EDITION
2

STATISTICS: LEARNING FROM DATA

Roxy Peck
California Polytechnic State University, San Luis Obispo

Tom Short
West Chester University of Pennsylvania

CENGAGE

Australia • Brazil • Mexico • Singapore • United Kingdom • United States

Statistics: Learning from Data,
Second Edition
Roxy Peck, Tom Short

Product Director: Mark Santee

Product Manager: Cassie Van Der Laan

Content Developer: Gabriela Carrascal

Product Assistant: Abby DeVeuve

Marketing Manager: Mike Saver

Senior Content Project Manager: Andrea Wagner

Manufacturing Planner: Doug Bertke

IP Analyst: Reba Frederics

IP Project Manager: Nick Barrows

Production Service: MPS Limited

Compositor: MPS Limited

Art Director: Vernon Boes

Text Designer: Diane Beasley

Cover Designer: Irene Morris

Cover Image: Color of the Rails by
Nicholas Rougeaux

For product information and technology assistance, contact us at
Cengage Customer & Sales Support, 1-800-354-9706.

For permission to use material from this text or product,
submit all requests online at **www.cengage.com/permissions.**
Further permissions questions can be emailed to
permissionrequest@cengage.com.

Library of Congress Control Number: 2017945324

Student Edition:
ISBN: 978-133-755808-2

Loose-leaf Edition:
ISBN: 978-133-755885-3

Cengage
20 Channel Center Street
Boston, MA 02210
USA

Cengage is a leading provider of customized learning solutions with employees residing in nearly 40 different countries and sales in more than 125 countries around the world. Find your local representative at **www.cengage.com.**

Cengage products are represented in Canada by Nelson Education, Ltd.

To learn more about Cengage platforms and services, visit **www.cengage.com.**

To register or access your online learning solution or purchase materials for your course, visit **www.cengagebrain.com.**

Printed in the United States of America
Print Number: 01   Print Year: 2017

To my friends and colleagues

in the Cal Poly Statistics Department

Roxy Peck

To Jerry Moreno and Jerry Senturia for inspiring

me to become a statistician

Tom Short

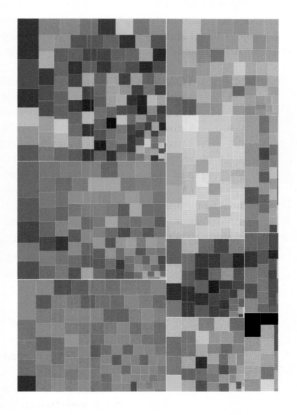

About the Cover

Artist Nicholas Rougeux used data extracted from the maps of metro systems from around the world to create a graphic image that grouped the colors used to represent the different transit lines into a single image. The sizes of the rectangles that make up the image are based on the number of stations on each transit line. For more information and for an interactive version of this image, see www.c82.net/work/?id=355.

Author Bios

 **ROXY PECK** is a professor emerita of statistics at California Polytechnic State University, San Luis Obispo. She was a faculty member in the Statistics Department for thirty years, serving for six years as Chair of the Statistics Department and thirteen years as Associate Dean of the College of Science and Mathematics. Nationally known in the area of statistics education, Roxy was made a Fellow of the American Statistical Association in 1998, and in 2003 she received the American Statistical Association's Founders Award in recognition of her contributions to K-12 and undergraduate statistics education. In 2009, she received the USCOTS Lifetime Achievement Award in Statistics Education. In addition to coauthoring the textbooks *Statistics: Learning from Data, Introduction to Statistics and Data Analysis*, and *Statistics: The Exploration and Analysis of Data*, she is also editor of *Statistics: A Guide to the Unknown*, a collection of expository papers that showcases applications of statistical methods. Roxy served from 1999 to 2003 as the Chief Faculty Consultant for the Advanced Placement Statistics exam, and she is a past chair of the joint ASA/NCTM Committee on Curriculum in Statistics and Probability for Grades K-12 and of the ASA Section on Statistics Education. Outside the classroom, Roxy enjoys travel and has visited all seven continents. She collects Navajo rugs and heads to Arizona and New Mexico whenever she can find the time.

 **TOM SHORT** is an Associate Professor in the Statistics Program within the Department of Mathematics at West Chester University of Pennsylvania. He previously held faculty positions at Villanova University, Indiana University of Pennsylvania, and John Carroll University. He is a Fellow of the American Statistical Association and received the 2005 Mu Sigma Rho Statistics Education Award. Tom is part of the leadership team for readings of the Advanced Placement (AP) Statistics Exam, and was a member of the AP Statistics Development Committee. He has also served on the Board of Directors of the American Statistical Association. Tom treasures the time he shares with his four children and the many adventures experienced with his wife, Darlene.

Brief Contents

Contents

SECTION III   **A FOUNDATION FOR INFERENCE: REASONING ABOUT PROBABILITY**

SECTION IV   LEARNING FROM SAMPLE DATA

Preface

Statistics is about learning from data and the role that variability plays in drawing conclusions from data. To be successful, it is not enough for students to master the computational aspects of descriptive and inferential statistics—they must also develop an understanding of the data analysis process at a conceptual level. The second edition of *Statistics: Learning from Data* is informed by careful and intentional thought about how the conceptual and the mechanical should be integrated in order to promote three key types of learning objectives for students:

- conceptual understanding
- mastery of the mechanics
- the ability to demonstrate conceptual understanding and mastery of the mechanics by "putting it into practice"

A Unique Approach

A number of innovative features distinguish this text from other introductory statistics books:

- **A New Approach to Probability**
  There is now quite a bit of research on how students develop an understanding of probability and chance. Using *natural frequencies* to reason about probability, especially conditional probability, is much easier for students to understand. The treatment of probability in this text is complete, including conditional probability and Bayes' Rule type probability calculations, but is done in a way that eliminates the need for the symbolism and formulas that are a roadblock for so many students. For those who also want to provide students with a more traditional coverage, there is an optional new section that introduces probability rules.

- **Chapter on Overview of Statistical Inference (Chapter 7)**
  This short chapter focuses on the things students need to think about in order to select an appropriate method of analysis. In most texts, this is "hidden" in the discussion that occurs when a new method is introduced. Considering this up front in the form of four key questions that need to be answered before choosing an inference method allows students to develop a general framework for inference and makes it easier for students to make correct choices.

- **An Organization That Reflects the Data Analysis Process**
  Students are introduced early to the idea that data analysis is a process that begins with careful planning, followed by data collection, data description using graphical and numerical summaries, data analysis, and finally interpretation of results. The ordering of topics in the text book mirrors this process: data collection, then data description, then statistical inference.

- **Inference for Proportions Before Inference for Means**
  Inference for proportions is covered before inference for means for the following reasons:
  - This makes it possible to develop the concept of a sampling distribution via simulation, an approach that is more accessible to students than a more formal, theoretical approach. Simulation is simpler in the context of proportions, where it is easy to construct a hypothetical population from which to sample (it is more complicated

to create a hypothetical population in the context of means because this requires making assumptions about shape and spread).
- Large-sample inferential procedures for proportions are based on the normal distribution and don't require the introduction of a new distribution (the *t* distribution). Students can focus on the new concepts of estimation and hypothesis testing without having to grapple at the same time with the introduction of a new probability distribution.
- **Parallel Treatments of Inference Based on Sample Data and Inference Based on Experiment Data**
Many statistical studies involve collecting data from a statistical experiment. The same inference procedures used to estimate or test hypotheses about population parameters also are used to estimate or test hypotheses about treatment effects. However, the necessary assumptions are slightly different (for example, random assignment replaces the assumption of random selection), and the wording of hypotheses and conclusions is also different. Trying to treat both cases together tends to confuse students. This text makes the distinction clear.

New in This Edition

- **New Sections on Randomization-Based Inference Methods**
Research indicates that randomization-based instruction in statistical inference may help learners to better understand the concepts of confidence and significance. The second edition includes new optional sections on randomization-based inference methods. These methods provide alternative analyses that can be used when the conditions required for normal distribution-based inference are not met. Each of the inference chapters (Chapters 9 through 13) now contains a new optional section on randomization-based inference that includes bootstrap methods for simulation-based confidence intervals and randomization tests of hypotheses. These new sections are accompanied by online Shiny apps, which can be used to construct bootstrap confidence intervals and to carry out randomization tests. The App collection that accompanies this text can be found at statistics.cengage.com/Peck2e/Apps.html.
- **Restructured Chapters on Statistical Inference**
The chapters on statistical inference have been restructured to include methods for learning from experiments in the same chapter as methods for learning from samples. While the coverage of inference based on data from statistical experiments (Chapter 14 in the first edition) has been integrated into earlier chapters, the important distinction between inferences based on data from experiments and inferences based on data from sampling is maintained in order to highlight the differences in how hypotheses are worded, in conditions, and in the wording of conclusions in these two situations. The sections of the chapter on inference for two means have also been reordered to put inference for paired samples before inference for independent samples, in order to better connect the paired samples structure with one sample inference for a mean in Chapter 12.
- **Expanded Treatment of Probability**
The second edition contains a new section titled "Calculating Probabilities—A More Formal Approach" for instructors who want to also provide a more traditional coverage of probability. For those who prefer the "hypothetical 1000" approach from the first edition, the newly added traditional section is optional and can be omitted without compromising any of the probability student learning objectives.
- **Updated Examples and Exercises**
In our continuing effort to keep things interesting and relevant, the second edition contains many updated examples and exercises on topics of interest to students that use data from recent journal articles, newspapers, and web posts.

Features That Support Student Engagement and Success

The text also includes a number of features that support conceptual understanding, mastery of mechanics, and putting ideas into practice.

- **Simple Design**

  There is now research showing that many of the "features" in current textbooks are not really helpful to students. In fact, cartoons, sidebars, historical notes, and the like, actually distract students and interfere with learning. The second edition of *Statistics: Learning from Data* has a simple, clean design in order to minimize clutter and maximize student understanding.

- **Chapter Learning Objectives—Keeping Students Informed About Expectations**

  Chapter learning objectives explicitly state the expected student outcomes. Learning objectives fall under three headings: Conceptual Understanding, Mastery of Mechanics, and Putting It into Practice.

- **Preview—Motivation for Learning**

  Each chapter opens with a *Preview* and *Preview Example* that provide motivation for studying the concepts and methods introduced in the chapter. They address why the material is worth learning, provide the conceptual foundation for the methods covered in the chapter, and connect to what the student already knows. A relevant and current example provides a context in which one or more questions are proposed for further investigation. This context is revisited in the chapter once students have the necessary understanding to more fully address the questions posed.

- **Real Data**

  Examples and exercises with overly simple settings do not allow students to practice interpreting results in authentic situations or give students the experience necessary to be able to use statistical methods in real settings. The exercises and examples are a particular strength of this text, and we invite you to compare the examples and exercises with those in other introductory statistics texts.

  Many students are skeptical of the relevance and importance of statistics. Contrived problem situations and artificial data often reinforce this skepticism. Examples and exercises that involve data extracted from journal articles, newspapers, and other published sources and that are of interest to today's students are used to motivate and engage students. Most examples and exercises in the book are of this nature; they cover a very wide range of disciplines and subject areas. These include, but are not limited to, health and fitness, consumer research, psychology and aging, environmental research, law and criminal justice, and entertainment.

- **Exercises Organized into a Developmental Structure—Structuring the Out-of-Class Experience**

  End-of-section exercises are organized into developmental sets. At the end of each section, there are two grouped problem sets. The exercises in each set work together to assess all of the learning objectives for that section. In addition to the two exercise sets, each section also has additional exercises for those who want more practice.

  Answers for the exercises of Exercise Set 1 in each section are included at the end of the book. In addition, many of the exercises in Exercise Set 1 include hints directing the student to a particular example or a relevant discussion that appears in the text. This feature provides direction for students who might need help getting started on a particular exercise. Instructors who prefer that students be more self-directed can assign Exercise Set 2. Answers and hints are not provided for the exercises in Exercise Set 2.

- **Are You Ready to Move On?—Students Test Their Understanding**

  Prior to moving to the next chapter, "Are You Ready to Move On?" exercises allow students to confirm that they have achieved the chapter learning objectives. Like the developmental problem sets of the individual sections, this collection of exercises is developmental in nature. These exercises assess all of the chapter learning objectives and serve as a comprehensive end-of-chapter review.

- **Explorations in Statistical Thinking—Real Data Algorithmic Sampling Exercises and Multivariable Thinking**

  Most chapters contain extended sampling-based, real-data exercises at the end of the chapter. Each student goes online to get a different random sample for the same exercise. These unique exercises are designed to develop conceptual understanding and to teach about sampling variability.

New guidelines from the American Statistical Association recommend that students in the introductory statistics course be provided with opportunities to develop multivariable thinking. To facilitate this, several chapters include an exploration that allows students to work with data sets that include more than two variables.

- **Data Analysis Software**
  JMP data analysis software may be bundled for free with the purchase of a new textbook. See Student Resources for more information.

- **Technology Notes**
  Technology Notes appear at the end of most chapters and give students helpful hints and guidance on completing tasks associated with a particular chapter. The following technologies are included in the notes: JMP, Minitab, SPSS, Microsoft Excel 2007, TI-83/84, and TI-nspire. They include display screens to help students visualize and better understand the steps. More complete technology manuals are also available on the text web site.

- **Chapter Activities—Engaging Students in Hands-On Activities**
  There is a growing body of evidence that students learn best when they are actively engaged. Chapter activities guide students' thinking about important ideas and concepts.

- **Support for Co-Requisite and Pre-Requisite Courses**
  In recognition of the emerging trend of placing students who might previously have been placed into a developmental mathematics sequence directly into the college-level introductory statistics course with co-requisite support, *Statistics Companion: The Math You Need to Know* provides a text companion for the co-requisite course. Also written by Peck and Short, this companion volume provides a just-in-time treatment of the mathematics needed for success in introductory statistics. While *Statistics Companion* can be adapted for use with any introductory statistics text book, it was written specifically with *Statistics: Learning from Data*, Second Edition, in mind and matches the terminology, notation and ordering of topics. The companion can also be adapted for use in a one-semester pre-statistics course for schools that prefer to have students complete their math preparation prior to beginning the statistics course. For more information or to receive a sample copy of *Statistics Companion: The Math You Need to Know*, contact your Cengage Learning Consultant.

Consistent with Recommendations for the Introductory Statistics Course Endorsed by the American Statistical Association

In 2005, the American Statistical Association endorsed the report "College Guidelines in Assessment and Instruction for Statistics Education (GAISE Guidelines)," which included the following six recommendations for the introductory statistics course:

1. Emphasize statistical literacy and develop statistical thinking.
2. Use real data.
3. Stress conceptual understanding rather than mere knowledge of procedures.
4. Foster active learning in the classroom.
5. Use technology for developing conceptual understanding and analyzing data.
6. Use assessments to improve and evaluate student learning.

In 2016, these guidelines were revised. The new guidelines reaffirmed the six recommendations and also included two new recommendations. The two new recommendations were:

- Teach statistics as an investigative process of problem-solving and decision making.
- Give students experience with multivariable thinking.

The second edition of *Statistics: Learning from Data* is consistent with these recommendations and supports the GAISE guidelines in the following ways:

1. **Emphasize Statistical Literacy and Develop Statistical Thinking.**
   Statistical literacy is promoted throughout the text in the many examples and exercises that are drawn from the popular press. In addition, a focus on the role of

variability, consistent use of context, and an emphasis on interpreting and communicating results in context work together to help students develop skills in statistical thinking.

2. **Use Real Data.**

The examples and exercises are context driven, and the reference sources include the popular press as well as journal articles.

3. **Stress Conceptual Understanding Rather Than Mere Knowledge of Procedures.**

Nearly all exercises in the text are multipart and ask students to go beyond just calculation, with a focus on interpretation and communication. The examples and explanations are designed to promote conceptual understanding. Hands-on activities in each chapter are also constructed to strengthen conceptual understanding. Which brings us to . . .

4. **Foster Active Learning in the Classroom.**

While this recommendation speaks more to pedagogy and classroom practice, the second edition of *Statistics: Learning from Data* provides more than 30 hands-on activities in the text and additional activities in the accompanying instructor resources that can be used in class or assigned to be completed outside of class.

5. **Use Technology for Developing Conceptual Understanding and Analyzing Data.**

The computer has brought incredible statistical power to the desktop of every investigator. The wide availability of statistical computer packages, such as JMP, Minitab, and SPSS, and the graphical capabilities of the modern microcomputer have transformed both the teaching and learning of statistics. To highlight the role of the computer in contemporary statistics, sample output is included throughout the book. In addition, numerous exercises contain data that can easily be analyzed using statistical software. JMP data analysis software can be bundled with new purchases of the text, and technology manuals for JMP and for other software packages, such as Minitab and SPSS, and for the graphing calculator are available in the online materials that accompany this text. The second edition of *Statistics: Learning from Data* also includes a number of Shiny web apps that can be used to illustrate statistical concepts and to implement the simulation-based inference methods covered in new optional sections. The App collection can be found at statistics.cengage.com/Peck2e/Apps.html.

6. **Use Assessments to Improve and Evaluate Student Learning.**

Comprehensive chapter review exercises that are specifically linked to chapter learning objectives are included at the end of each chapter. In addition, assessment materials in the form of a test bank, quizzes, and chapter exams are available in the instructor resources that accompany this text. The items in the test bank reflect the data-in-context philosophy of the text's exercises and examples.

7. **Teach Statistics as an Investigative Process of Problem-Solving and Decision Making.**

A systematic approach to inference helps students to see how data are used to answer questions and to learn about the world around them. Without such a foundation, students may see the methods they are learning in their statistics course as just a loose collection of tools and may not develop a real sense of the complete data analysis process. The organization of this text helps to highlight this process, addressing methods of data collection, followed by methods for summarizing data, followed by methods for learning from data. The data analysis process is also featured in Chapter 7, which provides an overview of statistical inference.

8. **Give Students Experience with Multivariable Thinking.**

Several new explorations have been included in the second edition as part of the Explorations in Statistical Thinking sections. These new explorations provide students with opportunities to work with data sets that include more than one variable in order to develop multivariable thinking.

Instructor and Student Resources

JMP Statistical Software

JMP is a statistics software for Windows and Macintosh computers from SAS, the market leader in analytics software and services for industry. JMP Student Edition is a streamlined, easy-to-use version that provides all the statistical methods and graphical displays covered in this textbook. Once data is imported, students will find that most procedures require just two or three mouse clicks. JMP can import data from a variety of formats, including Excel and other statistical packages, and you can easily copy and paste graphs and output into documents.

JMP also provides an interface to explore data visually and interactively, which will help your students develop a healthy relationship with their data, work more efficiently with data, and tackle difficult statistical problems more easily. Because its output provides both statistics and graphs together, the student will better see and understand the application of concepts covered in this book as well. JMP Student Edition also contains some unique platforms for student projects, such as mapping and scripting. JMP functions in the same way on both Windows and Mac platforms and instructions contained with this book apply to both platforms.

Access to this software is available for free with new copies of the book and available for purchase standalone at CengageBrain.com or http://www.jmp.com/getse. Find out more at www.jmp.com.

Student Resources

Digital

To access additional course materials and companion resources, please visit www.cengagebrain.com. At the CengageBrain.com home page, search for the ISBN of your title (from the back cover of your book) using the search box at the top of the page. This will take you to the product page where free companion resources can be found.

If your text includes a printed access card, you will have instant access to the following resources:

- Complete step-by-step instructions for JMP, TI-84 Graphing Calculators, Excel, Minitab, and SPSS.
- Data sets in JMP, TI-84, Excel, Minitab, SPSS, SAS, and ASCII file formats.
- Applets used in the Activities found in the text.

Prepare for class with confidence using WebAssign from Cengage *Statistics: Learning from Data*, Second Edition. This online learning platform fuels practice, so you truly absorb what you learn—and are better prepared come test time. Videos and tutorials walk you through concepts and deliver instant feedback and grading, so you always know where you stand in class. Focus your study time and get extra practice where you need it most. Study smarter with WebAssign!

Ask your instructor today how you can get access to WebAssign, or learn about self-study options at www.webassign.com

Print

Student Solutions Manual (ISBN: 9781337558389): Contains fully worked-out solutions to all of the Exercise Set 1 and odd-numbered additional exercises in the text, giving students a way to check their answers and ensure that they took the correct steps to arrive at an answer.

Instructor Resources

Digital

WebAssign from Cengage *Statistics: Learning from Data*, Second Edition, is a fully customizable online solution for STEM disciplines that empowers you to help your students learn, not just do homework. Insightful tools save you time and highlight exactly where your students are struggling. Decide when and what type of help students can access while working on assignments—and incentivize independent work so help features aren't abused. Meanwhile, your students get an engaging experience, instant feedback and better outcomes. A total win-win!

To try a sample assignment, learn about LMS integration or connect with our digital course support, visit http://www.webassign.com/cengage

Instructor Companion Website: Everything you need for your course in one place! Access the Instructor Solutions Manual, full lecture PowerPoints, and other support materials. This collection of book-specific lecture and class tools is available via http://www .cengage.com/login

Instructor Solutions Manual (ISBN: 9781337558396): This guide contains solutions to every exercise in the book. You can download the solutions manual from the Instructor Companion Site.

Print

Teacher's Resource Binder (ISBN: 9781337559263): The Teacher's Resource Binder is full of wonderful resources for both AP Statistics teachers and college professors. These include:

- Recommendations for instructors on how to teach the course, including sample syllabi, pacing guides, and teaching tips.
- Recommendations for what students should read and review for a particular class period or set of class periods.
- Extensive notes on preparing students to take the AP exam.
- Additional examples from published sources (with references), classified by chapter in the text. These examples can be used to enrich your classroom discussions.
- Model responses—examples of responses that can serve as a model for work that would be likely to receive a high mark on the AP exam.
- A collection of data explorations that can be used throughout the year to help students prepare for the types of questions that they may encounter on the investigative task on the AP Statistics Exam.
- Activity worksheets that can be duplicated and used in class.
- A test bank that includes assessment items, quizzes, and chapter exams.

Acknowledgments

We would like to express our thanks and gratitude to the following people who made this book possible:

Cassie Van Der Laan, our editor at Cengage, for her support of this project.

Spencer Arritt and Gabriela Carrascal, our content developers at Cengage, for their helpful suggestions and for keeping us on track.

Andrea Wagner, the content project manager.

Ed Dionne, our manager at MPS Limited.

Hunter Glanz and Alex Boyd for creating the Shiny Apps that accompany the text.

Stephen Miller, for his careful and complete work on the huge task of creating the student and instructor solutions manuals.

Roger Lipsett, for his attention to detail in checking the accuracy of examples and solutions.

Kathy Fritz, for creating the interactive PowerPoint presentations that accompany this text.

Melissa Sovak, for creating the Technology Notes sections.

Mike Saver, the Marketing Manager.

Chris Sabooni, the copy editor for the book.

MPS, for producing the artwork used in the book.

We would also like to give a special thanks to those who served on the Editorial Board for the book and those who class tested some of the chapters with their students:

Many people provided invaluable comments and suggestion as this text was being developed.

Reviewers

Susan Andrews, The College of Saint Rose
Melanie Autin, Western Kentucky University
Jordan Bertke, Central Piedmont Community College
Gregory Bloxom, Pensacola State College
Denise Brown, Collin College
Elena Buzaianu, University of North Florida
Andy Chang, Youngstown State University
Jerry Chen, Suffolk County Community College
Monte Cheney, Central Oregon Community College
Ivette Chuca, El Paso Community College
Mary Ann Connors, Westfield State University
George Davis, Georgia State University
Rob Eby, Blinn College, Bryan Campus
Karen Estes, St. Petersburg College
Larry Feldman, Indiana University of Pennsylvania
Kevin Fox, Shasta College
Leslie Hendrix, University of South Carolina, Columbia
Melinda Holt, Sam Houston State University
Brett Hunter, Colorado State University
Kelly Jackson, Camden County College
Keshav Jagannathan, Coastal Carolina University
Clarence Johnson, Cuyahoga Community College
Nancy Johnson, State College of Florida, Bradenton
Sue Ann Jones Dobbyn, Pellissippi State Technical Community College
Hoon Kim, Calfornia State Polytechnic University, Pomona
Cathy Lockwood, Sam Houston State University
Jackie MacLaughlin, Central Piedmont

Kim Massaro, University of Texas, San Antonio
Nola McDaniel, McNeese State
Glenn Miller, Borough of Manhattan Community College
Philip Miller, Indiana University SE
Sumona Mondal, Clarkson University
Kathy Mowers, Owensboro Community and Technical College
Linda Myers, Harrisburg Area Community College
Nguyet Nguyen, Youngstown State University
Ron Palcic, Johnson County Community College
Nishant Patel, Northwest Florida State College
Maureen Petkewich, Univiversity of South Carolina
Nancy Pevey, Pellissippi State Community College
Chandler Pike, University of Georgia
Blanche Presley, Macon State College
Daniel Rowe, Heartland Community College
Fary Sami, Harford Community College
Laura Sather, St. Cloud State University
Sean Simpson, Westchester Community College
Sam Soleymani, Santa Monica College
Cameron Troxell, Mount San Antonio College
Diane Van Deusen, Napa Valley College
Richard Watkins, Tidewater Community College
Jane-Marie Wright, Suffolk Community College
Cathy Zucco-Teveloff, Rider University

And last, but certainly not least, we thank our families, friends, and colleagues for their continued support.

Roxy Peck and Tom Short

EDITION
2

STATISTICS: LEARNING FROM DATA

1 Collecting Data in Reasonable Ways

Monkey Business Images/Shutterstock.com

PREVIEW

There is an old saying attributed to statistician Ed Deming, "without data, you are just another person with an opinion." Although anecdotes and coincidences may make for interesting stories, you wouldn't want to make important decisions on the basis of anecdotes alone. For example, just because a friend of a friend ate 16 apricots and then experienced relief from joint pain doesn't mean that this is all you would need to know to help one of your parents choose a treatment for arthritis. Before recommending apricots, you would definitely want to consider relevant data on the effectiveness of apricots as a treatment for arthritis.

Statistical methods help you to make sense of data and gain insight into the world around you. The ability to learn from data is critical for success in your personal and professional life. Data and conclusions based on data are everywhere—in newspapers, magazines, online resources, and professional publications. But should you believe what you read? For example, should you supplement your diet with black currant oil to stop hair loss? Will playing solitaire for 20 minutes each day help you feel less tired? If you eat proteins before carbohydrates when you eat a meal, will it lower your blood sugar? Should you donate blood twice a year to lower your risk of heart disease? These are just four recommendations out of many that appear in one issue of **Woman's World (April 4, 2016),** *a magazine with more than 1.3 million readers. In fact, if you followed all of the recommendations in that issue, you would also be loading up on prickly pear oil, hot chocolate, ginger tea, bread, bananas, sweet potatoes, bell peppers, tomatoes, and onions! Some of these recommendations are supported by evidence (data) from research studies, but how reliable is this evidence? Are the conclusions drawn reasonable, and do they apply to you? These important questions will be explored in this chapter.*

CHAPTER LEARNING OBJECTIVES

Conceptual Understanding

After completing this chapter, you should be able to

C1 Understand the difference between an observational study and an experiment.

C2 Understand that the conclusions that can be drawn from a statistical study depend on the way in which the data are collected.

C3 Explain the difference between a census and a sample.

C4 Explain the difference between a statistic and a population characteristic.

C5 Understand why random selection is an important component of a sampling plan.

C6 Understand why random assignment is important when collecting data in an experiment.

C7 Understand the difference between random selection and random assignment.

C8 Explain why volunteer response samples and convenience samples are unlikely to produce reliable information about a population.

C9 Understand the limitations of using volunteers as subjects in an experiment.

C10 Explain the purpose of a control group in an experiment.

C11 Explain the purpose of blinding in an experiment.

Mastering the Mechanics

After completing this chapter, you should be able to

M1 Create a sampling plan that could produce a simple random sample from a given population.

M2 Describe a procedure for randomly assigning experimental units to experimental conditions (for example, subjects to treatments) given a description of an experiment, the experimental conditions, and the experimental units.

Putting It into Practice

After completing this chapter, you should be able to

P1 Distinguish between an observational study and an experiment.

P2 Evaluate the design of an observational study.

P3 Evaluate the design of a simple comparative experiment.

P4 Evaluate whether conclusions drawn from a study are appropriate, given a description of the statistical study.

SECTION 1.1 Statistics—It's All About Variability

Statistical methods allow you to collect, describe, analyze, and draw conclusions from data. If you lived in a world where all measurements were identical for every individual, these tasks would be simple. For example, consider a population consisting of all of the students at a college. Suppose that every student is enrolled in the same number of courses, spent exactly the same amount of money on textbooks, and favors increasing student fees to support expanding library services. For this population, there is no variability in number of courses, amount spent on books, or student opinion on the fee increase. A person studying students from this population to draw conclusions about any of these three variables would have an easy task. It would not matter how many students were studied or how the students were selected. In fact, you could collect information on the number of courses, amount spent on books, and opinion on the fee increase by just stopping the next student who happened to walk by the library. Because there is no variability in the population, this one individual would provide complete and accurate information about the population, and you could draw conclusions with no risk of error.

The situation just described is obviously unrealistic. Populations with no variability are exceedingly rare, and they are of little statistical interest because they present no challenge. In fact, variability is almost universal. It is variability that makes life interesting. To be able to collect, describe, analyze, and draw conclusions from data in a sensible way, you need to develop an understanding of variability.

The following example illustrates how describing and understanding variability provide the foundation for learning from data.

Example 1.1   Monitoring Water Quality

As part of its regular water quality monitoring efforts, an environmental control board selects five water specimens from a particular well each day. The concentration of contaminants in parts per million (ppm) is measured for each of the five specimens, and then the average of the five measurements is calculated. The graph in Figure 1.1 is an example of a histogram. (You will learn how to construct and interpret histograms in Chapter 2.) This histogram summarizes the average contamination values for 200 days.

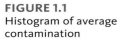
FIGURE 1.1
Histogram of average contamination

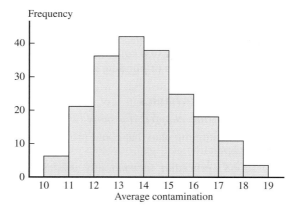

Suppose that a chemical spill has occurred at a manufacturing plant 1 mile from the well. It is not known whether a spill of this nature would contaminate groundwater in the area of the spill and, if so, whether a spill this distance from the well would affect the quality of well water.

One month after the spill, five water specimens are collected from the well, and the average contamination is 15.5 ppm. Considering the variation before the spill shown in the histogram, would you interpret this as evidence that the well water was affected by the spill? What if the calculated average was 17.4 ppm? How about 22.0 ppm?

Before the spill, the average contaminant concentration varied from day to day. An average of 15.5 ppm would not have been an unusual value, so seeing an average of 15.5 ppm after the spill isn't necessarily an indication that contamination has increased. On the other hand, an average as large as 17.4 ppm is less common, and an average as large as 22.0 ppm is not at all typical of the pre-spill values. In this case, you would probably conclude that the well contamination level has increased.

Reaching a conclusion requires an understanding of variability. Understanding variability allows you to distinguish between usual and unusual values. The ability to recognize unusual values in the presence of variability is an important aspect of many statistical methods and is also what enables you to quantify the chance of being incorrect when a conclusion is based on available data.

SECTION 1.2 Statistical Studies: Observation and Experimentation

If the goal is to make good decisions based on data, it should come as no surprise that the way you obtain the data is very important. It is also important to know what questions you hope to answer with data. Depending on what you want to learn, two types of statistical studies are common—*observational studies* and *experiments*.

Sometimes you are interested in answering questions about characteristics of a single **population** or in comparing two or more well-defined populations. To accomplish this, you select a **sample** from each population and use information from the samples to learn about characteristics of the populations.

DEFINITION

Population: The population is the entire collection of individuals or objects that you want to learn about.

Sample: A sample is a part of the population that is selected for study.

For example, many people, including the author of **"The 'CSI Effect': Does It Really Exist?" (*National Institute of Justice* [2008]: 1–7)**, have speculated that watching crime scene investigation TV shows (such as *CSI*, *Cold Case*, *Bones*, or *Numb3rs*) may be associated with the kind of high-tech evidence that jurors expect to see in criminal trials. Do people who watch such shows on a regular basis have higher expectations than those who do not watch them? To answer this question, you would want to learn about two populations, one consisting of people who watch crime scene investigation shows on a regular basis and the other consisting of people who do not. You could select a sample of people from each population and interview these people to determine their levels of expectation for high-tech evidence in a criminal case. This would be an example of an **observational study**. In an observational study, it is important to obtain samples that are representative of the corresponding populations.

Sometimes the questions you are trying to answer cannot be answered using data from an observational study. Such questions are often of the form, "What happens when …?" or "What is the effect of …?" For example, a teacher may wonder what happens to student test scores if the lab time for a chemistry course is increased from 3 hours to 6 hours per week. To answer this question, she could conduct an **experiment**. In such an experiment, the value of a *response* (test score) would be recorded under different *experimental conditions* (3-hour lab and 6-hour lab). The person carrying out the experiment creates the experimental conditions and also determines which people will be assigned to each experimental condition.

DEFINITION

An **observational study** is a study in which the person conducting the study observes characteristics of a sample selected from one or more existing populations. The goal of an observational study is to use data from the sample to learn about the corresponding population. In an observational study, it is important to obtain a sample that is representative of the population.

An **experiment** is a study in which the person conducting the study considers how a response behaves under different experimental conditions. The person carrying out the study determines who will be in each experimental group and what the experimental conditions will be. In an experiment, it is important to have comparable experimental groups.

Observational studies and experiments can both be used to compare groups. In an observational study, the person carrying out the study does not control who is in which population. However, in an experiment, the person conducting the study *does* control who is in which experimental group. For example, in the observational study to compare expectations of those who watch crime scene investigation shows and those who do not, the person conducting the study does not determine which people will watch crime scene investigation shows. However, in the chemistry experiment, the person conducting the study *does* determine which students will be in the 3-hour lab group and which students will be in the 6-hour lab group. This seemingly small difference is critical when it comes to drawing conclusions from a statistical study. This is why it is important to determine whether data are from an observational study or from an experiment. We will return to this important distinction in Section 1.5.

The following two examples illustrate how to determine whether a study is an observational study or an experiment.

Example 1.2   Chew More, Eat Less?

The article **"Increasing the Number of Chews before Swallowing Reduces Meal Size in Normal-Weight, Overweight and Obese Adults"** (*Journal of the Academy of Nutrition and Dietetics* **[2014]: 926–931**) describes a study that investigated whether chewing each bite of food more before swallowing would result in people eating less. Participants in the study were adults between the ages of 18 to 45 years. At the beginning of the study, each participant was observed as they ate five pizza rolls, and the number of chews made before swallowing was observed to determine a baseline for that participant.

Participants were then invited back for a second session on a different day. They were asked to eat their usual breakfast on that day and to not eat anything after breakfast. At the second session, the participants were assigned to one of three groups. All participants were provided with a platter of pizza rolls and were told to eat until they were comfortably full. They were also told they could request more pizza rolls if they wanted more. Each participant was also told how many times to chew each pizza roll before swallowing. The participants in group 1 were given a number of chews equal to their baselines. The participants in group 2 were given a number of chews that was 150% of (one and a half times as large as) their baselines. The participants in group 3 were given a number of chews that was 200% of (twice as large as) their baselines.

After analyzing data from this study, the researcher concluded that people ate about 10% less when they increased the number of chews by 50% (group 2) and about 15% less when they doubled the number of chews (group 3).

Is this study an observational study or an experiment? To answer this question, you need to consider how the three groups in the study were formed. Because the study participants were assigned to one of the three groups by the researchers conducting the study, the study is an experiment. As you will see in Section 1.4, the way the researchers decide which people go into each group is an important aspect of the study design.

Example 1.3   Caffeine and Sleep

The article **"Adolescents Living the 24/7 Lifestyle: Effects of Caffeine and Technology on Sleep Duration and Daytime Functioning" (*Pediatrics* [2009]: e1005–e1010)** describes a study in which researchers investigated whether there is a relationship between amount of sleep and caffeine consumption. They found that teenagers who usually get less than 8 hours of sleep on school nights were more likely to report falling asleep during school and consume more caffeine on average than teenagers who usually get 8 to 10 hours of sleep on school nights.

In the study described, two populations of teenagers were compared—teenagers who usually get less than 8 hours of sleep on school nights and teenagers who usually get 8 to 10 hours of sleep on school nights. Did the researchers determine which teenagers were in each group? The researchers had no control over how long the study participants slept, so the study is an observational study and not an experiment. It is still possible to make reasonable comparisons between the two populations, as long as the groups of teenagers in the study were chosen to be representative of the two populations of interest—*all* teenagers who usually get less than 8 hours of sleep on school nights and *all* teenagers who usually get 8 to 10 hours of sleep on school nights. The way in which the teenagers in the two study groups were chosen is an important aspect of the design of this observational study.

In the next sections, the design of observational studies and experiments will be considered in more detail.

Summing It Up—Section 1.2

The following learning objectives were addressed in this section:

Conceptual Understanding

C1: Understand the difference between an observational study and an experiment.

An **observational study** is a study in which the person conducting the study observes characteristics of a sample selected from a population. The goal of an observational study is to learn about a population.

An **experiment** is a study in which the person conducting the study considers how a response behaves under different experimental conditions. The person carrying out the study determines who will be in each experimental group.

Putting It into Practice

P1: Distinguish between an observational study and an experiment.

Once you understand the difference between an observational study and an experiment, if you are given a description of a study, you should be able to determine if it is an observational study or an experiment. See Examples 1.2 and 1.3.

SECTION 1.2   EXERCISES

Each Exercise Set assesses the following chapter learning objectives: C1, P1

SECTION 1.2 Exercise Set 1

For each of the statistical studies described in Exercises 1.1 to 1.5, indicate whether the study is an observational study or an experiment. Give a brief explanation for your choice. (Hint: See Examples 1.2 and 1.3.)

1.1 The following conclusion from a statistical study appeared in the article **"Smartphone Nation" (*AARP Bulletin,* September 2009)**: "If you love your smartphone, you're far from alone. Half of all boomers sleep with their cell phone within arm's length. Two of three people ages 50 to 64 use a cell phone to take photos, according to a 2010 Pew Research Center report."

1.2 The press release **"Men Need to Man Up, According to Ball Park Brand Survey" (*PR Newswire,* October 14, 2015)** describes the results of a study in which 1012 U.S. men were asked a number of questions about "life's tough conversations." One result from this survey was summarized in a *USA TODAY Snapshot* (*USA TODAY*, November 6, 2015) that said that "nearly 1 in 5 men would pay someone to handle their breakup for them."

1.3 An article in *USA TODAY* (October 19, 2010) describes a study of how young children learn. Sixty-four 18-month-old toddlers participated in the study. The toddlers were allowed to play in a lab equipped with toys, which also had a robot hidden behind a screen. The article states: "After allowing the infants play time, the team removed the screen and let the children see the robot. In some tests, an adult talked to the robot and played with it. In others the adult ignored the robot. After the adult left the room, the robot beeped and then turned its head to look at a toy to the side of the infant. In cases where the adult had played with the robot, the infant was four times more likely to follow the robot's gaze to the toy."

1.4 In a survey of 2500 U.S. adults, 69% responded that they were confident that "smart homes" will be a commonplace as smartphones within 10 years (*Intel Survey: Architecting the Future of the Smart Home 2025*, [2015]: download.intel.com /newsroom/kits/iot/pdfs/IntelSmartHomeSurveyBackgrounder .pdf, retrieved September 25, 2016).

1.5 A paper appearing in *The Journal of Pain* (March 2010, 199–209) described a study to determine if meditation has an effect on sensitivity to pain. Study participants were assigned to one of three groups. One group meditated for 20 minutes; one group performed a distraction task (working math problems!) for 20 minutes; and one group practiced a relaxation technique for 20 minutes. Sensitivity to pain was measured both before and after the 20-minute session.

SECTION 1.2 Exercise Set 2

For each of the statistical studies described in Exercises 1.6–1.10, indicate whether the study is an observational study or an experiment. Give a brief explanation for your choice.

1.6 A news release from Intel titled **"Intel's Security International Internet of Things Smart Home Survey Shows Many Respondents Sharing Personal Data for Money"** (March 30, 2016, newsroom.intel.com /news-releases/intel-securitys-international-internet-of -things-smart-home-survey/, retrieved September 25, 2016) described a survey conducted in 2015. The news release states "A total of 9,000 consumers were interviewed globally, including 2,500 from the United States, 1,000 from the United Kingdom, 1,000 from France, 1,000 from Germany, 1,000 from Brazil, 1,000 from India, 500 from Canada, 500 from Mexico and 500 from Australia." Among the findings from the survey were that 54% of the respondents worldwide would be willing to share personal data collected from devices in their homes with companies in exchange for money.

1.7 The paper **"Health Halos and Fast-Food Consumption"** (*Journal of Consumer Research* [2007]: 301–314) described a study in which 46 college students volunteered to participate. Half of the students were given a coupon for a McDonald's Big Mac sandwich and the other half were given a coupon for a Subway 12-inch Italian BMT sandwich. (For comparison, the Big Mac has 600 calories, and the Subway 12-inch Italian BMT sandwich has 900 calories.) The researchers were interested in how the perception of Subway as a healthy fast-food choice and McDonald's as an unhealthy fast-food choice would influence what additional items students would order with the sandwich. The researchers found that those who received the Subway coupon were less likely to order a diet soft drink, more likely to order a larger size drink, and more likely to order cookies than those who received the Big Mac coupon.

1.8 *USA TODAY* (August 25, 2015) reported that "American women favor Kate Middleton as a shopping buddy over Michelle Obama by 10 percentage points." This statement was based on a study in which 1001 adults were surveyed about their shopping preferences.

1.9 In a study of whether taking a garlic supplement reduces the risk of getting a cold, 146 participants were assigned to either a garlic supplement group or to a group that did not take a garlic supplement (**"Garlic for the Common Cold,"** Cochrane Database of Systematic Reviews, 2009). Based on the study, it was concluded that the proportion of people taking a garlic supplement who get a cold is lower than the proportion of those not taking a garlic supplement who get a cold.

1.10 The article **"Baby Scientists Experiment with Everything"** (*The Wall Street Journal*, April 18, 2015) describes a series of studies published in the journal *Science*. In one of these studies, 11-month old children were assigned to one of two groups. The children in one group were shown a ball behaving as expected, such as rolling into a wall or falling off an edge. The children in the other group were shown a ball behaving in an unexpected way, such as rolling through what appeared to be a solid wall or rolling off an edge and remaining suspended in the air. The children were then given a ball and another toy. The researchers found that the children in the group that saw the ball behaving as expected showed no preference for the ball over the other toy, but that the children who saw the ball behaving in an unexpected way tended to choose the ball, and that they also played with it differently and tested the ball's behavior by dropping it or rolling it.

ADDITIONAL EXERCISES

1.11 The article **"How Dangerous Is a Day in the Hospital?"** (*Medical Care* [2011]: 1068–1075) describes a study to determine if the risk of an infection is related to the length of a hospital stay. The researchers looked at a large number of hospitalized patients and compared the proportions who got an infection for two groups of patients—those who were hospitalized overnight and those who were hospitalized for more than one night. Indicate whether the study is an observational study or an experiment. Give a brief explanation for your choice.

1.12 The authors of the paper **"Fudging the Numbers: Distributing Chocolate Influences Student Evaluations of an Undergraduate Course"** (*Teaching in Psychology* [2007]: 245–247) carried out a study to see if events unrelated to an undergraduate course could affect student evaluations. Students enrolled in statistics courses taught by the same instructor participated in the study. All students attended the same lectures and one of six discussion sections that met once a week. At the end of the course, the researchers chose three of the discussion sections to be the "chocolate group." Students in these three sections were offered chocolate prior to having them fill out course evaluations. Students in the other three sections were not offered chocolate. The researchers concluded that "Overall, students offered chocolate gave more positive evaluations than students not offered chocolate." Indicate whether the study is an observational study or an experiment. Give a brief explanation for your choice.

1.13 The article **"Why We Fall for This"** (*AARP Magazine,* **May/June 2011**) described a study in which a business professor divided his class into two groups. He showed students a mug and then asked students in one of the groups how much they would pay for the mug. Students in the other group were asked how much they would sell the mug for if it belonged to them. Surprisingly, the average values assigned to the mug were quite different for the two groups! Indicate whether the study is an observational study or an experiment. Give a brief explanation for your choice.

1.14 The same article referenced in Exercise 1.13 also described a study which concluded that people tend to respond differently to the following questions:

> Question 1: Would you rather have $50 today or $52 in a week?
> Question 2: Imagine that you could have $52 in a week. Would you rather have $50 now?

The article attributes this to the question wording: the second question is worded in a way that makes you feel that you are "losing" $2 if you take the money now. Do you think that the study which led to the conclusion that people respond differently to these two questions was an observational study or an experiment? Explain why you think this.

SECTION 1.3   Collecting Data: Planning an Observational Study

In Section 1.2, two types of statistical studies were described—observational studies and experiments. In this section, you will look at some important considerations when planning an observational study or when deciding whether an observational study performed by others was well planned.

Planning an Observational Study—Collecting Data by Sampling

The purpose of an observational study is to collect data that will allow you to learn about a single population or about how two or more populations might differ. For example, you might want to answer the following questions about students at a college:

> What proportion of the students at the college support a proposed student fee for improved recreational facilities?

> What is the average number of hours per month that students at the college devote to community service?

In each case, the population of interest is all students at the college. The "ideal" study would involve carrying out a **census** of the population. A census collects data from everyone in the population, so that every student at the college would be included in the study. If you were to ask every student whether he or she supported the fee or how many hours per month he or she devotes to community service, you would be able to easily answer the questions above.

Unfortunately, very few observational studies involve a census of the population. It is usually not practical to get data from every individual in the population of interest. Instead, data are obtained from just a part of the population, called a sample. Then **statistics** calculated from the sample are used to answer questions about **population characteristics**.

Population characteristic (also sometimes called a population parameter): A population characteristic is a number that describes the entire population.

Statistic: A statistic is a number that describes a sample.

There are many reasons for studying a sample rather than the entire population. Sometimes the process of measuring what you are interested in is destructive. For example, if you are interested in studying the breaking strength of glass bottles or the lifetime of batteries used to power watches, it would be foolish to study the entire population. But the most common reasons for selecting a sample are limitations on the time or the money available to carry out the study.

If you hope to learn about a population by studying a sample selected from that population, it is important that the sample be representative of the population. To be reasonably sure of this, you must carefully consider the way the sample is selected. It is sometimes tempting to take the easy way out and gather data in the most convenient way. But if a sample is chosen on the basis of convenience alone, it is not clear that what you see in the sample can be generalized to the population.

For example, it might be easy to use the students in a statistics class as a sample of students at a particular college. But not all majors are required to take a statistics course, and so students from some majors may not be represented in the sample. Also, more students take statistics in the second year of college than in the first year, so using the students in a statistics class may over-represent second-year students and under-represent first-year students. It isn't clear how these factors (and others that you may not have thought about) will affect your ability to generalize from this sample to the population of all students at the college. *There isn't any way to tell just by looking at a sample whether it is representative of the population. The only assurance comes from the method used to select the sample.*

So how do you get a representative sample? One strategy is to use a sampling method called simple random sampling. One way to select a **simple random sample** is to make sure that every individual in the population has the same chance of being selected *each* time an individual is selected into the sample. Such a selection method then ensures that each different possible sample of the desired size has an equal chance of being the sample chosen. For example, suppose that you want a simple random sample of 10 employees chosen from all people who work at a large design firm. For the sample to be a simple random sample, each of the many different subsets of 10 employees must be equally likely to be the one selected. A sample taken from only full-time employees would not be a simple random sample of *all* employees, because someone who works part-time has no chance of being selected. Although a simple random sample may, by chance, include only full-time employees, it must be *selected* in such a way that each possible sample, and therefore *every* employee, has the same chance of being selected. *It is the selection process, not the final sample, which determines whether the sample is a simple random sample.*

The letter n is used to denote sample size, and it represents the number of individuals or objects in the sample. For the design firm example of the previous paragraph, $n = 10$.

Simple random sample of size n: A simple random sample of size n is a sample selected from a population in such a way that every different possible sample of this same size n has an equal chance of being selected.

The definition of a simple random sample implies that every individual member of the population has an equal chance of being selected. *However, the fact that every individual has an equal chance of selection, by itself, does not imply that the sample is a simple random sample.* For example, suppose that a class is made up of 100 students, 60 of whom are female. You decide to select 6 of the female students by writing all 60 names on slips of paper, mixing the slips, and then picking 6. You then select 4 male students from the class using a similar procedure. Even though every student in the class has an equal chance of being included in the sample (6 of 60 females are selected and 4 of 40 males are chosen), the resulting sample is *not* a simple random sample because not all different possible samples of 10 students from the class have an equal chance of selection. Many possible samples of 10 students—for example, a sample of 7 females and 3 males or a sample of all females—have no chance of being selected. The sample selection method described here is not necessarily a bad choice (in fact, it is an example of stratified sampling, to be discussed in more detail shortly), but it does not produce a simple random sample.

Selecting a Simple Random Sample

A number of different methods can be used to select a simple random sample. One way is to put the names or numbers of all members of the population on different but otherwise identical slips of paper. The process of thoroughly mixing the slips and then selecting *n* of them yields a simple random sample of size *n*. This method is easy to understand, but it has obvious drawbacks. The mixing must be adequate, and producing all the necessary slips of paper can be extremely tedious, even for relatively small populations.

A commonly used method for selecting a random sample is to first create a list, called a **sampling frame**, of the objects or individuals in the population. Each item on the list can then be identified by a number, and a table of random digits or a random number generator can be used to select the sample. Most statistics computer packages include a random number generator, as do many calculators. A small table of random digits can be found in Appendix A, Table 1.

For example, suppose a car dealership wants to learn about customer satisfaction. The dealership has a list containing the names of the 738 customers who purchased a new car from the dealership during 2017. The owner wants you to interview a sample of 20 customers. Because it would be tedious to write all 738 names on slips of paper, you can use random numbers to select the sample. To do this, you use three-digit numbers, starting with 001 and ending with 738, to represent the individuals on the list.

The random digits from rows 6 and 7 of Appendix A, Table 1 are shown here:

0 9 3   8 7 6   7 9 9   5 6 2   5 6 5   8 4 2   6 4

4 1 0   1 0 2   2 0 4   7 5 1   1 9 4   7 9 7   5 1

You can use blocks of three digits (underlined in the lists above) to identify the individuals who should be included in the sample. The first block of three digits is 093, so the 93rd person on the list will be included in the sample. The next two blocks of three digits (876 and 799) do not correspond to anyone on the list, so you ignore them. The next block that corresponds to a person on the list is 562. This process would continue until 20 people have been selected for the sample. You would ignore any three-digit repeats since a person should only be selected once for the sample.

Another way to select the sample would be to use computer software or a graphing calculator to generate 20 random numbers. For example, when 20 random numbers between 1 and 738 were requested, the statistics software package JMP produced the following:

71   193   708   582   64   62   624   336   187   455

349   204   619   683   658   183   39   606   178   610

These numbers could be used to determine which 20 customers to include in the sample.

When selecting a random sample, you can choose to do the sampling with or without replacement. **Sampling with replacement** means that after each successive item is selected for the sample, the item is "replaced" back into the population, so the same item may be selected again. In practice, sampling with replacement is rarely used. Instead, the more common method is **sampling without replacement**. This method does not allow the same item to be included in the sample more than once. After being included in the sample, an individual or object would not be considered for further selection.

> **DEFINITIONS**
>
> **Sampling with replacement:** Sampling in which an individual or object, once selected, is put back into the population before the next selection. A sample selected with replacement might include any particular individual from the population more than once.
>
> **Sampling without replacement:** Sampling in which an individual or object, once selected, is *not* put back into the population before the next selection. A sample selected without replacement always includes *n* distinct individuals from the population.
>
> **Although these two forms of sampling are different, when the sample size *n* is small relative to the population size, as is often the case, there is little practical difference between them. In practice, the two forms can be viewed as equivalent if the sample size is not more than 10% of the population size.**

The goal of random sampling is to produce a sample that is likely to be representative of the population. Although random sampling does not *guarantee* that the sample will be representative, you will see in later chapters that it does allow you to assess the risk of an unrepresentative sample. It is the ability to quantify this risk that enables you to generalize with confidence from a random sample to the corresponding population.

An Important Note Concerning Sample Size

Here is a common misconception: If the size of a sample is relatively small compared to the population size, the sample cannot possibly accurately reflect the population. Critics of polls often make statements such as, "There are 14.6 million registered voters in California. How can a sample of 1000 registered voters possibly reflect public opinion when only about 1 in every 14,000 people is included in the sample?"

These critics do not understand the power of random selection! Consider a population consisting of 5000 applicants to a state university, and suppose that you are interested in math SAT scores for this population. A graph of the values in this population is shown in Figure 1.2(a). Figure 1.2(b) shows graphs of the math SAT scores for individuals in each of five different random samples from the population, ranging in sample size from $n = 50$ to $n = 1000$. Notice that the samples all tend to reflect the distribution of scores in the population. If we were interested in using the sample to estimate the population average or the variability in math SAT scores, even the smallest of the samples pictured ($n = 50$) would provide reliable information.

Although it is possible to obtain a simple random sample that does not represent the population, this is likely only when the sample size is very small. Unless the population itself is small, this risk does not depend on what fraction of the population is sampled. The

FIGURE 1.2
(a) Math SAT scores for an entire population. (b) Math SAT scores for random samples of sizes 50, 100, 250, 500, and 1000.

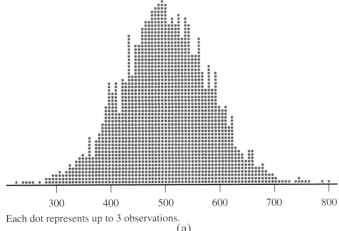

Each dot represents up to 3 observations.

(a)

(*Continued*)

FIGURE 1.2
(a) Math SAT scores for an entire population.
(b) Math SAT scores for random samples of sizes 50, 100, 250, 500, and 1000. (*Continued*)

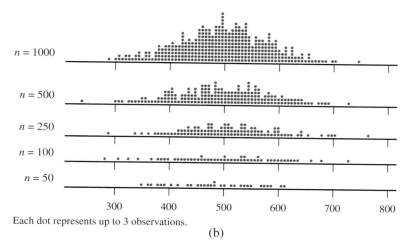

Each dot represents up to 3 observations.

(b)

random selection process allows you to be confident that the resulting sample will reflect the population, even when the sample consists of only a small fraction of the population.

Other Reasonable Sampling Strategies

Sometimes selecting a simple random sample can be difficult and impractical. For example, consider trying to take a simple random sample of leaves from a tree in order to learn about the proportion of leaves that show damage from cold weather. Or consider trying to take a simple random sample of all those attending a football game at a large stadium in order to learn about how far people traveled to get to the stadium. In situations like these, you might want to consider other strategies for selecting a sample. Above all, you want to select the sample in a way that makes it reasonable to believe that the sample will be representative of the population of interest. Two other common methods for selecting a sample from a population are stratified random sampling and systematic sampling.

When the population can be divided into a set of non-overlapping subgroups, **stratified random sampling** is often easier to implement and may be more cost-effective than simple random sampling. In stratified random sampling, separate simple random samples are independently selected from each subgroup. For example, to estimate the average cost of malpractice insurance, it might be convenient to view the population of all doctors practicing in a particular city as being made up of four subgroups: (1) surgeons, (2) internists and family practitioners, (3) obstetricians, and (4) all other doctors. Rather than taking a simple random sample from the population of *all* doctors, you could take four separate simple random samples—one from each subgroup. These four samples could be used to learn about the four subgroups as well as the overall population of doctors.

Systematic sampling is a procedure that can be used when the population of interest can be organized sequentially. A value k is specified (for example, $k = 50$ or $k = 200$). Then one of the first k individuals is selected at random, after which every k^{th} individual in the sequence is included in the sample. A sample selected in this way is called a **1 in k systematic sample**. For example, a sample of faculty members at a college might be selected from the faculty phone directory. One of the first $k = 20$ faculty members listed could be selected at random, and then every 20th faculty member after that on the list would also be included in the sample. This would result in a 1 in 20 systematic sample.

The value of k for a 1 in k systematic sample is generally chosen to achieve a desired sample size. If the population size is N and the desired sample size is n, you can determine k by dividing N by n. In the faculty directory scenario just described, if there are 900 faculty members at the college and a sample of size 45 is desired, the 1 in 20 systematic sample would achieve this result (because 900/45 = 20). If a sample size of 100 is desired, a 1 in 9 systematic sample could be used (because 900/100 = 9).

As long as there are no repeating patterns in the population list, systematic sampling works reasonably well. However, if there are patterns, systematic sampling can result in an unrepresentative sample. For example, suppose that workers at the entry station of a state park have recorded the number of visitors to the park each day for the past 10 years.

In a 1 in 70 systematic sample of days from this list, we would pick one of the first 70 days at random and then every 70th day after that. But if the first day selected happened to be a Wednesday, *every* day selected in the entire sample would also be a Wednesday (because there are 7 days in a week and 70 is a multiple of 7). Such a sample would probably not be representative of the entire population of days, because no weekend days (when the number of visitors is likely to be higher) would be included in the sample.

DEFINITION

Stratified random sampling divides a population into subgroups (strata) and then takes a separate random sample from each subgroup (stratum).

1 in k systematic sampling selects from an ordered arrangement of a population by choosing a starting point at random from the first k individuals and then selecting every k^{th} individual after that.

When you are considering how to select a sample or evaluating an observational study conducted by someone else, if the sampling plan does not result in a simple random sample, you should think about whether it might still result in a representative sample. Later in this text, you will see statistical methods that allow you to learn about a population using sample data. These methods rely on simple random sampling. If a study does not use a simple random sample, ask yourself the following question: Is the sampling strategy used likely to produce a sample that can be regarded as representative of the population?

Convenience Sampling—Don't Go There!

It may be tempting to resort to "convenience" sampling, which is using an easily available group to form a sample. This is often a recipe for disaster! Results from such samples are rarely informative, and it is a mistake to try to generalize from a convenience sample to any larger population.

One common form of convenience sampling is sometimes called **voluntary response sampling**. Such samples rely entirely on individuals who volunteer to be a part of the sample, often by responding to an advertisement, calling a publicized telephone number to register an opinion, or logging on to an Internet site to complete a survey. People who are motivated to volunteer responses often hold strong opinions, and it is extremely unlikely such individuals are representative of any larger population of interest.

One Other Consideration for Observational Studies—Avoiding Bias

Bias in sampling is the tendency for samples to differ from the corresponding population in some systematic way. Bias can result from the way in which the sample is selected or from the way in which information is obtained once the sample has been chosen. The most common types of bias encountered in sampling situations are selection bias, nonresponse bias, and measurement bias (sometimes also called response bias).

Selection bias is introduced when the way the sample is selected systematically excludes some part of the population of interest. For example, you may wish to learn about the population consisting of all residents of a particular city, but the method of selecting a sample from this population may exclude the homeless or those without telephones. If those who are excluded differ in some systematic way from those who are included, the sample is virtually guaranteed to be unrepresentative of the population. If the difference between those included and those excluded occurs consistently on a variable that is important to the study, conclusions based on the sample data may not be valid for the population of interest. Selection bias also occurs if only volunteers or self-selected individuals are used in a study, because those who choose to participate (for example, in a call-in telephone poll) may differ from those who do not choose to participate.

Measurement (response) bias occurs when the method of observation tends to produce values that systematically differ from the true value in some way. This might happen if an improperly calibrated scale is used to weigh items or if questions on a survey are worded in ways that tend to influence the response. David Wilson in his blog for *The Huffington Post* (**"Fox News Poll: What Does Bad Question Wording Say About Their Polling Data?," September 21, 2010, www.huffingtonpost.com/david-c-wilson/fox-news-poll-what-does-b_b_734101.html, retrieved September 25, 2016**) writes

> One of the main sources of error in interpreting poll results is poor question wording. In surveys questions are supposed to be accurate measures of one's attitudes, opinions, beliefs, and behaviors; just like the scales in bathrooms accurately measure one's weight. They should at least be well written and easily understood by respondents, and they should not be biased toward a particular viewpoint.

He goes on to give several examples of "bad" questions, including the following question, which was part of a Fox News poll conducted in September 2010:

> Do you agree or disagree with the following statement: the federal government has gotten totally out of control and threatens our basic liberties unless we clean house and commit to drastic change?

Wilson points out that this question makes a number of assertions, such as the government is out of control and basic liberties are threatened. If someone responded "agree" to this question, it would be unclear exactly which assertion they were in agreement with. In addition, words like *control* and *threaten* may lead people to respond in a particular way.

Other things that might contribute to response bias include the appearance or behavior of the person asking the question, the group or organization conducting the study, and the tendency for people to not be completely honest when asked about illegal behavior or unpopular beliefs.

Nonresponse bias occurs when responses are not obtained from all individuals selected into the sample. As with selection bias, nonresponse bias can distort results if those who respond differ in systematic ways from those who do not respond. Although some level of nonresponse is unavoidable in most surveys, the potential for bias is highest when the response rate is low. To minimize nonresponse bias, a serious effort should be made to follow up with individuals who do not respond to an initial request for information.

It is important to remember that bias is introduced by the way in which a sample is selected or by the way in which the data are collected. Increasing the size of the sample, although possibly desirable for other reasons, does nothing to reduce this bias if the method of selecting the sample is still flawed or if the nonresponse rate remains high.

As you consider the following examples, ask yourself these questions about the observational studies described:

What is the population of interest?

Was the sample selected in a reasonable way?

Is the sample likely to be representative of the population of interest?

Are there any obvious sources of bias?

Example 1.4 Telling Lies

The paper **"Deception and Design: The Impact of Communication Technology on Lying Behavior"** (*Computer-Human Interaction* [2009]: 130–136) describes a study designed to learn whether college students lie less often in face-to-face communication than in other forms of communication such as phone conversations or e-mail. Participants in this study were 30 students in an upper-division communications course at Cornell University who received course credit for participation. Participants were asked to record all of their social interactions for a week, making note of any lies they told. Based on data from these records, the authors of the paper concluded that students lie more often in phone conversations than in face-to-face conversations and more often in face-to-face conversations than in e-mail.

Let's consider the four questions posed just prior to this example.

Question	Response
What is the population of interest?	Because the authors wanted to learn about the behavior of college students, the population of interest is all college students.
Was the sample selected in a reasonable way?	In the study described, the authors used a convenience sample. This is problematic for two reasons—students taking the communications course may not be representative of *all* students enrolled at Cornell, and students at Cornell may not be representative of the population of *all* college students.
Is the sample likely to be representative of the population of interest?	Because the sample was a convenience sample, it is not likely to be representative of the population of interest. Even if the population of interest had been students at Cornell rather than all college students, it is still unlikely that the sample would be representative of the population.
Are there any obvious sources of bias?	The authors should be concerned about response bias. The fact that students were asked to monitor their behavior and record any lies that they told may have influenced their behavior during the week. Also, students may not have been completely truthful in reporting lies or may not have been thorough in recording their responses.

Example 1.5 **Think Before You Order That Burger!**

The article **"What People Buy from Fast-Food Restaurants: Caloric Content and Menu Item Selection" (*Obesity* [2009]: 1369–1374)** reported that the average number of calories consumed at lunch in New York City fast-food restaurants was 827. The researchers selected 267 fast-food locations at random. The paper states that at each of these locations "adult customers were approached as they entered the restaurant and asked to provide their food receipt when exiting and to complete a brief survey."

Let's again consider the four questions posed previously.

Question	Response
What is the population of interest?	Because the researchers wanted to learn about the behavior of people who eat at fast-food restaurants in New York City, this is the population of interest.
Was the sample selected in a reasonable way?	Because it would not be feasible to select a simple random sample from the population of interest, the researchers made an attempt to obtain a representative sample by randomly selecting the fast-food restaurants that would be used in the study. This means that every fast-food location in the city had the same chance of being included in the study. This is a reasonable strategy given the difficulty of sampling from this population.
Is the sample likely to be representative of the population of interest?	Because the locations were selected at random from all locations in the city, the researchers believed that it was reasonable to regard the people eating at these locations as representative of the population of interest.
Are there any obvious sources of bias?	Approaching customers as they entered the restaurant and before they ordered may have influenced their purchases. This introduces the potential for response bias. In addition, some people chose to not participate when approached. If those who chose to not participate differed from those who did participate, the researchers also need to be concerned about nonresponse bias. Both of these potential sources of bias limit the researchers' ability to generalize conclusions based on data from this study.

Summing It Up—Section 1.3

The following learning objectives were addressed in this section:

Conceptual Understanding

C3: Explain the difference between a census and a sample.

A **census** collects data from every individual in a population. A **sample** is a subset of the population.

C4: Explain the difference between a statistic and a population characteristic.

A **statistic** is a number that describes a sample, whereas a **population characteristic** is a number that describes the entire population. Population characteristics are also sometimes called **population parameters**.

C5: Understand why random selection is an important component of a sampling plan.

Random selection ensures that every individual in the population has an equal chance of being selected into the sample. This makes it likely that the resulting sample will be representative of the population.

C8: Explain why volunteer response samples and convenience samples are unlikely to produce reliable information about a population.

Although there are a number of ways to select a sample that should result in a representative sample (such as simple random sampling, stratified random sampling, and systematic sampling), volunteer response samples and convenience samples are usually not representative of the population. See the discussion in the subsections on convenience sampling and avoiding bias.

Mastering the Mechanics

M1: Create a sampling plan that could produce a simple random sample from a given population.

A simple random sample of size n is a sample that is selected in a way that ensures that every different sample of that same size from the population has an equal chance of being selected. This also implies that every individual in the population will have the same chance of being included in the sample. There are a number of methods that can be used to select a random sample. See the discussion in Section 1.3 on selecting a simple random sample.

Putting It into Practice

P2: Evaluate the design of an observational study.

Now that you have learned about observational studies, you should be able to decide whether an observational study has been conducted in a reasonable way. When evaluating an observational study, you should think about whether the sample was selected in a way that makes it likely that it will be representative of the population of interest and whether there are any obvious sources of bias. For examples of evaluating the design of an observational study, see Examples 1.4 and 1.5.

SECTION 1.3 EXERCISES

Each Exercise Set assesses the following chapter learning objectives: C3, C4, C5, C8, M1, P2

SECTION 1.3 Exercise Set 1

1.15 The article **"Bicyclists and Other Cyclists"** (*Annals of Emergency Medicine* [2010]: 426) reported that in 2008, 716 bicyclists were killed on public roadways in the United States and that the average age of the cyclists killed was 41 years. These figures were based on an analysis of the records of all traffic-related deaths of bicyclists on U.S. public roadways (this information is kept by the National Highway Traffic Safety Administration).

a. Does the group of 716 bicycle fatalities represent a census or a sample of the bicycle fatalities in 2008?

b. If the population of interest is bicycle traffic fatalities in 2008, is the given average age of 41 years a statistic or a population characteristic?

1.16 "San Fernando Valley residents OK with 1-cent transit tax, MTA poll says" is the headline of an article that appeared in the *LA Daily News* (April 12, 2016). This headline was based on responses from a sample of 100 San Fernando Valley residents. Describe the sample and the population of interest for this poll.

1.17 The article referenced in the previous exercise includes the following statement:

> Nearly three out of four San Fernando Valley residents support an upcoming sales tax measure to pay for traffic congestion relief, according to an MTA poll this week. The Los Angeles County Metropolitan Transportation Authority polled roughly 100 residents at a community forum in Van Nuys on Monday, of whom 72 percent said they would vote yes on a 1-cent sales tax to raise more than $120 billion for new highway, bus and rail projects and more, Metro officials reported Tuesday.

Based on the way this sample was selected, explain why the headline given in the previous exercise is not appropriate. Suggest a more appropriate headline.

1.18 The article **"Adolescents Living the 24/7 Lifestyle: Effects of Caffeine and Technology on Sleep Duration and Daytime Functioning" (*Pediatrics* [2009]: e1005–e1010)** reported that about 33% of teenagers have fallen asleep during school. Is this true for students at your school? Suppose that a list of all the students at your school is available. Describe the steps in a process that would use the list to select a simple random sample of 150 students. (Hint: See discussion on page 11 on selecting a random sample.)

1.19 The article **"Teenage Physical Activity Reduces Risk of Cognitive Impairment in Later Life" (*Journal of the American Geriatrics Society* [2010])** describes a study of more than 9000 women over 50 years old from Maryland, Minnesota, Oregon, and Pennsylvania. The women were asked about their physical activity as teenagers and at ages 30 and 50. A press release about this study (www.wiley.com/WileyCDA/PressRelease/pressReleaseId-77637.html, retrieved September 25, 2016) generalized the study results to all American women. In the press release, the researcher who conducted the study is quoted as saying

> Our study shows that women who are regularly physically active at any age have lower risk of cognitive impairment than those who are inactive, but that being physically active at teenage is most important in preventing cognitive impairment.

Answer the following four questions for this observational study. (Hint: Reviewing Examples 1.4 and 1.5 might be helpful.)
a. What is the population of interest?
b. Was the sample selected in a reasonable way?
c. Is the sample likely to be representative of the population of interest?
d. Are there any obvious sources of bias?

SECTION 1.3 Exercise Set 2

1.20 Data from a survey of 10,413 students and 588 teachers conducted for the Knight Foundation were used to calculate the following estimates: 65% of students and 40% of teachers chose freedom of speech as the most important of the rights guaranteed by the First Amendment (**USA TODAY, October 3, 2014**). Are the given percentages statistics or population characteristics?

1.21 The article **"Soldiers Hate Their Jobs" (USA TODAY, April 17, 2015)** reported that 52% of soldiers surveyed were pessimistic about their future in the military. This statement was based on data collected in 2014 in an annual assessment that all soldiers must take each year. If the population of interest is all soldiers serving in 2014, does the group of soldiers surveyed represent a census or a sample? Is the reported percentage (52%) a statistic or a population characteristic?

1.22 **"Should You Get a Flu Shot? Your Physical and Financial Health Is on the Line"** is the title of an article that appears in a blog on the WalletHub web site (**December 20, 2013, wallethub.com/blog/flu-shot-survey/1303/, retrieved September 25, 2016**). The author reported that an infectious disease expert from a top medical school in each of the 50 states was asked if he or she would recommend that the average person get a flu shot. Based on the 50 responses, it was reported that 94% would recommend a flu shot.
a. Suppose that the purpose of this survey was to estimate the percentage of all doctors who would recommend a flu shot. Would this sample be a simple random sample, a stratified sample, a systematic sample, or a convenience sample? Explain.
b. Explain why an estimate of the percentage who would recommend a flu shot that was based on data from this sample should not be generalized to all doctors.

1.23 A New York psychologist recommends that if you feel the need to check your e-mail in the middle of a movie or if you sleep with your cell phone next to your bed, it might be time to "power off" (**AARP Bulletin, September 2010**). Suppose that you want to learn about the proportion of students at your college who feel the need to check e-mail during the middle of a movie, and that you have access to a list of all students enrolled at your college. Describe the steps in a process that would use this list to select a simple random sample of 100 students.

1.24 A survey of Arizona drivers is described in the article **"Study Claims Safety Should Be Made Law" (*Red Rock News*, August 21, 2015).** The following statement is from the article:

> According to the annual survey, which aims to gauge the opinions and concerns of the motoring public across the state regarding traffic safety, Arizona drivers want better traffic safety laws. The survey showed that nine in 10 Arizonans—91 percent—favor a statewide ban on texting while driving for all drivers, and about two-thirds—64 percent—favor a primary seat belt law.

The article also describes how the data for this survey were collected. A survey was mailed to 2500 randomly selected AAA Arizona members (AAA Arizona is an automobile club that provides services to approximately 860,000 members in Arizona). The article did not indicate how many of the surveys were actually returned.

The results of this survey were generalized to all Arizona drivers. Answer the following four questions for this observational study. (Hint: Reviewing Examples 1.4 and 1.5 might be helpful.)
a. What is the population of interest?
b. Was the sample selected in a reasonable way?
c. Is the sample likely to be representative of the population of interest?
d. Are there any obvious sources of bias?

ADDITIONAL EXERCISES

1.25 The **SurveyMonkey Blog (February 11, 2015, retrieved September 25, 2016)** includes an article titled **"5 Common Survey Question Mistakes That'll Ruin Your Data."** Read this short article, which can be found at the following website, and then answer the following questions. www.surveymonkey.com/blog/2015/02/11/5 -common-survey-mistakes-ruin-your-data/
a. One of the recommendations in the article is "Don't write leading questions." Give an example of a leading question that is different from the two examples given in the article. Explain why you think the question is a leading question and then suggest a better way to word the question.
b. Select one of the other four recommendations and give an example of a bad question related to that recommendation. Then suggest a better way to word the question.

1.26 The supervisors of a rural county are interested in the proportion of property owners who support the construction of a sewer system. Because it is too costly to contact all 7000 property owners, a survey of 500 owners (selected at random) is undertaken. Describe the population and the sample for this problem.

1.27 A building contractor has a chance to buy an odd lot of 5000 used bricks at an auction. She is interested in determining the proportion of bricks in the lot that are cracked and therefore unusable for her current project, but she does not have enough time to inspect all 5000 bricks. Instead, she checks 100 bricks to determine whether each is cracked. Describe the population and the sample for this problem.

1.28 In 2000, the chairman of a California ballot initiative campaign to add "none of the above" to the list of ballot options in all candidate races was quite critical of a Field poll that showed his measure trailing by 10 percentage points. The poll was based on a random sample of 1000 registered voters in California. He is quoted by the **Associated Press (January 30, 2000)** as saying, "Field's sample in that poll equates to one out of $17,505$ voters." This was so dishonest, he added, that Field should get out of the polling business! If you worked on the Field poll, how would you respond to this criticism?

1.29 Whether or not to continue a Mardi Gras Parade through downtown San Luis Obispo, California, is a hotly debated topic. The parade is popular with students and many residents, but some celebrations have led to complaints and a call to eliminate the parade. The local newspaper conducted both an online survey and a telephone survey of its readers and was surprised by the results. The online survey received more than 400 responses, with more than 60% favoring continuing the parade, while the telephone response line received more than 120 calls, with more than 90% favoring banning the parade (*San Luis Obispo Tribune,* **March 3, 2004)**. What factors may have contributed to these very different results?

SECTION 1.4 # Collecting Data—Planning an Experiment

Experiments are usually conducted to answer questions such as "What happens when …?" or "What is the effect of …?" For example, tortilla chips are made by frying tortillas, and they are best when they are crisp, not soggy. One measure of crispness is moisture content. What happens to moisture content when the frying time of a chip is 30 seconds compared to frying times of 45 seconds or 60 seconds? An experiment could be designed to investigate the effect of frying time on moisture content.

It would be nice if you could just take three tortillas and fry one for 30 seconds, one for 45 seconds, and one for 60 seconds and then compare the moisture content of the three chips. However, even if we take two tortillas and fry each one for 30 seconds, there will be variability in moisture content. There are many reasons why this might occur: for example, small variations in environmental conditions, small changes in the temperature of the oil used to fry the chips, slight differences in the composition of the tortillas, and so on. This creates chance-like variability in the moisture content of chips—even for chips that are fried for the same amount of time. To be able to compare different frying times, you need to distinguish the variability in moisture content that is caused by differences in the frying time from the chance-like variability. A well-designed experiment produces data that allow you to do this.

In a **simple comparative experiment**, the value of a **response variable** (such as moisture content) is measured under different **experimental conditions** (for example, frying times). Experimental conditions are also sometimes called **treatments**, and the terms experimental conditions and treatments are used interchangeably. An **experimental unit** is the smallest unit to which a treatment is applied, such as individual people in an experiment to investigate the effect of room temperature on exam performance or the individual chips in the tortilla chip example.

The design of an experiment is the overall plan for conducting an experiment. A good design makes it possible to obtain unambiguous answers to the questions that the experiment was designed to answer. It does this by allowing you to distinguish the variability in the response that is attributable to differing experimental conditions from variability due to other sources.

In designing an experiment, you need to ensure that there are no systematic (as opposed to chance-like) sources of variation in the response that you can't isolate. For example, suppose that in the tortilla chip experiment you used three different batches of tortillas. One batch was used for all chips fried for 30 seconds, the second batch for all chips fried for 45 seconds, and the third for all chips fried for 60 seconds. This is not a good way to carry out the experiment because there might be differences in the composition of the tortillas in the three batches. This could produce variability in the response (moisture content) that is not distinguishable from variability in the response due to the experimental conditions (the frying times).

When you cannot distinguish between the effects of two variables on the response, the two variables are **confounded**. In the situation where three different batches of tortillas were used, the tortilla batch is called a **confounding variable**, and you would say that tortilla batch and frying time are confounded. If you observe a difference in moisture content for the three frying times, you can't tell if the difference is due to differences in frying time, differences in the tortilla batches, or some combination of both. A well-designed experiment will protect against potential confounding variables.

Design Strategies for Simple Comparative Experiments

The goal of a simple comparative experiment is to determine the effect of the experimental conditions (treatments) on the response variable of interest. To do this, you must consider other potential sources of variability in the response and then design your experiment to either eliminate them or ensure that they produce chance-like, as opposed to systematic, variability.

Eliminating Sources of Variability through Direct Control

An experiment can be designed to eliminate some sources of variability by using **direct control**. Direct control means holding a potential source of variability constant at some fixed level. For example, in the tortilla chip experiment, you might suspect that the oil temperature and the shape of the pan used for frying are possible sources of variability in the moisture content. You could eliminate these sources of variability through direct control by using just a single pan and by having a fixed oil temperature that is maintained for all chips. In general, if there are potential sources of variability that are easy to control, this should be incorporated into the experimental design.

Ensuring That Remaining Sources of Variability Produce Chance-Like Variability: Random Assignment

What about other sources of variability that can't be controlled, such as small differences in tortilla size in the tortilla chip example? These sources of variability are handled by **random assignment** to experimental groups. Random assignment ensures that an experiment does not favor one experimental condition over any other and tries to create "equivalent" experimental groups (groups that are as much alike as possible). Random assignment is an *essential* part of a good experimental design.

To see how random assignment tends to create similar groups, suppose 50 first-year college students participate in an experiment to investigate whether completing an online review of course material prior to an exam improves exam performance. The 50 subjects vary quite a bit in achievement, as shown in their math and verbal SAT scores, which are displayed in Figure 1.3.

FIGURE 1.3
Math SAT and Verbal SAT Scores for 50 First-Year Students

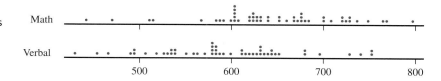

If these 50 students are to be assigned to the two experimental groups (one that will complete the online review and one that will not), you want to make sure that the assignment of students to groups does not tend to place the higher achieving students in one group and the lower achieving students in the other.

Random assignment makes it easy to create groups of students with similar achievement levels in both verbal and math SAT scores simultaneously. Figure 1.4(a) shows the verbal SAT scores of the students assigned to each of the two experimental groups (one shown in green and one shown in red) for each of three different random assignments of students to groups. Figure 1.4(b) shows the math SAT scores for the two experimental groups for each of the same three random assignments. Notice that each of the three random assignments produced groups that are quite similar with respect to *both* verbal and math SAT scores. So, if any of these three assignments were used and the two groups differed on exam performance, you could rule out differences in math or verbal SAT scores as possible competing explanations for the difference.

FIGURE 1.4
Three Different Random Assignments to Two Groups, One Shown in Green, One Shown in Red (a) Verbal SAT Score (b) Math SAT Score

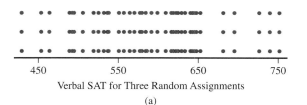

Verbal SAT for Three Random Assignments

(a)

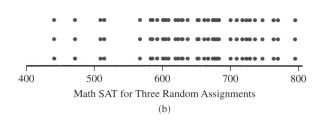

Math SAT for Three Random Assignments

(b)

Not only will random assignment tend to create groups that are similar in verbal and math SAT scores, but it will also even out the groups with respect to other potentially confounding variables. As long as the number of subjects is not too small, you can rely on the random assignment to produce comparable experimental groups. This is why random assignment is a part of all well-designed experiments.

As you have probably noticed, random assignment in an experiment is different from random selection of experimental units. Random assignment is a way of determining which experimental units go in each of the different experimental groups. Random selection is one possible way of choosing the experimental units. The ideal situation would be to have both random selection of experimental units and random assignment of these units to experimental conditions. This would allow conclusions from the experiment to be generalized to a larger population. However, for many experiments the random selection of experimental units is not possible. Even so, as long as experimental units are randomly assigned to experimental conditions, it is still possible to assess treatment effects.

Not all experiments require the use of human subjects to evaluate different experimental conditions. For example, a researcher might be interested in comparing the performance of three different gasoline additives as measured by gasoline mileage. The experiment might involve using a single car with an empty gas tank. One gallon of gas with one of the additives could be put in the tank and the car driven along a standard route at a constant speed until it runs out of gas. The total distance traveled could then be recorded. This could be repeated a number of times, 10 for example, with each additive.

This gas mileage experiment can be viewed as a sequence of trials. Because there are a number of potentially confounding variables that might have an effect on gas mileage (such as variations in environmental conditions, like wind speed or humidity, and small variations in the condition of the car), it would not be a good idea to use additive 1 for the first 10 trials, additive 2 for the next 10, and so on. An approach that would not unintentionally favor any of the additives would be to randomly assign additive 1 to 10 of the 30 planned trials, and then randomly assign additive 2 to 10 of the remaining 20 trials. The resulting plan might look as follows:

Trial	1	2	3	4	5	6	7	⋯	30
Additive	2	2	3	3	2	1	2	⋯	1

When an experiment can be viewed as a sequence of trials, you randomly assign treatments to trials. **Random assignment—either of subjects to treatments or of treatments to trials—is a critical component of a good experiment.**

There are several strategies that can be used to randomly assign subjects to treatments (experimental conditions). Two common strategies are:

- Write the name of each subject or a unique number that corresponds to a subject on a slip of paper. Place all of the slips in a container and mix well. Then draw out the desired number of slips to determine those who will be assigned to the first treatment group. This process of drawing slips of paper then continues until all treatment groups have been determined.
- Assign each subject a unique number from 1 to n, where n represents the total number of subjects. Use a random number generator or a table of random numbers to obtain numbers that will identify which subjects will be assigned to the first treatment group. This process would be repeated, ignoring any random numbers generated that correspond to subjects that have already been assigned to a treatment group, until all treatment groups have been formed.

The two strategies above work well and can be used for experiments in which the desired number of subjects in each treatment group has been predetermined.

Another strategy that is sometimes employed is to use a random mechanism (such as tossing a coin or rolling a die) to determine which treatment will be assigned to a particular subject. For example, in an experiment with two treatments, you might toss a coin to determine if the first subject is assigned to treatment 1 or treatment 2. This could continue for each subject—if the coin lands Heads, the subject is assigned to treatment 1, and if the coin lands Tails, the subject is assigned to treatment 2. This strategy is fine, but may result in treatment groups of unequal size. For example, in an experiment with 100 subjects, 53 might be assigned to treatment 1 and 47 to treatment 2. If this is acceptable, the coin flip strategy is a reasonable way to assign subjects to treatments.

Other Considerations When Planning an Experiment

The goal of an experimental design is to provide a method for collecting data that

1. Minimizes sources of variability in the response (other than the experimental conditions) so that any differences in response due to experimental conditions are easy to assess.

2. Creates experimental groups that are similar with respect to potential confounding variables that cannot be controlled.

Here are some additional considerations that you may need to think about when planning an experiment.

Use of a Control Group

If the purpose of an experiment is to determine whether some treatment has an effect, it is important to include an experimental group that does not receive the treatment. Such a group is called a **control group**. A control group allows the experimenter to assess how the response variable behaves when the treatment is not used. This provides a baseline against which the treatment groups can be compared to determine if the treatment has an effect.

Use of a Placebo

In experiments that use human subjects, the use of a control group may not be enough to determine if a treatment really does have an effect. People sometimes respond merely to the power of suggestion! For example, consider a study designed to determine if a particular herbal supplement is effective in promoting weight loss. Suppose the study is designed with an experimental group that takes the herbal supplement and a control group that takes nothing. It is possible that those taking the herbal supplement may believe that it will help them to lose weight. As a result, they may be more motivated and may unconsciously change their eating behavior or activity level, resulting in weight loss.

Although there is debate about the degree to which people respond, many studies have shown that people sometimes respond to treatments with no active ingredients, such as sugar pills or solutions that are nothing more than colored water. They often report that such "treatments" relieve pain or reduce symptoms such as nausea or dizziness. To determine if a treatment *really* has an effect, comparing a treatment group to a control group may not be enough. To address this problem, many experiments use what is called a placebo. A **placebo** is something that is identical (in appearance, taste, feel, etc.) to the treatment received by the treatment group, except that it contains no active ingredients.

In the herbal supplement study, rather than using a control group that receives *no* treatment, the researchers might want to use a placebo group. Individuals in the placebo group would take a pill identical in appearance to the herbal supplement but which contains no herb or any other active ingredient. Because the subjects don't know if they are taking the herb or the placebo, the placebo group provides a better basis for comparison. This allows the researchers to determine if the herbal supplement has any real effect over and above the "placebo effect." For an interesting discussion of the use of placebos and the placebo effect, see www.cancer.org/treatment/treatmentsandsideeffects/treatmenttypes /placebo-effect (retrieved April 10, 2017).

Single-Blind and Double-Blind Experiments

Because people often have their own beliefs about the effectiveness of various treatments, it is preferable to conduct experiments in which subjects do not know what treatment they are receiving. Such an experiment is called **single-blind**. Of course, not all experiments can be made single-blind. For example, in an experiment comparing the effect of two different types of exercise on blood pressure, it is not possible for participants to be unaware of whether they are in the swimming group or the jogging group! However, when it is possible, "blinding" the subjects in an experiment is generally a good strategy.

In some experiments, someone other than the subject is responsible for measuring the response. To ensure that they do not let personal beliefs influence their measurements, it is also preferable for this person to be unaware of which treatment each individual subject received. For example, in a medical experiment to determine if a new vaccine reduces the risk of getting the flu, doctors must decide whether a particular individual who is not feeling well actually has the flu or has some other unrelated illness. If the doctor making this assessment knows that a subject with flu-like symptoms has been vaccinated with the new flu vaccine, she might be less likely to determine that the subject has the flu and more likely to interpret the symptoms to be the result of some other illness.

There are two ways in which "blinding" might occur in an experiment. One involves blinding the subjects, while the other involves blinding the individuals who measure the response. If both subjects and those measuring the response do not know which treatment was received, the experiment is described as **double-blind**. If only one of the two types of blinding is present, the experiment is single-blind.

Using Volunteers as Subjects in an Experiment

Although using volunteers in a study that involves collecting data through sampling is never a good idea, it is common practice to use volunteers as subjects in an experiment. Even though using volunteers limits the researcher's ability to generalize to a larger population, random assignment of the volunteers to treatments should result in comparable groups, and so treatment effects can still be assessed.

Evaluating an Experiment

As you consider the following examples, ask yourself these questions about the experiments described:

1. What question is the experiment trying to answer?
2. What are the experimental conditions (treatments) for the experiment?
3. What is the response variable?
4. What are the experimental units and how were they selected?
5. Does the design incorporate random assignment of experimental units to the different experimental conditions? If not, are there potentially confounding variables that would make it difficult to draw conclusions based on data from the experiment?
6. Does the experiment incorporate a control group and/or a placebo group? If not, would the experiment be improved by including one or both of these?
7. Does the experiment involve blinding? If not, would the experiment be improved by making it single- or double-blind?

Example 1.6 | **Morality in the Morning**

The article **"The Morning Morality Effect: The Influence of Time of Day on Unethical Behavior"** (*Psychological Science* [2014]: 95–102) describes four studies that investigated whether people are more ethical in the morning than in the afternoon. In one of the studies (Experiment 1), volunteers selected either a morning or an afternoon session. During the session participants were paid to complete a task that presented opportunities to increase the amount earned by purposely providing incorrect responses. The researchers concluded that participants in the afternoon sessions cheated in order to increase the amount they would be paid significantly more often than those in the morning session.

Can you spot an obvious flaw in this study? You probably noticed that the researchers did not randomly assign participants to the two experimental conditions (morning time and afternoon time). This flaw was actually noted in the paper, and the authors stated

> An important limitation of the two previous experiments was that participants self-selected a morning or afternoon session. It is possible that unethical people, in general, are more likely to sign up for afternoon sessions than ethical people are; if true, this would provide an alternative explanation for our previous findings.

To address this limitation, two other experiments were also carried out. In one of these experiments, 70 volunteers were randomly assigned to a morning or afternoon session. In these sessions, participants were shown 20 sets each containing 12 numbers with three digits, such as 6.38. For each of these 20 sets, participants were instructed to look for a pair of numbers that added to 10 (such as 3.86 and 6.14). Each set of numbers was shown for 15 seconds and then participants indicated whether or not they had found a pair of numbers that added to 10. Participants did not have to specify what those numbers were, and were told that they would earn money for each pair that they found. Although the participants did not know this, only 10 of the 20 sets of numbers actually contained two numbers that added to 10. This allowed the researchers to assess whether people were cheating on the task by saying they had found the numbers in order to increase the amount they earned. The researchers found that those in the afternoon session reported finding significantly more pairs that added to 10 than the participants in the morning session. This led researchers to conclude that people are more honest in the morning.

This experiment incorporated random assignment to experimental conditions and allowed the researchers to conclude that the time of day was the cause of the difference in the way the two groups responded. Let's consider the seven questions posed just prior to this example:

Question	Response
What question is the experiment trying to answer?	The researchers wanted to find out if people are more likely to be dishonest in the afternoon than in the morning.
What are the experimental conditions (treatments) for the experiment?	There are two experimental conditions in this experiment. They are morning session and afternoon session.
What is the response variable?	The response variable is the number of times that a participant reported finding pairs of numbers that added to 10.
What are the experimental units and how were they selected?	The experimental units are the 70 volunteers.
Does the design incorporate random assignment of experimental units to the different experimental conditions? If not, are there potentially confounding variables that would make it difficult to draw conclusions based on data from the experiment?	Yes, the subjects were randomly assigned to one of the two experimental conditions.
Does the experiment incorporate a control group and/or a placebo group? If not, would the experiment be improved by including one or both of these?	No, but there is no need for a control group in this experiment.
Does the experiment involve blinding? If not, would the experiment be improved by making it single- or double-blind?	No. It would not be possible to include blinding of subjects in this experiment because they would know if they were in a morning session or an afternoon session, and there is no need to blind the person recording the response.

Example 1.7 Chilling Newborns? Then You Need a Control Group...

Researchers for the National Institute of Child Health and Human Development studied 208 infants whose brains were temporarily deprived of oxygen as a result of complications at birth **(*The New England Journal of Medicine*, October 13, 2005)**. These babies were subjects in an experiment to determine if reducing body temperature for three days after birth improved their chances of surviving without brain damage. The experiment was summarized in a paper that stated "infants were randomly assigned to usual care (control group) or whole-body cooling." Including a control group in the experiment provided a basis for comparison of death and disability rates for the proposed cooling treatment and those for usual care. Some variables that might also affect death and disability rates, such as duration of oxygen deprivation, could not be directly controlled. To ensure that the experiment did not unintentionally favor one experimental condition over the other, random assignment of the infants to the two groups was critical. Because this was a well-designed experiment, the researchers were able to use the resulting data and methods that you will see in Chapter 11 to conclude that cooling did reduce the risk of death and disability for infants deprived of oxygen at birth.

Consider the seven questions previously posed as a way of organizing your thinking about this experiment:

Question	Response
What question is the experiment trying to answer?	Does chilling newborns whose brains were temporarily deprived of oxygen at birth improve the chance of surviving without brain damage?
What are the experimental conditions (treatments) for the experiment?	The experiment compared two experimental conditions: usual care and whole-body cooling.
What is the response variable?	This experiment used two response variables—*survival* (yes, no) and *brain damage* (yes, no).
What are the experimental units and how were they selected?	The experimental units were the 208 newborns used in the study. The description of the experiment does not say how the newborns were selected, but it is unlikely that they were a random sample from some larger population.
Does the design incorporate random assignment of experimental units to the different experimental conditions? If not, are there potentially confounding variables that would make it difficult to draw conclusions based on data from the experiment?	Yes, the babies were randomly assigned to one of the two experimental conditions.
Does the experiment incorporate a control group and/or a placebo group? If not, would the experiment be improved by including one or both of these?	In this experiment, the usual care group serves as a control group. This group did not receive any cooling treatment.
Does the experiment involve blinding? If not, would the experiment be improved by making it single- or double-blind?	This experiment did not involve blinding. It might have been possible to blind the person who makes the assessment of whether or not there was brain damage, but this probably isn't necessary in this experiment.

Summing It Up—Section 1.4

The following learning objectives were addressed in this section:

Conceptual Understanding

C6: Understand why random assignment is important when collecting data in an experiment.

Random assignment ensures that an experiment does not favor one experimental condition (treatment) over any other and tries to create experimental groups that are as much alike as possible. Random assignment is an essential part of any good experimental design.

C10: Explain the purpose of a control group in an experiment.

A **control group** is a group that does not receive a treatment. This provides a baseline so that it is possible to determine if a treatment has an effect.

C11: Explain the purpose of blinding in an experiment.

Blinding occurs when either the subjects in an experiment or the person measuring the response in an experiment or both don't know which treatment was received. An experiment can be single-blind or double-blind. The purpose of blinding is to ensure that personal beliefs about subjects or the effectiveness of the treatments do not influence the results.

Mastering the Mechanics
M2: Describe a procedure for randomly assigning experimental units to experimental conditions given a description of an experiment, the experimental conditions, and the experimental units.

There are several reasonable methods for randomly assigning experimental units to experimental conditions. See the discussion in the subsection of Section 1.4, "Ensuring That Remaining Sources of Variability Produce Chance-Like Variability: Random Assignment."

Putting It into Practice
P3: Evaluate the design of a simple comparative experiment.

Now that you have learned about experiments, you should be able to decide whether an experiment has been conducted in a reasonable way. When evaluating an experiment, you should think about whether there is random assignment to experimental conditions (treatments) and whether there is appropriate use of a control group and blinding. For examples of evaluating the design of an experiment, see Examples 1.6 and 1.7.

SECTION 1.4 EXERCISES

Each Exercise Set assesses the following chapter learning objectives: C6, C10, C11, M2, P3

SECTION 1.4 Exercise Set 1

1.30 Does playing action video games provide more than just entertainment? The authors of the paper **"Action-Video-Game Experience Alters the Spatial Resolution of Vision"** (*Psychological Science* [2007]: 88–94) concluded that spatial resolution, an important aspect of vision, is improved by playing action video games. They based this conclusion on data from an experiment in which 32 volunteers who had not played action video games were "equally and randomly divided between the experimental and control groups." Subjects in each group played a video game for 30 hours over a period of 6 weeks. Those in the experimental group played *Unreal Tournament 2004*, an action video game. Those in the control group played *Tetris*, a game that does not require the user to process multiple objects at once. Explain why it was important for the researchers to randomly assign the subjects to the two groups.

1.31 In an experiment to assess the effect of wearing compression socks during a marathon, 20 runners in the 2013 Hartford Marathon were randomly assigned to two groups **("Compression and Clots in Athletes Who Travel,"** *Lower Extremities Review*, lermagazine.com/ler-archives/january -2016, retrieved July 2, 2017). Runners in one group wore a pair of compression socks during the marathon, while runners in the second group wore regular athletic socks. At the end of the marathon, blood samples were taken to measure variables related to preventing blood clots and speeding up recovery from exercise.
a. Describe why it was important for the researchers to assign participants to one of the two groups rather than letting the participants choose which group they wanted to be in.

b. The authors of the paper state that there is some evidence that suggests that wearing compression socks may result in a psychological advantage that might translate into performance gains. Suppose that instead of a response variable that was determined by a blood test, the response variable had been the time it took the runner to complete the marathon. Do you think it would be a good idea to have the runners be blind to the type of socks that they were given? Explain why or why not.

1.32 In an experiment to compare two different surgical procedures for hernia repair **("A Single-Blinded, Randomized Comparison of Laparoscopic Versus Open Hernia Repair in Children,"** *Pediatrics* [2009]: 332–336), 89 children were assigned at random to one of the two surgical methods. The methods studied were laparoscopic repair and open repair. In laparoscopic repair, three small incisions are made, and the surgeon works through these incisions with the aid of a small camera that is inserted through one of the incisions. In the open repair, a larger incision is used to open the abdomen. One of the response variables was the amount of medication given after the surgery to control pain and nausea. The paper states, "For postoperative pain, rescue fentanyl (1 mg/kg) and for nausea, ondansetron (0.1 mg/kg) were given as judged necessary by the attending nurse blinded to the operative approach."
a. Why do you think it was important that the nurse who administered the medications did not know which type of surgery was performed?
b. Explain why it was not possible for this experiment to be double-blind.

1.33 A study of college students showed a temporary gain of up to nine IQ points after listening to a Mozart piano sonata. This result, dubbed the Mozart effect, has since been criticized by a number of researchers who have been unable to confirm the result in similar studies. Suppose that you want to determine if there is really is a Mozart effect. You decide to carry out an experiment with three experimental groups. One group will listen to a Mozart piano sonata that lasts 24 minutes. The second group will listen to popular music for the same length of time, and the third group will relax for 24 minutes with no music playing. You will measure IQ before and after the 24 minute period. Suppose that you have 45 volunteers who have agreed to participate in the experiment. Describe the steps in a process you could use to randomly assign each of the volunteers to one of the experimental groups.

1.34 The paper **"Effect of a Nutritional Supplement on Hair Loss in Women"** (*Journal of Cosmetic Dermatology* [2015]: 76–82) describes an experiment to see if a dietary supplement consisting of Omega 3, Omega 6, and antioxidants could reduce hair loss in women with stage 1 hair loss. One hundred twenty women volunteered to participate in the study and were randomly assigned to either the supplement group or a control group. The women in the supplement group took the supplement for 6 months. Photos of the top of the head were taken of all the women at the beginning of the study and 6 months later at the end of the study. The two photos of each woman were evaluated by an independent expert who visually determined the change in hair density. The expert who determined the change in hair density did not know which of the women had taken the supplement.

Answer the following seven questions for the described experiment. (Hint: Reviewing Examples 1.6 and 1.7 might be helpful.)

1. What question is the experiment trying to answer?

2. What are the experimental conditions (treatments) for this experiment?

3. What is the response variable?

4. What are the experimental units and how were they selected?

5. Does the design incorporate random assignment of experimental units to the different experimental conditions? If not, are there potentially confounding variables that would make it difficult to draw conclusions based on data from the experiment?

6. Does the experiment incorporate a control group and/or a placebo group? If not, would the experiment be improved by including one or both of these?

7. Does the experiment involve blinding? If not, would the experiment be improved by making it single- or double-blind?

SECTION 1.4 Exercise Set 2

1.35 The Institute of Psychiatry at Kings College London found that dealing with "infomania" has a temporary, but significant, negative effect on IQ (*Discover,* **November 2005**). To reach this conclusion, researchers divided volunteers into two groups. Each subject took an IQ test. One group had to check e-mail and respond to instant messages while taking the test, and the other group took the test without any distraction. The distracted group had an average score that was 10 points lower than the average for the control group. Explain why it is important that the researchers use random assignment to create the two experimental groups.

1.36 The article **"Study Points to Benefits of Knee Replacement Surgery Over Therapy Alone"** (*New York Times,* **October 21, 2015**) describes a study to compare two treatments for people with knee pain. In the study, 50 people with arthritis received knee replacement surgery followed by a program of exercise. Another 50 people with arthritis did not have surgery but received the same program of exercise. After 1 year, 85% of the people who had surgery and 68% of the people who did not have surgery reported pain relief.

a. Why is it important to determine if the researchers randomly assigned the subjects to one of the two groups?

b. Explain why you think that the researchers might have wanted to include a control group in this study.

1.37 The article **"Doctor Dogs Diagnose Cancer by Sniffing It Out"** (*Knight Ridder Newspapers,* **January 9, 2006**) refers to an experiment described in the journal *Integrative Cancer Therapies*. In this experiment, dogs were trained to distinguish between people with breast and lung cancer and people without cancer by sniffing exhaled breath. Dogs were trained to lie down if they detected cancer in a breath sample. After training, the dogs' ability to detect cancer was tested using breath samples from people whose breath had not been used in training the dogs. The paper states, "The researchers blinded both the dog handlers and the experimental observers to the identity of the breath samples." Explain why this blinding is an important aspect of the design of this experiment.

1.38 Suppose that you would like to know if keyboard design has an effect on wrist angle, as shown in the accompanying figure.

You have 40 volunteers who have agreed to participate in an experiment to compare two different keyboards. Describe the steps in a process that you could use to randomly assign each of the volunteers to one of the experimental groups.

1.39 An advertisement for a sweatshirt that appeared in *SkyMall Magazine* (a catalog distributed by some airlines) stated the following:

> This is not your ordinary hoody! Why? Fact: Research shows that written words on containers of water can influence the water's structure for better or worse depending on the nature and intent of the word. Fact: The human body is 70% water. What if positive words were printed on the inside of your clothing?

For only $79, you could purchase a hooded sweatshirt that had over 200 positive words (such as hope, gratitude, courage, and love) in 15 different languages printed on the inside of the sweatshirt so that you could benefit from being surrounded by these positive words. The "fact" that written words on containers of water can influence the water's structure appears to be based on the work of Dr. Masaru Emoto, who typed words on paper and then pasted them on bottles of water. He noted how the water reacted to the words by observing crystals formed in the water. He describes several of his experiments in his self-published book, *The Message from Water* (2nd Edition, Hado Publishing, 1999). If you were going to interview Dr. Emoto, what questions would you want to ask him about his experiment?

ADDITIONAL EXERCISES

Use the following information to answer Exercises 1.40–1.43. Many surgeons play music in the operating room. Does the type of music played have an effect on the surgeons' performance? The report **"Death Metal in the Operating Room" (NPR, December 24, 2009, www.npr.org, retrieved April 8, 2017)** describes an experiment in which surgeons used a simulator to perform a surgery. Some of the surgeons listened to music with vocal elements while performing the surgery, and others listened to music that did not have vocal elements. The researchers concluded that the average time to complete the surgery was greater when music with vocal elements is played than when music without vocal elements is played.

1.40 What are the experimental conditions for the experiment described above? What is the response variable?

1.41 Explain why it is important to control each of the following variables in the experiment described above.
a. the type of surgery performed
b. operating room temperature
c. volume at which the music was played

1.42 Explain why it is important that the surgeons be assigned at random to the two music conditions in the experiment described above.

1.43 Could the experiment described above have been double-blind? Explain why or why not.

1.44 In an experiment comparing two different surgical procedures for hernia repair (**"A Single-Blinded, Randomized Comparison of Laparoscopic Versus Open Hernia Repair in Children,"** *Pediatrics* **[2009]: 332–336)**, 89 children were assigned at random to one of the two surgical methods. Because there were potentially confounding variables that they could not control, the researchers relied on the random assignment of subjects to treatments to create comparable groups. One such variable was age. After random assignment to treatments, the researchers looked at the age distribution of the children in each of the two experimental groups (laparoscopic repair [LR] and open repair [OR]). The figure below is similar to one in the paper. Based on this figure, has the random assignment of subjects to experimental groups been successful in creating groups that are similar in age? Explain.

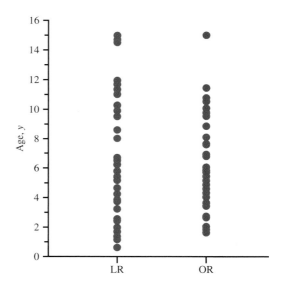

1.45 Pismo Beach, California, has an annual clam festival that includes a clam chowder contest. Judges rate clam chowders from local restaurants. The judges are not aware of which chowder is from which restaurant when they assign the ratings. One year, much to the dismay of the seafood restaurants on the waterfront, Denny's chowder was declared the winner! (When asked what the ingredients were, the cook at Denny's said he wasn't sure—he just had to add the right amount of nondairy creamer to the soup stock that he got from Denny's distribution center!)
a. Do you think that Denny's chowder would have won the contest if the judging had not been "blind"? Explain.
b. Although this was not an experiment, your answer to Part (a) helps to explain why those measuring the response in an experiment are often blinded. Using your answer in Part (a), explain why the results might have been different if the judges had known which restaurant—including Denny's—had prepared each of the clam chowders.

The Importance of Random Selection and Random Assignment: What Types of Conclusions Are Reasonable?

On September 25, 2009, results from a study of the relationship between spanking and a child's IQ were reported by a number of different news media. Some of the headlines that appeared that day were:

> **"Spanking lowers a child's IQ"** (*Los Angeles Times*)
> **"Do you spank? Studies indicate it could lower your kid's IQ"** (*SciGuy, Houston Chronicle*)
> **"Spanking can lower IQ"** (NBC4i, Columbus, Ohio)
> **"Smacking hits kids' IQ"** (*newscientist.com*)

In this study, the investigators followed 806 kids age 2 to 4 and 704 kids age 5 to 9 for 4 years. IQ was measured at the beginning of the study and again 4 years later. The researchers found that at the end of the study, the average IQ of the younger kids who were not spanked was 5 points higher than that of kids who were spanked. For the older group, the average IQ of kids who were not spanked was 2.8 points higher. These headlines all imply that spanking was the cause of the observed difference in IQ. Is this conclusion reasonable?

The type of conclusion that can be drawn from a statistical study depends on the study design. In an observational study, the goal is to use sample data to draw conclusions about one or more populations. In a well-designed observational study, the sample is selected in a way that you hope will make it representative of the population of interest. When this is the case, it is reasonable to generalize from the sample to the larger population. This is why good observational studies include random selection from the population of interest whenever possible.

A well-designed experiment can result in data that provide evidence for a cause-and-effect relationship. This is an important difference between an observational study and an experiment. In an observational study, it is not possible to draw clear cause-and-effect conclusions, because you cannot rule out the possibility that the observed effect is due to some variable other than the explanatory variable being studied. Consider the following example, which illustrates why it is not reasonable to draw cause-and-effect conclusions from an observational study.

Example 1.8 | **Noise Exposure and Heart Disease**

A health-related blog summarized a study of the relationship between exposure to loud noise and heart disease (**articles.mercola.com/sites/articles/archive/2015/11/11/loud -noise-exposure-increases-heart-disease-risk.aspx, November 11, 2015, retrieved September 25, 2016**). The title of the blog entry was "Long-Term Exposure to Loud Noise Raises Your Risk of Heart Disease." This title suggests a cause-and-effect relationship. But the study referenced in the blog entry—**"Exposure to Loud Noise, Bilateral High-Frequency Hearing Loss and Coronary Heart Disease," Occupational & Environmental Medicine, OnlineFirst, September 15, 2015, oem.bmj.com/content/early/2015/09/15 /oemed-2014-102778, retrieved September 25, 2016**—was an observational study that compared a group of people with hearing loss to a group of people without hearing loss. The researchers who conducted the study found that the percentage of people in the hearing loss group who had heart disease was higher than the percentage for the group that did not have hearing loss. Because this was an observational study, it is possible that the difference in percentages might be explained by other factors. For example, the two groups might have differed with respect to age, exercise habits, or general health, all of which might be alternative explanations for the observed difference. This makes it difficult to justify a cause-and-effect conclusion.

It is a good idea to be a bit cautious when you see headlines in newspapers and magazines that imply a cause-and-effect relationship between a treatment and a response, as illustrated in the following example.

Example 1.9 **Curbing Hair Loss**

The paper referenced in Exercise 1.34 (**"Effect of Nutritional Supplement on Hair Loss in Women," *Journal of Cosmetic Dermatology* [2015]: 76–82**) described a well-designed experiment to see if taking a dietary supplement could reduce hair loss in women with stage 1 hair loss. The subjects in the experiment were volunteers, and they were randomly assigned to either a supplement group or a control group. Because this was an experiment with random assignment to experimental conditions, it is reasonable for the researchers to conclude that the dietary supplement was the cause of the difference in the change in hair density for the two groups.

Is it reasonable to generalize this conclusion to all women with stage 1 hair loss? Because the women in the experiment were volunteers, before generalizing it is important that the researchers present a convincing argument that the women who participated in the experiment were representative of the population of interest—all women with stage 1 hair loss.

Although this study was well thought out, consider how it ended up being reported in the media. **Women's World (April 4, 2016)** summarized this study under the headline "Black Currant Oil Curbs Hair Loss." If you look at the original study, you will see that black current oil was only one part of the supplement taken by women in the study. According to the paper, the actual supplement consisted of "460 mg fish oil, 460 mg black currant seed oil, 5 mg vitamin E, 30 mg vitamin C and 1 mg lycopene." Based on this study, it is not reasonable to conclude that black currant oil alone is beneficial.

Let's return to the study on spanking and children's IQ described at the beginning of this section. Is this study an observational study or an experiment? Two groups were compared (children who were spanked and children who were not spanked), but the researchers did not randomly assign children to the spanking or no-spanking groups. The study is observational, and so cause-and-effect conclusions such as "spanking lowers IQ" are not justified. What you can say is that there is evidence that, as a group, children who are spanked tend to have a lower IQ than children who are not spanked. What you cannot say is that spanking is the *cause* of the lower IQ. It is possible that other variables—such as home or school environment, socioeconomic status, or parents' education levels—are related to both IQ and whether or not a child is spanked.

Fortunately, not everyone made the same mistake as the writers of the headlines given earlier in this section. Some examples of headlines that got it right are:

"Lower IQ's measured in spanked children" (*world-science.net*)
"Children who get spanked have lower IQs" (*livescience.com*)
"Research suggests an association between spanking and lower IQ in children" (*CBSnews.com*)

Drawing Conclusions from Statistical Studies

In this section, two different types of conclusions have been described. One type involves generalizing from what you have seen in a sample to some larger population, and the other involves reaching a cause-and-effect conclusion about the effect of an explanatory variable on a response. When is it reasonable to draw such conclusions? The answer depends on the way the data were collected. The following table summarizes when each of these types of conclusions is reasonable.

Type of Conclusion	Reasonable When
Generalize from sample to population	Random selection is used to obtain the sample
Difference in response is caused by experimental conditions (cause-and-effect conclusion)	There is random assignment of experimental units to experimental conditions

This table implies the following:

● For observational studies, cause-and-effect conclusions are not justified, but it is possible to generalize from the sample to the population of interest *if* the study design incorporated random selection or it can be argued that the sample is representative of the population.

● For experiments, it is possible to reach cause-and-effect conclusions *if* the study design incorporated random assignment to experimental conditions.

● If an experiment incorporates both random assignment to experimental conditions *and* random selection of experimental units from some population, it is possible to reach both cause-and-effect conclusions and to generalize these conclusions to the population from which the experimental units were selected. If an experiment does not include random selection of experimental units (for example, if volunteers are used as subjects), then it is not reasonable to generalize results unless a strong case can be made that the experimental units are representative of some larger group.

It is important to keep these three points in mind when drawing conclusions from a statistical study or when deciding if conclusions that others have made based on a statistical study are reasonable.

Summing It Up—Section 1.5

The following learning objectives were addressed in this section:

Conceptual Understanding

C2: Understand that the conclusions that can be drawn from a statistical study depend on the way in which the data are collected.

The type of conclusion that can be drawn from a statistical study depends on how the data are collected. If the individuals in a sample or the subjects in an experiment are randomly selected from some population, it is reasonable to generalize to that population. If there is random assignment in an experiment, the experiment can provide evidence of a cause-and-effect relationship. It is generally not reasonable to reach a cause-and-effect conclusion based on an observational study. See the table on page 31 to review drawing conclusions from statistical studies.

C7: Understand the difference between random selection and random assignment.

Random selection usually refers to how the individuals in a sample are selected. It is an important component of well-designed observational studies and is occasionally also used as a way to select subjects in an experiment. Random assignment refers to the way in which subjects are assigned to experimental conditions (treatments) in an experiment.

C9: Understand the limitations of using volunteers as subjects in an experiment.

Volunteers are often used in experiments. Even though the use of volunteers limits your ability to generalize to some larger population, random assignment of volunteers to experimental groups (treatments) should create comparable groups, which makes it possible to assess treatment effects.

Putting It into Practice

P4: Evaluate whether conclusions drawn from a study are appropriate given a description of the statistical study.

When evaluating conclusions drawn from an observational study, you should think about the way in which the sample was selected and whether it is reasonable to generalize to the population. When evaluating conclusions drawn from an experiment, you should think about whether the subjects were randomly selected and whether there was random assignment to treatments. If there was random assignment, a cause-and-effect conclusion may be reasonable.

SECTION 1.5   EXERCISES

Each Exercise Set assesses the following chapter learning objectives: C2, C7, C9, P4

SECTION 1.5   Exercise Set 1

1.46 A study described in **Food Network Magazine (January 2012)** concluded that people who push a shopping cart at a grocery store are less likely to purchase junk food than those who use a hand-held basket.

a. Do you think this study was an observational study or an experiment?

b. Is it reasonable to conclude that pushing a shopping cart causes people to be less likely to purchase junk food? Explain why or why not.

1.47 The article **"Heartfelt Thanks to Fido" (San Luis Obispo Tribune, July 5, 2003)** summarized a study that appeared in the **American Journal of Cardiology (March 15, 2003)**. In this study, researchers measured heart rate variability (a measure of the heart's ability to handle stress) in patients who had recovered from a heart attack. They found that heart rate variability was higher (which is good and means the heart can handle stress better) for those who owned a dog than for those who did not.

a. Based on this study, is it reasonable to conclude that owning a dog causes higher heart rate variability? Explain.

b. Is it reasonable to generalize the results of this study to all adult Americans? Explain why or why not.

1.48 The following is from an article titled **"After the Workout, Got Chocolate Milk?"** that appeared in the **Chicago Tribune (January 18, 2005)**:

> Researchers at Indiana University at Bloomington have found that chocolate milk effectively helps athletes recover from an intense workout. They had nine cyclists bike, rest four hours, then bike again, three separate times. After each workout, the cyclists downed chocolate milk or energy drinks Gatorade or Endurox (two to three glasses per hour); then, in the second workout of each set, they cycled to exhaustion. When they drank chocolate milk, the amount of time they could cycle until they were exhausted was similar to when they drank Gatorade and longer than when they drank Endurox.

For the experiment to have been well designed, it must have incorporated random assignment. Briefly explain where the researcher would have needed to use random assignment for the conclusion of the experiment to be valid.

1.49 "Pecans Lower Cholesterol" is a headline that appeared in the magazine **Woman's World (November 1, 2010)**. Consider the following five study descriptions. For each of the study descriptions, answer these five questions:

> Question 1: Is the described study an observational study or an experiment?
> Question 2: Did the study use random selection from some population?
> Question 3: Did the study use random assignment to experimental groups?

Question 4: Would the conclusion "pecans lower cholesterol" be appropriate given the study description? Explain.

Question 5: Would it be reasonable to generalize conclusions from this study to some larger population? If so, what population?

Study 1: Five hundred students were selected at random from those enrolled at a large college in Florida. Each student in the sample was asked whether they ate pecans more than once in a typical week, and their cholesterol levels were also measured. The average cholesterol level was significantly lower for the group who ate pecans more than once a week than for the group that did not.

Study 2: One hundred people who live in Los Angeles volunteered to participate in a statistical study. The volunteers were divided based on gender, with women in group 1 and men in group 2. Those in group 1 were asked to eat 3 ounces of pecans daily for 1 month. Those in group 2 were asked not to eat pecans for 1 month. At the end of the month, the average cholesterol level was significantly lower for group 1 than for group 2.

Study 3: Two hundred people volunteered to participate in a statistical study. Each person was asked how often he or she ate pecans, and their cholesterol levels were also measured. The average cholesterol level for those who ate pecans more than once a week was significantly lower than the average cholesterol level for those who did not eat pecans.

Study 4: Two hundred people volunteered to participate in a statistical study. For each volunteer, a coin was tossed. If the coin landed heads up, the volunteer was assigned to group 1. If the coin landed tails up, the volunteer was assigned to group 2. Those in group 1 were asked to eat 3 ounces of pecans daily for 1 month. Those in group 2 were asked not to eat pecans for 1 month. At the end of the month, the average cholesterol level was significantly lower for group 1 than for group 2.

Study 5: One hundred students were selected at random from those enrolled at a large college. Each of the selected students was asked to participate in a study, and all agreed to participate. For each student, a coin was tossed. If the coin landed heads up, the student was assigned to group 1. If the coin landed tails up, the student was assigned to group 2. Those in group 1 were asked to eat 3 ounces of pecans daily for 1 month. Those in group 2 were asked not to eat pecans for 1 month. At the end of the month, the average cholesterol level was significantly lower for group 1 than for group 2.

SECTION 1.5   Exercise Set 2

1.50 A survey of affluent Americans (those with incomes of $75,000 or more) indicated that 57% would rather have more time than more money (**USA TODAY, January 29, 2003**).

a. What condition on how the data were collected would make it reasonable to generalize this result to the population of affluent Americans?

b. Would it be reasonable to generalize this result to the population of *all* Americans? Explain why or why not.

1.51 Researchers at the University of Utah carried out a study to see if the size of the fork used to eat dinner has an effect on how much food is consumed (**Food Network Magazine, January 2012**). The researchers assigned people to one of two groups. One group ate dinner using a small fork, and the other group ate using a large fork. The researchers found that those who ate with a large fork ate less of the food on the plate than those who ate with the small fork. The title of the article describing this study was **"Dieters Should Use a Big Fork."** This title implies a cause-and-effect relationship between fork size and amount eaten and also generalizes this finding to the population of dieters. What would you need to know about the study design to determine if the conclusions implied by the headline are reasonable?

1.52 The paper **"Effect of Cell Phone Distraction on Pediatric Pedestrian Injury Risk"** (*Pediatrics* [2009]: e179–e185) describes an experiment examining whether people talking on a cell phone are at greater risk of an accident when crossing the street than when not talking on a cell phone. (No people were harmed in this experiment—a virtual interactive pedestrian environment was used.) One possible way of conducting such an experiment would be to have a person cross 20 streets in this virtual environment. The person would talk on a cell phone for some crossings and would not use the cell phone for others. Explain why it would be important to randomly assign the two treatments (talking on the phone, not talking on the phone) to the 20 trials (the 20 simulated street crossings).

1.53 "Strengthen Your Marriage with Prayer" is a headline that appeared in the magazine **Woman's World (November 1, 2010)**. The article went on to state that couples who attend religious services and pray together have happier, stronger marriages than those who do not. For each of the following study descriptions, answer these five questions:

Question 1: Is the study described an observational study or an experiment?

Question 2: Did the study use random selection from some population?

Question 3: Did the study use random assignment to experimental groups?

Question 4: Would the conclusion that you can "strengthen your marriage with prayer" be appropriate given the study description? Explain.

Question 5: Would it be reasonable to generalize conclusions from this study to some larger population? If so, what population?

Study 1: Married viewers of an afternoon television talk show were invited to participate in an online poll. A total of 1700 viewers chose to go online and participate in the poll. Each was asked whether they attended church regularly and whether they would rate their marriage as happy. The proportion of those attending church regularly who rated their marriage as happy was significantly higher than the proportion of those not attending church regularly who rated their marriage as happy.

Study 2: Two hundred women were selected at random from the membership of the American Association of University Women (a large professional organization for women). Each woman selected was asked if she was married. The 160 married women were asked questions about church attendance and whether or not they would rate their marriage as happy. The proportion of those attending church regularly who rated their marriage as happy was significantly higher than the proportion of those not attending church regularly who rated their marriage as happy.

Study 3: A researcher asked each man of the first 200 men arriving at a baseball game if he was married. If a man said he was married, the researcher also asked if he attended church regularly and if he would rate his marriage as happy. The proportion of those attending church regularly who rated their marriage as happy was significantly higher than the proportion of those not attending church regularly who rated their marriage as happy.

ADDITIONAL EXERCISES

1.54 An article titled **"Guard Your Kids Against Allergies: Get Them a Pet"** (*San Luis Obispo Tribune*, August 28, 2002) described a study that led researchers to conclude that "babies raised with two or more animals are about half as likely to have allergies by the time they turned six." Explain why it is not reasonable to conclude that being raised with two or more animals is the cause of the observed lower allergy rate.

1.55 Does living in the South cause high blood pressure? Data from a group of 6278 people questioned in the Third National Health and Nutritional Examination Survey between 1988 and 1994 indicate that a greater percentage of Southerners have high blood pressure than do people living in any other region of the United States (**"High Blood Pressure Greater Risk in U.S. South, Study Says," January 6, 2000, cnn.com**). This difference in rate of high blood pressure was found in every ethnic group, gender, and age category studied. What are two possible reasons we cannot conclude that living in the South causes high blood pressure?

Use the following information to answer Exercises 1.56–1.60.

The paper **"Turning to Learn: Screen Orientation and Reasoning from Small Devices"** (*Computers in Human Behavior* [2011]: 793–797) describes a study that investigated whether cell phones with small screens are useful for gathering information. The researchers wondered if the ability to reason using information read on a small screen was affected by the screen orientation. The researchers assigned 33 undergraduate students who were enrolled in a psychology course at a large public university to one of two groups at random.

One group read material that was displayed on a small screen in portrait orientation, and the other group read material on the same size screen but turned to display the information in landscape orientation (see the following figure).

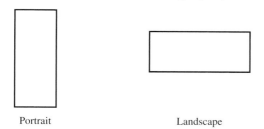

Portrait   Landscape

The researchers found that performance on a reasoning test based on the displayed material was better for the group that read material in the landscape orientation.

1.56 Is the study described above an observational study or an experiment?

1.57 Did the study described above use random selection from some population?

1.58 Did the study described above use random assignment to experimental groups?

1.59 Is the conclusion that reasoning using information displayed on a small screen is improved by turning the screen to landscape orientation appropriate, given the study design described above? Explain.

1.60 Is it reasonable to generalize the conclusions from the study described above to some larger population? If so, what population?

SECTION 1.6 Avoid These Common Mistakes

It is a big mistake to begin collecting data before thinking carefully about the questions you want to answer and then developing a plan. A poorly designed plan for data collection may result in data that will not allow you to answer key questions of interest or to generalize conclusions based on the data to the desired populations of interest.

Be sure to avoid these common mistakes:

1. Drawing a cause-and-effect conclusion from an observational study.
   Don't do this, and be wary when others do it!
2. Generalizing results of an experiment that uses volunteers as subjects.
   Only do this if it can be convincingly argued that the group of volunteers is representative of the population of interest.
3. Generalizing conclusions based on data from a poorly designed observational study.
   Though it is sometimes reasonable to generalize from a sample to a population, on other occasions it is not reasonable. Generalizing from a sample to a population is justified only when the sample is likely to be representative of the population. This would be the case if the sample was a random sample from the population, and there were no major potential sources of bias. If the sample was not selected at random or if potential sources of bias were present, these issues would have to be addressed before deciding if it is reasonable to generalize the study results.
4. Generalizing conclusions based on an observational study that used voluntary response or convenience sampling to a larger population.
   This is almost never reasonable!

CHAPTER ACTIVITIES

ACTIVITY 1.1 FACEBOOK FRIENDING

Background: The article **"Professors Prefer Face Time to Facebook"** appeared in the student newspaper at Cal Poly, San Luis Obispo (*Mustang Daily*, **August 27, 2009**). The article examined how professors and students felt about using Facebook as a means of faculty-student communication. The student who wrote this article got mixed opinions when she asked students whether they wanted to become Facebook friends with their professors. Two student comments included in the article were

> I think the younger the professor is, the more you can relate to them and the less awkward it would be if you were to become friends on Facebook. The older the professor, you just would have to wonder, "Why are they friending me?"

and

> I think becoming friends with professors on Facebook is really awkward. I don't want them being able to see into my personal life, and frankly, I am not really interested in what my professors do in their free time.

Even if the students interviewed had expressed a consistent opinion, it would still be unreasonable to think this represented how all students felt about this issue because only four students were interviewed, and it is not clear from the article how these students were selected.

In this activity, you will work with a partner to develop a plan to assess student opinion about being Facebook friends with professors at your school.

1. Suppose you will select a sample of 50 students at your school to participate in a survey. Write one or more questions to ask each student in the sample.

2. Discuss with your partner whether you think it would be easy or difficult to obtain a simple random sample

of 50 students at your school and to obtain the desired information from all the students selected for the sample. Write a summary of your discussion.

3. With your partner, decide how you might go about selecting a sample of 50 students from your school that reasonably could be considered representative of the population of interest, even if it may not be a simple random sample. Write a brief description of your sampling plan, and point out the aspects of your plan that you think will result in a representative sample.

4. Explain your plan to another pair of students. Ask them to critique your plan. Write a brief summary of the comments you received. Now reverse roles, and provide a critique of the plan devised by the other pair.

5. Based on the feedback you received in Step 4, would you modify your original sampling plan? If not, explain why this is not necessary. If so, describe how the plan would be modified.

ACTIVITY 1.2 McDONALD'S AND THE NEXT 100 BILLION BURGERS

Background: The article **"Potential Effects of the Next 100 Billion Hamburgers Sold by McDonald's"** (*American Journal of Preventative Medicine* [2005]: 379–381) estimated that 992.25 million pounds of saturated fat would be consumed as McDonald's sells its next 100 billion hamburgers. This estimate was based on the assumption that the average weight of a burger sold would be 2.4 oz. This is the average of the weight of a regular hamburger (1.6 oz.) and a Big Mac (3.2 oz.). The authors took this approach because

> McDonald's does not publish sales and profits of individual items. Thus, it is not possible to estimate how many of McDonald's first 100 billion beef burgers sold were 1.6 oz hamburgers, 3.2 oz. Big Macs (introduced in 1968), 4.0 oz. Quarter Pounders (introduced in 1973), or other sandwiches.

This activity can be completed by an individual or by a team. Your instructor will specify which approach (individual or team) you should use.

1. The authors of the article believe that the use of 2.4 oz. as the average size of a burger sold at McDonald's is "conservative," which would result in the estimate of 992.25 million pounds of saturated fat being lower than the actual amount that would be consumed. Explain why the authors' belief might be justified.

2. Do you think it would be possible to collect data that could lead to an estimate of the average burger size that would be better than 2.4 oz.? If so, explain how you would go about collecting such data. If not, explain why you think it is not possible.

ACTIVITY 1.3 VIDEO GAMES AND PAIN MANAGEMENT

Background: Video games have been used for pain management by doctors and therapists who believe that the attention required to play a video game can distract the player and thereby decrease the sensation of pain. The paper **"Video Games and Health"** (*British Medical Journal* [2005]: 122–123) states

> However, there has been no long term follow-up and no robust randomized controlled trials of such interventions. Whether patients eventually tire of such games is also unclear. Furthermore, it is not known whether any distracting effect depends simply on concentrating on an interactive task or whether the content of games is also an important factor as there have been no controlled trials comparing video games with other distracters.

Further research should examine factors within games such as novelty, users' preferences, and relative levels of challenge and should compare video games with other potentially distracting activities.

1. Working with a partner, select one of the areas of potential research suggested in the paper. Formulate a specific question that you could address by performing an experiment.

2. Propose an experiment that would provide data to address the question from Step 1. Be specific about how subjects might be selected, what the experimental conditions (treatments) would be, and what response would be measured.

CHAPTER 1 EXPLORATIONS IN STATISTICAL THINKING

EXPLORATION 1: UNDERSTANDING SAMPLING VARIABILITY

In the two exercises below, each student in your class will go online to select a random sample from a small population consisting of 300 adults between the ages of 18 and 64.

To learn about how the people in this population use text messaging, go online at statistics.cengage.com/Peck2e/Explore.html and click on the link for Chapter 1. This will take you to a web page where you can select a random sample of 20 people from the population.

Click on the sample button. This selects a random sample of 20 people from the population and will display information for these 20 people. You should see the following information for each person selected:

- An ID number that identifies the person selected.
- The number of text messages sent in a typical day.
- The response to the question "Have you ever sent or read a text message while driving?" The response is coded as 1 for yes and 2 for no.

Each student in your class will receive data from a different random sample.

Use the data from your sample to complete the following two exercises.

1. This exercise uses the number of text messages sent data.
   a. Below is a graph of the number of text messages sent data for the entire population.

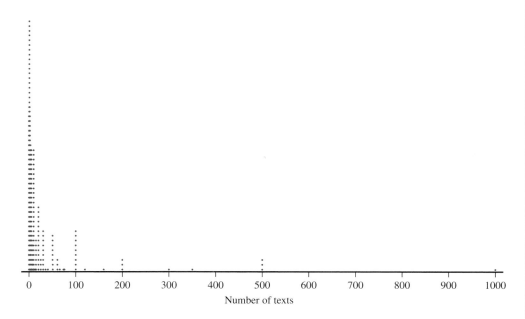

   b. Construct a similar graph with 20 dots for the number of text messages sent by the 20 people in your sample.
   c. How is the graph for your sample similar to the population graph? How is it different?
   d. If you had taken a random sample of 50 people from this population instead of 20, do you think your graph would have looked more like the population graph or less like the population graph? Why do you think this?

e. Calculate the average number of text messages sent in a typical day by the 20 people in your sample.

f. The average number of text messages sent for the entire population is 24.28. Did you get this same value for your sample average? Does this surprise you? Explain why or why not.

If asked to do so by your instructor, bring your graph with you to class. Your instructor will lead a class discussion of the following:

g. Did everyone in the class get the same value of the sample average? Is this surprising?

h. After comparing your graph to the graphs produced by several other students in your class, think about the following questions:

Why aren't all the sample graphs the same?

Even though not all of the sample graphs are the same, in what ways are they similar?

2. Now consider the data on text messaging while driving data.

a. Calculate the proportion of people in your sample who responded yes by counting the number of yes responses and dividing that number by 20.

b. For the entire population, the proportion who responded yes is $120/300 = 0.40$. Did you get this same value for your sample proportion? Does this surprise you? Explain why or why not.

If asked to do so by your instructor, bring your answers to parts (a) and (b) to class. Your instructor will lead a class discussion of the following:

c. Did everyone in the class get the same sample proportion? Why did this happen?

d. Of all the sample proportions, what was the smallest? What was the largest? Does this sample-to-sample variability surprise you?

e. Suppose everyone had selected a random sample of 50 people instead of 20 people. Do you think the sample proportions for these new samples would have varied more or varied less from sample to sample than what you saw for samples of size 20?

These two exercises explore the "big idea" of **sampling variability**. This important concept will be revisited often in the following chapters. Understanding sampling variability is the key to understanding how you can use sample data to learn about a population.

ARE YOU READY TO MOVE ON?   CHAPTER 1 REVIEW EXERCISES

All chapter learning objectives are assessed in these exercises. The learning objectives assessed in each exercise are given in parentheses.

1.61 **(C1, P1)**
For each of the following, determine whether the statistical study described is an observational study or an experiment. Give a brief explanation of your choice.

a. Can choosing the right music make wine taste better? This question was investigated by a researcher at a university in Edinburgh **(www.decanter.com/wine-news /montes-music-makes-wine-reach-parts-it-otherwise -couldnt-reach-82325/, retrieved April 8, 2017)**. Each of 250 volunteers was assigned at random to one of five rooms where they were asked to taste and rate a glass of wine. No music was playing in one of the rooms, and a different style of music was playing in each of the other four rooms. The researcher concluded that cabernet sauvignon is rated more highly when bold music is played than when no music is played.

b. The article **"Display of Health Risk Behaviors on MySpace by Adolescents"** (*Archives of Pediatrics and Adolescent Medicine* **[2009]: 27–34**) described a study of 500 publically accessible MySpace web profiles posted by 18-year-olds. The content of each profile was analyzed and the researchers concluded that those who indicated involvement in sports or a hobby were less likely to have references to risky behavior (such as sexual references or references to substance abuse or violence).

c. *USA TODAY* **(January 29, 2003)** reported that in a study of affluent Americans (defined as those with incomes of $75,000 or more per year) 57% indicated that they would rather have more time than more money.

d. The article **"Acupuncture for Bad Backs: Even Sham Therapy Works"** (*Time*, **May 12, 2009)** summarized a study conducted by researchers at the Group Health Center for Health Studies in Seattle. In this study, 638 adults with back pain were randomly assigned to one of four groups. People in group 1 received the usual care for back pain. People in group 2 received acupuncture at a set of points tailored specifically for each individual. People in group 3 received acupuncture at a standard set of points typically used in the treatment of back pain. Those in group 4 received fake acupuncture—they were poked with a toothpick at the same set of points used for the people in group 3. Two notable conclusions from the study were: (1) patients receiving real or fake acupuncture experienced a greater reduction in pain than those receiving usual care; and (2) there was no significant difference in pain reduction between those who received real acupuncture (groups 2 and 3) and those who received fake acupuncture toothpick pokes.

1.62 (C3)

The student senate at a college with 15,000 students is interested in the proportion of students who favor a change in the grading system to allow for plus and minus grades (for example, B+, B, B-, rather than just B). Two hundred students are interviewed to determine their attitude toward this proposed change. What is the population of interest? What group of students constitutes the sample in this problem?

1.63 (C4)

For each of the following statements, identify the number that appears in boldface type as the value of either a population characteristic or a statistic:

a. A department store reports that **84%** of all customers who use the store's credit plan pay their bills on time.

b. A sample of 100 students at a large university had a mean age of **24.1** years.

c. The Department of Motor Vehicles reports that **22%** of all vehicles registered in a particular state are imports.

d. A hospital reports that, based on the 10 most recent cases, the mean length of stay for surgical patients is **6.4** days.

e. A consumer group, after testing 100 batteries of a certain brand, reported an average life of **63** hours of use.

1.64 (C8)

According to the article **"Effect of Preparation Methods on Total Fat Content, Moisture Content, and Sensory Characteristics of Breaded Chicken Nuggets and Beef Steak Fingers"** (*Family and Consumer Sciences Research Journal* [1999]: 18–27), sensory tests were conducted using 40 college student volunteers at Texas Women's University. Give three reasons, other than the relatively small sample size, why it would not be a good idea to generalize any study results to the population of all college students.

1.65 (M1)

A petition with 500 signatures is submitted to a college's student council. The council president would like to determine what proportion of those who signed the petition are actually registered students at the college. There is not enough time to check all 500 names with the registrar, so the council president decides to select a simple random sample of 30 signatures. Describe the steps in a process she might use to do this.

1.66 (P2, P4)

The paper **"From Dr. Kildare to Grey's Anatomy"** (*Annals of Emergency Medicine* **[2010]: 21A–23A)** describes several studies of how the way in which doctors are portrayed on television might influence public perception of doctors. One study was described as follows:

> Rebecca Chory, Ph.D., now an associate professor of communication at West Virginia University, began studying the effect of such portrayals on patients' attitudes toward physicians. Using a survey of 300 undergraduate students, she compared perceptions of physicians in 1992—the end of the era when physicians were shown as all-knowing, wise father figures—with those in 1999, when shows such as *ER* and *Chicago Hope* (1994–2000) were continuing the transformation to showing the private side and lives of physicians, including vivid demonstrations of their weaknesses and insecurities.
>
> Dr. Chory found that, regardless of the respondents' personal experience with physicians, those who watched certain kinds of television had declining perceptions of physicians' composure and regard for others. Her results indicated that the more prime time physician shows that people watched in which physicians were the main characters, the more uncaring, cold, and unfriendly the respondents thought physicians were.

a. Answer the following four questions for the observational study described in this exercise. (Hint: Reviewing Examples 1.4 and 1.5 might be helpful.)

1. What is the population of interest?
2. Was the sample selected in a reasonable way?
3. Is the sample likely to be representative of the population of interest?
4. Are there any obvious sources of bias?

b. Based on the study design, do you think that the stated conclusions are reasonable?

1.67 (C6)

In many digital environments, users are allowed to choose how they are represented visually online. Does the way in which people are represented online affect online behavior? This question was examined by the authors of the paper **"The Proteus Effect: The Effect of Transformed Self-Representation on Behavior"** (*Human Communication Research* [2007]: 271–290).

Participants were randomly assigned either an attractive avatar (a graphical image that represents a person) to represent them or an unattractive avatar. The researchers concluded that when interacting with a person of the opposite gender in an online virtual environment, those assigned an attractive avatar moved significantly closer to the other person than those who had been assigned an unattractive avatar. This difference was attributed to the attractiveness of the avatar. Explain why the researchers would not have been able to reach this conclusion if participants had been allowed to choose one of the two avatars (attractive, unattractive) themselves.

1.68 **(C10)**
The article **"Yes that Miley Cyrus Biography Helps Learning"** (*The Globe and Mail*, **August 5, 2010**) describes an experiment investigating whether providing summer reading books to low-income children would affect school performance. Subjects in the experiment were 1300 children randomly selected from first and second graders at low-income schools in Florida. A group of 852 of these children were selected at random from the group of 1300 participants to be in the book group. The other 478 children were assigned to the control group. Children in the book group were invited to a book fair in the spring to choose any 12 reading books that they could then take home. Children in the control group were not given any reading books, but were given some activity and puzzle books. These children received books each year for three years until the children reached third and fourth grade. The researchers then compared reading test scores of the two groups.
a. Is randomly selecting 852 of the 1300 children to be in the book group is equivalent to random assignment of the two experimental conditions to subjects? Explain.
b. Explain the purpose of including a control group in this experiment.

1.69 **(C6, C11)**
The article **"Super Bowls: Serving Bowl Size and Food Consumption"** (*Journal of the American Medical Association* **[2005]: 1727–1728**) describes an experiment investigating how the size of serving bowls influences the amount a person eats. In this experiment, graduate students at a university were recruited to attend a Super Bowl party. The paper states that as the students arrived, they were

> ... led in an alternating order to one of two identical buffet tables on opposite sides of an adjoining room. The tables had identical amounts of snacks, such as nuts, pretzels, and chips. All of the snacks contained approximately the same number of calories per gram. On one of the tables the snacks were set out in large serving bowls and on the second table the snacks were set out in smaller serving bowls. The students were given a plate and invited to serve themselves before going to another room to watch the game. When they arrived at the game room, their plates were weighed and the number of calories in the food on the plate was estimated.

The researchers concluded that serving bowl size does make a difference, with those using large serving bowls tending to take more food.
a. Do you think that the alternate assignment to the experimental groups (large serving bowls, small serving bowls) based on arrival time is "close enough" to random assignment? That is, do you think it would tend to create comparable experimental groups?
b. In this study, the research assistant who weighed the plates and estimated the calorie content of the food on the plate was blinded as to which experimental group the plate belonged to and was also blinded as to the purpose of the experiment. Why do you think the researchers chose to incorporate this type of blinding?

1.70 **(M2)**
According to the article **"Rubbing Hands Together Under Warm Air Dryers Can Counteract Bacteria Reduction"** (*Infectious Disease News*, **September 22, 2010**), washing your hands isn't enough—good "hand hygiene" also includes drying hands thoroughly. The article described an experiment to compare bacteria reduction for three different hand-drying methods. In this experiment, subjects handled uncooked chicken for 45 seconds, then washed their hands with a single squirt of soap for 60 seconds, and then used one of the three hand-drying methods. The bacteria count on their hands was then measured. Suppose you want to carry out a similar experiment with 30 subjects who are willing to participate. Describe the steps in a process you might use to randomly assign each of the 30 subjects to one of the hand-drying methods.

1.71 **(P3, P4)**
Can moving their hands help children learn math? This is the question investigated by the authors of the paper **"Gesturing Gives Children New Ideas about Math"** (*Psychological Science* **[2009]: 267–272**). An experiment was conducted to compare two different methods for teaching children how to solve math problems of the form $3 + 2 + 8 = \underline{\quad} + 8$. One method involved having students point to the $3 + 2$ on the left side of the equal sign with one hand and then point to the blank on the right side of the equal sign before filling in the blank to complete the equation. The other method did not involve using these hand gestures. The paper states that the study used children ages 9 and 10 who were given a pretest containing six problems of the type described. Only children who answered all six questions incorrectly became subjects in the experiment. There were a total of 128 subjects. To compare the two methods, the 128 children were assigned at random to the two experimental conditions. Children in one group were taught a method that used hand gestures, and children in the other group were taught a similar strategy that did not involve hand gestures. Each child then took a test with six problems and received a score based on the number correct. From the resulting data, the researchers concluded that the average score for children who used hand gestures was significantly higher than the average score for children who did not use hand gestures.

a. Answer the following seven questions for the experiment described above. (Hint: Reviewing Examples 1.6 and 1.7 might be helpful.)

1. What question is the experiment trying to answer?
2. What are the experimental conditions (treatments) for this experiment?
3. What is the response variable?
4. What are the experimental units and how were they selected?
5. Does the design incorporate random assignment of experimental units to the different experimental conditions? If not, are there potentially confounding variables that would make it difficult to draw conclusions based on data from the experiment?
6. Does the experiment incorporate a control group and/or a placebo group? If not, would the experiment be improved by including them?
7. Does the experiment involve blinding? If not, would the experiment be improved by making it single- or double-blind?

b. Based on the study design, do you think that the conclusions are reasonable?

1.72 (C2)

The article **"Display of Health Risk Behaviors on MySpace by Adolescents"** (*Archives of Pediatrics and Adolescent Medicine* [2009]: 27–34) described a study in which researchers looked at a random sample of 500 publicly accessible MySpace web profiles posted by 18-year-olds. The content of each profile was analyzed. One of the conclusions reported was that displaying sport or hobby involvement was associated with decreased references to risky behavior (sexual references or references to substance abuse or violence).

a. Is it reasonable to generalize the stated conclusion to all 18-year-olds with a publicly accessible MySpace web profile? What aspect of the study supports your answer?
b. Not all MySpace users have a publicly accessible profile. Is it reasonable to generalize the stated conclusion to all 18-year-old MySpace users? Explain.
c. Is it reasonable to generalize the stated conclusion to *all* MySpace users with a publicly accessible profile? Explain.

1.73 (C2, C5)

The authors of the paper **"Popular Video Games: Quantifying the Presentation of Violence and Its Context"** (*Journal of Broadcasting & Electronic Media* [2003]: 58–76) investigated the relationship between video game rating—suitable for everyone (E), suitable for 13 years of age and older (T), and suitable for 17 years of age and older (M)—and the number of violent interactions per minute of play. The sample consisted of 60 video games—the 20 most popular (by sales) for each of three game systems. The researchers concluded that video games rated for older children had significantly more violent interactions per minute than video games rated for more general audiences.

a. Do you think that the sample of 60 games was selected in a way that makes it representative of the population of all video games?

b. Is it reasonable to generalize the researchers' conclusion to all video games? Explain why or why not.

1.74 (C6)

To examine the effect of exercise on body composition, healthy women aged 35 to 50 were classified as either active (nine or more hours of physical activity per week) or sedentary (**"Effects of Habitual Physical Activity on the Resting Metabolic Rates and Body Composition of Women Aged 35 to 50 Years,"** *Journal of the American Dietetic Association* [2001]: 1181–1191). Body fat percentage was measured, and the researchers found that this percentage was significantly lower for women who were active than for sedentary women.

a. Is the study described an experiment? If so, what is the explanatory variable and what is the response variable? If not, explain why it is not an experiment.
b. From this study alone, is it reasonable to conclude that physical activity is the cause of the observed difference in body fat percentage? Justify your answer.

1.75 (C7, C9)

The article **"Rethinking Calcium Supplements"** (*U.S. Airways Magazine*, October 2010) describes a study investigating whether taking calcium supplements increases the risk of heart attack. Consider the following four study descriptions. For each study, answer the following five questions:

Question 1: Is the study described an observational study or an experiment?
Question 2: Did the study use random selection from some population?
Question 3: Did the study use random assignment to experimental groups?
Question 4: Based on the study description, would it be reasonable to conclude that taking calcium supplements is the cause of the increased risk of heart attack?
Question 5: Would it be reasonable to generalize conclusions from this study to some larger population? If so, what population?

Study 1: Every heart attack patient and every patient admitted for an illness other than heart attack during the month of December, 2010, at a large urban hospital was asked if he or she took calcium supplements. The proportion of heart attack patients who took calcium supplements was significantly higher than the proportion of patients admitted for other illnesses who took calcium supplements.

Study 2: Two hundred people were randomly selected from a list of all people living in Minneapolis who receive Social Security. Each person in the sample was asked whether or not they took calcium supplements. These people were followed for 5 years, and whether or not they had had a heart attack during the 5-year period was noted. The proportion of heart attack victims in the group taking calcium supplements was significantly higher than the proportion of heart attack victims in the group not taking calcium supplements.

Study 3: Two hundred people were randomly selected from a list of all people living in Minneapolis who receive Social Security. Each person was asked to participate in a statistical study, and all agreed to participate. Those who had no previous history of heart problems were instructed not to take calcium supplements. Those with a previous history of heart problems were instructed to take calcium supplements. The participants were followed for 5 years, and whether or not they had had a heart attack during the 5-year period was noted. The proportion of heart attack victims in the calcium supplement group was significantly higher than the proportion of heart attack victims in the no calcium supplement group.

Study 4: Four hundred people volunteered to participate in a 10-year study. Each volunteer was assigned at random to either group 1 or group 2. Those in group 1 took a daily calcium supplement. Those in group 2 did not take a calcium supplement. The proportion who suffered a heart attack during the 10-year study period was noted for each group. The proportion of heart attack victims in group 1was significantly higher than the proportion of heart attack victims in group 2.

2 Graphical Methods for Describing Data Distributions

pollockg/Shutterstock.com

PREVIEW

When you carry out a statistical study, you hope to learn from the data you collect. But it is often difficult to "see" the information in data if it is presented as just a list of observations. An important step in the data analysis process involves summarizing data graphically and numerically. This makes it easier to see important characteristics of the data and is an effective way to communicate what you have learned.

CHAPTER LEARNING OBJECTIVES

Conceptual Understanding

After completing this chapter, you should be able to

C1 Distinguish between categorical and numerical data.

C2 Distinguish between discrete and continuous numerical data.

C3 Understand that selecting an appropriate graphical display depends on the number of variables in the data set, the data type, and the purpose of the graphical display.

Mastering the Mechanics

After completing this chapter, you should be able to

M1 Select an appropriate graphical display for a given data set.

M2 Construct and interpret bar charts and comparative bar charts.

M3 Construct and interpret dotplots and comparative dotplots.

M4 Construct and interpret stem-and-leaf and comparative stem-and-leaf displays.

M5 Construct and interpret histograms.

M6 Construct and interpret scatterplots.

M7 Construct and interpret time series plots.

Putting It into Practice

After completing this chapter, you should be able to

P1 Describe a numerical data distribution in terms of shape, center, variability, gaps, and outliers.

P2 Use graphical displays to compare groups on the basis of a categorical variable.

P3 Use graphical displays to compare groups on the basis of a numerical variable.

P4 Use a scatterplot to investigate the relationship between two numerical variables.

P5 Use a time series plot to investigate trends over time for a numerical variable.

P6 Critically evaluate graphical displays that appear in newspapers, magazines, and advertisements.

PREVIEW EXAMPLE

Data set available

Love Those New Cars?

Each year, marketing and information firm J.D. Power and Associates surveys new car owners 90 days after they have purchased their cars. The data collected are used to rate auto brands (Toyota, Ford, and others) on initial quality and initial customer satisfaction. *USA TODAY* (www.usatoday.com, March 29, 2016) reported both the number of manufacturing defects per 100 vehicles and a satisfaction score for all 30 brands sold in the United States. The data on number of defects are shown here:

76	80	55	76	76	67	65	58	63	64
69	70	67	68	74	64	60	72	58	74
66	63	58	71	72	65	38	78	43	51

There are several things that you might want to learn from these data. What is a typical value for number of defects per 100 vehicles? Is there a lot of variability in the number of defects for different brands, or are most brands about the same? Are some brands much better or much worse than most? If you were to look at the defect data and the customer satisfaction data together, would you find a relationship between the number of manufacturing defects and customer satisfaction? Questions like these are more easily answered if the data can be summarized in a graphical display. ■

In this chapter, you will learn how to summarize data graphically. The preview example will be revisited later in this chapter where you will see how graphical displays can help answer the questions posed.

SECTION 2.1   Selecting an Appropriate Graphical Display

The first step in learning from a data set is usually to construct a graph that will enable you to see the important features of the data distribution. This involves selecting an appropriate graphical display. The choice of an appropriate graphical display will depend on three things:

- The number of variables in the data set
- The data type
- The purpose of the graphical display

Let's consider each of these in more detail.

The Number of Variables in the Data Set

The individuals in any particular population typically possess many characteristics that might be studied. Consider the group of students currently enrolled at a college. One characteristic of the students in the population is political affiliation (Democrat, Republican, Independent, Green Party, Other). Another characteristic is the number of textbooks purchased this semester, and yet another is the distance (in miles) from home to the college. A **variable** is any characteristic whose value may change from one individual to another. For example, *political affiliation* is a variable, and so are *number of textbooks purchased* and *distance to the college*. **Data** result from making observations either on a single variable or simultaneously on two or more variables.

A **univariate data set** consists of observations on a single variable made on individuals in a sample or population. In some studies, however, two different characteristics are measured simultaneously. For example, both height (in inches) and weight (in pounds) might be recorded for each person. This results in a **bivariate data** set. It consists of pairs of numbers, such as (68, 146). **Multivariate data** result when we obtain a category or value for each of two or more attributes (so bivariate data are a special case of multivariate data). For example, a multivariate data set might include height, weight, pulse rate, and systolic blood pressure for each person in a group.

The Data Type

Variables in a statistical study can be classified as either categorical or numerical. For the population of students at a college, the variable *political affiliation* is categorical because each student's response to the question "What is your political affiliation?" is a category (such as Democrat or Republican). *Number of textbooks purchased* and *distance to the college* are both numerical variables. Sometimes you will see the term qualitative used in place of categorical and the term quantitative used in place of numerical.

> ### DEFINITION
>
> A data set consisting of observations on a single characteristic is a **univariate data set.**
>
> A univariate data set is **categorical** (or **qualitative**) if the individual observations are categorical responses.
>
> A univariate data set is **numerical** (or **quantitative**) if each observation is a number.

Numerical variables can be further classified into two types—*discrete* and *continuous*—depending on their possible values. Suppose that the variable of interest is the number of courses a student is taking. If no student is taking more than eight courses, the possible values are 1, 2, 3, 4, 5, 6, 7, and 8. These values are identified in Figure 2.1(a) by the isolated dots on the number line. On the other hand, the line segment in Figure 2.1(b) identifies a plausible set of possible values for the time (in seconds) it takes for the first kernel in a bag of microwave popcorn to pop. Here the possible values make up an entire interval on the number line. No possible value is isolated from other possible values.

FIGURE 2.1
Possible values of a variable:
(a) number of courses
(b) time to pop

Number of Courses

(a)

Time to Pop (seconds)

(b)

DEFINITION

A numerical variable is **discrete** if its possible values correspond to isolated points on the number line.

A numerical variable is **continuous** if its possible values form an entire interval on the number line.

Discrete data usually arise when you count (for example, the number of roommates a college student has or the number of petals on a certain type of flower). Usually, data are continuous when you measure. In practice, measuring instruments do not have infinite accuracy; so strictly speaking, the possible reported values do not form a continuum on the number line. For example, weight may be reported to the nearest pound. Even though reported weights are then whole numbers such as 140 and 141, an actual weight might be a value between 140 and 141. The distinction between discrete and continuous data is important when selecting an appropriate graphical display and also later when considering probability models.

Example 2.1   **Do U Txt?**

Consider the variable *number of text messages sent on a particular day*. Possible values for this variable are 0, 1, 2, 3… . These are isolated points on the number line, so this is an example of discrete numerical data.

Suppose that instead of the number of text messages sent, the *time spent texting* is recorded. This would result in continuous data. Even though time spent may be reported rounded to the nearest minute, the actual time spent could be 6 minutes, 6.2 minutes, 6.28 minutes, or any other value in an entire interval.

The Purpose of the Graphical Display

One last thing to consider when selecting an appropriate graphical display is the purpose of the graphical display. For univariate data, the purpose is usually to show the data distribution. For a categorical variable, this means showing how the observations are distributed among the different possible categories that might be observed. For a numerical variable, showing the data distribution means displaying how the observations are distributed along a numerical scale.

Some types of graphical displays are more informative than others when the goal is to compare two or more groups. For example, you might want to compare response times for two different Internet service providers or to compare full-time and part-time students with respect to where they purchase textbooks (a categorical variable with categories Campus bookstore, Off-campus bookstore, or Online bookseller).

For bivariate numerical data, the purpose of a constructing a graphical display is usually to see if there is a relationship between the two variables that define the data set. For example, if both a quality rating and a customer satisfaction score are available for the 30 brands of cars sold in the United States (as described in the chapter preview example), you might want to construct a display that would help you to explore the relationship between quality rating and satisfaction score.

The later sections of this chapter introduce a number of different graphical displays. Section 2.2 covers graphical displays for univariate categorical data, Section 2.3 covers graphical displays for univariate numerical data, and Section 2.4 covers the most common graphical display for bivariate numerical data.

Figure 2.2 shows how knowing the number of variables in the data set, the data type, and the purpose of the graphical display lead to an appropriate choice for a graphical display.

FIGURE 2.2
Choosing an appropriate graphical display

Graphical Display	Number of Variables	Data Type	Purpose
Bar Chart	1	Categorical	Display Data Distribution
Comparative Bar Chart	1 for two or more groups	Categorical	Compare 2 or More Groups
Dotplot	1	Numerical	Display Data Distribution
Comparative Dotplot	1 for two or more groups	Numerical	Compare 2 or More Groups
Stem-and-Leaf Display	1	Numerical	Display Data Distribution
Comparative Stem-and-Leaf Display	1 for two groups	Numerical	Compare 2 Groups
Histogram	1	Numerical	Display Data Distribution (Can also be used to compare groups, if done carefully)
Scatterplot	2	Numerical	Investigate Relationship Between 2 Numerical Variables
Time Series Plot	1, collected over time	Numerical	Investigate Trend Over Time

Summing It Up—Section 2.1

The following learning objectives were addressed in this section:

Conceptual Understanding

C1: Distinguish between categorical and numerical data.

Data are categorical if the observations are categorical responses. Data are numerical if the observations are numbers.

C2: Distinguish between discrete and continuous numerical data.

Numerical data are classified as discrete or continuous depending on whether the possible data values are isolated points along the number line (discrete) or an interval on the number line (continuous).

C3: Understand that selecting an appropriate graphical display depends on the number of variables in the data set, the data type, and the purpose of the graphical display.

There are many different types of graphical displays. Some are only appropriate for categorical data and others are only appropriate for numerical data. If a data set is univariate (observations on a single variable), the purpose of a graphical display is usually to show the data distribution. For bivariate data sets (observations on two variables) the purpose is usually to show the relationship between the variables.

Mastering the Mechanics

M1: Select an appropriate graphical display for a given data set.

When selecting a graphical display, you should think about the number of variables in the data set, whether the data are categorical or numerical, and the purpose of the graphical display.

SECTION 2.1 EXERCISES

Each Exercise Set assesses the following chapter learning objectives: C1, C2, C3, M1

SECTION 2.1 Exercise Set 1

2.1 Classify each of the following variables as either categorical or numerical. For those that are numerical, determine whether they are discrete or continuous.

a. Number of students in a class of 35 who turn in a term paper before the due date
b. Gender of the next baby born at a particular hospital
c. Amount of fluid (in ounces) dispensed by a machine used to fill bottles with soda pop
d. Thickness (in mm) of the gelatin coating of a vitamin E capsule
e. Birth order classification (only child, firstborn, middle child, lastborn) of a math major

2.2 For the following numerical variables, state whether each is discrete or continuous.

a. The number of insufficient-funds checks received by a grocery store during a given month
b. The amount by which a 1-pound package of ground beef decreases in weight (because of moisture loss) before purchase
c. The number of kernels in a bag of microwave popcorn that fail to pop
d. The number of students in a class of 35 who have purchased a used copy of the textbook

2.3 For each of the five data sets described, answer the following three questions and then use Figure 2.2 to select an appropriate graphical display.

> Question 1: How many variables are in the data set?
> Question 2: Are the variables in the data set categorical or numerical?
> Question 3: Would the purpose of a graphical display be to summarize the data distribution, to compare groups, or to investigate the relationship between two numerical variables?

Data Set 1: To learn about the reason parents believe their child is heavier than the recommended weight for children of the same age, each person in a sample of parents of overweight children was asked what they thought was the most important contributing factor. Possible responses were lack of exercise, easy access to junk food, unhealthy diet, medical condition, and other.

Data Set 2: To compare commute distances for full-time and part-time students at a large college, commute distance (in miles) was determined for each student in a random sample of 50 full-time students and for each student in a random sample of 50 part-time students.

Data Set 3: To learn about how number of years of education and income are related, each person in a random sample of 500 residents of a particular city was asked how many years of education he or she had completed and what his or her annual income was.

Data Set 4: To see if there is a difference between faculty and students at a particular college with respect to how they commute to campus (drive, walk, bike, and so on), each person in a random sample of 50 faculty members and each person in a random sample of 100 students was asked how he or she usually commutes to campus.

Data Set 5: To learn about how much money students at a particular college spend on textbooks, each student in a random sample of 200 students was asked how much he or she spent on textbooks for the current semester.

SECTION 2.1 Exercise Set 2

2.4 Classify each of the following variables as either categorical or numerical. For those that are numerical, determine whether they are discrete or continuous.

a. Brand of computer purchased by a customer
b. State of birth for someone born in the United States
c. Price of a textbook
d. Concentration of a contaminant (micrograms per cubic centimeter) in a water sample
e. Zip code (Think carefully about this one.)
f. Actual weight of coffee in a can labeled as containing 1 pound of coffee

2.5 For the following numerical variables, state whether each is discrete or continuous.

a. The length of a 1-year-old rattlesnake
b. The altitude of a location in California selected randomly by throwing a dart at a map of the state
c. The distance from the left edge at which a 12-inch plastic ruler snaps when bent far enough to break
d. The price per gallon paid by the next customer to buy gas at a particular station

2.6 For each of the five data sets described, answer the following three questions and then use Figure 2.2 to select an appropriate graphical display.

> Question 1: How many variables are in the data set?
> Question 2: Are the variables in the data set categorical or numerical?
> Question 3: Would the purpose of a graphical display be to summarize the data distribution, to compare groups, or to investigate the relationship between two numerical variables?

Data Set 1: To learn about the heights of five-year-old children, the height of each child in a sample of 40 five-year-old children was measured.

Data Set 2: To see if there is a difference in car color preferences of men and women, each person in a sample of

100 males and each person in a sample of 100 females was shown pictures of a new model car in five different colors and asked to select which color they would choose if they were to purchase the car.

Data Set 3: To learn how GPA at the end of the freshman year in college is related to high school GPA, both high school GPA and freshman year GPA were determined for each student in a sample of 100 students who had just completed their freshman year at a particular college.

Data Set 4: To learn how the amount of money spent on a fast-food meal might differ for men and women, the amount spent on lunch at a particular fast-food restaurant was determined for each person in a sample of 50 women and each person in a sample of 50 men.

Data Set 5: To learn about political affiliation (Democrat, Republican, Independent, and Other) of students at a particular college, each student in a random sample of 200 students was asked to indicate his or her political affiliation.

ADDITIONAL EXERCISES

2.7 Classify each of the following variables as either categorical or numerical.
a. Weight (in ounces) of a bag of potato chips
b. Number of items purchased by a grocery store customer
c. Brand of cola purchased by a convenience store customer
d. Amount of gas (in gallons) purchased by a gas station customer
e. Type of gas (regular, premium, diesel) purchased by a gas station customer

2.8 For the *numerical* variables in the previous exercise, which are discrete and which are continuous?

2.9 Classify each of the following variables as either categorical or numerical.
a. Color of an M&M candy selected at random from a bag of M&M's
b. Number of green M&M's in a bag of M&M's
c. Weight (in grams) of a bag of M&M's
d. Gender of the next person to purchase a bag of M&M's at a particular grocery store

2.10 For the *numerical* variables in the previous exercise, which are discrete and which are continuous?

2.11 Classify each of the following variables as either categorical or numerical.
a. Number of text messages sent by a college student in a typical day
b. Amount of time a high school senior spends playing computer or video games in a typical day
c. Number of people living in a house
d. A student's type of residence (dorm, apartment, house)
e. Dominant color on the cover of a book
f. Number of pages in a book
g. Rating (G, PG, PG-13, R) of a movie

Use the following instructions for Exercises 2.12–2.15.
Answer the following three questions and then use Figure 2.2 to select an appropriate graphical display.
Question 1: How many variables are in the data set?
Question 2: Are the variables in the data set categorical or numerical?
Question 3: Would the purpose of a graphical display be to summarize the data distribution, to compare groups, or to investigate the relationship between two numerical variables?

2.12 To learn about what super power middle school students would most like to have, each person in a sample of middle school students was asked to choose among invisibility, extreme strength, the ability to freeze time, and the ability to fly.

2.13 To compare the number of hours spent studying in a typical week for male and female students, data were collected from each person in a random sample of 50 female students and each person in a random sample of 50 male students.

2.14 To learn if there is a relationship between water consumption and headache frequency, people in a sample of young adults were asked how much water (in ounces) they drink in a typical day and how many days per month they experience a headache.

2.15 To learn about TV viewing habits of high school students, each person in a sample of students was asked how many hours he or she spent watching TV during the previous week.

SECTION 2.2 ## Displaying Categorical Data: Bar Charts and Comparative Bar Charts

In this section, you will see how bar charts and comparative bar charts can be used to summarize univariate categorical data. A bar chart is used when the purpose of the display is to show the data distribution, and a comparative bar chart (as the name implies) is used when the purpose of the display is to compare two or more groups.

Bar Charts

The first step in constructing a graphical display is often to summarize the data in a table and then use information in the table to construct the display. For categorical data, this table is called a *frequency distribution*.

A **frequency distribution for categorical data** is a table that displays the possible categories along with the associated frequencies and/or relative frequencies.

The **frequency** for a particular category is the number of times that category appears in the data set.

The **relative frequency** for a particular category is calculated as

$$\text{relative frequency} = \frac{\text{frequency}}{\text{number of observations in the data set}}$$

The relative frequency for a particular category is the proportion of the observations that belong to that category. If a table includes relative frequencies, it is sometimes referred to as a **relative frequency distribution**.

Example 2.2   **Motorcycle Helmets—Can You See Those Ears?**

The U.S. Department of Transportation established standards for motorcycle helmets. To comply with these standards, helmets should reach the bottom of the motorcyclist's ears. The report **"Motorcycle Helmet Use in 2014—Overall Results" (National Highway Traffic Safety Administration, January 2015)** summarized data collected by observing 806 motorcyclists nationwide at selected roadway locations. Each time a motorcyclist passed by, the observer noted whether the rider was wearing no helmet, a noncompliant helmet, or a compliant helmet. Using the coding

    N = no helmet
  NH = noncompliant helmet
  CH = compliant helmet

a few of the observations were

    CH   N   CH   NH   N   CH   CH   CH   N   N

There were 796 additional observations, which we didn't reproduce here. In total, there were 250 riders who wore no helmet, 40 who wore a noncompliant helmet, and 516 who wore a compliant helmet.

    The corresponding relative frequency distribution is given in Table 2.1.

TABLE 2.1 Relative Frequency Distribution for Helmet Use

Helmet Use Category	Frequency	Relative Frequency	
No helmet	250	0.310	← 250/806
Noncompliant helmet	40	0.050	← 40/806
Compliant helmet	516	0.640	
	806	1.000	

Total number of observations

Should add to 1, but in some cases may be slightly off due to rounding

From the relative frequency distribution, you can see that a large number of the riders (31%) were not wearing a helmet, but most of those who wore a helmet were wearing one that met the Department of Transportation safety standard.

A **bar chart** is a graphical display of categorical data. Each category in the frequency distribution is represented by a bar or rectangle, and the display is constructed so that the *area* of each bar is proportional to the corresponding frequency or relative frequency.

Bar Charts

When to Use   Number of variables: 1
Data type: categorical
Purpose: displaying data distribution

How to Construct
1. Draw a horizontal axis, and write the category names or labels below the line at regularly spaced intervals.
2. Draw a vertical axis, and label the scale using either frequency or relative frequency.
3. Place a rectangular bar above each category label. The height is determined by the category's frequency or relative frequency, and all bars should have the same width. With the same width, both the height and the area of the bar are proportional to the frequency or relative frequency of the corresponding category.

What to Look For
• Which categories occur frequently and which categories occur infrequently.

Example 2.3   Revisiting Motorcycle Helmets

Example 2.2 used data on helmet use from a sample of 806 motorcyclists to construct the frequency distribution in Table 2.1. A bar chart is an appropriate choice for displaying these data because

• There is one variable (helmet use).
• The variable is categorical.
• The purpose is to display the data distribution.

You can follow the three steps in the previous box to construct the bar chart.

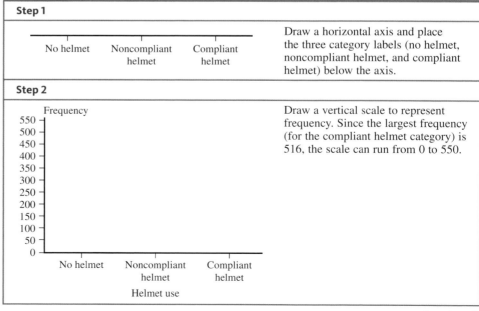

(continued)

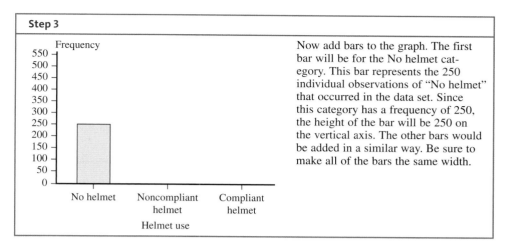

The completed bar chart is shown in Figure 2.3.

FIGURE 2.3
Bar chart of helmet use

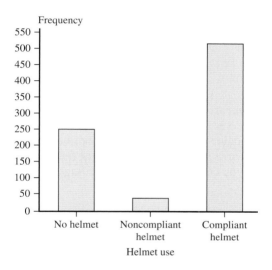

Relative frequency could also have been used for the vertical axis of the bar chart. Figure 2.4 shows a relative frequency bar chart. Notice that the only difference is in

FIGURE 2.4
Relative frequency bar chart
of helmet use

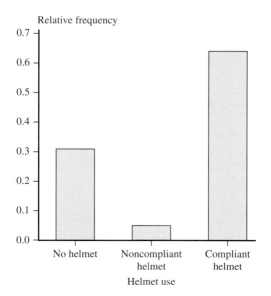

the scale on the vertical axis and that the overall picture of the distribution provided is unchanged. One advantage of the relative frequency bar chart is that the heights of the bars can be interpreted in terms of proportions or percentages.

The bar chart provides a visual representation of the distribution of the 806 values that make up the data set. From the bar chart, it is easy to see that the compliant helmet category occurred most often in the data set. The bar for compliant helmets is about two times as tall (and therefore has about two times the area) as the bar for no helmet because approximately two times as many motorcyclists wore compliant helmets than wore no helmet.

Comparative Bar Charts

Bar charts can also be used to provide a visual comparison of two or more groups. This is accomplished by constructing two or more bar charts that use the same set of horizontal and vertical axes.

Comparative Bar Chart

When to Use   Number of variables: 1 variable with observations for two
or more groups
Data type: categorical
Purpose: comparing two or more data distributions

A **comparative bar chart** is constructed by using the same horizontal and vertical axes for the bar charts of two or more groups. The bar charts are usually color-coded to indicate which bars correspond to each group. *For comparative bar charts, relative frequency should be used for the vertical axis so that you can make meaningful comparisons even if the sample sizes are not the same.*

Example 2.4   Is Education Worth the Cost?

Do people with college degrees think that their education was worth the cost? This question was posed to 2548 adults with an associate degree and to 30,151 adults with a bachelor's degree. The data, from the Gallup report **"Two-Year Grads Satisfied with Cost of Degree" (www.gallup.com, April 11, 2016)**, are summarized in the accompanying table.

	Frequency		Relative Frequency	
Response	**Associate Degree Holders**	**Bachelor's Degree Holders**	**Associate Degree Holders**	**Bachelor's Degree Holders**
Strongly disagree	178	1,508	0.07	0.05
Disagree	153	2,111	0.06	0.07
Neither agree or disagree	408	4,221	0.16	0.14
Agree	637	8,743	0.25	0.29
Strongly agree	1,172	13,568	0.46	0.45
Total	**2,548**	**30,151**	**1.00**	**1.00**

Because the two sample sizes are very different, it is important to use relative frequencies rather than frequency to construct the scale for the vertical axis in the comparative bar chart. The same set of steps that were used to construct a bar chart are used to construct a comparative bar chart, but in a comparative bar chart each category will have a bar for each group.

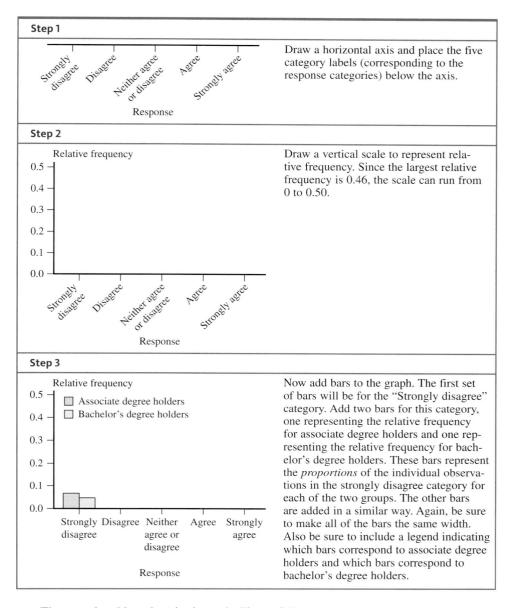

The completed bar chart is shown in Figure 2.5.

FIGURE 2.5
Comparative bar chart

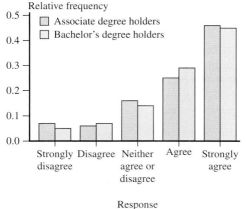

Looking at the comparative bar chart, it is easy to compare the response distributions of the two groups. Notice that the response distributions are similar, indicating that both associate degree holders and bachelor's degree holders generally agreed that their education was worth the cost, with more than 70% of each group in the Agree or Strongly agree categories.

To see why it is important to use relative frequencies rather than frequencies to compare groups of different sizes, consider the *inappropriate* bar chart constructed using the frequencies rather than the relative frequencies (Figure 2.6). Because there were so many more bachelor's degree holders than associate degree holders who participated in the poll (30,151 bachelor's degree holders and only 2548 associate degree holders), the inappropriate bar chart conveys a very different and misleading impression.

FIGURE 2.6

An *inappropriate* comparative bar chart using frequency rather than relative frequency for vertical axis

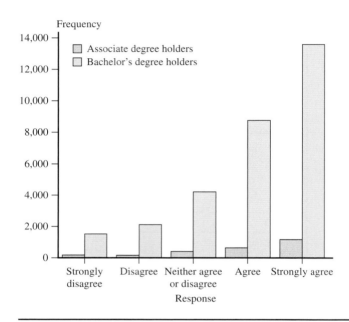

Summing It Up—Section 2.2

The following learning objectives were addressed in this section:

Mastering the Mechanics
M2: Construct and interpret bar charts and comparative bar charts.

Bar charts are appropriate when you have data on one variable and the data are categorical. To make a bar chart, first create a frequency distribution (see Example 2.2) and then follow the instructions given just prior to Example 2.3. A comparative bar chart is used when you have data on one categorical variable for two or more groups. Bars are drawn for each group (see Example 2.4).

Putting It into Practice
P2: Use graphical displays to compare groups on the basis of a categorical variable.

A comparative bar chart makes it easy to compare groups because bars for each group are drawn on the same set of axes. For an example, see Figure 2.5 and the discussion that follows.

SECTION 2.2 EXERCISES

Each Exercise Set assesses the following chapter learning objectives: M2, P2

SECTION 2.2 Exercise Set 1

2.16 The Gallup report **"More Americans Say Real Estate Is Best Long-Term Investment" (www.gallup.com, April 20, 2016, retrieved April 15, 2017)** included data from a poll of

1015 adults. The responses to the question "What do you think is the best long-term investment?" are summarized in the given relative frequency distribution.

Response	Relative Frequency
Real Estate	0.35
Stocks & Mutual Funds	0.22
Gold	0.17
Savings	0.15
Bonds	0.07
Other	0.04

a. Use this information to construct a bar chart for the response data.

b. Comment on how people responded to the question posed.

2.17 The report referenced in the previous exercise also gave responses to the question "What do you think is the best long-term investment?" by gender. Relative frequencies for the six response categories for men and for women are given in the accompanying table.

Response	Relative Frequency for Women	Relative Frequency for Men
Real Estate	0.33	0.32
Stocks & Mutual Funds	0.21	0.25
Gold	0.14	0.22
Savings	0.18	0.12
Bonds	0.08	0.05
Other	0.06	0.04

a. Construct a comparative bar chart that shows the distributions of responses for men and for women. (Hint: See Example 2.4. There should be two bars for each of the six response categories.)

b. Comment on similarities and differences in the response distributions for men and women.

SECTION 2.2 Exercise Set 2

2.18 The report **"Trends in Education 2010: Community Colleges" (www.collegeboard.com/trends)** included the accompanying information on student debt for students graduating with an AA degree from a public community college in 2008.

Debt	Relative Frequency
None	0.62
Less than $10,000	0.23
Between $10,000 and $20,000	0.10
More than $20,000	0.05

a. Use the given information to construct a bar chart.

b. Comment on student debt for public community college graduates.

2.19 Each year, *The Princeton Review* conducts surveys of high school students who are applying to college and of parents of college applicants. The report **"2016 College**

Hopes & Worries Survey Findings" (www.princetonreview .com/cms-content/final_cohowo2016survrpt.pdf, retrieved April 15, 2017) included a summary of how 8347 high school students responded to the question "Ideally how far from home would you like the college you attend to be?" Students responded by choosing one of four possible distance categories. Also included was a summary of how 2087 parents of students applying to college responded to the question "How far from home would you like the college your child attends to be?" The accompanying relative frequency table summarizes the student and parent responses.

	Frequency	
Ideal Distance	**Students**	**Parents**
Less than 250 miles	2,587	1,085
250 to 500 miles	2,671	626
500 to 1,000 miles	1,753	271
More than 1,000 miles	1,336	105
Total	**8,347**	**2,087**

a. Explain why you would want to use relative frequencies when constructing a comparative bar chart to compare ideal distance for students and parents.

b. Construct a comparative bar chart for these data.

c. Comment on similarities and differences in the distributions of ideal distance for parents and students.

ADDITIONAL EXERCISES

2.20 Heal the Bay is an environmental organization that releases an annual beach report card based on water quality **(Heal the Bay Beach Report Card, www.beachreportcard.org, retrieved May 7, 2016)**. The grades for 20 beaches in three counties in Washington (Whatcom, Snohomish, and Island counties) during dry weather were:

C   B   A+   A+   A+   A+   A+   A   A+   A+
A+   C   A+   A+   A+   C   A   F   F   B

a. Summarize the dry weather grades by constructing a relative frequency distribution and a bar chart.

b. The wet weather grades for these same beaches were:

A+   A+   A+   A+   A+   A+   A+   F   F   F
A+   A+   F   A+   A+   F   A+   A+   F   A+

Construct a bar chart for the wet weather grades.

c. Do the bar charts from Parts (a) and (b) support the statement that beach water quality tends to be better in dry weather conditions? Explain.

d. Construct a comparative bar chart for the grades for dry and wet weather and comment on the differences between the two grade distributions.

2.21 The report **"Findings from the 2014 College Senior Survey" (Higher Education Research Institute, December 2014)** summarizes data collected from more than 13,000 college seniors across the United States. One question in

Data set available

the survey asked students to rate themselves based on their critical thinking skills. For engineering majors, 60.4% rated critical thinking as "a major strength," whereas 39.6% did not see critical thinking as a major strength. Data were also provided for humanities majors, social science majors, biological sciences majors, and business majors, and these data are summarized in the accompanying table.

	Relative Frequency				
Response	**Engineering Majors**	**Humanities Majors**	**Social Science Majors**	**Biological Science Majors**	**Business Majors**
A major strength	0.604	0.513	0.470	0.461	0.391
Not a major strength	0.396	0.487	0.530	0.539	0.609

Construct a comparative bar chart and compare the responses over the five different majors.

2.22 The report "2013 International Bedroom Poll: Summary of Findings" describes a survey of 251 adult Americans conducted by the National Sleep Foundation **(www.sleep foundation.org/sites/default/files/RPT495a.pdf, retrieved April 15, 2017)**. Participants in the survey were asked how often they change the sheets on their bed and were asked to respond with one of the following categories: more than once a week, once a week, every other week, every three weeks, or less often than every three weeks. For this group, 10% responded more than once a week, 53% responded once a week, 26% responded every other week, 5% responded every three weeks, and 6% responded less often than every three weeks.

a. Use the given information to make a relative frequency distribution for the responses to the question.

b. Summarize the given information by constructing a bar chart.

c. The report also summarized data from 250 adults in Japan. For this group, 11% responded more than once a week, 30% responded once a week, 22% responded every other week, 9% responded every three weeks, and 28% responded less often than every three weeks. Construct a comparative bar chart that will allow you to compare the response distributions for the U.S. sample and the Japan sample. Comment on similarities and differences between the distributions for these two countries.

2.23 In the United States, movies are rated by the Motion Picture Association of America (MPAA). The accompanying table gives the MPAA rating of the 25 top money-making movies of 2015 (data from **www.boxofficemojo .com, retrieved October 10, 2016**).

Movie (Ordered from high to low based on amount of money made)	Rating
Star Wars: The Force Awakens	PG-13
Jurassic World	PG-13
Avengers: Age of Ultron	PG-13
Inside Out	PG
Furious 7	PG-13
Minions	PG
The Hunger Games: Mockingjay – Part 2	PG-13
The Martian	PG-13
Cinderella	PG
Spectre	PG-13
Mission Impossible: Rogue Nation	PG-13
Pitch Perfect 2	PG-13
The Revenant	R
Ant-Man	PG-13
Home	PG
Hotel Transylvania 2	PG
Fifty Shades of Grey	R
The SpongeBob Movie: Sponge Out of Water	PG
Straight Outta Compton	R
San Andreas	PG-13
Mad Max: Fury Road	R
Daddy's Home	PG-13
The Divergent Series: Insurgent	PG-13
Peanuts Movie	G
Kingsman: The Secret Service	R

Use the given information to construct a bar chart of the ratings for the top 25 movies of 2015. Describe the ratings distribution.

SECTION 2.3 **Displaying Numerical Data: Dotplots, Stem-and-Leaf Displays, and Histograms**

In this section, you will see three different types of graphical displays for univariate numerical data—dotplots, stem-and-leaf displays, and histograms.

Dotplots and Comparative Dotplots

A dotplot is a simple way to display numerical data when the data set is not too large. Each observation is represented by a dot above the location corresponding to its value on a number line. When a value occurs more than once in a data set, there is a dot for each occurrence, and these dots are stacked vertically in the plot.

Dotplots

When to Use   Number of variables: 1
Data Type: numerical
Purpose: display data distribution

How to Construct
1. Draw a horizontal line and mark it with an appropriate measurement scale.
2. Locate each value in the data set along the measurement scale, and represent it by a dot. If there are two or more observations with the same value, stack the dots vertically.

What to Look For   Dotplots convey information about
- A representative or typical value for the data set.
- The extent to which the data values vary.
- The nature of the distribution of values along the number line.
- The presence of unusual values in the data set.

Data set available

Example 2.5   Making it to Graduation

The article **"Keeping Score When It Counts: Graduation Success and Academic Progress Rates for 2016 NCAA Division I Basketball Tournament Teams" (The Institute for Diversity and Ethics in Sport, University of Central Florida, March 2016)** included data on graduation rates of basketball players for the universities and colleges that sent teams to the 2016 Division I playoffs. The following graduation rates are the average of the four six-year graduation percentages for basketball players starting college in 2005, 2006, 2007, and 2008.

Graduation Rates														
98	91	100	73	57	100	100	77	53	88	58	67	92	75	63
100	78	70	85	88	70	80	42	39	91	55	82	54	92	80
92	80	55	36	83	20	100	62	100	100	90	91	93	89	90
80	46	75	100	71	50	62	82	50	100	83	90	64	91	67
83	100	83	100	83	63	91	95							

One way to summarize these data graphically is to use a dotplot. A dotplot is an appropriate choice because the data set consists of one variable (graduation rate), the variable is numerical, and the purpose is to display the data distribution. The data set is not too large, with 68 observations. (If the data set had been much larger, a histogram, a graphical display to be introduced later in this section, might have been a better choice.)

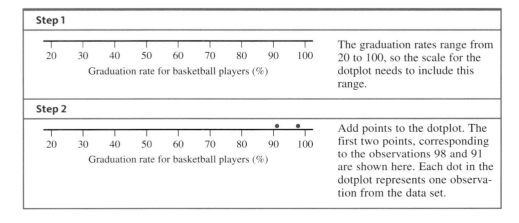

Step 1	
(dotplot scale: 20 30 40 50 60 70 80 90 100) Graduation rate for basketball players (%)	The graduation rates range from 20 to 100, so the scale for the dotplot needs to include this range.
Step 2	
(dotplot scale: 20 30 40 50 60 70 80 90 100 with two dots near 90 and 98) Graduation rate for basketball players (%)	Add points to the dotplot. The first two points, corresponding to the observations 98 and 91 are shown here. Each dot in the dotplot represents one observation from the data set.

The completed dotplot is shown in Figure 2.7.

FIGURE 2.7
Dotplot of graduation rates

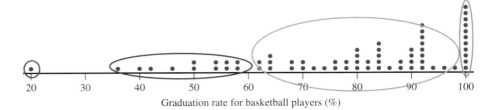

Graduation rate for basketball players (%)

The dotplot shows how the 68 graduation rates are distributed along the number line. You can see that basketball graduation rates vary from school to school, ranging from a low of 20% to a high of 100%. You can also see that the graduation rates seem to cluster in several groups, denoted by the colored ovals that have been added to the dotplot. There are 11 schools with graduation rates of 100% (excellent!). The majority of schools are in the large cluster, with graduation rates from about 62% to about 98%. And then there is a group of 12 schools with low graduation rates for basketball players and one school with an embarrassingly low graduation rate of 20% (University of Connecticut). ■

Most statistical software packages and some graphing calculators can be used to construct dotplots, so you may not need to construct dotplots by hand.

Comparative Dotplots

Dotplots can also be used for comparative displays. Comparative dotplots are constructed by using the same numerical scale for two or more dotplots.

Comparative Dotplots

When to Use   Number of variables: 1 variable with observations for two or more groups
Data type: Numerical
Purpose: Comparing two or more data distributions

Comparative dotplots are constructed using the same numerical scale for two or more dotplots. Be sure to include group labels for the dotplots in the display.

Data set available

Example 2.6   Making it to Graduation Revisited

The article referenced in Example 2.5 also gave graduation rates for *all* student athletes at the 68 schools in the 2016 Division I basketball playoffs. The data are listed below. Also listed are the differences between the graduation rate for all student athletes and the graduation rate for basketball players.

School	BB	ALL	Difference (ALL−BB)	School	BB	ALL	Difference (ALL−BB)
1	98	79	−19	10	88	97	9
2	91	88	−3	11	58	67	9
3	100	88	−12	12	67	87	20
4	73	71	−2	13	92	90	−2
5	57	74	17	14	75	80	5
6	100	98	−2	15	63	87	24
7	100	98	−2	16	100	87	−13
8	77	71	−6	17	78	82	4
9	53	69	16	18	70	91	21

(continued)

School	BB	ALL	Difference (ALL−BB)	School	BB	ALL	Difference (ALL−BB)
19	85	84	−1	44	89	89	0
20	88	92	4	45	90	85	−5
21	70	90	20	46	80	85	5
22	80	83	3	47	46	81	35
23	42	60	18	48	75	80	5
24	39	60	21	49	100	98	−2
25	91	81	−10	50	71	84	13
26	55	90	35	51	50	80	30
27	82	85	3	52	62	82	20
28	54	78	24	53	82	81	−1
29	92	79	−13	54	50	70	20
30	80	78	−2	55	100	85	−15
31	92	83	−9	56	83	87	4
32	80	78	−2	57	90	83	−7
33	55	79	24	58	64	86	22
34	36	79	43	59	91	92	1
35	83	86	3	60	67	85	18
36	20	85	65	61	83	93	10
37	100	95	−5	62	100	94	−6
38	62	78	16	63	83	76	−7
39	100	89	−11	64	100	69	−31
40	100	84	−16	65	83	82	−1
41	90	81	−9	66	63	80	17
42	91	85	−6	67	91	94	3
43	93	89	−4	68	95	98	3

How do the graduation rates of basketball players compare to those of all student athletes? Figure 2.8 shows comparative dotplots. Notice that the comparative dotplots use the same numerical scale.

There are some striking differences that are easy to see when the data are displayed in this way. The graduation rates for all student athletes tend to be higher and to vary less from school to school than the graduation rates for just basketball players.

The dotplots in Figure 2.8 are informative, but we can do even better. The data given here are *paired data*. Each basketball graduation rate can be paired with the graduation rate for all student athletes from the same school. When data are paired, it is usually more informative to look at the differences. These differences (All−Basketball) are also given in the data table. Figure 2.9 gives a dotplot of the 68 differences. Notice that one difference

FIGURE 2.8
Comparative dotplots of graduation rates for basketball players and for all athletes

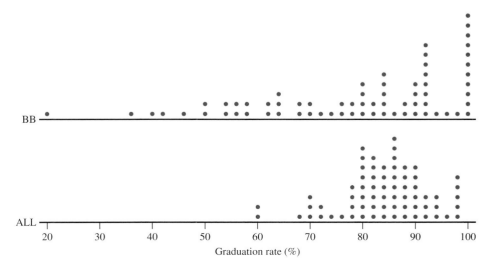

FIGURE 2.9
Dotplot of graduation rate
differences (ALL−BB)

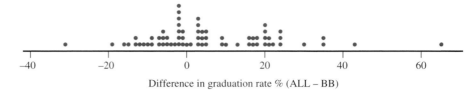

Difference in graduation rate % (ALL – BB)

is 0. This corresponds to a school where the basketball graduation rate is equal to the graduation rate of all student athletes. There are 30 schools for which the difference is negative. Negative differences correspond to schools that have a higher graduation rate for basketball players than for all student athletes. The most interesting feature of the difference dotplot is the spread in the positive differences. The positive differences correspond to schools with a higher graduation rate for all athletes than for basketball players. The positive differences range from 1% all the way up to 65%, and there were five schools in which the graduation rate for all athletes was 30 percentage points or more greater than the graduation rate for basketball players. (In case you were wondering, these schools were University of Oregon with a difference of 30%, Syracuse University and University of North Carolina Wilmington with differences of 35%, University of Cincinnati with a difference of 43%, and University of Connecticut with a difference of 65%.)

Stem-and-Leaf Displays (Optional)

A stem-and-leaf display is an effective way to summarize univariate numerical data when the data set is not too large. Each number in the data set is broken into two pieces, a stem and a leaf. The **stem** is the first part of the number and consists of the beginning digit(s). The **leaf** is the last part of the number and consists of the final digit(s). For example, the number 213 might be split into a stem of 2 and a leaf of 13 or a stem of 21 and a leaf of 3. The resulting stems and leaves are then used to construct the display.

Data set available

Example 2.7	Should Doctors Get Auto Insurance Discounts?

Many auto insurance companies give occupation-related discounts of 5 to 15%. The article **"Auto-Rate Discounts Seem to Defy Data"** (*San Luis Obispo Tribune*, **June 19, 2004**) included the accompanying data on the number of automobile accidents per year for every 1000 people in 40 occupations.

Occupation	Accidents per 1,000	Occupation	Accidents per 1,000
Student	152	Banking-finance	89
Physician	109	Customer service	88
Lawyer	106	Manager	88
Architect	105	Medical support	87
Real estate broker	102	Computer-related	87
Enlisted military	99	Dentist	86
Social worker	98	Pharmacist	85
Manual laborer	96	Proprietor	84
Analyst	95	Teacher, professor	84
Engineer	94	Accountant	84
Consultant	94	Law Enforcement	79
Sales	93	Physical therapist	78
Military Officer	91	Veterinarian	78
Nurse	90	Clerical, secretary	77
School administrator	90	Clergy	76
Skilled labor	90	Homemaker	76
Librarian	90	Politician	76
Creative arts	90	Pilot	75
Executive	89	Firefighter	67
Insurance agent	89	Farmer	43

Figure 2.10 shows a stem-and-leaf display for the accident rate data. The numbers in the vertical column on the left of the display are the **stems.** Each number to the right of the vertical line is a **leaf** corresponding to one of the observations in the data set. The legend

Stem: Tens

Leaf: Ones

tells you that the observation that had a stem of 4 and a leaf of 3 corresponds to the occupation with an accident rate of 43 per 1000 people. Similarly, the observation with the stem of 10 and leaf of 2 corresponds to 102 accidents per 1000 people.

FIGURE 2.10

Stem-and-leaf display for accident rate per 1000 people for 40 occupations

```
 4 | 3
 5 |
 6 | 7
 7 | 56667889
 8 | 444567788999
 9 | 0000013445689
10 | 2569
11 |
12 |
13 |
14 |              Stem: Tens
15 | 2            Leaf:  Ones
```

The display in Figure 2.10 suggests that a typical or representative value is in the stem 8 or 9 row, perhaps around 90. The observations are mostly concentrated in the 75 to 109 range, but there are a couple of values that stand out on the low end (43 and 67) and one observation (152) that is far removed from the rest of the data on the high end.

From the point of view of an auto insurance company, it might make sense to offer discounts to occupations with low accident rates—maybe farmers (43 accidents per 1000) or firefighters (67 accidents per 1000) or even some of the occupations with accident rates in the 70s. The "discounts seem to defy data" in the title of the article refers to the fact that some insurers provide discounts to doctors and engineers but not to homemakers, politicians, and other occupations with lower accident rates. Two possible explanations were offered for this apparent discrepancy. First, some occupations with higher accident rates could also have lower average costs per claim. Accident rates alone may not reflect the actual cost to the insurance company. Second, insurance companies may offer the discounted auto insurance in order to attract people who would also purchase other types of insurance, such as malpractice or liability insurance.

The leaves on each line of the display in Figure 2.10 have been arranged in order from smallest to largest. Most statistical software packages order the leaves this way, but even if this were not done, you could still see many important characteristics of the data set, such as shape and variability.

Stem-and-leaf displays can be useful for getting a sense of a typical value for the data set, as well as how much the values vary. It is also easy to spot data values that are unusually far from the rest. Such values are called outliers. The stem-and-leaf display of the accident rate data Figure 2.10 shows an outlier on the low end (43) and an outlier on the high end (152).

DEFINITION

An **outlier** is an unusually small or large data value.

A rule for deciding when an observation is an outlier is given in Chapter 3.

Stem-and-Leaf Display

When to Use   Number of variables: 1
Data type: Numerical
Purpose: Display data distribution

How to Construct

1. Select one or more leading digits for the stem values. The trailing digits (or sometimes just the first one of the trailing digits) become the leaves.
2. List possible stem values in a vertical column. Include *all* stem values between the stem associated with the smallest data value and the stem associated with the largest data value. Draw a line that will spearate the stems from the leaves in the display.
3. Record the leaf for every observation beside the corresponding stem value.
4. Indicate the units for stems and leaves someplace in the display.

What to Look For   The display conveys information about
- a representative or typical value for the data set
- the extent to which the data values vary
- the presence of any gaps and outliers
- the extent of symmetry in the data distribution
- the number and location of peaks

Example 2.8   Going Wireless

The U.S. Department of Health and Human Services reported the estimated percentage of households with only wireless phone service (no landline) in 2014 for the 50 U.S. states and the District of Columbia (**www.cdc.gov/nchs/data/nhis/earlyrelease/wireless_state _201602.pdf, retrieved January 20, 2017**). Data for the 19 Eastern states are given here.

State	Wireless %	State	Wireless %
CT	26.7	NY	31.1
DE	29.4	NC	42.9
DC	49.7	OH	45.8
FL	47.6	PA	30.0
GA	45.9	RI	34.6
ME	40.8	SC	49.5
MD	36.2	VA	41.1
MA	31.5	VT	37.2
NH	31.2	WV	37.2
NJ	25.1		

A stem-and-leaf display is an appropriate way to summarize these data because there is one variable (Wireless %), the variable is numerical, and the purpose is to display the data distribution. A dotplot would also have been a reasonable choice.

The sequence of steps given in the previous box can be used to construct a stem-and-leaf display for these data.

Step 1	
Stems: 2, 3, 4	The data values range from 25.1% to 49.7%.
	Using the first two digits would result in stems of 25, 26, ..., 49. Twenty-five stems are probably too many for a data set that has only 19 observations. Using the first digit as a stem results in three possible stems: 2, 3, 4; so a single digit will be used for the stem.
Step 2	
2 3 4	Getting ready to enter leaves, list all possible stems in a column and draw a line that will separate the stems from the leaves in the display.

(continued)

Step 3		
2 3 4	6.7, 9.4 9.7	The first data value is 26.7, which has a stem of 2 and a leaf of 6.7, so 6.7 is entered to the right of the 2 stem. The second data value is 29.4, which has a stem of 2 and a leaf of 9.4. The third data value is 49.7. These values have been added in to the display. Other data values are added in a similar way.
Step 4		
2 3 4	6.7, 9.4 9.7	The last step is to indicate the units for stems and leaves. The unit is determined by the place value of the rightmost digit. The stem is the tens digit, so the unit for the stem is tens. The rightmost digit of the leaves is in the tenths position, so the unit for the leaves is tenths.
Stem: tens Leaves: tenths		

The completed stem-and-leaf display is given in Figure 2.11.

FIGURE 2.11
Stem-and-leaf display of wireless percentage for eastern states

```
2 | 6.7, 9.4, 5.1
3 | 6.2, 1.5, 1.2, 1.1, 0.0, 4.6, 7.2, 7.2
4 | 9.7, 7.6, 5.9, 0.8, 2.9, 5.8, 9.5, 1.1
```

Stem: tens
Leaves: tenths

This display shows that for most eastern states, the percentage of households with only wireless phone service was in the 30% to 49% range. Three states had percentages that were smaller than the others: Connecticut at 26.7%, Delaware at 29.4%, and New Jersey at 25.1%.

An alternative display (Figure 2.12) results from dropping all but the first digit in each leaf. (This is called truncating, and is different than rounding.) This is what most statistical computer packages do when generating a stem-and-leaf display. The computer package Minitab was used to generate the display in Figure 2.12.

FIGURE 2.12
Minitab stem-and-leaf display of wireless percent using only the first leaf digit.

Stem-and-leaf of Wireless % East N = 19
Leaf Unit = 1.0

```
2    569
3    01114677
4    01255799
```

Repeated Stems to Stretch a Stem-and-Leaf Display

Sometimes a natural choice of stems results in a display in which many of the observations are concentrated on just a few stems. A more informative picture may be obtained by dividing the leaves at any given stem into two groups: those that begin with 0, 1, 2, 3, or 4 (the "low" leaves) and those that begin with 5, 6, 7, 8, or 9 (the "high" leaves"). Then each stem is listed twice when constructing the display, once for the low leaves and once for the high leaves. It is also possible to repeat a stem more than twice. For example, each stem might be repeated five times, once for each of the leaf groupings {0,1}, {2,3}, {4,5}, {6,7}, and {8,9}. The following example illustrates the use of repeated stems in a stem-and-leaf display.

Example 2.9 Seat Belt Use

The accompanying data on seat belt use for each of the 50 U.S. states and the District of Columbia are from **"Traffic Safety Facts,"** which was published in June 2015 by the **National Highway Traffic Safety Administration**. The observations represent the percentage of drivers wearing seat belts in a large nationwide observational survey.

95.7	88.4	87.2	74.4	97.1	82.4	85.1	91.9	93.2	88.8	97.3	93.5	80.2
94.1	90.2	92.8	85.7	86.1	84.1	85.0	92.1	76.6	93.3	94.7	78.3	78.8
74.0	70.0	94.0	70.4	87.6	92.1	90.6	90.6	81.0	85.0	86.3	97.8	83.6
87.4	90.0	68.9	87.7	90.7	83.4	84.1	77.3	94.5	87.8	84.7	79.2	

The values in the data set range from 68.9% to 97.8%. Using the first two digits in each data value for the stem results in a large number of stems, while using only the first digit results in a stem-and-leaf display with only four stems.

The stem-and-leaf display using single-digit stems and leaves truncated to a single digit is shown in Figure 2.13. A stem-and-leaf display that uses repeated stems is shown in Figure 2.14. Here each stem is listed twice, once for the low leaves (those beginning with 0, 1, 2, 3, or 4) and once for the high leaves (those beginning with 5, 6, 7, 8, or 9). This display is more informative than the one in Figure 2.13, but it is much more compact than a display based on two-digit stems.

```
6 | 8
7 | 004467889
8 | 01233444555566677777788
9 | 000001222333444445777
```

FIGURE 2.13
Stem-and-leaf display for the seat belt use data

```
6H | 8
7L | 0044
7H | 67889
8L | 01233444
8H | 5555667777788
9L | 000001222333444444
9H | 5777
```

FIGURE 2.14
Stem-and-leaf display for the seat belt use data using repeated stems

Comparative Stem-and-Leaf Displays (Optional)

It is also common to see a stem-and-leaf display used to provide a visual comparison of two groups. A comparative stem-and-leaf display, in which the leaves for one group are listed to the right of the stem values and the leaves for the second group are listed to the left, can show how the two groups are similar and how they differ. This display is also sometimes called a back-to-back stem-and-leaf display.

Comparative Stem-and-Leaf Displays

When to Use   Number of variables: 1 variable with observations for two groups
Data type: Numerical
Purpose: Compare two data distributions

A comparative stem-and-leaf display has a single column of stems, with the leaves for one data set listed to the right of the stems and the leaves for the second data set listed to the left of the stems. Be sure to include group labels to identify which group is on the left and which is on the right.

Data set available

Example 2.10   Seat Belt Use Revisited

The source referenced in Example 2.9 also gave information on whether or not each state currently enforces a seat belt law. Thirty-three states and the District of Columbia enforce a seat belt law and the other 17 states do not.

A comparative stem-and-leaf display (using only the first digit of each leaf) is shown in Figure 2.15. In this display, the leaves for data values corresponding to states that

FIGURE 2.15
Comparative stem-and-leaf display of the percentage of drivers using seat belts for states with and without enforcement of a seat belt law

```
   No Enforcement            Enforcement

              8 | 6 |
        9876400 | 7 | 48
       75433210 | 8 | 4455566777788
              4 | 9 | 000001222333444445777
```

enforce seat belt laws are listed to the right of the stems and the leaves corresponding to data values for states that do not enforce seat belt laws are listed to the left.

From the comparative stem-and-leaf display, you can see that although there was state-to-state variability in both the states with seat belt enforcement and the states without seat belt enforcement, the seat belt use percentages tended to be higher in states with enforcement.

Histograms

Dotplots and stem-and-leaf displays are not always effective ways to summarize numerical data. Both are awkward when the data set contains a large number of data values. Histograms are displays that may not work well for small data sets but do work well for larger numerical data sets. Histograms are constructed a bit differently, depending on whether the variable of interest is discrete or continuous.

Frequency Distributions and Histograms for Discrete Numerical Data

Discrete numerical data often result from counting. In such cases, each observation is a whole number. A frequency distribution for discrete numerical data lists each possible value (either individually or grouped into intervals), the associated frequency, and sometimes the corresponding relative frequency. Recall that relative frequencies are calculated by dividing each frequency by the total number of observations in the data set.

Data set available

Example 2.11 Promiscuous Queen Bees

Queen honey bees mate shortly after they become adults. During a mating flight, the queen usually takes multiple partners, collecting sperm that she will store and use throughout the rest of her life. The authors of the paper **"The Curious Promiscuity of Queen Honey Bees" (Annals of Zoology [2001]: 255–265)** studied the behavior of 30 queen honey bees to learn about the length of mating flights and the number of partners a queen takes during a mating flight. Data on number of partners on one mating flight (generated to be consistent with summary values and graphs given in the paper) are shown here.

Number of Partners

12	2	4	6	6	7	8	7	8	11
8	3	5	6	7	10	1	9	7	6
9	7	5	4	7	4	6	7	8	10

The corresponding relative frequency distribution is given in Table 2.2. The smallest value in the data set is 1 and the largest is 12, so the possible values from 1 to 12 are listed in the table, along with the corresponding frequencies and relative frequencies.

TABLE 2.2 Relative Frequency Distribution for Number of Partners

Number of Partners	Frequency	Relative Frequency	
1	1	0.033	$\frac{1}{30} = 0.033$
2	1	0.033	
3	1	0.033	
4	3	0.100	
5	2	0.067	
6	5	0.167	
7	7	0.233	
8	4	0.133	
9	2	0.067	
10	2	0.067	
11	1	0.033	
12	1	0.033	Differs from 1
Total	30	0.999	due to rounding

It is possible to create a more compact frequency distribution by grouping some of the possible values into intervals. For example, you might group 1, 2, and 3 partners to form an interval of 1–3, with a corresponding frequency of 3. The grouping of other values in a similar way results in the relative frequency distribution shown in Table 2.3.

TABLE 2.3 Relative Frequency Distribution for Number of Partners Using Intervals

Number of Partners	Frequency	Relative Frequency
1–3	3	0.100
4–6	10	0.333
7–9	13	0.433
10–12	4	0.133

A histogram for discrete numerical data is a graph that shows the data distribution. The following example illustrates the construction of a histogram for the queen bee data.

Data set available

Example 2.12 **Promiscuous Queen Bees Revisited**

Example 2.11 gave the following data on the number of partners during a mating flight for 30 queen honey bees.

Number of Partners

12	2	4	6	6	7	8	7	8	11
8	3	5	6	7	10	1	9	7	6
9	7	5	4	7	4	6	7	8	10

Let's start by looking at the dotplot of these data in Figure 2.16.

FIGURE 2.16
Dotplot of queen bee data

A histogram for this discrete numerical data set replaces the stacks of dots that appear above each possible value with a rectangle that is centered over the value and whose height corresponds to the number of observations at that value, as shown in Figure 2.17.

FIGURE 2.17
Histogram of queen bee data superimposed on dotplot

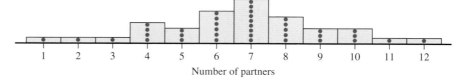

Notice that the area of each rectangle in the histogram is proportional to the frequency of the corresponding data value. For example, the rectangle corresponding to 4 partners is three times the area of the 1 partner rectangle because there were three times as many 4's in the data set as there were 1's.

In practice, histograms for discrete data don't show the dots corresponding to the actual data values—only the rectangles are shown, as in Figure 2.18 (at the bottom of the next page). We started with the dotplot and used it to construct the histogram to show that the rectangles in a histogram represent a collection of individual data values. The histogram could also have been constructed directly from the frequency distribution given in Example 2.11 using the following steps.

Histogram for Discrete Numerical Data

When to Use Number of variables: 1
Data Type: Discrete numerical
Purpose: Display data distribution

How to Construct
1. Draw a horizontal scale, and mark the possible values of the variable.
2. Draw a vertical scale, and add either a frequency or relative frequency scale.
3. Above each possible value, draw a rectangle centered at that value (so that the rectangle for 1 is centered at 1, the rectangle for 5 is centered at 5, and so on). The height of each rectangle is determined by the corresponding frequency or relative frequency. When possible values are consecutive whole numbers, the base width for each rectangle is 1.

What to Look For
- Center or typical value
- Amount of variability
- General shape
- Location and number of peaks
- Presence of gaps and outliers

Example 2.13 **Promiscuous Queen Bees One More Time**

Table 2.2 summarized the queen bee data of Examples 2.11 and 2.12 in a frequency distribution. The corresponding histogram and relative frequency histogram are shown in Figure 2.18. Notice that each rectangle in the histogram is centered over the corresponding value. When relative frequency is used for the vertical scale instead of frequency, the scale on the vertical axis is different, but all essential characteristics of the graph (shape, center, variability) are unchanged.

FIGURE 2.18
Histogram and relative frequency histogram of queen bee data

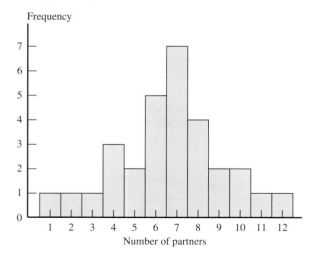

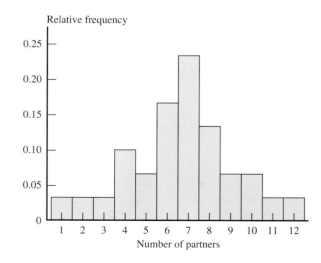

A histogram based on the grouped frequency distribution of Table 2.3 can be constructed in a similar fashion and is shown in Figure 2.19. A rectangle represents the frequency or relative frequency for each interval. For the interval 1–3, the rectangle extends from 0.5 to 3.5 so that there are no gaps between the rectangles of the histogram.

FIGURE 2.19
Histogram of queen bee data using intervals

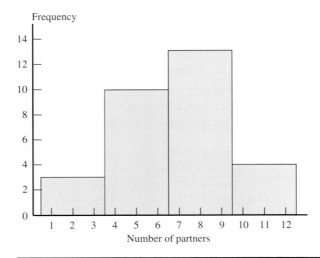

The difficulty in constructing tabular or graphical displays with continuous data, such as observations on reaction time or weight of airline passenger carry-on luggage, is that there are no natural intervals. The way out of this dilemma is to define your own intervals, as illustrated in the following example.

Data set available

Example 2.14 **Carry-on Luggage Weights**

Consider the following data on carry-on luggage weight (in pounds) for 25 airline passengers.

Carry-on Baggage Weight

25.0 17.9 10.1 27.6 30.0 18.0 28.7 28.2 27.8 28.0 31.4
20.9 33.8 27.6 21.9 19.9 20.8 28.5 22.4 24.9 26.4 22.0
34.5 22.7 25.3

A dotplot of these data is given in 2.20(a). Because the weights range from about 10 pounds to about 35 pounds, grouping the weights into 5-pound intervals is reasonable. The first interval would begin at 10 pounds and end at 15 pounds. Because you want each data value to fall in exactly one interval, you would include data values of 10 in this interval, along with any that are between 10 and 15 pounds. Data values of exactly 15 pounds would be put in the interval that starts at 15. The first interval could then be written as 10 to <15, where the less than symbol indicates that the interval will contain all data values that are greater than or equal to 10 and less than 15.

To see how the histogram is formed, take all of the dots in the interval 10 to <15 and stack them up at the center of the interval (at 12.5). Doing this for each of the intervals results in the graph shown in Figure 2.20(b).

FIGURE 2.20
Graphs of carry-on baggage weight: (a) dotplot; (b) dots stacked at midpoints of intervals.

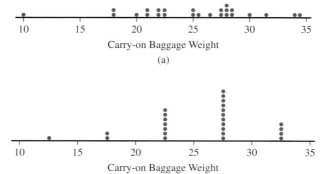

A histogram of these data is created by drawing a rectangle that spans the interval over each interval. The height of each rectangle is determined by the number of dots in the stack for that interval, as shown in Figure 2.21.

FIGURE 2.21
Histogram of carry-on baggage weight

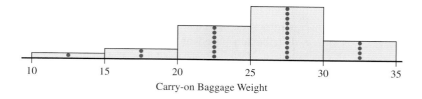

As with discrete data, histograms for continuous data don't show the "dots" and are usually constructed directly from a frequency distribution. This is illustrated in the examples that follow.

Frequency Distributions for Continuous Numerical Data

The first step in constructing a frequency distribution for continuous numerical data is to decide what intervals will be used to group the data. These intervals are called **class intervals**.

Data set available

Example 2.15   College from a Distance

States differ widely in the percentage of college students who are enrolled exclusively in distance education programs. The report **"Enrollment in Distance Education Courses, by State: Fall 2012" (National Center for Education Statistics, June 2014, nces.ed.gov /pubs2014/2014023.pdf, retrieved April 16, 2017)** included the accompanying data on the percentage of students at Title IV institutions in fall 2012 who were enrolled exclusively in distance education courses for the 50 U.S. states and the District of Columbia.

16.0	16.4	48.2	9.2	6.3	19.4	6.8	6.9	8.4	16.1	9.6	11.0	12.6
10.1	11.2	39.7	16.3	19.7	4.1	13.4	14.4	5.9	7.2	29.2	9.1	12.2
6.3	16.9	9.8	17.6	7.4	12.9	6.1	11.4	22.9	9.6	10.8	8.9	7.8
1.6	5.9	20.9	6.1	8.9	22.0	10.0	18.6	7.5	40.0	7.6	11.2	

The smallest observation is 1.6 (for Rhode Island) and the largest is 48.2 (for Arizona). It is reasonable to start the first class interval at 0 and let each interval have a width of 5. This results in class intervals starting with 0 to < 5 and continuing up to 45 to < 50.

Table 2.4 displays the resulting frequency distribution, along with the relative frequencies.

TABLE 2.4 Relative Frequency Distribution of Percentage of Students at Title IV Institutions Enrolled Exclusively in Distance Education Courses

Class Interval	Frequency	Relative Frequency
0 to < 5	2	0.039
5 to < 10	21	0.412
10 to < 15	12	0.235
15 to < 20	9	0.176
20 to < 25	3	0.059
25 to < 30	1	0.020
30 to < 35	0	0.000
35 to < 40	1	0.020
40 to < 45	1	0.020
45 to < 50	1	0.020
Total	**51**	**1.001**

There are no set rules for selecting either the number of class intervals or the length of the intervals. Using a few relatively wide intervals will bunch the data, whereas using a great many relatively narrow intervals may spread the data over too many intervals, so that no interval contains more than a few observations. Either way, interesting features of the data set may be missed. In general, with a small amount of data, relatively few intervals, perhaps between 5 and 10, should be used. With a large amount of data, a distribution based on 15 to 20 (or even more) intervals is often recommended. The quantity

$$\sqrt{\text{number of observations}}$$

is sometimes used as an estimate of an appropriate number of intervals: 5 intervals for 25 observations, 10 intervals when the number of observations is 100, and so on.

Histograms for Continuous Numerical Data

When the class intervals in a frequency distribution are all of equal width, you construct a histogram in a way that is very similar to what is done for discrete data.

Histogram for Continuous Numerical Data When Class Intervals Are Equal Width

When to Use   Number of variables: 1
Data Type: Continuous numerical
Purpose: Displaying data distribution

How to Construct
1. Mark the boundaries of the class intervals on a horizontal axis.
2. Use either frequency or relative frequency on the vertical axis.
3. Draw a rectangle for each class interval directly above that interval (so that the edges are at the class interval boundaries). The height of each rectangle is the frequency or relative frequency of the corresponding class interval.

What to Look For
• Center or typical value
• Amount of variability
• General shape
• Location and number of peaks
• Presence of gaps and outliers

Data set available

Example 2.16   Sleep Deficit and School Start Time

The authors of the paper **"The Influence of School Time on Sleep Patterns of Children and Adolescents"** (*Sleep Medicine* [2016]: 33–39) were interested in determining if early school start time had an effect on the amount of sleep school-age children get. They studied students in Brazil in a region that offered both a morning school start (with classes from 7:30 a.m. to noon) time and an afternoon school start (with classes from 1:30 p.m. to 5:30 p.m.). One variable of interest in the study was sleep deficit, which they defined as the difference in sleep duration on weekends and the sleep duration on nights with school the next day. Sleep deficit was measured in hours, so a student who typically slept 9 hours a night on weekends and only 7 hours a night on weeknights would have a sleep deficit of 2 hours. A student with a sleep deficit of −3 hours would be one who typically slept 3 hours longer on weeknights than on weekends, resulting in a negative difference.

Table 2.5 gives frequencies (approximate values based on a graph that appears in the paper) and relative frequencies for various sleep deficit categories for students with the morning start time.

TABLE 2.5 Relative Frequency Distribution of Sleep Deprivation for Morning Start Students

Sleep Deficit (in hours)	Frequency	Relative Frequency
−6 to < −4	4	0.007
−4 to < −2	15	0.028
−2 to < 0	35	0.065
0 to < 2	238	0.442
2 to < 4	196	0.364
4 to < 6	42	0.078
6 to < 8	8	0.015
Total	**538**	**0.999**

You can use the steps in the previous box to construct a histogram for the data summarized in Table 2.5.

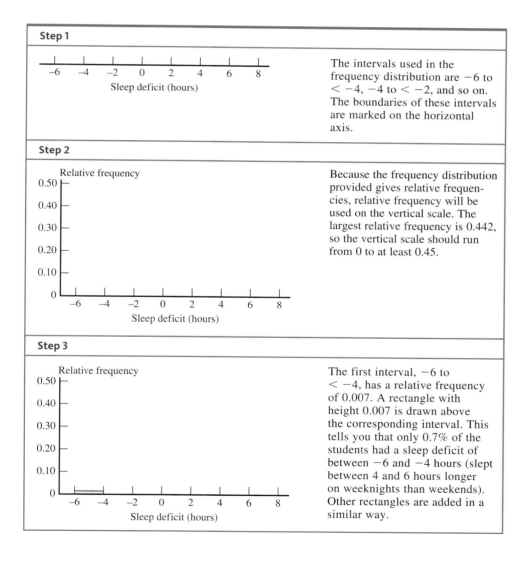

Step 1

Sleep deficit (hours)

The intervals used in the frequency distribution are −6 to < −4, −4 to < −2, and so on. The boundaries of these intervals are marked on the horizontal axis.

Step 2

Relative frequency

Sleep deficit (hours)

Because the frequency distribution provided gives relative frequencies, relative frequency will be used on the vertical scale. The largest relative frequency is 0.442, so the vertical scale should run from 0 to at least 0.45.

Step 3

Relative frequency

Sleep deficit (hours)

The first interval, −6 to < −4, has a relative frequency of 0.007. A rectangle with height 0.007 is drawn above the corresponding interval. This tells you that only 0.7% of the students had a sleep deficit of between −6 and −4 hours (slept between 4 and 6 hours longer on weeknights than weekends). Other rectangles are added in a similar way.

Figure 2.22 shows the completed relative frequency histogram. Notice that the histogram has a single peak and that the majority of the students had a sleep deficit that was greater than 0, which means that they get less sleep on school nights. Many of the students had a sleep deficit that was between 0 and 4 hours.

FIGURE 2.22

Histogram of sleep deficit for students with morning school start time

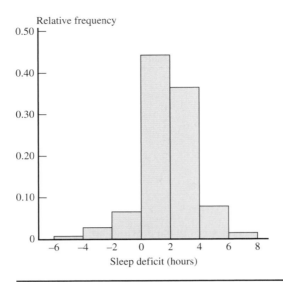

Using Histograms to Compare Groups

Histograms can be used to compare groups, but keep two things in mind:

1. Always use relative frequency rather than frequency on the vertical axis since the number of observations in each group might be different.
2. Use the same scales on both the horizontal and vertical axes to make drawing comparisons easier.

The use of histograms to compare distributions is illustrated in Example 2.17.

Data set available

Example 2.17 Sleep Deficit and School Start Time Revisited

The paper referenced in Example 2.16 also reported data on sleep deficit for students enrolled in school with an afternoon start time. Because there were 538 students in the morning start group and only 101 students in the afternoon start group, it is important to use relative frequencies rather than frequencies if you want to compare the distributions of sleep deficits for the two groups of students.

Table 2.6 gives relative frequencies (approximate values based on a graph that appears in the paper) for various sleep deficit categories for students with the morning start time and for students with the afternoon start time.

TABLE 2.6 Relative Frequency Distribution for Sleep Deficit for Students with Morning Start Time and Students with Afternoon Start Time

Sleep Deficit (in hours)	Morning Start Relative Frequency	Afternoon Start Relative Frequency
−6 to < −4	0.007	0.020
−4 to < −2	0.028	0.050
−2 to < 0	0.065	0.190
0 to < 2	0.442	0.570
2 to < 4	0.364	0.120
4 to < 6	0.078	0.040
6 to < 8	0.015	0.010

Figure 2.23(a) is the relative frequency histogram for the morning school start time group, and Figure 2.23(b) is the relative frequency histogram for the afternoon school start time

FIGURE 2.23
Histograms of sleep deficit
(a) morning school start time
(b) afternoon school start time

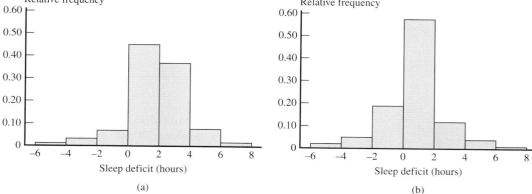

Morning Start Group

Afternoon Start Group

Sleep deficit (hours)

(a)

Sleep deficit (hours)

(b)

group. Notice that both histograms have a single peak, with the majority of students in both groups having positive values for sleep deficit. However, students in the afternoon school start time group tended to have smaller deficits than the students in the morning start time group, indicating that difference between sleep duration on weekends and sleep duration on weekdays tended to be smaller for students with an afternoon start time. This is one reason that many people are recommending later school start times.

Histograms with Unequal Width Intervals

If the intervals in the frequency distribution are not all the same width, histograms must be constructed using a different scale on the vertical axis, called the *density scale*. Using density ensures that the areas of the rectangles in the histogram will be proportional to the frequencies or relative frequencies.

> ### CONSTRUCTING A HISTOGRAM FOR CONTINUOUS DATA WHEN CLASS INTERVAL WIDTHS ARE UNEQUAL
>
> When class intervals are not of equal width, frequencies or relative frequencies should not be used on the vertical axis. Instead, the height of each rectangle, called the **density** for the corresponding class interval, is given by
>
> $$\text{density} = \text{rectangle height} = \frac{\text{relative frequency of class interval}}{\text{class interval width}}$$
>
> The vertical axis is called the **density scale**.

The formula for density can also be used when interval widths are equal. However, when the intervals are of equal width, the extra arithmetic required to obtain the densities is unnecessary.

Data set available

Example 2.18   Student Debt on the Rise

At many U.S. colleges and universities, low-income students often need to take on a large debt burden in order to pay for their education. The article **"Poor Feel the Bite of Rising College Costs" (*The Wall Street Journal*, February 20, 2016)** classified 1319 colleges into intervals based on the median yearly debt for students from families with an annual income of $30,000 or less. The data is summarized in the frequency distribution given in Table 2.7

TABLE 2.7 Frequency Distribution for Median Student Debt

Class Interval	Frequency	Relative Frequency	Interval Width	Density
$0 to < $10,000	166	0.126	10,000	0.000013
$10,000 to < $15,000	389	0.295	5,000	0.000059
$15,000 to < $20,000	456	0.346	5,000	0.000069
$20,000 to < $40,000	308	0.234	20,000	0.000012

Looking at the frequency distribution, you can see that the median student debt for low-income students was between $20,000 and $40,000 for 308 of the colleges and was between $15,000 and $20,000 for 456 of the colleges. Notice also that the income intervals used in the article are not of equal width. Two intervals have a width of $5000 (for example, the interval from $10,000 to < $15,000) and the other two intervals have widths of $10,000 and $20,000.

Figure 2.24 displays two histograms based on this frequency distribution. The histogram in Figure 2.24(a) is correctly drawn, with density used to determine the height of each bar. The histogram in Figure 2.24(b) has height equal to relative frequency and is therefore *not correct*. In particular, this second histogram exaggerates the proportion of colleges with low median debt and the proportion of colleges with very high median debt—the areas of the two most extreme rectangles are much too large. The eye is naturally drawn to large areas, so it is important that the areas correctly represent the relative frequencies.

FIGURE 2.24
Histograms for median student debt
(a) correct histogram (height = density)
(b) incorrect histogram (height = frequency)

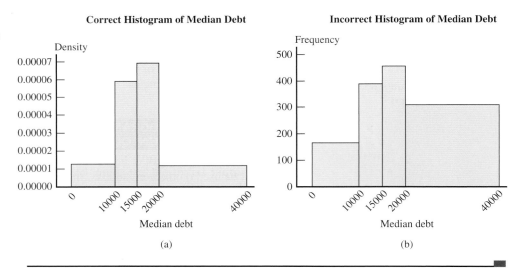

(a)

(b)

Histogram Shapes

General shape is an important characteristic of a histogram. In describing various shapes, it is convenient to approximate the histogram with a smooth curve (called a *smoothed histogram*). This is illustrated in Figure 2.25.

FIGURE 2.25
Approximating a histogram with a smooth curve

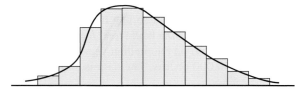

One description of general shape relates to the number of peaks, or **modes**.

A histogram is said to be **unimodal** if it has a single peak, **bimodal** if it has two peaks, and **multimodal** if it has more than two peaks.

These shapes are illustrated in Figure 2.26.

FIGURE 2.26
Smoothed histograms with various numbers of modes: (a) unimodal; (b) bimodal; (c) multimodal

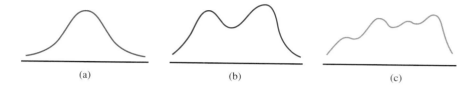

(a) (b) (c)

Bimodality sometimes occurs when the data set consists of observations on two quite different kinds of individuals or objects. For example, consider a large data set consisting of driving times for automobiles traveling between San Luis Obispo and Monterey, California. This histogram would show two peaks, one for those cars traveling on the inland route (roughly 2.5 hours) and another for those cars traveling up the coast highway (3.5 to 4 hours). However, bimodality does not automatically follow in such situations. Bimodality will occur in the histogram of the combined groups only if the centers of the two separate histograms are far apart (relative to the variability in the two data sets). For example, a large data set consisting of heights of college students would probably not produce a bimodal histogram because the typical height for males (about 69 inches) and the typical height for females (about 66 inches) are not very far apart.

A histogram is **symmetric** if there is a vertical line of symmetry such that the part of the histogram to the left of the line is a mirror image of the part to the right. Several different symmetric unimodal smoothed histograms are shown in Figure 2.27.

FIGURE 2.27
Several symmetric unimodal smoothed histograms

Proceeding to the right from the peak of a unimodal histogram, you move into what is called the **upper tail** of the histogram. Going in the opposite direction moves you into the **lower tail**.

A unimodal histogram that is not symmetric is said to be **skewed**. If the upper tail of the histogram stretches out much farther than the lower tail, then the distribution of values is **positively skewed** or **right skewed**. If the lower tail is much longer than the upper tail, the histogram is **negatively skewed** or **left skewed**.

These two types of skewness are illustrated in Figure 2.28. Positive skewness is much more frequently encountered than negative skewness. An example of positive skewness would be the distribution of single-family home prices in Los Angeles County. Most homes are moderately priced (at least for California), whereas the relatively few homes in Beverly Hills and Malibu have much higher price tags.

FIGURE 2.28
Two examples of skewed smoothed histograms: (a) positive skew; (b) negative skew

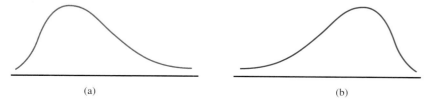

(a) (b)

FIGURE 2.29
A normal curve

One shape, a **normal curve,** arises frequently in statistical applications. A normal curve is symmetric and bell-shaped (See Figure 2.29). Many histograms of characteristics such as arm span or the weight of an apple can be well approximated by a normal curve. A more detailed discussion of normal curves is found in Chapter 6.

Summing It Up—Section 2.3

The following learning objectives were addressed in this section:

Mastering the Mechanics

M3: Construct and interpret dotplots and comparative dotplots.
Dotplots are appropriate when you have data on one variable and the data are numerical. A dotplot is a graph of the data points along a number line. The box just prior to Example 2.5 provides instructions for how to make a dotplot and points out things that you should look for when interpreting dotplots. A comparative dotplot is two (or more) dotplots using the same numerical scale (see Example 2.6).

M4: Construct and interpret stem-and-leaf-displays and comparative stem-and-leaf displays.
Stem-and-leaf displays are appropriate when you have data on one variable and the data are numerical. To make a stem-and-leaf display, each data value is divided into a stem and a leaf. The box just prior to Example 2.8 provides instructions for how to make a stem-and-leaf display and points out things you should look for when interpreting stem-and-leaf displays. A comparative stem-and-leaf display is two stem-and-leaf displays constructed using the same set of stems (see Example 2.10).

M5: Construct and interpret histograms.
Histograms are appropriate when you have data on one variable and the data are numerical. Histograms are constructed a bit differently for discrete numerical data and continuous numerical data. See the box just prior to Example 2.13 for instructions for how to make a histogram and things you should look for when interpreting a histogram for discrete numerical data and the box just prior to Example 2.14 for continuous numerical data.

Putting It into Practice

P1: Describe a numerical data distribution in terms of shape, center, variability, gaps, and outliers.
Data distributions differ in terms of where they are centered and how much they vary. The center of a numerical data distribution is often used to describe a typical value for the distribution. A data distribution that is spread out indicates that there is substantial variability in the observations in the data set. Numerical data distributions are also described by their shape using terms like symmetric, skewed, unimodal (one peak), and bimodal (two peaks). When looking at a data distribution, it is also common to note any gaps, outliers (extreme values), and any other unusual features of the distribution, as this often reveals something interesting about the variable being studied.

P3: Use graphical displays to compare groups on the basis of a numerical variable.
Well-constructed graphical displays are often used to compare groups. To make comparisons easier, graphs that are used for comparisons should be constructed using the same numerical scales.

SECTION 2.3 EXERCISES

Each Exercise Set assesses the following chapter learning objectives: M3, M4, M5, P1, P3

SECTION 2.3 Exercise Set 1

2.24 An article in the **San Luis Obispo *New Times* (February 4, 2016)** reported the accompanying concussion rates for different high school sports. The given data are concussion rates per 10,000 athletes participating in high school sports in 2012.

Sport	Concussion Rate (Concussions per 10,000 athletes)
Football	11.2
Lacrosse (Boys)	6.9
Lacrosse (Girls)	5.2
Wrestling	6.2
Basketball (Girls)	5.6
Basketball (Boys)	2.8
Soccer (Girls)	6.7
Soccer (Boys)	4.2
Field Hockey	4.2
Volleyball	2.4
Softball	1.6
Baseball	1.2

a. Construct a dotplot for the concussion rate data.

b. In addition to the three girls' sports indicated in the table (lacrosse, basketball, and soccer), the article also reported concussion rates for field hockey, volleyball, and softball, which are girls' sports. Locate the points on the dotplot that correspond to concussion rates for girls' sports and highlight them in a different color. Based on the dotplot, would you say that the concussion rates tend to be lower for girls' sports? (Hint: See Example 2.5.)

2.25 Box Office Mojo **(www.boxofficemojo.com)** tracks movie ticket sales. Ticket sales (in millions of dollars) for each of the top 20 movies in 2014 and 2015 are shown in the accompanying tables.

Movie	2014 Sales (millions of dollars)
American Sniper	350.1
The Hunger Games: Mockingjay—Part 1	337.1
Guardians of the Galaxy	333.1
Captain America: The Winter Soldier	259.8
The LEGO Movie	257.8
The Hobbit: The Battle of the Five Armies	255.1
Transformers: Age of Extinction	245.4
Maleficent	241.4
X-Men: Days of Future Past	233.9
Big Hero 6	222.5
Dawn of the Planet of the Apes	208.5
The Amazing Spider-Man 2	202.9
Godzilla	200.7
22 Jump Street	191.7
Teenage Mutant Ninja Turtles	191.2
Interstellar	188.0
How to Train Your Dragon 2	177.0
Gone Girl	167.8
Divergent	150.9
Neighbors	150.2

Movie	2015 Sales (millions of dollars)
Star Wars: The Force Awakens	936.7
Jurassic World	652.3
Avengers: Age of Ultron	459.0
Inside Out	356.5
Furious 7	353.0
Minions	336.0
The Hunger Games: Mockingjay—Part 2	281.7
The Martian	228.4
Cinderella	201.2
Spectre	200.1
Mission Impossible—Rogue Nation	195.0
Pitch Perfect 2	184.3
The Revenant	183.6
Ant-Man	180.2
Home	177.4
Hotel Transylvania 2	169.7
Fifty Shades of Grey	166.2
The SpongeBob Movie: Sponge Out of Water	163.0
Straight Outta Compton	161.2
San Andreas	155.2

Construct comparative dotplots of the 2014 and 2015 ticket sales data. Comment on any interesting features of the dotplots. In what ways are the distributions of the 2014 and 2015 ticket sales observations similar? In what ways are they different? (Hint: See Example 2.6.)

2.26 *USA TODAY* **(June 11, 2010)** gave the following data on median age for each of the 50 U.S. states and the District of Columbia (DC). Construct a stem-and-leaf display using stems 28, 29, …, 42. Comment on shape, center, and variability of the data distribution. Are there any unusual values in the data set that stand out? (Hint: See Example 2.8.)

State	Median Age	State	Median Age
Alabama	37.4	Maine	42.2
Alaska	32.8	Maryland	37.7
Arizona	35.0	Massachusetts	39.0
Arkansas	37.0	Michigan	38.5
California	34.8	Minnesota	37.3
Colorado	35.7	Mississippi	35.0
Connecticut	39.5	Missouri	37.6
Delaware	38.4	Montana	39.0
DC	35.1	Nebraska	35.8
Florida	40.0	Nevada	35.5
Georgia	34.7	New Hampshire	40.4
Hawaii	37.5	New Jersey	38.8
Idaho	34.1	New Mexico	35.6
Illinois	36.2	New York	38.1
Indiana	36.8	North Carolina	36.9
Iowa	38.0	North Dakota	36.3
Kansas	35.9	Ohio	38.5
Kentucky	37.7	Oklahoma	35.8
Louisiana	35.4	Oregon	38.1

(continued)

State	Median Age	State	Median Age
Pennsylvania	39.9	Vermont	41.2
Rhode Island	39.2	Virginia	36.9
South Carolina	37.6	Washington	37.0
South Dakota	36.9	West Virginia	40.5
Tennessee	37.7	Wisconsin	38.2
Texas	33.0	Wyoming	35.9
Utah	28.8		

2.27 A report from **Texas Transportation Institute (Texas A&M University System, 2005)** titled **"Congestion Reduction Strategies"** included the following data on extra travel time during rush hour for very large and for large urban areas.

Very Large Urban Areas	Extra Hours per Year per Traveler
Los Angeles, CA	93
San Francisco, CA	72
Washington DC-VA-MD	69
Atlanta, GA	67
Houston, TX	63
Dallas, Fort Worth, TX	60
Chicago, IL-IN	58
Detroit, MI	57
Miami, FL	51
Boston, MA-NH-RI	51
New York, NY-NJ-CT	49
Phoenix, AZ	49
Philadelphia, PA-NJ-DE-MD	38

Large Urban Areas	Extra Hours per Year per Traveler
Riverside, CA	55
Orlando, FL	55
San Jose, CA	53
San Diego, CA	52
Denver, CO	51
Baltimore, MD	50
Seattle, WA	46
Tampa, FL	46
Minneapolis, St Paul, MN	43
Sacramento, CA	40
Portland, OR-WA	39
Indianapolis, IN	38
St Louis, MO-IL	35
San Antonio, TX	33
Providence, RI-MA	33
Las Vegas, NV	30
Cincinnati, OH-KY-IN	30
Columbus, OH	29
Virginia Beach, VA	26
Milwaukee, WI	23
New Orleans, LA	18
Kansas City, MO-KS	17
Pittsburgh, PA	14
Buffalo, NY	13
Oklahoma City, OK	12
Cleveland, OH	10

a. Construct a back-to-back stem-and-leaf display for the two different sizes of urban areas. (Hint: See Example 2.10.)

b. Is the display constructed in Part (a) consistent with the following statement? Explain. Statement: The larger the urban area, the greater the extra travel time during peak period travel.

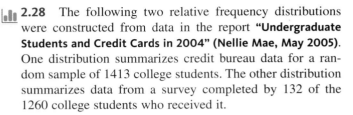 **2.28** The following two relative frequency distributions were constructed from data in the report **"Undergraduate Students and Credit Cards in 2004" (Nellie Mae, May 2005)**. One distribution summarizes credit bureau data for a random sample of 1413 college students. The other distribution summarizes data from a survey completed by 132 of the 1260 college students who received it.

Credit Card Balance (dollars)— Credit Bureau Data	Relative Frequency
0 to < 100	0.18
100 to < 500	0.19
500 to < 1,000	0.14
1,000 to < 2,000	0.16
2,000 to < 3,000	0.10
3,000 to < 7,000	0.16
7,000 or more	0.07

Credit Card Balance (dollars)— Survey Data	Relative Frequency
0 to < 100	0.18
100 to < 500	0.22
500 to < 1,000	0.17
1,000 to < 2,000	0.22
2,000 to < 3,000	0.07
3,000 to < 7,000	0.14
7,000 or more	0.00

a. Construct a histogram for the credit bureau data. Assume that no one had a balance greater than 15,000 and that the last interval is 7000 to < 15,000. Be sure to use the density scale. (Hint: See Example 2.18.)

b. Construct a histogram for the survey data. Use the same scales that you used for the histogram in Part (a) so that it will be easy to compare the two.

c. Comment on the similarities and differences in the histograms from Parts (a) and (b).

d. Do you think the high nonresponse rate for the survey may have contributed to the observed differences in the two histograms? Explain.

2.29 An exam is given to students in an introductory statistics course. Comment on the expected shape of the histogram of scores if:

a. the exam is very easy

b. the exam is very difficult

c. half the students in the class have had calculus, the other half have had no prior college math courses, and the exam emphasizes higher-level math skills

Explain your reasoning in each case.

 Data set available

2.30 The following data on violent crime on Florida college campuses during 2014 are from the FBI web site.

University/College	Student Enrollment	Number of Violent Crimes Reported (2014)
Edison State College	15,423	3
Florida A&M University	10,743	10
Florida Atlantic University	30,759	5
Florida Gulf Coast University	14,077	5
Florida International University	47,663	9
Florida State University	40,909	15
New College of Florida	793	2
Pensacola State College	11,235	0
Santa Fe College	15,113	1
Tallahassee Community College	13,509	4
University of Central Florida	59,589	29
University of Florida	49,878	15
University of North Florida	16,083	4
University of South Florida	46,088	15
University of West Florida	12,588	6

a. Construct a dotplot using the 15 observations on number of violent crimes reported. Which schools stand out from the rest?

b. One of the Florida schools only has 793 students, and a few of the schools are quite a bit larger than the rest. Because of this, it might make more sense to consider a crime *rate* by calculating the number of violent crimes reported *per 1000 students*. For example, for Florida A&M University the violent crime rate would be

$$\frac{10}{10,743}(1000) = (0.000931)(1000) = 0.931$$

Calculate the violent crime rate for the other 14 schools. Use these values to construct a second dotplot. Do the same schools stand out as unusual in this dotplot?

c. Based on your answers from Parts (a) and (b), comment on violent crimes reported at Florida universities and colleges in 2014.

2.31 The report **"Seat Belt Use in 2014" (National Highway Traffic Safety Administration)** included the estimated percentages of drivers who wear seat belts for the 50 states and the District of Columbia. In the accompanying data table, each state was also classified into one of three geographical regions—West (W), Middle states (M), or East (E).

SeatBelt%	Region	State	SeatBelt%	Region	State
95.7	M	AL	82.4	W	CO
88.4	W	AK	85.1	E	CT
87.2	W	AZ	91.9	E	DE
74.4	M	AR	93.2	E	DC
97.1	W	CA	88.8	E	FL

SeatBelt%	Region	State	SeatBelt%	Region	State
97.3	E	GA	87.6	E	NJ
93.5	W	HI	92.1	W	NM
80.2	W	ID	90.6	E	NY
94.1	M	IL	90.6	E	NC
90.2	M	IN	85.0	E	OH
92.8	M	IA	86.3	M	OK
85.7	M	KS	97.8	W	OR
86.1	M	KY	83.6	E	PA
84.1	M	LA	87.4	E	RI
85.0	E	ME	90.0	E	SC
92.1	E	MD	68.9	M	SD
76.6	E	MA	87.7	M	TN
93.3	M	MI	90.7	M	TX
94.7	M	MN	83.4	W	UT
78.3	M	MS	77.3	E	VA
78.8	M	MO	84.1	E	VT
74.0	W	MT	94.5	W	WA
79.0	M	NE	87.8	E	WV
94.0	W	NV	84.7	M	WI
81.0	M	ND	79.2	W	WY
70.4	E	NH			

a. Display the data in comparative dotplots that make it possible to compare seat belt use for the three geographical regions.

b. Does the graphical display in Part (a) reveal any striking differences in seat belt use for the three geographical regions or are the distributions of seat belt use similar for the three regions?

2.32 Credit card fraud is a growing problem for both consumers and merchants. The data below on the percentage of credit card holders who have been impacted by fraud between 2009 and 2014 for 20 countries appeared in the article **"Credit Card & Debit Card Fraud Statistics" (www.cardhub.com/edu/credit-debit-card-fraud-statistics/, retrieved April 16, 2017)**.

Country	Cardholders Affected (%)
Australia	27
Brazil	27
Canada	18
China	31
France	24
Germany	14
India	32
Indonesia	13
Italy	16
Mexico	28
New Zealand	15
Poland	15
Russia	19
Singapore	25
South Africa	25
Sweden	8
The Netherlands	10
United Arab Emirates	39
United Kingdom	25
United States	36

(continued)

a. The smallest value in the data set is 8 and the largest value is 39. One possible choice of stems for a stem-and-leaf display would be to use the tens digit. Thinking of each data value as having two digits before the decimal place, the smallest value of 8 would be represented as 08 and would have a tens digit of 0. This would result in stems of 0, 1, 2, and 3. Construct a stem-and-leaf display using these four stems.

b. Do any countries stand out as having unusually high or low percentages of cardholders affected by fraud?

2.33 Wikipedia gives the following data on percentage increase in population between 2010 and 2015 for the 50 U.S. states and the District of Columbia (DC) **(en.wikipedia.org /wiki/List_of_U.S._states_by_population_growth_rate, retrieved October 16, 2016)**. Each state is also classified as belonging to the eastern or western part of the United States:

State	Percent Change	East/ West	State	Percent Change	East/ West
Alabama	1.7	E	Missouri	1.6	E
Alaska	4.0	W	Montana	4.4	W
Arizona	6.8	W	Nebraska	3.8	W
Arkansas	2.1	E	Nevada	7.1	W
California	5.1	W	New Hampshire	1.1	E
Colorado	8.5	W	New Jersey	1.9	E
Connecticut	0.5	E	New Mexico	1.3	W
Delaware	5.4	E	New York	2.2	E
DC	11.7	E	North Carolina	5.3	E
Florida	7.8		North Dakota	12.5	W
Georgia	5.4	E	Ohio	0.7	E
Hawaii	5.2	E	Oklahoma	4.3	W
Idaho	5.6	W	Oregon	5.2	W
Illinois	0.2	W	Pennsylvania	0.8	E
Indiana	2.1	E	Rhode Island	0.4	E
Iowa	2.6	E	South Carolina	5.9	E
Kansas	2.1	E	South Dakota	5.4	W
Kentucky	2.0	W	Tennessee	4.0	E
Louisiana	3.0	E	Texas	9.2	W
Maine	0.1	E	Utah	8.4	W
Maryland	4.0	E	Vermont	0.1	E
Massachusetts	3.8	E	Virginia	4.8	E
Michigan	0.4	E	Washington	6.6	W
Minnesota	3.5	E	West Virginia	0.0	E
Mississippi	0.8	E	Wisconsin	1.5	E
		E	Wyoming	4.0	W

a. Construct a stem-and-leaf display for the entire data set.

b. Comment on any interesting features of the display. Do any of the observations appear to be outliers?

c. Now construct a comparative stem-and-leaf display for the Eastern and Western states. Write a few sentences comparing the two distributions.

2.34 The accompanying relative frequency table is based on data from the **2015 College Bound Seniors** report for California (**College Board, 2016**).

Score on SAT Critical Reading Exam	Relative Frequency for Females	Relative Frequency for Males
200 to < 300	0.040	0.049
300 to < 400	0.144	0.160
400 to < 500	0.294	0.290
500 to < 600	0.330	0.305
600 to < 700	0.158	0.153
700 to < 800	0.035	0.043

a. Construct a relative frequency histogram for males.

b. Using the same scale as the histogram from Part (a), construct a relative frequency histogram for females.

c. Based on the histograms from Parts (a) and (b), write a few sentences commenting on the similarities and differences in the two distributions.

2.35 Using the five class intervals 100 to < 120, 120 to < 140, … , 180 to < 200, construct a frequency distribution based on 70 observations whose histogram could be described as follows:

a. symmetric

b. bimodal

c. positively skewed

d. negatively skewed

ADDITIONAL EXERCISES

2.36 Example 2.9 provided data on seat belt use for each of the 50 U.S. states and the District of Columbia (**"Traffic Safety Facts," June, 2015, National Highway Traffic Safety Administration**). The same report also gave data on traffic fatality rate (traffic deaths per 100,000 population) for 2014. A stem-and-leaf display for the fatality rate data using repeated stems is shown here:

Stem-and-leaf display for the fatality rate data

```
0 | 3 4
0 | 5 6 6 7 7 7 8 8 8 9 9 9 9 9 9 9
1 | 0 0 0 0 0 1 1 2 2 2 3 3 3 3 3 4 4 4 4
1 | 5 5 5 5 6 6 7 7 7 8
2 | 0 2 3
2 | 5
```

Stems: tens
Leaves: ones

Some, but not all, states enforce seat belts laws. Below are the fatality rate data divided into two groups—states that enforce seat belts laws and those that do not.

States with Seat Belt Law Enforcement

17	9	18	8	7	13	3	15	14	7	8	12	10
12	17	16	12	9	10	7	23	6	14	6	14	16
11	4	20	15	13	8	15	10					

States without Seat Belt Law Enforcement

13	10	13	5	14	22	13	11	9	17	10	9	15
9	9	9	25									

 Data set available

a. Construct a comparative stem-and-leaf display of traffic fatality rate for states with seat belt law enforcement and states without seatbelt law enforcement using the same repeated stems as in the display above.

b. Comment on similarities or differences in the fatality rate distributions for states with seat belt enforcement and states without seat belt enforcement.

2.37 The paper **"Simpson's Paradox: A Data and Discrimination Case Study Exercise" (Journal of Statistics Education [2014])** included data on the ages (in years) of clients of the California Department of Developmental Services. The authors grouped the age data using intervals based on typical costs for services, using the age groups in the frequency distribution below.

Age Interval	Frequency
0 to < 6	82
6 to < 13	175
13 to < 18	212
18 to < 22	199
22 to < 51	226
51 to < 100	106

Below is an **incorrectly** drawn histogram that uses frequency on the vertical scale.

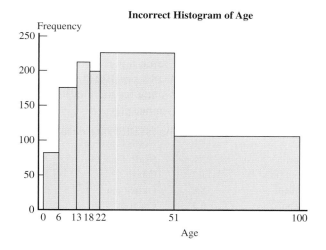

Incorrect Histogram of Age

Construct a correctly drawn histogram that uses density to determine the heights of the rectangles in the histogram. Compare the correct histogram to the one given here and comment on how the incorrectly drawn histogram is misleading. (Hint: See Example 2.18.)

2.38 The Bloomberg Visual Data web site included the following data on the total number of federal firearm background checks performed in 2014 and also the number of background checks per 1000 residents in 2014 for each of the 50 U.S. states (**www.bloomberg.com/graphics/best-and-worst/#most-nics-firearm-background-checks-per-capita-states, retrieved April 16, 2017**).

State	Number of Background Checks	Background Checks per 1,000 Residents
Alabama	621,305	128
Alaska	87,623	119
Arizona	310,672	46
Arkansas	234,282	79
California	1,474,616	38
Colorado	413,284	77
Connecticut	270,297	75
Delaware	42,950	46
Florida	1,034,546	52
Georgia	484,580	48
Hawaii	15,225	11
Idaho	131,742	81
Illinois	1,344,096	104
Indiana	647,550	98
Iowa	127,022	41
Kansas	172,167	59
Kentucky	2,492,184	565
Louisiana	315,357	68
Maine	83,085	62
Maryland	142,207	24
Massachusetts	179,344	27
Michigan	424,091	43
Minnesota	481,122	88
Mississippi	214,829	72
Missouri	517,063	85
Montana	121,836	119
Nebraska	68,568	36
Nevada	116,735	41
New Hampshire	124,677	94
New Jersey	92,320	10
New Mexico	139,780	67
New York	365,427	19
North Carolina	1,182,349	119
North Dakota	70,548	95
Ohio	596,389	51
Oklahoma	348,495	90
Oregon	243,044	61
Pennsylvania	899,241	70
Rhode Island	20,400	19
South Carolina	289,764	60
South Dakota	83,659	98
Tennessee	533,394	81
Texas	1,465,992	54
Utah	263,812	90
Vermont	31,502	50
Virginia	419,764	50
Washington	482,115	68
West Virginia	221,847	120
Wisconsin	334,308	58
Wyoming	63,063	108

a. The dotplot on the next page was constructed using the data on total number of background checks performed. Comment on the interesting features of this dotplot. Which states stand out as being unusual in terms of total number of background checks performed?

 Data set available

FIGURE FOR EXERCISE 2.38

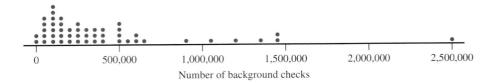

Number of background checks

b. Some states are much larger than others in terms of population (for example, California, Texas, Florida, and New York). You may expect the total number of background checks performed to be related to population size, so it might be more informative to compare states based on number of background checks per 1000 residents. Construct a dot plot using the given data on number of background checks per 1000 residents.

c. Comment on the differences between the dotplot you constructed in Part (b) and the one given earlier in this exercise. What does this tell you about how states differ?

2.39 The accompanying data on annual maximum wind speed (in meters per second) in Hong Kong for each year in a 45-year period are from an article that appeared in the journal **Renewable Energy (March 2007)**. Use the data to construct a histogram. Is the histogram approximately symmetric, positively skewed, or negatively skewed? Would you describe the histogram as unimodal, bimodal, or multimodal?

30.3	39.0	33.9	38.6	44.6	31.4	26.7	51.9	31.9
27.2	52.9	45.8	63.3	36.0	64.0	31.4	42.2	41.1
37.0	34.4	35.5	62.2	30.3	40.0	36.0	39.4	34.4
28.3	39.1	55.0	35.0	28.8	25.7	62.7	32.4	31.9
37.5	31.5	32.0	35.5	37.5	41.0	37.5	48.6	28.1

2.40 The accompanying frequency distribution summarizes data on the number of times smokers attempted to quit before their final successful attempt **("Demographic Variables, Smoking Variables, and Outcome Across Five Studies,"** *Health Psychology* **[2007]: 278–287).**

Number of Attempts	Frequency
0	778
1	306
2	274
3–4	221
5 or more	238

Assume that no one had made more than 10 unsuccessful attempts, so that the last entry in the frequency distribution can be regarded as 5–10 attempts. Summarize this data set

using a histogram. Because the class intervals are not all the same width, you will need to use a density scale for the histogram. Also remember that for a discrete variable, the bar for 1 will extend from 0.5 to 1.5. Think about what this will mean for the bars for the 3–4 group and the 5–10 group.

2.41 The accompanying data are the percentage of drivers who are uninsured in each of the 50 U.S. states and the District of Columbia (DC) as reported in the article **"2015's Most and Least Risky States for Drivers' Wallets" (www.wallethub.com, retrieved April 16, 2017)**. Construct a graphical display that shows the distribution of percent of uninsured drivers, and write a few sentences commenting on what the display reveals about the distribution.

State	Percent Uninsured Drivers	State	Percent Uninsured Drivers
Alabama	19.6	Missouri	13.5
Alaska	13.2	Montana	14.1
Arizona	10.6	Nebraska	6.7
Arkansas	15.9	Nevada	12.2
California	14.7	New Hampshire	9.3
Colorado	16.2	New Jersey	10.3
Connecticut	8.0	New Mexico	21.6
Delaware	11.5	New York	5.3
DC	11.9	North Carolina	9.1
Florida	23.8	North Dakota	5.9
Georgia	11.7	Ohio	13.5
Hawaii	8.9	Oklahoma	25.9
Idaho	6.7	Oregon	9.0
Illinois	13.3	Pennsylvania	6.5
Indiana	14.2	Rhode Island	17.0
Iowa	9.7	South Carolina	7.7
Kansas	9.4	South Dakota	7.8
Kentucky	15.8	Tennessee	20.1
Louisiana	13.9	Texas	13.3
Maine	4.7	Utah	5.8
Maryland	12.2	Vermont	8.5
Massachusetts	3.9	Virginia	10.1
Michigan	21.0	Washington	16.1
Minnesota	10.8	West Virginia	8.4
Mississippi	22.9	Wisconsin	11.7
		Wyoming	8.7

SECTION 2.4 **Displaying Bivariate Numerical Data: Scatterplots and Time Series Plots**

A bivariate data set consists of measurements or observations on two variables, x and y. For example, x might be the weight of a car and y the gasoline mileage rating of the car. When both x and y are numerical variables, each observation consists of a pair of numbers, such as (14, 5.2) or (27.63, 18.9). The first number in a pair is the value of x, and the second number is the value of y.

An unorganized list of bivariate data doesn't tell you much about the distribution of the x values or the distribution of the y values, and tells you even less about how the two variables might be related to one another. Just as graphical displays are used to summarize univariate data, they can also be used to summarize bivariate data. An important graph for bivariate numerical data is a **scatterplot**.

In a scatterplot, each observation (pair of numbers) is represented by a point on a rectangular coordinate system, like the one shown in Figure 2.30(a). Figure 2.30(b) shows the point representing the observation (4.5, 15).

FIGURE 2.30
Constructing a scatterplot:
(a) rectangular coordinate system; (b) point corresponding to (4.5, 15)

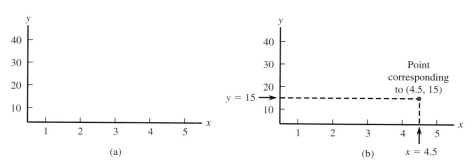

Scatterplot

When to Use   Number of variables: 2
Data type: Numerical
Purpose: Investigate the relationship between two variables

How to Construct
1. Draw horizontal and vertical axes. Label the horizontal axis and include an appropriate scale for the x variable. Label the vertical axis and include an appropriate scale for the y variable.
2. For each (x, y) pair in the data set, add a dot at the appropriate location in the display.

What to Look For   Any relationship between x and y

In many data sets, the range of the x values, the range of the y values, or both may be far away from 0. For example, a study of how air conditioner efficiency is related to maximum daily outdoor temperature might involve observations at temperatures of 80°, 82°, . . . , 98°, 100°. In such cases, the plot will be more informative if the axes intersect at some point other than (0, 0) and are marked accordingly.

Data set available

Example 2.19   **Worth the Price You Pay?**

Consumer Reports rated 29 fitness trackers (such as Fitbit and Jawbone) on factors such as ease of use and accuracy of step count to obtain an overall score **(www.consumerreports .org, retrieved October 13, 2016)**. The accompanying table gives price and overall score for these 29 fitness trackers.

Price	Overall Score	Price	Overall Score	Price	Overall Score
260	87	125	71	100	56
150	83	120	70	70	56
200	82	100	69	100	55
200	82	90	66	120	54
150	82	200	66	100	54
150	79	220	65	125	51
130	78	60	61	100	50
100	77	100	61	100	44
176	74	100	60	22	41
180	74	150	57		

Is there a relationship between x = price and y = overall score? A scatterplot can help answer this question.

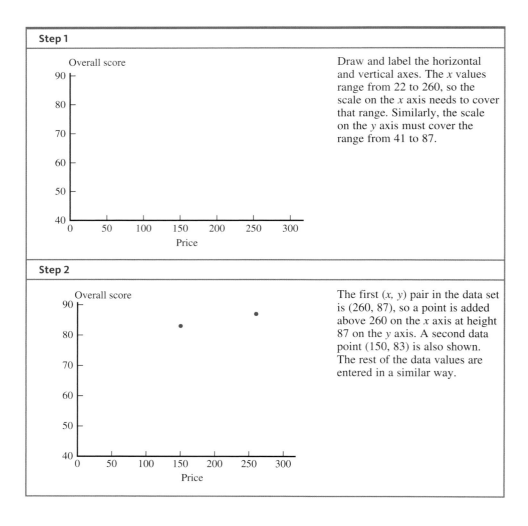

Figure 2.31 shows the completed scatterplot. There does appear to be a general pattern to the points in the scatterplot. The fitness trackers with higher prices tended to have higher overall scores. But also notice that not all high-priced fitness trackers had high overall scores, and that some of the fitness trackers in the $100 to $150 price range also had high overall scores.

FIGURE 2.31
Scatterplot of overall score versus price for 29 fitness trackers

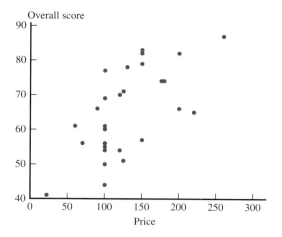

Time Series Plots

Data sets often consist of measurements collected over time at regular intervals so that you can learn about change over time. For example, stock prices, sales figures, and other socio-economic indicators might be recorded on a weekly or monthly basis. A **time series plot** (sometimes also called a time plot) is a simple graph of data collected over time that can help you see interesting trends or patterns.

A time series plot can be constructed by thinking of the data set as a bivariate data set, where *y* is the variable observed and *x* is the time at which the observation was made. These (*x*, *y*) pairs are plotted as in a scatterplot. Consecutive observations are then connected by a line segment.

Data set available

Example 2.20   **Exercise on the Rise?**

Gallup conducts frequent polls in which large samples of adult Americans are asked how often they exercise. The article **"So Far in 2015, More Americans Exercising Frequently" (www.gallup.com/poll/184403/far-2015-americans-exercising-frequently.aspx?g_source, retrieved April 17, 2017)** used information from these polls to estimate that during the first half of 2015, on average 52.5% of Americans exercised for 30 minutes or more at least three days a week. The article also provided estimates for the years 2008 to 2015, as shown in the accompanying table.

Year	Percentage
2008	51.9
2009	49.3
2010	51.1
2011	50.8
2012	53.0
2013	51.4
2014	51.2
2015	52.5

Figure 2.32 shows a time series plot of these data. Notice that the eight (year, percentage) pairs have been plotted and that these points have been connected by line segments. This makes it easier to see any trend over time. You can see from the time series plot that the percentage has not steadily increased year to year, although there does appear to be a general upward trend following the drop that occurred in 2009.

The article also included a time series plot that was based on monthly estimates of the percentage exercising 30 minutes or more at least three times per week. Figure 2.33 is similar to the plot that appeared in the article. In addition to the general increasing

FIGURE 2.32
Time series plot of percentage of Americans who exercise for 30 minutes or more at least three times per week

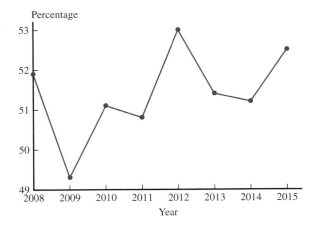

FIGURE 2.33
Time series plot of percentage exercising based on monthly data

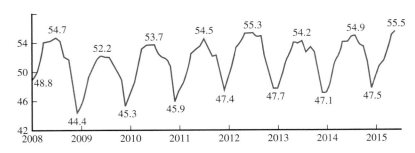

Monthly averages
Gallup-Healthways Well-Being Index

trend from year to year, you can also see a pattern that repeats each year, with the percentage tending to increase during the first half of each year and decrease in the second half of each year.

■

Example 2.21   **The Cost of Christmas**

The Christmas Price Index is calculated each year by PNC Advisors. It is a humorous look at the cost of giving all of the gifts described in the popular Christmas song "The 12 Days of Christmas." The year 2015 was the most costly year since the index began in 1984, with the "cost of Christmas" at $34,131. A plot of the Christmas Price Index over time appears on the PNC web site (**www.pncchristmaspriceindex.com**), and the data given there were used to construct the time series plot of Figure 2.34. The plot shows an upward trend

FIGURE 2.34
Time series plot for the Christmas Price Index data

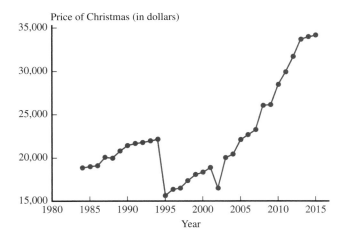

in the index from 1984 until 1993. There has also been a clear upward trend in the index since 1995. You can visit the web site to see individual time series plots for each of the 12 gifts that are used to determine the Christmas Price Index (a partridge in a pear tree, two turtle doves, and so on). See if you can figure out what caused the dramatic decline from 1994 to 1995.

Example 2.22 | **Education and Income—Stay in School!**

The time series plot in Figure 2.35 is similar to one appearing on the web site of the National Center for Education Statistics **(nces.ed.gov/programs/coe/indicator_cba.asp, retrieved April 17, 2017)**. It shows the change over time in median annual earnings by education level. (The median annual earnings is the value for which half of the population earns less and half earns more. For example, in the year 2000, about half of those with bachelor's degrees earned less than $55,000 per year and half of those with bachelor's degrees earned more than $55,000.) From this plot, you can see that the median annual earnings varied more from year to year for those with master's or higher degrees, but that the large gap between the median earnings for those who only completed high school and those with a college degree has remained about the same since the year 2000.

FIGURE 2.35
Time series plots of median annual earnings by education level

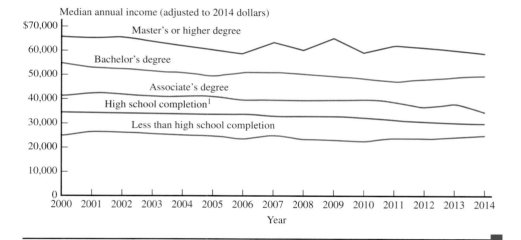

Summing It Up—Section 2.4

The following learning objectives were addressed in this section:

Mastering the Mechanics
M6: Construct and interpret scatterplots.
A scatterplot is a graph of bivariate numerical data. The box just prior to Example 2.19 provides instructions for how to make a scatterplot.

M7: Construct and interpret time series plots.
A time series plot is a graph of data collected over time. Time is plotted on the horizontal axis and the points in the plot are connected by line segments in order to make it easier to see trends and other patterns over time.

Putting It into Practice
P4: Use a scatterplot to investigate the relationship between two numerical variables.
When looking at a scatterplot, the usual question of interest is whether or not there is a relationship between the two variables represented in the scatterplot. In order to decide if there appears to be a relationship, look for a pattern in the scatterplot.

P5: Use a time series plot to investigate trends over time for a numerical variable.
When looking at a time series plot, you often want to see how the value of some variable is changing over time.

EXERCISES

Each Exercise Set assesses the following chapter learning objectives: M6, M7, P4, P5

Exercise Set 1

 2.42 *Consumer Reports Health* (www.consumerreports.org) gave the following data on saturated fat (in grams), sodium (in mg), and calories for 36 fast-food items.

a. Construct a scatterplot using y = calories and x = fat. Does it look like there is a relationship between fat and calories? Is the relationship what you expected? Explain.

b. Construct a scatterplot using y = calories and x = sodium. Write a few sentences commenting on the difference between this scatterplot and the scatterplot from Part (a).

c. Construct a scatterplot using y = sodium and x = fat. Does there appear to be a relationship between fat and sodium?

d. Add a vertical line at x = 3 and a horizontal line at y = 900 to the scatterplot in Part (c). This divides the scatterplot into four regions, with some points falling into each region. Which of the four regions corresponds to healthier fast-food choices? Explain.

Fat	Sodium	Calories	Fat	Sodium	Calories
2	1,042	268	0	520	140
5	921	303	2.5	1,120	330
3	250	260	1	240	120
2	770	660	3	650	180
1	635	180	1	1,620	340
6	440	290	4	660	380
4.5	490	290	3	840	300
5	1,160	360	1.5	1,050	490
3.5	970	300	3	1,440	380
1	1,120	315	9	750	560
2	350	160	1	500	230
3	450	200	1.5	1,200	370
6	800	320	2.5	1,200	330
3	1,190	420	3	1,250	330
2	1,090	120	0	1,040	220
5	570	290	0	760	260
3.5	1,215	285	2.5	780	220
2.5	1,160	390	3	500	230

 2.43 The report **"Daily Cigarette Use: Indicators on Children and Youth"** (Child Trends Data Bank, www.childtrends.org/wp-content/uploads/2012/11/03_Smoking_new.pdf, retrieved April 17, 2017) included the accompanying data on the percentage of students who report smoking cigarettes daily, for students in grades 8, 10, and 12.

	Percentage of Students Who Smoke		
Year	Grade 8	Grade 10	Grade 12
2000	7.4	14.0	20.6
2001	5.5	12.2	19.0
2002	5.1	10.1	16.9
2003	4.5	8.9	15.8
2004	4.4	8.3	15.6
2005	4.0	7.5	13.6
2006	4.0	7.6	12.2
2007	3.0	7.2	12.3
2008	3.1	5.9	11.4
2009	2.7	6.3	11.2
2010	2.9	6.6	10.7
2011	2.4	5.5	10.3
2012	1.9	5.0	9.3
2013	1.8	4.4	8.5
2014	1.4	3.2	6.7

a. Construct a time series plot for students in grade 12, and comment on any trend over time.

b. Construct a time series plot that shows trend over time for each of the three grade levels. Graph each of the three time series on the same set of axes, using different colors to distinguish the different grade levels. Either label the time series in the plot or include a legend to indicate which time series corresponds to which grade level. (Hint: See Example 2.22.)

c. Write a paragraph based on your plot from Part (b). Discuss the similarities and differences for the three different grade levels.

Exercise Set 2

 2.44 The accompanying table provides data on a measure of emotional health and a measure of the quality of the work environment for 12 different occupations. These data are from an article titled **"U.S. Teachers Love Their Lives, but Struggle in the Workplace"** on the Gallup web site (www.gallup.com/poll/161516/teachers-love-lives-struggle-workplace.aspx?g_source, retrieved April 17, 2017).

Occupation	Emotional Health Score	Work Environment Score
Physician	83.7	61.5
Teacher	82.7	49.9
Farming, fishing or forestry	82.7	53.3
Professional	82.3	54.0

Occupation	Emotional Health Score	Work Environment Score
Nurse	81.8	50.1
Business owner	81.7	67.8
Manager, executive or official	81.5	56.5
Construction or mining	80.9	50.3
Installation or repair	80.7	44.8
Clerical or office	80.3	44.4
Sales	80.2	45.5
Manufacturing or production	80.1	41.5
Transportation	80.0	39.2

a. Construct a scatterplot using y = Emotional health score and x = Work environment score.

b. Based on the scatterplot from part (a), does there appear to be a relationship between Emotional health score and Work environment score for these 12 occupations? Does the scatterplot indicate that occupations with higher Work environment scores tend to have higher emotional health scores?

c. Identify the point in the scatterplot that corresponds to teachers. Is the location of this point in the scatterplot consistent with the title of the article? Explain.

2.45 The Census Bureau collects data on the percentage of households in the United States that have Internet access in the home. The accompanying table shows this percentage for the years 1997 to 2013 (**census.gov/hhes/computer /publications/2012.html, retrieved August 1, 2016**).

Year	Percentage of Households with Internet
1997	18.0
2000	41.5
2001	50.4
2003	54.7
2007	61.7
2009	68.7
2010	71.1
2011	71.7
2012	74.8
2013	74.4

a. Construct a time series plot that shows how this percentage has changed over time. (Note that data were not collected for every year, so the points in the time series plot will not be equally spaced along the x axis.)

b. Has the percentage of households with Internet access increased at a fairly steady rate? How can you tell from the time series plot?

2.46 The report **"Credit Card & Debit Card Fraud Statistics"** (**www.cardhub.com/edu/credit-debit-card-fraud-statistics/, retrieved April 17, 2017**) included the data in the accompany-

ing table on the amount of money lost to online fraud for each of the years from 2001 to 2012.

Year	Money Lost to Online Fraud (in billions of dollars)
2001	1.7
2002	2.1
2003	1.9
2004	2.6
2005	2.8
2006	3.1
2007	3.7
2008	4.0
2009	3.3
2010	2.7
2011	3.4
2012	3.5

Construct a time series plot for these data and describe the trend over time. Has the amount of money lost to online fraud increased at a fairly steady rate? Explain.

ADDITIONAL EXERCISES

2.47 The Census Bureau collects data on the percentage of households in the United States that have a computer. The accompanying table shows this percentage for the years 1984 to 2013 (**census.gov/hhes/computer/publications/2012.html, retrieved August 1, 2016**).

Year	Percentage of Households with a Computer
1984	8.2
1989	15.0
1993	22.9
1997	36.6
2000	51.0
2001	56.3
2003	61.8
2007	69.7
2009	74.1
2010	76.7
2011	75.6
2012	78.9
2013	83.8

a. Construct a time series plot for these data. Be careful—the observations are not equally spaced in time. The points in the plot should not be equally spaced along the x axis.

b. Comment on any trend over time.

2.48 *Consumer Reports* (**www.consumerreports.org**) rated 37 different models of laptops that were for sale in 2015. An overall score was assigned to each model based on considering a number of factors, including performance, portability, and battery life. Data on price and overall score were used to construct the following scatterplot.

 Data set available

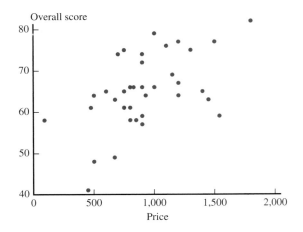

Write a few sentences commenting on this scatterplot. Would you describe the relationship between price and overall score as positive (overall score tends to increase as price increases) or negative (overall score tends to decrease as price increases)? Explain.

2.49 The National Center for Education Statistics included the following data on the average cost per year for tuition, fees, and room and board for four-year public institutions in the United States **(nces.ed.gov/fastfacts/display.asp?id=76, retrieved April 17, 2017)**. Construct a time series plot of these data and comment on the trend over time.

Year	Average Cost
2002	$14,439
2003	$15,505
2003	$16,510
2005	$17,451
2006	$18,471

(*continued*)

Year	Average Cost
2007	$19,363
2008	$20,409
2009	$21,126
2010	$22,074
2011	$23,011
2012	$23,872
2013	$24,706

2.50 One cause of tennis elbow, a malady that strikes fear into the hearts of all serious tennis players, is the impact-induced vibration of the racket-and-arm system at ball contact. The likelihood of getting tennis elbow depends on various properties of the racket used. Consider the accompanying scatterplot of x = racket resonance frequency (in hertz) and y = sum of peak-to-peak accelerations (a characteristic of arm vibration, in meters per second per second) for 23 different rackets **("Transfer of Tennis Racket Vibrations into the Human Forearm,"** *Medicine and Science in Sports and Exercise* **[1992]: 1134–1140)**. Discuss interesting features of the scatterplot.

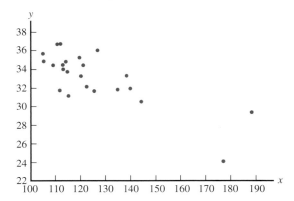

SECTION 2.5 ## Graphical Displays in the Media

There are several types of graphical displays that appear frequently in newspapers and magazines. In this section, you will see some alternative uses of bar charts as well as two other types of graphical displays—pie charts and segmented bar charts. The details on how to construct these graphs aren't covered here, but they pop up often enough to justify taking a quick look at them.

Pie Charts

A pie chart is another way of displaying the distribution of a categorical data set. A circle is used to represent the whole data set, with "slices" of the pie representing the categories. The area of a particular category's slice is proportional to its frequency or relative frequency. Pie charts are most effective for summarizing data sets when there are not too many categories.

Example 2.23 | **Life Insurance for Cartoon Characters?**

The article **"Fred Flintstone, Check Your Policy" (***The Washington Post***, October 2, 2005)** summarized a survey of 1014 adults conducted by the Life and Health Insurance Foundation for Education. Each person surveyed was asked to select which of five fictional characters had the greatest need for life insurance: Spider-Man, Batman, Fred Flintstone, Harry Potter, and Marge Simpson. The data are summarized in the pie chart of Figure 2.36.

FIGURE 2.36
Pie chart of which fictional character most needs life insurance

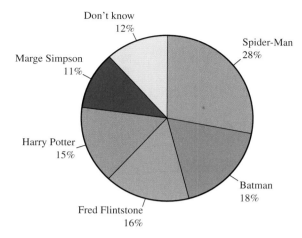

The survey results were quite different from the assessment of an insurance expert. His opinion was that Fred Flintstone, a married father with a young child, was by far the one with the greatest need for life insurance. Spider-Man, unmarried with an elderly aunt, would need life insurance only if his aunt relied on him to supplement her income. Batman, a wealthy bachelor with no dependents, doesn't need life insurance, in spite of his dangerous job!

A Different Type of "Pie" Chart: Segmented Bar Charts

A pie chart can be difficult to construct by hand, and the circular shape sometimes makes it difficult to compare areas for different categories, particularly when the relative frequencies are similar. The **segmented bar chart** (also called a stacked bar chart) avoids these difficulties by using a rectangular bar rather than a circle to represent the entire data set. The bar is divided into segments, with different segments representing different categories. As with pie charts, the area of the segment for a particular category is proportional to the relative frequency for that category. Example 2.24 illustrates the use of a segmented bar chart.

Example 2.24   **How College Seniors Spend Their Time**

Each year, the Higher Education Research Institute conducts a survey of college seniors. In 2008, approximately 23,000 seniors participated in the survey **("Findings from the 2008 Administration of the College Senior Survey," Higher Education Research Institute, June, 2009)**. The accompanying relative frequency table summarizes student responses to the question: "During the past year, how much time did you spend studying and doing homework in a typical week?"

Studying/Homework	
Amount of Time	**Relative Frequency**
2 hours or less	0.074
3 to 5 hours	0.227
6 to 10 hours	0.285
11 to 15 hours	0.181
16 to 20 hours	0.122
Over 20 hours	0.111

The corresponding segmented bar chart is shown in Figure 2.37.

FIGURE 2.37
Segmented bar chart for the
study time data of Example 2.24

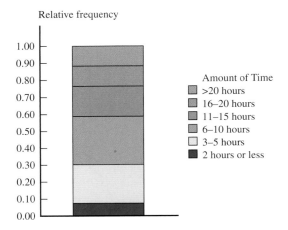

The same report also gave data on the amount of time spent on exercise or sports in a typical week. Figure 2.38 shows horizontal segmented bar charts (segmented bar charts can be displayed either vertically or horizontally) for both time spent studying and time spent exercising. Viewing these displays side-by-side makes it easy to see differences in the distributions of time spent on these two types of activities.

FIGURE 2.38
Segmented bar charts for time
spent studying and time spent
exercising per week

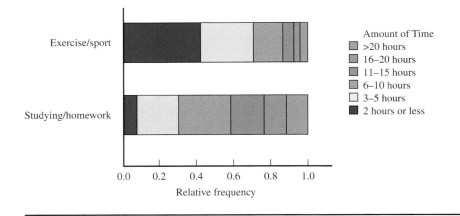

Other Uses of Bar Charts and Pie Charts

Bar charts and pie charts are used to summarize categorical data sets. However, they are also occasionally used for other purposes, as illustrated in Examples 2.25 and 2.26.

Example 2.25   **Grape Production**

The 2015 Grape Crush Report for California gave the following information on grape production for each of four different types of grapes **(www.nass.usda.gov/Statistics_by_State /California/Publications/Grape_Crush/Prelim/, retrieved April 17, 2017)**:

Type of Grape	Tons Produced
Red Wine Grapes	2,037,000
White Wine Grapes	1,662,000
Raisin Grapes	92,000
Table Grapes	71,000
Total	**3,862,000**

Although this table is not a frequency distribution, it is common to represent information of this type graphically using a pie chart, as shown in Figure 2.39. The pie represents the total grape production, and the slices show the proportion of the total production for each of the four types of grapes.

FIGURE 2.39
Pie chart for grape production data

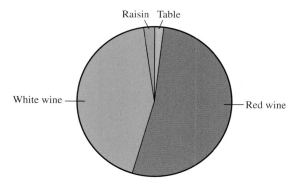

The 2015 report also included segmented bar charts like the ones shown in Figure 2.40 that represent how grape production has changed over time. It is easier to see the change over time from these segmented bar charts that can be displayed side-by-side than it would be from 10 pie charts displaying the same information. From the segmented bar charts, you can see that total grape production is down a bit from its high in 2013, with the biggest change over the years being in the production of table grapes and grapes produced for raisins.

FIGURE 2.40
Segmented bar charts showing change in grape production over time

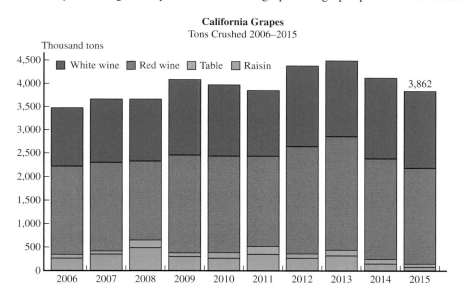

Data set available

Example 2.26   Back-to-College Spending

The National Retail Federation's 2015 Back to College Consumer Intentions and Actions Survey (**www.nrf.com, retreived August 1, 2016**) asked college students how much they planned to spend in various categories during the upcoming academic year. The average amounts of money (in dollars) that men and women planned to spend for five different types of purchases are shown in the accompanying table.

Type of Purchase	Average for Men	Average for Women
Clothing and accessories	$195.03	$180.62
Dorm of apartment furnishings	$242.95	$248.78
Electronics or computers	$403.65	$429.40
School supplies	$83.45	$74.03
Shoes	$116.91	$93.27

Even though this table is not a frequency distribution, this type of information is often represented graphically in the form of a bar chart, as illustrated in Figure 2.41. Here you can see that the average amounts are similar for all of the types of purchases except for electronics and computers, where the average for women is a bit greater than for men and the average for shoes is a bit higher for men than for women.

FIGURE 2.41
Comparative bar chart of back-to-college spending for men and women

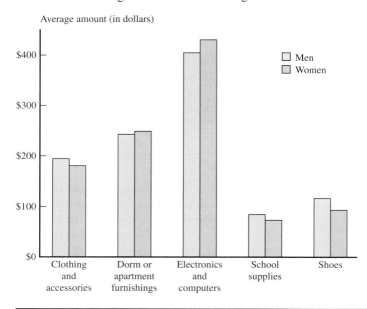

Summing It Up—Section 2.5

The following learning objective was addressed in this section:

Putting It into Practice
P6: Critically evaluate graphical displays that appear in newspapers, magazines, and advertisements.
There are some types of graphical displays that are fairly common in the media that were not covered in the earlier sections of this chapter, including pie charts and segmented bar charts. Graphical displays that appear in the media are not always constructed correctly and are sometimes misleading. When you are evaluating a graphical display, you should also be watching for the common mistakes that are discussed in Section 2.6

SECTION 2.5 **EXERCISES**

Each Exercise Set assesses the following chapter learning objectives: P6

SECTION 2.5 Exercise Set 1

2.51 The following display is similar to one that appeared in **USA TODAY (June 29, 2009)**. It is meant to be a bar graph of responses to the question shown in the display.
a. Is response to the question a categorical or numerical variable?
b. Explain why a bar chart rather than a dotplot was used to display the response data.
c. There must have been an error made in constructing this display. How can you tell that it is not a correct representation of the response data?

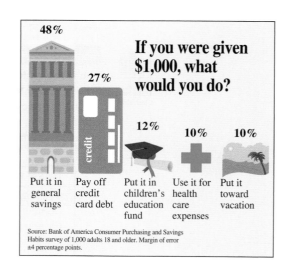

▌▐▐ **2.52** The Center for Science in the Public Interest evaluated school cafeterias in 20 school districts across the United States. Each district was assigned a numerical score on the basis of rigor of food codes, frequency of food safety inspections, access to inspection information, and the results of cafeteria inspections. Based on the score assigned, each district was also assigned one of four grades. The scores and grades are summarized in the accompanying table, which appears in the report **"Making the Grade: An Analysis of Food Safety in School Cafeterias" (cspi.us/new/pdf/makingthegrade.pdf, 2007)**.

■ **Top of the Class** ■ **Passing** ■ **Barely Passing** ■ **Failing**

School District	Overall Score (out of 100)
City of Fort Worth, TX	80
King County, WA	79
City of Houston, TX	78
Maricopa County, AZ	77
City and County of Denver, CO	75
Dekalb County, GA	73
Farmington Valley Health District, CT	72
State of Virginia	72
Fulton County, GA	68
City of Dallas, TX	67
City of Philadelphia, PA	67
City of Chicago, IL	65
City and County of San Francisco, CA	64
Montgomery County, MD	63
Hillsborough County, FL	60
City of Minneapolis, MN	60
Dade County, FL	59
State of Rhode Island	54
District of Columbia	46
City of Hartford, CT	37

a. Two variables are summarized in the table: grade and overall score. Is overall score a numerical or categorical variable? Is grade (indicated by the different colors in the table) a numerical or categorical variable?
b. Explain how the accompanying table is equivalent to a segmented bar chart of the grade data.
c. Construct a dotplot of the overall score data. Based on the dotplot, suggest an alternate assignment of grades (top of class, passing, and so on) to the 20 school districts. Explain the reasoning you used to make your assignment.

2.53 The accompanying comparative segmented bar charts are similar to ones appearing in the report **"The Future of the First Amendment: 2014 Survey of High School Students and Teachers" (www.knightfoundation.org/future-first-amendment-survey/, retrieved April 17, 2017)**. The segmented bar charts summarize responses from a large sample of high school students and a large sample of high school teachers.

Suppose that you plan to include these displays in an article that you are writing for your school newspaper. Write a few sentences that could accompany the displays. Be sure to comment on how responses to the two questions are similar or different and on the differences between how teachers and students responded.

SECTION 2.5 Exercise Set 2

2.54 The graphical display at the top left of the next page is similar to one that appeared in **USA TODAY (October 22, 2009)**. It summarizes survey responses to a question about visiting social networking sites while at work. Which of the graph types introduced in this section is used here? (*USA TODAY* frequently adds artwork and text to their graphs to make them look more interesting.)

FIGURE FOR EXERCISE 2.53

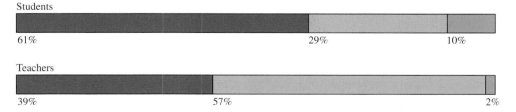
"Students should be allowed to express their opinions about teachers and school administrators on Facebook without worrying about being punished by school administrators for what they say."

■ Agree ■ Disagree ■ Don't know

Students
61%   29%   10%

Teachers
39%   57%   2%

"High school students should be allowed to report on controversial issues in their student newspapers without the approval of school authorities."

■ Agree ■ Disagree ■ Don't know

Students
61%   34%   5%

Teachers
29%   67%   4%

▌▐▐ Data set available

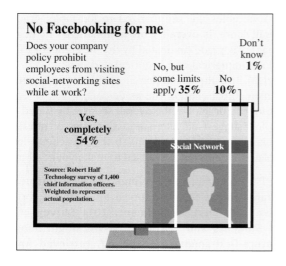

2.55 The accompanying graph is similar to one that appeared in **USA TODAY (August 5, 2008)**. This graph is a modified comparative bar graph. The modifications (incorporating hands and the earth) were most likely made to construct a display that readers would find more interesting.

a. Use the information in the graph to construct a traditional comparative bar chart.

b. Explain why the modifications made in the graph may make interpretation more difficult than for the traditional comparative bar chart.

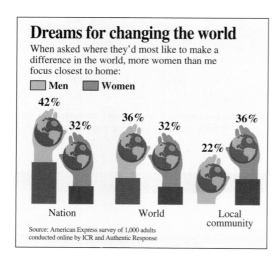

2.56 The accompanying graph is similar to one that appears in the **USA TODAY Snapshots** collection **(www.usatoday.com/picture-gallery/news/2015/04/07/usa-today-snapshots/6340793/, retrieved October 13, 2016)**. This graph was intended to show the change in the average U.S. household expenditure on gas and motor oil over time.

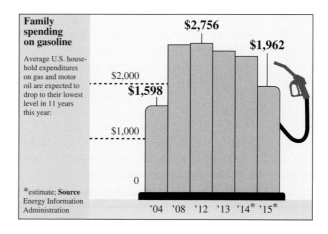

In order to make the graph look like a gasoline pump, a bar chart was used, but this is not the best choice for showing change over time, especially since the years for which averages are plotted are not equally spaced.

a. Construct a time series plot for the average expenditure on gas and motor oil. Three of the observations (for 2004, 2012, and 2015) are identified in the given graph, and you can approximate the averages for the other years from the graph. Make sure to locate the points in your time series plot appropriately along the time (x) axis.

b. Write a few sentences commenting on how the graph and your time series plot give different impressions of the change over time. Which graph do you think more accurately represents the change over time?

ADDITIONAL EXERCISES

2.57 The following display is similar to one that appeared in **USA TODAY (May 16, 2011)**. It is meant to display responses to the question shown.

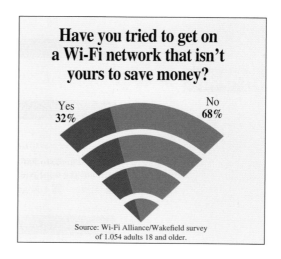

a. Explain how this display may be misleading.
b. Construct a pie chart or a bar chart that accurately summarizes the responses.

2.58 The following graphical display is similar to one that appeared in *USA TODAY* **(August 3, 2009)** and is meant to be a comparative bar graph. Do you think it is an effective summary of the data? If so, explain why. If not, explain why not and construct a display that makes it easier to compare the ice cream preferences of men and women.

2.59 The accompanying graphical display is similar to one that appeared in the *USA TODAY Snapshots* collection **(www.usatoday.com/picture-gallery/news/2015/04/07/usa-today-snapshots/6340793/, retrieved October 13, 2016)**. The graph is meant to represent responses to a question asking Millennials (people born between 1983 and 2001) whether they thought that student debt should be given equal weight with other debt when applying for a loan to buy a home.

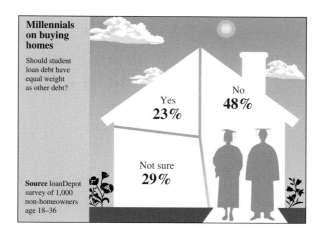

a. Explain how this graph is similar to a pie chart.
b. Which of the following segmented bar charts (Chart I, Chart II, or Chart III) is a graph of the data used to create the graph? Explain.

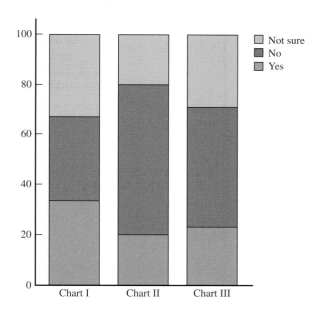

SECTION 2.6   Avoid These Common Mistakes

When constructing or evaluating graphical displays, you should keep the following in mind:

1. ***Areas should be proportional to frequency, relative frequency, or magnitude of the number being represented.*** The eye is naturally drawn to large areas in graphical displays, and it is natural for the observer to make informal comparisons based on area. Correctly constructed graphical displays, such as pie charts, bar charts, and histograms, are designed so that the areas of the pie slices or the bars are proportional to frequency or relative frequency. Sometimes, in an effort to make graphical displays more interesting, designers lose sight of this important principle, and the resulting graphs are misleading. For example, consider the following graph, which is similar to one that appeared in *USA TODAY* **(October 3, 2002)**:

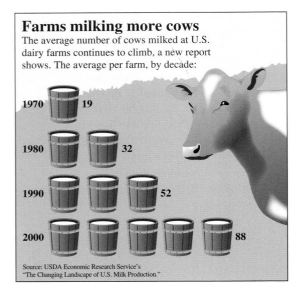

In trying to make the graph more visually interesting by replacing the bars of a bar chart with milk buckets, areas are distorted. For example, the two buckets for 1980 represent 32 cows, whereas the one bucket for 1970 represents 19 cows. This is misleading because 32 is not twice as big as 19. Other areas are distorted as well.

Another common distortion occurs when a third dimension is added to bar charts or pie charts. For example, the following pie chart is similar to one that appeared in *USA TODAY* (September 17, 2009).

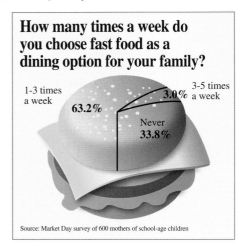

Adding the third dimension distorts the areas and makes it much more difficult to interpret correctly. A correctly drawn pie chart follows.

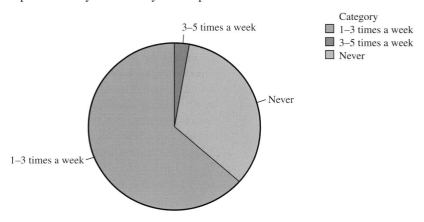

2. ***Be cautious of graphs with broken axes or axes that don't start at 0.*** Although it is common to see scatterplots with broken axes, be extremely cautious of time series plots, bar charts, or histograms with broken axes. The use of broken axes in a scatterplot does not result in a misleading picture of the relationship in the bivariate data set used to construct the display. On the other hand, in time series plots, broken axes can sometimes exaggerate the magnitude of change over time. Although it is not always a bad idea to break the vertical axis in a time series plot, it is something you should watch for. If you see a time series plot with a broken axis, as in the accompanying time series plot of the number of people employed by the U.S. Secret Service (similar to a graph that appeared in **USA TODAY, October 3, 2014**), you should pay particular attention to the scale on the vertical axis and take extra care in interpreting the graph. This time series plot appears to exaggerate the change over time.

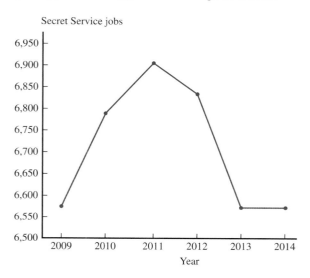

In bar charts and histograms, the vertical axis (which represents frequency, relative frequency, or density) should *never* be broken. If the vertical axis is broken in this type of graph, the resulting display will violate the "proportional area" principle, and the display will be misleading. For example, the following bar chart is similar to one in an advertisement for a software product designed to raise student test scores. By starting the vertical axis at 50, the gain from using the software is exaggerated. Areas of the bars are not proportional to the magnitude of the numbers represented—the area for the rectangle representing 68 is more than three times the area of the rectangle representing 55!

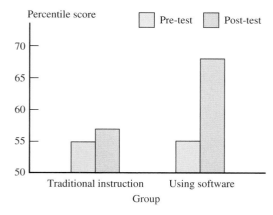

3. ***Watch out for unequal time spacing in time series plots.*** If observations over time are not made at regular time intervals, special care must be taken in constructing the time series plot. Consider the accompanying time series plot, which is similar to one in a *San Luis Obispo Tribune* **(September 22, 2002)** article on online banking:

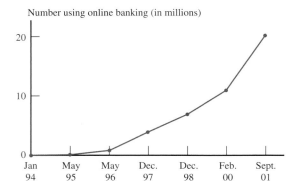

Notice that the intervals between observations are irregular, yet the points in the plot are equally spaced along the time axis. This makes it difficult to assess the rate of change over time. This could have been remedied by spacing the observations appropriately along the time axis, as shown here:

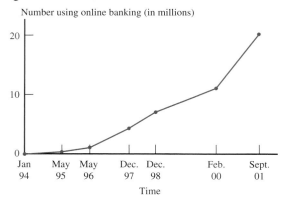

4. ***Be careful how you interpret patterns in scatterplots.*** A strong pattern in a scatterplot means that the two variables tend to vary together in a predictable way, but it does not mean that there is a cause-and-effect relationship. This point will be considered further in Chapter 4, but in the meantime, when describing patterns in scatterplots, you can't say that changes in one variable *cause* changes in the other.

5. ***Make sure that a graphical display creates the right first impression.*** For example, consider the following graph, which is similar to one that appeared in **USA TODAY (June 25, 2001)**. Although this graph does not violate the proportional area principle, the way the "bar" for the none category is displayed makes this graph difficult to read, and a quick glance at this graph may leave the reader with an incorrect impression.

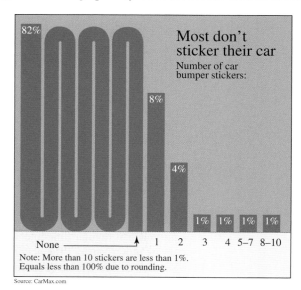

CHAPTER ACTIVITIES

ACTIVITY 2.1 BEAN COUNTERS!

Materials needed: A large bowl of dried beans (or marbles, plastic beads, or any other small, fairly regular objects) and a coin.

In this activity, you will investigate whether people can hold more in their right hand or in their left hand.

1. Flip a coin to determine which hand you will measure first. If the coin lands heads side up, start with the right hand. If the coin lands tails side up, start with the left hand. With the designated hand, reach into the bowl and grab as many beans as possible. Raise the hand over the bowl and count to four. If no beans drop during the count to four, drop the beans onto a piece of paper and record the number of beans grabbed. If any beans drop during the count, restart the count. Repeat the process with the other hand, and then record the following information: (1) right-hand number, (2) left-hand number, and (3) dominant hand (left or right, depending on whether you are left- or right-handed).

2. Create a class data set by recording the values of the three variables listed in Step 1 for each student in your class.

3. Using the class data set, construct a comparative dot-plot or a comparative stem-and-leaf display with the right-hand numbers displayed on the right and the left-hand numbers displayed on the left. Comment on the interesting features of the display and include a comparison of the right-hand number and left-hand number distributions.

4. Now construct a comparative display that allows you to compare dominant-hand count to nondominant-hand count. Does the display support the theory that dominant-hand count tends to be higher than nondominant-hand count?

5. For each person, calculate the difference dominant number − nondominant number. Construct a display of these differences. Comment on the interesting features of this display.

6. Explain why looking at the distribution of the differences (Step 5) provides more information than the comparative display (Step 4). What information is lost in the comparative display?

CHAPTER 2 EXPLORATIONS IN STATISTICAL THINKING

EXPLORATION 1: UNDERSTANDING SAMPLING VARIABILITY

In the two exercises below, each student in your class will go online to select a random sample from a small population consisting of 300 adults between the ages of 18 and 64.

To learn about the age distribution of the people in this population, go online at statistics.cengage.com/Peck2e/Explore.html and click on the link for Chapter 2. When you click on this link, it will take you to a web page where you can select a random sample of 50 people from the population.

Click on the sample button. This selects a random sample and will display information for the 50 people in your sample. You should see the following information for each person selected:

- An ID number that identifies the person selected
- Sex
- Age

Each student in your class will receive data from a different random sample.

Use the data from your random sample to complete the following two exercises.

1. This exercise uses the age data.

   a. Flip a coin two times and record the outcome (H or T) for each toss. These outcomes will determine which of the following class intervals you will use to create a frequency distribution and histogram of your age data.

Coin Toss Outcomes	Use Class Intervals
HH	15 to < 25, 25 to < 35, 35 to < 45, 45 to < 55, 55 to < 65
HT	18 to < 28, 28 to < 38, 38 to < 48, 48 to < 58, 58 to < 68
TH	15 to < 20, 20 to < 25, 25 to < 30, ..., 55 to < 60, 60 to < 65
TT	18 to < 23, 23 to < 28, 28 to < 33, ..., 58 to < 63, 63 to < 68

What class intervals will you be using?
- **b.** Using the age data for the people in your sample, construct a frequency distribution.
- **c.** Draw a histogram of the sample age data.
- **d.** Describe the shape of the age histogram from Part (c).
- **e.** Here is a histogram that displays the age distribution of the entire population. How is your histogram similar to the population histogram? How is it different?

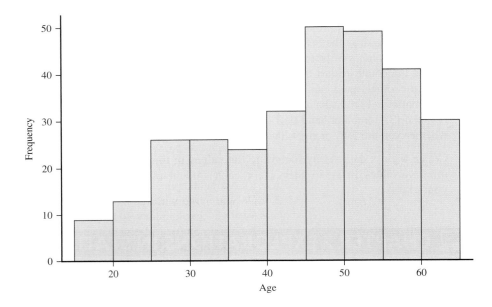

If asked to do so by your instructor, bring your histogram with you to class. Your instructor will lead a class discussion of the following:
- **f.** Compare your histogram to a histogram produced by another student who used different class intervals. How are the two histograms similar? How are the two histograms different?
- **g.** How do the histograms produced by the class support the following statement: Histograms based on random samples from a population tend to look like the population histogram.

2. In this exercise, you will compare the age distribution of the males and the age distribution of the females in your sample.
- **a.** Flip a coin to determine what type of comparative graphical display you will construct. If your flip results in a head, you will use a comparative dotplot. If your flip results in a tail, you will use a comparative stem-and-leaf display. Which type of graphical display will you be constructing?
- **b.** Using the data in your sample, construct either a comparative dotplot or a comparative stem-and-leaf display—depending on the outcome of Part (a)—that allows you to compare the age distributions of males and females. You will probably have different numbers of males and females in your sample, but this is not something to worry about.
- **c.** Based on your graphical display, write a few sentences comparing the two age distributions.

If asked to do so by your instructor, bring your graphical display to class. Your instructor will lead a class discussion of the following:

d. Compare your graphical display to that of another student who used a different type of display. Did you both make similar statements when you compared the two age distributions in Part (c), even though you each used different random samples and constructed different types of displays?

EXPLORATION 2: EXPLORING MULTIVARIABLE THINKING

How is the price of gas at the pump related to the price of oil? Oil prices and gas prices tend to rise and fall together, but this isn't always the case. In this set of exercises, you will work with data on the average price of oil (in dollars per barrel) and the average price of a gallon of gas (in dollars) for the years 1970 to 2014.

When looking at the change in price of something over time, it is common to also look at 2015 inflation-adjusted prices. Because the purchasing power of a dollar has decreased over the years, it is hard to compare the actual price of a gallon of gas in 1970 ($0.36) with the actual price in 2014 ($3.30). The 2015 inflation-adjusted price is calculated by economists and represents what the price would have been if you were buying it in 2015 dollars. For example, the 1970 price of $0.36 per gallon, would have been equivalent to a price of $2.19 in 2015.

The web site **InflationData.com** provides actual prices and inflation-adjusted prices for many products. The data used in this task are from this web site (retrieved September 19, 2016).

Go online at statistics.cengage.com/Peck2e/Explore.html and click on the link for Chapter 2. This will take you to a page where you can download the oil and gas prices.

1. Use the actual prices of oil and the actual prices of gas to construct two time series plots.
2. Do you agree with the statement that oil prices and gas prices tend to rise and fall together? Explain how you arrived at your opinion based on the time series plots.
3. An article on the InflationData.com web site titled **"Are Oil Companies Ripping Us Off with Gas Prices?" (December 12, 2015, www.inflationdata.com/Inflation/Inflation_Rate/Gasoline_vs_Oil_Price_Chart.asp, retrieved September 19, 2016)** includes the following statement: "But as we can see from the chart gas and oil prices are fairly closely related. They tend to rise and fall in tandem but at some extremes oil rose faster while at others gas seems to rise faster. Up until 1972, prices of crude and gasoline tracked very closely but since then there has been some more divergence." Do the time series plots support this statement? Explain.
4. Looking at the time series plots of actual gas and oil prices, it is clear that the actual prices of gas and the price of oil have increased over time. Now construct time series plots of the inflation-adjusted prices for oil and for gas.
5. In what ways are the time series plots of the inflation-adjusted prices different from the plots of the actual prices?
6. Based on what you have learned from the time series plots, write a paragraph commenting on how gas and oil prices have changed over time and whether or not gas price and oil prices appear to be related.

ARE YOU READY TO MOVE ON? CHAPTER 2 REVIEW EXERCISES

All chapter learning objectives are assessed in these exercises. The learning objectives assessed in each exercise are given in parentheses.

2.60 (C1, C2)
In a survey of 100 people who had recently purchased motorcycles, data on the following variables were recorded:
Gender of purchaser
Brand of motorcycle purchased
Number of previous motorcycles owned by purchaser
Telephone area code of purchaser
Weight of motorcycle as equipped at purchase

a. Which of these variables are categorical?
b. Which of these variables are discrete numerical?

c. Which type of graphical display would be an appropriate choice for summarizing the gender data, a bar chart or a dotplot?

d. Which type of graphical display would be an appropriate choice for summarizing the weight data, a bar chart or a dotplot?

2.61 (C3, M1)

For each of the five data sets described, answer the following three questions and then use Figure 2.2 to choose an appropriate graphical display for summarizing the data.

> Question 1: How many variables are in the data set?
> Question 2: Is the data set categorical or numerical?
> Question 3: Would the purpose of the graphical display be to summarize the data distribution, to compare groups, or to investigate the relationship between two numerical variables?

Data Set 1: To learn about credit card debt of students at a college, the financial aid office asks each student in a random sample of 75 students about his or her amount of credit card debt.

Data Set 2: To learn about how number of hours worked per week and number of hours spent watching television in a typical week are related, each person in a sample of size 40 was asked to keep a log of hours worked and hours spent watching television for one week. At the end of the week, each person reported the total number of hours spent on each activity.

Data Set 3: To see if satisfaction level differs for airline passengers based on where they sit on the airplane, all passengers on a particular flight were surveyed at the end of the flight. Passengers were grouped based on whether they sat in an aisle seat, a middle seat, or a window seat. Each passenger was asked to indicate his or her satisfaction with the flight by selecting one of the following choices: very satisfied, satisfied, dissatisfied, and very dissatisfied.

Data Set 4: To learn about where students purchase textbooks, each student in a random sample of 200 students at a particular college was asked to select one of the following responses: campus bookstore, off-campus bookstore, purchased all books online, or used a combination of online and bookstore purchases.

Data Set 5: To compare the amount of money men and women spent on their most recent haircut, each person in a sample of 20 women and each person in a sample of 20 men was asked how much was spent on his or her most recent haircut.

2.62 (M2)

The article **"Where College Students Buy Textbooks" (*USA TODAY*, October 14, 2010)** gave data on where students purchased books. The accompanying frequency table summarizes data from a sample of 1152 full-time college students.

Where Books Purchased	Frequency
Campus bookstore	576
Campus bookstore web site	48
Online bookstore other than campus bookstore	240
Off campus bookstore	168
Rented textbooks	36
Purchased mostly eBooks	12
Didn't buy any textbooks	72

Construct a bar chart to summarize the data distribution. Write a few sentences commenting on where students are buying textbooks.

2.63 (M2, P2)

An article about college loans **("New Rules Would Protect Students," *USA TODAY*, June 16, 2010)** reported the percentage of students who had defaulted on a student loan within 3 years of when they were scheduled to begin repayment. Information was given for public colleges, private nonprofit colleges, and for-profit colleges.

	Relative Frequency		
Loan Status	Public Colleges	Private Nonprofit Colleges	For-Profit Colleges
Good Standing	0.928	0.953	0.833
In Default	0.072	0.047	0.167

a. Construct a comparative bar chart that would allow you to compare loan status for the three types of colleges.

b. The article states "those who attended for-profit schools were more likely to default than those who attended public or private non-profit schools." What aspect of the comparative bar chart supports this statement?

2.64 (M3)

The article **"Fliers Trapped on Tarmac Push for Rules on Release" (*USA TODAY*, July 28, 2009)** gave the following data for 17 airlines on the number of flights that were delayed on the tarmac for at least 3 hours for the period between October 2008 and May 2009:

Airline	Number of Delays	Rate per 10,000 Flights
ExpressJet	93	4.9
Continental	72	4.1
Delta	81	2.8
Comair	29	2.7
American Eagle	44	1.6
US Airways	46	1.6
JetBlue	18	1.4
American	48	1.3

(continued)

Data set available

Airline	Number of Delays	Rate per 10,000 Flights
Northwest	24	1.2
Mesa	17	1.1
United	29	1.1
Frontier	5	0.9
SkyWest	29	0.8
Pinnacle	13	0.7
Atlantic Southeast	11	0.6
AirTran	7	0.4
Southwest	11	0.1

State	Gasoline Tax (cents per gallon)
Hawaii	18.5
Idaho	33.0
Illinois	33.1
Indiana	29.0
Iowa	31.8
Kansas	25.0
Kentucky	26.0
Louisiana	20.9
Maine	31.4
Maryland	32.8
Massachusetts	26.7
Michigan	30.9
Minnesota	30.6
Mississippi	18.4
Missouri	17.3
Montana	27.8
Nebraska	27.7
Nevada	23.8
New Hampshire	23.8
New Jersey	14.6
New Mexico	18.9
New York	33.8
North Carolina	35.3
North Dakota	23.0
Ohio	28.0
Oklahoma	17.0
Oregon	30.0
Pennsylvania	51.4
Rhode Island	34.1
South Carolina	16.8
South Dakota	30.0
Tennessee	21.4
Texas	20.0
Utah	30.1
Vermont	30.5
Virginia	16.8
Washington	44.6
West Virginia	33.2
Wisconsin	32.9
Wyoming	24.0

The figure at the bottom of the page shows two dotplots: one displays the number of delays data, and one displays the rate per 10,000 flights data.

a. If you were going to rank airlines based on flights delayed on the tarmac for at least 3 hours, would you use the *total number of flights* data or the *rate per 10,000 flights* data? Explain the reason for your choice.

b. Write a short paragraph that could be part of a newspaper article on flight delays that could accompany the dotplot of the *rate per 10,000 flights* data.

2.65 (M4, P1)

The following gasoline tax per gallon data for each of the 50 U.S. states and the District of Columbia (DC) in 2015 were obtained from the **U.S. Energy Information Administration (www.eia.gov/tools/faqs/faq.cfm?id=10&t=10,** retrieved April 17, 2017).

a. Construct a stem-and-leaf display of these data.

State	Gasoline Tax (cents per gallon)
Alabama	19.0
Alaska	9.0
Arizona	19.0
Arkansas	21.8
California	37.2
Colorado	23.3
Connecticut	25.0
Delaware	23.0
DC	23.5
Florida	30.6
Georgia	26.5

(*continued*)

FIGURE FOR EXERCISE 2.64

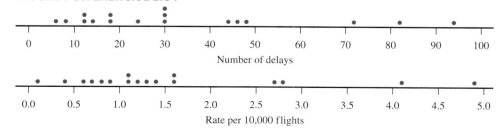

Number of delays

Rate per 10,000 flights

Data set available

b. Based on the stem-and-leaf display, what do you notice about the center and variability of the data distribution?

c. Do any values in the data set stand out as unusual? If so, which states correspond to the unusual observations and how do these values differ from the rest?

2.66 (M5, P1, P3)

The report **"Trends in College Pricing 2015" (trends.college board.org/sites/default/files/2015-trends-college-pricing -final-508.pdf, retrieved April 17, 2017)** included the information in the accompanying relative frequency distributions for public and private nonprofit four-year college students.

a. Construct a relative frequency histogram for tuition and fees for students at public four-year colleges. Write a few sentences describing the distribution of tuition and fees, commenting on shape, center, and variability.

b. Construct a relative frequency histogram for tuition and fees for students at private nonprofit four-year colleges. Be sure to use the same scale for the vertical and horizontal axes as you used for the histogram in Part (a). Write a few sentences describing the distribution of tuition and fees for students at private nonprofit four-year colleges.

c. Write a few sentences describing the differences in the two distributions.

Tuition and Fees	Proportion of Students (Relative Frequency)	
	Public Four-Year College Students	Private Nonprofit Four-Year College Students
$3,000 to < $6,000	0.04	0.02
$6,000 to < $9,000	0.38	0.06
$9,000 to < $12,000	0.27	0.02
$12,000 to < $15,000	0.12	0.02
$15,000 to < $18,000	0.04	0.03
$18,000 to < $21,000	0.04	0.03
$21,000 to < $24,000	0.02	0.05
$24,000 to < $27,000	0.02	0.07
$27,000 to < $30,000	0.02	0.09
$30,000 to < $33,000	0.02	0.09
$33,000 to < $36,000	0.01	0.08
$36,000 to < $39,000	0.00	0.09
$39,000 to < $42,000	0.00	0.07
$42,000 to < $45,000	0.00	0.07
$45,000 to < $48,000	0.00	0.21

2.67 (M3, M5, M6, P1, P4)

The chapter preview example introduced data from a survey of new car owners conducted by the J.D. Power and Associates marketing firm (*USA TODAY*, www.usaatoday. com, March 29, 2016). For each brand of car sold in the United States, data on a quality rating (number of defects per 100 vehicles) and a customer satisfaction rating (called the APEAL rating) are given in the accompanying table. The APEAL rating is a score between 0 and 1000, with higher values indicating greater satisfaction. In this exercise, you will use graphical displays of these data to answer the questions posed in the preview example.

Brand	Quality Rating	APEAL Rating
Acura	64	814
Audi	80	858
BMW	76	849
Buick	74	792
Cadillac	58	826
Chevrolet	64	791
Chrysler	58	783
Dodge	58	790
Fiat	38	768
Ford	66	789
GMC	60	791
Honda	71	783
Hyundai	70	804
Infiniti	63	826
Jeep	43	762
Kia	72	791
Land Rover	55	853
Lexus	76	844
Lincoln	65	835
Mazda	74	790
Mercedes-Benz	67	842
MINI	68	795
Mitsubishi	51	748
Nissan	63	786
Porsche	76	882
Scion	65	779
Subaru	78	766
Toyota	72	783
Volkswagen	67	796
Volvo	69	812

a. Construct a dotplot of the quality rating data.

b. Based on the dotplot, what is a typical value for quality rating?

c. Describe the variability in the quality rating for the different brands? What aspect of the dotplot supports your answer?

d. Are some brands much better or much worse than most in terms of quality rating? Write a few sentences commenting on the brands that appear to be different from most.

e. Construct a histogram of the APEAL ratings and comment on shape, center, and variability of the data distribution.

f. Construct a scatterplot of x = quality rating and y = APEAL rating. Does customer satisfaction (as measured

Data set available

by the APEAL rating) appear to be related to car quality? Explain.

2.68 (M7, P5)

The Solid Waste Management section of the Environmental Protection Agency *Report on the Environment* **(www.epa .gov/roe/, retrieved April 17, 2017)** included a graph similar to the accompanying graph. The report also included the following statement:

> The last several decades have seen steady growth in recycling and composting, while the total amounts landfilled peaked in 1990 (145 MT) and have generally declined since then (134 MT in 2013).

Explain how the time series plot is consistent or is not consistent with the given statement.

Exhibit 1. Municipal solid waste generated and managed in the U.S., 1960—2013

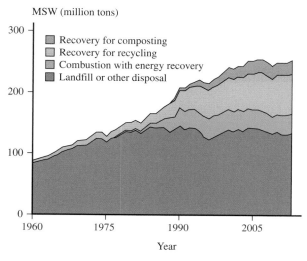

2.69 (P6)

The accompanying graphical display is from the **Fall 2008 Census Enrollment Report at Cal Poly, San Luis Obispo**. It uses both a pie chart and a segmented bar graph to summarize ethnicity data for students enrolled in Fall 2008.

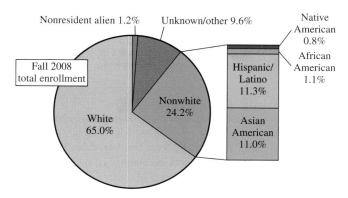

a. Use the information in the graphical display to construct a single segmented bar graph for the ethnicity data.

b. Do you think that the original graphical display or the one you created in Part (a) is more informative? Explain your choice.

c. Why do you think that the original graphical display format (combination of pie chart and segmented bar graph) was chosen over a single pie chart with seven slices?

d. After 2008, Cal Poly changed the format of the graph it used in the yearly enrollment reports and now uses just a pie chart. The graph below shows the pie chart for 2015. Which type of graph (the single pie chart, a segmented bar graph, or the combination of a pie chart and segmented bar graph that was used in 2008) would you recommend the university use in future years?

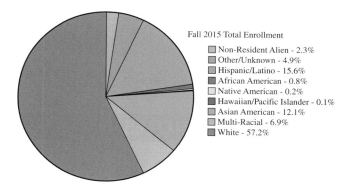

2.70 (P6)

The following two graphical displays are similar to ones that appeared in **USA TODAY (June 1, 2009 and July 28, 2009)**. One is an appropriate representation and the other is not. For each of the two, explain why it is or is not drawn appropriately.

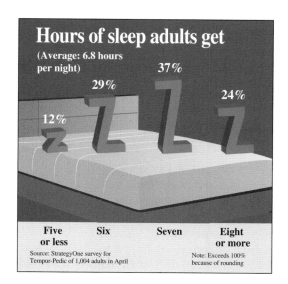

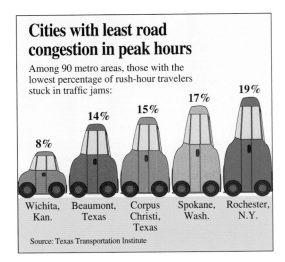

Cities with least road congestion in peak hours

Among 90 metro areas, those with the lowest percentage of rush-hour travelers stuck in traffic jams:

8% — Wichita, Kan.
14% — Beaumont, Texas
15% — Corpus Christi, Texas
17% — Spokane, Wash.
19% — Rochester, N.Y.

Source: Texas Transportation Institute

2.71   (P6)

Explain why the following graphical display (similar to a graph that appeared in *USA TODAY,* **September 17, 2009**) is misleading.

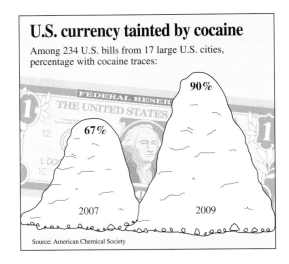

U.S. currency tainted by cocaine

Among 234 U.S. bills from 17 large U.S. cities, percentage with cocaine traces:

67% — 2007
90% — 2009

Source: American Chemical Society

TECHNOLOGY NOTES

Bar Graphs

JMP

1. Enter the raw data into a column (**Note:** To open a new data table, click **File** then select **New** then **Data Table**)
2. Click **Graph** and select **Chart**
3. Click and drag the column name containing your stored data from the box under **Select Columns** to the box next to **Categories, X, Levels**
4. Click **OK**

MINITAB

1. Enter the raw data into C1
2. Select **Graph** and choose **Ba̱r Chart...**
3. Highlight Simple
4. Click **OK**
5. Double click C1 to add it to the **Categorical Variables** box
6. Click **OK**

Note: You may add or format titles, axis titles, legends, etc., by clicking on the **Labels...** button prior to performing step 6 above.

SPSS

1. Enter the raw data into a column
2. Select **Graph** and choose **Chart Builder...**
3. Under **C̱hoose from** highlight **Bar**
4. Click and drag the first bar chart in the right box (Simple Bar) to the Chart preview area

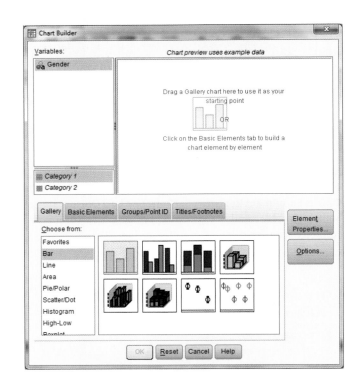

5. Click and drag the variable from the **Variables** box to the **X-Axis?** box in the chart preview area
6. Click **OK**

Note: By default, SPSS displays the count of each category in a bar chart. To display the percentage for each category, follow the above instructions to Step 5. Then, in the Element Properties dialog box, select Percentage from the drop-down box under Statistic. Click **Apply** then click **OK** on the Chart Builder dialog box.

Excel 2007

1. Enter the category names into column A (you may input the title for the variable in cell A1)
2. Enter the count or percent for each category into column B (you may put the title "Count" or "Percentage" in cell B1)
3. Select all data (including column titles if used)
4. Click on the **Insert** Ribbon

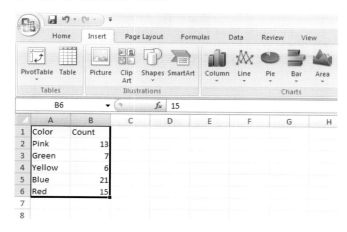

5. Choose **Column** and select the first chart under 2-D Column (Clustered Column)
6. The chart will appear on the same worksheet as your data

Note: You may add or format titles, axis titles, legends, and so on, by right-clicking on the appropriate piece of the chart.

Note: Using the Bar Option on the Insert Ribbon will produce a horizontal bar chart.

TI-83/84

The TI-83/84 does not have the functionality to produce bar charts.

TI-Nspire

1. Enter the category names into a list (to access data lists select the spreadsheet option and press **enter**)

Note: In order to correctly enter category names, you will input them as text. Begin by pressing ?!> and then select " and press **enter**. Now type the category name and press **enter**.

Note: Be sure to title the list by selecting the top row of the column and typing a title.

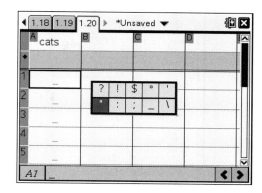

2. Enter the counts for each category into a second list (be sure to title this list as well!)
3. Press the **menu** key and navigate to **3:Data** and then **5:Frequency Plot** and select this option
4. For **Data List** select the list containing your category names from the drop-down menu
5. For **Frequency List** select the list containing the counts for each category from the drop-down menu
6. Press **OK**

Comparative Bar Charts

JMP

1. Enter the raw data into one column
2. In a separate column enter the group information

		Column 1	Column 2
	11	Pink	F
	12	Pink	F
	13	Pink	F
	14	Yellow	F
	15	Red	F
	16	Red	F
	17	Blue	F
	18	Blue	F
	19	Blue	F
	20	Blue	F
	21	Blue	M
	22	Blue	M
	23	Blue	M

3. Click **Graph** then select **Chart**
4. Click and drag the column containing the raw data from the box under **Select Columns** to the box next to **Categories, X, Levels**
5. Click and drag the column containing the group information from the box under **Select Columns** to the box next to **Categories, X, Labels**
6. Click **OK**

MINITAB

1. Input the group information into C1
2. Input the raw data into C2 (be sure to match the response from the first column with the appropriate group)

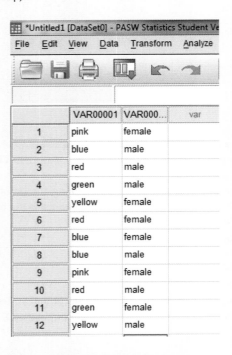

3. Select **Graph** then choose **Bar Chart...**
4. Highlight Cluster
5. Click **OK**
6. Double click C1 to add it to the **Categorical Variables** box
7. Then double click C2 to add it to the **Categorical Variables** box
8. Click **OK**

Note: Be sure to list the grouping variable first in the Categorical Variables box.

Note: You may add or format titles, axis titles, legends, and so on, by clicking on the **Labels...** button prior to performing Step 8 above.

SPSS
1. Enter the raw data into one column
2. Enter the group information into a second column (be sure to match the response from the first column with the appropriate group)

3. Select **Graph** and choose **Chart Builder...**
4. Under **Choose from** highlight Bar

5. Click and drag the second bar chart in the first column (Clustered Bar) to the Chart preview area
6. Click and drag the group variable (second column) into the **Cluster on X** box in the upper right corner of the chart preview area
7. Click and drag the data variable (first column) into the **X-Axis?** box in the chart preview area

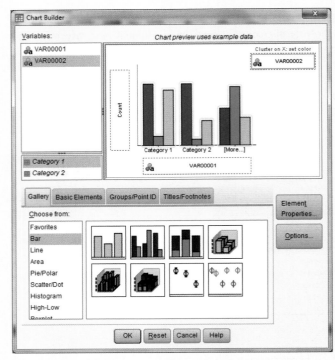

8. Click **Ok**

Excel 2007
1. Enter the category names into column A (you may input the title for the variable in cell A1)
2. Enter the count or percent for different groups into a new column, starting with column B
3. Select all data (including column titles if used)

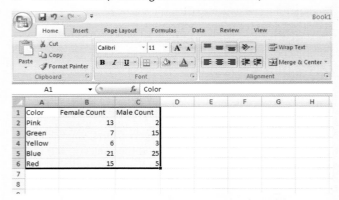

4. Click on the **Insert** Ribbon
5. Choose **Column** and select the first chart under 2-D Column (Clustered Column)
6. The chart will appear on the same worksheet as your data

Note: You may add or format titles, axis titles, legends, and so on, by right-clicking on the appropriate piece of the chart.

Note: Using the Bar Option on the Insert Ribbon will produce a horizontal bar chart.

TI-83/84

The TI-83/84 does not have the functionality to produce comparative bar charts.

TI-Nspire

The TI-Nspire does not have the functionality to produce comparative bar charts.

Dotplots

JMP

1. Input the raw data into a column
2. Click **Graph** and select **Chart**
3. Click and drag the column name containing the data from the box under **Select columns** to the box next to **Categories, X, Levels**
4. Select **Point Chart** from the second drop-down box in the **Options** section

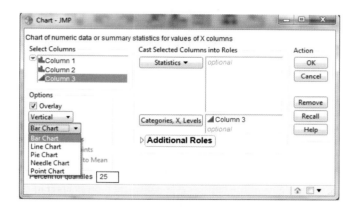

5. Click **OK**

MINITAB

1. Input the raw data into C1
2. Select **Graph** and choose **Dotplot**
3. Highlight Simple under One Y
4. Click **OK**
5. Double click C1 to add it to the **Graph Variables** box
6. Click **OK**

Note: You may add or format titles, axis titles, legends, and so on, by clicking on the **Labels...** button prior to performing Step 6 above.

SPSS

1. Input the raw data into a column
2. Select **Graph** and choose **Chart Builder...**
3. Under **Choose from** highlight Scatter/Dot
4. Click and drag the second option in the second row (Simple Dot Plot) to the Chart preview area
5. Click and drag the variable name from the **Variables** box into the **X-Axis?** box
6. Click **OK**

Excel 2007

Excel 2007 does not have the functionality to create dotplots.

TI-83/84

The TI-83/84 does not have the functionality to create dotplots.

TI-Nspire

1. Enter the data into a data list (to access data lists select the spreadsheet option and press **enter**)

Note: Be sure to title the list by selecting the top row of the column and typing a title.

2. Press the **menu** key then select **3:Data** then select **6:QuickGraph** and press **enter** (a dotplot will appear)

Stem-and-leaf plots

JMP

1. Enter the raw data into a column
2. Click **Analyze** then select **Distribution**
3. Click and drag the column name containing the data from the box under **Select Columns** to the box next to **Y, Columns**
4. Click **OK**
5. Click the red arrow next to the column name

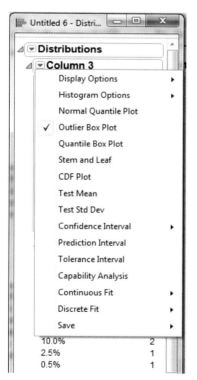

6. Select **Stem and Leaf**

MINITAB

1. Input the raw data into C1
2. Select **Graph** and choose **Stem-and-leaf...**
3. Double click C1 to add it to the **Graph Variables** box
4. Click **OK**

Note: You may add or format titles, axis titles, legends, and so on, by clicking on the **Labels...** button prior to performing Step 4 above.

SPSS

1. Input the raw data into a column
2. Select **Analyze** and choose **Descriptive Statistics** then **Explore**
3. Highlight the variable name from the box on the left
4. Click the arrow to the right of the Dependent List box to add the variable to this box

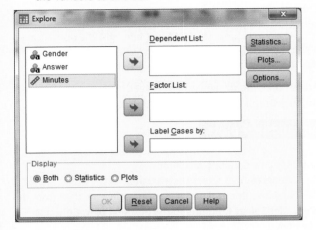

5. Click **OK**
6. **Note:** The stem-and-leaf plot is one of the plots produced along with several other descriptive statistics and plots.

Excel 2007

Excel 2007 does not have the functionality to create stem-and-leaf plots automatically.

TI-83/84

The TI-83/84 does not have the functionality to create stem-and-leaf plots.

TI-Nspire

The TI-Nspire does not have the functionality to create stem-and-leaf plots.

Histograms

JMP

1. Enter the raw data into a column
2. Click **Analyze** then select **Distribution**
3. Click and drag the column name containing the data from the box under **Select Columns** to the box next to **Y, Columns**
4. Click **OK**
5. Click the red arrow next to the column name
6. Select **Histogram Options** and click **Vertical**

MINITAB

1. Input the raw data into C1
2. Select **Graph** and choose **Histogram…**
3. Highlight Simple
4. Click **OK**
5. Double click C1 to add it to the **Graph Variables** box
6. Click **OK**

Note: You may add or format titles, axis titles, legends, and so on, by clicking on the **Labels…** button prior to performing Step 5 above.

SPSS

1. Input the raw data into a column
2. Select **Graph** and choose **Chart Builder…**
3. Under **Choose from** highlight Histogram
4. Click and drag the first option (Simple Histogram) to the Chart preview area
5. Click and drag the variable name from the **Variables** box into the **X-Axis?** box
6. Click **OK**

Excel 2007

Note: In order to produce histograms in Excel 2007, we will use the Data Analysis Add-On. For instructions on installing this add-on, please see the note at the end of this chapter's Technology Notes.

1. Input the raw data in column A (you may use a column heading in cell A1)
2. Click on the **Data** ribbon
3. Choose the **Data Analysis** option from the **Analysis** group
4. Select **Histogram** from the Data Analysis dialog box and click **OK**
5. Click in the **Input Range:** box and then click and drag to select your data (including the column heading)
   a. If you used a column heading, click the checkbox next to **Labels**
6. Click the checkbox next to **Chart Output**
7. Click **OK**

Note: If you have specified bins manually in another column, click in the box next to **Bin Range:** and select your bin assignments.

TI-83/84

1. Enter the data into **L1** (In order to access lists press the **STAT** key, highlight the option called **Edit…** then press **ENTER**)
2. Press the **2ⁿᵈ** key then the **Y =** key
3. Highlight the option for Plot1 and press **ENTER**
4. Highlight **On** and press **ENTER**
5. For **Type**, select the histogram shape and press **ENTER**

6. Press **GRAPH**

Note: If the graph window does not display appropriately, press the **WINDOW** button and reset the scales appropriately.

TI-Nspire

1. Enter the data into a data list (In order to access data lists select the spreadsheet option and press **enter**)

Note: Be sure to title the list by selecting the top row of the column and typing a title.

2. Press the **menu** key then select **3:Data** then select **6:QuickGraph** and press **enter** (a dotplot will appear)
3. Press the **menu** key and select **1:Plot Type**

4. Select **3:Histogram** and press **enter**

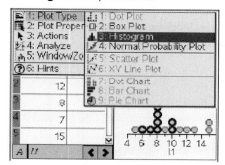

Scatterplots

JMP
1. Input the data for the independent variable into the first column
2. Input the data for the dependent variable into the second column
3. Click **Analyze** then select **Fit Y by X**
4. Click and drag the column name for the independent data from the box under **Select Columns** to the box next to **Y, Response**
5. Click and drag the column name for the dependent data from the box under **Select Columns** to the box next to **X, Factor**
6. Click **OK**

MINITAB
1. Input the raw data for the independent variable into C1
2. Input the raw data for the dependent variable into C2
3. Select **Graph** and choose **Scatterplot**
4. Highlight Simple
5. Click **OK**
6. Double click on C2 to add it to the **Y Variables** column of the spreadsheet
7. Double click on C1 to add it to the **X Variables** column of the spreadsheet
8. Click **OK**

Note: You may add or format titles, axis titles, legends, and so on, by clicking on the **Labels...** button prior to performing Step 7 above.

SPSS
1. Input the raw data into two columns
2. Select **Graph** and choose **Chart Builder...**
3. Under **Choose from** highlight Scatter/Dot
4. Click and drag the first option (Simple Scatter) to the Chart preview area
5. Click and drag the variable name representing the independent variable to the **X-Axis?** box
6. Click and drag the variable name representing the dependent variable to the **Y-Axis?** box
7. Click **OK**

Excel 2007
1. Input the raw data into two columns (you may enter column headings)
2. Select both columns of data

3. Click on the **Insert** ribbon
4. Click **Scatter** and select the first option (Scatter with Markers Only)

Note: You may add or format titles, axis titles, legends, and so on, by right-clicking on the appropriate piece of the chart.

TI-83/84
1. Input the data for the dependent variable into L2 (In order to access lists press the **STAT** key, highlight the option called **Edit...** then press **ENTER**)
2. Input the data for the dependent variable into L1
3. Press the **2nd** key then press the **Y=** key
4. Select the **Plot1** option and press **ENTER**
5. Select **On** and press **ENTER**
6. Select the scatterplot option and press **ENTER**

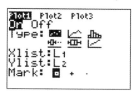

7. Press the **GRAPH** key

Note: If the graph window does not display appropriately, press the **WINDOW** button and reset the scales appropriately.

TI-Nspire
1. Enter the data for the independent variable into a data list (In order to access data lists select the spreadsheet option and press **enter**)

Note: Be sure to title the list by selecting the top row of the column and typing a title.

2. Enter the data for the dependent variable into a separate data list
3. Highlight both columns of data (by scrolling to the top of the screen to highlight one list then press shift and use the arrow key to highlight the second list)
4. Press the **menu** key and select **3:Data** then select **6:QuickGraph** and press **enter**

Time series plots

JMP
1. Input the raw data into one column
2. Input the time increment data into a second column

Column 4	Column 5
1	1900
2	1901
3	1902
4	1903
5	1904
6	1905
7	1906
8	1907
9	1908
1	1909
2	1910
3	1911
4	1912

3. Click **Graph** then select **Control Chart** then select **Run Chart**
4. Click and drag the name of the column containing the raw data from the box under **Select Columns** to the box next to Process
5. Click and drag the name of the column containing the time increment data from the box under **Select Columns** to the box next to **Sample Label**
6. Click **OK**

MINITAB
1. Input the raw data into C1
2. Select **Graph** and choose **Time Series Plot...**
3. Highlight Simple
4. Click **OK**
5. Double click C1 to add it to the **Series** box
6. Click the Time/Scale button
7. Choose the appropriate time scale under Time Scale (Choose Calendar if you want to use Days, Years, Months, Quarters, etc.)
8. Click to select one set for each variable
9. In the spreadsheet, fill in the start value for the time scale
10. Input the increment value into the box next to **Increment** (i.e., 1 to move by one year or one month, etc.)
11. Click **OK**
12. Click **OK**

Note: You may add or format titles, axis titles, legends, and so on, by clicking on the **Labels...** button prior to performing Step 12 above.

SPSS
1. Input the raw data into two columns: one representing the data and one representing the time increments
2. Select **Analyze** then choose **Forecasting** then **Sequence Charts...**
3. Highlight the column name containing the raw data and add it to the **Variables** box
4. Highlight the column name containing the time data and add it to the **Time Axis Labels** box
5. Click **OK**

Excel 2007
1. Input the data into two columns: one representing the data and one representing the time increments
2. Select an empty cell
3. Click on the **Insert** ribbon
4. Click **Line** and select the first option under 2-D Line (Line)
5. Right-click on the empty chart area that appears and select **Select Data...**
6. Under **Legend Entries (Series)** click **Add**
7. In the **Series name:** box select the column title for the data
8. In the **Series values:** box select the data values

9. Click **OK**
10. Under **Horizontal (Category) Axis Labels** click **Edit**
11. In the **Axis label range:** select the time increment data (do NOT select the column title)
12. Click **OK**
13. Click **OK**

TI-83/84
1. Input the time increment data into **L1**
2. Input the data values for each time into **L2**
3. Press the **2nd** key then the **Y=** key
4. Select **1:Plot1** and press **ENTER**
5. Highlight **On** and press **ENTER**
6. Highlight the second option in the first row and press **ENTER**
7. Press **GRAPH**

Note: If the graph window does not display appropriately, press the **WINDOW** button and reset the scales appropriately.

TI-Nspire
1. Enter the data for the time increments into a data list (In order to access data lists select the spreadsheet option and press **enter**)

Note: Be sure to title the list by selecting the top row of the column and typing a title.

2. Enter the data for each time increment into a separate data list
3. Highlight both columns of data (by scrolling to the top of the screen to highlight one list then press shift and use the arrow key to highlight the second list)
4. Press the **menu** key and select **3:Data** then select **6:QuickGraph** and press **enter**
5. Press the **menu** key and select **1:Plot Type** then select **6:XY Line Plot** and press **enter**

Installing Excel 2007's Data Analysis Add-On

1. Click the **Microsoft Office Button** and then Click **Excel Options**
2. Click **Add-Ins**, then in the **Manage** box, select **Excel Add-ins**.
3. Click **Go**
4. In the **Add-Ins available** box, select the **Analysis ToolPak** checkbox, and then click **OK**

Note: If this option is not listed, click **Browse** to locate it.

Note: If you get prompted that the Analysis ToolPak is not currently installed on your computer, click **Yes** to install it.

5. After you load the Analysis ToolPak, the **Data Analysis** command is available in the **Analysis** group on the **Data** ribbon.

3

Numerical Methods for Describing Data Distributions

Blend Images/Alamy Stock Photo

PREVIEW

In Chapter 2, graphical displays were used to summarize data. By creating a visual display of the data distribution, it is easier to see and describe important characteristics, such as shape, center, and variability. In this chapter, you will see how numerical measures are also used to describe characteristics of a data distribution.

CHAPTER LEARNING OBJECTIVES

Conceptual Understanding

After completing this chapter, you should be able to

C1 Understand how numerical summary measures are used to describe characteristics of a numerical data distribution.

C2 Understand how the variance and standard deviation describe variability in a data distribution.

C3 Understand the impact that outliers can have on measures of center and variability.

C4 Understand how the shape of a data distribution is considered when selecting numerical summary measures.

C5 Understand how the relative values of the mean and median may be related to the shape of a data distribution.

Mastering the Mechanics

After completing this chapter, you should be able to

M1 Select appropriate summary statistics for describing center and variability of a numerical data distribution.

M2 Calculate and interpret the value of the sample mean.

M3 Calculate and interpret the value of the sample standard deviation.

M4 Calculate and interpret the value of the sample median.

M5 Calculate and interpret the value of the sample interquartile range.

M6 Calculate and interpret the values in the five-number summary.

M7 Given a numerical data set, construct a boxplot.

M8 Identify outliers in a numerical data set.

Putting It into Practice

After completing this chapter, you should be able to

P1 Interpret measures of center in context.

P2 Interpret measures of variability in context.

P3 Use boxplots to make comparisons between two or more groups.

P4 Use *z*-scores and percentiles to describe relative standing.

PREVIEW EXAMPLE

Just Thinking About Exercise??

Studies have shown that in some cases people overestimate the extent to which physical exercise can compensate for food consumption. When this happens, people increase food intake more than what is justified based on the exercise performed. The authors of the paper **"Just Thinking About Exercise Makes Me Serve More Food: Physical Activity and Calorie Compensation" (*Appetite* [2011]: 332–335)** wondered if even *thinking* about exercise would lead to increased food consumption. They carried out an experiment in which people were offered snacks as a reward for participating in the experiment. People read a short essay and then answered a few questions about the essay. Some participants read an essay that was unrelated to exercise (the control group), some read an essay that described listening to music while taking a 30 minute walk (the fun group), and some read an essay that described strenuous exercise (the exercise group). Participants were then provided with two plastic bags and invited to help themselves to two types of snacks—Chex Mix and M&M's. After the participants served themselves, the bags were weighed so that the researchers could determine the number of calories in snacks taken.

Data on number of calories, consistent with summary values in the paper, were used to construct the comparative dotplot shown in Figure 3.1. From the dotplots, it is clear that the number of calories tends to be quite a bit higher for those who read about exercise than for those in the control group! If you want to compare the distributions with more precision, the first step is to describe them numerically. ■

FIGURE 3.1
Comparative dotplot of calories

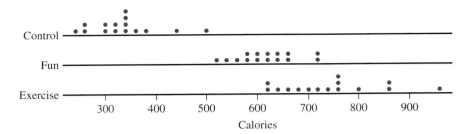

SECTION 3.1 Selecting Appropriate Numerical Summaries

Suppose that you have just received your score on an exam in one of your classes. What would you want to know about the distribution of scores for this exam? Like many students, you would probably want to know two things: "What was the class average?" and "What were the high and the low scores?"

These questions are related to important characteristics of the data distribution. By asking for the class average, you want to know what a "typical" exam score was. What number best describes the entire set of scores? By asking for the high and low scores on the exam, you want to know about the variability in the data set. Were the scores similar or did they differ quite a bit from student to student?

When describing a numerical data set, it is common to report both a value that describes where the data distribution is centered along the number line and a value that describes how much variability there is in the data distribution.

Measures of Center describe where the data distribution is located along the number line. A measure of center provides information about what is "typical."

Measures of Variability describe how much variability there is in a data distribution. A measure of variability provides information about how much individual values tend to differ from one another.

There are multiple ways to describe center and variability in a data distribution. Two common choices are to either use the mean and standard deviation or the median and interquartile range. These terms will be defined in Sections 3.2 and 3.3, where you will see how these summary measures are calculated and interpreted. For now, we will focus on which of these two options is the best choice in a given situation.

The key factor in deciding how to measure center and variability is the shape of the data distribution, as described in Table 3.1. Because this decision is based on the shape of the distribution, the best place to start is with a graphical display of the data.

TABLE 3.1 Choosing Appropriate Measures for Describing Center and Variability

If the Shape of the Data Distribution Is...	Describe Center and Variability Using...
Approximately symmetric	Mean and standard deviation
Skewed or has outliers	Median and interquartile range

Data set available

Example 3.1 **Medical Errors**

The stress of the final years of medical training can sometimes contribute to depression and burnout. The authors of the paper **"Rates of Medication Errors Among Depressed and Burnt Out Residents" (*British Medical Journal* [2008]: 488)** studied 24 residents in pediatrics who were classified as depressed based on their score on the Harvard national depression screening scale. Medical records of patients treated by these residents during a fixed time period were examined for errors in ordering or administering medications. The accompanying data are the total number of medication-related errors for each of the 24 residents.

Number of Errors

0	0	0	0	0	0	0	0	0	0	0	0
0	0	0	0	0	1	1	1	2	3	5	11

A dotplot of these data is shown in Figure 3.2.

FIGURE 3.2
Dotplot of number of medication-related errors

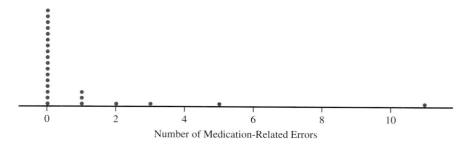

From the dotplot, you can see that the data distribution is not symmetric and that there are outliers. For these reasons, you would choose the median and interquartile range to describe center and variability for this data set.

Before leaving this section, let's look at why the shape of the data distribution affects the choice of the summary measures that would be used to describe center and variability.

Data set available

Example 3.2 Baseball Salaries

ESPN publishes professional baseball players' salaries at the beginning of each season. The following are the 2016 salaries for the Kansas City Royals players in millions of dollars (**espn.go.com/mlb/team/salaries/_/name/kc/kansas-city-royals, retrieved August 4, 2016**).

Player	2016 Salary (in millions of dollars)	Player	2016 Salary (in millions of dollars)
Alcides Escobar	3.00	Kelvin Herrera	1.60
Alex Gordon	13.75	Kendrys Morales	6.50
Chris Young	0.68	Kris Medlen	2.00
Christian Colon	0.51	Lorenzo Cain	2.73
Danny Duffy	2.42	Luke Hochevar	4.00
Dillon Gee	5.30	Mike Moustakas	2.64
Drew Butera	0.99	Omar Infante	7.50
Edinson Volquez	7.50	Paulo Orlando	0.51
Eric Hosmer	5.65	Salvador Perez	1.75
Ian Kennedy	9.85	Wade Davis	7.00
Jarrod Dyson	1.23	Yordano Ventura	0.75
Joakim Soria	7.00		

A dotplot of these data is given in Figure 3.3.

FIGURE 3.3
Dotplot of 2016 salaries for players on the Kansas City Royals

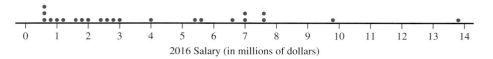

Suppose that you calculated the average salary for this team's players to describe a "typical" value. The average salary is $4.124 million. Does this seem typical for players on this team? The 13 players (57% of the team) with salaries of $3 million or less probably don't think so. Because this distribution is quite skewed and has outliers, the average is pulled up by the small number of very large salaries. When this happens, the average may not be the best choice to describe the center. The median (to be introduced in Section 3.3) may be a better choice.

Summing It Up—Section 3.1

The following learning objectives were addressed in this section:

Conceptual Understanding

C1: Understand how numerical summary measures are used to describe characteristics of a numerical data distribution.

Two important characteristics of a numerical data distribution are center (which gives you a sense of a typical value for the data) and variability (which gives you an idea of how much the individual data values tend to differ from one another). Numerical summary measures are often used to describe center and variability.

C4: Understand how the shape of a data distribution is considered when selecting numerical summary measures.

Depending on whether a data distribution is approximately symmetric or if it is skewed or has outliers, different summary measures may be used to describe center and variability.

Mastering the Mechanics

M1: Select appropriate summary statistics for describing center and variability of a numerical data distribution.

When a data distribution is approximately symmetric, the usual choice for a measure of center is the mean and the usual choice for a measure of variability is the standard deviation. When a data distribution is skewed or has outliers, it is usually preferable to use the median as a measure of center and the interquartile range as a measure of variability.

SECTION 3.1 EXERCISES

Each Exercise Set assesses the following chapter learning objectives: C1, C4, M1

SECTION 3.1 Exercise Set 1

For each data set described in Exercises 3.1–3.3, construct a graphical display of the data distribution and then indicate what summary measures you would use to describe center and variability. (Hint: Consider the shape of the data distribution.)

3.1 The accompanying data are consistent with summary statistics from the paper **"Shape of Glass and Amount of Alcohol Poured: Comparative Study of Effect of Practice and Concentration"** (*British Medical Journal* [2005]: 1512–1514). They represent the actual amount (in milliliters) poured into a short, wide glass for individuals asked to pour 1.5 ounces.

89.2	68.6	32.7	37.4	39.6	46.8	66.1	79.2
66.3	52.1	47.3	64.4	53.7	63.2	46.4	63.0
92.4	57.8						

3.2 Data on tipping percent for 20 restaurant tables, consistent with summary statistics given in the paper **"Beauty and the Labor Market: Evidence from Restaurant Servers" (unpublished manuscript by Matt Parrett, 2007)**, are:

0.0	5.0	45.0	32.8	13.9	10.4	55.2	50.0
10.0	14.6	38.4	23.0	27.9	27.9	105.0	19.0
10.0	32.1	11.1	15.0				

3.3 Data on manufacturing defects per 100 cars for the 30 brands of cars sold in the United States (*USA TODAY,* **March 29, 2016)** are:

97	134	198	142	95	135	132	145	136	129	152
158	169	155	106	125	120	153	208	163	204	173
165	126	113	167	171	166	181	161			

3.4 The accompanying data on number of cell phone minutes used in one month are consistent with summary

Data set available

statistics published in a report of a marketing study of San Diego residents **(Tele-Truth, March 2009)**:

189	0	189	177	106	201	0	212	0	306
0	0	59	224	0	189	142	83	71	165
236	0	142	236	130					

Explain why the average may not be the best measure of a typical value for this data set.

SECTION 3.1 Exercise Set 2

For each data set described in Exercises 3.5–3.7, construct a graphical display of the data distribution and then indicate what summary measures you would use to describe center and variability. (Hint: Consider the shape of the data distribution.)

3.5 The accompanying data are a subset of data that appeared in the paper **"Ladies First? A Field Study of Discrimination in Coffee Shops"** (*Applied Economics* [April 2008]). The data are the times (in seconds) between ordering and receiving coffee for 19 male customers at a Boston coffee shop.

40	60	70	80	85	90	100	100	110	120
125	125	140	140	160	160	170	180	200	

3.6 For each brand of car sold in the United States, data on a customer satisfaction rating (called the APEAL rating) are given (*USA TODAY,* **March 29, 2016**). The APEAL rating is a score between 0 and 1,000, with higher values indicating greater satisfaction.

882	858	853	849	844	842	835	826	826	814	812
804	796	795	792	791	791	791	790	790	789	786
783	783	783	779	768	766	762	748			

3.7 Data on weekday exercise time for 20 males, consistent with summary quantities given in the paper **"An Ecological Momentary Assessment of the Physical Activity and Sedentary Behaviour Patterns of University Students"** (*Health Education Journal* [2010]: 116–125), are:

Male—Weekday

43.5	91.5	7.5	0.0	0.0	28.5	199.5	57.0	142.5
8.0	9.0	36.0	0.0	78.0	34.5	0.0	57.0	151.5
8.0	0.0							

3.8 The report **"State of the News Media 2015"** (**Pew Research Center, April 29, 2015**) published the accompanying

circulation numbers for 15 news magazines (such as *Time* and *The New Yorker*) for 2014:

3,284,012	1,469,223	1,214,590	1,046,977	993,043
931,228	905,755	843,914	783,353	574,370
483,360	412,062	147,808	119,297	41,518

Explain why the average may not be the best measure of a typical value for this data set.

ADDITIONAL EXERCISES

For each data set described in Exercises 3.9–3.10, construct a graphical display of the data distribution and then indicate what summary measures you would use to describe center and variability.

3.9 Data on weekend exercise time for 20 males, consistent with summary quantities given in the paper **"An Ecological Momentary Assessment of the Physical Activity and Sedentary Behaviour Patterns of University Students"** (*Health Education Journal* [2010]: 116–125), are:

Male—Weekend

0	0	15	0	5	0	1	1	57	95
73	155	125	7	27	97	61	0	73	155

3.10 Data on weekday exercise time for 20 females, consistent with summary quantities given in the paper referenced in the previous exercise, are:

Female—Weekday

10.0	90.6	48.5	50.4	57.4	99.6
0.0	5.0	0.0	0.0	5.0	2.0
10.5	5.0	47.0	0.0	5.0	54.0
0.0	48.6				

3.11 Increasing joint extension is one goal of athletic trainers. In a study to investigate the effect of therapy that uses ultrasound and stretching (**Trae Tashiro, Masters Thesis, University of Virginia, 2004**), passive knee extension was measured after treatment. Passive knee extension (in degrees) is given for each of 10 study participants.

59	46	64	49	56	70	45	52	63	52

Which would you choose to describe center and variability—the mean and standard deviation or the median and interquartile range? Justify your choice.

SECTION 3.2 Describing Center and Variability for Data Distributions That Are Approximately Symmetric

When the shape of a numerical data distribution is approximately symmetric, the mean and standard deviation are appropriate choices for describing center and variability.

Data set available

The Mean

The mean of a numerical data set is just the familiar arithmetic average—the sum of all the observations in the data set divided by the total number of observations. Before we introduce a formula for calculating the mean, it is helpful to have some simple notation for the variable of interest and for the data set.

Notation

x = the variable of interest
n = the number of observations in the data set (the sample size)
x_1 = the first observation in the data set
x_2 = the second observation in the data set
...
x_n = the nth (last) observation in the data set

For example, data consisting of the number of hours spent doing community service during the fall semester for $n = 10$ students might be

$x_1 = 25$	$x_2 = 4$	$x_3 = 0$	$x_4 = 90$	$x_5 = 0$
$x_6 = 10$	$x_7 = 20$	$x_8 = 0$	$x_9 = 16$	$x_{10} = 36$

Notice that the value of the subscript on x doesn't tell you anything about how small or large the data value is. In this example, x_1 is just the first observation in the data set and not necessarily the smallest observation. Similarly, x_n is the last observation but not necessarily the largest.

The sum of $x_1, x_2, \ldots, x_n$ is written $x_1 + x_2 + \ldots x_n$, but this can be shortened by using the Greek letter Σ, which represents "sum." In particular Σx is used to denote the sum of all of the x values in a data set.

DEFINITION

The **sample mean** is the arithmetic average of the values in a sample. It is denoted by the symbol $\bar{x}$ (pronounced as *x-bar*).

$$\bar{x} = \frac{\text{sum of all observations in the sample}}{\text{number of observations in the sample}} = \frac{\Sigma x}{n}$$

Data set available

Example 3.3 **Thinking About Exercise Again**

The example in the Chapter Preview described a study that investigated how thinking about exercise might affect food intake. Data on calories in snacks taken for people in each of three groups (control, read about fun walk, and read about strenuous exercise) are given in Table 3.2. Dot plots of these three data sets were given in the Chapter Preview and are reproduced here as Figure 3.4.

FIGURE 3.4
Dotplot of calories

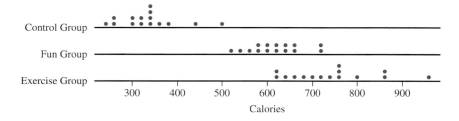

TABLE 3.2 Calories

Control Group	Fun Group	Exercise Group
340	668	626
300	585	802
329	637	768
381	588	738
331	529	751
445	597	715
320	719	670
256	612	630
252	622	862
342	553	761
332	600	648
296	658	956
357	542	671
505	711	854
242	641	703

Because each of the three data distributions is approximately symmetric, the mean is a reasonable choice for describing center. For the Control Group, the sum of the sample data values is

$$\Sigma x = 340 + 300 + \cdots + 242 = 5030$$

and the sample mean number of calories is

$$\bar{x} = \frac{\Sigma x}{n} = \frac{5030}{15} = 335.33$$

The value of the sample mean describes where number of calories for the Control Group is centered along the number line. It can be interpreted as a typical number of calories for people in the Control Group. For the Fun Group, the sample mean number of calories is

$$\bar{x} = \frac{\Sigma x}{n} = \frac{9261}{15} = 617.40$$

and for the Exercise Group the sample mean number of calories is

$$\bar{x} = \frac{\Sigma x}{n} = \frac{14,291}{15} = 952.73$$

Notice that the sample mean number of calories for the Control Group was much smaller than the sample means for the other two groups.

It is customary to use Roman letters to denote sample statistics, as we have done with the sample mean $\bar{x}$. Population characteristics (also known as parameters) are usually denoted by Greek letters. The population mean is denoted by the Greek letter μ.

DEFINITION

The **population mean** is denoted by μ. It is the arithmetic average of all the x values in an entire population.

For example, the mean fuel efficiency for *all* 600,000 cars of a certain make and model might be $\mu = 27.5$ miles per gallon. A particular sample of 5 cars might have efficiencies of 27.3, 26.2, 28.4, 27.9, and 26.5, which results in $\bar{x} = 27.26$ (somewhat smaller than μ). However, a second sample of 5 cars might result in $\bar{x} = 28.52$, a third $\bar{x} = 26.85$, and so on. The value of $\bar{x}$ varies from sample to sample, whereas there is just one value of μ. In later chapters, you will see how the value of $\bar{x}$ from a particular sample can be used to draw conclusions about the value of μ.

Measuring Variability

Reporting a measure of center gives only partial information about a data set. It is also important to describe how much the observations in the data set differ from one another. For example, consider the three sets of six exam scores displayed in Figure 3.5. Each of these data sets has a mean of 75, but as you go from Set 1 to Set 2 to Set 3, they decrease in variability.

FIGURE 3.5
Three samples with the same mean but different amounts of variability

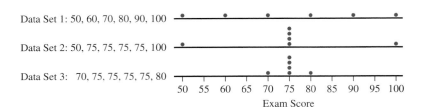

A simple numerical measure of variability is the **range**, which is just the difference between the largest and smallest observations in the data set. While the range is easy to calculate, it is not a very good measure of variability. For the data sets shown in Figure 3.5, both Data Set 1 and Data Set 2 have a range of $100 - 50 = 50$, but Data Set 1 has more variability than Data Set 2. For this reason, measures of variability that are based on all of the observations in the data set (not just the two most extreme values) are usually preferred over the range.

Deviations from the Mean

Among the most widely used measures of variability are those based on how far each observation deviates from the sample mean. Subtracting $\bar{x}$ from each observation gives the set of **deviations from the mean**.

DEFINITION

There is a deviation from the mean for each observation x in a data set. To calculate a **deviation from the mean**, subtract the sample mean $\bar{x}$ from the observation to get $(x - \bar{x})$.

A deviation from the mean is positive if the corresponding observation is greater than $\bar{x}$ and negative if the observation is less than $\bar{x}$.

Data set available

Example 3.4   The Big Mac Index

McDonald's fast-food restaurants are found in many countries around the world. Table 3.3 shows the cost of a Big Mac for 12 European Union countries (converted to U.S. dollars) from the article **"Big Mac Index 2015" (www.bigmacindex.org, January 22, 2015, retrieved April 18, 2017).**

TABLE 3.3 Big Mac Prices for Twelve European Union Countries

Country	2015 Big Mac Price in U.S. Dollars
Austria	3.93
Belgium	4.29
Estonia	3.36
Finland	4.75
France	4.52
Germany	4.25
Greece	3.53
Ireland	4.04
Italy	4.46
Netherlands	4.00
Portugal	3.48
Spain	4.23

Notice that there is quite a bit of variability in the Big Mac prices. For this data set, $\Sigma x = 48.84$ and $\bar{x} = \$4.07$. Table 3.4 displays the data along with the corresponding deviations, formed by subtracting $\bar{x} = \$4.07$ from each observation. Six of the deviations are positive because six of the observations are larger than $\bar{x}$. Some of the deviations are quite large in magnitude (-0.71 and 0.68, for example), indicating observations that are far from the sample mean.

TABLE 3.4 Deviations from the Mean for the Big Mac Data

2015 Big Mac Price in U.S. Dollars	Deviation from the Mean ($x - \bar{x}$)
3.93	−0.14
4.29	0.22
3.36	−0.71
4.75	0.68
4.52	0.45
4.25	0.18
3.53	−0.54
4.04	−0.03
4.46	0.39
4.00	−0.07
3.48	−0.59
4.23	0.16

In general, when there is more variability in the sample, the observations will tend to fall farther away from the mean. This will be reflected in the deviations from the mean, and this is why the deviations from the mean can be combined to obtain an overall measure of variability. You might think that a reasonable way to combine the deviations from the mean into a measure of variability is to find the average deviation from the mean. The problem with this approach is that some of the deviations are positive and some are negative, and when the deviations are added together, the positive and negative deviations offset each other. In fact, except for small differences due to rounding, the sum of the deviations from the mean is always equal to 0. One way around this is to use a measure of variability that is based on the squared deviations.

Variance and Standard Deviation

When the data distribution is approximately symmetric, two widely used measures of variability are the variance and the standard deviation. Both of these measures of variability are based on the squared deviations from the mean.

DEFINITION

The **sample variance**, denoted by s^2, is the sum of the squared deviations from the mean divided by $n - 1$.

$$s^2 = \frac{\Sigma (x - \bar{x})^2}{n - 1}$$

The **sample standard deviation**, denoted by s, is the square root of the sample variance.

$$s = \sqrt{s^2} = \sqrt{\frac{\Sigma (x - \bar{x})^2}{n - 1}}$$

In calculating the value of the variance or the standard deviation, notice that the deviations from the mean are first squared and then added together. When the values in a data set are spread out, some will fall far from the mean. For these values, the squared deviations from the mean will be large, resulting in a large sample variance. This is why data distributions that are quite spread out will have a large variance and standard deviation.

People find the standard deviation to be a more natural measure of variability than the variance because it is expressed in the same units as the original data values. For example, if the observations in a data set represent the price of a Big Mac in dollars, the mean and the deviations from the mean are also in dollars. But when the deviations are squared and then combined to obtain the variance, the units are dollars squared—not something that is familiar to most people! This makes it difficult to interpret the value of the variance and to decide whether the variance is large or small. Taking the square root of the variance to obtain the standard deviation results in a measure of variability that is expressed in the same units as the original data values, making it easier to interpret.

Example 3.5   Big Mac Revisited

Let's use the Big Mac data and the deviations from the mean calculated in Example 3.4 to calculate the values of the sample variance and the sample standard deviation. Table 3.5 shows the original observations and the deviations from the mean, along with the squared deviations.

TABLE 3.5 Deviations and Squared Deviations for the Big Mac Data

2015 Big Mac Price in U.S. Dollars	Deviation from the Mean $(x - \bar{x})$	Squared Deviation from the Mean $(x - \bar{x})^2$
3.93	−0.14	0.0196
4.29	0.22	0.0484
3.36	−0.71	0.5041
4.75	0.68	0.4624
4.52	0.45	0.2025
4.25	0.18	0.0324
3.53	−0.54	0.2916
4.04	−0.03	0.0009
4.46	0.39	0.1521
4.00	−0.07	0.0049
3.48	−0.59	0.3481
4.23	0.16	0.0256
		$\Sigma(x-\bar{x})^2 = 2.0926$

Combining the squared deviations to calculate the values of s^2 and s gives

$$s^2 = \frac{\Sigma(x - \bar{x})^2}{n - 1} = \frac{2.0926}{12 - 1} = \frac{2.0926}{11} = 0.1902$$

$$s = \sqrt{0.1902} = 0.436$$

The computation of s can be a bit tedious, especially if the sample size is large. Fortunately, many calculators and computer software packages can easily compute the standard deviation. One commonly used statistical computer package is JMP. The output resulting from using the JMP Analyze command with the Big Mac data is shown here.

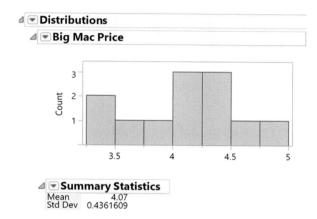

Distributions

Big Mac Price

Summary Statistics

Mean	4.07
Std Dev	0.4361609

The standard deviation can be informally interpreted as the size of a "typical" or "representative" deviation from the mean. In Example 3.5, a typical deviation from $\bar{x}$ is about 0.436. Some observations are closer to $\bar{x}$ than 0.436, and others are farther away. The standard deviation is often used to compare variability in data sets. For example, if the standard deviation of Big Mac prices for a larger group of 56 countries was $s = 1.153$, we would conclude that the original sample of European Union countries has much less variability than the data set consisting of all 56 countries.

The **population variance** and the **population standard deviation** are measures of variability for the entire population. They are denoted by σ^2 (sigma squared) and σ (sigma), respectively. (Again, lowercase Greek letters are used for population characteristics.)

Notation

s^2 sample variance

σ^2 population variance

s sample standard deviation

σ population standard deviation

In many statistical procedures, you would like to use the value of σ, but you don't usually know this value. In this case, the sample standard deviation s is used as an estimate of σ. Using $n - 1$ rather than n (which might seem to be a more natural choice) as the divisor in the calculation of the sample variance and standard deviation results in estimates that tend to be better estimates of the corresponding population values.

Putting It Together

There are three simple steps to complete when summarizing a numerical data distribution:

1. **Select** appropriate measures of center and variability. This will involve looking at the shape of the distribution.
2. **Calculate** the values of the selected measures.
3. **Interpret** the values in context.

Let's look at one more example that illustrates this process.

Data set available

Example 3.6	Thirsty Bats

The article **"How to Confuse Thirsty Bats"** (*Nature*, **November 11, 2010**) summarized a study that was published in the journal ***Nature Communications*** **("Innate Recognition of Water Bodies in Echolocating Bats," November 2, 2010)**. The article states

> Echolocating bats have a legendary ability to find prey in the dark—so you'd think they would be able to tell the difference between water and a sheet of metal. Not so, report Greif and Siemers in *Nature Communications*. They have found that bats identify any extended, echo-acoustically smooth surface as water, and will try to drink from it.

This conclusion was based on a study where bats were placed in a room that had two large plates on the floor. One plate was made of wood and had an irregular surface. The other plate was made of metal and had a smooth surface. The researchers found that the bats never attempted to drink from the irregular surface, but that they made repeated attempts to drink from the smooth, metal surface. The number of attempts to drink from the smooth metal surface for 11 bats are shown here:

$$66 \quad 144 \quad 13 \quad 26 \quad 94 \quad 163 \quad 8 \quad 125 \quad 1 \quad 64 \quad 56$$

These data will be used to select, calculate, and interpret appropriate summary measures of center and variability.

Step	Explanation
Select 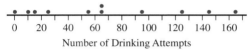 Number of Drinking Attempts The data distribution is approximately symmetric, so use the sample mean and sample standard deviation to describe center and variability.	Although there is a lot of variability from bat to bat in number of drinking attempts, the data distribution does not appear to be skewed, and there are no observations that seem to be outliers. So, the sample mean and sample standard deviation are appropriate for describing center and variability of this data set.
Calculate For these data $$\Sigma x = 760$$ $$\bar{x} = 69.09$$ $$\Sigma(x-\bar{x})^2 = 31{,}754.9$$ $$s^2 = \frac{\Sigma(x-\bar{x})^2}{n-1} = \frac{31{,}754.9}{10} = 3175.49$$ $$s = \sqrt{s^2} = \sqrt{3175.49} = 56.35$$	Calculation of the values of $\bar{x}$ and s can be done using a calculator with statistics functions, a computer software package, or by hand. If calculating by hand, first find the value of the sample mean, $\bar{x}$. Then use $\bar{x}$ to calculate the deviations from the mean and the sum of the squared deviations. Finally, use this sum to calculate the variance and the standard deviation.
Interpret On average, the bats in the sample made 69.09 attempts to drink from the smooth, metal surface. The sample standard deviation is 56.35. This is relatively large compared to the mean of the values in the data set, indicating a lot of variability from bat to bat in number of drinking attempts.	Always finish with an interpretation of the sample mean and sample standard deviation in context. The mean represents a typical or representative value for the data set, and the sample standard deviation describes how much the values in the data set vary around the mean. It can be informally interpreted as a representative distance from the mean.

Summing It Up—Section 3.2

The following learning objectives were addressed in this section:

Conceptual Understanding
C2: Understand how the variance and standard deviation describe variability in a data distribution.

The variance and standard deviation measure variability in a data set by considering variability around the mean, as measured by the sum of the squared deviations from the mean.

C4: Understand how the shape of a data distribution is considered when selecting numerical summary measures.

When a data distribution is approximately symmetric, an appropriate choice for a measure of center is the mean and an appropriate choice for a measure of variability is the standard deviation.

Mastering the Mechanics
M2: Calculate and interpret the value of the sample mean.

The sample mean is the arithmetic average of the values in the sample. For an example of how the mean is calculated and interpreted, see Example 3.3.

M3: Calculate and interpret the value of the sample standard deviation.

The sample standard deviation is a measure of variability in a data set, and it can be loosely interpreted as a typical deviation from the mean for the values in the sample. For an example of how the sample standard deviation is calculated and interpreted, see Example 3.5 and the discussion following that example.

Putting It into Practice
P1: Interpret measures of center in context.

For a data distribution that is approximately symmetric, the mean describes where the data distribution is centered along the number line and can be interpreted as a typical value for the data set.

P2: Interpret measures of variability in context.

The standard deviation describes how far observations in a data set tend to vary from the mean. For a data distribution that is approximately symmetric, the standard deviation can be loosely interpreted as a typical deviation from the mean.

SECTION 3.2   EXERCISES

Each Exercise Set assesses the following chapter learning objectives: C2, C4, M2, M3, P1, P2

SECTION 3.2   Exercise Set 1

3.12 The accompanying data are consistent with summary statistics in the paper **"Shape of Glass and Amount of Alcohol Poured: Comparative Study of Effect of Practice and Concentration"** (*British Medical Journal* [2005]: 1512–1514). The data are the actual amount (in ml) poured into a tall, slender glass for individuals asked to pour 1.5 ounces (44.3 ml). Calculate and interpret the values of the mean and standard deviation.

44.0	49.6	62.3	28.4	39.1	39.8	60.5	73.0
57.5	56.5	65.0	56.2	57.7	73.5	66.4	32.7
40.4	21.4						

3.13 The paper referenced in the previous exercise also gave data on the actual amount (in ml) poured into a short, wide glass for individuals asked to pour 1.5 ounces (44.3 ml).

89.2	68.6	32.7	37.4	39.6	46.8	66.1	79.2
66.3	52.1	47.3	64.4	53.7	63.2	46.4	63.0
92.4	57.8						

a. Calculate and interpret the values of the mean and standard deviation.
b. What do the values of the means for the two types of glasses (the mean for a tall, slender glass was calculated in Exercise 3.12) suggest about the shape of glasses used for alcoholic drinks?

3.14 The **2015 Urban Mobility Scorecard (Texas A&M Transportation Institute, mobility.tamu.edu/ums/report/, retrieved April 19, 2017)** included data on the estimated cost (in millions of dollars) resulting from traffic congestion for different urban areas. The following are the data for the 13 largest U.S. urban areas.

Urban Area	Total Cost (millions of dollars)
New York	15
Los Angeles	13
Chicago	7
Washington, D.C.	5
Houston	5
Dallas, Fort Worth	4
Detroit	4
Miami	4
Phoenix	4
Philadelphia	4
San Francisco	3
Boston	3
Atlanta	3

 Data set available

a. Calculate the mean and standard deviation for this data set.

b. Delete the observations for New York and Los Angeles and recalculate the mean and standard deviation. Compare this mean and standard deviation to the values calculated in Part (a). What does this suggest about using the mean and standard deviation as measures of center and variability for a data set with outliers?

3.15 Children going back to school can be expensive for parents—second only to the Christmas holiday season in terms of spending. **Forbes.com (August 12, 2015)** estimated that in 2015, parents had an average of $1269 in back-to-school expenses. However, not every parent spent the same amount of money. Imagine a data set consisting of the amount spent on back-to-school items for each student at a particular elementary school. Do you think that the standard deviation for this data set would be closer to $3 or closer to $30? Explain.

SECTION 3.2 Exercise Set 2

3.16 The U.S. Department of Transportation reported the number of speed-related crash fatalities for the 15 states that had the highest number of these fatalities in 2012 (*2012 Speeding Traffic Safety Facts*, May 2014).

a. Calculate and interpret the mean and standard deviation for this data set.

State	Speed-Related Traffic Fatalities
Texas	1,247
California	916
Pennsylvania	614
North Carolina	440
Illinois	387
Florida	361
New York	360
Ohio	356
Missouri	326
South Carolina	316
Arizona	297
Alabama	272
Virginia	271
Michigan	250
Oklahoma	218

b. Explain why it is not reasonable to generalize from this sample of 15 states to the population of all 50 states.

3.17 Cost per serving (in cents) for 15 high-fiber cereals rated very good or good by *Consumer Reports* are shown below.

46 49 62 41 19 77 71 30 53 53 67 43 48 28 54

Calculate and interpret the mean and standard deviation for this data set.

Data set available

3.18 The accompanying data are a subset of data read from a graph in the paper **"Ladies First? A Field Study of Discrimination in Coffee Shops"** (*Applied Economics* **[April, 2008]**). The data are the waiting times (in seconds) between ordering and receiving coffee for 19 female customers at a Boston coffee shop.

60   80   80   100   100   100   120   120
120   140   140   150   160   180   200   200
220   240   380

a. Calculate the mean and standard deviation for this data set.

b. Delete the observation of 380 and recalculate the mean and standard deviation. How do these values compare to the values calculated in Part (a)? What does this suggest about using the mean and standard deviation as measures of center and variability for a data set with outliers?

3.19 Morningstar is an investment research firm that publishes some online educational materials. The materials for an online course called **"Looking at Historical Risk"** (news.morningstar.com/classroom2/course .asp?docId =2927&page=2&CN=com, retrieved August 3, 2016) included the following paragraph referring to annual return (in percent) for investment funds:

> Using standard deviation as a measure of risk can have its drawbacks. It's possible to own a fund with a low standard deviation and still lose money. In reality, that's rare. Funds with modest standard deviations tend to lose less money over short time frames than those with high standard deviations. For example, the one-year average standard deviation among ultrashort-term bond funds, which are among the lowest-risk funds around (other than money market funds), is a mere 0.64%.

a. Explain why the standard deviation of percent return is a reasonable measure of unpredictability and why a smaller standard deviation for a funds percent return means less risk.

b. Explain how a fund with a small standard deviation can still lose money. (Hint: Think about the average percent return.)

ADDITIONAL EXERCISES

3.20 The article **"Caffeinated Energy Drinks—A Growing Problem"** (*Drug and Alcohol Dependence* **[2009]: 1–10**) gave the accompanying data (on the next page) on caffeine concentration (mg/ounce) for eight top-selling energy drinks.

a. What is the mean caffeine concentration for this set of energy drinks?

b. Coca-Cola has 2.9 mg/ounce of caffeine and Pepsi Cola has 3.2 mg/ounce of caffeine. Write a sentence explaining how the caffeine concentration of top-selling energy drinks compares to that of these colas.

Energy Drink	Caffeine Concentration (mg/ounce)
Red Bull	9.6
Monster	10.0
Rockstar	10.0
Full Throttle	9.0
No Fear	10.9
Amp	8.9
SoBe Adrenaline Rush	9.5
Tab Energy	9.1

3.21 Acrylamide, a possible cancer-causing substance, forms in high-carbohydrate foods cooked at high temperatures. Acrylamide levels can vary widely even within the same type of food. An article appearing in the journal *Food Chemistry* **(March 2014, 204–211)** included the following acrylamide content (in nanograms/gram) for five brands of bisquits:

$$345 \quad 292 \quad 334 \quad 276 \quad 248$$

a. Calculate the mean acrylamide level. For each data value, calculate the deviation from the mean.
b. Verify that, except for the effect of rounding, the sum of the five deviations from the mean is equal to 0 for this data set. (If you rounded the sample mean or the deviations, your sum may not be exactly zero, but it should still be close to zero.)
c. Use the deviations from Part (a) to calculate the variance and standard deviation for this data set.

3.22 Although bats are not known for their eyesight, they are able to locate prey (mainly insects) by emitting high-pitched sounds and listening for echoes. A paper appearing in *Animal Behaviour* **("The Echolocation of Flying Insects by Bats" [1960]: 141–154)** gave the following distances (in centimeters) at which a bat first detected a nearby insect:

$$62 \quad 23 \quad 27 \quad 56 \quad 52 \quad 34 \quad 42 \quad 40 \quad 68 \quad 45 \quad 83$$

a. Calculate and interpret the mean distance at which the bat first detects an insect.
b. Calculate the sample variance and standard deviation for this data set. Interpret these values.

3.23 For the data in Exercise 3.22, subtract 10 from each sample observation. For the new set of values, calculate the mean and all the deviations from the mean. How do these deviations compare to the deviations from the mean for the original sample? How will the new s^2 compare to s^2 for the old values? In general, what effect does subtracting the same number from (or adding the same number to) every value have on s^2 and s? Explain.

3.24 For the data of Exercise 3.22, multiply each data value by 10, then calculate the standard deviation. How does this value compare to s for the original data? More generally, what happens to s if each observation is multiplied by the same positive constant c?

SECTION 3.3 # Describing Center and Variability for Data Distributions That Are Skewed or Have Outliers

In this section, you will learn about measures of center and variability that are more appropriate when the data distribution is noticeably skewed or when there are outliers (unusual observations).

Describing Center

One potential drawback to the mean as a measure of center is that if the data distribution is skewed or has outliers, its value can be greatly affected. Consider the following example.

Data set available

Example 3.7	Number of Visits to a Class Web Site

Forty students were enrolled in a statistical reasoning course at a California college. The instructor made course materials, grades, and lecture notes available to students on a class web site, and course management software kept track of how often each student accessed any of these web pages. One month after the course began, the instructor requested a report on how many times each student had accessed a class web page. The 40 observations were:

20	37	4	20	0	84	14	36	5	331	19	0
0	22	3	13	14	36	4	0	18	8	0	26
4	0	5	23	19	7	12	8	13	16	21	7
13	12	8	42								

The sample mean for this data set is $\bar{x} = 23.10$. Figure 3.6 is a dotplot of the data. Many would argue that 23.10 is not a very representative value for this sample, because only 7 of 40 observations are larger than 23.10. The two outlying values of 84 and 331 (no, that was not a typo!) have a big impact on the value of $\bar{x}$.

FIGURE 3.6
Dotplot of the web site visit data of Example 3.7

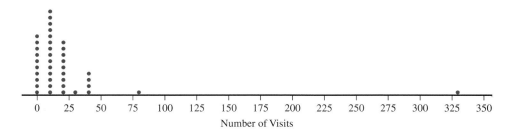

Number of Visits

Because the data distribution is skewed and there are outliers, the mean is not the best choice for describing center. A better choice is a measure called the median.

The Median

The median strip of a highway divides the highway in half. The median of a numerical data set works the same way. Once the data values have been listed in order from smallest to largest, the **median** is the middle value in the list, dividing it into two equal parts.

The process of determining the median of a sample is slightly different depending on whether the sample size n is even or odd. When n is an odd number, the sample median is the single middle value. But when n is even, there are two middle values in the ordered list, and you average these two middle values to obtain the sample median.

DEFINITION

The **sample median** is obtained by first ordering the n observations from smallest to largest (with any repeated values included, so that every sample observation appears in the ordered list). Then

$$\text{sample median} = \begin{cases} \text{the single middle value} & \text{if } n \text{ is odd} \\ \text{the average of the two middle values} & \text{if } n \text{ is even} \end{cases}$$

Data set available

Example 3.8   Web Site Data Revisited

The sample size for the web site visit data of Example 3.7 was $n = 40$, an even number. The median is the average of the middle two values in the ordered data set (the 20th and 21st values in the ordered list of the data). Arranging the data from smallest to largest produces the following list (with the two middle values highlighted).

0	0	0	0	0	0	3	4	4	4	5	5
7	7	8	8	8	12	12	**13**	**13**	13	14	14
16	18	19	19	20	20	21	22	23	26	36	36
37	42	84	331								

The median can now be calculated.

$$median = \frac{13 + 13}{2} = 13$$

Usually, half of the values in a data set are smaller than the median and half are greater. This isn't quite the case here because the value 13 occurs three times. Even so, it is common to interpret the median as the value that divides the data set in half. With a median of 13, you would say that half of the students in the class visited the web site fewer than 13 times, and half of the students visited the web site more than 13 times. Looking at the dotplot (Figure 3.6), you can see that 13 is more typical of the values in the data set than the sample mean of $\bar{x} = 23.10$.

The sample mean can be sensitive to even a single outlier. The median, on the other hand, is quite insensitive to outliers. For example, the largest observations in Example 3.8 can be increased by any amount without changing the value of the median. Similarly, a decrease in several of the smallest observations would not affect the value of the median.

Because the median is not sensitive to unusual observations in a data set, it is considered to be a better choice for describing a typical value for data distributions with unusual observations and also for data distributions that are skewed. This is why the median is often what is reported when describing salary distributions or house price distributions, which often have a long upper tail. An interesting discussion of the use of the median to describe a typical value appeared in the article **"$115K! The 13 Best Paying U.S. Companies"** (***USA TODAY*, August 11, 2015**). This article reported that the median salary for employees of the 13 companies with the highest salaries (which included companies like Jupiter Networks, Netflix, Yahoo, Microsoft, and eBay) was $115,068. They go on to say

> Getting paid $115,000 is certainly not the norm. The median pay—or the amount where half the workers earn more and half earn less—in the S&P 500 on average is $68,000 a year among the 459 companies in the S&P 500 for which Glassdoor has valid pay statistics.

Median salary was used for these comparisons because the mean salary, which would include the very high salaries of the company CEO and upper management, would not be representative of a typical salary.

Measuring Variability—The Interquartile Range

The value of s can also be greatly affected by the presence of even one outlier. A better choice for describing variability when the data distribution is noticeably skewed or has outliers is the *interquartile range*, a measure of variability that is resistant to the effects of outliers. The interquartile range (often abbreviated as *iqr*) is based on quantities called *quartiles*. The *lower quartile* separates the smallest 25% of the observations from the largest 75%, and the *upper quartile* separates the largest 25% from the smallest 75%. The *middle quartile* is the median, and it separates the smallest 50% from the largest 50%. Figure 3.7 illustrates the locations of these quartiles for a smoothed histogram.

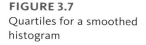

FIGURE 3.7
Quartiles for a smoothed histogram

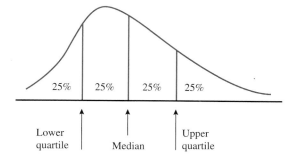

The quartiles for sample data are obtained by dividing the *n* ordered observations into a lower half and an upper half. If *n* is odd, the median is excluded from both halves. The lower and upper quartiles are then the medians of the two halves. (The median is only temporarily excluded when calculating quartiles. It is not excluded from the data set.)

lower quartile = median of the lower half of the data set

upper quartile = median of the upper half of the data set

If n is odd, the median of the entire data set is excluded from both halves when calculating quartiles.

The **interquartile range (iqr)** is defined by

iqr = upper quartile − lower quartile

Note: There are several other sensible ways to define quartiles. Some calculators and software packages use alternative definitions.

The standard deviation measures variability in terms of how far the observations in a data set tend to fall from the center of the data set (the mean). The interquartile range also measures variability in a data set, but it does so by looking at how spread out the middle half of the data set is. If the interquartile range is small, the values that make up the middle half of the data set are tightly clustered, indicating less variability. On the other hand, if the interquartile range is large, the values in the middle half of the data set are more spread out. By focusing on the middle half of the data set rather than all of the data values, the value of the interquartile range is not influenced by extreme values.

Data set available

Example 3.9 Web Site Visits One More Time

In Example 3.8, the median was used to describe the center of the web site visit data distribution. The interquartile range can be used to describe variability—how much the number of web site visits vary from student to student. You start by dividing the ordered data list from Example 3.8 into two parts at the median. The lower half of the data set consists of the 20 smallest observations and the upper half consists of the 20 largest observations, as shown below.

Lower Half

0 0 0 0 0 0 3 4 4 **4**

5 5 7 7 8 8 8 12 12 13

Upper Half

13 13 14 14 16 18 19 19 20 **20**

21 22 23 26 36 36 37 42 84 331

The lower quartile is the median of the lower half. There are 20 observations (an even number) in the lower half, so the median is the average of the two middle observations (highlighted in the preceding data set):

$$\text{lower quartile} = \frac{4 + 5}{2} = 4.5$$

The upper quartile is found in the same way, using the upper half of the data set:

$$\text{upper quartile} = \frac{20 + 21}{2} = 20.5$$

The interquartile range is the difference between the upper quartile and the lower quartile:

$$\text{iqr} = 20.5 - 4.5 = 16.0$$

Because the lower quartile separates the smallest 25% of the data from the rest, you can say that 25% of the students visited the web site fewer than 4.5 times. You can also say that 50% of the students visited the web site between 4.5 and 20.5 times, and 25% visited more than

20.5 times. An iqr of 16.0 tells you there was quite a bit of variability in the data set, but not nearly as much as suggested by the value of the standard deviation $s = 52.33$, which was influenced by the two outliers.

Data set available

Example 3.10   Some Strange Things Can Happen with Quartiles…

Example 3.1 gave data on the number of medication errors made by 24 medical residents who were classified as depressed (**"Rates of Medication Errors Among Depressed and Burnt Out Residents," British Medical Journal [2008]: 488**). This data set is interesting because some data values appear multiple times. For example, there are 17 observations that all have the value 0.

Because the distribution is skewed and has outliers, the median and interquartile range would be an appropriate choice for describing center and variability. The ordered data values are

0 0 0 0 0 0 0 0 0 0 0 0 0 0 0 0 0 1 1 1 2 3 5 11

The median is the average of the two middle values. Here, both middle values are 0, and so the median is also 0.

Dividing the data set into the lower and upper halves and then finding the median of each half leads to the following quartiles.

Lower half

0 0 0 0 0 0 0 0 0 0 0 0

lower quartile = 0

Upper half

0 0 0 0 0 1 1 1 2 3 5 11

upper quartile = 1

You can now calculate the value of the interquartile range.

$$\text{iqr} = \text{upper quartile} - \text{lower quartile} = 1 - 0 = 1$$

Notice here that the lower quartile and the median are both 0. This is because more than half of the observations in the data set have a value of 0. For this data set, the smallest 25% are all 0, and even the next 25% in the ordered list are all 0. One way to see this is to take another look at the ordered data list:

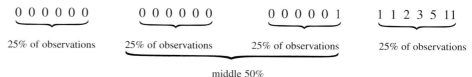

The small value of the interquartile range indicates that there is relatively little variability in the middle half of the data distribution.

Putting It Together

When describing center and variability of a data distribution that is skewed or has outliers, the following three steps can be used.

1. **Select** appropriate measures of center and variability. This will involve taking the shape of the data distribution into account.
2. **Calculate** the values of the selected measures.
3. **Interpret** the values in context.

These are the same steps that were used when describing center and variability of a data distribution that is approximately symmetric. The key difference is that in this case, you choose the median and interquartile range as the appropriate measures of center and variability in the first step.

Let's look at one more example that illustrates this process.

Data set available

Example 3.11   **Educational Attainment by State**

Data on the 2014 percentage of the adult population that have a bachelor's degree or higher for each of the 50 states and the District of Columbia are shown below. The data for the states appeared in the article **"America's Most and Least Educated States: A Survey of All 50 States" (24/7 Wall Street, September 23, 2015)** and the data for Washington, D.C. appeared in the **Wall Street Journal (January 20, 2016)**. The 51 data values are shown below.

```
23.5  28.0  27.6  21.4  31.7  38.3  38.0  30.6  27.3  29.1  31.0  25.0  32.8  24.7
27.7  31.5  22.2  22.9  29.4  38.2  41.2  27.4  34.3  21.1  27.5  29.3  29.5  23.1
35.0  37.4  26.4  34.5  28.7  27.4  26.6  24.2  30.8  29.0  30.4  26.3  27.8  25.3
27.8  31.1  34.9  36.7  33.1  49.3  19.2  28.4  26.6
```

These data will be used to select, calculate, and interpret appropriate summary measures of center and variability.

Step	Explanation
Select	To select the most appropriate measures of center and variability. The best place to start is with a graph of the data. From the histogram, you can see that the distribution is skewed, and that there is an outlier. It would be best to use measures of center and variability that are not heavily influenced by extreme values.

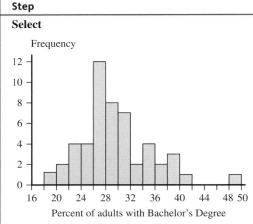

The data distribution is skewed, and there is an outlier. Use the median and interquartile range to describe center and variability.

Calculate	
To calculate the quartiles and the interquartile range, first order the data and use the median to divide the data into a lower half and an upper half. Because there are an odd number of observations ($n = 51$), the median is excluded from both the upper and lower halves when calculating the quartiles.	Calculation of the median, quartiles, and interquartile range can be done using a calculator with statistics functions, using a computer package, or by hand. If calculating by hand, you would arrange the data in order from smallest to largest and then find the median, the quartiles and the interquartile range, as shown to the left.
Ordered Data	An alternative to hand calculation would be to use a software package such as JMP. JMP produces the accompanying output:
Lower Half:	

```
19.2  21.1  21.4  22.2  22.9  23.1
23.5  24.2  24.7  25.0  25.3  26.3
26.4  26.6  26.6  27.3  27.4  27.4
27.5  27.6  27.7  27.8
27.8  28.0  28.4
```

(continued)

Step	Explanation
Median: 28.7 **Upper Half:** 29.0 29.1 29.3 29.4 29.5 30.4 30.6 30.8 31.0 31.1 31.5 31.7 **32.8** 33.1 34.3 34.5 34.9 35.0 36.7 37.4 38.0 38.2 38.3 41.2 49.3	◢ ▾ **Distributions** ◢ ▾ **Percent with Bachelor's Degree** ◢ **Quantiles** | 100.0%	maximum	49.3
99.5%		49.3
97.5%		46.87
90.0%		37.88
75.0%	quartile	32.8
50.0%	median	28.7
25.0%	quartile	26.4
10.0%		22.94
2.5%		19.77
0.5%		19.2
0.0%	minimum	19.2	
Each half of the sample contains 25 observations. The lower quartile, 26.4, is the median of the lower half. The upper quartile, 32.8, is the median of the upper half. Then $iqr = 32.8 - 26.4 = 6.4$	Notice that the median and quartiles agree with the calculations shown to the left. In the JMP output, the upper quartile is the one labeled 75% (because 75% of the data values are below the upper quartile) and the lower quartile is the one labeled 25%.
Interpret The median for this data set is 28.7. For half of the states, the percentage of the population with a bachelor's degree or higher is less than 28.7%. For the other half, more than 28.7% of the population have a bachelor's degree or higher. The interquartile range of 6.4 indicates that the percentages for the states in the middle half were spread out over an interval of 6.4 percentage points.	Always finish with an interpretation of the sample median and sample interquartile range in context. The median represents a typical or representative value for the data set in the sense that half of the values in the data set are smaller and half are larger. The sample interquartile range describes how much the values in the middle half of the data set spread out.

The region with the largest percentage, at 49.3%, was the District of Columbia. The second highest, at 41.2%, was Massachusetts, and the smallest, at 19.2%, was West Virginia.

One final note. When a data distribution is approximately symmetric, we have recommended using the mean and standard deviation to describe center and variability because these are the most commonly used summary measures. However, the median and interquartile range can also be used for approximately symmetric data distributions. In fact, when the data distribution is approximately symmetric, the values of the mean and median will be similar and the interquartile range still describes variability. But it may be a misleading to use the mean and standard deviation to describe center and variability for data distributions that are very skewed or that have notable outliers. Because the values of the mean and standard deviation can be greatly affected by extreme values, it can be misleading to use them in these situations.

Summing It Up—Section 3.3

The following learning objectives were addressed in this section:

Conceptual Understanding

C3: Understand the impact that outliers can have on measures of center and variability.
The values of both the mean and the standard deviation are influenced by extreme values in the data set. This is why the median and interquartile range are usually used to describe center and variability for data distributions that have outliers.

C4: Understand how the shape of a data distribution is considered when selecting numerical summary measures.
When a data distribution is skewed or has outliers, an appropriate choice for a measure of center is the median and an appropriate choice for a measure of variability is the interquartile range.

C5: Understand how the relative values of the mean and median may be related to the shape of a data distribution.
If a data distribution is skewed or has outliers, the value of the mean is pulled toward the extreme values compared to the median, which is not generally affected by a few unusual values in a data set. For distributions that are approximately symmetric, the values of the mean and median tend to be similar.

Mastering the Mechanics
M4: Calculate and interpret the value of the sample median.
The sample median is the middle value (or average of the middle two values if there is an even number of data values) when the data have been arranged from smallest to largest. Half of the data values are less than or equal to the median and half are greater than or equal to the median. For an example of how the median is calculated and interpreted, see Example 3.8.

M5: Calculate and interpret the value of the sample interquartile range.
The interquartile range (iqr) is a measure of variability in a data set, and it can be interpreted as the length of the interval that captures the middle half of the data distribution. For an example of how the interquartile range is calculated and interpreted, see Example 3.9.

Putting It into Practice
P1: Interpret measures of center in context.
The median describes a typical value using a value for which half of the data are less than and half of the data are greater than this value. It is an appropriate choice for describing center for a data distribution that is skewed or has outliers.

P2 Interpret measures of variability in context.
The interquartile range (iqr) describes how spread out the middle half of the data distribution is. It is an appropriate choice for describing variability in a data distribution that is skewed or has outliers.

SECTION 3.3 EXERCISES

Each Exercise Set assesses the following chapter learning objectives: C3, C4, C5, M4, M5, P1, P2

SECTION 3.3 Exercise Set 1

3.25 The report titled **"State of the News Media 2013" (Pew Research Center, May 7, 2013)** included the weekday circulation numbers for the top 20 newspapers in the country. Here are the data for the 6 months ending September 2012:

2,293798	1,713,833	1,613,866	641,369	535,875
529,999	522,868	462,228	432,455	412,669
411,960	410,130	392,989	325,814	313,003
311,504	300,277	296,427	293,139	285,088

a. Calculate and interpret the value of the median of this data set.
b. Explain why the median is preferable to the mean for describing center for this data set.
c. Explain why it would be unreasonable to generalize from this sample of 20 newspapers to the population of all daily newspapers in the United States.

3.26 The paper **"Can Pizza Fit in to the Renal Diet? A Review of the Phosphorus, Potassium and Sodium Content of Selected Frozen and Delivery Options"** (*Journal of Renal Nutrition* [2015]: e15–e18) gave information on the sodium content (in mg per slice) for different types of pizzas. Data on sodium content for 13 different brands of cheese pizzas are shown here.

Sodium Content (mg/slice)

565	440	404	441	218	676	459
658	624	541	708	650	677	

Calculate and interpret the values of the quartiles and the interquartile range. (Hint: See Example 3.9.)

3.27 The accompanying data on total amount of time per day (in minutes) spent using a cell phone are consistent with summary statistics in the paper **"The Relationship Between Cell Phone Use and Academic Performance in a Sample of U.S. College Students"** (*SAGE Open* [2015]: 1–9).

225	318	468	0	236	601	144	196	374
0	424	198	156	734	331	502	0	492
563	195	237	110	516	422	740		

Calculate and interpret the values of the median and the interquartile range.

3.28 Data on tipping percent for 20 restaurant tables, consistent with summary statistics given in the paper **"Racial and**

Ethnic Differences in Tipping: The Role of Perceived Descriptive and Injunctive Tipping Norms" (*Restaurant Management* [2015]: 68–79), are:

14.0	16.5	15.4	13.3	27.6	1.2	5.2
21.8	19.1	17.3	17.2	11.8	10.2	8.5
12.8	19.5	19.2	18.8	14.8	14.9	

Calculate and interpret the values of the median and the interquartile range.

SECTION 3.3 Exercise Set 2

3.29 The **Insurance Institute for Highway Safety (www.iihs .org, June 11, 2009)** published data on repair costs for cars involved in different types of accidents. In one study, seven different 2009 models of mini- and micro-cars were driven at 6 mph straight into a fixed barrier. The following table gives the cost of repairing damage to the bumper for each of the seven models.

Model	Repair Cost
Smart Fortwo	$1,480
Chevrolet Aveo	$1,071
Mini Cooper	$2,291
Toyota Yaris	$1,688
Honda Fit	$1,124
Hyundai Accent	$3,476
Kia Rio	$3,701

a. Calculate and interpret the value of the median for this data set.
b. Explain why the median is preferable to the mean for describing center in this situation.

3.30 The **2015 Urban Mobility Scorecard (Texas A&M Transportation Institute, mobility.tamu.edu/ums/report/, retrieved April 19, 2017)** included data on the estimated cost (in millions of dollars) resulting from traffic congestion for different urban areas. The following data are for the 13 largest U.S. urban areas that had a population of over 3 million.

Urban Area	Total Cost (millions of dollars)
New York	15
Los Angeles	13
Chicago	7
Washington, D.C.	5
Houston	5
Dallas/Fort Worth	4
Detroit	4
Miami	4
Phoenix	4
Philadelphia	4
San Francisco	3
Boston	3
Atlanta	3

Calculate and interpret the values of the quartiles and the interquartile range.

3.31 Data on weekday exercise time for 20 males, consistent with summary quantities given in the paper **"An Ecological Momentary Assessment of the Physical Activity and Sedentary Behaviour Patterns of University Students"** (*Health Education Journal* [2010]: 116–125), are shown below. Calculate and interpret the values of the median and interquartile range.

Male—Weekday

43.5	91.5	7.5	0.0	0.0	28.5	199.5	57.0	142.5
8.0	9.0	36.0	0.0	78.0	34.5	0.0	57.0	151.5
8.0	0.0							

3.32 Data on weekday exercise time for 20 females, consistent with summary quantities given in the paper **"An Ecological Momentary Assessment of the Physical Activity and Sedentary Behaviour Patterns of University Students"** (*Health Education Journal* [2010]: 116–125), are shown below.

Female—Weekday

10.0	90.6	48.5	50.4	57.4	99.6
0.0	5.0	0.0	0.0	5.0	2.0
10.5	5.0	47.0	0.0	5.0	54.0
0.0	48.6				

a. Calculate and interpret the values of the median and interquartile range.
b. How do the values of the median and interquartile range for women compare to those for men calculated in the previous exercise?

ADDITIONAL EXERCISES

3.33 The article **"The Wedding Industry's Pricey Little Secret" (June 12, 2013, www.slate.com, retrieved April 19, 2017)** stated that the widely reported average wedding cost is grossly misleading. The article reports that in 2012, the average wedding cost was $27,427 and the median cost was $18,086.

a. What does the large difference between the mean cost and the median cost tell you about the distribution of wedding costs in 2012?
b. Do you agree that the average wedding cost is misleading? Explain why or why not.
c. The article also states "the proportion of couples who spent the 'average' or more was actually a minority." Do you agree with this statement? Explain why or why not using the reported values of the mean and median wedding cost.

3.34 The state of California defines family income groups in terms of median county income as follows:

Extremely low income: below 30% of county median income

Very low income: between 30% and 50% of county median income

Low income: between 50% and 80% of county median income

Moderate income: between 80% and 120% of county median income

For San Luis Obispo County, the median income as of May 24, 2016 for single person households was $53,950 (**www.slocounty.ca.gov/Assets/PL/Housing/AHS/AHS.pdf, August 1, 2016, retrieved April 19, 2017**).

a. Interpret the value of the median income for a single-person household in San Luis Obispo County.
b. Each of the following statements is incorrect. For each statement, use the given information to explain why it is incorrect.

Statement 1: 30% of the single-person households in San Luis Obispo County would be classified as extremely low income.
Statement 2: More than 50% of the single-person households in San Luis Obispo County would be classified as extremely low income or very low income.
Statement 3: There cannot be any single-person households in San Luis Obispo County that would be

classified as having an income that was greater than those in the moderate income category.

3.35 A sample of 26 offshore oil workers took part in a simulated escape exercise, resulting in the following data on time (in seconds) to complete the escape (**"Oxygen Consumption and Ventilation During Escape from an Offshore Platform,"** Ergonomics [1997]: 281–292):

389	356	359	363	375	424	325	394	402
373	373	370	364	366	364	325	339	393
392	369	374	359	356	403	334	397	

a. Construct a dotplot of the data. Will the mean or the median be larger for this data set? Explain your reasoning.
b. Calculate the values of the mean and median.
c. By how much could the largest time be increased without affecting the value of the sample median? By how much could this value be decreased without affecting the value of the median?

SECTION 3.4 Summarizing a Data Set: Boxplots

In Sections 3.2 and 3.3, you saw ways to describe center and variability of a data distribution. It is sometimes also helpful to have a method of summarizing data that gives more information than just a measure of center and a measure of variability. A boxplot is a compact display that provides information about the center, variability, and symmetry or skewness of a data distribution.

The construction of a boxplot is illustrated in the following example.

Data set available

Example 3.12 **Beating That High Score**

The authors of the paper **"Striatal Volume Predicts Level of Video Game Skill Acquisition"** (*Cerebral Cortex* [2010]: 2522–2530) studied a number of factors that affect performance in a complex video game. One factor was practice strategy. Forty college students who all reported playing video games less than 3 hours per week over the past two years and who had never played the game *Space Fortress* were assigned at random to one of two groups. Each person completed 20 two-hour practice sessions. Those in the fixed priority group were told to work on improving their total score at each practice session. Those in the variable priority group were told to focus on a different aspect of the game, such as improving speed score, in each practice session. The investigators were interested in whether practice strategy makes a difference. They measured the improvement in total score from the first practice session to the last.

Improvement scores (approximated from a graph in the paper) for the 20 people in each practice strategy group are given here.

Fixed Priority	1,100	1,900	2,000	2,400	2,600	2,600	2,550	3,500
Practice Group	3,500	3,500	3,700	3,800	3,900	4,000	4,000	4,100
	4,100	4,500	5,000	6,000				
Variable Priority	1,200	1,300	2,300	3,200	3,300	3,800	4,000	4,100
Practice Group	4,300	4,800	5,500	5,700	5,700	5,800	6,000	6,300
	6,800	6,800	6,900	7,700				

Let's begin by looking at the fixed priority improvement scores. From the accompanying dotplot of these data, you can locate the median and the lower and upper quartiles.

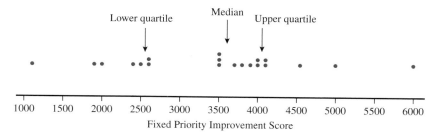

Notice that the lower quartile, the median, and the upper quartile divide this data set into four parts, each with five observations.

To construct a boxplot, first draw a box around the middle 50% of the data, as shown.

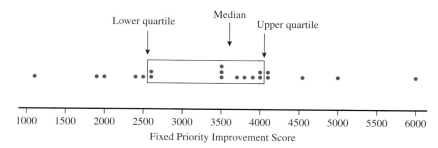

To complete the boxplot, draw a vertical line at the median. Then draw two lines—one that extends from the box to the smallest observation and one that extends from the box to the largest observation. These lines are called "whiskers."

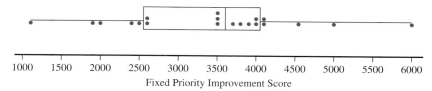

This is the boxplot for the fixed priority improvement scores. You can think of the boxplot as being made up of four parts—the lower whisker, the lower part of the box, the upper part of the box, and the upper whisker. Notice that each of these four parts represents the same number of observations (25% of the data values). If a part of the box is narrow or a whisker is short, it means the data values fall close together in this region. If a part of the box is wide or a whisker is long, then the data values are more spread out.

Starting with the dotplot illustrates the relationship between the parts of the boxplot and the data distribution. However, boxplots are usually constructed without a dotplot and the underlying data values (the dots). You will return to this example to construct a boxplot for the variable priority group improvement scores after looking at how boxplots are usually constructed.

To construct a boxplot without first drawing a dotplot, you need the following information: the smallest observation, the lower quartile, the median, the upper quartile, and the largest observation. This collection of measures is called the **five-number summary**.

DEFINITION

The **five-number summary** consists of the following:

1. Smallest observation in the data set (also called the Minimum)
2. Lower quartile
3. Median
4. Upper quartile
5. Largest observation in the data set (also called the Maximum)

A boxplot is a graph of the five-number summary. To construct a boxplot, follow the steps in the accompanying box.

Constructing a Boxplot

1. Calculate the values in the five-number summary.
2. Draw an axis and add an appropriate scale.
3. Draw a box above the line that extends from the lower quartile to the upper quartile.
4. Draw a line segment inside the box at the location of the median.
5. Draw two line segments, called whiskers, which extend from the box to the smallest observation and from the box to the largest observation.

Data set available

Example 3.13   **Video Game Practice Strategies**

Let's return to the video game improvement scores of Example 3.12. For the variable priority practice strategy, the ordered improvement scores are

| 1200 | 1300 | 2300 | 3200 | 3300 | 3800 | 4000 | 4100 | 4300 | 4800 |

| 5500 | 5700 | 5700 | 5800 | 6000 | 6300 | 6800 | 6800 | 6900 | 7700 |

The median is the average of the middle two observations, so

$$\text{median} = \frac{4800 + 5500}{2} = 5150$$

The lower half of the data set is

| 1200 | 1300 | 2300 | 3200 | 3300 | 3800 | 4000 | 4100 | 4300 | 4800 |

so the lower quartile is

$$\text{lower quartile} = \frac{3300 + 3800}{2} = 3550$$

The upper half of the data set is

| 5500 | 5700 | 5700 | 5800 | 6000 | 6300 | 6800 | 6800 | 6900 | 7700 |

so the upper quartile is

$$\text{upper quartile} = \frac{6000 + 6300}{2} = 6150$$

You now have what you need for the five-number summary and are ready to construct the boxplot.

Step	Explanation
The five-number summary smallest observation = 1200 lower quartile = 3550 median = 5150 upper quartile = 6150 largest observation = 7700	
Draw an axis and add an appropriate scale 1000 2000 3000 4000 5000 6000 7000 8000 Variable Priority Improvement Score	Since the values in the data set range from 1200 to 7700, a scale from 1000 to 8000 is a reasonable choice.

(continued)

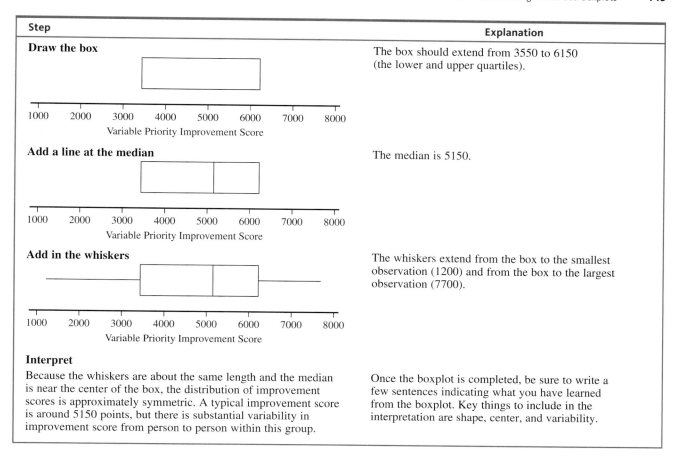

Step	Explanation
Draw the box	The box should extend from 3550 to 6150 (the lower and upper quartiles).
Add a line at the median	The median is 5150.
Add in the whiskers	The whiskers extend from the box to the smallest observation (1200) and from the box to the largest observation (7700).
Interpret Because the whiskers are about the same length and the median is near the center of the box, the distribution of improvement scores is approximately symmetric. A typical improvement score is around 5150 points, but there is substantial variability in improvement score from person to person within this group.	Once the boxplot is completed, be sure to write a few sentences indicating what you have learned from the boxplot. Key things to include in the interpretation are shape, center, and variability.

Boxplots can be tedious to construct by hand, especially when the data set is large. Most graphing calculators and statistical software packages will construct boxplots. Figure 3.8 shows a boxplot for the data of this example that was made using JMP. Boxplots can be displayed either vertically or horizontally, and JMP displays them vertically.

FIGURE 3.8
JMP boxplot for the variable priority group improvement scores

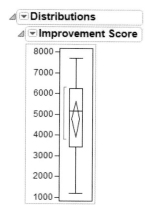

Using Boxplots for Comparing Groups

With two or more data sets consisting of observations on the same variable (for example, the video game improvement scores for the two different practice strategies described in Examples 3.12 and 3.13), a **comparative boxplot** is often used to compare the data sets.

A **comparative boxplot** is two or more boxplots drawn using the same numerical scale.

Example 3.14 **Comparing Practice Strategies**

The video game improvement scores from Examples 3.12 and 3.13 were used to construct the comparative boxplot shown in Figure 3.9.

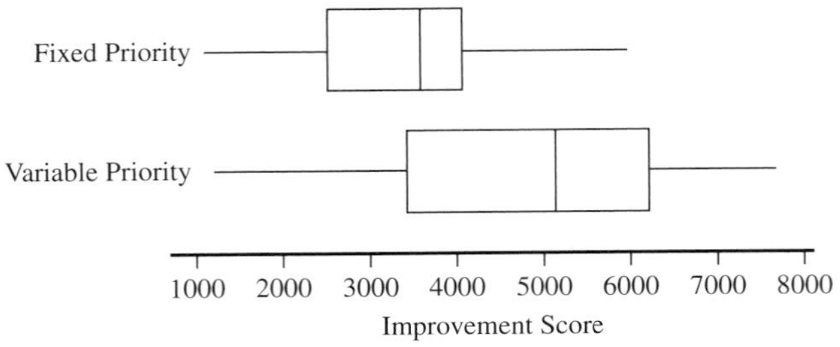

FIGURE 3.9
Comparative boxplot for improvement scores of fixed priority practice group and variable priority practice group.

FIGURE 3.9
Comparative boxplot for improvement scores of fixed priority practice group and variable priority practice group.

From the comparative boxplot, you can see that both data distributions are approximately symmetric. The improvement scores tend to be higher for the variable priority group. However, there is more consistency in the improvement scores for the fixed priority group. Improvement scores in the variable priority group are more spread out, indicating more variability in improvement scores.

Outliers and Modified Boxplots

Up to this point, the term outlier has been used informally to describe an observation that stands out as unusual. However, there is a formal definition of an outlier. The rules in the accompanying box are used to determine if an observation is "unusual enough" to officially qualify as an outlier.

DEFINITION

An observation is an **outlier** if it is

$$\textbf{greater than } \text{upper quartile} + 1.5(\text{iqr})$$

or

$$\textbf{less than } \text{lower quartile} - 1.5(\text{iqr})$$

A **modified boxplot** is a boxplot that shows outliers. The steps for constructing a modified boxplot are described in the following box. The first steps are the same as those for constructing a regular boxplot. The differences are in how the whiskers are added and in the identification of outliers.

Constructing a Modified Boxplot

1. Calculate the values in the five-number summary.
2. Draw an axis and add an appropriate scale.
3. Draw a box above the line that extends from the lower quartile to the upper quartile.
4. Draw a line segment inside the box at the location of the median.
5. Determine if there are any outliers in the data set.
6. Add a whisker to the plot that extends from the box to the smallest observation in the data set *that is not an outlier.* Add a second whisker that extends from the box to the largest observation *that is not an outlier.*
7. If there are outliers, add dots to the plot to indicate the positions of the outliers.

The following examples illustrate the construction of a modified boxplot and the use of modified boxplots to compare groups.

Example 3.15 Another Look at Big Mac Prices

Big Mac prices in U.S. dollars for 56 different countries were given in the article **"Big Mac Index 2015,"** first introduced in Example 3.4. The following 56 Big Mac prices are arranged in order from the lowest price (Ukraine at $1.20) to the highest (Switzerland at $7.54). The price for the United States was $4.79.

1.20	1.36	1.89	2.11	2.22	2.24	2.30	2.43	2.48	2.51
2.53	2.65	2.77	**2.81**	**2.92**	2.93	2.98	3.04	3.14	3.17
3.25	3.32	3.34	3.35	3.35	3.36	3.48	**3.53**	**3.53**	3.54
3.67	3.78	3.93	3.96	4.00	4.01	4.04	4.23	4.25	4.26
4.29	**4.32**	**4.37**	4.45	4.46	4.49	4.52	4.63	4.64	4.75
4.79	4.97	5.21	5.38	6.30	7.54				

The Big Mac price data will be used to construct a modified boxplot.

Step	Explanation
The five-number summary smallest observation $= 1.20$ lower quartile $= \dfrac{2.81 + 2.92}{2} = 2.865$ median $= \dfrac{3.53 + 3.53}{2} = 4.345$ upper quartile $= \dfrac{4.32 + 4.37}{2} = 4.345$ largest observation $= 7.54$	The median and quartiles can be calculated using the highlighted information in the ordered data list.
Draw a horizontal line and add an appropriate scale 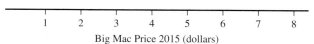	Since the values in the data set range from 1.20 to 7.54, a scale from 1 to 8 is a reasonable choice.
Draw the box 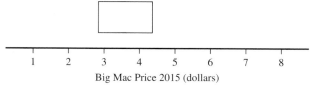	The box should extend from 2.865 to 4.345 (the lower and upper quartiles).
Add a line at the median 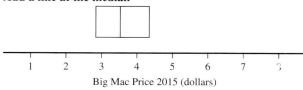	The median is 3.530.

(continued)

Step	Explanation
Determine if there are outliers iqr = 4.345 − 2.865 = 1.480 1.5(iqr) = 1.5(1.480) = 2.220 lower quartile − 1.5(iqr) = 2.865 − 2.220 = 0.645 upper quartile + 1.5(iqr) = 4.345 + 2.220 = 6.565 There are no outliers on the low end because no country had a Big Mac price less than \$0.645. On the high end, one country had Big Mac price greater than \$6.565, so 7.54 is an outlier.	To decide if there are outliers, you need to calculate the two outlier boundaries. In this example, the two boundaries are 0.645 and 6.565, so to identify outliers you are looking for values in the data set that are less than 0.645 or that are greater than 6.565.
Add in the whiskers 	The whiskers extend from the box to the smallest observation that is not an outlier and from the box to the largest observation that is not an outlier. There are no outliers on the low side, so the whisker extends to the smallest observation, which is 1.20. The largest observation in the data set that is not an outlier is 6.30, so the whisker on the high end extends from the box to 6.30.
Add the outliers 	One dot is added to the plot to indicate the positions of the outlier 7.54.
Interpret The most noticeable aspect of the boxplot is the presence of the outlier, indicating that the price of a Big Mac in Switzerland is much greater than the rest. Because the whiskers are about the same length and the median is near the center of the box, the rest of the data distribution (excluding Switzerland) is approximately symmetric. A typical Big Mac price is around \$3.53, but there is substantial variability in Big Mac price from country to country. The price of a Big Mac in the United States was listed as \$4.79, which is above the upper quartile, indicating that 75% or more of the 56 countries had lower prices than the United States.	Once the boxplot is completed, be sure to write a few sentences indicating what you have learned from the boxplot. Key things to include in the interpretation are shape, center, variability, and outliers.

Data set available

Example 3.16 **An Odd Looking Boxplot...**

Example 3.10 gave the following data on the number of medication errors made by 24 medical residents who were classified as depressed.

Number of Errors

0 0 0 0 0 0 0 0 0 0 0 0 0 0 0

0 0 1 1 1 2 3 5 11

Figure 3.10 shows a modified boxplot constructed using these data.

FIGURE 3.10
Boxplot of the medication error data

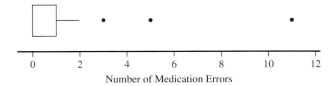

Three outliers are clearly shown in the boxplot, corresponding to residents that made 3, 5, and 11 medication errors. The boxplot looks a little odd. It appears that the median line and the lower whisker have been left out. But there were 17 observations in this data set that all had the value 0. As a result, the smallest value, the lower quartile, and the median are all equal to 0. The median line is actually at the left edge of the box, and there is no lower whisker shown.

Example 3.17 **NBA Player Salaries**

The web site **HoopsHype** (**hoopshype.com/salaries**) publishes salaries of NBA players. The 2016–2017 salaries of players on three teams were used to construct the comparative boxplot shown in Figure 3.11.

FIGURE 3.11
Comparative boxplot for salaries of three NBA teams

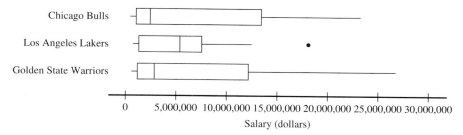

The comparative boxplot reveals some interesting similarities and differences in the salary distributions of the three teams. The minimum salary was lower for the Warriors, but the minimum salary was not that different for the three teams. The median salary was greater for the Lakers, but the other two teams both had players with salaries that were much greater than the highest Laker salary. The Lakers were also the only team with an outlier salary, indicating that there was one player on this team that had a salary that was noticeably higher than the other players on the team. Salaries are more variable from player to player for the Warriors and for the Bulls than for the Lakers. In fact, the upper whisker for the Lakers ends at a value that is about equal to the upper quartile for the other two teams, indicating that about 25% of the players on the Bulls and the Warriors have salaries that are greater than at most one of the Lakers (Kobe Bryant).

Summing It Up—Section 3.4

The following learning objectives were addressed in this section:

Mastering the Mechanics

M6: Calculate and interpret the values in the five-number summary.
The five-number summary for a data set consists of the values of the minimum, the lower quartile, the median, the upper quartile, and the maximum. These five numbers divide a data set into quarters and are used to construct a boxplot. For an example of calculating the numbers in the five-number summary, see Example 3.13.

M7: Given a numerical data set, construct a boxplot.
A boxplot uses the numbers in the five-number summary to create a graph that provides information about center, variability, and symmetry or skewness in a data distribution. Modified boxplots also show outliers. For the steps in constructing a simple boxplot, see the box just prior to Example 3.13. For an example that illustrates the identification of outliers and construction of a boxplot that shows outliers, see Example 3.15 and the discussion just prior to that example.

M8: Identify outliers in a numerical data set.
For a univariate numerical data set, outliers are unusually small or unusually large data values. A data value is considered an outlier if it is more than 1.5 times the iqr away from the nearest quartile. For an example of identifying outliers in a data set, see Example 3.15.

Putting It into Practice

P3: Use boxplots to make comparisons between two or more groups.
A comparative boxplot can be used to compare two or more groups. A comparative boxplot is two or more boxplots drawn using the same numerical scale. For an example of using boxplots to compare groups, see Example 3.14.

SECTION 3.4   EXERCISES

Each Exercise Set assesses the following chapter learning objectives: M6, M7, M8, P3

SECTION 3.4   Exercise Set 1

3.36   The **U.S. Environmental Protection Agency reported** the following sulphur dioxide emissions (in tons) for the 48 states in the continental United States **(www3.epa .gov/airmarkets/progress/reports/emissions_reductions_so2 .html#figure2, retrieved April 20, 2017).**

State	SO$_2$ Emissions 2013	State	SO$_2$ Emissions 2013
AL	106,155	NC	48,154
AR	73,578	ND	55,203
AZ	23,689	NE	65,824
CA	227	NH	3,167
CO	38,287	NJ	2,432
CT	1,107	NM	17,735
DE	2,240	NV	7,427
FL	89,065	NY	17,797
GA	80,949	OH	281,986
IA	76,844	OK	74,425
ID	7	OR	14,004
IL	135,866	PA	252,078
IN	268,217	RI	16
KS	30,021	SC	26,779
KY	188,115	SD	15,342
LA	80,133	TN	56,405
MA	10,841	TX	365,507
MD	25,117	UT	21,144
ME	873	VA	38,778
MI	194,390	VT	2
MN	24,366	WA	2,859
MO	141,430	WI	62,434
MS	77,486	WV	86,201
MT	16,216	WY	40,671

Use these data to calculate the values in the five-number summary. (Hint: See Example 3.13.)

3.37   The data below are manufacturing defects per 100 cars for the 30 brands of cars sold in the United States **(USA TODAY, March 29, 2016).** Many of these values are larger than 100 because one car might have many defects.

97	134	198	142	95	135	132	145
136	129	152	158	169	155	106	125
120	153	208	163	204	173	165	126
113	167	171	166	181	161		

Use these data to construct a boxplot. Write a few sentences describing the important characteristics of the boxplot.

3.38   Data on the gasoline tax per gallon (in cents) in 2015 for the 50 U.S. states and the District of Columbia (DC) are shown below **(www.eia.gov/tools/faqs/faq.cfm?id=10&t=10, retrieved September 1, 2016).**

State	Gasoline Tax (cents per gallon)	State	Gasoline Tax (cents per gallon)
Alabama	19.0	Montana	27.8
Alaska	9.0	Nebraska	27.7
Arizona	19.0	Nevada	23.8
Arkansas	21.8	New Hampshire	23.8
California	37.2	New Jersey	14.6
Colorado	23.3	New Mexico	18.9
Connecticut	25.0	New York	33.8
Delaware	23.0	North Carolina	35.3
DC	23.5	North Dakota	23.0
Florida	30.6	Ohio	28.0
Georgia	26.5	Oklahoma	17.0
Hawaii	18.5	Oregon	30.0
Idaho	33.0	Pennsylvania	51.4
Illinois	33.1	Rhode Island	34.1
Indiana	29.0	South Carolina	16.8
Iowa	31.8	South Dakota	30.0
Kansas	25.0	Tennessee	21.4
Kentucky	26.0	Texas	20.0
Louisiana	20.9	Utah	30.1
Maine	31.4	Vermont	30.5
Maryland	32.8	Virginia	16.8
Massachusetts	26.7	Washington	44.6
Michigan	30.9	West Virginia	33.2
Minnesota	30.6	Wisconsin	32.9
Mississippi	18.4	Wyoming	24.0
Missouri	17.3		

a. The smallest value in the data set is 9.0 (Alaska), and the largest value is 51.4 (Pennsylvania). Are these values outliers? Explain. (Hint: See Example 3.15.)

b. Construct a boxplot for this data set and comment on the interesting features of the plot.

Data set available

3.39   The National Climate Data Center gave the accompanying annual rainfall (in inches) for Medford, Oregon, from 1950 to 2014 (**www.ncdc.noaa.gov/cdo-web/datasets#ANNUAL, retrieved April 20, 2017):**

20.0 18.7 25.6 16.3 19.9 28.8 20.5

23.3 10.4 20.6 19.6 23.1 18.9 29.1

18.1 21.0 18.7 15.3 20.5 23.3 19.4

20.6 18.9 18.6 19.4 12.3 22.2 15.7

20.2 15.9 21.9 19.2 30.2 18.6 10.7

17.1 14.8 13.7 15.1 13.5 14.5 15.0

16.6 12.1 21.8 31.4 17.9 28.7 16.5

18.8 15.0 18.0 19.8 19.1 23.4 21.8

17.1 13.8 11.8 21.3 16.4 26.9 9.0

20.2

a. Calculate the quartiles and the interquartile range.
b. Are there outliers in this data set? If so, which observations are outliers?
c. Draw a modified boxplot for this data set and comment on the interesting features of this plot.

3.40   The accompanying data are consistent with summary statistics that appeared in the paper **"Shape of Glass and Amount of Alcohol Poured: Comparative Study of Effect of Practice and Concentration"** (*British Medical Journal* [2005]: 1512–1514). Data represent the actual amount (in ml) poured into a glass for individuals asked to pour 1.5 ounces (44.3 ml) into either a tall, slender glass or a short, wide glass.

Tall, slender glass

44.0 49.6 62.3 28.4 39.1 39.8 60.5 73.0 57.5

56.5 65.0 56.2 57.7 73.5 66.4 32.7 40.4 21.4

Short, wide glass

89.2 68.6 32.7 37.4 39.6 46.8 66.1 79.2

66.3 52.1 47.3 64.4 53.7 63.2 46.4 63.0

92.4 57.8

Construct a comparative boxplot and comment on the differences and similarities in the two data distributions. (Hint: See Examples 3.14 and 3.16.)

SECTION 3.4   **Exercise Set 2**

3.41   The accompanying data are median ages for residents of each of the 50 U.S. states and the District of Columbia (DC) in 2014 **(The Statistics Portal, retrieved May 18, 2016 from www.statista.com/statistics/208048/median -age-of-population-in-the-usa-by-state/).**

State	Median Age	State	Median Age
Alabama	38.6	Montana	39.6
Alaska	33.3	Nebraska	36.2
Arizona	36.9	Nevada	37.4
Arkansas	37.8	New Hampshire	42.5
California	36.0	New Jersey	39.4
Colorado	36.3	New Mexico	37.2
Connecticut	40.5	New York	38.2
Delaware	39.6	North Carolina	38.3
DC	33.8	North Dakota	35.1
Florida	41.6	Ohio	39.4
Georgia	36.1	Oklahoma	36.2
Hawaii	38.1	Oregon	39.3
Idaho	35.9	Pennsylvania	40.7
Illinois	37.5	Rhode Island	39.8
Indiana	37.4	South Caroline	38.8
Iowa	38.2	South Dakota	36.6
Kansas	36.2	Tennessee	38.6
Kentucky	38.5	Texas	34.3
Louisiana	36.1	Utah	30.5
Maine	44.1	Vermont	42.8
Maryland	38.3	Virginia	37.7
Massachusetts	39.4	Washington	37.5
Michigan	39.6	West Virginia	41.9
Minnesota	37.8	Wisconsin	39.2
Mississippi	36.7	Wyoming	36.6
Missouri	38.5		

Use these data to calculate the values in the five-number summary.

3.42   The accompanying data are a subset of data read from a graph in the paper **"Ladies First? A Field Study of Discrimination in Coffee Shops"** (*Applied Economics* [April, 2008]). The data are the waiting time (in seconds) between ordering and receiving coffee for 19 male customers at a Boston coffee shop.

40   60   70   80   85   90 100 100 110 120

125 125 140 140 160 160 170 180 200

Use these data to construct a boxplot. Write a few sentences describing the important characteristics of the boxplot.

3.43   The accompanying data are the percentage of babies born prematurely in 2014 for the 50 U.S. states and the District of Columbia (DC) **(The Henry J. Kaiser Family Foundation, kff.org/other/state-indicator/preterm-births-by -raceethnicity/, retrieved April 20, 2017).**

Data set available

State	Premature Percent	State	Premature Percent
Alabama	11.7	Montana	9.3
Alaska	8.5	Nebraska	9.1
Arizona	9.0	Nevada	10.1
Arkansas	10.0	New Hampshire	8.2
California	8.3	New Jersey	9.6
Colorado	8.4	New Mexico	9.2
Connecticut	9.2	New York	8.9
Delaware	9.3	North Carolina	9.7
DC	9.6	North Dakota	8.4
Florida	9.9	Ohio	10.3
Georgia	10.8	Oklahoma	10.3
Hawaii	10.0	Oregon	7.7
Idaho	8.2	Pennsylvania	9.4
Illinois	10.1	Rhode Island	8.6
Indiana	9.7	South Carolina	10.8
Iowa	9.3	South Dakota	8.5
Kansas	8.7	Tennessee	10.8
Kentucky	10.7	Texas	10.4
Louisiana	12.3	Utah	9.1
Maine	8.4	Vermont	7.9
Maryland	10.1	Virginia	9.2
Massachusetts	8.6	Washington	8.1
Michigan	9.8	West Virginia	10.8
Minnesota	8.7	Wisconsin	9.2
Mississippi	12.9	Wyoming	11.2
Missouri	9.8		

a. The smallest value in the data set is 7.7 (Oregon), and the largest value is 12.9 (Mississippi). Are these values outliers? Explain.

b. Construct a boxplot for this data set and comment on the interesting features of the plot.

3.44 The article **"The Best—and Worst—Places to Be a Working Woman" (The Economist, Graphic Detail for March 3, 2016)** reported values of what it calls the glass-ceiling index, which is designed to rate countries based on women's chances of equal treatment at work. The index weights factors that include participation of women in higher education, participation in the workforce by women, pay, child-care cost, and maternity benefits. The best possible value for this index is 100. Data for 29 countries are shown in the accompanying table.

Country	Glass-Ceiling Index
Iceland	82.8
Norway	79.3
Sweden	79.0
Finland	73.8
Hungary	70.4
Poland	70.0
France	68.3
Denmark	66.6
New Zealand	63.8
Belgium	63.3
Canada	62.3
Portugal	60.8
Spain	59.1
Israel	58.3
Slovakia	57.8
Austria	57.1
Germany	57.0
Australia	56.2
United States	55.9
Czech Republic	55.1
Italy	53.7
Netherlands	51.8
Greece	50.8
Great Britain	50.5
Ireland	48.4
Switzerland	40.6
Japan	28.8
Turkey	27.2
South Korea	25.0

a. Are there outliers in this data set? If so, which observations are outliers?

b. Draw a modified boxplot for this data set.

c. The article points out that Nordic countries (Iceland, Sweden, Norway, and Finland) come out on top on this index. Where are the values for the Nordic countries located in terms of the box plot?

3.45 Like many people, dogs can also suffer from back problems. Animal researchers were interested in learning whether laser therapy after back surgery would shorten recovery time for dogs that needed back surgery (**"Low-Level Laser Therapy Reduces Time to Ambulation in Dogs After Hemilaminectomy: A Preliminary Study," Journal of Small Animal Practice [2012]: 465–469**). Thirty-six dogs that needed back surgery were assigned to one of two experimental groups. One group had surgery only and the other group had surgery followed by laser therapy daily for five days after surgery. The researchers recorded the time (in days) required to reach a specified level of movement for each dog. The following summary statistics were reported for the two groups:

Experimental Group	Minimum	Lower Quartile	Median	Upper Quartile	Maximum
Surgery Only	2	6	14	16	29
Surgery + Laser	2	2	3.5	9	19

a. What does the fact that the minimum and the lower quartile are equal for the Surgery + Laser group indicate?
b. Construct a comparative boxplot and comment on any differences and similarities in the recovery times for the two groups.
c. Based on the comparative boxplot, does it look like the laser therapy is effective in reducing recovery time? Explain.

ADDITIONAL EXERCISES

3.46   A graph similar to the one below appeared in the paper **"Does the Weight of the Dish Influence our Perception of Food?"** (*Food Quality and Preference*, [2011]: 753–756). The graph was constructed using data from an experiment in which people rated yogurt served in three different bowls. The bowls looked the same but were of different weights. The yogurt in each bowl was actually the same yogurt. Subjects held each bowl in their hands while they tasted the yogurt and provided a rating.

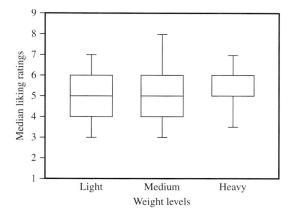

This graph summarizes the ratings for the light, medium, and heavy bowls. The graph is described as follows: "The boundary of the box closest to 0 indicates the 25th percentile, the thicker line indicates the median rating, and the boundary of the box farthest from zero indicates the 75th percentile. Error bars indicate the 90th and 10th percentiles of the response."

Explain how this graph is similar to and how it is different from a comparative boxplot.

3.47   The U.S. Department of Health and Human Services reported the estimated percentage of households with only wireless phone service (no landline) in 2014 for the 50 states and the District of Columbia (DC) **(cdc.gov/nchs/data /nhis/earlyrelease/wireless_state_201602.pdf,** retrieved **April 20, 2017).** In the accompanying data table, each state was also classified into one of three geographical regions— West (W), Middle (M), and East (E).

Wireless %	Region	State	Wireless %	Region	State
43.4	M	AL	41.0	W	MT
39.7	W	AK	46.5	M	NE
49.4	W	AZ	48.4	W	NV
56.2	M	AR	43.6	M	ND
42.8	W	CA	31.2	E	NH
50.5	W	CO	25.1	E	NJ
26.7	E	CT	47.0	W	NM
29.4	E	DE	31.1	E	NY
49.7	E	DC	42.9	E	NC
47.6	E	FL	45.8	E	OH
45.9	E	GA	50.4	M	OK
38.3	W	HI	47.0	W	OR
56.1	W	ID	30.0	E	PA
45.7	M	IL	34.6	E	RI
47.7	M	IN	49.5	E	SC
50.7	M	IA	41.4	M	SD
51.6	M	KS	46.6	M	TN
47.1	M	KY	54.6	M	TX
40.9	M	LA	52.2	W	UT
40.8	E	ME	41.1	E	VA
36.2	E	MD	37.2	E	VT
31.5	E	MA	48.3	W	WA
47.8	M	MI	37.2	E	WV
43.1	M	MN	46.6	M	WI
55.1	M	MS	51.8	W	WY
51.5	M	MO			

a. Construct a comparative boxplot that makes it possible to compare wireless percent for the three geographical regions.
b. Does the graphical display in Part (a) reveal any striking differences, or are the distributions similar for the three regions?

3.48   The Bloomberg web site included the data in the accompanying table on the number of movies made by 25 *Saturday Night Live* cast members as of 2014 **(www.bloomberg.com/graphics/best-and-worst/#top-grossing-saturday -night-live-alumni,** retrieved April 20, 2017). Also given was the top-grossing movie made by each and the gross income for that movie adjusted for inflation.

Star	Number of Movies	Top-Grossing Movie	Inflation-Adjusted Gross of Top Movie (millions of dollars)
Eddie Murphy	38	Shrek 2	550
Dan Aykroyd	56	Ghostbusters	514
Robert Downey, Jr.	57	The Avengers	646
Ben Stiller	50	Meet the Fockers	352
Bill Murray	46	Ghostbusters	514
Adam Sandler	36	The Waterboy	236
Chris Rock	33	Beverly Hills Cop II	322
Will Ferrell	40	Austin Powers: The Spy Who Shagged Me	293
Mike Myers	16	Shrek 2	550
Joan Cusack	48	Toy Story 3	453
Bill Hader	29	Monsters University	274
Rob Schneider	29	Home Alone 2: Lost in New York	294
Brian Doyle-Murray	32	As Good as It Gets	220
Laurie Metcalf	28	Toy Story 3	453
Kristen Wiig	29	Despicable Me 2	376
Molly Shannon	33	How the Grinch Stole Christmas	359

(continued)

Star	Number of Movies	Top-Grossing Movie	Inflation-Adjusted Gross of Top Movie (millions of dollars)
David Koechner	45	Austin Powers: The Spy Who Shagged Me	293
Jon Lovitz	34	Big	231
Amy Poehler	23	Shrek the Third	370
Chevy Chase	32	National Lampoon's Vacation	147
Billy Crystal	26	Monsters University	274
Harry Shearer	27	The Simpsons Movie	210
Randy Quaid	38	Independence Day	464
David Spade	25	Grown Ups	177
Martin Short	28	Madagascar 3: Europe	224

Construct a boxplot for the number of movies data and comment on what the boxplot tells you about the data distribution for the number of movies.

 3.49   Refer to the data given in the previous exercise.
a. Are there any outliers in the inflation-adjusted gross movie income data? If so, which data values are outliers?
b. Construct a boxplot for the inflation-adjusted gross movie income data.
c. For the inflation-adjusted gross movie income data, the mean is $351.8 million and the median is $322.0 million. What characteristic of the boxplot explains why the mean is greater than the median for this data set?

SECTION 3.5   Measures of Relative Standing: z-scores and Percentiles

When you obtain your score after taking an exam, you probably want to know how it compares to the scores of others. Is your score above or below the mean, and by how much? Does your score place you among the top 5% of the class or only among the top 25%? Answering these questions involves measuring the position, or relative standing, of a particular value in a data set. One measure of relative standing is a *z-score*.

DEFINITION

The **z-score** corresponding to a particular data value is

$$z\text{-score} = \frac{\text{data value} - \text{mean}}{\text{standard deviation}}$$

The *z-score* tells you how many standard deviations the data value is from the mean. The *z-score* is positive when the data value is greater than the mean and negative when the data value is less than the mean.

The process of subtracting the mean and then dividing by the standard deviation is sometimes referred to as *standardizing*. A *z-score* is one example of a *standardized score*.

Example 3.18   **Relatively Speaking, Which Is the Better Offer?**

Suppose that two graduating seniors, one a marketing major and one an accounting major, are comparing job offers. The accounting major has an offer for $55,000 per year, and the marketing student has an offer for $53,000 per year. Summary information about the distribution of offers is shown here:

Major	Mean	Standard Deviation
Accounting	$56,000	$1,500
Marketing	$52,500	$1,000

Then,

$$\text{accounting } z\text{-score} = \frac{55,000-56,000}{1500} = -0.67$$

so $55,000 is 0.67 standard deviations below the mean, whereas

$$\text{marketing } z\text{-score} = \frac{53,000-52,500}{1000} = 0.5$$

which is 0.5 standard deviations above the mean. Relative to their distributions, the marketing offer is actually more attractive than the accounting offer.

The z-score is particularly useful when the data distribution is mound shaped and approximately symmetric. In this case, a z-score outside the interval from −2 to +2 occurs in about 5% of all cases, whereas a z-score outside the interval from −3 to +3 occurs only about 0.3% of the time. This means that in this type of distribution, data values with z-scores less than −2 or greater than +2 can be thought of as being among the smallest or largest values in the data set.

A rule called the **Empirical Rule** allows you to use the mean and standard deviation to make statements about the data values. You can use this rule when the shape of the data distribution is mound shaped (has a single peak) and is approximately symmetric.

The Empirical Rule

If a data distribution is mound shaped and approximately symmetric, then

Approximately 68% of the observations are within 1 standard deviation of the mean.
Approximately 95% of the observations are within 2 standard deviations of the mean.
Approximately 99.7% of the observations are within 3 standard deviations of the mean.

Figure 3.12 illustrates the percentages given by the Empirical Rule.

FIGURE 3.12
Approximate percentages implied by the Empirical Rule

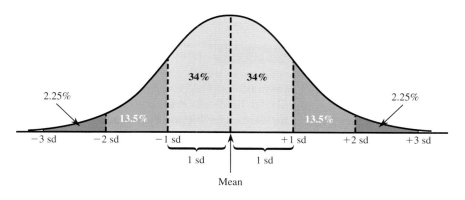

Example 3.19 | **Heights of Mothers and the Empirical Rule**

One of the earliest articles to promote using a distribution called the normal distribution was **"On the Laws of Inheritance in Man. I. Inheritance of Physical Characters" (*Biometrika* [1903]: 375–462)**. One of the data sets discussed in this article contained 1052 measurements of the heights of mothers. The mean and standard deviation were

$$\bar{x} = 62.484 \text{ inches} \qquad s = 2.390 \text{ inches}$$

The data distribution was described as being mound shaped and approximately symmetric. Table 3.6 contrasts actual percentages from the data set along with those suggested by the Empirical Rule. Notice how closely the Empirical Rule approximates the percentages of observations falling within 1, 2, and 3 standard deviations of the mean.

TABLE 3.6 Summarizing the Distribution of Mothers' Heights

Number of Standard Deviations	Interval	Actual	Empirical Rule
1	60.094 to 64.874	72.1%	Approximately 68%
2	57.704 to 67.264	96.2%	Approximately 95%
3	55.314 to 69.654	99.2%	Approximately 99.7%

Percentiles

A particular observation can be located even more precisely by giving the percentage of the data that fall at or below that value. For example, if 95% of all test scores are at or below 650, which also means only 5% are above 650, then 650 is called the *95th percentile* of the data set (or of the distribution of scores). Similarly, if 10% of all scores are at or below 400, then the value 400 is the 10th percentile.

DEFINITION

For a number r between 0 and 100, the ***r*th percentile** is a value such that r percent of the observations in the data distribution fall at or below that value.

Figure 3.13 illustrates the 90th percentile. You have already met several percentiles in disguise. The median is the 50th percentile, and the lower and upper quartiles are the 25th and 75th percentiles, respectively.

FIGURE 3.13
90th percentile for a smoothed histogram

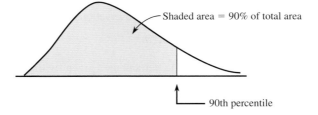

Shaded area = 90% of total area

90th percentile

Example 3.20 | **Head Circumference at Birth**

In addition to weight and length, head circumference is another measure of health in newborn babies. The National Center for Health Statistics reports the following summary values for head circumference (cm) at birth for boys (approximate values from graphs on the **Centers for Disease Control and Prevention** web site at **www.cdc.gov/growthcharts /data/set1clinical/cj41l019.pdf, retrieved April 20, 2017**):

Percentile	5	10	25	50	75	90	95
Head Circumference (cm)	32.2	33.2	34.5	35.8	37.0	38.2	38.6

Interpreting these percentiles, half of newborn boys have head circumferences of 35.8 cm or less, because 35.8 is the 50th percentile (the median). The middle 50% have head circumferences between 34.5 cm and 37.0 cm, with 25% 34.5 cm or less and 25% greater than 37.0 cm. You can tell that this distribution for newborn boys is not symmetric, because the 5th percentile is 3.6 cm below the median, whereas the 95th percentile is only 2.8 cm above the median. This means that the bottom part of the distribution stretches out more than the top part, which creates a negatively skewed distribution, like the one shown in Figure 3.14.

FIGURE 3.14
Negatively skewed distribution

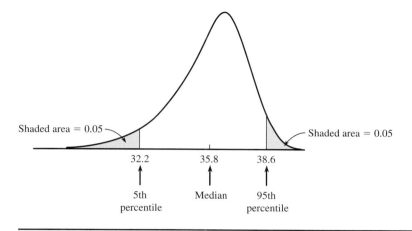

Summing It Up—Section 3.5

The following learning objectives were addressed in this section:

Putting It into Practice
P4: Use percentiles and z-scores to describe relative standing.
Percentiles and z-scores are both measures of relative standing. Measures of relative standing describe where a particular data value is relative to the other values in the data set.

A z-score indicates how far away an observation is from the mean in terms of the standard deviation. For example, a z-score of 1.5 corresponds to an observation that is 1.5 standard deviations above the mean. Negative z-scores correspond to values that are below the mean. Example 3.18 illustrates the calculation of z-scores.

The Empirical Rule is a result that links percentiles and z-scores for data distributions that are approximately symmetric and mound shaped. Use of the Empirical Rule is illustrated in Example 3.19.

Percentiles also describe relative standing. If a data value is at the rth percentile, r percent of the observations in the data set are less than or equal to that value. Percentiles are discussed in Example 3.20.

SECTION 3.5 EXERCISES

Each Exercise Set assesses the following chapter learning objectives: P4

SECTION 3.5 Exercise Set 1

3.50 A student took two national aptitude tests. The mean and standard deviation were 475 and 100, respectively, for the first test, and 30 and 8, respectively, for the second test. The student scored 625 on the first test and 45 on the second test. Use z-scores to determine on which exam the student performed better relative to the other test takers. (Hint: See Example 3.18.)

3.51 The mean playing time for a large collection of compact discs is 35 minutes, and the standard deviation is 5 minutes.
a. What value is 1 standard deviation above the mean? One standard deviation below the mean? What values are 2 standard deviations away from the mean?
b. Assuming that the distribution of times is mound shaped and approximately symmetric, approximately what

percentage of times are between 25 and 45 minutes? Less than 20 minutes or greater than 50 minutes? Less than 20 minutes? (Hint: See Example 3.19.)

3.52 The report **"Who Borrows Most? Bachelor's Degree Recipients with High Levels of Student Debt"** (trends .collegeboard.org/content/who-borrows-most-bachelors -degree-recipients-high-levels-student-debt-april-2010, **retrieved April 20, 2017)** included the following percentiles for the amount of student debt for students graduating with a bachelor's degree in 2010:

10th percentile = $0   25th percentile = $0
50th percentile = $11,000   75th percentile = $24,600
90th percentile = $39,300

For each of these percentiles, write a sentence interpreting the value of the percentile. (Hint: See Example 3.20.)

3.53 The paper **"Study of the Flying Ability of *Rhynchophorus ferrugineus* Adults Using a Computer-Monitored Mill"** (*Bulletin of Entomological Research* **[2014]: 462–467)** summarized data from a study of red palm weevils, a pest that is a threat to palm trees. The following frequency distribution from the paper was constructed using the longest flight (in meters) observed for 132 weevils.

Longest Flight (m)	Frequency (Number of Weevils)
0 to < 100	71
100 to < 2,000	32
2000 to < 5,000	18
5000 to < 10,000	8
10,000 or more	3

Estimate the approximate values of the following percentiles:
a. 54th
b. 80th
c. 92nd

SECTION 3.5 Exercise Set 2

3.54 Suppose that your statistics professor returned your first midterm exam with only a z-score written on it. She also told you that a histogram of the scores was mound shaped and approximately symmetric. How would you interpret each of the following z-scores?
a. 2.2   d. 1.0
b. 0.4   e. 0
c. 1.8

3.55 In a study investigating the effect of car speed on accident severity, the vehicle speed at impact was recorded for 5000 fatal accidents. For these accidents, the mean speed was 42 mph and the standard deviation was 15 mph. A histogram revealed that the vehicle speed distribution was mound shaped and approximately symmetric.

a. Approximately what percentage of the vehicle speeds were between 27 and 57 mph?
b. Approximately what percentage of the vehicle speeds exceeded 57 mph?

3.56 Suppose that your younger sister is applying to college and has taken the SAT exam. She scored at the 83rd percentile on the verbal section of the test and at the 94th percentile on the math section. Because you have been studying statistics, she asks you for an interpretation of these values. What would you tell her?

3.57 The following data values are the 2014 per capita operating expenditures on public libraries for each of the 50 U.S. states and Washington, D.C. (from **www.imls.gov /research-evaluation/data-collection/public-libraries-survey /explore-pls-data, retrieved April 20, 2017**):

16.48	16.66	16.76	18.77	18.87	19.95	20.43
21.92	24.21	25.05	25.77	26.01	26.07	26.50
26.55	26.60	26.80	28.25	30.70	31.69	31.77
33.23	33.33	33.51	33.74	34.59	35.53	37.09
37.25	37.37	37.61	38.19	38.47	41.34	42.65
43.02	44.16	45.02	48.22	49.39	50.21	51.81
53.39	54.01	54.97	55.88	56.36	58.52	59.62
59.95	67.44					

a. Summarize this data set with a frequency distribution. Construct the corresponding histogram.
b. Use the histogram from Part (a) to find the approximate values of the following percentiles:
   i. 50th   iv. 90th
   ii. 70th   v. 40th
   iii. 10th

ADDITIONAL EXERCISES

3.58 The mean number of text messages sent per month by customers of a cell phone service provider is 1650, and the standard deviation is 750. Find the z-score associated with each of the following numbers of text messages sent.
a. 0
b. 10,000
c. 4500
d. 300

3.59 Suppose that the distribution of weekly water usage for single-family homes in a particular city is mound shaped and approximately symmetric. The mean is 1400 gallons, and the standard deviation is 300 gallons.
a. What is the approximate value of the 16th percentile?
b. What is the approximate value of the median?
c. What is the approximate value of the 84th percentile?

3.60 The mean reading speed of students completing a speed-reading course is 450 words per minute (wpm). If the standard deviation is 70 wpm, find the *z*-score associated with each of the following reading speeds.

a. 320 wpm   **c.** 420 wpm
b. 475 wpm   **d.** 610 wpm

3.61 Suppose that the distribution of scores on an exam is mound shaped and approximately symmetric. The exam scores have a mean of 100 and the 16th percentile is 80.

a. What is the 84th percentile?
b. What is the approximate value of the standard deviation of exam scores?
c. What is the *z*-score for an exam score of 90?
d. What percentile corresponds to an exam score of 140?
e. Do you think there were many scores below 40? Explain.

SECTION 3.6   Avoid These Common Mistakes

When calculating or interpreting summary measures, keep the following in mind.

1. Watch out for categorical data that look numerical! Often, categorical data are coded numerically. For example gender might be coded as 0 = female and 1 = male, but this does not make gender a numerical variable. Also, sometimes a categorical variable can "look" numerical, as is the case for variables like zip code or telephone area code. But zip codes and area codes are really just labels for geographic regions and don't behave like numbers. Categorical data should not be summarized using the mean and standard deviation or the median and interquartile range.

2. Measures of center don't tell all. Although measures of center, such as the mean and the median, do give you a sense of what might be considered a typical value for a variable, this is only one characteristic of a data set. Without additional information about variability and distribution shape, you don't really know much about the behavior of the variable.

3. Data distributions with different shapes can have the same mean and standard deviation. For example, consider the following two histograms:

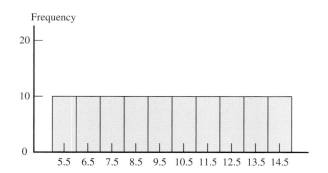
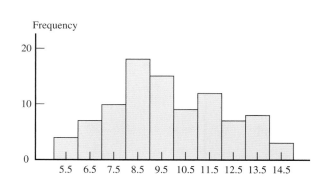

Both histograms summarize data sets that have a mean of 10 and a standard deviation of 2, yet they have different shapes.

4. Both the mean and the standard deviation can be sensitive to extreme values in a data set, especially if the sample size is small. If a data distribution is markedly skewed or if the data set has outliers, the median and the interquartile range are often a better choice for describing center and variability.

5. Measures of center and measures of variability describe values of a variable, not frequencies in a frequency distribution or heights of the bars in a histogram. For example, consider the following two frequency distributions and histograms shown in Figure 3.20.

There is more variability in the data summarized by Frequency Distribution and Histogram A than in the data summarized by Frequency Distribution and Histogram B. This is because the values of the variable described by Histogram and Frequency Distribution B are more concentrated near the mean. Don't be misled by the fact that

FIGURE 3.20
Two distributions with different standard deviations

Frequency Distribution A		Frequency Distribution B	
Value	Frequency	Value	Frequency
1	10	1	5
2	10	2	10
3	10	3	20
4	10	4	10
5	10	5	5

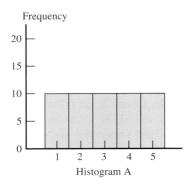

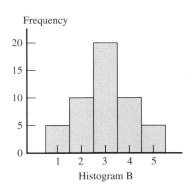

there is no variability in the frequencies in Frequency Distribution A or the heights of the bars in Histogram A.

6. Be careful with boxplots based on small sample sizes. Boxplots convey information about center, variability, and shape, but interpreting shape information is problematic when the sample size is small. For example, it is not really possible to decide whether a data distribution is symmetric or skewed from a boxplot based on a small sample.

7. Not all distributions are mound shaped. Using the Empirical Rule in situations where you are not convinced that the data distribution is mound shaped and approximately symmetric can lead to incorrect statements.

8. Watch for outliers! Unusual observations in a data set often provide important information about the variable under study, so it is important to consider outliers in addition to describing what is typical. Outliers can also be problematic because the values of some summaries are influenced by outliers and because some methods for drawing conclusions from data are not appropriate if the data set has outliers.

CHAPTER ACTIVITIES

ACTIVITY 3.1   COLLECTING AND SUMMARIZING NUMERICAL DATA

In this activity, you will work in groups to collect data on how many hours per week students at your school spend engaged in a particular activity.

1. With your group, pick one of the following activities to be the focus of your study (or you may choose a different activity with the approval of your instructor):
   i. Using the Internet
   ii. Studying or doing homework
   iii. Watching TV
   iv. Exercising
   v. Sleeping

2. Develop a plan to collect data on the variable you have chosen.

3. Summarize your data using both numerical and graphical summaries. Be sure to address both center and variability.

4. Write a short article for your school paper summarizing your findings regarding student behavior. Your article should include both numerical and graphical summaries.

ACTIVITY 3.2 AIRLINE PASSENGER WEIGHTS

The article **"Airlines Should Weigh Passengers, Bags, NTSB Says"** (*USA TODAY*, February 27, 2004) reported on a recommendation from the National Transportation Safety Board that airlines weigh passengers and their bags to prevent planes from being overloaded. This recommendation was the result of an investigation into the crash of a small commuter plane in 2003, which determined that too much weight contributed to the crash.

Rather than weighing passengers, airlines currently use estimates of mean passenger and luggage weights. After the 2003 accident, this estimate was increased by 10 pounds for passengers and 5 pounds for luggage. Although a somewhat overweight airplane can fly if all systems are working properly, if one of its engines fails, it becomes difficult for the pilot to control.

Assuming that the new estimate of the mean passenger weight is accurate, discuss the following questions with a partner and then write a paragraph that answers these questions.

1. What role does variability in passenger weights play in creating a potentially dangerous situation for an airline?

2. Would an airline have a lower risk of a potentially dangerous situation if the variability in passenger weight is large? What if the variabiity in passenger weight is small?

ACTIVITY 3.3 BOXPLOT SHAPES

In this activity, you will investigate the relationship between a boxplot and its corresponding five-number summary. The following table gives five-number summaries labeled I–IV. Also shown are four boxplots labeled A–D. Notice that scales have been intentionally omitted from the boxplots. Match each five-number summary to its boxplot.

Five-Number Summaries

	I	II	III	IV
Minimum	40	4	0.0	10
Lower quartile	45	8	0.1	34
Median	71	16	0.9	44
Upper quartile	88	25	2.2	82
Maximum	106	30	5.1	132

A 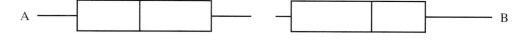 B

C  D

💻 CHAPTER 3 EXPLORATIONS IN STATISTICAL THINKING

🖱 EXPLORATION 1: UNDERSTANDING SAMPLING VARIABILITY

In Exercise 1, each student in your class will go online to select two random samples, one of size 10 and one of size 30, from a small population consisting of 300 adults.

1. To learn about how the values of sample statistics vary from sample to sample, go online at statistics.cengage.com/Peck2e/Explore.html and click on the link for Chapter 3. Then click on the Exercise 1 link. This link will take you to a web page where you can select two random samples from the population.

Click on the samples button. This selects a random sample of size 10 and a random sample of size 30 and will display the ages of the people in your samples. Each student in your class will receive data from different random samples.

Use the data from your random samples to complete the following.
   a. Calculate the mean age of the people in your sample of size 10.
   b. The mean age for the entire population is 44.4 years. Did you get 44.4 years for your sample mean? How far was your sample mean from 44.4?

c. Now calculate the mean age for the people in your sample of size 30. Is the mean for your sample of size 30 closer to 44.4 years or farther away than the mean from your sample of size 10? Does this surprise you? Explain why or why not.

If asked to do so by your instructor, bring your answers to Parts (a)–(c) to class. Your instructor will lead the class through the rest of this exercise.

Have one student draw a number line on the board or on a piece of poster-size chart paper. The scale on the number line should range from 20 to 60. The class should now create a dotplot of the sample means for the samples of size 10 by having each student enter a dot that corresponds to his or her sample mean. Label this dotplot "Means of Random Samples of Size 10."

d. Based on the class dotplot for samples of size 10, how would you describe the sample-to-sample variability in the sample means? Was there substantial variability, or were the sample means fairly similar?

e. Locate 44.4 on the scale for the class dotplot. Does it look like distribution of the sample means is centered at about 44.4?

f. Here is a dotplot of the ages for the entire population. How is the dotplot of sample means for samples of size 10 similar to the population dotplot? How is it different?

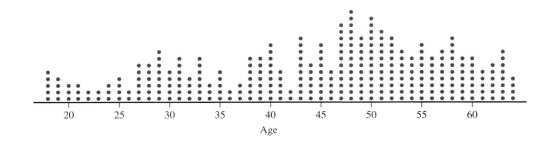

g. Now create a class dotplot of the sample means for the samples of size 30. Label this dotplot "Means of Random Samples of Size 30." Based on the class dotplots for samples of size 10 and samples of size 30, was there more sample-to-sample variability in the values of the sample mean for samples of size 10 or for samples of size 30?

h. How do the dotplots of the sample means for samples of size 10 and the sample means for samples of size 30 compare in terms of shape, center, and variability?

i. How do the dotplots support the statement that the sample mean will tend to be closer to the actual value of the population mean for a random sample of size 30 than for a random sample of size 10?

In Exercise 2, each student in your class will go online to select a random sample of 25 airplanes from a population of airplanes.

2. The cost of jet fuel is one factor that contributes to the rising cost of airfares. In a study of aircraft fuel consumption that was part of an environmental impact report, different types of planes that use the Phoenix airport were studied, and the amount of fuel consumed from the time the plane left the gate until the time the plane reached an altitude of 3000 feet was determined. Go online at statistics.cengage.com/Peck2e/Expore.html and click on the link for Chapter 3. Then click on the Exercise 2 link. This link will take you to a web page where you can select a random sample of 25 planes. Click on the sample button. This

selects the sample and displays the amount of fuel consumed (in gallons) for each plane in your sample.

a. Calculate the mean, median, and standard deviation for this sample.

b. Construct a dotplot of the fuel consumption values.

c. Write a few sentences commenting on the dotplot. What does the dotplot tell you about the distribution of fuel consumption values in the sample?

d. Which of the following best describes the distribution of fuel consumption?

 i. The distribution is approximately symmetric. A typical value for fuel consumption is approximately 130,000 gallons, and there is relatively little variability in fuel consumption for the different plane types.

 ii. The distribution is mound shaped. A typical value for fuel consumption is approximately 130,000 gallons, and there is substantial variability in fuel consumption for the different plane types.

 iii. The distribution is markedly skewed, with a long upper tail (on the right-hand side) indicating that there are some plane types with a fuel consumption that is much higher than for most plane types.

 iv. The distribution is markedly skewed, with a long lower tail (on the left-hand side) indicating that there are some plane types with a fuel consumption that is much lower than for most plane types.

e. Which of the following do you think would have the biggest impact on overall fuel consumption at this airport?

 i. Modifications were made that would reduce the fuel consumption of every type of plane by 20 gallons.

 ii. Changes were implemented that would result in a 10% reduction in the fuel consumption of the 10 types of planes with the largest consumption.

f. Write a few sentences explaining your choice in Part (e).

EXPLORATION 2: EXPLORING MULTIVARIABLE THINKING

As people age, they tend to lose muscle mass. The authors of the paper **"Relationship Between Muscle Mass and Muscle Strength, and the Impact of Comorbidities: A Population-Based, Cross-Sectional Study of Older Adults in the United States" (July 2013, www.researchgate.net/publication/250307435, retrieved September 27, 2016)** carried out a study exploring the relationship between muscle mass (kg/m^2) and age for both men and women. Below is a graph that is similar to one that appeared in the paper.

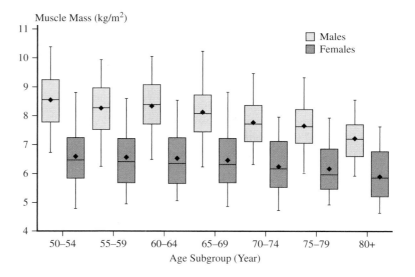

The following explanation of this graph is from the paper:

> **Figure 1 Distribution of muscle mass by age and gender.** Lower half of box-plot depicts 25th percentile, upper half of boxplot represents 75th percentile, horizontal line dividing upper and lower half of boxplot represents median, upper whisker represents 95th percentile, lower whisker represents 5th percentile; and solid diamond in boxplot represents mean.

Use this graphical display to answer the following questions.

1. The boxplots in this graphical display are a bit different from the boxplots introduced in this chapter. In what ways are these boxplots different?
2. Do the boxplots support the following statement? Justify your answer based on the boxplots.
   Statement 1: Muscle mass tends to decrease with age.
3. Do the boxplots support the following statement? Justify your answer based on the boxplots.
   Statement 2: Starting at about age 50, muscle mass tends to decrease at a fairly steady rate.
4. Do the boxplots support the following statement? Justify your answer based on the boxplots.
   Statement 3: For any of the given age groups, muscle mass for women tends to be about the same as muscle mass for men.
5. Do the boxplots support the following statement? Justify your answer based on the boxplots.
   Statement 4: After age 65, muscle mass tends to decline more quickly for men than for women.
6. Do the boxplots support the following statement? Justify your answer based on the boxplots.
   Statement 5: Muscle mass declines with age at about the same rate for men and for women.
7. For which age group is the difference in muscle mass for men and women the greatest? For which age group is it the smallest?

ARE YOU READY TO MOVE ON? CHAPTER 3 REVIEW EXERCISES

All chapter learning objectives are assessed in these exercises. The learning objectives assessed in each exercise are given in parentheses.

3.62 (C1, C2, M1)
For each of the following data sets, construct a graphical display of the data distribution and then indicate what summary measures you would use to describe center and variability.

a. The following are data on weekend exercise time for 20 females consistent with summary quantities given in the paper **"An Ecological Momentary Assessment of the Physical Activity and Sedentary Behaviour Patterns of University Students" (Health Education Journal [2010]: 116–125).**

Female—Weekend

84.0	27.0	82.5	0.0	5.0	13.0	44.5
3.0	0.0	14.5	45.5	39.5	6.5	34.5
0.0	14.5	40.5	44.5	54.0	0.0	

b. The accompanying data are consistent with summary statistics that appeared in the paper **"Shape of Glass and**

Amount of Alcohol Poured: Comparative Study of Effect of Practice and Concentration" (British Medical Journal [2005]: 1512–1514). Data represent the actual amount (in ml) poured into a tall, slender glass for individuals asked to pour 1.5 ounces (44.3 ml).

44.0	49.6	62.3	28.4	39.1	39.8
60.5	73.0	57.5	56.5	65.0	56.2
57.7	73.5	66.4	32.7	40.4	21.4

c. The accompanying data are from a graph that appeared in the paper **"Ladies First? A Field Study of Discrimination in Coffee Shops" (Applied Economics [April, 2008]).** The data are the wait times (in seconds) between ordering and receiving coffee for 19 female customers at a Boston coffee shop.

60	80	80	100	100	100	120
120	120	140	140	150	160	180
200	200	220	240	380		

3.63 (M2, M3, P1, P2)

Data on a customer satisfaction rating (called the APEAL rating) are given for each brand of car sold in the United States (*USA TODAY,* **March 29, 2016**). The APEAL rating is a score between 0 and 1,000, with higher values indicating greater satisfaction.

```
748 762 766 768 779 783 783 783 786 789
790 790 791 791 791 792 795 796 804 812
814 826 826 835 842 844 849 853 858 882
```

Calculate and interpret the mean and standard deviation for this data set.

3.64 (C3, M3, P2)

The paper **"Caffeinated Energy Drinks—A Growing Problem"** (*Drug and Alcohol Dependence* **[2009]: 1–10**) reported caffeine per ounce for 8 top-selling energy drinks and for 11 high-caffeine energy drinks:

Top Selling Energy Drinks

```
9.6   10.0   10.0   9.0   10.9   8.9   9.5   9.1
```

High-Caffeine Energy Drinks

```
21.0   25.0   15.0   21.5   35.7   15.0
33.3   11.9   16.3   31.3   30.0
```

The mean caffeine per ounce is clearly higher for the high-caffeine energy drinks, but which of the two groups of energy drinks is the most variable? Compare the standard deviations of the caffeine contents for the two types of energy drinks in order to determine which is more variable.

3.65 (C4)

Data on tipping percent for 20 restaurant tables, consistent with summary statistics given in the paper **"Beauty and the Labor Market: Evidence from Restaurant Servers"** (unpublished manuscript by Matt Parrett, 2007), are:

```
0.0    5.0    45.0   32.8   13.9   10.4   55.2
50.0   10.0   14.6   38.4   23.0   27.9   27.9
105.0  19.0   10.0   32.1   11.1   15.0
```

a. Calculate the mean and standard deviation for this data set.
b. Delete the observation of 105.0 and recalculate the mean and standard deviation. How do these values compare to the values from Part (a)? What does this suggest about using the mean and standard deviation as measures of center and variability for a data set with outliers?

3.66 (C5)

The **Insurance Institute for Highway Safety** (**www.iihs.org, June 11, 2009**) published data on repair costs for cars involved in different types of accidents. In one study, seven different 2009 models of mini- and micro-cars were driven at 6 miles per hour straight into a fixed barrier. The following table

gives the cost of repairing damage to the bumper for each of the seven models.

Model	Repair Cost
Smart Fortwo	$1,480
Chevrolet Aveo	$1,071
Mini Cooper	$2,291
Toyota Yaris	$1,688
Honda Fit	$1,124
Hyundai Accent	$3,476
Kia Rio	$3,701

Calculate the values of the mean and median. Why are these values so different? Which of the two (the mean or the median) appears to be a better description of a typical value for this data set?

3.67 (M4, M5, P1, P2)

The accompanying data are a subset of data read from a graph in the paper **"Ladies First? A Field Study of Discrimination in Coffee Shops"** (*Applied Economics,* **April 2008**). The data are wait times (in seconds) between ordering and receiving coffee for 19 female customers at a Boston coffee shop.

```
60    80    80    100   100   100   120
120   120   140   140   150   160   180
200   200   220   240   380
```

a. Calculate and interpret the values of the median and interquartile range.
b. Explain why the median and interquartile range is an appropriate choice of summary measures to describe center and variability for this data set.

3.68 (M4, M5, M6, M7, M8)

The report **"Most Licensed Drivers Age 85+: States"** (**www. bloomberg.com/graphics/best-and-worst/#most-licensed-drivers-age-85-plus-states, retrieved April 20, 2017**) gives the percentage of drivers in each state and the District of Columbia in 2011 who were over 85 years of age.

State	Percentage of Drivers Over Age 85
Alaska	0.72
Nevada	0.96
New Mexico	0.98
Utah	1.05
Georgia	1.11
Hawaii	1.13
Texas	1.13
California	1.17
Montana	1.20
Virginia	1.22
Arizona	1.25

Data set available

(continued)

State	Percentage of Drivers Over Age 85
Mississippi	1.26
Maryland	1.28
Kentucky	1.30
Idaho	1.33
Washington	1.33
Indiana	1.34
North Carolina	1.37
South Carolina	1.37
District of Columbia	1.43
Tennessee	1.48
Wyoming	1.48
Missouri	1.49
New Hampshire	1.50
Wisconsin	1.61
Ohio	1.62
West Virginia	1.64
Illinois	1.66
Michigan	1.69
Colorado	1.72
Delaware	1.72
Oklahoma	1.72
Arkansas	1.75
Iowa	1.78
Massachusetts	1.78
Kansas	1.81
New Jersey	1.85
Oregon	1.90
Louisiana	1.92
North Dakota	1.96
Rhode Island	1.97
Florida	1.98
Pennsylvania	1.98
South Dakota	1.99
Nebraska	2.02
New York	2.09
Minnesota	2.11
Vermont	2.22
Alabama	2.23
Maine	2.34
Connecticut	5.10

a. Find the values of the median, the lower quartile, and the upper quartile.

b. The largest value in the data set is 5.10% (Connecticut). Is this state an outlier?

c. Construct a modified boxplot for this data set and comment on the interesting features of the plot. How would you describe the shape of the distribution if you don't consider the outlier?

3.69 (M4, M5, M6, M7, M8, P3)
Fiber content (in grams per serving) and sugar content (in grams per serving) for 18 high-fiber cereals (www .consumerreports.com) are shown.

Fiber Content

7	10	10	7	8	7	12	12	8
13	10	8	12	7	14	7	8	8

Sugar Content

11	6	14	13	0	18	9	10
19	6	10	17	10	10	0	9
5	11						

a. Find the median, quartiles, and interquartile range for the fiber content data set.

b. Find the median, quartiles, and interquartile range for the sugar content data set.

c. Are there any outliers in the sugar content data set?

d. Explain why the minimum value and the lower quartile are equal for the fiber content data set.

e. Construct a comparative boxplot and use it to comment on the differences and similarities in the fiber and sugar distributions.

3.70 (P4)
The accompanying table gives the mean and standard deviation of reaction times (in seconds) for each of two different stimuli:

	Stimulus 1	Stimulus 2
Mean	6.0	3.6
Standard deviation	1.2	0.8

Suppose your reaction time is 4.2 seconds for the first stimulus and 1.8 seconds for the second stimulus. Compared to other people, to which stimulus are you reacting more quickly?

3.71 (P4)
The report **"Education Pays: How College Shapes Lives"** (trends.collegeboard.org/sites/default/files/education -pays-how-college-shapes-lives-report-022714.pdf, retrieved April 20, 2017) gave the following percentiles for annual earnings in 2011 for full-time female workers age 25 and older with an associate degree:

25th percentile = $26,900   50th percentile = $39,300
75th percentile = $53,400

a. For each of these percentiles, write a sentence interpreting the value of the percentile.

b. The report also gave percentiles for men age 25 and older with an associate degree:

25th percentile = $36,000   50th percentile = $50,900
75th percentile = $71,900

Write a few sentences commenting on how these values compare to the percentiles for women.

TECHNOLOGY NOTES

Mean

JMP
1. Input the raw data into a column
2. Click **Analyze** and select **Distribution**
3. Click and drag the column name containing the data from the box under **Select Columns** to the box next to **Y, Columns**
4. Click **OK**

Note: These commands also produce the following statistics: standard deviation, minimum, quartile 1, median, quartile 3, and maximum.

Minitab
1. Input the raw data into C1
2. Select **Stat** and choose **Basic Statistics** then choose **Display Descriptive Statistics...**
3. Double-click C1 to add it to the **Variables** list
4. Click **OK**

Note: These commands also produce the following statistics: standard deviation, minimum, quartile 1, median, quartile 3, and maximum.

SPSS
1. Input the raw data into the first column
2. Select **Analyze** and choose **Descriptive Statistics** then choose **Explore...**
3. Highlight the column name for the variable
4. Click the arrow to move the variable to the **Dependent List** box
5. Click **OK**

Note: These commands also produce the following statistics: median, variance, standard deviation, minimum, maximum, interquartile range, and several plots.

Excel 2007
1. Input the raw data into the first column
2. Click on the **Data** ribbon and select **Data Analysis**

   Note: If you do not see **Data Analysis** listed on the Ribbon, see the Technology Notes for Chapter 2 for instructions on installing this add-on.

3. Select **Descriptive Statistics** from the dialog box
4. Click **OK**
5. Click in the box next to **Input Range:** and select the data (if you used and selected column titles, check the box next to **Labels in First Row**)
6. Check the box next to **Summary Statistics**
7. Click **OK**

Note: These commands also produce the following statistics: standard error, median, mode, standard deviation, sample variance, range, minimum, and maximum.

Note: You can also find the mean using the Excel function average.

TI-83/84
1. Input the raw data into **L1** (To access lists press the **STAT** key, highlight the option called **Edit...** then press **ENTER**)
2. Press the **STAT** key
3. Use the arrows to highlight **CALC**
4. Highlight **1-Var Stats** and press **ENTER**
5. Press the **2nd** key and then the **1** key
6. Press **ENTER**

Note: You may need to scroll to view all of the statistics. This procedure also produces the standard deviation, minimum, Q1, median, Q3, maximum.

TI-Nspire
1. Enter the data into a data list (To access data lists select the spreadsheet option and press **enter**)

   Note: Be sure to title the list by selecting the top row of the column and typing a title.

2. Press the **menu** key and select **4:Statistics** then select **1:Stat Calculations** then **1:One-Variable Statistics...**
3. Press **OK**
4. For **X1 List,** select the title for the column containing your data from the drop-down menu
5. Press **OK**

Note: You may need to scroll to view all of the statistics. This procedure also produces the standard deviation, minimum, Q1, median, Q3, maximum.

Median

JMP
1. Input the raw data into a column
2. Click **Analyze** and select **Distribution**
3. Click and drag the column name containing the data from the box under **Select Columns** to the box next to **Y, Columns**
4. Click **OK**

Note: These commands also produce the following statistics: mean, standard deviation, minimum, quartile 1, quartile 3, and maximum.

Minitab
1. Input the raw data into C1
2. Select **Stat** and choose **Basic Statistics** then choose **Display Descriptive Statistics...**
3. Double-click C1 to add it to the **Variables** list
4. Click **OK**

Note: These commands also produce the following statistics: mean, standard deviation, minimum, quartile 1, quartile 3, maximum.

SPSS
1. Input the raw data into the first column
2. Select **Analyze** and choose **Descriptive Statistics** then choose **Explore...**

3. Highlight the column name for the variable
4. Click the arrow to move the variable to the **Dependent List** box
5. Click **OK**

Note: These commands also produce the following statistics: mean, variance, standard deviation, minimum, maximum, interquartile range, and several plots.

Excel 2007

1. Input the raw data into the first column
2. Click on the **Data** ribbon and select **Data Analysis**
3. **Note:** If you do not see **Data Analysis** listed on the Ribbon, see the Technology Notes for Chapter 2 for instructions on installing this add-on.
4. Select **Descriptive Statistics** from the dialog box
5. Click **OK**
6. Click in the box next to **Input Range:** and select the data (if you used and selected column titles, check the box next to **Labels in First Row**)
7. Check the box next to **Summary Statistics**
8. Click **OK**

Note: These commands also produce the following statistics: mean, standard error, mode, standard deviation, sample variance, range, minimum, and maximum.

Note: You can also find the median using the Excel function **median**.

TI-83/84

1. Input the raw data into **L1** (To access lists press the **STAT** key, highlight the option called **Edit...** then press **ENTER**)
2. Press the **STAT** key
3. Use the arrows to highlight **CALC**
4. Highlight **1-Var Stats** and press **ENTER**
5. Press the **2nd** key and then the **1** key
6. Press **ENTER**

Note: You may need to scroll to view all of the statistics. This procedure also produces the mean, standard deviation, minimum, Q1, Q3, maximum.

TI-Nspire

1. Enter the data into a data list (To access data lists select the spreadsheet option and press **enter**)

   Note: Be sure to title the list by selecting the top row of the column and typing a title.

2. Press the **menu** key and select **4:Statistics** then select **1:Stat Calculations** then **1:One-Variable Statistics...**
3. Press **OK**
4. For **X1 List**, select the title for the column containing your data from the drop-down menu
5. Press **OK**

Note: You may need to scroll to view all of the statistics. This procedure also produces the mean, standard deviation, minimum, Q1, Q3, maximum.

Variance

JMP

1. Input the raw data into a column
2. Click **Analyze** and select **Distribution**
3. Click and drag the column name containing the data from the box under **Select Columns** to the box next to **Y, Columns**
4. Click **OK**
5. Click the red arrow next to the column name
6. Click **Display Options** then select **More Moments**

Note: These commands also produce the following statistics: standard deviation, minimum, quartile 1, median, quartile 3, and maximum.

Minitab

1. Input the raw data into C1
2. Select **Stat** and choose **Basic Statistics** then choose **Display Descriptive Statistics...**
3. Double-click C1 to add it to the **Variables** list
4. Click the **Statistics** button
5. Check the box next to **Variance**
6. Click **OK**
7. Click **OK**

SPSS

1. Input the raw data into the first column
2. Select **Analyze** and choose **Descriptive Statistics** then choose **Explore...**
3. Highlight the column name for the variable
4. Click the arrow to move the variable to the **Dependent List** box
5. Click **OK**

Note: These commands also produce the following statistics: mean, median, standard deviation, minimum, maximum, interquartile range, and several plots.

Excel 2007

1. Input the raw data into the first column
2. Click on the **Data** ribbon and select **Data Analysis**
3. **Note:** If you do not see **Data Analysis** listed on the Ribbon, see the Technology Notes for Chapter 2 for instructions on installing this add-on.
4. Select **Descriptive Statistics** from the dialog box
5. Click **OK**
6. Click in the box next to **Input Range:** and select the data (if you used and selected column titles, check the box next to **Labels in First Row**)
7. Check the box next to **Summary Statistics**
8. Click **OK**

Note: These commands also produce the following statistics: mean, standard error, median, mode, standard deviation, range, minimum, and maximum.

Note: You can also find the variance using the Excel function **var**.

TI-83/84

The TI-83/84 does not automatically produce the variance; however, this can be determined by finding the standard deviation and squaring it.

TI-Nspire

The TI-Nspire does not automatically produce the variance; however, this can be determined by finding the standard deviation and squaring it.

Standard Deviation

JMP

1. Input the raw data into a column
2. Click **Analyze** and select **Distribution**
3. Click and drag the column name containing the data from the box under **Select Columns** to the box next to **Y, Columns**
4. Click **OK**

Note: These commands also produce the following statistics: mean, minimum, quartile 1, median, quartile 3, and maximum.

Minitab

1. Input the raw data into C1
2. Select **Stat** and choose **Basic Statistics** then choose **Display Descriptive Statistics...**
3. Double-click C1 to add it to the **Variables** list
4. Click **OK**

Note: These commands also produce the following statistics: standard deviation, minimum, quartile 1, median, quartile 3, maximum.

SPSS

1. Input the raw data into the first column
2. Select **Analyze** and choose **Descriptive Statistics** then choose **Explore...**
3. Highlight the column name for the variable
4. Click the arrow to move the variable to the **Dependent List** box
5. Click **OK**

Note: These commands also produce the following statistics: mean, median, variance, minimum, maximum, interquartile range, and several plots.

Excel 2007

1. Input the raw data into the first column
2. Click on the **Data** ribbon and select **Data Analysis**
3. **Note:** If you do not see **Data Analysis** listed on the Ribbon, see the Technology Notes for Chapter 2 for instructions on installing this add-on.
4. Select **Descriptive Statistics** from the dialog box
5. Click **OK**
6. Click in the box next to **Input Range:** and select the data (if you used and selected column titles, check the box next to **Labels in First Row**)

7. Check the box next to **Summary Statistics**
8. Click **OK**

Note: These commands also produce the following statistics: mean, standard error, median, mode, sample variance, range, minimum, and maximum.

Note: You can also find the standar deviation using the Excel function **sd**.

TI-83/84

1. Input the raw data into **L1** (To access lists press the **STAT** key, highlight the option called **Edit...** then press **ENTER**)
2. Press the **STAT** key
3. Use the arrows to highlight **CALC**
4. Highlight **1-Var Stats** and press **ENTER**
5. Press the **2nd** key and then the **1** key
6. Press **ENTER**

Note: You may need to scroll to view all of the statistics. This procedure also produces the mean, minimum, Q1, median, Q3, maximum.

TI-Nspire

1. Enter the data into a data list (To access data lists select the spreadsheet option and press **enter**)

   Note: Be sure to title the list by selecting the top row of the column and typing a title.

2. Press the **menu** key and select **4:Statistics** then select **1:Stat Calculations** then **1:One-Variable Statistics...**
3. Press **OK**
4. For **X1 List**, select the title for the column containing your data from the drop-down menu
5. Press **OK**

Note: You may need to scroll to view all of the statistics. This procedure also produces the mean, minimum, Q1, median, Q3, maximum.

Quartiles

JMP

1. Input the raw data into a column
2. Click **Analyze** and select **Distribution**
3. Click and drag the column name containing the data from the box under **Select Columns** to the box next to **Y, Columns**
4. Click **OK**

Note: These commands also produce the following statistics: mean, standard deviation, minimum, median, and maximum.

Minitab

1. Input the raw data into C1
2. Select **Stat** and choose **Basic Statistics** then choose **Display Descriptive Statistics...**
3. Double-click C1 to add it to the **Variables** list
4. Click **OK**

Note: These commands also produce the following statistics: standard deviation, minimum, quartile 1, median, quartile 3, maximum.

SPSS

1. Input the raw data into the first column
2. Select **Analyze** and choose **Descriptive Statistics** then choose **Frequencies...**
3. Highlight the column name for the variable
4. Click the arrow to move the variable to the **Dependent List** box
5. Click the **Statistics** button
6. Check the box next to Quartiles
7. Click **Continue**
8. Click **OK**

Excel 2007

1. Input the raw data into the first column
2. Select the cell where you would like to place the first quartile results
3. Click the **Formulas** Ribbon
4. Click **Insert Function**
5. Select the category **Statistical** from the drop-down menu
6. In the **Select a function:** box click **Quartile**
7. Click **OK**
8. Click in the box next to **Array** and select the data
9. Click in the box next to **Quart** and type 1
10. Click **OK**

Note: To find the third quartile, type 3 into the box next to Quart in Step 9.

TI-83/84

1. Input the raw data into **L1** (To access lists press the **STAT** key, highlight the option called **Edit...** then press **ENTER**)
2. Press the **STAT** key
3. Use the arrows to highlight **CALC**
4. Highlight **1-Var Stats** and press **ENTER**
5. Press the **2nd** key and then the **1** key
6. Press **ENTER**

Note: You may need to scroll to view all of the statistics. This procedure also produces the mean, standard deviation, minimum, median, maximum.

TI-Nspire

1. Enter the data into a data list (To access data lists select the spreadsheet option and press **enter**)
   Note: Be sure to title the list by selecting the top row of the column and typing a title.
2. Press the **menu** key and select **4:Statistics** then select **1:Stat Calculations** then **1:One-Variable Statistics...**
3. Press **OK**
4. For **X1 List**, select the title for the column containing your data from the drop-down menu
5. Press **OK**

Note: You may need to scroll to view all of the statistics. This procedure also produces the mean, standard deviation, minimum, median, maximum.

IQR

JMP

JMP does not have the functionality to produce the IQR automatically.

Minitab

1. Input the raw data into C1
2. Select **Stat** and choose **Basic Statistics** then choose **Display Descriptive Statistics...**
3. Double-click C1 to add it to the **Variables** list
4. Click the **Statistics** button
5. Check the box next to **Interquartile range**
6. Click **OK**
7. Click **OK**

SPSS

1. Input the raw data into the first column
2. Select **Analyze** and choose **Descriptive Statistics** then choose **Explore...**
3. Highlight the column name for the variable
4. Click the arrow to move the variable to the **Dependent List** box
5. Click **OK**

Note: These commands also produce the following statistics: mean, median, variance, standard deviation, minimum, maximum, and several plots.

Excel 2007

1. Use the steps under the **Quartiles** section to find both the first and third quartiles
2. Click on an empty cell where you would like the result for IQR to appear
3. Type = into the cell
4. Click on the cell containing the third quartile
5. Type −
6. Click on the cell containing the first quartile
7. Press **Enter**

TI-83/84

The TI-83/84 does not have the functionality to produce the IQR automatically.

TI-Nspire

The TI-Nspire does not have the functionality to produce the IQR automatically.

Boxplot

JMP

1. Input the raw data into a column
2. Click **Analyze** and select **Distribution**
3. Click and drag the column name containing the data from the box under **Select Columns** to the box next to **Y, Columns**
4. Click **OK**

Minitab

1. Input the raw data into C1
2. Select **Graph** then choose **Boxplot...**
3. Highlight Simple under One Y
4. Click **OK**
5. Double-click C1 to add it to the **Graph Variables** box
6. Click **OK**

Note: You may add or format titles, axis titles, legends, and so on by clicking on the **Labels**... button prior to performing Step 6 above.

SPSS

1. Enter the raw data the first column
2. Select **Graph** and choose **Chart Builder...**
3. Under **Choose from** highlight Boxplot
4. Click and drag the first boxplot (Simple Boxplot) to the Chart preview area
5. Click and drag the data variable into the **Y-Axis?** box in the chart preview area
6. Click **OK**

Note: Boxplots are also be produced when summary statistics such as mean, median, standard deviation, and so on are produced.

Excel 2007

Excel 2007 does not have the functionality to create boxplots.

TI-83/84

1. Input the raw data into **L1** (To access lists press the **STAT** key, highlight the option called **Edit...** then press **ENTER**)
2. Press the **2ⁿᵈ** key and then press the **Y** = key
3. Select **Plot1** and press **ENTER**
4. Highlight **On** and press **ENTER**
5. Highlight the graph option in the second row, second column, and press **ENTER**
6. Press **GRAPH**

Note: If the graph window does not display appropriately, press the **WINDOW** button and reset the scales appropriately.

TI-Nspire

1. Enter the data into a data list (To access data list select the spreadsheet option and press **enter**)

   Note: Be sure to title the list by selecting the top row of the column and typing a title.

2. Press **menu** and select **3:Data** then select **6:QuickGraph**
3. Press **menu** and select **1:Plot Type** then select **2:Box Plot** and press **enter**

Side-by-side Boxplots

JMP

1. Enter the raw data for both groups into a column
2. Enter the group information into a second column

Column 2	Column 3
F	8
F	3
F	6
F	3
F	9
F	2
F	3
F	4
F	8
M	3
M	5
M	4
M	6
M	7
M	2

3. Click **Analyze** then select **Fit Y by X**
4. Click and drag the column name containing the raw data from the box under **Select Columns** to the box next to **Y, Response**
5. Click and drag the column name containing the group information from the box under **Select Columns** to the box next to **X, Factor**
6. Click **OK**
7. Click the red arrow next to **Oneway Analysis of...**
8. Click **Quantiles**

Minitab

1. Input the raw data for the first group into C1
2. Input the raw data for the second group into C2
3. Continue to input data for each group into a separate column
4. Select **Graph** then choose **Boxplot...**
5. Highlight Simple under Multiple Y's
6. Click **OK**
7. Double-click the column names for each column to be graphed to add it to the **Graph Variables** box
8. Click **OK**

SPSS

1. Enter the raw data the first column
2. Enter the group data into the second column

*Untitled1 [DataSet0] - PASW Statistics
File   Edit   View   Data   Transform

	VAR00001	VAR00002
1	3	male
2	6	female
3	5	male
4	7	female
5	5	male
6	8	female
7	2	male
8	3	female
9	6	male
10	5	female

3. Select **<u>G</u>raph** and choose **<u>C</u>hart Builder...**
4. Under **<u>C</u>hoose from** highlight Boxplot
5. Click and drag the first boxplot (Simple Boxplot) to the Chart preview area
6. Click and drag the data variable into the **Y-Axis?** box in the chart preview area
7. Click and drag the group variable into the **X-Axis?** box in the chart preview area
8. Click **OK**

Excel 2007
Excel 2007 does not have the functionality to create side-by-side boxplots.

TI-83/84
The TI-83/84 does not have the functionality to create side-by-side boxplots.

TI-Nspire
The TI-Nspire does not have the functionality to create side-by-side boxplots.

4 Describing Bivariate Numerical Data

Pablo Paul/Alamy Stock Photo

PREVIEW

What can you learn from bivariate numerical data? A good place to start is with a scatterplot of the data. If two variables are related, it may be possible to describe the relationship in a way that allows you to predict the value of one variable based on the value of the other. For example, if there is a relationship between a blood test measure and age and you could describe that relationship mathematically, it might be possible to predict the age of a crime victim. If you can describe the relationship between fuel efficiency and the weight of a car, you could predict the fuel efficiency of a car based on its weight. In this chapter, you will see how this can be accomplished.

CHAPTER LEARNING OBJECTIVES

Conceptual Understanding

After completing this chapter, you should be able to

C1 Understand that the relationship between two numerical variables can be described in terms of form, direction, and strength.

C2 Understand that the correlation coefficient is a measure of the strength and direction of a linear relationship.

C3 Understand the difference between a statistical relationship and a causal relationship (the difference between correlation and causation).

C4 Understand how a line might be used to describe the relationship between two numerical variables.

C5 Understand the meaning of *least squares* in the context of fitting a regression line.

C6 Understand how a least squares regression line can be used to make predictions.

C7 Explain why it is risky to use the least squares regression line to make predictions for values of the predictor variable that are outside the range of the data.

C8 Explain why it is important to consider both the standard deviation about the least squares regression line, s_e, and the value of r^2 when assessing the usefulness of the least squares regression line.

C9 Explain why it is desirable to have a small value of s_e and a large value of r^2 in a linear regression setting.

C10 Understand the role of a residual plot in assessing whether a line is the most appropriate way to describe the relationship between two numerical variables.

C11 Describe the effect of an influential observation on the equation of the least squares regression line.

Mastering the Mechanics

After completing this chapter, you should be able to

M1 Identify linear and nonlinear patterns in scatterplots.

M2 Distinguish between positive and negative linear relationships.

M3 Informally describe form, direction, and strength of a linear relationship, given a scatterplot.

M4 Calculate and interpret the value of the correlation coefficient.

M5 Know the properties of the correlation coefficient r.

M6 Identify the response variable and the predictor variable in a linear regression setting.

M7 Find the equation of the least squares regression line.

M8 Interpret the slope of the least squares regression line in context.

M9 Interpret the intercept of the least squares regression line in context, when appropriate.

M10 Use the least squares regression line to make predictions.

M11 Calculate and interpret the value of s_e, the standard deviation about the least squares regression line.

M12 Calculate and interpret the value of r^2, the coefficient of determination.

M13 Calculate and interpret residuals, given a bivariate numerical data set and the equation of the least squares regression line.

M14 Construct a residual plot in a linear regression setting.

M15 Use a residual plot to comment on the appropriateness of a line for summarizing the relationship between two numerical variables.

M16 Identify outliers and potentially influential observations in a linear regression setting.

Putting it into Practice

After completing this chapter, you should be able to

P1 Describe the relationship between two numerical variables (using a scatterplot and the correlation coefficient).

P2 Investigate the usefulness of the least squares regression line for describing the relationship between two numerical variables (using s_e, r^2, and a residual plot).

P3 Use the least squares regression line to make predictions, when appropriate.

P4 Describe the anticipated accuracy of predictions based on the least squares regression line.

PREVIEW EXAMPLE

Help for Crime Scene Investigators

Forensic scientists must often estimate the age of an unidentified crime victim. Prior to 2010, this was usually done by analyzing teeth and bones, and the resulting estimates were not very reliable. A groundbreaking study described in the paper **"Estimating Human Age from T-Cell DNA Rearrangements"** (*Current Biology* **[2010]**) examined the relationship between age and a measure based on a blood test. Age and the blood test measure were recorded for 195 people ranging in age from a few weeks to 80 years. A scatterplot of the data appears in Figure 4.1.

FIGURE 4.1
Scatterplot of age versus blood test measure

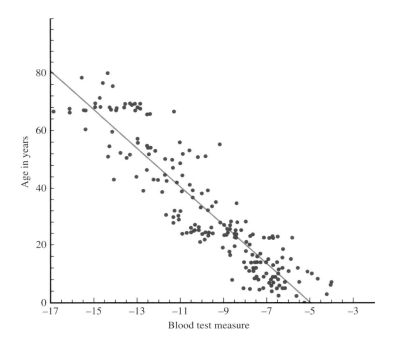

Based on the scatterplot, it does look like there is a relationship between the blood test measure and age. Smaller values of the blood test measure tend to be paired with larger values of age. Because the pattern in the scatterplot is linear, the line shown in Figure 4.1 was proposed as a way to summarize the relationship between age and the blood test measure. This line can be used to estimate the age of a crime victim from a blood sample. ■

Later in this chapter, you will revisit the preview example to see how a line that summarizes the relationship between two numerical variables can be used to make predictions and what can be said about the accuracy of those predictions.

SECTION 4.1   Correlation

A bivariate data set consists of measurements or observations on two variables, x and y. When both x and y are numerical variables, each observation is an (x, y) pair, such as (14, 5.2) or (27.63, 18.9). When investigating the relationship between two numerical variables, looking at a scatterplot of the data is the best place to start.

Look at the scatterplots of Figure 4.2 and consider these questions:

1. Does it look like there is a relationship between the two variables represented in each scatterplot?
2. If there is a relationship, is it linear?

To answer the first question, look for a pattern in each scatterplot. Notice that in the scatterplot of Figure 4.2(e), there is no obvious pattern. The points appear to be scattered at random. In this case, you would conclude that there does not appear to be a relationship between the two variables.

FIGURE 4.2
Scatterplots of five bivariate data sets

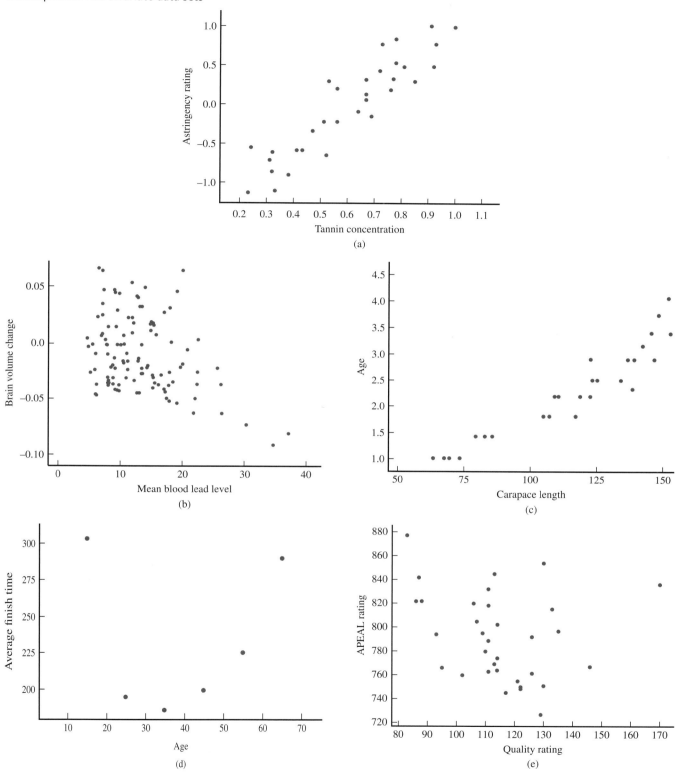

All of the other scatterplots in Figure 4.2 show some sort of pattern, although the pattern is more pronounced in some of the scatterplots than in others. Notice that the patterns in the scatterplots of Figures 4.2(a) and (b) look roughly linear, whereas the patterns in the scatterplots of Figures 4.2(c) and (d) are curved. The accompanying table summarizes how the two relationship questions would be answered for each of the five scatterplots of Figure 4.2.

Scatterplot	Variables	Does It Look Like There Is a Relationship?	Does the Relationship Look Linear?
4.2 (a)	x = Tannin concentration y = Astringency rating	Yes	Yes
4.2 (b)	x = Mean blood lead level y = Brain volume change	Yes	Yes
4.2 (c)	x = Carapace length y = Age	Yes	No
4.2 (d)	x = Age y = Average finish time	Yes	No
4.2 (e)	x = Quality rating y = APEAL rating	No	

The direction of a linear relationship can be either positive or negative. In a positive linear relationship, larger values of x tend to be paired with larger values of y. When this is the case, the linear pattern in the scatterplot slopes upward as you move from left to right. The scatterplot in Figure 4.2(a) shows a positive linear relationship between x = Tannin concentration and y = Astringency rating. The scatterplot of Figure 4.2(b) shows a negative linear relationship between x = Blood lead level and y = Brain volume change. How would you describe a negative linear relationship? Do larger values of x tend to be paired with larger or smaller values of y? Does the linear pattern in the scatterplot slope upward or downward as you move from left to right?

Next, consider the four scatterplots of Figure 4.3. These scatterplots were constructed using data from graphs in the paper **"Simple Memory Test Predicts Intelligence"** (*Archives*

FIGURE 4.3

Four scatterplots showing positive linear relationships

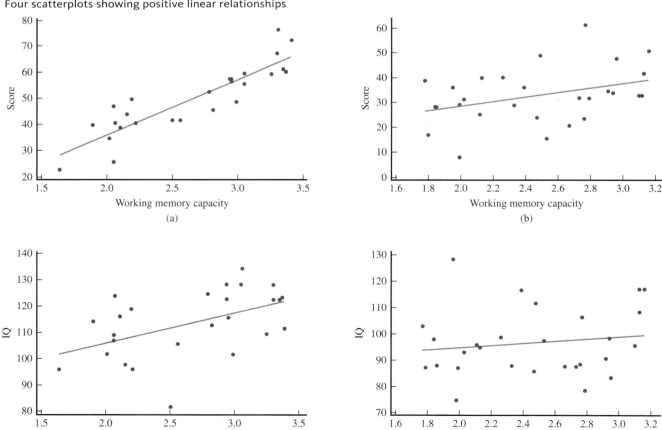

of General Psychiatry [2010]: 570–577). The variables used to create these scatterplots will be described in more detail in Example 4.14. For now, just look at the general patterns in these plots. All four of these scatterplots were described as showing a positive linear relationship. The line shown in each scatterplot was used to describe the relationship between *x* and *y*.

When the points in a scatterplot tend to cluster tightly around a line, the relationship is described as strong. Try to order the scatterplots in Figure 4.3 from the one that shows the strongest relationship to the one that shows the weakest. For which of the four scatterplots does the line provide the most information about how *x* and *y* are related? For which one is the line least informative?

Did you order the scatterplots (a), (c), (b), (d)? Scatterplots (b) and (c) are pretty close, so don't worry if you had the order of those two reversed. It can sometimes be difficult to judge the strength of a linear relationship from a scatterplot. In these cases, it is helpful to have a way to measure the strength of a linear relationship. The correlation coefficient is a statistic that can be used to assess the strength and direction of a linear relationship.

Pearson's Sample Correlation Coefficient

Pearson's sample correlation coefficient, denoted by *r*, measures the strength and direction of a linear relationship between two numerical variables. It is usually referred to as just the correlation coefficient. Let's first investigate some general properties of *r*.

Consider Figure 4.4, which shows eight scatterplots and the associated values of the correlation coefficient. Keeping in mind that *r* measures the strength and direction of a linear relationship, look at the figure and try to answer the following questions:

- When is the value of *r* positive?
- When is the value of *r* negative?
- For positive linear relationships, as the value of *r* increases, is the linear relationship getting stronger or weaker?
- What is the largest possible value of *r*? Why?
- What is the smallest possible value of *r*? Why?
- If $r = -0.7$ for one data set and $r = 0.7$ for a different data set, which data set has the stronger linear relationship?

Let's look at one more scatterplot, shown in Figure 4.5. The correlation coefficient for this data set is $r = 0.038$, which is close to zero. Because *r* is a measure of the strength of *linear* relationship, this indicates that there is not a linear relationship between Age and Average finish time. The scatterplot, however, shows a definite *nonlinear* pattern. This is an important point—when the value of *r* is near 0, you should not conclude that there is no relationship whatsoever between the two variables. Be sure to look at the scatterplot of the data to see if there might be a nonlinear relationship.

Properties of *r*

By looking at the scatterplots and correlation coefficients in Figure 4.4, you should now have discovered most of the general properties of *r* that follow.

1. The sign of *r* matches the direction of the linear relationship: *r* is positive when the linear relationship is positive and negative when the linear relationship is negative.
2. *The value of r is always greater than or equal to −1 and less than or equal to +1.* A value near the upper limit, +1, indicates a strong positive relationship, whereas a value close to the lower limit, −1, suggests a strong negative relationship. Figure 4.6 shows a useful way to describe the strength of a linear relationship based on the value of *r*. It may seem surprising that a value as far away from 0 as −0.5 or +0.5 is in the weak category. An explanation for this is given later in the chapter.

FIGURE 4.4
Scatterplots and correlation coefficients

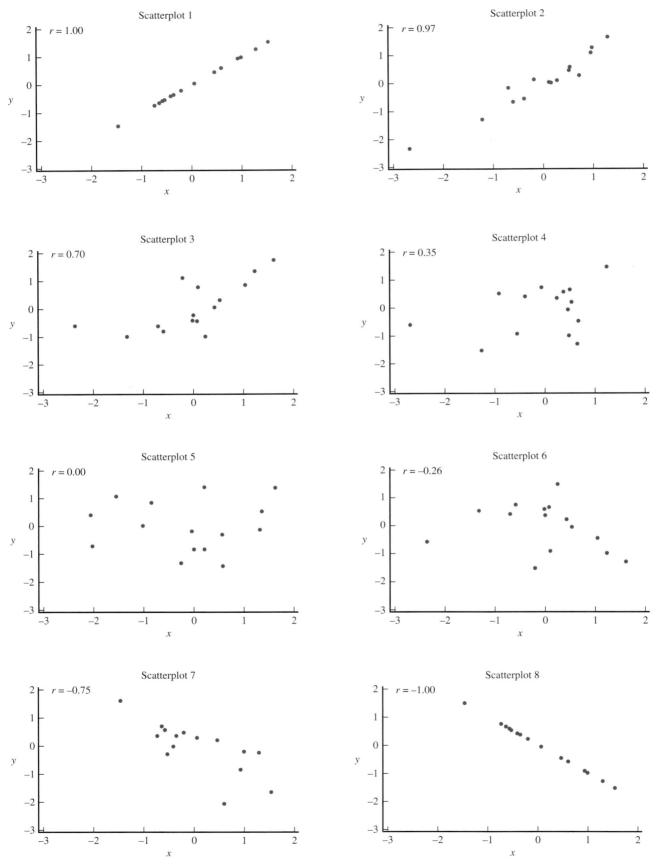

FIGURE 4.5
Scatterplot showing a strong but nonlinear relationship

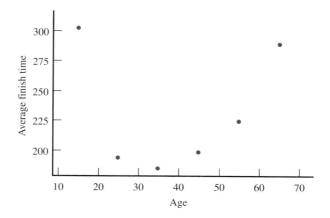

FIGURE 4.6
Describing the strength of a linear relationship

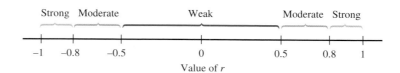

3. $r = 1$ only when all the points in the scatterplot fall on a straight line that slopes upward. Similarly, $r = -1$ only when all the points fall on a downward-sloping line.
4. *r is a measure of the extent to which x and y are linearly related*—that is, the extent to which the points in the scatterplot fall close to a straight line. Even if the value of r is close to 0, there could still be a strong nonlinear relationship.
5. *The value of r does not depend on the unit of measurement for either variable.* For example, if x is height, the value of r is the same whether height is expressed in inches, meters, or miles.

Calculating and Interpreting the Value of the Correlation Coefficient

The first step in calculating the correlation coefficient is to calculate a z-score for each x value (by subtracting $\bar{x}$ and then dividing by s_x):

$$z_x = \frac{x - \bar{x}}{s_x}$$

Next, calculate a z-score for each y value (by subtracting $\bar{y}$ and then dividing by s_y):

$$z_y = \frac{y - \bar{y}}{s_y}$$

Then use the formula in the following box to calculate the value of r.

The correlation coefficient *r* is calculated using the following formula:

$$r = \frac{\Sigma z_x z_y}{n - 1}$$

$\Sigma z_x z_y$ is found by multiplying z_x and z_y for each observation in the data set and then adding the $z_x z_y$ values.

Example 4.1 Missing Teeth

The article **"More Americans Turning to Dentures to Get, Keep Jobs" (*USA TODAY*, August 10, 2015)** includes the following statement:

> Some areas that top the nation for unemployment also fare worst in a health measure that can keep people from getting jobs—missing teeth.

Is there a relationship between the unemployment rate and the percentage of the population with no natural teeth? The table below gives the 2015 unemployment rate for the eight states in the South Atlantic region of the United States (from the report **"Regional and State Unemployment – 2015," Bureau of Labor Statistics, www.bls.gov/news .release/pdf/srgune.pdf, retrieved September 1, 2016)**. Also given in the table is the percentage of seniors with no remaining natural teeth (from the report **"Worst Dental Health: States," www.bloomberg.com/graphics/best-and-worst/#worst-dental-health -states, retrieved April 21, 2017)**.

State	Percentage with No Natural Teeth	2015 Unemployment Rate (%)
Delaware	16.4	4.9
Florida	13.3	5.4
Georgia	21.0	5.9
Maryland	13.6	5.2
North Carolina	21.5	5.7
South Carolina	21.6	6.0
Virginia	15.0	4.4
West Virginia	36.0	6.7

Figure 4.7 is a scatterplot of these data. Based on the scatterplot, it appears that there is a positive relationship that is approximately linear between unemployment rate and percentage with no natural teeth.

FIGURE 4.7
Scatterplot of 2015 unemployment rate versus percentage with no natural teeth

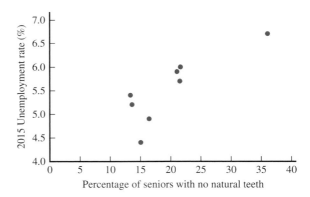

If you are calculating the value of the correlation coefficient by hand, it is a good idea to keep several decimal places in the intermediate calculations to avoid compounding rounding errors. In this example, three decimal places are used in the intermediate calculations.

Using x to denote the percentage with no natural teeth and y to denote 2015 unemployment rate, you can calculate the following values:

$$\bar{x} = 19.800 \qquad s_x = 7.419 \qquad \bar{y} = 5.525 \qquad s_y = 0.713$$

To calculate the value of the correlation coefficient, begin by calculating z-scores for each (x, y) pair in the data set. For example, the first observation (for Delaware) is (16.4, 4.9). The corresponding z-scores are

$$z_x = \frac{16.4 - 19.800}{7.419} = -0.458 \qquad z_y = \frac{4.9 - 5.525}{0.713} = -0.877$$

The following table shows the z-scores and the product $z_x z_y$ for each observation.

State	x	y	z_x	z_y	$z_x z_y$
Delaware	16.4	4.9	−0.458	−0.877	0.402
Florida	13.3	5.4	−0.876	−0.175	0.154
Georgia	21.0	5.9	0.162	0.526	0.085
Maryland	13.6	5.2	−0.836	−0.456	0.381
North Carolina	21.5	5.7	0.229	0.245	0.056
South Carolina	21.6	6.0	0.243	0.666	0.162
Virginia	15.0	4.4	−0.647	−1.578	1.021
West Virginia	36.0	6.7	2.184	1.648	3.599

Then

$$\Sigma z_x z_y = 5.859$$

and

$$r = \frac{\Sigma z_x z_y}{n - 1} = \frac{5.859}{8 - 1} = \frac{5.859}{7} = 0.837$$

Based on the scatterplot and the value of the correlation coefficient, you would conclude that for these eight states, there is a strong positive linear relationship between percentage of seniors with no natural teeth and 2015 unemployment rate.

Hand calculation of the value of the correlation coefficient is quite tedious. Fortunately, all statistical software packages and most scientific and graphing calculators can compute the value of r. For example, SPSS, a widely used statistics package, was used to compute the correlation coefficient, producing the following computer output.

Correlations

		Unemployment Rate	Percentage with No Teeth
Unemployment Rate	Pearson Correlation	1	.837**
	Sig. (2-tailed)		.009
	N	8	8
Percentage with No Teeth	Pearson Correlation	.837**	1
	Sig. (2-tailed)	.009	
	N	8	8

**Correlation is significant at the 0.01 level (2-tailed).

Example 4.2 A Face You Can Trust

The article **"How to Tell if a Guy Is Trustworthy"** *(LiveScience,* **March 8, 2010)** described an interesting research study published in the journal *Psychological Science* **("Valid Facial Cues to Cooperation and Trust: Male Facial Width and Trustworthiness,"** *Psychological Science* **[2010]: 349–354).** This study investigated whether there is a relationship between facial characteristics and peoples' assessment of trustworthiness. Sixty-two students were told that they would play a series of two-person games in which they could earn real money. In each game, the student could choose between two options:

1. They could end the game immediately and a total of $10 would be paid out, with each player receiving $5.
2. They could "trust" the other player. In this case, $3 would be added to the money available, but the other player could decide to either split the money fairly with $6.50 to each player or could decide to keep $10 and only give $3 to the first player.

Each student was shown a picture of the person he or she would be playing with. The student then indicated whether they would end the game immediately and take $5 or trust the other player to do the right thing in hopes of increasing the payout to $6.50. This process was repeated for a series of games, each with a photo of a different second player. For each photo, the researchers recorded the width-to-height ratio of the face in the photo and the percentage of the students who chose the trust option when shown that photo. A representative subset of the data (from a graph that appeared in the paper) is given here:

Face Width-to-Height Ratio	Percentage Choosing Trust Option	Face Width-to-Height Ratio	Percentage Choosing Trust Option
1.75	58	2.10	62
1.80	80	2.15	42
1.90	30	2.15	31
1.95	50	2.15	15
1.95	65	2.15	65
2.00	10	2.20	42
2.00	28	2.20	22
2.00	62	2.20	14
2.00	70	2.25	9
2.00	63	2.25	20
2.05	8	2.25	35
2.05	35	2.30	43
2.05	50	2.30	8
2.05	58	2.35	41
2.10	64	2.40	53
2.10	25	2.40	35
2.10	35	2.60	55
2.10	50	2.68	20
2.10	66	2.70	11

A scatterplot of these data is shown in Figure 4.8.

FIGURE 4.8
Scatterplot of percent choosing trust option versus facial width-to-height ratio

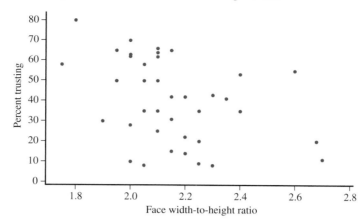

JMP was used to compute the value of the correlation coefficient, with the following result:

Correlation					
Variable	Mean	Std Dev	Correlation	Signif. Prob	Number
Percentage Choosing Trust Option	40.26316	20.58691	-0.39169	0.0150*	38
Face Width-to-Height Ratio	2.153421	0.209598			

The value of the correlation coefficient (-0.392) is found in the JMP output under the heading "Correlation." This indicates a weak negative linear relationship between face width-to-height ratio and the percentage of people who chose to trust the face when playing the game. The relationship is negative, indicating that those with larger face width-to-height ratios (wider faces) tended to be trusted less. Based on this observation, the researchers concluded, "It is clear that male facial width ratio is a cue to male trustworthiness and that it predicts trust placed in male faces."

An interesting side note: In another study (described in the same article) where subjects were randomly assigned to the player 1 and player 2 roles, the researchers found that those with larger face width-to-height ratios *were* less trustworthy. They found a positive correlation between player 2 face width-to-height ratio and the percentage of the time that player 2 decided to keep $10 and only give $3 to player 1.

Example 4.3   Does It Pay to Pay More for a Bike Helmet?

Are more expensive bike helmets safer than less expensive ones? Data on x = Price (in dollars) and y = Quality rating for 35 different brands of bike helmets were used to construct the scatterplot in Figure 4.9. The data are from the *Consumer Reports* web site **(www.consumerreports.org /products/bike-helmets/ratings-overview/, retrieved August 28, 2016)**. Quality rating was a number from 0 (the worst possible rating) to 100 and was determined using factors that included how well the helmet absorbed the force of an impact, the strength of the helmet, ventilation, and ease of use.

FIGURE 4.9
Scatterplot for the bike helmet data of Example 4.3

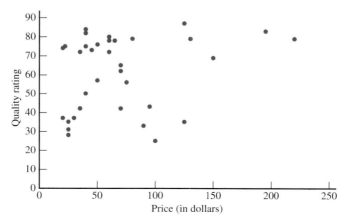

From the scatterplot, it appears that there is only a weak positive relationship between price and quality rating. The correlation coefficient, obtained using Minitab, is

Correlations: Price, Quality Rating

Pearson correlation of Price and Quality Rating = 0.221

A correlation coefficient of $r = 0.221$ indicates that although there is a weak tendency for higher quality ratings to be associated with higher priced helmets, the relationship is not very strong. In fact, some inexpensive helmets had high-quality ratings.

How the Correlation Coefficient Measures the Strength of a Linear Relationship

The correlation coefficient measures the strength of a linear relationship by using z-scores. Because $z_x = \dfrac{x - \bar{x}}{s_x}$, x values that are larger than $\bar{x}$ will have positive z-scores and those smaller than $\bar{x}$ will have negative z-scores. Also y values larger than $\bar{y}$ will have positive z-scores and those smaller will have negative z-scores. Keep this in mind as you look at the scatterplots in Figure 4.10. The scatterplot in Figure 4.10(a)

FIGURE 4.10
Viewing a scatterplot according to the signs of z_x and z_y;
(a) a positive relationship;
(b) a negative relationship,
(c) no strong relationship

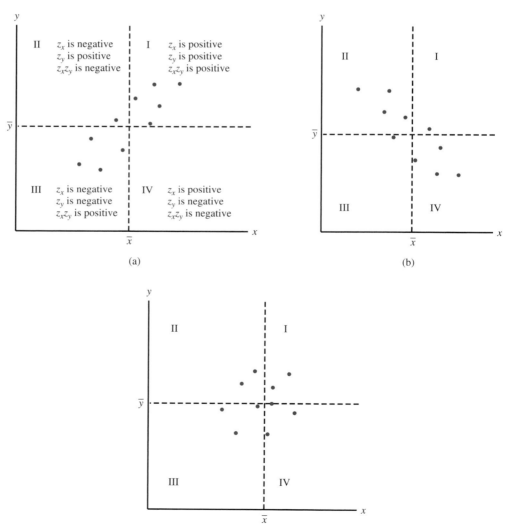

indicates a strong positive relationship. A vertical line through $\bar{x}$ and a horizontal line through $\bar{y}$ divide the plot into four regions. In Region I, x is greater than $\bar{x}$ and y is greater than $\bar{y}$. So, in this region, the z-score for x and the z-score for y are both positive. It follows that the product $z_x z_y$ is positive. The product of the z-scores is also positive for any point in Region III, because in Region III both z-scores are negative, and multiplying two negative numbers results in a positive number. In each of the other two regions, one z-score is positive and the other is negative, so $z_x z_y$ is negative. But because the points in Figure 4.10(a) generally fall in Regions I and III, the products of z-scores tend to be positive, and the *sum* of the products will be a relatively large positive number.

Similar reasoning for the data displayed in Figure 4.10(b), which exhibits a strong negative relationship, implies that $\Sigma z_x z_y$ will be a relatively large (in magnitude) negative number. When there is no strong relationship, as in Figure 4.10(c), positive and negative products tend to counteract one another, producing a value of $\Sigma z_x z_y$ that is close to zero. The sum $\Sigma z_x z_y$ can be a large positive number, a large negative number, or a number close to 0, depending on whether there is a strong positive, a strong negative, or no strong linear relationship. Dividing this sum by $n-1$ results in a value between -1 and $+1$.

Correlation and Causation

A value of r close to 1 indicates that the larger values of one variable tend to be associated with the larger values of the other variable. This does not mean that a large value of one variable *causes* the value of the other variable to be large. Correlation measures the extent of association, but *association does not imply causation*. It frequently happens that two variables are highly correlated not because a change in one causes a change in the other but because they are both strongly related to a third variable. For example, among all elementary school children, the relationship between the number of cavities in a child's teeth and the size of his or her vocabulary is strong and positive. Yet no one advocates eating foods that result in more cavities to increase vocabulary size. Number of cavities and vocabulary size are both strongly related to age, so older children tend to have higher values of both variables. In the ABCNews.com series **"Who's Counting?" (February 1, 2001)**, John Allen Paulos reminded readers that correlation does not imply causation and gave the following example: Consumption of hot chocolate is negatively correlated with crime rate (high values of hot chocolate consumption tend to be paired with lower crime rates), but both are responses to cold weather.

Scientific experiments can frequently make a strong case for causality by carefully controlling the values of all variables that might be related to the ones under study. Then, if y is observed to change in a "smooth" way as the experimenter changes the value of x, there may be a causal relationship between x and y. In the absence of such control and ability to manipulate values of one variable, there is always the possibility that an unidentified third variable is influencing both variables under investigation. A high correlation in many uncontrolled studies carried out in different settings can also marshal support for causality—as in the case of cigarette smoking and cancer—but proving causality is an elusive task.

Summing It Up—Section 4.1

The following learning objectives were addressed in this section:

Conceptual Understanding
C1: Understand that the relationship between two numerical variables can be described in terms of form, direction, and strength.
A useful way to explore whether there is a relationship between two numerical variables is by looking at a scatterplot. An obvious pattern in the scatterplot indicates that there is a relationship. The terms linear and nonlinear can be used to describe the form of a relationship. If the points in a scatterplot appear to be scattered around a line, the relationship is described as approximately linear. If the points appear to be scattered around a curve rather than a line, the relationship is described as nonlinear.

The direction of a relationship is positive if larger values of one variable tend to be paired with larger values of the other variable. If larger values of one variable tend to be paired with smaller values of the other variable, the direction of the relationship is negative.

Relationships are also described as weak, moderate, or strong, depending on how tightly the points in the scatterplot tend to cluster around a line or a curve. Figure 4.3 and the discussion of the scatterplots in that figure provide examples of linear relationships of varying strengths.

C2: Understand that the correlation coefficient is a measure of the strength and direction of a linear relationship.

The correlation coefficient is a statistic that is calculated using bivariate numerical data. The value of the correlation coefficient is always greater than or equal to -1 and less than or equal to $+1$, and it is used to represent the direction and strength of a linear relationship between two numerical variables. For an explanation of how the value of the correlation coefficient is related to strength and direction of a linear relationship, see the discussion of Figure 4.4.

C3: Understand the difference between a statistical relationship and a causal relationship (the difference between correlation and causation).

There is a statistical relationship between two numerical variables if they tend to vary together in a predictable way. This does not necessarily mean that there is a cause-and-effect relationship between the two variables. For a discussion of this important idea, see the subsection "Correlation and Causation" that appears at the end of Section 4.1.

Mastering the Mechanics

M1: Identify linear and nonlinear patterns in scatterplots.

Figure 4.2 and the discussion that follows the figure illustrate how to identify linear and nonlinear patterns in scatterplots.

M2: Distinguish between positive and negative linear relationships.

Figure 4.2 and the discussion that follows the figure illustrate the difference between positive and negative relationships.

M3: Informally describe form, direction, and strength of a linear relationship, given a scatterplot.

When describing a linear relationship, it is common to address form (linear or nonlinear), direction (positive or negative), and strength (weak, moderate, or strong).

M4: Calculate and interpret the value of the correlation coefficient.

For examples of the calculation and interpretation of the value of the correlation coefficient, see Examples 4.1 and 4.2.

M5: Know the properties of the correlation coefficient, *r*.

Important properties of the correlation coefficient include (1) r is negative when the relationship is negative and positive when the relationship is positive; (2) the value of r is always greater than or equal to -1 and less than or equal to $+1$; and (3) a value of $r = 1$ or $r = -1$ indicates a perfect linear relationship. For a discussion of these and other properties of the correlation coefficient, see the discussion that follows Figure 4.5.

Putting It into Practice

P1: Describe the relationship between two numerical variables (using a scatterplot and the correlation coefficient).

Based on a scatterplot and the value of the correlation coefficient, relationships between two numerical variables are usually described in terms of form, direction, and strength. For examples that illustrate how relationships are described, see Examples 4.2 and 4.3.

SECTION 4.1 EXERCISES

Each Exercise Set assesses the following chapter learning objectives: C1, C2, C3, M1, M2, M3, M4, M5, P1

SECTION 4.1 Exercise Set 1

4.1 For each of the scatterplots shown, answer the following questions:

- **i.** Does there appear to be a relationship between x and y?
- **ii.** If so, does the relationship appear to be linear?
- **iii.** If so, would you describe the linear relationship as positive or negative?

Scatterplot 1:

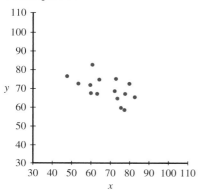

Scatterplot 2:

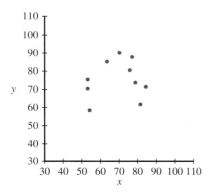

Scatterplot 3:

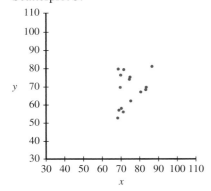

Scatterplot 4:

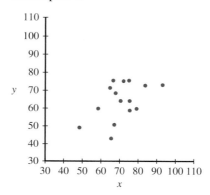

4.2 For each of the following pairs of variables, indicate whether you would expect a positive correlation, a negative correlation, or a correlation close to 0. Explain your choice.
- **a.** Interest rate and number of loan applications
- **b.** Height and IQ
- **c.** Height and shoe size
- **d.** Minimum daily temperature and cooling cost

4.3 The paper **"The Relationship Between Cell Phone Use, Academic Performance, Anxiety, and Satisfaction with Life in College Students"** (*Computers in Human Behavior* **[2014]: 343–350)** described a study of cell phone use among undergraduate college students at a large, Midwestern public university. The paper reported that the value of the correlation coefficient between x = Cell phone use (measured as total amount of time (in hours) spent using a cell phone on a typical day) and y = GPA (cumulative grade point average (GPA) determined from university records) was $r = -0.203$.

- **a.** Interpret the given value of the correlation coefficient. Does the value of the correlation coefficient suggest that students who use a cell phone for more hours per day tend to have higher GPAs or lower GPAs?
- **b.** The study also investigated the correlation between texting (measured as the total number of texts sent and texts received per day) and GPA. The direction of the relationship between texting and GPA was the same as the direction of the relationship between cell phone use and GPA, but the relationship between texting and GPA was not as strong. Which of the following possible values for the correlation coefficient between texting and GPA could have been the one observed?

$$r = -0.30 \qquad r = -0.10 \qquad r = 0.10 \qquad r = 0.30$$

- **c.** The paper included the following statement: "Participants filled in two blanks—one for texts sent and one for texts received. These two texting items were nearly perfectly correlated." Do you think that the value of the correlation coefficient for texts sent and texts received was close to -1, close to 0, or close to $+1$? Explain your reasoning.

4.4 The article **"$115K! The 13 Best Paying U.S. Companies"** (*USA TODAY*, **August 11, 2015**) gave the following data on median worker pay (in thousands of dollars) and the 1-year percent change in stock price for the 13 highest paying companies in the United States.

Company	Median Worker Pay	Percent Change in Stock Price
Jupiter Networks	134.7	21.7
Netflix	132.2	92.2
Equinix	125.0	33.9
Altera	122.8	52.2
Visa	122.5	41.9
Yahoo	121.2	2.8
Xilinx	121.2	5.2
VeriSign	119.0	30.1
Microsoft	118.0	8.0
Broadcom	118.0	37.3
F5 Networks	117.6	15.9
Adobe Systems	117.4	22.0
eBay	115.1	−44.7

a. Construct a scatterplot for these data.
b. Calculate and interpret the value of the correlation coefficient.
c. The article states that companies that pay more are seeing a payoff in their stock performance. Is this conclusion justified based on these data? Explain.
d. Is it reasonable to generalize conclusions based on these data to the population of all companies in the United States? Explain why or why not.

4.5 Each year, marketing firm **J.D. Power and Associates** surveys new car owners 90 days after they purchase their cars. These data are used to rate auto brands (Toyota, Ford, etc.) on quality and customer satisfaction. Data for 2015 on a quality rating (number of defects per 100 vehicles) and a satisfaction score for 30 brands sold in the United States are given in the accompanying table.
a. Construct a scatterplot of y = Satisfaction rating versus x = Quality rating. How would you describe the relationship between x and y?
b. Calculate and interpret the value of the correlation coefficient.

Brand	Quality Rating	Satisfaction Rating
Acura	129	814
Audi	134	858
BMW	142	849
Buick	106	792
Cadillac	145	826
Chevrolet	125	791
Chrysler	165	783
Dodge	208	790
Fiat	171	768
Ford	204	789
GMC	120	791
Honda	126	783
Hyundai	158	804
Infiniti	136	826
Jeep	181	762
Kia	153	791
Land Rover	198	853
Lexus	95	844
Lincoln	132	835
Mazda	163	790
Mercedes-Benz	135	842
MINI	155	795
Mitsubishi	161	748
Nissan	173	786
Porsche	97	882
Scion	167	779
Subaru	166	766
Toyota	113	783
Volkswagen	169	796
Volvo	152	812

4.6 Is the following statement correct? Explain why or why not.

A correlation coefficient of 0 implies that there is no relationship between two variables.

4.7 The article **"That's Rich: More You Drink, More You Earn"** (*Calgary Herald*, **April 16, 2002**) reported that there was a positive correlation between alcohol consumption and income. Is it reasonable to conclude that increasing alcohol consumption will increase income? Explain why or why not.

4.8 The paper **"Noncognitive Predictors of Student Athletes' Academic Performance"** (*Journal of College Reading and Learning* **[2000]: e167)** summarizes a study of 200 Division I athletes. It was reported that the correlation coefficient for college grade point average (GPA) and a measure of academic self-worth was $r = 0.48$. Also reported were the correlation coefficient for college GPA and high school GPA ($r = 0.46$) and the correlation coefficient for college GPA and a measure of tendency to procrastinate ($r = -0.36$). Write a few sentences summarizing what these correlation coefficients tell you about GPA for the 200 athletes in the sample.

SECTION 4.1 **Exercise Set 2**

4.9 For each of the scatterplots shown, answer the following questions:
  i. Does there appear to be a relationship between x and y?
  ii. If so, does the relationship appear to be linear?
  iii. If so, would you describe the linear relationship as positive or negative?

Scatterplot 1:

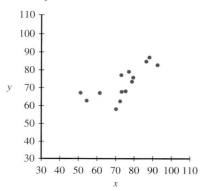

Scatterplot 2:

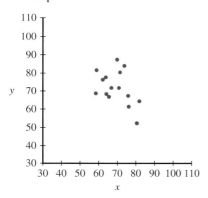

Scatterplot 3:

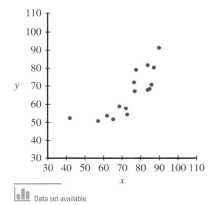

Data set available

Scatterplot 4:

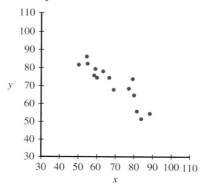

4.10 For each of the following pairs of variables, indicate whether you would expect a positive correlation, a negative correlation, or a correlation close to 0. Explain your choice.
a. Weight of a car and gas mileage
b. Size and selling price of a house
c. Height and weight
d. Height and number of siblings

4.11 The paper **"Digit Ratio as an Indicator of Numeracy Relative to Literacy in 7-Year-Old British Schoolchildren" (British Journal of Psychology [2008]: 75–85)** investigated a possible relationship between x = Digit ratio (the ratio of the length of the second finger to the length of the fourth finger) and y = Difference between numeracy score and literacy score on a national assessment. (The digit ratio is thought to be inversely related to the level of prenatal testosterone exposure.) The authors concluded that children with smaller digit ratios tended to have larger differences in test scores, meaning that they tended to have a higher numeracy score than literacy score. This conclusion was based on a correlation coefficient of $r = -0.22$. Does the value of the correlation coefficient indicate that there is a strong linear relationship? Explain why or why not.

4.12 Draw two scatterplots, one for which $r = 1$ and a second for which $r = -1$.

4.13 Data from the **U.S. Federal Reserve Board (www.federalreserve.gov/releases/housedebt/, retrieved April 21, 2017)** on consumer debt (as a percentage of personal income) and mortgage debt (also as a percentage of personal income) for the 10 years from 2006 to 2015 are shown in the following table:

Consumer Debt	Household Debt
6.73	5.97
7.07	5.97
6.96	5.87
6.69	5.60
6.18	5.16
5.58	5.06
5.14	4.97
4.95	5.22
4.73	5.31
4.59	5.45

a. What is the value of the correlation coefficient for this data set?
b. Is it reasonable to conclude that there is no strong relationship between the variables (linear or otherwise)? Use a graphical display to support your answer.

4.14 The paper **"Religiosity and Teen Birth Rate in the United States"** (*Reproductive Health* [2009]: 14–20) included data on teen birth rate and on a measure of conservative religious beliefs for each of 49 U.S. states. Birth rate was measured as births per 1000 teenage females. The authors of the paper created a "religiosity" score for each state by using responses to a large national survey conducted by the Pew Forum on Religion and Public Life. Higher religiosity scores correspond to more conservative religious beliefs. The authors reported that "teen birth rate is very highly correlated with religiosity at the state level, with more religious states having a higher rate of teen birth." The following data for 49 states are from a graph in the paper.
a. Construct a scatterplot of $y =$ Birth rate versus $x =$ Religiosity score. How would you describe the relationship between x and y? What aspects of the scatterplot support what the authors concluded about teen birth rate and religiosity?
b. Calculate and interpret the value of the correlation coefficient.

Religiosity	Teen Birth Rate	Religiosity	Teen Birth Rate
26.5	10.7	42.1	24.5
28.9	8.5	42.6	37.5
29.7	16.2	42.8	58.3
30.1	36.7	43.3	32.0
30.5	11.2	43.3	31.3
32.1	13.7	44.1	26.4
32.7	18.3	44.4	32.4
34.8	16.1	44.7	35.6
35.0	36.1	45.5	38.0
35.9	21.7	46.1	34.0
36.8	15.1	47.5	30.9
36.9	27.1	49.1	37.1
37.1	31.9	50.1	57.2
38.3	18.5	51.1	47.2
38.6	24.5	51.1	53.3
39.4	49.1	51.3	47.7
39.5	56.1	53.5	42.4
39.8	24.0	54.2	56.2
39.9	31.5	54.7	25.0
40.7	17.0	55.3	46.9
40.7	25.0	55.5	47.7
40.9	21.8	55.7	45.9
41.2	34.0	58.1	46.5
41.4	24.7	62.3	62.8
41.4	31.9		

ADDITIONAL EXERCISES

4.15 Each year the Harris Poll surveys Americans on a number of issues. It uses responses to several questions to calculate a "Happiness" index that measures overall happiness. The article **"Latest Happiness Index Reveals American Happiness at All-Time Low"** (www.theharrispoll.com/health-and-life/American-Happiness_at-All-Time-Low.html, retrieved April 21, 2017) included the happiness index for the seven years between 2008 and 2016. Also included in the article were the percentages of people who responded "Somewhat Agree" or "Strongly Agree" to the following statements:

Statement 1 (Happy with Life) 1: At this time, I'm generally happy with my life.
Statement 2 (Won't Benefit): I won't get much benefit from the things that I do anytime soon.

Year	Happiness Index	Happy with Life Statement (percentage who Somewhat or Strongly Agree)	Won't Benefit Statement (percentage who Somewhat or Strongly Agree)
2008	35	83	32
2009	35	81	38
2010	33	80	36
2011	33	80	38
2013	33	77	42
2015	34	82	36
2016	31	81	41

a. Calculate the value of the correlation coefficient for Happiness index and the response to the Happy with Life statement.
b. Calculate the value of the correlation coefficient for Happiness index and the response to the Won't Benefit statement.
c. Is there a stronger relationship between Happiness index and the response to Statement 1 or between Happiness index and the response to Statement 2?
d. Write a few sentences describing the relationships between Happiness index and the responses to the two statements.

4.16 The amount of money spent each year on science, space, and technology in the United States (in millions of dollars) and the amount of money spent on pets in the United States (in billions of dollars) for the years 2000 to 2009 were used to construct the following graph. (The data are from the web site www.tylervigen.com/spurious-correlations, retrieved August 28, 2016.)

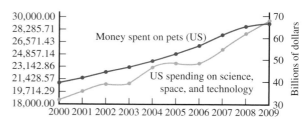

Based on these time series plots, would the correlation coefficient between amount spent on science, space, and technology and the amount of money spent on pets be positive or negative? Weak or strong? What aspect of the time series plots support your answer?

4.17 Below are the data used to construct the time series plots in the previous exercise. Calculate the value of the correlation coefficient for the amount spent on science, space, and technology and the amount spent on pets. Explain how this value is consistent with your answer to the previous exercise.

Year	Amount Spent on Science, Space, and Technology (millions of dollars)	Amount Spent on Pets (billions of dollars)
2000	18,594	39.7
2001	19,753	41.9
2002	20,734	44.6
2003	20,831	48.8
2004	23,029	49.8
2005	23,597	53.1
2006	23,584	56.9
2007	25,525	61.8
2008	27,731	65.7
2009	29,449	67.1

4.18 An auction house released a list of 25 recently sold paintings. The artist's name and the sale price of each painting appear on the list. Would the correlation coefficient be an appropriate way to summarize the relationship between artist and sale price? Why or why not?

4.19 A sample of automobiles traveling on a particular segment of a highway is selected. Each one travels at roughly a constant rate of speed, although speed does vary from auto to auto. Let x = Speed and y = Time needed to travel this segment. Would the sample correlation coefficient be closest to 0.9, 0.3, -0.3, or -0.9? Explain.

4.20 It may seem odd, but biologists can tell how old a lobster is by measuring the concentration of pigment in the lobster's eye. The authors of the paper **"Neurolipofuscin Is a Measure of Age in *Panulirus argus*, the Caribbean Spiny Lobster, in Florida" (*Biological Bulletin* [2007]: 55–66)** wondered if it was sufficient to measure the pigment in just one eye, which would be the case if there is a strong relationship between the concentration in the right eye and the concentration in the left eye. Pigment concentration (as a percentage of tissue sample) was measured in both eyes for 39 lobsters, resulting in the following summary quantities (based on data from a graph in the paper):

$$n = 39 \qquad \Sigma x = 88.8 \qquad \Sigma y = 86.1$$
$$\Sigma xy = 281.1 \qquad \Sigma x^2 = 288.0 \qquad \Sigma y^2 = 286.6$$

An alternative formula for calculating the correlation coefficient that doesn't involve calculating the z-scores is

$$r = \frac{\Sigma xy - \frac{(\Sigma x)(\Sigma y)}{n}}{\sqrt{\Sigma x^2 - \frac{(\Sigma x)^2}{n}}\sqrt{\Sigma y^2 - \frac{(\Sigma y)^2}{n}}}$$

Use this formula to calculate the value of the correlation coefficient, and interpret this value.

SECTION 4.2   Linear Regression: Fitting a Line to Bivariate Data

When there is a relationship between two numerical variables, you can use information about one variable to learn about the value of the second variable. For example, you might want to predict y = Product sales for a month when x = Amount spent on advertising is $10,000. The letter y is used to denote the variable you would like to predict, and this variable is called the **response variable** (also sometimes called the **dependent variable**). The other variable, denoted by x, is the **predictor variable** (also sometimes called the **independent or explanatory variable**).

Many scatterplots exhibit linear patterns. When this is the case, the relationship can be summarized by finding a line that is as close as possible to the points in the scatterplot. Before seeing how this is done, let's review some elementary facts about lines and linear relationships.

Lines and Linear Relationships

The equation of a line is

$$y = a + bx$$

A particular line is specified by choosing values of a and b. For example, one line is $y = 10 + 2x$ and another is $y = 10 - 5x$. If we choose some x values and calculate

$y = a + bx$ for each value, the points corresponding to the resulting (x, y) pairs will fall exactly on a straight line.

> ### DEFINITION
>
> The equation of a line is
>
> $$y = a + bx$$
>
> The value of b, called the **slope** of the line, is the amount by which y increases when x increases by 1 unit.
>
> The value of a, called the **intercept** (also called the **y-intercept** or **vertical intercept**) of the line, is the height of the line above the value $x = 0$.

The line $y = 10 + 2x$ has slope $b = 2$, so y will increase by 2 units for each 1-unit increase in x. When $x = 0$, $y = 10$, so the line crosses the vertical axis (where $x = 0$) at a height of 10. This is illustrated in Figure 4.11(a). The slope of the line $y = 100 - 5x$ is -5, so y decreases by 5 when x increases by 1. The height of this line above $x = 0$ is $a = 100$. This line is pictured in Figure 4.11(b).

FIGURE 4.11
Graphs of two lines
(a) slope $b = 2$, intercept $a = 10$;
(b) slope $b = -5$, intercept $a = 100$

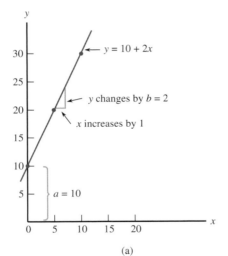
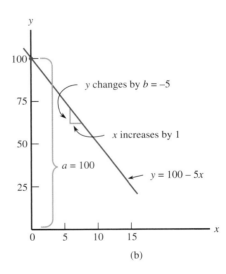

(a)   (b)

It is easy to draw the graph of any particular line. Choose any two x values and substitute them into the equation to obtain the corresponding y values. Plot the resulting two (x, y) pairs and then draw the line that passes through these points.

Choosing a Line to Summarize a Linear Relationship: The Principle of Least Squares

Consider the scatterplot shown in Figure 4.12(a). This scatterplot shows a positive linear relationship between x and y. You can think about using a line to summarize this relationship, but what line should you use? Two possible lines are shown in Figure 4.12(b). It is easy to choose between these two lines—Line 1 provides a better "fit" for these data.

But now consider the two lines shown in Figure 4.12(c). Which of these two lines is a better choice for describing the relationship? Here the choice isn't obvious. You need a way to judge how well a line fits the data in order to find the "best" line.

To measure how well a particular line fits a given data set, you can use the vertical deviations from the line. There is a deviation for each point in the scatterplot. To calculate

FIGURE 4.12

Comparing the fit of two lines

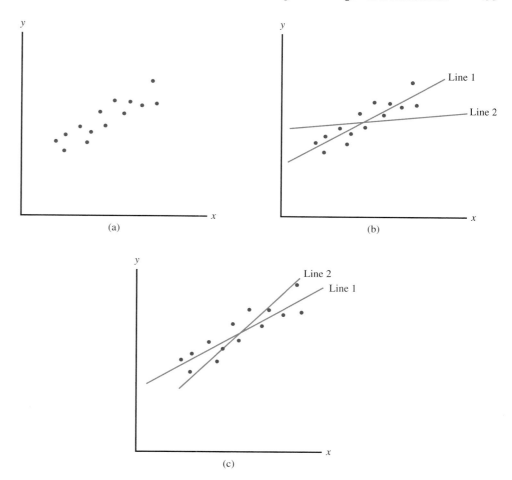

the deviation for a point (x, y), find the height of the line for that x value (by substituting x into the equation of the line) and then subtract this value from y.

For example, Figure 4.13 shows a scatterplot that also includes the line

$$y = 10 + 2x$$

The labeled points in the plot are the fifth and sixth points from the left. These two points are $(x_5, y_5) = (15, 44)$ and $(x_6, y_6) = (20, 45)$. For these two points, the vertical deviations from the line are

$$\text{5th deviation} = y_5 - \text{height of the line at } x_5$$
$$= 44 - [10 + 2(15)]$$
$$= 4$$

and

$$\text{6th deviation} = y_6 - \text{height of the line at } x_6$$
$$= 45 - [10 + 2(20)]$$
$$= -5$$

Notice that deviations are positive for data points that fall above the line and negative for data points that fall below the line.

A particular line is a good fit if the data points tend to be close to the line. This means that the deviations from the line will be small in magnitude. Looking back at Figure 4.12(b), Line 2 does not fit as well as Line 1 because the deviations from Line 2 tend to be larger than the deviations from Line 1.

To assess the overall fit of a line, the n deviations can be combined into a single measure of fit. The standard approach is to square the deviations (to obtain nonnegative numbers) and then to sum these squared deviations.

FIGURE 4.13
Scatterplot with line and two deviations shown

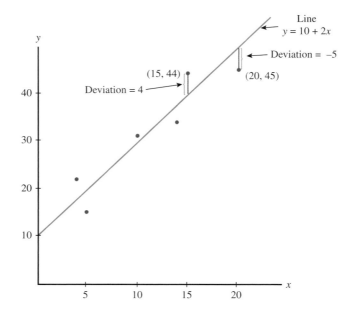

DEFINITION

A widely used measure of the fit of a line $y = a + bx$ to bivariate data $(x_1, y_1), (x_2, y_2), \ldots, (x_n, y_n)$ is the **sum of squared deviations** about the line:

$$\Sigma[y - (a + bx)]^2 = [y_1 - (a + bx_1)]^2 + [y_2 - (a + bx_2)]^2 + \ldots + [y_n - (a + bx_n)]^2$$

You can use the sum of squared deviations to judge the fit of a line. For a given data set, you would like to find the line that makes this sum as small as possible. This line is called the **least squares regression line**.

DEFINITION

The **least squares regression line** is the line that minimizes the sum of squared deviations.

Fortunately, the equation of the least squares regression line can be obtained without having to calculate deviations from any particular line. The accompanying box gives relatively simple formulas for the slope and intercept of the least squares regression line.

DEFINITION

The slope of the least squares regression line is

$$b = \frac{\Sigma(x - \bar{x})(y - \bar{y})}{\Sigma(x - \bar{x})^2}$$

and the y intercept is

$$a = \bar{y} - b\bar{x}$$

The equation of the least squares regression line is

$$\hat{y} = a + bx$$

where $\hat{y}$ (read as y-hat) is the predicted value of y that results from substituting a particular x value into the equation.

Statistical software packages and graphing calculators can compute the slope and intercept of the least squares regression line. If the slope and intercept are to be calculated by hand, a calculating formula for the slope can be used to reduce the amount of time required (see Exercise 4.33).

Example 4.4   Pomegranate Juice and Tumor Growth

Pomegranate, a fruit native to Persia, has been used in the folk medicines of many cultures to treat various ailments. Researchers have investigated if pomegranate's antioxidant properties are useful in the treatment of cancer. One study, described in the paper **"Pomegranate Fruit Juice for Chemoprevention and Chemotherapy of Prostate Cancer"** (*Proceedings of the National Academy of Sciences* [October 11, 2005]: 14,813–14,818), investigated whether pomegranate fruit extract (PFE) was effective in slowing the growth of prostate cancer tumors. In this study, 24 mice were injected with cancer cells. The mice were then randomly assigned to one of three treatment groups. One group of eight mice received normal drinking water, the second group of eight mice received drinking water supplemented with 0.1% PFE, and the third group received drinking water supplemented with 0.2% PFE. The average tumor volume for the mice in each group was recorded at several points in time. The following data (approximated from a graph in the paper) are for the mice that received plain drinking water. Here, y = Average tumor volume (in mm^3) and x = Number of days after injection of cancer cells.

x	11	15	19	23	27
y	150	270	450	580	740

A scatterplot of these data (Figure 4.14) shows that the relationship between these two variables can reasonably be summarized by a straight line.

FIGURE 4.14
Scatterplot for the tumor volume data of Example 4.4

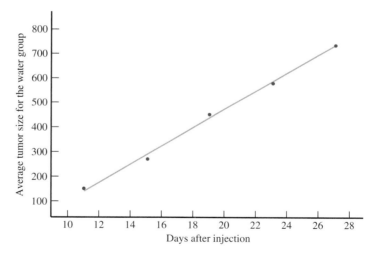

The summary quantities necessary to calculate the equation of the least squares regression line are

$$\Sigma x = 95 \qquad \Sigma y = 2190 \qquad \Sigma(x - \bar{x})(y - \bar{y}) = 5960 \qquad \Sigma(x - \bar{x})^2 = 160$$

From these quantities,

$$\bar{x} = 19 \qquad \bar{y} = 438$$

$$b = \frac{\Sigma(x - \bar{x})(y - \bar{y})}{\Sigma(x - \bar{x})^2} = \frac{5960}{160} = 37.25$$

and

$$a = \bar{y} - b\bar{x} = 438 - (37.25)(19) = -269.75$$

The least squares regression line is then

$$\hat{y} = -269.75 + 37.25x$$

Computer software or a graphing calculator could also be used to find the equation of the least squares regression line. For example, JMP was used to produce the following output. The equation of the least squares regression line appears at the bottom of the output.

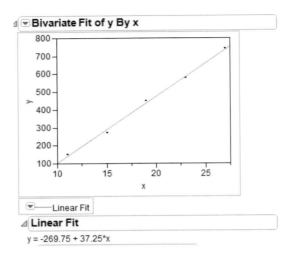

Bivariate Fit of y By x

Linear Fit
y = -269.75 + 37.25*x

To predict average tumor volume 20 days after injection of cancer cells for mice who drink plain water, use the y value of the point on the least squares regression line above $x = 20$:

$$\hat{y} = -269.75 + 37.25(20) = 475.25$$

You could predict average tumor volumes for a different numbers of days in the same way.

Figure 4.15 shows a scatterplot for both the group of mice that drank plain water and the group that drank water supplemented by 0.2% PFE. Notice that the tumor growth seems to be much slower for the 0.2% PFE group. In fact, for this group, the relationship between average tumor volume and number of days after injection of cancer cells appears to be curved rather than linear.

FIGURE 4.15

Scatterplot of average tumor volume versus number of days after injection of cancer cells for the water group and for the 0.2% PFE group

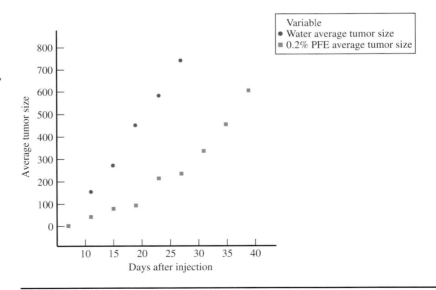

A word of caution: Think twice about using the least squares regression line to predict y for an x value that is outside the range of x values in the data set, because you don't know whether the observed linear pattern continues outside this range. Making predictions outside the range of the x values in the data set can produce misleading predictions if the pattern does not continue. This is sometimes referred to as the **danger of extrapolation**.

Consider the least squares regression line in Example 4.4. You can see that using this line to predict average tumor volume for fewer than 10 days after injection of cancer cells can lead to nonsensical predictions. For example, if the number of days after injection is five, the predicted average tumor volume is negative:

$$\hat{y} = -269.75 + 37.25(5) = -83.5$$

Because it is impossible for average tumor volume to be negative, it is a clear that the pattern observed for x values from 11 to 27 days does not apply to values less than 11 days, and it may not apply to values greater than 27 days either. Even so, the least squares regression line can be a useful tool for making predictions within the 11- to 27-day range.

The Danger of Extrapolation

The least squares regression line should not be used to make predictions outside the range of the x values in the data set because the linear pattern may not continue outside this range.

Example 4.5 Ankle Motion and Balance

Good balance is important in preventing falls, especially in older adults. The paper **"Correlation Between Ankle Range of Motion and Balance in Community-Dwelling Elderly Population"** (*Indian Journal of Physiotherapy and Occupational Therapy* **[2012]: 127–129**) describes a study of the relationship between ankle range of motion and balance. The accompanying data on x = Ankle range of motion (in degrees) and y = Balance test score for 30 people ages 65 to 85 are consistent with a scatterplot and summary measures given in the paper.

x = Ankle Range of Motion	y = Balance Score	x = Ankle Range of Motion	y = Balance Score
85	12.1	82	13.6
74	14.0	78	14.7
76	10.4	86	14.2
88	11.8	86	13.4
49	6.0	101	16.8
75	9.6	75	12.7
70	8.2	52	10.3
60	9.9	70	10.9
81	14.2	77	9.2
54	6.8	53	8.1
50	7.0	87	13.7
96	15.5	65	6.6
60	8.3	78	8.6
65	9.5	66	9.5
61	11.7	74	11.6

Minitab was used to construct a scatterplot of the data (Figure 4.16) and to compute the value of the correlation coefficient.

Correlations: Ankle Range of Motion, Balance Score

Pearson correlation of Ankle Range of Motion and Balance Score = 0.807 P-Value = 0.000

FIGURE 4.16
Minitab scatterplot for the data
of Example 4.5

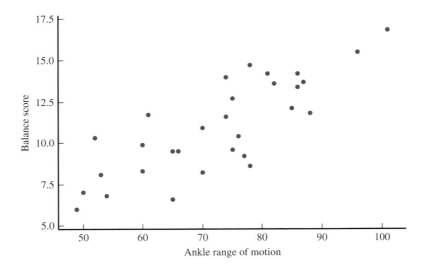

Notice that there is a linear pattern in the scatterplot, suggesting that a line is a reasonable way to summarize the relationship between balance score and ankle range of motion. The correlation coefficient of $r = 0.807$ is consistent with the pattern in the scatterplot and indicates a strong positive linear relationship.

Figure 4.17 shows part of the Minitab regression output. Instead of x and y, the labels "Balance Score" and "Ankle Range of Motion" are used. In the table just below the equation of the regression line, the first row gives information about the intercept, a, and the second row gives information about the slope, b. Notice that the coefficient column, labeled "Coef", contains more precise values for a and b than those that appear in the equation.

FIGURE 4.17
Partial Minitab output for
Example 4.5

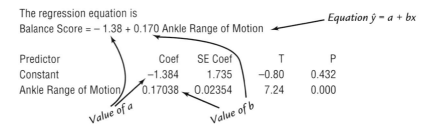

Figure 4.18 shows the scatterplot with the least squares regression line included. The equation of the least squares regression line

$$\hat{y} = -1.38 + 0.170x$$

can be used to predict the balance score for a given ankle range of motion. For example, for a person with ankle range of motion of 85 degrees, you would predict a balance score of

$$\hat{y} = -1.38 + 0.170(85) = 13.07$$

The least squares regression line should not be used to predict the balance score for people with ankle ranges of motion such as $x = 30$ or $x = 120$. These x values are well outside the range of the data, and you do not know if a linear relationship is appropriate outside the observed range.

FIGURE 4.18

Scatterplot and least squares
regression line for the range
of motion and balance data of
Example 4.5

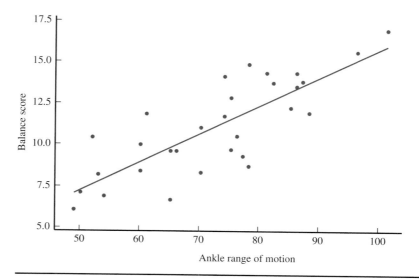

Regression

Why is the line used to summarize a linear relationship called the least squares regression line? The "least squares" part of the name means the line has a smaller sum of squared deviations than any other line. But why do we include "regression" in the name? This terminology comes from the relationship between the least squares regression line and Pearson's sample correlation coefficient. To understand this relationship, alternative expressions for the slope b and the equation of the least squares regression line are needed. With s_x and s_y denoting the sample standard deviations of the x's and y's, respectively, a bit of algebra results in

$$b = r\left(\frac{s_y}{s_x}\right)$$

$$\hat{y} = \bar{y} + r\left(\frac{s_y}{s_x}\right)(x - \bar{x})$$

You do not need to use these formulas in any computations, but several of their implications are important for appreciating what the least squares regression line does.

1. When $x = \bar{x}$ is substituted into the equation of the line, $\hat{y} = \bar{y}$ results. That is, the least squares regression line passes through the point of averages $(\bar{x}, \bar{y})$.

2. Suppose for the moment that $r = 1$. Then all the data points fall on the line whose equation is

$$\hat{y} = \bar{y} + \left(\frac{s_y}{s_x}\right)(x - \bar{x})$$

Now substitute $x = \bar{x} + s_x$, which is an x value that is one standard deviation above $\bar{x}$:

$$\hat{y} = \bar{y} + \frac{s_y}{s_x}(\bar{x} + s_x - \bar{x}) = \bar{y} + s_y$$

This means that with $r = 1$, when x is one standard deviation above its mean, you predict that the associated y value will be one standard deviation above its mean. Similarly, if $x = \bar{x} - 2s_x$ (two standard deviations below its mean), then

$$\hat{y} = \bar{y} + \frac{s_y}{s_x}(\bar{x} - 2s_x - \bar{x}) = \bar{y} - 2s_y$$

which is also two standard deviations below the mean. If $r = -1$, then $x = \bar{x} + s_x$ results in $\hat{y} = \bar{y} - s_y$, so the predicted y is also one standard deviation from its mean but on the opposite side of $\bar{y}$ from where x is relative to $\bar{x}$. In general, if x and y are perfectly correlated, the predicted y value will be the same number of standard deviations from its mean, $\bar{y}$, as x is from its mean, $\bar{x}$.

3. Now suppose that x and y are not perfectly correlated. For example, suppose $r = 0.5$. Then the equation of the least squares regression line is

$$\hat{y} = \bar{y} + 0.5 \left(\frac{s_y}{s_x}\right)(x - \bar{x})$$

Substituting $x = \bar{x} + s_x$ gives

$$\hat{y} = \bar{y} + 0.5 \frac{s_y}{s_x}(\bar{x} + s_x - \bar{x}) = \bar{y} + 0.5 s_y$$

That is, for $r = 0.5$, when x is one standard deviation above its mean, you predict that y will be only 0.5 standard deviation above its mean. Similarly, if $r = -0.5$ then the predicted y value will be only half the number of standard deviations from $\bar{y}$ than x is from $\bar{x}$, but in the opposite direction.

Consider using the least squares regression line to predict the value of y associated with an x value that is a certain number of standard deviations away from $\bar{x}$. The predicted y value will be only r times this number of standard deviations from $\bar{y}$. Except when $r = 1$ or $r = -1$, the predicted y will always be closer to $\bar{y}$ (in standard deviations) than x is to $\bar{x}$.

Using the least squares regression line for prediction results in a predicted y that is pulled back in, or *regressed*, toward the mean of y compared to where x is relative to its mean (this is the meaning of the phrase "regression to the mean"). This regression effect was first noticed by Sir Francis Galton (1822–1911), a famous biologist, when he was studying the relationship between the heights of fathers and their sons. He found that sons whose fathers were taller than average were predicted to be taller than average (because r is positive here) but not as tall as their fathers. He found a similar relationship for sons whose fathers were shorter than average. This "regression effect" led to the term **regression analysis**, which involves fitting lines, curves, and more complicated functions to bivariate and multivariate data.

The alternative form of the least squares regression line emphasizes that predicting y from knowledge of x is not the same problem as predicting x from knowledge of y. The slope of the least squares regression line for predicting x is $r\left(\frac{s_x}{s_y}\right)$ rather than $r\left(\frac{s_y}{s_x}\right)$ and the intercepts of the lines are almost always different. For purposes of prediction, it makes a difference whether y is regressed on x, as we have done, or x is regressed on y. *The least squares regression line of y on x should not be used to predict x, because it is not the line that minimizes the sum of squared deviations in the x direction.*

Summing It Up—Section 4.2

The following learning objectives were addressed in this section:

Conceptual Understanding
C4: Understand how a line might be used to describe the relationship between two numerical variables.
When there is a linear pattern in a scatterplot of bivariate data, it is possible to use a line to describe how one variable tends to change as the other variable increases.

C5: Understand the meaning of *least squares* in the context of fitting a regression line.
The least squares regression line is the line that minimizes the sum of the squared deviations (the differences between the actual y values and the y values predicted by the line). No other line has a smaller sum of squared deviations.

C6: Understand how a least squares regression line can be used to make predictions.
The equation of the least squares regression line has the form $\hat{y} = a + bx$. This line can be used to obtain a predicted y value ($\hat{y}$) for a given x value by substituting that x value into the equation and then calculating the value of $\hat{y}$.

C7: Explain why it is risky to use the least squares regression line to make predictions for values of the predictor variable that are outside the range of the data.

The least squares regression line should not be used to make predictions outside the range of the x values in the data set because there is no evidence that the linear pattern continues outside this range.

Mastering the Mechanics

M6: Identify the response variable and the predictor variable in a linear regression setting.

In a regression setting, the response variable is the variable you would like to predict and the predictor is the variable that will be used to make the prediction. The response variable is also sometimes called the dependent variable, and the predictor variable is also sometimes called the independent variable or the explanatory variable.

M7: Find the equation of the least squares regression line.

The equation of the least squares regression line can be found using technology or by using the formulas for the slope and intercept given in this section. The calculation of the equation of the least squares regression line is illustrated in Examples 4.4 and 4.5.

M8: Interpret the slope of the least squares regression line in context.

The slope of the least squares regression line is interpreted as the change in the predicted value of y (the predicted value of the response variable) when the value of the predictor variable increases by one unit.

M9: Interpret the intercept of the least squares regression line in context, when appropriate.

The intercept of the least squares regression line is interpreted as the predicted value of y when the value of x is 0. It is only appropriate to interpret the intercept in this way if 0 is in the range of the x values in the data set, because predicting the value of y for x values that are outside the range of the data set can be misleading and should be avoided.

M10: Use the least squares regression line to make predictions.

The equation of the least squares regression line has the form $\hat{y} = a + bx$. This line can be used to obtain a predicted y value ($\hat{y}$) for a given x value by substituting that x value into the equation and then calculating the value of $\hat{y}$.

Putting It into Practice

P3: Use the last squares regression line to make predictions, when appropriate.

When there is a linear relationship between two numerical variables, the least squares regression line can be used to make a prediction of a y value for a given x value. However, such predictions should be limited to values of x that are within the range of the x values in the data set used to obtain the equation of the least squares regression line.

SECTION 4.2   EXERCISES

Each Exercise Set assesses the following chapter learning objectives: C4, C5, C6, C7, M6, M7, M8, M9, M10, P3

SECTION 4.2   Exercise Set 1

4.21 Two scatterplots are shown below. Explain why it makes sense to use the least squares regression line to summarize the relationship between x and y for one of these data sets but not the other.

Scatterplot 1

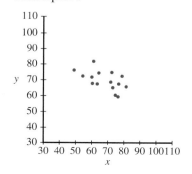

Scatterplot 2

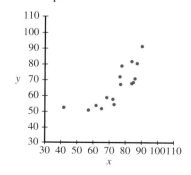

4.22 Data on y = Time to complete a task (in minutes) and x = Number of hours of sleep on the previous night were used to find the least squares regression line.

The equation of the line was $\hat{y} = 12 - 0.36x$. For this data set, would the sum of squared deviations from the line $y = 12.5 - 0.5x$ be greater or less than the sum of squared deviations from the least squares regression line? Explain your choice. (Hint: Think about the definition of the least squares regression line.)

4.23 Data on x = Size of a house (in square feet) and y = Amount of natural gas used (therms) during a specified period were used to fit the least squares regression line. The slope was 0.017 and the intercept was −5.0. Houses in this data set ranged in size from 1000 to 3000 square feet.
a. What is the equation of the least squares regression line?
b. What would you predict for gas usage for a 2100 sq. ft. house?
c. What is the approximate change in gas usage associated with a 1 sq. ft. increase in size?
d. Would you use the least squares regression line to predict gas usage for a 500 sq. ft. house? Why or why not?

4.24 Medical researchers have noted that adolescent females are much more likely to deliver low-birth-weight babies than are adult females. Because low-birth-weight babies have a higher mortality rate, a number of studies have examined the relationship between birth weight and mother's age. One such study is described in the article **"Body Size and Intelligence in 6-Year-Olds: Are Offspring of Teenage Mothers at Risk?"** (*Maternal and Child Health Journal* [2009]: 847–856). The following data on maternal age (in years) and birth weight of baby (in grams) are consistent with summary values given in the article and also with data published by the **National Center for Health Statistics**.

Mother's age	15	17	18	15	16	19
Birth weight	2289	3393	3271	2648	2897	3327

Mother's age	17	16	18	19
Birth weight	2970	2535	3138	3573

a. If the goal is to learn about how birth weight is related to mother's age, which of these two variables is the response variable and which is the predictor variable?
b. Construct a scatterplot of these data. Would it be reasonable to use a line to summarize the relationship between birth weight and mother's age?
c. Find the equation of the least squares regression line.
d. Interpret the slope of the least squares regression line in the context of this study.
e. Does it make sense to interpret the intercept of the least squares regression line? If so, give an interpretation. If not, explain why it is not appropriate for this data set. (Hint: Think about the range of the x values in the data set.)
f. What would you predict for birth weight of a baby born to an 18-year-old mother?

g. What would you predict for birth weight of a baby born to a 15-year-old mother?
h. Would you use the least squares regression equation to predict birth weight for a baby born to a 23-year-old mother? If so, what is the predicted birth weight? If not, explain why.

4.25 Acrylamide is a chemical that is sometimes found in cooked starchy foods and which is thought to increase the risk of certain kinds of cancer. The paper **"A Statistical Regression Model for the Estimation of Acrylamide Concentrations in French Fries for Excess Lifetime Cancer Risk Assessment"** (*Food and Chemical Toxicology* [2012]: 3867–3876) describes a study to investigate the effect of Frying time (in seconds) and Acrylamide concentration (in micrograms per kilogram) in french fries. The data in the accompanying table are approximate values read from a graph that appeared in the paper.

Frying Time	Acrylamide Concentration
150	155
240	120
240	190
270	185
300	140
300	270

a. If the goal is to learn how Acrylamide concentration is related to Frying time, which of these two variables is the response variable and which is the predictor variable?
b. Construct a scatterplot of these data. Describe any interesting features of the scatterplot.
c. Find the equation of the least squares regression line for predicting Acrylamide concentration using Frying time.
d. Is the slope of the least squares line positive or negative? Is this consistent with the scatterplot in Part (b)?
e. Do the scatterplot and the equation of the least squares regression line support the conclusion that longer frying times tend to be paired with higher acrylamide concentrations? Explain.
f. What is the predicted acrylamide concentration for a frying time of 225 seconds?
g. Would you use the least squares regression equation to predict acrylamide concentration for a frying time of 500 seconds? If so, what is the concentration? If not, explain why.

4.26 The authors of the paper **"Statistical Methods for Assessing Agreement Between Two Methods of Clinical Measurement"** (*International Journal of Nursing Studies* [2010]: 931–936) compared two different instruments for measuring a person's ability to breathe out air. (This measurement is helpful in diagnosing various lung disorders.) The two instruments considered were a Wright peak flow meter and a mini-Wright peak flow meter. Seventeen people

Data set available

participated in the study, and for each person air flow was measured once using the Wright meter and once using the mini-Wright meter.

Subject	Mini-Wright Meter	Wright Meter	Subject	Mini-Wright Meter	Wright Meter
1	512	494	4	428	434
2	430	395	5	500	476
3	520	516	6	600	557
7	364	413	13	260	267
8	380	442	14	477	478
9	658	650	15	259	178
10	445	433	16	350	423
11	432	417	17	451	427
12	626	656			

a. Suppose that the Wright meter is considered to provide a better measure of air flow, but the mini-Wright meter is easier to transport and to use. If the two types of meters produce different readings but there is a strong relationship between the readings, it would be possible to use a reading from the mini-Wright meter to predict the reading that the larger Wright meter would have given. Use the given data to find an equation to predict Wright meter reading using a reading from the mini-Wright meter.

b. What would you predict for the Wright meter reading for a person whose mini-Wright meter reading was 500?

c. What would you predict for the Wright meter reading for a person whose mini-Wright meter reading was 300?

SECTION 4.2 Exercise Set 2

4.27 Two scatterplots follow. Explain why it makes sense to use the least squares regression line to summarize the relationship between x and y for one of these data sets but not the other.

Scatterplot 1

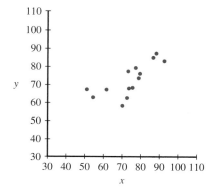

 Data set available

Scatterplot 2

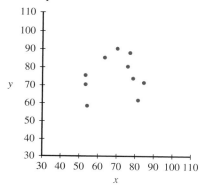

4.28 What does it mean when we say that the regression line is the *least squares* regression line?

4.29 The accompanying data are a subset of data from the report **"Great Jobs, Great Lives" (Gallup-Purdue Index 2015 Report, www.gallup.com/reports/197144/gallup-purdue -index-report-2015.aspx, retrieved April 22, 2017)**. The values are approximate values read from a scatterplot. Students at a number of universities were asked if they agreed that their education was worth the cost. One variable in the table is the percentage of students at the university who responded that they strongly agree. The other variable in the table is the *U.S. News & World Report* ranking of the university.

Ranking	Percentage of Alumni Who Strongly Agree
28	53
29	58
30	62
37	55
45	54
47	62
52	55
54	62
57	70
60	58
65	66
66	55
72	65
75	57
82	67
88	59
98	75

a. Construct a scatterplot for these data.

b. What is the value of the correlation coefficient?

c. Find the equation of the least squares regression line.

d. Predict the percentage of alumni who would strongly agree that their education was worth the cost for a university with a ranking of 50.

e. Explain why it would not be a good idea to use the least squares regression line to predict the percentage of alumni who strongly agree that their education was worth the cost for a university that had a ranking of 10.

4.30 The authors of the paper **"Evaluating Existing Movement Hypotheses in Linear Systems Using Larval Stream Salamanders"** (*Canadian Journal of Zoology* **[2009]: 292–298**) investigated whether water temperature was related to how far a salamander would swim and to whether it would swim upstream or downstream. Data for 14 streams with different mean water temperatures appear below (from a graph in the paper). The two variables of interest are mean water temperature (°C) and net directionality, which was defined as the difference in the relative frequency of released larvae moving upstream and the relative frequency of released larvae moving downstream. A positive value of net directionality means a higher proportion were moving upstream than downstream. A negative value means a higher proportion were moving downstream.

Mean Temperature (x)	Net Directionality (y)	Mean Temperature (x)	Net Directionality (y)
6.17	−0.08	17.98	0.29
8.06	0.25	18.29	0.23
8.62	−0.14	19.89	0.24
10.56	0.00	20.25	0.19
12.45	0.08	19.07	0.14
11.99	0.03	17.73	0.05
12.50	−0.07	19.62	0.07

a. If the goal is to learn about how net directionality is related to mean temperature, which of these two variables is the response variable and which is the predictor variable?

b. Construct a scatterplot of the data. How would you describe the relationship between these two variables?

c. Find the equation of the least squares regression line describing the relationship between y = Net directionality and x = Mean water temperature.

d. Interpret the slope of the least squares regression line in the context of this study.

e. Does it make sense to interpret the intercept of the least squares regression line? If so, give an interpretation. If not, explain why.

f. What value of net directionality would you predict for a stream that had a mean water temperature of 15°C?

g. The authors state that "when temperatures were warmer, more larvae were captured moving upstream, but when temperatures were cooler, more larvae were captured moving downstream." Do the scatterplot and least squares regression line support this statement?

4.31 The **California State Park System Statistical Report** for the 2014/2015 Fiscal Year **(www.parks.ca.gov/pages/795 /files/14-15%20Statistical%20Report%20-%20INTERNET .pdf, retrieved April 22, 2017)** gave the accompanying data on

x = Amount of money collected in user fees (in thousands of dollars) and y = Operating cost (in thousands of dollars) for nine state parks in the North Coast Redwoods District.

User Fees (thousands of dollars)	Operating Costs (thousands of dollars)
17	100
74	359
811	3,582
380	1,377
33	241
427	760
500	1,044
734	2,205
760	1,620

a. Use a statistical software package or a graphing calculator to construct a scatterplot of these data. Describe any interesting features of the scatterplot.

b. Find the equation of the least squares regression line (use software or a graphing calculator).

c. Is the slope of the least squares regression line positive or negative? Is this consistent with your description in Part (a)?

d. Based on the scatterplot, do you think that the correlation coefficient for this data set would be less than 0.5 or greater than 0.5? Explain.

4.32 The report **"Airline Quality Rating 2016"** **(www .airlinequalityrating.com/reports/2016_AQR_Final.pdf, retrieved April 22, 2017)** included the accompanying data on the on-time arrival percentage and the number of complaints files per 100,000 passengers for U.S. airlines.

The report did not include data on the number of complaints for two of the airlines. Use the given data from the other airlines to fit the least squares regression line and use it to predict the number of complaints per 100,000 passengers for Spirit and for Virgin America.

Airline	On-Time Arrival Percentage	Complaints per 100,000 Passengers
Alaska	86	0.42
American	80	2.12
Delta	86	0.72
Envoy Air	74	1.59
Express Jet	78	1.01
Frontier	73	3.91
Hawaiian	88	0.89
JetBlue	76	1.17
SkyWest	80	0.84
Southwest	80	0.53
Spirit	69	Not reported
United	78	2.71
Virgin America	80	Not reported

Data set available

ADDITIONAL EXERCISES

4.33 The following table gives data on age, number of cell phone calls made in a typical day, and number of text messages sent in a typical day for a random sample of 10 people selected from those enrolled in adult education classes offered by a school district.

Age	Number of Cell Phone Calls	Number of Text Messages Sent	Age	Number of Cell Phone Calls	Number of Text Messages Sent
55	6	20	19	15	14
33	5	35	30	10	30
60	1	3	33	3	2
38	2	1	37	3	1
55	4	2	52	5	20

An alternative formula for calculating the slope of the least squares regression line that doesn't involve calculating the deviations from mean for the x values and the y values is

$$b = \frac{\Sigma xy - \frac{(\Sigma x)(\Sigma y)}{n}}{\Sigma x^2 - \frac{(\Sigma x)^2}{n}}$$

Use this formula to find the equation of the least squares regression line for predicting y = Number of cell phone calls using x = Age as a predictor.

4.34 Use the data given in the previous exercise to find the equation of the least squares regression line for predicting y = Number of text messages sent using x = Age as a predictor.

4.35 Use the data given in Exercise 4.33 to construct two scatterplots—one of number of cell phone calls versus age and the other of number of text messages sent versus age. Based on the scatterplots, do you think age is a better predictor of number of cell phone calls or number of text messages sent? Explain why you think this.

4.36 In a study of the relationship between TV viewing and eating habits, a sample of 548 ethnically diverse students from Massachusetts was followed over a 19-month period (*Pediatrics* [2003]: 1321–1326). For each additional hour of television viewed per day, the number of fruit and vegetable servings per day was found to decrease on average by 0.14 serving.

a. For this study, what is the response variable? What is the predictor variable?
b. Would the least squares regression line for predicting number of servings of fruits and vegetables using number of hours spent watching TV have a positive or negative slope? Justify your choice.

4.37 An article on the cost of housing in California (*San Luis Obispo Tribune*, March 30, 2001) included the following statement: "In Northern California, people from the San Francisco Bay area pushed into the Central Valley, benefiting from home prices that dropped on average $4000 for every mile traveled east of the Bay." If this statement is correct, what is the slope of the least squares regression line, $\hat{y} = a + bx$, where y = House price (in dollars) and x = Distance east of the Bay (in miles)? Justify your answer.

4.38 The data below on runoff sediment concentration for plots with varying amounts of grazing damage are representative values from a graph in the paper **"Effect of Cattle Treading on Erosion from Hill Pasture: Modeling Concepts and Analysis of Rainfall Simulator Data"** (*Australian Journal of Soil Research* [2002]: 963–977). Damage was measured by the percentage of bare ground in the plot. Data are given for gradually sloped and for steeply sloped plots.

Gradually Sloped Plots

Bare ground (%)	5	10	15	25	30	40
Concentration	50	200	250	500	600	500

Steeply Sloped Plots

Bare ground (%)	5	5	10	15	20	25	20
Concentration	100	250	300	600	500	500	900

Bare ground (%)	30	35	40	35
Concentration	800	1100	1200	1000

a. *Using the data for steeply sloped plots*, find the equation of the least squares regression line for predicting y = Runoff sediment concentration using x = Percentage of bare ground.
b. What would you predict runoff sediment concentration to be for a steeply sloped plot with 18% bare ground?
c. Would you recommend using the least squares regression line from Part (a) to predict runoff sediment concentration for gradually sloped plots? Explain.

SECTION 4.3 Assessing the Fit of a Line

Once the equation of the least squares regression line has been obtained, the next step is to investigate how effectively the line summarizes the relationship between x and y. Important questions to consider are:

1. Is a line an appropriate way to summarize the relationship between the two variables?
2. Are there any unusual aspects of the data set that you need to consider before using the least squares regression line to make predictions?
3. If you decide that it is reasonable to use the least squares regression line as a basis for prediction, how accurate can you expect the predictions to be?

Data set available

This section looks at graphical and numerical methods that will allow you to answer these questions. Most of these methods are based on the vertical deviations of the data points from the least squares regression line, which represent the differences between actual y values and the corresponding predicted $\hat{y}$ values.

Predicted Values and Residuals

The predicted value corresponding to the first observation in a data set, (x_1, y_1), is obtained by substituting x_1 into the regression equation to obtain $\hat{y}_1$, so

$$\hat{y}_1 = a + bx_1$$

The difference between the actual y value, y_1, and the corresponding predicted value is

$$y_1 - \hat{y}_1$$

This difference, called a *residual*, is the vertical deviation of a point in the scatterplot from the least squares regression line. A point that is above the line results in a positive residual, whereas a point that is below the line results in a negative residual. This is shown in Figure 4.19.

FIGURE 4.19
Positive and negative residuals

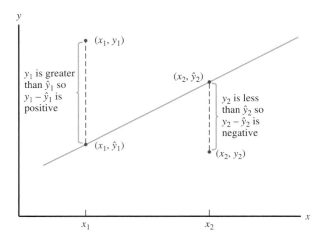

DEFINITION

The **predicted values** result from substituting each x value into the equation for the least squares regression line. This gives

$$\hat{y}_1 = \text{first predicted value} = a + bx_1$$
$$\hat{y}_2 = \text{second predicted value} = a + bx_2$$
$$\vdots$$
$$\hat{y}_n = n^{th} \text{ predicted value} = a + bx_n$$

The **residuals** are the n quantities

$$y_1 - \hat{y}_1 = \text{first residual}$$
$$y_2 - \hat{y}_2 = \text{second residual}$$
$$\vdots$$
$$y_n - \hat{y}_n = n^{th} \text{ residual}$$

Each residual is the difference between an observed y value and the corresponding predicted y value.

Data set available

| **Example 4.6** | **It May Be a Pile of Debris to You, but It Is Home to a Mouse** |

The accompanying data is a subset of data from a scatterplot that appeared in the paper **"Small Mammal Responses to fine Woody Debris and Forest Fuel Reduction in Southwest Oregon"** (*Journal of Wildlife Management* **[2005]: 625–632**). The authors of the paper were interested in how the distance a deer mouse will travel for food is related to the distance from the food to the nearest pile of fine woody debris that could provide a hiding place for the mouse. Distances were measured in meters. The data are given in Table 4.1.

TABLE 4.1 Data, Predicted Values, and Residuals for Example 4.6

Distance from Debris (x)	Distance Traveled (y)	Predicted Distance Traveled ($\hat{y}$)	Residual ($y - \hat{y}$)
6.94	0.00	14.76	−14.76
5.23	6.13	9.23	−3.10
5.21	11.29	9.16	2.13
7.10	14.35	15.28	−0.93
8.16	12.03	18.70	−6.67
5.50	22.72	10.10	12.62
9.19	20.11	22.04	−1.93
9.05	26.16	21.58	4.58
9.36	30.65	22.59	8.06

JMP was used to fit the least squares regression line. The resulting output is shown in Figure 4.20.

Rounding the slope and intercept, the equation of the least squares regression line is

$$\hat{y} = -7.69 + 3.234x.$$

FIGURE 4.20
JMP output for the data of Example 4.6

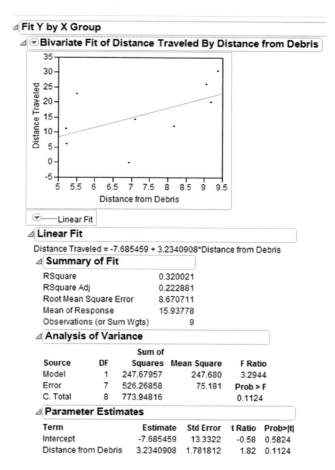

A scatterplot of the data showing the least squares regression line is included in the JMP output of Figure 4.20. The residuals for this data set are the signed vertical distances from the points to the line.

For the mouse with the smallest x value ($x_3 = 5.21$ and $y_3 = 11.29$), the corresponding predicted value and residual are

$$\text{predicted value} = \hat{y}_3 = -7.69 + 3.234(x_3) = -7.69 + 3.234(5.21) = 9.16$$
$$\text{residual} = y_3 - \hat{y}_3 = 11.29 - 9.16 = 2.13$$

The other predicted values and residuals are calculated in a similar manner and are included in Table 4.1.

Calculating the predicted values and residuals by hand can be tedious, but JMP and other statistical software packages, as well as many graphing calculators, can compute and save the residuals. When requested, JMP adds the predicted values and residuals to the data table, as shown here:

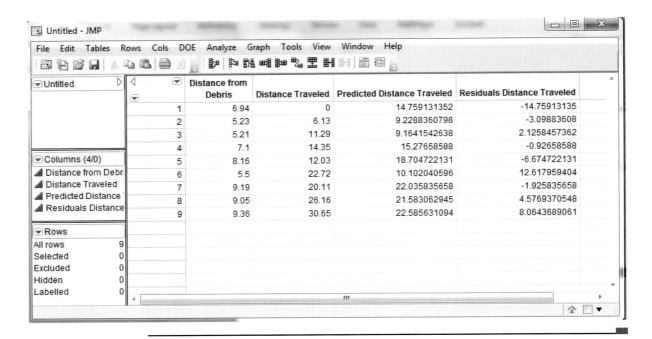

Plotting the Residuals

A careful look at the residuals can reveal many potential problems. To assess the appropriateness of the least squares regression line, a *residual plot* is a good place to start.

A **residual plot** is a scatterplot of the (x, residual) pairs.

Isolated points or a pattern of points in the residual plot indicate potential problems.

Example 4.7   **Revisiting the Deer Mice**

In the previous example, the nine residuals for the deer mice data were calculated. From Table 4.1 (or the JMP output), the nine (x, residual) pairs are

(6.94, −14.76)   (5.23, −3.10)   (5.21, 2.13)   (7.10, −0.93)

(8.16, −6.67)   (5.50, 12.62) (9.19, −1.93)   (9.05, 4.58) (9.36, 8.06)

Figure 4.21 shows the corresponding residual plot.

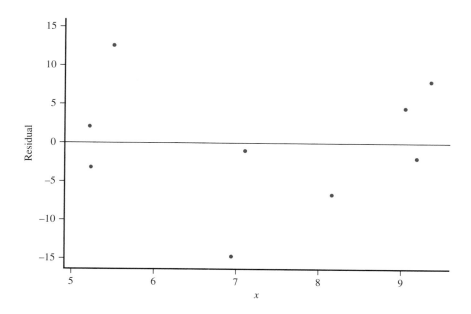

FIGURE 4.21
Residual plot for the deer mice data

There is no strong pattern in the residual plot. The points in the residual plot appear scattered at random.

A desirable residual plot is one that exhibits no particular pattern, such as curvature. Curvature in the residual plot indicates that the relationship between x and y is not linear. In this case, a curve would be a better choice for describing the relationship. It is sometimes easier to see curvature in a residual plot than in a scatterplot, as illustrated in Example 4.8.

Data set available

Example 4.8 Record Times

Consider the accompanying data on x = Distance (in meters) and y = Record time (in seconds) for men for races of various length in international track and field competitions (from *The World Almanac and Book of Facts 2016*).

Distance (in meters)	Record Time (in seconds)
100	10
200	19
400	43
800	101
1,000	132
1,500	206
2,000	285
3,000	441
5,000	757

The scatterplot, displayed in Figure 4.22(a), appears quite straight. However, even though the value of the correlation coefficient is very close to 1, when the residuals from the least squares regression line are plotted (see Figure 4.22(b)), there is a definite curved pattern.

Because of this, it is not accurate to say that men's record times increase linearly with distance.

FIGURE 4.22

Plots for the data of Example 4.8
(a) scatterplot; (b) residual plot

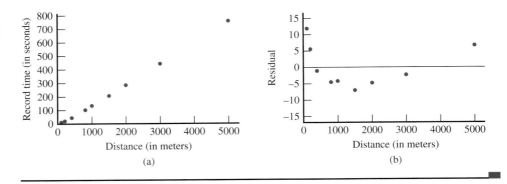

It is also important to look for unusual values in the scatterplot and in the residual plot. In the residual plot, a point falling far above or below the horizontal line at height 0 corresponds to a large residual. This may indicate some type of unusual behavior, such as a recording error. A point whose x value differs greatly from others in the data set may have excessive influence in determining the equation of the least squares regression line. To assess the impact of such an isolated point, delete it from the data set and then recalculate the equation of the least squares regression line. You can then evaluate the extent to which the equation has changed.

Data set available

Example 4.9   **Older Than Your Average Bear**

The accompanying data on x = Age (in years) and y = Weight (in kg) for 12 black bears appeared in the paper **"Habitat Selection by Black Bears in an Intensively Logged Boreal Forest"** (*Canadian Journal of Zoology* **[2008]: 1307–1316**). A scatterplot and residual plot are shown in Figures 4.23(a) and 4.23(b), respectively. One bear in the sample was much older than the others (Bear 3 with an age of x = 28.5 years and a weight of y = 62 kg). This results in a point in the scatterplot that is far to the right of the other points. Because the least squares regression line minimizes the sum of squared residuals, the line is pulled toward this observation. As a result, this single observation plays a big role in determining the slope of the least squares regression line, and it is therefore called an *influential observation*. Notice that this influential observation does not have a large residual, because the least squares regression line actually passes near this point. Figure 4.24 shows what happens when the influential observation is removed from the data set. Both the slope and intercept of the new least squares regression line are quite different from the slope and intercept of the original line.

Bear	Age	Weight	Bear	Age	Weight
1	10.5	54	7	6.5	62
2	6.5	40	8	5.5	42
3	28.5	62	9	7.5	40
4	10.5	51	10	11.5	59
5	6.5	55	11	9.5	51
6	7.5	56	12	5.5	50

FIGURE 4.23
Minitab plots for the bear data of Example 4.9. (a) scatterplot; (b) residual plot

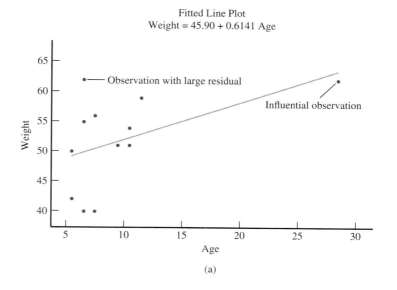

(a)

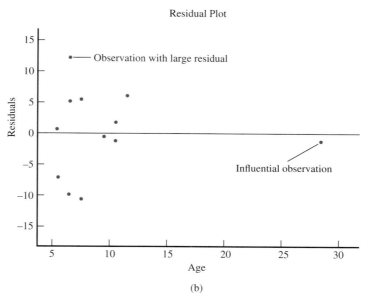

(b)

FIGURE 4.24
Scatterplot and least squares regression line with Bear 3 removed from the data set

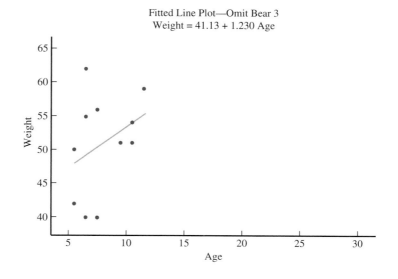

Some points in the scatterplot may fall far from the least squares regression line in the y direction, resulting in a large residual. These points are sometimes referred to as *outliers*. In this example, the observation with the largest residual is Bear 7 with an age of $x =$ 6.5 years and a weight of $y = 62$ kg (labeled in Figure 4.23). Even though this observation has a large residual, it is not influential. The equation of the least squares regression line for the complete data set is $\hat{y} = 45.90 + 0.6141x$, which is not much different from $\hat{y} = 43.81 + 0.7131x$, the equation that results when Bear 7 is omitted.

Unusual points in a bivariate data set are those that are far away from most of the other points in the scatterplot in either the x direction or the y direction.

An observation is potentially an **influential observation** if it has an x value that is far away from the rest of the data (separated from the rest of the data in the x direction). To decide if an observation *is* influential, determine if removal of the observation has a large impact on the value of the slope or intercept of the least squares regression line. (The decision about whether the impact is large is subjective and depends on the context of the given problem.)

An observation is an **outlier** if it has a large residual. Outliers fall far away from the least squares regression line in the y direction.

Careful examination of a scatterplot and a residual plot can help you evaluate whether a line is an appropriate way to summarize a relationship. If you decide that a line is appropriate, the next step is to assess the accuracy of predictions based on the least squares regression line and whether these predictions tend to be better than those made without knowledge of the value of x. Two numerical measures that help with this assessment are the coefficient of determination, r^2, and the standard deviation about the least squares regression line, s_e.

Coefficient of Determination, r^2

Suppose that you would like to predict the price of houses in a particular city. A random sample of 20 houses that are for sale is selected, and y = Price and x = Size (in square feet) are recorded for each house. There will be variability in house price, and it is this variability that makes accurate price prediction a challenge. But if you know that differences in house size account for a large proportion of the variability in house price, then knowing the size of a house will help you predict its price. The proportion of variability in house price that can be accounted for, or "explained" by, a linear relationship between house size and house price is called the **coefficient of determination**.

DEFINITION

The **coefficient of determination**, denoted by r^2, is the proportion of variability in y that can be explained by the linear relationship between x and y.

The value of r^2 is often converted to a percentage (by multiplying by 100).

To understand how r^2 is calculated, consider variability in the y values. Variability in y can be effectively explained by an approximate straight-line relationship when the points in the scatterplot tend to fall close to the least squares regression line—that is, when the residuals are small. A natural measure of variability about the least squares regression line is the sum of the squared residuals. (Squaring before combining prevents negative and positive residuals from offsetting one another.) A second sum of squares assesses the total amount of variability in the observed y values by considering how spread out the y values are around the mean y value.

DEFINITION

The **residual sum of squares** (sometimes also referred to as the error sum of squares), denoted by **SSResid**, is defined as

$$\text{SSResid} = (y_1 - \hat{y}_1)^2 + (y_2 - \hat{y}_2)^2 + \cdots + (y_n - \hat{y}_n)^2 = \Sigma(y - \hat{y})^2$$

The **total sum of squares**, denoted by **SSTotal**, is defined as

$$\text{SSTotal} = (y_1 - \bar{y})^2 + (y_2 - \bar{y})^2 + \cdots + (y_n - \bar{y})^2 = \Sigma(y - \bar{y})^2$$

These sums of squares can be found as part of the regression output from most statistical packages.

Note: It is also possible to calculate these sums of squares using the following computational formulas:

$$\text{SSTotal} = \Sigma y^2 - \frac{(\Sigma y)^2}{n}$$
$$\text{SSResid} = \Sigma y^2 - a\Sigma y - b\Sigma xy$$

Example 4.10 Deer Mice Revisited

Figure 4.25 displays part of the Minitab output that results from fitting the least squares regression line to the data on $y =$ Distance traveled for food and $x =$ Distance to nearest woody debris pile from Example 4.6. From the output,

$$\text{SSTotal} = 773.95 \text{ and SSResid} = 526.27$$

FIGURE 4.25
Minitab output for the deer mice data

Regression Analysis: Distance Traveled versus Distance to Debris

The regression equation is
Distance Traveled = − 7.7 + 3.23 Distance to Debris

Predictor	Coef	SE Coef	T	P
Constant	−7.69	13.33	−0.58	0.582
Distance to Debris	3.234	1.782	1.82	0.112

S = 8.67071   R-Sq = 32.0%   R-Sq(adj) = 22.3%

Analysis of Variance

Source	DF	SS	MS	F	P
Regression	1	247.68	247.68	3.29	0.112
Residual Error	7	526.27	75.18		
Total	8	773.95			

SSTotal   SSResid

The residual sum of squares is the sum of the squared vertical deviations from the least squares regression line. As Figure 4.26 illustrates, SSTotal is also a sum of squared vertical deviations from a line—the horizontal line at height $\bar{y}$. The least squares regression

FIGURE 4.26
Interpreting sums of squares:
(a) SSResid = sum of squared vertical deviations from the least squares regression line
(b) SSTotal = sum of squared vertical deviations from the horizontal line at height $\bar{y}$

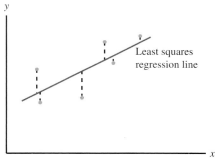
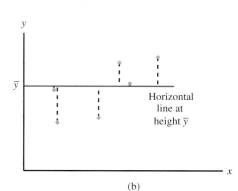

(a)   (b)

line passes through the point $(\bar{x}, \bar{y})$ and is, by definition, the line that has the smallest sum of squared deviations. It follows that SSResid $\leq$ SSTotal. These two sums of squares are equal only when the least squares regression line *is* the horizontal line $y = \bar{y}$.

SSResid is often referred to as a measure of "unexplained" variability—the variability in y that cannot be explained by the linear relationship between x and y. The more the points in the scatterplot deviate from the least squares regression line, the larger the value of SSResid. Similarly, SSTotal is interpreted as a measure of total variability. The greater the amount of variability in the y values, the larger the value of SSTotal.

The ratio SSResid/SSTotal is the fraction or proportion of total variability that can't be explained by a linear relationship. Subtracting this ratio from 1 gives the proportion of total variability that is explained, which is the coefficient of determination.

The coefficient of determination is calculated as

$$r^2 = 1 - \frac{\text{SSResid}}{\text{SSTotal}}$$

Multiplying r^2 by 100 gives the percentage of y variability that can be explained by the approximate linear relationship. The closer this percentage is to 100%, the more successful the relationship is in explaining variability in y.

Example 4.11 r^2 for the Deer Mice Data

For the data on distance traveled for food and distance to nearest debris pile from Example 4.10, SSTotal $= 773.95$ and SSResid $= 526.27$. From these values,

$$r^2 = 1 - \frac{\text{SSResid}}{\text{SSTotal}} = 1 - \frac{526.27}{773.95} = 0.32$$

This means that 32% of the observed variability in distance traveled for food can be explained by an approximate linear relationship between distance traveled for food and distance to nearest debris pile. Notice that the r^2 value can be found in the Minitab output of Figure 4.25, labeled "R-Sq" and in the JMP output of Figure 4.20 labeled "RSquare."

The symbol r was used in Section 4.1 to denote the sample correlation coefficient. This notation suggests how the correlation coefficient and the coefficient of determination are related:

$$(\text{correlation coefficient})^2 = \text{coefficient of determination}$$

If $r = 0.8$ or $r = -0.8$, then $r^2 = 0.64$, so 64% of the observed variability in y can be explained by the linear relationship. When $r = 0.5$, $r^2 = 0.25$, so only 25% of the observed variability in y is explained by the linear relationship. This is why a value of r between -0.5 and 0.5 is interpreted as indicating only a weak linear relationship. Because the value of r does not depend on which variable is labeled x, the same is true of r^2. The coefficient of determination is one of the few quantities calculated in a regression analysis with a value that remains the same when the roles of response and predictor variables are interchanged.

Example 4.12 Lead Exposure and Brain Volume

The authors of the paper **"Decreased Brain Volume in Adults with Childhood Lead Exposure"** (*Public Library of Science Medicine* **[May 27, 2008]: e112)** studied the relationship between childhood environmental lead exposure and volume change in a particular region of the brain. Data on $x =$ Mean childhood blood lead level (mg/dL) and $y =$ Brain volume change (percent) from a graph in the paper were used to produce the scatterplot in Figure 4.27. The least squares regression line is also shown on the scatterplot.

Notice that although there is a slight tendency for smaller y values (corresponding to a brain volume decrease) to be paired with higher values of mean blood lead levels, the relationship is weak. The points in the plot are widely scattered around the least squares regression line.

FIGURE 4.27
Scatterplot and least squares regression line for the data of Example 4.12

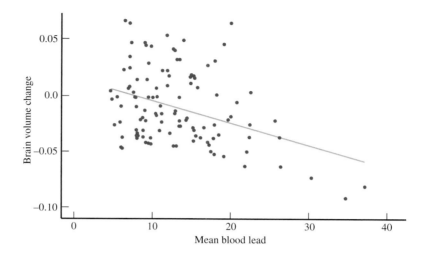

Figure 4.28 displays part of the Minitab output that results from fitting the least squares regression line to the data.

FIGURE 4.28
Minitab output for the data of Example 4.12

Regression Analysis: Brain Volume Change versus Mean Blood Lead

The regression equation is
Brain Volume Change = 0.01559 − 0.001993 Mean Blood Lead

S = 0.0310931   R-Sq = 13.6%   R-Sq(adj) = 12.9%

Analysis of Variance

Source	DF	SS	MS	F	P
Regression	1	0.016941	0.0169410	17.52	0.000
Error	111	0.107313	0.0009668		
Total	112	0.124254			

From the computer output, $r^2 = 0.136$. This means that differences in childhood mean blood lead level explain only 13.6% of the variability in adult brain volume change. You can calculate the value of the correlation coefficient by taking the square root of r^2. In this case, you know that the correlation coefficient is negative (because there is a negative relationship between x and y), so you want the negative square root:

$$r = -\sqrt{0.136} = -0.369$$

Based on the values of r and r^2, you would conclude that there is a weak negative linear relationship and that childhood mean blood lead level explains only about 13.6% of adult change in brain volume.

Standard Deviation About the Least Squares Regression Line

The coefficient of determination (r^2) measures the extent of variability about the least squares regression line *relative* to overall variability in y. A large value of r^2 does not necessarily imply that the deviations from the line are small in an absolute sense. A typical observation could deviate from the line by quite a bit.

Recall that the sample standard deviation

$$s = \sqrt{\frac{\Sigma(x - \bar{x})^2}{n - 1}}$$

is a measure of variability in a single sample. The value of s can be interpreted as a typical amount by which a sample observation deviates from the mean. There is an analogous measure of variability around the least squares regression line.

DEFINITION

The **standard deviation about the least squares regression line**, denoted by s_e, is

$$s_e = \sqrt{\frac{\text{SSResid}}{n - 2}}$$

The value of s_e can be interpreted as the typical amount by which an observation deviates from the least squares regression line.

Data set available

Example 4.13 **Predicting Graduation Rates**

Consider the accompanying data from 2014 on six-year graduation rate (%), instructional expenditure per full-time student (in dollars), and median SAT score for 9 primarily undergraduate public universities and colleges in the western United States with enrollments between 10,000 and 20,000 **(Source: College Results Online, The Education Trust)**.

Graduation Rate	Instructional Expenditures	Median SAT
75.0	6,960	1,242
71.5	7,274	1,114
59.3	5,361	1,014
56.4	5,374	1,070
52.4	5,070	920
48.0	5,226	888
45.8	5,927	970
42.7	5,600	937
41.1	5,073	871

Figure 4.29 displays scatterplots of graduation rate versus instructional expenditure and graduation rate versus median SAT score. The least squares regression lines and the values of r^2 and s_e are also shown.

FIGURE 4.29

Scatterplots for the data of Example 4.13: (a) graduation rate versus instructional expenditure (b) graduation rate versus median SAT

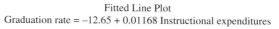

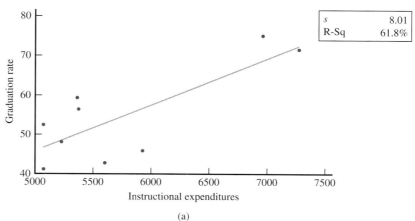

(a)

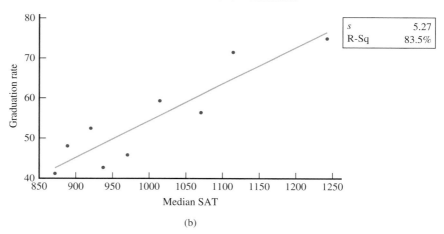

(b)

Notice that while there is a positive linear relationship between instructional expenditure and graduation rate, the relationship is not as strong as the relationship between median SAT and graduation rate. The value of r^2 is 0.618, indicating that about 61.8% of the variability in graduation rate from university to university is explained by differences in instructional expenditures. The standard deviation about the least squares regression line is $s_e = 8.01$, which is larger than s_e for median SAT. This means that the points in the first scatterplot tend to fall farther from the least squares regression line than the points in the second scatterplot. The value of r^2 for the second scatterplot is 0.835 and $s_e = 5.27$, indicating that median SAT does a better job of explaining variability in graduation rates. You would expect the least squares regression line that uses median SAT as a predictor to produce more accurate estimates of graduation rates than the line that uses instructional expenditure as a predictor. Based on the values of r^2 and s_e, median SAT would be a better choice for predicting graduation rates than instructional expenditures.

Also, take a second look at the scatterplot of graduation rate versus instructional expenditures. There are two schools that stand out in the scatterplot as potentially influential observations. The two points in the upper right-hand corner of the plot are far removed from the others in the x direction, indicating that these two universities had noticeably higher expenditures than the other seven universities. If these two data points are removed, the equation of the least squares regression line changes dramatically. For this smaller data set, the equation of the least squares regression line is

$$\text{Graduation rate} = 67.0 - 0.0033 \text{ Instructional expenditures}$$

and r^2 is only 2.12%. This is another reason that Median SAT would be preferred as a predictor of graduation rate.

When evaluating the usefulness of the least squares regression line for making predictions, it is important to consider both the value of r^2 *and* the value of s_e. These two measures assess different aspects of the fit of the line. In general, you would like to have a small value for s_e (which indicates that deviations from the line tend to be small) and a large value for r^2 (which indicates that the linear relationship explains a large proportion of the variability in the y values).

Interpreting the Values of s_e and r^2

A small value of s_e indicates that residuals tend to be small. Because a residual represents the difference between a predicted y value and an observed y value, the value of s_e tells you about the accuracy you can expect when using the least squares regression line to make predictions.

A large value of r^2 indicates that a large proportion of the variability in y can be explained by the approximate linear relationship between x and y. This tells you that knowing the value of x is helpful for predicting y.

A useful least squares regression line will have a reasonably small value of s_e and a reasonably large value of r^2.

Example 4.14 **Predicting IQ**

On May 25, 2010, LiveScience published an article titled **"Simple Memory Test Predicts Intelligence" (www.livescience.com/6519-simple-memory-test-predicts-intelligence.html, retrieved April 22, 2017).** The article summarized a study that found the score on a test of working memory capacity was correlated with a number of different measures of intelligence. The actual study **(*Archives of General Psychiatry* [2010]: 570–577)** looked at how working memory capacity was related to scores on a test of cognitive functioning and to scores on an IQ test. Two groups were studied—one group consisted of patients diagnosed with schizophrenia and the other group consisted of healthy control subjects.

There are several interesting things to note about the linear relationships in Figure 4.30. Looking first at the cognitive functioning score scatterplots, you can see that the relationship between cognitive functioning score and working memory capacity is much stronger for the healthy control group (Figure 4.30(b)) than for the schizophrenic patient group (Figure 4.30(a)). For the patient group, $s_e = 10.74$ and $r^2 = 0.140$. If you were to use the least squares regression line $\hat{y} = 10.24 + 9.23x$ to predict $y = $ Cognitive functioning score based on $x = $ Working memory capacity, a typical prediction error would be around 10.74. Also, for this group, the approximate linear relationship between the variables explains only about 14% of the variability in cognitive functioning score. The value of the correlation coefficient is

$$r = \sqrt{r^2} = \sqrt{0.140} = 0.374$$

There is a weak positive linear relationship between working memory capacity and cognitive functioning score for the patient group.

FIGURE 4.30

Scatterplots for Example 4.14. (a) Cognitive functioning score versus Working memory capacity for patients, (b) Cognitive functioning score versus Working memory capacity for controls, (c) IQ versus Working memory capacity for patients, (d) IQ versus Working memory capacity for controls

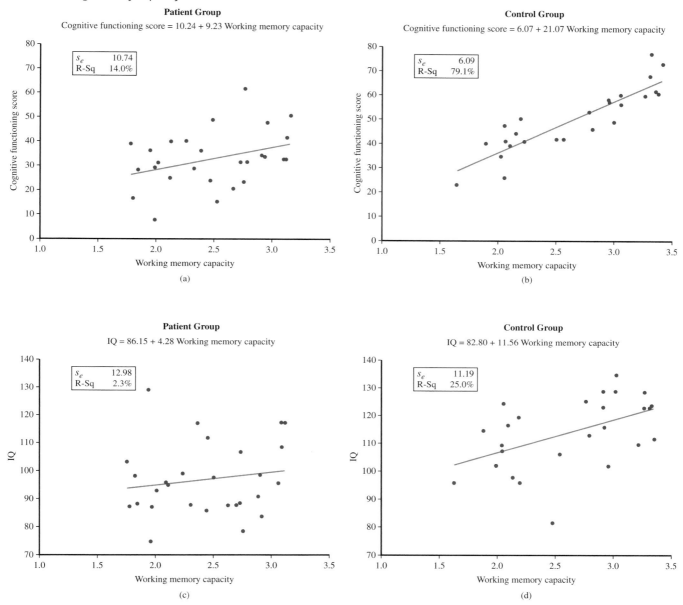

For the healthy control group, $s_e = 6.09$ and $r^2 = 0.791$. This means that using the least squares regression line $\hat{y} = 6.07 + 21.07x$ to predict y = Cognitive functioning score for healthy controls would tend to result in smaller prediction errors, typically around 6.09. For the healthy control group, the approximate linear relationship between the two variables explains about 79.1% of the variability in cognitive functioning score, a much larger percentage than for the patient group.

Working memory capacity is not nearly as good a predictor of IQ as it is for cognitive functioning score. For both groups, the linear relationship between working memory capacity and IQ is much weaker than the relationship between working memory capacity and cognitive functioning score. Also, notice that the relationship between working memory capacity and IQ is stronger for the healthy control group than for the patient group.

Summing It Up—Section 4.3

The following learning objectives were addressed in this section:

Conceptual Understanding

C8: Explain why it is important to consider both the standard deviation about the least squares regression line, s_e and the value of r^2 when assessing the usefulness of the least squares regression line.

The standard deviation about the least squares regression line (s_e) and r^2 assess different aspects of the fit of the line. See the box titled "Interpreting the Values of s_e and r^2" for a discussion of the information provided by each of these measures.

C9: Explain why it is desirable to have a small value of s_e and a large value of r^2 in a linear regression setting.

A small value of s_e indicates that the residuals tend to be small, indicating that predictions are expected to be relatively accurate. A large value of r^2 indicates that a large proportion of the variability in y can be explained by the linear relationship between x and y.

C10: Understand the role of a residual plot in assessing whether a line is the most appropriate way to describe the relationship between two numerical variables.

A residual plot is a scatterplot of the (x, residual) pairs. Isolated points or a pattern of points in the residual plot indicate potential problems that should be investigated before a line is used to describe the relationship between two numerical variables.

C11: Describe the effect of an influential observation on the equation of the least squares regression line.

An influential observation is one that has excessive influence in determining the equation of the least squares regression line. Data points that have x values that differ greatly from the other x values in the data set might be influential.

Mastering the Mechanics

M11: Calculate and interpret the value of s_e, the standard deviation about the least squares regression line.

s_e is interpreted as the typical amount by which an observation deviates from the least squares line (which represents the prediction error). The calculation and interpretation of the value of s_e are illustrated in Example 4.14.

M12: Calculate and interpret the value of r^2, the coefficient of determination.

r^2 is interpreted as the proportion of variability in y that can be explained by a linear relationship with x. The calculation and interpretation of the value of r^2 are illustrated in Example 4.12.

M13: Calculate and interpret residuals, given a bivariate numerical data set and the equation of the least squares regression line.

The residuals are the differences between the observed y values and the corresponding predicted values. The calculation and interpretation of residuals are illustrated in Example 4.6.

M14: Construct a residual plot in a linear regression setting.

A residual plot is a scatterplot of the (x, residual) pairs. For examples of the construction of a residual plot, see Examples 4.6 and 4.7.

M15: Use a residual plot to comment on the appropriateness of a line for summarizing the relationship between two numerical variables.

Isolated points or a pattern of points in the residual plot indicate potential problems that should be investigated before a line is used to describe the relationship between two numerical variables. Example 4.8 shows how the residual plot can be used to evaluate the appropriateness of a line for summarizing the relationship between two numerical variables.

M16: Identify outliers and potentially influential observations in a linear regression setting.

An outlier is an observation with a large residual, which corresponds to a point in the residual plot that is far from the least squares regression line in the y direction. An observation has an x value that is far from the other x values in the data set might be an influential observation. In order to determine if an observation is influential, delete it from the data set and then recalculate the equation of the least squares regression line.

Putting It into Practice

P2: Investigate the usefulness of the least squares regression line for describing the relationship between two numerical variables (using s_e, r^2, and a residual plot).
Example 4.14 in this section and Example 4.15 of the next section (which looks at the entire regression analysis process developed in this chapter) both provide good examples of investigating the usefulness of a least squares regression line.

P4: Describe the anticipated accuracy of predictions based on the least squares regression line.
The value of s_e can be interpreted as a typical prediction error and provides information about the accuracy of predictions based on the least squares regression line. The use of s_e to describe accuracy of predictions is illustrated in Example 4.13.

SECTION 4.3 EXERCISES

Each Exercise Set assesses the following chapter learning objectives: C8, C9, C10, C11, M11, M12, M13, M14, M15, M16, P2, P4

SECTION 4.3 Exercise Set 1

 4.39 The accompanying data are a subset of data from the report **"Great Jobs, Great Lives" (Gallup-Purdue Index 2015 Report, www.gallup.com/reports/197144/gallup-purdue -index-report-2015.aspx , retrieved April 22, 2017)**. The values are approximate values read from a scatterplot. Students at a number of universities were asked if they agreed that their education was worth the cost. One variable in the table is the percentage of students at the university who responded that they strongly agree. The other variable in the table is the **U.S. News & World Report** ranking of the university.

Ranking	Percentage of Alumni Who Strongly Agree
28	53
29	58
30	62
37	55
45	54
47	62
52	55
54	62
57	70
60	58
65	66
66	55
72	65
75	57
82	67
88	59
98	75

a. What is the value of r^2 for this data set? Write a sentence interpreting this value in context. (Hint: See Example 4.13.)
b. What is the value of s_e for this data set? Write a sentence interpreting this value in context.
c. Is the linear relationship between the percentage of alumni who think that their education was worth the cost and university ranking positive or negative? Is it weak, moderate, or strong? Justify your answer.

4.40 Briefly explain why a small value of s_e is desirable in a regression setting.

4.41 Briefly explain why it is important to consider the value of r^2 in addition to the value of s_e when evaluating the usefulness of the least squares regression line.

4.42 The data in the accompanying table are from the paper **"Six-Minute Walk Test in Children and Adolescents" (The Journal of Pediatrics [2007]: 395–399)**. Two hundred and eighty boys completed a test that measures the distance that the boy can walk on a flat, hard surface in 6 minutes. For each age group shown in the table, the median distance walked by the boys in that age group is given.

Age Group	Representative Age (midpoint of age group)	Median Six-minute Walk Distance (meters)
3–5	4.0	544.3
6–8	7.0	584.0
9–11	10.0	667.3
12–15	13.5	701.1
16–18	17.0	727.6

 Data set available

a. With x = Representative age and y = Median distance walked in 6 minutes, construct a scatterplot. Does the pattern in the scatterplot look linear?
b. Find the equation of the least squares regression line.
c. Calculate the five residuals and construct a residual plot. Are there any unusual features in the residual plot? (Hint: See Examples 4.6 and 4.7.)

4.43 The paper referenced in the previous exercise also gave the 6-minute walk distances for 248 girls ages 3 to 18 years. The median distances for the five age groups were

492.4   578.3   655.8   657.6   660.9

a. With x = Representative age and y = Median distance walked in 6 minutes, construct a scatterplot. How does the pattern for girls differ from the pattern for boys from Exercise 4.42?
b. Find the equation of the least squares regression line that describes the relationship between median distance walked in 6 minutes and representative age for girls.
c. Calculate the five residuals and construct a residual plot. The authors of the paper decided to use a curve rather than a straight line to describe the relationship between median distance walked in 6 minutes and age for girls. What aspect of the residual plot supports this decision?

 4.44 Acrylamide is a chemical that is sometimes found in cooked starchy foods and which is thought to increase the risk of certain kinds of cancer. The paper **"A Statistical Regression Model for the Estimation of Acrylamide Concentrations in French Fries for Excess Lifetime Cancer Risk Assessment"** (*Food and Chemical Toxicology* [2012]: 3867–3876) describes a study to investigate the effect of x = Frying time (in seconds) and y = Acrylamide concentration (in micrograms per kg) in french fries. The data in the accompanying table are approximate values read from a graph that appeared in the paper.

Frying Time	Acrylamide Concentration
150	155
240	120
240	190
270	185
300	140
300	270

a. Construct a scatterplot of these data.
b. Find the equation of the least squares regression line. Based on this line, what would you predict Acrylamide concentration to be for a frying time of 270 seconds? What is the residual associated with the observation (270, 185)?
c. Look again at the scatterplot from Part (a). Which observation is potentially influential? Explain the reason for your choice. (Hint: See Example 4.9.)
d. When the potentially influential observation is deleted from the data set, the equation of the least squares regression line

fit to the remaining five observations is $\hat{y} = -44 + 0.83x$. Use this equation to predict acrylamide concentration for a frying time of 270 seconds. Compare this prediction to the prediction made in Part (b).

 4.45 Researchers have observed that bears hunting salmon in a creek often carry the salmon away from the creek before eating it. The relationship between x = Total number of salmon in a creek and y = Percentage of salmon killed by bears that were transported away from the stream prior to being eaten by the bear was examined in the paper **"Transportation of Pacific Salmon Carcasses from Streams to Riparian Forests by Bears"** (*Canadian Journal of Zoology* [2009]: 195–203). Data for the 10 years from 1999 to 2008 are given in the accompanying table.

Total Number	Percentage Transported
19,504	77.8
3,460	28.7
1,976	28.9
8,439	27.9
11,142	55.3
3,467	20.4
3,928	46.8
20,440	76.3
7,850	40.3
4,134	24.1

a. Construct a scatterplot of these data. Does there appear to be a relationship between the total number of salmon and the percentage of transported salmon?
b. Find the equation of the least squares regression line.
c. The residuals from the least squares regression line are shown in the accompanying table. The observation (3928, 46.8) has a large residual. Is this data point also an influential observation?

Total Number	Percentage Transported	Residual
19,504	77.8	3.43
3,460	28.7	0.30
1,976	28.9	4.76
8,439	27.9	−14.76
11,142	55.3	4.89
3,467	20.4	−8.02
3,928	46.8	17.06
20,440	76.3	−0.75
7,850	40.3	−0.68
4,134	24.1	−6.23

d. The two points with unusually large x values (19,504 and 20,440) were not thought to be influential observations even though they are far removed from the rest of the points. Explain why these two points are not influential.
e. Partial Minitab output resulting from fitting the least squares regression line is shown here. What is the value of s_e? Write a sentence interpreting this value.

Regression Analysis: Percent Transported versus Total Number

The regression equation is
Percent Transported = 18.5 + 0.00287 Total Number

Predictor	Coef
Constant	18.483
Total Number	0.0028655

S = 9.16217   R-Sq = 83.2%   R-Sq(adj) = 81.1%

f. What is the value of r^2? Write a sentence interpreting this value.

SECTION 4.3 Exercise Set 2

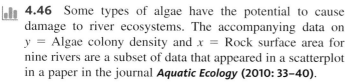 **4.46** Some types of algae have the potential to cause damage to river ecosystems. The accompanying data on y = Algae colony density and x = Rock surface area for nine rivers are a subset of data that appeared in a scatterplot in a paper in the journal **Aquatic Ecology (2010: 33–40)**.

x	50	55	50	79	44	37	70	45	49
y	152	48	22	35	38	171	13	185	25

a. Calculate the equation of the least squares regression line.
b. What is the value of r^2 for this data set? Write a sentence interpreting this value in context.
c. What is the value of s_e for this data set? Write a sentence interpreting this value in context.
d. Is the linear relationship between rock surface area and algae colony density positive or negative? Is it weak, moderate, or strong? Justify your answer.

4.47 Briefly explain why a large value of r^2 is desirable in a regression setting.

4.48 Briefly explain why it is important to consider the value of s_e in addition to the value of r^2 when evaluating the usefulness of the least squares regression line.

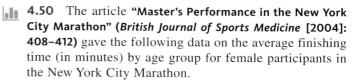 **4.49** The report **"Airline Quality Rating 2016"** (airlinequalityrating.com/reports/2016_AQR_Final.pdf, retrieved April 22, 2017) included the data for 13 U.S. airlines given in the table below.

Airline	On-Time Arrival Percentage	Airline Quality Rating
Alaska	86	5
American	80	10
Delta	86	3
Envoy Air	74	12
Express Jet	78	9
Frontier	73	11
Hawaiian	88	4
JetBlue	76	2
SkyWest	80	7
Southwest	80	6
Spirit	69	13
United	78	8
Virgin America	80	1

a. With x = Airline quality rating and y = On-time arrival percentage, construct a scatterplot. Does the pattern in the scatterplot look linear?
b. Find the equation of the least squares regression line.
c. Calculate the residuals and construct a residual plot. Are there any unusual features in the residual plot?

4.50 The article **"Master's Performance in the New York City Marathon"** (*British Journal of Sports Medicine* [2004]: 408–412) gave the following data on the average finishing time (in minutes) by age group for female participants in the New York City Marathon.

Age Group	Representative Age	Average Finish Time
10–19	15	302.38
20–29	25	193.63
30–39	35	185.46
40–49	45	198.49
50–59	55	224.30
60–69	65	288.71

a. Find the equation of the least squares regression line that describes the relationship between y = Average finish time and x = Representative age.
b. Calculate the six residuals and construct a residual plot. The authors of the paper decided to use a curve rather than a straight line to describe the relationship between average finish time and representative age. What aspect of the residual plot supports this decision?

4.51 The first Batman movie was made over 50 years ago in 1966. Over the years, Batman has been played on screen by a number of actors and even by a Lego figure in the Lego Batman movies. In the original comic books, Batman was described as being 188 cm tall (about 6′2″) and as weighing 95 kg (about 210 pounds). The article **"50 Years of Batman on Film: How Has His Physique Changed?"** (economist.com, March 28, 2016, retrieved April 22, 2017) included the heights and weights of all of the onscreen Batmen in the table below.

Batman	Height (cm)	Weight (kg)
Comic book	188	95
Lego Batman	4	4
Adam West	188	91
Michael Keaton	178	72
Val Kilmer	183	93
George Clooney	180	78
Christian Bale	183	82
Ben Affleck	193	98

With x = Height and y = Weight, the equation of the least squares regression line is $\hat{y} = 1.052 + 0.466x$.

 Data set available

a. Calculate the residuals.
b. Construct a residual plot. Are there any unusual features of the residual plot?
c. The observation for Lego Batman (4, 4) is far removed from the other values in the data set (not surprising!). Is this observation influential in determining either the slope of the least squares regression line, or the intercept, or both? Justify your answer.

4.52 The article **"Examined Life: What Stanley H. Kaplan Taught Us About the SAT"** (*The New Yorker* [December 17, 2001]: 86–92) included a summary of findings regarding the use of SAT I scores, SAT II scores, and high school grade point average (GPA) to predict first-year college GPA. The article states that "among these, SAT II scores are the best predictor, explaining 16 percent of the variance in first-year college grades. GPA was second at 15.4 percent, and SAT I was last at 13.3 percent."
a. If the data from this study were used to fit a least squares regression line with y = First-year college GPA and x = High school GPA, what would be the value of r^2?
b. The article stated that SAT II was the best predictor of first-year college grades. Do you think that predictions based on a least squares regression line with y = First-year college GPA and x = SAT II score would be very accurate? Explain why or why not.

4.53 The paper **"Effects of Age and Gender on Physical Performance"** (*Age* [2007]: 77–85) describes a study investigating the relationship between age and swimming performance. Data on age and 1-hour swim distance for over 10,000 men participating in a national long-distance 1-hour swimming competition are summarized in the accompanying table.

Age Group	Representative Age (midpoint of age group)	Average Swim Distance (m)
20–29	25	3,913.5
30–39	35	3,728.8
40–49	45	3,579.4
50–59	55	3,361.9
60–69	65	3,000.1
70–79	75	2,649.0
80–89	85	2,118.4

a. Find the equation of the least squares regression line using x = Representative age and y = Average swim distance.
b. Calculate the seven residuals and use them to construct a residual plot. What does the residual plot suggest about the appropriateness of using a line to describe the

relationship between representative age and average swim distance?
c. Would it be reasonable to use the least squares regression line from Part (a) to predict the average swim distance for women ages 40 to 49 by substituting the representative age of 45 into the equation of the least squares regression line? Explain why or why not.

ADDITIONAL EXERCISES

4.54 The **California State Park System Statistical Report** for the 2014/2015 Fiscal Year (**www.parks.ca.gov/pages/795/files/14-15%20Statistical%20Report%20-%20INTERNET.pdf, retrieved April 22, 2017**) gave the accompanying data on x = Amount of money collected in user fees (in thousands of dollars) and y = Operating costs (in thousands of dollars) for nine state parks in the North Coast Redwoods District.

User Fees (thousands of dollars)	Operating Costs (thousands of dollars)
17	100
74	359
811	3,582
380	1,377
33	241
427	760
500	1,044
734	2,205
760	1,620

a. Construct a scatterplot of these data.
b. Find the equation of the least squares regression line. Do you think this line would give accurate predictions? Explain.
c. Delete the observation with the largest x value from the data set and recalculate the equation of the least squares regression line. Does this observation have a big effect on the equation of the line?

4.55 The paper **"Depression, Body Mass Index, and Chronic Obstructive Pulmonary Disease—A Holistic Approach"** (*International Journal of COPD* [2016]:239–249) gave data on change in Body Mass Index (BMI in kg/m²) and change in a measure of depression for patients suffering from depression who participated in a pulmonary rehabilitation program. The data in the accompanying table are a subset of the data given in the paper and are approximate values read from a scatterplot in the paper.

BMI Change (kg/m²)	Depression Score Change
0.5	−1
−0.5	9
0	4
0.1	4
0.7	6
0.8	7
1	13
1.5	14
1.1	17
1	18
0.3	12
0.3	14

a. Find the equation of the least squares regression line that would allow you to predict change in depression score based on change in BMI.

b. Find the values of r^2 and s_e. Based on these values, do you think that the least squares regression line does a good job of describing the relationship between change in depression score and change in BMI? Explain.

c. The following graph is a scatterplot of these data. The least squares regression line is also shown. Which observations are outliers? Do the observations with the largest residuals correspond to the patients with the largest change in BMI?

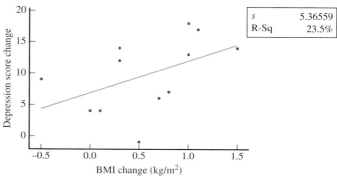

Fitted Line Plot
Depression score change = 6.873 + 5.078 BMI change

4.56 The following quote is from the paper **"The Weight of the Bottle as a Possible Extrinsic Cue with Which to Estimate the Price (and Quality) of the Wine? Observed Correlations"** (*Food Quality and Preference* [2012]: 41–45):

> The weight of the wine bottles was positively correlated ($r = .12; p < .001$) with the price of the wines

a. Does the value of the correlation coefficient indicate that the relationship is weak, moderate, or strong?

b. What is the value of r^2? Write a sentence that gives an interpretation of this value.

SECTION 4.4 Describing Linear Relationships and Making Predictions—Putting It All Together

Now that you have considered all of the parts of a linear regression analysis, you can put these parts together. The steps in a linear regression analysis are summarized in the accompanying box.

Steps in a Linear Regression Analysis

Given a bivariate numerical data set consisting of observations on a response variable y and a predictor variable x:

Step 1 Summarize the data graphically by constructing a scatterplot.

Step 2 Based on the scatterplot, decide if it looks like the relationship between x and y is approximately linear. If so, proceed through the next steps.

Step 3 Find the equation of the least squares regression line.

Step 4 Construct a residual plot and look for any patterns or unusual features that may indicate that a line is not the best way to summarize the relationship between x and y. If none are found, proceed to the next steps.

Step 5 Calculate the values of s_e and r^2 and interpret them in context.

Step 6 Based on what you have learned from the residual plot and the values of s_e and r^2, decide whether the least squares regression line is useful for making predictions. If so, proceed to the last step.

Step 7 Use the least squares regression line to make predictions.

Let's return to the Chapter Preview example to see how following these steps can help you learn about the age of an unidentified crime victim.

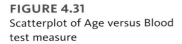

Example 4.15 **Revisiting Help for Crime Scene Investigators**

One of the tasks that forensic scientists face is estimating the age of an unidentified crime victim. Prior to 2010, this was usually done by analyzing teeth and bones, and the resulting estimates were not very reliable. In a groundbreaking study described in the paper **"Estimating Human Age from T-Cell DNA Rearrangements"** (*Current Biology* [2010]), scientists examined the relationship between age and a blood test measure. They recorded age and the blood test measure for 195 people ranging in age from a few weeks to 80 years. Because the scientists were interested in predicting age using the blood test measure, the response variable is y = Age (in years), and the predictor variable is x = Blood test measure.

Step 1: The scientists first constructed a scatterplot of the data, which is shown in Figure 4.31.

Step 2: Based on the scatterplot, it does appear that there is a reasonably strong negative linear relationship between age and the blood test measure. The scientists also reported that the correlation coefficient for this data set was $r = -0.92$, which is consistent with the strong negative linear pattern in the scatterplot.

Step 3: The scientists calculated the equation of the least squares regression line to be

$$\hat{y} = -33.65 - 6.74x$$

FIGURE 4.31
Scatterplot of Age versus Blood test measure

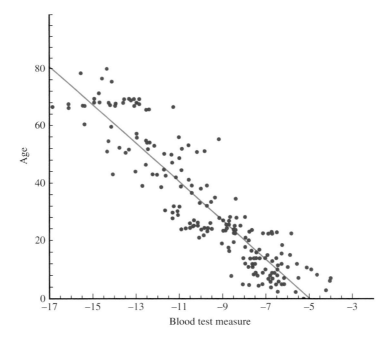

Step 4: A residual plot constructed from these data showed a few observations with large residuals, but these observations were not far removed from the rest of the data in the x direction. These observations were not judged to be influential. Also, there were no unusual patterns in the residual plot that would suggest a nonlinear relationship between age and the blood test measure.

Step 5: The values of s_e and r^2 were

$$s_e = 8.9 \qquad \text{and} \qquad r^2 = 0.835$$

This means that approximately 83.5% of the variability in age can be explained by the approximate linear relationship between age and blood test measure. The value of s_e tells you that if the least squares regression line is used to predict age from the blood test measure, a typical difference between the predicted age and the actual age would be about 9 years.

Step 6: Based on the residual plot, the large value of r^2, and the relatively small value of s_e, the scientists proposed using the blood test measure and the least squares regression line as a way to estimate ages of crime victims.

Step 7: To illustrate predicting age, suppose that a blood sample is taken from an unidentified crime victim and that the value of the blood test measure is determined to be −10. The predicted age of the victim would be

$$\hat{y} = -33.65 - 6.74\,(-10)$$

$$= -33.65 - (-67.4)$$

$$= 33.75 \text{ years}$$

Remember that this is just an estimate of the actual age, and a typical prediction error is about 9 years.

SECTION 4.5 Avoid These Common Mistakes

There are a number of ways you can get into trouble when analyzing bivariate numerical data! Here are some things to keep in mind when conducting your own analyses or when reading reports of such analyses:

1. Correlation does not imply causation. A common media blunder is to infer a cause-and-effect relationship between two variables simply because there is a strong correlation between them. Don't fall into this trap! A strong correlation implies only that the two variables tend to vary together in a predictable way, but there are many possible explanations for why this is occurring other than one variable causing change in the other.

   For example, the article **"Ban Cell Phones? You May as Well Ban Talking Instead"** (*USA TODAY,* **April 27, 2000**) provided data that showed a strong negative correlation between the number of cell phone subscribers and traffic fatality rates. During the years from 1985 to 1998, the number of cell phone subscribers increased from 200,000 to 60,800,000, and the number of traffic deaths per 100 million miles traveled decreased from 2.5 to 1.6 over the same period. However, based on this correlation alone, the conclusion that cell phone use is the cause of improved road safety is not reasonable!

   Similarly, the *Calgary Herald* (**April 16, 2002**) reported that heavy and moderate drinkers earn more than light drinkers or those who do not drink. Based on the correlation between number of drinks consumed and income, the author of the study concluded that moderate drinking "causes" higher earnings. This is obviously a misleading statement, but at least the article goes on to state that "there are many possible reasons for the correlation. It could be because better-off men simply choose to buy more alcohol. Or it might have something to do with stress: Maybe those with stressful jobs tend to earn more after controlling for age, occupation, etc., and maybe they also drink more in order to deal with the stress."

2. A correlation coefficient near 0 does not necessarily imply that there is no relationship between two variables. Before drawing such a conclusion, carefully examine a scatterplot of the data. Although the variables may be unrelated, it is also possible that there is a strong but nonlinear relationship.

3. The least squares regression line for predicting y from x is not the same line as the least squares regression line for predicting x from y. The least squares regression line is, by definition, the line that has the smallest possible sum of squared deviations of points from the line *in the y direction* (it minimizes $\Sigma(y - \hat{y})^2$). This is not generally the same line as the line that minimizes the sum of the squared deviations in the x direction. It is not appropriate, for example, to fit a line to data using $y =$ House price and $x =$ House size and then use the resulting least squares regression line Price $= a + b$(Size) to predict the size of a house by substituting in a price and then solving for size. Make sure that the response and predictor variables are clearly identified and that the appropriate line is fit.

4. Beware of extrapolation. You can't assume that a least squares regression line is valid over a wider range of x values. Using the least squares regression line to make predictions outside the range of x values in the data set often leads to poor predictions.

5. Be careful in interpreting the value of the intercept of the least squares regression line. In many instances interpreting the intercept as the value of y that would be predicted when $x = 0$ represents extrapolation beyond the range of the x values in the data set. You should not interpret the intercept unless $x = 0$ is within the range of the data.

6. Remember that the least squares regression line may be the "best" line (in that it has a smaller sum of squared deviations than any other line), but that doesn't necessarily mean that the line will produce good predictions. Be cautious of predictions based on a least squares regression line without any information about the appropriateness of the least squares regression line, such as s_e and r^2.

7. It is not enough to look at just r^2 or just s_e when evaluating the least squares regression line. Remember to consider both values. These two measures address different aspects of the fit of the line. In general, you would like to have both a small value for s_e (which indicates that deviations from the line tend to be small) and a large value for r^2 (which indicates that the linear relationship explains a large proportion of the variability in the y values).

8. The value of the correlation coefficient, as well as the values for the intercept and slope of the least squares regression line, can be sensitive to influential observations in the data set, particularly if the sample size is small. Because potentially influential observations are those whose x values are far away from most of the x values in the data set, it is important to *always* start with a plot of the data.

CHAPTER ACTIVITIES

ACTIVITY 4.1   AGE AND FLEXIBILITY

Materials needed: Yardsticks.

In this activity, you will investigate the relationship between age and a measure of flexibility. Flexibility will be measured by asking a person to bend at the waist as far as possible, extending his or her arms toward the floor. Using a yardstick, measure the distance from the floor to the finger tip closest to the floor.

1. Age and the measure of flexibility just described will be recorded for a group of individuals. Your goal is to determine if there is a relationship between age and this measure of flexibility. What are two reasons why it would not be a good idea to use just the students in your class as the subjects for your study?

2. Working as a class, decide on a reasonable way to collect data.

3. After your class has collected appropriate data, use these data to construct a scatterplot. Comment on the interesting features of the plot. Does it look like there is a relationship between age and flexibility?

4. If there appears to be a linear relationship between age and flexibility, calculate the equation of the least squares regression line.

5. *In the context of this activity*, write a brief description of the danger of extrapolation.

🖥 CHAPTER 4 EXPLORATIONS IN STATISTICAL THINKING

🖱 EXPLORATION 1: EXPLORING RELATIONSHIPS AND UNDERSTANDING SAMPLING VARIABILITY

In the following exercise, each student in your class will go online to select a random sample of size 20 from a small population consisting of 300 adults.

To learn about the relationship between age and the number of text messages sent in a typical day for this population, go online at statistics.cengage.com/Peck2e/Explore.html and click on the link for Chapter 4. This link will take you to a web page where you can select a random sample from the population.

Click on the sample button. This selects a random sample of size 20 and will display the age and number of text messages sent for each of the people in your sample. Each student in your class will receive data from a different random sample.

Use the data from your random sample to complete the following.
a. Calculate the value of the correlation coefficient.
b. Based on the value of the correlation coefficient, which of the following best describes the linear relationship between age and the number of text messages sent?
 i. Strong and negative
 ii. Moderate and negative
 iii. Weak and negative
 iv. Weak and positive
 v. Moderate and positive
 vi. Strong and positive
c. What is the value of the slope of the least squares regression line?
d. What is the value of the intercept of the least squares regression line?
e. What is the value of r^2 for this data set? Express your answer as a decimal and round it to three decimal places.
f. Write a sentence interpreting the value of r^2.
g. What is the value of s_e?
h. Based on the values of r^2 and s_e, would you recommend using this regression equation to predict number of text messages sent? Explain why or why not.

If asked to do so by your instructor, bring the equation of the least squares regression line for your sample to class. Your instructor will lead the class through the rest of this exercise.

After students have had a chance to compare regression equations, consider the following questions.
i. Why didn't every student get the same regression equation?
j. If two different students were to use their regression equations to predict the number of text messages sent for a person who is 28 years old, would the predictions be the same? Do you think they would be similar? Explain.
k. If every student had obtained a random sample of 50 people instead of 20, do you think the regression equations would have differed more or less from student to student than they did for samples of size 20? Explain.

🖱 EXPLORATION 2: EXPLORING MULTIVARIABLE THINKING

Hill racing is a popular sport in Scotland. A hill race is one that is run off road in hilly terrain where changes in elevation contribute to the difficulty of the race. In this Exploration, you will work with data from 44 Scottish hill races run between January 1, 2016, and June 22, 2016 (data from **scottishhillracing.co.uk, retrieved June 30, 2016**). The variables in the data set are race name, race distance (in km), race climb (in m), fastest time for a male runner (in minutes), and fastest time for a female runner (in minutes).

Go online at statistics.cengage.com/Peck2e/Explore.html and click on the link for Chapter 4. This will take you to a web page where you can download the Scottish Hill Races data.

1. Use the data from these 44 races to construct graphical displays and regression analyses that would allow you to answer the following questions.
2. **a.** Does the fastest time for a female runner appear to be related to race distance?
   b. Does the fastest time for a female runner appear to be related to race climb?
   c. Which of race distance and race climb is a better predictor of the fastest time for a female runner?
   d. Which of race distance and race climb is a better predictor of the fastest time for a male runner?
3. Think of another question that could be investigated using this data set. Use the data to answer the question.
4. Based on what you have learned from the data, write a paragraph commenting on how fastest times in Scottish hill races are related to the sex of the runner, race distance, and race climb.

ARE YOU READY TO MOVE ON? CHAPTER 4 REVIEW EXERCISES

All chapter learning objectives are assessed in these exercises. The learning objectives assessed in each exercise are given in parentheses for each exercise.

4.57 (M1, M2)

For each of the four scatterplots shown, answer the following questions:
i. Does there appear to be a relationship between x and y?
ii. If so, does the relationship appear to be linear?
iii. If so, would you describe the linear relationship as positive or negative?

Scatterplot 1:

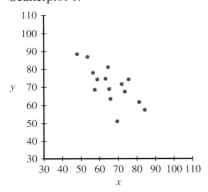

Scatterplot 2:

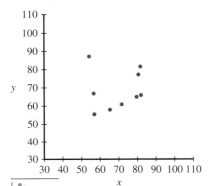

Scatterplot 3:

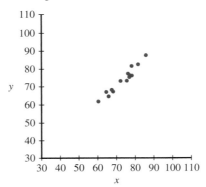

Scatterplot 4:

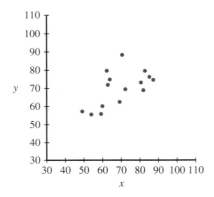

4.58 (M1, M2)

For each of the following pairs of variables, indicate whether you would expect a positive correlation, a negative correlation, or a correlation close to 0. Explain your choice.
a. Price and weight of an apple
b. A person's height and the number of pets he or she has

c. Time spent studying for an exam and score on the exam
d. A person's weight and the time it takes him or her to run one mile

4.59   (C2, M5)

The authors of the paper **"Flat-Footedness Is Not a Disadvantage for Athletic Performance in Children Aged 11 to 15 Years"** (*Pediatrics* **[2009]: e386–e392**) studied the relationship between y = Arch height and scores on five different motor ability tests. They reported the following correlation coefficients:

Motor Ability Test	Correlation Between Test Score and Arch Height
Height of counter movement jump	−0.02
Hopping: average height	−0.10
Hopping: average power	−0.09
Balance, closed eyes, one leg	0.04
Toe flexion	0.05

a. Interpret the value of the correlation coefficient between average hopping height and arch height. What does the fact that the correlation coefficient is negative say about the relationship? Do larger arch heights tend to be paired with larger or smaller average hopping heights?

b. The title of the paper suggests that having a small value for Arch height (flat-footedness) is not a disadvantage when it comes to motor skills. Do the given correlation coefficients support this conclusion? Explain.

4.60   (C1, C2, M3, M4)

The accompanying data are x = Cost (cents per serving) and y = Fiber content (grams per serving) for 18 high-fiber cereals rated by *Consumer Reports* (**www.consumerreports.org /health**).

Cost per Serving	Fiber per Serving	Cost per Serving	Fiber per Serving
33	7	53	13
46	10	53	10
49	10	67	8
62	7	43	12
41	8	48	7
19	7	28	14
77	12	54	7
71	12	27	8
30	8	58	8

a. Construct a scatterplot of y = Fiber content versus Cost. Based on the scatterplot, how would you describe the relationship between fiber content and cost?

b. Calculate and interpret the value of the correlation coefficient.

c. The serving size differed for the different cereals, with serving sizes varying from a half cup to one and a quarter cups. Converting price and fiber content to "per cup" rather than "per serving" results in the following data. Is the correlation coefficient for the per cup data greater than or less than the correlation coefficient for the per-serving data?

Cost per Cup	Fiber per Cup	Cost per Cup	Fiber per Cup
44.0	9.3	53.0	13.0
46.0	10.0	53.0	10.0
49.0	10.0	67.0	8.0
62.0	7.0	43.0	12.0
32.8	6.4	48.0	7.0
19.0	7.0	56.0	28.0
77.0	12.0	54.0	7.0
56.8	9.6	54.0	16.0
30.0	8.0	77.3	10.7

4.61   (C3)

Based on data from six countries, the paper **"A Cross-National Relationship Between Sugar Consumption and Major Depression?"** (*Depression and Anxiety* **[2002]: 118–120**) concluded that there was a correlation between refined sugar consumption (calories per person per day) and annual rate of major depression (cases per 100 people). The following data are from a graph in the paper:

Country	Sugar Consumption	Depression Rate
Korea	150	2.3
United States	300	3.0
France	350	4.4
Germany	375	5.0
Canada	390	5.2
New Zealand	480	5.7

a. Calculate and interpret the value of the correlation coefficient for this data set.

b. Is it reasonable to conclude that increasing sugar consumption leads to higher rates of depression? Explain.

c. What concerns do you have about this study that would make you hesitant to generalize these conclusions to other countries?

4.62   (P1)

The article **"Air Pollution and Medical Care Use by Older Americans"** (*Health Affairs* **[2002]: 207– 214**) gave data on a measure of pollution (in micrograms of particulate matter per cubic meter of air) and the cost of medical care per person over age 65 for six geographical regions of the United States:

Data set available

Region	Pollution	Cost of Medical Care
North	30.0	915
Upper South	31.8	891
Deep South	32.1	968
West South	26.8	972
Big Sky	30.4	952
West	40.0	899

a. Construct a scatterplot for these data.
b. Calculate and interpret the value of the correlation coefficient.
c. Write a few sentences summarizing what the scatterplot and value of the correlation coefficient tell you about the relationship between pollution and medical cost per person over age 65.

4.63 (C4)

The relationship between hospital patient-to-nurse ratio and various characteristics of job satisfaction and patient care has been the focus of a number of research studies. Suppose x = Patient-to-nurse ratio is the predictor variable. For each of the following response variables, indicate whether you expect the slope of the least squares regression line to be positive or negative, and give a brief explanation for your choice.

a. y = Measure of nurse's job satisfaction (higher values indicate higher satisfaction)
b. y = Measure of patient satisfaction with hospital care (higher values indicate higher satisfaction)
c. y = Measure of quality of patient care (higher values indicate higher quality)

4.64 (C5)

For a given data set, the sum of squared deviations from the line $y = 40 + 6x$ is 529.5. For this same data set, which of the following could be the sum of squared deviations from the least squares regression line? Explain your choice.

i. 308.6
ii. 529.6
iii. 617.4

4.65 (C7)

Explain why it can be misleading to use the least squares regression line to obtain predictions for x values that are substantially larger or smaller than the x values in the data set.

4.66 (C6, M7, M8, M9, M10)

Is living in a large high-rise apartment building a disadvantage in a medical emergency? This question was investigated in the paper **"Impact of Building Height and Volume on Cardiac Arrest Response Time"** (*Prehospital Emergency Care* [2016]: 212–219). The accompanying data on the median time (in minutes) it took an emergency team responding to

a cardiac arrest call to get from their vehicle at a building entrance to the patient and administer a defibrillator shock (y) and a measure of building size (x, the natural logarithm of the building volume) are approximate values read from a scatterplot that appeared in the paper.

Measure of Building Size	Median Curb-to-Defib Time
8	1.9
9	2
10	2
11	2.3
12	2.5
13	2.6
14	2.9
15	3.1

a. Find the equation of the least squares regression line.
b. Interpret the slope of the least squares regression line in the context of this study.
c. Does it make sense to interpret the intercept of the least squares regression line? If so, give an interpretation. If not, explain why it is not appropriate for this data set.
d. Use the least squares regression line to predict the median curb to defib time for people who live in buildings with a building size of 12.5.

4.67 (M6, M7)

The following data on sale price, size, and land-to-building ratio for 10 large industrial properties appeared in the paper **"Using Multiple Regression Analysis in Real Estate Appraisal"** (*Appraisal Journal* [2002]: 424–430):

Property	Sale Price (millions of dollars)	Size (thousands of sq. ft.)	Land-to-Building Ratio
1	10.6	2,166	2.0
2	2.6	751	3.5
3	30.5	2,422	3.6
4	1.8	224	4.7
5	20.0	3,917	1.7
6	8.0	2,866	2.3
7	10.0	1,698	3.1
8	6.7	1,046	4.8
9	5.8	1,108	7.6
10	4.5	405	17.2

a. If you wanted to predict sale price and you could use either size or land-to-building ratio as the basis for making predictions, which would you use? Explain.
b. Based on your choice in Part (a), find the equation of the least squares regression line for predicting sale price.

4.68 (P3)
Does it pay to stay in school? The report *Trends in Higher Education* **(The College Board, 2010)** looked at the median hourly wage gain per additional year of schooling. The report states that workers with a high school diploma had a median hourly wage that was 10% higher than those who had only completed 11 years of school. Workers who had completed 1 year of college (13 years of education) had a median hourly wage that was 11% higher than that of the workers who had completed only 12 years of school. The gain in median hourly wage for each additional year of school is shown in the accompanying table. The entry for 15 years of schooling has been intentionally omitted from the table.

Years of schooling	Median hourly wage gain for the additional year (percent)
12	10
13	11
14	13
16	16
17	18
18	19

a. Use the given data to predict the median hourly wage gain for the fifteenth year of schooling.
b. The actual wage gain for fifteenth year of schooling was 14%. How close was the predicted wage gain percent from Part (a) to the actual value?

4.69 (C10, M13, M14, M15)
The following table gives the number of heart transplants performed in the United States each year from 2006 to 2015 **(U.S. Department of Health and Human Services, optn.transplant.hrsa.gov/data/view-data-reports/national -data/, retrieved April 22, 2017):**

Year	Number of Heart Transplants
1 (2006)	2,193
2	2,209
3	2,163
4	2,211
5	2,332
6	2,322
7	2,378
8	2,531
9	2,655
10 (2015)	2,804

a. Construct a scatterplot of these data, and then find the equation of the least squares regression line that describes the relationship between y = Number of heart transplants performed and x = Year. Describe how the number of heart transplants performed has changed over time from 2006 to 2015.
b. Calculate the 10 residuals, and construct a residual plot. Are there any features of the residual plot that indicate that the relationship between year and number of heart transplants performed would be better described by a curve rather than a line? Explain.

4.70 (C10, M14, M15)
Can you tell how old a lobster is by its size? This question was investigated by the authors of a paper that appeared in the *Biological Bulletin* **(August 2007)**. Researchers measured carapace (the exterior shell) length of 27 laboratory-raised lobsters of known age. The data on x = Carapace length (in mm) and y = Age (in years) in the following table were read from a graph that appeared in the paper.

Age	Carapace Length	Age	Carapace Length
1.00	63.32	2.33	138.47
1.00	67.50	2.50	133.95
1.00	69.58	2.51	125.25
1.00	73.41	2.50	123.51
1.42	79.32	2.93	146.82
1.42	82.80	2.92	139.17
1.42	85.59	2.92	136.73
1.82	105.07	2.92	122.81
1.82	107.16	3.17	142.30
1.82	117.25	3.41	152.73
2.18	109.24	3.42	145.78
2.18	110.64	3.75	148.21
2.17	118.99	4.08	152.04
2.17	122.81		

a. Construct a scatterplot of these data, and then find the equation of the least squares regression line that describes the relationship between y = Age and x = Carapace length. Describe how age varies with carapace length.
b. Using computer software or a graphing calculator, construct a residual plot. Are there any features of the residual plot that indicate that the relationship between age and carapace length would be better described by a curve rather than a line? Explain.

4.71 (C11, M16)
The article **"$115K! The 13 Best Paying U.S. Companies"** (*USA TODAY*, **August 11, 2015**) gave the following data on median worker pay (in thousands of dollars) and the 1-year percent change in stock price for the 13 highest paying companies in the U.S. A scatterplot of these data is also shown.

Company	Median Worker Pay	Percent Change in Stock Price
Jupiter Networks	134.7	21.7
Netflix	132.2	92.2
Equinix	125.0	33.9
Altera	122.8	52.2
Visa	122.5	41.9
Yahoo	121.2	2.8
Xilinx	121.2	5.2
VeriSign	119.0	30.1
Microsoft	118.0	8.0
Broadcom	118.0	37.3
F5 Networks	117.6	15.9
Adobe Systems	117.4	22.0
eBay	115.1	−44.7

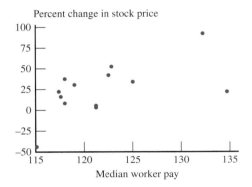

The equation of the least squares regression line with $y =$ Percent change in stock price and $x =$ Median worker pay is $\hat{y} = -358 + 3.14x$.

a. The observations for Jupiter Networks and Netflix are both far removed from the other points in the scatterplot in the x direction. Which of these two observations would have the greatest impact on the equation of the least squares regression line if it were to be omitted from the data set?

b. Explain the difference between an influential observation and an outlier in a bivariate data set.

4.72   **(C6, C8, C9, M11, M12, P2, P4)**

The accompanying data on $x =$ Average energy density (calories per 100 grams) and $y =$ Average cost (in dollars) for eight different food groups are from the paper **"The Cost of U.S. Foods as Related to Their Nutritional Value"** (*American Journal of Clinical Nutrition* [2010]: 1181–1188). The authors were interested in the relationship between average cost and the average energy density of foods.

Food Group	Average Cost	Average Energy Density
Milk and milk products	0.40	182
Eggs	0.32	171
Dry beans, legumes, nuts, and seeds	0.50	330
Grain products	0.47	337
Fruit	0.28	67
Vegetables	0.33	83
Fats, oils, and salad dressings	0.37	390
Sugars, sweets and beverages	0.40	242

a. Construct a scatterplot of these data. Does the relationship between cost and energy density look approximately linear?

b. Find the equation of the least squares regression line.

c. Construct a residual plot and comment on any unusual features of the residual plot.

d. Calculate and interpret the value of r^2 in the context of this study.

e. Explain why it would also be informative to look at the value of s_e in addition to the value of r^2 when evaluating the usefulness of the least squares regression line.

f. Calculate and interpret the value of s_e.

g. Based on you answers to Parts (c) – (f), do you think it is appropriate to use the least squares regression line to predict average cost for a food group based on its average energy density? Explain.

h. The meat, poultry, and fish food group has an average energy density of 196. What average cost would you predict for this food group?

TECHNOLOGY NOTES

Correlation

JMP

1. Input the data for the dependent variable into one column
2. Input the data for the independent variable into another column
3. Click **Analyze** then select **Multivariate Methods** then select **Multivariate**
4. Click and drag the column name containing the dependent variable from the box under **Select Columns** to the box next to **Y, Columns**

5. Click and drag the column name containing the independent variable from the box under **Select Columns** to the box next to **Y, Columns**
6. Click **OK**

Note: This produces a table of correlations. The correlation between the two variables can be found in the first row, second column.

MINITAB

1. Input the data for the dependent variable into the first column
2. Input the data for the independent variable into the second column
3. Select **Stat** then **Basic Statistics** then **Correlation...**
4. Double-click each column name in order to move it to the box under **Variables:**
5. Click **OK**

SPSS

1. Input the data for the dependent variable into the first column
2. Input the data for the independent variable into the second column
3. Select **Analyze** then choose **Correlate** then choose **Bivariate...**
4. Highlight the name of both columns by holding the **ctrl** key and clicking on each name
5. Click the arrow button to move both variables to the **Variables** box
6. Click **OK**

Note: This produces a table of correlations. The correlation between the two variables can be found in the first row, second column.

Note: The correlation can also be produced by following the steps to produce the regression equation.

Excel 2007

1. Input the data into two separate columns
2. Click the **Data** ribbon and select **Data Analysis**
3. **Note:** If you do not see **Data Analysis** listed on the Ribbon, see the Technology Notes for Chapter 2 for instructions on installing this add-on.
4. Select **Correlation** from the dialog box and click **OK**
5. Click in the box next to **Input Range:** and select BOTH columns of data (if you input and selected titles for the columns, click the box next to **Labels in First Row**)
6. Click **OK**

Note: The correlation between the variables can be found in the first column, second row of the table that is output.

TI-83/84

Note: Before beginning this chapter, press the **2nd** key then the **0** key and scroll down to the entry DiagnosticOn. Press **ENTER** twice. After doing this, the regression correlation coefficient r will appear as output with the linear regression line.

1. Enter the data for the independent variable into **L1** (to access lists press the **STAT** key, highlight the option called **Edit...** then press **ENTER**)
2. Input the data for the dependent variable into **L2**
3. Press the **STAT** key
4. Highlight **CALC** then select **LinReg(a+bx)** and press **ENTER**
5. Press the **2nd** key then the **1** key
6. Press,
7. Press the **2nd** key then the **2** key
8. Press **ENTER**

TI-Nspire

1. Enter the data for the independent variable into a data list (to access data lists select the spreadsheet option and press **enter**)

   Note: Be sure to title the list by selecting the top row of the column and typing a title.

2. Enter the data for the dependent variable into a separate data list
3. Press the **menu** key and select **4:Statistics** then **1:Stat Calculations** then **3:Linear Regression(mx+b)...** and press **enter**
4. For **X List:** select the column with the independent variable data from the drop-down menu
5. For **Y List:** select the column with the dependent variable data from the drop-down menu
6. Press **OK**

Note: You may need to scroll to view the correlation coefficient in the list of output.

Regression

JMP

1. Enter the data for the dependent variable into the first column
2. Input the data for the independent variable into the second column
3. Click **Analyze** then select **Fit Y by X**
4. Click and drag the column name containing the dependent data from the box under **Select Columns** to the box next to **Y, Response**
5. Click and drag the column name containing the independent data from the box under **Select Columns** to the box next to **X, Factor**
6. Click **OK**
7. Click the red arrow next to **Bivariate Fit...**
8. Click **Fit Line**

MINITAB

1. Input the data for the dependent variable into the first column
2. Input the data for the independent variable into the second column
3. Select **Stat** then **Regression** then **Regression...**
4. Highlight the name of the column containing the dependent variable and click **Select**
5. Highlight the name of the column containing the independent variable and click **Select**
6. Click **OK**

Note: You may need to scroll up in the Session window to view the regression equation.

SPSS

1. Input the data for the dependent variable into the first column
2. Input the data for the independent variable into the second column
3. Select **Analyze** then choose **Regression** then choose **Linear...**
4. Highlight the name of the column containing the dependent variable

5. Click the arrow button next to the **Dependent** box to move the variable to this box
6. Highlight the name of the column containing the independent variable
7. Click the arrow button next to the **Independent** box to move the variable to this box
8. Click **OK**

Note: The regression coefficients can be found in the **Coefficients** table. The intercept value can be found in the first column of the (Constant) row. The value of the slope can be found in the first column of the row labeled with the independent variable name.

Excel 2007

1. Input the data into two separate columns
2. Click the **Data** ribbon and select **Data Analysis**
3. **Note:** If you do not see Data Analysis listed on the Ribbon, see the Technology Notes for Chapter 2 for instructions on installing this add-on.
4. Select **Regression** from the dialog box and click **OK**
5. Click in the box next to **Y:** and select the dependent variable data
6. Click in the box next to **X:** and select the independent variable data (if you input and selected titles for BOTH columns, check the box next to **Labels**)
7. Click **OK**

Note: The regression coefficients can be found in the third table under the **Coefficients** column.

TI-83/84

1. Enter the data for the independent variable into **L1** (to access lists press the **STAT** key, highlight the option called **Edit...** then press **ENTER**)
2. Input the data for the dependent variable into **L2**
3. Press the **STAT** key
4. Highlight **CALC** then select **LinReg(a+bx)** and press **ENTER**
5. Press the **2nd** key then the **1** key
6. Press,
7. Press the **2nd** key then the **2** key
8. Press **ENTER**

TI-Nspire

1. Enter the data for the independent variable into a data list (to access data lists select the spreadsheet option and press **enter**)

    Note: Be sure to title the list by selecting the top row of the column and typing a title.

2. Enter the data for the dependent variable into a separate data list
3. Press the **menu** key and select **4:Statistics** then **1:Stat Calculations** then **3:Linear Regression(mx+b)...** and press **enter**
4. For **X List:** select the column with the independent variable data from the drop-down menu
5. For **Y List:** select the column with the dependent variable data from the drop-down menu
6. Press **OK**

Residuals

JMP

1. Enter the data for the dependent variable into the first column
2. Input the data for the independent variable into the second column
3. Click **Analyze** then select **Fit Y by X**
4. Click and drag the column name containing the dependent data from the box under **Select Columns** to the box next to **Y, Response**
5. Click and drag the column name containing the independent data from the box under **Select Columns** to the box next to **X, Factor**
6. Click **OK**
7. Click the red arrow next to **Bivariate Fit...**
8. Click **Fit Line**
9. Click the red arrow next to **Linear Fit**
10. Click **Save Residuals**

MINITAB

1. Input the data for the dependent variable into the first column
2. Input the data for the independent variable into the second column
3. Select **Stat** then **Regression** then **Regression...**
4. Highlight the name of the column containing the dependent variable and click **Select**
5. Highlight the name of the column containing the independent variable and click **Select**
6. Click **Storage...**
7. Check the box next to **Residuals**
8. Click **OK**
9. Click **OK**

SPSS

1. Input the data for the dependent variable into the first column
2. Input the data for the independent variable into the second column
3. Select **Analyze** then choose **Regression** then choose **Linear...**

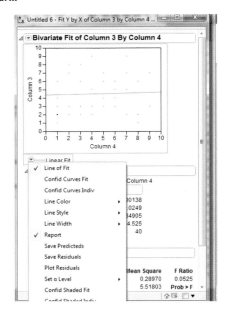

4. Highlight the name of the column containing the dependent variable
5. Click the arrow button next to the **Dependent** box to move the variable to this box
6. Highlight the name of the column containing the independent variable
7. Click the arrow button next to the **Independent** box to move the variable to this box
8. Click on the **Save...** button
9. Click the check box next to **Unstandardized** under the **Residuals** section
10. Click **Continue**
11. Click **OK**

Note: The residuals will be saved in the SPSS worksheet in a new column.

Excel 2007

1. Input the data into two separate columns
2. Click the **Data** ribbon and select **Data Analysis**
   Note: If you do not see Data Analysis listed on the Ribbon, see the Technology Notes for Chapter 2 for instructions on installing this add-on.
3. Select **Regression** from the dialog box and click **OK**
4. Click in the box next to **Y:** and select the dependent variable data
5. Click in the box next to **X:** and select the independent variable data (if you input and selected titles for BOTH columns, check the box next to **Labels**)
6. Check the box next to **Residuals** under the **Residuals** section of the dialog box
7. Click **OK**

TI-83/84

The TI-83/84 does not have the functionality to produce the residuals automatically. After using Linreg(a+bx), select 2nd, Lists, Names, and select RESID. This is a list of the residuals.

TI-Nspire

1. Enter the data for the independent variable into a data list (to access data lists select the spreadsheet option and press **enter**)
   Note: Be sure to title the list by selecting the top row of the column and typing a title.
2. Enter the data for the dependent variable into a separate data list
3. Press the menu key and select **4:Statistics** then **1:Stat Calculations** then **3:Linear Regression(mx+b)...** and press **enter**
4. For **X List:** select the column with the independent variable data from the drop-down menu
5. For **Y List:** select the column with the dependent variable data from the drop-down menu
6. Press **OK**

Residual Plot

JMP

1. Begin by saving the residuals as described in the previous section
2. Form a scatterplot of the independent variable versus the residuals using the procedures described in Chapter 2 for scatterplots

MINITAB

1. Input the data for the dependent variable into the first column
2. Input the data for the independent variable into the second column
3. Select **Stat** then **Regression** then **Regression...**
4. Highlight the name of the column containing the dependent variable and click **Select**
5. Highlight the name of the column containing the independent variable and click **Select**
6. Click **Graphs...**
7. Click the box under **Residuals versus the variables**:
8. Double-click the name of the independent variable
9. Click **OK**
10. Click **OK**

SPSS

1. Begin by saving the residuals as described in the previous section
2. Form a scatterplot of the independent variable versus the residuals using the procedures described in Chapter 2 for scatterplots

Excel 2007

1. Input the data into two separate columns
2. Click the **Data** ribbon and select **Data Analysis**

   Note: If you do not see **Data Analysis** listed on the Ribbon, see the Technology Notes for Chapter 2 for instructions on installing this add-on.

3. Select **Regression** from the dialog box and click **OK**
4. Click in the box next to **Y:** and select the dependent variable data
5. Click in the box next to **X:** and select the independent variable data (if you input and selected titles for BOTH columns, check the box next to **Labels**)
6. Check the box next to **Residuals Plots** under the **Residuals** section of the dialog box
7. Click **OK**

TI-83/84

The TI-83/84 does not have the functionality to produce a residual plot automatically. After using Linreg(a+bx), select 2nd, Statplot. Set the first plot as a scatter with XList: L1 and YList: RESID. Select Zoom, Stats, and a residual plot is displayed.

TI-Nspire

The TI-Nspire does not have the functionality to produce a residual plot automatically.

5 Probability

iStock.com/Baris Simsek

PREVIEW

In many situations, you need to assess risks in order to make an informed decision. Should you purchase an extended warranty for a new laptop? An extended warranty can be expensive, but it would result in significant savings if the laptop failed during the period covered by the extended warranty. How likely is it that your laptop will fail during this period? Suppose that the deadline to apply for a summer internship is five days away, and you are just mailing your application. Should you pay for priority mail service, which promises two-day delivery, or is it likely that your application will arrive on time by regular mail? Suppose that a prescription drug is available that will reduce pain caused by arthritis, but there are possible side effects. How serious are the side effects, and how likely are they to occur? Each of the situations just described involves making a decision in the face of uncertainty. Making good decisions in situations that involve uncertainty is easier when you can be more precise about the meaning of terms such as "likely" or "unlikely."

CHAPTER LEARNING OBJECTIVES

Conceptual Understanding

After completing this chapter, you should be able to

C1 Interpret a probability as a long-run relative frequency of occurrence.

C2 Understand what it means for two events to be mutually exclusive.

C3 Understand what it means for two events to be independent.

C4 Understand the difference between an unconditional probability and a conditional probability.

C5 (Optional) Understand how probabilities can be estimated using simulation.

Mastering the Mechanics

After completing this chapter, you should be able to

M1 Interpret probabilities in context.

M2 Calculate the probability of an event when outcomes in the sample space are equally likely.

M3 Use information in a two-way table to calculate probabilities of events, unions of two events, and intersections of two events.

M4 Given the probabilities of two events E and F and the probability of the intersection $E \cap F$, construct a hypothetical 1000 table, and use the table to calculate other probabilities of interest.

M5 Given the probabilities of two events E and F and the probability of the union $E \cup F$, construct a hypothetical 1000 table, and use the table to calculate other probabilities of interest.

M6 Calculate probabilities of unions for mutually exclusive events.

M7 Calculate probabilities of intersections for independent events.

M8 Given the probabilities of two independent events E and F, construct a hypothetical 1000 table, and use the table to calculate other probabilities of interest.

M9 Use information in a two-way table to calculate conditional probabilities.

M10 Given probability and conditional probability information, construct a hypothetical 1000 table, and use the table to calculate other probabilities of interest.

M11 (Optional) Use probability formulas to calculate probabilities of unions and intersections and to calculate conditional probabilities.

M12 (Optional) Carry out a simulation to estimate a probability.

Putting It into Practice

After completing this chapter, you should be able to

P1 Distinguish between questions that can be answered by calculating an unconditional probability and questions that can be answered by calculating a conditional probability.

P2 Given a question that can be answered by calculating a probability, calculate and interpret an appropriate probability to answer the question.

P3 Use probability to make decisions and justify conclusions.

PREVIEW EXAMPLE

Should You Paint the Nursery Pink?

Ultrasound is a medical imaging technique routinely used to assess the health of a baby prior to birth. It is sometimes possible to determine the baby's sex during an ultrasound examination. How accurate are sex identifications made during the first trimester (3 months) of pregnancy?

The paper **"The Use of Three-Dimensional Ultrasound for Fetal Gender Determination in the First Trimester"** (*The British Journal of Radiology* [2003]: 448–451) describes a study of ultrasound sex predictions. An experienced radiologist looked at 159 first trimester ultrasound images and made a sex prediction for each one. When each baby was born, the ultrasound sex prediction was compared to the baby's actual sex. The following table summarizes the resulting data.

	Radiologist 1	
	Predicted Male	**Predicted Female**
Baby is Male	74	12
Baby is Female	14	59

Notice that sex prediction based on the ultrasound image is not always correct. Several questions come to mind:

1. How likely is it that a predicted sex is correct?
2. Is a predicted sex more likely to be correct when the baby is male than when the baby is female?
3. If the predicted sex is female, should you paint the nursery pink? If you do, how likely is it that you will need to repaint?

The paper also included sex predictions made by a second radiologist, who looked at 154 first trimester ultrasound images. The data are summarized in the following table.

	Radiologist 2	
	Predicted Male	**Predicted Female**
Baby is Male	81	8
Baby is Female	7	58

In addition to the questions posed previously, you could also compare the accuracy of sex predictions for the two radiologists. Does the skill of the radiologist make a difference?

All of these questions can be answered using the methods introduced in this chapter, and this example will be revisited in Section 5.4.

From its roots in the analysis of games of chance, probability has evolved into a science that enables you to make informed decisions with confidence. In this chapter, you will be introduced to the basic ideas of probability, explore strategies for calculating probabilities, and consider ways to estimate probabilities when it is difficult to calculate them directly. ■

SECTION 5.1 Interpreting Probabilities

People often find themselves in situations where the outcome is uncertain. For example, when a ticketed passenger shows up at the airport, she faces two possible outcomes: (1) she is able to take the flight, or (2) she is denied a seat as a result of overbooking by the airline, and must take a later flight. Based on her past experience with this particular flight, the passenger may know that one outcome is more likely than the other. She may believe that the chance of being denied a seat is quite small. Although this outcome is possible, she views it as unlikely.

To quantify the likelihood of its occurrence, a number between 0 and 1 can be assigned to an outcome. This number is called a *probability*. Assigning a probability to an outcome is an attempt to quantify what is meant by "*likely*" or "*unlikely*."

A **probability** is a number between 0 and 1 that reflects the likelihood of occurrence of some outcome.

There are several ways to interpret a probability. One is a subjective interpretation, in which a probability is interpreted as a personal measure of the strength of belief that an outcome will occur. A probability of 1 represents a belief that the outcome will certainly occur. A probability of 0 represents a belief that the outcome will certainly *not* occur—that it is impossible. All other probabilities fall between these two extremes. This interpretation is common in ordinary speech. For example, you might say, "There's about a 50–50 chance," or "My chances are nil."

The subjective interpretation, however, presents some difficulties. Because different people may have different subjective beliefs, they may assign different probabilities to the same outcome. Whenever possible, we will use an objective *relative frequency* approach to probability. With this approach, a probability specifies the *long-run proportion* of the time that an outcome will occur. A probability of 1 corresponds to an outcome that occurs 100% of the time—a certain outcome. A probability of 0 corresponds to an outcome that occurs 0% of the time—an impossible outcome.

Relative Frequency Interpretation of Probability

The probability of an outcome, denoted by *P*(outcome), is interpreted as the proportion of the time that the outcome occurs *in the long run*.

Consider the following situation. A package delivery service promises 2-day delivery between two cities but is often able to deliver packages in just 1 day. The company reports that the probability of next-day delivery is 0.3. This implies that in the long run, 30% of all packages arrive in 1 day. An equivalent way to interpret this probability would be to say that in the long run, about 30 out of 100 packages shipped would arrive in 1 day.

Suppose that you track the delivery of packages shipped with this company. With each new package shipped, you could calculate the relative frequency of packages shipped so far that have arrived in 1 day:

$$\frac{\text{number of packages that arrived in 1 day}}{\text{total number of packages shipped}}$$

The results for the first 15 packages might be as follows:

Package Number	Did the Package Arrive in 1 Day?	Relative Frequency of Packages Shipped so Far That Arrived in 1 Day
1	Yes	$\frac{1}{1} = 1.00$
2	Yes	$\frac{2}{2} = 1.00$
3	No	$\frac{2}{3} = 0.67$
4	Yes	$\frac{3}{4} = 0.75$
5	No	$\frac{3}{5} = 0.60$
6	No	$\frac{3}{6} = 0.50$
7	Yes	$\frac{4}{7} = 0.57$
8	No	$\frac{4}{8} = 0.50$
9	No	$\frac{4}{9} = 0.44$
10	No	$\frac{4}{10} = 0.40$
11	No	$\frac{4}{11} = 0.36$
12	Yes	$\frac{5}{12} = 0.42$
13	No	$\frac{5}{13} = 0.38$
14	No	$\frac{5}{14} = 0.36$
15	No	$\frac{5}{15} = 0.33$

FIGURE 5.1
Relative frequency of packages
delivered in 1 day for 15 packages

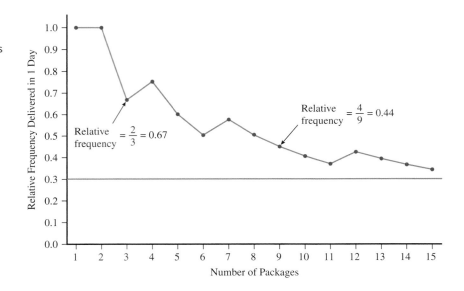

FIGURE 5.2
Relative frequency of packages
delivered in 1 day for 50 packages

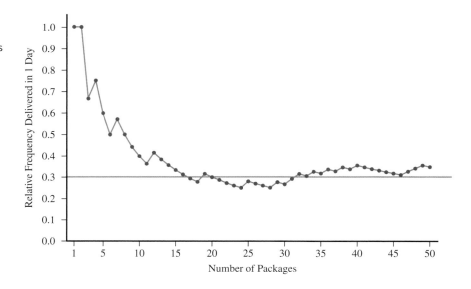

FIGURE 5.3
Relative frequency of
packages delivered in
1 day for 1000 packages

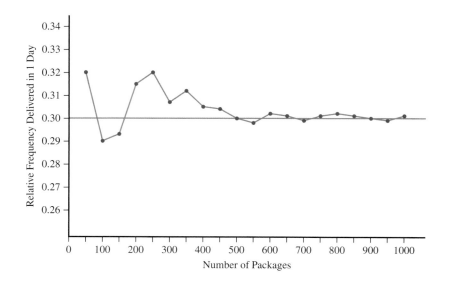

Figure 5.1 shows how this relative frequency of packages arriving in 1 day changes over the first 15 packages.

Figure 5.2 illustrates how this relative frequency fluctuates during a sequence of 50 shipments. As the number of packages in the sequence increases, the relative frequency does not continue to fluctuate wildly but instead settles down and approaches a specific value, which is the probability of interest. Figure 5.3 illustrates how the relative frequency settles down over a sequence of 1000 shipments. What you see happening in this figure is not unique to this particular example. This "settling down" is a consequence of the Law of Large Numbers.

Law of Large Numbers

As the number of observations increases, the proportion of the time that an outcome occurs gets close to the actual value of the probability of that outcome.

The Law of Large Numbers is the basis for the relative frequency interpretation of probabilities.

Some Basic Properties of Probabilities

Keep the relative frequency interpretation of probability in mind as you consider the following basic properties of probabilities:

1. *The probability of any outcome is a number between 0 and 1.* A relative frequency, which is the number of times the outcome occurs divided by the total number of observations, cannot be less than 0 (there can't be a negative number of occurrences) or greater than 1 (the outcome can't occur more often than the total number of observations).

2. *If outcomes can't occur at the same time, then the probability that any one of them will occur is the sum of their individual probabilities.* For example, **The Chronicle of Higher Education Almanac (2016)** reports that of students enrolled full time at 4-year colleges in 2015, 65% were enrolled in public colleges, 26% were enrolled in private nonprofit colleges, and 9% were enrolled in private for-profit colleges. Imagine selecting a student at random from this group of students. If you observe the type of college the selected student attends, there are three possible outcomes: (1) public, (2) private nonprofit, and (3) private for-profit. Interpreting the given percentages as probabilities, you can say that

   $$P(\text{selected student attends a private nonprofit}) = 0.26$$

   and

   $$P(\text{selected student attends a private for-profit}) = 0.09$$

   Assuming that no person is enrolled *full-time* at more than one college, the probability that the selected student attends a private nonprofit *or* a private for-profit college is

   $$P(\text{private nonprofit } or \text{ private for-profit}) = 0.26 + 0.09 = 0.35$$

   About 35 in 100 full-time 4-year college students attend a private college (nonprofit or for-profit).

3. *The probability that an outcome will not occur is equal to 1 minus the probability that the outcome will occur.* Continuing with the 4-year college student example, you know that about 35 in 100 full-time students attend a private college. This means that the others do not attend a private college, so about 65 in 100 full-time students are not at private colleges. Here

   $$P(not \text{ private}) = 1 - P(\text{private}) = 1 - 0.35 = 0.65$$

   Notice that this is equal to $P(\text{public})$, because there are only three possible outcomes. Also notice that because there are only three possible outcomes that might be observed,

   $$P(\text{public } or \text{ private nonprofit } or \text{ private for-profit}) = 0.65 + 0.26 + 0.09 = 1$$

Because a probability represents a long-run relative frequency, in situations where exact probabilities are not known, it is common to estimate probabilities based on observation. For example, a shipping company's reported 1-day delivery probability of 0.3 is most likely an estimate based on historical data by calculating the proportion of packages delivered in 1 day over a large number of shipments.

Summing It Up—Section 5.1

The following learning objectives were addressed in this section:

Conceptual Understanding
C1: Interpret a probability as a long-run relative frequency of occurrence.
A probability is a number between 0 and 1 that reflects the likelihood of occurrence of a particular outcome. The probability of an outcome is interpreted as the proportion of the time that the outcome occurs in the long run (a long-run relative frequency).

Mastering the Mechanics
M1: Interpret probabilities in context.
The probability of an outcome is interpreted as the proportion of the time that the outcome occurs in the long run (a long-run relative frequency). For an example of interpreting a probability as a long-run relative frequency in context, see the discussion of on-time delivery rates and Figures 5.1–5.3.

SECTION 5.1 EXERCISES

Each Exercise Set assesses the following chapter learning objectives: C1, M1

SECTION 5.1 Exercise Set 1

5.1 The article **"Scrambled Statistics: What Are the Chances of Finding Multi-Yolk Eggs?"** (*Significance* **[August 2016]: 11)** gives the probability of a double-yolk egg as 0.001.
a. Give a relative frequency interpretation of this probability.
b. If 5000 eggs were randomly selected, about how many double-yolk eggs would you expect to find?

5.2 An airline reports that for a particular flight operating daily between Phoenix and Atlanta, the probability of an on-time arrival is 0.86. Give a relative frequency interpretation of this probability.

5.3 For a monthly subscription fee, a video download site allows people to download and watch up to five movies per month. Based on past download histories, the following table gives the estimated probabilities that a randomly selected subscriber will download 0, 1, 2, 3, 4 or 5 movies in a particular month.

Number of downloads	0	1	2	3	4	5
Estimated probability	0.03	0.45	0.25	0.10	0.10	0.07

If a subscriber is selected at random, what is the estimated probability that this subscriber downloads
a. three or fewer movies?
b. at most three movies?
c. four or more movies?
d. zero or one movie?
e. more than one movie?

SECTION 5.1 Exercise Set 2

5.4 What does it mean to say that the probability that a coin toss will land head side up is 0.5?

5.5 In a particular state, automobiles that are more than 10 years old must pass a vehicle inspection in order to be registered. This state reports the probability that a car more than 10 years old will fail the vehicle inspection is 0.09. Give a relative frequency interpretation of this probability.

5.6 A bookstore sells books in several formats—hardcover, paperback, digital, and audio.

Based on past sales, the table below gives the estimated probabilities that a randomly selected purchase will be of particular types.

Hardcover	Paperback	Digital	Audio
0.16	0.36	0.40	0.08

If a purchase is selected at random, what is the probability that this purchase is for a book that is
a. digital or audio?
b. not digital?
c. a printed book?

ADDITIONAL EXERCISES

5.7 If you were to roll a fair die 1000 times, about how many sixes do you think you would observe? What is the probability of observing a six when a fair die is rolled?

5.8 The probability of getting a king when a card is selected at random from a standard deck of 52 playing cards is $\frac{1}{13}$.

a. Give a relative frequency interpretation of this probability.

b. Express the probability as a decimal rounded to three decimal places. Then complete the following statement: If a card is selected at random, I would expect to see a king about _____ times in 1000.

5.9 At a particular college, students have several options for purchasing textbooks. The options and the proportion of students choosing each option are shown in the accompanying table.

If a student at this college is selected at random, what is the probability that the student

a. purchased all print books online?

b. purchased all print books?

c. purchased some e-books or all e-books?

d. did not purchase all print books from the campus bookstore?

5.10 Give a relative frequency interpretation for each probability calculated in the previous exercise.

TABLE FOR EXERCISE 5.9

Option	All Print Books From Campus Bookstore	All Print Books From Off-Campus Bookstore	All Print Books From Online Booksellers	All Print Books, but Purchased Both From Campus Bookstore and Online	All e-Books Downloaded From Publishers	Mix of Print and e-Books
Proportion Choosing Option	0.25	0.12	0.17	0.31	0.04	0.11

SECTION 5.2 Calculating Probabilities

In the previous section, you saw how probabilities are interpreted and learned some basic properties. In this section, you will learn two ways that probabilities can be calculated. Before you begin, a few key terms need to be introduced.

Chance Experiments and Sample Spaces

In most probability settings, there is uncertainty about what the outcome will be. Such situations are called *chance experiments*.

> **DEFINITION**
>
> A **chance experiment** is the process of making an observation when there is uncertainty about which of two or more possible outcomes will result.

For example, in an opinion poll, there is uncertainty about whether an individual selected at random from some population supports a school bond, and when a die is rolled there is uncertainty about which side will land face up. Observing outcomes in either of these situations are examples of chance experiments. Notice that a *chance* experiment is different from the kind of experiment considered in Chapter 1, where the term experiment was used to describe a type of statistical study.

Consider a chance experiment to investigate whether men or women are more likely to choose a hybrid engine over a traditional internal combustion engine when purchasing a Honda Civic at a particular car dealership. A customer will be selected at random from those who purchased a Honda Civic. The type of vehicle purchased (hybrid or traditional) will be determined, and the customer's sex will be recorded. Before the customer is selected, there is uncertainty about the outcome, so this is a chance experiment. You do know, however, what the possible outcomes are. This set of possible outcomes is called the *sample space*.

> **DEFINITION**
>
> The collection of all possible outcomes of a chance experiment is the **sample space** for the experiment. The sample space is usually denoted by the letter *S*.

The sample space of a chance experiment can be represented in many ways. One representation is a simple list of all the possible outcomes. For the chance experiment where sex and engine type were observed, the possible outcomes are the following:

1. male, hybrid
2. female, hybrid
3. male, traditional
4. female, traditional

To simplify, you could use M and F to indicate sex and H and T to indicate engine type. Using set notation, the sample space could be written as

$$S = \{MH, FH, MT, FT\}$$

Events

In the car-purchase chance experiment, you might be interested in a particular outcome. Or you might focus on the group of outcomes that involve the purchase of a hybrid—the group consisting of MH and FH. When you combine one or more individual outcomes into a collection, you are creating what is known as an *event*.

DEFINITION

An **event** is any collection of outcomes from the sample space of a chance experiment.

A **simple event** is an event consisting of exactly one of the outcomes in the sample space.

An event can be represented by a name, such as *hybrid*, or by an uppercase letter, such as A, B, or C. Sometimes different events are denoted using the same letter with different subscripts, such as E_1, E_2, and E_3.

Example 5.1 **Car Preferences**

Reconsider the situation in which a person who purchased a Honda Civic was categorized by sex (M or F) and type of car purchased (H or T). The sample space is

$$S = \{MH, FH, MT, FT\}$$

Because there are four outcomes, there are four simple events. One event of interest consists of all outcomes in which a hybrid is purchased. The event *hybrid* is

$$hybrid = \{MH, FH\}$$

Another event is the event that the purchaser is male,

$$male = \{MH, MT\}$$

Because it consists of only one outcome, one of the different possible simple events occurs whenever a chance experiment is performed. You also say that a given event occurs whenever one of the outcomes making up the event occurs. For example, if the outcome in Example 5.1 is MH, then the simple event *male purchasing a hybrid* has occurred, and so has the non-simple event *hybrid*.

Example 5.2 **Simple Coin Toss Game**

Consider a game in which two players each toss a coin. If both coins land heads up, Player 1 is declared the winner. If both coins land tails up, Player 2 is declared the winner. If the coins land with one showing heads and the other showing tails, the game is considered a tie. This game can be viewed as a chance experiment. Before the coins are tossed, there is uncertainty about what outcome will occur.

For this chance experiment, one possible outcome is that Player 1 tosses a head and Player 2 tosses a tail. This outcome can be abbreviated as HT. Using this notation, the sample space is

$$S = \{HH, HT, TH, TT\}$$

Using W_1 to denote the event *player 1 wins*,

$$W_1 = \{HH\}$$

For the event *game results in a tie*, denoted by T,

$$T = \{HT, TH\}$$

Calculating Probabilities

Some chance experiments have outcomes that are equally likely to occur. This would be the case when, for example, a fair coin is tossed (two equally likely outcomes, H and T) or when a fair die is rolled (six equally likely outcomes). When outcomes are equally likely, it is easy to calculate the probability of an event using what is known as the classical approach.

The Classical Approach to Calculating Probabilities

When a chance experiment has equally likely outcomes, in the long run you would expect each of these outcomes to occur the same proportion of the time. For example, if there are 10 equally likely outcomes, each one would occur about $\frac{1}{10}$ or 10% of the time in the long run. If an event consisted of 3 of these 10 outcomes, this event would occur about $\frac{3}{10}$ or 30% of the time.

In general, if there are N equally likely outcomes, the probability of each outcome is $\frac{1}{N}$, and the probability of an event can be determined if you know how many of the possible outcomes are included in the event.

Classical Approach to Calculating Probabilities for Equally Likely Outcomes

When the outcomes in the sample space of a chance experiment are equally likely, the **probability of an event** E, denoted by $P(E)$, is calculated by dividing the number of outcomes in the event E by the total number of outcomes in the sample space:

$$P(E) = \frac{\text{number of outcomes in the event E}}{\text{number of outcomes in the sample space}}$$

Example 5.3 Calling the Toss

On some football teams, the honor of calling the toss at the beginning of a football game is determined by random selection. Suppose that, this week, a member of the offense will call the toss. There are 5 linemen on the 11-player offense. Because a player will be selected at random, each of these 11 players is equally likely to be chosen. If you define the event L as the event that a lineman is selected to call the toss, 5 of the 11 possible outcomes are included in L. The probability that a lineman will be selected is then

$$P(L) = \frac{5}{11} = 0.455$$

In Section 5.1, probabilities were interpreted as long-run relative frequencies. This interpretation is appropriate even when probabilities are calculated using the classical approach. The calculated probability indicates that in a long sequence of selections, a lineman will be chosen about 45.5% of the time.

Example 5.4   Math Contest

Four students (Adam, Bettina, Carlos, and Debra) submitted correct solutions to a math contest that had two prizes. The contest rules specify that if more than two correct responses are submitted, the winners will be selected at random from those submitting correct responses. You can use AB to denote the outcome that Adam and Bettina are the two selected. The other possible outcomes can be denoted in a similar way. Then the sample space for selecting the two winners from the four correct responses is

$$S = \{AB, AC, AD, BC, BD, CD\}$$

Because the winners are selected at random, the six possible outcomes are equally likely and the probability of each individual outcome is $\frac{1}{6}$.

If E is the event that both selected winners are the same sex,

$$E = \{AC, BD\}$$

Because E contains two outcomes, $P(E) = \frac{2}{6} = 0.333$. In the long run, if two winners are selected at random from this group, both winners will be the same sex about 33.3% of the time. If F denotes the event that at least one of the selected winners is female, then F consists of all outcomes except AC and $P(F) = \frac{5}{6} = 0.833$. In the long-run, about 83.3% of the selections would include at least one female winner.

The classical approach to probability works well for chance experiments that have a finite set of outcomes that are equally likely. However, many chance experiments do not have equally likely outcomes. For example, consider the chance experiment of selecting a student from those enrolled at a particular school and observing whether the student is a freshman, sophomore, junior, or senior. If there are more seniors than sophomores at the school, the four possible outcomes for this chance experiment are not equally likely. In this situation, it would be a mistake to think that the probability of each outcome is $\frac{1}{4}$. To calculate or estimate probabilities in situations where outcomes are not equally likely, you need an alternate approach.

Relative Frequency Approach to Estimating Probabilities

When a chance experiment is performed, some events may be likely to occur, whereas others may not be as likely. For a specified event E, its probability indicates how frequently E occurs when the chance experiment is performed many times. For example, recall the package delivery example from Section 5.1. Figures 5.1–5.3 showed that the relative frequency (proportion) of packages delivered in 1 day fluctuated quite a bit over the short run, but in the long run this relative frequency settled down and stayed close to 0.3, the reported probability of a next-day delivery.

The Law of Large Numbers tells you that, as the number of repetitions of a chance experiment increases, the proportion of the time an event occurs gets close to the actual probability of the event, *even if the value of this probability is not known*. This means that you can observe outcomes from a chance experiment and then use the observed outcomes to estimate probabilities.

Relative Frequency Approach to Estimating Probabilities

The **probability of an event** E, denoted by $P(E)$, is defined to be the value approached by the relative frequency of occurrence of E in a very long series of observations from a chance experiment. If the number of observations is large,

$$P(E) \approx \frac{\text{number of times } E \text{ occurs}}{\text{number of observations}}$$

The relative frequency definition of probability depends on being able to repeat a chance experiment many times. For example, suppose that you perform a chance experiment that consists of flipping a cap from a 20-ounce bottle of soda and noting whether the cap lands with the top up or down. Unlike tossing a coin, there is no particular reason to believe the cap is equally likely to land top up or top down. You can flip the cap many times and calculate the relative frequency of the event $T = top\ up$

$$\frac{\text{number of times the event } top\ up \text{ occurs}}{\text{total number of flips}}$$

This relative frequency is an estimate of the probability of the event T. If the bottle cap was flipped 1000 times and it landed top up 694 times, the estimate of $P(T)$ would be

$$P(T) \approx \frac{694}{1000} = 0.694$$

In some situations, probabilities are estimated based on past history. For example, an insurance company may use past claims to estimate the probability that a 20-year-old male driver will submit a claim for a car accident in a given year. If 45,000 20-year-old males were insured in the previous year and 3200 of them submitted a car accident claim, an estimate of this probability would be $\frac{3200}{45,000} = 0.071$. This probability is interpreted as meaning that for 20-year-old males insured by this company, about 71 out of 1000 will submit a car accident claim in a given year.

The relative frequency approach to probability is based on observation. By observing many outcomes, you can obtain stable long-run relative frequencies that provide reasonable estimates of the probabilities of different events. The relative frequency approach to probability is intuitive, and it can be used in situations where the classical approach is not appropriate. In situations where outcomes *are* equally likely, however, either the classical or the relative frequency approach can be used.

Summing It Up—Section 5.2

The following learning objectives were addressed in this section:

Mastering the Mechanics
M1: Interpret probabilities in context.
The probability of an outcome is interpreted as the proportion of the time that the outcome occurs in the long run (a long-run relative frequency). Examples 5.3 and 5.4 illustrate the interpretation of probabilities in context.

M2: Calculate the probability of an event when outcomes in the sample space are equally likely.
When the outcomes in the sample space are equally likely, the probability of an event can be calculated by dividing the number of outcomes in the event by the total number of outcomes in the sample space. This is illustrated in Example 5.3.

SECTION 5.2 EXERCISES

Each Exercise Set assesses the following chapter learning objectives: M1, M2

SECTION 5.2 Exercise Set 1

5.11 Phoenix is a hub for a large airline. Suppose that on a particular day, 8000 passengers arrived in Phoenix on this airline. Phoenix was the final destination for 1800 of these passengers. The others were all connecting to flights to other cities. On this particular day, several inbound flights were late, and 480 passengers missed their connecting flight. Of these 480 passengers, 75 were delayed overnight and had to spend the night in Phoenix. Consider the chance experiment of choosing a passenger at random from these 8000 passengers. Calculate the following probabilities:

a. the probability that the selected passenger had Phoenix as a final destination.

b. the probability that the selected passenger did not have Phoenix as a final destination.

c. the probability that the selected passenger was connecting and missed the connecting flight.

d. the probability that the selected passenger was a connecting passenger and did not miss the connecting flight.

e. the probability that the selected passenger either had Phoenix as a final destination or was delayed overnight in Phoenix.

f. An independent customer satisfaction survey is planned. Fifty passengers selected at random from the 8000 passengers who arrived in Phoenix on the day described above will be contacted for the survey. The airline knows that the survey results will not be favorable if too many people who were delayed overnight are included. Write a few sentences explaining whether or not you think the airline should be worried, using relevant probabilities to support your answer.

5.12 A professor assigns five problems to be completed as homework. At the next class meeting, two of the five problems will be selected at random and collected for grading. You have only completed the first three problems.

a. What is the sample space for the chance experiment of selecting two problems at random? (Hint: You can think of the problems as being labeled A, B, C, D, and E. One possible selection of two problems is A and B. If these two problems are selected and you did problems A, B, and C, you will be able to turn in both problems. There are nine other possible selections to consider.)

b. Are the outcomes in the sample space equally likely?

c. What is the probability that you will be able to turn in both of the problems selected?

d. Does the probability that you will be able to turn in both problems change if you had completed the last three problems instead of the first three problems? Explain.

e. What happens to the probability that you will be able to turn in both problems selected if you had completed four of the problems rather than just three?

5.13 Suppose you want to estimate the probability that a patient will develop an infection while hospitalized at a particular hospital. In the past year, this hospital had 6450 patients, and 712 of them developed an infection. What is the estimated probability that a patient at this hospital will develop an infection?

SECTION 5.2 Exercise Set 2

5.14 A college job placement center has requests from five students for employment interviews. Three of these students are math majors, and the other two students are statistics majors. Unfortunately, the interviewer has time to talk to only two of the students. These two will be randomly selected from among the five.

a. What is the sample space for the chance experiment of selecting two students at random? (Hint: You can think of the students as being labeled A, B, C, D, and E. One possible selection of two students is A and B. There are nine other possible selections to consider.)

b. Are the outcomes in the sample space equally likely?

c. What is the probability that both selected students are statistics majors?

d. What is the probability that both students are math majors?

e. What is the probability that at least one of the students selected is a statistics major?

f. What is the probability that the selected students have different majors?

5.15 Roulette is a game of chance that involves spinning a wheel that is divided into 38 segments of equal size, as shown in the accompanying picture.

Anna Baburkina/Shutterstock.com

A metal ball is tossed into the wheel as it is spinning, and the ball eventually lands in one of the 38 segments. Each segment has an associated color. Two segments are green. Half of the other 36 segments are red, and the others are black. When a balanced roulette wheel is spun, the ball is equally likely to land in any one of the 38 segments.

a. When a balanced roulette wheel is spun, what is the probability that the ball lands in a red segment?

b. In the roulette wheel shown, black and red segments alternate. Suppose instead that all red segments were grouped together and that all black segments were together. Does this increase the probability that the ball will land in a red segment? Explain.

c. Suppose that you watch 1000 spins of a roulette wheel and note the color that results from each spin. What would be an indication that the wheel was not balanced?

5.16 Suppose you want to estimate the probability that a randomly selected customer at a particular grocery store will pay by credit card. Over the past 3 months, 80,500 purchases were made, and 37,100 of them were paid for by credit card. What is the estimated probability that a randomly selected customer will pay by credit card?

ADDITIONAL EXERCISES

5.17 According to *The Chronicle for Higher Education Almanac* **(2016)**, there were 1,003,329 Associate degrees awarded by U.S. community colleges in the 2013–2014 academic year. A total of 613,034 of these degrees were awarded to women.

a. If a person who received an Associate degree in 2013–2014 is selected at random, what is the probability that the selected person will be female?

b. What is the probability that the selected person will be male?

5.18 The same issue of *The Chronicle for Higher Education Almanac* referenced in the previous exercise also reported the following information for Ph.D. degrees awarded by U.S. colleges in the 2013–2014 academic year:

- A total of 54,070 Ph.D. degrees were awarded.
- 12,504 of these degrees were in the life sciences.
- 9859 of these degrees were in the physical sciences.
- The remaining degrees were in majors other than life or physical sciences.

What is the probability that a randomly selected Ph.D. student who received a degree in 2013–2014

a. received an degree in the life sciences?

b. received a degree that was not in a life or a physical science?

c. did not receive a degree in the physical sciences?

5.19 The **National Center for Health Statistics (www.cdc .gov/nchs/data/nvsr/nvsr64/nvsr64_12.pdf, retrieved April 25, 2017)** gave the following information on births in the United States in 2014:

Type of Birth	Number of Births
Single birth	3,848,214
Twins	135,336
Triplets	4,233
Quadruplets	246
Quintuplets or higher	47

Use this information to estimate the probability that a randomly selected pregnant woman who gave birth in 2014

a. delivered twins

b. delivered quadruplets

c. gave birth to more than a single child

5.20 A deck of 52 cards is mixed well, and 5 cards are dealt.

a. It can be shown that (disregarding the order in which the cards are dealt) there are 2,598,960 possible hands, of which only 1287 are hands consisting entirely of spades. What is the probability that a hand will consist entirely of spades? What is the probability that a hand will consist entirely of a single suit?

b. It can be shown that exactly 63,206 of the possible hands contain only spades and clubs, with both suits represented. What is the probability that a hand consists entirely of spades and clubs with both suits represented?

5.21 Six people hope to be selected as a contestant on a TV game show. Two of these people are younger than 25 years old. Two of these six will be chosen at random to be on the show.

a. What is the sample space for the chance experiment of selecting two of these people at random? (Hint: You can think of the people as being labeled A, B, C, D, E, and F. One possible selection of two people is A and B. There are 14 other possible selections to consider.)

b. Are the outcomes in the sample space equally likely?

c. What is the probability that both the chosen contestants are younger than 25?

d. What is the probability that both the chosen contestants are not younger than 25?

e. What is the probability that one is younger than 25 and the other is not?

SECTION 5.3 Probabilities of More Complex Events: Unions, Intersections, and Complements

In many situations, two or more different events are of interest. For example, consider a chance experiment that consists of selecting a student at random from those enrolled at a particular college. If there are 9000 students enrolled at the college, the sample space would consist of 9000 different possible outcomes, each corresponding to a student who might be selected.

In this situation, here are some possible events:

F = event that the selected student is female
O = event that the selected student is older than 30
A = event that the selected student favors expanding the athletics program
S = event that the selected student is majoring in one of the lab sciences

Because a student is to be selected at random, the outcomes in the sample space are equally likely. This means that you can use the classical approach to calculate probabilities. For example, if 6000 of the 9000 students at the college are female,

$$P(F) = \frac{6000}{9000} = 0.67$$

Similarly, if 4300 of the 9000 students favor expanding the athletics program,

$$P(A) = \frac{4300}{9000} = 0.48$$

Once a number of events have been specified, it is possible to use these events to create new events. For example, you might be interested in the event that the selected student is *not* majoring in a lab science or the event that the selected student is female *and* older than 30.

Complements

The complement of an event E is a new event denoted by E^C. E^C is the event that E does *not* occur. For the event A described previously, the complement of A is

$A^C = not\ A =$ event that the selected student does *not* favor expanding
the athletics program

In terms of outcomes, the event A^C includes all the possible outcomes in the sample space that are not in the event A. This implies

$$P(A^C) = 1 - P(A)$$

Since $P(A)$ was calculated to be $P(A) = 0.48$, you know $P(A^C) = 1.0 - 0.48 = 0.52$.

Complement Probabilities

If E is an event, the **complement of E**, denoted by E^C, is the event that E does *not* occur.

The probability of E^C can be calculated from the probability of E as follows:

$$P(E^C) = 1 - P(E)$$

Intersections

The intersection of two events E and F is denoted by $E \cap F$. The intersection $E \cap F$ is the event that E *and* F both occur. For example, consider the two events previously described

$O =$ event that the selected student is over 30

and

$S =$ event that the selected student is majoring in one of the lab sciences

The event $O \cap S$ is the event that the selected student is over 30 *and* is majoring in one of the lab sciences. In the following table, the 9000 students have been classified into one of four cells. The rows of the table correspond to whether or not the event O occurs and the columns correspond to whether or not the event S occurs.

	S (Majoring in Lab Science)	Not S (Not Majoring in Lab Science)	Total
O (Over 30)	400	1,700	2,100
Not O (Not Over 30)	1,100	5,800	6,900
Total	1,500	7,500	9,000

From this table, you can see that

$$P(O) = \frac{2100}{9000} = 0.23$$

$$P(S) = \frac{1500}{9000} = 0.17$$

$$P(O \cap S) = \frac{400}{9000} = 0.04$$

The 400 in the numerator of the fraction used to calculate $P(O \cap S)$ comes from the upper-left cell of the table—these 400 students are both over 30 and majoring in a lab science.

Intersection Events

If E and F are events, the **intersection of E and F** is denoted by $E \cap F$ and is the new event that *both E and F* occur.

Unions

The union of two events E and F is denoted by $E \cup F$. The event $E \cup F$ is the event that either *E or F* (or both) occur. In the chance experiment of selecting a student at random, you might be interested in the event that the selected student favors expanding the athletics program or is over 30 years old. This event is the union of the events A and O described earlier, and would be denoted by $A \cup O$.

Suppose that the following table describes the students at the college:

	A (Favors Expanding Athletics)	Not A (Does Not Favor Expanding Athletics)	Total
O (Over 30)	1,600	500	2,100
Not O (Not Over 30)	2,700	4,200	6,900
Total	4,300	4,700	9,000

From this table you can see that

$$P(O) = \frac{2100}{9000} = 0.23$$

$$P(A) = \frac{4300}{9000} = 0.48$$

$$P(O \cup A) = \frac{1600 + 500 + 2700}{9000} = \frac{4800}{9000} = 0.53$$

Notice that the 4800 in the numerator of the fraction used to calculate $P(O \cup A)$ comes from adding the numbers in three cells of the table. These cells represent outcomes for which at least one of the events O or A occurs. Also notice that just adding the total for event O (2100) and the total for event A (4300) does not result in the correct numerator because the 1600 students who are both over 30 and favor expanding athletics are counted twice.

Union Events

If E and F are events, the **union** of these events is denoted by $E \cup F$. The event $E \cup F$ is the new event that *E or F* (or both) occurs.

Working with Hypothetical 1000 Tables to Calculate Probabilities

In the previous discussion, you were able to use tables to calculate the probability of an intersection of two events and the probability of a union of two events. This was possible because a student was to be selected at random (making each of the 9000 possible outcomes equally likely) and because the numbers of students falling into each of the cells of the appropriate table were given. In many situations, you may only know the probabilities of some events. In this case, it is often possible to create a *hypothetical 1000 table* and then use the table to calculate probabilities. The following examples show how this is done.

Example 5.5   Health Information From TV Shows

The report **"TV Drama/Comedy Viewers and Health Information" (cdc.gov/healthcommunication /pdf/healthstyles_2005.pdf, retrieved April 25, 2017)** describes a large survey that was conducted for the Centers for Disease Control and Prevention (CDC). The CDC believed that the sample used was representative of adult Americans. One question on the survey asked respondents if they had learned something new about a health issue or disease from a TV show in the previous 6 months.

Consider the following events:

L = event that a randomly selected adult American reports learning something new about a health issue or disease from a TV show in the previous 6 months

and

F = event that a randomly selected adult American is female

Data from the survey were used to estimate the following probabilities:

$$P(L) = 0.58 \quad P(F) = 0.50 \quad P(L \cap F) = 0.31$$

From the given information, you can create a table for a hypothetical 1000 people. You would start by labeling rows and columns of the table as follows:

	F (Female)	Not F (Not Female)	Total
L (Learned from TV)			
Not L (Did Not Learn from TV)			
Total			

Because you are creating a table of 1000 individuals, $P(L) = 0.58$ tells you that 58% of the 1000 people should be in the L row: $(0.58)(1000) = 580$. This means that $1000 - 580 = 420$ should be in the *not L* row. You can add these row totals to the table:

	F (Female)	Not F (Not Female)	Total
L (Learned from TV)			580
Not L (Did Not Learn from TV)			420
Total			1,000

Similarly, you can figure out what the column totals should be. $P(F) = 0.50$, so the total for the F column should be $(0.50)(1000) = 500$. This means that the column total for the *not F* column should also be 500 because $1000 - 500 = 500$. Adding these column totals to the table gives the following:

	F (Female)	Not F (Not Female)	Total
L (Learned from TV)			580
Not L (Did Not Learn from TV)			420
Total	500	500	1,000

One other probability was given:

$$P(L \cap F) = 0.31$$

This means that you would expect there to be $(0.31)(1000) = 310$ people who were female *and* who learned something about a health issue or disease from TV in the previous 6 months. This number would go in the upper-left cell of the table. This cell and the marginal totals can then be used to calculate values for the remaining cells of the hypothetical 1000 table.

	F (Female)	Not F (Not Female)	Total
L (Learned from TV)	310	270	**580**
Not L (Did Not Learn from TV)	190	230	**420**
Total	**500**	**500**	**1,000**

This table can now be used to calculate other probabilities from those that were given. For example,

$$P(L \cup F) = \frac{310 + 270 + 190}{1000} = \frac{770}{1000} = 0.770$$

Example 5.6 Tattoos

The article **"Chances Are You Know Someone with a Tattoo, and He's Not a Sailor" (*Associated Press*, June 11, 2006)** reported the following approximate probabilities based on a survey of adults ages 18 to 50:

$$P(\text{age 18 to 29}) = 0.50$$

$$P(\text{tattoo}) = 0.24$$

$$P(\text{tattoo } or \text{ age 18 to 29}) = 0.56$$

Consider a chance experiment that consists of selecting a person at random from a population of adults age 18 to 50. Here is some notation for events of interest:

$$T = \text{the event that the selected person has a tattoo}$$

and

$$A = \text{the event that the selected person is age 18 to 29}$$

Assuming that the survey participants are representative of the population of adults age 18 to 50, you know

$$P(T) = 0.24 \quad P(A) = 0.50 \quad P(T \cup A) = 0.56$$

Using the given probability information, you can now create a hypothetical 1000 table. Setting up the table and labeling the rows and columns gives

	T (Tattoo)	Not T (No Tattoo)	Total
A (Age 18 to 29)			
Not A (Age 30 to 50)			
Total			**1,000**

Notice that "not age 18 to 29" is equivalent to age 30 to 50 because everyone in the population of interest is age 18 to 50. $P(18 \text{ to } 29) = 0.50$ tells you that the hypothetical 1000 people are divided evenly between the two age groups. Because $P(tattoo) = 0.24$, you expect $(0.24)(1000) = 240$ of the hypothetical 1000 people to be in the tattoo column and the rest $(1000 - 240 = 760)$ to be in the no tattoo column, as shown in the following table.

	T (Tattoo)	*Not T* (No Tattoo)	Total
A (Age 18 to 29)	Cell 1	Cell 2	**500**
Not A (Age 30 to 50)	Cell 3		**500**
Total	**240**	**760**	**1,000**

The next step is to try to fill in the cells of the table using the given information. Because you know that $P(T \cup A) = 0.56$, you know that the sum of the counts in the cells labeled Cell 1, Cell 2, and Cell 3 in the preceding table must total $(0.56)(1000) = 560$ because these cells correspond to the union of the events *tattoo* and *age 18 to 29*. Adding the total for the *tattoo* column and the *age 18 to 29* row gives

$$240 + 500 = 740$$

Because this counts the people in Cell 1 twice, the difference between 740 and the union count of 560 must be the number of people in Cell 1. This means that the Cell 1 count is $740 - 560 = 180$. Once you know the marginal totals and this cell count, you can fill in the rest of the cells to get the following table:

	T (Tattoo)	*Not T* (No Tattoo)	Total
A (Age 18 to 29)	180	320	**500**
Not A (Age 30 to 50)	60	440	**500**
Total	**240**	**760**	**1,000**

This table can now be used to calculate other probabilities. For example,

$$P(T \cap A) = \frac{180}{1000} = 0.180$$

Completing a hypothetical 1000 table for two events E and F usually requires that you know $P(E)$, $P(F)$, and either $P(E \cup F)$ or $P(E \cap F)$. However, there are two special cases where it is possible to complete a hypothetical 1000 table from just $P(E)$ and $P(F)$. These special cases are when the events E and F are *mutually exclusive* or when the events E and F are *independent*.

Mutually Exclusive Events

Two events E and F are *mutually exclusive* if they can't occur at the same time. For example, consider the chance experiment that involves selecting a student at random from those enrolled at a particular college. The two events

$$F = \text{event that the selected student is a freshman}$$

and

$$S = \text{event that the selected student is a sophomore}$$

are mutually exclusive because no outcome could result in both of these events occurring. This implies that in the following hypothetical 1000 table, the number in the upper-left cell must be 0.

	S (Sophomore)	*Not S* (Not Sophomore)	Total
F (Freshman)	0		
Not F (Not Freshman)			
Total			**1,000**

In this case, knowing $P(F)$ and $P(S)$ would provide enough information to complete the table.

Notice that because the upper-left cell of the table has a value of 0, $P(E \cap F) = 0$. Also, if E and F are mutually exclusive, $P(E \cup F)$ will be equal to the sum of the two individual event probabilities $P(E)$ and $P(F)$.

Example 5.7 Calls to 9-1-1

Sometimes people call the emergency 9-1-1 number to report situations that are not considered emergencies (such as to report a lost dog). Suppose that 30% of the calls to the emergency 9-1-1 number in a particular county are for medical emergencies and that 20% are calls that are not considered emergencies. Consider the two events

M = event that the next call to the emergency line is for a medical emergency

and

N = event that the next call to the emergency line is not considered an emergency

Based on the given information, you know $P(M) = 0.30$ and $P(N) = 0.20$. In this situation, M and N are mutually exclusive because the next call to the emergency line can't be both a medical emergency *and* a call that is not considered an emergency.

Setting up the hypothetical 1000 table for these two events gives

	N (Non-Emergency Call)	Not N (Emergency Call)	Total
M (Medical Emergency)	0		300
Not M (Not Medical Emergency)			700
Total	200	800	1,000

Completing the table by filling in the remaining cells gives

	N (Non-Emergency Call)	Not N (Emergency Call)	Total
M (Medical Emergency)	0	300	300
Not M (Not Medical Emergency)	200	500	700
Total	200	800	1,000

From the table, you can see that

$$P(M \cap N) = 0$$

and

$$P(M \cup N) = \frac{300 + 200}{1000} = 0.500$$

which is equal to $P(M) + P(N) = 0.3 + 0.2 = 0.5$.

As you probably noticed in working through Example 5.7, in the case of mutually exclusive events, you don't really need to complete the hypothetical 1000 table to calculate the probability of a union or an intersection.

Addition Rule for Mutually Exclusive Events

If E and F are mutually exclusive events, then

$$P(E \cap F) = 0$$

and

$$P(E \cup F) = P(E) + P(F)$$

Independent Events

Suppose that the incidence rate for a particular disease in a certain population is known to be 1 in 1000. Then the probability that a randomly selected individual from this population has the disease is 0.001. Further suppose that a diagnostic test for the disease is given to the selected individual, and the test result is positive. Even if the diagnostic test sometimes returns a positive result for someone who does not have the disease, a positive result will make you want to revise the probability that this person has the disease upward from 0.001. Knowing that one event has occurred (the selected person has a positive test result) can change your assessment of the probability of another event (the selected person has the disease).

As another example, consider a university's course registration process, which divides students into 12 registration priority groups. Suppose that overall, only 10% of all students receive all requested classes, but 75% of those in the first priority group receive all requested classes. Interpreting these figures as probabilities, you would say that the probability that a randomly selected student at this university receives all requested classes is 0.10. However, if you know that the selected student is in the first priority group, you revise the probability that the selected student receives all requested classes to 0.75. Knowing that the event *selected person is in the first priority group* has occurred changes your assessment of the probability of the event *selected person received all requested classes*. These two events are said to be **dependent**.

It is sometimes the case, however, that the likelihood of one event is not affected by the occurrence of another event. For example, suppose that you purchase a desktop computer system with a separate monitor and keyboard. Two possible events of interest are

Event 1: The monitor needs service while under warranty

Event 2: The keyboard needs service while under warranty

Because the two components operate independently of one another, learning that the monitor has needed warranty service would not affect your assessment of the likelihood that the keyboard will need repair. If you know that 1% of all keyboards need repair while under warranty, you say that the probability that a keyboard needs warranty service is 0.01. Knowing that the monitor needed warranty service does not affect this probability. This means that *monitor failure* (Event 1) and *keyboard failure* (Event 2) are **independent** events.

DEFINITION

Independent events: Two events are **independent** if knowing that one of the events has occurred does not change your assessment of the probability that the other event has occurred.

Dependent events: Two events are **dependent** if knowing that one event has occurred changes your assessment of the probability that the other event occurs.

In the previous examples,
- the events *has disease* and *tests positive* are dependent.
- the events *receives all requested classes* and *is in the first priority group* are dependent.
- the events *monitor needs warranty service* and *keyboard needs warranty service* are independent.

Multiplication Rule for Independent Events

An individual who purchases a computer system might wonder how likely it is that *both* the monitor and the keyboard will need service while under warranty. A student who must take either a chemistry course or a physics course to fulfill a science requirement might be concerned about the chance that both courses are full before the student has a chance to register. In each of these cases, the focus is on the probability that two different events occur together. For independent events, a simple multiplication rule relates the individual event probabilities to the probability that the events occur together.

Multiplication Rule for Independent Events

If two events are independent, the probability that *both* events occur is the product of the individual event probabilities. Denoting the events as E and F,

$$P(E \cap F) = P(E)P(F)$$

More generally, if there are k independent events, the probability that all the events occur is the product of all the individual event probabilities.

Example 5.8 Nuclear Power Plant Warning Sirens

The Diablo Canyon nuclear power plant in Avila Beach, California, has a warning system that includes a network of sirens. Sirens are located approximately 0.5 miles from each other. When the system is tested, individual sirens sometimes fail. The sirens operate independently of one another, so knowing that a particular siren has failed does not change the probability that any other siren fails.

Imagine that you live near Diablo Canyon and that there are two sirens (Siren 1 and Siren 2) that can be heard from your home. You might be concerned about the probability that *both* Siren 1 *and* Siren 2 fail. Suppose that when the siren system is activated, about 5% of the individual sirens fail. Then

$$P(\text{Siren 1 fails}) = 0.05$$

$$P(\text{Siren 2 fails}) = 0.05$$

and the two events *Siren 1 fails* and *Siren 2 fails* are independent. Using the multiplication rule for independent events,

$$P(\text{Siren 1 fails} \cap \text{Siren 2 fails}) = P(\text{Siren 1 fails})P(\text{Siren 2 fails})$$

$$= (0.05)(0.05)$$

$$= 0.0025$$

Even though the probability that any individual siren will fail is 0.05, the probability that *both* of the two sirens will fail is much smaller. This would happen in the long run only about 2.5 times for every 1000 times the system is activated.

If two events E and F are known to be independent, it is possible to complete a hypothetical 1000 table using just $P(E)$ and $P(F)$. This is illustrated in the following example.

Example 5.9 Satellite Backup Systems

A satellite has both a main and a backup solar power system. (It is common to design such redundancy into products to increase their reliability.) Suppose that the probability of failure during the 10-year lifetime of the satellite is 0.05 for the main power system and 0.08 for the backup system. What is the probability that both systems will fail? Because one system failing has no effect on whether the other one does, it is reasonable to assume that the following two events are independent:

M = event that main system fails
B = event that backup system fails

From the given information, you know

$P(M) = 0.05$
$P(B) = 0.08$

Because M and B are independent,

$$P(M \cap B) = P(M)P(B) = (0.05)(0.08) = 0.004$$

This gives the information needed to complete a hypothetical 1000 table:

	B (Backup Fails)	**Not B** (Backup Does Not Fail)	**Total**
M (Main Fails)	4	46	**50**
Not M (Main Does Not Fail)	76	874	**950**
Total	**80**	**920**	**1,000**

Using this table allows you to calculate $P(M \cup B)$ as

$$P(M \cup B) = \frac{4 + 46 + 76}{1000} = \frac{126}{1000} = 0.126$$

You can also calculate the probability of the satellite having at least one functioning system as

P(at least one system works) = P(main system does not fail $\cup$ backup system does not fail)

$$= P(M^C \cup B^C)$$

$$= \frac{76 + 874 + 46}{1000}$$

$$= \frac{996}{1000}$$

$$= 0.996$$

The probability that the main system fails but the backup system does not fail is

$$P(M \cap B^C) = \frac{46}{1000} = 0.046$$

Summing It Up—Section 5.3

The following learning objectives were addressed in this section:

Conceptual Understanding
C2: Understand what it means for two events to be mutually exclusive.
Two events are mutually exclusive if they can't both occur at the same time. For mutually exclusive events, $P(E \cap F) = 0$.

C3: Understand what it means for two events to be independent.
Two events are independent if knowing that one of the events has occurred does not change your assessment of the probability that the other event has occurred.

Mastering the Mechanics
M3: Use the information in a two-way table to calculate probabilities of events, unions of two events, and intersections of two events.
When data are summarized in a two-way table, the entries in the table can be used to calculate probabilities. This process is illustrated in the discussion that appears at the beginning of this section.

M4: Given the probabilities of two events E and F and the probability of the intersection $E \cap F$, construct a hypothetical 1000 table, and use the table to calculate other probabilities of interest.
When you know the probabilities of some events, you can create a hypothetical 1000 table and then use the table to calculate other probabilities. Example 5.5 illustrates how this is done when you know the probabilities of two events and the probability of the intersection of those two events.

M5: Given the probabilities of two events E and F and the probability of the union $E \cup F$, construct a hypothetical 1000 table, and use the table to calculate other probabilities of interest.

When you know the probabilities of some events, you can create a hypothetical 1000 table and then use the table to calculate other probabilities. Example 5.6 illustrates how this is done when you know the probabilities of two events and the probability of the union of those two events.

M6: Calculate probabilities of unions for mutually exclusive events.

When you know the probabilities of two events and also know that the events are mutually exclusive, the probability of the union of the two events is just the sum of the probabilities of the two events. That is, if E and F are mutually exclusive, $P(E \cup F) = P(E) + P(F)$.

M7: Calculate probabilities of intersections of independent events.

When you know the probabilities of two events and also know that the events are independent, the probability of the intersection of the two events is just the product of the probabilities of the two events. That is, if E and F are independent, $P(E \cap F) = P(E) + P(F)$.

M8: Given the probabilities of two independent events E and F, construct a hypothetical 1000 table, and use the table to calculate other probabilities of interest.

When you know the probabilities of two events and also know that the events are independent, you can create a hypothetical 1000 table and then use the table to calculate other probabilities. Example 5.9 illustrates how this is done when you know the probabilities of two events and also know that the two events are independent.

Putting It into Practice

P2: Given a question that can be answered by calculating a probability, calculate and interpret an appropriate probability to answer the question.

Many interesting questions can be answered by calculating probabilities. This process is illustrated in Examples 5.8 and 5.9.

SECTION 5.3 | EXERCISES

Each Exercise Set assesses the following chapter learning objectives: C2, C3, M3, M4, M5, M6, M7, M8, P2

SECTION 5.3   Exercise Set 1

5.22   A small college has 2700 students enrolled. Consider the chance experiment of selecting a student at random. For each of the following pairs of events, indicate whether or not you think they are mutually exclusive and explain your reasoning.

a. the event that the selected student is a senior and the event that the selected student is majoring in computer science.

b. the event that the selected student is female and the event that the selected student is majoring in computer science.

c. the event that the selected student's college residence is more than 10 miles from campus and the event that the selected student lives in a college dormitory.

d. the event that the selected student is female and the event that the selected student is on the college football team.

5.23   A Gallup survey found that 64% of women and 55% of men said that they favor affirmative action programs for women **(Gallup Poll Social Series, July 28, 2016)**. Suppose that this information is representative of U.S. adults. If a U.S. adult is selected at random, are the events *selected adult is male* and *selected adult favors affirmative action programs for women* independent or dependent? Explain.

5.24   A bank offers both adjustable-rate and fixed-rate mortgage loans on residential properties, which are classified into three categories: single-family houses, condominiums, and multifamily dwellings. Suppose each loan made in 2016 was classified according to type of mortgage and type of property, resulting in the following table. Consider the chance experiment of selecting one of these 3750 loans at random.

	Single-Family	Condo	Multifamily	Total
Adjustable	1,500	788	337	**2,625**
Fixed-Rate	375	377	373	**1,125**
Total	**1,875**	**1,165**	**710**	**3,750**

a. What is the probability that the selected loan will be for an adjustable rate mortgage?

b. What is the probability that the selected loan will be for a multifamily property?

c. What is the probability that the selected loan will not be for a single-family property?

d. What is the probability that the selected loan will be for a single-family property or a condo?

e. What is the probability that the selected loan will be for a multifamily property or for an adjustable rate loan?

f. What is the probability that the selected loan will be a fixed-rate loan for a condo?

5.25   There are two traffic lights on Shelly's route from home to work. Let E denote the event that Shelly must stop at the first light, and define the event F in a similar manner for the second light. Suppose that $P(E) = 0.4$, $P(F) = 0.3$, and $P(E \cap F) = 0.15$.

a. Use the given probability information to set up a hypothetical 1000 table with columns corresponding to E and *not E* and rows corresponding to F and *not F*.

b. Use the table from Part (a) to find the following probabilities:

   i. the probability that Shelly must stop for at least one light (the probability of $E \cup F$).

   ii. the probability that Shelly does not have to stop at either light.

   iii. the probability that Shelly must stop at exactly one of the two lights.

   iv. the probability that Shelly must stop only at the first light.

5.26 A large cable company reports that 80% of its customers subscribe to its cable TV service, 42% subscribe to its Internet service, and 97% subscribe to at least one of these two services.

a. Use the given probability information to set up a hypothetical 1000 table.

b. Use the table from Part (a) to find the following probabilities:

   i. the probability that a randomly selected customer subscribes to both cable TV and Internet service.

   ii. the probability that a randomly selected customer subscribes to exactly one of these services.

5.27 a. Suppose events E and F are mutually exclusive with $P(E) = 0.41$ and $P(E) = 0.23$.

   i. What is the value of $P(E \cap F)$?

   ii. What is the value of $P(E \cup F)$?

b. Suppose that for events A and B, $P(A) = 0.26$, $P(B) = 0.34$, and $P(A \cup B) = 0.47$. Are A and B mutually exclusive? How can you tell?

5.28 Each time a class meets, the professor selects one student at random to explain the solution to a homework problem. There are 40 students in the class, and no one ever misses class. Luke is one of these students. What is the probability that Luke is selected both of the next two times that the class meets?

5.29 A large retail store sells headphones. A customer who purchases headphones can pay either by cash or credit card. An extended warranty is also available for purchase. Suppose that the events

   M = event that the customer pays by cash

   E = event that the customer purchases an extended warranty

are independent with $P(M) = 0.47$ and $P(E) = 0.16$.

a. Construct a hypothetical 1000 table with columns corresponding to cash or credit card and rows corresponding to whether or not an extended warranty is purchased.

b. Use the table to calculate $P(M \cup E)$. Give a long-run relative frequency interpretation of this probability.

SECTION 5.3 Exercise Set 2

5.30 Consider a chance experiment that consists of selecting a student at random from a high school with 3000 students.

a. In the context of this chance experiment, give an example of two events that would be mutually exclusive.

b. In the context of this chance experiment, give an example of two events that would not be mutually exclusive.

5.31 False positive results are not uncommon with mammograms, a test used to screen for breast cancer. For a woman who has a positive mammogram, the probability that she actually has breast cancer is less than 0.05 if she is under 40 years old, and ranges from 0.050 to 0.109 if she is over 40 years old (**"Breast Cancer Screenings: Does the Evidence Support the Recommendations?,"** *Significance* **[August 2016]: 24–37).** If a woman with a positive mammogram is selected at random, are the two events

   B = event that selected woman has breast cancer

and

   A = event that selected woman is over 40 years old

independent events? Justify your answer using the given information.

5.32 The accompanying data are from the article **"Characteristics of Buyers of Hybrid Honda Civic IMA: Preferences, Decision Process, Vehicle Ownership, and Willingness-to-Pay" (Institute for Environmental Decisions, November 2006).** Each of 311 people who purchased a Honda Civic was classified according to sex and whether or not the car purchased had a hybrid engine.

	Hybrid	Not Hybrid
Male	77	117
Female	34	83

Suppose one of these 311 individuals is to be selected at random.

a. What is the probability that the selected individual purchased a hybrid?

b. What is the probability that the selected individual is male?

c. What is the probability that the selected individual is male and purchased a hybrid?

d. What is the probability that the selected individual is female or purchased a hybrid?

e. What is the probability that the selected individual is female and purchased a hybrid?

f. Explain why the probabilities calculated in Parts (d) and (e) are not equal.

5.33 The paper **"Predictors of Complementary Therapy Use Among Asthma Patients: Results of a Primary Care Survey"** (*Health and Social Care in the Community* **[2008]: 155–164)**

described a study in which each person in a large sample of asthma patients responded to two questions:

> Question 1: Do conventional asthma medications usually help your symptoms?
>
> Question 2: Do you use complementary therapies (such as herbs, acupuncture, aroma therapy) in the treatment of your asthma?

Suppose that this sample is representative of asthma patients. Consider the following events:

E = event that the patient uses complementary therapies

F = event that the patient reports conventional medications usually help

The data from the sample were used to estimate the following probabilities:

$$P(E) = 0.146 \quad P(F) = 0.879 \quad P(E \cap F) = 0.122$$

a. Use the given probability information to set up a hypothetical 1000 table with columns corresponding to E and *not E* and rows corresponding to F and *not F*.

b. Use the table from Part (a) to find the following probabilities:

i. The probability that an asthma patient responds that conventional medications do not help and that the patient uses complementary therapies.

ii. The probability that an asthma patient responds that conventional medications do not help and that the patient does not use complementary therapies.

iii. The probability that an asthma patient responds that conventional medications usually help or the patient uses complementary therapies.

c. Are the events E and F independent? Explain.

5.34 An appliance manufacturer offers extended warranties on its washers and dryers. Based on past sales, the manufacturer reports that of customers buying both a washer and a dryer, 52% purchase the extended warranty for the washer, 47% purchase the extended warranty for the dryer, and 59% purchase at least one of the two extended warranties.

a. Use the given probability information to set up a hypothetical 1000 table.

b. Use the table from Part (a) to find the following probabilities:

i. the probability that a randomly selected customer who buys a washer and a dryer purchases an extended warranty for both the washer and the dryer.

ii. the probability that a randomly selected customer purchases an extended warranty for neither the washer nor the dryer.

5.35 **a.** Suppose events E and F are mutually exclusive with $P(E) = 0.14$ and $P(F) = 0.76$.

i. What is the value of $P(E \cap F)$?

ii. What is the value of $P(E \cup F)$?

b. Suppose that for events A and B, $P(A) = 0.24$, $P(B) = 0.24$, and $P(A \cup B) = 0.48$. Are A and B mutually exclusive? How can you tell?

5.36 In some states, such as Iowa and Nevada, the presidential primaries are decided by caucuses rather than a primary election. The caucuses determine winners at the precinct level, and turnout is often low. As a result, it is not uncommon in a close race to have some caucuses end in a tie. The article **"A Nevada Tie to Be Decided by Cards"** (*The Wall Street Journal*, **February 20, 2016**) reported that in Nevada a tie is decided by having each side draw a card, with the high card winning. In Iowa, a tie is decided by a coin toss. In 2016, in the primary race between Hillary Clinton and Bernie Sanders, some Democratic caucuses were in fact decided by coin tosses.

a. Suppose two caucuses resulted in a tie between Bernie Sanders and Hillary Clinton. What is the probability that both would be decided in favor of Hillary Clinton?

b. Suppose two caucuses resulted in a tie between Bernie Sanders and Hillary Clinton. What is the probability that both would be decided in favor of Bernie Sanders?

c. Suppose two caucuses resulted in a tie between Bernie Sanders and Hillary Clinton. What is the probability that both would be decided in favor of the same candidate?

d. Suppose that three caucuses resulted in a tie between Bernie Sanders and Hillary Clinton. What is the probability that all three caucuses would be decided in favor of the same candidate?

5.37 A rental car company offers two options when a car is rented. A renter can choose to pre-purchase gas or not and can also choose to rent a GPS device or not. Suppose that the events

A = event that gas is pre-purchased

B = event that a GPS is rented

are independent with $P(A) = 0.20$ and $P(B) = 0.15$.

a. Construct a hypothetical 1000 table with columns corresponding to whether or not gas is pre-purchased and rows corresponding to whether or not a GPS is rented.

b. Use the table to find $P(A \cup B)$. Give a long-run relative frequency interpretation of this probability.

ADDITIONAL EXERCISES

5.38 The table at the top of the next page describes the approximate distribution of students by sex and college at a midsize public university in the West. Suppose you were to randomly select one student from this university.

a. What is the probability that the selected student is a male?

b. What is the probability that the selected student is in the College of Agriculture?

c. What is the probability that the selected student is a male in the College of Agriculture?

d. What is the probability that the selected student is a male who is not from the College of Agriculture?

TABLE FOR EXERCISE 5.38

Sex	College						
	Education	Engineering	Liberal Arts	Science and Math	Agriculture	Business	Architecture
Male	200	3,200	2,500	1,500	2,100	1,500	200
Female	300	800	1,500	1,500	900	1,500	300

5.39 The report **"Teens, Social Media & Technology Overview 2015" (Pew Research Center, April 9, 2015)** summarized data from a large survey of teens age 13–17. Of those surveyed, 71% use Facebook and 52% use Instagram. Use these percentages to explain why the two events

F = event that a randomly selected survey participant uses Facebook

and

I = event that a randomly selected survey participant uses Instagram

cannot be mutually exclusive events.

5.40 The report **"Improving Undergraduate Learning" (Social Science Research Council, 2011)** summarizes data from a survey of several thousand college students. These students were thought to be representative of the population of all college students in the United States. When asked about an upcoming semester, 68% said they would be taking a class that is reading-intensive (requires more than 40 pages of reading per week). Only 50% said they would be taking a class that is writing-intensive (requires more than 20 pages of writing over the course of the semester). The percentage who said that they would be taking both a reading-intensive course and a writing-intensive course was 42%.
a. Use the given information to set up a hypothetical 1000 table.
b. Use the table to find the following probabilities:
  i. the probability that a randomly selected student would be taking at least one reading-intensive or writing-intensive course.
  ii. the probability that a randomly selected student would be taking a reading-intensive course or a writing-intensive course, but not both.

  iii. the probability that a randomly selected student would be taking neither a reading-intensive nor a writing-intensive course.

5.41 Airline tickets can be purchased online, by telephone, or by using a travel agent. Passengers who have a ticket sometimes don't show up for their flights. Suppose a person who purchased a ticket is selected at random. Consider the following events:

O = event selected person purchased ticket online
N = event selected person did not show up for flight

Suppose $P(O) = 0.70$, $P(N) = 0.07$, and $P(O \cap N) = 0.04$.
a. Are the events N and O independent? How can you tell?
b. Construct a hypothetical 1000 table with columns corresponding to N and *not N* and rows corresponding to O and *not O*.
c. Use the table to find $P(O \cup N)$. Give a relative frequency interpretation of this probability.

5.42 The following statement is from a letter to the editor that appeared in **USA TODAY (September 3, 2008)**: "Among Notre Dame's current undergraduates, our ethnic minority students (21%) and international students (3%) alone equal the percentage of students who are children of alumni (24%). Add the 43% of our students who receive need-based financial aid (one way to define working-class kids), and more than 60% of our student body is composed of minorities and students from less affluent families."

Do you think that the statement that more than 60% of the student body is composed of minorities and students from less affluent families is likely to be correct? Explain.

SECTION 5.4 Conditional Probability

In the previous section, you saw that sometimes the knowledge that one event has occurred changes the assessment of the chance that another event occurs. In this section, we begin by revisiting one of the examples from Section 5.3 in more detail.

Consider a population in which 0.1% of all individuals have a certain disease. The presence of the disease cannot be discerned from outward appearances, but there is a diagnostic test available. Unfortunately, the test is not always correct. Of those with positive test results, 80% actually have the disease and the other 20% who show positive test results are false-positives. To put this in probability terms, consider the chance experiment in which an individual is randomly selected from the population. Define the following events:

$$E = \text{event that the individual has the disease}$$
$$F = \text{event that the individual's diagnostic test is positive}$$

We will use $P(E|F)$ to denote the probability of the event E given that the event F is known to have occurred. This new notation is used to indicate that a probability calculation has been made conditional on the occurrence of another event. The standard symbol for this is a vertical line, and it is read "given." You would say, "the probability that an individual has the disease *given* that the diagnostic test is positive," and represent this symbolically as $P(\text{has disease}|\text{positive test})$ or $P(E|F)$. This probability is called a conditional probability.

The information provided above implies that

$$P(E) = 0.001$$

$$P(E|F) = 0.8$$

Before you have diagnostic test information, you would think that the occurrence of E is unlikely, but once it is known that the test result is positive, your assessment of the likelihood of the disease increases dramatically. (Otherwise the diagnostic test would not be very useful!)

DEFINITION

Conditional Probability

If E and F are events, the conditional probability of E given F is denoted by $P(E|F)$.

Example 5.10 **After-School Activities**

After-school activities at a large high school can be classified into three types: athletics, fine arts, and other. In this example, suppose that *every* student at the school is in *exactly* one of these after-school activities. The following table gives the number of students participating in each of these types of activities by grade:

	9th grade	10th grade	11th grade	12th grade	Total
Athletics	150	160	140	150	**600**
Fine Arts	100	90	120	125	**435**
Other	125	140	150	150	**656**
Total	**375**	**300**	**410**	**425**	**1,600**

Look to see how each of the following statements follows from the information in the table:

1. There are 160 10th-grade students participating in athletics.
2. The number of 12th graders participating in fine arts activities is 125.
3. There are 435 students participating in fine arts activities.
4. There are 410 11th graders.
5. The total number of students at the school is 1600.

Each week during the school year, the principal plans to select a student at random and invite this student to have lunch with her to discuss various student concerns. She feels that random selection will give her the greatest chance of hearing from a representative cross-section of the student body. Because the student will be selected at random, each of the 1600 students is equally likely to be selected. What is the probability that a randomly selected student is an athlete? You can calculate this probability as follows:

$$P(\text{Athlete}) = \frac{\text{number of athletes}}{\text{total number of students}} = \frac{600}{1600} = 0.375$$

Now suppose that the principal's secretary records not only the student's name but also the student's grade level. The secretary has indicated that the selected student is a senior. Does this information change the assessment of the likelihood that the selected student is an athlete? Because 150 of the 425 seniors participate in athletics, this suggests that

$$P(\text{Athlete}|\text{Senior}) = \frac{\text{number of senior athletes}}{\text{total number of seniors}} = \frac{150}{425} = 0.353$$

The probability is calculated in this way because you know that the selected student is one of 425 seniors, each of whom is equally likely to have been selected. If you were to repeat the chance experiment of selecting a student at random, about 35.3% of the times *that resulted in a senior being selected* would also result in the selection of someone who participates in athletics.

Notice that the unconditional probability $P(\text{Athlete})$ is not equal to the conditional probability $P(\text{Athlete}|\text{Senior})$. Because the assessment of the probability that the selected student is an athlete changes if it is known that the selected student is a senior, the two events

A = event selected student is an athlete

and

S = event selected student is a senior

are not independent events.

Example 5.11 **GFI Switches**

A GFI (ground fault interrupt) switch turns off power to a system in the event of an electrical malfunction. A spa manufacturer currently has 25 spas in stock, each equipped with a single GFI switch. Suppose that two different companies supplied the switches, and that some of the switches are defective, as summarized in the following table:

	Not Defective	Defective	Total
Company 1	10	5	15
Company 2	8	2	10
Total	18	7	25

A spa is randomly selected for testing. Consider the following events:

E = event that GFI switch in the selected spa is from Company 1

F = event that GFI switch in the selected spa is defective

Using the table, you can calculate the following probabilities:

$$P(E) = \frac{15}{25} = 0.60$$

$$P(F) = \frac{7}{25} = 0.28$$

$$P(E \cap F) = \frac{5}{25} = 0.20$$

Now suppose that testing reveals a defective switch. This means that the chosen spa is one of the seven in the "defective" column. How likely is it that the switch came from the first company? Because five of the seven defective switches are from Company 1,

$$P(E|F) = P(\text{company 1}|\text{defective}) = \frac{5}{7} = 0.714$$

Notice that this is larger than the unconditional probability $P(E)$. This is because Company 1 has a much higher defective rate than Company 2.

Example 5.12 Surviving a Heart Attack

Medical guidelines recommend that a hospitalized patient who suffers cardiac arrest should receive defibrillation (an electric shock to the heart) within 2 minutes. The paper **"Delayed Time to Defibrillation After In-Hospital Cardiac Arrest"** (***The New England Journal of Medicine*** **[2008]: 9–17)** describes a study of the time to defibrillation for hospitalized patients in hospitals of various sizes. The authors examined medical records of 6716 patients who suffered cardiac arrest while hospitalized, recording the size of the hospital and whether or not defibrillation occurred within 2 minutes. Data from this study are summarized here:

Hospital Size	Time to Defibrillation		Total
	Within 2 Minutes	More Than 2 Minutes	
Small (Fewer Than 250 Beds)	1,124	576	1,700
Medium (250–499 Beds)	2,178	886	3,064
Large (500 or More Beds)	1,387	565	1,952
Total	4,689	2,027	6,716

In this example, assume that these data are representative of the larger group of all hospitalized patients who suffer a cardiac arrest. Suppose that a hospitalized patient who suffered a cardiac arrest is selected at random. The following events are of interest:

S = event that the selected patient is at a small hospital
M = event that the selected patient is at a medium-sized hospital
L = event that the selected patient is at a large hospital
D = event that the selected patient receives defibrillation within 2 minutes

Using the information in the table, you can calculate

$$P(D) = \frac{4689}{6716} = 0.698$$

This probability is interpreted as the proportion of hospitalized patients suffering cardiac arrest that receive defibrillation within 2 minutes. This means that 69.8% of these patients receive timely defibrillation.

Now suppose it is known that the selected patient was at a small hospital. How likely is it that this patient received defibrillation within 2 minutes? To answer this question, you need to calculate $P(D|S)$, the probability of defibrillation within 2 minutes, given that the patient is at a small hospital. From the information in the table, you know that of the 1700 patients in small hospitals, 1124 received defibrillation within 2 minutes, so

$$P(D|S) = \frac{1124}{1700} = 0.661$$

Notice that this is smaller than the unconditional probability, $P(D) = 0.698$. This tells you that there is a smaller probability of timely defibrillation at a small hospital.

Two other conditional probabilities of interest are

$$P(D|M) = \frac{2178}{3064} = 0.711$$

and

$$P(D|L) = \frac{1387}{1952} = 0.711$$

From this, you see that the probability of timely defibrillation is the same for patients at medium-sized and large hospitals, and that this probability is higher than that for patients at small hospitals.

It is also possible to calculate $P(L|D)$, the probability that a patient is at a large hospital given that the patient received timely defibrillation:

$$P(L|D) = \frac{1387}{4689} = 0.296$$

Let's look carefully at the interpretation of some of these probabilities:

1. $P(D) = 0.698$ is interpreted as the proportion of *all* hospitalized patients suffering cardiac arrest who receive timely defibrillation. Approximately 69.8% of these patients receive timely defibrillation.
2. $P(D \cap L) = \frac{1387}{6716} = 0.207$ is the proportion of *all* hospitalized patients who suffer cardiac arrest who are at a large hospital *and* who receive timely defibrillation.
3. $P(D|L) = 0.711$ is the proportion *of patients at large hospitals* suffering cardiac arrest who receive timely defibrillation.
4. $P(L|D) = 0.296$ is the proportion *of patients who receive timely defibrillation* who were at large hospitals.

Notice the difference between the unconditional probabilities in Interpretations 1 and 2 and the conditional probabilities in Interpretations 3 and 4. The unconditional probabilities involve the entire group of interest (all hospitalized patients suffering cardiac arrest), whereas the conditional probabilities only involve a subset of the entire group defined by the "given" event.

As mentioned in the opening paragraphs of this section, one of the most important applications of conditional probability is in making diagnoses. Your mechanic diagnoses your car by hooking it up to a machine and reading the pressures and speeds of the various components. A meteorologist diagnoses the weather by looking at temperatures, isobars, and wind speeds. Medical doctors diagnose the state of a person's health by performing various tests and gathering information, such as weight and blood pressure. Doctors also observe characteristics of their patients in an attempt to determine whether or not their patients have a certain disease. Many diseases are not actually observable—or at least not easily so—and often the doctor must make a probabilistic judgment. As you will see, conditional probability plays a large role in evaluating diagnostic techniques.

Sometimes, a criterion exists to determine with certainty whether or not a person has a disease. This criterion is colloquially known as the "gold standard." The gold standard might be an invasive surgical procedure, which can be expensive or dangerous. Or, the gold standard might be a test that takes a long time to perform in the lab or that is costly. For a diagnostic test to be preferred over a gold standard, it would need to be faster, less expensive, or less invasive and yet still produce results that agree with the gold standard. In this context, *agreement* means (1) that the test generally comes out positive when the patient has the disease and (2) that the test generally comes out negative when the patient does not have the disease. These statements involve conditional probabilities, as illustrated in the following example.

Example 5.13 **Diagnosing Tuberculosis**

To illustrate the calculations involved in evaluating a diagnostic test, consider the case of tuberculosis (TB), an infectious disease that typically attacks lung tissue. Before 1998, culturing was the existing gold standard for diagnosing TB. This method took 10 to 15 days to get a result. In 1998, investigators evaluated a DNA technique that turned out to be much faster **("LCx: A Diagnostic Alternative for the Early Detection of *Mycobacterium tuberculosis* Complex," *Diagnostic Microbiology and Infectious Diseases* [1998]: 259–264).** The DNA technique for detecting tuberculosis was evaluated by comparing results from the test to the existing gold standard, with the following results for 207 patients exhibiting symptoms:

	Has Tuberculosis (Gold Standard)	Does Not Have Tuberculosis (Gold Standard)	Total
DNA Test Positive	14	0	**14**
DNA Test Negative	12	181	**193**
Total	**26**	**181**	**207**

Looking at the table, the DNA technique seems to be effective as a diagnostic test. Patients who tested positive with the DNA test were also positive with the gold standard in every case. Patients who tested negative with the DNA test generally tested negative with the gold standard, but the table also indicates some false-negative results for the DNA test.

Consider a randomly selected individual who is tested for tuberculosis. The following events are of interest:

$$T = \text{event that the individual has tuberculosis}$$

$$N = \text{event that the DNA test is negative}$$

What is the probability that a person has tuberculosis, given that the DNA test was negative? To answer this question, you need to calculate $P(T|N)$. From the table, you see that of the 193 people with negative DNA tests, 12 actually did have tuberculosis. So,

$$P(T|N) = P(\text{tuberculosis}|\text{negative DNA test}) = \frac{12}{193} = 0.062$$

Notice that 12.6% $\left(\frac{26}{207} = 0.126\right)$ of those tested had tuberculosis. The added information provided by the diagnostic test has altered the probability—and provided some measure of relief for the patients who test negative. Once it is known that the test result is negative, the estimated probability of having TB is only about half as large.

Pay Attention to the Denominator

A probability calculated from information in a table is a fraction. The key to calculating these probabilities is having the right numerator and the right denominator when you put your fraction together. The denominator for a conditional probability will be a row or column total, whereas the denominator for an unconditional probability is the total for the entire table.

In the examples considered so far, you have started with information about a population that was already in table form. When this is not the case, it is sometimes possible to use given probability information to construct a hypothetical 1000 table. This approach usually makes calculating conditional probabilities relatively easy, as illustrated in the following two examples.

Example 5.14 Internet Addiction

Internet addiction has been defined by researchers as a disorder characterized by excessive time spent on the Internet, impaired judgment and decision-making ability, social withdrawal, and depression. The paper **"The Association between Aggressive Behaviors and Internet Addiction and Online Activities in Adolescents"** (*Journal of Adolescent Health* **[2009]: 598–605**) describes a study of a large number of adolescents. Each participant in the study was assessed using the Chen Internet Addiction Scale to determine if he or she suffered from Internet addiction. The following statements are based on the survey results:

1. 51.8% of the study participants were female and 48.2% were male.
2. 13.1% of the females suffered from Internet addiction.
3. 24.8% of the males suffered from Internet addiction.

Consider the chance experiment that consists of selecting a study participant at random, and define the following events:

$$F = \text{the event that the selected participant is female}$$

$$M = \text{the event that the selected participant is male}$$

$$I = \text{the event that the selected participant suffers from Internet addiction}$$

The three statements from the paper define the following probabilities:

$$P(F) = 0.518$$

$$P(M) = 0.482$$

$$P(I|F) = 0.131$$

$$P(I|M) = 0.248$$

Let's create a hypothetical 1000 table based on the given probabilities. You can set up a table with rows that correspond to whether or not the selected person is female and columns that correspond to whether or not the selected person is addicted to the Internet.

	I (Internet Addiction)	Not *I* (No Internet Addiction)	Total
F (Female)			
Not F (Male)			
Total			1,000

Because $P(F) = 0.518$, you would expect $(0.518)(1000) = 518$ females and $(0.482)(1000) = 482$ males in the hypothetical 1000. This gives the row totals for the table:

	I (Internet Addiction)	Not *I* (No Internet Addiction)	Total
F (Female)			518
Not F (Male)			482
Total			1,000

How do you use the information provided by the given conditional probabilities? The probability $P(I|F) = 0.131$ tells you that 13.1% *of the females* are addicted to the Internet. So, you would expect the 518 females to be divided into

$$(0.131)(518) \approx 68$$

who are addicted and $518 - 68 = 450$ who are not. For males, you would expect

$$(0.248)(482) \approx 120$$

who are addicted and the remaining $482 - 120 = 362$ who are not. This allows you to complete the hypothetical 1000 table as shown here:

	I (Internet Addiction)	Not *I* (No Internet Addiction)	Total
F (Female)	68	450	518
Not F (Male)	120	362	482
Total	188	812	1,000

You can now use this table to calculate some probabilities of interest that were not given. For example,

$$P(I) = \frac{188}{1000} = 0.188$$

means that 18.8% of the study participants were addicted to the Internet. Also, the probability that a selected participant is female given that the selected person is an Internet addict is

$$P(F|I) = \frac{68}{188} = 0.362$$

Notice that this probability is *not* the same as $P(I|F)$. $P(I|F)$ is the proportion *of females* who are addicted, whereas $P(F|I)$ is the proportion *of Internet addicts* who are female.

Example 5.15   Should You Paint the Nursery Pink? Revisited

The example in the Chapter Preview section dealt with sex predictions based on ultrasounds performed during the first trimester of pregnancy (**"The Use of Three-Dimensional Ultrasound for Fetal Gender Determination in the First Trimester,"** *The British Journal of Radiology* **[2003]: 448–451**). Radiologist 1 looked at 159 first trimester ultrasound images and made a sex prediction for each one. The predictions were then compared with the actual sex when the baby was born. The table below summarizes the data for Radiologist 1.

	Radiologist 1		
	Predicted Male	Predicted Female	Total
Baby is Male	74	12	86
Baby is Female	14	59	73
Total	88	71	159

Several questions were posed in the Preview Example:

1. How likely is it that a predicted sex is correct?
2. Is a predicted sex more likely to be correct when the baby is male than when the baby is female?
3. If the predicted sex is female, should you paint the nursery pink? If you do, how likely is it that you will need to repaint?

The first question posed can be answered by looking for cells in the table where the sex prediction is correct—the upper-left and the lower-right cells. Then

$$P(\text{predicted sex is correct}) = \frac{74 + 59}{159} = \frac{133}{159} = 0.836$$

Assuming that these 159 ultrasound images are representative of all first trimester ultrasound images, you can interpret this probability as a long-run relative frequency. For Radiologist 1, the sex prediction is correct about 83.6% of the time.

The second question posed asks if Radiologist 1 is more likely to be correct when the baby is male than when the baby is female. You can answer this question by comparing two conditional probabilities,

$$P(\text{predicted sex is male}|\text{baby is male})$$

and

$$P(\text{predicted sex is female}|\text{baby is female})$$

From the table, you can see that when the baby was male, 74 of the 86 predictions were correct, so

$$P(\text{predicted sex is male}|\text{baby is male}) = \frac{74}{86} = 0.86$$

Also

$$P(\text{predicted sex is female}|\text{baby is female}) = \frac{59}{73} = 0.81$$

Radiologist 1 is correct slightly more often when the baby is male than when the baby is female (about 86% of the time for males compared with about 81% of the time for females).

The third question posed is an interesting one: If Radiologist 1 predicts that the sex is female and you paint the nursery pink, how likely is it that you will need to repaint? To answer this question, you need to look at a conditional probability that is different from the one calculated above. There you looked at the probability that the predicted sex is female given that the baby was female. To answer question 3, the "given" part is that the prediction is female. That is, you want to know the probability that the baby is female given that the prediction is female:

$$P(\text{baby is female}|\text{prediction is female})$$

From the table, you can see that for the 71 babies that were predicted to be female, 59 were actually female. This gives

$$P(\text{baby is female}|\text{prediction is female}) = \frac{59}{71} = 0.831$$

For Radiologist 1, when the predicted sex is female, about 83.1% of the time the baby is actually female. This means that if the nursery was painted pink based on a predicted first trimester ultrasound image, the probability that it would need to be repainted is $1 - 0.831 = 0.169$, or about 17% of the time.

The preview example also gave the following data for a second radiologist who made sex predictions based on 154 first trimester ultrasound images.

	Radiologist 2		
	Predicted Male	**Predicted Female**	**Total**
Baby is Male	81	8	**89**
Baby is Female	7	58	**65**
Total	**88**	**66**	**154**

An additional question was posed: Does the chance of a correct sex prediction differ for the two radiologists? For Radiologist 2

$$P(\text{correct prediction}) = \frac{81 + 58}{154} = \frac{139}{154} = 0.903$$

Notice that this is quite a bit greater than the corresponding probability of 0.836 for Radiologist 1.

Let's take this example a little further. Suppose that these two radiologists both work in the same clinic and that the probability of a correct sex prediction from a first trimester ultrasound image for each of the two radiologists is equal to the probability previously calculated—0.836 for Radiologist 1 and 0.903 for Radiologist 2. Further, suppose that Radiologist 1 works part-time and handles 30% of the ultrasounds for the clinic and that Radiologist 2 handles the remaining 70%. You can then ask the following questions:

1. What is the probability that a sex prediction based on a first trimester ultrasound at this clinic is correct?
2. If a first trimester ultrasound sex prediction is incorrect, what is the probability that the prediction was made by Radiologist 2?

To answer these questions, you can translate the given probability information into a hypothetical 1000 table. You know the following for first trimester sex predictions made at this clinic:

$$P(\text{prediction is made by Radiologist 1}) = 0.30$$

$$P(\text{prediction is made by Radiologist 2}) = 0.70$$

$$P(\text{prediction is correct}|\text{prediction made by Radiologist 1}) = 0.836$$

$$P(\text{prediction is correct}|\text{prediction made by Radiologist 2}) = 0.903$$

Setting up the table with row totals included gives

	Prediction Correct	Prediction Incorrect	Total
Radiologist 1			300
Radiologist 2			700
Total			1,000

The row totals are calculated from the probability that a sex prediction was made by a particular radiologist—you would expect 30% or 300 of the hypothetical 1000 predictions to have been made by Radiologist 1 and 70% or 700 of the hypothetical 1000 to have been made by Radiologist 2.

Next consider the 300 predictions made by Radiologist 1. Because Radiologist 1 is correct about 83.6% of the time (from $P(\text{prediction is correct}|\text{prediction made by Radiologist 1}) = 0.836$), you would expect

$$(0.836)(300) \approx 251$$

to be correct and the remaining $300 - 251 = 49$ to be incorrect. For Radiologist 2, you would expect

$$(0.903)(700) \approx 632$$

to be correct and the remaining $700 - 632 = 68$ to be incorrect.

Adding these numbers to the hypothetical 1000 table gives

	Prediction Correct	Prediction Incorrect	Total
Radiologist 1	251	49	300
Radiologist 2	632	68	700
Total	883	117	1,000

You can now calculate the two desired probabilities:

$$P(\text{Prediction is correct}) = \frac{883}{1000} = 0.883$$

and

$$P(\text{Radiologist 2}|\text{prediction incorrect}) = \frac{68}{117} = 0.581$$

This last probability tells you that about 58.1% of the incorrect sex predictions at this clinic are made by Radiologist 2. This may seem strange at first, given that Radiologist 2 is correct a greater proportion of the time than Radiologist 1. But remember that Radiologist 2 does more than twice as many predictions as Radiologist 1. So, even though Radiologist 2 has a lower error rate, more of the errors are made by Radiologist 2 because many more sex predictions are made by Radiologist 2.

Summing It Up—Section 5.4
The following learning objectives were addressed in this section:

Conceptual Understanding
C4: Understand the difference between an unconditional probability and a conditional probability.
A conditional probability is a probability of an event that is calculated conditional on the occurrence of another event. For example, consider the chance experiment of selecting a patient at random from all patients that had surgery at a particular hospital during 2016. The probability P(patient had hip replacement surgery) is an unconditional probability and represents the proportion of *all* surgery patients who had hip replacement surgery. The probability P(patient had hip replacement surgery given that the patient was over 60 years old) is a conditional probability and represents the proportion of surgery *patients over 60 years* old who had hip replacement surgery. For another example, see Example 5.10.

Mastering the Mechanics
M9: Use information in a two-way table to calculate conditional probabilities.
When data are summarized in a two-way table, the entries in the table can be used to calculate both unconditional probabilities and conditional probabilities. Calculations of conditional probabilities are illustrated in Examples 5.10, 5.11, and 5.12.

M10: Given probability and conditional probability information, construct a hypothetical 1000 table and use the table to calculate other probabilities of interest.
You can create a hypothetical 1000 table using known unconditional and conditional probabilities and then use the table to calculate other probabilities. The second half of Example 5.15 provides an example that illustrates this process.

Putting It into Practice
P1: Distinguish between questions that can be answered by calculating an unconditional probability and questions that can be answered by calculating a conditional probability.
For a discussion of the different interpretations of unconditional and conditional probabilities, see Example 5.12.

SECTION 5.4 EXERCISES

Each Exercise Set assesses the following chapter learning objectives: C4, M9, M10, P1

SECTION 5.4 Exercise Set 1

5.43 Suppose that 20% of all teenage drivers in a certain county received a citation for a moving violation within the past year. Assume in addition that 80% of those receiving such a citation attended traffic school so that the citation would not appear on their permanent driving record. Consider the chance experiment that consists of randomly selecting a teenage driver from this county.
a. One of the percentages given in the problem specifies an unconditional probability, and the other percentage specifies a conditional probability. Which one is the conditional probability, and how can you tell?
b. Suppose that two events E and F are defined as follows:

   E = selected driver attended traffic school
   F = selected driver received such a citation

Use probability notation to translate the given information into two probability statements of the form P(___) = probability value.

5.44 The accompanying data are from the article **"Characteristics of Buyers of Hybrid Honda Civic IMA: Preferences, Decision Process, Vehicle Ownership, and Willingness-to-Pay" (Institute for Environmental Decisions, November 2006).** Each of 311 people who purchased a Honda Civic was classified according to sex and whether the car purchased had a hybrid engine or not.

	Hybrid	Not Hybrid
Male	77	117
Female	34	83

a. Suppose that one of these 311 people will be selected at random. Find the following probabilities:

   i. $P(male)$

   ii. $P(hybrid)$

   iii. $P(hybrid|male)$

   iv. $P(hybrid|female)$

   v. $P(female|hybrid)$

b. For each of the probabilities calculated in Part (a), write a sentence interpreting the probability.

c. Are the probabilities $P(hybrid|male)$ and $P(male|hybrid)$ equal? If not, write a sentence or two explaining the difference.

5.45 The paper **"Action Bias among Elite Soccer Goalkeepers: The Case of Penalty Kicks" (** *Journal of Economic Psychology* **[2007]: 606–621)** presents an interesting analysis of 286 penalty kicks in televised championship soccer games from around the world. In a penalty kick, the only players involved are the kicker and the goalkeeper from the opposing team. The kicker tries to kick a ball into the goal from a point located 11 meters away. The goalkeeper tries to block the ball from entering the goal. For each penalty kick analyzed, the researchers recorded the direction that the goalkeeper moved (jumped to the left, stayed in the center, or jumped to the right) and whether or not the penalty kick was successfully blocked. Consider the following events:

   L = the event that the goalkeeper jumps to the left

   C = the event that the goalkeeper stays in the center

   R = the event that the goalkeeper jumps to the right

   B = the event that the penalty kick is blocked

Based on their analysis of the penalty kicks, the authors of the paper gave the following probability estimates:

$$P(L) = 0.493 \qquad P(C) = 0.063 \qquad P(R) = 0.444$$
$$P(B|L) = 0.142 \quad P(B|C) = 0.333 \quad P(B|R) = 0.126$$

a. For each of the given probabilities, write a sentence giving an interpretation of the probability in the context of this problem.

b. Use the given probabilities to construct a hypothetical 1000 table with columns corresponding to whether or not a penalty kick was blocked and rows corresponding to whether the goalkeeper jumped left, stayed in the center, or jumped right.

c. Use the table to calculate the probability that a penalty kick is blocked.

d. Based on the given probabilities and the probability calculated in Part (c), what would you recommend to a goalkeeper as the best strategy when trying to defend against a penalty kick? How does this compare to what goalkeepers actually do when defending against a penalty kick?

5.46 The accompanying table summarizes data from a Gallup Survey of 3594 parents with school-aged children **("Five Insights into U.S. Parents' Satisfaction with Education," August 25, 2016, www.gallup.com, retrieved April 25, 2017).** In this survey, parents were asked if they were completely satisfied with the education their oldest child receives.

School Type	Percentage of Parents Completely Satisfied
Public	28%
Private	62%

Of the 3594 parents surveyed, 608 were parents whose oldest child attended a private school. The parents participating in this survey were thought to be representative of U.S. parents of school-aged children.

a. Use the given information to determine the *number of parents surveyed* falling into each of the cells in the table below.

	Completely Satisfied	Not Completely Satisfied	Total
Public School			
Private School			608
Total			3,594

b. Estimate the probability that a randomly selected parent of school-aged children is completely satisfied with his or her oldest child's education.

c. Estimate the probability that a randomly selected parent of school-aged children has an oldest child who attends a private school.

d. Estimate the probability that a randomly selected parent of school-aged children is not completely satisfied with his or her oldest child's education given that the oldest child attends a private school.

e. Estimate the probability that a randomly selected parent of school-aged children is completely satisfied with his or her oldest child's education and the oldest child attends public school.

f. Consider the event E = event that a randomly selected parent of school-aged children is completely satisfied and the event F = event that the selected parent's oldest child attends a private school. Are these independent events? Explain.

SECTION 5.4 Exercise Set 2

5.47 An electronics store sells two different brands of DVD players. The store reports that 30% of customers purchasing a DVD choose Brand 1. Of those that choose Brand 1, 20% purchase an extended warranty. Consider the chance experiment of randomly selecting a customer who purchased a DVD player at this store.

a. One of the percentages given in the problem specifies an unconditional probability, and the other percentage specifies a conditional probability. Which one is the conditional probability, and how can you tell?

b. Suppose that two events B and E are defined as follows:

B = selected customer purchased Brand 1

E = selected customer purchased an extended warranty

Use probability notation to translate the given information into two probability statements of the form $P(\_\_\_\_)$ = probability value.

5.48 The article **"Americans Growing More Concerned About Head Injuries in Football" (www.theharrispoll.com, December 21, 2015, retrieved April 25, 2017)** describes a survey of 2096 adult Americans. Survey participants were asked if they were football fans and also if they agreed or disagreed that the rules that the National Football League adopted in 2010 designed to limit head injuries have been effective. Data from the survey are summarized in the table below.

	Agree	Disagree	Total
Football Fan	693	523	1,216
Not a Football Fan	229	651	880
Total	922	1,174	2,096

Suppose that a survey participant is to be selected at random. Consider the following events:

A = event selected participant agreed that the rules have been effective

D = event selected participant disagreed that the rules have been effective

F = event selected participant was a football fan

a. Calculate the following probabilities
   i. $P(A)$
   ii. $P(D)$
   iii. $P(A|F)$
   iv. $P(A|F^C)$
b. Are the events F and A independent events? Justify your answer using relevant probabilities.

5.49 Lyme disease is transmitted by infected ticks. Several tests are available for people with symptoms of Lyme disease. One of these tests is the EIA/IFA test. The paper **"Lyme Disease Testing by Large Commercial Laboratories in the United States" (*Clinical Infectious Disease* [2014]: 676–681)** found that 11.4% of those tested actually had Lyme disease.

Consider the following events:

+ represents a positive result on the blood test
− represents a negative result on the blood test
L represents the event that the patient actually has Lyme disease
L^C represents the event that the patient actually does not have Lyme disease

The following probabilities are based on percentages given in the paper:

$$P(L) = 0.114$$
$$P(L^C) = 0.886$$

$$P(+|L) = 0.933$$
$$P(-|L) = 0.067$$
$$P(+|L^C) = 0.039$$
$$P(-|L^C) = 0.961$$

a. For each of the given probabilities, write a sentence giving an interpretation of the probability in the context of this problem.
b. Use the given probabilities to construct a hypothetical 1000 table with columns corresponding to whether or not a person has Lyme disease and rows corresponding to whether the blood test is positive or negative.
c. Notice the form of the known conditional probabilities; for example, $P(+|L)$ is the probability of a positive test given that a person selected at random from the population actually has Lyme disease. Of more interest is the probability that a person has Lyme disease, given that the test result is positive. Use information from the table constructed in Part (b) to calculate this probability.

5.50 The article **"U.S. Investors Split Between Digital and Traditional Banking" (www.gallup.com, August 5, 2016, retrieved April 25, 2017)** summarized data from a Gallup survey of a random sample of 1019 U.S. adults with investments of $10,000 or more. Based on the survey data, it was estimated that 31% of investors manage their investments by doing everything they possibly can online. But the authors of the article also noted that there was a quite a difference between younger investors (ages 18 to 49) and older investors (ages 50 and older). For younger investors, 43% said they do everything they possibly can online, while the percentage for older investors was 23%.

a. Use the given information to estimate $P(O)$, $P(O|Y)$, and $P(O|F)$ where O = event that a randomly selected investor does everything possible online, Y = event that a randomly selected investor is age 18 to 49, and F = event that a randomly selected investor is 50 years old or older.
b. Suppose that 40% of investors are between the ages of 18 and 49. Use the probabilities from Part (a) and the estimate $P(Y) = 0.40$ to construct a hypothetical 1000 table. Then use information in the table to calculate $P(Y|O)$ and write a sentence interpreting this value in the context of this exercise.

ADDITIONAL EXERCISES

5.51 Suppose that an individual is randomly selected from the population of all adult males living in the United States. Let A be the event that the selected individual is over 6 feet in height, and let B be the event that the selected individual is a professional basketball player. Which do you think is greater, $P(A|B)$ or $P(B|A)$? Why?

5.52 The paper **"Good for Women, Good for Men, Bad for People: Simpson's Paradox and the Importance of Sex-Specific Analysis in Observational Studies" (*Journal of Women's Health**

and Gender-Based Medicine **[2001]: 867–872)** described the results of a medical study in which one treatment was shown to be better for men and better for women than a competing treatment. However, if the data for men and women are combined, it appears as though the competing treatment is better. To see how this can happen, consider the accompanying data tables constructed from information in the paper. Subjects in the study were given either Treatment A or Treatment B, and their survival was noted. Let *S* be the event that a patient selected at random survives, *A* be the event that a patient selected at random received Treatment A, and *B* be the event that a patient selected at random received Treatment B.

a. The following table summarizes data for men and women combined:

	Survived	Died	Total
Treatment A	215	85	300
Treatment B	241	59	300
Total	456	144	600

   i. Find $P(S)$.
   ii. Find $P(S|A)$.
   iii. Find $P(S|B)$.
   iv. Which treatment appears to be better?

b. Now consider the summary data for the men who participated in the study:

	Survived	Died	Total
Treatment A	120	80	200
Treatment B	20	20	40
Total	140	100	240

   i. Find $P(S)$.
   ii. Find $P(S|A)$.
   iii. Find $P(S|B)$.
   iv. Which treatment appears to be better?

c. Now consider the summary data for the women who participated in the study:

	Survived	Died	Total
Treatment A	95	5	100
Treatment B	221	39	260
Total	316	44	360

   i. Find $P(S)$.
   ii. Find $P(S|A)$.
   iii. Find $P(S|B)$.
   iv. Which treatment appears to be better?

d. You should have noticed from Parts (b) and (c) that for both men and women, Treatment A appears to be better. But in Part (a), when the data for men and women are combined, it looks like Treatment B is better. This is an example of what is called Simpson's Paradox. Write a brief explanation of why this apparent inconsistency occurs for this data set. (Hint: Do men and women respond similarly to the two treatments?)

5.53 A large cable TV company reports the following:
  80% of its customers subscribe to its cable TV service
  42% of its customers subscribe to its Internet service
  32% of its customers subscribe to its telephone service
  25% of its customers subscribe to both its cable TV and Internet service
  21% of its customers subscribe to both its cable TV and phone service
  23% of its customers subscribe to both its Internet and phone service
  15% of its customers subscribe to all three services

Consider the chance experiment that consists of selecting one of the cable company customers at random. Find and interpret the following probabilities:
a. $P(\text{cable TV only})$
b. $P(\text{Internet}|\text{cable TV})$
c. $P(\text{exactly two services})$
d. $P(\text{Internet and cable TV only})$

5.54 A man who works in a big city owns two cars, one small and one large. Three-quarters of the time he drives the small car to work, and one-quarter of the time he takes the large car. If he takes the small car, he usually has little trouble parking and so is at work on time with probability 0.9. If he takes the large car, he is on time to work with probability 0.6. Given that he was at work on time on a particular morning, what is the probability that he drove the small car?

5.55 Students at a particular university use an online registration system to select their courses for the next term. There are four different priority groups, with students in Group 1 registering first, followed by those in Group 2, and so on. Suppose that the university provided the accompanying information on registration for the fall semester. The entries in the table represent the proportion of students falling into each of the priority-unit combinations.

Priority Group	Number of Units Secured During First Attempt to Register				
	0–3	4–6	7–9	10–12	More Than 12
1	0.01	0.01	0.06	0.10	0.07
2	0.02	0.03	0.06	0.09	0.05
3	0.04	0.06	0.06	0.06	0.03
4	0.04	0.08	0.07	0.05	0.01

a. What proportion of students at this university got 10 or more units during the first attempt to register?
b. Suppose that a student reports receiving 11 units during the first attempt to register. Is it more likely that he or she is in the first or the fourth priority group? Explain.
c. If you are in the third priority group next term, is it likely that you will get more than 9 units during the first attempt to register? Explain.

5.56   Consider the following events:

C = event that a randomly selected driver is observed to be texting or using a hand-held cell phone

A = event that a randomly selected driver is observed driving a car

V = event that a randomly selected driver is observed driving a van or SUV

T = event that a randomly selected driver is observed driving a pickup truck

Based on data from the **National Highway Traffic Safety Administration ("Traffic Safety Facts," February 2014),** the following probability estimates are reasonable:

$$P(C) = 0.015 \quad P(C|A) = 0.017 \quad P(C|V) = 0.014$$
$$P(C|T) = 0.010$$

Explain why $P(C)$ is not just the average of the three given conditional probabilities.

SECTION 5.5   **Calculating Probabilities—A More Formal Approach (Optional)**

In the previous two sections, you saw how tables could be used to calculate event probabilities and conditional probabilities. Sometimes you started with a data table and used the data in the table to estimate probabilities. In other situations, you started with probability information and used that information to construct a hypothetical 1000 table that could be used to calculate probabilities of interest. In this section, you will see that it is possible to bypass the construction of a hypothetical 1000 table and use a few key probability formulas to calculate probabilities directly.

Probability Formulas

The Complement Rule (for calculating complement probabilities)
For any event E,

$$P(E^c) = 1 - P(E)$$

The Addition Rule (for calculating union probabilities)
For any two events E and F,

$$P(E \cup F) = P(E) + P(F) - P(E \cap F)$$

For **mutually exclusive events**, this simplifies to

$$P(E \cup F) = P(E) + P(F)$$

The Multiplication Rule (for calculating intersection probabilities)
For any two events E and F,

$$P(E \cap F) = P(F)P(E|F) = P(E|F)P(F)$$

For **independent events,** this simplifies to

$$P(E \cap F) = P(E)P(F)$$

Conditional Probabilities
For any two events E and F, with $P(F) \neq 0$,

$$P(E \mid F) = \frac{P(E \cap F)}{P(F)}$$

Three of the probability formulas in the preceding box were introduced in Section 5.3 (the Complement Rule, the Addition Rule for mutually exclusive events, and the Multiplication Rule for independent events). In the following examples, you will see how the other formulas follow from the work you have already done with hypothetical 1000 tables.

Union Probabilities: The Addition Rule

To see how the Addition Rule can be used to calculate the probability of a union event, let's revisit the health information study first introduced in Example 5.5.

| **Example 5.16** | **Health Information from TV Shows Revisited** |

The report **"TV Drama/Comedy Viewers and Health Information" (cdc.gov/healthcommunication/pdf/healthstyles_2005.pdf, retrieved April 25, 2017)** describes a large survey that was conducted for the Centers for Disease Control and Prevention (CDC). The CDC believed that the sample used was representative of adult Americans. One question on the survey asked respondents if they had learned something new about a health issue or disease from a TV show in the previous 6 months. Consider the following events:

L = event that a randomly selected adult American reports learning something new about a health issue or disease from a TV show in the previous 6 months

and

F = event that a randomly selected adult American is female

Data from the survey were used to estimate the following probabilities:

$$P(L) = 0.58 \quad P(F) = 0.50 \quad P(L \cap F) = 0.31$$

In Example 5.5, the given probability information was used to construct the following hypothetical 1000 table:

	F (female)	*not F* (not female)	Total
L(learned from TV)	310	270	**580**
not L(did not learn from TV)	190	230	**420**
Total	**500**	**500**	**1,000**

The table was used to calculate

$$P(L \cup F) = \frac{310 + 270 + 190}{1000} = \frac{770}{1000} = 0.770$$

This probability could have been calculated directly using the Addition Rule as follows:
$$P(L \cup F) = P(L) + P(F) - P(L \cap F)$$

$$= 0.58 + 0.50 - 0.31$$

$$= 0.77$$

If you are comfortable working with formulas, this is much quicker than setting up the hypothetical 1000 table.

To see why the Addition Rule works, look back at the calculation based on the hypothetical 1000 table:

$$P(L \cup F) = \frac{310 + 270 + 190}{1000} \qquad 0$$

This can be rewritten as

$$P(L \cup F) = \frac{310 + 270 + 190 + (310 - 310)}{1000}$$

$$= \frac{(310 + 270)}{1000} + \frac{(190 + 310)}{1000} - \frac{310}{1000}$$

$$= 0.580 + 0.500 - 0.310$$

$$= P(L) + P(F) - P(L \cap F)$$

Subtracting $P(L \cap F)$ from $P(L) + P(F)$ corrects for the fact that outcomes in $L \cap F$ are included both in L and in F.

Conditional Probabilities

A probability that takes into account whether or not another event has occurred is called a "conditional" probability. The probability of the event E given that the event F is known to have occurred is denoted by $P(E|F)$. To see how the formula for calculating a conditional probability is used, let's revisit the heart attack study of Example 5.12.

Example 5.17 Surviving a Heart Attack Revisited

Medical guidelines recommend that a hospitalized patient who suffers cardiac arrest should receive defibrillation (an electric shock to the heart) within 2 minutes. The paper **"Delayed Time to Defibrillation After In-Hospital Cardiac Arrest" (*The New England Journal of Medicine* [2008]: 9–17)** describes a study of the time to defibrillation for hospitalized patients in hospitals of various sizes. The authors examined medical records of 6716 patients who suffered cardiac arrest while hospitalized, recording the size of the hospital and whether or not defibrillation occurred within 2 minutes. Data from this study are reproduced here:

Hospital Size	Time to Defibrillation		Total
	Within 2 Minutes	More Than 2 Minutes	
Small (Fewer Than 250 Beds)	1,124	576	**1,700**
Medium (250–499 Beds)	2,178	886	**3,064**
Large (500 or More Beds)	1,387	565	**1,952**
Total	**4,689**	**2,027**	**6,716**

In Example 5.12, it was assumed that these data were representative of the larger group of all hospitalized patients who suffer a cardiac arrest. The following events were defined:

S = event that a randomly selected patient is at a small hospital
M = event that a randomly selected patient is at a medium-sized hospital
L = event that a randomly selected patient is at a large hospital
D = event that a randomly selected patient receives defibrillation within 2 minutes

Information in the table was used to calculate $P(D)$, the probability that a randomly selected patient receives defibrillation within 2 minutes. Because 4689 out of the 6716 patients received defibrillation within 2 minutes: $P(D) = \dfrac{4689}{6716} = 0.698$. Conditional probabilities involving event D were also calculated:

$$P(D \mid S) = \frac{1124}{1700} = 0.661 \quad P(D \mid M) = \frac{2178}{3064} = 0.711 \quad P(D \mid L) = \frac{1387}{1952} = 0.711$$

These probabilities are interpreted as the proportions of patients at small, medium, and large hospitals, respectively, who received timely defibrillations.

The reasoning used to calculate these conditional probabilities based on information in the given table is the same as using the formula for calculating a conditional probability. To see this, let's use the formula to calculate $P(D \mid S)$:

$$P(D \mid S) = \frac{P(D \cap S)}{P(S)}$$
$$= \frac{1124/6716}{1700/6716}$$
$$= \frac{1124}{1700}$$
$$= 0.661 \qquad \longleftarrow \text{\textit{Same as calculation based on the table.}}$$

Notice that when you reasoned directly from the table, you calculated this probability by looking only at patients in small hospitals, and then calculating the proportion of those patients who received timely defibrillation. The formula for calculating a conditional probability is based on this same reasoning.

In Example 5.17, using the formula to calculate a conditional probability isn't any quicker than using the information in the table because the data table was given. However, if you are starting with probability information rather than a data table, it is possible to use the formula to calculate conditional probabilities without having to set up a hypothetical 1000 table. This is illustrated in the following example.

Example 5.18 **Tattoos**

The article **"Chances Are You Know Someone with a Tattoo, and He's Not a Sailor" (Associated Press, June 11, 2006)** summarized data from a representative sample of adults ages 18 to 50. Consider the following events:

T = the event that a randomly selected person in the targeted age group has a tattoo

A = the event that a randomly selected person in the targeted age group is between 18 and 29 years old

The following probabilities were estimated based on data from the sample:

$$P(T) = 0.24$$

$$P(A) = 0.50$$

$$P(T \cap A) = 0.18$$

Suppose you are interested in the probability that a person who is between 18 and 29 years old has a tattoo. This is the conditional probability $P(T \mid A)$. One way to estimate this probability is to construct a hypothetical 1000 table using the given information. This involves completing the following table:

	T	Not T	Total
A	180		500
Not A			
Total	240		1,000

You could then use the hypothetical 1000 table to calculate $P(T \mid A)$. Using the hypothetical 1000 table would give you a correct answer, but you could calculate $P(T \mid A)$ directly from the conditional probability formula, as shown here:

$$P(T \mid A) = \frac{P(T \cap A)}{P(A)} = \frac{0.18}{0.50} = 0.36$$

You could also calculate $P(A \mid T)$:

$$P(A \mid T) = \frac{P(T \cap A)}{P(T)} = \frac{0.18}{0.24} = 0.75$$

Notice the difference between the three probabilities

$$P(T) = 0.24$$

$$P(T \mid A) = 0.36$$

$$P(A \mid T) = 0.75$$

These three probabilities have different interpretations.

Probability	Interpretation
$P(T) = 0.24$	About 24% of *adults ages 18 to 50* have a tattoo.
$P(T \mid A) = 0.36$	About 36% of *adults ages 18 to 29* have a tattoo.
$P(A \mid T) = 0.75$	About 75% of *adults ages 18 to 50 who have a tattoo* are between 18 and 29 years old.

Intersection Probabilities: The Multiplication Rule

To see how the Multiplication Rule can be used to calculate the probability of an intersection event, consider the following example.

Example 5.19 **DVD Player Warranties**

A large electronics store sells two different portable DVD players, Brand 1 and Brand 2. Based on past sales records, the store sales manager reports that 70% of the DVD players sold are Brand 1 and 30% are Brand 2. The manager also reports that 20% of the people who buy Brand 1 also purchase an extended warranty, and 40% of the people who buy Brand 2 purchase an extended warranty. Consider selecting a person at random from those who purchased a DVD player from this store, and define the following events:

B_1 = event that the selected customer bought Brand 1
B_2 = event that the selected customer bought Brand 2
E = event that the selected customer purchased an extended warranty

Suppose you would like to know the probability that the selected customer bought Brand 1 and purchased an extended warranty, $P(B_1 \cap E)$.

There are three different ways you might calculate this probability.

Method 1: Hypothetical 1000 Table
One way to calculate $P(B_1 \cap E)$ is to use the given information to complete a hypothetical 1000 table. This would result in the accompanying table.

	E	*Not E*	Total
Brand 1	140	560	**700**
Brand 2	120	180	**300**
Total	**260**	**740**	**1,000**

The table could then be used to calculate $P(B_1 \cap E)$, as shown.

$$P(B_1 \cap E) = \frac{140}{1000} = 0.14$$

Method 2: The Multiplication Rule
Translating the given information into probability statements results in

$P(B_1) = 0.70$ (70% buy Brand 1)
$P(B_2) = 0.30$ (30% buy Brand 2)
$P(E \mid B_1) = 0.20$ (20% *of those who buy Brand 1* purchase an extended warranty)
$P(E \mid B_2) = 0.40$ (40% *of those who buy Brand 2* purchase an extended warranty)

Using the Multiplication Rule to calculate $P(B_1 \cap E)$, you get

$$P(B_1 \cap E) = P(E \cap B_1) = P(B_1)\, P(E \mid B_1) = (0.70)\,(0.20) = 0.14$$

Notice that this is equal to the probability calculated from the hypothetical 1000 table and can be calculated without having to construct the table.

Method 3: Using a Tree Diagram
Another way to calculate the probability of an intersection event when some conditional probabilities are known is to use a tree diagram. Many people prefer this more visual approach. The tree diagram of Figure 5.4 gives a nice display that is based on the Multiplication Rule.

FIGURE 5.4
A tree diagram for the probability calculations of Example 5.19.

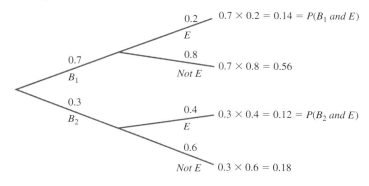

The two first-generation branches are labeled with events B_1 and B_2 along with their probabilities. Two second-generation branches extend from each first-generation branch, corresponding to the two events E and *not E*. The *conditional* probabilities $P(E \mid B_1)$, $P(not\ E \mid B_1)$, $P(E \mid B_2)$, and $P(not\ E \mid B_2)$, appear on these branches. Application of the Multiplication Rule then consists of multiplying probabilities across the branches of the tree diagram. For example,

$$P(B_2 \cap E) = P(B_2)\,P(E \mid B_2)$$
$$= (0.3)\,(0.4)$$
$$= 0.12$$

and this probability is displayed to the right of the E branch that comes from the B_2 branch. You can now also easily calculate $P(E)$, the probability that an extended warranty is purchased. The event E can occur in two different ways: Buy Brand 1 *and* warranty, or buy Brand 2 *and* warranty. Symbolically, these events are $B_1 \cap E$ and $B_2 \cap E$. Furthermore, if each customer purchased a single DVD player, he or she could not have simultaneously purchased both Brand 1 and Brand 2, so the two events $B_1 \cap E$ and $B_2 \cap E$ are mutually exclusive.
Then

$$P(E) = P(B_1 \cap E) + P(B_2 \cap E)$$
$$= P(B_1)\,P(E \mid B_1) + P(B_2)\,P(E \mid B_2)$$
$$= (0.7)(0.2) + (0.3)(0.4)$$
$$= 0.14 + 0.12$$
$$= 0.26$$

This probability is the sum of two of the probabilities shown on the right-hand side of the tree diagram. This means that 26% of all DVD player purchasers selected an extended warranty.

The general multiplication rule can be extended to give an expression for the probability that several events occur together. In the case of three events E, F, and G,

$$P(E \cap F \cap G) = P(G)P(F \mid G)P(E \mid F \cap G)$$

When the events are all independent, $P(E \mid F \cap G) = P(E)$ and $P(F \mid G) = P(F)$ so the right-hand side of the equation for $P(E \cap F \cap G)$ is simply the product of the three unconditional probabilities.

Example 5.20   Lost Luggage

Twenty percent of all passengers who fly from Los Angeles (LA) to New York (NY) do so on Airline G. This airline misplaces luggage for 10% of its passengers, and 90% of this lost luggage is subsequently recovered. If a passenger who has flown from LA to NY is randomly selected, what is the probability that the selected individual flew on Airline G (event G), had luggage misplaced (event F), and subsequently recovered the misplaced luggage (event E)? The given information implies that

$$P(G) = 0.20 \quad P(F \mid G) = 0.10 \quad P(E \mid F \cap G) = 0.90$$

Then

$$P(E \cap F \cap G) = P(G)P(F \mid G)\,P(E \mid F \cap G) = (0.20)(0.10)(0.90) = 0.018$$

That is, about 1.8% of passengers flying from LA to NY fly on Airline G, have their luggage misplaced, and subsequently recover the lost luggage.

More Complex Probability Calculations: The Law of Total Probability and Bayes' Rule (Optional)

Law of Total Probability

Let's reconsider the information on DVD player sales from Example 5.19. In this example, the following events were defined:

B_1 = event that Brand 1 is purchased
B_2 = event that Brand 2 is purchased
E = event that an extended warranty is purchased

Based on the information given in Example 5.19, the following probabilities are known:

$$P(B_1) = 0.7 \qquad P(B_2) = 0.3 \qquad P(E \mid B_1) = 0.2 \qquad P(E \mid B_2) = 0.4$$

Notice that the conditional probabilities $P(E \mid B_1)$ and $P(E \mid B_2)$ are known but that the unconditional probability $P(E)$ is not known.

To find $P(E)$ notice that the event E can occur in two ways: (1) A customer purchases an extended warranty *and* buys Brand 1 (which can be written as $E \cap B_1$); or (2) a customer purchases an extended warranty *and* buys Brand 2 (which can be written as $E \cap B_2$). Because these are the only ways in which E can occur, you can write the event E as

$$E = (E \cap B_1) \cup (E \cap B_2)$$

The two events $(E \cap B_1)$ and $(E \cap B_2)$ are mutually exclusive (since B_1 and B_2 are mutually exclusive), so using the addition rule for mutually exclusive events gives

$$P(E) = P((E \cap B_1) \cup (E \cap B_2))$$
$$= P(E \cap B_1) + P(E \cap B_2)$$

Finally, using the general multiplication rule to evaluate $P(E \cap B_1)$ and $P(E \cap B_2)$ results in

$$P(E) = P(E \cap B_1) + P(E \cap B_2)$$
$$= P(B_1) P(E \mid B_1) + P(B_2) P(E \mid B_2)$$

Substituting in the known probabilities gives

$$P(E) = P(B_1) P(E \mid B_1) + P(B_2) P(E \mid B_2)$$
$$= (0.7)(0.2) + (0.3)(0.4)$$
$$= 0.26$$

You can conclude that 26% of the DVD player customers purchased an extended warranty.

As this example illustrates, when conditional probabilities are known, they can sometimes be used to calculate unconditional probabilities. The **law of total probability** formalizes this use of conditional probabilities.

The Law of Total Probability

If B_1 and B_2 are mutually exclusive events with $P(B_1) + P(B_2) = 1$, then for any event E

$$P(E) = P(E \cap B_1) + P(E \cap B_2)$$
$$= P(B_1) P(E \mid B_1) + P(B_2) P(E \mid B_2)$$

More generally, if $B_1, B_2, \cdots, B_k$ are mutually exclusive events with $P(B_1) + P(B_2) + \cdots + P(B_k) = 1$, then for any event E

$$P(E) = P(E \cap B_1) + P(E \cap B_2) + \cdots + P(E \cap B_k)$$
$$= P(B_1) P(E \mid B_1) + P(B_2) P(E \mid B_2) + \cdots + P(B_k) P(E \mid B_k)$$

<hr>

Example 5.21 **Which Way to Jump?**

The paper **"Action Bias among Elite Soccer Goalkeepers: The Case of Penalty Kicks" (*Journal of Economic Psychology* [2007]: 606–621)** presents an interesting analysis of 286 penalty kicks in televised championship soccer games from around the world. In a penalty kick, the only players involved are the kicker and the goalkeeper from the opposing team. The kicker tries to kick a ball into the goal from a point located 11 meters away. The goalkeeper tries to block the ball from reaching the goal. For each penalty kick analyzed, the researchers recorded the direction that the goalkeeper moved (jumped to the left, stayed in the center, or jumped to the right) and whether or not the penalty kick was successfully blocked.

Consider the following events:

L = the event that the goalkeeper jumps to the left
C = the event that the goalkeeper stays in the center
R = the event that the goalkeeper jumps to the right
B = the event that the penalty kick is blocked

Based on their analysis of the penalty kicks, the authors of the paper gave the following probability estimates:

$$P(L) = 0.493 \qquad P(C) = 0.063 \qquad P(R) = 0.444$$
$$P(B \mid L) = 0.142 \qquad P(B \mid C) = 0.333 \qquad P(B \mid R) = 0.126$$

What proportion of penalty kicks were blocked? You can use the law of total probability to answer this question. Here, the three events L, C, and R play the role of B_1, B_2, and B_3 and B plays the role of E in the formula for the law of total probability.

Substituting into the formula, you get

$$P(B) = P(B \cap L) + P(B \cap C) + P(B \cap R)$$
$$= P(L)P(B \mid L) + P(C)P(B \mid C) + P(R)P(B \mid R)$$
$$= (0.493)(0.142) + (0.063)(0.333) + (0.444)(0.126)$$
$$= 0.070 + 0.021 + 0.056$$
$$= 0.147$$

A tree diagram could also have been used here. Figure 5.5 is a visual display of the given probability information. Following the branches in the tree that lead to a blocked kick and adding these three probabilities gives

$$P(B) = 0.070 + 0.021 + 0.056 = 0.147$$

FIGURE 5.5
Tree diagram for the soccer example

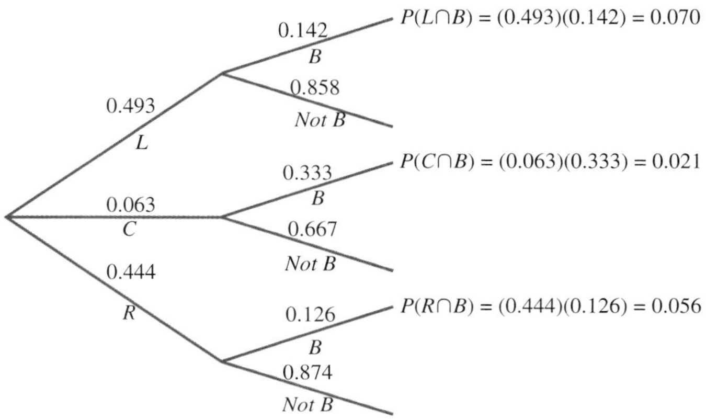

This means that only 14.7% of penalty kicks were successfully blocked. Two other interesting findings of this study were

1. The direction that the goalkeeper moves appears to be independent of whether the kicker kicked the ball to the left, center, or right of the goal. This was attributed to the fact that goalkeepers have to choose their action before they can clearly observe the direction of the kick.

2. The three conditional probabilities are $P(B \mid L) = 0.142$, $P(B \mid C) = 0.333$, and $P(B \mid R) = 0.126$. The optimal strategy for a goalkeeper appears to be to stay in the center of the goal. However, staying in the center was only chosen 6.3% of the time—much less often that jumping left or right. The authors believe that this is because a goalkeeper does not feel as bad about not successfully blocking a kick if some action (jumping left or right) is taken compared to if no action (staying in the center) is taken. This is the "action bias" referred to in the title of the paper.

Bayes' Rule

In this subsection, you will see a formula discovered by the Reverend Thomas Bayes (1702–1761), an English Presbyterian minister. He discovered what is now known as Bayes' Rule. This rule provides a solution to what Bayes called the "converse problem." A converse problem is one where some conditional probabilities are known, such as the probability of a positive test result given cancer, but you would like to know the probability that has the conditioning "reversed." That is, you would like to know the probability of cancer given a positive test result.

You have already seen how to find such probabilities using hypothetical 1000 tables. Now you will see how Bayes' Rule might be used to calculate these probabilities without constructing a hypothetical 1000 table. Although many people find the hypothetical 1000 table to be the easier approach, either method can be used.

Bayes' Rule

If B_1 and B_2 are mutually exclusive events with $P(B_1) + P(B_2) = 1$, then for any event E

$$P(B_1 \mid E) = \frac{P(B_1)P(E \mid B_1)}{P(B_1)P(E \mid B_1) + P(B_2)P(E \mid B_2)}$$

More generally, if $B_1, B_2, \cdots, B_k$ are mutually exclusive events with $P(B_1) + P(B_2) + \cdots + P(B_k) = 1$ then for any event E,

$$P(B_i \mid E) = \frac{P(B_i)P(E \mid B_i)}{P(B_1)P(E \mid B_1) + P(B_2)P(E \mid B_2) + \ldots + P(B_k)P(E \mid B_k)}$$

To see how Bayes' Rule can be used to calculate a conditional probability, let's revisit the Internet addiction study first described in Example 5.14.

Example 5.22 **Internet Addiction Revisited**

Internet addiction has been defined by researchers as a disorder characterized by excessive time spent on the Internet, impaired judgment and decision-making ability, social withdrawal, and depression. The paper **"The Association between Aggressive Behaviors and Internet Addiction and Online Activities in Adolescents"** (*Journal of Adolescent Health* **[2009]: 598–605**) describes a study of a large number of adolescents. Each participant in the study was assessed using the Chen Internet Addiction Scale to determine if he or she suffered from Internet addiction. The following statements are based on the survey results:

1. 51.8% of the study participants were female and 48.2% were male.
2. 13.1% of the females suffered from Internet addiction.
3. 24.8% of the males suffered from Internet addiction.

Consider the chance experiment that consists of selecting a study participant at random and define the following events:

F = the event that the selected participant is female
M = the event that the selected participant is male
I = the event that the selected participant suffers from Internet addiction

The three statements from the paper define the following probabilities:

$$P(F) = 0.518$$
$$P(M) = 0.482$$
$$P(I \mid F) = 0.131$$
$$P(I \mid M) = 0.248$$

In Example 5.14, the given information was used to construct the following hypothetical 1000 table:

	I (Internet addiction)	not I (no Internet addiction)	Total
F (female)	68	450	**518**
not F (male)	120	362	**482**
Total	**188**	**812**	**1,000**

This table was then used to calculate

$$P(F \mid I) = \frac{68}{188} = 0.362$$

Notice that $P(I \mid F)$ was given, and then the table was used to calculate $P(F \mid I)$.

Bayes' Rule can also be used to calculate this probability. With F and M playing the role of B_1 and B_2 and I playing the role of E in the Bayes' Rule formula, you get

$$P(F \mid I) = \frac{P(F)(I \mid F)}{P(F)P(I \mid F) + P(M)P(I \mid M)}$$

$$= \frac{(0.518)(0.131)}{(0.518)(0.131) + (0.482)(0.248)}$$

$$= \frac{0.068}{0.068 + 0.120}$$

$$= \frac{0.068}{0.188}$$

$$= 0.362$$

Notice that this probability is equal to the one calculated using the hypothetical 1000 table, but was calculated without having to construct the table.

The formulas for calculating probabilities introduced in this section provide an alternative to using a hypothetical 1000 table. Either a table or the formulas can be used, so you can choose the method you prefer.

Summing It Up—Section 5.5

The following learning objectives were addressed in this section:

Mastering the Mechanics
M11: (Optional) Use probability formulas to calculate probabilities of unions and intersections and to calculate conditional probabilities.
When you know the probabilities of some events, probability formulas for complements, unions, intersections, and conditional probabilities provide a way to calculate probabilities that can be used as an alternative to the hypothetical 1000 table approach used in the previous sections. A summary of the probability formulas can be found in the box at the beginning of this section. The examples of this section revisit examples from previous sections to illustrate the use of probability formulas to calculate probabilities. Because the formulas and the hypothetical 1000 table approaches are equivalent, you can choose the method you prefer when calculating probabilities.

SECTION 5.5 EXERCISES

Each Exercise Set assesses the following chapter learning objectives: M11

SECTION 5.5 Exercise Set 1

5.57 There are two traffic lights on Shelly's route from home to work. Let E denote the event that Shelly must stop at the first light, and define the event F in a similar manner for the second light. Suppose that $P(E) = 0.4$, $P(F) = 0.3$ and $P(E \cap F) = 0.15$. In Exercise 5.25, you constructed a hypothetical 1000 table to calculate the following probabilities. Now use the probability formulas of this section to find these probabilities.

a. The probability that Shelly must stop for at least one light (the probability of the event $E \cup F$).

b. The probability that Shelly does not have to stop at either light.

c. The probability that Shelly must stop at exactly one of the two lights.

d. The probability that Shelly must stop only at the first light.

5.58 A large cable company reports that 80% of its customers subscribe to cable TV, 42% subscribe to Internet service, and 97% subscribe to at least one of cable TV and Internet service. In Exercise 5.26, you constructed a hypothetical 1000 table to calculate the following probabilities. Now use the probability formulas of this section to find these probabilities.

a. The probability that a randomly selected customer subscribes to both cable TV and Internet service.

b. The probability that a randomly selected customer subscribes to exactly one of cable TV or Internet service.

5.59 Two different airlines have a flight from Los Angeles to New York that departs each weekday morning at a certain time. Let E denote the event that the first airline's flight is fully booked on a particular day, and let F denote the event that the second airline's flight is fully booked on that same day. Suppose that $P(E) = 0.7$, $P(F) = 0.6$ and $P(E \cap F) = 0.54$.

a. Calculate $P(E \mid F)$, the probability that the first airline's flight is fully booked given that the second airline's flight is fully booked.

b. Calculate $P(F \mid E)$.

5.60 A certain university has 10 vehicles available for use by faculty and staff. Six of these are vans and four are cars. On a particular day, only two requests for vehicles have been made. Suppose that the two vehicles to be assigned are chosen at random from among the 10 vehicles.

a. Let E denote the event that the first vehicle assigned is a van. What is the value of $P(E)$?

b. Let F denote the event that the second vehicle assigned is a van. What is the value of $P(F \mid E)$?

c. Use the results of Parts (a) and (b) to calculate $P(E \cap F)$.

SECTION 5.5 Exercise Set 2

5.61 The paper **"Predictors of Complementary Therapy Use Among Asthma Patients: Results of a Primary Care Survey"** (*Health and Social Care in the Community* [2008]: 155–164) described a study in which each person in a large sample of asthma patients responded to two questions:

　Question 1: Do conventional asthma medications usually help your asthma symptoms?

　Question 2: Do you use complementary therapies (such as herbs, acupuncture, aroma therapy) in the treatment of your asthma?

Suppose that this sample is representative of asthma patients. Consider the following events:

　E = event that an asthma patient uses complementary therapies

　F = event that an asthma patient reports that conventional medications usually help

The data from the sample were used to estimate the following probabilities:

$$P(E) = 0.146 \quad P(F) = 0.879 \quad P(E \cap F) = 0.122$$

In Exercise 5.33, you constructed a hypothetical 1000 table to calculate the following probabilities. Now use the probability formulas of this section to find these probabilities.

a. The probability that an asthma patient responds that conventional medications do not help and uses complementary therapies.

b. The probability that an asthma patient responds that conventional medications do not help and does not use complementary therapies.

c. The probability that an asthma patient responds that conventional medications usually help or the patient uses complementary therapies.

5.62 An appliance manufacturer offers extended warranties on its washers and dryers. Based on past sales, the manufacturer reports that of customers buying both a washer and a dryer, 52% purchase the extended warranty for the washer, 47% purchase the extended warranty for the dryer, and 59% purchase at least one of the two extended warranties. In Exercise 5.34, you constructed a hypothetical 1000 table to calculate the following probabilities. Now use the probability formulas of this section to find these probabilities.

a. The probability that a randomly selected customer who buys a washer and a dryer purchases an extended warranty for both the washer and the dryer.

b. The probability that a randomly selected customer does not purchase an extended warranty for either the washer or dryer.

5.63 The report **"Great Jobs, Great Lives. The Relationship Between Student Debt, Experiences and Perceptions of**

College Worth" (Gallup-Purdue Index 2015 Report) gave information on the percentage of recent college graduates (those graduating between 2006 and 2015, inclusive) who strongly agree with the statement "My college education was worth the cost." Suppose that a college graduate will be selected at random, and consider the following events:

A = event that the selected graduate strongly agrees that education was worth the cost

N = event that the selected graduate finished college with no student debt

H = event that the selected graduate finished college with high student debt (over \$50,000)

The following probability estimates were given in the report:

$$P(A) = 0.38 \qquad P(A \mid N) = 0.49 \qquad P(A \mid H) = 0.18$$

a. Interpret the value of $P(A \mid N)$.
b. Interpret the value of $P(A \mid H)$.
c. Are the events A and H independent? Justify your answer.

5.64 A construction firm bids on two different contracts. Let E_1 be the event that the bid on the first contract is successful, and define E_2 analogously for the second contract. Suppose that $P(E_1) = 0.4$ and $P(E_2) = 0.3$ and that E_1 and E_2 are independent events.
a. Calculate the probability that both bids are successful (the probability of the event E_1 *and* E_2).
b. Calculate the probability that neither bid is successful (the probability of the event *not* E_1 *and not* E_2).
c. What is the probability that the firm is successful in at least one of the two bids?

5.65 The authors of the paper **"Do Physicians Know When Their Diagnoses Are Correct?"** (*Journal of General Internal Medicine* **[2005]: 334–339**) presented detailed case studies to medical students and to faculty at medical schools. Each participant was asked to provide a diagnosis in the case and also to indicate whether his or her confidence in the correctness of the diagnosis was high or low. Define the events C, I, and H as follows:

C = event that diagnosis is correct

I = event that diagnosis is incorrect

H = event that confidence in the correctness of the diagnosis is high

a. Data appearing in the paper were used to estimate the following probabilities for medical students:

$$P(C) = 0.261$$
$$P(I) = 0.739$$
$$P(H \mid C) = 0.375$$
$$P(H \mid I) = 0.073$$

Use Bayes' Rule to calculate the probability of a correct diagnosis given that the student's confidence level in the correctness of the diagnosis is high.
b. Data from the paper were also used to estimate the following probabilities for medical school faculty:

$$P(C) = 0.495$$
$$P(I) = 0.505$$
$$P(H \mid C) = 0.537$$
$$P(H \mid I) = 0.252$$

Calculate $P(C \mid H)$ for medical school faculty. How does the value of this probability compare to the value of $P(C \mid H)$ for students calculated in Part (a)?

ADDITIONAL EXERCISES

5.66 The report **"Improving Undergraduate Learning"** (**Social Science Research Council, 2011**) summarizes data from a survey of several thousand college students. These students were thought to be representative of the population of all U.S. college students. When asked about a typical semester, 68% said they would be taking a class that is reading intensive (requires more than 40 pages of reading per week). Only 50% said they would be taking a class that is writing intensive (requires more than 20 pages of writing over the course of the semester). The percentage who said that they would be taking both a reading intensive course and a writing intensive course in a typical semester was 42%. In Exercise 5.40, you constructed a hypothetical 1000 table to calculate the following probabilities. Now use the probability formulas of this section to find these probabilities.
a. The probability that a randomly selected student would be taking at least one reading intensive or writing intensive course.
b. The probability that a randomly selected student would be taking a reading intensive course or a writing intensive course, but not both.
c. The probability that a randomly selected student is taking neither a reading intensive nor a writing intensive course.

5.67 A large cable company reports the following:
80% of its customers subscribe to cable TV service
42% of its customers subscribe to Internet service
32% of its customers subscribe to telephone service
25% of its customers subscribe to both cable TV and Internet service
21% of its customers subscribe to both cable TV and phone service
23% of its customers subscribe to both Internet and phone service
15% of its customers subscribe to all three services
Consider the chance experiment that consists of selecting one of the cable company customers at random. In Exercise 5.53, you constructed a hypothetical 1000 table to calculate the following probabilities. Now use the probability formulas of this section to find these probabilities.
a. $P(\text{cable TV only})$
b. $P(\text{Internet} \mid \text{cable TV})$
c. $P(\text{exactly two services})$
d. $P(\text{Internet and cable TV only})$

5.68 In a January 2016 Harris Poll, each of 2252 American adults was asked the following question: "If you had to choose, which ONE of the following sports would you say is your favorite?" **("Pro Football is Still America's Favorite Sport," www.theharrispoll.com/sports/Americas_Fav_Sport_2016. html, retrieved April 25, 2017)**. Of the survey participants, 33% chose pro football as their favorite sport. The report also included the following statement, "Adults with household incomes of $75,000 − < $100,000 (48%) are especially likely to name pro football as their favorite sport, while love of this particular game is especially low among those in $100,000 + households (21%)."

Suppose that the percentages from this poll are representative of American adults in general. Consider the following events:

F = event that a randomly selected American adult names pro football as his or her favorite sport
L = event that a randomly selected American has a household income of $75,000 − < $100,000
H = event that a randomly selected American has a household income of $100,000 +

a. Use the given information to estimate the following probabilities:
 i. $P(F)$
 ii. $P(F \mid L)$
 iii. $P(F \mid H)$

b. Are the events F and L mutually exclusive? Justify your answer.

c. Are the events H and L mutually exclusive? Justify your answer.

d. Are the events F and H independent? Justify your answer.

5.69 Consider the following events:

T = event that a randomly selected adult trusts credit card companies to safeguard his or her personal data
M = event that a randomly selected adult is between the ages of 19 and 36
O = event that a randomly selected adult is 37 or older

Based on a June 9, 2016, Gallup survey **("Data Security: Not a Big Concern for Millennials," www.gallup.com, retrieved April 25, 2017)**, the following probability estimates are reasonable:

$$P(T \mid M) = 0.27 \qquad P(T \mid O) = 0.22$$

Explain why $P(T)$ is not just the average of the two given probabilities.

SECTION 5.6 Probability as a Basis for Making Decisions

Probability plays an important role in drawing conclusions from data. This is illustrated in Examples 5.23 and 5.24.

Example 5.23 Age and Sex of College Students

Table 5.1 shows the proportions of the students at a college in various age–sex combinations at a college. These proportions can be interpreted as probabilities if a student is to be selected at random from the students at the college. For example, the probability (long-run proportion of the time) that a male, age 21–24, will be selected is 0.16.

TABLE 5.1 Age and Sex Distribution

Sex	Under 17	17–20	21–24	25–30	31–39	40 and Older
Male	0.006	0.15	0.16	0.08	0.04	0.01
Female	0.004	0.18	0.14	0.10	0.04	0.09

Suppose that you are told that a 43-year-old student from this college is waiting to meet you. Do you think the student is male or female. What if the student was 27 years old? What about 33 years old? Are you equally confident in all three of your choices?

Reasonable responses to these questions can be based on the probabilities in Table 5.1. You would probably think that the 43-year-old student was female. You cannot be certain that this is correct, but you can see that someone in the 40-and-over age group is much less likely to be male than female. You would probably also think that the 27-year-old was female, but you would be less confident in this conclusion. For the age group 31–39, the proportion of males and the proportion of females are equal, so you would think it equally likely that a 33-year-old student would be male or female. You could decide in favor of male (or female), but with little confidence in your decision. In other words, there is a good chance that you would be incorrect.

Example 5.24 Can You Pass by Guessing?

A professor planning to give a quiz that consists of 20 true–false questions is interested in knowing how someone who answers by guessing would do on such a test. To investigate, he asks the 500 students in his introductory psychology course to write the numbers from 1 to 20 on a piece of paper and then to write T or F next to each number. The students are forced to guess at the answer to each question, because they are not even told what the questions are! These answer sheets are then collected and graded using the key for the quiz. The results are summarized in Table 5.2.

TABLE 5.2 Quiz "Guessing" Distribution

Number of Correct Responses	Number of Students	Proportion of Students	Number of Correct Responses	Number of Students	Proportion of Students
0	0	0.000	11	79	0.158
1	0	0.000	12	61	0.122
2	1	0.002	13	39	0.078
3	1	0.002	14	18	0.036
4	2	0.004	15	7	0.014
5	8	0.016	16	1	0.002
6	18	0.036	17	1	0.002
7	37	0.074	18	0	0.000
8	58	0.116	19	0	0.000
9	81	0.162	20	0	0.000
10	88	0.176			

Because probabilities are long-run proportions, an entry in the "Proportion of Students" column of Table 5.2 can be considered an estimate of the probability of correctly guessing the answers to a specific number of questions. For example, a proportion of 0.122 (or 12.2%) of the 500 students got 12 of the 20 correct. You could estimate the probability of a student correctly guessing 12 answers to be (approximately) 0.122.

You can use the information in Table 5.2 to answer the following questions.

1. Would you be surprised if someone guessing on a 20-question true–false quiz got only three correct? The approximate probability of someone getting only three correct by guessing is 0.002. This means that, in the long run, only about 2 in 1000 guessers would get exactly three correct. Because it is so unlikely, this outcome would be considered surprising.

2. If a score of 15 or more correct is a passing grade on the quiz, is it likely that someone who is guessing will pass? The long-run proportion of people who are guessing who would pass is the sum of the proportions for all the passing scores (15, 16, 17, 18, 19, and 20). It follows that

$$\text{probability of passing} \approx 0.014 + 0.002 + 0.002 + 0.000 + 0.000 + 0.000$$
$$= 0.018$$

It would be unlikely that a student who is guessing would be able to pass.

3. The professor actually gives the quiz, and a student scores 16 correct. Do you think that the student was just guessing? You could begin by assuming that the student was guessing and determine whether a score at least as high as 16 is a likely or an

unlikely occurrence. Table 5.2 tells you that the approximate probability of getting a score at least as high as this student's score is

$$\text{probability of scoring 16 or higher} \approx 0.002 + 0.002 + 0.000 + 0.000 + 0.000$$
$$= 0.004$$

That is, in the long run, only about 4 times in 1000 would someone score 16 or higher by guessing. There are two possible explanations for a score of 16: (1) The student was guessing and was *really* lucky, or (2) the student was not just guessing. Given that the first possibility is highly unlikely, you could conclude that a student who scored 16 was not just guessing at the answers. Although you cannot be certain that you are correct, the evidence is compelling.

4. What score on the quiz would it take to convince you that a student was not just guessing? You should be convinced by a score that is high enough to be unlikely for a guesser. Consider the following approximate probabilities (calculated from the entries in Table 5.2):

Score	Approximate Probability
20	0.000
19 or better	$0.000 + 0.000 = 0.000$
18 or better	$0.000 + 0.000 + 0.000 = 0.000$
17 or better	$0.002 + 0.000 + 0.000 + 0.000 = 0.002$
16 or better	$0.002 + 0.002 + 0.000 + 0.000 + 0.000 = 0.004$
15 or better	$0.014 + 0.002 + 0.002 + 0.000 + 0.000 + 0.000 = 0.018$
14 or better	$0.036 + 0.014 + 0.002 + 0.002 + 0.000 + 0.000 + 0.000 = 0.054$
13 or better	$0.078 + 0.036 + 0.014 + 0.002 + 0.002 + 0.000 + 0.000 + 0.000 = 0.132$

You might say that a score of 14 or higher is reasonable evidence that someone is not just guessing, because the approximate probability that a guesser would score this high is only 0.054. Of course, if you conclude that a student is not guessing based on a quiz score of 14 or higher, then there is a risk that you are incorrect (about 1 in 20 guessers would score this high by chance). A guesser will score 13 or more correct about 13.2% of the time, which is often enough that most people would not rule out guessing.

Examples 5.23 and 5.24 illustrate how probabilities can be used to make decisions. Later chapters look more formally at drawing conclusions based on available but often incomplete information and at assessing the reliability of such conclusions.

Summing It Up—Section 5.6

The following learning objectives were addressed in this section:

Putting It into Practice
P3: Use probability to make decisions and justify conclusions.
Probability plays an important role when drawing conclusions based on data. The two examples in this section (Examples 5.23 and 5.24) illustrate how probabilities can be used to make decisions.

SECTION 5.6 EXERCISES

Each Exercise Set assesses the following chapter learning objectives: P3

SECTION 5.6 Exercise Set 1

5.70 The paper **"Accuracy and Reliability of Self-Reported Weight and Height in the Sister Study" (***Public Health Nutrition*** [2012]: 989–999)** investigates whether women provide an accurate value when reporting weight. The table below is based on comparing actual weight to self-reported weight for women participating in a large-scale medical study. Each participant was classified into a category describing accuracy of reported weight and also by age.

	Accuracy Category			
Age	Under-Reported by More Than 7 lbs	Under-Reported by Between 4 and 7 lbs	Reported Weight Within 3 lbs	Over-Reported by 4 or More lbs
45 years	297	373	1594	207
45–54 years	763	902	4130	510
55–64 years	677	966	4383	545
65 + years	263	444	2285	300

Assume that it is reasonable to consider these data representative of adult women in the United States. Consider the following conclusion:

> Most women reported their weight to within 3 pounds of their actual weight. Older women were less likely to under report their weight and more likely to over report their weight than younger women.

Provide a justification for this conclusion. Use the information in the table to calculate relevant probabilities.

5.71 The report **"2015 Utah Seat Belt Use Survey" (Utah Department of Public Safety—Highway Safety Office, September 14, 2015)** stated that based on observing a large number of vehicle occupants, the estimated percentage of Utah drivers and passengers who wear seat belts is 87.2%. The report also gave information on seat belt use by sex and by whether the vehicle is traveling in an urban or rural area. The information in the following table is consistent with summary values given in the report.

	Urban Areas		Rural Areas	
	Male	Female	Male	Female
Seat belt	871	928	770	837
No Seat Belt	129	72	230	163

Assume that these data are representative of drivers and passengers in Utah. Consider the following conclusion:

> Females are more likely to wear seat belts than males in both urban and rural areas. The difference in the percentage of females and the percentage of males who wear seat belts is greater for rural areas than for urban areas.

Provide a justification for this conclusion. Use the information in the table to calculate relevant probabilities.

SECTION 5.7 Estimating Probabilities Empirically and Using Simulation (Optional)

In the examples presented in the previous section, reaching conclusions required knowledge of the probabilities of various events. This is reasonable when you know the actual long-run proportion of the time that each event will occur. However, in some situations, these probabilities are not known and must be determined by using hypothetical 1000 tables or probability formulas, including the basic ones introduced in this chapter.

In this section, we shift gears a bit and focus on an empirical approach to probability. When an analytical approach is impossible, impractical, or requires more than the limited probability tools of an introductory course, you can *estimate* probabilities empirically through observation or by using simulation.

Estimating Probabilities Empirically

It is fairly common practice to use observed long-run proportions to estimate probabilities. The process used to estimate probabilities is simple:

1. Observe a large number of chance outcomes under controlled circumstances.
2. Interpreting probability as a long-run relative frequency, estimate the probability of an event by using the observed proportion of occurrence.

This process is illustrated in Examples 5.25 and 5.26.

Example 5.25 **Fair Hiring Practices**

To recruit a new faculty member, a university biology department intends to advertise for someone with a Ph.D. in biology and at least 10 years of college-level teaching experience. A member of the department expresses the belief that requiring at least 10 years of teaching experience will exclude most potential applicants and will exclude far more female applicants than male applicants. The biology department would like to determine the probability an applicant would be excluded because of the experience requirement.

A similar university just completed a search in which there was no requirement for prior teaching experience but the information about prior teaching experience was recorded. Data from the 410 applications received is summarized in the following table:

	Number of Applicants		
	Less Than 10 Years of Experience	10 Years of Experience or More	Total
Male	178	112	**290**
Female	99	21	**120**
Total	**277**	**133**	**410**

Let's assume that the populations of applicants for the two positions is the same. The available information can be used to approximate the probability that an applicant will fall into each of the four sex–experience combinations. The estimated probabilities (obtained by dividing the number of applicants in each sex–experience combination by 410) are given in Table 5.3.

TABLE 5.3 Estimated Probabilities for Example 5.25

	Less Than 10 Years of Experience	10 Years of Experience or More
Male	0.434	0.273
Female	0.242	0.051

From Table 5.3,

P(candidate excluded because of experience requirement) $\approx 0.434 + 0.242 = 0.676$

You can also assess the impact of the experience requirement separately for male and female applicants. From the given information, the proportion of male applicants who have less than 10 years of experience is $\frac{178}{290} = 0.614$, whereas the corresponding proportion for females is $\frac{99}{120} = 0.825$. Therefore, approximately 61% of the male and about 83% of the female applicants would be excluded.

These subgroup proportions—0.614 for males and 0.825 for females—are conditional probabilities. In this example, the probability that a potential candidate has less than 10 years of experience is 0.676, but this probability changes to 0.825 if you know that a candidate is female. These probabilities can be expressed as an unconditional probability

P(less than 10 years of experience) $= 0.676$

and a conditional probability

P(less than 10 years of experience|female) $= 0.825$

Example 5.26 **Who Has the Upper Hand?**

Men and women frequently express intimacy through touch. A common instance of this is holding hands. Some researchers have suggested that hand-holding might not be just an expression of intimacy but also might communicate status differences. For two people to hold hands, one must assume an "overhand" grip and one an "underhand" grip. Research in this area has shown that it is the male who usually assumes the overhand grip. In the view of some investigators, the overhand grip implies status or superiority. The authors of the paper **"Men and Women Holding Hands: Whose Hand Is Uppermost?"** (*Perceptual and Motor Skills* **[1999]: 537–549)** investigated an alternative explanation: Perhaps the positioning of hands is a function of the heights of the individuals. Because men tend to be taller than women, maybe comfort, not status, dictates the positioning. Investigators at two universities observed hand-holding male–female pairs, resulting in the following data:

	Sex of Person with Uppermost Hand		
	Men	Women	Total
Man Taller	2,149	299	**2,448**
Equal Height	780	246	**1,026**
Woman Taller	241	205	**446**
Total	**3,170**	**750**	**3,920**

Assuming that these hand-holding couples are representative of hand-holding couples in general, you can use the available information to estimate various probabilities. For example, if a hand-holding couple is selected at random, then

$$P(\text{man's hand uppermost}) = \frac{3170}{3920} \approx 0.809$$

For a randomly selected hand-holding couple, if the man is taller, the probability that the male has the uppermost hand is $\frac{2149}{2448} = 0.878$. On the other hand—so to speak—if the woman is taller, the probability that the female has the uppermost hand is $\frac{205}{446} = 0.460$. Notice that these probabilities, P(male uppermost *given* male taller) and P(female uppermost *given* female taller), are conditional. Also, because P(male uppermost *given* male taller) = 0.878 is not equal to P(male uppermost) = 0.809, the events *male uppermost* and *male taller* are not independent events. And even when the female is taller, the male is still more likely to have the upper hand!

Estimating Probabilities by Using Simulation

Simulation provides a way to estimate probabilities when you are unable (or do not have the time or resources) to determine probabilities analytically and when it is impractical to estimate them empirically by observation. Simulation involves generating "observations" in a situation that is similar to the real situation of interest.

To illustrate the idea of simulation, consider the situation in which a professor wishes to estimate the probabilities of scores resulting from students guessing on a 20-question true–false quiz. Observations could be collected by having 500 students actually guess at the answers to 20 questions and then scoring the resulting papers. This approach requires considerable time and effort. Simulation provides an alternative approach.

Because each question on the quiz is a true–false question, a person who is guessing should be equally likely to answer correctly or incorrectly. Rather than asking a student to select *true* or *false* and then comparing the choice to the correct answer, an equivalent process would be to pick a ball at random from a box that contains equal numbers of red and blue balls, with a blue ball representing a correct answer. Making 20 selections from the box (replacing each ball selected before picking the next one) and then counting the number of correct choices (the number of times a blue ball is selected) is a physical substitute for an observation from a student who has guessed at the answers to 20 true–false questions. The number of blue balls in 20 selections should have the same probability as the corresponding number of correct responses to the quiz when a student is guessing.

For example, 20 selections of balls might yield the following results (R, red ball; B, blue ball):

Selection	1	2	3	4	5	6	7	8	9	10
Result	R	R	B	R	B	B	R	R	R	B

Selection	11	12	13	14	15	16	17	18	19	20
Result	R	R	B	R	R	B	B	R	R	B

This corresponds to a quiz with eight correct responses, and it provides you with one observation for estimating the probabilities of interest. This process can then be repeated a large number of times to generate additional observations. For example, you might find the following:

Repetition	Number of Correct Responses
1	8
2	11
3	10
4	12
⋮	⋮
1,000	11

The 1000 simulated quiz scores could then be used to construct a table of estimated probabilities.

Taking this many balls out of a box and writing down the results would be cumbersome and tedious. The process can be simplified by using random digits to substitute for drawing balls from the box. For example, a single digit could be selected at random from the 10 digits 0, 1, 2, 3, 4, 5, 6, 7, 8, 9. When using random digits, each of the 10 possibilities is equally likely to occur, so you can use the even digits (including 0) to indicate a correct response and the odd digits to indicate an incorrect response. This would maintain the important property that a correct response and an incorrect response are equally likely, because correct and incorrect are each represented by 5 of the 10 digits.

To aid in carrying out such a simulation, tables of random digits (such as Appendix A Table 1) or computer-generated random digits can be used. The numbers in Appendix A Table 1 were produced using a computer's random number generator. (Imagine repeatedly drawing a chip from a box containing 10 chips numbered 0, 1, . . ., 9. After each selection, the result is recorded, the chip returned to the box, and the chips mixed. Thus, any one of the digits is equally likely to occur on any of the selections.)

To see how a table of random numbers can be used to carry out a simulation, let's reconsider the quiz example. A random digit is used to represent the guess on a single question, with an even digit representing a correct response. A sequence of 20 digits could represent the answers to the 20 quiz questions. To use the random digit table, start by picking an arbitrary starting point in Appendix A Table 1. Suppose that you start at Row 10 and take the 20 digits in a row to represent one quiz. The first five "quizzes" and the corresponding number correct are

Quiz	Random Digits	Number Correct
1 (Table 1 Row 10)	9 4 6 0 6 9 7 8 8 2 5 2 9 6 0 1 4 6 0 5	13
2 (Table 1 Row 11)	6 6 9 5 7 4 4 6 3 2 0 6 0 8 9 1 3 6 1 8	12
3 (Table 1 Row 12)	0 7 1 7 7 7 2 9 7 8 7 5 8 8 6 9 8 4 1 0	9
4 (Table 1 Row 13)	6 1 3 0 9 7 3 3 6 6 0 4 1 8 3 2 6 7 6 8	11
5 (Table 1 Row 14)	2 2 3 6 2 1 3 0 2 2 6 6 9 7 0 2 1 2 5 8	13

denotes correct response

This process can be repeated to generate a large number of observations, which could be used to construct a table of estimated probabilities.

The simulation method for generating observations must preserve the important characteristics of the actual process being considered. For example, consider a multiple-choice quiz in which each of the 20 questions on the quiz has 5 possible responses, only 1 of which is correct. For any particular question, you would expect a student to be able to guess the correct answer only one-fifth of the time in the long run. To simulate this situation, you could select at random from a box that contained four times as many red balls as blue balls. If you are using random digits for the simulation, you could use 0 and 1 to represent a correct response and 2, 3, …, 9 to represent an incorrect response.

Using Simulation to Approximate a Probability

1. Propose a method that uses a random mechanism (such as a random number generator or table, the selection of a ball from a box, or the toss of a coin) to represent an observation. Be sure that the important characteristics of the actual process are preserved.
2. Generate an observation using the method from Step 1, and determine whether the event of interest has occurred.
3. Repeat Step 2 a large number of times.
4. Calculate the estimated probability by dividing the number of observations for which the event of interest occurred by the total number of observations generated.

The simulation process is illustrated in Examples 5.27–5.29.

Example 5.27   Building Permits

Many California cities limit the number of building permits that are issued each year. Because of limited water resources, one such city plans to issue permits for only 10 dwelling units in the upcoming year. The city will decide who is to receive permits by holding a lottery. Suppose that you are 1 of 39 individuals applying for a permit. Thirty people are requesting a permit for a single-family home, 8 are requesting a permit for a duplex (which counts as two dwelling units), and 1 person is requesting a permit for a small apartment building with eight units (which counts as eight dwelling units). Each request will be entered into the lottery. Requests will be selected at random one at a time, and if there are enough permits remaining, the request will be granted. This process will continue until all 10 permits have been issued. If your request is for a single-family home, what are your chances of receiving a permit? You can use simulation to estimate this probability. (It is not easy to determine analytically.)

To carry out the simulation, you can view the requests as being numbered from 1 to 39 as follows:

01–30: Requests for single-family homes
31–38: Requests for duplexes
39: Request for eight-unit apartment

For ease of discussion, let's assume that your request is Request 01.
One method for simulating the permit lottery consists of these three steps:

1. Choose a random number between 01 and 39 to indicate which permit request is selected first, and grant this request.
2. Select a random number between 01 and 39 to indicate which permit request is considered next. (If a number that has already been selected is chosen, ignore it and select again.) Determine the number of dwelling units for the selected request. Grant the request only if there are enough permits remaining to satisfy the request.
3. Repeat Step 2 until permits for 10 dwelling units have been granted.

Minitab was used to generate random numbers between 01 and 39 to imitate the lottery drawing. (The random number table in Appendix A Table 1 could also be used, by selecting 2 digits and ignoring 00 and any value over 39.) The first sequence generated by Minitab is:

Random Number	Type of Request	Total Number of Units So Far
25	Single-family home	1
07	Single-family home	2
38	Duplex	4
31	Duplex	6
26	Single-family home	7
12	Single-family home	8
33	Duplex	10

You would stop at this point, because permits for 10 units would have been issued. In this simulated lottery, Request 01 was not selected, so you would not have received a permit.

The next simulated lottery (using Minitab to generate the selections) is as follows:

Random Number	Type of Request	Total Number of Units So Far
38	Duplex	2
16	Single-family home	3
30	Single-family home	4
39	Apartment, not granted because there are not 8 permits remaining	4
14	Single-family home	5
26	Single-family home	6
36	Duplex	8
13	Single-family home	9
15	Single-family home	10

Again, Request 01 was not selected, so you would not have received a permit in this simulated lottery either.

Now that a strategy for simulating a lottery has been devised, the tedious part of the simulation begins. You now have to simulate a large number of lottery drawings, determining for each whether Request 01 is granted. Five hundred such drawings were simulated, and Request 01 was selected in 85 of the lotteries. Based on this simulation,

$$\text{estimated probability of receiving a building permit} = \frac{85}{500} = 0.17$$

Example 5.28 One-Boy Family Planning

Suppose that couples who wanted children were to continue having children until a boy is born. Assuming that each newborn child is equally likely to be a boy or a girl, would this behavior change the proportion of boys in the population? This question was posed in an article that appeared in *The American Statistician* **(1994: 290–293)**, and many people answered the question incorrectly. Simulation will be used to estimate the long-run proportion of boys in the population. This proportion is an estimate of the probability that a randomly selected child from this population is a boy. Note that every sibling group would have exactly one boy.

You can use a single-digit random number to represent a child. The odd digits (1, 3, 5, 7, 9) could represent a male birth, and the even digits could represent a female birth. An observation is constructed by selecting a sequence of random digits. If the first random number obtained is odd (a boy), the observation is complete. If the first selected number is even (a girl), another digit is chosen. You would continue in this way until an odd digit is obtained. For example, reading across Row 15 of the random number table Appendix Table 1, the first 10 digits are

$$0 \quad 7 \quad 1 \quad 7 \quad 4 \quad 2 \quad 0 \quad 0 \quad 0 \quad 1$$

Using these numbers to simulate sibling groups, results in the following:

Sibling group 1	0 7	girl, boy
Sibling group 2	1	boy
Sibling group 3	7	boy
Sibling group 4	4 2 0 0 0 1	girl, girl, girl, girl, girl, boy

Continuing along Row 15 of the random number table,

Sibling group 5	3	boy
Sibling group 6	1	boy

Sibling group 7   2 0 4 7   girl, girl, girl, boy

Sibling group 8   8 4 1   girl, girl, boy

After simulating eight sibling groups, there were 8 boys among 19 children. The proportion of boys is $\frac{8}{19}$, which is close to 0.5. Continuing the simulation to obtain a large number of observations leads to the conclusion that the long-run proportion of boys in the population would still be 0.5.

Example 5.29   ESP?

Can a close friend read your mind? Try the following experiment. Write the word *blue* on one piece of paper and the word *red* on another, and place the two slips of paper in a box. Select one slip of paper from the box, look at the word written on it, then try to convey the word by sending a mental message to a friend who is seated in the same room. Ask your friend to select either red or blue, and record whether the response is correct. Repeat this 10 times and count the number of correct responses. How did your friend do? Is your friend receiving your mental messages or just guessing?

Let's investigate by using simulation to get the approximate probabilities for someone who *is* guessing. You can use a random digit to represent a response, with an even digit representing a correct response (C) and an odd digit representing an incorrect response (X). A sequence of 10 digits can be used to generate one observation. For example, using the last 10 digits in Row 25 of the random number table (Appendix A Table 1) gives

5	2	8	3	4	3	0	7	3	5
X	C	C	X	C	X	C	X	X	X

which results in four correct responses. Minitab was used to generate 150 sequences of 10 random digits and the following results were obtained:

Sequence Number	Digits	Number Correct
1	3996285890	5
2	1690555784	4
3	9133190550	2
⋮	⋮	⋮
149	3083994450	5
150	9202078546	7

Table 5.4 summarizes the results of the simulation. The estimated probabilities in Table 5.4 are based on the assumption that a correct and an incorrect response are equally likely. Evaluate your friend's performance in light of this information. Is it likely that someone guessing would have been able to get as many correct as your friend did? Do you think your friend was receiving your mental messages? How are the estimated probabilities in Table 5.4 used to support your answer?

TABLE 5.4 Estimated Probabilities for Example 5.29

Number Correct	Number of Sequences	Estimated Probability
0	0	0.000
1	1	0.007
2	8	0.053
3	16	0.107
4	30	0.200
5	36	0.240

(continued)

Number Correct	Number of Sequences	Estimated Probability
6	35	0.233
7	17	0.113
8	7	0.047
9	0	0.000
10	0	0.000
Total	**150**	**1.000**

Summing It Up—Section 5.7

The following learning objectives were addressed in this section:

Conceptual Understanding
C5: (Optional) Understand how probabilities can be estimated using simulation.
Simulation provides a way to estimate probabilities when it is difficult to calculate them analytically or to estimate them by making a long sequence of actual observations. In a simulation to estimate a probability, you generate "observations" in a situation that is similar to the real situation of interest.

Mastering the Mechanics
M12: (Optional) Carry out a simulation to estimate a probability.
The steps for carrying out a simulation to estimate a probability are described in the box just prior to Example 5.27. For examples of how these steps are implemented in several different settings, see Examples 5.27, 5.28, and 5.29.

SECTION 5.7 EXERCISES

Each Exercise Set assesses the following chapter learning objectives: C5, M12

SECTION 5.7 Exercise Set 1

5.72 The report **"Airline Quality Rating 2016" (www.airlinequalityrating.com/reports/2016_AQR_Final.pdf, retrieved April 25, 2017)** provided an overview of the complaints about airlines received by the U.S. Department of Transportation. The table below gives the number of complaints received by type of complaint for the years 2014 and 2015.

Type of Complaint	2014 Number of Complaints Received	2015 Number of Complaints Received
Flight Problems	4,304	5,506
Baggage Handling	1,628	2,050
Reservations, Ticketing, Boarding	1,276	1,807
Customer Service	1,201	1,728
Fares	699	1,300
Other	2,257	2,869

Use the given information to estimate the following probabilities:
a. The probability that a complaint made in 2014 was about baggage handling.
b. The probability that a complaint made in 2015 was not about flight problems.
c. The probability that two independent complaints made in 2015 were both about flight problems.
d. The probability that a complaint made in 2014 was either about flight problems or customer service.

5.73 Many cities regulate the number of taxi licenses, and there is a great deal of competition for both new and existing licenses. Suppose that a city has decided to sell 10 new licenses for $25,000 each. A lottery will be held to determine who gets the licenses, and no one may request more than 3 licenses. Twenty individuals and taxi companies have entered the lottery. Six of the 20 entries are requests for 3 licenses, nine are requests for 2 licenses, and the rest are requests for a single license. The city will select requests at random, filling as much of the request as possible. For example, if there were only one license left, any request selected would only receive this single license.
a. An individual who wishes to be an independent driver has put in a request for a single license. Use simulation to approximate the probability that the request will be granted. Perform at least 20 simulated lotteries (more is better!).
b. Do you think that this is a fair way of distributing licenses? Can you propose an alternative procedure for distribution?

5.74 Five hundred first-year students at a state university were classified according to both high school grade point average (GPA) and whether they were on academic probation at the end of their first semester. The data are summarized in the accompanying table.

	High School GPA			
Probation	2.5 to <3.0	3.0 to <3.5	3.5 and Above	Total
Yes	50	55	30	**135**
No	45	135	185	**365**
Total	**95**	**190**	**215**	**500**

a. Use the given table to approximate the probability that a randomly selected first-year student at this university will be on academic probation at the end of the first semester.

b. What is the estimated probability that a randomly selected first-year student at this university had a high school GPA of 3.5 or above?

c. Are the two events *selected student has a high school GPA of 3.5 or above* and *selected student is on academic probation at the end of the first semester* independent events? How can you tell?

d. Estimate the proportion of first-year students with high school GPAs between 2.5 and 3.0 who are on academic probation at the end of the first semester.

e. Estimate the proportion of those first-year students with high school GPAs of 3.5 and above who are on academic probation at the end of the first semester.

5.75 Four students must work together on a group project. They decide that each will take responsibility for a particular part of the project, as follows:

Person	Maria	Alex	Juan	Jacob
Task	Survey design	Data collection	Analysis	Report writing

Because of the way the tasks have been divided, one student must finish before the next student can begin work. To ensure that the project is completed on time, a time line is established, with a deadline for each team member. If any one of the team members is late, the timely completion of the project is jeopardized. Assume the following probabilities:

1. The probability that Maria completes her part on time is 0.8.

2. If Maria completes her part on time, the probability that Alex completes on time is 0.9, but if Maria is late, the probability that Alex completes on time is only 0.6.

3. If Alex completes his part on time, the probability that Juan completes on time is 0.8, but if Alex is late, the probability that Juan completes on time is only 0.5.

4. If Juan completes his part on time, the probability that Jacob completes on time is 0.9, but if Juan is late, the probability that Jacob completes on time is only 0.7.

Use simulation (with at least 20 trials) to estimate the probability that the project is completed on time. Think carefully about this one. For example, you might use a random digit to represent each part of the project (four in all). For the first digit (Maria's part), 1–8 could represent *on time*, and 9 and 0 could represent *late*. Depending on what happened with Maria (late or on time), you would then look at the digit representing Alex's part. If Maria was on time, 1–9 would represent *on time* for Alex, but if Maria was late, only 1–6 would represent *on time*. The parts for Juan and Jacob could be handled similarly.

ADDITIONAL EXERCISES

5.76 In Exercise 5.75, the probability that Maria completes her part on time was 0.8. Suppose that this probability is really only 0.6. Use simulation (with at least 20 trials) to estimate the probability that the project is completed on time.

5.77 Suppose that the probabilities of timely completion are as in Exercise 5.75 for Maria, Alex, and Juan, but that Jacob has a probability of completing on time of 0.7 if Juan is on time and 0.5 if Juan is late.

a. Use simulation (with at least 20 trials) to estimate the probability that the project is completed on time.

b. Compare the probability from Part (a) to the ones calculated in Exercises 5.75 and 5.76. Which decrease in the probability of on-time completion (Maria's or Jacob's) resulted in the bigger change in the probability that the project is completed on time?

5.78 A medical research team wishes to evaluate two different treatments for a disease. Subjects are selected two at a time, and one is assigned to Treatment 1 and the other to Treatment 2. The treatments are applied, and each is either a success (S) or a failure (F). The researchers keep track of the total number of successes for each treatment. They plan to continue the experiment until the number of successes for one treatment exceeds the number of successes for the other by 2. For example, based on the results in the accompanying table, the experiment would stop after the sixth pair, because Treatment 1 has two more successes than Treatment 2. The researchers would conclude that Treatment 1 is preferable to Treatment 2.

Pair	Treatment 1	Treatment 2	Cumulative Number of Successes for Treatment 1	Cumulative Number of Successes for Treatment 2
1	S	F	1	0
2	S	S	2	1
3	F	F	2	1
4	S	S	3	2
5	F	F	3	2
6	S	F	4	2

Suppose that Treatment 1 has a success rate of 0.7 and Treatment 2 has a success rate of 0.4. Use simulation to estimate the probabilities requested in Parts (a) and (b). (Hint: Use a pair of random digits to simulate one pair of subjects. Let the first digit represent Treatment 1 and use 1–7 as an indication of a success and 8, 9, and 0 to indicate a failure. Let the second digit represent Treatment 2, with 1–4 representing a success. For example, if the two digits selected to represent a pair were 8 and 3, you would record failure for Treatment 1 and success for Treatment 2. Continue to select pairs, keeping track of the cumulative number of successes for each treatment. Stop the trial as soon as the number of successes for one treatment exceeds that for the other by 2. This would complete one trial. Now repeat this whole process until you have results for at least 20 trials [more is better]. Finally, use the simulation results to estimate the desired probabilities.)

a. Estimate the probability that more than five pairs must be treated before a conclusion can be reached. (Hint: $P(\text{more than } 5) = 1 - P(5 \text{ or fewer}).$)

b. Estimate the probability that the researchers will incorrectly conclude that Treatment 2 is the better treatment.

5.79 A single-elimination tournament with four players is to be held. A total of three games will be played. In Game 1, the players seeded (rated) first and fourth play. In Game 2, the players seeded second and third play. In Game 3, the winners of Games 1 and 2 play, with the winner of Game 3 declared the tournament winner. Suppose that the following probabilities are known:

$$P(\text{Seed 1 defeats Seed 4}) = 0.8$$
$$P(\text{Seed 1 defeats Seed 2}) = 0.6$$
$$P(\text{Seed 1 defeats Seed 3}) = 0.7$$
$$P(\text{Seed 2 defeats Seed 3}) = 0.6$$
$$P(\text{Seed 2 defeats Seed 4}) = 0.7$$
$$P(\text{Seed 3 defeats Seed 4}) = 0.6$$

a. How would you use random digits to simulate Game 1 of this tournament?

b. How would you use random digits to simulate Game 2 of this tournament?

c. How would you use random digits to simulate the third game in the tournament? (This will depend on the outcomes of Games 1 and 2.)

d. Simulate one complete tournament, giving an explanation for each step in the process.

e. Simulate 10 tournaments, and use the resulting information to estimate the probability that the first seed wins the tournament.

f. Ask four classmates for their simulation results. Along with your own results, this should give you information on 50 simulated tournaments. Use this information to estimate the probability that the first seed wins the tournament.

g. Why do the estimated probabilities from Parts (e) and (f) differ? Which do you think is a better estimate of the actual probability? Explain.

CHAPTER ACTIVITIES

ACTIVITY 5.1 KISSES

Background: The paper **"What Is the Probability of a Kiss? (It's Not What You Think)"** (found in the online *Journal of Statistics Education* [2002]) posed the following question: What is the probability that a Hershey's Kiss candy will land on its base (as opposed to its side) if it is flipped onto a table? Unlike flipping a coin, there is no reason to believe that this probability is 0.5.

Working as a class, develop a plan that would enable you to estimate this probability. Once you have an acceptable plan, use it to produce an estimate of the desired probability.

ACTIVITY 5.2 A CRISIS FOR EUROPEAN SPORTS FANS?

Background: The *New Scientist* (January 4, 2002) reported on a controversy surrounding the Euro coins that have been introduced as a common currency across most of Europe. Each country mints its own coins, but these coins are accepted in any of the countries that have adopted the Euro as its currency.

A group in Poland claims that the Belgium-minted Euro does not have an equal chance of landing heads or tails. This claim was based on 250 tosses of the Belgium Euro, of which 140 (56%) came up heads. Should this be cause for alarm for European sports fans, who know that "important" decisions are made by the flip of a coin?

In this activity, you will investigate whether this difference should be cause for alarm by examining whether observing 140 heads out of 250 tosses is an unusual outcome if the coin is fair.

1. For this first step, you can either (a) flip a U.S. penny 250 times, keeping a tally of the number of heads and tails observed (this won't take as long as you think) or (b) simulate 250 coin tosses by using a random number table, your calculator, or a statistics software package to generate random numbers (if you choose this option, give a brief description of how you carried out the simulation).

2. For your sequence of 250 tosses, calculate the proportion of heads observed.

3. Form a data set that consists of the values for the proportion of heads observed in 250 tosses of a fair coin for the entire class. Summarize this data set by constructing a graphical display.

4. Working with a partner, write a paragraph explaining why European sports fans should or should not be worried by the results of the Polish experiment. Your explanation should be based on the observed proportion of heads from the Polish experiment simulation results.

ACTIVITY 5.3 THE "HOT HAND" IN BASKETBALL

Background: Consider a mediocre basketball player who has consistently made only 50% of his free throws over several seasons. If you were to examine his free throw record over the last 50 free throw attempts, is it likely that you would see a "streak" of 5 in a row where he is successful in making the free throw? In this activity, you will investigate this question. Assume that the outcomes of successive free throw attempts are independent and that the probability that the player is successful on any particular attempt is 0.5.

1. Begin by simulating a sequence of 50 free throws for this player. Because this player has a probability of success of 0.5 for each attempt and the attempts are independent, you can model a free throw by tossing a coin. Using heads to represent a successful free throw and tails to represent a missed free throw, simulate 50 free throws by tossing a coin 50 times, recording the outcome of each toss.

2. For your sequence of 50 tosses, identify the longest streak by looking for the longest string of heads in your sequence. Determine the length of this streak.

3. Combine your longest streak value with those from the rest of the class, and construct a histogram or dotplot of all the longest streak values.

4. Based on the graph from Step 3, does it appear likely that a player of this skill level would have a streak of 5 or more successes at some point during a sequence of 50 free throw attempts? Justify your answer based on the graph from Step 3.

5. Use the combined class data to estimate the probability that a player of this skill level has a streak of at least 5 somewhere in a sequence of 50 free throw attempts.

6. Using the multiplication rule for independent events, the probability of a player of this skill level being successful on the *next* 5 free throw attempts is

$$P(SSSSS) = \left(\frac{1}{2}\right)\left(\frac{1}{2}\right)\left(\frac{1}{2}\right)\left(\frac{1}{2}\right)\left(\frac{1}{2}\right) = \left(\frac{1}{2}\right)^5 = 0.031$$

which is relatively small. At first, this value might seem inconsistent with your answer in Step 5, but the estimated probability from Step 5 and the calculated probability of 0.031 are really considering different situations. Explain how these situations are different.

7. Do you think that the assumption that the outcomes of successive free throws are independent is reasonable? Explain. (This is a hotly debated topic among both sports fans and statisticians!)

ARE YOU READY TO MOVE ON? CHAPTER 5 REVIEW EXERCISES

All chapter learning objectives are assessed in these exercises. The learning objectives assessed in each exercise are given in parentheses.

5.80 (C1)
The article "A Crash Course in Probability" from **The Economist** (www.economist.com/blogs/gulliver/2015/01/air-safety, January 29, 2015, retrieved April 25, 2017) included the following information: The chance of being involved in an airplane crash when flying on an Airbus 330 from London to New York City on Virgin Atlantic Airlines is 1 in 5,371,369. This was interpreted as meaning that you "would expect to go down if you took this flight every day for 14,716 years." The

article also states that a person could "expect to fly on the route for 14,716 years before plummeting into the Atlantic."

Comment on why these statements are misleading.

5.81 (M1)
A company that offers roadside assistance to drivers reports that the probability that a call for assistance will be to help someone who is locked out of his or her car is 0.18. Give a relative frequency interpretation of this probability.

5.82 (P1)

Eighty-six countries won medals at the 2016 Olympics in Rio de Janeiro. Based on results posted on **www.bbc.com /sport/olympics/rio-2016/medals/countries (retrieved April 25, 2017)**,

> 1 country won more than 100 medals
> 2 countries won between 51 and 100 medals
> 3 countries won between 31 and 50 medals
> 4 countries won between 21 and 30 medals
> 15 countries won between 11 and 20 medals
> 15 countries won between 6 and 10 medals
> 46 countries won between 1 and 5 medals

Suppose one of the 86 countries winning medals at the 2016 Olympics is selected at random.

a. What is the probability that the selected country won more than 50 medals?
b. What is the probability that the selected country did not win more than 100 medals?
c. What is the probability that the selected country won 10 or fewer medals?
d. What is the probability that the selected country won between 11 and 50 medals?

5.83 (M2)

The student council for a school of science and math has one representative from each of five academic departments: Biology (B), Chemistry (C), Mathematics (M), Physics (P), and Statistics (S). Two of these students are to be randomly selected for inclusion on a university-wide student committee.

a. What are the 10 possible outcomes?
b. From the description of the selection process, all outcomes are equally likely. What is the probability of each outcome?
c. What is the probability that one of the committee members is the statistics department representative?
d. What is the probability that both committee members come from laboratory science departments?

5.84 (M1, P3)

Consider the following two lottery-type games:

> Game 1: You pick one number between 1 and 50. After you have made your choice, a number between 1 and 50 is selected at random. If the selected number matches the number you picked, you win.
> Game 2: You pick two numbers between 1 and 10. After you have made your choices, two different numbers between 1 and 10 are selected at random. If the selected numbers match the two you picked, you win.

a. The cost to play either game is $1, and if you win you will be paid $20. If you can only play one of these games, which game would you pick and why? Use relevant probabilities to justify your choice.
b. For either of these games, if you plan to play the game 100 times, would you expect to win money or lose money overall? Explain.

5.85 (C2)

Consider a chance experiment that consists of selecting a customer at random from all people who purchased a car at a large car dealership during 2016.

a. In the context of this chance experiment, give an example of two events that would be mutually exclusive.
b. In the context of this chance experiment, give an example of two events that would not be mutually exclusive.

5.86 (C3)

The article **"Obesity, Smoking Damage U.S. Economy,"** which appeared in the **Gallup online Business Journal (www.gallup. com, September 7, 2016, retrieved Arpil 25, 2017)**, reported that based on a large representative sample of adult Americans, 52.7% claimed that they exercised at least 30 minutes on three or more days per week during 2015. It also reported that the percentage for millennials (people age 19–35) was 57.1%, and for those over 35 it was 51.1%. If an adult American were to be selected at random, are the events *selected adult exercises at least 30 minutes three times per week* and *selected adult is a millennial* independent or dependent events? Justify your answer using the given information.

5.87 (M3)

The following table summarizes data on smoking status and age group, and is consistent with summary quantities obtained in a Gallup Poll published in the online article **"In U.S., Young Adults' Cigarette Use Is Down Sharply" (www .gallup.com, December 10, 2015, retrieved April 25, 2017)**.

Age Group	Smoking Status	
	Smoker	Nonsmoker
18 to 29	174	618
30 to 49	333	1,115
50 to 64	384	1,445
65 and older	211	1,707

Assume that it is reasonable to consider these data as representative of the American adult population. Consider the chance experiment or randomly selecting an adult American.

a. What is the probability that the selected adult is a smoker?
b. What is the probability that the selected adult is under 50 years of age?
c. What is the probability that the selected adult is a smoker that is 65 or older?
d. What is the probability that the selected adult is a smoker or is age 65 or older?

5.88 (M4)

A study of the impact of seeking a second opinion about a medical condition is described in the paper **"Evaluation of Outcomes from a National Patient-Initiated Second-Opinion Program" (The American Journal of Medicine [2015]: 1138e (25–1138e33)**. Based on a review of 6791 patient-initiated

second opinions, the paper states the following: "Second opinions often resulted in changes in diagnosis (14.8%), treatment (37.4%), or changes in both (10.6%)."
Consider the following two events:

D = event that second opinion results in a change in diagnosis

T = event that second opinion results in a change in treatment

a. What are the values of $P(D)$, $P(T)$, and $P(D \cap T)$?

b. Use the given probability information to set up a hypothetical 1000 table with columns corresponding to D and D^C and rows corresponding to T and T^C.

c. What is the probability that a second opinion results in neither a change in diagnosis nor a change in treatment?

d. What is the probability that a second opinion results is a change in diagnosis or a change in treatment?

5.89 (M5)
A large cable company reports that 42% of its customers subscribe to its Internet service, 32% subscribe to its phone service, and 51% subscribe to its Internet service or its phone service (or both).

a. Use the given probability information to set up a hypothetical 1000 table.

b. Use the table to find the following:
  i. the probability that a randomly selected customer subscribes to both the Internet service and the phone service.
  ii. the probability that a randomly selected customer subscribes to exactly one of the two services.

5.90 (M6)
a. Suppose events E and F are mutually exclusive with $P(E) = 0.64$ and $P(F) = 0.17$.
  i. What is the value of $P(E \cap F)$?
  ii. What is the value of $P(E \cup F)$?

b. Suppose that A and B are events with $P(A) = 0.3, P(B) = 0.5$, and $P(A \cap B) = 0.15$. Are A and B mutually exclusive? How can you tell?

c. Suppose that A and B are events with $P(A) = 0.65$ and $P(B) = 0.57$. Are A and B mutually exclusive? How can you tell?

5.91 (M7)
In a small city, approximately 15% of those eligible are called for jury duty in any one calendar year. People are selected for jury duty at random from those eligible, and the same individual cannot be called more than once in the same year. What is the probability that an eligible person in this city is selected in both of the next 2 years? All of the next 3 years?

5.92 (M8)
An online store offers two methods of shipping—regular ground service and an expedited 2-day shipping. Customers

may also choose whether or not to have the purchase gift wrapped. Suppose that the events

E = event that the customer chooses expedited shipping

G = event that the customer chooses gift wrap

are independent with $P(E) = 0.26$ and $P(G) = 0.12$.

a. Construct a hypothetical 1000 table with columns corresponding to whether or not expedited shipping was chosen and rows corresponding to whether or not gift wrap was selected.

b. Use the information in the table to calculate $P(E \cup G)$. Give a long-run relative frequency interpretation of this probability.

5.93 (M9)
The National Highway Traffic Safety Administration requires each U.S. state to carry out an observational study to assess the level of seat belt use in the state. The report **"2015 Utah Seat Belt Use Survey" (Utah Department of Public Safety, September 14, 2015)** summarized data from the study done in Utah. The proportions in the accompanying table are based on observations of over 25,000 drivers and passengers.

	Uses Seatbelt	Does Not Use Seat Belt
Male	0.423	0.077
Female	0.452	0.048

Assume that these proportions are representative of drivers and passengers in Utah and that an adult from Utah is selected at random.

a. What is the probability that the selected adult uses a seat belt?

b. What is the probability that the selected adult uses a seat belt, given that the individual selected is male?

c. What is the probability that the selected adult does not use a seat belt, given that the selected individual is female?

d. What is the probability that the selected individual is female, given that the selected individual does not use a seat belt?

e. Are the probabilities from Parts (c) and (d) equal? Write a couple of sentences explaining why or why not.

5.94 (C4, M1, M10)
The report **"Twitter in Higher Education: Usage Habits and Trends of Today's College Faculty" (Magna Publications, September 2009)** describes a survey of nearly 2000 college faculty. The report indicates the following:

30.7% reported that they use Twitter, and 69.3% said that they do not use Twitter.

Of those who use Twitter, 39.9% said they sometimes use Twitter to communicate with students.

Of those who use Twitter, 27.5% said that they sometimes use Twitter as a learning tool in the classroom.

Consider the chance experiment that selects one of the study participants at random.

a. Two of the percentages given in the problem specify unconditional probabilities, and the other two percentages specify conditional probabilities. Which are conditional probabilities? How can you tell?

b. Suppose the following events are defined:

T = event that selected faculty member uses Twitter

C = event that selected faculty member sometimes uses Twitter to communicate with students

L = event that selected faculty member sometimes uses Twitter as a learning tool in the classroom

Use the given information to determine the following probabilities:

  i. $P(T)$ **iii.** $P(C|T)$
  ii. $P(T^C)$ **iv.** $P(L|T)$

c. Construct a hypothetical 1000 table using the given probabilities and use the information in the table to calculate $P(C)$, the probability that the selected study participant sometimes uses Twitter to communicate with students.

d. Construct a hypothetical 1000 table using the given probabilities and use the information in the table to calculate the probability that the selected study participant sometimes uses Twitter as a learning tool in the classroom.

5.95   **(M10, P2)**
In an article that appears on the website of the **American Statistical Association (www.amstat.org/meetings/jsm/2000 /usei/gunn.pdf, retrieved April 25, 2017)**, Carlton Gunn, a public defender in Seattle, Washington, wrote about how he uses statistics in his work as an attorney. He states:

> I personally have used statistics in trying to challenge the reliability of drug testing results. Suppose the chance of a mistake in the taking and processing of a urine sample for a drug test is just 1 in 100. And your client has a "dirty" (i.e., positive) test result. Only a 1 in 100 chance that it could be wrong? Not necessarily. If the vast majority of all tests given—say 99 in 100—are truly clean, then you get one false dirty and one true dirty in every 100 tests, so that half of the dirty tests are false.

Define the following events as

TD = event that the test result is dirty

TC = event that the test result is clean

D = event that the person tested is actually dirty

C = event that the person tested is actually clean

a. Using the information in the quote, what are the values of
  i. $P(TD | D)$ **iii.** $P(C)$
  ii. $P(TD | C)$ **iv.** $P(D)$

b. Use the probabilities from Part (a) to construct a hypothetical 1000 table.

c. What is the value of $P(TD)$?

d. Use the information in the table to calculate the probability that a person is clean given that the test result is dirty, $P(C|TD)$. Is this value consistent with the argument given in the quote? Explain.

5.96   **(M10, M11, P2)**
Are people more confident in their answers when the answer is actually correct than when it is not? The article **"Female Students Less Confident, More Accurate Than Male Counterparts" (American Academy of Family Physicians News, March 5, 2015)** described a study that measured medical students' confidence and the accuracy of their responses. Participants categorized their confidence levels using either "sure," "feeling lucky," or "no clue."
Define the following events:

C = event that a response is correct
S = event that confidence level is "sure"
L = event that confidence level is "feeling lucky"
N = event that confidence level is "no clue"

a. Data from the article were used to estimate the following probabilities for males:

  $P(S) = 0.442$ $P(L) = 0.422$ $P(N) = 0.136$
$P(C|S) = 0.783$ $P(C|L) = 0.498$ $P(C|N) = 0.320$

Use the given probabilities to construct a hypothetical 1000 table with rows corresponding to confidence level and columns corresponding to whether the response was correct or not.

b. Calculate the probability that a male student's confidence level is "sure" given that the response is correct.

c. Calculate the probability that a male student's confidence level is "no clue" given that the response is incorrect.

d. Calculate the probability that a male student's response is correct.

e. Data from the article were also used to estimate the following probabilities for females:

  $P(S) = 0.395$ $P(L) = 0.444$ $P(N) = 0.161$
$P(C|S) = 0.805$ $P(C|L) = 0.535$ $P(C|N) = 0.320$

Use the given probabilities to construct a hypothetical 1000 table with rows corresponding to confidence level and columns corresponding to whether the response was correct or not.

f. Calculate the probability that a female student's confidence level is "sure" given that the response is correct.

g. Calculate the probability that a female student's confidence level is "no clue" given that the response is incorrect.

h. Calculate the probability that a female student's response is correct.

i. Do the given probabilities and the probabilities that you calculated support the statement in the title of the article? Explain.

5.97   **(P3, P4)**
When treating patients in the emergency room, it is important to be able to make a quick decision about whether a

female patient is pregnant or not. The usual laboratory test for pregnancy uses a urine sample. But in an emergency room situation, it may be easier to obtain a whole blood sample. The authors of the paper **"Substituting Whole Blood for Urine in a Bedside Pregnancy Test"** (*Clinical Laboratory in Emergency Medicine* [2012]: 478–482) carried out an evaluation of a pregnancy test based on a whole blood sample. Data from this study are summarized in the accompanying table.

	Test Positive	Test Negative
Pregnant	202	9
Not Pregnant	0	416

If this whole blood test is used to predict whether or not a woman is pregnant, can you be equally confident in a positive test result and a negative test result? Justify your answer using relevant probabilities. (Hint: Consider the error probabilities—the probability that a woman is not pregnant given a positive test result and the probability that a woman is pregnant given a negative test result.)

5.98 (M11)

To help ensure the safety of school classrooms, the local fire marshal does an inspection at Thomas Jefferson High School each month to check for faulty wiring, overloaded circuits, and other fire code violations. Each month, one room is selected for inspection. Suppose that the probability that the selected room is a science classroom (biology, chemistry, or physics) is 0.6 and the probability that the selected room is a chemistry room is 0.4. Use probability formulas to find the following probabilities.

a. The probability that the selected room is not a science room.

b. The probability that the selected room is a chemistry room and a science room.

c. The probability that the selected room is a chemistry room given that the room selected was a science room.

d. The probability that the selected room was a chemistry room or a science room.

5.99 (M11)

Three of the most common types of pets are cats, dogs, and fish. Many families have more than one type of pet. Suppose that a family is selected at random and consider the following events and probabilities:

F = event that the selected family has at least one fish

D = event that the selected family has at least one dog

C = event that the selected family has at least one cat

$$P(F) = 0.20 \qquad P(D) = 0.32 \qquad P(C) = 0.35$$
$$P(F \cap D) = 0.18 \quad P(F \cap C) = 0.07 \quad P(D \mid C) = 0.30$$

Calculate the following probabilities.

a. $P(F \mid D)$

b. $P(F \cup D)$

c. $P(C \cap D)$

6 Random Variables and Probability Distributions

Kathy deWitt/Alamy Stock Photo

PREVIEW

*One way to learn from data is to use information from a sample to learn about a population distribution. In this situation, you are usually interested in the distribution of one or more variables. For example, an environmental scientist who obtains an air sample from a specified location might be interested in the concentration of ozone. Before selection of the air sample, the value of the ozone concentration is uncertain. Because the value of a variable quantity such as ozone concentration is subject to uncertainty, such variables are called **random variables**. In this Chapter, you will learn how probability models are used to describe the behavior of random variables.*

CHAPTER LEARNING OBJECTIVES

Conceptual Understanding

After completing this chapter, you should be able to

C1 Understand that a probability distribution describes the long-run behavior of a random variable.

C2 Understand how the long-run behavior of a random variable is described by its mean and standard deviation.

C3 Understand that areas under a density curve for a continuous random variable are interpreted as probabilities.

Mastering the Mechanics

After completing this chapter, you should be able to

M1 Distinguish between discrete and continuous random variables.

M2 Given a probability distribution for a discrete random variable, calculate and interpret probabilities.

M3 Construct the probability distribution of a discrete random variable.

M4 Calculate and interpret the mean and standard deviation of a discrete random variable.

M5 Given the probability density curve for a continuous random variable, identify areas corresponding to various probabilities.

M6 Calculate probabilities for continuous random variables whose density curves have a simple form.

M7 Interpret an area under a normal curve as a probability.

M8 For a normal random variable x, use technology or tables to calculate probabilities of the form $P(a < x < b)$, $P(x < b)$, and $P(x > a)$ and to find percentiles.

M9 Use technology or given normal scores to construct a normal probability plot.

M10 Given a normal probability plot, assess whether it is reasonable to think that a population distribution is approximately normal.

M11 **(Optional)** Distinguish between binomial and geometric random variables.

M12 **(Optional)** Calculate binomial probabilities using technology or tables.

M13 **(Optional)** Calculate probabilities using the geometric distribution.

M14 **(Optional)** Given a binomial distribution, determine if it is appropriate to use a normal approximation to estimate binomial probabilities. If appropriate, approximate binomial probabilities using a normal distribution.

Putting It into Practice

After completing this chapter, you should be able to

P1 Use information provided by a probability distribution to draw conclusions in context.

PREVIEW EXAMPLE

iPod Shuffles

The paper **"Does Your iPod *Really* Play Favorites?" (*The American Statistician* [2009]: 263–268)** investigated the shuffle feature of the iPod. The shuffle feature takes a group of songs, called a playlist, and plays them in a random order. Some users have questioned the "randomness" of the shuffle, citing situations where several songs from the same artist were played in close proximity to each other. One such example appeared in a *Newsweek* article in 2005, where the author states that Steely Dan songs always seem to pop up in the first hour of play. (For readers unfamiliar with Steely Dan, see www.steelydan.com.) Is this consistent with a "random" shuffle? You will return to this example in Section 6.2 to answer this question. ∎

SECTION 6.1   Random Variables

In most chance experiments, an investigator focuses attention on one or more variable quantities. For example, consider the chance experiment of randomly selecting a customer who is leaving a store. One numerical variable of interest to the store manager might be the number of items purchased by the customer. The letter x can be used to denote this variable. Possible values of x are 0 (a person who left without making a purchase), 1, 2, 3, and so on. In this example, the values of x are isolated points on the number s line. Until a customer is selected and the number of items counted, the value of x is uncertain. Another variable of interest might be y = number of minutes spent in a checkout line. One possible value of y is 3.0 minutes and another is 4.0 minutes, but other numbers between 3.0 and 4.0 are also possible. The possible values of y form an entire interval on the number line.

> **DEFINITION**
>
> **Random variable:** a numerical variable whose value depends on the outcome of a chance experiment. A random variable associates a numerical value with each outcome of a chance experiment.
>
> A random variable is **discrete** if its possible values are isolated points along the number line.
>
> A random variable is **continuous** if its possible values are *all* points in some interval.

Lowercase letters, such as x and y, will be used to represent random variables.* Figure 6.1 shows a set of possible values for each type of random variable. In practice, a discrete random variable almost always involves counting (for example, the number of items purchased, the number of gas pumps in use, or the number of broken eggs in a carton). A continuous random variable is one whose value is typically obtained by measuring (for example, the temperature in a freezer, the weight of a pineapple, or the amount of time spent in a store). Because there is a limit to the accuracy of any measuring instrument, such as a watch or a scale, it may seem that any variable should be regarded as discrete. For example, when weight is measured to the nearest pound, the observed values appear to be isolated points along the number line, such as 2 pounds, 3 pounds, and so on. But this is just a function of the accuracy with which weight is recorded and not because a weight between 2 and 3 pounds is impossible. In this case, the variable's behavior is continuous.

FIGURE 6.1
Two different types of random variables

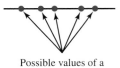
Possible values of a discrete random variable

Possible values of a continuous random variable

Example 6.1   Book Sales

Consider a chance experiment in which the type of book, print (P) or digital (D), chosen by each of three successive customers making a purchase from on an online bookstore is noted. Define a random variable x as

$$x = \text{number of customers purchasing a digital book}$$

The outcome in which the first and third customers purchase a digital book and the second customer purchases a print book can be abbreviated DPD. The associated x value is 2, because two of the three customers selected a digital book. Similarly, the x value for the

---

*In some books, uppercase letters are used to name random variables, with lowercase letters representing a particular value that the variable might assume. This text uses a simpler and less formal notation.

outcome DDD (all three purchased a digital book) is 3. The following table displays each of the eight possible outcomes of the chance experiment and the corresponding value of *x*.

Outcome	PPP	DPP	PDP	PPD	DDP	DPD	PDD	DDD
x value	0	1	1	1	2	2	2	3

There are only four possible *x* values—0, 1, 2, and 3. Because these possible values are isolated points on the number line, *x* is a discrete random variable.

In some situations, the random variable of interest is discrete, but the number of possible values is not finite. This is illustrated in Example 6.2.

Example 6.2   This Could Be a Long Game

Two friends agree to play a game that consists of a sequence of trials. The game continues until one player wins two trials in a row. One random variable of interest might be

$$x = \text{number of trials required to complete the game}$$

Let A denote a win for Player 1 and B denote a win for Player 2. The simplest possible outcomes are AA (Player 1 wins the first two trials and the game ends) and BB (Player 2 wins the first two trials). With either of these two outcomes, *x* = 2. There are also two outcomes for which *x* = 3: ABB and BAA. Some other possible outcomes and associated *x* values are

Outcomes	*x* Value
AA, BB	2
BAA, ABB	3
ABAA, BABB	4
ABABB, BABAA	5
⋮	⋮
ABABABABAA, BABABABABB	10

Any positive integer that is 2 or greater is a possible value for *x*. Because the values 2, 3, 4, … are isolated points on the number line, *x* is a discrete random variable even though there is no upper limit to the number of possible values.

Example 6.3   Stress

In an engineering stress test, pressure is applied to a thin 1-foot-long bar until the bar snaps. The precise location where the bar will snap is uncertain. You can use *x* to denote the distance (in feet) from the left end of the bar to the break. Then *x* = 0.25 is one possibility, *x* = 0.9 is another, and in fact any number between 0 and 1 is a possible value of *x*. Figure 6.2 shows the outcome *x* = 0.6. Because the set of possible values is an entire interval on the number line, *x* is a continuous random variable.

Depending on the accuracy of the measuring instrument, you may only be able to measure the distance to the nearest hundredth of a foot or thousandth of a foot. The *actual* distance, though, could be any number between 0 and 1, which indicates that the variable is continuous.

FIGURE 6.2
The bar for Example 6.3 and the outcome *x* = 0.6

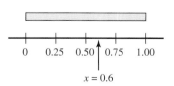

In data analysis, random variables often arise in the context of summarizing data from a sample that has been selected from some population. This is illustrated in Example 6.4.

Example 6.4   **College Plans**

Suppose that a high school counselor plans to select a random sample of 50 seniors. She will ask each student in the sample if he or she plans to attend college after graduation. The process of selecting a sample is a chance experiment. The sample space for this experiment consists of all the different possible random samples of size 50 that might result (there are a very large number of them), and for random sampling, each of these outcomes is equally likely. Suppose

$$x = \text{number of successes in the sample}$$

where a success is defined as a student who plans to attend college. Then x is a random variable, because it associates a numerical value with each of the possible outcomes (random samples) that might occur. The possible values of x are 0, 1, 2, . . . , 50, so x is a discrete random variable.

Summing It Up—Section 6.1

The following learning objective was addressed in this section:

Mastering the Mechanics
M1: Distinguish between discrete and continuous random variables.
The possible values of a discrete random variable are isolated points along the number line. The possible value of a continuous random variable are all points in an interval.

SECTION 6.1   EXERCISES

Each Exercise Set assesses the following chapter learning objective: M1

SECTION 6.1   Exercise Set 1

6.1   State whether each of the following random variables is discrete or continuous:
a. The number of defective tires on a car
b. The body temperature of a hospital patient
c. The number of pages in a book
d. The number of draws (with replacement) from a deck of cards until a heart is selected
e. The lifetime of a light bulb

6.2   Starting at a particular time, each car entering an intersection is observed to see whether it turns left (L), turns right (R), or goes straight ahead (S). The experiment terminates as soon as a car is observed to go straight. Let x denote the number of cars observed. What are possible x values? List five different outcomes and their associated x values. (Hint: See Example 6.2.)

6.3   A box contains four slips of paper marked 1, 2, 3, and 4. Two slips are selected *without* replacement. List the possible values for each of the following random variables:
a. x = sum of the two numbers
b. y = difference between the first and second numbers (first − second)
c. z = number of slips selected that show an even number
d. w = number of slips selected that show a 4

SECTION 6.1   Exercise Set 2

6.4   Classify each of the following random variables as either discrete or continuous:
a. The fuel efficiency (miles per gallon) of an automobile
b. The amount of rainfall at a particular location during the next year
c. The distance that a person throws a baseball
d. The number of questions asked during a 1-hour lecture
e. The tension (in pounds per square inch) at which a tennis racket is strung
f. The amount of water used by a household during a given month
g. The number of traffic citations issued by the highway patrol in a particular county on a given day

6.5   A point is randomly selected from the interior of the square pictured here:

Let x denote the distance from the lower left-hand corner A to the selected point. What are possible values of x? Is x a discrete or a continuous variable?

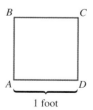

1 foot

6.6   A person stands at the corner marked A of the square pictured in the previous exercise and tosses a coin. If it lands heads up, the person moves one corner clockwise, to

B. If the coin lands tails up, the person moves one corner counterclockwise, to *D.* This process is then repeated until the person arrives back at *A.* Let *y* denote the number of coin tosses. What are possible values of *y*? Is *y* discrete or continuous?

ADDITIONAL EXERCISES

6.7   State whether each of the following random variables is discrete or continuous.
a. The number of courses a student is enrolled in
b. The time spent completing a homework assignment

c. The length of a person's forearm
d. The number of times out of 10 throws that a dog catches a Frisbee

6.8   A person is asked to draw a line segment that they think is 3 inches long. The length of the line segment drawn will be measured and the value of $x =$ (actual length $- 3$) will be calculated.
a. What is the value of x for a person who draws a line segment that is 3.1 inches long?
b. Is x a discrete or continuous random variable? Explain.

6.9   Two six-sided dice, one red and one white, will be rolled. List the possible values for each of the following random variables.
a. $x =$ sum of the two numbers showing
b. $y =$ difference between the number on the red die and the number on the white die (red $-$ white)
c. $w =$ largest number showing

SECTION 6.2   # Probability Distributions for Discrete Random Variables

The probability distribution of a random variable is a model that describes the long-run behavior of that variable. For example, suppose that the Department of Animal Regulation in a particular county is interested in studying the variable

$x =$ number of licensed dogs or cats at a randomly selected house

County regulations prohibit more than five dogs or cats per house. Consider the chance experiment of randomly selecting a house in this county. In this case, x is a discrete random variable because it associates a numerical value (0, 1, 2, 3, 4, or 5) with each of the possible outcomes (houses) in the sample space. Although you know what the possible values for x are, it would also be useful to know how this variable would behave if it were observed for many houses. What would be the most common value? What proportion of the time would $x = 5$ be observed? $x = 3$? A probability distribution provides answers to questions like these.

> **DEFINITION**
>
> The **probability distribution of a discrete random variable** x gives the probability associated with each possible x value. Each probability is the long-run proportion of the time that the corresponding x value will occur.
>
> Common ways to display a probability distribution for a discrete random variable are a table, a probability histogram, or a formula.
>
> If one possible value of x is 2, it is common to write $p(2)$ in place of $P(x = 2)$. Similarly, $p(5)$ denotes the probability that $x = 5$, and so on.

Example 6.5   **Energy-Efficient Refrigerators**

Suppose that each of four randomly selected customers purchasing a refrigerator at a large appliance store chooses either an energy-efficient model (E) or one from a less expensive group of models (G) that do not have an energy-efficient rating. Assume that these customers make their choices independently of one another and that 40% of all customers select an energy-efficient model. This implies that for any particular one of the four customers, $P(E) = 0.4$ and $P(G) = 0.6$. One possible outcome is EGGE, where the first and fourth customers select energy-efficient models and the other two choose less expensive models.

Because the customers make their choices independently, the multiplication rule for independent events implies that

$$P(\text{EGGE}) = P(\text{1st chooses E } and \text{ 2nd chooses G } and \text{ 3rd chooses G } and \text{ 4th chooses E})$$
$$= P(E)P(G)P(G)P(E)$$
$$= (0.4)(0.6)(0.6)(0.4)$$
$$= 0.0576$$

Similarly,

$$P(\text{EGEG}) = P(E)P(G)P(E)P(G)$$
$$= (0.4)(0.6)(0.4)(0.6)$$
$$= 0.0576 \text{ (which is equal to } P(\text{EGGE}))$$

and

$$P(\text{GGGE}) = (0.6)(0.6)(0.6)(0.4) = 0.0864$$

The number among the four customers who purchase energy-efficient models is a random variable. Using x to denote this random variable,

x = the number of energy efficient refrigerators purchased by a group of four customers.

Table 6.1 displays the 16 possible outcomes that might be observed, the probability of each outcome, and the value of the random variable x that is associated with each outcome.

TABLE 6.1 Outcomes and Probabilities for Example 6.5

Outcome	Probability	x Value	Outcome	Probability	x Value
GGGG	0.1296	0	GEEG	0.0576	2
EGGG	0.0864	1	GEGE	0.0576	2
GEGG	0.0864	1	GGEE	0.0576	2
GGEG	0.0864	1	GEEE	0.0384	3
GGGE	0.0864	1	EGEE	0.0384	3
EEGG	0.0576	2	EEGE	0.0384	3
EGEG	0.0576	2	EEEG	0.0384	3
EGGE	0.0576	2	EEEE	0.0256	4

The probability distribution of x is easily obtained from this information. Consider the smallest possible x value, 0. The only outcome for which $x = 0$ is GGGG, so

$$p(0) = P(x = 0) = P(\text{GGGG}) = 0.1296$$

There are four different outcomes for which $x = 1$, so $p(1)$ results from summing the four corresponding probabilities:

$$p(1) = P(x = 1) = P(\text{EGGG } or \text{ GEGG } or \text{ GGEG } or \text{ GGGE})$$
$$= P(\text{EGGG}) + P(\text{GEGG}) + P(\text{GGEG}) + P(\text{GGGE})$$
$$= 0.0864 + 0.0864 + 0.0864 + 0.0864$$
$$= 4(0.0864)$$
$$= 0.3456$$

Similarly,

$$p(2) = P(\text{EEGG}) + \cdots + P(\text{GGEE}) = 6(0.0576) = 0.3456$$
$$p(3) = 4(0.0384) = 0.1536$$
$$p(4) = 0.0256$$

The probability distribution of x is summarized in the following table:

x Value	0	1	2	3	4
$p(x)$	0.1296	0.3456	0.3456	0.1536	0.0256

To interpret $p(3) = 0.1536$, think of performing the chance experiment repeatedly, each time with a new group of four customers. In the long run, 15.36% of these groups of four will have exactly three customers who purchase an energy-efficient refrigerator.

The probability distribution can be used to determine probabilities of various events involving x. For example, the probability that at least two of the four customers choose energy-efficient models is

$$P(x \geq 2) = P(x = 2 \ or \ x = 3 \ or \ x = 4)$$
$$= p(2) + p(3) + p(4)$$
$$= 0.5248$$

This means that in the long run, a group of four refrigerator purchasers will include at least two who select energy-efficient models about 52.48% of the time.

A probability distribution table for a discrete variable x shows the possible x values and the value of $p(x)$ for each one. Because $p(x)$ is a probability, it must be a number between 0 and 1 for each possible value of x, and because the probability distribution includes all possible x values, the sum of all the $p(x)$ values must equal 1. These properties of discrete probability distributions are summarized in the following box.

> **Properties of Discrete Probability Distributions**
> 1. For every possible x value, $0 \leq p(x) \leq 1$.
> 2. $\displaystyle\sum_{\text{all } x \text{ values}} p(x) = 1$

A *probability histogram* is a graphical representation of a discrete probability distribution. The graph has a rectangle centered above each possible value of x, and the area of each rectangle is proportional to the probability of the corresponding value. Figure 6.3 displays the probability histogram for the probability distribution of Example 6.5.

FIGURE 6.3
Probability histogram for the distribution of number of energy-efficient refrigerators purchased

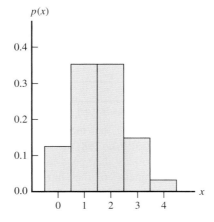

In Example 6.5, the probability distribution was determined by starting with a simple chance experiment and applying basic probability rules. Sometimes this approach is not feasible because of the complexity of the chance experiment. In this case, the long-run behavior of a random variable is often described by an approximate probability distribution based on empirical evidence and prior knowledge. This distribution must still follow rules of probability. Specifically,

1. $p(x) \geq 0$ for every x value

2. $\displaystyle\sum_{\text{all } x \text{ values}} p(x) = 1$

Example 6.6 | Paint Flaws

In automobile manufacturing, painting is one of the last steps in the process of assembling a new car. Some minor blemishes in the paint surface are considered acceptable, but if there are too many, it becomes noticeable to a potential customer, and the car must be repainted. Cars coming off the assembly line are carefully inspected, and the number of minor blemishes in the paint surface is determined. Let x denote the number of minor blemishes on a randomly selected car from a particular manufacturing plant. Suppose that a large number of automobiles are evaluated, leading to the following approximate probability distribution:

x	0	1	2	3	4	5	6	7	8	9	10
$p(x)$	0.041	0.130	0.209	0.223	0.178	0.114	0.061	0.028	0.011	0.004	0.001

The corresponding probability histogram appears in Figure 6.4. The probabilities in this distribution reflect the car manufacturer's experience. For example, $p(3) = 0.223$ indicates that 22.3% of new automobiles had exactly 3 minor paint blemishes. The probability that the number of minor paint blemishes is between 2 and 5 inclusive is

$$P(2 \leq x \leq 5) = p(2) + p(3) + p(4) + p(5) = 0.724$$

If many cars of this type were examined, about 72.4% would have 2, 3, 4, or 5 minor paint blemishes.

FIGURE 6.4
Probability histogram for the distribution of the number of minor paint blemishes on a randomly selected car

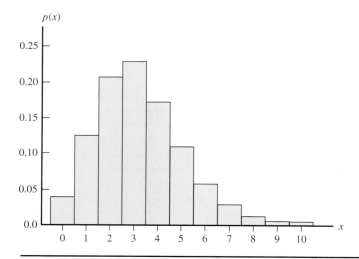

Example 6.7 | iPod Shuffles Revisited

Recall from the Preview Example that the paper **"Does Your iPod *Really* Play Favorites?"** (*The American Statistician* **[2009]: 263–268**) investigated the shuffle feature of the iPod. The shuffle feature is supposed to play the songs in a playlist in a random order. Some users have questioned the "randomness" of the shuffle, citing situations where several songs from the same artist were played in close proximity to each other. One such example appeared in a *Newsweek* article in 2005, where the author states that Steely Dan songs always seem to pop up in the first hour of play. Is this consistent with a "random" shuffle?

To investigate the claim that the shuffle might not be random, suppose you create a playlist of 3000 songs that includes 50 songs by Steely Dan (this is the situation considered by the authors of the *The American Statistician* paper). You could then carry out a simulation by creating a shuffle of 20 songs (about 1 hour of playing time) from the playlist and noting the number of songs by Steely Dan that were among the 20. This could be done by thinking of the songs in the playlist as being numbered from 1 to 3000, with numbers 1 to 50 representing the Steely Dan songs. A random number generator could then be used to select 20 random numbers between 1 and 3000. Counting how many times a number between 1 and 50 was included in this list would give you the number of Steely Dan songs in this particular shuffle of 20 songs. Repeating this process a large number of times would enable you to estimate the probabilities needed for the probability distribution of

x = number of Steely Dan songs in a random shuffle consisting of 20 songs

The accompanying probability distribution is from the *The American Statistician* paper. Even though the possible values of x are 0, 1, 2, ... , 20, the probabilities of x values of 4 or greater are very small (0 when rounded to three decimal places). For this reason, 4, 5, ... , 20 have been grouped into a single entry in the table.

x	0	1	2	3	4 or more
$p(x)$	0.714	0.244	0.039	0.004	0.000

Notice that $P(x \geq 1) = 1 - 0.714 = 0.286$. This means that about 28.6% of the time, at least one Steely Dan song would occur in a random shuffle of 20 songs. Given that there are only 50 Steely Dan songs in the playlist of 3000 songs, this result surprises many people!

You have seen examples in which the probability distribution of a discrete random variable has been given as a table and as a probability (relative frequency) histogram. It is sometimes also possible to use a formula that specifies the probability for each possible value of the random variable. Examples of this approach can be found in Section 6.6.

Summing It Up—Section 6.2

The following learning objectives were addressed in this section:

Conceptual Understanding
C1: Understand that a probability distribution describes the long-run behavior of a random variable.
The probability distribution of a discrete random variable specifies the possible values that the variable might take and gives the probability associated with each of these values. These probabilities describe what proportion of the time, in the long run, that the variable will take on each of the possible values.

Mastering the Mechanics
M2: Given a probability distribution for a discrete random variable, calculate and interpret probabilities.
It is possible to calculate the probability of an event defined in terms of a random variable by identifying the possible values that satisfy the event and then adding the corresponding probabilities given in the probability distribution. For an example, see Example 6.6.

M3: Construct the probability distribution of a discrete random variable.
Random variables assign a numerical value to the possible outcomes of a chance experiment. To construct a probability distribution for a discrete random variable, you can start by listing out all the possible outcomes of the chance experiment and determining the probability associated with each outcome. Then, by considering the value that the random variable assigns to each outcome, you will be able to determine the possible values of the random variable and the probability of each one. For an example, see Example 6.5.

SECTION 6.2 EXERCISES

Each Exercise Set assesses the following chapter learning objectives: C1, M2, M3

SECTION 6.2 Exercise Set 1

6.10 Consider the random variable y = the number of broken eggs in a randomly selected carton of one dozen eggs. Suppose the probability distribution of y is as follows:

y	0	1	2	3	4
$p(y)$	0.65	0.20	0.10	0.04	?

a. Only y values of 0, 1, 2, 3, and 4 have probabilities greater than 0. What is $p(4)$?
b. How would you interpret $p(1) = 0.20$?
c. Calculate $P(y \leq 2)$, the probability that the carton contains at most two broken eggs, and interpret this probability.
d. Calculate $P(y < 2)$, the probability that the carton contains *fewer than* two broken eggs. Why is this smaller than the probability in Part (c)?
e. What is the probability that the carton contains exactly 10 unbroken eggs?
f. What is the probability that at least 10 eggs are unbroken?

6.11 Suppose that fund-raisers at a university call recent graduates to request donations for campus outreach programs. They report the following information for last year's graduates:

Size of donation	$0	$10	$25	$50
Proportion of calls	0.45	0.30	0.20	0.05

Three attempts were made to contact each graduate. A donation of $0 was recorded both for those who were contacted but declined to make a donation and for those who were not reached in three attempts. Consider the variable x = amount of donation for a person selected at random from the population of last year's graduates of this university.

a. Write a few sentences describing what donation amounts you would expect to see if the value of x was observed for each of 1000 graduates.
b. What is the most common value of x in this population?
c. What is $P(x \geq 25)$?
d. What is $P(x > 0)$?

6.12 A restaurant has four bottles of a certain wine in stock. The wine steward does not know that two of these bottles (Bottles 1 and 2) are bad. Suppose that two bottles are ordered, and the wine steward selects two of the four bottles at random. Consider the random variable x = the number of good bottles among these two.

a. When two bottles are selected at random, one possible outcome is (1,2) (Bottles 1 and 2 are selected) and another is (2,4). List all possible outcomes.

b. What is the probability of each outcome in Part (a)?

c. The value of x for the (1,2) outcome is 0 (neither selected bottle is good), and $x = 1$ for the outcome (2,4). Determine the x value for each possible outcome. Then use the probabilities in Part (b) to determine the probability distribution of x. (Hint: See Example 6.5.)

6.13 Suppose that 20% of all homeowners in an earthquake-prone area of California are insured against earthquake damage. Four homeowners are selected at random. Define the random variable x as the number among the four who have earthquake insurance.

a. Find the probability distribution of x. (Hint: Let S denote a homeowner who has insurance and F one who does not. Then one possible outcome is SFSS, with probability $(0.2)(0.8)(0.2)(0.2)$ and associated x value of 3. There are 15 other outcomes.)
b. What is the most likely value of x?
c. What is the probability that at least two of the four selected homeowners have earthquake insurance?

SECTION 6.2 Exercise Set 2

6.14 Suppose x = the number of courses a randomly selected student at a certain university is taking. The probability distribution of x appears in the following table:

x	1	2	3	4	5	6	7
$p(x)$	0.02	0.03	0.09	0.25	0.40	0.16	0.05

a. What is $P(x = 4)$?
b. What is $P(x \leq 4)$?
c. What is the probability that the selected student is taking at most five courses?
d. What is the probability that the selected student is taking at least five courses? More than five courses?

e. Calculate $P(3 \leq x \leq 6)$ and $P(3 < x < 6)$. Explain in words why these two probabilities are different.

6.15 A pizza shop sells pizzas in four different sizes. The 1000 most recent orders for a single pizza resulted in the following proportions for the various sizes:

Size	12 in.	14 in.	16 in.	18 in.
Proportion	0.20	0.25	0.50	0.05

With x = the size of a pizza in a single-pizza order, the given table is an approximation to the population distribution of x.

a. Write a few sentences describing what you would expect to see for pizza sizes over a long sequence of single-pizza orders.
b. What is the approximate value of $P(x < 16)$?
c. What is the approximate value of $P(x \leq 16)$?

6.16 Of all airline flight requests received by a certain ticket broker, 70% are for domestic travel (D) and 30% are for international flights (I). Define x to be the number that are for domestic flights among the next three requests received. Assuming independence of successive requests, determine the probability distribution of x. (Hint: One possible outcome is DID, with the probability $(0.7)(0.3)(0.7) = 0.147$.)

6.17 Suppose that a computer manufacturer receives computer boards in lots of five. Two boards are selected from each lot for inspection. You can represent possible outcomes of the selection process by pairs. For example, the pair $(1, 2)$ represents the selection of Boards 1 and 2 for inspection.
a. List the 10 different possible outcomes.
b. Suppose that Boards 1 and 2 are the only defective boards in a lot of five. Two boards are to be chosen at random. Let x = the number of defective boards observed among those inspected. Find the probability distribution of x.

ADDITIONAL EXERCISES

6.18 A business has six customer service telephone lines. Let x denote the number of lines in use at any given time. Suppose that the probability distribution of x is as follows:

x	0	1	2	3	4	5	6
$p(x)$	0.10	0.15	0.20	0.25	0.20	0.06	0.04

Write each of the following events in terms of x, and then calculate the probability of each one:

a. At most three lines are in use
b. Fewer than three lines are in use
c. At least three lines are in use
d. Between two and five lines (inclusive) are in use
e. Between two and four lines (inclusive) are not in use
f. At least four lines are not in use

6.19 A contractor is required by a county planning department to submit anywhere from one to five forms (depending on the nature of the project) when applying for a building permit. Let y be the number of forms required of the next applicant. Suppose the probability that y forms are required is known to be proportional to y; that is, $p(y) = ky$ for $y = 1, \ldots, 5$.
a. What is the value of k? (Hint: $\Sigma p(y) = 1$.)
b. What is the probability that at most three forms are required?
c. What is the probability that between two and four forms (inclusive) are required?

6.20 A new battery's voltage may be acceptable (A) or unacceptable (U). A certain flashlight requires two batteries, so batteries will be independently selected and tested until two acceptable ones have been found. Suppose that 80% of all batteries have acceptable voltages, and let y denote the number of batteries that must be tested.
a. What are the possible values of y?
b. What is $p(2) = P(y = 2)$?
c. What is $p(3)$? (Hint: There are two different outcomes that result in $y = 3$.)

SECTION 6.3 Probability Distributions for Continuous Random Variables

The possible values of a continuous random variable form an entire interval on the number line. One example of a continuous random variable is the weight x (in pounds) of a full-term newborn baby. Suppose for the moment that weight is recorded only to the nearest pound. Then a reported weight of 7 pounds would be used for any weight greater than or equal to 6.5 pounds and less than 7.5 pounds. If you observe the weights of a large number of newborns, you could construct a density histogram with rectangles centered at 4, 5, and so on. In this histogram, the area of each rectangle is the approximate probability of the corresponding weight value. The total area of all the rectangles is 1, and the probability that a weight (to the nearest pound) is between two values, such as 6 and 8, is the sum of the corresponding rectangular areas. Figure 6.5(a) illustrates this.

FIGURE 6.5
Probability distribution for birth weight: (a) weight measured to the nearest pound; (b) weight measured to the nearest tenth of a pound; (c) limiting curve as measurement accuracy increases: shaded area = $P(6 \leq \text{weight} \leq 8)$

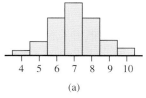
(a)

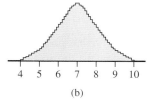
(b)

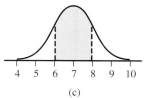
(c)

Now suppose that weight is measured to the nearest tenth of a pound. There are many more possible reported weight values than before, such as 5.0, 5.1, 5.7, 7.3, and 8.9. As shown in Figure 6.5(b), the rectangles in the density histogram are much narrower, giving this histogram a much smoother appearance. As before, the area of each rectangle is an approximate probability, and the total area of all the rectangles is 1.

Figure 6.5(c) shows what happens as weight is measured to a greater degree of accuracy. The resulting histogram approaches a smooth curve. The curve cannot go below the horizontal measurement scale, and the total area under the curve is 1 (because this is true of every probability histogram). The probability that x falls in an interval such as $6 \leq x \leq 8$ is represented by the area under the curve and above that interval.

DEFINITION

A **probability distribution for a continuous random variable** x is specified by a curve called a **density curve**. The function that defines this curve is denoted by $f(x)$ and is called the **density function**.

The following are properties of all continuous probability distributions:

1. $f(x) \geq 0$ (so that the curve never drops below the horizontal axis).
2. The total area under the density curve is equal to 1.

The probability that x falls in any particular interval is the area under the density curve and above the interval.

Many probability calculations for continuous random variables involve the following three types of events:

1. $a < x < b$, the event that the random variable x takes a value between two given numbers, a and b
2. $x < a$, the event that the random variable x takes a value less than a given number a
3. $x > b$, the event that the random variable x takes a value greater than a given number b (this can also be written as $b < x$)

Figure 6.6 shows the probabilities of these events as areas under a density curve.

FIGURE 6.6
Probabilities as areas under a probability density curve

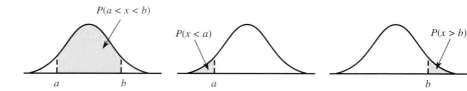

$P(a < x < b)$ $P(x < a)$ $P(x > b)$

a b a b

Example 6.8 **Application Processing Times**

Let x represent a continuous random variable defined as the amount of time (in minutes) required to process a credit application. Suppose that x has a probability distribution with density function

$$f(x) = \begin{cases} 0.5 & 4 < x < 6 \\ 0 & \text{otherwise} \end{cases}$$

The graph of $f(x)$, the density curve, is shown in Figure 6.7(a). It is especially easy to use this density curve to calculate probabilities, because it just requires finding the areas of rectangles using the formula

$$\text{area} = (\text{base})(\text{height})$$

The curve has positive height, 0.5, only between $x = 4$ and $x = 6$. The total area under the curve is just the area of the rectangle with base extending from 4 to 6 and with height 0.5. This gives

$$\text{area} = (6 - 4)(0.5) = 1$$

as required.

FIGURE 6.7

The uniform distribution for Example 6.8

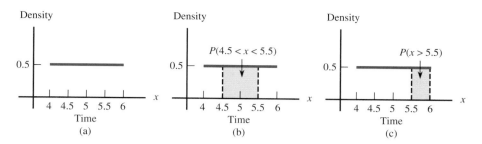

When the density function is constant over an interval (resulting in a horizontal density curve), the probability distribution is called a **uniform distribution**.

As illustrated in Figure 6.7(b), the probability that x is between 4.5 and 5.5 is

$$P(4.5 < x < 5.5) = \text{area of shaded rectangle}$$
$$= (\text{base})(\text{height})$$
$$= (5.5 - 4.5)(0.5)$$
$$= 0.5$$

Figure 6.7(c) shows the probability that x is greater than 5.5. Because, in this context, $x > 5.5$ is equivalent to $5.5 < x < 6$,

$$P(x > 5.5) = (6 - 5.5)(0.5) = 0.25$$

According to this model, in the long run, about 25% of all credit applications will have processing times that exceed 5.5 min.

The probability that a *discrete* random variable x lies in the interval between two endpoints a and b depends on whether either endpoint is included in the interval. Suppose, for example, that x is the number of major defects on a new automobile. Then

$$P(3 \le x \le 7) = p(3) + p(4) + p(5) + p(6) + p(7)$$

whereas

$$P(3 < x < 7) = p(4) + p(5) + p(6)$$

However, if x is a *continuous* random variable, such as task completion time, then

$$P(3 \le x \le 7) = P(3 < x < 7)$$

because the area under a density curve and above a single value such as 3 or 7 is 0. Geometrically, you can think of finding the area above a single point as finding the area of a rectangle with base = 0. The area above an interval of values, therefore, does not depend on whether or not either endpoint is included.

If x is a continuous random variable, then for any two numbers a and b with $a < b$,

$$P(a \le x \le b) = P(a < x \le b) = P(a \le x < b) = P(a < x < b)$$

Example 6.9   Priority Mail Package Weights

Two hundred packages shipped using the Priority Mail rate for packages less than 2 pounds were weighed, resulting in a sample of 200 observations of the variable

$$x = \text{package weight (in pounds)}$$

from the population of all Priority Mail packages under 2 pounds. A histogram (using the density scale, where height = (relative frequency)/(interval width)) of the 200 weights is shown in Figure 6.8(a). Because the histogram is based on a sample of 200 packages, it provides only an approximation to the actual distribution of x. The shape of this histogram suggests that a reasonable model for the distribution of x might be the triangular distribution shown in Figure 6.8(b).

FIGURE 6.8
Graphs for Example 6.9:
(a) histogram of package weights;
(b) continuous probability
distribution for x = package
weight

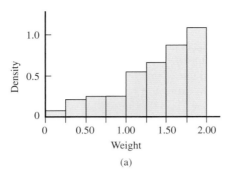
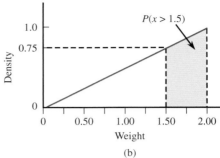

(a) (b)

Notice that the total area under the triangular probability density curve is equal to

$$\text{total area of triangle} = \frac{1}{2}(\text{base})(\text{height}) = \frac{1}{2}(2)(1) = 1$$

This probability distribution can be used to calculate the proportion of packages over 1.5 pounds, $P(x > 1.5)$. This corresponds to the area of the shaded region in Figure 6.8(b). In this case, it is easier to calculate the area of the unshaded triangular region (which corresponds to $P(x \leq 1.5)$):

$$P(x \leq 1.5) = \frac{1}{2}(1.5)(0.75) = 0.563$$

Because the total area under a probability density curve is 1,

$$P(x > 1.5) = 1 - 0.563 = 0.437$$

It is also the case that

$$P(x \geq 1.5) = 0.437$$

and that

$$P(x = 1.5) = 0$$

(because the area under the density curve above a single x value is 0).

Example 6.10 Service Times

An airline's toll-free reservation number recorded the length of time (in minutes) required to provide service to each of 500 callers. This resulted in 500 observations of the continuous numerical variable

$$x = \text{service time}$$

A density histogram is shown in Figure 6.9(a).

FIGURE 6.9
Graphs for Example 6.10:
(a) histogram of service time;
(b) continuous probability
distribution for service time

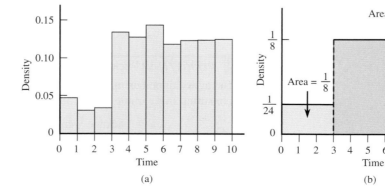

(a) (b)

The population of interest is all callers to the reservation line. Based on the density histogram, it appears that a reasonable model for the population distribution would be flat over the interval from 0 to 3 and higher but also flat over the interval from 3 to 10. Because service requests were usually one of two types: (1) requests to make a flight reservation and (2) requests to cancel a reservation, this type of model seems reasonable. Canceling a reservation, which accounted for about one-eighth of the calls to the reservation line, could usually be accomplished fairly quickly, whereas making a reservation (seven-eighths of the calls) required more time.

Figure 6.9(b) shows the probability distribution curve proposed as a model for the variable x = service time. The height of the curve for each of the two segments was chosen so that the total area under the curve would be 1 and so that $P(x \leq 3) = \frac{1}{8}$ (these were thought to be cancellation calls) and $P(x > 3) = \frac{7}{8}$.

Once the model has been developed, it can be used to calculate probabilities. For example,

$$P(x > 8) = \text{area under curve and above interval from 8 to 10}$$

$$= 2\left(\frac{1}{8}\right) = \frac{2}{8} = \frac{1}{4} = 0.250$$

In the long run, about one-fourth of all service requests will require more than 8 minutes.
Similarly,

$$P(2 < x < 4) = \text{area under curve and above interval from 2 to 4}$$

$$= (\text{area under curve and above interval from 2 to 3})$$

$$+ (\text{area under curve and above interval from 3 to 4})$$

$$= 1\left(\frac{1}{24}\right) + 1\left(\frac{1}{8}\right)$$

$$= \frac{1}{24} + \frac{3}{24}$$

$$= \frac{4}{24}$$

$$= \frac{1}{6} = 0.167$$

In each of the previous examples, the model for the probability distribution was simple enough that it was possible to calculate probabilities (evaluate areas under the density curve) using simple area formulas from geometry. Example 6.11 shows that this is not always the case.

Example 6.11 Online Registration Times

Students at a university use an online registration system to register for courses. The variable

$$x = \text{length of time (in minutes) required for a student to register}$$

was recorded for a large number of students using the system, and the resulting values were used to construct a density histogram (see Figure 6.10). The general form of the histogram can be described as bell shaped and symmetric, and a smooth curve has been superimposed. This smooth curve is a reasonable model for the probability distribution of x. Although this is a common probability model for many variables, it is not obvious how you could use it to calculate probabilities, because it is not clear how to calculate areas under such a curve.

FIGURE 6.10

Histogram and continuous probability distribution for time to register for Example 6.11

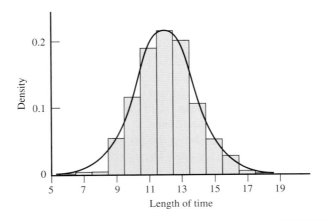

The symmetric, bell-shaped probability model of Example 6.11 is known as a *normal probability distribution*. Normal distributions are widely used, and they are investigated in more detail in Section 6.5.

Probabilities for continuous random variables are often calculated using cumulative areas. A cumulative area is all of the area under the density curve to the left of a particular value. Figure 6.11 illustrates the cumulative area to the left of 0.5, which is $P(x < 0.5)$.

FIGURE 6.11

A cumulative area under a probability density curve

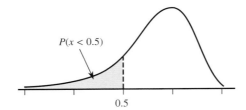

The probability that x is in any particular interval, $P(a < x < b)$, can be calculated as the difference between two cumulative areas.

The probability that a continuous random variable x lies between a lower endpoint a and an upper endpoint b is

$$P(a < x < b) = \text{(cumulative area to the left of } b) - \text{(cumulative area to the left of } a)$$

$$= P(x < b) - P(x < a)$$

For continuous random variables, the area above an interval of values does not depend on whether or not either endpoint is included, so this is also the probability of

$$P(a \leq x \leq b), P(a < x \leq b), \text{ and } P(a \leq x < b).$$

This property is illustrated in Figure 6.12 for the case of $a = 0.25$ and $b = 0.75$. You will use this result often in Section 6.5 when you calculate probabilities using the normal distribution.

FIGURE 6.12

Calculation of $P(a < x < b)$ using cumulative areas

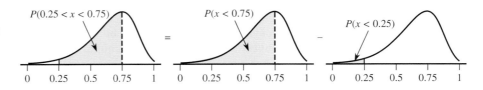

For some continuous distributions, cumulative areas can be calculated using methods from the branch of mathematics called integral calculus. However, because this text does not assume knowledge of calculus, you will rely on tables that have been constructed for the commonly encountered continuous probability distributions. Many graphing calculators and statistical software packages will also compute areas for the most widely used continuous probability distributions.

Summing It Up—Section 6.3

The following learning objectives were addressed in this section:

Conceptual Understanding

C1: Understand that a probability distribution describes the long-run behavior of a random variable.

The probability distribution of a continuous random variable is specified by a smooth curve called a density curve. For a continuous random variable, the probability associated with a particular interval of possible values represents the proportion of the time, in the long run, that the variable will take on a value in the corresponding interval.

C3: Understand that areas under a density curve for a continuous random variable are interpreted as probabilities.

Areas under the density curve for a continuous random variable represent probabilities. The area under the curve and above a given interval represents the probability of observing a value in that interval.

Mastering the Mechanics

M5: Given the probability density curve for a continuous random variable, identify areas corresponding to various probabilities.

The probability that a continuous random variable takes on a value in an interval from a to b is the area under the probability density curve and above the interval.

M6: Calculate probabilities for continuous random variables whose density curves have a simple form.

When the density curve has a simple form, it is possible to calculate probabilities (areas under the curve) using simple area formulas from geometry. For examples of calculating probabilities, see Examples 6.9 and 6.10.

SECTION 6.3 EXERCISES

Each Exercise Set assesses the following chapter learning objectives: C1, C3, M5, M6

SECTION 6.3 Exercise Set 1

6.21 Let x denote the lifetime (in thousands of hours) of a certain type of fan used in diesel engines. Suppose the density curve of x is as pictured.

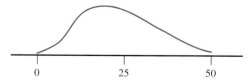

Shade the area under the curve corresponding to each of the following probabilities (draw a new curve for each part):
a. $P(10 < x < 25)$
b. $P(10 \leq x \leq 25)$
c. $P(x < 30)$
d. The probability that the lifetime is at least 25,000 hours

6.22 Suppose that the random variable

x = waiting time for service at a bank (in minutes)

has the probability distribution described by the density curve pictured below.

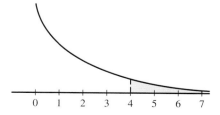

a. What probability is represented by the shaded area?

b. Suppose the shaded area = 0.26. Interpret this probability in the context of this problem.

6.23 Let x denote the time (in seconds) necessary for an individual to react to a certain stimulus. Suppose the probability distribution of x is specified by the accompanying density curve.

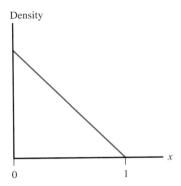

Density

0   1 x

a. What is the height of the density curve above $x = 0$? (Hint: Total area = 1.)
b. What is the probability that reaction time exceeds 0.5 sec? (Hint: See Example 6.9.)
c. What is the probability that reaction time is at most 0.25 sec?

6.24 Consider the population that consists of all soft contact lenses made by a particular manufacturer, and define the random variable x = thickness (mm). Suppose that a reasonable model for the probability distribution of x is the one shown in the following figure.

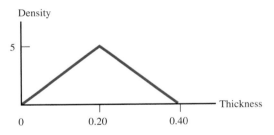

Density

5

0   0.20   0.40   Thickness

a. Verify that the total area under the density curve is equal to 1. [Hint: The area of a triangle is equal to (1/2)(base)(height).]
b. What is the probability that x is less than 0.20? Less than 0.10? More than 0.30?
c. What is the probability that x is between 0.10 and 0.20? (Hint: First find the probability that x is *not* between 0.10 and 0.20.)

SECTION 6.3 Exercise Set 2

6.25 An online store charges for shipping based on the weight of the items in an order. Define the random variable

x = weight of a randomly selected order (in pounds)

The density curve of x is shown here:

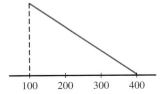

100   200   300   400

Shade the area under the curve corresponding to each of the following probabilities (draw a new curve for each part):
a. $P(x > 200)$
b. $P(100 < x < 300)$
c. $P(x \leq 250)$
d. The probability that the selected order has a weight greater than 350 pounds.

6.26 Suppose that the random variable

x = weekly water usage (in gallons) for a randomly selected studio apartment in Los Angeles

has the probability distribution described by the following density curve.

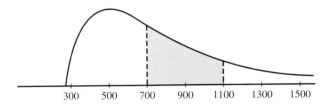

300   500   700   900   1100   1300   1500

a. What probability is represented by the shaded area?
b. Suppose the shaded area = 0.35. Interpret this probability in the context of this problem.

6.27 The article **"Probabilistic Risk Assessment of Infrastructure Networks Subjected to Hurricanes"** (*12th International Conference on Applications of Statistics and Probability in Civil Engineering*, **2015**) suggests a uniform distribution as a model for the actual landfall position of the eye of a hurricane. Consider the random variable x = *distance of actual landfall from predicted landfall*. Suppose that a uniform distribution on the interval that ranges from 0 km to 400 km is a reasonable model for x.
a. Draw the density curve for x.
b. What is the height of the density curve?
c. What is the probability that x is at most 100?
d. What is the probability that x is between 200 and 300? Between 50 and 150? Why are these two probabilities equal?

ADDITIONAL EXERCISES

6.28 The continuous random variable x has the probability distribution shown here:

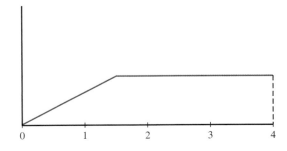

0   1   2   3   4

Shade the area under the curve corresponding to each of the following probabilities. (Draw a new curve for each part.)
a. $P(x < 1)$
b. $P(x > 3)$
c. $P(1 < x < 2)$
d. $P(2 < x < 3)$

6.29 Refer to the probabilities given in Parts (a)–(d) of the previous exercise. Which of these probabilities is smallest? Which is largest?

6.30 A particular professor never dismisses class early. Let x denote the amount of additional time (in minutes) that elapses before the professor dismisses class. Suppose that x has a uniform distribution on the interval from 0 to 10 minutes. The density curve is shown in the following figure:

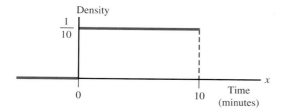

a. What is the probability that at most 5 minutes elapse before dismissal?
b. What is the probability that between 3 and 5 minutes elapse before dismissal?

6.31 Refer to the probability distribution given in the previous exercise. Put the following probabilities in order, from smallest to largest:

$$P(2 < x < 3) \quad P(2 \le x \le 3) \quad P(x < 2) \quad P(x > 7)$$

Explain your reasoning.

SECTION 6.4 The Mean and Standard Deviation of a Random Variable

We study a random variable x, such as the number of insurance claims made by a homeowner (a discrete variable) or the birth weight of a baby (a continuous variable), to learn something about how its values are distributed along the measurement scale. For sample data, the sample mean $\bar{x}$ and sample standard deviation s summarize center and variability. Similarly, the mean value and standard deviation of a random variable describe where its probability distribution is centered and the extent to which it spreads out about the center.

> The **mean value of a random variable x,** denoted by μ_x, describes where the probability distribution of x is centered. The mean is interpreted as the long-run average value for observed x values.
>
> The **standard deviation of a random variable x,** denoted by σ_x, describes variability in the probability distribution. When the value of σ_x is small, observed values of x will tend to be close to the mean value. The larger the value of σ_x, the more variability there will be in observed x values.

Figure 6.13(a) shows two discrete probability distributions with the same standard deviation (variability) but different means (centers). One distribution has a mean of $\mu_x = 6$ and the other has $\mu_x = 10$. Which is which? Figure 6.13(b) shows two continuous probability distributions that have the same mean but different standard deviations. Which distribution—(i) or (ii)—has the larger standard deviation? Finally, Figure 6.13(c) shows three continuous distributions with different means and standard deviations. Which of the three distributions has the largest mean? Which has a mean of about 10? Which distribution has the smallest standard deviation? (The correct answers to these questions are the following: Figure 6.13(a)(ii) has a mean of 6, and Figure 6.13(a)(i) has a mean

FIGURE 6.13
Some probability distributions:
(a) different values of μ_x with the same value of σ_x;
(b) different values of σ_x with the same value of μ_x;
(c) different values of μ_x and σ_x

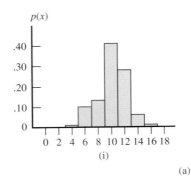
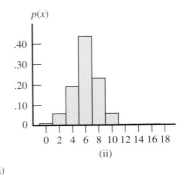

(a)

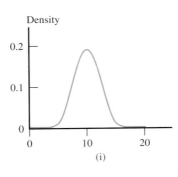
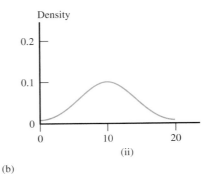

(b)

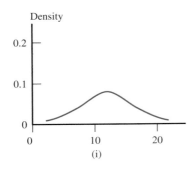
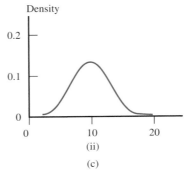
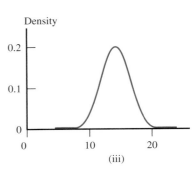

(c)

of 10; Figure 6.13(b)(ii) has the larger standard deviation; Figure 6.13(c)(iii) has the largest mean, Figure 6.13(c)(ii) has a mean of about 10, and Figure 6.13(c)(iii) has the smallest standard deviation.)

It is common to use the terms *mean of the random variable x* and *mean of the probability distribution of x* interchangeably. Similarly, the standard deviation of the random variable x and the standard deviation of the probability distribution of x refer to the same thing. Although the mean and standard deviation are calculated differently for discrete and continuous random variables, the interpretation is the same in both cases.

The Mean Value of a Discrete Random Variable

Consider an experiment consisting of randomly selecting a car licensed in a particular state. Define the discrete random variable x to be the number of low-beam headlights on the selected car that are in need of adjustment. Possible x values are 0, 1, and 2, and the probability distribution of x might be as follows:

x value	0	1	2
probability	0.5	0.3	0.2

The corresponding probability histogram appears in Figure 6.14.

FIGURE 6.14
Probability histogram for the distribution of the number of headlights needing adjustment

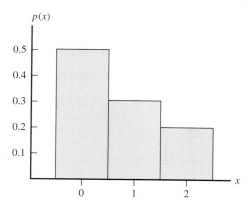

In a sample of 100 cars, the sample relative frequencies might differ somewhat from the given probabilities (which are the long-run relative frequencies). For example, we might see:

x value	0	1	2
frequency	46	33	21

The sample average value of x for these 100 observations is the sum of 46 zeros, 33 ones, and 21 twos, all divided by 100:

$$\bar{x} = \frac{(0)(46) + (1)(33) + (2)(21)}{100}$$

$$= (0)\left(\frac{46}{100}\right) + (1)\left(\frac{33}{100}\right) + (2)\left(\frac{21}{100}\right)$$

$$= (0)(\text{rel. freq. of } 0) + (1)(\text{rel. freq. of } 1) + (2)(\text{rel. freq. of } 2)$$

$$= 0.75$$

As the sample size increases, each relative frequency will approach the corresponding probability, and the value of $\bar{x}$ will approach

$$(0)P(x=0) + (1)P(x=1) + (2)P(x=2) = (0)(0.5) + (1)(0.3) + (2)(0.2)$$

$$= 0.70$$

$$= \text{mean value of } x$$

Notice that the expression for $\bar{x}$ is a weighted average of the possible x values (the weight for each possible x value is the observed relative frequency). Similarly, the mean value of the random variable x is a weighted average, but now the weights are the probabilities from the probability distribution.

DEFINITION

The **mean value of a discrete random variable x,** denoted by μ_x, is calculated by first multiplying each possible x value by the probability of observing that value and then adding the resulting quantities. Symbolically,

$$\mu_x = \sum_{\text{all possible } x \text{ values}} x \cdot p(x)$$

The term **expected value** is sometimes used in place of mean value, and $E(x)$ is another way to denote μ_x.

Example 6.12 **Exam Attempts**

Individuals applying for a certain license are allowed up to four attempts to pass the licensing exam. Consider the random variable

$$x = \text{the number of attempts made by a randomly selected applicant}$$

The probability distribution of x is as follows:

x	1	2	3	4
$p(x)$	0.10	0.20	0.30	0.40

Then x has mean value

$$
\begin{aligned}
\mu_x &= \Sigma x \cdot p(x) \\
&= (1)p(1) + (2)p(2) + (3)p(3) + (4)p(4) \\
&= (1)(0.10) + (2)(0.20) + (3)(0.30) + (4)(0.40) \\
&= 0.10 + 0.40 + 0.90 + 1.60 \\
&= 3.00
\end{aligned}
$$

Example 6.13 **Apgar Scores**

At 1 minute after birth and again at 5 minutes, each newborn child is given a numerical rating called an Apgar score. Possible values of this score are 0, 1, 2, ... , 9, 10. A child's score is determined by five factors: muscle tone, skin color, respiratory effort, strength of heartbeat, and reflex, with a high score indicating a healthy infant. Consider the random variable

$$x = \text{Apgar score of a randomly selected newborn infant at a particular hospital}$$

and suppose that x has the following probability distribution:

x	0	1	2	3	4	5	6	7	8	9	10
$p(x)$	0.002	0.001	0.002	0.005	0.02	0.04	0.17	0.38	0.25	0.12	0.01

The mean value of x is

$$
\begin{aligned}
\mu_x &= \Sigma x \cdot p(x) \\
&= (0)p(0) + (1)p(1) + \cdots + (9)p(9) + (10)p(10) \\
&= (0)(0.002) + (1)(0.001) + \cdots + (9)(0.12) + (10)(0.01) \\
&= 7.16
\end{aligned}
$$

The mean Apgar score for a *sample* of newborn children born at this hospital may be $\bar{x} = 7.05$, $\bar{x} = 8.30$, or any one of a number of other possible values between 0 and 10. However, as child after child is born and rated, the mean score will approach the value 7.16. This value can be interpreted as the mean Apgar score for the population of all babies born at this hospital.

The Standard Deviation of a Discrete Random Variable

The mean value μ_x provides only a partial summary of a probability distribution. Two different distributions can have the same value of μ_x, but observations from one distribution might vary more than observations from the other distribution.

Example 6.14 **Glass Panels**

Flat screen TVs require high-quality glass with very few flaws. A television manufacturer receives glass panels from two different suppliers. Consider

$$x = \text{number of flaws in a randomly selected glass panel from Supplier 1}$$
$$y = \text{number of flaws in a randomly selected glass panel from Supplier 2}$$

Suppose that the probability distributions for x and y are as follows:

x	0	1	2	3		y	0	1	2	3
$p(x)$	0.4	0.3	0.2	0.1		$p(y)$	0.2	0.6	0.2	0

Probability histograms for x and y are shown in Figure 6.15.

FIGURE 6.15
Probability histograms for the number of flaws in a glass panel for Example 6.14: (a) Supplier 1; (b) Supplier 2

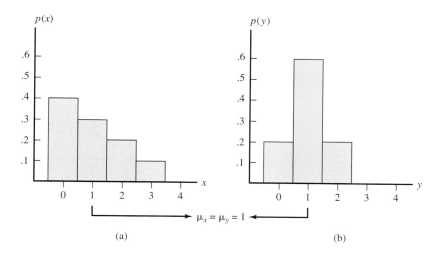

It is easy to verify that the mean values of both x and y are 1, so for either supplier the long-run average number of flaws per panel is 1. However, the two probability histograms show that the probability distribution for the first supplier is more spread out than that of the second supplier.

The greater variability of the first distribution means there will be more variability in observed x values than in observed y values. For example, none of the y observations will be 3, but about 10% of x observations will be 3.

The variance and standard deviation are used to summarize variability in a probability distribution.

DEFINITION

The **variance of a discrete random variable x,** denoted by σ_x^2, is calculated by first subtracting the mean from each possible x value to obtain the deviations, then squaring each deviation and multiplying the result by the probability of the corresponding x value, and finally adding these quantities. Symbolically,

$$\sigma_x^2 = \sum_{\text{all possible } x \text{ values}} (x - \mu_x)^2 p(x)$$

The **standard deviation of x,** denoted by σ_x, is the square root of the variance.

As with s^2 and s (the variance and standard deviation of a sample), the variance and standard deviation of a random variable x are based on squared deviations from the mean. A value far from the mean results in a large squared deviation. However, such a value contributes substantially to variability in x only if the probability associated with that value is not too small. For example, if $\mu_x = 1$ and $x = 25$ is a possible value, then the squared deviation is $(25 - 1)^2 = 576$. But if $P(x = 25) = .000001$, the value 25 will hardly ever be observed, so it won't contribute much to variability in observed x values. This is why each squared deviation is multiplied by its probability to obtain a measure of variability.

Example 6.15 **Glass Panels Revisited**

For x = number of flaws in a glass panel from the first supplier in Example 6.14

$$\sigma_x^2 = (0 - 1)^2\, p(0) + (1 - 1)^2\, p(1) + (2 - 1)^2\, p(2) + (3 - 1)^2\, p(3)$$
$$= (1)(0.4) + (0)(0.3) + (1)(0.2) + (4)(0.1)$$
$$= 1.0$$

Then $\sigma_x = \sqrt{1.0} = 1.0$. For y = the number of flaws in a glass panel from the second supplier,

$$\sigma_y^2 = (0 - 1)^2\, p(0) + (1 - 1)^2\, p(1) + (2 - 1)^2\, p(2)$$
$$= (0 - 1)^2(0.2) + (1 - 1)^2(0.6) + (2 - 1)^2\,(0.2)$$
$$= 0.4$$

Then $\sigma_y = \sqrt{0.4} = 0.632$. The fact that $\sigma_x > \sigma_y$ confirms what you saw in Figure 6.15.

The Mean and Standard Deviation When x Is Continuous

For continuous probability distributions, μ_x and σ_x are defined and calculated using methods from calculus, so you don't need to worry about how to calculate them. However, it is important to know that μ_x and σ_x play the same role for continuous distributions as they do for discrete distributions. The mean value μ_x describes the center of the continuous distribution and gives the approximate long-run average for observed x values. The standard deviation σ_x measures the extent that the continuous distribution (density curve) spreads out about μ_x and describes the amount of variability that can be expected in observed x values.

Example 6.16 **Body Mass Index**

Body mass index (BMI) is a continuous variable calculated from height and weight that is measured in kg/m². The authors of the paper **"Concordance of Self-Report and Measured Height and Weight of College Students"** (*Journal of Nutrition Education and Behavior* **[2015]: 94–98**) collected data on BMI from a large number of students at eight different colleges. For these students, the authors reported that the continuous random variables

$$x = \text{BMI based on self-reported height and weight}$$

and

$$y = \text{BMI based on measured height and weight}$$

have the following means and standard deviations:

$$\mu_x = 24.2 \qquad\qquad \mu_y = 24.5$$
$$\sigma_x = 2.6 \qquad\qquad \sigma_y = 3.9$$

This suggests that if you were to observe self-reported BMI and measured BMI for a large number of these college students, you would not see much difference in the averages of the observations for these variables. But because the standard deviation of y = measured BMI is larger than the standard deviation of x = self-reported BMI, there would be more variability in the values of the measured BMI than the BMI that was based on self-reported height and weight.

Summing It Up—Section 6.4

The following learning objectives were addressed in this section:

Conceptual Understanding

C2: Understand how the long-run behavior of a random variable is described by its mean and standard deviation.

The mean of a random variable identifies where its probability distribution is centered and is interpreted as the average value in a large number of observations of the variable. The standard deviation describes how much the probability distribution spreads out around the center and is an indication of how much variability there would be in a large number of observations of the variable.

Mastering the Mechanics

M4: Calculate and interpret the mean and standard deviation of a discrete random variable.

The mean and standard deviation of a discrete random variable are calculated using the possible values and probabilities specified by its probability distribution. The formula for the mean can be found in the box just prior to Example 6.12, and the formula for the standard deviation is given in the box just prior to Example 6.15.

SECTION 6.4 EXERCISES

Each Exercise Set assesses the following chapter learning objectives: C2, M4

SECTION 6.4 Exercise Set 1

6.32 Consider the random variable y = the number of broken eggs in a randomly selected carton of one dozen eggs. Suppose the probability distribution of y is as follows:

y	0	1	2	3	4
$p(y)$	0.65	0.20	0.10	0.04	0.01

a. Calculate and interpret μ_y. (Hint: See Example 6.13.)
b. In the long run, for what percentage of cartons is the number of broken eggs less than μ_y? Does this surprise you?
c. Explain why μ_y is not equal to $\dfrac{0 + 1 + 2 + 3 + 4}{5} = 2.0$.

6.33 An appliance dealer sells three different models of freezers having 13.5, 15.9, and 19.1 cubic feet of storage space. Consider the random variable x = the amount of storage space purchased by the next customer to buy a freezer. Suppose that x has the following probability distribution:

x	13.5	15.9	19.1
$p(x)$	0.2	0.5	0.3

a. Calculate the mean and standard deviation of x. (Hint: See Example 6.15.)
b. Give an interpretation of the mean and standard deviation of x in the context of observing the outcomes of many purchases.

6.34 A grocery store has an express line for customers purchasing at most five items. Consider the random variable x = the number of items purchased by a randomly selected customer using this line. Make two tables that represent two different possible probability distributions for x that have the same mean but different standard deviations.

SECTION 6.4 Exercise Set 2

6.35 A local television station sells 15-second, 30-second, and 60-second advertising spots. Consider the random variable x = length of a randomly selected commercial appearing on this station, and suppose that the probability distribution of x is given by the following table:

x	15	30	60
$p(x)$	0.1	0.3	0.6

What is the mean length for commercials appearing on this station?

6.36 Suppose that for a given computer salesperson, the probability distribution of

x = the number of systems sold in 1 week

is given by the following table:

x	1	2	3	4	5	6	7	8
$p(x)$	0.05	0.10	0.12	0.30	0.30	0.11	0.01	0.01

a. Find the mean value of x (the mean number of systems sold).
b. Find the variance and standard deviation of x. How would you interpret these values?
c. What is the probability that the number of systems sold is within 1 standard deviation of its mean value?
d. What is the probability that the number of systems sold is more than 2 standard deviations from the mean?

6.37 A grocery store has an express line for customers purchasing at most five items. Consider the random variable x = the number of items purchased by a randomly selected customer using this line. Make two tables that represent two different possible probability distributions for x that have the same standard deviation but different means.

ADDITIONAL EXERCISES

6.38 The probability distribution of x, the number of tires needing replacement on a randomly selected automobile checked at a certain inspection station, is given in the following table:

x	0	1	2	3	4
$p(x)$	0.54	0.16	0.06	0.04	0.20

The mean value of x is $\mu_x = 1.2$. Calculate the values of σ_x^2 and σ_x.

6.39 A company makes hardwood flooring, which it sells in boxes that will cover 500 square feet of floor. Let $x =$ the number of boxes ordered by a randomly chosen customer.

Suppose the probability distribution of x is as follows:

x	1	2	3	4
$p(x)$	0.2	0.4	0.3	0.1

a. Calculate and interpret the mean value of x.
b. Calculate and interpret the variance and standard deviation of x.

6.40 A business has six customer service telephone lines. Consider the random variable $x =$ number of lines in use at a randomly selected time. Suppose that the probability distribution of x is as follows:

x	0	1	2	3	4	5	6
$p(x)$	0.10	0.15	0.20	0.25	0.20	0.06	0.04

a. Calculate the mean value and standard deviation of x.
b. What is the probability that the number of lines in use is farther than 3 standard deviations from the mean value?

SECTION 6.5 Normal Distributions

Normal distributions are continuous probability distributions that formalize the mound-shaped distributions introduced in Section 6.3. Normal distributions are widely used for two reasons. First, they approximate the distributions of many different variables. Second, they also play a central role in many of the inferential procedures that will be discussed in later chapters.

Normal distributions are bell shaped and symmetric, as shown in Figure 6.16. They are sometimes referred to as *normal curves*.

FIGURE 6.16
A normal distribution

There are many different normal distributions. They are distinguished from one another by their mean μ and standard deviation σ. The mean μ of a normal distribution describes where the corresponding curve is centered, and the standard deviation σ describes how much the curve spreads out around that center. As with all continuous probability distributions, the total area under any normal curve is equal to 1.

Three normal distributions are shown in Figure 6.17. Notice that the smaller the standard deviation, the taller and narrower the corresponding curve. There will also be a larger area concentrated around μ at the center of the curve. Because areas under a continuous probability distribution curve represent probabilities, the chance of observing a value near the mean is much greater when the standard deviation is small.

The value of μ is the number on the measurement axis lying directly below the top of the bell. The value of σ can also be approximated from a picture of the curve. It is the distance to either side of μ at which a normal curve changes from turning downward to turning upward (these are sometimes called inflection points). Consider the normal curve in Figure 6.18. Starting at the top of the bell (above $\mu = 100$) and moving to the right, the curve turns downward until it is above the value 110. After that point, it continues to decrease in height but begins to turn upward rather than downward. Similarly, to the left of $\mu = 100$, the curve turns downward until it reaches 90 and then begins to turn upward. Because the curve changes from turning downward to turning upward at a distance of 10 on either side of μ, $\sigma = 10$ for this normal curve.

To use a particular normal distribution to describe the behavior of a random variable, a mean and a standard deviation must be specified. For example, you might use a normal distribution with mean 7 pounds and standard deviation 1 pound as a model for

FIGURE 6.17
Three normal distributions

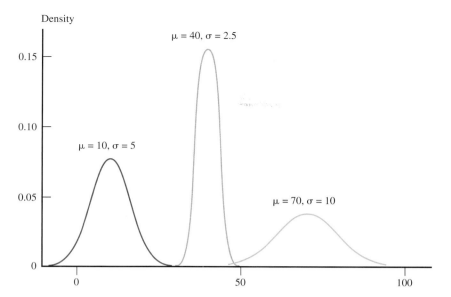

Density

μ = 40, σ = 2.5

μ = 10, σ = 5

μ = 70, σ = 10

FIGURE 6.18
Mean μ and standard deviation σ for a normal curve

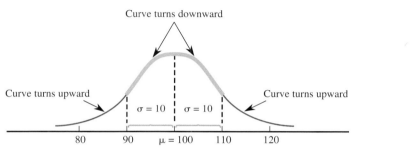

Curve turns downward

Curve turns upward

Curve turns upward

σ = 10   σ = 10

80   90   μ = 100   110   120

the distribution of x = birth weight of a baby. If this model is a reasonable description of the probability distribution, you could then use areas under the normal curve with $\mu = 7$ and $\sigma = 1$ to approximate various probabilities related to birth weight. For example, the probability that a birth weight is over 8 pounds (expressed as $P(x > 8)$) corresponds to the shaded area in Figure 6.19(a). The shaded area in Figure 6.19(b) represents the probability of a birth weight falling between 6.5 and 8 pounds, $P(6.5 < x < 8)$.

FIGURE 6.19
Normal distribution for birth weight:
(a) shaded area = $P(x > 8)$;
(b) shaded area = $P(6.5 < x < 8)$

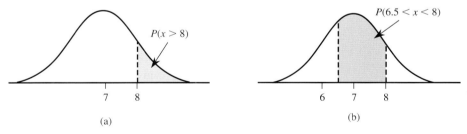

$P(x > 8)$

$P(6.5 < x < 8)$

7   8

6   7   8

(a)

(b)

Unfortunately, calculating such probabilities (areas under a normal curve) is not simple. To overcome this difficulty, you can rely on technology or a table of areas for a reference normal distribution, called the *standard normal distribution*.

DEFINITION

The **standard normal distribution** is the normal distribution with

$$\mu = 0 \text{ and } \sigma = 1$$

The corresponding density curve is called the *standard normal curve*. It is customary to use the letter z to represent a variable whose distribution is described by the standard normal curve. The standard normal curve is also sometimes called the z **curve**.

Few naturally occurring variables have distributions that are well described by the standard normal distribution, but this distribution is important because it is used in probability calculations for other normal distributions. When you are interested in finding a probability based on some other normal curve, you either rely on technology or translate the problem into an equivalent problem that involves finding an area under the standard normal curve. A table for the standard normal distribution is then used to find the desired area. To be able to do this, you must first learn to work with the standard normal distribution.

Standard Normal Distribution

In working with normal distributions, you need two general skills:

1. You must be able to use the normal distribution to calculate probabilities, which are areas under a normal curve and above given intervals.
2. You must be able to characterize extreme values in the distribution, such as the largest 5%, the smallest 1%, and the most extreme 5% (which would include the largest 2.5% and the smallest 2.5%).

Let's begin by looking at how to accomplish these tasks for the standard normal distribution.

The standard normal or z curve is shown in Figure 6.20(a). It is centered at $\mu = 0$, and the standard deviation, $\sigma = 1$, is a measure of how much it spreads out about its mean. Notice that this picture is consistent with the Empirical Rule of Chapter 3: About 95% of the area is associated with values that are within 2 standard deviations of the mean (between -2 and 2), and almost all of the area is associated with values that are within 3 standard deviations of the mean (between -3 and 3).

FIGURE 6.20
(a) The standard normal curve;
(b) a cumulative area

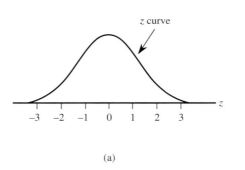

(a)

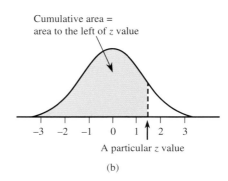

(b)

Appendix A Table 2 gives cumulative z curve areas like the one shown in Figure 6.20(b) for many different values of z. The smallest value for which the cumulative area is given is $z = -3.89$, a value far out in the lower tail of the z curve. The next smallest value for which an area is given is -3.88, then -3.87, and so on in increments of 0.01 all the way up to 3.89.

Using the Table of Standard Normal Curve Areas

For any number z^* between -3.89 and 3.89 and rounded to two decimal places, Appendix A Table 2 gives

$$\text{(area under } z \text{ curve to the left of } z^*) = P(z < z^*) = P(z \le z^*)$$

where the letter z is used to represent a random variable whose distribution is the standard normal distribution.

(continued)

To find this probability using the table, locate the following:

1. The row labeled with the sign of $z*$ and the digit on either side of the decimal point (for example, -1.7 or 0.5)
2. The column identified with the second digit to the right of the decimal point in $z*$ (for example, .06 if $z* = -1.76$)

The number at the intersection of this row and column is the desired probability, $P(z < z*)$.

A portion of the table of standard normal curve areas appears in Figure 6.21. To find the area under the z curve to the left of 1.42, look in the row labeled 1.4 and the column labeled .02 (the highlighted row and column in Figure 6.21). From the table, the corresponding cumulative area is 0.9222. So

$$z \text{ curve area to the left of } 1.42 = 0.9222$$

FIGURE 6.21
Portion of the table of standard normal curve areas

$z*$	.00	.01	.02	.03	.04	.05
0.0	.5000	.5040	.5080	.5120	.5160	.5199
0.1	.5398	.5438	.5478	.5517	.5557	.5596
0.2	.5793	.5832	.5871	.5910	.5948	.5987
0.3	.6179	.6217	.6255	.6293	.6331	.6368
0.4	.6554	.6591	.6628	.6664	.6700	.6736
0.5	.6915	.6950	.6985	.7019	.7054	.7088
0.6	.7257	.7291	.7324	.7357	.7389	.7422
0.7	.7580	.7611	.7642	.7673	.7704	.7734
0.8	.7881	.7910	.7939	.7967	.7995	.8023
0.9	.8159	.8186	.8212	.8238	.8264	.8289
1.0	.8413	.8438	.8461	.8485	.8508	.8531
1.1	.8643	.8665	.8686	.8708	.8729	.8749
1.2	.8849	.8869	.8888	.8907	.8925	.8944
1.3	.9032	.9049	.9066	.9082	.9099	.9115
1.4	.9192	.9207	.9222	.9236	.9251	.9265
1.5	.9332	.9345	.9357	.9370	.9382	.9394
1.6	.9452	.9463	.9474	.9484	.9495	.9505
1.7	.9554	.9564	.9573	.9582	.9591	.9599
1.8	.9641	.9649	.9656	.9664	.9671	.9678

$P(z < 1.42)$

You can also use the table to find the area to the right of 1.42. Because the total area under the z curve is 1, it follows that

$$(z \text{ curve area to the right of } 1.42) = 1 - (z \text{ curve area to the left of } 1.42)$$

$$= 1 - 0.9222$$

$$= 0.0778$$

These probabilities can be interpreted to mean that about 92.22% of observed z values will be smaller than 1.42, and about 7.78% will be larger than 1.42.

Example 6.17 **Finding Standard Normal Curve Areas**

The probability $P(z < -1.76)$ appears in the z table at the intersection of the -1.7 row and the .06 column. The result is

$$P(z < -1.76) = 0.0392$$

as shown in the following figure:

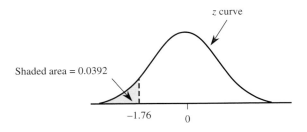

In other words, about 3.9% of observed z values will be smaller than -1.76. Similarly,

$$P(z \leq 0.58) = \text{entry in } 0.5 \text{ row and } .08 \text{ column of Table 2} = 0.7190$$

as shown in the following figure:

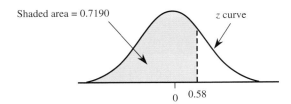

Now consider $P(z < -4.12)$. This probability does not appear in Appendix A Table 2; there is no -4.1 row. However, it must be less than $P(z < -3.89)$, the smallest z value in the table, because -4.12 is farther out in the lower tail of the z curve. Since $P(z < -3.89) \approx .0000$ (that is, zero to four decimal places), it follows that

$$P(z < -4.12) \approx 0$$

Similarly,

$$P(z < 4.12) > P(z < 3.89) \approx 1.0000$$

so

$$P(z < 4.12) \approx 1$$

As illustrated in Example 6.17, you can use Appendix A Table 2 to calculate other probabilities involving z. The probability that z is larger than a value c is

$$P(z > c) = \text{area under the } z \text{ curve to the right of } c = 1 - P(z \leq c)$$

In other words, the area to the right of a value (a right-tail area) is 1 minus the corresponding cumulative area. This is illustrated in Figure 6.22.

FIGURE 6.22
The relationship between an upper-tail area and a cumulative area

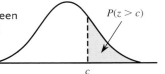 = 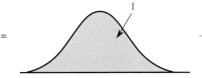 −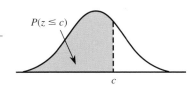

Similarly, the probability that z falls in the interval between a lower endpoint a and an upper endpoint b is

$$P(a < z < b) = \text{area under the } z \text{ curve and above the interval from } a \text{ to } b$$
$$= P(z < b) - P(z < a)$$

That is, $P(a < z < b)$ is the difference between two cumulative areas, as illustrated in Figure 6.23.

FIGURE 6.23
$P(a < z < b)$ as the difference between two cumulative areas

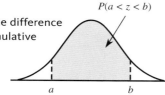 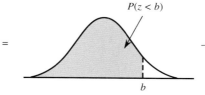 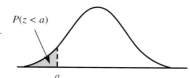

Example 6.18   **More About Standard Normal Curve Areas**

The probability that z is between -1.76 and 0.58 is

$$P(-1.76 < z < 0.58) = P(z < 0.58) - P(z < -1.76)$$
$$= 0.7190 - 0.0392$$
$$= 0.6798$$

as shown in the following figure:

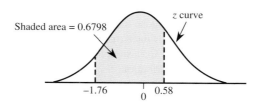

The probability that z is between -2 and $+2$ (within 2 standard deviations of its mean, since $\mu = 0$ and $\sigma = 1$) is

$$P(-2.00 < z < 2.00) = P(z < 2.00) - P(z < -2.00)$$
$$= 0.9772 - 0.0228$$
$$= 0.9544$$

as shown in the following figure:

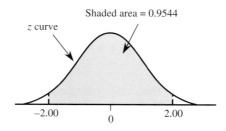

This last probability is the basis for one part of the Empirical Rule, which states that when a data distribution is well approximated by a normal curve, approximately 95% of the values are within 2 standard deviations of the mean.

The probability that the value of z exceeds 1.96 is

$$P(z > 1.96) = 1 - P(z < 1.96)$$
$$= 1 - 0.9750$$
$$= 0.0250$$

as shown in the following figure:

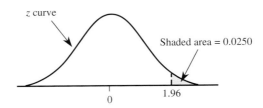

That is, 2.5% of the area under the z curve lies to the right of 1.96 in the upper tail. Similarly,

$$P(z > -1.28) = \text{area to the right of } -1.28$$
$$= 1 - P(z < -1.28)$$
$$= 1 - 0.1003$$
$$= 0.8997$$
$$\approx 0.90$$

Identifying Extreme Values

Suppose that you want to describe the values included in the smallest 2% of a distribution or the values making up the most extreme 5% (which includes the largest 2.5% and the smallest 2.5%). Examples 6.19 and 6.20 show how to identify extreme values in the standard normal distribution.

Example 6.19 Identifying Extreme Values

Suppose that you want to describe the values that make up the smallest 2% of the standard normal distribution. Symbolically, you are trying to find a value (call it z^*), such that

$$P(z < z^*) = 0.02$$

This is illustrated in Figure 6.24, which shows that the cumulative area for z^* is 0.02.

FIGURE 6.24
The smallest 2% of the standard normal distribution

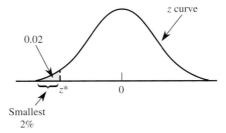

This means you should look for a cumulative area of 0.0200 in the body of Appendix A Table 2. The closest cumulative area in the table is 0.0202 (in the -2.0 row and .05 column), so $z^* = -2.05$. Variable values less than -2.05 make up the smallest 2% of the standard normal distribution. Notice that, because of the symmetry of the normal curve, variable values greater than 2.05 make up the largest 2% of the standard normal distribution.

Now suppose that you are interested in the largest 5% of all z values. You would then be trying to find a value of z^* for which

$$P(z > z^*) = 0.05$$

as illustrated in Figure 6.25. Because Appendix A Table 2 always works with cumulative area (area to the left), the first step is to determine

$$\text{area to the left of } z^* = 1 - 0.05 = 0.95$$

FIGURE 6.25
The largest 5% of the standard normal distribution

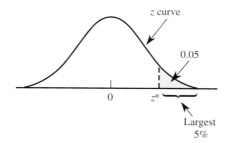

Looking for the cumulative area closest to 0.95 in Appendix A Table 2, you find that 0.95 falls exactly halfway between 0.9495 (corresponding to a z value of 1.64) and 0.9505 (corresponding to a z value of 1.65). Because 0.9500 is exactly halfway between the two areas, use a z value that is halfway between 1.64 and 1.65. (If one value had been closer to 0.9500 than the other, you could just use the z value corresponding to the closest area.) This gives

$$z^* = \frac{1.64 + 1.65}{2} = 1.645$$

Values greater than 1.645 make up the largest 5% of the standard normal distribution. By symmetry, -1.645 separates the smallest 5% of all z values from the others.

Example 6.20 **More Extremes**

Sometimes you are interested in identifying the most extreme (unusually large *or* small) values in a distribution. How can you describe the values that make up the most extreme 5% of the standard normal distribution? Here you want to separate the middle 95% from the most extreme 5%. This is illustrated in Figure 6.26.

FIGURE 6.26
The most extreme 5% of the standard normal distribution

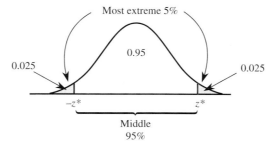

Because the standard normal distribution is symmetric, the most extreme 5% is equally divided between the high side and the low side of the distribution, resulting in an area of 0.025 for each of the tails of the z curve. To find z^*, first determine the cumulative area for z^*, which is

area to the left of $z^* = 0.95 + 0.025 = 0.975$

The cumulative area 0.9750 appears in the 1.9 row and .06 column of Appendix A Table 2, so $z^* = 1.96$. Symmetry about 0 implies that if z^* denotes the value that separates the largest 2.5%, the value that separates the smallest 2.5% is $-z^*$, or -1.96. For the standard normal distribution, values that are either less than -1.96 or greater than 1.96 are the most extreme 5%.

Other Normal Distributions

Areas under the z curve can be used to calculate probabilities and to describe values for *any* normal distribution. Remember that the letter z is usually reserved for variables that have a standard normal distribution. Other letters, such as x or y, are used to denote a variable whose distribution is described by a normal curve with mean μ and standard deviation σ.

Suppose that you want to calculate $P(a < x < b)$, the probability that the variable x lies in a particular range. This probability corresponds to an area under a normal curve and above the interval from a to b, as shown in Figure 6.27(a).

FIGURE 6.27
Equality of nonstandard and standard normal curve areas

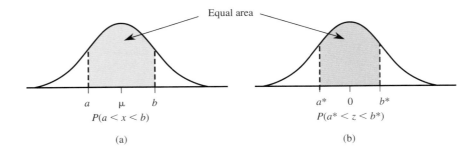

The strategy for obtaining this probability is to find an equivalent problem involving the standard normal distribution. Finding an equivalent problem means determining an interval (a^*, b^*) that has the same probability for z (same area under the z curve) as the interval (a, b) in the original normal distribution (see Figure 6.27(b)). The asterisk is used to distinguish a and b, the values from the original normal distribution from a^* and b^*, the values from the z distribution.

To find a^* and b^*, simply calculate z-scores for a and b. This process is called **standardizing** the endpoints. For example, suppose that the variable x has a normal distribution with mean $\mu = 100$ and standard deviation $\sigma = 5$. To find

$$P(98 < x < 107)$$

first translate this problem into an equivalent one for the standard normal distribution. Recall from Chapter 3 that a z-score is calculated by first subtracting the mean and then dividing by the standard deviation. Converting the lower endpoint $a = 98$ to a z-score gives

$$a^* = \frac{98 - 100}{5} = \frac{-2}{5} = -0.40$$

and converting the upper endpoint yields

$$b^* = \frac{107 - 100}{5} = \frac{7}{5} = 1.40$$

Then

$$P(98 < x < 107) = P(-0.40 < z < 1.40)$$

The probability $P(-0.40 < z < 1.40)$ can now be evaluated using technology or Appendix A Table 2.

Finding Probabilities

To calculate probabilities for any normal distribution, standardize the relevant values and then use the table of z curve areas. More specifically, if x is a variable whose behavior is described by a normal distribution with mean μ and standard deviation σ, then

$$P(x < b) = P(z < b^*)$$
$$P(x > a) = P(z > a^*)$$
$$P(a < x < b) = P(a^* < z < b^*)$$

where z is a variable whose distribution is standard normal and

$$a^* = \frac{a - \mu}{\sigma} \qquad b^* = \frac{b - \mu}{\sigma}$$

Example 6.21 Newborn Birth Weights

Data from the paper **"Birth Weight Curves Tailored to Maternal World Region"** (*Journal of Obstetrics and Gynaecology Canada* **[2012]: 159–171**) suggest that a normal distribution with mean $\mu = 3500$ grams and standard deviation $\sigma = 550$ grams is a reasonable model for the probability distribution of $x =$ birth weight of a randomly selected full-term baby born in Canada. What proportion of birth weights in Canada are between 2900 and 4700 grams?

To answer this question, you must find

$$P(2900 < x < 4700)$$

First, translate the interval endpoints to equivalent endpoints for the standard normal distribution:

$$a^* = \frac{a - \mu}{\sigma} = \frac{2900 - 3500}{550} = -1.09$$

$$b^* = \frac{b - \mu}{\sigma} = \frac{4700 - 3500}{550} = 2.18$$

Then

$$P(2900 < x < 4700) = P(-1.09 < z < 2.18)$$
$$= (z \text{ curve area to the left of } 2.18)$$
$$- (z \text{ curve area to the left of } -1.09)$$
$$= 0.9854 - 0.1379$$
$$= 0.8475$$

The probabilities for x and z are shown in Figure 6.28. If birth weights were observed for many babies born in Canada, about 85% of them would be between 2900 and 4700 grams.

FIGURE 6.28
$P(2900 < x < 4700)$ and corresponding z curve area for the birth weight problem of Example 6.21

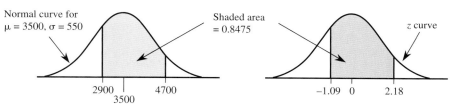

What is the probability that a randomly chosen baby will have a birth weight greater than 4500? To evaluate $P(x > 4500)$. First calculate

$$a^* = \frac{a - \mu}{\sigma} = \frac{4500 - 3500}{550} = 1.82$$

Then (see Figure 6.29)

$$P(x > 4500) = P(z > 1.82)$$
$$= z \text{ curve area to the right of } 1.82$$
$$= 1 - (z \text{ curve area to the left of } 1.82)$$
$$= 1 - 0.9656$$
$$= 0.0344$$

FIGURE 6.29
$P(x > 4500)$ and corresponding z curve area for the birth weight problem of Example 6.21

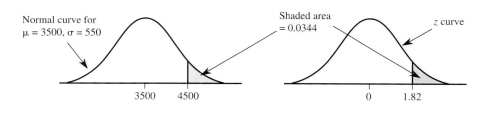

Example 6.22 **IQ Scores**

Although there is some controversy regarding the appropriateness of IQ scores as a measure of intelligence, they are still used for a variety of purposes. One commonly used IQ scale has a mean of 100 and a standard deviation of 15, and these IQ scores are approximately normally distributed. (Because this IQ score is based on the number of correct responses on a test, it is actually a discrete variable. However, the population distribution of IQ scores closely resembles a normal curve.) If we define the random variable

$$x = \text{IQ score of a randomly selected individual}$$

then x has a distribution that is approximately normal with $\mu = 100$ and $\sigma = 15$.

One way to become eligible for membership in Mensa, an organization for those of high intelligence, is to have an IQ score above 130. What proportion of the population would qualify for Mensa membership? Answering this question requires evaluating $P(x > 130)$. This probability is shown in Figure 6.30.

FIGURE 6.30
Normal distribution and desired proportion for Example 6.22

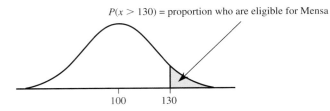

$P(x > 130) = $ proportion who are eligible for Mensa

With $a = 130$,

$$a^* = \frac{a - \mu}{\sigma} = \frac{130 - 100}{15} = 2.00$$

So (see Figure 6.31)

$$P(x > 130) = P(z > 2.00)$$
$$= z \text{ curve area to the right of } 2.00$$
$$= 1 - (z \text{ curve area to the left of } 2.00)$$
$$= 1 - 0.9772$$
$$= 0.0228$$

Only about 2.28% of the population would qualify for Mensa membership.

FIGURE 6.31
$P(x > 130)$ and corresponding z curve area

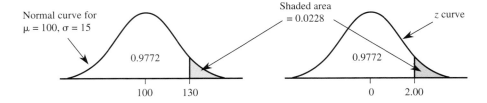

Suppose that you are interested in the proportion of the population with IQ scores below 80—that is, $P(x < 80)$. With $b = 80$,

$$b^* = \frac{b - \mu}{\sigma} = \frac{80 - 100}{15} = -1.33$$

So

$$P(x < 80) = P(z < -1.33)$$
$$= z \text{ curve area to the left of } -1.33$$
$$= 0.0918$$

as shown in Figure 6.32. This probability tells you that just a little over 9% of the population has an IQ score below 80.

FIGURE 6.32

$P(x < 80)$ and corresponding z curve area for the IQ problem of Example 6.22

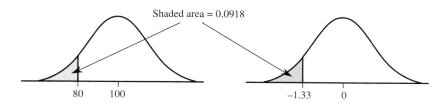

Now consider the proportion of the population with IQs between 75 and 125. Using $a = 75$ and $b = 125$, you obtain

$$a^* = \frac{75 - 100}{15} = -1.67 \qquad b^* = \frac{125 - 100}{15} = 1.67$$

so

$$P(75 < x < 125) = P(-1.67 < z < 1.67)$$
$$= z \text{ curve area between } -1.67 \text{ and } 1.67$$
$$= (z \text{ curve area to the left of } 1.67)$$
$$\quad - (z \text{ curve area to the left of } -1.67)$$
$$= 0.9525 - 0.0475$$
$$= 0.9050$$

This is illustrated in Figure 6.33. This probability tells you that about 90.5% of the population has an IQ score between 75 and 125. Of the 9.5% with an IQ score that is not between 75 and 125, about half of them (4.75%) have scores greater than 125, and about half have scores less than 75.

FIGURE 6.33

$P(75 < x < 125)$ and corresponding z curve area

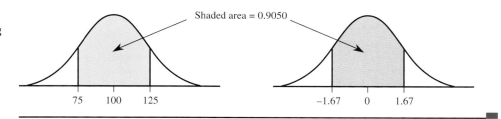

When translating a problem involving a normal distribution with mean μ and standard deviation σ to a problem involving the standard normal distribution, you convert to z-scores:

$$z = \frac{x - \mu}{\sigma}$$

Recall from Chapter 3 that a z-score can be interpreted as the distance of an x value from the mean in units of the standard deviation. A z-score of 1.4 corresponds to an x value that is 1.4 standard deviations above the mean, and a z-score of -2.1 corresponds to an x value that is 2.1 standard deviations below the mean.

Suppose that you are trying to evaluate $P(x < 60)$ for a variable whose distribution is normal with $\mu = 50$ and $\sigma = 5$. Converting the endpoint 60 to a z-score gives

$$z = \frac{60 - 50}{5} = 2$$

which tells you that the value 60 is two standard deviations above the mean. You then have

$$P(x < 60) = P(z < 2)$$

where z is a standard normal variable. Notice that for the standard normal distribution, which has $\mu = 0$ and $\sigma = 1$, the value 2 is two standard deviations above the mean. The value $z = 2$ is located the same distance (measured in standard deviations) from the mean of the standard normal distribution as is the value $x = 60$ from the mean in the normal distribution with $\mu = 50$ and $\sigma = 5$. This is why the translation using z-scores results in an equivalent problem involving the standard normal distribution.

Describing Extreme Values in a Normal Distribution

To describe the extreme values for a normal distribution with mean μ and standard deviation σ, first solve the corresponding problem for the standard normal distribution and then translate your answer back to the normal distribution of interest. This process is illustrated in Example 6.23.

Example 6.23 **Registration Times**

Data on the length of time (in minutes) required to register for classes using an online system suggest that the distribution of the variable

$$x = \text{time to register}$$

for students at a particular university can be well approximated by a normal distribution with mean $\mu = 12$ minutes and standard deviation $\sigma = 2$ minutes. Because some students do not log off properly, the university would like to log off students automatically after some amount of time has elapsed. The university has decided to choose this time so that only 1% of the students are logged off while they are still attempting to register.

To determine the amount of time that should be allowed before disconnecting a student, you need to describe the largest 1% of the distribution of time to register. These are the individuals who will be mistakenly disconnected. This is illustrated in Figure 6.34(a). To determine the value of x^*, first solve the corresponding problem for the standard normal distribution, as shown in Figure 6.34(b).

FIGURE 6.34

Capturing the largest 1% in a normal distribution

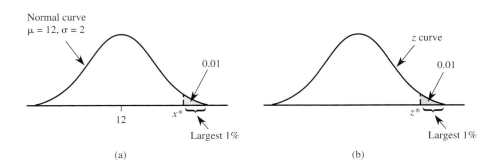

(a)   (b)

By looking for a cumulative area of 0.99 in Appendix A Table 2, you find the closest entry (0.9901) in the 2.3 row and the .03 column, so $z^* = 2.33$. For the standard normal distribution, the largest 1% of the distribution is made up of those values greater than 2.33. This implies that in the distribution of time to register x (or any other normal distribution), the largest 1% are those values with z-scores greater than 2.33 or, equivalently, those x values more than 2.33 standard deviations above the mean. Here, the standard deviation is 2, so 2.33 standard deviations is 2.33(2), and it follows that

$$x^* = 12 + 2.33(2) = 12 + 4.66 = 16.66$$

The largest 1% of the distribution for time to register is made up of values that are greater than 16.66 minutes. If the system was set to log off students after 16.66 minutes, only about 1% of the students registering would be logged off before completing their registration.

A general formula for converting a z-score back to an x value results from solving

$$z^* = \frac{x^* - \mu}{\sigma}$$

for x^*, as shown in the accompanying box.

To convert a z score, denoted by z^* back to an x value, use

$$x^* = \mu + z^*\sigma$$

Example 6.24 **Garbage Truck Processing Times**

Garbage trucks entering a particular waste management facility are weighed before offloading garbage into a landfill. Data from the paper **"Estimating Waste Transfer Station Delays Using GPS" (*Waste Management* [2008]: 1742–1750)** suggest that a normal distribution with mean $\mu = 13$ minutes and $\sigma = 3.9$ minutes is a reasonable model for the probability distribution of the random variable $x = $ *total processing time* for a garbage truck at this waste management facility (total processing time includes waiting time as well). Suppose that you want to describe the total processing times of the trucks making up the 10% with the longest processing times. These trucks would be the 10% with times corresponding to the shaded region in the accompanying illustration.

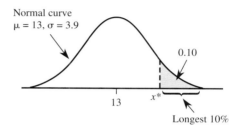

For the standard normal distribution, the longest 10% are those with z-scores greater than $z^* = 1.28$ (from Appendix A Table 2, based on a cumulative area of 0.90).
    Then

$$x^* = \mu + z^*\sigma$$
$$= 13 + 1.28(3.9)$$
$$= 13 + 4.992$$
$$= 17.992$$

About 10% of the garbage trucks using this facility would have a total processing time of more than about 18 minutes.
    The 5% with the fastest processing times would be those with z-scores less than $z^* = -1.645$ (from Appendix A Table 2, based on a cumulative area of 0.05). Then

$$x^* = \mu + z^*\sigma$$
$$= 13 + (-1.645)(3.9)$$
$$= 13 - 6.416$$
$$= 6.584$$

About 5% of the garbage trucks processed at this facility will have total processing times of less than about 6.6 minutes.

Summing It Up—Section 6.5

The following learning objectives were addressed in this section:

Mastering the Mechanics
M7: Interpret an area under a normal curve as a probability.
Because normal distributions are continuous probability distributions, an area under the normal curve and above a given interval represents the probability of observing a value in that interval.

M8: For a normal random variable x**, use technology or tables to calculate probabilities of the form** $P(a < x < b)$, $P(x < b)$, **and** $P(x > a)$ **and to find percentiles.** Technology or tables can be used to find areas (which represent probabilities) under a normal curve with a specified mean and standard deviation. If using a table, the endpoints of the interval of interest are converted to z-scores, which results in an interval that has the same area for the standard normal distribution as the original interval for the given normal distribution. This is illustrated in Example 6.19. It is also possible to use technology or tables to find percentiles, which can be used to describe extreme values in a normal distribution. For an example, see Example 6.24.

Putting It into Practice
P1: Use information provided by a probability distribution to draw conclusions in context. Because the long-run behavior of many different random variables is well described by a normal probability distribution, it is common to use the normal distribution to answer questions in a variety of contexts. This is illustrated in many of the examples in this section.

SECTION 6.5 EXERCISES
Each Exercise Set assesses the following chapter learning objectives: M7, M8, P1

SECTION 6.5 Exercise Set 1

6.41 Determine the following standard normal (z) curve areas:
a. The area under the z curve to the left of 1.75
b. The area under the z curve to the left of -0.68
c. The area under the z curve to the right of 1.20
d. The area under the z curve to the right of -2.82
e. The area under the z curve between -2.22 and 0.53
f. The area under the z curve between -1 and 1
g. The area under the z curve between -4 and 4

6.42 Let z denote a random variable that has a standard normal distribution. Determine each of the following probabilities:
a. $P(z < 2.36)$
b. $P(z \leq 2.36)$
c. $P(z < -1.23)$
d. $P(1.14 < z < 3.35)$
e. $P(-0.77 \leq z \leq -0.55)$
f. $P(z > 2)$
g. $P(z \geq -3.38)$
h. $P(z < 4.98)$

6.43 Consider the population of all one-gallon cans of dusty rose paint manufactured by a particular paint company. Suppose that a normal distribution with mean $\mu = 5$ ml and standard deviation $\sigma = 0.2$ ml is a reasonable model for the distribution of the variable

x = amount of red dye in the paint mixture

Use the normal distribution to calculate the following probabilities. (Hint: See Example 6.21.)
a. $P(x < 5.0)$
b. $P(x < 5.4)$
c. $P(x \leq 5.4)$
d. $P(4.6 < x < 5.2)$
e. $P(x > 4.5)$
f. $P(x > 4.0)$

6.44 Suppose that the distribution of typing speed in words per minute (wpm) for experienced typists using a new type of split keyboard can be approximated by a normal curve with mean 60 wpm and standard deviation 15 wpm (**"The Effects of Split Keyboard Geometry on Upper Body Postures, Ergonomics [2009]: 104–111**).
a. What is the probability that a randomly selected typist's speed is at most 60 wpm? Less than 60 wpm?
b. What is the probability that a randomly selected typist's speed is between 45 and 90 wpm?
c. Would you be surprised to find a typist in this population whose speed exceeded 105 wpm?
d. Suppose that two typists are independently selected. What is the probability that both their speeds exceed 75 wpm?
e. Suppose that special training is to be made available to the slowest 20% of the typists. What typing speeds would qualify individuals for this training? (Hint: See Example 6.23.)

6.45 A machine that cuts corks for wine bottles operates in such a way that the distribution of the diameter of the corks produced is well approximated by a normal distribution with mean 3 cm and standard deviation 0.1 cm. The specifications call for corks with diameters between 2.9 and 3.1 cm. A cork not meeting the specifications is considered defective. (A cork that is too small leaks and causes the wine to deteriorate; a cork that is too large doesn't fit in the bottle.) What proportion of corks produced by this machine are defective?

6.46 Refer to the previous exercise. Suppose that there are two machines available for cutting corks. The machine described in the preceding problem produces corks with diameters that are approximately normally distributed with mean 3 cm and standard deviation 0.1 cm. The second machine produces corks with diameters that are approximately normally distributed with mean 3.05 cm and standard deviation 0.01 cm. Which machine would you recommend? (Hint: Which machine would produce fewer defective corks?)

SECTION 6.5 Exercise Set 2

6.47 Determine each of the following areas under the standard normal (z) curve:
a. To the left of -1.28
b. To the right of 1.28
c. Between -1 and 2
d. To the right of 0
e. To the right of -5
f. Between -1.6 and 2.5
g. To the left of 0.23

6.48 Let z denote a random variable having a normal distribution with $\mu = 0$ and $\sigma = 1$. Determine each of the following probabilities:
a. $P(z < 0.10)$
b. $P(z < -0.10)$
c. $P(0.40 < z < 0.85)$
d. $P(-0.85 < z < -0.40)$
e. $P(-0.40 < z < 0.85)$
f. $P(z > -1.25)$
g. $P(z < -1.50 \ or \ z > 2.50)$

6.49 The article **"New York City's Graffiti-Removal Response Time Rises" (***The Wall Street Journal,*** September 16, 2016, www.wsj.com/articles/new-york-citys-graffiti-removal-response-time-rises-1473287392, retrieved May 1, 2017)** states that the city took an average of 114 days to handle graffiti complaints in 2015. Suppose that the response time is approximately normally distributed with a mean of 114 days and a standard deviation of 20 days.
a. Approximately what proportion of graffiti removal requests are handled within 60 days?
b. Approximately what proportion of graffiti removal requests take more than 120 days?

6.50 Purchases made at small "corner stores" were studied by the authors of the paper **"Changes in Quantity, Spending, and Nutritional Characteristics of Adult, Adolescent and Child Urban Corner Store Purchases After an Environmental Intervention" (***Preventive Medicine*** [2015]: 81–85)**. Corner stores were defined as stores that are less than 200 square feet in size, have only one cash register, and primarily sell food. After observing a large number of corner store purchases in Philadelphia, the authors reported that the average number of grams of fat in a corner store purchase was 21.1. Suppose that the variable

x = number of grams of fat in a corner store purchase

has a distribution that is approximately normal with a mean of 21.1 grams and a standard deviation of 7 grams.
a. What is the probability that a randomly selected corner store purchase has more than 30 grams of fat?
b. What is the probability that a randomly selected corner store purchase has between 15 and 25 grams of fat?
c. If two corner store purchases are randomly selected, what it the probability that both of these purchases will have more than 25 grams of fat?

6.51 The paper referenced in Example 6.21 suggested that a normal distribution with mean 3500 grams and standard deviation 550 grams is a reasonable model for birth weights of babies born in Canada.
a. One common medical definition of a large baby is any baby that weighs more than 4000 grams at birth. What is the probability that a randomly selected Canadian baby is a large baby?
b. What is the probability that a randomly selected Canadian baby weighs either less than 2000 grams or more than 4000 grams at birth?
c. What birth weights describe the 10% of Canadian babies with the greatest birth weights?

6.52 The time that it takes a randomly selected job applicant to perform a certain task has a distribution that can be approximated by a normal distribution with a mean of 120 seconds and a standard deviation of 20 seconds. The fastest 10% are to be given advanced training. What task times qualify individuals for such training?

ADDITIONAL EXERCISES

6.53 The paper **"Examining Communication- and Media-Based Recreational Sedentary Behaviors Among Canadian Youth: Results from the COMPASS Study" (***Preventive Medicine*** [2015]: 74–80)** estimated that the time spent playing video or computer games by high school boys had a mean of 123.4 minutes per day and a standard deviation of 117.1 minutes per day. Based on this mean and standard deviation, explain why it is not reasonable to think that the distribution of the random variable x = *time spent playing video or computer games* is approximately normal.

6.54 A machine that produces ball bearings has initially been set so that the mean diameter of the bearings it produces is 0.500 inches. A bearing is acceptable if its diameter is within 0.004 inches of this target value. Suppose, however, that the setting has changed during the course of production, so that the distribution of the diameters produced is now approximately normal with mean 0.499 inch and standard deviation 0.002 inch. What percentage of the bearings produced will not be acceptable?

6.55 The paper referenced in Example 6.24 **("Estimating Waste Transfer Station Delays Using GPS,"** *Waste Management* **[2008]: 1742–1750)** describing processing times for garbage trucks also provided information on processing times at a second facility. At this second facility, the mean total processing time was 9.9 minutes and the standard deviation of the processing times was 6.2 minutes. Explain why a normal distribution with mean 9.9 and standard deviation 6.2 would not be an appropriate model for the probability distribution of the variable x = *total processing time* of a randomly selected truck entering this second facility.

6.56 The number of vehicles leaving a highway at a certain exit during a particular time period has a distribution that is approximately normal with mean value 500 and standard deviation 75. What is the probability that the number of cars exiting during this period is
a. At least 650?
b. Strictly between 400 and 550? (*Strictly* means that the values 400 and 550 are not included.)
c. Between 400 and 550 (inclusive)?

6.57 A pizza company advertises that it puts 0.5 pound of real mozzarella cheese on its medium-sized pizzas. In fact, the amount of cheese on a randomly selected medium pizza is normally distributed with a mean value of 0.5 pound and a standard deviation of 0.025 pound.
a. What is the probability that the amount of cheese on a medium pizza is between 0.525 and 0.550 pounds?
b. What is the probability that the amount of cheese on a medium pizza exceeds the mean value by more than 2 standard deviations?
c. What is the probability that three randomly selected medium pizzas each have at least 0.475 pounds of cheese?

6.58 Suppose that fuel efficiency (miles per gallon, mpg) for a particular car model under specified conditions is normally distributed with a mean value of 30.0 mpg and a standard deviation of 1.2 mpg.
a. What is the probability that the fuel efficiency for a randomly selected car of this model is between 29 and 31 mpg?
b. Would it surprise you to find that the efficiency of a randomly selected car of this model is less than 25 mpg?
c. If three cars of this model are randomly selected, what is the probability that each of the three have efficiencies exceeding 32 mpg?
d. Find a number x^* such that 95% of all cars of this model have efficiencies exceeding x^* (i.e., $P(x > x^*) = 0.95$).

6.59 Suppose that the amount of time spent by a statistical consultant with a client at their first meeting is a random variable that has a normal distribution with a mean value of 60 minutes and a standard deviation of 10 minutes.
a. What is the probability that more than 45 minutes is spent at the first meeting?
b. What amount of time is exceeded by only 10% of all clients at a first meeting?

6.60 When used in a particular DVD player, the lifetime of a certain brand of battery is normally distributed with a mean value of 6 hours and a standard deviation of 0.8 hour. Suppose that two new batteries are independently selected and put into the player. The player ceases to function as soon as one of the two batteries fails.

a. What is the probability that the DVD player functions for at least 4 hours?
b. What is the probability that the DVD player functions for at most 7 hours?
c. Find a number x^* such that only 5% of all DVD players will function without battery replacement for more than x^* hours.

6.61 A machine producing vitamin E capsules operates so that the actual amount of vitamin E in each capsule is normally distributed with a mean of 5 mg and a standard deviation of 0.05 mg. What is the probability that a randomly selected capsule contains less than 4.9 mg of vitamin E? At least 5.2 mg of vitamin E?

6.62 *The Wall Street Journal* **(February 15, 1972)** reported that General Electric was sued in Texas for sex discrimination over a minimum height requirement of 5 feet, 7 inches. The suit claimed that this restriction eliminated more than 94% of adult females from consideration. Let x represent the height of a randomly selected adult woman. Suppose that x is approximately normally distributed with mean 66 inches (5 ft. 6 in.) and standard deviation 2 inches.
a. Is the claim that 94% of all women are shorter than 5 ft. 7 in. correct?
b. What proportion of adult women would be excluded from employment as a result of the height restriction?

6.63 Suppose that your statistics professor tells you that the scores on a midterm exam were approximately normally distributed with a mean of 78 and a standard deviation of 7. The top 15% of all scores have been designated A's. Your score is 89. Did you earn an A? Explain.

6.64 Suppose that the pH of soil samples taken from a certain geographic region is normally distributed with a mean of 6.00 and a standard deviation of 0.10. Suppose the pH of a randomly selected soil sample from this region will be determined.
a. What is the probability that the resulting pH is between 5.90 and 6.15?
b. What is the probability that the resulting pH exceeds 6.10?
c. What is the probability that the resulting pH is at most 5.95?
d. What value will be exceeded by only 5% of all such pH values?

6.65 The light bulbs used to provide exterior lighting for a large office building have an average lifetime of 700 hours. If lifetime is approximately normally distributed with a standard deviation of 50 hours, how often should all the bulbs be replaced so that no more than 20% of the bulbs will have already burned out?

6.66 Let x denote the duration of a randomly selected pregnancy (the time elapsed between conception and birth). Accepted values for the mean value and standard deviation of *x* are 266 days and 16 days, respectively. Suppose that the probability distribution of *x* is (approximately) normal.

a. What is the probability that the duration of a randomly selected pregnancy is between 250 and 300 days?

b. What is the probability that the duration is at most 240 days?

c. What is the probability that the duration is within 16 days of the mean duration?

d. A "Dear Abby" newspaper column dated January 20, 1973, contained a letter from a woman who stated that the duration of her pregnancy was exactly 310 days. (She wrote that the last visit with her husband, who was in the navy, occurred 310 days before the birth of her child.) What is the probability that the duration of pregnancy is at least 310 days? Does this probability make you skeptical of the claim?

e. Some insurance companies will pay the medical expenses associated with childbirth only if the insurance has been in effect for more than 9 months (275 days). This restriction is designed to ensure that benefits are only paid if conception occurred during coverage. Suppose that conception occurred 2 weeks after coverage began. What is the probability that the insurance company will refuse to pay benefits because of the 275-day requirement?

SECTION 6.6   Checking for Normality

Some of the most frequently used statistical methods are valid only when a sample is from a population distribution that is at least approximately normal. One way to see if the population distribution is approximately normal is to construct a **normal probability plot** of the data. This plot uses quantities called **normal scores**. The values of the normal scores depend on the sample size *n*. For example, the normal scores when *n* = 10 are as follows:

$$-1.547 \quad -1.000 \quad -0.655 \quad -0.375 \quad -0.123 \quad 0.123 \quad 0.375 \quad 0.655 \quad 1.000 \quad 1.547$$

To interpret these numbers, think of selecting sample after sample from a normal distribution, each one consisting of *n* = 10 observations. Then −1.547 is the long-run average *z*-score of the smallest observation from each sample, −1.000 is the long-run average *z*-score of the second smallest observation from each sample, and so on.

Tables of normal scores for many different sample sizes are available. Alternatively, many software packages (such as JMP and Minitab) and some graphing calculators can compute these scores and then use them to construct a normal probability plot. Not all calculators and software packages use the same algorithm to compute normal scores. However, this does not change the overall character of a normal probability plot, so either tabulated values or those given by the computer or calculator can be used.

To construct a normal probability plot, first arrange the sample observations from smallest to largest. The smallest normal score is then paired with the smallest observation, the second smallest normal score with the second smallest observation, and so on. The first number in a pair is the normal score, and the second number in the pair is the corresponding data value. A normal probability plot is just a scatterplot of the (normal score, observed value) pairs.

If the sample has been selected from a *standard* normal distribution, the second number in each pair should be reasonably close to the first number (ordered observation ≈ corresponding mean value). If this is the case, the *n* plotted points will fall near a line with slope equal to 1 (a 45° line) passing through (0, 0). In general, when the sample is from a normal population distribution, the plotted points should be close to *some* straight line (but not necessarily one with slope 1 and intercept 0).

DEFINITION

A **normal probability** plot is a scatterplot of the (normal score, observed value) pairs. A strong linear pattern in a normal probability plot suggests that it is reasonable to think that the population distribution is approximately normal. On the other hand, systematic departure from a straight-line pattern (such as curvature in the plot) suggests that the population distribution is not normal.

Some software packages and graphing calculators plot the (observed value, normal score) pairs instead of the (normal score, observed value) pairs. This is also an acceptable way to construct a normal probability plot.

Data set available

Example 6.25 **Egg Weights**

The following data represent egg weights (in grams) for a sample of 10 eggs. These data are consistent with summary quantities in the paper **"Evaluation of Egg Quality Traits of Chickens Reared under Backyard System in Western Uttar Pradesh"** *(Indian Journal of Poultry Science*, 2009)*.

53.04 53.50 52.53 53.00 53.07 52.86 52.66 53.23 53.26 53.16

Arranging the sample observations in order from smallest to largest results in

52.53 52.66 52.86 53.00 53.04 53.07 53.16 53.23 53.26 53.50

Pairing these ordered observations with the normal scores for a sample of size 10 (given at the beginning of this section) results in the following 10 ordered pairs that can be used to construct the normal probability plot:

$(-1.547, 52.53)$	$(-1.000, 52.66)$	$(-0.655, 52.86)$	$(-0.375, 53.00)$
$(-0.123, 53.04)$	$(0.123, 53.07)$	$(0.375, 53.16)$	$(0.655, 53.23)$
$(1.000, 53.26)$	$(1.547, 53.50)$		

The normal probability plot is shown in Figure 6.35. The linear pattern in the plot suggests that it is reasonable to think that the population egg-weight distribution is approximately normal.

FIGURE 6.35

A normal probability plot for the egg weight data of Example 6.25

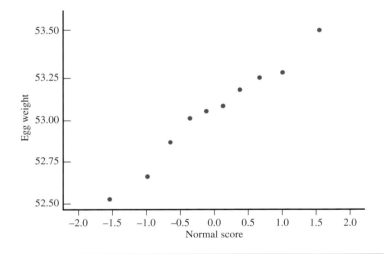

How do you know if a normal probability plot shows a strong linear pattern? Sometimes it is a matter of opinion. Particularly when n is small, normality should not be ruled out unless the pattern is clearly nonlinear. Figure 6.36 displays several plots for population distributions that are not normal. Figure 6.36(a) shows a pattern that results when the population distribution is skewed. The pattern in Figure 6.36(b) is one that occurs when the population distribution is roughly symmetric but flatter and not as bell shaped as the normal distribution. Figure 6.36(c) shows what the normal probability plot might look like if there is an outlier in a small data set. Any of these patterns suggest that the population distribution may not be normal.

FIGURE 6.36

Plots suggesting a population distribution is not normal:
(a) indication that the population distribution is skewed;
(b) indication that the population distribution has heavier tails than a normal distribution;
(c) indication of an outlier.

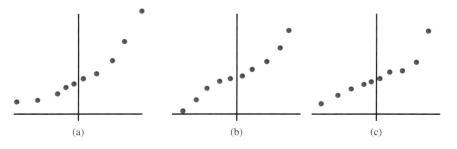

(a)　　　　(b)　　　　(c)

Using the Correlation Coefficient to Check Normality

The correlation coefficient r was introduced in Chapter 4 as a way to measure how closely the points in a scatterplot fall to a straight line. Consider calculating the value of r using the n (normal score, observed value) pairs:

(smallest normal score, smallest observation)

$$\vdots$$

(largest normal score, largest observation)

The normal probability plot always slopes upward (because it is based on values ordered from smallest to largest), so r will be a positive number. A value of r quite close to 1 indicates a very strong linear relationship in the normal probability plot. If r is much smaller than 1, the population distribution is not likely to be normal.

How much smaller than 1 does r have to be before you should begin to seriously doubt normality? The answer depends on the sample size n. If n is small, an r value somewhat below 1 is not surprising, even when the population distribution is normal. However, if n is large, only an r value very close to 1 supports normality. For selected values of n, Table 6.2 gives critical values that can be used to check for normality. If the sample size is in between two tabulated values of n, use the critical value for the larger sample size. (For example, if $n = 46$, use the value 0.966 given for sample size 50.)

TABLE 6.2 Critical Values for r*

n	5	10	15	20	25	30	40	50	60	75
Critical r	0.832	0.880	0.911	0.929	0.941	0.949	0.960	0.966	0.971	0.976

*Source: *"Normal Probability Plots and Tests for Normality," www.minitab.com/uploadedFiles/Content/News /Published_Articles/normal_probability_plots.pdf.*

Checking for Normality

If

$r <$ critical r for the corresponding n

it is not reasonable to think that the population distribution is normal.

How were the critical values in Table 6.2, such as the critical value 0.941 for $n = 25$, obtained? Consider selecting a large number of different random samples, each one consisting of 25 observations from a normally distributed population, then using the 25 (normal score, observed value) pairs to calculate the value of r for each sample. Only about 1% of these samples would result in an r value less than the critical value 0.941. That is, 0.941 was chosen to guarantee a 1% error rate: In only 1% of all cases will you think normality is not reasonable when the distribution really is normal. The critical values for other sample sizes were also chosen to yield a 1% error rate.

Another type of error is also possible: obtaining a large value of r and concluding that normality is reasonable when the distribution is actually not normal. This type of error is more difficult to control than the type mentioned previously, but the procedure described here generally does a good job in controlling for both types of error.

Example 6.26   Egg Weights Continued

The sample size for the egg-weight data of Example 6.25 is $n = 10$. The critical r for $n = 10$ from Table 6.2 is 0.880. From Minitab, the correlation coefficient calculated using the (normal score, observed value) pairs is $r = 0.986$. Because r is larger than the critical r for a sample of size 10, it is reasonable to think that the population distribution of egg weights is approximately normal.

Correlations: Egg Weight, Normal Score

Pearson correlation of Egg Weight and Normal Score = 0.986

Summing It Up—Section 6.6

The following learning objectives were addressed in this section:

Mastering the Mechanics

M9: Use technology or given normal scores to construct a normal probability plot.
To construct a normal probability plot, normal scores are paired with observed data values that have been arranged in order from smallest to largest. A normal probability plot is a scatterplot of the (normal score, observed value) pairs. See Example 6.25.

M10: Given a normal probability plot, assess whether it is reasonable to think that a population distribution is approximately normal.
If the normal probability plot shows a strong linear pattern, it is reasonable to think that the population distribution may be approximately normal. The value of the correlation coefficient calculated using the (normal score, observed value) pairs can also be used to assess whether it is reasonable to think that the population distribution is approximately normal. See Example 6.26.

SECTION 6.6 EXERCISES

Each Exercise Set assesses the following chapter learning objectives: M9, M10

SECTION 6.6 Exercise Set 1

6.67 The authors of the paper **"Development of Nutritionally At-Risk Young Children Is Predicted by Malaria, Anemia, and Stunting in Pemba, Zanzibar"** (*The Journal of Nutrition* **[2009]: 763–772**) studied factors that might be related to dietary deficiencies in children. Children were observed for a fixed period of time, and the amount of time spent in various activities was recorded. One variable of interest was the amount of time (in minutes) a child spent fussing. Data for 15 children, consistent with summary quantities in the paper, are given in the accompanying table. Normal scores for a sample size of 15 are also given.

Fussing Time (x)	Normal Score
0.05	−1.739
0.10	−1.245
0.15	−0.946
0.40	−0.714
0.70	−0.515
1.05	−0.335
1.95	−0.165
2.15	0.000
3.70	0.165
3.90	0.335
4.50	0.515
6.00	0.714
8.00	0.946
11.00	1.245
14.00	1.739

a. Construct a normal probability plot for the fussing time data. Does the plot look linear? Do you agree with the authors of the paper that the fussing time distribution is not normal? (Hint: See Example 6.25.)

b. Calculate the correlation coefficient for the (normal score, *x*) pairs. Compare this value to the appropriate critical *r* value from Table 6.2 to determine if it is reasonable to think that the fussing time distribution is approximately normal. (Hint: See Example 6.26.)

6.68 The paper **"Risk Behavior, Decision Making, and Music Genre in Adolescent Males" (Marshall University, May 2009)** examined the effect of type of music playing and performance on a risky, decision-making task.

a. Participants in the study responded to a questionnaire used to assign a risk behavior score. Risk behavior scores (from a graph in the paper) for 15 participants follow. Use these data to construct a normal probability plot (the normal scores for a sample of size 15 appear in the previous exercise).

102 105 113 120 125 127 134 135
139 141 144 145 149 150 160

b. Participants also completed a positive and negative affect scale (PANAS) designed to measure emotional response to music. PANAS values (from a graph in the paper) for 15 participants follow. Use these data to construct a normal probability plot (the normal scores for a sample of size 15 appear in the previous exercise).

36 40 45 47 48 49 50 52
53 54 56 59 61 62 70

 Data set available

c. The author of the paper stated that it was reasonable to consider both risk behavior scores and PANAS scores to be approximately normally distributed. Do the normal probability plots from Parts (a) and (b) support this conclusion? Explain.

SECTION 6.6 Exercise Set 2

6.69 Consider the following 10 observations on the lifetime (in hours) for a certain type of power supply:

152.7   172.0   172.5   173.3   193.0
204.7   216.5   234.9   262.6   422.6

Construct a normal probability plot, and comment on whether it is reasonable to think that the distribution of power supply lifetime is approximately normal. (The normal scores for a sample of size 10 are -1.547, -1.000, -0.655, -0.375, -0.123, 0.123, 0.375, 0.655, 1.000, and 1.547.)

6.70 Fat contents (x, in grams) for seven randomly selected hot dog brands that were rated as very good by **Consumer Reports (www.consumerreports.org)** are shown below.

14   15   11   10   6   15   16

The normal scores for a sample of size 7 are

-1.364 -0.758 -0.353 0 0.353 0.758 1.364

a. Construct a normal probability plot for the fat content data. Does the plot look linear?
b. Calculate the correlation coefficient for the (normal score, x) pairs. Compare this value to the appropriate critical r value from Table 6.2 to determine if it is reasonable to think that the fat content distribution is approximately normal.

ADDITIONAL EXERCISES

6.71 An automobile manufacturer is interested in the fuel efficiency of a proposed new car design. Six nonprofessional drivers were selected, and each one drove a prototype of the new car from Phoenix to Los Angeles. The resulting fuel efficiencies (x, in miles per gallon) are:

27.2   29.3   31.2   28.4   30.3   29.6

The normal scores for a sample of size 6 are

-1.282 -0.643 -0.202 0.202 0.643 1.282

a. Construct a normal probability plot for the fuel efficiency data. Does the plot look linear?
b. Calculate the correlation coefficient for the (normal score, x) pairs. Compare this value to the appropriate critical r value from Table 6.2 to determine if it is reasonable to think that the fuel efficiency distribution is approximately normal.

6.72 The accompanying data on x = student-teacher ratio is for a random sample of 20 high schools in Maine selected from a population of 85 high schools. The data are consistent

 Data set available

with summary values for the state of Maine that appeared in an article in the *Bangor Daily News* **(September 22, 2016, bangordailynews.com/2016/09/22/mainefocus/we-discovered -a-surprise-when-we-looked-deeper-into-our-survey-of -maine-principals/?ref=morelnmidcoast,** retrieved May 2, **2017).** The corresponding normal scores are also shown.

Student-Teacher Ratio (x)	Normal Score
9.0	-1.868
10.0	-1.403
11.0	-1.128
11.2	-0.919
11.6	-0.744
11.7	-0.589
11.8	-0.448
11.9	-0.315
12.0	-0.187
12.1	-0.062
12.5	0.062
12.6	0.187
13.0	0.315
13.2	0.448
13.6	0.589
13.7	0.744
14.0	0.919
14.5	1.128
14.9	1.403
15.0	1.868

a. Construct a normal probability plot.
b. Calculate the correlation coefficient for the (normal score, x) pairs. Compare this value to the appropriate critical r value from Table 6.2 to determine if it is reasonable to think that the distribution of student-teacher ratios for high schools in Maine is approximately normal.

6.73 Consider the following sample of 25 observations on x = diameter (in centimeters) of CD disks produced by a particular manufacturer:

15.66 15.78 15.82 15.84 15.89 15.92 15.94 15.95
15.99 16.01   16.04 16.05 16.06 16.07   16.08 16.10
16.11 16.13 16.13 16.15 16.15 16.19 16.22 16.27
16.29

The 13 largest normal scores for a sample of size 25 are 1.964, 1.519, 1.259, 1.064, 0.903, 0.763, 0.636, 0.519, 0.408, 0.302, 0.200, 0.099, and 0. The 12 smallest scores result from placing a negative sign in front of each of the given nonzero scores. Construct a normal probability plot. Is it reasonable to think that the disk diameter distribution is approximately normal? Explain.

Binomial and Geometric Distributions (Optional)

In this section, two discrete probability distributions are introduced: the binomial distribution and the geometric distribution. These distributions arise when a chance experiment consists of a sequence of trials, each with only two possible outcomes. For example, one characteristic of blood type is the Rh factor, which can be either positive or negative. You can think of a chance experiment that consists of recording the Rh factor for each of 25 blood donors as a sequence of 25 trials, where each trial consists of observing the Rh factor (positive or negative) of a single donor.

You could also conduct a different chance experiment that consists of observing the Rh factor of blood donors until a donor who is Rh-negative is encountered. This second experiment can also be viewed as a sequence of trials, but here the total number of trials in this experiment is not predetermined, as it was in the previous experiment. Chance experiments of the two types just described are characteristic of those leading to the binomial probability distribution and the geometric probability distributions, respectively.

Binomial Distributions

Suppose that you decide to record the sex of each of the next 25 newborn children at a particular hospital. What is the chance that at least 15 are female? What is the chance that between 10 and 15 are female? How many among the 25 can you expect to be female? These and other similar questions can be answered by studying the *binomial probability distribution*. This distribution arises when the experiment of interest is a *binomial experiment*. Binomial experiments have the properties listed in the following box.

Properties of a Binomial Experiment

A **binomial experiment** consists of a sequence of trials with the following conditions:

1. There is a fixed number of trials.
2. Each trial can result in one of only two possible outcomes, labeled success (S) and failure (F).
3. Outcomes of different trials are independent.
4. The probability of success is the same for each trial.

The **binomial random variable x** is defined as

   x = number of successes observed when a binomial experiment is performed

The probability distribution of x is called the **binomial probability distribution**.

The term *success* here does not necessarily mean something positive. It is simply the "counted" outcome of a trial. For example, if the random variable counts the number of female births among the next 25 births at a particular hospital, then a female birth would be labeled a success (because this is what the variable counts). If male births were counted instead, a male birth would be labeled a success and a female birth a failure.

One example of a binomial random variable was given in Example 6.5. In that example, you considered x = *number among four customers who selected an energy efficient refrigerator* (rather than a less expensive model). This is a binomial experiment with four trials, where the purchase of an energy-efficient refrigerator is considered a success and $P(\text{success}) = P(E) = 0.4$. The 16 possible outcomes, along with the associated probabilities, were displayed in Table 6.1.

Consider now the case of five customers, a binomial experiment with five trials. The possible values of

   x = number who purchase an energy-efficient refrigerator

are 0, 1, 2, 3, 4, and 5. There are 32 possible outcomes of the binomial experiment, each one a sequence of five successes and failures. Five of these outcomes result in $x = 1$:

   SFFFF, FSFFF, FFSFF, FFFSF, and *FFFFS*

Because the trials are independent, the first of these outcomes has probability

$$P(SFFFF) = P(S)P(F)P(F)P(F)P(F)$$
$$= (0.4)(0.6)(0.6)(0.6)(0.6)$$
$$= (0.4)(0.6)^4$$
$$= 0.052$$

The probability calculation will be the same for any outcome with only one success ($x = 1$). It does not matter where in the sequence the single success occurs. This means that

$$p(1) = P(x = 1)$$
$$= P(SFFFF \text{ or } FSFFF \text{ or } FFSFF \text{ or } FFFSF \text{ or } FFFFS)$$
$$= 0.052 + 0.052 + 0.052 + 0.052 + 0.052$$
$$= 5(0.052)$$
$$= 0.260$$

Similarly, there are 10 outcomes for which $x = 2$, because there are 10 ways to select two from among the five trials to be the S's: $SSFFF$, $SFSFF$, ... , and $FFFSS$. The probability of each results from multiplying together (0.4) two times and (0.6) three times. For example,

$$P(SSFFF) = (0.4)(0.4)(0.6)(0.6)(0.6)$$
$$= (0.4)^2(0.6)^3$$
$$= 0.0346$$

and so

$$p(2) = P(x = 2)$$
$$= P(SSFFF) + \cdots + P(FFFSS)$$
$$= (10)(0.4)^2(0.6)^3$$
$$= 0.346$$

Other probabilities can be calculated in a similar way, A general formula for calculating the probabilities associated with the different possible values of x is

$$p(x) = P(x\ S's \text{ among the five trials})$$
$$= (\text{number of outcomes with } x\ S\text{'s}) \cdot (\text{probability of any given outcome with } x\ S\text{'s})$$

To use this general formula, you need to determine the number of different outcomes that have x successes. Let n denote the number of trials in the experiment. Then the number of outcomes with $x\ S$'s is the number of ways of selecting x success trials from among the n trials. A simple expression for this quantity is

$$\text{number of outcomes with } x \text{ successes} = \frac{n!}{x!(n-x)!}$$

where, for any positive whole number m, the symbol $m!$ (read "m factorial") is defined by

$$m! = m(m-1)(m-2)\cdots(2)(1)$$

and $0! = 1$.

The Binomial Distribution

Suppose

n = number of independent trials in a binomial experiment
p = constant probability that any particular trial results in a success

Then

$$p(x) = P(x \text{ successes among the } n \text{ trials})$$
$$= \frac{n!}{x!(n-x)!} p^x (1-p)^{n-x} \qquad x = 0,1,2, \ldots, n$$

The expressions $\binom{n}{x}$ or $_nC_x$ are sometimes used in place of $\frac{n!}{x!(n-x)!}$. Both are read as "n choose x" and represent the number of ways of choosing x items from a set of n.

Notice that here the probability distribution is specified by a formula rather than a table or a probability histogram.

Example 6.27 **Recognizing Your Roommate's Scent**

An interesting experiment was described in the paper **"Sociochemosensory and Emotional Functions" (*Psychological Science* [2009]: 1118–1123)**. The authors of this paper wondered if a female college student could recognize her roommate by scent. They carried out an experiment in which the participants used fragrance-free soap, deodorant, shampoo, and laundry detergent for a period of time. Their bedding was also laundered using a fragrance-free detergent. Each person was then given a new t-shirt that she slept in for one night. The shirt was then collected and sealed in an airtight bag. Later, the roommate was presented with three identical t-shirts (one worn by her roommate and two worn by other women) and asked to pick the one that smelled most like her roommate. (Yes, hard to believe, but people really do research like this!) This process was repeated, with the shirts refolded and rearranged before the second trial. The researchers recorded how many times (0, 1, or 2) that the shirt worn by the roommate was correctly identified.

This can be viewed as a binomial experiment consisting of $n = 2$ trials. Each trial results in either a correct identification or an incorrect identification. Because the researchers counted the number of correct identifications, a correct identification is considered a success. You can then define

$$x = \text{number of correct identifications}$$

Suppose that a participant is not able to identify her roommate by smell. In this case, she is essentially just picking one of the three shirts at random and so the probability of success (picking the correct shirt) is $\frac{1}{3}$. Here, it would also be reasonable to regard the two trials as independent. In this case, the experiment satisfies the conditions of a binomial experiment, and x is a binomial random variable with $n = 2$ and $p = \frac{1}{3}$.

You can use the binomial distribution formula to calculate the probability associated with each of the possible x values:

$$p(0) = \frac{2!}{0!2!}\left(\frac{1}{3}\right)^0\left(\frac{2}{3}\right)^2 = (1)(1)\left(\frac{2}{3}\right)^2 = 0.444$$

$$p(1) = \frac{2!}{1!1!}\left(\frac{1}{3}\right)^1\left(\frac{2}{3}\right)^1 = (2)\left(\frac{1}{3}\right)^1\left(\frac{2}{3}\right)^1 = 0.444$$

$$p(2) = \frac{2!}{2!0!}\left(\frac{1}{3}\right)^2\left(\frac{2}{3}\right)^0 = (1)\left(\frac{1}{3}\right)^2(1) = 0.111$$

Summarizing in table form gives

x	p(x)
0	0.444
1	0.444
2	0.111

This means that about 44.4% of the time, a person who is just guessing would pick the correct shirt on neither trial, about 44.4% of the time the correct shirt would be identified on one of the two trials, and about 11.1% of the time the correct shirt would be identified on both trials.

The authors actually performed this experiment with 44 subjects. They reported that 47.7% of the subjects identified the correct shirt on neither trial, 22.7% identified the correct shirt on one trial, and 31.7% identified the correct shirt on both trials. These results differed quite a bit from the expected "just guessing" results. This difference was interpreted as evidence that some women could identify their roommates by smell.

Example 6.28   Computer Sales

Suppose that 60% of all computers sold by a large computer retailer are laptops and 40% are desktop models. The type of computer purchased by each of the next 12 customers will be recorded. Define a random variable x as

$$x = \text{number of laptops among these } 12$$

Because x counts the number of laptops, you use S to denote the sale of a laptop. Then x is a binomial random variable with $n = 12$ and $p = P(S) = 0.60$. The probability distribution of x is given by

$$p(x) = \frac{12!}{x!(12-x)!}(0.6)^x(0.4)^{12-x} \qquad x = 0,1,2,\ldots,12$$

The probability that exactly four computers are laptops is

$$p(4) = P(x = 4)$$
$$= \frac{12!}{4!8!}(0.6)^4(0.4)^8$$
$$= (495)(0.6)^4(0.4)^8$$
$$= 0.042$$

If many groups of 12 purchases are examined, about 4.2% of them will include exactly four laptops.

The probability that between four and seven (inclusive) are laptops is

$$P(4 \le x \le 7) = P(x = 4 \text{ or } x = 5 \text{ or } x = 6 \text{ or } x = 7)$$

Since these outcomes are mutually exclusive, this is equal to

$$P(4 \le x \le 7) = p(4) + p(5) + p(6) + p(7)$$
$$= \frac{12!}{4!8!}(0.6)^4(0.4)^8 + \cdots + \frac{12!}{7!5!}(0.6)^7(0.4)^5$$
$$= 0.042 + 0.101 + 0.177 + 0.227$$
$$= 0.547$$

Notice that

$$P(4 < x < 7) = P(x = 5 \text{ or } x = 6)$$
$$= p(5) + p(6)$$
$$= 0.278$$

so the probability depends on whether $<$ or $\le$ appears. (This is typical of *discrete* random variables.)

The binomial distribution formula can be tedious to use unless n is small. Appendix A Table 9 gives binomial probabilities for selected values of n in combination with selected values of p. Statistical software packages and most graphing calculators can also be used to compute binomial probabilities. The following box explains how Appendix A Table 9 can be used.

Using Appendix A Table 9 to Calculate Binomial Probabilities

To find $p(x)$ for any particular value of x,

1. Locate the part of the table corresponding to the value of n (5, 10, 15, 20, or 25).
2. Move down to the row labeled with the value of x.
3. Go across to the column headed by the specified value of p.

The desired probability is at the intersection of the designated x row and p column. For example, when $n = 20$ and $p = 0.8$,

$$p(15) = P(x = 15) = (\text{entry at intersection of } n = 15 \text{ row and } p = 0.8 \text{ column}) = 0.175$$

Although $p(x)$ is positive for each possible x value, many probabilities are zero when rounded to three decimal places, so they appear as 0.000 in the table.

Sampling Without Replacement

If sampling is done by selecting an element at random from the population, observing whether it is a success or a failure, and then returning it to the population before the next selection is made, the variable x = *number of successes* observed in the sample would fit all the requirements of a binomial random variable. However, sampling is usually carried out without replacement. That is, once an element has been selected for the sample, it is not a candidate for future selection. When sampling is done without replacement, the trials (individual selections) are not independent. In this case, the number of successes observed in the sample does not have a binomial distribution but rather a different type of distribution called a *hypergeometric distribution.* The probability calculations for this distribution are even more tedious than for the binomial distribution. Fortunately, when the sample size n is small relative to N, the population size, probabilities calculated using the binomial distribution and probabilities calculated using the hypergeometric distribution are nearly equal. They are so close, in fact, that you can ignore the difference and use the binomial probabilities in place of the hypergeometric probabilities. The following guideline can be used to determine if it is reasonable to use the binomial probability distribution to calculate probabilities when sampling without replacement.

Let x denote the number of S's in a sample of size n selected at random and without replacement from a population consisting of N individuals or objects. If $\frac{n}{N} \leq 0.05$, meaning that no more than 5% of the population is sampled, then the binomial distribution gives a good approximation to the probability distribution of x.

Example 6.29 Saying I Do...

A survey of 505 American women in 2016 found that only about 25% favor preserving the tradition of having the bride promise to obey her husband as part of wedding vows **(www .yahoo.com/style/women-want-the-word-obey-dropped-from-wedding-162058081.html, retrieved May 2, 2017)**. Suppose that exactly 25% of American women favor preserving this tradition. Consider a random sample of $n = 20$ American women (much less than 5% of the population). Then

x = the number in the sample who favor preserving the promise to obey

has (approximately) a binomial distribution with $n = 20$ and $p = 0.25$. The probability that five of those sampled favor preserving the promise to obey is (from Appendix A Table 9)

$$p(5) = P(x = 5)$$
$$= 0.202$$

The probability that at least half of those in the sample (10 or more) favor preserving the promise to obey is

$$P(x \geq 10) = p(10) + p(11) + \cdots + p(20)$$
$$= 0.010 + 0.003 + 0.001 + \cdots + 0.000$$
$$= 0.014$$

If $p = 0.25$, only about 1.4% of all samples of size 20 would contain at least 10 people who favor preserving the promise to obey. Because $P(x \geq 10)$ is so small when $p = 0.25$, if $x \geq 10$ were actually observed, you would have to wonder whether the reported value of $p = 0.25$ is correct. In Chapter 10, you will see how hypothesis-testing methods can be used to decide between two contradictory claims about a population (such as $p = 0.25$ and $p > 0.25$).

Technology, Appendix A Table 9, or hand calculations using the formula for binomial probabilities can be used to calculate binomial probabilities. Figure 6.37 shows the probability histogram for the binomial distribution with $n = 20$ and $p = 0.25$. Notice that the distribution is slightly skewed to the right. (The binomial distribution is symmetric only when $p = 0.5$.)

FIGURE 6.37
The binomial probability histogram when $n = 20$ and $p = 0.25$

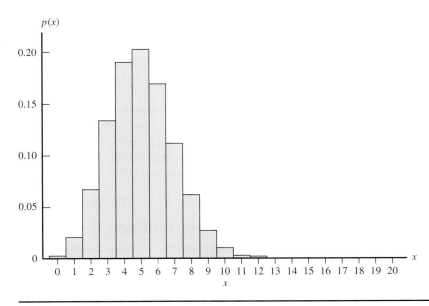

Mean and Standard Deviation of a Binomial Random Variable

A binomial random variable x based on n trials has possible values 0, 1, 2, … , n, so the mean value is

$$\mu_x = \Sigma x p(x) = (0)p(0) + (1)\, p(1) + \ldots + (n)p(n)$$

and the variance of x is

$$\sigma_x^2 = \Sigma(x - \mu_x)^2 p(x)$$
$$= (0 - \mu_x)^2 p(0) + (1 - \mu_x)^2 p(1) + \cdots (n - \mu_x)^2 p(n)$$

These expressions would be very tedious to evaluate for any particular values of n and p. Fortunately, they can be simplified to the expressions given in the following box.

The mean value and the standard deviation of a binomial random variable x are

$$\mu_x = np \qquad \text{and} \qquad \sigma_x = \sqrt{np(1 - p)}$$

Example 6.30   **Budgets and Tracking Spending**

The report **"The 2016 Consumer Financial Literacy Survey" (The National Foundation for Credit Counseling, nfcc.org/wp-content/uploads/2016/04/NFCC_BECU_2016-FLS _datasheet-with-key-findings_041516.pdf, retrieved May 2, 2017)** estimates that 40% of U.S. adults say that they have a budget and keep close track of their spending. This estimate was based on a representative sample of more than 1500 people. Suppose that 40% of all U.S. adults have a budget and keep close track of their spending. A random sample of $n = 25$ U.S. adults is to be selected. Consider the random variable

$x = $ number in the sample who have a budget and keep track of spending

Even if sampling here is done without replacement, the sample size $n = 25$ is very small compared to the total number of U.S. adults, so you can approximate the probability distribution of x using a binomial distribution with $n = 25$ and $p = 0.4$. Having a budget and keeping track of spending is identified as a success because this is the outcome counted by the random variable x. The mean value of x is then

$$\mu_x = np = 25(0.40) = 10.0$$

and the standard deviation is

$$\sigma_x = \sqrt{np(1 - p)} = \sqrt{25(0.4)(0.6)} = \sqrt{6} = 2.45$$

There are two cases where there is no variability in x. When $p = 0$, x will always equal 0. When $p = 1$, x will always equal n. In both cases, the value of σ_x is 0. It is also easily verified that $p(1 - p)$ is largest when $p = 0.5$. This means that the binomial distribution spreads out the most when sampling from a population that consists of half successes and half failures. The binomial distribution is less spread out and more skewed the farther p is from 0.5.

Geometric Distributions

A binomial random variable is defined as the number of successes in n independent trials, where each trial can result in either a success or a failure and the probability of success is the same for each trial. Suppose, however, that you are not interested in the number of successes in a *fixed* number of trials but rather in the number of trials that must be carried out before a success occurs. Counting the number of boxes of cereal that must be purchased before finding one with a rare toy and counting the number of games that a professional bowler must play before achieving a score over 250 are two examples of this.

The variable

$$x = \text{number of trials to first success}$$

is called a *geometric random variable*, and the probability distribution that describes its behavior is called a *geometric probability distribution*.

Properties of a Geometric Experiment

Suppose an experiment consists of a sequence of trials with the following conditions:

1. The trials are independent.
2. Each trial can result in one of two possible outcomes, success or failure.
3. The probability of success is the same for all trials.

A **geometric random variable** is defined as

$x = $ number of trials until the first success is observed (including the success trial)

The probability distribution of x is called the **geometric probability distribution**.

For example, suppose that 40% of the students who drive to campus at your college carry jumper cables. Your car has a dead battery and you don't have jumper cables, so you decide to ask students who are headed to the parking lot whether they have a pair of jumper cables. You might be interested in the number of students you would have to ask in order to find one who has jumper cables. Defining a success as a student with jumper cables, a trial would consist of asking an individual student for help. The random variable

$$x = \text{number of students asked in order to find one with jumper cables}$$

is an example of a geometric random variable, because it can be viewed as the number of trials to the first success in a sequence of independent trials.

The probability distribution of a geometric random variable is easy to construct. As before, p is used to denote the probability of success on any given trial. Possible outcomes can be denoted as follows:

Outcome	x = Number of Trials to First Success
S	1
FS	2
FFS	3
⋮	⋮
$FFFFFFS$	7
⋮	⋮

Each possible outcome consists of 0 or more failures followed by a single success.
This means that

$$p(x) = P(x \text{ trials to first success})$$
$$= P(FF \ldots FS)$$

x − 1 failures followed by a success on trial x

The probability of success is p for each trial, so the probability of failure for each trial is $1 - p$. Because the trials are independent,

$$p(x) = P(x \text{ trials to first success}) = P(FF \ldots FS)$$
$$= P(F)P(F) \cdots P(F)P(S)$$
$$= (1 - p)(1 - p) \cdots (1 - p)p$$
$$= (1 - p)^{x-1} p$$

This leads to the formula for the geometric probability distribution.

The Geometric Distribution

If x is a geometric random variable with probability of success $= p$ for each trial, then

$$p(x) = (1 - p)^{x-1}p \qquad x = 1, 2, 3, \ldots$$

Example 6.31   Jumper Cables

Consider the jumper cable problem described previously. Because 40% of the students who drive to campus carry jumper cables, $p = 0.4$. The probability distribution of

$$x = \text{number of students asked in order to find one with jumper cables}$$

is

$$p(x) = (0.6)^{x-1}(0.4) \qquad x = 1, 2, 3, \ldots$$

The probability distribution can now be used to calculate various probabilities. For example, the probability that the first student asked has jumper cables ($x = 1$) is

$$p(1) = (0.6)^{1-1}(0.4) = (0.6)^0(0.4) = 0.4$$

The probability that three or fewer students must be asked is

$$P(x \leq 3) = p(1) + p(2) + p(3)$$
$$= (0.6)^0(0.4) + (0.6)^1(0.4) + (0.6)^2(0.4)$$
$$= 0.4 + 0.24 + 0.144$$
$$= 0.784$$

Summing It Up—Section 6.7

The following learning objectives were addressed in this section:

Mastering the Mechanics

M11: Distinguish between binomial and geometric random variables.

Some experiments consist of observing the outcomes of independent trials where there are two possible outcomes for each trial (success and failure), and the probability of success is the same for all trials. If there are a fixed number of trials and the random variable counts the number of successes in those trials, it is a binomial random variable. If there is not a predetermined number of trials and the random variable counts the number of trials until the first success is observed, it is a geometric random variable.

M12: Calculate binomial probabilities using technology or tables.

Examples 6.27 and 6.28 illustrate calculations of binomial probabilities.

M13: Calculate probabilities using the geometric distribution.

Example 6.31 illustrates the calculation of geometric probabilities.

SECTION 6.7 EXERCISES

Each Exercise Set assesses the following chapter learning objectives: (Optional) M11, (Optional) M12, (Optional) M13

SECTION 6.7 Exercise Set 1

6.74 Example 6.27 described a study in which a person was asked to determine which of three t-shirts had been worn by her roommate by smelling the shirts **("Sociochemosensory and Emotional Functions,"** *Psychological Science* **[2009]: 1118–1123)**. Suppose that instead of three shirts, each participant was asked to choose among four shirts and that the process was performed five times. If a person can't identify her roommate by smell and is just picking a shirt at random, then x = *number of correct identifications* is a binomial random variable with $n = 5$ and $p = \frac{1}{4}$.
a. What are the possible values of x?
b. For each possible value of x, find the associated probability $p(x)$ and display the possible x values and $p(x)$ values in a table. (Hint: See Example 6.27.)
c. Construct a histogram displaying the probability distribution of x.

6.75 Suppose that in a certain metropolitan area, 90% of all households have cable TV. Let x denote the number among four randomly selected households that have cable TV. Then x is a binomial random variable with $n = 4$ and $p = 0.9$.
a. Calculate $p(2) = P(x = 2)$, and interpret this probability.
b. Calculate $p(4)$, the probability that all four selected households have cable TV.
c. Calculate $P(x \leq 3)$.

6.76 Twenty-five percent of the customers of a grocery store use an express checkout. Consider five randomly selected customers, and let x denote the number among the five who use the express checkout.
a. Calculate $p(2)$, that is, $P(x = 2)$.
b. Calculate $P(x \leq 1)$.
c. Calculate $P(x \geq 2)$. (Hint: Make use of your answer to Part (b).)
d. Calculate $P(x \neq 2)$.

6.77 Industrial quality control programs often include inspection of incoming materials from suppliers. If parts are purchased in large lots, a typical plan might be to select 20 parts at random from a lot and inspect them. Suppose that a lot is judged acceptable if one or fewer of these 20 parts are defective. If more than one part is defective, the lot is rejected and returned to the supplier. Find the probability of accepting lots that have each of the following (Hint: Identify success with a defective part.):
a. 5% defective parts
b. 10% defective parts
c. 20% defective parts

6.78 Suppose a playlist on an MP3 music player consisting of 100 songs includes 8 by a particular artist. Suppose that songs are played by selecting a song at random (with replacement) from the playlist. The random variable x represents the number of songs until a song by this artist is played.
a. Explain why the probability distribution of x is not binomial.
b. Find the following probabilities. (Hint: See Example 6.31.)
   i. $p(4)$
   ii. $P(x \leq 4)$
   iii. $P(x > 4)$
   iv. $P(x \geq 4)$
c. Interpret each of the probabilities in Part (b) and explain the difference between them.

6.79 *Women's Health Magazine* surveyed 1187 readers to find out how often people wash their sheets **(March 26, 2015, www.womenshealthmag.com/health/dirty-sheets, retrieved May 2, 2017)**. They found that even though microbiologists recommend that you wash your sheets at least once a week, only 44% said that they wash their sheets that often. Suppose this group is representative of adult Americans and define the random variable x to be the number of adult Americans you would have to ask before you found someone that washes his or her sheets at least once a week.

a. Is the probability distribution of x binomial or geometric? Explain.

b. What is the probability that you would have to ask three people before finding one who washes sheets at least once a week?

c. What is the probability that fewer than four people would have to be asked before finding one who washes sheets at least once a week?

d. What is the probability that more than three people would have to be asked before finding one who washes sheets at least once a week?

SECTION 6.7 Exercise Set 2

6.80 Information Security Buzz provides news for the information security community. In an article published on September 24, 2016, it reported that based on a large international survey of Internet users, 60% of Internet users have installed security solutions on all of the devices they use to access the Internet (**www.informationsecuritybuzz.com /articles/21-29-60-kaspersky-lab-presents-first-cybersecurity -index/, retrieved May 2, 2017**).

a. Suppose that the true proportion of Internet users who have security solutions on all of the devices they use to access the Internet is 0.60. If 20 Internet users are selected at random, what is the probability that more than 10 have security solutions installed on all devices used to access the Internet?

b. Suppose that a random sample of Internet users is selected. Which is more likely—that more than 15 have security solutions on all devices used to access the Internet or that fewer than 5 have security solutions on all devices used to access the Internet? Justify your answer based on probability calculations.

6.81 FlightView surveyed 2600 North American airline passengers and reported that approximately 80% said that they carry a smartphone when they travel (**www.flightview.com /TravelersSurvey/downloads/survey_infographic_poster.pdf, retrieved May 2, 2017**). Suppose that the actual percentage is 80%. Consider randomly selecting six passengers and define the random variable x to be the number of the six selected passengers who travel with a smartphone. The probability distribution of x is the binomial distribution with $n = 6$ and $p = 0.8$.

a. Calculate $p(4)$, and interpret this probability.

b. Calculate $p(6)$, the probability that all six selected passengers travel with a smartphone.

c. Calculate $P(x \geq 4)$.

6.82 Refer to the previous exercise, and suppose that 10 rather than 6 passengers are selected ($n = 10$, $p = 0.8$). Use Appendix A Table 9 to find the following:

a. $p(8)$

b. $P(x \leq 7)$

c. The probability that more than half of the selected passengers travel with a smartphone.

6.83 Thirty percent of all automobiles undergoing an emissions inspection at a certain inspection station fail the inspection.

a. Among 15 randomly selected cars, what is the probability that at most 5 fail the inspection?

b. Among 15 randomly selected cars, what is the probability that between 5 and 10 (inclusive) fail the inspection?

c. Among 25 randomly selected cars, what is the mean value of the number that pass inspection, and what is the standard deviation?

d. What is the probability that among 25 randomly selected cars, the number that pass is within 1 standard deviation of the mean value?

6.84 Sophie is a dog who loves to play catch. Unfortunately, she isn't very good at this, and the probability that she catches a ball is only 0.1. Let x be the number of tosses required until Sophie catches a ball.

a. Does x have a binomial or a geometric distribution?

b. What is the probability that it will take exactly two tosses for Sophie to catch a ball?

c. What is the probability that more than three tosses will be required?

6.85 Suppose that 5% of cereal boxes contain a prize and the other 95% contain the message, "Sorry, try again." Consider the random variable x, where x = number of boxes purchased until a prize is found.

a. What is the probability that at most two boxes must be purchased?

b. What is the probability that exactly four boxes must be purchased?

c. What is the probability that more than four boxes must be purchased?

ADDITIONAL EXERCISES

6.86 An experiment was conducted to investigate whether a graphologist (a handwriting analyst) could distinguish a normal person's handwriting from that of a psychotic. A well-known expert was given 10 files, each containing handwriting samples from a normal person and from a person diagnosed as psychotic, and asked to identify the psychotic's handwriting. The graphologist made correct identifications in 6 of the 10 trials (data taken from *Statistics in the Real World*, by R. J. Larsen and D. F. Stroup [New York: Macmillan, 1976]). Does this indicate that the graphologist has an ability to distinguish the handwriting of psychotics? (Hint: What is the probability of correctly guessing 6 or more times out of 10? Your answer should depend on whether this probability is relatively small or relatively large.)

6.87 A breeder of show dogs is interested in the number of female puppies in a litter. If a birth is equally likely to result in a male or a female puppy, give the probability distribution of the variable x = *number of female puppies in a litter of size 5*.

6.88 You are to take a multiple-choice exam consisting of 100 questions with five possible responses to each question. Suppose that you have not studied and so must guess (randomly select one of the five answers) on each question. Let x represent the number of correct responses on the test.
a. What kind of probability distribution does x have?
b. What is your expected score on the exam? (Hint: Your expected score is the mean value of the x distribution.)
c. Calculate the variance and standard deviation of x.
d. Based on your answers to Parts (b) and (c), is it likely that you would score over 50 on this exam? Explain the reasoning behind your answer.

6.89 Suppose that 20% of the 10,000 signatures on a certain recall petition are invalid. Would the number of invalid signatures in a sample of size 2000 have (approximately) a binomial distribution? Explain.

6.90 A soft-drink machine dispenses only regular Coke and Diet Coke. Sixty percent of all purchases from this machine are diet drinks. The machine currently has 10 cans of each type. If 15 customers want to purchase drinks before the machine is restocked, what is the probability that each of the 15 is able to purchase the type of drink desired? (Hint: Let x denote the number among the 15 who want a diet drink. For which possible values of x is everyone satisfied?)

6.91 A coin is flipped 25 times. Let x be the number of flips that result in heads (H). Consider the following rule for deciding whether or not the coin is fair:

Judge the coin fair if $8 \le x \le 17$.
Judge the coin biased if either $x \le 7$ or $x \ge 18$.

a. What is the probability of judging the coin biased when it is actually fair?
b. Suppose that a coin is not fair and that $P(\text{H}) = 0.9$. What is the probability that this coin would be judged fair? What is the probability of judging a coin fair if $P(\text{H}) = 0.1$?
c. What is the probability of judging a coin fair if $P(\text{H}) = 0.6$? if $P(\text{H}) = 0.4$? Why are these probabilities large compared to the probabilities in Part (b)?
d. What happens to the "error probabilities" of Parts (a) and (b) if the decision rule is changed so that the coin is judged fair if $7 \le x \le 18$ and unfair otherwise? Is this a better rule than the one first proposed? Explain.

6.92 The longest "run" of S's in the 10-trial sequence $SSFSSSSFFS$ has length 4, corresponding to the S's on the fourth, fifth, sixth, and seventh trials. Consider a binomial experiment with $n = 4$, and let y be the length (number of trials) in the longest run of S's.
a. When $p = 0.5$, the 16 possible outcomes are equally likely. Determine the probability distribution of y in this case (first list all outcomes and the y value for each one). Then calculate μ_y.
b. Repeat Part (a) for the case $p = 0.6$.

SECTION 6.8 Using the Normal Distribution to Approximate a Discrete Distribution (Optional)

In this section, you will see how probabilities for some discrete random variables can be approximated using a normal curve. The most important case of this is the approximation of binomial probabilities.

The Normal Curve and Discrete Variables

The probability distribution of a discrete random variable x is represented graphically by a probability histogram. The probability of a particular value is the area of the rectangle centered at that value. Possible values of x are isolated points on the number line, usually whole numbers. For example, if $x = $ IQ of a randomly selected 8-year-old child, then x is a discrete random variable, because an IQ score must be a whole number.

Often, a probability histogram can be well approximated by a normal curve, as illustrated in Figure 6.38. In such cases, the distribution of x is said to be approximately normal. A normal distribution can then be used to calculate approximate probabilities of events involving x.

FIGURE 6.38
A normal curve approximation to a probability histogram

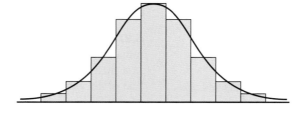

Example 6.32 | **Express Mail Packages**

The number of express mail packages mailed at a certain post office on a randomly selected day is approximately normally distributed with mean 18 and standard deviation 6. Suppose you want to know the probability that $x = 20$. Figure 6.39(a) shows a portion of the probability histogram for x with the approximating normal curve superimposed. The area of the shaded rectangle is $P(x = 20)$. The left edge of this rectangle is at 19.5 on the horizontal scale, and the right edge is at 20.5. Therefore, the desired probability is approximately the area under the normal curve between 19.5 and 20.5. Standardizing these endpoints gives

$$\frac{20.5 - 18}{6} = 0.42 \qquad \frac{19.5 - 18}{6} = 0.25$$

so

$$P(x = 20) \approx P(0.25 < z < 0.42) = 0.6628 - 0.5987 = 0.0641$$

Figure 6.39(b) shows that $P(x \le 10)$ is approximately the area under the normal curve to the left of 10.5. It follows that

$$P(x \le 10) \approx P\left(z \le \frac{10.5 - 18}{6}\right) = P(z \le -1.25) = 0.1056$$

FIGURE 6.39
The normal approximation for Example 6.32

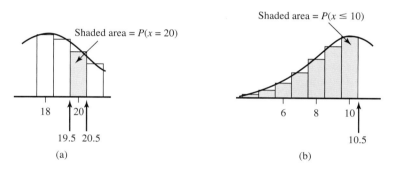

(a)

(b)

The calculation of probabilities in Example 6.32 illustrates the use of what is known as a **continuity correction**. Because the rectangle for $x = 10$ extends to 10.5 on the right, the normal curve area to the left of 10.5 rather than 10 is used. In general, if possible x values are consecutive whole numbers, then $P(a \le x \le b)$ will be approximately the normal curve area between limits $a - \frac{1}{2}$ and $b + \frac{1}{2}$. Similarly, $P(a < x < b)$ will be approximately the area between $a + \frac{1}{2}$ and $b - \frac{1}{2}$.

Normal Approximation to a Binomial Distribution

Figure 6.40 shows the probability histograms for two binomial distributions, one with $n = 25$, $p = 0.4$, and the other with $n = 25$, $p = 0.1$. For each distribution, $\mu = np$ and

FIGURE 6.40
Normal approximations to binomial distributions

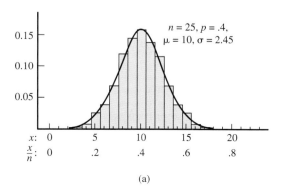

(a)

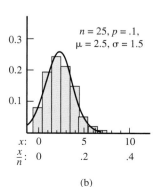

(b)

$\sigma = \sqrt{np(1-p)}$ were calculated and a normal curve with this μ and σ was superimposed. A normal curve fits the probability histogram well in the first case (Figure 6.40(a)). When this happens, binomial probabilities can be accurately approximated by areas under the normal curve. In the second case (Figure 6.40(b)), the normal curve does not give a good approximation because the probability histogram is skewed, whereas the normal curve is symmetric.

APPROXIMATING BINOMIAL PROBABILITIES

Suppose x is a binomial random variable based on n trials and success probability p, so that

$$\mu = np \text{ and } \sigma = \sqrt{np(1-p)}$$

If n and p are such that

$$np \geq 10 \text{ and } n(1-p) \geq 10$$

Then the distribution of x is approximately normal.

Combining this result with the continuity correction implies that

$$P(a \leq x \leq b) \approx P\left(\frac{a - \frac{1}{2} - \mu}{\sigma} \leq z \leq \frac{b + \frac{1}{2} - \mu}{\sigma}\right)$$

That is, the probability that x is between a and b inclusive is approximately the area under the approximating normal curve between $a - \frac{1}{2}$ and $b + \frac{1}{2}$.

Similarly,

$$P(x \leq b) \approx P\left(z \leq \frac{b + \frac{1}{2} - \mu}{\sigma}\right)$$

$$P(a \leq x) \approx P\left(\frac{a - \frac{1}{2} - \mu}{\sigma} \leq z\right)$$

When either $np < 10$ or $n(1-p) < 10$, the binomial distribution is too skewed for the normal approximation to give reasonably accurate probability estimates.

Example 6.33 | Premature Babies

Premature babies are those born before 37 weeks, and those born before 34 weeks are most at risk. The paper **"Some Thoughts on the True Value of Ultrasound"** (*Ultrasound in Obstetrics and Gynecology* **[2007]: 671–674**) reported that 2% of births in the United States occur before 34 weeks. Suppose that 1000 births will be randomly selected and that the value of

$$x = \text{the number of these births that occurred prior to 34 weeks}$$

is to be determined. Because

$$np = 1000\,(0.02) = 20 \geq 10$$
$$n(1-p) = 1000\,(0.98) = 980 \geq 10$$

the distribution of x is approximately normal with

$$\mu = np = 1000(0.02) = 20$$
$$\sigma = \sqrt{np(1-p)} = \sqrt{1000(0.02)(0.98)} = \sqrt{19.60} = 4.427$$

You can now calculate the probability that the number of babies born prior to 34 weeks in a sample of 1,000 will be between 10 and 25 (inclusive):

$$P(10 \leq x \leq 25) = P\left(\frac{9.5 - 20}{4.427} \leq z \leq \frac{25.5 - 20}{4.427}\right)$$

$$= P(-2.37 \leq z \leq 1.24)$$

$$= 0.8925 - 0.0089$$

$$= 0.8836$$

as shown in the following figure:

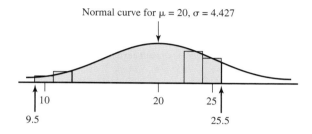

Normal curve for $\mu = 20$, $\sigma = 4.427$

10   20   25
9.5   25.5

Summing It Up—Section 6.8

The following learning objective was addressed in this section:

Mastering the Mechanics

M14: Given a binomial distribution, determine if it is appropriate to use a normal approximation to estimate binomial probabilities. If appropriate, approximate binomial probabilities using a normal distribution.

For binomial probability distributions that are not too skewed, the normal distribution can be used to approximate binomial probabilities. It is appropriate to use a normal approximation if both np and $n(1 - p)$ are greater than or equal to 10. Example 6.33 illustrates how binomial probabilities can be approximated by an area under a normal curve.

SECTION 6.8 EXERCISES

Each Exercise Set assesses the following chapter learning objectives: (Optional) M14

SECTION 6.8 Exercise Set 1

6.93 Let x denote the IQ of an individual selected at random from a certain population. The value of x must be a whole number. Suppose that the distribution of x can be approximated by a normal distribution with mean value 100 and standard deviation 15. Approximate the following probabilities. (Hint: See Example 6.32.)
a. $P(x = 100)$
b. $P(x \leq 110)$
c. $P(x < 110)$ (Hint: $x < 110$ is the same as $x \leq 109$.)
d. $P(75 \leq x \leq 125)$

6.94 Seventy percent of the bicycles sold by a certain store are mountain bikes. Among 100 randomly selected bike purchases, what is the approximate probability that
a. At most 75 are mountain bikes? (Hint: See Example 6.33.)

b. Between 60 and 75 (inclusive) are mountain bikes?
c. More than 80 are mountain bikes?
d. At most 30 are not mountain bikes?

6.95 Suppose that 65% of all registered voters in a certain area favor a seven-day waiting period before purchase of a handgun. Among 225 randomly selected registered voters, what is the approximate probability that
a. At least 150 favor such a waiting period?
b. More than 150 favor such a waiting period?
c. Fewer than 125 favor such a waiting period?

SECTION 6.8 Exercise Set 2

6.96 The distribution of the number of items produced by an assembly line during an 8-hour shift can be approximated by a normal distribution with mean value 150 and standard deviation 10.

a. What is the approximate probability that the number of items produced is at most 120?
b. What is the approximate probability that at least 125 items are produced?
c. What is the approximate probability that between 135 and 160 (inclusive) items are produced?

6.97 Suppose that 25% of the fire alarms in a large city are false alarms. Let x denote the number of false alarms in a random sample of 100 alarms. Approximate the following probabilities:
a. $P(20 \leq x \leq 30)$
b. $P(20 < x < 30)$
c. $P(x \geq 35)$
d. The probability that x is farther than 2 standard deviations from its mean value.

6.98 A company that manufactures mufflers for cars offers a lifetime warranty on its products, provided that ownership of the car does not change. Only 20% of its mufflers are replaced under this warranty.
a. In a random sample of 400 purchases, what is the approximate probability that between 75 and 100 (inclusive) mufflers are replaced under warranty?
b. Among 400 randomly selected purchases, what is the approximate probability that at most 70 mufflers are replaced under warranty?
c. If you were told that fewer than 50 among 400 randomly selected purchases were replaced under warranty, would you question the 20% figure? Explain.

ADDITIONAL EXERCISES

6.99 A symptom validity test (SVT) is sometimes used to confirm diagnosis of psychiatric disorders. The paper **"Developing a Symptom Validity Test for Posttraumatic Stress Disorder: Application of the Binomial Distribution"** (*Journal of Anxiety Disorders* [2008]: 1297–1302) investigated the use of SVTs in the diagnosis of post-traumatic stress disorder. One SVT proposed is a 60-item test (called the MENT test), where each item has only a correct or incorrect response. The MENT test is designed so that responses to the individual questions can be considered independent of one another. For this reason, the authors of the paper believe that the score on the MENT test can be viewed as a binomial random variable with $n = 60$. The MENT test is designed to help in distinguishing fictitious claims of post-traumatic stress disorder. The items on the test are written so that the correct response to an item should be relatively obvious, even to people suffering from stress disorders. Researchers have found that a patient with a fictitious claim of stress disorder will try to "fake" the test, and that the probability of a correct response to an item for these patients is 0.7 (compared to 0.96 for other patients).

The authors used a normal approximation to the binomial distribution with $n = 60$ and $p = 0.7$ to calculate various probabilities of interest, where $x = $ *number of correct responses on the MENT test* for a patient who is trying to fake the test.
a. Verify that it is appropriate to use a normal approximation to the binomial distribution in this situation.
b. Approximate the following probabilities:
   i. $P(x = 42)$
   ii. $P(x < 42)$
   iii. $P(x \leq 42)$
c. Explain why the probabilities calculated in Part (b) are not all equal.
d. The authors calculated the exact binomial probability of a score of 42 or less for someone who is not faking the test. Using $p = 0.96$, they found

$$P(x \leq 42) = .000000000013$$

Explain why the authors calculated this probability using the binomial formula rather than using a normal approximation.
e. The authors propose that someone who scores 42 or less on the MENT exam is faking the test. Explain why this is reasonable, using some of the probabilities from Parts (b) and (d) as justification.

6.100 Studies have found that women diagnosed with cancer in one breast also sometimes have cancer in the other breast that was not initially detected by mammogram or physical examination **("MRI Evaluation of the Contralateral Breast in Women with Recently Diagnosed Breast Cancer,"** *The New England Journal of Medicine* **[2007]: 1295–1303).** To determine if magnetic resonance imaging (MRI) could detect missed tumors in the other breast, 969 women diagnosed with cancer in one breast had an MRI exam. The MRI detected tumors in the other breast in 30 of these women.
a. Use $p = \dfrac{30}{969} = 0.031$ as an estimate of the probability that a woman diagnosed with cancer in one breast has an undetected tumor in the other breast. Consider a random sample of 500 women diagnosed with cancer in one breast. Explain why it is reasonable to think that the random variable $x = $ *number in the sample who have an undetected tumor in the other breast* has a binomial distribution with $n = 500$ and $p = 0.031$.
b. Is it reasonable to use the normal distribution to approximate probabilities for the random variable x defined in Part (a)? Explain why or why not.
c. Approximate the following probabilities:
   i. $P(x < 10)$
   ii. $P(10 \leq x \leq 25)$
   iii. $P(x > 20)$
d. For each of the probabilities calculated in Part (c), write a sentence interpreting the probability.

CHAPTER ACTIVITIES

ACTIVITY 6.1   IS IT REAL?

Background: Three students were asked to complete the following:

a. Flip a coin 30 times, and note the number of heads observed in the 30 flips.

b. Repeat Step (a) 99 more times, to obtain 100 observations of the random variable x = *number of heads in 30 flips*.

c. Construct a dotplot of the 100 values of x.

Because this was a tedious assignment, one or more of the three did not really carry out the coin flipping and just made up 100 values of x that they thought would look "real." The three dotplots produced by these students are shown here.

1. Which student(s) made up x values? What about the dotplot makes you think the student did not actually do the coin flipping?

2. Working as a group, each student in your class should flip a coin 30 times and note the number of heads in the 30 tosses. If there are fewer than 50 students in the class, each student should repeat this process until there

are a total of at least 50 observations of x = *number of heads in 30 flips*. Using the data from the entire class, construct a dotplot of the x values.

3. After looking at the dotplot in Step 2 that resulted from actually flipping a coin 30 times and observing number of heads, review your answers in Question 1. For each of the three students, explain why you now think that he or she did or did not actually do the coin flipping.

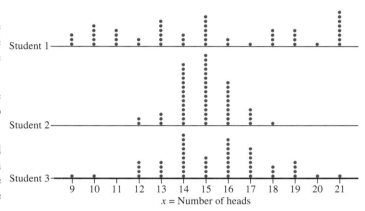

ARE YOU READY TO MOVE ON?   CHAPTER 6 REVIEW EXERCISES

All chapter learning objectives are assessed in these exercises. The learning objectives assessed in each exercise are given in parentheses.

6.101 (M1)

A point is randomly selected on the surface of a lake that has a maximum depth of 100 feet. Let x be the depth of the lake at the randomly chosen point. What are possible values of x? Is x discrete or continuous?

6.102 (M3)

A box contains five slips of paper, marked $1, $1, $1, $10, and $25. The winner of a contest selects two slips of paper at random and then gets the larger of the dollar amounts on the two slips. Define a random variable w by w = *amount awarded*. Determine the probability distribution of w. (Hint: Think of the slips as numbered 1, 2, 3, 4, and 5. An outcome of the experiment will consist of two of these numbers.)

6.103 (M2, P1)

Airlines sometimes overbook flights. Suppose that for a plane with 100 seats, an airline takes 110 reservations. Define the random variable x as

x = the number of people who actually show up for a sold-out flight on this plane

From past experience, the probability distribution of x is given in the following table:

x	p(x)
95	0.05
96	0.10
97	0.12
98	0.14
99	0.24
100	0.17
101	0.06
102	0.04
103	0.03
104	0.02
105	0.01
106	0.005
107	0.005
108	0.005
109	0.0037
110	0.0013

a. What is the probability that the airline can accommodate everyone who shows up for the flight?
b. What is the probability that not all passengers can be accommodated?
c. If you are trying to get a seat on such a flight and you are number 1 on the standby list, what is the probability that you will be able to take the flight? What if you are number 3?

6.104 (C1, C2)
Suppose that the random variable

x = actual weight (in ounces) of a randomly selected package of cereal

has the probability distribution described by the density curve pictured here.

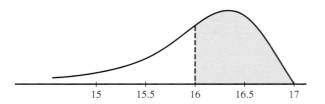

a. What probability is represented by the shaded area?
b. Suppose the shaded area = 0.50. Interpret this probability in the context of this problem.

6.105 (M6)
Let x be the amount of time (in minutes) that a particular San Francisco commuter will have to wait for a BART train. Suppose that the density curve is as pictured (a uniform distribution):

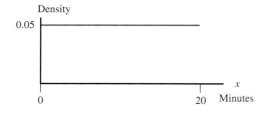

a. What is the probability that x is less than 10 minutes? More than 15 minutes?
b. What is the probability that x is between 7 and 12 minutes?
c. Find the value c for which $P(x < c) = 0.9$.

6.106 (M5)
Referring to the previous exercise, let x and y be waiting times on two independently selected days. Define a new random variable w by $w = x + y$, the sum of the two waiting times. The set of possible values for w is the interval from 0 to 40 (because both x and y can range from 0 to 20). It can be shown that the density curve of w is as pictured (this curve is called a triangular distribution, for obvious reasons!):

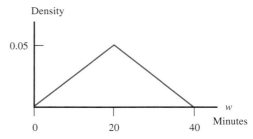

a. Shade the area under the density curve that corresponds to $P(w > 35)$.
b. Put the following probabilities in order from smallest to largest:

$P(w < 10)$ $P(20 < w < 25)$ $P(25 < w < 30)$ $P(w > 30)$

Explain your reasoning.

6.107 (C2)
A company receives light bulbs from two different suppliers. Define the variables x and y as
x = lifetime of a bulb from Supplier 1
y = lifetime of a bulb from Supplier 2
Five hundred bulbs from each supplier are tested, and the lifetime of each bulb (in hours) is recorded. The density histograms below are constructed from these two sets of observations. Although these histograms are constructed using data from only 500 bulbs, they can be considered approximations to the corresponding probability distributions.
a. Which probability distribution has the larger mean?
b. Which probability distribution has the larger standard deviation?
c. Assuming that the cost of the light bulbs is the same for both suppliers, which supplier would you recommend? Explain.
d. One of the two distributions pictured has a mean of approximately 1000, and the other has a mean of about 900. Which

HISTOGRAMS FOR EXERCISE 6.107

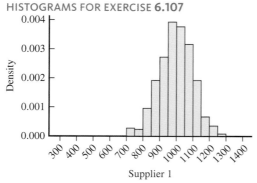
Supplier 1

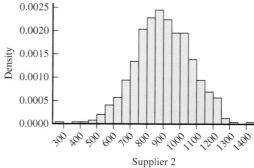
Supplier 2

Data set available

of these is the mean of the distribution for the variable x (lifetime for a bulb from Supplier 1)?

e. One of the two distributions pictured has a standard deviation of approximately 100, and the other has a standard deviation of about 175. Which of these is the standard deviation of the distribution for the variable x (lifetime for a bulb from Supplier 1)?

6.108 (M4)

The probability distribution of x, the number of defective tires on a randomly selected automobile checked at a certain inspection station, is given in the following table:

x	0	1	2	3	4
$p(x)$	0.54	0.16	0.06	0.04	0.20

a. Calculate the mean value of x.
b. Interpret the mean value of x in the context of a long sequence of observations of number of defective tires.
c. What is the probability that x exceeds its mean value?
d. Calculate the standard deviation of x.

6.109 (M7, M8)

The size of the left upper chamber of the heart is one measure of cardiovascular health. When the upper left chamber is enlarged, the risk of heart problems is increased. The paper **"Left Atrial Size Increases with Body Mass Index in Children"** (*International Journal of Cardiology* [2009]: 1–7) described a study in which the left atrial size was measured for a large number of children ages 5 to 15 years. Based on these data, the authors concluded that for healthy children, left atrial diameter was approximately normally distributed with a mean of 26.4 mm and a standard deviation of 4.2 mm.

a. Approximately what proportion of healthy children have left atrial diameters less than 24 mm?
b. Approximately what proportion of healthy children have left atrial diameters greater than 32 mm?
c. Approximately what proportion of healthy children have left atrial diameters between 25 and 30 mm?
d. For healthy children, what is the value for which only about 20% have a larger left atrial diameter?

6.110 (M7, M8)

The paper referenced in the previous exercise also included data on left atrial diameter for children who were considered overweight. For these children, left atrial diameter was approximately normally distributed with a mean of 28 mm and a standard deviation of 4.7 mm.

a. Approximately what proportion of overweight children have left atrial diameters less than 25 mm?
b. Approximately what proportion of overweight children have left atrial diameters greater than 32 mm?
c. Approximately what proportion of overweight children have left atrial diameters between 25 and 30 mm?
d. What proportion of overweight children have left atrial diameters greater than the mean for healthy children?

6.111 (M7, M8)

Consider the variable x = *time required for a college student to complete a standardized exam*. Suppose that for the population of students at a particular university, the distribution of x is well approximated by a normal curve with mean 45 minutes and standard deviation 5 minutes.

a. If 50 minutes is allowed for the exam, what proportion of students at this university would be unable to finish in the allotted time?
b. How much time should be allowed for the exam if you wanted 90% of the students taking the test to be able to finish in the allotted time?
c. How much time is required for the fastest 25% of all students to complete the exam?

6.112 (C2, P1)

Example 6.14 gave the probability distributions shown below for

$\quad$ x = number of flaws in a randomly selected glass panel
$\quad\quad$ from Supplier 1
$\quad$ y = number of flaws in a randomly selected glass panel
$\quad\quad$ from Supplier 2

for two suppliers of glass used in the manufacture of flat screen TVs. If the manufacturer wanted to select a single supplier for glass panels, which of these two suppliers would you recommend? Justify your choice based on consideration of both center and variability.

x	0	1	2	3	y	0	1	2	3
$p(x)$	0.4	0.3	0.2	0.1	$p(y)$	0.2	0.6	0.2	0

6.113 (M9)

The following normal probability plot was constructed using data on the price of seven 2015 Honda Accords with automatic transmissions that were listed for sale within 25 miles of the zip code 19383 **(www.autotrader.com, search conducted on September 24, 2016)**. For purposes of this exercise, you may assume that this sample is representative of 2015 Honda Accord prices in this area. Based on the normal probability plot, is it reasonable to think that the distribution of 2015 Honda Accord prices in this area is approximately normal? Explain.

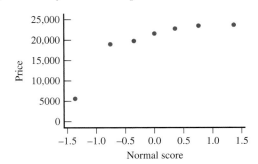

6.114 (M10)

Macular degeneration is the most common cause of blindness in people older than 60 years. One variable thought to be related to a type of inflammation associated with this disease is level of a substance called soluble Fas Ligand (sFasL) in the blood. The accompanying data are representative values of $x = sFasL\ level$ for 10 patients with age-related macular degeneration. These data are consistent with summary quantities and descriptions of the data given in the paper **"Associations of Plasma-Soluble Fas Ligand with Aging and Age-Related Macular Degeneration"** (*Investigative Ophthalmology and Visual Science* **[2008]: 1345–1349**).

x	Normal Score
0.069	−1.547
0.074	−1.000
0.176	−0.655
0.185	−0.375
0.216	−0.123
0.287	0.123
0.343	0.375
0.343	0.655
0.512	1.000
0.729	1.547

a. Construct a normal probability plot. Does the normal probability plot appear linear or curved?

b. Calculate the correlation coefficient for the (normal score, x) pairs. Compare this value to the appropriate critical r value from Table 6.2 to determine if it is reasonable to consider the distribution of sFasL levels to be approximately normal.

TECHNOLOGY NOTES

Finding Normal Probabilities

JMP

1. After opening a new data table, click **Rows** then select **Add rows**
2. Type **1** in the box next to **How many rows to add:**
3. Click **OK**
4. Double-click on the **Column 1** heading
5. Click **Column Properties** and select **Formula**
6. Click **Edit Formula**
7. In the box under **Functions (grouped)** click **Probability** then select **Normal Distribution**
8. In the white box at the bottom half of the screen, double-click in the red box around x

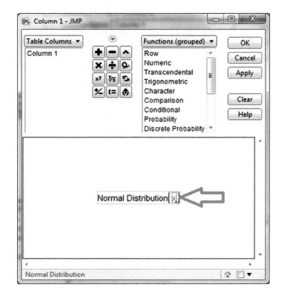

9. Type the value that you would like to find a probability for and click **OK**
10. Click **OK**

Note: This procedure outputs the value for $P(X \leq x)$. If you want to find the value for $P(X \geq x)$ you will need to subtract the output from one.

Minitab

1. Click **Calc** then click **Probability Distributions** then click **Normal...**
2. In the box next to **Mean:** type the mean of the normal distribution that you are working with
3. In the box next to **Standard deviation:** type the standard deviation of the normal distribution that you are working with
4. Click the radio button next to **Input constant:**
5. Click in the box next to **Input constant:** and type the value that you are finding the probability for
6. Click **OK**

Note: This procedure outputs the value for $P(X \leq x)$. If you want to find the value for $P(X \geq x)$ you will need to subtract the output from one.

Note: You may also type a column of values for which you would like to find probabilities (these must ALL be from the SAME distribution) and use the **Input Column** option to find probabilities for each value in the column selected.

SPSS

1. Type in the values that you would like to find the probabilities for in one column (this column will be automatically titled VAR00001)

2. Click <u>T</u>ransform and select <u>C</u>ompute Variable...
3. In the box under **Target Variable:** type NormProb (this will be the title of a new column that lists the cumulative probabilities for each value in VAR00001)
4. Under **Function group:** select **CDF & Noncentral CDF**
5. Under **Functions and Special Variables:** double-click **Cdf. Normal**
6. In the box under **Num<u>e</u>ric Expression:** you should see CDF. NORMAL(?,?,?)
7. Highlight the first ? and double-click VAR00001
8. Highlight the second ? and type the mean of the normal distribution that you are working with
9. Highlight the third ? and type the standard deviation of the normal distribution that you are working with

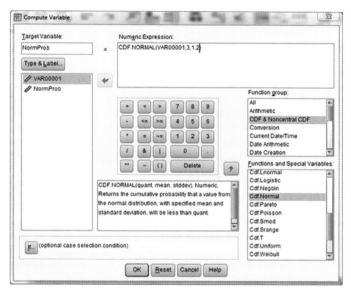

10. Click **OK**

Note: This procedure outputs the value for $P(X \leq x)$. If you want to find the value for $P(X \geq x)$ you will need to subtract the output from one.

Excel
1. Click in an empty cell
2. Click the **Formulas** ribbon
3. Select **Insert Function**
4. Select **Statistical** from the drop-down menu for category
5. Select **NORMDIST** from the **Select a function:** box
6. Click **OK**
7. Click in the box next to **X** and type the data value that you would like to find the probability for
8. Click in the box next to **Mean** and type the mean of the Normal distribution that you are working with
9. Click in the box next to **Standard_dev** and type the standard deviation of the Normal distribution that you are working with
10. Click in the box next to **Cumulative** and type **TRUE**
11. Click **OK**

Note: This procedure outputs the value for $P(X \leq x)$. If you want to find the value for $P(X \geq x)$ you will need to subtract the output from one.

TI-83/84
$P(x_1 < X < x_2)$
1. Press the **2nd** key then press the **VARS** key
2. Highlight **normalcdf(** and press **ENTER**
3. Type in the lower bound x_1
4. Type,
5. Type in the upper bound x_2
6. Type,
7. Type in the value of the mean
8. Type,
9. Type in the value of the standard deviation
10. Type **)**
11. Press **ENTER**

Note: If you are looking for $P(X < x)$, use a lower bound of $-10{,}000{,}000$. If you are looking for $P(X > x)$, use an upper bound of $10{,}000{,}000$.

TI-Nspire
$P(x_1 < X < x_2)$
1. Enter the Scratchpad
2. Press the **menu** key and select **5:Probablity** then **5:Distributions** then **2:Normal Cdf...** then press **enter**
3. In the box next to **Lower Bound** type the value for x_1
4. In the box next to **Upper Bound** type the value for x_2
5. In the box next to **μ** type the value of the mean
6. In the box next to **σ** type the value of the standard deviation
7. Press **OK**

Note: If you are looking for $P(X < x)$, use a lower bound of $-10{,}000{,}000$. If you are looking for $P(X > x)$, use an upper bound of $10{,}000{,}000$.

Finding Binomial Probabilities

JMP
$P(X \leq x)$
1. After opening a new data table, click **Rows** then select **Add rows**
2. Type **1** in the box next to **How many rows to add:**
3. Click **OK**
4. Double-click on the **Column 1** heading
5. Click **Column Properties** and select **Formula**
6. Click **Edit Formula**
7. In the box under **Functions (grouped)** click **Discrete Probability** then select **Binomial Distribution**
8. In the white box at the bottom half of the screen, double-click in the red box around p and type the value for the success probability, p and press **enter** on your keyboard
9. Double-click in the red box around n and type the value for the number of trials, n and press **enter** on your keyboard
10. Double-click in the red box around **k** and type the value for the number of successes for which you would like to find the probability
11. Click **OK**
12. Click **OK**

P(X = x)

1. After opening a new data table, click **Rows** then select **Add rows**
2. Type **1** in the box next to **How many rows to add:**
3. Click **OK**
4. Double-click on the **Column 1** heading
5. Click **Column Properties** and select **Formula**
6. Click **Edit Formula**
7. In the box under **Functions (grouped)** click **Discrete Probability** then select **Binomial Probability**
8. In the white box at the bottom half of the screen, double-click in the red box around **p** and type the value for the success probability, *p,* and press **enter** on your keyboard
9. Double-click in the red box around **n** and type the value for the number of trials, *n,* and press **enter** on your keyboard
10. Double-click in the red box around **k** and type the value for the number of successes for which you would like to find the probability
11. Click **OK**
12. Click **OK**

Minitab

P(X ≤ x)

1. Click **Calc** then click **Probability Distributions** then click **Binomial...**
2. In the box next to **Number of Trials:** input the value for *n,* the total number of trials
3. In the box next to **Probability of Success:** input the value for *p,* the success probability
4. Click the radio button next to **Input Constant**
5. In the box next to **Input Constant:** type the value that you want to find the probability for
6. Click **OK**

Note: You may also type a column of values for which you would like to find probabilities (these must ALL be from the SAME distribution) and use the **Input Column** option to find probabilities for each value in the column selected.

P(X = x)

1. Click **Calc** then click **Probability Distributions** then click **Binomial...**
2. Click the radio button next to **Probability**
3. In the box next to **Number of Trials:** input the value for *n,* the total number of trials
4. In the box next to **Probability of Success:** input the value for *p,* the success probability
5. Click the radio button next to **Input Constant**
6. In the box next to **Input Constant:** type the value that you want to find the probability for
7. Click **OK**

Note: You may also type a column of values for which you would like to find probabilities (these must ALL be from the SAME distribution) and use the **Input Column** option to find probabilities for each value in the column selected.

SPSS

P(X = x)

1. Type in the values that you would like to find the probabilities for in one column (this column will be automatically titled VAR00001)
2. Click **Transform** and select **Compute Variable...**
3. In the box under **Target Variable:** type **BinomProb** (this will be the title of a new column that lists the cumulative probabilities for each value in VAR00001)
4. Under **Function group:** select **CDF & Noncentral CDF**
5. Under **Functions and Special Variables:** double-click **Cdf. Binom**
6. In the box under **Numeric Expression:** you should see CDF. BINOM(?,?,?)
7. Highlight the first ? and double-click VAR00001
8. Highlight the second ? and type the value for *n,* the total number of trials
9. Highlight the third ? and type success probability, *p*
10. Click **OK**

P(X = x)

1. Type in the values that you would like to find the probabilities for in one column (this column will be automatically titled VAR00001)
2. Click **Transform** and select **Compute Variable...**
3. In the box under **Target Variable:** type BinProb (this will be the title of a new column that lists the cumulative probabilities for each value in VAR00001)
4. Under **Function group:** select **PDF & Noncentral PDF**
5. Under **Functions and Special Variables:** double-click Pdf. Binom
6. In the box under **Numeric Expression:** you should see PDF. BINOM(?,?,?)
7. Highlight the first ? and double-click VAR00001
8. Highlight the second ? and type the value for *n,* the total number of trials
9. Highlight the third ? and type success probability, *p*
10. Click **OK**

Excel

1. Click in an empty cell
2. Click the **Formulas** ribbon
3. Select **Insert Function**
4. Select **Statistical** from the drop-down menu for category
5. Select **BINOMDIST** from the **Select a function:** box
6. Click **OK**
7. Click in the box next to **Number_s** and type the number of successes that you are finding a probability for
8. Click in the box next to **Trials** and type the value for *n,* the total number of trials
9. Click in the box next to **Probability_s** and type success probability, *p*
10. Click in the box next to **Cumulative** and type **TRUE** if you are finding *P(X ≤ x)* or type **FALSE** if you are finding *P(X = x)*
11. Click **OK**

Note: This procedure outputs the value for $P(X \leq x)$ when **TRUE** is used as input for **Cumulative**. If you want to find the value for $P(X \geq x)$ you will need to subtract this output from one.

TI-83/84
$P(X \leq x)$
1. Press the **2nd** key then press the **VARS** key
2. Highlight the **binomcdf(** option and press the **ENTER** key
3. Type in the number of trials, n
4. Type,
5. Type in the success probability, p
6. Type,
7. Type the value for x
8. Type **)**
9. Press **ENTER**

$P(X = x)$
1. Press the **2nd** key then press the **VARS** key
2. Highlight the **binompdf(** option and press the **ENTER** key
3. Type in the number of trials, n
4. Type,
5. Type in the success probability, p
6. Type,
7. Type the value for x
8. Type **)**
9. Press **ENTER**

TI-Nspire
$P(x_1 \leq X \leq x_2)$
1. Enter the Scratchpad
2. Press the **menu** key and select **5:Probability** then select **5:Distributions** then select **E:Binomial Cdf...** and press the **enter** key
3. In the box next to **Num Trials, n** type in the number of trials, n
4. In the box next to **Prob Success, p** type in the success probability, p
5. In the box next to **Lower Bound** type in the value for x_1
6. In the box next to **Upper Bound** type in the value for x_2
7. Press **OK**

Note: In order to find $P(X \leq x)$ input a 0 for the lower bound.

$P(X = x)$
1. Enter the Calculate Scratchpad
2. Press the **menu** key and select **5:Probability** then select **5:Distributions** then select **D:Binomial Pdf...** and press the **enter** key
3. In the box next to **Num Trials, n** type in the number of trials, n
4. In the box next to **Prob Success, p** type in the success probability, p
5. In the box next to **X Value** type the value for x
6. Press **OK**

Normal Probability Plots

JMP
1. Enter the raw data into a column
2. Click **Analyze** then select **Distribution**

3. Click and drag the column name containing the data from the box under **Select Columns** to the box next to **Y, Columns**
4. Click **OK**
5. Click the red arrow next to the column name
6. Select **Normal Quantile Plot**

Minitab
1. Input the data for which you would like to check Normality into a column
2. Click **Graph** then click **Probability Plot...**
3. Highlight the **Single** plot
4. Click **OK**
5. Double click the column name for the column that contains your data to move it into the **Graph Variables:** box
6. Click **OK**

SPSS
1. Input the data for which you would like to check Normality into a column
2. Click **Analyze** then select **Descriptive Statistics** then select **Q-Q Plots...**
3. Highlight the column name for the variable
4. Click the arrow to move the variable to the **Variables:** box
5. Click **OK**

Note: The normal probability plot is output with several other plots and statistics. This plot can be found under the title **Normal Q-Q Plot.**

Excel
Excel does not have the functionality to automatically produce Normal Probability Plots.

However, Excel can produce Normal Probability Plots in the course of running a regression using the Analysis ToolPak.

TI-83/84
1. Input the raw data into **L1** (In order to access lists press the **STAT** key, highlight the option called **Edit...** then press **ENTER**)
2. Press the **2nd** key then press the **Y =** key
3. Highlight **Plot1** and press **ENTER**
4. Highlight **On** and press **ENTER**
5. Highlight the plot type on the second row, third column and press **ENTER**
6. Press **GRAPH**

TI-Nspire
1. Enter the data into a data list (In order to access data lists select the spreadsheet option and press **ENTER**)

Note: Be sure to title the list by selecting the top row of the column and typing a title.

2. Press **menu** and select **3:Data** then **6:QuickGraph** then press **enter**
3. Press **menu** and select **1:Plot Type** then select **4:Normal Probability Plot** and press **ENTER**

7

An Overview of Statistical Inference— Learning from Data

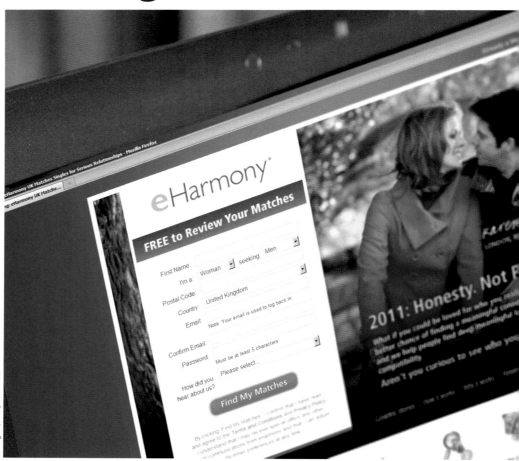

Ian Dagnall/Alamy Stock Photo

PREVIEW

Whether data are collected by sampling from populations or result from an experiment to compare treatments, the ultimate goal is to learn from the data.

This chapter introduces the inferential process shared by all of the methods for learning from data that are covered in the chapters that follow. You will also see how the answers to four key questions guide the selection of an appropriate inference method.

Conceptual Understanding

After completing this chapter, you should be able to

C1   Understand the difference between questions that can be answered by using sample data to estimate population characteristics and those that can be answered by testing hypotheses about population characteristics.

C2   Understand that there is risk involved in drawing conclusions from sample data—that when generalizing from sample data, the sample may not always provide an accurate picture of the population.

C3   Understand that there is risk involved with drawing conclusions from experiment data—that when generalizing from experiment data, the observed difference in treatment effects may sometimes be due to variability in the response variable and the random assignment to treatments.

C4   Understand that data type distinguishes between questions that involve proportions and those that involve means.

C5   Know that different methods are used to draw conclusions based on categorical data and to draw conclusions based on numerical data.

C6   Know the four key questions that help identify an appropriate inferential method.

C7   Know the five-step process for estimation problems.

C8   Know the five-step process for hypothesis testing problems.

Mastering the Mechanics

After completing this chapter, you should be able to

M1   Distinguish between estimation problems and hypothesis testing problems.

M2   Distinguish between problems that involve proportions and those that involve means.

Putting it into Practice

After completing this chapter, you should be able to

P1   Given a scenario, answer the four key questions that help identify an appropriate inference method.

Deception in Online Dating Profiles

With the increasing popularity of online dating services, the truthfulness of information in the personal profiles provided by users is a topic of interest. The authors of the paper **"Self-Presentation in Online Personals: The Role of Anticipated Future Interaction, Self-Disclosure, and Perceived Success in Internet Dating" (*Communication Research* [2006]: 152–177)** designed a statistical study to investigate misrepresentation of personal characteristics. The researchers hoped to answer three questions:

1.   What proportion of online daters believe they have misrepresented themselves in an online profile?

2.   What proportion of online daters believe that others frequently misrepresent themselves?

3.   Are people who place a greater importance on developing a long-term, face-to-face relationship more honest in their online profiles?

What did the researchers learn? Based on the data, they estimated that only about 6% of online daters believe that they have intentionally misrepresented themselves in online profiles. In spite of the fact that most users believe themselves to be honest, about 86% believe that others frequently misrepresented characteristics such as physical appearance in the online profile. The researchers also found that the data supported the claim that those who place greater importance on developing a long-term, face-to-face relationship are more honest in their online profiles.

How were these researchers able to reach these conclusions? The estimates (the 6% and the 86% in the previous statements) are based on sample data. Do these estimates provide an accurate picture of the entire population of online daters? The researchers concluded that the data supported the claim that those who place greater importance on developing a long-term, face-to-face relationship are more honest in the way they represented themselves online, but how did they reach this conclusion, and should you be convinced? These are important questions. In this chapter and those that follow, you will see how questions like these can be answered. ■

SECTION 7.1 Statistical Inference—What You Can Learn from Data

Statistical inference is all about learning from data. Inferential methods involve using sample data to learn about a population or using experiment data to learn about treatment effects.

Learning from Sample Data

When you obtain information from a sample selected from some population, it is usually because

1. You want to learn something about characteristics of the population. This results in an **estimation problem**. It involves using sample data to estimate population characteristics.

OR

2. You want to use the sample data to decide whether there is support for some claim or statement about the population. This results in a **hypothesis testing problem**. It involves testing a claim (a hypothesis) about the population.

Example 7.1 Deception in Online Dating Profiles Revisited

Let's revisit the online dating example of the chapter preview. In that example, the population of interest was all online daters. Three questions about this population were identified:

1. What proportion of online daters believe they have misrepresented themselves in an online profile?
2. What proportion of online daters believe that others frequently misrepresent themselves?
3. Are people who place a greater importance on developing a long-term, face-to-face relationship more honest in their online profiles?

The first two of these questions are estimation problems because they involve using sample data to learn something about a population characteristic. The population characteristic of interest in the first question is the proportion of all online daters who believe they have misrepresented themselves online. In the second question, the population characteristic of interest is the proportion of all online daters who believe that others frequently misrepresent themselves. The third question is a hypothesis testing problem because it involves determining if sample data support a claim about the population of online daters.

■

An **estimation problem** involves using sample data to estimate the value of a population characteristic.

A **hypothesis testing problem** involves using sample data to test a claim about a population.

| Example 7.2 | Whose Reality? |

The article **"Who's Afraid of Reality Shows?"** (*Communication Research* **[2008]:382–397**) considers social concern over reality television shows. Researchers conducted telephone interviews with 606 individuals in a sample selected to represent the adult population of Israel. One of the things that the researchers hoped to learn was whether the data supported the theory that a majority of Israeli adults believed they are much less affected by reality shows than other people. They concluded that the sample data did provide support for this theory. This study involves generalizing from the sample to the population of Israeli adults and it is a hypothesis testing problem because it uses sample data to test a claim (that a majority consider themselves less affected than others).

Learning from Sample Data When There Are Two or More Populations

Sometimes sample data are obtained from two or more populations of interest, and the goal is to learn about differences between the populations. Consider the following two examples.

| Example 7.3 | Electronic Devices in the Bedroom |

In 2014, the National Sleep Foundation conducted a survey of 1103 adult Americans who have a child between the age of 6 and 17 living with them. Data from this survey were published in the report **"Sleep in the Modern Family" (2014 Sleep in America Poll, www.sleepfoundation.org/sleep-polls-data/sleep-in-america-poll/2014-sleep-and-family, retrieved May 3, 2017)**. Because it is thought that having electronic devices such as televisions, computers, and smartphones in the bedroom has the potential to disturb sleep, the National Sleep Foundation was interested in learning about how many electronic devices parents and children have in their bedrooms. Based on data from the survey, they were able to estimate the proportion of parents who have 2 or more electronic devices in their bedrooms (0.54 or 54%) and the proportion of children who have 2 or more electronic devices in their bedrooms (0.45 or 45%). From this information, it is possible to estimate the difference in the proportion with two or more devices for the two populations of interest (parents and children). The proportion with two or more electronic devices in the bedroom is 0.09, or 9 percentage points, greater for parents than children. This study illustrates statistical inference because it involves generalizing from samples to corresponding populations. It is an estimation problem because sample data were used to estimate the values of population characteristics. Because there were two samples, one from each of two different populations, it is also possible to learn something about how the two populations differ with respect to a population characteristic.

| Example 7.4 | Does School Start Time Make a Difference? |

If the school day were to start later, would kids get more sleep? The authors of the paper **"Adolescent Sleep, School Start Times, and Teen Motor Vehicle Crashes"** (*Journal of Clinical Sleep Medicine* **[2008]: 533–535**) collected data from a large sample of students in grades 6 to 12 during the year prior to a change in school start time and from a second sample of students in grades 6 to 12 during the year after the change. The students in both samples attended a school district that implemented a one-hour delay in the school start time for the second year of the study. One of the variables that the researchers studied was the number of hours spent sleeping in a typical day. The researchers concluded that the mean time spent sleeping was 0.5 hours greater after the change to a delayed start time than before the change. This study involves generalizing from samples. It is a hypothesis testing problem because it involves testing a claim (the mean number of hours of sleep on a typical night is greater when students have a delayed start). This claim is about the difference between two groups (students with an early start time schedule and students on a delayed start schedule).

Learning from Experiment Data

Statistical inference methods are also used to learn from experiment data. When data are obtained from an experiment, it is usually because

1. You want to learn about the effect of the different experimental conditions (treatments) on the measured response. This is an **estimation problem** because it involves using sample data to estimate a characteristic of the treatments, such as the mean response for a treatment or the difference in mean response for two treatments.

OR

2. You want to determine if experiment data provide support for a claim about how the effects of two or more treatments differ. This is a **hypothesis testing problem** because it involves testing a claim (hypothesis) about treatment effects.

The following two examples illustrate an estimation problem and a hypothesis testing problem in the context of learning from experiment data.

Example 7.5   Physical Therapy or Surgery?

The article **"Study Points to Benefits of Knee Replacement Surgery Over Therapy Alone"** (*The New York Times*, **October 21, 2015**) describes a study in which researchers randomly assigned adults with moderate to severe osteoarthritis to either a group who received a knee replacement followed by physical therapy or a group who did not have knee replacement surgery and who received extensive nonsurgical therapy. After one year, the people participating in the study were each asked if they had less pain than before surgery or treatment. Using data from the study, the researchers estimated that the proportion who experience pain relief is higher by 0.17 (17 percentage points) for those who have knee replacement surgery than for those who have nonsurgical therapy. This is an example of an estimation problem. It involves generalizing from experiment data to treatment characteristics—in this case, the difference in the proportion who experience pain relief for the two treatments (surgery and nonsurgical therapy).

Example 7.6   Use a White Mug When You Drink Coffee ...

Have you ever wondered why most restaurants serve coffee in white mugs? The paper **"Does the Colour of the Mug Influence the Taste of the Coffee?"** (*Flavour* [2014]: 1–7) describes an experiment in which researchers investigated whether people rated flavor characteristics of coffee differently depending on the color of the mug that the coffee was served in. The researchers wanted to test the claim that people rate coffee differently depending on whether it is served in a clear glass mug or in a white mug. Using methods you will encounter in Chapter 13, the researchers concluded that the data supported the claim that the color of the mug makes a difference, finding that people tended to rate the coffee flavor as more intense and as less sweet when served in a white mug. This is an example of a hypothesis testing problem. It involves using experiment data to test a claim about treatment effect—in this case, about the difference in mean coffee ratings for the white mug treatment and for the clear glass mug treatment.

Statistical Inference Involves Risk

Examples 7.1–7.6 illustrate learning from sample data and from experiment data in a variety of settings. However, one important aspect has not yet been addressed. *Learning from data involves risk.* To be fully informed, you need to recognize what these risks are and consider how you can assess and describe the risks involved. Even when a reasonable sampling plan is used or a well-designed experiment is carried out, the resulting data provide only incomplete information about the population or about treatment effects.

The risks associated with statistical inference arise because you are attempting to draw conclusions on the basis of data that provide partial rather than complete information. But you can still make good decisions, even with incomplete information. The statistical inference methods that you will learn in the chapters that follow provide ways to make effective use of available data, but it is important to recognize that there are still risks involved.

What are the risks? In estimation problems, where sample data are used to estimate population or treatment characteristics, you run the risk that these estimates will be inaccurate. Even when you accompany an estimate with a description of the anticipated accuracy, sometimes estimates aren't as accurate as you say. When you produce estimates and evaluate estimates produced by others, you need to understand that there is a risk that the estimates are not as accurate as claimed. It is important to understand how likely it is that the method used to produce the estimates and accompanying measures of accuracy might mislead in this way.

In hypothesis testing situations, where you use partial rather than complete information to determine whether there is support for a claim about population or treatment characteristics, you run the risk of an incorrect conclusion. Based on the available data you might conclude that there is support for a claim that is actually not true. Or, you may decide that the data do not support a claim that really is true. When you carry out a hypothesis test, you will need to understand the consequences of reaching an incorrect conclusion. You also must understand how likely it is that the method used to decide whether or not a claim is supported might lead to an incorrect conclusion.

Variability in Data

Sampling

The most important factor in evaluating risk is understanding variability in the data. When a sample is selected from a population, the sample provides only a partial picture of the population. When there is variability in the population, you need to consider whether this partial picture is representative of the population. Even when a sample is selected in a reasonable way, if there is a lot of variability in the population, the partial pictures provided by different samples might be quite different. This sample-to-sample variability should be considered when you assess the risk associated with drawing conclusions about the population from sample data.

Experimentation

Variability also plays an important role in determining the risks involved when generalizing from experiment data. Experiments are usually conducted to investigate the effect of experimental conditions (treatments) on a response. For example, an experiment might be designed to determine if noise level has an effect on the time required to perform a task requiring concentration. Suppose there are 20 individuals available to serve as subjects in this experiment with two treatments conditions (quiet environment and noisy environment). The response variable is the time required to complete the task.

If noise level has no effect on completion time, the time observed for each of the 20 subjects would be the same whether they are in the quiet group or the noisy group. Any observed differences in the completion times for the two treatments would not be due to noise level. These differences would only be due to person-to-person variability in completion times and the random assignment of individuals to experimental groups. That is, the distribution of completion times for the two treatment groups might differ, but the difference can be explained by person-to-person variability in completion times and by the random assignment (different assignments will result in different people being assigned to the quiet and noisy conditions). You must understand how differences might result from variability in the response and the random assignment to treatment groups in order to distinguish them from differences created by a treatment effect. This important idea will be explored more fully in Chapter 13.

Summing It Up—Section 7.1

The following learning objectives were addressed in this section:

Conceptual Understanding

C1: Understand the difference between questions that can be answered by using sample data to estimate population characteristics and those that can be answered by testing hypotheses about population characteristics.

Sometimes data are collected in order to learn about the value of a population characteristic. This is an estimation problem. In other situations, data are collected in order to decide if there is support for a claim about a population. This is a hypothesis testing problem.

C2: Understand that there is risk involved in drawing conclusions from sample data—that when generalizing from sample data, the sample may not always provide an accurate picture of the population.

When using sample data to estimate a population characteristic, there is a risk that your estimate will be inaccurate. It is important to understand how likely it is that estimate and any associated measures of accuracy that accompany the estimate might be misleading. It is also possible that using sample data to test a claim about a population might lead to an incorrect conclusion. Again, it is important to understand what the risk of error is when using sample data to reach a decision. For a more complete discussion of risk, see the subsection "Statistical Inference Involves Risk."

C3: Understand that there is risk involved with drawing conclusions from experiment data—that when generalizing from experiment data, the observed difference in treatment effects may sometimes be due to variability in the response variable and the random assignment to treatments.

When using data from an experiment to determine if there are differences in treatment effects, there is a risk that an incorrect conclusion might be reached. As with situations that involve drawing a conclusion based on data from samples, it is important to recognize this risk and to understand how likely it is that an incorrect conclusion might be reached. For a more complete discussion of risk, see the subsection "Statistical Inference Involves Risk."

Mastering the Mechanics

M1: Distinguish between estimation problems and hypothesis testing problems.

Estimation problems involve using data to estimate the value of a population characteristic. Hypothesis testing problems involve using data to test a claim about a population.

SECTION 7.1 EXERCISES

Each Exercise Set assesses the following chapter learning objectives: C1, C2, C3, M1

SECTION 7.1 Exercise Set 1

7.1 The report **"Teens and Distracted Driving: Texting, Talking and Other Uses of the Cell Phone Behind the Wheel" (Pew Research Center, November 16, 2009)** summarizes data from a survey of a representative sample of 800 teens between the ages of 12 and 17. The following statements were made on the basis of the resulting data:

75% of all American teens own a cell phone.

66% of all American teens use a cell phone to send and receive text messages.

26% of American teens ages 16–17 have used a cell phone to text while driving.

Are the inferences made ones that involve estimation or ones that involve hypothesis testing? (Hint: See Example 7.1.)

7.2 For the study described in the previous exercise, answer the following questions.

a. What is the population of interest?

b. What population characteristics are being estimated?

c. Do you think that the actual percentage of all American teens who own a cell phone is exactly 75%? Explain why or why not.

d. Two of the estimates of population characteristics from this study were that 75% of teens own a cell phone and that 26% of teens ages 16–17 have used their phones to text while driving. Which of these two estimates do you think is more accurate and why? (Hint: What do you know about the number of teens surveyed?)

7.3 The article **"More Communities Banning 'Television on a Stick'" (USA TODAY, March 23, 2010)** describes an ongoing controversy over the distraction caused by digital billboards along highways. One study mentioned in the newspaper article is described in **"Effects of Advertising Billboards During Simulated Driving" (Applied Ergonomics [2010]: 1–8)**. In this study, 48 people made a 9 km drive in a driving simulator. Drivers were instructed to change lanes according to roadside lane change signs. Some of the lane changes occurred near digital billboards. What was displayed on the digital billboard changed once during the time that the billboard was visible by the driver to simulate the changing digital billboards that appear along highways. Data from this study supported the theory that the time required to respond to road signs was greater when digital billboards were present.

Is the inference made one that involves estimation or one that involves hypothesis testing? (Hint: See Example 7.1.)

7.4   For the study described in the previous exercise, answer the following questions.
a. What is the population of interest?
b. What claim was tested?
c. What additional information would you want before deciding if it is reasonable to generalize the conclusions of this study to the population of interest?
d. Assuming that the people who participated in the study are representative of the population of interest, do you think that the risk of an incorrect conclusion would have been lower, about the same, or higher if 100 people had participated instead of 48?

7.5   Consider the population that consists of all students enrolled at a college.
a. Give an example of a question about this population that could be answered by collecting data and using it to estimate a population characteristic.
b. Give an example of a question about this population that could be answered by collecting data and using it to test a claim about this population.

SECTION 7.1   Exercise Set 2

7.6   Do people better remember what they learned if they are in the same physical space where they first learned it? The authors of the paper **"The Dynamics of Memory: Context-Dependent Updating"** (*Learning & Memory* (2008): 574–579) asked people to learn a set of 20 unrelated objects. Two days later, these people were asked to recall the objects learned on the first day. Some of the people were asked to recall the objects in the same room where they originally learned the objects. The others were asked to recall the objects in a different room. People were assigned at random to one of these two recall conditions. The authors found that the data on the number of objects recalled supported the claim that recall is better when people return to the original learning context. Is the inference made one that involves estimation or one that involves hypothesis testing?

7.7   The article **"The Largest Last Supper: Depictions of Food Portions and Plate Size Increase Over the Millennium"** (*International Journal of Obesity* [2010]: 1–2) describes a study in which each painting in a sample of 52 paintings of *The Last Supper* was analyzed by comparing the size of the food plates in the painting to the head sizes of the people in the painting. For paintings that were painted prior to the year 1500, the estimated average plate-to-head size ratio was smaller than this ratio for the paintings that were painted after the year 1500. Is the inference made one that involves estimation or one that involves hypothesis testing?

7.8   For the study described in the previous exercise, answer the following questions.

a. The original sample consisted of 52 paintings. These paintings were then divided into two samples consisting of 30 painted before 1500 and 22 painted after 1500. What are the two populations of interest?
b. What population characteristics are being estimated?
c. Suppose that the paintings selected for analysis were selected at random from all paintings that portray *The Last Supper*. Do you think that the estimate produced for average plate-to-head size ratio for paintings made before 1500 is likely to be less accurate than the corresponding estimate for paintings made after 1500? Explain.

7.9   Consider the population that consists of all employees of a large computer manufacturer.
a. Give an example of a question about this population that could be answered by collecting data and using it to estimate a population characteristic.
b. Give an example of a question about this population that could be answered by collecting data and using it to test a claim about this population.

ADDITIONAL EXERCISES

7.10   Fans of professional soccer are probably aware that players sometimes fake injuries (called dives or flops). But how common is this practice? The articles **"A Field Guide to Fakers and Floppers"** (*The Wall Street Journal*, June 28, 2010) and **"Red Card for Faking Footballers"** (*Science Daily*, October 10, 2009) describe a study of deceptive behavior in soccer. Based on this study, it was possible to categorize injuries as real or fake based on movements that were characteristic of fake injuries (such as an arched back with hands raised, which is meant to attract the attention of a referee but which is not characteristic of the way people fall naturally). Data from an analysis of a sample of soccer games were then used to make the following statements:

On average, referees stop a soccer game to deal with apparent injuries 11 times per game.

On average, there is less than one "real" injury per soccer game.

Are the inferences made ones that involve estimation or ones that involve hypothesis testing?

7.11   **"Want to Lose More Fat? Skip Breakfast Before Workout"** (*The Tribune*, June 4, 2010) is the headline of a newspaper article describing a study comparing men who did endurance training without eating before training and men who ate before training. Twenty men were assigned at random to one of two 6-week diet and exercise programs. Both groups followed a similar diet and performed the same daily morning exercise routine. Men in one group did the exercise routine prior to eating, and those in the other group ate first and then exercised. The resulting data supported the claim that those who do not eat prior to exercising burn a higher proportion of fat than those who eat before exercising. Is the inference made one that involves estimation or one that involves hypothesis testing?

7.12 For the study described in the previous exercise, answer the following questions.
a. What is the population of interest?
b. What claim was tested?
c. What additional information would you want before deciding if it is reasonable to generalize the conclusions of this study to the population of interest?

d. Assuming that the people who participated in the study are representative of the people in the population of interest, do you think that the risk of an incorrect conclusion would have been lower, about the same, or higher if 10 men had participated instead of 20?

Selecting an Appropriate Method—Four Key Questions

This chapter together with Chapter 8 lay the conceptual groundwork for the methods that are covered in the rest of the book. There are quite a few methods for learning from data, and choosing a method that is appropriate for a particular problem is one of the biggest challenges of learning statistics. Identifying characteristics of a problem that determine an appropriate method will help you develop a systematic approach.

For each new situation that you encounter, begin by answering the four key questions that follow. The answers to these questions will guide your decision about which methods to consider.

Question Type (Q): Is the question you are trying to answer an estimation problem or a hypothesis testing problem?

In Section 7.1, you learned how to distinguish between estimation problems and hypothesis testing problems. It is important to make this distinction because you will choose different methods depending on the answer to this question.

Study Type (S): Does the situation involve generalizing from a sample to learn about a population (an observational study or survey), or does it involve generalizing from an experiment to learn about treatment effects?

The answer to this question affects the choice of method as well as the type of conclusion that can be drawn. For example, the study investigating concern over reality television shows described in Example 7.2 was an observational study and used data from a sample to learn about the population of Israeli adults. The study of Example 7.5, which examined whether knee replacement surgery or nonsurgical therapy was more effective for pain releif in people with osteoarthritis, used data from an experiment to learn about treatment effects.

Type of Data (T): What type of data will be used to answer the question? Is the data set univariate (one variable) or bivariate (two variables)? Are the data categorical or numerical?

Univariate data are used to answer questions about a single variable. For example, the goal of the study in Example 7.3 was to learn how the proportion with two or more electronic devices in the bedroom differed for parents and children. There were two samples (one consisting of parents and one consisting of children). The two groups were compared on the basis of one categorical variable (*more than two electronic devices in the bedroom*, with possible values of yes or no). Bivariate data are used to learn about the relationship between two variables. For example, the study of deception in online dating profiles (see Example 7.1) investigated whether people who place a greater importance on developing a long-term, face-to-face relationship are more honest in their online profiles. Answering this question involves looking at two variables (*importance placed on developing a long-term face-to-face relationship* and a measure of *honesty in the online profile*) for each person in the sample.

Whether you are working with categorical data or numerical data is also an important consideration in selecting an appropriate method. For example, if you have a single variable and the data are categorical, the question of interest is probably about a population proportion. But if the data are numerical, the question of interest is probably about a population mean.

The methods used to learn about proportions are different from those used to learn about means. The easiest way to distinguish between a situation involving proportions and one involving means is to determine whether the data are categorical or numerical. For example, suppose the variable of interest is whether there are two or more electronic devices in a child's bedroom. Possible values for this variable are yes and no, so the variable is categorical. These data can be used to estimate a proportion or to test a hypothesis about a proportion. On the other hand, if the variable of interest is the number of hours per day a child spends watching TV, possible values are numbers, so the variable is numerical. Data on this variable can be used to estimate a mean or test a hypothesis about a mean. In Chapter 2, you learned that the graphical displays used for categorical data (such as bar charts and pie charts) are different from those used for numerical data (such as dotplots, histograms, and boxplots). The same is true for statistical inference methods—different methods are used for different types of data.

Number of Samples or Treatments (N): How many samples are there, or, if the data are from an experiment, how many treatments are being compared?

For situations that involve sample data, different methods are used depending on whether there are one, two, or more than two samples. Also, you may choose a different method to analyze data from an experiment with only two treatments than you would for an experiment with more than two treatments.

One way to remember these four questions is to use the acronym **QSTN**—think of this as the word "question" without the vowels. A brief version of the four key questions is given here:

Q **Question Type**	Estimation or hypothesis testing?
S **Study Type**	Sample data or experiment data?
T **Type of Data**	One variable or two? Categorical or numerical?
N **Number of Samples or Treatments**	How many samples or treatments?

Table 7.1 shows how answering these questions can help you to identify a data analysis method for consideration. You will learn the methods identified in the table's "Method to Consider" column in the remaining chapters of this text.

The following examples illustrate how the four key questions (QSTN) are answered for three different studies.

Example 7.7 My Funny Valentine

It probably wouldn't surprise you to know that Valentine's Day means big business for florists, jewelry stores, and restaurants. But would it surprise you to know that it is also a big day for pet stores? In January 2015, the National Retail Federation conducted a survey of 6375 consumers that were selected in a way that the federation believed would produce a representative sample of U.S. adults **("Cupid to Shower Americans with Jewelry, Candy This Valentine's Day," www.nrf.com, retrieved November 7, 2016)**. One question in the survey asked, "Do you plan to spend money on a Valentine's Day gift for your pet this year?" The Federation hoped to learn about the proportion of U.S. adults that planned to buy a gift for their pet. Let's answer the four key questions that would help identify an appropriate analysis method.

TABLE 7.1 Answering Four Key Questions to Identify An Appropriate Method

Table Row	Q Question Type Estimation or Hypothesis Test?	S Study Type Sample Data or Experimental Data?	T Type of Data 1 Variable or 2? Categorical or Numerical?	N Number How Many Samples or Treatments?	Method to Consider	Chapter
1	Estimation	Sample	1 categorical variable	1	One-Sample z Confidence Interval for a Proportion	9
2	Hypothesis Test	Sample	1 categorical variable	1	One-Sample z Test for a Proportion	10
3	Estimation	Sample or Experiment	1 categorical variable	2	Two-Sample z Confidence Interval for a Difference in Proportions	11
4	Hypothesis Test	Sample or Experiment	1 categorical variable	2	Two-Sample z Test for a Difference in Proportions	11
5	Estimation	Sample	1 numerical variable	1	One-Sample t Confidence Interval for a Mean	12
6	Hypothesis Test	Sample	1 numerical variable	1	One-Sample t Test for a Mean	12
7	Estimation	Sample or Experiment	1 numerical variable	2	Two-Sample t or Paired t Confidence Interval for a Difference in Means	13
8	Hypothesis Test	Sample or Experiment	1 numerical variable	2	Two-Sample t or Paired t Test for a Difference in Means	13
9	Hypothesis Test	Sample	1 numerical variable	More than 2	ANOVA F Test	16
10	Estimation	Sample	1 numerical variable	More than 2	Multiple Comparisons	16

Chapter 14 includes methods for comparing **more than two group proportions** and for learning about **relationships between categorical variables**. **Chapter 15** expands on what you learned in Chapter 4, providing additional methods for learning about **relationships between numerical variables.**

Q Question Type	Estimation or hypothesis testing?	Estimation. The researchers wanted to estimate a population characteristic—the proportion of U.S. adults who plan to buy a gift for their pet. There is no claim made about this proportion that would lead to a hypothesis test.
S Study Type	Sample data or experiment data?	Sample data. The study involved a survey that was given to people in a sample selected from the population of U.S. adults.
T Type of Data	One variable or two? Categorical or numerical?	One categorical variable. Data were collected on one variable—people were asked whether or not they planned to purchase a gift for their pet. This variable is categorical—possible values are yes and no.
N Number of Samples or Treatments	How many samples or treatments?	One. There is only one sample, consisting of 6375 adults.

The answers to these four questions lead to a suggested method (see row 1 of Table 7.1). In case you are curious, the researchers estimated the proportion of U.S. adults that planned to purchase a gift for a pet to be 0.212, or 21.2%. Using methods that you will learn in Chapter 9, they were also able to say that they believed that this estimate was accurate to within 1.3 percentage points of the actual population percentage.

Example 7.8 Coffee Ratings Revisited

Example 7.6 described a study that investigated whether the color of the mug has an effect on how people rate flavor characteristics of coffee (**"Does the Colour of the Mug Influence the Taste of the Coffee?," *Flavour* [2014]: 1–7**). In the study, subjects were assigned at random to the experimental conditions (white mug or clear glass mug). After tasting the coffee, they were asked to rate the coffee flavor, using a scale of 0 to 100. The researchers were interested in answering the following question: Does the color of the mug make a difference in how people rate the flavor of coffee?

Before you look at the answers to the four key questions that follow, think about how you would answer these questions based on the study description.

Q Question Type	Estimation or hypothesis testing?	Hypothesis test. The researchers planned to use the data to test a claim about treatment effects. Specifically, they wanted to test the claim that flavor rating differs for the two mug colors.
S Study Type	Sample data or experiment data?	Experiment data. This was a study to investigate the effect of mug color on flavor rating. Subjects were randomly assigned to the two conditions considered.
T Type of Data	One variable or two? Categorical or numerical?	One numerical variable. The claim that is to be tested is about a single variable—*flavor rating*. Flavor rating was measured on a scale from 0 to 100, resulting in numerical data.
N Number of Samples or Treatments	How many samples or treatments?	Two. This experiment has two treatments (experimental conditions). In this example, the two treatments are white mug and clear glass mug.

The answers to these questions direct you to a method listed in Table 7.1. This method can be used to confirm the researchers' conclusion that mean flavor rating is not the same when coffee is served in a white mug as when it is served in a clear glass mug.

Example 7.9 | Red Wine or Onions?

Flavonols are compounds found in some foods that are thought to have beneficial health-related properties when absorbed into the blood. Just because a food contains flavanols, however, doesn't mean that it will be absorbed into the blood. A study described in the paper **"Red Wine Is a Poor Source of Bioavailable Flavonols in Men"** (*Journal of Nutrition* **[2001]: 745–748**) investigated the absorption of one type of flavonol called quercetin. In this study, one group of healthy men consumed 750 mL of red wine daily for four days and another group consumed 50 g of fried onions daily for four days. This amount of wine contains the same amount of quercetin as the onions. Blood quercetin concentration was measured at the end of the four-day period. The resulting data were then used to learn about the difference in blood quercetin concentration between the two sources of quercetin.

How would you answer the four key questions for this study? The answer to Question 1 (Q) isn't obvious, but try to answer the other questions. Then take a look at the discussion that follows.

Q **Question Type**	Estimation or hypothesis testing?	From the given description, it is hard to tell. If the objective was to estimate the difference in mean concentration for the two treatments, then this would be an estimation problem. On the other hand, if the goal was to test a claim about the difference—for example, to test the claim that the mean blood quercetin concentration was not the same for wine and onions—this would be a hypothesis testing problem. The answer here will depend on the research objective. This would need to be clarified before analyzing the data.
S **Study Type**	Sample data or experiment data?	Experiment data. Two different experimental conditions (wine and onions) were investigated, with the goal of learning about their effect on blood quercetin concentration.
T **Type of Data**	One variable or two? Categorical or numerical?	One numerical variable. The variable of interest is blood quercetin concentration, and this is a numerical variable.
N **Number of Samples or Treatments**	How many samples or treatments?	Two. This experiment involved two treatments (red wine and onions).

A football playbook is a book containing all of the plays that a team may use during a game. Players must learn how to run each of these plays, and the person calling the plays must learn which plays are appropriate in a given situation. You can think of your statistics course as helping you to develop your statistics "playbook." Each new method that you learn is like a new play that you get to add to your playbook as you master it. The four key questions of this section will help you to determine which plays in your playbook might be appropriate in a particular situation and give you a systematic way of thinking about each new method as you add it to your statistics playbook.

Summing It Up—Section 7.2

The following learning objectives were addressed in this section:

Conceptual Understanding

C4: Understand that data type distinguishes between questions that involve proportions and those that involve means.

When you have data on a single variable and the data are categorical, the question of interest is usually about a population proportion. When the data are numerical, the question of interest is usually about a population mean.

C5: Know that different methods are used to draw conclusions based on categorical data and to draw conclusions based on numerical data.

It is important to determine if you are working with categorical data or numerical data because the methods used to learn about proportions are different from the methods used to learn about means.

C6: Know the four key questions that help identify an appropriate inferential method.

Answering four key questions can help you identify an appropriate method. The four key questions are summarized below, and a more complete discussion of these questions can be found at the beginning of this section.

Q Question Type	Estimation or hypothesis testing?
S Study Type	Sample data or experiment data?
T Type of Data	One variable or two? Categorical or numerical?
N Number of Samples or Treatments	How many samples or treatments?

Mastering the Mechanics

M2: Distinguish between problems that involve proportions and those that involve means.

To distinguish between problems that involve proportions and those that involve means, consider whether you are working with categorical data or numerical data. When the data are categorical, the question of interest is usually about a population proportion. When the data are numerical, the question of interest is usually about a population mean.

Putting It into Practice

P1: Given a scenario, answer the four key questions that help identify an appropriate inference method.

The answers to the four key questions can be used to identify an appropriate inference method. Once these questions have been answered, you can use Table 7.1 to identify a potential method. Examples 7.7, 7.8, and 7.9 illustrate the process of answering the four key questions.

SECTION 7.2 EXERCISES

Each Exercise Set assesses the following chapter learning objectives: C4, C5, C6, M2, P1

SECTION 7.2 Exercise Set 1

7.13 Suppose that a study was carried out in which each person in a random sample of students at a particular college was asked how much money he or she spent on textbooks for the current semester. Would you use these data to estimate a population mean or to estimate a population proportion? How did you decide?

7.14 When you collect data to learn about a population, why do you worry about whether the data collected are categorical or numerical?

7.15 Consider the four key questions that guide the choice of an inference method. Two of these questions are

T: Type of data. One variable or two? Categorical or numerical?

N: Number of samples or treatments. How many samples or treatments?

What are the other two questions that make up the four key questions?

For each of the studies described in Exercises 7.16 to 7.18, answer the four key questions:

Q Question Type	Estimation or hypothesis testing?
S Study Type	Sample data or experiment data?
T Type of Data	One variable or two? Categorical or numerical?
N Number of Samples or Treatments	How many samples or treatments?

7.16 Refer to the instructions prior to this exercise. The article **"Smartphone Nation"** (*AARP Bulletin*, **September 2009**) describes a study of how people ages 50 to 64 years use cell phones. In this study, each person in a sample of adults thought to be representative of this age group was asked about whether he or she kept a cell phone by the bed at night. The researchers conducting this study hoped to use the resulting data to learn about the proportion of people in this age group who sleep with their cell phone nearby. (Hint: See Example 7.7.)

7.17 Refer to the instructions prior to Exercise 7.16. Do children diagnosed with attention deficit/hyperactivity disorder (ADHD) have smaller brains than children without this condition? This question was the topic of a research study described in the paper **"Developmental Trajectories of Brain Volume Abnormalities in Children and Adolescents with Attention Deficit/Hyperactivity Disorder"** (*Journal of the American Medical Association* [2002]: 1740–1747). Brain scans were completed for 152 children with ADHD and 139 children of similar age without ADHD. The researchers wanted to see if the resulting data supported the claim that the mean brain volume of children with ADHD is smaller than the mean for children without ADHD. (Hint: See Example 7.7.)

7.18 Refer to the instructions prior to Exercise 7.16. The article **"Arctic Sea Ice Is Slipping Away—and You're to Blame"** (*USA TODAY*, **November 4, 2016**) describes a study that appeared in the journal *Science*. In this study, researchers looked at carbon pollution levels and the amount of sea ice (frozen ocean water that melts each summer) each year over a period of years. The resulting data were used to learn about how the amount of snow ice might be related to carbon pollution level.

SECTION 7.2 Exercise Set 2

7.19 Suppose that a study is carried out in which each student in a random sample selected from students at a particular college is asked whether or not he or she would purchase a recycled paper product even if it cost more than the same product that was not made with recycled paper. Would you use the resulting data to estimate a population mean or to estimate a population proportion? How did you decide?

7.20 Comment on the following statement: The same statistical inference methods are used for learning from categorical data and for learning from numerical data.

7.21 Consider the four key questions that guide the choice of an inference method. Two of these questions are

   Q: Question type. Estimation or hypothesis testing?
   S: Study type. Sample data or experiment data?

What are the other two questions that make up the four key questions?

For each of the studies described in Exercises 7.22 to 7.24, answer the four key questions:

Q Question Type	Estimation or hypothesis testing?
S Study Type	Sample data or experiment data?
T Type of Data	One variable or two? Categorical or numerical?
N Number of Samples or Treatments	How many samples or treatments?

7.22 Refer to the instructions prior to this exercise. A study of fast-food intake is described in the paper **"What People Buy From Fast-Food Restaurants"** (*Obesity* [2009]: 1369–1374). Adult customers at three hamburger chains (McDonald's, Burger King, and Wendy's) at lunchtime in New York City were approached as they entered the restaurant and were asked to provide their receipt when exiting. The receipts were then used to determine what was purchased and the number of calories consumed. The sample mean number of calories consumed was 857, and the sample standard deviation was 677. This information was used to learn about the mean number of calories consumed in a New York fast-food lunch.

7.23 Refer to the instructions prior to Exercise 7.22. Common Sense Media surveyed 1000 teens and 1000 parents of teens to learn about how teens are using social networking sites such as Facebook and MySpace (**"Teens Show, Tell Too Much Online,"** *San Francisco Chronicle*, **August 10, 2009**). The two samples were independently selected and were chosen in a way that makes it reasonable to regard them as representative of American teens and parents of American teens. When asked if they check their online social networking sites more than 10 times a day, 220 of the teens surveyed said yes. When parents of teens were asked if their teen checks his or her site more than 10 times a day, 40 said yes. The researchers used these data to conclude that there was evidence that the proportion of all parents who think their teen checks a social networking site more than 10 times a day is less than the proportion of all teens who report that they check the sites more than 10 times a day.

7.24 Refer to the instructions prior to Exercise 7.22. Researchers at the Medical College of Wisconsin studied 2121 children between the ages of 1 and 4 (*Milwaukee Journal Sentinel*, **November 26, 2005**). For each child in the study, a measure of iron deficiency and the length of time the child was bottle-fed were recorded. The resulting data were used to learn about whether there was a relationship between iron deficiency and the length of time a child is bottle fed.

ADDITIONAL EXERCISES

For each of the studies described in Exercises 7.25 to 7.29, answer the following key questions:

Q Question Type	Estimation or hypothesis testing?
S Study Type	Sample data or experiment data?
T Type of Data	One variable or two? Categorical or numerical?
N Number of Samples or Treatments	How many samples or treatments?

7.25   Refer to the instructions prior to this exercise. The concept of a "phantom smoker" was introduced in the paper **"I Smoke but I Am Not a Smoker: Phantom Smokers and the Discrepancy Between Self-Identity and Behavior"** (*Journal of American College Health* [2010]: 117– 125). Previous studies of college students found that how students respond when asked to identify themselves as either a smoker or a nonsmoker was not always consistent with how they respond to a question about how often they smoked cigarettes. A phantom smoker is defined to be someone who self-identifies as a nonsmoker but who admits to smoking cigarettes when asked about frequency of smoking. This prompted researchers to wonder if asking college students to self-identify as being a smoker or nonsmoker might be resulting in an underestimate of the actual percentage of smokers. The researchers planned to use data from a sample of 899 students to estimate the percentage of college students who are phantom smokers.

7.26   Refer to the instructions prior to Exercise 7.25. An article in **USA TODAY (October 19, 2010)** described a study to investigate how young children learn. Sixty-four toddlers age 18 months participated in the study. The toddlers were allowed to play in a lab equipped with toys and which had a robot that was hidden behind a screen. The article states: "After allowing the infants playtime, the team removed the screen and let the children see the robot. In some tests, an adult talked to the robot and played with it. In others the adult ignored the robot. After the adult left the room, the robot beeped and then turned its head to look at a toy to the side of the infant." The researchers planned to see if the resulting data supported the claim that children are more likely to follow the robot's gaze to the toy when they see an adult play with the robot than when they see an adult ignore the robot.

7.27   Refer to the instructions prior to Exercise 7.25. In a study of whether taking a garlic supplement reduces the risk of getting a cold, 146 participants were assigned to either a garlic supplement group or to a group that did not take a garlic supplement (**"Garlic for the Common Cold," Cochrane Database of Systematic Reviews, 2009**). Researchers planned to see if there is evidence that the proportion of people taking a garlic supplement who get a cold is lower than the proportion of those not taking a garlic supplement who get a cold.

7.28   Refer to the instructions prior to Exercise 7.25. **"Spending on Favorite Drinks"** is the title of a graph that appeared as part of the USA Snapshot series in the newspaper **USA TODAY (November 4, 2016)**. This graph summarized data from a survey of adult Americans. Each survey participant reported how much they spend each month on various beverages, such as coffee, wine, and beer. These data were used to learn about the mean monthly spending for each type of beverage.

7.29   Refer to the instructions prior to Exercise 7.25. Can moving their hands help children learn math? This question was investigated by the authors of the paper **"Gesturing Gives Children New Ideas about Math"** (*Psychological Science* [2009]: 267–272). A study was conducted to compare two different methods for teaching children how to solve math problems of the form $3 + 2 + 8 =$ ____$+ 8$. One method involved having students point to the $3 + 2$ on the left side of the equal sign with one hand and then point to the blank on the right side of the equal sign before filling in the blank to complete the equation. The other method did not involve using these hand gestures. To compare the two methods, 128 children were assigned at random to one of the methods. Each child then took a test with six problems, and the number correct was determined for each child. The researchers planned to see if the resulting data supported the theory that the mean number correct for children who use hand gestures is higher than the mean number correct for children who do not use hand gestures.

SECTION 7.3   # A Five-Step Process for Statistical Inference

In the chapters that follow, you will see quite a few different methods for learning from data. Even though you will choose different methods in different situations, all of the methods used for estimation problems are carried out using the same overall process, and all of the hypothesis testing methods follow a similar overall process.

A Five-Step Process for Estimation Problems (EMC^3)

Once a problem has been identified as an estimation problem (the E in EMC^3), the following steps will be used:

Step	What Is This Step?
E	**Estimate**: Explain what population characteristic you plan to estimate.
M	**Method**: Select a potential method. This step involves using the answers to the four key questions (QSTN) to identify a potential method.
C	**Check**: Check to make sure that the method selected is appropriate. Many methods for learning from data only provide reliable information under certain conditions. It will be important to verify that any such conditions are met before proceeding.
C	**Calculate**: Sample data are used to perform any necessary calculations. This step is often accomplished using technology, such as a graphing calculator or computer software.
C	**Communicate Results**: This is a critical step in the process. In this step, you will answer the question of interest, explain what you have learned from the data, and acknowledge potential risks.

A Five-Step Process for Hypothesis Testing Problems (HMC^3)

Once a problem has been identified as a hypothesis testing problem (the H in HMC^3), the first step in the five-step process is

H	**Hypotheses**: Define the hypotheses that will be tested.

This is really the only step in the process that is different from the steps described for estimation—the MC^3 steps are the same, resulting in

Step	What Is This Step?
H	**Hypotheses**: Define the hypotheses that will be tested.
M	**Method**: Select a potential method. This step involves using the answers to the four key questions (QSTN) to identify a potential method.
C	**Check**: Check to make sure that the method selected is appropriate. Many methods for learning from data only provide reliable information under certain conditions. It will be important to verify that any such conditions are met before proceeding.
C	**Calculate**: Sample data are used to perform any necessary calculations. This step is often accomplished using technology, such as a graphing calculator or computer software.
C	**Communicate Results**: This is a critical step in the process. In this step, you will answer the question of interest, explain what you have learned from the data, and acknowledge potential risks.

These steps provide a common systematic approach to learning from data in a variety of situations. Returning to the playbook analogy, you can think of the four key questions of the last section as providing a strategy for selecting a play (statistical method). The five-step processes of this section describe the structure of each play in the playbook. So, as you move forward in Chapters 9 through 16, just remember—you will answer four questions to select a "play," and then each "play" has five steps.

Summing It Up—Section 7.3

The following learning objectives were addressed in this section:

Conceptual Understanding

C7: Know the five-step process for estimation problems.

The five-step EMC^3 process provides a systematic approach for solving estimation problems. The steps in the process are Estimate, Method, Check, Calculate, and Communicate. These steps are described in more detail in the table at the beginning of this section.

C8: Know the five-step process for hypothesis testing problems.

The five-step HMC^3 process provides a systematic approach for solving hypothesis testing problems. The steps in the process are Hypotheses, Method, Check, Calculate, and Communicate. These steps are described in more detail in the table at the end of this section.

CHAPTER ACTIVITIES

ACTIVITY 7.1 TONY'S APOLOGY

In 2010, one of the worst oil spills in history occurred along the Gulf Coast of the United States when a BP offshore oil well failed. The online article **"Are Men More Sympathetic Than Women to BP CEO Tony Hayward?" (www.fastcompany.com/1671707/are-men-more-sympathetic-to-bp-ceo-tony-hayward, retrieved May 3, 2017)** describes a study conducted by Innerscope. The article states

> BP's marketing response to the Gulf oil disaster has been an unquestionable failure, thanks in large part to CEO Tony Hayward and his many callous remarks ("I'd like my life back," anyone?). But according to neuromarketing research company Innerscope, men and women don't hate Hayward's halfhearted apologies equally. Innerscope measured the biometric response (skin response, heart rate, body movement, and breathing) of 54 volunteers while they watched BP's now-infamous apology advertisements. The 27 men and 27 women were all college-educated, but otherwise had nothing in common.
>
> The biometrics, visible in the video, reveal that men and women had wildly different biometric responses at two different points during the ad—when Tony Hayward talks about how BP is engaging in the largest environmental response in the country's history, and when the camera zooms in on an apologetic Hayward.

Take a look at the video referred to in the quote. The video can be found at **www.youtube.com/watch?v=eirWV0Z63YQ&feature=player_embedded, retrieved May 3, 2017**.

After discussing the following questions with a partner, give a brief answer to each question.

a. The video was intended to show the estimated responses of men and women at various points in time during Tony Hayward's apology. Do you think that the display does a good job of conveying this information visually? Explain.

b. Are the two points where the responses of men and women were "wildly different" obvious in the graphical display?

c. Was this study an observational study or an experiment? Did the study incorporate random selection or random assignment?

d. If the goal of the study was to generalize the estimates of the mean biometric responses for men and women to the population of American adult males and American adult females, do you have any concerns about how the samples were selected?

e. Can you suggest a better way to carry out a study so that the generalization described in Part (d) would be more reasonable?

ARE YOU READY TO MOVE ON? CHAPTER 7 REVIEW EXERCISES

All chapter learning objectives are assessed in these exercises. The learning objectives assessed in each exercise are given in parentheses.

7.30 (C1, M1)
Should advertisers worry about people with digital video recorders (DVRs) fast-forwarding through their TV commercials? Recent studies by MillwardBrown and Innerscope Research indicate that when people are fast-forwarding through commercials they are actually still quite engaged and paying attention to the screen to see when the commercials end and the show they were watching starts again. If a commercial goes by that the viewer has seen before, the impact of the commercial may be equivalent to viewing the commercial at normal speed. One study of DVR viewing behavior is described in the article **"Engaging at Any Speed? Commercials Put to the Test" (*The New York Times*, July 3, 2007)**. For each person in a sample of adults, physical responses (such as respiratory rate and heart rate) were recorded while watching commercials at normal speed and while watching commercials at fast-forward speed. These responses were used to calculate an engagement score. Engagement scores ranged from 0 to 100 (higher values indicate greater engagement). The researchers concluded that the mean engagement score for people watching at regular speed was 66, and for people watching at fast-forward speed it was 68. Is the described inference one that resulted from estimation or one that resulted from hypothesis testing?

7.31 (C1, M1)
Are people willing to eat blemished produce? An article that described the result of a survey of 2025 adult Americans was titled **"Eight in Ten Americans Say Appearance Is at Least Somewhat Important When Shopping for Fresh Produce" (www.theharrispoll.com/business/Appearance-is-Important-When-Shopping-for-Produce.html, September 22, 2016, retrieved July 24, 2017)**. Is the inference described in the title of this article one that resulted from estimation or one that resulted from hypothesis testing? Explain.

7.32 (C1, M1)
The article **"Display of Health Risk Behaviors on MySpace by Adolescents" (*Archives of Pediatrics and Adolescent Medicine* [2009]: 27–34)** describes a study of 500 publically

accessible MySpace Web profiles posted by 18-year-olds. The content of each profile was analyzed, and the researchers used the resulting data to conclude that there was support for the claim that those involved in sports or a hobby were less likely to have references to risky behavior (such as sexual references or references to substance abuse or violence). Is the described inference one that resulted from estimation or one that resulted from hypothesis testing?

7.33 (C1)
Consider the population that consists of all people who purchased season tickets for home games of the New York Yankees.
a. Give an example of a question about this population that could be answered by collecting data and using the data to estimate a population characteristic.
b. Give an example of a question about this population that could be answered by collecting data and using the data to test a claim about this population.

7.34 (C2)
Data from a poll of working women conducted in 2016 by Gallup led to the following estimates: Approximately 48% of working women are actively looking for a different job and 60% of working women rate greater work-life balance and well-being as a very important attribute in a new job (**"Women in America: Work and a Life-Well Lived," www.gallup.com, retrieved November 8, 2016**).
a. What additional information about the survey would you need in order to decide if it is reasonable to generalize these estimates to the population of all American adult working women?
b. Assuming that the given estimates were based on a representative sample, do you think that the estimates would more likely be closer to the actual population values if the sample size had been 1000 or if the sample size had been 2000? Explain.

7.35 (C3)
In a study of whether taking a garlic supplement reduces the risk of getting a cold, 146 participants were randomly assigned to either a garlic supplement group or to a group that did not take a garlic supplement (**"Garlic for the Common Cold,"** *Cochrane Database of Systematic Reviews*, **2009**). Based on the study, it was concluded that the proportion of people taking a garlic supplement who get a cold is lower than the proportion of those not taking a garlic supplement who get a cold.
a. What claim about the effect of taking garlic is supported by the data from this study?
b. Is it possible that the conclusion that the proportion of people taking garlic who get a cold is lower than the proportion for those not taking garlic is incorrect? Explain.
c. If the number of people participating in the study had been 50, do you think that the chance of an incorrect

conclusion would be greater than, about the same as, or lower than for the study described?

7.36 (C4, M2)
Suppose that a study was carried out in which each student in a random sample of students at a particular college was asked if he or she was registered to vote. Would these data be used to estimate a population mean or to estimate a population proportion? How did you decide?

7.37 (C5)
Explain why the question

T: Type of data—one variable or two? Categorical or numerical?

is one of the four key questions used to guide decisions about what inference method should be considered.

For each of the studies described in Exercises 7.38 to 7.40, answer the four key questions:

Q Question Type	Estimation or hypothesis testing?
S Study Type	Sample data or experiment data?
T Type of Data	One variable or two? Categorical or numerical?
N Number of Samples or Treatments	How many samples or treatments?

7.38 (C6, M2, P1)
Refer to the instructions prior to this exercise. A study of adult Americans conducted by the polling organization Ipsos (**"One in Five Americans Consider Themselves 'Entrepreneurs,'" November 7, 2016, www.ipsos-na.com /news-polls/pressrelease.aspx?id=7462, retrieved November 8, 2016**) asked each person in a sample whether he or she self-identified as an entrepreneur. The responses to this question were used to learn about the proportion of adult Americans who self-identify as an entrepreneur.

7.39 (C6, M2, P1)
Refer to the instructions prior to Exercise 7.38. In a study to determine if using low-intensity laser therapy reduces pain for orthodontic patients who are fitted with new braces, patients were randomly assigned to either a control group (who did not receive laser treatment) and an experimental group who did receive laser treatment (**"Low-Level Laser Therapy for Alleviation of Pain from Fixed Orthodontic Appliance Therapy,"** *Journal of Advanced Clinical and Research Insights* **[2016]: 43–46**). The researchers found that the mean pain rating was lower for those in the laser treatment group than for those in the control group both two days and six days after receiving braces. The researchers used these data to determine if there was evidence to support the claim that

the mean pain rating is lower for patients who receive the laser treatment.

7.40   (C6, M2, P1)
Refer to the instructions prior to Exercise 7.38. The authors of the paper **"Flat-Footedness Is Not a Disadvantage for Athletic Performance in Children Aged 11 to 15 Years"** (*Pediatrics* **[2009]: e386–e392)** collected data from 218 children on foot arch height and motor ability. The resulting data were used to investigate the relationship between arch height and motor ability.

7.41   (C7)
The process for statistical inference described in Section 7.3 consists of five steps:
a. What are the five steps in the process for estimation problems?
b. Explain how the first step differs for estimation problems and hypothesis testing problems.

8   Sampling Variability and Sampling Distributions

Chabruken/Getty Images

PREVIEW

*When you collect sample data, it is usually to learn something about the
population from which the sample was selected. To do this, you need to
understand **sampling variability**—the chance differences that occur from one
random sample to another as a result of random selection. The following example
illustrates the key ideas of this chapter in a simple setting. Even though this
example is a bit unrealistic because it considers a very small population, it will
help you understand what is meant by sampling variability.*

CHAPTER LEARNING OBJECTIVES

Conceptual Understanding

After completing this chapter, you should be able to

C1 Understand that the value of a sample statistic varies from sample to sample.

C2 Understand that a sampling distribution describes the sample-to-sample variability of a statistic.

C3 Understand how the standard deviation of the sampling distribution of a sample proportion is related to sample size.

C4 Understand how the sampling distribution of a sample proportion enables you to use sample data to learn about a population proportion.

Mastering the Mechanics

After completing this chapter, you should be able to

M1 Define the terms statistic and sampling variability.

M2 Distinguish between a sample proportion and a population proportion and use correct notation to denote a sample proportion and a population proportion.

M3 Know that the sampling distribution of a sample proportion is centered at the actual value of the population proportion.

M4 Know the formula for the standard deviation of the sampling distribution of a sample proportion.

M5 Know when the sampling distribution of a sample proportion will be approximately normal.

Putting It into Practice

After completing this chapter, you should be able to

P1 Determine the mean and standard deviation of the sampling distribution of a sample proportion and interpret the mean and standard deviation in context.

P2 Determine if the sampling distribution of a sample proportion is approximately normal.

P3 Use sample data and properties of the sampling distribution of a sample proportion to reason informally about a population proportion.

PREVIEW EXAMPLE

Suppose that you are interested in learning about the proportion of women in the group of students pictured in Figure 8.1 and that this group is the entire population of interest. Because 19 of the 34 students in the group are female, the proportion of women in the population is $\frac{19}{34} = 0.56$. You can calculate this proportion because the picture provides complete information on sex for the entire population (a census).

FIGURE 8.1
A population of students

Digital Vision/Getty Images

But, suppose that the population information is not available. To learn about the proportion of women in the population, you decide to select a sample from the population by choosing 5 of the students at random. One possible sample is shown in Figure 8.2(a). The proportion of women in this sample (a sample statistic) is $\frac{3}{5} = 0.6$. A different sample, also resulting from random selection, is shown in Figure 8.2(b). For this second sample, the proportion of women is $\frac{2}{5} = 0.4$.

FIGURE 8.2
Two random samples of size 5

(a)

(b)

Notice the following:

1. The sample proportion is different for the two different random samples. Figure 8.2 shows only two samples, but there are many other possible samples that might result when five students are selected from this population. *The value of the sample proportion varies from sample to sample*.
2. For both of the samples considered, the value of the sample proportion was not equal to the value of the corresponding population proportion, which was 0.56.

These observations raise the following important questions:

1. How much will the value of a sample proportion tend to differ from sample to sample?
2. How much will a sample proportion tend to differ from the actual value of the population proportion?
3. Based on what you see in a sample, can you provide an accurate estimate of a population proportion? In the context of this example, based on a sample, what can you say about the proportion of students in the population who are female?
4. What conclusions can you draw based on the value of a sample proportion? For instance, in the context of this example, you might want to know if it is reasonable to conclude that a majority (more than half) of the students in the population are female.

In this chapter, you will see that these questions can be answered by looking at the *sampling distribution* of the sample proportion. ∎

Just as the distribution of a numerical variable describes the long-run behavior of that variable, the sampling distribution of the sample proportion provides information about the long-run behavior of the sample proportion when many different random samples are selected. In this chapter, you will see how sampling distributions provide a foundation for using sample data to learn about a population proportion.

Statistics and Sampling Variability

A number calculated from the values in a sample is called a **statistic**. Statistics, such as the sample mean $\bar{x}$, the sample median, the sample standard deviation s, or the proportion of individuals in a sample that possess a particular property $\hat{p}$ (read as p-hat), provide information about population characteristics.

The most common way to learn about the value of a population characteristic is to study a sample from the population. For example, suppose you want to learn about the proportion of students at a particular college who participated in community service projects during the previous semester. A random sample of 100 students from the college could be selected. Each student in the sample could be asked whether he or she had participated in one or more community service projects in the previous semester. This would result in an observation for the variable x = *community service* (with possible values "yes" and "no") for each student in the sample. The sample proportion (denoted by $\hat{p}$) could be calculated as an estimate of the value of the corresponding population proportion (denoted by p).

NOTATION

A **sample proportion** is denoted by $\hat{p}$.

A **population proportion** is denoted by p.

It would be nice if the value of $\hat{p}$ was equal to the value of the population proportion p, but this is an unusual occurrence. For example, $\hat{p}$ might be 0.26 for your random sample of 100 students, but this doesn't mean that exactly 26% of the students at the college participated in a community service project. The value of the sample proportion depends on which 100 students are randomly selected to be in the sample. Not only will the value of $\hat{p}$ for a particular sample differ from the population proportion p, but $\hat{p}$ values from different samples can also differ from one another. This sample-to-sample variability makes it challenging to generalize from a sample to the population from which it was selected. To meet this challenge, you need to understand sample-to-sample variability.

DEFINITION

Any quantity calculated from values in a sample is called a **statistic**.

The observed value of a statistic varies from sample to sample depending on the particular sample selected. This variability is called **sampling variability**.

Example 8.1   Opposition to Fracking

Fracking is a process of injecting liquid into the ground in order to fracture rocks and release natural gas. While this has increased the production of natural gas, some people have expressed concerns that this process may have a negative impact on the environment. Consider a small population of 100 adult Americans and suppose that when asked if he or she supported or opposed fracking as a way to increase production of natural gas and oil, 40 indicated that they were opposed to fracking. Although this is not a real population, the proportion opposed to fracking is consistent with a large survey of adult Americans conducted in 2015 by the Gallup organization **("Americans Split on Support for Fracking in Oil, Natural Gas," March 13, 2015, www.gallup.org, retrieved October 8, 2016)**. So that you have an easy way to identify the adults in this population, you can assign each one a number from 1 to 100, starting with those who indicated opposition to fracking, as shown here.

Opposed to Fracking	1, 2, 3, ... , 39, 40
Not Opposed to Fracking	41, 42, 43, ... , 99, 100

For this population,

$$p = \frac{40}{100} = 0.40$$

indicating that 40% were opposed to fracking.

Suppose that you don't know the value of this population proportion. You decide to estimate p by taking a random sample of people from this population. Each person in the sample is asked about his or her opposition to fracking, and you determine how many in the sample are opposed to fracking. You then calculate $\hat{p}$, the sample proportion of people opposed to fracking. Is the value of $\hat{p}$ likely to be close to 0.40, the corresponding population proportion?

To answer this question, consider a simple investigation that examines the behavior of the statistic $\hat{p}$ when random samples of size 20 are selected. (This scenario is not realistic. Most populations of interest consist of many more than 100 individuals. However, this small population size is easier to work with as the idea of sampling variability is developed.)

Begin by selecting a random sample of size 20 from this population. One way to do this is to write the numbers from 1 to 100 on slips of paper, mix them well, and then select 20 slips without replacement. The numbers on the 20 slips identify which of the 100 people in the population are included in the sample. Another approach is to use either a table of random digits or a random number generator to determine which 20 people should be selected. Using JMP to obtain 20 random numbers between 1 and 100 resulted in the following:

78	**16**	**31**	86	**5**	67	**39**	**28**	97	70
34	65	47	89	**26**	79	52	**4**	82	**6**

The nine highlighted numbers correspond to people who were identified as opposing fracking (remember, these were the people numbered 1 to 40). So, for this sample

$$\hat{p} = \frac{9}{20} = 0.45$$

This value is 0.05 larger than the actual population proportion of 0.40. Is this difference typical, or is this particular sample proportion unusually far away from p? Looking at some additional samples will provide some insight.

Four more random samples (Samples 2–5) from this same population are shown in the following table, along with the value of $\hat{p}$ for each sample.

Sample	People Selected	$\hat{p}$
2	89, 63, 19, 53, 71, 62, 42, 60, 69, 43, 40, 36, 45, 90, 73, 32, 48, 56, 44, 5	$\hat{p} = \dfrac{5}{20} = 0.25$
3	73, 6, 49, 38, 21, 92, 45, 10, 93, 66, 64, 5, 59, 3, 85, 75, 81, 18, 57, 60	$\hat{p} = \dfrac{7}{20} = 0.35$
4	31, 53, 6, 64, 38, 22, 62, 92, 84, 100, 60, 17, 61, 27, 34, 55, 49, 80, 66, 8	$\hat{p} = \dfrac{8}{20} = 0.40$
5	93, 37, 25, 68, 21, 79, 44, 2, 81, 58, 17, 1, 99, 49, 3, 52, 89, 18, 11, 20	$\hat{p} = \dfrac{10}{20} = 0.50$

Remember that $p = 0.40$. You can now see the following:

1. The value of the sample proportion $\hat{p}$ varies from one random sample to another (sampling variability).

2. Some samples produced $\hat{p}$ values that were greater than p (Samples 1 and 5), some produced values less than p (Samples 2 and 3), and there was one sample with $\hat{p}$ equal to p.

3. Samples 1, 3, and 4 resulted in sample proportions that were equal to or relatively close to the value of the population proportion, but Sample 2 resulted in a value that was much farther away from p.

Continuing the investigation, 45 more random samples (each of size $n = 20$) were selected. The resulting sample proportions are shown here:

Sample	$\hat{p}$	Sample	$\hat{p}$	Sample	$\hat{p}$
6	0.40	21	0.25	36	0.65
7	0.55	22	0.25	37	0.35
8	0.55	23	0.40	38	0.50
9	0.55	24	0.35	39	0.45
10	0.30	25	0.45	40	0.30
11	0.45	26	0.35	41	0.25
12	0.50	27	0.50	42	0.30
13	0.20	28	0.40	43	0.50
14	0.35	29	0.40	44	0.35
15	0.25	30	0.30	45	0.35
16	0.40	31	0.30	46	0.55
17	0.50	32	0.55	47	0.35
18	0.35	33	0.40	48	0.50
19	0.50	34	0.45	49	0.45
20	0.25	35	0.20	50	0.35

Figure 8.3 is a relative frequency histogram of the 50 sample proportions. This histogram describes the behavior of $\hat{p}$. Many of the samples resulted in $\hat{p}$ values that are equal to or relatively close to $p = 0.40$, but occasionally the $\hat{p}$ values were as small as 0.20 or as large as 0.65. This tells you that if you were to take a sample of size 20 from this population and use $\hat{p}$ as an estimate of the population proportion p, you should not necessarily expect $\hat{p}$ to be close to p.

FIGURE 8.3
Histogram of $\hat{p}$ values from
50 random samples for Example 8.1

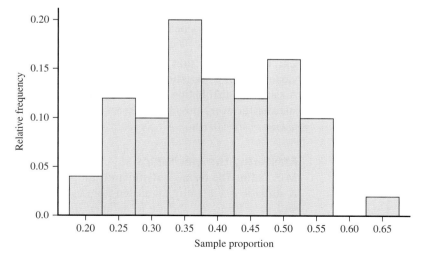

The histogram of Figure 8.3 shows the sampling variability in the statistic $\hat{p}$. It also provides an approximation to the distribution of $\hat{p}$ values that would have been observed if you had considered every different possible sample of size 20 from this population.

In the example just considered, the approximate sampling distribution of the statistic $\hat{p}$ was based on just 50 different samples. The actual sampling distribution results from considering *all* possible samples.

DEFINITION

The distribution formed by the values of a sample statistic for every possible different sample of a given size is called its **sampling distribution**.

The sampling distribution of a statistic, such as $\hat{p}$, provides important information about variability in the values of the statistic. The histogram of Figure 8.3 is an *approximation* of the sampling distribution of the statistic $\hat{p}$ for samples of size 20 from the population described in Example 8.1. To determine the actual sampling distribution of $\hat{p}$, you would need to consider every possible sample of size 20 from the population of 100 students, calculate the proportion for each sample, and then construct a relative frequency histogram of these $\hat{p}$ values. But you wouldn't really want to do this—there are more than a quintillion (that's a 1 followed by 18 zeros!) different possible samples of size 20 that can be selected from a population of size 100. And, for more realistic situations with larger population and sample sizes, the situation becomes even worse! Fortunately, as you consider more examples in Section 8.2, patterns emerge that allow you to describe the sampling distribution of $\hat{p}$ without actually having to look at all possible samples.

In this section and the sections that follow in this chapter, the focus is on the sample proportion. There are many other sample statistics that might be of interest, such as the sample mean and the sample median. These sample statistics also have sampling distributions that describe their long-run behavior. You will see other sampling distributions in the chapters that follow.

Summing It Up—Section 8.1

The following learning objectives were addressed in this section:

Conceptual Understanding

C1: Understand that the value of a sample statistic varies from sample to sample.
The observed value of a sample statistic, such as the sample proportion, will vary from one sample to another depending on the particular individuals that are included in the sample. This variability is called **sampling variability**.

C2: Understand that a sampling distribution describes the sample-to-sample variability of a statistic.
A sampling distribution for a statistic is the distribution formed by the values of a sample statistic for every possible different sample of a given size. It describes the long-run behavior of the statistic.

Mastering the Mechanics

M1: Define the terms statistic and sampling variability.
A **statistic** is a quantity calculated from values in a sample. The value of a sample statistic will vary from one sample to another, depending on the particular individuals that are included in the sample. This variability is called **sampling variability**.

M2: Distinguish between a sample proportion and a population proportion and use correct notation to denote a sample proportion and a population proportion.
A sample proportion is the proportion of successes observed in a sample. A sample proportion is denoted by $\hat{p}$. A population proportion is the proportion of successes in an entire population. A population proportion is denoted by p.

SECTION 8.1 EXERCISES

Each Exercise Set assesses the following chapter learning objectives: C1, C2, M1, M2

SECTION 8.1 Exercise Set 1

8.1 A random sample of 1000 students at a large college included 428 who had one or more credit cards. For this sample, $\hat{p} = \frac{428}{1000} = 0.428$. If another random sample of 1000 students from this university were selected, would you expect that $\hat{p}$ for that sample would also be 0.428? Explain why or why not.

8.2 Consider the two relative frequency histograms at the bottom of this page. The histogram on the left was constructed by selecting 100 different random samples of size 50 from a population consisting of 55% females and 45% males. For each sample, the sample proportion of females, $\hat{p}$, was calculated. The 100 $\hat{p}$ values were used to construct the histogram. The histogram on the right was constructed in a similar way, except that the samples were of size 75 instead of 50.
a. Which of the two histograms shows more sample-to-sample variability? How can you tell?
b. For which of the two sample sizes, $n = 50$ or $n = 75$, do you think the value of $\hat{p}$ is more likely to be close to 0.55? What about the given histograms supports your choice?

8.3 Consider the following statement:
The Department of Motor Vehicles reports that the proportion of all vehicles registered in California that are imports is **0.22**.
a. Is the number that appears in boldface in this statement a sample proportion or a population proportion?
b. Which of the following use of notation is correct, $p = 0.22$ or $\hat{p} = 0.22$? (Hint: See definitions and notation on page 403.)

8.4 Consider the following statement:
A sample of size 100 was selected from those admitted to a particular college in fall 2017. The proportion of these 100 who were transfer students is **0.38**.
a. Is the number that appears in boldface in this statement a sample proportion or a population proportion?
b. Which of the following use of notation is correct, $p = 0.38$ or $\hat{p} = 0.38$?

8.5 Explain the difference between p and $\hat{p}$.

SECTION 8.1 Exercise Set 2

8.6 A random sample of 100 employees of a large company included 37 who had worked for the company for less than one year. For this sample, $\hat{p} = \frac{37}{100} = 0.37$. If a different random sample of 100 employees were selected, would you expect that $\hat{p}$ for that sample would also be 0.37? Explain why or why not.

8.7 Consider the two relative frequency histograms at the top of the next page. The histogram on the left was constructed by selecting 100 different random samples of size 40 from a population consisting of 20% part-time students and 80% full-time students. For each sample, the sample proportion of part-time students, $\hat{p}$, was calculated. The 100 $\hat{p}$ values were used to construct the histogram. The histogram on the right was constructed in a similar way, but using samples of size 70.

HISTOGRAMS FOR EXERCISE **8.2**

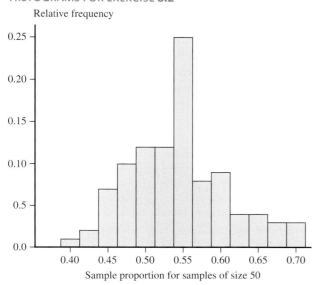

Sample proportion for samples of size 50

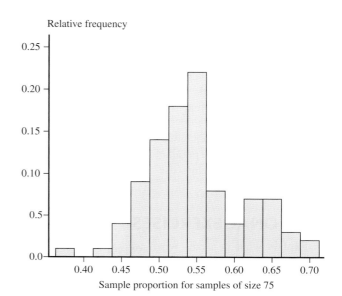

Sample proportion for samples of size 75

HISTOGRAMS FOR EXERCISE **8.7**

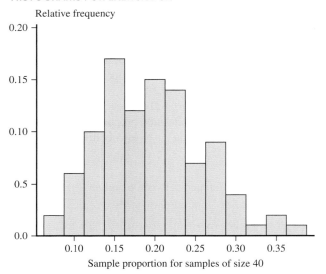

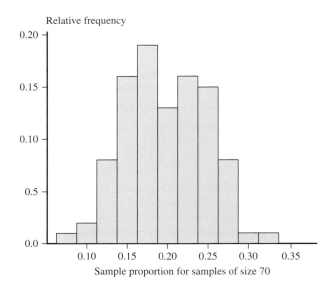

a. Which of the two histograms indicates that the value of $\hat{p}$ has smaller sample-to-sample variability? How can you tell?

b. For which of the two sample sizes, $n = 40$ or $n = 70$, do you think the value of $\hat{p}$ would be less likely to be close to 0.20? What about the given histograms supports your choice?

8.8 Explain what the term sampling variability means in the context of using a sample proportion to estimate a population proportion.

8.9 Consider the following statement:
Fifty people were selected at random from those attending a football game. The proportion of these 50 who made a food or beverage purchase while at the game was **0.83**.
a. Is the number that appears in boldface in this statement a sample proportion or a population proportion?
b. Which of the following use of notation is correct, $p = 0.83$ or $\hat{p} = 0.83$?

8.10 Consider the following statement:
The proportion of all calls made to a county 9-1-1 emergency number during the year 2017 that were nonemergency calls was **0.14**.
a. Is the number that appears in boldface in this statement a sample proportion or a population proportion?
b. Which of the following use of notation is correct, $p = 0.14$ or $\hat{p} = 0.14$?

ADDITIONAL EXERCISES

8.11 Explain what it means when we say the value of a sample statistic varies from sample to sample.

8.12 Consider the two relative frequency histograms at the top of the next page. The histogram on the left was constructed by selecting 100 different random samples of size 35 from a population in which 17% donated to a nonprofit organization. For each sample, the sample proportion $\hat{p}$ was calculated and then the 100 $\hat{p}$ values were used to construct the histogram. The histogram on the right was constructed in a similar way but used samples of size 110.
a. Which of the two histograms indicates that the value of $\hat{p}$ has more sample-to-sample variability? How can you tell?
b. For which of the two sample sizes, $n = 35$ or $n = 110$, do you think the value of $\hat{p}$ would be more likely to be close to the actual population proportion of $p = 0.17$? What about the given histograms supports your choice?

8.13 Consider the following statement:
In a sample of 20 passengers selected from those who flew from Dallas to New York City in April 2017, the proportion who checked luggage was **0.45**.
a. Is the number that appears in boldface in this statement a sample proportion or a population proportion?
b. Which of the following use of notation is correct, $p = 0.45$ or $\hat{p} = 0.45$?

8.14 Consider the following statement:
A county tax assessor reported that the proportion of property owners who paid 2016 property taxes on time was **0.93**.
a. Is the number that appears in boldface in this statement a sample proportion or a population proportion?
b. Which of the following use of notation is correct, $p = 0.93$ or $\hat{p} = 0.93$?

HISTOGRAMS FOR EXERCISE **8.12**

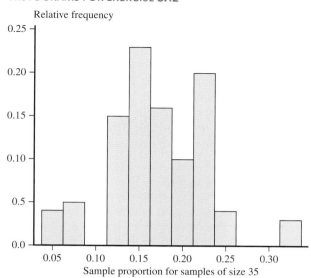

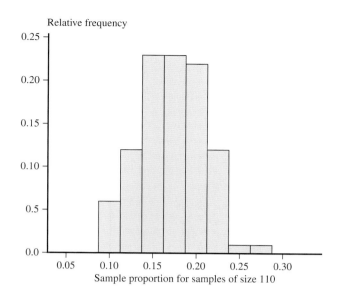

SECTION 8.2   The Sampling Distribution of a Sample Proportion

Sometimes you want to learn about the proportion of individuals or objects in a population that possess a particular characteristic. For example, you might be interested in the proportion of smartphones that possess a particular manufacturing flaw, the proportion of registered voters who support a particular political candidate, or the proportion of coffee drinkers who prefer decaffeinated coffee. Because proportions are of interest in so many different settings, it helps to introduce some general terminology and notation that can be used in any context. An individual or object that possesses the characteristic of interest is called a *success*. One that does not possess the characteristic is called a *failure*. The terms success and failure mean different things in different settings. For example, in a setting where you are interested in learning about the proportion of smartphones that have a particular manufacturing flaw, a flawed smartphone is a "success." The letter p denotes the proportion of successes in the population. The value of p is a number between 0 and 1, and $100p$ is the percentage of successes in the population. For example, if $p = 0.75$, the population consists of 75% successes and 25% failures. If $p = 0.01$, the population consists of 1% successes and 99% failures.

The value of the population proportion p is usually unknown. When a random sample of size n is selected from the population, some of the individuals in the sample are successes, and the others are failures. The statistic that is used to draw conclusions about the population proportion is $\hat{p}$, the **sample proportion of successes**:

$$\hat{p} = \frac{\text{number of successes in the sample}}{n}$$

To learn about the sampling distribution of the statistic $\hat{p}$, begin by considering the results of some sampling investigations. In the examples that follow, a sample size n was specified, and 500 different random samples of this size were selected from the population. The value of $\hat{p}$ is calculated for each sample and a histogram of these 500 $\hat{p}$ values is constructed. Because 500 is a reasonably large number of samples, the histogram of the $\hat{p}$ values should resemble the actual sampling distribution of $\hat{p}$ (which would be obtained by considering *all* possible samples). This process is repeated for several different values of n to see how the choice of sample size affects the sampling distribution. Careful examination of the resulting histograms will help you to understand some general properties to be stated shortly.

Example 8.2 **STEM College Students**

In fall 2015, there were 20,944 students enrolled at California Polytechnic State University, San Luis Obispo. Of these students, 9082 (43.4%) were enrolled in a science, technology, engineering, or mathematics (STEM) major. What would you expect to see for the sample proportion of STEM majors if you were to take a random sample of size 10 from this population? To investigate, you can simulate sampling from this student population. With success denoting a STEM major student, the proportion of successes in the population is $p = 0.434$. A statistical software package was used to select 500 random samples of size $n = 10$ from the population. The sample proportion of STEM majors $\hat{p}$ was calculated for each sample, and these 500 values of $\hat{p}$ were used to construct the relative frequency histogram shown in Figure 8.4.

FIGURE 8.4

Histogram of 500 $\hat{p}$ values based on random samples of size $n = 10$ from a population with $p = 0.434$

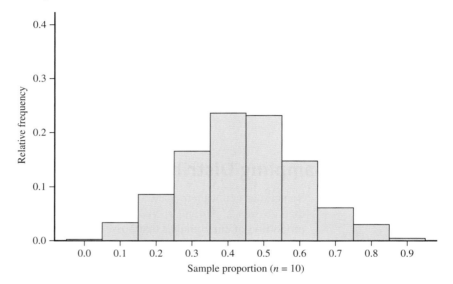

The histogram in Figure 8.4 describes the behavior of the sample proportion $\hat{p}$ for samples of size $n = 10$ from this population. Notice that there is a lot of sample-to-sample variability in the value of $\hat{p}$. For some samples, $\hat{p}$ was as small as 0 or 0.1, and for other samples, $\hat{p}$ was as large as 0.9! This tells you that a sample of size 10 from this population of students may not provide very accurate information about the proportion of STEM majors in the population. What if a larger sample were selected? To investigate the effect of sample size on the behavior of $\hat{p}$, 500 samples of size 25, 500 samples of size 50, and 500 samples of size 100 were selected. Histograms of the resulting $\hat{p}$ values, along with the histogram for the samples of size 10, are displayed in Figure 8.5.

The most noticeable feature of the histogram shapes in Figure 8.5 is that they are all relatively symmetric. All four histograms appear to be centered at roughly 0.434, the value of p, the population proportion. Had the histograms been based on all possible samples rather than just 500, each histogram would have been centered at exactly 0.434. Finally, notice that the histograms spread out more for small sample sizes than for large sample sizes. Not surprisingly, the value of $\hat{p}$ based on a large sample tends to be closer to p, the population proportion of successes, than does the value of $\hat{p}$ from a small sample.

Example 8.3 **Contracting Hepatitis from Blood Transfusion**

The development of viral hepatitis after a blood transfusion can cause serious complications for a patient. The article **"An Assessment of Hepatitis E Virus in U.S. Blood Donors and Recipients" (*Transfusion* [2013]: 2505–2511)** reported that in a sample of 342 blood donors age 18 to 45, about 7% tested positive for the Hepatitis E virus. Suppose that the actual proportion that test positive for Hepatitis E in the population of all blood donors in this age group is 0.07. You can simulate sampling from this population of blood donors. For this example, a blood donor who tests positive for Hepatitis E will be considered a success, so

FIGURE 8.5

Histogram of 500 $\hat{p}$ values (population proportion $p = 0.434$) (a) $n = 10$ (b) $n = 25$ (c) $n = 50$ (d) $n = 100$

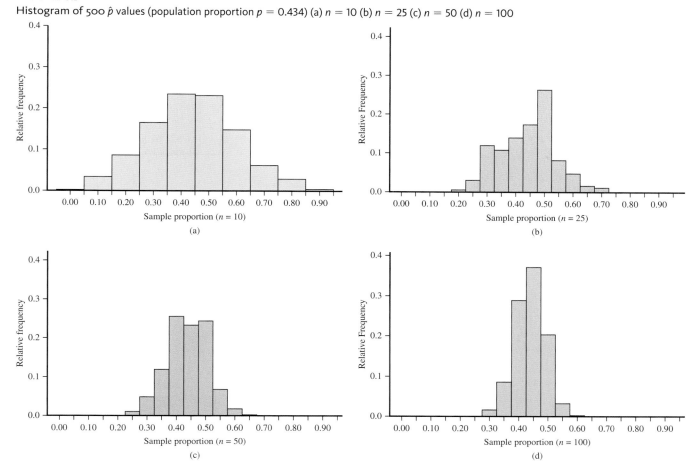

$p = 0.07$. Figure 8.6 displays histograms of 500 values of $\hat{p}$ for each of the four sample sizes $n = 10, 25, 50,$ and 100.

As was the case in Example 8.2, all four histograms are centered at approximately the value of p, the population proportion. If the histograms had been based on all possible samples, they would all have been centered at exactly $p = 0.07$. Notice that the scales on the axes are not the same for the four histograms in Figure 8.6. This was done so that it would be easier to see the behavior of the sample proportion for each sample size. Taking the differences in scales into account, you can see that the sample-to-sample variability decreases as the sample size increases. For example, the distribution is much less spread out in the histogram for $n = 100$ than for $n = 25$. The larger the value of n, the closer the sample proportion $\hat{p}$ tends to be to the value of the population proportion p.

Another thing to notice about the histograms in Figure 8.6 is the progression toward the shape of a normal curve as n increases. The histograms for $n = 10$ and $n = 25$ are quite skewed, and the skew of the histogram for $n = 50$ is still moderate (compare Figure 8.6(c) with Figure 8.5(c)). Only the histogram for $n = 100$ looks approximately normal in shape. It appears that whether a normal curve provides a good approximation to the sampling distribution of $\hat{p}$ depends on the values of both n and p. Knowing only that $n = 50$ is not enough to guarantee that the shape of the $\hat{p}$ histogram will be approximately normal.

General Properties of the Sampling Distribution of $\hat{p}$

Examples 8.2 and 8.3 suggest that the center of the sampling distribution of $\hat{p}$ (that is, the mean value of $\hat{p}$) is equal to the value of the population proportion for any sample size. Also, the variability of the sampling distribution of $\hat{p}$ decreases as n increases. The sample histograms of Figure 8.6 also suggest that in some cases, the sampling distribution of $\hat{p}$ is

FIGURE 8.6

Histogram of 500 $\hat{p}$ values
(population proportion $p = 0.07$)
(a) $n = 10$ (b) $n = 25$ (c) $n = 50$
(d) $n = 100$

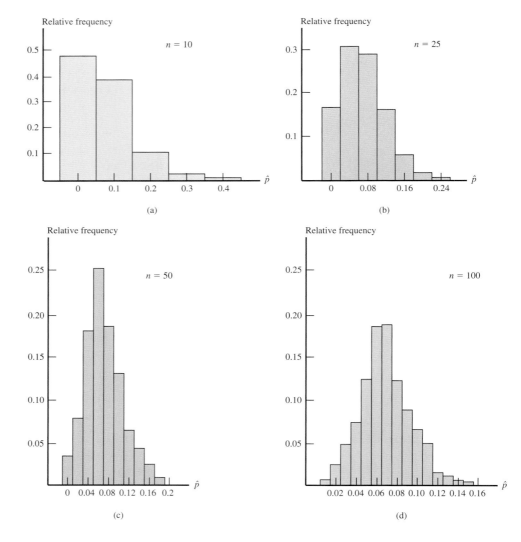

approximately normal in shape. These observations are stated more formally in the following general rules.

General Properties of the Sampling Distribution of $\hat{p}$

$\hat{p}$ is the proportion of successes in a random sample of size n from a population where the proportion of successes is p. The mean value of $\hat{p}$ is denoted by $\mu_{\hat{p}}$, and the standard deviation of $\hat{p}$ is denoted by $\sigma_{\hat{p}}$.

The following rules hold.

Rule 1. $\mu_{\hat{p}} = p$

Rule 2. $\sigma_{\hat{p}} = \sqrt{\dfrac{p(1 - p)}{n}}$. This rule is exact if the population is infinite and is approximately correct if the population is finite and no more than 10% of the population is included in the sample.

Rule 3. When n is large and p is not too near 0 or 1, the sampling distribution of $\hat{p}$ is approximately normal.

Rule 1, $\mu_{\hat{p}} = p$, states that the sampling distribution of $\hat{p}$ is always centered at the value of the population success proportion p. This means that the $\hat{p}$ values from many different random samples will tend to cluster around the actual value of the population proportion. This also means that the sample proportion does not consistently tend to overestimate or to underestimate the value of the population proportion.

Rule 2, $\sigma_{\hat{p}} = \sqrt{\dfrac{p(1-p)}{n}}$, implies that sample-to-sample variability in $\hat{p}$ decreases as n increases because the sample size n is in the denominator of the expression. This means that for larger samples, $\hat{p}$ values tend to cluster more tightly around the actual value of the population proportion. This rule also gives a precise relationship between the standard deviation of $\hat{p}$ and both the population proportion p and the sample size n. Knowing that the standard deviation of the sample proportion decreases as the sample size increases implies that if the value of the sample proportion is used as an estimate of the population proportion, then the estimate is more likely to be close to the actual value if the estimate is based on a larger sample size than if it is based on a smaller sample size.

Rule 3 states that in some cases the sampling distribution is approximately normal. Examples 8.2 and 8.3 suggest that both p and n must be considered when deciding if the sampling distribution of $\hat{p}$ is approximately normal.

When is the sampling distribution of $\hat{p}$ approximately normal?

The farther the value of p is from 0.5, the larger n has to be in order for $\hat{p}$ to have a sampling distribution that is approximately normal.

A conservative rule is that if both $np \geq 10$ and $n(1-p) \geq 10$, then the sampling distribution of $\hat{p}$ is approximately normal.

An equivalent way of stating this rule is to say that the sampling distribution of $\hat{p}$ is approximately normal if the sample size is large enough to expect at least 10 successes and at least 10 failures in the sample.

A sample size of $n = 100$ is not by itself enough to say that the sampling distribution of $\hat{p}$ is approximately normal. If $p = 0.01$, the sampling distribution is still quite skewed, so a normal curve is not a good approximation. Similarly, if $n = 100$ and $p = 0.99$ (so that $n(1-p) = 1 < 10$), the distribution of $\hat{p}$ is negatively skewed. The conditions $np \geq 10$ and $n(1-p) \geq 10$ ensure that the sampling distribution of $\hat{p}$ is not too skewed. If $p = 0.5$, the sampling distribution is approximately normal for n as small as 20, whereas for $p = 0.05$ or 0.95, n must be at least 200.

In the next section, you will see how properties of the sampling distribution of $\hat{p}$ allow you to use sample data to draw conclusions about a population proportion.

Summing It Up—Section 8.2

The following learning objectives were addressed in this section:

Mastering the Mechanics

M3: Know that the sampling distribution of a sample proportion is centered at the actual value of the population proportion.
The sampling distribution of the sample proportion $\hat{p}$ is centered at the value of the population proportion p. This means that $\hat{p}$ values from many different random samples will tend to cluster around the actual value of the population proportion.

M4: Know the formula for the standard deviation of the sampling distribution of a sample proportion.
The standard deviation of the sampling distribution of a sample proportion describes how much the value of the sample proportion tends to vary from sample to sample. The formula for the standard deviation of the sampling distribution of a sample proportion is

$$\sigma_{\hat{p}} = \sqrt{\dfrac{p(1-p)}{n}}.$$

M5: Know when the sampling distribution of a sample proportion will be approximately normal.
Whether or not the sampling distribution of $\hat{p}$ is approximately normal depends on both the value of the population proportion p and the sample size n. If both $np \geq 10$ and

$n(1 - p) \geq 10$, then it is reasonable to regard the sampling distribution of the sample proportion as approximately normal. An equivalent condition is that the sample size is large enough so that the sample will include at least 10 successes and at least 10 failures.

Putting It into Practice

P1: Determine the mean and standard deviation of the sampling distribution of a sample proportion and interpret them in context.

The general properties of the sampling distribution of $\hat{p}$ given in this section can be used to calculate the mean and standard deviation of the sampling distribution of a sample proportion. These values describe where the sampling distribution is centered and how much it spreads out around that center.

P2: Determine if the sampling distribution of a sample proportion is approximately normal.

To determine if the sampling distribution of $\hat{p}$ is approximately normal, check one of the following conditions: (1) both $np \geq 10$ and $n(1 - p) \geq 10$, or (2) the sample includes at least 10 successes and at least 10 failures.

SECTION 8.2 EXERCISES

Each Exercise Set assesses the following chapter learning objectives: M3, M4, M5, P1, P2

SECTION 8.2 Exercise Set 1

8.15 A random sample is to be selected from a population that has a proportion of successes $p = 0.65$. Determine the mean and standard deviation of the sampling distribution of $\hat{p}$ for each of the following sample sizes:

a. $n = 10$ **d.** $n = 50$
b. $n = 20$ **e.** $n = 100$
c. $n = 30$ **f.** $n = 200$

8.16 For which of the sample sizes given in the previous exercise would the sampling distribution of $\hat{p}$ be approximately normal if $p = 0.65$? If $p = 0.2$?

8.17 The article **"Younger Adults More Likely Than Their Elders to Prefer Reading the News"** (October 6, 2016, www .pewresearch.org/fact-tank/2016/10/06/younger-adults -more-likely-than-their-elders-to-prefer-reading-news/) estimated that only 3% of those age 65 and older who prefer to watch the news, rather than to read or listen, watch the news online. This estimate was based on a survey of a large sample of adult Americans conducted by the Pew Research Center. Consider the population consisting of all adult Americans age 65 and older who prefer to watch the news and suppose that for this population the actual proportion who prefer to watch online is 0.03.

a. A random sample of $n = 100$ people will be selected from this population and $\hat{p}$, the proportion of people who prefer to watch online, will be calculated. What are the mean and standard deviation of the sampling distribution of $\hat{p}$?

b. Is the sampling distribution of $\hat{p}$ approximately normal for random samples of size $n = 100$? Explain.

c. Suppose that the sample size is $n = 400$ rather than $n = 100$. Does the change in sample size affect the mean and standard deviation of the sampling distribution of $\hat{p}$? If so, what are the new values for the mean and standard deviation? If not, explain why not.

d. Is the sampling distribution of $\hat{p}$ approximately normal for random samples of size $n = 400$? Explain.

8.18 The article referenced in the previous exercise also reported that for people age 18 to 29 who prefer to watch the news, the proportion that prefer to watch online is 0.37. Answer the questions posed in Parts (a)–(d) of the previous exercise for the population of people age 18 to 29 who prefer to watch the news.

SECTION 8.2 Exercise Set 2

8.19 A random sample is to be selected from a population that has a proportion of successes $p = 0.25$. Determine the mean and standard deviation of the sampling distribution of $\hat{p}$ for each of the following sample sizes:

a. $n = 10$ **d.** $n = 50$
b. $n = 20$ **e.** $n = 100$
c. $n = 30$ **f.** $n = 200$

8.20 For which of the sample sizes given in the previous exercise would the sampling distribution of $\hat{p}$ be approximately normal if $p = 0.25$? If $p = 0.6$?

8.21 A certain chromosome defect occurs in only 1 in 200 adult Caucasian males. A random sample of 100 adult Caucasian males will be selected. The proportion of men in this sample who have the defect, $\hat{p}$, will be calculated.

a. What are the mean and standard deviation of the sampling distribution of $\hat{p}$?

b. Is the sampling distribution of $\hat{p}$ approximately normal? Explain.

c. What is the smallest value of n for which the sampling distribution of $\hat{p}$ would be approximately normal?

8.22 The U.S. Census Bureau reported that in 2015 the proportion of adult Americans age 25 and older who have a bachelor's degree or higher is 0.325 (**"Educational Attainment in the United States: 2015," www.census.gov, retrieved January 22, 2017**). Consider the population of all adult Americans age 25 and over in 2015 and define $\hat{p}$ to the proportion of people in a random sample from this population who have a bachelor's degree or higher.

a. Would $\hat{p}$ based on a random sample of only 10 people from this population have a sampling distribution that is approximately normal? Explain why or why not.

b. What are the mean and standard deviation of the sampling distribution of $\hat{p}$ if the sample size is 400?

c. Suppose that the sample size is $n = 200$ rather than $n = 400$. Does the change in sample size affect the mean and standard deviation of the sampling distribution of $\hat{p}$? If so, what are the new values for the mean and standard deviation? If not, explain why not.

ADDITIONAL EXERCISES

8.23 For which of the following combinations of sample size and population proportion would the standard deviation of $\hat{p}$ be smallest?

$n = 40 \qquad p = 0.3$
$n = 60 \qquad p = 0.4$
$n = 100 \qquad p = 0.5$

8.24 Explain why the standard deviation of $\hat{p}$ is equal to 0 when the population proportion is equal to 1.

8.25 For which of the following sample sizes would the sampling distribution of $\hat{p}$ be approximately normal when $p = 0.2$? When $p = 0.8$? When $p = 0.6$?

$n = 10 \qquad n = 25$
$n = 50 \qquad n = 100$

8.26 A random sample of size 300 is to be selected from a population. Determine the mean and standard deviation of the sampling distribution of $\hat{p}$ for each of the following population proportions.

a. $p = 0.20$
b. $p = 0.45$
c. $p = 0.70$
d. $p = 0.90$

8.27 For which of the population proportions given in the previous exercise would the sampling distribution of $\hat{p}$ be approximately normal if $n = 40$? If $n = 75$?

8.28 The article **"Facebook Etiquette at Work" (USA TODAY, March 24, 2010)** reported that 56% of people participating in a survey of social network users said it was not OK for someone to "friend" his or her boss. Let p denote the proportion of all social network users who feel this way and suppose that $p = 0.56$.

a. Would $\hat{p}$ based on a random sample of 50 social network users have a sampling distribution that is approximately normal?

b. What are the mean and standard deviation of the sampling distribution of $\hat{p}$ if the sample size is 100?

SECTION 8.3 ## How Sampling Distributions Support Learning from Data

In this section, you will take a quick and informal look at the role that sampling distributions play in learning about population characteristics. The two examples in this section show how a sampling distribution provides important information in a estimation setting and in a hypothesis testing setting.

In an estimation situation, you need to understand sampling variability to assess how close an estimate is likely to be to the actual value of the corresponding population characteristic. Published reports often include statements about a margin of error. For example, a report based on a survey of randomly selected registered voters might state that the proportion of registered voters in California who support increasing state funding for community colleges is 0.55 (or 55%) with a margin of error of 0.03 (or 3%). The reported margin of error acknowledges that the actual population proportion is not likely to be exactly 0.55. It indicates that the estimate of 55% support is likely to be within 3 percentage points of the actual population percentage. This margin of error is based on an assessment of sampling variability as described by a sampling distribution. This is illustrated in Example 8.4.

| Example 8.4 | Will Cash Become a Thing of the Past? |

The article **"Most Americans Foresee Death of Cash in Their Lifetime" (July 15, 2016, www .gallup.com, retrieved October 8, 2016)** used data from a random sample of 1024 adults to estimate the proportion of all adults in the United States who think it is likely that within their lifetime, the United States will become a cashless society with all purchases being made by credit card, debit card, or some other form of electronic payment. Of the 1024 people surveyed, 635 indicated that they thought this was likely, resulting in a sample proportion of $\hat{p} = \frac{635}{1024} = 0.62$. The population proportion who think that a cashless society is likely probably isn't exactly 0.62. How accurate is this estimate likely to be?

To answer this question, you can use what you know about the sampling distribution of $\hat{p}$ for random samples of size $n = 1024$. Using the general results given in Section 8.2, you know three things about the sampling distribution of $\hat{p}$:

What You Know	How You Know It
The sampling distribution of $\hat{p}$ is centered at p. This means that the $\hat{p}$ values from random samples cluster around the actual value of the population proportion.	Rule 1 states that $\mu_{\hat{p}} = p$. This is true for random samples, and the description of the study says that the sample was a random sample.
Values of $\hat{p}$ will cluster fairly tightly around the actual value of the population proportion. The standard deviation of $\hat{p}$, which describes how much the $\hat{p}$ values spread out around the population proportion p, is $\sqrt{\frac{p(1-p)}{n}}$. An estimate of this standard deviation is $\sqrt{\frac{\hat{p}(1-\hat{p})}{1024}} = \sqrt{\frac{0.62(1-0.62)}{1024}} = 0.015$	Rule 2 states that $\sigma_{\hat{p}} = \sqrt{\frac{p(1-p)}{n}}$. In this example, $n = 1024$. The actual value of the population proportion p is not known. However, $\hat{p}$ provides an estimate of p that can be used to estimate the standard deviation of the sampling distribution. The estimated standard deviation provides information about how tightly the $\hat{p}$ values from different random samples will cluster around p.
The sampling distribution of $\hat{p}$ is approximately normal.	Rule 3 states that the sampling distribution of $\hat{p}$ is approximately normal if n is large and p is not too close to 0 or 1. Here the sample size is 1024. The sample includes 635 successes and 389 failures, which are both much greater than 10. So, the sampling distribution of $\hat{p}$ is approximately normal.

Summarizing, you know that the $\hat{p}$ distribution is centered at the actual population proportion, has a standard deviation of about 0.015, and is approximately normal. By using this information and what you know about normal distributions, you can now get a sense of the accuracy of the estimate $\hat{p} = 0.62$.

For any variable described by a normal distribution, about 95% of the values are within two standard deviations of the center. Since the sampling distribution of $\hat{p}$ is approximately normal and is centered at the actual population proportion p, you now know that about 95% of all possible random samples will produce a sample proportion that is within $2(0.015) = 0.030$ of the actual value of the population proportion. So, a margin of error of 0.030 can be reported. This tells you that the sample estimate $\hat{p} = 0.62$ is likely to be within 0.030 of the actual proportion of U.S. adults who think that the United States will become a cashless society in their lifetime. You could also say that plausible values for the actual proportion of U.S. adults who think this are those between 0.590 and 0.650. This example shows that the sampling distribution of the sample proportion is the key to assessing the accuracy of the estimate.

Sample data can also be used to evaluate whether or not a claim about a population is believable. For example, the study **"Digital Footprints" (Pew Internet & American Life Project, www.pewinternet.org, 2007, retrieved May 3, 2017)** reported that 47% of Internet users have searched for information about themselves online. The value of 47% was based on a random sample of Internet users. Is it reasonable to conclude that fewer than half of *all* Internet users have searched online for information about themselves? This requires careful thought. There are two reasons why the sample proportion might be less than 0.50. One reason is sampling variability—the value of the sample proportion varies from sample to sample and won't usually be exactly equal to the value of the population proportion. Maybe the population proportion is 0.50 (or even greater) and a sample proportion of 0.47 is "explainable" just due to sampling variability. If this is the case, you can't interpret the sample proportion of 0.47 as convincing evidence that fewer than half of *all* Internet users have searched online for information about themselves. Another reason that the sample proportion might be less than 0.50 is that the actual population proportion is less than 0.50. Is 0.47 enough less than 0.50 that the difference can't be explained by sampling variability alone? If so, the sample data provide convincing evidence that fewer than half of Internet users have performed such a search. This determination can be based on an assessment of sampling variability, which is illustrated in Example 8.5.

Example 8.5   Blood Transfusions Continued

In the article referenced in Example 8.3, the proportion of blood donors testing positive for Hepatitis E was given as 0.07. Suppose that a new screening procedure is implemented and it is hoped that this will reduce the number of donors who test positive. In a random sample of $n = 200$ blood donors, only 6 of the 200 test positive. This appears to be a favorable result, because $\hat{p} = \frac{6}{200} = 0.03$. The question of interest is whether this result indicates that the actual proportion of donors who test positive for Hepatitis E when the new screening procedure is used is less than 0.07, or if the smaller value of the sample proportion could be attributed to just sampling variability.

If the screening procedure is not effective, here is what you know about sample-to-sample variability in $\hat{p}$:

What You Know	How You Know It
The sampling distribution of $\hat{p}$ is centered at p. This means that if the screening procedure is not effective, the $\hat{p}$ values from random samples cluster around 0.07.	Rule 1 states that $\mu_{\hat{p}} = p$. This is true for random samples, and the description of the study says that the sample was selected at random. If the screening procedure is not effective, the proportion who get hepatitis will be the same as it was without screening, which was 0.07.
Values of $\hat{p}$ will cluster fairly tightly around the actual value of the population proportion. If the screening procedure is not effective, the standard deviation of $\hat{p}$, which describes how much the $\hat{p}$ values spread out around the population proportion p, is $\sqrt{\frac{p(1-p)}{n}} = \sqrt{\frac{(0.07)(1-0.07)}{200}} = 0.018.$	Rule 2 states that $\sigma_{\hat{p}} = \sqrt{\frac{p(1-p)}{n}}$. In this example, $n = 200$. If the screening is not effective, the value of p is 0.07. These values of n and p can be used to calculate the standard deviation of the $\hat{p}$ distribution. This standard deviation provides information about how tightly the $\hat{p}$ values from different random samples will cluster around p.
The sampling distribution of $\hat{p}$ is approximately normal.	Rule 3 states that the sampling distribution of $\hat{p}$ is approximately normal if n is large and p is not too close to 0 or 1. Here the sample size is 200 and if the screening procedure is not effective, $p = 0.07$. Because $np = 200(0.07) = 14 \geq 10$ and $n(1 - p) = 200(0.93) = 186 \geq 10$, the sampling distribution of $\hat{p}$ is approximately normal.

Summarizing, you know that if the screening is not effective, the $\hat{p}$ distribution is centered at 0.07, has a standard deviation of 0.018, and is approximately normal. By using this information and what you know about normal distributions, you can now determine how likely it is that a sample proportion as small as 0.03 would be observed if the screening is not effective.

Step	Explanation
Determine what probability will answer the question posed: $P(\hat{p} \leq 0.03)$	You are interested in whether a sample proportion as small as 0.03 would be unusual if the screening is not effective. This can be evaluated by calculating the probability of observing a sample proportion of 0.03 or smaller just by chance (due to sampling variability alone).
$P(\hat{p} \leq 0.03) = P\left(z \leq \dfrac{0.03 - 0.07}{0.018}\right)$ $= P(z \leq -2.22)$	You know that if the screening is not effective, the $\hat{p}$ distribution for random samples of size $n = 200$ is centered at 0.07, has standard deviation 0.018, and is approximately normal. Since the $\hat{p}$ distribution is approximately normal, but not *standard* normal, to calculate the desired probability, translate the problem into an equivalent problem for the standard normal by using a z-score. (See Chapter 6 if you don't remember how to do this.)
$P(z \leq -2.22) = 0.0132$	The standard normal table, a calculator with statistical functions, or computer software can then be used to obtain the desired probability.

Based on the probability calculation, you now know that if the screening is *not* effective, it would be very unlikely (probability 0.0132) that a sample proportion as small as 0.03 would be observed for a sample of size $n = 200$. The new screening procedure does appear to lead to a smaller proportion of blood donors testing positive for Hepatitis E. The sampling distribution of $\hat{p}$ played a key role in reaching this conclusion.

Examples 8.4 and 8.5 used informal reasoning based on the sampling distribution of $\hat{p}$ to learn about a population proportion. This type of reasoning is considered more formally in Chapters 9 and 10.

Summing It Up—Section 8.3

The following learning objectives were addressed in this section:

Conceptual Understanding
C4: Understand how the sampling distribution of a sample proportion enables you to use sample data to learn about a population proportion.
The sampling distribution of a sample proportion provides information about the long-run behavior of the sample proportion. This information is the key to assessing the accuracy of an estimate of a population proportion and to drawing a conclusion about a claim about a population proportion.

Putting It into Practice
P3: Use sample data and properties of the sampling distribution of a sample proportion to reason informally about a population proportion.
Using properties of the sampling distribution of a population proportion to assess the accuracy of an estimate of a population proportion is illustrated in Example 8.4. Drawing a conclusion about a claim involving a population proportion based on information provided by the sampling distribution is illustrated in Example 8.5.

SECTION 8.3 EXERCISES

Each Exercise Set assesses the following chapter learning objectives: C4, P3

SECTION 8.3 Exercise Set 1

8.29 Suppose that a particular candidate for public office is favored by 48% of all registered voters in the district. A polling organization will take a random sample of 500 of these voters and will use $\hat{p}$, the sample proportion, to estimate p.

a. Show that $\sigma_{\hat{p}}$, the standard deviation of $\hat{p}$, is equal to 0.022.

b. If for a different sample size, $\sigma_{\hat{p}} = 0.071$, would you expect more or less sample-to-sample variability in the sample proportions than when $n = 500$?

c. Is the sample size that resulted in $\sigma_{\hat{p}} = 0.071$ larger than 500 or smaller than 500? Explain your reasoning.

8.30 The article **"The Average American Is in Credit Card Debt, No Matter the Economy"** (*Money Magazine,* February 9, 2016) reported that only 35% of credit card users pay off their bill every month. Suppose that the reported percentage was based on a random sample of 1000 credit card users. Suppose you are interested in learning about the value of p, the proportion of all credit card users who pay off their bill every month.

The following table is similar to the table that appears in Examples 8.4 and 8.5, and is meant to summarize what you know about the sampling distribution of $\hat{p}$ in the situation just described. The "What You Know" information has been provided. Complete the table by filling in the "How You Know It" column.

What You Know	How You Know It
The sampling distribution of $\hat{p}$ is centered at the actual (but unknown) value of the population proportion.	
An estimate of the standard deviation of $\hat{p}$, which describes how much the $\hat{p}$ values spread out around the population proportion p, is 0.015.	
The sampling distribution of $\hat{p}$ is approximately normal.	

8.31 The article **"Long-Term Effects of Tongue Piercing—A Case-Control Study"** (*Clinical Oral Investigations* [2012]: 231–237) describes a study of 46 males with pierced tongues. Suppose that it is reasonable to regard this sample as a random sample from the population of all males with pierced tongues. The researchers found receding gums, which can lead to tooth loss, in 27 of the study participants.
a. Suppose you are interested in learning about the value of p, the proportion of all males with pierced tongues who have receding gums. This proportion can be estimated using the sample proportion, $\hat{p}$. What is the value of $\hat{p}$ for this sample?
b. Based on what you know about the sampling distribution of $\hat{p}$, is it reasonable to think that this estimate is within 0.05 of the actual value of the population proportion? Explain why or why not. (Hint: See Example 8.4.)

8.32 The report **"The Role of Two-Year Institutions in Four-Year Success"** (**National Student Clearinghouse Research Center, 2015, nscresearchcenter.org/wp-content/uploads/SnapshotReport17-2YearContributions.pdf, retrieved May 3, 2017**) states that nationwide, 46% of students graduating with a four-year degree in the 2013–2014 academic year had been enrolled in a two-year college sometime in the previous 10 years. The proportion of students graduating with a four-year degree in California with previous two-year college enrollment was estimated to be 0.62 (62%) for that year.

Suppose that this estimate was based on a random sample of 1500 California four-year degree graduates. Is it reasonable to conclude that the proportion of California four-year degree graduates who attended a two-year college in the previous 10 years is different from the national figure? (Hint: Use what you know about the sampling distribution of $\hat{p}$. You might also refer to Example 8.5.)

SECTION 8.3 **Exercise Set 2**

8.33 Suppose that 20% of the customers of a cable television company watch the Shopping Channel at least once a week. The cable company does not know the actual proportion of all customers who watch the Shopping Channel at least once a week and is trying to decide whether to replace this channel with a new local station. The company plans to take a random sample of 100 customers and to use $\hat{p}$ as an estimate of the population proportion.
a. Show that $\sigma_{\hat{p}}$, the standard deviation of $\hat{p}$, is equal to 0.040.
b. If for a different sample size, $\sigma_{\hat{p}} = 0.023$, would you expect more or less sample-to-sample variability in the sample proportions than when $n = 100$?
c. Is the sample size that resulted in $\sigma_{\hat{p}} = 0.023$ larger than 100 or smaller than 100? Explain your reasoning.

8.34 The article **"CSI Effect Has Juries Wanting More Evidence"** (*USA TODAY,* August 5, 2004) examines how the popularity of crime-scene investigation television shows is influencing jurors' expectations of what evidence should be produced at a trial. In a random sample of 500 potential jurors, one study found that 350 were regular watchers of at least one crime-scene forensics television series. Suppose you are interested in learning about the value of p, the proportion of all potential jurors who are regular watchers of crime-scene shows. The following table is similar to the table that appears in Examples 8.4 and 8.5 and is meant to summarize what you know about the sampling distribution of $\hat{p}$ in the situation just described. The "What You Know" information has been provided. Complete the table by filling in the "How You Know It" column.

What You Know	How We Know It
The sampling distribution of $\hat{p}$ is centered at the actual (but unknown) value of the population proportion.	
An estimate of the standard deviation of $\hat{p}$, which describes how much the $\hat{p}$ values spread out around the population proportion p, is 0.021.	
The sampling distribution of $\hat{p}$ is approximately normal.	

8.35 The article **"60% of Employers Are Peeking into Candidates' Social Media Profiles"** (CareerBuilders, April 28, 2016, nscresearchcenter.org/wp-content/uploads/SnapshotReport17**

-2YearContributions.pdf, retrieved October 9, 2016) included data from a survey of 2186 hiring managers and human resource professionals. The article noted that many employers are using social networks to screen job applicants and that this practice is becoming more common. Of the 2186 people who participated in the survey, 1312 indicated that they use social networking sites (such as Facebook, MySpace, and LinkedIn) to research job applicants. For the purposes of this exercise, assume that the sample can be regarded as a random sample of hiring managers and human resource professionals.

a. Suppose you are interested in learning about the value of p, the proportion of all hiring managers and human resource managers who use social networking sites to research job applicants. This proportion can be estimated using the sample proportion, $\hat{p}$. What is the value of $\hat{p}$ for this sample?

b. Based on what you know about the sampling distribution of $\hat{p}$, is it reasonable to think that this estimate is within 0.02 of the actual value of the population proportion? Explain why or why not.

8.36 The report **"A Crisis in Civic Education" (January 2016, goacta.org/images/download/A_Crisis_in_Civic_Education .pdf, retrieved May 3, 2017)** indicated that in a survey of a random sample of 1000 recent college graduates, 96 indicated that they believed that Judith Sheindlin (also known on TV as "Judge Judy") was a member of the U.S. Supreme Court. Is it reasonable to conclude that the proportion of recent college graduates who have this incorrect belief is greater than 0.09 (9%)? (Hint: Use what you know about the sampling distribution of $\hat{p}$. You might also refer to Example 8.5.)

ADDITIONAL EXERCISES

8.37 Some colleges now allow students to rent textbooks for a semester. Suppose that 38% of all students enrolled at a particular college would rent textbooks if that option were available to them. If the campus bookstore uses a random sample of size 100 to estimate the proportion of students at the college who would rent textbooks, is it likely that this estimate would be within 0.05 of the actual population proportion? Use what you know about the sampling distribution of $\hat{p}$ to support your answer.

8.38 In a study of pet owners, it was reported that 24% celebrate their pet's birthday (**Pet Statistics**, **Bissell Homecare, Inc., 2010**). Suppose that this estimate was based on a random sample of 200 pet owners. Is it reasonable to conclude that the proportion of all pet owners who celebrate their pet's birthday is less than 0.25? Use what you know about the sampling distribution of $\hat{p}$ to support your answer.

8.39 The article **"Facebook Etiquette at Work" (USA TODAY, March 24, 2010)** reported that 56% of 1200 social network users surveyed indicated that they thought it was not OK for someone to "friend" his or her boss. Suppose that this sample can be regarded as a random sample of social network users. Is it reasonable to conclude that more than half of social network users feel this way? Use what you know about the sampling distribution of $\hat{p}$ to support your answer.

8.40 The article referenced in the previous exercise also reported that 38% of the 1200 social network users surveyed said it was OK to ignore a coworker's friend request. If $\hat{p} = 0.38$ is used as an estimate of the proportion of *all* social network users who believe this, is it likely that this estimate is within 0.05 of the actual population proportion? Use what you know about the sampling distribution of $\hat{p}$ to support your answer.

CHAPTER ACTIVITIES

ACTIVITY 8.1 **DEFECTIVE M&MS**

Materials needed: 100 plain M&M's for each group of four students

In this activity, you will work in a group with three other students as M&M's inspectors. You will investigate the distribution of the proportion of "defective" M&M's in a sample of size 100.

1. Some M&M's are defective. Examples of possible defects include a broken M&M, a misshapen M&M, and an M&M that does not have the "m" on the candy coating.

Suppose that a claim has been made that 10% of plain M&M's are defective. Assuming this claim is correct, describe the sampling distribution of $\hat{p}$, the proportion of defective M&M's in a sample of size 100. (That is, if $p = 0.10$ and $n = 100$, what is the sampling distribution of $\hat{p}$?)

2. Data collection: With your teammates, carefully inspect 100 M&M's. Sort them into two piles, with nondefective M&M's in one pile and defective M&M's in a second pile. Calculate the sample proportion of defective M&M's.

3. If the claim of 10% defective is true, would a sample proportion as extreme as what you observed in Step 2 be unusual, or is your sample proportion from Step 2 consistent with a 10% population defective rate? Justify your answer based on what you said about the sampling distribution of $\hat{p}$ in Step 1.

🖥 CHAPTER 8 EXPLORATIONS IN STATISTICAL THINKING

🖱 EXPLORATION 1: UNDERSTANDING SAMPLING VARIABILITY

In the exercise below, you will go online to select random samples from a population of adults between the ages of 18 and 64.

Each person in this population was asked if they ever slept with their cell phone. The proportion of people who responded yes to this question was $p = 0.64$. This is a population proportion. Suppose that you didn't know the value of this proportion and that you planned to take a random sample of $n = 20$ people in order to estimate p.

Go online at **statistics.cengage.com/Peck2e/Explore.html** and click on the link for Chapter 8. This link will take you to a web page where you can select random samples of 20 people from the population.

Click on the Select Sample button. This selects a random sample and will display the following information:

1. The ID number that identifies the person selected.
2. The response to the question "Do you sometimes sleep with your cell phone?" These responses were coded numerically—a 1 indicates a yes response and a 2 indicates a no response.

The value of $\hat{p}$ for this sample will also be computed and displayed.

Record the value of $\hat{p}$ for your first sample. Then click the Select Sample button again to get another random sample. Record the value of $\hat{p}$ for your second sample. Continue to select samples until you have recorded $\hat{p}$ values for 25 different random samples.

Use the 25 $\hat{p}$ values to complete the following:

a. Construct a dotplot of the 25 $\hat{p}$ values. Is the dotplot centered at about 0.64, the value of the population proportion?
b. One way to estimate $\sigma_{\hat{p}}$, the standard deviation of the sampling distribution of $\hat{p}$, is to calculate the standard deviation of the 25 $\hat{p}$ values. Calculate this estimate.
c. Because $p = 0.64$, the standard deviation of the sampling distribution of $\hat{p}$ is
$$\sigma_{\hat{p}} = \sqrt{\frac{p(1-p)}{n}} = \sqrt{\frac{(0.64)(0.36)}{20}} = 0.11.$$ Was your estimate of $\sigma_{\hat{p}}$ from Part (b) close to this value?
d. Explain how the dotplot you constructed in Part (a) is or is not consistent with the following statement: Because $p = 0.64$ and $n = 20$, it is not reasonable to think that the sampling distribution of $\hat{p}$ is approximately normal.

ARE YOU READY TO MOVE ON? CHAPTER 8 REVIEW EXERCISES

All chapter learning objectives are assessed in these exercises. The learning objectives assessed in each exercise are given in parentheses.

8.41 (C1)
A random sample of 50 registered voters in a particular city included 32 who favored using city funds for the construction of a new recreational facility. For this sample, $\hat{p} = \frac{32}{50} = 0.64$. If a second random sample of 50 registered voters was selected, would it surprise you if $\hat{p}$ for that sample was not equal to 0.64? Why or why not?

8.42 (C1, C2)
Consider the following three relative frequency histograms. Each histogram was constructed by selecting 100 random samples from a population composed of 40% women and 60% men. For each sample, the sample proportion of women, $\hat{p}$, was calculated and the 100 $\hat{p}$ values were used to construct the histogram. For each histogram, a different sample size was used. One histogram was constructed using 100 samples of size 20, one was constructed using 100 samples of size 40, and one was constructed using 100 samples of size 100.

Histogram I

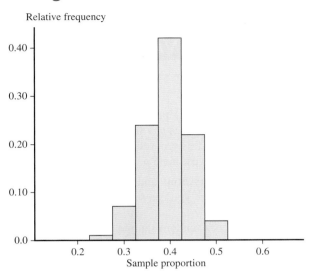

Histogram II

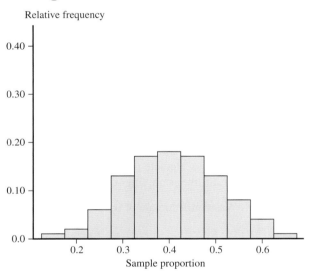

Histogram III

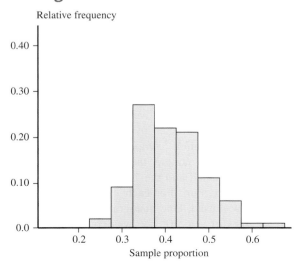

a. These three histograms are approximate sampling distributions of $\hat{p}$ for the three different sample sizes. Which histogram corresponds to the sample size that would produce sample proportions that varied the most from sample to sample?

b. Match the three sample sizes to the three histograms. That is, which histogram is the approximate sampling distribution for sample size = 20, which is for sample size = 40, and which is for sample size = 100? Explain how you made your choices.

8.43 (M1)
Explain why there is sample-to-sample variability in $\hat{p}$ but not in p.

8.44 (M1, M2)
Consider the following statement:
An inspector selected 20 eggs at random from the eggs processed at a large egg production facility. These 20 eggs were tested for salmonella, and the proportion of the eggs that tested positive for salmonella was **0.05**.

a. Is the number that appears in boldface in this statement a sample proportion or a population proportion?

b. Which of the following use of notation is correct, $p = 0.05$ or $\hat{p} = 0.05$?

8.45 (M1, M2)
Consider the following statement:
The proportion of all students enrolled at a particular university during 2017 who lived on campus was **0.21**.

a. Is the number that appears in boldface in this statement a sample proportion or a population proportion?

b. Which of the following use of notation is correct, $p = 0.21$ or $\hat{p} = 0.21$?

8.46 (M2, M3)
A random sample will be selected from a population that has a proportion of successes $p = 0.70$. Determine the mean and standard deviation of the sampling distribution of $\hat{p}$ for each of the following sample sizes:

a. $n = 10$ d. $n = 50$
b. $n = 20$ e. $n = 100$
c. $n = 30$ f. $n = 200$

8.47 (M4)
For which of the sample sizes given in the previous exercise would the sampling distribution of $\hat{p}$ be approximately normal if $p = 0.70$? If $p = 0.30$?

8.48 (P1, P2)
The article **"Fewer Americans Are Reading, But Don't Blame the Millennials" (*Los Angeles Times*, October 9, 2016)** indicates that 80% of millennials (those age 18–29) have read a book in the last year. Suppose that this is the actual percentage for the population of all millennials. Consider a sample proportion $\hat{p}$ that is based on a random sample of 225 millennials. If $p = 0.80$, what are the mean value and standard deviation

of the sampling distribution of $\hat{p}$? Answer this question for $p = 0.70$. Is the sampling distribution of $\hat{p}$ approximately normal in both cases? Explain.

8.49 (P1, P2)

The company Digital Trends reported that 48% of Americans have shared passwords for TV and movie streaming (**"Americans Know They Shouldn't Share Their Passwords, But Do It Anyway," February 18, 2016, www.digitaltrends.com /computing/everyone-shares-passwords-lastpass-survey/, retrieved October 9, 2016**). For purposes of this exercise, assume that the 48% figure is correct for the population of adult Americans.

a. A random sample of size $n = 200$ will be selected from this population and $\hat{p}$, the proportion who have shared TV and movie streaming passwords, will be calculated. What are the mean and standard deviation of the sampling distribution of $\hat{p}$?

b. Is the sampling distribution of $\hat{p}$ approximately normal for random samples of size $n = 200$? Explain.

c. Suppose that the sample size is $n = 50$ rather than $n = 200$. Does the change in sample size affect the mean and standard deviation of the sampling distribution of $\hat{p}$? If so, what are the new values of the mean and standard deviation? If not, explain why not.

d. Is the sampling distribution of $\hat{p}$ approximately normal for random samples of size $n = 50$? Explain.

8.50 (C4)

Suppose that the actual proportion of students at a particular college who use public transportation to travel to campus is 0.15. In a study of parking needs at the campus, college administrators would like to estimate this proportion. They plan to take a random sample of 75 students and use the sample proportion who use public transportation, $\hat{p}$, as an estimate of the population proportion.

a. Show that the standard deviation of $\hat{p}$ is equal to $\sigma_{\hat{p}} = 0.041$.

b. If for a different sample size, $\sigma_{\hat{p}} = 0.032$, would you expect more or less sample-to-sample variability in the sample proportions than for when $n = 75$?

c. Is the sample size that resulted in $\sigma_{\hat{p}} = 0.032$ larger than 75 or smaller than 75? Explain your reasoning.

8.51 (C4)

In a survey of a representative sample of adults in the United States, participants were asked if they agreed or disagreed with the statement "I can't imagine my life without my smartphone" (**July 13, 2015, www.gallup.com, retrieved October 9, 2016**). In response, 46% said that they agreed with this statement. Suppose that this estimate of the

population proportion was based on a random sample of 1000 adults and that you are interested in learning about the value of p, the population proportion of all adults who agree with this statement. The following table is similar to the table that appears in Examples 8.4 and 8.5 and is meant to summarize what you know about the sampling distribution of $\hat{p}$ in the situation just described. The "What You Know" information has been provided. Complete the table by filling in the "How You Know It" column.

What You Know	How You Know It
The sampling distribution of $\hat{p}$ is centered at the actual (but unknown) value of the population proportion.	
An estimate of the standard deviation of $\hat{p}$, which describes how much the $\hat{p}$ values spread out around the population proportion p, is 0.016.	
The sampling distribution of $\hat{p}$ is approximately normal.	

8.52 (P3)

If a hurricane were headed your way, would you evacuate? The headline of a press release issued January 21, 2009, by the survey research company **International Communications Research (icrsurvey.com)** states, "Thirty-one Percent of People on High-Risk Coast Will Refuse Evacuation Order, Survey of Hurricane Preparedness Finds." This headline was based on a survey of 5046 adults who live within 20 miles of the coast in high hurricane risk counties of eight southern states. The sample was selected to be representative of the population of coastal residents in these states, so assume that it is reasonable to regard the sample as if it were a random sample.

a. Suppose you are interested in learning about the value of p, the proportion of adults who would refuse to evacuate. This proportion can be estimated using the sample proportion, $\hat{p}$. What is the value of $\hat{p}$ for this sample?

b. Based on what you know about the sampling distribution of $\hat{p}$, is it reasonable to think that the estimate is within 0.03 of the actual value of the population proportion? Explain why or why not.

8.53 (P3)

In a national sample of 1907 American adults, 1297 indicated that they own a smartphone (**"Technology Device Ownership: 2015," Pew Research Center**). Assume that it is reasonable to regard this sample as a random sample of adult Americans. Is it reasonable to conclude that the proportion of adults who own a smartphone is greater than 0.60?

9 Estimating a Population Proportion

iStock.com/Bill Manning

PREVIEW

When a sample is selected from a population, it is usually because you hope it will provide information about the population. For example, you might want to use sample data to learn about the value of a population characteristic such as the proportion of students enrolled at a college who purchase textbooks online or the mean number of hours that students at the college spend studying each week. This chapter considers how sample data can be used to estimate the value of a population proportion.

Conceptual Understanding

After completing this chapter, you should be able to

C1 Understand what makes a statistic a "good" estimator of a population characteristic.

C2 Know what it means to say that a statistic is an unbiased estimator of a population characteristic.

C3 Understand how the standard deviation of the sample proportion, $\hat{p}$, is related to sample size and to the value of the population proportion.

C4 Understand the relationship between sample size and margin of error.

C5 Understand how an interval is used to estimate a population characteristic.

C6 Know what factors affect the width of a confidence interval estimate of a population proportion.

C7 Understand the meaning of the confidence level associated with a confidence interval estimate.

Mastering the Mechanics

After completing this chapter, you should be able to

M1 Use a sample proportion $\hat{p}$ to estimate a population proportion p and calculate the associated margin of error.

M2 Know the conditions for appropriate use of the margin of error and confidence interval formulas when estimating a population proportion.

M3 Know the key characteristics that lead to selection of the z confidence interval for a population proportion as an appropriate method.

M4 Use the five-step process for estimation problems (EMC³) to calculate and interpret a confidence interval for a population proportion.

M5 Calculate the sample size necessary to achieve a desired margin of error when estimating a population proportion.

M6 (Optional) Calculate and interpret a bootstrap confidence interval for a population proportion.

Putting It in to Practice

After completing this chapter, you should be able to

P1 Interpret a margin of error in context.

P2 Interpret a confidence interval for a population proportion in context and interpret the associated confidence level.

P3 Determine the required sample size, given a description of a proposed study and a desired margin of error.

PREVIEW EXAMPLE

Hurricane Evacuation

If a hurricane were headed your way, would you evacuate? The headline of a press release issued January 21, 2009, by the survey research company **International Communications Research (icrsurvey.com)** states "31 Percent of People on High Risk Coast Will Refuse Evacuation Order, Survey of Hurricane Preparedness Finds." This headline was based on a survey of a representative sample of 5046 coastal residents in eight southern states.

The survey was conducted so that officials involved in emergency planning could learn what proportion of residents might refuse to evacuate. For an estimate to be useful, these officials would need to know something about its accuracy. Because the given estimate of 0.31 (corresponding to 31%) is a sample proportion and the sample proportion will vary from sample to sample, you don't expect the actual *population* proportion to be exactly 0.31. The sample size was large and the sample was selected to be representative of the population of coastal residents, so you might think that the sample proportion would be "close" to the actual value of the population proportion. But how close? Could the actual value of the population proportion be as large as 0.40? As small as 0.25? Being able to evaluate the accuracy of this estimate is important to the people responsible for developing emergency response plans.

This example will be revisited in Section 9.3 after you have learned how sampling distributions enable you to evaluate the accuracy of an estimate of a population proportion. ■

Selecting an Estimator

The first step in estimating a population characteristic, such as a population proportion or a population mean, is to select an appropriate sample statistic. When the goal is to estimate a population proportion, p, the usual choice is the sample proportion $\hat{p}$. But what makes $\hat{p}$ a reasonable choice? More generally, what makes any particular statistic a good choice for estimating a population characteristic?

Because of sampling variability, the value of any statistic you might choose as a potential estimator will vary from one random sample to another. Taking this variability into account, two questions are of interest:

1. Will the statistic consistently tend to overestimate or to underestimate the value of the population characteristic?
2. Will the statistic tend to produce values that are close to the actual value of the population characteristic?

A statistic that does not consistently tend to underestimate or to overestimate the value of a population characteristic is said to be an **unbiased** estimator of that characteristic. Ideally, you would choose a statistic that is unbiased or for which the bias is small (meaning, for example, that if there is a tendency for the statistic to underestimate, the underestimation tends to be small). You can tell if this is the case by looking at the statistic's sampling distribution. If the sampling distribution is centered at the actual value of the population characteristic, then the statistic is unbiased.

The sampling distribution of a statistic also has a standard deviation, which describes how much the values of the statistic vary from sample to sample. If a sampling distribution is centered very close to the actual value of the population characteristic, a small standard deviation ensures that values of the statistic will cluster tightly around the actual value of the population characteristic. This means that the value of the statistic will tend to be close to the population value. Because the standard deviation of a sampling distribution is so important in evaluating the accuracy of an estimate, it is given a special name: the **standard error** of the statistic. For example, the standard error of the sample proportion $\hat{p}$ is

$$\text{standard error of } \hat{p} = \sqrt{\frac{p(1 - p)}{n}}$$

where p is the value of the population proportion and n is the sample size (see Section 8.2).

You can now use this new terminology to describe what makes a statistic a good estimator of a population characteristic.

A statistic with a sampling distribution that is centered at the actual value of a population characteristic is an **unbiased estimator** of that population characteristic. In other words, a statistic is unbiased if the mean of its sampling distribution is equal to the actual value of the population characteristic.

The standard deviation of the sampling distribution of a statistic is called the **standard error** of the statistic.

What makes a statistic a good estimator of a population characteristic?

1. Unbiased (or nearly unbiased)
2. Small standard error

A statistic that is unbiased and has a small standard error is likely to result in an estimate that is close to the actual value of the population characteristic.

As an example of a statistic that is biased, consider the sample range as an estimator of the population range. Recall that the range of a population is defined as the difference between its smallest and largest values. The sample range will be equal to the population range *only* if the sample includes both the largest and the smallest values in the entire population. In any other case, the sample range will be smaller than the population range, so the sample range tends to underestimate the population range. The sampling distribution of the sample range is not centered at the value of the population range.

The sampling distribution of a statistic provides information about the accuracy of estimation. Figure 9.1 displays the sampling distributions of three different statistics. The value of the population characteristic, which is labeled *population value* in the figure, is marked on the measurement axis.

FIGURE 9.1

Sampling distributions of three different statistics for estimating a population value

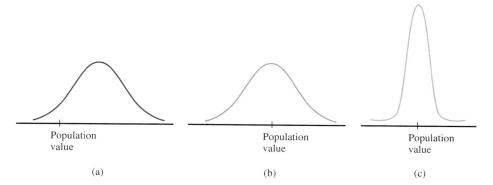

Population value
(a)

Population value
(b)

Population value
(c)

The sampling distribution in Figure 9.1(a) is for a statistic that is not likely to result in an estimate close to the population value. The distribution is centered to the right of the population value, making it very likely that an estimate (the value of the statistic for a particular sample) will be larger than the population value. This statistic is not unbiased and will consistently tend to overestimate the population value.

The sampling distribution in Figure 9.1(b) is centered at the population value. Although one estimate may be smaller than the population value and another may be larger, with this statistic there will be no long-run tendency to over- or underestimate the population value. However, even though the statistic is unbiased, the sampling distribution spreads out quite a bit around the population value. Because of this, some estimates will be far above or far below the population value.

The mean value of the statistic with the distribution shown in Figure 9.1(c) is also equal to the population value, implying no systematic tendency to over- or underestimate. The narrow shape of the distribution curve implies that the statistic's standard error (standard deviation) is relatively small. Estimates based on this statistic will almost always be quite close to the population value—certainly more often than estimates based on the statistic with the sampling distribution shown in Figure 9.1(b). This statistic has the characteristics of a good estimator—it is unbiased and has a small standard error.

Now consider the problem of using sample data to estimate a population proportion. In Chapter 8, properties of the sampling distribution of the statistic $\hat{p}$ (the sample proportion) were introduced. Two of the properties considered there were:

1. $\mu_{\hat{p}} = p$. This tells you that the mean of the sampling distribution of $\hat{p}$ is equal to the value of the population proportion p. The $\hat{p}$ values from all the different possible random samples of size n will center around the actual value of the population proportion. This means that $\hat{p}$ is an unbiased estimator of the population proportion, p.

2. $\sigma_{\hat{p}} = \sqrt{\dfrac{p(1 - p)}{n}}$. This specifies the standard error (standard deviation) of the statistic $\hat{p}$. The standard error describes how much $\hat{p}$ values will spread out around the actual value of the population proportion. If the sample size is large, the standard error will tend to be small.

These two observations suggest that the statistic $\hat{p}$ is a reasonable estimator of the corresponding population proportion as long as the sample size is large enough to ensure a relatively small standard error.

Example 9.1 **State Supreme Court Decisions**

The authors of the paper **"A Comparison of the Criminal Appellate Decisions of Appointed State Supreme Courts: Insights, Questions, and Implications for Judicial Independence" (*The Fordham Urban Law Journal* [April 4, 2007]: 343–362)** examined *all* criminal cases heard by the Supreme Courts of 11 states from 2000 to 2004. Of the 1488 cases considered, 391 were decided in favor of the defendant. Using this information, you can calculate

$$p = \frac{391}{1488} = 0.263$$

where p is a population proportion because *all* cases were considered.

Suppose that the population proportion $p = 0.263$ was not known. To estimate this proportion, you plan to select 25 of the 1488 cases at random and calculate $\hat{p}$, the sample proportion that were decided in favor of the defendant. Because $\hat{p}$ is an unbiased estimator of p, there is no consistent tendency for $\hat{p}$ to over- or underestimate the population proportion. Also, because $p = 0.263$ and $n = 25$,

$$\text{standard error of } \hat{p} = \sqrt{\frac{p(1-p)}{n}} = \sqrt{\frac{(0.263)(1-0.263)}{25}} = \sqrt{\frac{0.1938}{25}}$$
$$= \sqrt{0.0078} = 0.088$$

In practice, the value of p is not known, and the standard error can only be estimated. You will see how to do this in Section 9.2. Even though this example is a bit unrealistic, it can help you to understand what the standard error tells you about an estimator.

If the sample size is 100 rather than 25, then

$$\text{standard error of } \hat{p} = \sqrt{\frac{p(1-p)}{n}} = \sqrt{\frac{(0.263)(1-0.263)}{100}} = \sqrt{\frac{0.1938}{100}}$$
$$= \sqrt{0.0019} = 0.044$$

Notice that the standard error of $\hat{p}$ is much smaller for a sample size of 100 than it is for a sample size of 25. This tells you that $\hat{p}$ values from random samples of size 100 will cluster more tightly around the actual value of the population proportion than the $\hat{p}$ values from random samples of size 25.

In Example 9.1, you saw that the standard error of $\hat{p}$ is smaller when $n = 100$ than when $n = 25$. The fact that the standard error of $\hat{p}$ becomes smaller as the sample size increases makes sense because a large sample provides more information about the population than a small sample. What may not be as evident is how the standard error is related to the actual value of the population proportion.

In Example 9.1, p was 0.263 and the standard error of $\hat{p}$ for random samples of size 25 was 0.088. What if the population proportion had been 0.4 instead of 0.263? Then

$$\text{standard error of } \hat{p} = \sqrt{\frac{p(1-p)}{n}} = \sqrt{\frac{(0.4)(1-0.4)}{25}} = 0.098$$

For the same sample size ($n = 25$), the standard error of $\hat{p}$ is larger when the population proportion is 0.4 than when it is 0.263. A larger standard error corresponds to less accurate estimates of the population proportion. For a fixed sample size, the standard error of $\hat{p}$ is greatest when $p = 0.5$.

It may initially surprise you that $\hat{p}$ tends to produce more accurate estimates the farther the population proportion is from 0.5. To see why this actually makes sense, consider the extreme case where $p = 0$. If $p = 0$, this means that *no* individual in the population possesses the characteristic of interest. *Every* possible sample from this population will result in $\hat{p} = 0$, since every sample will include zero individuals with the characteristic of interest. There is no sample-to-sample variability in the values of $\hat{p}$, so it makes sense that the standard error of $\hat{p}$ is 0. The same would be true in the other extreme case where $p = 1$. Every sample would result in $\hat{p} = 1$, and again there would be no sample-to-sample variability in the value of $\hat{p}$.

For populations where p is close to 0 or to 1, it is fairly easy to get a good estimate of the population proportion, because most samples will tend to produce similar results. The case where there will be the most sample-to-sample variability is where the population is most diverse. For a categorical variable with just two possible values, this is when $p = 0.5$ (half of the population possesses the characteristic of interest and half does not). When $p = 0.5$, the values of $\hat{p}$ will vary more from sample to sample.

These ideas will be developed more formally in Sections 9.2 and 9.3.

Summing It Up—Section 9.1

The following learning objectives were addressed in this section:

Conceptual Understanding

C1: Understand what makes a statistic a "good" estimator of a population characteristic.
A statistic is a "good" estimator of a population characteristic if it tends to produce estimates that are reasonably close to the actual value. This is the case when the statistic is unbiased and has a small standard error (that is, the sample-to-sample variability is small).

C2: Know what it means to say that a statistic is an unbiased estimator of a population characteristic.
A statistic is an unbiased estimator of a population characteristic if it does not consistently tend to underestimate or to overestimate the value of the population characteristic. This is the case when the sampling distribution of the statistic is centered at the value of the population characteristic.

C3: Understand how the standard deviation of the sample proportion, $\hat{p}$, is related to sample size and to the value of the population proportion.
The standard deviation of $\hat{p}$ is equal to $\sqrt{\dfrac{p(1-p)}{n}}$. The standard deviation decreases as the sample size n increases. For a fixed sample size, the closer the value of the population proportion is to 0.5, the larger the standard deviation of $\hat{p}$.

SECTION 9.1 EXERCISES

Each Exercise Set assesses the following chapter learning objectives: C1, C2, C3

SECTION 9.1 Exercise Set 1

9.1 For estimating a population characteristic, why is an unbiased statistic with a small standard error preferred over an unbiased statistic with a larger standard error?

9.2 Three different statistics are being considered for estimating a population characteristic. The sampling distributions of the three statistics are shown in the following illustration:

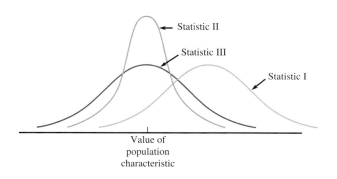

Which of these statistics are unbiased estimators of the population characteristic? (Hint: See the discussion of Figure 9.1.)

9.3 Three different statistics are being considered for estimating a population characteristic. The sampling distributions of the three statistics are shown in the following illustration:

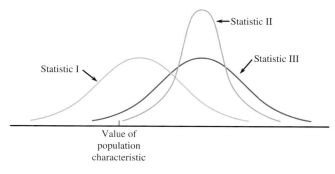

Which statistic would you recommend? Explain your choice.

9.4 A researcher wants to estimate the proportion of students enrolled at a university who eat fast food more than three times in a typical week. Would the standard error of the sample proportion $\hat{p}$ be smaller for random samples of size $n = 50$ or random samples of size $n = 200$?

9.5 Use the formula for the standard error of $\hat{p}$ to explain why
a. the standard error is greater when the value of the population proportion p is near 0.5 than when it is near 1.
b. the standard error of $\hat{p}$ is the same when the value of the population proportion is $p = 0.2$ as it is when $p = 0.8$.

9.6 A random sample will be selected from the population of all adult residents of a particular city. The sample proportion $\hat{p}$ will be used to estimate p, the proportion of all adult residents who are employed full time. For which of the following situations will the estimate tend to be closest to the actual value of p?
i. $n = 500, p = 0.6$
ii. $n = 450, p = 0.7$
iii. $n = 400, p = 0.8$

SECTION 9.1 **Exercise Set 2**

9.7 For estimating a population characteristic, why is an unbiased statistic generally preferred over a biased statistic? Does unbiasedness alone guarantee that the estimate will be close to the actual value of the population characteristic? Explain.

9.8 Three different statistics are being considered for estimating a population characteristic. The sampling distributions of the three statistics are shown in the following illustration:

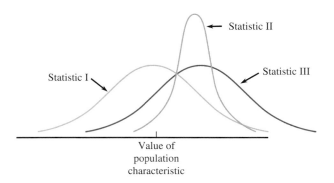

Which of these statistics are unbiased estimators of the population characteristic?

9.9 Three different statistics are being considered for estimating a population characteristic. The sampling distributions of the three statistics are shown in the following illustration:

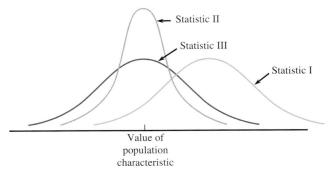

Which statistic would you recommend? Explain your choice.

9.10 A researcher wants to estimate the proportion of students enrolled at a university who are registered to vote. Would the standard error of the sample proportion $\hat{p}$ be larger if the actual population proportion was $p = 0.4$ or $p = 0.8$?

9.11 Use the formula for the standard error of $\hat{p}$ to explain why increasing the sample size decreases the standard error.

9.12 A random sample will be selected from the population of all adult residents of a particular city. The sample proportion $\hat{p}$ will be used to estimate p, the proportion of all adult residents who do not own a car. For which of the following situations will the estimate tend to be closest to the actual value of p?
i. $n = 500, p = 0.1$
ii. $n = 1000, p = 0.2$
iii. $n = 1200, p = 0.3$

ADDITIONAL EXERCISES

9.13 If two statistics are available for estimating a population characteristic, under what circumstances might you choose a biased statistic over an unbiased statistic?

9.14 A random sample will be selected from the population of all students enrolled at a large college. The sample proportion $\hat{p}$ will be used to estimate p, the proportion of all students who use public transportation to travel to campus. For which of the following situations will the estimate tend to be closest to the actual value of p?

i. $n = 300, p = 0.3$
ii. $n = 700, p = 0.2$
iii. $n = 1000, p = 0.1$

9.15 A researcher wants to estimate the proportion of city residents who favor spending city funds to promote tourism. Would the standard error of the sample proportion $\hat{p}$ be smaller for random samples of size $n = 100$ or random samples of size $n = 200$?

9.16   A researcher wants to estimate the proportion of property owners who would pay their property taxes one month early if given a $50 reduction in their tax bill. Would the standard error of the sample proportion $\hat{p}$ be larger if the actual population proportion were $p = 0.2$ or if it were $p = 0.4$?

SECTION 9.2   Estimating a Population Proportion—Margin of Error

When sample data are used to estimate a population proportion, understanding sampling variability will help you assess how close the estimate is likely to be to the actual value of the population proportion. Newspapers and journals often include statements about the *margin of error* associated with an estimate. For example, based on survey of a representative sample of 1000 adult Americans, YouGov estimated that the proportion of adult Americans who have less than $1000 in savings is $\hat{p} = 0.430$ **("People More Likely to Save with an Opt-Out System," today.yougov.com/news/2016/04/25/savings/, retrieved May 5, 2017)**. The report also gave a margin of error of 0.042 (or 4.2 percentage points) for this estimate and for other estimated proportions included in the report. By reporting a margin of error, the people who conducted the survey acknowledge that the actual population proportion of *all* adult Americans who have less than $1000 in savings is not likely to be *exactly* 0.430. The margin of error indicates that plausible values for the actual population proportion include values between $0.430 - 0.042 = 0.388$ and $0.430 + 0.042 = 0.472$.

The value of the sample proportion $\hat{p}$ provides an estimate of the population proportion p. If $\hat{p} = 0.426$ and $p = 0.484$, the estimate is "off" by 0.058, which is the difference between the actual value and the estimate. This difference represents the error in the estimate. A different sample might produce an estimate of $\hat{p} = 0.498$, resulting in an estimation error (the difference between the value of the statistic and the value of the population characteristic being estimated) of 0.014. For a given sampling method, statistic, and sample size, the maximum likely estimation error is called the **margin of error**.

The **margin of error** of a statistic is the maximum likely estimation error. It is unlikely that an estimate will differ from the actual value of the population characteristic by more than the margin of error.

In Section 9.1, the standard error (the standard deviation of the sampling distribution) of a statistic was introduced. The standard error and information about the shape of the sampling distribution of the statistic are used to find the margin of error.

With p denoting the proportion in the population possessing some characteristic of interest, recall that three things are known about the sampling distribution of $\hat{p}$ when a random sample of size n is selected:

1.   The mean of the $\hat{p}$ sampling distribution is p. This means that $\mu_{\hat{p}} = p$.

2.   The standard error (standard deviation) of $\hat{p}$ is $\sqrt{\dfrac{p(1-p)}{n}}$.

3.   If n is large, the $\hat{p}$ sampling distribution is approximately normal.

The sampling distribution describes the behavior of the statistic $\hat{p}$. If the sample size is large, the sampling distribution of $\hat{p}$ is approximately normal and you can make use of what you know about normal distributions. Consider the following line of reasoning:

Step	What You Know	How You Know It
1	If a variable has a standard normal distribution, about 95% of the time the value of the variable will be between -1.96 and 1.96.	The table of z curve areas (Appendix A Table 2) can be used to show that for the standard normal distribution, a central area of 0.95 is captured between -1.96 and 1.96 (see Section 6.5). This tells you that for a variable that has a standard normal distribution, observed values will be between -1.96 and 1.96 about 95% of the time. Most extreme 5% 0.95 0.025 0.025 -1.96 1.96 Middle 95%
2	For any normal distribution, about 95% of the observed values will be within 1.96 standard deviations of the mean.	Because the mean of the standard normal distribution is 0 and the standard deviation is 1, the statement in Step 1 is equivalent to saying that 95% of the values are within 1.96 standard deviations of the mean. This is true for *all* normal distributions, not just the standard normal distribution.
3	If n is large, the $\hat{p}$ sampling distribution is approximately normal with mean p and standard error $\sqrt{\dfrac{p(1-p)}{n}}$.	These are the three properties of the sampling distribution of $\hat{p}$.
4	If n is large, 95% of all possible random samples will produce a value of $\hat{p}$ that is within $1.96\sqrt{\dfrac{p(1-p)}{n}}$ of the value of the population proportion p.	This is just a rewording of the statement in Step 2 in terms of the sampling distribution of $\hat{p}$. This is reasonable because you know that if the sample size is large, the sampling distribution of $\hat{p}$ is approximately normal, and so the statement in Step 2 also applies to the sampling distribution of $\hat{p}$.
5	If n is large, it is unlikely that $\hat{p}$ will be farther than $$1.96\sqrt{\dfrac{p(1-p)}{n}}$$ from the actual value of the population proportion p. In other words, if n is large, it is unlikely that $\hat{p}$ will differ from p by more than 1.96 times the standard error of $\hat{p}$.	This is rewording the statement in Step 4 in terms of the standard error, which is the standard deviation of $\hat{p}$.

The margin of error of a statistic is the maximum likely estimation error expected when the statistic is used as an estimator. Based on the reasoning described in Steps 1–5, you know that if the sample size is large, about 95% of the time a random sample will result in a sample proportion that is closer than $1.96\sqrt{\dfrac{p(1-p)}{n}}$ to the actual value of the population proportion. This leads to the following definition of margin of error when $\hat{p}$ from a large sample is used to estimate a population proportion:

$$\text{margin of error} = 1.96\sqrt{\frac{p(1-p)}{n}} = (1.96)(\text{standard error of } \hat{p})$$

There are just two more questions that need to be answered before looking at an example.

Question	Answer
How large does the sample need to be?	The formula for margin of error is appropriate when the sampling distribution of $\hat{p}$ is approximately normal. Recall from Chapter 8 that whether this is the case depends on the population proportion. If both $np \geq 10$ and $n(1-p) \geq 10$, it is reasonable to think that the sampling distribution of $\hat{p}$ is approximately normal. Since the value of the population proportion p will be unknown, you check the sample size conditions using the sample proportion, $\hat{p}$.

(continued)

Question	Answer
How large does the sample need to be? *(continued)*	The sample size is large enough if 1. $n\hat{p} \geq 10$ and $n(1 - \hat{p}) \geq 10$ OR (equivalently) 2. The sample includes at least 10 successes (individuals that have the characteristic of interest) and at least 10 failures.
The formula for margin of error is margin of error $= 1.96 \sqrt{\dfrac{p(1 - p)}{n}}$ Notice that p is the value of the population proportion, which is unknown. How can the margin of error be calculated if you don't know the value of p?	Since the value of p is unknown, you can estimate the margin of error by using $\hat{p}$, the sample proportion, in place of p. The estimated standard error is then $\sqrt{\dfrac{\hat{p}(1 - \hat{p})}{n}}$ and the estimated margin of error is $1.96 \sqrt{\dfrac{\hat{p}(1 - \hat{p})}{n}}$.

The following box summarizes what is needed to calculate and interpret a margin of error when estimating a population proportion.

> ### Margin of Error When the Sample Proportion $\hat{p}$ is Used to Estimate a Population Proportion p.
>
> **Appropriate when the following conditions are met**
>
> 1. The sample is a random sample from the population of interest
>   OR
>   the sample is selected in a way that makes it reasonable to think the sample is representative of the population.
>
> 2. The sample size is large enough. This condition is met if
> $n\hat{p} \geq 10$ and $n(1 - \hat{p}) \geq 10$
>   OR (equivalently)
>   the sample includes at least 10 successes and at least 10 failures.
>
> **When these conditions are met**
> $$\text{margin of error} = 1.96 \sqrt{\frac{\hat{p}(1 - \hat{p})}{n}}$$
>
> **Interpretation of margin of error**
> It would be unusual for the sample proportion to differ from the actual value of the population proportion by more than the margin of error. For about 95% of all random samples, the estimation error will be less than the margin of error.
>
> **Notes**
>
> 1. The formula given for margin of error is actually for the *estimated* margin of error, but it is common to refer to it without including "estimated." Any time a margin of error is reported, it will be an estimated margin of error.
> 2. The value 2 is often used in place of 1.96 in the margin of error calculation. This simplifies the calculations and doesn't usually change the result by much. When calculating a margin of error, it is acceptable to use either 2 or 1.96.

The following examples illustrate the calculation and interpretation of margin of error when estimating a population proportion.

Example 9.2   Saving for a Rainy Day

Earlier in this section, a survey to estimate the proportion of adult Americans who have less than $1000 in savings was described. Based on a representative sample of 1000 adult Americans, **YouGov (today.yougov.com/news/2016/04/25/savings/, retrieved May 5, 2017)** the proportion of adult Americans who have less than $1000 in savings was estimated to be $\hat{p} = 0.430$. You now have enough information to calculate a margin of error for this estimate.

Step	Explanation
Check Conditions It is appropriate to use the margin of error formula because 1. The sample was representative of the population. 2. The sample size is large enough.	The sample was described as representative of the population of adult Americans. Although no information about how the sample was selected was given, it is reasonable to think that YouGov took appropriate steps to ensure a representative sample. Since $\hat{p} = 0.430$ and $n = 1000$, you can verify that the sample size is large enough: $n\hat{p} = 1000(0.430) = 430$ and $n(1 - \hat{p}) = 1000(1 - 0.430) = 570.$ Both $n\hat{p}$ and $n(1 - \hat{p})$ are at least 10.
Calculate margin of error margin of error $= 1.96\sqrt{\dfrac{\hat{p}(1 - \hat{p})}{n}}$ $\quad = 1.96\sqrt{\dfrac{(0.430)(1 - 0.430)}{1000}}$ $\quad = 1.96(0.016)$ $\quad = 0.031$	Substitute $n = 1000$ and $\hat{p} = 0.430$ into the formula for margin of error.

Interpret margin of error

An estimate of the proportion of adult Americans who have less than $1000 in savings is 0.430. Although this is only an estimate, it is unlikely that this estimate differs from the actual population proportion by more than 0.031.

The margin of error given in the YouGov report was 0.042. The reported margin of error is greater than the one calculated here because several different proportions were estimated in the report, and so a conservative margin of error that could be applied to all of the estimates was provided. For example, one other proportion estimated was the proportion of adult American males who have over $5000 in savings. This proportion was based on responses from $n = 486$ men, and the estimate was $\hat{p} = 0.30$. Using the formula for margin of error for this estimate gives

$$1.96\sqrt{\frac{(0.30)(1 - 0.30)}{486}} = 1.96(0.021) = 0.041$$

and if 2 is used in place of the 1.96 in the calculation of the margin of error, the margin of error is the reported value of 0.042.

Example 9.3 **Support for Solar Power**

The article **"Americans Strongly Favor Expanding Solar Power to Help Address Costs and Environmental Concerns" (Pew Research Center, October 5, 2016)** reported that in a survey of a representative sample of 1534 adult Americans, 89% indicated that they favored expanding solar panel farms in the United States. Based on the survey results, $\hat{p} = 0.89$ provides an estimate of the proportion of all adult Americans who favor the expansion of solar farms. How accurate is this estimate? You can describe the accuracy by calculating the margin of error.

Check Conditions

It is appropriate to use the margin of error formula if

1. The sample was selected in a way that makes it reasonable to think the sample is representative of the population.
2. The sample size is large.

The given information states that the sample was a representative sample of American adults. With $\hat{p} = 0.89$ and $n = 1534$,

$$n\hat{p} = 1534(0.89) = 1365.3 \text{ and } n(1 - \hat{p}) = 1534(1 - 0.89) = 168.7$$

Because both $n\hat{p}$ and $n(1 - \hat{p})$ are at least 10, the sample size is large enough to use the margin of error formula.

Calculate margin of error

$$\text{margin of error} = 1.96\sqrt{\frac{\hat{p}(1 - \hat{p})}{n}}$$

$$= 1.96\sqrt{\frac{(0.890)(1 - 0.890)}{1534}}$$

$$= 1.96(0.008)$$

$$= 0.016$$

Interpretation
An estimate of the proportion of adult Americans who favor the expansion of solar farms is 0.890, or 89%. Although this is only an estimate, it is unlikely that this estimate differs from the actual population proportion by more than 0.016 (or 1.6 percentage points).

If the margin of error is large, it indicates that you don't have very accurate information about a population characteristic. On the other hand, if the margin of error is small, you would think that the sample resulted in an estimate that is relatively accurate.

In the chapters that follow, you will see that the same reasoning used to develop the margin of error for $\hat{p}$ can also be used in other situations.

In general, for an unbiased statistic that has a sampling distribution that is approximately normal,

$$\text{margin of error} = (1.96)(\text{standard error})$$

The estimation error, which is the difference between the value of the statistic and the value of the population characteristic being estimated, will be less than the margin of error for about 95% of all possible random samples.

In the next section, you will see another approach that also conveys information about the accuracy of an estimate. This approach uses an interval of plausible values rather than just a single number estimate.

Summing It Up—Section 9.2

The following learning objectives were addressed in this section:

Conceptual Understanding
C4: Understand the relationship between sample size and margin of error.
As the sample size increases, the margin of error decreases. This indicates that it is more likely that an estimate will be close to the actual value of the population characteristic when the sample size is larger.

Mastering the Mechanics
M1: Use a sample proportion $\hat{p}$ to estimate a population proportion p and calculate the associated margin of error.
The sample proportion provides an estimate of the population proportion. When appropriate, the margin of error can be calculated using the formula given in this section. The margin of error describes the maximum likely estimation error. For examples of estimating a population proportion and calculating the associated margin of error, see Examples 9.2 and 9.3.

M2: Know the conditions for appropriate use of the margin of error and confidence interval formulas when estimating a population proportion.
There are two conditions that need to be reasonably met in order for it to be appropriate to use the margin of error formula given in this section. These conditions are (1) that the sample is a random sample from the population of interest or the sample is selected in a way that makes it reasonable to think that the sample is representative of the population,

and (2) the sample size is large. The sample size is considered to be large if $n\hat{p}$ and $n(1 - \hat{p})$ are at least 10 (or equivalently, if the sample includes at least 10 successes and at least 10 failures).

Putting It into Practice

P1: Interpret a margin of error in context.

The margin of error is interpreted at the maximum likely estimation error. For examples of interpreting margin of error in context, see Examples 9.2 and 9.3.

SECTION 9.2 EXERCISES

Each Exercise Set assesses the following chapter learning objectives: C4, M1, M2, P1

SECTION 9.2 Exercise Set 1

9.17 A large online retailer is interested in learning about the proportion of customers making a purchase during a particular month who were satisfied with the online ordering process. A random sample of 600 of these customers included 492 who indicated they were satisfied. For each of the three following statements, indicate if the statement is correct or incorrect. If the statement is incorrect, explain what makes it incorrect.

Statement 1: It is unlikely that the estimate $\hat{p} = 0.82$ differs from the value of the actual population proportion by more than 0.016.

Statement 2: It is unlikely that the estimate $\hat{p} = 0.82$ differs from the value of the actual population proportion by more than 0.031.

Statement 3: The estimate $\hat{p} = 0.82$ will never differ from the value of the actual population proportion by more than 0.031.

9.18 Consider taking a random sample from a population with $p = 0.40$.

a. What is the standard error of $\hat{p}$ for random samples of size 100?

b. Would the standard error of $\hat{p}$ be greater for samples of size 100 or samples of size 200?

c. If the sample size were doubled from 100 to 200, by what factor would the standard error of $\hat{p}$ decrease?

9.19 The report **"2007 Electronic Monitoring & Surveillance Survey" (American Management Association)** summarized a survey of 304 U.S. businesses. The report stated that 91 of the 304 businesses had fired workers for misuse of the Internet. Assume that this sample is representative of businesses in the United States.

a. Estimate the proportion of all businesses in the U.S. that have fired workers for misuse of the Internet. What statistic did you use?

b. Use the sample data to estimate the standard error of $\hat{p}$.

c. Calculate and interpret the margin of error associated with the estimate in Part (a). (Hint: See Example 9.3.)

9.20 The use of the formula for margin of error requires a large sample. For each of the following combinations of n and $\hat{p}$, indicate whether the sample size is large enough for use of this formula to be appropriate.

a. $n = 50$ and $\hat{p} = 0.30$

b. $n = 50$ and $\hat{p} = 0.05$

c. $n = 15$ and $\hat{p} = 0.45$

d. $n = 100$ and $\hat{p} = 0.01$

9.21 The paper **"Sleeping with Technology: Cognitive, Affective and Technology Usage Predictors of Sleep Problems Among College Students"** (*Sleep Health* [2016]: 49–56) summarized data from a survey of a sample of college students. Of the 734 students surveyed, 125 reported that they sleep with their cell phones near the bed and check their phones for something other than the time at least twice during the night. For purposes of this exercise, assume that this sample is representative of college students in the United States.

a. Use the given information to estimate the proportion of college students who check their cell phones for something other than the time at least twice during the night.

b. Verify that the conditions needed in order for the margin of error formula to be appropriate are met.

c. Calculate the margin of error.

d. Interpret the margin of error in the context of this problem.

9.22 The report **"The 2016 Consumer Financial Literacy Survey" (The National Foundation for Credit Counseling, www.nfcc.org, retrieved October 28, 2016)** summarized data from a representative sample of 1668 adult Americans. When asked if they typically carry credit card debt from month to month, 584 of these people responded "yes."

a. Use the given information to estimate the proportion of adult Americans who carry credit card debt from month to month.

b. Verify that the conditions needed in order for the margin of error formula to be appropriate are met.

c. Calculate the margin of error.

d. Interpret the margin of error in the context of this problem.

SECTION 9.2 Exercise Set 2

9.23 A car manufacturer is interested in learning about the proportion of people purchasing one of its cars who plan to purchase another car of this brand in the future. A random sample of 400 of these people included 267 who said they would purchase this brand again. For each of the three statements below, indicate if the statement is correct or incorrect. If the statement is incorrect, explain what makes it incorrect.

> **Statement 1:** The estimate $\hat{p} = 0.668$ will never differ from the value of the actual population proportion by more than 0.046.
> **Statement 2:** It is unlikely that the estimate $\hat{p} = 0.668$ differs from the value of the actual population proportion by more than 0.024.
> **Statement 3:** It is unlikely that the estimate $\hat{p} = 0.668$ differs from the value of the actual population proportion by more than 0.046.

9.24 Consider taking a random sample from a population with $p = 0.70$.
a. What is the standard error of $\hat{p}$ for random samples of size 100?
b. Would the standard error of $\hat{p}$ be smaller for samples of size 100 or samples of size 400?
c. Does decreasing the sample size by a factor of 4, from 400 to 100, result in a standard error of $\hat{p}$ that is four times as large?

9.25 The report **"Parents, Teens and Digital Monitoring"** **(Pew Research Center, January 7, 2016, www.pewinternet .org/2016/01/07/parents-teens-and-digital-monitoring, retrieved May 5, 2017)** reported that 61% of parents of teens aged 13 to 17 said that they had checked which web sites their teens had visited. The 61% figure was based on a representative sample of 1060 parents of teens in this age group.
a. Use the given information to estimate the proportion of parents of teens age 13 to 17 who have checked which web sites their teen has visited. What statistic did you use?
b. Use the sample data to estimate the standard error of $\hat{p}$.
c. Calculate and interpret the margin of error associated with the estimate in Part (a).

9.26 The use of the formula for margin of error requires a large sample. For each of the following combinations of n and $\hat{p}$, indicate whether the sample size is large enough for use of this formula to be appropriate.
a. $n = 100$ and $\hat{p} = 0.70$
b. $n = 40$ and $\hat{p} = 0.25$
c. $n = 60$ and $\hat{p} = 0.25$
d. $n = 80$ and $\hat{p} = 0.10$

9.27 The *USA Snapshot* titled **"Big Bang Theory"** (**USA TODAY, October 14, 2016)** summarized data from a sample of 1003 American parents of children age 6 to 11 years. It reported that 53% of these parents view science-oriented TV shows as a good way to expose their kids to science outside of school. Assume that this sample is representative of American parents of children age 6 to 11 years.
a. Use the given information to estimate the proportion of American parents of children age 6 to 11 years who view

science-oriented TV shows as a good way to expose their children to science.
b. Verify that the conditions needed in order for the margin of error formula to be appropriate are met.
c. Calculate the margin of error.
d. Interpret the margin of error in the context of this problem.

9.28 Suppose that 935 smokers each received a nicotine patch, which delivers nicotine to the bloodstream at a much slower rate than cigarettes do. Dosage was decreased to 0 over a 12-week period. Of these 935 people, 245 were still not smoking 6 months after treatment. Assume this sample is representative of all smokers.
a. Use the given information to estimate the proportion of all smokers who, when given this treatment, would refrain from smoking for at least 6 months.
b. Verify that the conditions needed in order for the margin of error formula to be appropriate are met.
c. Calculate the margin of error.
d. Interpret the margin of error in the context of this problem.

ADDITIONAL EXERCISES

9.29 *USA TODAY* reported that the proportion of Americans who prefer cheese on their burgers is 0.84 (**USA TODAY, September 7, 2016**). This estimate was based on a survey of a representative sample of 1000 adult Americans. Calculate and interpret a margin of error for the reported proportion.

9.30 The article **"Most Americans Don't Understand the Cloud, But They Should" (foxbusiness.com, October 17, 2016, retrieved November 12, 2016)** reported that in a sample of 1000 people, 22% said they have pretended to know what the cloud is or how it works. Assuming that it is reasonable to regard the sample as representative of adult Americans, an estimate of the proportion who have pretended to know what the cloud is or how it works is 0.22. The margin of error associated with this estimate is 0.026. Interpret the value of this margin of error.

9.31 The *USA Snapshot* titled **"Have a Nice Trip"** (**USA TODAY, November 17, 2015**) summarized data from a survey of 1000 U.S. adults who had traveled by air at least once in the previous year. The Snapshot includes the following statement: "38% admit to yelling at a complete stranger while traveling."
a. Assuming that the sample was selected to be representative of the population of U.S. adults who have traveled by air at least once in the previous year, what is an estimate of the population proportion who have yelled at a complete stranger while traveling?
b. Calculate and interpret the margin of error associated with your estimate in Part (a).

9.32 *Business Insider* reported that a study commissioned by eBay Motors found that nearly 40% of millennials who drive a car that is more than 5 years old have named their cars

("**Millennials Have an Odd Habit When It Comes to Their Cars,**"
April 14, 2016).

a. Assuming that the sample was selected to be representative of the population of millennials who drive a car that is more than 5 years old, what is an estimate of the population proportion who have named their car?

b. Suppose that the sample size for the study described was 800. Calculate and interpret the margin of error associated with your estimate in Part (a).

9.33 An article in the *Chicago Tribune* (August 29, 1999) reported that in a poll of residents of the Chicago suburbs, 43% felt that their financial situation had improved during the past year. The following statement is from the article: "The findings of this Tribune poll are based on interviews with 930 randomly selected suburban residents. The sample included suburban Cook County plus DuPage, Kane, Lake, McHenry, and Will Counties. In a sample of this size, one can say with 95% certainty that results will differ by no more than 3% from results obtained if all residents had been included in the poll." Give a statistical argument to justify the claim that the estimate of 43% is within 3% of the actual percentage of all residents who feel that their financial situation has improved.

SECTION 9.3 A Large Sample Confidence Interval for a Population Proportion

The preview example of this chapter described how data from a survey were used to estimate the proportion of coastal residents who would refuse to evacuate if a hurricane was approaching. Based on a representative sample of 5046 adults living within 20 miles of the coast in high hurricane risk counties of eight southern states, it was estimated that the proportion of residents who would refuse to evacuate was 0.31 (or 31%).

When you estimate that 31% of residents would refuse to evacuate, you don't really believe that *exactly* 31% would refuse. The estimate $\hat{p} = 0.31$ is a sample proportion, and the value of $\hat{p}$ varies from sample to sample. What the estimate tells you is that the actual population proportion is "around" 0.31. In Section 9.2 you saw one way to be more specific about what is meant by "around" by reporting both the estimate and a margin of error. For example, you could say that the estimated proportion who would refuse to evacuate is 0.31, with a margin of error of 0.013. This implies that "around 0.31" means "within 0.013 of 0.31."

Another common way to be more specific about what is meant by "around" is to report an interval of reasonable values for the population proportion rather than a single number estimate. For example, you could report that, based on data from the sample of coastal residents, you believe the actual proportion who would refuse to evacuate is a value in the interval from 0.297 to 0.323. By reporting an interval, you acknowledge that you don't know the exact value of the population proportion. The width of the interval conveys information about accuracy. The fact that the interval here is narrow implies that you have fairly precise information about the value of the population proportion. On the other hand, if you say that you think the population proportion is between 0.21 and 0.41, it would be clear that you don't know as much about the actual value. In this section, you will learn how to use sample data to construct an interval estimate, called a confidence interval.

What Is a Confidence Interval?

A **confidence interval** specifies a set of plausible values for a population characteristic. For example, using sample data and what you know about the behavior of the sample proportion, it is possible to construct an interval that you think should include the actual value of the population proportion. Because a sample provides only incomplete information about the population, there is some risk involved with a confidence interval estimate. Occasionally, but hopefully not very often, a sample will lead you to an interval that does not include the actual value of the population characteristic. If this were to happen and you make a statement such as "the proportion of coastal residents who would refuse to evacuate is between 0.30 and 0.32," you would be wrong. It is important to know how likely it is that the method used to calculate a confidence interval estimate will lead to a correct statement. So, associated with every confidence interval is a **confidence level** which specifies the "success rate" of the method used to produce the confidence interval.

> **DEFINITION**
>
> **Confidence Interval**
> A confidence interval is an interval that you think includes the value of the population characteristic.
>
> **Confidence Level**
> The confidence level associated with a confidence interval is the success rate of the *method* used to construct the interval.

What Does the Confidence Level Tell You About a Confidence Interval?

The confidence level associated with a confidence interval tells you how much "confidence" you can have in the *method* used to construct the interval (but *not* your confidence in any *one* particular interval). Usual choices for confidence levels are 90%, 95%, and 99%, although other levels are also possible. If you construct a 95% confidence interval using the method to be described shortly, you would be using a method that is "successful" in capturing the value of the population characteristic 95% of the time. Different random samples will lead to different intervals, but 95% of these intervals will include the value of the population characteristic. Similarly, a 99% confidence interval is one constructed using a method that captures the actual value of the population characteristic for 99% of all possible random samples.

Margin of Error and a 95% Confidence Interval

The margin of error when the sample proportion $\hat{p}$ is used to estimate a population proportion was introduced in Section 9.2. When the following conditions are met,

1. The sample size is large ($n\hat{p} \geq 10$ and $n(1 - \hat{p}) \geq 10$)
2. The sample is a random sample or the sample is selected in a way that should result in a representative sample from the population

the margin of error was defined as

$$\text{margin of error} = (1.96)(\text{standard error of } \hat{p}) = 1.96\sqrt{\frac{\hat{p}(1 - \hat{p})}{n}}$$

The margin of error formula was based on the following reasoning:

1. When the conditions are met, the sampling distribution of $\hat{p}$ is approximately normal, is centered at the actual value of the population proportion, and has a standard error that is estimated by $\sqrt{\dfrac{\hat{p}(1 - \hat{p})}{n}}$.
2. For any variable that has a standard normal distribution, about 95% of the values will be between -1.96 and 1.96. This also means that for any normal distribution, about 95% of the values are within 1.96 standard deviations of the mean.

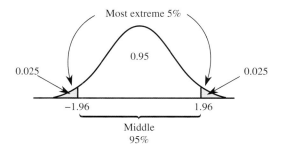

3. Because $\hat{p}$ has a normal distribution (approximately) with mean p, about 95% of all random samples have a sample proportion that is within $1.96\sqrt{\dfrac{\hat{p}(1 - \hat{p})}{n}}$ of the population proportion p.

Based on this reasoning, you are able to say that for 95% of all random samples, the estimation error (the difference between the value of the sample proportion $\hat{p}$ and the value of the population proportion p) will be less than the margin of error. This means that the interval

$$(\hat{p} - \text{margin of error}, \hat{p} + \text{margin of error})$$

which is often abbreviated as $\hat{p} \pm$ margin of error, should include the value of the population proportion for 95% of all possible random samples. The interval

$$\hat{p} \pm \text{margin of error}$$

is the 95% confidence interval (meaning the confidence level is 95%) for a population proportion.

A Large-Sample 95% Confidence Interval for a Population Proportion

Appropriate when the following conditions are met:

1. The sample is a random sample from the population of interest or the sample is selected in a way that should result in a representative sample from the population.
2. The sample size is large. This condition is met when both $n\hat{p} \geq 10$ and $n(1 - \hat{p}) \geq 10$ or (equivalently) the sample includes at least 10 successes and at least 10 failures.

When these conditions are met, a 95% confidence interval for the population proportion is

$$\hat{p} \pm 1.96 \sqrt{\frac{\hat{p}(1 - \hat{p})}{n}}$$

Note: This formula corresponds to a confidence level of 95%. Other confidence levels will be considered later in this section.

Interpretation of Confidence Interval

You can be 95% confident that the actual value of the population proportion is included in the calculated interval.

Interpretation of 95% Confidence Level

A method has been used to produce the confidence interval that is successful in capturing the actual population proportion approximately 95% of the time.

Before looking at an example of calculating and interpreting a 95% confidence interval, here is a quick review of some important ideas from Chapter 7.

Chapter 7 Review: Four Key Questions and the Five-Step Process for Estimation Problems

In Section 7.2, you learned four key questions that guide the decision about what statistical inference method to consider in any particular situation. In Section 7.3, a five-step process for estimation problems was introduced. In the playbook analogy, the four key questions provide a way to select an appropriate play from the playbook (that is, to select an appropriate statistical inference method). The five-step process provides a structure for carrying out each play in the playbook.

The four key questions of Section 7.2 were

Q **Question Type**	Estimation or hypothesis testing?
S **Study Type**	Sample data or experiment data?
T **Type of Data**	One variable or two? Categorical or numerical?
N **Number of Samples or Treatments**	How many samples or treatments?

When the answers to these questions are

   Q: Estimation
   S: Sample data
   T: One categorical variable
   N: One sample

the method you will want to consider is the large-sample confidence interval for a population proportion.

Because this is an estimation problem, you would proceed by following the five-step process for estimation problems (EMC^3). The steps in this process are (see Section 7.3):

Step	What Is This Step?
E	**Estimate:** Explain what population characteristic you plan to estimate.
M	**Method:** Select a potential method. This step involves using the answers to the four key questions (QSTN) to identify a potential method.
C	**Check:** Check to make sure that the method selected is appropriate. Many methods for learning from data only provide reliable information under certain conditions. It will be important to verify that any such conditions are met before proceeding.
C	**Calculate:** Use sample data to perform any necessary calculations. This step is often accomplished using technology, such as a graphing calculator or computer software for statistical analysis.
C	**Communicate Results:** This is a critical step in the process. In this step, you answer the research question of interest, explain what you have learned from the data, and acknowledge potential risks.

You will see how this process is used in the examples that follow. Be sure to include all five steps when you use a confidence interval to estimate a population characteristic.

Example 9.4 Moving Home

When unemployment rises and the economy struggles, some people make the decision to move in with friends or family. Researchers studying relocation patterns were interested in learning about the proportion of U.S. adults 21 years of age or older who moved back home or in with friends during the previous year. In a survey of 743 U.S. adults age 21 or older, 52 reported that in the previous year they had made the change from living independently to living with friends or relatives (*USA TODAY*, July 29, 2008). Based on these data, what can you learn about the proportion of all adults who moved home or in with friends during the previous year?

Begin by answering the four key questions for this problem.

Q **Question Type**	Estimation or hypothesis testing?	Estimation
S **Study Type**	Sample data or experiment data?	Sample data
T **Type of Data**	One variable or two? Categorical or numerical?	One categorical variable
N **Number of Samples or Treatments**	How many samples or treatments?	One sample

The answers are *estimation, sample data, one categorical variable, one sample*. It is this combination of answers that indicates you should consider a confidence interval for a population proportion. This is the only statistical inference method you have considered so far, but as the number of available methods grows, the answers to the four key questions will help you decide which method to use.

Now you can use the five-step process to learn from the data.

Step	
Estimate: Explain what population characteristic you plan to estimate.	In this example, you will estimate the value of p, **the proportion of U.S. adults age 21 and older who have moved home or in with friends during the past year.**
Method: Select a potential method.	Because the answers to the four key questions are estimation, sample data, one categorical variable, one sample, consider constructing a **confidence interval for a population proportion** (see Table 7.1).
	Associated with every confidence interval is a confidence level. When the method selected is a confidence interval, you will also need to specify a confidence level.
	For now, you only know how to construct an interval corresponding to a 95% confidence level. You will see how to construct intervals for other confidence levels after this example.
Check: Check to make sure that the method selected is appropriate.	There are **two conditions** that need to be met in order to use the confidence interval of this section.
	The sample size is large enough because there are at least 10 successes (52 people who have moved home or in with friends) and at least 10 failures (the other $743 - 52 = 691$ people in the sample).
	The requirement of a random or representative sample is more difficult. No information was provided regarding how the sample was selected. To proceed, you need to assume that the sample was selected in a reasonable way. You will need to keep this in mind when you get to the step that involves interpretation.
Calculate: Use the sample data to perform any necessary calculations.	$n = 743$
	$\hat{p} = \dfrac{52}{743} = 0.070$
	95% confidence interval
	$$\hat{p} \pm 1.96\sqrt{\dfrac{\hat{p}(1 - \hat{p})}{n}}$$
	$$0.07 \pm 1.96\sqrt{\dfrac{(0.070)(1 - 0.070)}{743}}$$
	$0.07 \pm 1.96\sqrt{0.0001}$
	$0.07 \pm 1.96(0.010)$
	0.07 ± 0.020
	$(0.050, 0.090)$

(continued)

Step

| **Communicate Results:** Answer the research question of interest, explain what you have learned from the data, and acknowledge potential risks. | **Confidence interval:** Assuming that the sample was selected in a reasonable way, you can be 95% confident that the actual proportion of U.S. adults age 21 or older who have moved home or in with friends during the previous year is somewhere between 0.050 and 0.090. **Confidence level:** The method used to construct this interval estimate is successful in capturing the actual value of the population proportion about 95% of the time. In this example, no information was given about how the sample was selected. Because of this, it is important to also state that the interpretation is only valid if the sample was selected in a reasonable way. |

Be Careful...

The 95% confidence interval for p calculated in Example 9.4 is (0.050, 0.090). It is tempting to say that there is a "probability" of 0.95 that p is between 0.050 and 0.090, but you can't make a chance (probability) statement concerning the value of p. The 95% refers to the percentage of *all* possible random samples that would result in an interval that includes p. In other words, if you take sample after sample from the population and use each one separately to calculate a 95% confidence interval, in the long run about 95% of these intervals will capture p. Figure 9.2 illustrates this concept for intervals generated from 100 different random samples; 93 of the intervals include p, whereas 7 do not. Any specific interval, and the interval (0.050, 0.090) in particular, either includes p or it does

FIGURE 9.2
One hundred 95% confidence intervals for p calculated from 100 different random samples (asterisks identify intervals that do not include p)

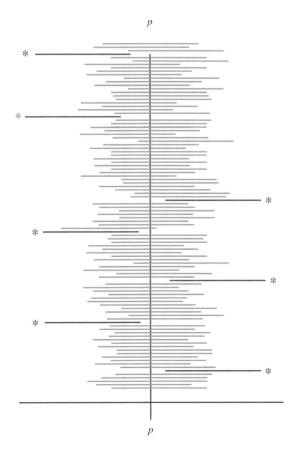

not (remember, the value of p is fixed but not known). *The confidence level of 95% refers to the method used to construct the interval rather than to any particular interval, such as the one you obtained.*

Other Confidence Levels

The formula given for a 95% confidence interval can easily be adapted for other confidence levels. The choice of a 95% confidence level led to the use of the z value 1.96 (chosen to capture a central area of 0.95 under the standard normal curve) in the formula. Any other confidence level can be obtained by using an appropriate z critical value in place of 1.96. For example, suppose that you wanted to achieve a confidence level of 99%. To obtain a central area of 0.99, the appropriate z critical value would have a cumulative area (area to the left) of 0.995, as illustrated in Figure 9.3. From Appendix A Table 2, the corresponding z critical value is 2.58.

FIGURE 9.3
Finding the z critical value for a 99% confidence level

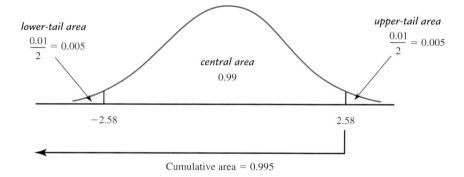

For other confidence levels, the appropriate z critical values are found in a similar way. The desired confidence level specifies a central area under the standard normal curve and then Appendix A Table 2 or technology can be used to determine the corresponding z critical value.

A Large-Sample Confidence Interval for a Population Proportion p

Appropriate when the following conditions are met:

1. The sample is a random sample from the population of interest or the sample is selected in a way that should result in a representative sample from the population.
2. The sample size is large. This condition is met when $n\hat{p} \geq 10$ and $n(1 - \hat{p}) \geq 10$ or (equivalently) the sample includes at least 10 successes and at least 10 failures.

When these conditions are met, a confidence interval for the population proportion is

$$\hat{p} \pm (z \text{ critical value})\sqrt{\frac{\hat{p}(1 - \hat{p})}{n}}$$

The desired confidence level determines which z critical value is used. The three most common confidence levels use the following z critical values:

Confidence Level	z Critical Value
90%	1.645
95%	1.96
99%	2.58

(continued)

Interpretation of Confidence Interval

You can be confident that the actual value of the population proportion is included in the calculated interval. In any given problem, this statement should be worded in context.

Interpretation of Confidence Level

The confidence level specifies the approximate percentage of the time that this method is expected to be successful in capturing the actual population proportion.

A Note on Confidence Level

Why settle for 95% confidence when 99% confidence is possible? Because the higher confidence level comes at a price. The 99% confidence interval is wider than the 95% confidence interval. The width of the 95% interval is $2\left(1.96\sqrt{\dfrac{p(1-p)}{n}}\right)$ whereas the 99% interval has width $2\left(2.58\sqrt{\dfrac{p(1-p)}{n}}\right)$. The wider interval provides less precise information about the value of the population proportion. On the other hand, a 90% confidence interval would be narrower than the 95% confidence interval, but the associated risk of being incorrect is higher. In the opinion of many investigators, a 95% confidence level is a reasonable compromise between confidence and precision, and this is a widely used confidence level.

Example 9.5   Babies on Social Media

The article **"Have a Social Media Account for Your Baby? 40% of Millennial Moms Do" (www .today.com/parents/have-social-media-account-your-baby-40-percent-millennial-moms -1D80224937, October 18, 2014, retrieved May 5, 2017)** reported on a survey conducted by Gerber of 1000 women age 18 to 34 years who had children under the age of 2 years. Of those surveyed, 400 had created a social media account for their babies before their first birthday. Assuming that it is reasonable to regard this sample of 1000 as representative of the population of millennial moms, you can use this information to construct an estimate of all millennial moms who have created a social media account for their babies before their first birthday.

Answers to the four key questions for this problem are:

Q Question Type	Estimation or hypothesis testing?	Estimation
S Study Type	Sample data or experiment data?	Sample data. The data are from a sample of 1000 millennial moms.
T Type of Data	One variable or two? Categorical or numerical?	One categorical variable. The data are responses to the question "Have you created a social media account for your baby before his or her first birthday?" so there is one variable. The variable is categorical with two possible values—yes and no.
N Number of Samples or Treatments	How many samples or treatments?	One sample.

The answers are *estimation, sample data, one categorical variable, one sample*. This combination of answers indicates that you should consider a confidence interval for a population proportion.

You can now use the five-step process for estimation problems to construct a 90% confidence interval.

E Estimate:

The proportion of millennial moms who set up a social media account for her baby before the baby's first birthday, p, will be estimated.

M Method:

Because the answers to the four key questions are estimation, sample data, one categorical variable, one sample, a confidence interval for a population proportion will be considered. A confidence level of 90% was specified for this example.

C Check:

There are two conditions that need to be met for the confidence interval of this section to be appropriate. The large sample condition is easily verified:

$$n\hat{p} = 1000 \left(\frac{400}{1000} \right) = 1000(0.400) = 400 \geq 10$$

$$\text{and} \quad n(1 - \hat{p}) = 1000(0.600) = 600 \geq 10$$

The requirement of a random sample or representative sample is more difficult. You do not know how the sample was selected. In order to proceed, you need to assume that the sample was selected in a reasonable way. You will need to keep this in mind when you get to the step that involves interpretation.

C Calculate:

The appropriate z critical value is found in the box on page 444 or by using Appendix A Table 2 to find the z critical value that captures a central area of 0.90.

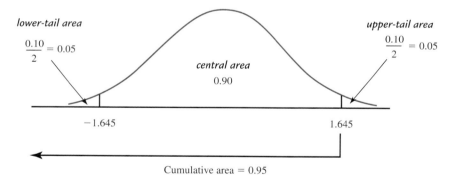

$n = 1000$

$\hat{p} = \dfrac{400}{1000} = 0.400$

z critical value $= 1.645$

90% confidence interval

$$\hat{p} \pm 1.645 \sqrt{\frac{\hat{p}(1 - \hat{p})}{n}}$$

$$0.400 \pm 1.645 \sqrt{\frac{(0.400)(1 - 0.400)}{1000}}$$

$0.400 \pm 1.645(0.015)$

0.400 ± 0.025

$(0.375, 0.425)$

C Communicate Results:

Interpret Confidence interval:
Assuming that the sample was selected in a reasonable way, you can be 90% confident that the actual proportion of millennial moms who have created a social media account for their baby before the baby's first birthday is somewhere between 0.375 and 0.425.

Interpret Confidence level:
The method used to construct this interval estimate is successful in capturing the actual value of the population proportion about 90% of the time.

In this example, you do not know how the sample was selected. Because of this, it is also important to state that the given interpretation is only valid if the sample was selected in a reasonable way.

Notice that the confidence interval in Example 9.5 is relatively narrow—it ranges from 0.375 to 0.425. In general, a narrow interval is desirable because it conveys more information about the actual value of the population proportion than a wider interval. Saying that the value of the population proportion is between 0.375 and 0.425 is more informative than saying that the value of the population proportion is between 0.300 and 0.500.

There are three things that affect the width of the interval. These are the confidence level, the sample size, and the value of the sample proportion.

What You Know	How You Know It
The greater the confidence level, the wider the interval.	The formula for the confidence interval is $\hat{p} \pm (z \text{ critical value}) \sqrt{\dfrac{\hat{p}(1 - \hat{p})}{n}}$. The greater the confidence level, the greater the z critical value will be, making the interval wider.
The greater the sample size, the narrower the interval.	The sample size n appears in the denominator of the standard error of $\hat{p}$. The greater the sample size, the smaller the standard error $\sqrt{\dfrac{\hat{p}(1 - \hat{p})}{n}}$ will be. When the standard error is smaller, the interval will be narrower.
The closer $\hat{p}$ is to 0.5, the wider the interval.	The numerator of the standard error of $\hat{p}$ is $\hat{p}(1 - \hat{p})$. This value is greatest when $\hat{p}$ is 0.5. This means that the interval will be wider for values of $\hat{p}$ close to 0.5. This is the case where the population is very diverse, making it most difficult to accurately estimate the population proportion.

A Few Final Things to Consider

1. What About Small Samples?

The confidence level associated with a z confidence interval for a population proportion is only approximate. When you report a 95% confidence interval for a population proportion, the 95% confidence level implies that you have used a method that produces an interval that includes the actual value of the population proportion about 95% of the time in the long run. However, the normal distribution is only an approximation to the sampling distribution of $\hat{p}$, and so the true confidence level may differ somewhat from the reported value. If $n\hat{p} \geq 10$ and $n(1 - \hat{p}) \geq 10$, the approximation is reasonable and the actual confidence level is usually quite close to the reported level. This is why it is important to check the sample size conditions before calculating and reporting a z confidence interval for a population proportion.

What should you do if these conditions are not met? If the sample size is too small to satisfy the sample size conditions, an alternative procedure can be used. One alternative is a bootstrap confidence interval, which is introduced in Section 9.5.

2. General Form of a Confidence Interval

Many confidence intervals have the same general form as the large-sample z interval for a population proportion. The confidence interval

$$\hat{p} \pm (z \text{ critical value}) \sqrt{\frac{\hat{p}(1-\hat{p})}{n}}$$

is one example of a more general form for confidence intervals, which is

$$(\text{statistic}) \pm (\text{critical value}) \left(\begin{array}{c}\text{standard error}\\ \text{of the statistic}\end{array}\right)$$

For a population characteristic other than p, a statistic for estimating the characteristic is selected. Then the standard error of the statistic and an appropriate critical value are used to construct a confidence interval with this same general structure.

An Alternative to the Large-Sample z Interval

Investigators have shown that in some instances, even when the sample size conditions of the large sample z confidence interval for a population proportion are met, the actual confidence level associated with the method may be noticeably different from the reported confidence level. One way to achieve an actual confidence level that is closer to the reported confidence level is to use a modified sample proportion, $\hat{p}_{\text{mod}}$, the proportion of successes after adding two successes and two failures to the sample. This means that $\hat{p}_{\text{mod}}$ is

$$\hat{p}_{\text{mod}} = \frac{\text{number of successes} + 2}{n + 4}$$

$\hat{p}_{\text{mod}}$ is used in place of $\hat{p}$ in the usual confidence interval formula. Properties of this modified confidence interval are investigated in Activity 9.2 at the end of this chapter.

Summing It Up—Section 9.3

The following learning objectives were addressed in this section:

Conceptual Understanding

C5: Understand how an interval is used to estimate a population characteristic.

A confidence interval provides an estimate of the value of a population characteristic by specifying an interval of plausible values. The width of the interval provides information about the accuracy of the estimate.

C6: Know what factors affect the width of a confidence interval estimate of a population proportion.

The width of a confidence interval for a population proportion is affected by the confidence level, the sample size, and the value of the sample proportion, $\hat{p}$. The greater the confidence level, the wider the interval. The larger the sample size, the narrower the interval. The closer $\hat{p}$ is to 0.5, the wider the interval.

C7: Understand the meaning of the confidence level associated with a confidence interval estimate.

The confidence level associated with a confidence interval is the success rate for the method used to construct the interval. It specifies the percentage of all possible random samples that would result in an interval that includes the actual value of the population characteristic.

Mastering the Mechanics

M2: Know the conditions for appropriate use of the margin of error and confidence interval formulas when estimating a population proportion.

There are two conditions that need to be reasonably met in order for it to be appropriate to use the confidence interval formula given in this section. These conditions are (1) that the sample is a random sample from the population of interest or the sample is selected in a way that makes it reasonable to think that the sample is representative of the population, and (2) the sample size is large. The sample size is considered to be large if $n\hat{p}$ and $n(1 - \hat{p})$ are both at least 10 (or equivalently, if the sample includes at least 10 successes and at least 10 failures).

M3: Know the key characteristics that lead to selection of the z confidence interval for a population proportion as an appropriate method.
The z confidence interval for a population proportion is the method that you would consider if the answers to the four key questions are

> Q (question type): estimation
>
> S (study type): sample data
>
> T (type of data): one categorical variable
>
> N (number of samples or treatments): one sample

Example 9.4 illustrates how these questions are answered in a particular context.

M4: Use the five-step process for estimation problems (EMC³) to calculate and interpret a confidence interval for a population proportion.
The five steps for estimation problems are

> E: Estimate
>
> M: Method
>
> C: Check
>
> C: Calculate
>
> C: Communicate

The five-step process is illustrated in Examples 9.4 and 9.5.

Putting It into Practice
P2: Interpret a confidence interval for a population proportion in context and interpret the associated confidence level.
A confidence interval for a population proportion is interpreted as an interval of plausible values for the population proportion. The confidence level specifies the approximate proportion of the time that the method used to construct the interval is successful in capturing the actual value of the population proportion. For examples of interpreting a confidence interval for a population proportion and the associated confidence level in context, see Examples 9.4 and 9.5.

SECTION 9.3 EXERCISES

Each Exercise Set assesses the following chapter learning objectives: C5, C6, C7, M2, M3, M4, P2

SECTION 9.3 Exercise Set 1

9.34 Suppose that a city planning commission wants to know the proportion of city residents who support installing streetlights in the downtown area. Two different people independently selected random samples of city residents and used their sample data to construct the following confidence intervals for the population proportion:

> Interval 1: (0.28, 0.34)
> Interval 2: (0.31, 0.33)

(Hint: Consider the formula for the confidence interval given on page 444.)
a. Explain how it is possible that the two confidence intervals are not centered in the same place.

b. Which of the two intervals conveys more precise information about the value of the population proportion?
c. If both confidence intervals have a 95% confidence level, which confidence interval was based on the smaller sample size? How can you tell?
d. If both confidence intervals were based on the same sample size, which interval has the higher confidence level? How can you tell?

9.35 For each of the following choices, explain which one would result in a wider large-sample confidence interval for p:
a. 90% confidence level or 95% confidence level
b. $n = 100$ or $n = 400$

9.36 Based on data from a survey of 1200 randomly selected Facebook users **(USA TODAY, March 24, 2010)**, a 95% confidence interval for the proportion of all Facebook users who say it is OK for someone to "friend" his or her boss is (0.41, 0.47). What is the meaning of the confidence level of 95% that is associated with this interval? (Hint: See Example 9.5.)

9.37 Appropriate use of the interval

$$\hat{p} \pm (z \text{ critical value})\sqrt{\frac{\hat{p}(1-\hat{p})}{n}}$$

requires a large sample. For each of the following combinations of n and $\hat{p}$, indicate whether the sample size is large enough for this interval to be appropriate.
a. $n = 50$ and $\hat{p} = 0.30$
b. $n = 50$ and $\hat{p} = 0.05$
c. $n = 15$ and $\hat{p} = 0.45$
d. $n = 100$ and $\hat{p} = 0.01$

9.38 The formula used to calculate a large-sample confidence interval for p is

$$\hat{p} \pm (z \text{ critical value})\sqrt{\frac{\hat{p}(1-\hat{p})}{n}}$$

What is the appropriate z critical value for each of the following confidence levels?
a. 90%
b. 99%
c. 80%

9.39 The article **"Career Expert Provides DOs and DON'Ts for Job Seekers on Social Networking" (CareerBuilder.com, August 19, 2009)** included data from a survey of 2667 hiring managers and human resource professionals. The article noted that more employers are now using social networks to screen job applicants. Of the 2667 people who participated in the survey, 1200 indicated that they use social networking sites such as Facebook, MySpace, and LinkedIn to research job applicants. Assume that the sample is representative of hiring managers and human resource professionals. Answer the four key questions (QSTN) to confirm that the suggested method in this situation is a confidence interval for a population proportion.

9.40 For the study described in the previous exercise, use the five-step process for estimation problems (EMC³) to construct and interpret a 95% confidence interval for the proportion of hiring managers and human resource professionals who use social networking sites to research job applicants. Identify each of the five steps in your solution. (Hint: See Example 9.4.)

9.41 The *Princeton Review 2016 College Hopes and Worries Survey Report* (www.princetonreview.com/cms-content/final_cohowo2016survrpt.pdf, retrieved May 6, 2017) reported that 31% of students applying to college

wanted to attend a college that was within 250 miles of their home and that 51% of parents of students applying to college wanted their child to attend a college that was within 250 miles from home. Suppose that the reported percentages were based on random samples of 8347 students applying to college and of 2087 parents of students applying to college. (Hint: See Example 9.5.)
a. Construct and interpret a 90% confidence interval for the proportion of students applying to college who want to attend a college within 250 miles of their home.
b. Construct and interpret a 90% confidence interval for the proportion of parents of students applying to college who want their child to attend a college within 250 miles of home.
c. Explain why the two 90% confidence intervals are not the same width.

9.42 In a survey of 800 college students in the United States, 576 indicated that they believe that a student or faculty member on campus who uses language considered racist, sexist, homophobic, or offensive should be subject to disciplinary action **("Listening to Dissenting Views Part of Civil Debate," USA TODAY, November 17, 2015).** Assuming that the sample is representative of college students in the United States, construct and interpret a 95% confidence interval for the proportion of college students who have this belief.

9.43 It probably wouldn't surprise you to know that Valentine's Day means big business for florists, jewelry stores, and restaurants. But did you know that it is also a big day for pet stores? In January 2015, the National Retail Federation conducted a survey of consumers in a representative sample of adult Americans **("Survey of Online Shopping for Valentine's Day 2015," nrf.com/news/delivering-customer-delight-valentines-day, retrieved November 14, 2016).** One of the questions in the survey asked if the respondent planned to spend money on a Valentine's Day gift for his or her pet.
a. The proportion who responded that they did plan to purchase a gift for their pet was 0.212. Suppose that the sample size for this survey was $n = 200$. Construct and interpret a 95% confidence interval for the proportion of all adult Americans who planned to purchase a Valentine's Day gift for their pet.
b. The actual sample size for the survey was much larger than 200. Would a 95% confidence interval calculated using the actual sample size have been narrower or wider than the confidence interval calculated in Part (a)?

SECTION 9.3 Exercise Set 2

9.44 Suppose that a campus bookstore manager wants to know the proportion of students at the college who purchase some or all of their textbooks online. Two different people independently selected random samples of students at the college and used their sample data to construct the following confidence intervals for the population proportion:

> Interval 1: (0.54, 0.57)
> Interval 2: (0.46, 0.62)

a. Explain how it is possible that the two confidence intervals are not centered in the same place.

b. Which of the two intervals conveys more precise information about the value of the population proportion?

c. If both confidence intervals have a 95% confidence level, which confidence interval was based on the smaller sample size? How can you tell?

d. If both confidence intervals were based on the same sample size, which interval has the higher confidence level? How can you tell?

9.45 For each of the following choices, explain which would result in a narrower large-sample confidence interval for *p:*

a. 95% confidence level or 99% confidence level

b. $n = 200$ or $n = 500$

9.46 Based on data from a survey of 1200 randomly selected Facebook users (*USA TODAY*, March 24, 2010), a 98% confidence interval for the proportion of all Facebook users who say it is OK to ignore a coworker's "friend" request is (0.35, 0.41). What is the meaning of the confidence level of 98% that is associated with this interval?

9.47 Appropriate use of the interval

$$\hat{p} \pm (z \text{ critical value})\sqrt{\frac{\hat{p}(1 - \hat{p})}{n}}$$

requires a large sample. For each of the following combinations of n and $\hat{p}$, indicate whether the sample size is large enough for this interval to be appropriate.

a. $n = 100$ and $\hat{p} = 0.70$

b. $n = 40$ and $\hat{p} = 0.25$

c. $n = 60$ and $\hat{p} = 0.25$

d. $n = 80$ and $\hat{p} = 0.10$

9.48 The formula used to calculate a large-sample confidence interval for *p* is

$$\hat{p} \pm (z \text{ critical value})\sqrt{\frac{\hat{p}(1 - \hat{p})}{n}}$$

What is the appropriate *z* critical value for each of the following confidence levels?

a. 95%

b. 98%

c. 85%

9.49 The *USA Snapshot* titled **"Baby's First Photo Reveal"** (*USA TODAY*, October 17, 2014) summarized data from a survey of 1001 mothers with children under the age of 2. The Snapshot includes the following statement: "83% of moms post new baby photos from the delivery room." This

information could be used to provide an estimate of the proportion of new mothers who post pictures on social media from the delivery room. Assume that the sample is representative of the population of mothers with children under the age of 2 years. Answer the four key questions (QSTN) to confirm that the suggested method in this situation is a confidence interval for a population proportion.

9.50 For the situation described in the previous exercise, use the five-step process for estimation problems (EMC[3]) to construct and interpret a 99% confidence interval for the proportion of mothers of children under the age of 2 years who post pictures of their new baby on social media from the delivery room. Identify each of the five steps in your solution.

9.51 The *USA Snapshot* titled **"Social Media Jeopardizing Your Job?"** (*USA TODAY*, November 12, 2014) summarized data from a survey of 1855 recruiters and human resource professionals. The Snapshot indicted that 53% of the people surveyed had reconsidered a job candidate based on his or her social media profile. Assume that the sample is representative of the population of recruiters and human resource professionals in the United States.

a. Use the given information to estimate the proportion of recruiters and human resource professionals who have reconsidered a job candidate based on his or her social media profile using a 95% confidence interval. Give an interpretation of the interval in context and an interpretation of the confidence level of 95%.

b. Would a 90% confidence interval be wider or narrower than the 95% confidence interval from Part (a)?

9.52 The report **"Job Seeker Nation Study 2016"** (www .jobvite.com/wp-content/uploads/2016/03/Jobvite _Jobseeker_Nation_2016.pdf, retrieved May 6, 2017) summarized a survey of 2305 working adults. The report indicates that 484 of the working adults surveyed said they were very concerned that their job will be automated, outsourced, or otherwise made obsolete in the next five years. The sample was selected in a way designed to produce a representative sample of working adults. Construct and interpret a 95% confidence interval for the proportion of working adults who are very concerned that their job will be automated, outsourced, or otherwise made obsolete in the next five years.

9.53 The article **"Most Dog Owners Take More Pictures of Their Pet Than Their Spouse"** (August 22, 2016, news .fastcompany.com/most-dog-owners-take-more-pictures-of-their-pet-than-their-spouse-4017458, retrieved May 6, 2017) indicates that in a sample of 1000 dog owners, 650 said that they take more pictures of their dog than their significant others or friends, and 460 said that they are more likely to complain to their dog than to a friend. Suppose that it is reasonable to consider this sample as representative of the population of dog owners.

a. Construct and interpret a 90% confidence interval for the proportion of dog owners who take more pictures of their dog than of their significant others or friends.

b. Construct and interpret a 95% confidence interval for the proportion of dog owners who are more likely to complain to their dog than to a friend.

c. Give two reasons why the confidence interval in Part (b) is wider than the interval in Part (a).

ADDITIONAL EXERCISES

9.54 In 2010, the National Football League adopted new rules designed to limit head injuries. In a survey conducted in 2015 by the Harris Poll, 1216 of 2096 adults indicated that they were football fans and followed professional football. Of these football fans, 692 said they thought that the new rules were effective in limiting head injuries **(December 21, 2015, www.theharrispoll.com/sports/Football-Injuries.html, retrieved May 6, 2017)**.

a. Assuming that the sample is representative of adults in the United States, construct and interpret a 95% confidence interval for the proportion of U.S. adults who consider themselves to be football fans.

b. Construct and interpret a 95% confidence interval for the proportion *of football fans* who think that the new rules have been effective in limiting head injuries.

c. Explain why the confidence intervals in Parts (a) and (b) are not the same width even though they both have a confidence level of 95%.

9.55 One thousand randomly selected adult Americans participated in a survey conducted by the **Associated Press (June 2006)**. When asked "Do you think it is sometimes justified to lie, or do you think lying is never justified?" 52% responded that lying was never justified. When asked about lying to avoid hurting someone's feelings, 650 responded that this was often or sometimes OK.

a. Construct and interpret a 90% confidence interval for the proportion of adult Americans who would say that lying is never justified.

b. Construct and interpret a 90% confidence interval for the proportion of adult Americans who think that it is often or sometimes OK to lie to avoid hurting someone's feelings.

c. Using the confidence intervals from Parts (a) and (b), comment on the apparent inconsistency in the responses.

9.56 The article **"Write It by Hand to Make It Stick"** (*Advertising Age*, **July 27, 2016**) summarizes data from a survey of 1001 students age 13 to 19 years. Of the students surveyed, 851 reported that they learn best using a mix of digital and nondigital tools. Construct and interpret a 95% confidence interval for the proportion of students age 13 to 19 who would say that they learn best using a mix of digital and nondigital tools. In order for the method used

to construct the interval to be valid, what assumption about the sample must be reasonable?

9.57 The article referenced in the previous exercise also indicated that 811 of the 1001 students surveyed said that they would feel restricted if they could only work on digital devices. Would a 95% confidence interval for the proportion of students age 13 to 19 years who would say that they would feel restricted if they could only use digital devices be narrower or wider than the interval constructed in the previous exercise for the proportion who would say that they learn best using a mix of digital and nondigital tools? Explain your reasoning—you should be able to answer this question without constructing the second confidence interval.

9.58 The report **"The 2016 Consumer Financial Literacy Survey" (The National Foundation for Credit Counseling, www.nfcc.org, retrieved October 28, 2016)** summarized data from a representative sample of 1668 adult Americans. Based on data from this sample, it was reported that over half of U.S. adults would give themselves a grade of A or B on their knowledge of personal finance. This statement was based on observing that 934 people in the sample would have given themselves a grade of A or B.

a. Construct and interpret a 95% confidence interval for the proportion of all adult Americans who would give themselves a grade of A or B on their financial knowledge of personal finance.

b. Is the confidence interval from Part (a) consistent with the statement that a majority of adult Americans would give themselves a grade of A or B? Explain why or why not.

9.59 The report **"The Politics of Climate" (Pew Research Center, October 4, 2016, www.pewinternet.org/2016/10/04 /the-politics-of-climate, retrieved May 6, 2017)** summarized data from a survey on public opinion of renewable and other energy sources. It was reported that 52% of the people in a sample from western states said that they have considered installing solar panels on their homes. This percentage was based on a representative sample of 369 homeowners in the western United States. Use the given information to construct and interpret a 90% confidence interval for the proportion of all homeowners in western states who have considered installing solar panels.

9.60 The report referenced in the previous exercise also indicated that 33% of those in a representative sample of 533 homeowners in southern states said that they had considered installing solar panels.

a. Use the given information to construct and interpret a 90% confidence interval for the proportion of all homeowners in the southern states who have considered installing solar panels.

b. Give two reasons why the confidence interval in Part (a) is narrower than the confidence interval calculated in the previous exercise.

SECTION 9.4 **Choosing a Sample Size to Achieve a Desired Margin of Error**

In the previous two sections, you saw that sample size plays a role in determining both the margin of error and the width of the confidence interval. Before collecting any data, you might want to determine a sample size that ensures a particular value for the margin of error. For example, with p representing the actual proportion of students at a college who purchase textbooks online, you may want to estimate p to within 0.03. The required sample size n is found by setting the expression for margin of error equal to 0.03 to obtain

$$1.96\sqrt{\frac{p(1-p)}{n}} = 0.03$$

and then solving for n.

In general, if M is the desired margin of error, finding the necessary sample size requires solving the equation

$$M = 1.96\sqrt{\frac{p(1-p)}{n}}$$

Solving this equation for n results in

$$n = p(1-p)\left(\frac{1.96}{M}\right)^2$$

To use this formula to determine a sample size, you need the value of p, which is unknown. One solution is to carry out a preliminary study and use the resulting data to get a rough estimate of p. In other cases, prior knowledge may suggest a reasonable estimate of p. If there is no prior information about the value of p and a preliminary study is not feasible, a more conservative solution makes use of the fact that the maximum value of $p(1-p)$ is 0.25 (its value when $p = 0.5$). Replacing $p(1-p)$ with 0.25 yields

$$n = 0.25\left(\frac{1.96}{M}\right)^2$$

Using this formula to determine n results in a sample size for which $\hat{p}$ will be within M of the population proportion at least 95% of the time for any value of p.

Using a 95% confidence level, the sample size required to estimate a population proportion p with a margin of error M is

$$n = p(1-p)\left(\frac{1.96}{M}\right)^2$$

The value of p may be estimated using prior information. If prior information is not available, using $p = 0.5$ in this formula results in a conservatively large value for the required sample size.

Example 9.6 **Sniffing Out Cancer**

Dogs have a sense of smell that is much more powerful than humans. Because of this, dogs can be trained to identify the presence of odors unique to specific types of cancer. The article **"Meet the Dogs Who Can Sniff Out Cancer Better Than Some Lab Tests" (February 3, 2016, www .cnn.com/2015/11/20/health/cancer-smelling-dogs, retrieved May 6, 2017)** describes a large-scale study that is being planned by researchers in England to investigate whether dogs really can identify the presence of cancer by sniffing urine samples. Suppose you want to collect data that would allow you to estimate the long-run proportion of accurate identifications for a particular dog that has completed training. The dog has been trained to lie down when presented with a urine specimen from a cancer patient and to remain standing when presented with a specimen from a person who does not have cancer. How many different urine specimens should be used if you want to estimate the long-run proportion of correct identifications for this dog with a margin of error of 0.10?

Using the conservative value of $p = 0.5$ in the formula for required sample size gives

$$n = p(1 - p)\left(\frac{1.96}{M}\right)^2 = 0.25\left(\frac{1.96}{0.10}\right)^2 = 96.04$$

A sample of at least 97 urine specimens should be used. Notice that, in sample size calculations, you always round up.

Summing It Up—Section 9.4

The following learning objectives were addressed in this section:

Mastering the Mechanics
M5: Calculate the sample size necessary to achieve a desired margin of error when estimating a population proportion.
The formula

$$n = p(1 - p)\left(\frac{1.96}{M}\right)^2$$

can be used to calculate the sample size required to estimate a population proportion with a given margin of error, M. The value of p may be estimated using prior information, or if prior information is not available, $p = 0.5$ may be used to obtain a conservatively large value for the required sample size.

Putting It into Practice
P3: Determine the required sample size, given a description of a proposed study and a desired margin of error.
Example 9.6 illustrates the calculation of a required sample size in a particular context.

SECTION 9.4 EXERCISES

Each Exercise Set assesses the following chapter learning objectives: M5, P3

SECTION 9.4 Exercise Set 1

9.61 A discussion of digital ethics appears in the article **"Academic Cheating, Aided by Cell Phones or Web, Shown to be Common" (Los Angeles Times, June 17, 2009)**. One question posed in the article is: What proportion of college students have used cell phones to cheat on an exam? Suppose you have been asked to estimate this proportion for students enrolled at a large college. How many students should you include in your sample if you want to estimate this proportion with a margin of error of 0.02? (Hint: See Example 9.6.)

9.62 In 2010, the online security firm Symantec estimated that 63% of computer users don't change their passwords very often **(www.cnet.com/news/survey-63-dont -change-passwords-very-often, retrieved November 19, 2016)**. Because this estimate may be outdated, suppose that you want to carry out a new survey to estimate the proportion of students at your school who do not change their password. You would like to determine the sample size required to estimate this proportion with a margin of error of 0.05.
a. Using 0.63 as a preliminary estimate, what is the required sample size if you want to estimate this proportion with a margin of error of 0.05?

b. How does the the sample size in part (a) compare to the sample size that would result from using the conservative value of 0.5?
c. What sample size would you recommend? Justify your answer.

9.63 A manufacturer of small appliances purchases plastic handles for coffeepots from an outside vendor. If a handle is cracked, it is considered defective and can't be used. A large shipment of plastic handles is received. How many handles from the shipment should be inspected in order to estimate p, the proportion of defective handles in the shipment, with a margin of error of 0.1?

SECTION 9.4 Exercise Set 2

9.64 The article **"Should Canada Allow Direct-to-Consumer Advertising of Prescription Drugs?" (Canadian Family Physician [2009]: 130–131)** calls for the legalization of advertising of prescription drugs in Canada. Suppose you wanted to conduct a survey to estimate the proportion of Canadians who would allow this type of advertising. How large a random sample would be required to estimate this proportion with a margin of error of 0.02?

9.65 The 1991 publication of the book *Final Exit*, which includes chapters on doctor-assisted suicide, caused a great deal of controversy in the medical community. The Society for the Right to Die and the American Medical Association quoted very different figures regarding the proportion of primary-care physicians who have participated in some form of doctor-assisted suicide for terminally ill patients (*USA TODAY*, July 1991). Suppose that a survey of physicians will be carried out to estimate this proportion with a margin of error of 0.05. How many primary-care physicians should be included in a random sample?

ADDITIONAL EXERCISES

9.66 In spite of the potential safety hazards, some people would like to have an Internet connection in their car. A preliminary survey of adult Americans has estimated the proportion of adult Americans who would like Internet access in their car to be somewhere around 0.30 (*USA TODAY*, May 1, 2009). Use the given preliminary estimate to determine the sample size required to estimate this proportion with a margin of error of 0.02.

9.67 Data from a representative sample were used to estimate that 32% of all computer users in 2011 had tried to get on a Wi-Fi network that was not their own in order to save money (*USA TODAY*, May 16, 2011). You decide to conduct a survey to estimate this proportion for the current year. What is the required sample size if you want to estimate this proportion with a margin of error of 0.05? Calculate the required sample size first using 0.32 as a preliminary estimate of p and then using the conservative value of 0.5. How do the two sample sizes compare? What sample size would you recommend for this study?

9.68 *USA TODAY* (January 24, 2012) reported that ownership of tablet computers and e-readers is soaring. Suppose you want to estimate the proportion of students at your college who own at least one tablet or e-reader. What sample size would you use in order to estimate this proportion with a margin of error of 0.03?

SECTION 9.5

Bootstrap Confidence Intervals for a Population Proportion (Optional)

In Section 9.3, you learned how to construct a confidence interval for a population proportion, p, using data from a random sample. That interval has the form $\hat{p}$ (the sample proportion) plus and minus a margin of error. The margin of error used in that confidence interval is calculated by multiplying the standard error of $\hat{p}$ by a z critical value. To use this approach, you need to know that the sample is a random sample from a population (or is selected in a way that makes it reasonable to think that the sample is representative of the population). In addition, in order for the margin of error used in that confidence interval to provide a reasonable description of sample-to-sample variability in $\hat{p}$, you need to know that the sampling distribution of $\hat{p}$ is approximately normal. This is reasonable when the sample size is large ($n\hat{p} \geq 10$ and $n(1 - \hat{p}) \geq 10$) but isn't necessarily the case when the sample size is small. In this section, you will see an alternative method that doesn't require a large sample size but that can still be used to obtain a confidence interval for a population proportion.

A Bootstrap Confidence Interval for a Population Proportion

When you use the sample proportion $\hat{p}$ to estimate a population proportion, you know that the value of $\hat{p}$ is not likely to be exactly equal to the value of the population proportion. But if the sample is selected in a reasonable way, you expect that the value of $\hat{p}$ will be somewhere around the value of the population proportion. A confidence interval quantifies what is meant by "around the value of the population proportion." When the assumptions of the large-sample confidence interval are reasonable, the margin of error is based on knowing that the sampling distribution of $\hat{p}$ is approximately normal. The margin of error tells you the largest value that the difference between the observed sample proportion and the actual value of the population proportion is likely to be for a given confidence level. This margin of error can then be used to construct a confidence interval.

But what can you do if you don't know that the sampling distribution of $\hat{p}$ is approximately normal? To construct a confidence interval, you would still need to know something about how far away you think your sample proportion is likely to be from the value of the population proportion. For example, suppose that your sample proportion is

$\hat{p} = 0.32$ and that you knew that it was unlikely to be smaller than the population proportion by more than 0.04. Then you could say that you think that the population proportion is less than 0.36 (from $0.32 + 0.04$). If you also knew that it was unlikely that your sample proportion was greater than the population proportion by more than 0.05, you could say that you think that the population proportion is greater than 0.27 (from $0.32 - 0.05$). A reasonable interval estimate of the population proportion based on your sample would then be (0.27, 0.36). **Bootstrapping** is a way to determine what number you should add and what number you should subtract from the sample proportion in order to form a confidence interval.

To understand how bootstrapping works, think about how you could figure out how far sample proportions from random samples of size n tend to be from the value of the population proportion if you happened to know the value of the population proportion. You could take many different random samples of size n from such a population and calculate the sample proportion for each one. Looking at the distribution of these sample proportions and thinking about how they cluster around the value of the population proportion would tell you what you need to know. But of course, in practice this won't work because you don't know the value of the population proportion (if you did, then you wouldn't need to estimate it!).

What you can do instead is to think of a hypothetical population that you expect to be very similar to the population that your sample is actually from. You can see what happens with sample proportions from this hypothetical population and use that information to tell you about variability in $\hat{p}$ values. To do this, bootstrapping uses the observed sample proportion as the proportion for the hypothetical population.

To create a bootstrap confidence interval, many random samples of size n are taken from this hypothetical population to form a **bootstrap distribution**. The variability in this distribution indicates how far $\hat{p}$ values for samples from this hypothetical population might be from the original observed value of $\hat{p}$. Knowing how far these simulated $\hat{p}$ values tend to fall from the observed value of $\hat{p}$ gives you an idea of how far the observed value of $\hat{p}$ is likely to be from the actual value of p in the population.

For a 95% confidence level, using the boundaries that capture the middle 95% of the simulated bootstrap distribution is equivalent to determining the endpoints of a confidence interval, which is represented as a $\hat{p}$ minus a number and $\hat{p}$ plus a number. For bootstrap confidence intervals, the number subtracted from $\hat{p}$ and the number added to $\hat{p}$ won't always be the same since the bootstrap distribution may not be symmetric. This interval is called a **bootstrap confidence interval** estimate for p, and it is based on simulation rather than on knowing that the sampling distribution of $\hat{p}$ is approximately normal.

Taking a random sample from a hypothetical population that has a proportion of successes equal to the original sample proportion is equivalent to selecting a random sample with replacement from your original sample. The process of drawing new samples from an original representative sample is called **resampling**, because you are taking new samples directly from the original sample you collected. This method is also called "bootstrapping," because it is like "pulling yourself up by the bootstraps" in the sense that you are using nothing more than the original data you collected in one sample to generate information about sample-to-sample variability in the sample proportion, $\hat{p}$.

Example 9.7 | **Generating a Bootstrap Distribution**

Suppose that you take a random sample of 10 students at your school and ask each one if they spend more than four hours a day online. You would like to use the sample to estimate the proportion of students at your school who would respond "yes" to this question. If the responses for the people in your sample were those shown in the table below, the value of the sample proportion of "yes" responses would be $\hat{p} = 5/10 = 0.5$.

Student Number	1	2	3	4	5	6	7	8	9	10
Response	No	Yes	Yes	No	No	Yes	No	No	Yes	Yes

To create a bootstrap distribution, you think of a hypothetical population in which the proportion of successes is 0.5, and then take samples of size 10 from this hypothetical population. One way to do this is to select a random sample of size 10 with replacement from the original sample using random numbers. Ten random numbers, selected with replacement from the numbers from 1 to 10, are shown in the table below. Also shown in the table are the responses from the original sample associated with each of these random numbers.

Random Number	10	9	3	8	3	2	8	3	6	9
Response	Yes	Yes	Yes	No	Yes	Yes	No	Yes	Yes	Yes

The proportion of "yes" responses in this simulated sample is $\hat{p} = 8/10 = 0.8$, even though the proportion in the original sample was 0.5.

To form a bootstrap distribution, you resample from the original sample many times. These simulated samples provide information about the sample-to-sample variability in the sample proportions, $\hat{p}$, and the bootstrap distribution can then be used to determine an interval of plausible values for p associated with a specified confidence level.

Bootstrap Confidence Intervals for One Proportion

This section explains how to generate a bootstrap distribution and construct a bootstrap confidence interval for a population proportion using one of the Shiny web apps that accompany this text. You can find these web apps at statistics.cengage.com/Peck2e/Apps .html. There are also a number of other resources for constructing bootstrap confidence intervals (for example, see The StatKey apps at www.lock5stat.com/StatKey/).

Example 9.8 | **Generating a Bootstrap Confidence Interval for a Population Proportion (Example 9.7 continued)**

Recall that in the previous example, a random sample of 10 students produced a sample proportion who spend more than four hours a day online of $\hat{p} = 0.50$. To use the Shiny app to construct a bootstrap confidence interval for the proportion of students at the school who spend more than four hours a day online, follow the instructions below.

Enter the number of observations and the number of successes into the Shiny app titled "Bootstrap Confidence Interval for One Proportion." In this example, the number of observations, also known as the sample size, is $n = 10$, and five successes were observed in the original sample. The confidence level in this example will be 95%, so you should enter this value as well, or recognize that 95% is the default.

Bootstrap Confidence Interval for One Proportion

Select number of observations

10

Select the number of successes

5

Select confidence level (in %):

95

Select the number of simulated samples to generate:

● 1   ○ 10   ○ 100   ○ 1,000   ○ 10,000

Generate Simulated Sample(s)   Reset

Select the number of simulated bootstrap samples you want to generate. In this example, choose 1000 simulated samples. Then, click "Generate Simulated Sample(s)."

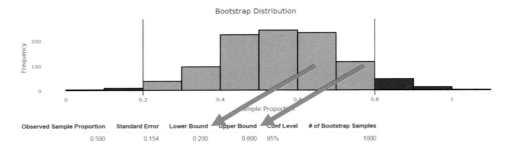

Bootstrap Confidence Interval for One Proportion

Select number of observations

10

Select the number of successes

5

Select confidence level (in %):

95

Select the number of simulated samples to generate:

○ 1 ○ 10 ○ 100 ◉ 1,000 ○ 10,000

[Generate Simulated Sample(s)] [Reset]

The Shiny app identifies the "Lower Bound" to be the value of the sample proportion that has 2.5% of the simulated proportions below and the "Upper Bound" to be the value that has 2.5% of the simulated proportions above:

Bootstrap Distribution

Observed Sample Proportion	Standard Error	Lower Bound	Upper Bound	Conf Level	# of Bootstrap Samples
0.500	0.154	0.200	0.800	95%	1000

Different sets of simulated samples may produce slightly different results. Based on the output from the Shiny app for the simulation shown here, 2.5% of the simulated sample proportions fall at or below 0.2, and 2.5% of the simulated sample proportions fall at or above 0.8. A 95% bootstrap confidence interval for the proportion of students at the school who would respond that they spend more than four hours a day online is (0.2, 0.8). You can be 95% confident that the actual proportion of "yes" responses falls between 0.2 and 0.8. Notice that this interval is very wide, with plausible values for the population proportion ranging from 0.2 all the way to 0.8. This is a function of the very small sample size. It is difficult to estimate population proportions accurately with a small sample, no matter what method you use!

Because a bootstrap confidence interval is based on a distribution of simulated proportions, the interval that is generated may vary from one simulation to another. But if the confidence interval is based on a large number of simulated proportions, different bootstrap confidence intervals based on the same sample won't differ substantially from one another.

Now let's take a look at a more realistic example.

Example 9.9 **Moving Home Revisited**

Recall that in Example 9.4 you used the large-sample confidence interval formula to find a 95% confidence interval for p, the population proportion of U.S. adults 21 and older who have moved home or in with friends during the past year. The observed value of the sample proportion was $\hat{p} = 52/743 = 0.07$, and the resulting confidence interval was (0.050, 0.090).

You can use the bootstrap method as an alternative way to find a 95% confidence interval for p.

In the Shiny app "Bootstrap Confidence Interval for One Proportion," enter the sample size and the number of successes. In Example 9.4 the sample size is $n = 743$, and 52 successes were observed in the original sample. Select a 95% confidence level and then choose 1000 simulated samples. Click on "Generate Simulated Sample(s)."

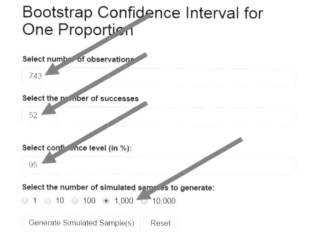

For the simulation shown here, the Shiny app identifies the value that has 2.5% of the simulated proportions below and the value that has 2.5% of the simulated proportions above:

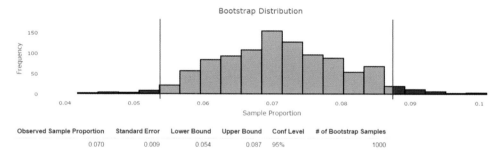

Observed Sample Proportion	Standard Error	Lower Bound	Upper Bound	Conf Level	# of Bootstrap Samples
0.070	0.009	0.054	0.087	95%	1000

For this simulation, 2.5% of the simulated sample proportions fall at or below 0.054, and that 2.5% of the simulated sample proportions fall at or above 0.087. A bootstrap confidence interval for the actual proportion of U.S. adults 21 and older who have moved home or in with friends during the past year is (0.054, 0.087). This is a set of plausible values for the actual proportion of U.S. adults 21 and older who have moved home or in with friends during the past year. Assuming that the sample is representative of the population, you can be 95% confident that the actual proportion falls between 0.054 and 0.087 (between 5.4% and 8.7%).

Notice that this bootstrap confidence interval is slightly narrower than the interval provided by the large-sample method in Section 9.3, which was (0.050, 0.090). The conditions for the large-sample confidence interval are satisfied, and the large-sample confidence interval is very close to the bootstrap interval.

What Happens When the Conditions for the Large-Sample Interval Are Not Met?

In the previous example, the bootstrap confidence interval turned out to be slightly narrower than the large-sample confidence interval. It is not surprising here that the two intervals are similar, because the sample size was large enough to justify the use of the large-sample interval and so both methods provide appropriate ways to obtain a confidence interval. But the bootstrap method can still be used even if the sample size is not large enough to satisfy the conditions for the large-sample confidence interval. This is illustrated in the following example.

Example 9.10   Liver Injuries in Newborns

The article **"Severe Liver Injury While Using Umbilical Venous Catheter: Case Series and Literature Review" (*American Journal of Perinatology* [2014]: 965–974)** describes a study of newborns who were placed in intensive care and required insertion of an umbilical vein catheter so that fluids could be administered. Researchers found that 9 out of the 1081 newborns studied developed catheter-associated liver injury. The authors were interested in estimating the proportion of newborns who suffer liver injury as a result of the use of umbilical vein catheters.

The researchers considered this sample of 1081 infants to be representative of the population of newborns who required use of the catheter. The sample proportion with liver injury is $\hat{p} = \dfrac{9}{1081} = 0.00833$. Notice that although the sample size is 1081, because $n\hat{p} = (1081)(0.00833) = 9$, the sample size is not large enough to justify the use of the large-sample confidence interval for a population proportion.

The Shiny app "Bootstrap Confidence Interval for One Proportion" was used to generate a bootstrap distribution based on 1000 simulated sample proportions.

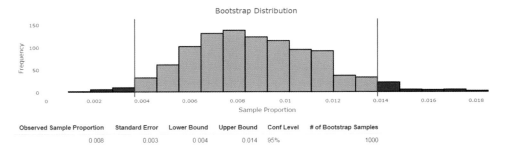

Observed Sample Proportion	Standard Error	Lower Bound	Upper Bound	Conf Level	# of Bootstrap Samples
0.008	0.003	0.004	0.014	95%	1000

Based on this output, a 95% bootstrap confidence interval for the population proportion, p, is (0.004, 0.014). Based on this sample, you can be 95% confident that the actual proportion of newborns with umbilical catheters who suffer liver injury is somewhere between 0.004 and 0.014.

Although the conditions for the large-sample interval are not met, if you were to compute a large-sample 95% confidence interval, you would get (0.003, 0.014). Because the conditions for the large-sample interval are not met, the bootstrap interval is a more appropriate choice.

Summing It Up—Section 9.5

The following learning objectives were addressed in this section:

Mastering the Mechanics
M6: Calculate and interpret a bootstrap confidence interval for a population proportion. A bootstrap confidence interval is an alternative method for calculating a confidence interval for a population proportion. This method can be used even in situations when the conditions necessary for the large-sample z confidence interval are not met. A bootstrap confidence interval is interpreted in the same way as the large-sample z confidence interval. Examples 9.9 and 9.10 illustrate the calculation and interpretation of a bootstrap confidence interval for a population proportion.

SECTION 9.5 EXERCISES

Each Exercise Set assesses the following learning objectives: M6

SECTION 9.5 Exercise Set 1

9.69 A survey on SodaHead (**www.sodahead.com/survey /featured/anonymous-advice/?results51, retrieved May 13, 2016**) reported that 603 out of 753 respondents replied "no" to the question "Should you be friends with your boss on Facebook?"

a. Use the accompanying output from the "Bootstrap Confidence Interval for One Proportion" Shiny app to report a 95% bootstrap confidence interval for the population proportion who would reply "no" to the question. Interpret the confidence interval in context.

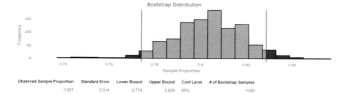

b. SodaHead provides summaries for anonymous and voluntary responses to survey questions. Do you believe that the proportion of respondents who reply "no" to the question in an anonymous and voluntary situation would tend to underestimate or overestimate the actual population proportion of interest? Explain your reasoning.

9.70 The article **"Report: More Than Half of DC-Area Millennials Are Using Ride-Hailing Apps" (June 23, 2016, www .washingtonian.com/2016/06/23/report-half-dc-area -millennials-using-ride-hailing-apps/, retrieved May 4, 2017)** refers to a study summarized at the following site: www.wbaresearch.com/wp-content/uploads/2016/06 /Transportation-MarkeTrak-Spring-20161.pdf (retrieved May 4, 2017). The study indicates that 21% of Washington-area adults who are 55 years old and older have used transportation apps such as Uber or Lyft at least once.

Suppose that a small local transportation service for older residents is monitoring usage of app-based transportation. The service conducted a survey of a random sample of 21 of its regular customers who are 55 or older and found that 3 of them had tried Uber or Lyft at least once.

a. Would it be appropriate to use the large-sample confidence interval for a population proportion to estimate the proportion of the transportation services customers who have tried Uber or Lyft at least once? Explain.

b. Would it be appropriate to use a bootstrap confidence interval for a population proportion to estimate the proportion of the transportation services customers who have tried Uber or Lyft at least once? Explain.

c. Use the accompanying output from the Shiny app to report a 95% bootstrap confidence interval for the population proportion of customers 55 or older who have used Uber or Lyft at least once. Interpret the confidence interval in context.

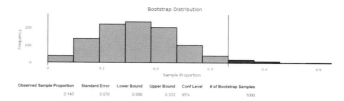

d. Is the value obtained in the study for Washington-area adults who are over 55 years old who have used Uber or Lyft at least once (21%) in the bootstrap confidence interval? What does this tell you?

9.71 An article titled **"The Latest on Workplace Monitoring and Surveillance" (American Management Association, November 17, 2014)** referred to the "2007 Electronic Monitoring & Surveillance Survey." In a summary of survey results submitted by 304 U.S. businesses, 85 of these businesses had fired workers for e-mail misuse.

Suppose that it is reasonable to regard these 304 businesses as a representative sample of businesses in the United States. Use the "Bootstrap Confidence Interval for One Proportion" Shiny app to generate a bootstrap confidence interval for the proportion of U.S. businesses who have fired workers for e-mail misuse. Interpret the interval in context.

9.72 During the 2016 NBA Finals, Kevin Love of the Cleveland Cavaliers successfully made 5 three-point shots out of 19 attempts. Assume that these attempts comprise a sample that is representative of his ability during the entire 2016 season.

a. Explain why it would not be appropriate to use a large-sample confidence interval for one proportion to estimate Kevin Love's success rate for three-point shots during the 2016 season.

b. Use the "Bootstrap Confidence Interval for One Proportion" Shiny app to generate a 90% bootstrap confidence interval for Kevin Love's three-point shot success rate during the 2016 NBA season. Interpret the interval in context.

SECTION 9.5 Exercise Set 2

9.73 A survey of a representative sample of 478 U.S. employers found that 359 ranked stress as their top health and productivity concern (**June 29, 2016, www.globenewswire .com/news-release/2016/06/29/852338/0/en/Seventy-five -percent-of-U-S-employers-say-stress-is-their-number-one -workplace-health-concern.html?print=1, retrieved May 4, 2017**).

a. Use the accompanying output from the "Bootstrap Confidence Interval for One Proportion" Shiny app to report a 95% bootstrap confidence interval for the proportion of all U.S. employers who would rank stress at their top health and productivity concern. Interpret the confidence interval in context.

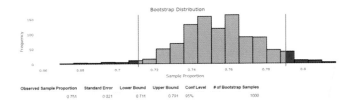

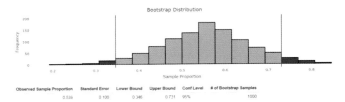

sold at auction in the post-Brexit UK. Interpret the confidence interval in context.

b. A number of international employers were also surveyed. If the international employers had a similar rate of identifying stress as their top health and productivity concern, and if the results from international employers were included in the sample, would the width of the resulting confidence interval remain the same, decrease, or increase? Explain your reasoning.

9.74 In mid-2016 the United Kingdom (UK) withdrew from the European Union (an event known as "Brexit"), causing economic concerns throughout the world. One indicator that economists use to monitor the health of the economy is the proportion of residential properties offered for sale at auction that are successfully sold.

An article titled **"Going, going, gone through the roof–sky's the limit at auction" (October 22, 2016, www.estateagenttoday .co.uk/features/2016/10/going-going-gone-through-the-roof—the-skys-the-limit-at-auction, retrieved May 4, 2017)** reported the success rate of a sample of 26 residential properties offered for sale at auctions in the UK in the summer of 2016. For this sample of properties, 14 of the 26 residential properties were successfully sold. Suppose it is reasonable to consider these 26 properties as representative of residential properties offered at auction in the post-Brexit UK.

a. Would it be appropriate to use the large-sample confidence interval for a population proportion to estimate the proportion of residential properties successfully sold at auction in the post-Brexit UK? Explain.

b. Would it be appropriate to use a bootstrap confidence interval for a population proportion to estimate the proportion of residential properties successfully sold at auction in the post-Brexit UK? Explain.

c. Use the accompanying output from the "Bootstrap Confidence Interval for One Proportion" Shiny app to report a 95% bootstrap confidence interval for the population proportion of residential properties successfully

d. The success rate for properties sold at auction throughout the UK during one stretch a year earlier—in 2015—was 72%. Does this value fall within the bootstrap confidence interval you reported in Part (c)? What does this tell you?

9.75 The report titled **"One in Three American Households Are Stuck in a Relationship with a Financial Services Provider They Don't Trust" (June 29, 2016, www.businesswire.com/news /home/20160629005198/en/American-Households-Stuck -Relationship-Financial-Services-Provider, retrieved May 4, 2017)** estimated that 31% of American households feel obliged to do business with one or more financial services companies they distrust. This estimate is based on a representative sample of 1056 consumers age 18 and older.

Use the "Bootstrap Confidence Interval for One Proportion" Shiny app to generate a 95% bootstrap confidence interval for the proportion of all U.S. households that feel obliged to do business with one or more financial services companies they distrust. Interpret the interval in context.

9.76 A 2016 study of 120 U.S. brand-name products found that 70% were active on Snapchat **(June 15, 2016, www.businessinsider.com/what-exactly-are-brands-posting -on-snapchat-2016-6, retrieved May 4, 2017)**. The researchers conducting the study used bootstrap methods to determine a confidence interval.

Suppose that it is reasonable to consider this sample of brand-name products as representative of all brand-name products. Use the "Bootstrap Confidence Interval for One Proportion" Shiny app to find a 95% confidence interval for the proportion of all brand-name products that are active on Snapchat, and interpret the interval in context.

SECTION 9.6 Avoid These Common Mistakes

When using sample data to estimate a population characteristic, either a single number estimate or a confidence interval estimate might be used. Confidence intervals are generally preferred because a single number estimate, by itself, does not convey any information about its accuracy. For this reason, whenever you report the value of a single number estimate, it is a good idea to also include a margin of error.

Reporting and interpreting a confidence interval estimate requires a bit of care. First, always report both the confidence interval and the associated confidence level. Also remember that both the confidence interval and the confidence level should be interpreted. A good strategy is to begin with an interpretation of the confidence interval in the context of the problem and to follow that with an interpretation of the confidence level. For example, if a 90% confidence interval for p, the proportion of students at a particular college who own a car, is (0.56, 0.78), you might say

Interpretation of interval → $\left\{ \begin{array}{l} \text{You can be 90\% confident that between} \\ \text{56\% and 78\% of the students at this} \\ \text{college own a car.} \end{array} \right.$

explanation of "90% confidence" → $\left\{ \begin{array}{l} \text{You have used a method to produce} \\ \text{this estimate that is successful in} \\ \text{capturing the actual population} \\ \text{proportion about 90\% of the time.} \end{array} \right.$
interpretation of confidence level

Unfortunately, there is no customary way of reporting interval estimates of population characteristics in published sources. Possibilities include

confidence interval
estimate $\pm$ margin of error
estimate $\pm$ standard error

If the reported interval is described as a confidence interval, a confidence level should accompany it. These intervals can be interpreted just as you have interpreted the confidence intervals in this chapter, and the confidence level specifies the long-run success rate associated with the method used to construct the interval.

A form particularly common in news articles is estimate $\pm$ margin of error. The margin of error reported is usually two times the standard error of the estimate. This method of reporting is a little more informal than a confidence interval, but if the sample size is reasonably large, it is roughly equivalent to reporting a 95% confidence interval. You can interpret these intervals as you would a confidence interval with an approximate confidence level of 95%.

Be careful when interpreting intervals reported in the form of an estimate $\pm$ standard error. Recall from Section 9.3 that the general form of a confidence interval is

$$\text{statistic} \pm (\text{critical value})(\text{standard error of the statistic})$$

The critical value in the confidence interval formula is determined by the sampling distribution of the statistic and by the confidence level. Note that the form estimate $\pm$ standard error is equivalent to a confidence interval with the critical value set equal to 1. For a statistic whose sampling distribution is approximately normal (such as a large-sample proportion), a critical value of 1 corresponds to a confidence level of about 68%. Because a confidence level of 68% is rather low, you may want to use the given information and the confidence interval formula to convert to an interval with a higher confidence level.

When working with single number estimates and confidence interval estimates, here are a few things you should keep in mind:

1. In order for an estimate to be useful, you must know something about its accuracy. You should beware of single number estimates that are not accompanied by a margin of error or some other measure of accuracy.
2. A confidence interval estimate that is wide indicates that you don't have very precise information about the population characteristic being estimated. Don't be fooled by a high confidence level if the resulting interval is wide. High confidence, while desirable, is not the same thing as saying you have precise information about the value of a population characteristic.
3. The width of a confidence interval is affected by the confidence level, the sample size, and the standard deviation of the statistic used to construct the interval. The best strategy for decreasing the width of a confidence interval is to take a larger sample. It is far better to think about this before collecting data. Use the sample size formula to determine a sample size that will result in a small margin of error or that will result in a confidence interval estimate that is narrow enough to provide useful information.
4. The accuracy of an estimate depends on the sample size, not the population size. This may be counter to intuition, but as long as the sample size is small relative to the population size (n is less than 10% of the population size), the margin of error for

estimating a population proportion with 95% confidence is approximately $2\sqrt{\frac{p(1-p)}{n}}$. Notice that the approximate margin of error involves the sample size n, and decreases as n increases. However, the approximate margin of error does not depend on the population size, N.

The size of the population does need to be considered if sampling is without replacement and the sample size is more than 10% of the population size. In this case, the margin of error is adjusted by multiplying it by a **finite population correction factor** $\sqrt{\frac{N-n}{N-1}}$. Since this correction factor is always less than 1, the adjusted margin of error will be smaller.

5. Conditions are important. Appropriate use of the margin of error formula and the large-sample confidence interval of this chapter requires certain conditions be met:

   i. The sample is a random sample from the population of interest or is selected in a way that should result in a representative sample.
   ii. The sample size is large enough for the sampling distribution of $\hat{p}$ to be approximately normal.

   If these conditions are met, the large-sample confidence interval provides a method for using sample data to estimate the population proportion with confidence, and the confidence level is a good approximation of the success rate for the method.

   Whether the random or representative sample condition is plausible will depend on how the sample was selected and the intended population. Conditions for the sample size are the following:

   $$n\hat{p} \geq 10 \text{ and } n(1 - \hat{p}) \geq 10$$

   (so that the sampling distribution of $\hat{p}$ will be approximately normal).

   If the conditions for the large-sample confidence interval are not met, the bootstrap confidence interval of Section 9.5 provides an alternate way to estimate a population proportion.

6. When reading published reports, don't fall into the trap of thinking "confidence interval" every time you see a $\pm$ in an expression. As was discussed earlier in this section, published reports are not consistent. In addition to confidence intervals, it is common to see both estimate $\pm$ margin of error and estimate $\pm$ standard error reported.

CHAPTER ACTIVITIES

ACTIVITY 9.1 GETTING A FEEL FOR CONFIDENCE LEVEL

Technology Activity Open the "Confidence Interval Visualization" app that can be found in the app collection at statistics.cengage.com/Peck2e/Apps.html. You should see a screen like the one shown.

Confidence Interval Visualization

Select population proportion:

| 0.40 |

Select sample size (n):

| 50 |

Select number of intervals:

| 1 |

Select confidence level (in %):

| 95 |

| Draw Sample(s) from Population | Reset |

This app will select a random sample from a population with a specified value for the population proportion. It will then use that sample to construct a confidence interval for the population proportion and plot that interval on a graph. From that graph you will be able to see if the calculated confidence interval includes the value of the population proportion.

For purposes of this activity, you will sample from a population with $p = 0.4$ and begin with a sample size of 50. Enter 0.40 in the box for the population proportion and 50 in the box for the sample size, as shown in the previous figure. Make sure that the confidence level is set to be 95%, and leave the number of intervals set at 1.

Click on the "Draw Sample(s) from Population" button. You should now see a confidence interval appear on the display on the right hand side. If the interval contains the population proportion of 0.40, the interval will be drawn in green. If 0.40 is not in the confidence interval, the interval will be drawn in red. The graph on your screen should now look something like the one shown here.

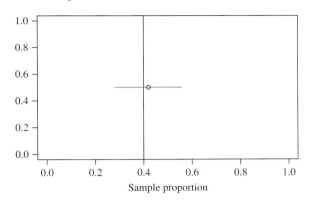

Click on the "Draw Sample(s) from Population" button a few more times and watch what happens.

To look at more than one interval at a time, change the number of intervals from 1 to 100, and then click on the "Draw Sample(s) from Population" button. You should now see a graph similar to the following, with 100 intervals in the display. Again, intervals containing 0.40 (the value of p in this population) will be green and those that do not include 0.40 will be red.

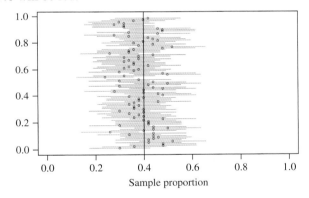

In the output shown here, 4 of the 100 confidence intervals do not contain the value of the population proportion and 96 of them contain the value of the population proportion. For these 100 intervals, 96% capture the actual value of the population proportion. Because these are 95% confidence intervals, you would expect that in the long run, about 95% of the intervals would include the value of the population proportion.

Now use the app to answer the following questions:

1. Generate 500 confidence intervals based on samples of size 50 from a population with a population proportions of 0.40. How does the percentage of these intervals that contain the population proportion of 0.40 compare to the confidence level of 95%?

2. Experiment with three other confidence levels of your choice. In general, is the proportion of calculated confidence intervals that contain the value of the population proportion close to the stated confidence level?

ACTIVITY 9.2   AN ALTERNATIVE CONFIDENCE INTERVAL FOR A POPULATION PROPORTION

Technology Activity: This activity presumes that you have completed Activity 9.1.

Background: As an alternative to the usual large-sample z confidence interval, consider the confidence interval

$$\hat{p}_{mod} \pm (z \text{ critical value})\sqrt{\frac{\hat{p}_{mod}(1-\hat{p}_{mod})}{n}}$$

where $\hat{p}_{mod} = \dfrac{\text{successes} + 2}{n + 4}$. This alternative interval is preferred by many statisticians because, in repeated sampling, the proportion of intervals constructed that include the actual value of the population proportion, p, tends to be closer to the stated confidence level. In this activity, you will compare the success rates for the two different confidence interval methods.

Open the apps "Confidence Interval Visualization" and "Alternative Confidence Interval for a Population Proportion." These two apps can be found in the app collection at statistics.cengage.com/Peck2e/Apps.html.

The "Confidence Interval Visualization" app uses the large-sample confidence interval for a population proportion to construct confidence intervals. The app "Alternative Confidence Interval for a Population Proportion" uses the alternative method based on p_{mod} to construct confidence intervals.

1. Consider sampling from a population with $p = 0.3$ using a sample size of 40. Notice that this sample size is large enough to meet the sample size conditions for the large-sample confidence interval for a population

proportion. Use the "Confidence Interval Visualization" app to generate 300 95% confidence intervals. How does the percentage of intervals constructed that include $p = 0.3$ compare to the confidence level of 95%? Does this surprise you?

2. Now use the "Alternative Confidence Interval for a Population Proportion" app to generate 300 95% confidence intervals using samples of size 40 from a population with $p = 0.3$. How does the percentage of intervals constructed that include $p = 0.3$ compare to the confidence level of 95%? Is this percentage closer to

95% than was the case for the large-sample confidence interval?

3. Experiment with different combinations of values of sample size and population proportion p. Can you find a combination for which the large-sample z interval has a success rate that is close to 95%? Can you find a combination for which it has a success rate that is even farther from 95% than it was for $n = 40$ and $p = 0.3$? How does the modified interval perform in each of these cases?

ACTIVITY 9.3   VERIFYING SIGNATURES ON A RECALL PETITION

Background: In 2003, petitions were submitted to the California Secretary of State calling for the recall of Governor Gray Davis. Each of California's 58 counties reported the number of valid signatures on the petitions from that county so that the state could determine whether there were enough to certify the recall and set a date for the recall election. The following paragraph appeared in the *San Luis Obispo Tribune* (**July 23, 2003**):

> In the campaign to recall Gov. Gray Davis, the secretary of state is reporting 16,000 verified signatures from San Luis Obispo County. In all, the County Clerk's Office received 18,866 signatures on recall petitions and was instructed by the state to check a random sample of 567. Out of those, 84.48% were good. The verification

process includes checking whether the signer is a registered voter and whether the address and signature on the recall petition match the voter registration.

1. Use the data from the random sample of 567 San Luis Obispo County signatures to construct a 95% confidence interval for the proportion of petition signatures that are valid.

2. How do you think that the reported figure of 16,000 verified signatures for San Luis Obispo County was obtained?

3. Based on your confidence interval from Part 1, explain why you think that the reported figure of 16,000 verified signatures is or is not reasonable.

🖥 CHAPTER 9 EXPLORATIONS IN STATISTICAL THINKING

🖱 EXPLORATION 1: UNDERSTANDING SAMPLING VARIABILITY AND THE MEANING OF CONFIDENCE LEVEL

In the exercise below, each student in your class will go online to select a random sample of 30 animated movies produced between 1980 and 2011. You will use this sample to estimate the proportion of animated movies made between 1980 and 2011 that were produced by Walt Disney Studios.

Go online at statistics.cengage.com/Peck2e/Explore.html and click on the link for Chapter 9. This link will take you to a web page where you can select a random sample from the animated movie population.

Click on the "sample" button. This selects a random sample of 30 movies and will display the movie name and the production studio for each movie in your sample. Each student in your class will receive a different random sample.

Use the data from your random sample to complete the following:
a. Calculate the proportion of movies in your sample that were produced by Disney.
b. Is the proportion you calculated in Part (a) a sample proportion or a population proportion?

c. Construct a 90% confidence interval for the proportion of animated movies made between 1980 and 2011 that were produced by Disney.

Write a few sentences interpreting the confidence interval from Part (c) and the associated confidence level.

d. The actual population proportion is 0.41. Did your confidence interval from Part (c) include this value?

e. Which of the following is a correct interpretation of the 90% confidence level?

   1. The probability that the actual population proportion is contained in the calculated interval is 0.90.

   2. If the process of selecting a random sample of movies and then calculating a 90% confidence interval for the proportion of all animated movies made between 1980 and 2011 that were produced by Disney is repeated 100 times, exactly 90 of the 100 intervals will include the actual population proportion.

   3. If the process of selecting a random sample of movies and then calculating a 90% confidence interval for the proportion of all animated movies made between 1980 and 2011 that were produced by Disney is repeated a very large number of times, approximately 90% of the intervals will include the actual population proportion.

If asked to do so by your instructor, bring your confidence interval estimate of the proportion of animated movies made between 1980 and 2011 that were produced by Disney to class. Your instructor will lead the class through a discussion of the questions that follow.

Compare your confidence interval to the confidence interval obtained by another student in your class.

f. Are the two confidence intervals the same?

g. Did both intervals contain the actual population proportion of 0.41?

h. How many people in your class have a confidence interval that does not include the actual value of the population proportion? Is this surprising, given the 90% confidence level associated with the confidence intervals?

ARE YOU READY TO MOVE ON? CHAPTER 9 REVIEW EXERCISES

All chapter learning objectives are assessed in these exercises. The learning objectives assessed in each exercise are given in parentheses.

9.77 (C1)
Two statistics are being considered for estimating the value of a population characteristic. The sampling distributions of the two statistics are shown here. Explain why Statistic II would be preferred over Statistic I.

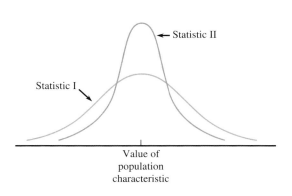

9.78 (C2)
Two statistics are being considered for estimating the value of a population characteristic. The sampling distributions of the two statistics are shown here.

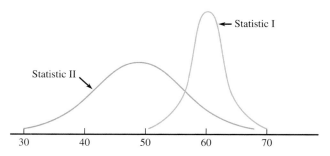

a. Suppose the actual value of the population characteristic is 50. Is Statistic I or Statistic II unbiased? If so, which one?

b. Suppose the actual value of the population characteristic is 54. In this case, neither Statistic I nor Statistic II is unbiased. Which of these two statistics would you recommend as the better estimator of the population characteristic? Explain the reasoning for your choice.

9.79 (C3)
Will $\hat{p}$ from a random sample from a population with 60% successes tend to be closer to 0.6 for a sample size of $n = 400$ or a sample size of $n = 800$? Provide an explanation for your choice.

9.80 (C3)
Will $\hat{p}$ from a random sample of size 400 tend to be closer to the actual value of the population proportion when $p = 0.4$ or when $p = 0.7$? Provide an explanation for your choice.

9.81 (C3)
A random sample will be selected from the population of all adult residents of a particular city. The sample proportion $\hat{p}$ will be used to estimate p, the proportion of all adult residents who are registered to vote. For which of the following situations will the estimate tend to be closest to the actual value of p?
I. $n = 1000, p = 0.5$
II. $n = 200, p = 0.6$
III. $n = 100, p = 0.7$

9.82 (P1)
In response to budget cuts, county officials are interested in learning about the proportion of county residents who favor closure of a county park rather than closure of a county library. In a random sample of 500 county residents, 198 favored closure of a county park. For each of the three statements below, indicate if the statement is correct or incorrect. If the statement is incorrect, explain what makes it incorrect.

Statement 1: It is unlikely that the estimate $\hat{p} = 0.396$ differs from the value of the actual population proportion by more than 0.043.
Statement 2: The estimate $\hat{p} = 0.396$ will never differ from the value of the actual population proportion by more than 0.043.
Statement 3: It is unlikely that the estimate $\hat{p} = 0.396$ differs from the value of the actual population proportion by more than 0.022.

9.83 (C4)
Consider taking a random sample from a population with $p = 0.25$.
a. What is the standard error of $\hat{p}$ for random samples of size 400?
b. Would the standard error of $\hat{p}$ be smaller for random samples of size 200 or samples of size 400?

c. Does cutting the sample size in half from 400 to 200 double the standard error of $\hat{p}$?

9.84 (C5, C6, M1, M2, P1, P2)
In a survey on supernatural experiences, 722 of 4013 adult Americans reported that they had seen a ghost (**"What Supernatural Experiences We've Had," USA TODAY, February 8, 2010**). Assume that this sample is representative of the population of adult Americans.
a. Use the given information to estimate the proportion of adult Americans who would say they have seen a ghost.
b. Verify that the conditions for use of the margin of error formula to be appropriate are met.
c. Calculate the margin of error.
d. Interpret the margin of error in context.
e. Construct and interpret a 90% confidence interval for the proportion of all adult Americans who would say they have seen a ghost.
f. Would a 99% confidence interval be narrower or wider than the interval calculated in Part (e)? Justify your answer.

9.85 (C6)
Suppose that county planners are interested in learning about the proportion of county residents who would pay a fee for a curbside recycling service if the county were to offer this service. Two different people independently selected random samples of county residents and used their sample data to construct the following confidence intervals for the proportion who would pay for curbside recycling:

Interval 1: (0.68, 0.74)
Interval 2: (0.68, 0.72)

a. Explain how it is possible that the two confidence intervals are not centered in the same place.

b. Which of the two intervals conveys more precise information about the value of the population proportion?
c. If both confidence intervals are associated with a 95% confidence level, which confidence interval was based on the smaller sample size? How can you tell?

d. If both confidence intervals were based on the same sample size, which interval has the higher confidence level? How can you tell?

9.86 (C6)
Describe how each of the following factors affects the width of the large-sample confidence interval for p:
a. The confidence level
b. The sample size
c. The value of $\hat{p}$

9.87 (C7)
Based on data from a survey of 1200 randomly selected Facebook users (**USA TODAY, March 24, 2010**), a 90% confidence interval for the proportion of all Facebook users who

say it is not OK to "friend" someone who reports to you at work is (0.60, 0.64). What is the meaning of the 90% confidence level associated with this interval?

9.88 (M3)

The study **"The Demographics of Social Media Users" (Pew Research Center, August 19, 2015)** reported that 72% of adult American Internet users use Facebook. The 72% figure was based on a representative sample of $n = 1602$ adult American Internet users. Suppose that you would like to use the data from this survey to estimate the proportion of all adult American Internet users who use Facebook. Answer the four key questions (QSTN) to confirm that the suggested method in this situation is a large-sample confidence interval for a population proportion.

9.89 (M4)

For the study described in the previous exercise, use the five-step process for estimation problems (EMC³) to construct and interpret a 90% confidence interval for the proportion of adult American Internet users who use Facebook. Identify each of the five steps in your solution.

9.90 (C6, P2)

The article **"Kids Digital Day: Almost 8 Hours" (*USA TODAY*, January 20, 2010)** summarized a national survey of 2002 Americans age 8 to 18. The sample was selected to be representative of Americans in this age group.

a. Of those surveyed, 1321 reported owning a cell phone. Use this information to construct and interpret a 90% confidence interval for the proportion of all Americans ages 8 to 18 who owned a cell phone in 2010.

b. Of those surveyed, 1522 reported owning an MP3 music player. Use this information to construct and interpret a 90% confidence interval for the proportion of all Americans ages 8 to 18 who owned an MP3 music player in 2010.

c. Explain why the confidence interval from Part (b) is narrower than the confidence interval from Part (a) even though the confidence levels and the sample sizes used to calculate the two intervals were the same.

9.91 (M5, P3)

A consumer group is interested in estimating the proportion of packages of ground beef sold at a particular store that have an actual fat content exceeding the fat content stated on the label. How many packages of ground beef should be tested in order to have a margin of error of 0.05?

TECHNOLOGY NOTES

Confidence Intervals for Proportions

JMP

Summarized data

1. Enter the data table into the JMP data table with categories in the first column and counts in the second column

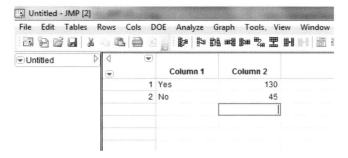

2. Click **Analyze** and select **Distribution**
3. Click and drag the first column name from the box under **Select Columns** to the box next to **Y, Columns**
4. Click and drag the second column name from the box under **Select Columns** to the box next to **Freq**
5. Click **OK**
6. Click the red arrow next to the column name and click **Confidence Interval** then select the appropriate level or select **Other** to input a level that is not listed

Raw Data

1. Enter the raw data into a column
2. Click **Analyze** and select **Distribution**
3. Click and drag the first column name from the box under **Select Columns** to the box next to **Y, Columns**
4. Click **OK**
5. Click the red arrow next to the column name and click **Confidence Interval** then select the appropriate level or select **Other** to input a level that is not listed

Minitab

Summarized data

1. Click **Stat** then click **Basic Statistics** then click **1 Proportion...**
2. Click the radio button next to **Summarized data**
3. In the box next to **Number of Trials:** type the value for n, the total number of trials
4. In the box next to **Number of events:** type the value for the number of successes
5. Click **Options...**
6. Input the appropriate confidence level in the box next to **Confidence Level**
7. Check the box next to **Use test and interval based on normal distribution**

8. Click **OK**
9. Click **OK**

Raw data

1. Input the raw data into a column
2. Click **Stat** then click **Basic Statistics** then click **1 Proportion...**
3. Click in the box under **Samples in columns:**
4. Double click the column name where the raw data is stored
5. Click **Options...**
6. Input the appropriate confidence level in the box next to **Confidence Level**
7. Check the box next to **Use test and interval based on normal distribution**
8. Click **OK**
9. Click **OK**

SPSS

SPSS does not have the functionality to automatically produce confidence intervals for a population proportion.

Excel

Excel does not have the functionality to automatically produce confidence intervals for a population proportion. However, you can manually type in the formula for the lower and upper limits separately into two different cells to have Excel calculate the result for you.

TI-83/84

1. Press the **STAT** key
2. Highlight **TESTS**
3. Highlight **1-PropZInterval** and press **ENTER**
4. Next to x type the number of successes
5. Next to n type the number of trials, n
6. Next to **C-Level** type the value for the confidence level
7. Highlight **Calculate** and press **ENTER**

TI-Nspire

1. Enter the Calculate Scratchpad
2. Press the **menu** key then select **6:Statistics** then select **6:Confidence Intervals** then **5:1-Prop z Interval...** then press **enter**
3. In the box next to **Successes, x** type the number of successes
4. In the box next to n type the number of trials, n
5. In the box next to **C Level** type the confidence level
6. Press **OK**

10 Asking and Answering Questions About a Population Proportion

spass/Shutterstock.com

PREVIEW

Two types of inference problems are considered in this text. In estimation problems, sample data are used to learn about the value of a population characteristic. In hypothesis testing problems, sample data are used to decide if some claim about the value of a population characteristic is plausible. In Chapter 9, you saw how to use sample data to estimate a population proportion. In this chapter, you will see how sample data can also be used to decide whether a claim, called a hypothesis, about a population proportion is believable.

CHAPTER LEARNING OBJECTIVES

Conceptual Understanding

After completing this chapter, you should be able to

C1 Understand how a research question is translated into hypotheses.

C2 Understand that rejecting the null hypothesis implies strong support for the alternative hypothesis.

C3 Understand why failing to reject the null hypothesis does not imply strong support for the null hypothesis.

C4 Know the two types of errors that are possible in a hypothesis test.

C5 Understand how the significance level of a test relates to the probability of a Type I error.

C6 Understand the reasoning used to reach a decision in a hypothesis test.

Mastering the Mechanics

After completing this chapter, you should be able to

M1 Translate a research question or claim about a population proportion into null and alternative hypotheses.

M2 Describe Type I and Type II errors in context.

M3 Know the key characteristics that lead to selection of a large-sample test for a population proportion as an appropriate method.

M4 Know the conditions for appropriate use of the large-sample test for a population proportion.

M5 Select an appropriate significance level based on consideration of the consequences of Type I and Type II errors.

M6 Use the five-step process for hypothesis testing (HMC^3) to carry out a large-sample test for a population proportion.

M7 **(Optional)** Carry out a randomization test for a population proportion.

M8 **(Optional)** Carry out an exact binomial test for a population proportion.

Putting It into Practice

After completing this chapter, you should be able to

P1 Recognize when a situation calls for testing hypotheses about a population proportion.

P2 Carry out a large-sample test for a population proportion and interpret the conclusion in context.

PREVIEW EXAMPLE

Choosing Baby's Sex

The article **"Boy or Girl: Which Gender Baby Would You Pick?" (*LiveScience*, March 23, 2005, www.livescience.com, retrieved May 8, 2017)** summarized a study that was published in *Fertility and Sterility*. The *LiveScience* article makes the following statements:

> "When given the opportunity to choose the sex of their baby, women are just as likely to choose pink socks as blue, a new study shows."

> "Of the 561 women who participated in the study, 229 said they would like to choose the sex of a future child. Among these 229 there was no greater demand for boys or girls."

These statements are equivalent to the claim that the proportion of women who would choose a girl is 0.50 or 50%,

Is this claim supported by the study data? The paper referenced in the *LiveScience* article **("Preimplantation Sex Selection Demand and Preferences in an Infertility Population,"** *Fertility and Sterility* **[2005]: 649–658)** provides the following information about the study:

- A survey with 19 questions was mailed to 1385 women who had visited the Center for Reproductive Medicine at Brigham and Women's Hospital.
- A total of 561 women returned the survey.

- Of the 561 women who responded, 229 (or 40.8%) wanted to select the sex of their next child.
- Of the 229 who wanted to select the baby's sex, 89 (or 38.9%) wanted a boy and 140 (or 61.1%) wanted a girl.

You should have a few questions at this point. What is the population of interest, and are the 561 women who responded to the survey representative of this population? There was a high nonresponse rate—only 561 of 1385 surveys were returned. Is it possible that women who returned the survey differ in some important way from women who did not return the survey? It will be important to address these questions before deciding if it is reasonable to generalize from this sample to a larger population.

Moreover, the proportion of women wanting to choose the baby's sex who would select a girl was $\frac{140}{229} = 0.611$. However, the *LiveScience* summary indicates that there is no preference for girls over boys, implying that the proportion who would choose a girl is 0.50. Is a sample proportion of 0.611 consistent with a population proportion of 0.50? There are two possibilities to consider:

1. Even when the population proportion who would choose a girl is 0.50, the sample proportion will tend to differ from 0.50. So, a sample proportion as large as 0.611 might just be due to sampling variability. If this is the case, the sample data are considered consistent with the claim that the population proportion is 0.50.

or

2. When the population proportion who would choose a girl is 0.50, observing a sample proportion as large as 0.611 might be very unlikely. In this case, the sample data would provide evidence against the claim that the population proportion is 0.50.

How should you decide between these two possibilities? This example will be revisited later in this chapter where you will apply what you learned about sampling distributions in Chapter 8 to decide which of these two possibilities is the more reasonable choice.

This chapter is the first of several that cover tests of hypotheses. In this chapter, you will see how to use sample data to choose between two competing claims about a population proportion. ∎

SECTION 10.1   Hypotheses and Possible Conclusions

In Chapter 7, you learned to distinguish estimation problems from hypothesis testing problems. Recall that a hypothesis testing problem involves using sample data to test a claim (called a hypothesis) about a population. Later in this chapter, you will learn how to test hypotheses about a population proportion. But first some general terminology is needed and you need to consider the risks involved in this type of inference.

In its simplest form, a hypothesis is a claim or statement about the value of a single population characteristic. (You will see more complicated hypotheses in later chapters.) For example, consider a population consisting of all e-mail messages sent using a campus e-mail system during a particular month. The following are examples of hypotheses about population proportions:

Hypothesis	Population Proportion of Interest	The Hypothesis Says...
$p < 0.01$	where p is the proportion of e-mail messages that were returned as undeliverable	Less than 1% of the e-mail messages sent were returned as undeliverable.
$p < 0.25$	where p is the proportion of e-mail messages that included an attachment	Less than 25% of the e-mail messages sent included an attachment.

(continued)

Hypothesis	Population Proportion of Interest	The Hypothesis Says...
$p > 0.8$	where p is the proportion of e-mail messages that were longer than 500 characters	More than 80% of the e-mail messages sent were longer than 500 characters.
$p = 0.3$	where p is the proportion of e-mail messages that were sent to more than one person.	30% of the e-mail messages sent were sent to more than one person.

Notice that hypotheses are *always* statements about population characteristics and *never* about sample statistics. Statements like $\hat{p} < 0.50$ or $\hat{p} = 0.72$ are not hypotheses because $\hat{p}$ is a sample proportion, not a population proportion.

In a hypothesis testing situation, you want to know if the data provide support for a claim about a population characteristic. For example, suppose that a particular community college claims that the majority (more than half) of students completing an associate's degree transfer to a 4-year college. To determine if you should believe this claim, you might begin by selecting a random sample of 100 graduates of this community college. Each graduate could be asked if he or she transferred to a 4-year college. You would then want to determine if the sample data provide convincing evidence in support of the hypothesis $p > 0.5$, where p is the population proportion of graduates who transfer to a 4-year college. This hypothesis states that the population proportion is greater than 0.5, or equivalently that more than 50% (a majority) of the graduates transfer to a 4-year college.

Although it will seem a bit odd at first, the way to decide if the sample data provide convincing evidence in support of a particular hypothesis is to set up a pair of competing hypotheses that includes the hypothesis of interest. In the community college transfer example, the two competing hypotheses might be

$$p \le 0.5 \text{ and } p > 0.5$$

The hypothesis $p > 0.5$ says that a majority of graduates transfer to a 4-year college, and the hypothesis $p \le 0.5$ says that the proportion of graduates who transfer is less than or equal to 0.5. If it were possible to survey *every* graduate (a census), you would know the actual value of p and you would know which of these two hypotheses was correct. Unfortunately, you must usually decide between the two competing hypotheses using data from a sample, which means that you will not have complete information about the population.

A **hypothesis test** uses sample data to choose between two competing hypotheses about a population characteristic.

A criminal trial is a familiar situation in which a choice between two contradictory claims must be made. A person accused of a crime must be judged either guilty or not guilty. Under the U.S. system of justice, the individual on trial is initially presumed not guilty. Only strong evidence to the contrary causes the not guilty verdict to be rejected in favor of a guilty verdict. The burden is put on the prosecution to *convince* the jurors that a defendant is guilty.

As in a judicial proceeding, in a hypothesis test you initially assume that a particular hypothesis, called the *null hypothesis*, is the correct one. You then consider the evidence (data) and reject the null hypothesis in favor of a competing hypothesis, called the *alternative hypothesis*, only if there is convincing evidence against the null hypothesis.

The **null hypothesis,** denoted by H_0, is a claim about a population characteristic that is initially assumed to be true.

The **alternative hypothesis,** denoted by H_a, is a competing claim.

In carrying out a test of H_0 versus H_a, the null hypothesis H_0 is rejected in favor of the alternative hypothesis H_a only if the data provide convincing evidence that H_0 is false. If the data do not provide such evidence, H_0 is not rejected.

The two possible conclusions in a hypothesis test are *reject H_0* and *fail to reject H_0.*

In choosing between the null and alternative hypotheses, notice that you initially assume that the null hypothesis is true. When you observe data that would be very unlikely to occur if the null hypothesis were true, you have convincing evidence against the null hypothesis. In this case, you would reject the null hypothesis in favor of the alternative hypothesis. This is a strong conclusion because you have *convincing* evidence against the null hypothesis.

If you are not convinced that the null hypothesis is false, you fail to reject the null hypothesis. It is important to note that this does *not* mean that you have strong or convincing evidence that the null hypothesis is true. You can only say that you were not convinced that the null hypothesis is false.

Hypothesis Test Conclusion	This Means That...
Reject H_0	There is convincing evidence against the null hypothesis. If H_0 were true, the data would be very surprising.
Fail to reject H_0	There is not convincing evidence against the null hypothesis. If H_0 were true, the data would not be considered surprising.

A statistical hypothesis test can only demonstrate strong support for the alternative hypothesis. If the goal of a study is to decide whether data support a particular claim about a population, this claim will determine the alternative hypothesis. In a hypothesis testing situation, you usually begin by translating the claim of interest into the alternative hypothesis.

The **alternative hypothesis** will have one of the following three forms:

H_a: population characteristic $>$ hypothesized value

H_a: population characteristic $<$ hypothesized value

H_a: population characteristic $\neq$ hypothesized value

The hypothesized value and whether $>$, $<$, or $\neq$ appears in the alternative hypothesis is determined by the study context and the question of interest. Some examples illustrating selection of the alternative hypothesis follow.

Example 10.1 Living at Home

The article **"For First Time in Modern Era Living with Parents Edges Out Other Living Arrangements for 18- to 34-Year-Olds"** (*Pew Research Center Social and Demographic Trends,* **May 24, 2016)** summarizes a survey of a representative sample of 18- to 34-year-olds. Suppose that you are interested in determining if the sample data support the claim that fewer than one-third of adults in this age group are living with parents. With p = the proportion of adults age 18 to 34 who live with their parents, the question of interest becomes "Is $p < 0.333$?" The hypothesized value of 0.333 is used because "one third" is equivalent to $\frac{1}{3} \approx 0.333$. The appropriate alternative hypothesis is H_a: $p < 0.333$.

Example 10.2 | **Personal Finance**

A national survey of 1012 adult Americans conducted by Gallup **("Americans Still Generally Upbeat About Personal Finances," www.gallup.com, January 25, 2016, retrieved November 16, 2016)** asked survey participants if they thought they were in better financial shape than they were one year ago. Suppose that you want to determine if the survey data provide convincing evidence that a majority of adult Americans believe they are in better financial shape than one year ago. With p = the population proportion of adult Americans who believe they are better off, the claim of interest (a majority, or more than 50%) is "p greater than 0.50." The appropriate alternative hypothesis is H_a: $p > 0.50$.

Example 10.3 | **M&M Color Mix**

The manufacturer of M&Ms claims that 40% of plain M&M's are brown. A sample of M&M's will be used to determine if the population proportion of brown M&M's is different from what the manufacturer claims. With p = the population proportion of plain M&M's that are brown, the manufacturer's claim is $p = 0.40$. The question is whether there is evidence that the manufacturer's claim is false, so the alternative hypothesis would be H_a: $p \neq 0.40$.

Example 10.4 | **Whose Reality Revisited**

Example 7.2 described a study in which researchers wanted to learn about how people view reality TV shows **("Who's Afraid of Reality Shows?"** *Communications Research* **[2008]: 382–397)**. The researchers hoped to learn if the study data supported the theory that a majority of adults in Israel believe they are much less affected by reality TV shows than other people. With p = the population proportion of Israeli adults who believe they are much less affected than other people, the researchers' claim is $p > 0.5$. To see if there is evidence to support this claim, the researchers could carry out a hypothesis test with an alternative hypothesis of H_a: $p > 0.5$.

Once an alternative hypothesis has been selected, the next step is to formulate a competing null hypothesis. The null hypothesis can be formed by replacing the inequality that appears in the alternative hypothesis ($<$, $>$, or $\neq$) with an equal sign. For the alternative hypothesis H_a: $p > 0.5$ from Example 10.4, the corresponding null hypothesis would be H_0: $p = 0.5$.

Because the alternative hypothesis in Example 10.4 was H_a: $p > 0.5$, it might seem sensible to state H_0 as the inequality $p \leq 0.5$ rather than $p = 0.5$. While either is acceptable, H_0 is usually stated as a claim of equality. There are several reasons for this. First, the development of a test statistic is most easily understood if there is only a single hypothesized value in the null hypothesis. Second, suppose that the sample data provide convincing evidence that H_0: $p = 0.5$ should be rejected in favor of H_a: $p > 0.5$. This means that you are convinced by the sample data that the population proportion is greater than 0.5. It follows that you must also be convinced that the population proportion is not 0.4 or 0.3 or any other value that is less than 0.5. As a consequence, the conclusion when testing H_0: $p = 0.5$ versus H_a: $p > 0.5$ is always the same as the conclusion for a test where the null hypothesis is H_0: $p \leq 0.5$.

Because it is customary to state the null hypothesis as a claim of equality, this convention will be followed in the examples and exercises in this text. However, it is also acceptable to write the null hypothesis as $\leq$ when the inequality in the alternative hypothesis is $>$ or as $\geq$ when the inequality in the alternative hypothesis is $<$. *Just be sure to remember that the null hypothesis must include the "equal to" case.*

The form of the null hypothesis is

$$H_0\text{: population characteristic} = \text{hypothesized value}$$

where the hypothesized value is a specific number determined by the problem context.

Example 10.5 illustrates how the selection of H_0 (the claim initially believed true) and H_a depend on the objectives of a study.

Example 10.5 | **Evaluating a New Medical Treatment**

A medical research team has been given the task of evaluating a new laser treatment for certain types of tumors. Consider the following two scenarios:

> Scenario 1: The current standard treatment is considered reasonable and safe by the medical community, is not costly, has no major side effects, and has a known success rate of 0.85 (or 85%).

> Scenario 2: The current standard treatment sometimes has serious side effects, is costly, and has a known success rate of 0.30 (or 30%).

In the first scenario, the research question of interest would probably be "Does the new treatment have a higher success rate than the standard treatment?" Unless there is convincing evidence that the new treatment has a higher success rate, it is unlikely that current medical practice would change. With p representing the actual proportion of success for the laser treatment, the following hypotheses would be tested:

$$H_0: p = 0.85 \text{ versus } H_a: p > 0.85$$

In this case, rejecting the null hypothesis would require convincing evidence that the success rate is higher for the new treatment.

In the second scenario, the current standard treatment does not have much to recommend it. The new laser treatment may be considered preferable because of cost or because it has fewer or less serious side effects, as long as the success rate for the new procedure is no worse than that of the standard treatment. Here, researchers might decide to test

$$H_0: p = 0.30 \text{ versus } H_a: p < 0.30$$

If the null hypothesis is rejected, the new treatment would not be put forward as an alternative to the standard treatment, because there is strong evidence that the laser method has a lower success rate. On the other hand, if the null hypothesis is not rejected, there is not convincing evidence that the success rate for the laser treatment is lower than that for the standard treatment. This is *not* the same as saying that there is evidence that the laser treatment is as good as the standard treatment. If medical practice were to embrace the new procedure, it would not be because it has a higher success rate but rather because it costs less or because it has fewer side effects.

One last reminder—you should be careful in setting up the hypotheses for a test. *Remember that a statistical hypothesis test is only capable of demonstrating strong support for the alternative hypothesis (by rejecting the null hypothesis). When the null hypothesis is not rejected, it does not mean strong support for H_0—only lack of strong evidence against it.* When deciding what alternative hypothesis to use, be sure to keep the research objectives in mind.

Summing It Up—Section 10.1

The following learning objectives were addressed in this section:

Conceptual Understanding

C1: Understand how a research question is translated into hypotheses.

In a hypothesis testing situation, data are used to decide if there is support for a claim about a population characteristic. When data are collected to answer a question about a population characteristic, that question will determine what hypotheses are tested.

C2: Understand that rejecting the null hypothesis implies strong support for the alternative hypothesis.

In a hypothesis test, the null hypothesis is assumed to be true, and then it is only rejected if there is compelling evidence against it. This means that if the null hypothesis is rejected, there is strong support for the alternative hypothesis.

C3: Understand why failing to reject the null hypothesis does not imply strong support for the null hypothesis.

The conclusion in a hypothesis test will be "fail to reject the null hypothesis" unless there is strong evidence against the null hypothesis. Failure to reject the null hypothesis does not mean that there is strong support for the null hypothesis. It only means that there was not convincing evidence against it.

Mastering the Mechanics

M1: Translate a research question or claim about a population proportion into null and alternative hypotheses.

Example 10.5 illustrates how a research question is translated into null and alternative hypotheses.

SECTION 10.1 EXERCISES

Each Exercise Set assesses the following chapter learning objectives: C1, C2, C3, M1

SECTION 10.1 Exercise Set 1

10.1 Explain why the statement $\hat{p} = 0.40$ is not a legitimate hypothesis.

10.2 **CareerBuilder.com** conducted a survey to learn about the proportion of employers who perform background checks when evaluating a candidate for employment **("Majority of Employers Background Check Employees…Here's Why," November 17, 2016, retrieved November 19, 2016)**. Suppose you are interested in determining if the resulting data provide strong evidence in support of the claim that more than two-thirds of employers perform background checks. To answer this question, what null and alternative hypotheses should you test? (Hint: See Example 10.4.)

10.3 The article **"Facebook Use and Academic Performance Among College Students,"** *Computers in Human Behavior* **[2015]: 265–272)** estimated that 70% of students at a large public university in California who are Facebook users log into their Facebook profiles at least six times a day. Suppose that you plan to select a random sample of 400 students at your college. You will ask each student in the sample if they are a Facebook user and if they log into their Facebook profile at least six times a day. You plan to use the resulting data to decide if there is evidence that the proportion for your college is different from the proportion reported in the article for the college in California. What hypotheses should you test? (Hint: See Example 10.3.)

10.4 The article **"Poll Finds Most Oppose Return to Draft, Wouldn't Encourage Children to Enlist" (Associated Press, December 18, 2005)** reports that in a random sample of 1000 American adults, 430 answered "yes" to the following question: "If the military draft were reinstated, would you favor drafting women as well as men?" The data were used to test $H_0: p = 0.5$ versus $H_a: p < 0.5$, and the null hypothesis was rejected. (Hint: See discussion following Example 10.5.)
a. Based on the result of the hypothesis test, what can you conclude about the proportion of American adults who favor drafting women if a military draft were reinstated?

b. Is it reasonable to say that the data provide strong support for the alternative hypothesis?
c. Is it reasonable to say that the data provide strong evidence against the null hypothesis?

10.5 According to an article in *Science Daily* **("Sill No Strong Evidence That Adjunct Treatment with HGH in IVF Improves Results," sciencedaily.com, July 4, 2016, retrieved November 26, 2016)**, women who are having difficulty becoming pregnant sometimes use human growth hormone (HGH) in addition to in-vitro fertilization (IVF) to try to have a baby. A large study found that "there was no strong evidence" that the proportion of women who became pregnant while taking HGH along with IVF was greater than the success rate for IVF alone.
a. Is this consistent with testing

H_0: HGH in addition to IVF increases the chance of getting pregnant
versus
H_a: HGH in addition to IVF does not increase the chance of getting pregnant
or with testing
H_0: HGH in addition to IVF does not increase the chance of getting pregnant
versus
H_a: HGH in addition to IVF increases the chance of getting pregnant
Explain.
b. Does the stated conclusion of "no strong evidence" indicate that the null hypothesis was rejected? Explain.

10.6 In a hypothesis test, what does it mean to say that the null hypothesis was rejected?

SECTION 10.1 Exercise Set 2

10.7 Explain why the statement $\hat{p} > 0.50$ is not a legitimate hypothesis.

10.8 *USA TODAY* **(March 4, 2010)** described a survey of 1000 women age 22 to 35 who work full time. Each woman who participated in the survey was asked if she would be willing to give up some personal time in order to make more money. To determine if the resulting data provided convincing evidence that the majority of women age 22 to 35 who work full time would be willing to give up some personal time for more money, what hypotheses should you test?

10.9 The article **"Facebook Use and Academic Performance Among College Students"** (*Computers in Human Behavior* [2015]: 265–272) estimated that 87% percent of students at a large public university in California who are Facebook users update their status at least two times a day. Suppose that you plan to select a random sample of 400 students at your college. You will ask each student in the sample if they are a Facebook user and if they update their status at least two times a day. You plan to use the resulting data to decide if there is evidence that the proportion for your college is different from the proportion reported in the article for the college in California. What hypotheses should you test?

10.10 The article **"Public Acceptability in the UK and the USA of Nudging to Reduce Obesity: The Example of Reducing Sugar-Sweetened Beverages"** (*PLOS One*, June 8, 2016) describes a survey in which each person in a representative sample of 1082 adult Americans was asked about whether they would find different types of interventions acceptable in an effort to reduce consumption of sugary beverages. When asked about a tax on sugary beverages, 459 of the people in the sample said they thought that this would be an acceptable intervention. These data were used to test $H_0: p = 0.5$ versus $H_a: p < 0.5$ and the null hypothesis was rejected.
a. Based on the hypothesis test, what can you conclude about the proportion of adult Americans who think that taxing sugary beverages is an acceptable intervention in an effort to reduce consumption of sugary beverages?
b. Is it reasonable to say that the data provide strong support for the alternative hypothesis?
c. Is it reasonable to say that the data provide strong evidence against the null hypothesis?

10.11 A press release about a paper that appeared in *The Journal of Youth and Adolescence* (www.springer.com/about1springer/media/springer1select?SGWID50-11001-6-1433942-0, August 26, 2013, retrieved May 8, 2017) was titled **"Video Games Do Not Make Vulnerable Teens More Violent."** The press release includes the following statement about the study described in the paper: "Study finds no evidence that violent video games increase antisocial behavior in youths with pre-existing psychological conditions." In the context of a hypothesis test with the null hypothesis being that video games do not increase antisocial behavior, explain why the title of the press release is misleading.

10.12 In a hypothesis test, what does it mean to say that the null hypothesis was not rejected?

ADDITIONAL EXERCISES

10.13 Which of the following are legitimate hypotheses?
a. $p = 0.65$
b. $\hat{p} = 0.90$
c. $\hat{p} = 0.10$
d. $p = 0.45$
e. $p > 4.30$

10.14 Which of the following specify legitimate pairs of null and alternative hypotheses?
a. $H_0: p = 0.25$ $H_a: p > 0.25$
b. $H_0: p < 0.40$ $H_a: p > 0.40$
c. $H_0: p = 0.40$ $H_a: p < 0.65$
d. $H_0: p \neq 0.50$ $H_a: p = 0.50$
e. $H_0: p = 0.50$ $H_a: p > 0.50$
f. $H_0: \hat{p} = 0.25$ $H_a: \hat{p} > 0.25$

10.15 A college has decided to introduce the use of plus and minus with letter grades, as long as there is convincing evidence that more than 60% of the faculty favor the change. A random sample of faculty will be selected, and the resulting data will be used to test the relevant hypotheses. If p represents the proportion of *all* faculty who favor a change to plus–minus grading, which of the following pairs of hypotheses should be tested?
$$H_0: p = 0.6 \text{ versus } H_a: p < 0.6$$
or
$$H_0: p = 0.6 \text{ versus } H_a: p > 0.6$$
Explain your choice.

10.16 A television station has been providing live coverage of a sensational criminal trial. The station's program director wants to know if more than half of potential viewers prefer a return to regular daytime programming. A survey of randomly selected viewers is conducted. With p representing the proportion of all viewers who prefer regular daytime programming, what hypotheses should the program director test?

SECTION 10.2 Potential Errors in Hypothesis Testing

Once hypotheses have been formulated, a hypothesis test is carried out to decide whether H_0 should be rejected. Just as a jury may reach the wrong verdict in a trial, there is some chance that sample data may lead you to the wrong conclusion about a population characteristic in a hypothesis test. In this section, the kinds of errors that can occur are considered.

One incorrect conclusion in a criminal trial is for a jury to convict an innocent person, and another is for a guilty person to be found not guilty. Similarly, there are two different types of errors that might be made when reaching a decision in a hypothesis testing problem. One type of error involves rejecting H_0 even though the null hypothesis is true. The second type of error results from failing to reject H_0 when it is false. These errors are known as Type I and Type II errors, respectively.

DEFINITION

A **Type I error** is the error of rejecting H_0 when H_0 is true

A **Type II error** is the error of failing to reject H_0 when H_0 is false

---

When testing hypotheses about a population characteristic, the only way to guarantee that neither type of error occurs is to base the decision on a census of the entire population. When you base a decision on sample data (which provide incomplete information abut the population), you run the risk of making the wrong decision—there is some chance that you will make a Type I error and some chance that you will make a Type II error.

Example 10.6 **On-Time Arrivals**

The **U.S. Department of Transportation** reports that for 2015, 79.9% of all domestic passenger flights arrived within 15 minutes of the scheduled arrival time (***Air Travel Consumer Report*, February 2016)**. Suppose that an airline with a poor on-time record decides to offer its employees a bonus if the airline's proportion of on-time flights exceeds the overall industry rate of 0.799 in an upcoming month. In this context, p will represent the actual proportion of the airline's flights that are on time during the month of interest. A random sample of flights could be selected and used as a basis for choosing between

$$H_0: p = 0.799 \text{ and } H_a: p > 0.799$$

In this context, a Type I error (rejecting a true H_0) is concluding that the airline on-time rate exceeds the overall industry rate, when in fact the airline does not have a better on-time record. This Type I error would result in the airline rewarding its employees when the proportion of on-time flights was not actually greater than 0.799. A Type II error (not rejecting a false H_0) is not concluding that the airline's on-time proportion is better than the industry proportion when the airline really did have a better on-time record. A Type II error would result in the airline employees *not* receiving a reward that they deserved. Notice that the consequences associated with Type I and Type II errors are different.

Example 10.7 **Treating Blocked Arteries**

The article **"Boston Scientific Stent Study Flawed" (*The Wall Street Journal*, August 14, 2008)** described a research study conducted by Boston Scientific Corporation. Boston Scientific developed a new heart stent used to treat arteries blocked by heart disease. The new stent, called the "Liberte," is made of thinner metal than heart stents that were currently in use, including a stent called the "Express" that is also made by Boston Scientific. Because the new stent was made of thinner metal, the manufacturer thought it would be easier for doctors to direct through a patient's arteries to get to a blockage.

One type of problem that can sometimes occur after a stent is used is that the artery can become blocked again. In order to obtain approval to manufacture and sell the new Liberte stent in the United States, the Food and Drug Administration (FDA) required Boston Scientific to provide evidence that the proportion of patients receiving the Liberte stent who experienced a re-blocked artery was less than 0.10. The proportion of 0.10 (or 10%) was determined by the FDA based on the re-block proportions for currently approved stents, including the Express stent, which had a historical re-block proportion of 0.07.

Because Boston Scientific was asked to provide evidence that the re-block proportion for the new Liberte stent was less than 0.10, an alternative hypothesis of $H_a: p < 0.10$ was selected. This resulted in the following null and alternative hypotheses:

$$H_0: p = 0.10 \quad H_a: p < 0.10$$

The null hypothesis states that the re-block proportion for the new stent is 0.10 (or equivalently that 10% of patients receiving the new stent experience a re-blocked artery). The alternative hypothesis states that the re-block proportion is less than 0.10.

Consider Type I and Type II errors in this context:

Error	Definition of Error	Description of Error in Context	Consequence of Error
Type I error	Reject a true H_0	Conclude that the re-block proportion for the new stent is less than 0.10 when it really is 0.10 (or greater).	The new stent is approved for sale. A greater proportion of patients will experience re-blocked arteries.
Type II error	Fail to reject a false H_0	You are not convinced that the re-block proportion is less than 0.10 when it really is less than 0.10.	The new stent is not approved for sale. Patients and doctors will not benefit from the new design that makes the new stent easier to use.

Notice that the consequences associated with Type I and Type II errors are very different. This is something to consider when evaluating the acceptable error risks.

Examples 10.6 and 10.7 illustrate the two different types of errors that might occur when testing hypotheses. Type I and Type II errors—and the associated consequences of making such errors—are different. The accompanying box introduces the terminology and notation used to describe error probabilities.

The **probability of a Type I error** is denoted by α.

The **probability of a Type II error** is denoted by β.

In a hypothesis test, the probability of a Type I error, α, is also called the **significance level**. For example, a hypothesis test with $\alpha = 0.01$ is said to have a significance level of 0.01.

Example 10.8 **Early Detection of Lung Cancer**

Early detection has been shown to increase the chance of survival for patients with lung cancer. The paper **"Urinary Protein Biomarkers in the Early Detection of Lung Cancer"** (*Cancer Prevention Research* **[2015]: 111–117**) includes data from a study to determine if it is possible to accurately diagnose lung cancer using biomarkers found in a patient's urine. The researchers developed a screening method and then tested the method with a group of people who were known to have lung cancer and a group of patients who were known to not have lung cancer. The paper included the following information:

- The test correctly identified lung cancer in 39 of 54 patients known to have lung cancer.
- The test correctly identified as cancer free all of the 49 people tested who were known to not have lung cancer.

You can think of using this test to choose between two hypotheses:

H_0: patient has lung cancer

H_a: patient does not have lung cancer

Although these are not "statistical hypotheses" (statements about a population characteristic), the possible decision errors are analogous to Type I and Type II errors.

In this situation, believing that a patient with lung cancer is cancer-free would be a Type I error—rejecting the hypothesis of lung cancer when it is, in fact, true. Believing that a cancer-free patient may have lung cancer is a Type II error—not rejecting the null hypothesis when it is, in fact, false. Based on the preliminary study results, you can estimate the error probabilities. The probability of a Type I error, α, is approximately $(54 - 39)/54 = 15/54 = 0.278$. The probability of a Type II error, β, is approximately $0/49 = 0$.

The ideal hypothesis test procedure would have both $\alpha = 0$ and $\beta = 0$. However, if you must base your decision on incomplete information—a sample rather than a census— this is impossible to achieve. Standard test procedures allow you to specify a desired value for α, but they provide no direct control over β. Because α represents the probability of rejecting a true null hypothesis, selecting a significance level $\alpha = 0.05$ results in a test procedure that, used over and over with different samples, rejects a *true* null hypothesis about 5 times in 100. Selecting $\alpha = 0.01$ results in a test procedure with a Type I error rate of 1% in long-term repeated use. Choosing a smaller value for α implies that the user wants a procedure with a smaller Type I error rate.

One question arises naturally at this point: If you can select α (the probability of making a Type I error), why would you ever select $\alpha = 0.05$ rather than $\alpha = 0.01$? Why not always select a very small value for α? To achieve a small probability of making a Type I error, the corresponding test procedure must require very strong evidence against H_0 before the null hypothesis can be rejected. Although this makes a Type I error unlikely, it increases the risk of a Type II error (*not* rejecting H_0 when it should have been rejected). You must balance the consequences of Type I and Type II errors. If a Type II error has serious consequences, it may be a good idea to select a somewhat larger value for α.

In general, there is a compromise between small α and small β, leading to the following widely accepted principle for selecting the significance level to be used in a hypothesis test.

After assessing the consequences of Type I and Type II errors, identify the largest α that is acceptable. Then carry out the hypothesis test using this maximum acceptable value as the level of significance (because using any smaller value for α will result in a larger value for β).

Example 10.9 | Heart Stents Revisited

In Example 10.7, the following hypotheses were selected:

$$H_0: p = 0.10 \qquad H_a: p < 0.10$$

where p represents the proportion of all patients receiving a new Liberte heart stent who experience a re-blocked artery. For these hypotheses, a consequence of a Type I error is that the new stent would be approved for sale even though the re-block proportion is actually 0.10 or greater, and that more patients will experience re-blocked arteries. A consequence of a Type II error is that the new stent would not be approved for sale even though the actual re-block proportion is less than 0.10, so patients and doctors would not benefit from the design that makes the new stent easier to use. A Type I error would result in the approval of a heart stent for which 10% or more of the patients receiving the stent would experience a re-blocked artery. Because this represents an unnecessary risk to patients (given that other stents with lower re-block proportions are available), a small value for α (Type I error probability), such as $\alpha = 0.01$, would be selected. Of course, selecting a small value for α increases the risk of a Type II error (not approving the new stent when the re-block proportion is actually acceptable).

Summing It Up—Section 10.2

The following learning objectives were addressed in this section:

Conceptual Understanding

C4: Know the two types of errors that are possible in a hypothesis test.
A Type I error is the error of rejecting H_0 when H_0 is true. A Type II error is the error of failing to reject H_0 when H_0 is false. Examples 10.6 and 10.7 illustrate identifying Type I and Type II errors in context.

C5: Understand how the significance level of a test relates to the probability of a Type I error.
The significance level for a test is the probability of a Type I error, α.

Mastering the Mechanics

M2: Describe Type I and Type II errors in context.
Examples 10.6 and 10.7 illustrate describing Type I and Type II errors in context.

M5: Select an appropriate significance level based on consideration of the consequences of Type I and Type II errors.
Choosing a small value for the significance level will make the probability of making a Type I error small, but it can increase the probability of a Type II error. For this reason, you need to assess the consequences of making each type of error before selecting a significance level. The selection of an appropriate significance level in a given context is illustrated in Example 10.9.

SECTION 10.2 EXERCISES

Each Exercise Set assesses the following chapter learning objectives: C4, C5, M2, M5

SECTION 10.2 Exercise Set 1

10.17 One type of error in a hypothesis test is rejecting the null hypothesis when it is true. What is the other type of error that might occur when a hypothesis test is carried out?

10.18 Suppose that for a particular hypothesis test, the consequences of a Type I error are very serious. Would you want to carry out the test using a small significance level α (such as 0.01) or a larger significance level (such as 0.10)? Explain the reason for your choice.

10.19 Occasionally, warning flares of the type contained in most automobile emergency kits fail to ignite. A consumer group wants to investigate a claim that the proportion of defective flares made by a particular manufacturer is higher than the advertised value of 0.10. A large number of flares will be tested, and the results will be used to decide between $H_0: p = 0.10$ and $H_a: p > 0.10$, where p represents the actual proportion of defective flares made by this manufacturer. If H_0 is rejected, charges of false advertising will be filed against the manufacturer.
a. Explain why the alternative hypothesis was chosen to be $H_a: p > 0.10$.
b. Complete the last two columns of the following table. (Hint: See Example 10.7 for an example of how this is done.)

Error	Definition of Error	Description of Error in Context	Consequence of Error
Type I error	Reject a true H_0		
Type II error	Fail to reject a false H_0		

10.20 A television manufacturer states that at least 90% of its TV sets will not need service during the first 3 years of operation. A consumer group wants to investigate this statement. A random sample of $n = 100$ purchasers is selected and each person is asked if the set purchased needed repair during the first 3 years. Let p denote the proportion of all sets made by this manufacturer that will not need service in the first 3 years. The consumer group does not want to claim false advertising unless there is strong evidence that $p < 0.90$. The appropriate hypotheses are then $H_0: p = 0.90$ versus $H_a: p < 0.90$.
a. In the context of this problem, describe Type I and Type II errors, and discuss the possible consequences of each.
b. Would you recommend a test procedure that uses $\alpha = 0.01$ or one that uses $\alpha = 0.10$? Explain. (Hint: See Example 10.9.)

10.21 The paper **"Breast MRI as an Adjunct to Mammography for Breast Cancer Screening in High-Risk Patients"** (*American Journal of Roentgenology* **[2015]: 889–897**) describes a study that investigated the usefulness of using MRI (magnetic resonance imaging) to diagnose breast cancer. MRI exams from 650 women were reviewed. Of the 650 women, 13 had breast cancer, and the MRI exam detected breast cancer in 12 of these women. Of the 637 women who did not have breast cancer, the MRI correctly identified that no cancer was present for 547 of them. The accompanying table summarizes this information.

	Breast Cancer Present	Breast Cancer Not Present	Total
MRI Indicated Breast Cancer	12	90	102
MRI Did Not Indicate Breast Cancer	1	547	548
Total	13	637	650

Suppose that an MRI exam is used to decide between the two hypotheses

H_0 : A woman does not have breast cancer
H_a : A woman has breast cancer

(Although these are not hypotheses about a population characteristic, this exercise illustrates the definitions of Type I and Type II errors.)

a. One possible error would be deciding that a woman who has breast cancer is cancer free. Is this a Type I error or a Type II error? Use the information in the table to approximate the probability of this type of error. (Hint: See Example 10.8.)
b. There is a second type of error that is possible in this context. Describe this error and use the information in the table to approximate the probability of this type of error.

SECTION 10.2 Exercise Set 2

10.22 One type of error in a hypothesis test is failing to reject a false null hypothesis. What is the other type of error that might occur when a hypothesis test is carried out?

10.23 Suppose that for a particular hypothesis test, the consequences of a Type I error are not very serious, but there are serious consequences associated with making a Type II error. Would you want to carry out the test using a small significance level α (such as 0.01) or a larger significance level (such as 0.10)? Explain the reason for your choice.

10.24 A manufacturer of handheld calculators receives large shipments of printed circuits from a supplier. It is too costly and time-consuming to inspect all incoming circuits, so when each shipment arrives, a sample is selected for inspection. A shipment is defined to be of inferior quality if it contains more than 1% defective circuits. Information from the sample is used to test $H_0: p = 0.01$ versus $H_a: p > 0.01$, where p is the actual proportion of defective circuits in the shipment. If the null hypothesis is not rejected, the shipment is accepted, and the circuits are used in the production of calculators. If the null hypothesis is rejected, the entire shipment is returned to the supplier because of inferior quality.
a. Complete the last two columns of the following table. (Hint: See Example 10.7 for an example of how this is done.)

Error	Definition of Error	Description of Error in Context	Consequence of Error
Type I error	Reject a true H_0		
Type II error	Fail to reject a false H_0		

b. From the calculator manufacturer's point of view, which type of error would be considered more serious? Explain.
c. From the printed circuit supplier's point of view, which type of error would be considered more serious? Explain.

10.25 An automobile manufacturer is considering using robots for part of its assembly process. Converting to robots is expensive, so it will be done only if there is strong evidence that the proportion of defective installations is less for the robots than for human assemblers. Let p denote the actual proportion of defective installations for the robots. It is known that the proportion of defective installations for human assemblers is 0.02.
a. Which of the following pairs of hypotheses should the manufacturer test?

$$H_0: p = 0.02 \text{ versus } H_a: p < 0.02$$

or

$$H_0: p = 0.02 \text{ versus } H_a: p > 0.02$$

Explain your choice.
b. In the context of this exercise, describe Type I and Type II errors.
c. Would you prefer a test with $\alpha = 0.01$ or $\alpha = 0.10$? Explain your reasoning.

10.26 Medical personnel are required to report suspected cases of child abuse. Because some diseases have symptoms that are similar to those of child abuse, doctors who see a child with these symptoms must decide between two competing hypotheses:

H_0: symptoms are due to child abuse
H_a: symptoms are not due to child abuse

(Although these are not hypotheses about a population characteristic, this exercise illustrates the definitions of Type I and Type II errors.) The article **"Blurred Line Between Illness, Abuse Creates Problem for Authorities"** (*Macon Telegraph,* **February 28, 2000)** included the following quote from a doctor in Atlanta regarding the consequences of making an incorrect decision: "If it's disease, the worst you have is an angry family. If it is abuse, the other kids (in the family) are in deadly danger."

a. For the given hypotheses, describe Type I and Type II errors.

b. Based on the quote regarding consequences of the two kinds of error, which type of error is considered more serious by the doctor quoted? Explain.

ADDITIONAL EXERCISES

10.27 Describe the two types of errors that might be made when a hypothesis test is carried out.

10.28 Give an example of a situation where you would want to select a small significance level.

10.29 Give an example of a situation where you would not want to select a very small significance level.

10.30 How accurate are DNA paternity tests? By comparing the DNA of the baby and the DNA of a man that is being tested, one maker of DNA paternity tests claims that their test is 100% accurate if the man is not the father and 99.99% accurate if the man is the father **(IDENTIGENE, www.dnatesting .com/paternity-test-questions/paternity-test-accuracy/, retrieved November 16, 2016).**

a. Consider using the result of this DNA paternity test to decide between the following two hypotheses:

H_0: a particular man is not the father
H_a: a particular man is the father

In the context of this problem, describe Type I and Type II errors. (Although these are not hypotheses about a population characteristic, this exercise illustrates the definitions of Type I and Type II errors.)

b. Based on the information given, what are the values of α, the probability of a Type I error, and β, the probability of a Type II error?

10.31 **The United States Elections Project (www.electproject .org/2016g, retrieved January 22, 2017)** reported that 57.8% of registered voters in California voted in the 2016 presidential election and that this was less than the national percentage of 60.0%. Explain why it is not necessary to carry out a hypothesis test to determine if the proportion of registered voters in California who voted in the election is less than the national proportion of 0.600.

SECTION 10.3 The Logic of Hypothesis Testing—An Informal Example

Now that the necessary foundation is in place, you are ready to see how knowing about the sampling distribution of a sample proportion makes it possible to reach a conclusion in a hypothesis test. A simple example will be used to illustrate the reasoning that leads to a choice between a null and an alternative hypothesis.

With p denoting the proportion in the population that possess some characteristic of interest, recall that three things are known about the sampling distribution of the sample proportion $\hat{p}$ when a random sample of size n is selected:

1. The mean of the sampling distribution of $\hat{p}$ is p. This means $\mu_{\hat{p}} = p$.

2. The standard error (standard deviation) of $\hat{p}$ is $\sqrt{\frac{p(1-p)}{n}}$.

3. If n is large, then the sampling distribution of $\hat{p}$ is approximately normal.

The sampling distribution describes the behavior of the statistic $\hat{p}$. If the sampling distribution of $\hat{p}$ is approximately normal, you can use what you know about normal distributions.

An Informal Example

When data are used to choose between a null and an alternative hypothesis, there are two possible conclusions. You either decide to reject the null hypothesis in favor of the alternative hypothesis or you decide that the null hypothesis cannot be rejected. The fundamental idea behind hypothesis testing procedures is this: *You reject the null hypothesis H_0 if the observed sample is very unlikely to have occurred when H_0 is true.* Example 10.10 illustrates this in the context of testing hypotheses about a population proportion when the sample size is large.

Example 10.10	Impact of Food Labels

An **Associated Press** survey was conducted to investigate how people use the nutritional information provided on food package labels (*Spartansburg Herald*, **March 19, 2016**). Interviews were conducted with 1003 randomly selected adult Americans, and each participant was asked a series of questions, including the following two:

> Question 1: When purchasing packaged food, how often do you check the nutrition labeling on the package?
>
> Question 2: How often do you purchase foods that are bad for you, even after you've checked the nutrition labels?

It was reported that 582 responded "frequently" to Question 1 and 441 responded "very often or somewhat often" to Question 2.

Start by considering the responses to the first question. Based on these data, is it reasonable to conclude that a majority of adult Americans (more than 50%) frequently check nutrition labels? With p denoting the proportion of all American adults who frequently check nutrition labels (a population proportion), this question can be answered by testing the following hypotheses:

$$H_0: p = 0.50$$
$$H_a: p > 0.50$$

This alternative hypothesis states that the population proportion of adult Americans who frequently check nutrition labels is greater than 0.50. This would mean that more than half (a majority) frequently check nutrition labels. The null hypothesis will be rejected only if there is convincing evidence against $p = 0.50$.

For this sample,

$$\hat{p} = \frac{582}{1003} = 0.580$$

The observed sample proportion is certainly greater than 0.50, but could this just be due to sampling variability? That is, when $p = 0.50$ (meaning H_0 is true), the sample proportion $\hat{p}$ usually differs somewhat from 0.50 simply because of chance variability from one sample to another. Is it plausible that a sample proportion of $\hat{p} = 0.580$ occurred only as a result of this sampling variability, or is it unusual to observe a sample proportion this large when $p = 0.50$?

If the null hypothesis, $H_0: p = 0.50$, is true, you know the following:

What You Know	How You Know It...
The $\hat{p}$ values will center around 0.50—that is, the mean of the $\hat{p}$ distribution is 0.50.	The sample is described as 1003 randomly selected adult Americans. When random samples are selected from a population, $\mu_{\hat{p}} = p$. If the null hypothesis is true, $p = 0.50$, so $\mu_{\hat{p}} = 0.50$.
The standard deviation of $\hat{p}$, which describes how much the $\hat{p}$ values tend to vary from sample to sample, is $\sqrt{\dfrac{(0.50)(1 - 0.50)}{1003}}$.	When random samples of size n are selected from a population, $\sigma_{\hat{p}} = \sqrt{\dfrac{p(1 - p)}{n}}$. The sample size is $n = 1003$ and, if the null hypothesis is true, $p = 0.50$, so $$\sigma_{\hat{p}} = \sqrt{\frac{(0.50)(1 - 0.50)}{1003}}.$$
The sampling distribution of $\hat{p}$ is approximately normal.	When random samples of size n are selected from a population and the sample size is large, the sampling distribution of $\hat{p}$ is approximately normal. When $p = 0.50$, the sample size of $n = 1003$ is large enough for the $\hat{p}$ distribution to be approximately normal.

(continued)

What You Know	How You Know It...
The statistic $$z = \frac{\hat{p} - 0.50}{\sqrt{\dfrac{(0.50)(1-0.50)}{1003}}}$$ has (approximately) a standard normal distribution.	Transforming a statistic that has a normal distribution to a z statistic (by subtracting its mean and then dividing by its standard deviation) results in a statistic that has a standard normal distribution. Since $\hat{p}$ has (approximately) a normal distribution with mean 0.50 and standard deviation $\sqrt{\dfrac{(0.50)(1-0.50)}{1003}}$ when the null hypothesis is true, the statistic $z = \dfrac{\hat{p} - 0.50}{\sqrt{\dfrac{(0.50)(1-0.50)}{1003}}}$ will have (approximately) a standard normal distribution.

You can now use what you know about normal distributions to judge whether the observed sample proportion would be surprising if the null hypothesis were true. For the sample in the study, $\hat{p} = \frac{582}{1003} = 0.580$. Using this $\hat{p}$ value in the z statistic formula

$$z = \frac{\hat{p} - 0.50}{\sqrt{\dfrac{(0.50)(1-0.50)}{1003}}}$$

results in

$$z = \frac{p - 0.50}{\sqrt{\dfrac{(0.50)(0.50)}{1003}}} = \frac{0.580 - 0.50}{\sqrt{\dfrac{(0.50)(0.50)}{1003}}} = \frac{0.080}{0.016} = 5.00$$

This means that $\hat{p} = 0.580$ is 5 standard deviations greater than what you would expect it to be if the null hypothesis H_0: $p = 0.50$ were true. This would be *very* unusual for a normal distribution and can be regarded as inconsistent with the null hypothesis. You can evaluate just how unusual this would be by calculating the following probability:

P(observing a value of z at least as unusual as 5.00 when H_0 is true)

$= P(z \geq 5.00$ when H_0 is true)

$=$ area under the z curve to the right of 5.00

≈ 0

When H_0 is true, there is virtually no chance of seeing a sample proportion and corresponding z value this extreme as a result of sampling variability alone. In this case, the evidence for rejecting H_0 in favor of H_a is very compelling.

Interestingly, in spite of the fact that there is strong evidence that a majority of American adults frequently check nutrition labels, the responses to the second question suggest that the percentage of people who then ignore the information on the label and purchase "bad" foods anyway is not small—the sample proportion of adults who gave a response of very often or somewhat often was 0.440.

___

The preceding example illustrates the logic of hypothesis testing. It is worth reading a second time, because understanding the reasoning illustrated is critical to mastering the hypothesis testing methods you will see in the next section. You begin by assuming that the null hypothesis is correct. The sample is then examined in light of this assumption. If the observed sample proportion would not be unusual when H_0 is true, then chance variability from one sample to another is a plausible explanation for what has been observed, and H_0 would not be rejected.

On the other hand, if the observed sample proportion would have been very unlikely when H_0 is true, you would interpret this as convincing evidence against the null hypothesis and you would reject H_0. The decision to reject or to fail to reject the null hypothesis is based on an assessment of how unlikely the observed sample data would be if the null hypothesis were true.

Summing It Up—Section 10.3

The following learning objective was addressed in this section:

Conceptual Understanding

C6: Understand the reasoning used to reach a decision in a hypothesis test.

In a hypothesis test, you begin by assuming that the null hypothesis is true. You then consider whether the observed data would have been surprising or not surprising given that the null hypothesis is true. If the observed data would have been unlikely to occur if the null hypothesis is true, this is evidence against the null hypothesis, and the null hypothesis would be rejected. This reasoning is illustrated in Example 10.10.

SECTION 10.3 EXERCISES

Each Exercise Set assesses the following chapter learning objectives: C6

SECTION 10.3 Exercise Set 1

10.32 In the report **"Healthy People 2020 Objectives for the Nation,"** The Centers for Disease Control and Prevention (CDC) set a goal of 0.341 for the proportion of mothers who will still be breastfeeding their babies one year after birth **(www.cdc.gov/breastfeeding/policy/hp2020.htm, April 11, 2016, retrieved November 28, 2016)**. The CDC also estimated the proportion who were still being breastfed one year after birth to be 0.307 for babies born in 2013 **(www.cdc.gov/breastfeeding/pdf/2016breastfeedingreportcard.pdf, retrieved November 28, 2016)**. This estimate was based on a survey of women who had given birth in 2013. Suppose that the survey used a random sample of 1000 mothers and that you want to use the survey data to decide if there is evidence that the goal is not being met. Let p denote the population proportion of all mothers of babies born in 2013 who were still breast-feeding at 12 months. (Hint: See Example 10.10.)

a. Describe the shape, center, and variability of the sampling distribution of $\hat{p}$ for random samples of size 1000 if the null hypothesis H_0: $p = 0.341$ is true.

b. Would you be surprised to observe a sample proportion as small as $\hat{p} = 0.333$ for a sample of size 1000 if the null hypothesis H_0: $p = 0.341$ were true? Explain why or why not.

c. Would you be surprised to observe a sample proportion as small as $\hat{p} = 0.310$ for a sample of size 1000 if the null hypothesis H_0: $p = 0.341$ were true? Explain why or why not.

d. The actual sample proportion observed in the study was $\hat{p} = 0.307$. Based on this sample proportion, is there convincing evidence that the goal is not being met, or is the observed sample proportion consistent with what you would expect to see when the null hypothesis is true? Support your answer with a probability calculation.

10.33 The paper **"Bedtime Mobile Phone Use and Sleep in Adults"** (*Social Science and Medicine* **[2016]: 93–101**) describes a study of 844 adults living in Belgium. Suppose that it is reasonable to regard this sample as a random sample of adults living in Belgium. You want to use the survey data to decide if there is evidence that a majority of adults living in Belgium take their cell phones to bed with them. Let p denote the population proportion of all adults living in Belgium who take their cell phones to bed with them. (Hint: See Example 10.10.)

a. Describe the shape, center, and variability of the sampling distribution of $\hat{p}$ for random samples of size 844 if the null hypothesis H_0: $p = 0.50$ is true.

b. Would you be surprised to observe a sample proportion as large as $\hat{p} = 0.52$ for a sample of size 844 if the null hypothesis H_0: $p = 0.50$ were true? Explain why or why not.

c. Would you be surprised to observe a sample proportion as large as $\hat{p} = 0.54$ for a sample of size 844 if the null hypothesis H_0: $p = 0.50$ were true? Explain why or why not.

d. The actual sample proportion observed in the study was $\hat{p} = 0.59$. Based on this sample proportion, is there convincing evidence that the null hypothesis H_0: $p = 0.50$ is not true, or is $\hat{p}$ consistent with what you would expect to see when the null hypothesis is true? Support your answer with a probability calculation.

e. Do you think it would be reasonable to generalize the concusion of this test to adults living in the United States? Explain why or why not.

SECTION 10.3 Exercise Set 2

10.34 The article **"Most Customers OK with New Bulbs"** (*USA TODAY*, February 18, 2011) describes a survey of 1016 randomly selected adult Americans. Each person in the sample was asked if they have replaced standard light bulbs in their home with the more energy efficient compact fluorescent (CFL) bulbs. Suppose you want to use the survey data to determine if there is evidence that more than 70% of adult Americans have replaced standard bulbs with CFL bulbs. Let p denote the population proportion of all adult Americans who have replaced standard bulbs with CFL bulbs.

a. Describe the shape, center, and variability of the sampling distribution of $\hat{p}$ for random samples of size 1016 if the null hypothesis $H_0: p = 0.70$ is true.

b. Would you be surprised to observe a sample proportion as large as $\hat{p} = 0.72$ for a sample of size 1016 if the null hypothesis $H_0: p = 0.70$ were true? Explain why or why not.

c. Would you be surprised to observe a sample proportion as large as $\hat{p} = 0.75$ for a sample of size 1016 if the null hypothesis $H_0: p = 0.70$ were true? Explain why or why not.

d. The actual sample proportion observed in the study was $\hat{p} = 0.71$. Based on this sample proportion, is there convincing evidence that more than 70% have replaced standard bulbs with CFL bulbs, or is this sample proportion consistent with what you would expect to see when the null hypothesis is true? Support your answer with a probability calculation.

10.35 The report **"Robot, You Can Drive My Car: Majority Prefer Driverless Technology" (Transportation Research Institute University of Michigan, www.umtri.umich.edu/what-were-doing/news/robot-you-can-drive-my-car-majority-prefer-driverless-technology, July 22, 2015, retrieved May 8, 2017)** describes a survey of 505 licensed drivers. Each driver in the sample was asked if they would prefer to keep complete control of the car while driving, to use a partially self-driving car that allowed partial driver control, or to turn full control over to a driverless car. Suppose that it is reasonable to regard this sample as a random sample of licensed drivers in the United States, and that you want to use the data from this survey to decide if there is evidence that fewer than half of all licensed drivers in the United States prefer to keep complete control of the car while driving.

a. Describe the shape, center, and variability of the sampling distribution of $\hat{p}$ for random samples of size 505 if the null hypothesis $H_0: p = 0.50$ is true.

b. Would you be surprised to observe a sample proportion as small as $\hat{p} = 0.48$ for a sample of size 505 if the null hypothesis $H_0: p = 0.50$ were true? Explain why or why not.

c. Would you be surprised to observe a sample proportion as small as $\hat{p} = 0.46$ for a sample of size 505 if the null hypothesis $H_0: p = 0.50$ were true? Explain why or why not.

d. The actual sample proportion observed in the study was $\hat{p} = 0.44$. Based on this sample proportion, is there convincing evidence that fewer than 50% of licensed drivers prefer to keep complete control of the car when driving, or is the sample proportion consistent with what you would expect to see when the null hypothesis is true? Support your answer with a probability calculation.

ADDITIONAL EXERCISES

10.36 Every year on Groundhog Day (February 2), the famous groundhog "Punxsutawney Phil" tries to predict whether there will be 6 more weeks of winter. The article **"Groundhog Has Been Off Target" (USA TODAY, February 1, 2011)** states that "based on weather data, there is no predictive skill for the groundhog." Suppose that you plan to take a random sample of 20 years and use weather data to determine the proportion of these years the groundhog's prediction was correct.

a. Describe the shape, center, and variability of the sampling distribution of $\hat{p}$ for samples of size 20 if the groundhog has only a 50–50 chance of making a correct prediction.

b. Based on your answer to Part (a), what sample proportion values would convince you that the groundhog's predictions have a better than 50–50 chance of being correct?

10.37 **USA TODAY, (February 17, 2011)** described a survey of 1008 American adults. One question on the survey asked people if they had ever sent a love letter using e-mail. Suppose that this survey used a random sample of adults and that you want to decide if there is evidence that more than 20% of American adults have written a love letter using e-mail.

a. Describe the shape, center, and variability of the sampling distribution of $\hat{p}$ for random samples of size 1008 if the null hypothesis $H_0: p = 0.20$ is true.

b. Based on your answer to Part (a), what sample proportion values would convince you that more than 20% of adults have sent a love letter via e-mail?

SECTION 10.4 A Procedure for Carrying Out a Hypothesis Test

The first three sections of this chapter provide what is needed to develop a systematic strategy for testing hypotheses. In Section 10.1, you saw how research questions or claims about a population are translated into two competing hypotheses—the null hypothesis and the alternative hypothesis. You also considered the two possible conclusions in a hypothesis test—rejecting the null hypothesis and failing to reject the null hypothesis. In Section 10.2, you considered the two types of errors (Type I and Type II) that are possible. You also saw that evaluating the consequences of Type I and Type II errors helps in the selection of an appropriate significance level α. Finally, in Section 10.3, you saw that the reasoning process used to reach a decision in a hypothesis test is straightforward and intuitive. You first assume that the null hypothesis is true. You then determine if the sample data are consistent with the null hypothesis (the data would not be surprising if the null hypothesis were true) or if the sample data convinces you that the null hypothesis should be rejected (the data would be unlikely if the null hypothesis were true).

In Example 10.10 (Section 10.3), sample data from a survey of 1003 adult Americans were used to decide between

$$H_0: p = 0.50$$

and

$$H_a: p > 0.50$$

where p represented the population proportion of adult Americans who frequently check nutritional labels when purchasing packaged foods. The decision in Example 10.10 was based on calculating the probability of seeing sample data as unusual as what was observed if the null hypothesis is true. This calculated probability was approximately 0, so a decision was made to reject the null hypothesis.

The probability calculated in Example 10.10 is called a **P-value**. A P-value specifies how likely it is that a sample would be as or more extreme than the one observed if H_0 were true. To find the P-value, a z statistic was calculated:

$$z = \frac{\hat{p} - hypothesized\ value}{standard\ deviation\ of\ \hat{p}}$$

This z is called a **test statistic**. Knowing the value of the test statistic (for example $z = 5.00$) allows calculation of the corresponding P-value (for example, P-value ≈ 0). The P-value is used to make a decision in a hypothesis test. If the P-value is small (meaning that sample data as or more extreme as what was observed would be very unlikely if the null hypothesis were true), you reject the null hypothesis. If the P-value is not small, you fail to reject the null hypothesis.

A **test statistic** is calculated using sample data. The value of the test statistic is used to determine the P-value associated with the test.

The **P-value** (also sometimes called the **observed significance level**) is a measure of inconsistency between the null hypothesis and the observed sample. It is the probability, assuming that H_0 is true, of obtaining a test statistic value at least as inconsistent with H_0 as what actually resulted.

You reject the null hypothesis when the P-value is small.

The test statistic appropriate for a particular hypothesis testing situation will depend on the answers to the four key questions that are used to select a method. Recall that in the playbook analogy, these are the questions you need to answer in order to select the right statistics "play."

The null hypothesis will be rejected if the P-value is small, but just how small does the P-value have to be in order to reject H_0? The answer depends on the significance level α (the probability of a Type I error) selected for the test. For example, suppose you select $\alpha = 0.05$. This implies that the probability of rejecting a true null hypothesis should be 0.05. To obtain a test procedure with this probability of a Type I error, you would reject the null hypothesis if the sample result is among the most unusual 5% of all samples when H_0 is true. This means that H_0 is rejected if the calculated P-value is less than 0.05. If you select $\alpha = 0.01$, H_0 will be rejected only if you observe a sample result so extreme that it would be among the most unusual 1% if H_0 is true. This would be the case if P-value < 0.01.

A decision in a hypothesis test is based on comparing the P-value to the chosen significance level α.

H_0 is rejected if P-value $< \alpha$.

H_0 is not rejected if P-value $\geq \alpha$.

For example, suppose that P-value $= 0.035$ and that a significance level of 0.05 is chosen. Then, because

$$P\text{-value} = 0.035 < 0.05 = \alpha$$

H_0 would be rejected. This would not be the case, though, for $\alpha = 0.01$, because then P-value $\geq \alpha$.

The five-step process for hypothesis testing (HMC³) first described in Chapter 7 is used to carry out a hypothesis test. The five steps are:

Step	What Is This Step?
H	**Hypotheses:** Define the hypotheses that will be tested.
M	**Method:** Use the answers to the four key questions (QSTN) to identify a potential method.
C	**Check:** Check to make sure that the method selected is appropriate. Many methods for learning from data only provide reliable information under certain conditions. Verify that these conditions are met before proceeding.
C	**Calculate:** Use the sample data to perform any necessary calculations.
C	**Communicate Results:** This is a critical step in the process. In this step, you will answer the research question of interest, explain what you have learned from the data, and acknowledge potential risks.

The accompanying table details what each of the steps entails when carrying out a hypothesis test.

Step	This Step Includes...
H Hypotheses	1. Describe the population and the population characteristic of interest. 2. Translate the research question or claim made about the population into null and alternative hypotheses.
M Method	1. Identify the appropriate test and test statistic. 2. Select a significance level for the test.
C Check	1. Verify that any conditions for the selected test are met.
C Calculate	1. Find the values of any sample statistics needed to calculate the value of the test statistic. 2. Calculate the value of the test statistic. 3. Determine the P-value for the test.
C Communicate Results	1. Compare the P-value to the selected significance level and make a decision to either reject H_0 or fail to reject H_0. 2. Provide a conclusion in words that is in context and that addresses the question of interest.

A statistical software program or a graphing calculator is often used to complete the calculate step in this process.

You will see how this process is used to test hypotheses about a population proportion in Section 10.5.

Summing It Up—Section 10.4

The following learning objectives were addressed in this section:

Conceptual Understanding

C5: Understand how the significance level of a test relates to the probability of a Type I error.

The significance level for a test is the probability of a Type I error, α.

C6: Understand the reasoning used to reach a decision in a hypothesis test.

In a hypothesis test, you begin by assuming that the null hypothesis is true. You then consider whether what was observed in the data would have been surprising or not surprising given that the null hypotheses were true. If what was observed in the data would have been unlikely to occur if the null hypothesis were true, this is evidence against the null hypothesis, and the null hypothesis would be rejected. The determination of whether the data would have been unlikely to occur if the null hypothesis were true is based on the P-value, which is the probability of observing data at least as extreme as what was observed if the null hypothesis were true. If the P-value is less than α, the significance level selected for the test, the null hypothesis is rejected.

SECTION 10.4 EXERCISES

Each Exercise Set assesses the following chapter learning objectives: C5, C6

SECTION 10.4 Exercise Set 1

10.38 Use the definition of the P-value to explain the following:
a. Why H_0 would be rejected if P-value $= 0.003$
b. Why H_0 would not be rejected if P-value $= 0.350$

10.39 According to a survey of a random sample of 2278 adult Americans conducted by the Harris Poll (**"Do Americans Prefer Name Brands or Store Brands? Well, That Depends" (theharrispoll.com, February 11, 2015, retrieved November 29, 2016)**, 1162 of those surveyed said that they prefer name brands to store brands when purchasing frozen vegetables. Suppose that you want to use this information to determine if there is convincing evidence that a majority of adult Americans prefer name-brand frozen vegetables over store brand frozen vegetables.
a. What hypotheses should be tested in order to answer this question?
b. The P-value for this test is 0.173. What conclusion would you reach if $\alpha = 0.05$?

10.40 Step 2 of the five-step process for hypothesis testing is selecting an appropriate method. What is involved in completing this step?

SECTION 10.4 Exercise Set 2

10.41 For which of the following P-values will the null hypothesis be rejected when performing a test with a significance level of 0.05?
a. 0.001   **d.** 0.047
b. 0.021   **e.** 0.148
c. 0.078

10.42 The article **"Euthanasia Still Acceptable to Solid Majority in U.S." (www.gallup.com, June 24, 2016, retrieved November 29, 2016)** summarized data from a survey of 1025 adult Americans. When asked if doctors should be able to end a terminally ill patient's life by painless means if requested to do so by the patient, 707 of those surveyed responded yes. For proposes of this exercise, assume that it is reasonable to regard this sample as a random sample of adult Americans. Suppose that you want to use the data from this survey to decide if there is convincing evidence that more than two-thirds of adult Americans believe that doctors should be able to end a terminally ill patient's life if requested to do so by the patient.
a. What hypotheses should be tested in order to answer this question?
b. The P-value for this test is 0.058. What conclusion would you reach if $\alpha = 0.05$?
c. Would you have reached a different conclusion if $\alpha = 0.10$? Explain.

10.43 Step 5 of the five-step process for hypothesis testing is communication of results. What is involved in completing this step?

ADDITIONAL EXERCISES

10.44 For which of the following combinations of *P*-value and significance level would the null hypothesis be rejected?
a. *P*-value = 0.426 $\alpha = 0.05$
b. *P*-value = 0.033 $\alpha = 0.01$
c. *P*-value = 0.046 $\alpha = 0.10$
d. *P*-value = 0.026 $\alpha = 0.05$
e. *P*-value = 0.004 $\alpha = 0.01$

10.45 Explain why a *P*-value of 0.002 would be interpreted as strong evidence against the null hypothesis.

10.46 Explain why you would not reject the null hypothesis if the *P*-value were 0.370.

SECTION 10.5 Large-Sample Hypothesis Test for a Population Proportion

In this section, you will see how the five-step process for hypothesis testing (HMC³) is used to test hypotheses about a population proportion. It may be helpful to start with a quick review of the four key questions (QSTN) introduced in Section 7.2:

Q **Question Type**	Estimation or hypothesis testing?
S **Study Type**	Sample data or experiment data?
T **Type of Data**	One variable or two? Categorical or numerical?
N **Number of Samples or Treatments**	How many samples or treatments?

When the answers to these questions are *hypothesis testing, sample data, one categorical variable, and one sample,* a method to consider is the large-sample hypothesis test for a population proportion.

Test Statistic

The test statistic for a large-sample hypothesis test for a population proportion is

$$z = \frac{\hat{p} - p_0}{\sqrt{\dfrac{p_0(1 - p_0)}{n}}}$$

where p_0 is the hypothesized value from the null hypothesis.

When the null hypothesis is true and the sample size is large, this test statistic has a distribution that is approximately standard normal.

The sample size is large enough for the distribution of the test statistic to be approximately standard normal if both np_0 and $n(1 - p_0)$ are greater than or equal to 10.

Calculating the *P*-value

When the null hypothesis is true and the sample size is large, the test statistic

$$z = \frac{\hat{p} - p_0}{\sqrt{\dfrac{p_0(1 - p_0)}{n}}}$$

has a distribution that is approximately standard normal. Because the *P*-value is the probability of observing a test statistic value that is at least as inconsistent with the null hypothesis as what was observed in the sample, calculating the *P*-value involves finding an area under the standard normal curve.

The way you find the *P*-value depends on the form of the inequality ($<$, $>$, or $\neq$) in the alternative hypothesis, H_a. Suppose, for example, that you want to test

$$H_0: p = 0.60 \qquad \text{versus} \qquad H_a: p > 0.60$$

using a large sample. The appropriate test statistic is

$$z = \frac{\hat{p} - 0.60}{\sqrt{\dfrac{(0.60)(1 - 0.60)}{n}}}$$

Values of $\hat{p}$ inconsistent with H_0 and more consistent with H_a are those that are much greater than 0.60. In fact, to reject H_0 you would need to see a value of $\hat{p}$ that is enough greater than 0.60 that it can't just be explained by sample-to-sample variability. These values of $\hat{p}$ will correspond to z values that are much greater than 0. For example, if $n = 400$ and $\hat{p} = 0.679$, then

$$z = \frac{0.679 - 0.60}{\sqrt{\dfrac{(0.60)(1 - 0.60)}{400}}} = \frac{0.079}{0.024} = 3.29$$

If H_0 is true, the value $\hat{p} = 0.679$ is more than 3 standard deviations larger than what you would have expected. This means that

P-value = *P*(*z* at least as inconsistent with H_0 as 3.29 when H_0 is true)
= *P*($z \geq 3.29$ when H_0 is true)
= area under the *z* curve to the right of 3.29
= $1 - 0.9995$
= 0.0005

This *P*-value is illustrated in Figure 10.1. If H_0 is true, in the long run only about 5 out of 10,000 samples would result in a *z* value as or more extreme than what actually resulted. This would be quite unusual. Using a significance level of $\alpha = 0.01$, you would reject the null hypothesis because the *P*-value is less than α (0.0005 $<$ 0.01).

FIGURE 10.1
Calculating a *P*-value

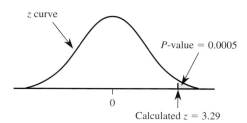

Now consider testing $H_0: p = 0.30$ versus $H_a: p \neq 0.30$. A value of $\hat{p}$ *either* much greater than 0.30 *or* much less than 0.30 is inconsistent with H_0 and provides support for H_a. This corresponds to *z* values far out in *either* tail of the *z* curve. For example, if $z = 1.75$, then (as shown in Figure 10.2)

P-value = *P*(*z* value at least as inconsistent with H_0 as 1.75 when H_0 is true)
= *P*($z \geq 1.75$ or $z \leq -1.75$ when H_0 is true)
= (*z* curve area to the right of 1.75) + (*z* curve area to the left of -1.75)
= $(1 - 0.9599) + 0.0401$
= 0.0802

FIGURE 10.2
P-value as the sum of two tail areas

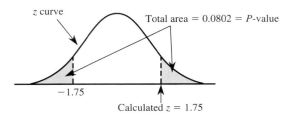

Determining the *P*-value When the Test Statistic Has a Standard Normal Distribution

1. **Upper-tailed test**
   $H_a: p >$ hypothesized value
   P-value calculated as illustrated:

   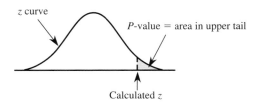

2. **Lower-tailed test**
   $H_a: p <$ hypothesized value
   P-value calculated as illustrated:

   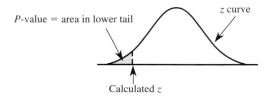

3. **Two-tailed test**
   $H_a: p \neq$ hypothesized value
   P-value calculated as illustrated:

   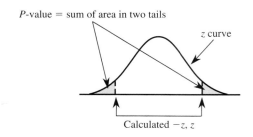

The symmetry of the *z* curve implies that when the test is two-tailed (the "not equal" alternative), it is not necessary to add two curve areas. Instead,

If *z* is positive, *P*-value = 2(area to the right of *z*).
If *z* is negative, *P*-value = 2(area to the left of *z*).

The following box summarizes the large-sample test for a population proportion.

Large-Sample Test for a Population Proportion

Appropriate when the following conditions are met:
1. The sample is a random sample from the population of interest or the sample is selected in a way that should result in a representative sample.
2. The sample size is large. This condition is met when both $np_0 \geq 10$ and $n(1 - p_0) \geq 10$.

When these conditions are met, the following test statistic can be used:

$$z = \frac{\hat{p} - p_0}{\sqrt{\dfrac{p_0(1 - p_0)}{n}}}$$

where p_0 is the hypothesized value from the null hypothesis.

Form of the null hypothesis: $H_0\!: p = p_0$

Associated *P*-value:

When the Alternative Hypothesis Is...	The *P*-value Is...
$H_a\!: p > p_0$	Area under the z curve to the right of the calculated value of the test statistic
$H_a\!: p < p_0$	Area under the z curve to the left of the calculated value of the test statistic
$H_a\!: p \neq p_0$	2(area to the right of z) if z is positive or 2(area to the left of z) if z is negative

---

To carry out the test, follow the five-step process (HMC³) described in Section 10.4.

Example 10.11   Love Those Cell Phones...

The article **"You Now Have a Shorter Attention Span Than a Goldfish" (*Time*, May 14, 2015)** describes a study of 2000 Canadians over the age of 18 that was carried out by Microsoft. Study participants were asked whether the following statement described them: "When nothing is occupying my attention, the first thing I do is reach for my phone." Of the study participants in the age group 18 to 24 years old, 77% responded "yes" to this question. Suppose that this group of 18- to 24-year-olds can be considered as a representative sample of Canadians in this age group, and also suppose that 800 of the study participants were in this age group.

Does this sample support the claim that more than 75% of all Canadians in this age group would respond "yes" to the given statement?

Begin by answering the four key questions (QSTN) for this problem.

Q Question Type	Estimation or hypothesis testing?	Hypothesis testing. You are asked to test a claim about a population (more than 75% of Canadians age 18 to 24 years would say that the given statement describes them).
S Study Type	Sample data or experiment data?	Sample data. The data are from a sample of Canadians age 18 to 24.
T Type of Data	One variable or two? Categorical or numerical?	One categorical variable. The variable here (response to statement) is categorical, with two possible values—yes and no.
N Number of Samples or Treatments	How many samples or treatments?	One.

Because the answers are *hypothesis testing, sample data, one categorical variable, and one sample*, you should consider a large-sample test for a population proportion.

Now you can use the five-step process for hypothesis testing problems (HMC³) to carry out the test.

Process Step	
H Hypotheses	The claim is about the proportion of Canadians age 18 to 24 years who would say that the given statement describes them, so the population characteristic of interest is a population proportion: p = proportion of Canadians age 18 to 24 years who would say that the given statement describes them The claim (more than 75%) translates into a hypothesis of $p > 0.75$. You want to know if there is evidence to support this claim, so this will be the alternative hypothesis. The null hypothesis is found by replacing the inequality ($>$) in the alternative hypothesis by an equal sign. **Hypotheses:** *Null hypothesis*: H_0: $p = 0.75$ *Alternative hypothesis*: H_a: $p > 0.75$
M Method	Because the answers to the four key questions are hypothesis testing, sample data, one categorical variable, and one sample, consider a large-sample hypothesis test for a population proportion. **Potential method:** Large-sample test for a population proportion. The test statistic for this test is $$z = \frac{\hat{p} - 0.75}{\sqrt{\dfrac{0.75(1 - 0.75)}{n}}}$$ The value of 0.75 in the test statistic is the hypothesized value from the null hypothesis. You also need to select a significance level for the test. In some cases, a significance level will be specified. If not, you should choose a significance level based on a consideration of the consequences of Type I and Type II errors. In this example, a Type I error would be deciding that more than 75% of Canadians age 18 to 24 years would say that the given statement describes them, when in fact the actual percentage is 75% or less. A Type II error is thinking that 75% or fewer would respond "yes" when the actual percentage is greater than 75%. In this situation, because neither type of error is much more serious than the other, you might choose a value for α of 0.05 (as opposed to something much smaller or much larger). **Significance level:** $\alpha = 0.05$
C Check	There are two conditions that need to be met in order for the large sample test for a population proportion to be appropriate. The large-sample condition is easily verified. The sample size is large enough because $np_0 = 800(0.75) = 600$ and $n(1 - p_0) = 800(0.25) = 200$ are both greater than or equal to 10. The requirement of a random or representative sample is more difficult. The researchers conducting the study indicated that they believed that the sample was representative of Canadians in this age group. If this is the case, it is reasonable to proceed.
C Calculate	Substituting the values for n and $\hat{p}$ into the test statistic formula results in **Test statistic:** $$z = \frac{\hat{p} - 0.75}{\sqrt{\dfrac{0.75(1 - 0.75)}{800}}} = \frac{0.77 - 0.75}{\sqrt{\dfrac{0.75(1 - 0.75)}{800}}} = \frac{0.02}{0.015} = 1.33$$ **Associated *P*-value:** This is an upper-tailed test (the inequality in the null hypothesis is $>$), so the *P*-value is the area under the z curve and to the right of the calculated z value. $$P\text{-value} = \text{area to the right of } 1.33$$ $$= P(z > 1.33)$$ $$= 0.0918$$
C Communicate Results	Because the *P*-value is greater than or equal to the selected significance level, you fail to reject the null hypothesis. **Decision:** $0.0918 \geq 0.05$, fail to reject H_0. The final conclusion for the test should be stated in context and answer the question posed. **Conclusion:** The sample does not provide convincing evidence that more than 75% of Canadians ages 18 to 24 think that the statement "When nothing is occupying my attention, the first thing I do is reach for my phone" describes them.

| Example 10.12 | Choosing Baby's Sex Revisited |

The preview example at the beginning of the chapter described a study published in the journal **Fertility and Sterility** and summarized in the article **"Boy or Girl: Which Gender Baby Would You Pick?" (LiveScience, March 23, 2005)**. The *LiveScience* summary states that, of 229 women surveyed who said that they would like to choose the sex of a future child, "there was no greater demand for boys or girls." However, the journal article states that, of the 229 who wanted to select the baby's sex, 89 wanted a boy and 140 wanted a girl. Does this provide evidence against the statement of no preference in the *LiveScience* summary?

Considering the four key questions (QSTN), this situation can be described as *hypothesis testing, sample data, one categorical variable* (gender preference), and *one sample*. This combination suggests a large-sample test for a population proportion.

H Hypotheses

In this example, you want to use sample data to determine if there is evidence against a claim that has been made. Before you can proceed, you need to think about the appropriate population. In this study, all of the women who participated in the survey were patients at a particular fertility clinic—the Center for Reproductive Medicine at Brigham and Women's Hospital in Boston. Because of this, it is wise to limit any conclusions to the population of all female patients of this fertility clinic who would like to choose a future baby's sex.

The claim is about the proportion of all women using this fertility clinic who want to choose a future baby's sex and who would choose a particular sex, so the population characteristic of interest is a population proportion. You can focus on the proportion who would choose a boy.

Population characteristic of interest:

p = population proportion of female patients of the Center for Reproductive Medicine who would choose a boy, out of all whose who want to choose a future baby's sex.

The claim of no preference translates into $p = 0.50$ because if there is no preference, you would expect half to choose boy and half to choose girl. Since you are asked if there is evidence against this claim, the alternative hypothesis is $p \neq 0.50$.

Hypotheses:
Null hypothesis: H_0: $p = 0.50$
Alternative hypothesis: H_a: $p \neq 0.50$

M Method

Because the answers to the four key questions are hypothesis testing, sample data, one categorical variable, and one sample, a large-sample hypothesis test for a population proportion will be considered. The test statistic for this test is

$$z = \frac{\hat{p} - 0.50}{\sqrt{\dfrac{0.50(1 - 0.50)}{n}}}$$

The value of 0.50 in the test statistic is the hypothesized value from the null hypothesis.

You also need to select a significance level for the test. Because no significance level was specified, you should choose a significance level after considering the consequences of Type I and Type II errors. In this situation, neither type of error is much more serious than the other, so you might choose a value of 0.05 for α.

Significance level: $\alpha = 0.05$

C Check

There are two conditions that must be met before using the large-sample test for a population proportion. The large-sample condition is easily verified. The sample size is large enough because $np_0 = 229(0.50) = 114.5$ and $n(1 - p_0) = 229(1 - 0.50) = 114.5$ are both greater than or equal to 10.

The requirement of a random or representative sample is more difficult. Although a random sample of patients from the clinic was selected, there was a very large non-response rate. Only about 40% of those who received the survey returned it (see the study description in the Chapter Preview). In order for the sample to be considered representative of the population, you need to be willing to assume that those who returned the survey did not differ in any important way from those that did not return the survey.

C Calculate   To calculate the value of the test statistic, you need to find the value of the sample proportion:

$$n = 229$$

$$\hat{p} = \frac{89}{229} = 0.389$$

Next, you calculate the value of the test statistic:

$$z = \frac{\hat{p} - 0.50}{\sqrt{\dfrac{0.50(1 - 0.50)}{n}}} = \frac{0.389 - 0.50}{\sqrt{\dfrac{(0.50)(1 - 0.50)}{229}}} = \frac{-0.111}{0.033} = -3.36$$

This is a two-tailed test (the inequality in H_a is $\neq$), so the P-value is twice the area under the z curve to the left of -3.36. This area can be found using a graphing calculator, statistical software, or the table of normal curve areas.
Associated P-value:

$$
\begin{aligned}
P\text{-value} &= 2(\text{area under } z \text{ curve to the left of } -3.36) \\
&= 2P(z < -3.36) \\
&= 2(0.0004) \\
&= 0.0008
\end{aligned}
$$

C Communicate Results

Because the P-value is less than the selected significance level, the null hypothesis is rejected.

Decision: $0.0008 < 0.05$, Reject H_0.

Interpretation:
There is convincing evidence that the proportion of women who would choose a boy is not equal to 0.50. This means that there is evidence that the *LiveScience* summary of the study findings is not correct when it states that there is no gender preference among female patients at the clinic who would like to choose the sex of a future baby. This conclusion is only valid if the sample is representative of the population—nonresponse is a concern here.

Most statistical computer packages and graphing calculators can calculate and report P-values for a variety of hypothesis tests, including the large-sample test for a population proportion. Minitab was used to carry out the test of Example 10.12, and the resulting computer output follows:

Test and CI for One Proportion
Test of p = 0.5 vs. p not = 0.5

Sample	X	N	Sample p	95% CI	Z-Value	P-Value
1	89	229	0.388646	(0.325514, 0.451779)	−3.37	0.001

From the Minitab output, $z = -3.37$, and the associated P-value is 0.001. The small differences in the values of the test statistic and P-value from Example 10.12 are the result of differences in rounding.

Example 10.13   **Be Careful What You Put on Your Facebook Page!**

The article **"60% of Employers Are Peeking Into Candidates Social Media Profiles"** **(CareerBuilder.com, April 28, 2016, retrieved November 29, 2016)** included data from a survey of 2186 hiring managers and human resource professionals. The article noted that many employers are using social networks to screen job applicants and that this practice is becoming more common. Of the 2186 people who participated in the survey, 1312 indicated that they use social networking sites (such as Facebook, MySpace, and LinkedIn) to research job applicants. Based on the survey data, is there convincing evidence that fewer than two-thirds of hiring managers and human resource professionals use social networking sites in this way?

In terms of the four key questions, this study can be classified as *hypothesis testing, sample data, one categorical variable* (use of social networking sites to screen job applicants, with possible values use or do not use), and *one sample*. This combination suggests a large-sample test for a population proportion.

H Hypotheses In this example, the claim is about the population proportion of hiring managers and human resource professionals who use social networking sites to screen job applicants, so the population characteristic of interest is

> p = population proportion of hiring managers and human resource professionals who use social networking sites to screen job applicants

The question of interest is whether fewer than two-thirds use social networking sites to screen job applicants, so the alternative hypothesis is $p < 0.667$. The null hypothesis is then $p = 0.667$.

Hypotheses:
Null hypothesis: H_0: $p = 0.667$
Alternative hypothesis: H_a: $p < 0.667$

M Method Because the answers to the four key questions are hypothesis testing, sample data, one categorical variable, and one sample, a large-sample hypothesis test for a population proportion is considered.

Test statistic:

$$z = \frac{\hat{p} - 0.667}{\sqrt{\dfrac{(0.667)(1 - 0.667)}{n}}}$$

The value of 0.667 in the test statistic is the hypothesized value from the null hypothesis.

Next, you need to select a significance level for the test. As a future job applicant, you might consider a Type I error (incorrectly believing that fewer than two-thirds screen applicants using social networking sites) to be a more serious error than a Type II error. In this situation, it would make sense to choose a relatively small value for α, such as 0.01.

Significance level: $\alpha = 0.01$

C Check There are two conditions that must be met before using the large-sample test for a population proportion. The large sample condition is easily verified. The sample size is large enough because $np_0 = 2186(0.667) = 1458$ and $n(1 - p_0) = 2186(1 - 0.667) = 728$ are both greater than or equal to 10.

The study authors indicate that they believe that the sample of hiring managers and human resource professionals was selected in a way that would result in a representative sample. Because both conditions required for the test are met, it is reasonable to proceed.

C Calculate To calculate the value of the test statistic, you need to find the value of the sample proportion.

$$n = 2186$$

$$\hat{p} = \frac{1312}{2186} = 0.600$$

Substituting the values for n and $\hat{p}$ into the test statistic formula results in the following:

$$z = \frac{\hat{p} - 0.667}{\sqrt{\dfrac{(0.667)(1 - 0.667)}{2186}}} = \frac{0.600 - 0.667}{\sqrt{\dfrac{(0.667)(1 - 0.667)}{2186}}} = \frac{-0.067}{0.010} = -6.70$$

This is a lower-tailed test (the inequality in H_a is $<$), so the P-value is the area under the z curve and to the left of -6.70.

Associated *P*-value:

$$P\text{-value} = \text{area under } z \text{ curve to the left of } -6.70$$
$$= P(z < -6.70)$$
$$\approx 0$$

C Communicate Results   Because the *P*-value is less than the selected significance level, the null hypothesis is rejected. There is convincing evidence that the proportion of hiring managers and human resource professionals who use social networking sites to screen job applicants is less than 0.667.

Even though the survey data provide convincing evidence that less than two-thirds currently use social networking sites to screen job applicants, it was noted that the practice is becoming more common (only 45% of those surveyed in 2009, compared to 60% in 2016). Just in case you are applying for a job in the near future, here are a few of the things that hiring managers said that they found on social networking sites that caused them not to hire a candidate: provocative or inappropriate photos, references to drinking or drugs, negative comments about a previous employer or coworker, and poor communication.

A Few Final Things to Consider

1. What About Small Samples?

If $np_0 \geq 10$ and $n(1 - p_0) \geq 10$, the standard normal distribution is a reasonable approximation to the distribution of the *z* test statistic when the null hypothesis is true. In this case, *P*-values based on the standard normal distribution can be used to reach a conclusion in the hypothesis test. If the sample size is not large enough to satisfy the large sample conditions, the distribution of the test statistic may be quite different from the standard normal distribution. In this case, you can't use the standard normal distribution to judge whether what was observed in the sample is consistent or inconsistent with the null hypothesis. If you have a small sample, you can consider one of the two methods introduced in Section 10.6—Randomization Tests and Exact Binomial Tests for One Proportion.

2. Choosing a Potential Method

Take a look back at Table 7.1 (on page 390 and also on the inside back cover of the text). You have now seen how the answers to the four key questions lead you to the potential methods indicated in the first two rows of the table. You now have these two "plays" in your statistics playbook. As you get into the habit of answering the four key questions for each new situation that you encounter, it will become easier to use this table to select an appropriate method in a given situation.

Summing It Up—Section 10.5

The following learning objectives were addressed in this section:

Conceptual Understanding

C6: Understand the reasoning used to reach a decision in a hypothesis test.

In a hypothesis test, you begin by assuming that the null hypothesis is true. You then consider whether what was observed in the data would have been surprising or not surprising given that the null hypotheses were true. If what was observed in the data would have been unlikely to occur if the null hypothesis is true, this is evidence against the null hypothesis, and the null hypothesis would be rejected. The determination of whether the data would have been unlikely to occur if the null hypothesis were true is based on the *P*-value, which is the probability of observing data at least as inconsistent with H_0 as what was observed if the null hypothesis were true. If the *P*-value is less than α, the significance level selected for the test, the null hypothesis is rejected.

Mastering the Mechanics

M3: Know the key characteristics that lead to selection of a large-sample test for a population proportion as an appropriate method.

When the answers to the four key questions (QSTN) are hypothesis testing, sample data, one categorical variable, and one sample, the large-sample test for a population proportion is a method to consider. This method would be an appropriate choice if the conditions required for use of this method are met.

M4: Know the conditions for appropriate use of the large-sample test for a population proportion.

There are two conditions that should be verified before a large-sample test for a population proportion is used. These are:

(1) The sample is a random sample from the population of interest or the sample is selected in a way that should result in a representative sample.
(2) The sample size is large. This condition is met when both np_0 and $n(1 - p_0)$ are greater than or equal to 10.

M6: Use the five-step process for hypothesis testing (HMC³) to carry out a large-sample test for a population proportion.

The five steps in the process for hypothesis testing are Hypotheses, Method, Check, Calculate, and Communicate results. Examples 10.11, 10.12, and 10.13 all illustrate how the steps in the process are followed to reach a conclusion in a hypothesis testing situation.

Putting It into Practice

P1: Recognize when a situation calls for testing hypotheses about a population proportion.

A test of hypotheses about a population proportion is used when you want to see if there is evidence to support a claim about a population proportion. The data in the sample used to test such a claim will be categorical.

P2: Carry out a large-sample test for a population proportion and interpret the conclusion in context.

Examples 10.11, 10.12, and 10.13 all illustrate hypothesis tests for a population proportion, including interpreting the result of the test in context.

SECTION 10.5 EXERCISES

Each Exercise Set assesses the following chapter learning objectives: C6, M3, M4, M6, P1, P2

SECTION 10.5 Exercise Set 1

For questions 10.47–10.49, answer the following four key questions and indicate whether the method that you would consider would be a large-sample hypothesis test for a population proportion.

Q Question Type	Estimation or hypothesis testing?
S Study Type	Sample data or experiment data?
T Type of Data	One variable or two? Categorical or numerical?
N Number of Samples or Treatments	How many samples or treatments?

10.47 Refer to the instructions given prior to this exercise. The paper **"College Students' Social Networking Experiences on Facebook"** (*Journal of Applied Developmental*

Psychology **[2009]: 227–238)** summarized a study in which 92 students at a private university were asked how much time they spent on Facebook on a typical weekday. You would like to estimate the average time spent on Facebook by students at this university.

10.48 Refer to the instructions given prior to Exercise 10.47. The article **"iPhone Can Be Addicting, Says New Survey" (www.msnbc.com, March 8, 2010)** described a survey administered to 200 college students who owned an iPhone. One of the questions on the survey asked students if they slept with their iPhone in bed with them. You would like to use the data from this survey to determine if there is convincing evidence that a majority of college students with iPhones sleep with their phones.

10.49 Refer to the instructions given prior to Exercise 10.47. *USA TODAY* **(February 17, 2011)** reported that 10% of 1008 American adults surveyed about their use of e-mail said that they had ended a relationship by e-mail. You would like to use this information to estimate the proportion of all adult Americans who have used e-mail to end a relationship.

10.50 Assuming a random sample from a large population, for which of the following null hypotheses and sample sizes is the large-sample z test appropriate?
a. $H_0: p = 0.2$, $n = 25$
b. $H_0: p = 0.6$, $n = 200$
c. $H_0: p = 0.9$, $n = 100$
d. $H_0: p = 0.05$, $n = 75$

10.51 Let p denote the proportion of students at a large university who plan to purchase a campus meal plan in the next academic year. For a large-sample z test of $H_0: p = 0.20$ versus $H_a: p < 0.20$, find the P-value associated with each of the following values of the z test statistic. (Hint: See page 496.)
a. -0.55
b. -0.92
c. -1.99
d. -2.24
e. 1.40

10.52 The paper **"Debt Literacy, Financial Experiences and Over-Indebtedness"** (*Social Science Research Network, Working paper W14808, 2008*) included data from a survey of 1000 Americans. One question on the survey was: "You owe $3000 on your credit card. You pay a minimum payment of $30 each month. At an Annual Percentage Rate of 12% (or 1% per month), how many years would it take to eliminate your credit card debt if you made no additional charges?" Answer options for this question were: (a) less than 5 years; (b) between 5 and 10 years; (c) between 10 and 15 years; (d) never—you will continue to be in debt; (e) don't know; and (f) prefer not to answer.
a. Only 354 of the 1000 respondents chose the correct answer of "never." Assume that the sample is representative of adult Americans. Is there convincing evidence that the proportion of adult Americans who can answer this question correctly is less than 0.40 (40%)? Use the five-step process for hypothesis testing (HMC[3]) described in this section and $\alpha = 0.05$ to test the appropriate hypotheses. (Hint: See Example 10.13.)
b. The paper also reported that 37.8% of those in the sample chose one of the wrong answers (a, b, or c) as their response to this question. Is it reasonable to conclude that more than one-third of adult Americans would select a wrong answer to this question? Use $\alpha = 0.05$.

10.53 The paper **"Teens and Distracted Driving" (Pew Internet & American Life Project, 2009)** reported that in a representative sample of 283 American teens age 16 to 17, there were 74 who indicated that they had sent a text message while driving. For purposes of this exercise, assume that this sample is a random sample of 16- to 17-year-old Americans. Do these data provide convincing evidence that more than a quarter of Americans age 16 to 17 have sent a text message while driv-

ing? Test the appropriate hypotheses using a significance level of 0.01. (Hint: See Example 10.11.)

10.54 The article **"Streaming Overtakes Live TV Among Consumer Viewing Preferences"** (*Variety*, **April 22, 2015**) states that "U.S. consumers are more inclined to stream entertainment from an internet service than tune in to live TV." This statement is based on a survey of a representative sample of 2076 U.S. consumers. Of those surveyed, 1100 indicated that they prefer to stream TV shows rather than watch TV programs live. Do the sample data provide convincing evidence that a majority of U.S. consumers prefer to stream TV shows rather than to watch them live? Test the relevant hypotheses using a 0.05 significance level.

10.55 The report **"Digital Democracy Survey" (Deloitte Development LLC, 2016, www2.deloitte.com/us/en.html, retrieved November 30, 2016)** describes a large national survey. In a representative sample of Americans ages 14 to 18 years, 45% indicated that they usually use social media while watching TV. Suppose that the sample size was 500.
a. Is there convincing evidence that less than half of Americans ages 14 to 18 years usually use social media while watching TV? Use a significance level of 0.05.
b. Suppose that the sample size had been 100 rather than 500 and that 45% of those in the sample indicated that they usually use social media while watching TV. Based on this sample of 100, is there convincing evidence that less than half of Americans ages 14 to 18 years usually use social media while watching TV? Use a significance level of 0.05.
c. Explain why different conclusions were reached in the hypothesis tests of Parts (a) and (b).

10.56 In a survey of 1005 adult Americans, 46% indicated that they were somewhat interested or very interested in having web access in their cars (*USA TODAY*, **May 1, 2009**). Suppose that the marketing manager of a car manufacturer claims that the 46% is based only on a sample and that 46% is close to half, so there is no reason to believe that the proportion of all adult Americans who want car web access is less than 0.50. Is the marketing manager correct in his claim? Provide statistical evidence to support your answer. For purposes of this exercise, assume that the sample can be considered representative of adult Americans.

SECTION 10.5 Exercise Set 2

For questions 10.57–10.59, answer the following four key questions and indicate whether the method that you would consider would be a large-sample hypothesis test for a population proportion.

Q	Estimation or hypothesis testing?
Question Type	
S	Sample data or experiment data?
Study Type	
T	One variable or two? Categorical
Type of Data	or numerical?
N	How many samples or treatments?
Number of Samples or Treatments	

10.57 Refer to the instructions given prior to this exercise. The paper **"College Students' Social Networking Experiences on Facebook"** (*Journal of Applied Developmental Psychology* **[2009]: 227–238**) summarized a study in which 92 students at a private university were asked whether they used Facebook just to pass the time. Twenty-three responded "yes" to this question. The researchers were interested in estimating the proportion of students at this college who use Facebook just to pass the time.

10.58 Refer to the instructions given prior to Exercise 10.57. The paper **"Pathological Video-Game Use Among Youth Ages 8 to 18: A National Study"** (*Psychological Science* **[2009]: 594–601**) summarizes data from a random sample of 1178 students age 8 to 18. The paper reported that for the students in the sample, the mean amount of time spent playing video games was 13.2 hours per week. The researchers were interested in using the data to estimate the mean amount of time spent playing video games for students age 8 to 18.

10.59 Refer to the instructions given prior to Exercise 10.57. The paper referenced in the previous exercise also reported that when each of the 1178 students who participated in the study was asked if he or she played video games at least once a day, 271 responded "yes." The researchers were interested in using this information to decide if there is convincing evidence that more than 20% of students age 8 to 18 play video games at least once a day.

10.60 Let p denote the proportion of students living on campus at a large university who plan to move off campus in the next academic year. For a large sample z test of $H_0: p = 0.70$ versus $H_a: p > 0.70$, find the P-value associated with each of the following values of the z test statistic.
a. 1.40
b. 0.92
c. 1.85
d. 2.18
e. −1.40

10.61 Assuming a random sample from a large population, for which of the following null hypotheses and sample sizes is the large-sample z test appropriate?
a. $H_0: p = 0.8$, $n = 40$
b. $H_0: p = 0.4$, $n = 100$
c. $H_0: p = 0.1$, $n = 50$
d. $H_0: p = 0.05$, $n = 750$

10.62 A representative sample of 1000 likely voters in the United States included 440 who indicated that they think that women should not be required to register for the military draft (**"Most Women Oppose Having to Register for the Draft,"** www.rasmessenreports.com, February 10, 2016, retrieved November 30, 2016). Using the five-step process for hypothesis testing (HMC³) and a 0.05 significance level, determine if there is convincing evidence that less than half of likely voters in the United States think that women should not be required to register for the military draft.

10.63 The report **"2007 Electronic Monitoring and Surveillance Survey: Many Companies Monitoring, Recording, Videotaping—and Firing—Employees"** (American Management Association, 2007) summarized a survey of 304 U.S. businesses. Of these companies, 201 indicated that they monitor employees' web site visits. Assume that it is reasonable to regard this sample as representative of businesses in the United States.
a. Is there sufficient evidence to conclude that more than 75% of U.S. businesses monitor employees' web site visits? Test the appropriate hypotheses using a significance level of 0.01.
b. Is there sufficient evidence to conclude that a majority of U.S. businesses monitor employees' web site visits? Test the appropriate hypotheses using a significance level of 0.01.

10.64 Duck hunting in populated areas faces opposition on the basis of safety and environmental issues. In a survey to assess public opinion regarding duck hunting on Morro Bay (located along the central coast of California), a random sample of 750 local residents included 560 who strongly opposed hunting on the bay. Does this sample provide convincing evidence that a majority of local residents oppose hunting on Morro Bay? Test the relevant hypotheses using $\alpha = 0.01$.

10.65 The report **"A Crisis in Civic Education"** (American Council of Trustees and Alumni, January 2016, www.goacta.org /images/download/A_Crisis_in_Civic_Education.pdf, retrieved November 30, 2016) summarizes data from a survey of a representative sample of 1000 adult Americans regarding their understanding of U.S. government. Only 459 of the adults in the sample were able to give a correct response to a question asking them to choose a correct definition of the Bill of Rights from a list of five possible answers. Using a significance level of 0.01, determine if there is convincing evidence that less than half of adult Americans could identify the correct definition of the Bill of Rights.

ADDITIONAL EXERCISES

10.66 In a survey conducted by CareerBuilder.com, employers were asked if they had ever fired an employee for holiday shopping online while at work **("Cyber Monday Shopping at Work? You're Not Alone," November 22, 2016, retrieved November 30, 2016).** Of the 2379 employers responding to the survey, 262 said they had fired an employee for shopping online while at work. Suppose that it is reasonable to assume that the sample is representative of employers in the United States. Do the sample data provide convincing evidence that more than 10% of employers have fired an employee for shopping online at work? Test the relevant hypotheses using $\alpha = 0.01$.

10.67 In a survey of 1000 women age 22 to 35 who work full-time, 540 indicated that they would be willing to give up some personal time in order to make more money **(USA TODAY, March 4, 2010).** The sample was selected to be representative of women in the targeted age group.
a. Do the sample data provide convincing evidence that a majority of women age 22 to 35 who work full-time would be willing to give up some personal time for more money? Test the relevant hypotheses using $\alpha = 0.01$.
b. Would it be reasonable to generalize the conclusion from Part (a) to all working women? Explain why or why not.

10.68 According to a large national survey conducted by the Pew Research Center **("What Americans Think About NSA Surveillance, National Security and Privacy," May 2, 2015, www.pewresearch.org, retrieved December 1, 2016),** 54% of adult Americans disapprove of the National Security Agency collecting records of phone and Internet data. Suppose that this estimate was based on a random sample of 1000 adult Americans.
a. Is there convincing evidence that a majority of adult Americans feel this way? Test the relevant hypotheses using a 0.05 significance level.
b. The actual sample size was much larger than 1000. If you had used the actual sample size when doing the calculations for the test in Part (a), would the P-value have been larger than, the same as, or smaller than the P-value you obtained in Part (a)? Provide a justification for your answer.

10.69 In a representative sample of adult Americans ages 26 to 32 years, 27% indicated that they owned a fitness band that kept track of the number of steps walked each day and their daily activity levels **("Digital Democracy Survey", Deloitte Development LLC, 2016, www2.deloitte.com/us/en.html, retrieved November 30, 2016).** Suppose that the sample size was 500. Is there convincing evidence that more than one-quarter of all adult Americans in this age group own a fitness band?

Test the relevant hypotheses using a significance level of 0.05.

10.70 The article **"Facebook Use and Academic Performance Among College Students" (Computers in Human Behavior [2015]: 265–272)** estimated that 87% percent of students at a large public university in California who are Facebook users update their status at least two times a day. This estimate was based on a random sample of 261 students at this university.
a. Does this sample provide convincing evidence that more than 80% of the students at this college who are Facebook users update their status at least two times a day? Test the relevant hypotheses using $\alpha = 0.05$.
b. Would it be reasonable to generalize the conclusion from the test in Part (a) to all college students in the United States? Explain why or why not.

10.71 A number of initiatives on the topic of legalized gambling have appeared on state ballots. A political candidate has decided to support legalization of casino gambling if he is convinced that more than two-thirds of American adults approve of casino gambling. Suppose that 1035 of the people in a random sample of 1523 American adults said they approved of casino gambling. Is there convincing evidence that more than two-thirds approve?

10.72 In 2016, the **National Foundation for Credit Counseling** released a report titled **"The 2016 Consumer Financial Literacy Survey" (www.nfcc.org, retrieved December 1, 2016).** In a nationally representative sample of 1668 adult Americans, 965 indicated that they had checked their credit score within the last 12 months. Is there convincing evidence that a majority of adult Americans have checked their credit scores within the last 12 months? Test the relevant hypotheses using $\alpha = 0.05$.

10.73 The survey described in the previous exercise also noted that of the 965 people that had checked their credit report within the last 12 months, 38% had done so as part of their regular financial planning. Does this provide convincing evidence that more than one-third of adult Americans who have checked their credit scores within the past 12 months did so as part of regular financial planning? Carry out a test using a significance level of 0.01.

10.74 The article titled **"13% of Americans Don't Use the Internet. Who Are They?"** describes a study conducted by the **Pew Research Center (pewrearch.org, September 7, 2016, retrieved December 1, 2016).** Suppose that the title of this article is based on a representative sample of 600 adult Americans. Does this support the claim that the proportion of adult Americans who do not use the Internet is greater than 0.10 (10%)?

Randomization Tests and Exact Binomial Tests for One Proportion (Optional)

In Section 10.5, you learned how to use data from a random sample to carry out a large-sample z test for one proportion. To use that test, you need to know that the sample is a random sample from a population (or that it is selected in a way that makes it reasonable to think that the sample is representative of a population). In addition, you need to know that the distribution of the sample proportion, $\hat{p}$, is aproximately normal. It is reasonable to think that the distribution of $\hat{p}$ is approximately normal when the sample size is large ($np \geq 10$ and $n(1 - p) \geq 10$), but this isn't necessarily true when the sample size is small.

When the sample size is large, you can carry out a z test and use the normal distribution to calculate a P-value. But even when the sampling distribution of $\hat{p}$ is well approximated by a normal distribution, the resulting P-value for the large-sample z test is still just an approximation to the actual value. When you want to carry out a hypothesis test for a population proportion but the sample size is not large enough to assume that the sampling distribution of $\hat{p}$ is normal, there are other methods that do not require a large sample size and that can be used to approximate a P-value for the hypothesis test. Two of these methods are a randomization test and an "exact" test that is based on the binomial distribution.

A Randomization Test for One Proportion

In a hypothesis test for one proportion, the null hypothesis is of the form H_0: $p =$ hypothesized value, where p is the population proportion and the hypothesized value is determined by the question of interest. For example, the null hypothesis might be H_0: $p = 0.50$.

Recall that in a hypothesis test, the P-value is a measure of how likely it would be to see something as extreme or more extreme as what was observed in the sample data if the null hypothesis were true. One way to approximate a P-value is to assume that the population proportion is in fact equal to the value specified in the null hypothesis and then simulate random samples from such a population. The distribution of these simulated sample proportions would give you a sense of what values for the sample proportion you would expect to see when the null hypothesis is true and what values would be unlikely to occur. The distribution of simulated proportions also allows you to find an approximate P-value. The distribution of simulated sample proportions is called a **randomization distribution**.

Example 10.14	**Using a Shiny App to Create a Randomization Distribution**

Suppose that you are interested in deciding if there is evidence that a majority of the students at your school are registered to vote. You take a random sample of 10 students and find that 8 of the 10 students in the sample are registered to vote. A majority of the students in the sample are registered, but does this mean it is reasonable to conclude that there is convincing evidence that a majority of *all* students are registered to vote? To answer this question, you would test the null hypothesis H_0: $p = 0.50$ against the alternative hypothesis H_a: $p > 0.50$.

If you assume that H_0: $p = 0.50$ is true, then you can generate values of $\hat{p}$ for many different simulated samples of size $n = 10$. Go online at statistics.cengage.com/Peck2e /Apps.html and open the Shiny app called "Randomization Test for One Proportion." Enter 10 for the number of observations, and 8 for the number of successes. The default hypothesized value for p is already 0.5. Choose "Upper-Tailed ($>$)" for the form of the alternative hypothesis, and request 1000 simulated samples. Click "Generate Simulated Samples."

Randomization Test for One Proportion

Select number of observations

10

Select the number of successes

8

Enter hypothesized value:

0.5

Select form of alternative hypothesis:
○ Two-Tailed (Not Equal)
◉ Upper-Tailed (>)
○ Lower-Tailed (<)

Select number of simulated samples to generate:
○ 1 ○ 10 ○ 100 ◉ 1,000 ○ 10,000

[Generate Simulated Samples] [Reset]

The Shiny app generates 1000 simulated random samples of size 10 from a population with $p = 0.50$, and provides a histogram of the values of the 1000 simulated sample proportions. The Shiny app also indicates where the observed value of the sample proportion, $\hat{p} = 0.8$ falls in the randomization distribution. The output for one set of 1000 simulated sample proportions is shown below.

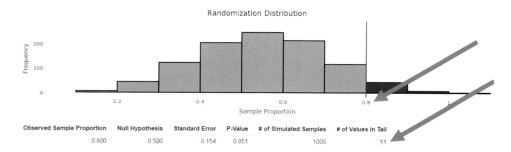

Randomization Distribution

Observed Sample Proportion	Null Hypothesis	Standard Error	P-Value	# of Simulated Samples	# of Values in Tail
0.800	0.500	0.154	0.051	1000	51

The app reports that for this particular simulation, 51 of the 1000 simulated samples, or 5.1%, were as or more extreme than what was observed in the actual sample. This value, $\dfrac{51}{1000} = 0.051$, is an approximation for the P-value for the one-sided test. The approximate P-value is also shown in the output produced by the app. Although different simulations will give different sets of simulated sample proportions, and therefore different values for the approximate P-value, the results from different simulations should be similar as long as a large number of sample proportions are used to create the randomization distribution.

Based on this randomization distribution of 1000 simulated values of $\hat{p}$ when $n = 10$ and the null hypothesis $H_0: p = 0.5$ is true, the approximate P-value is $51/1000 = 0.051$. You can then compare this P-value to a selected significance level in order to make a decision about whether the null hypothesis should be rejected. For example, if you have selected a significance level of 0.05, you will fail to reject H_0 because the P-value of 0.051 is close to, but greater than $\alpha = 0.05$. This means that even though 8 out of 10 students in the sample were registered to vote, it would not have been surprising to have seen this just by chance when the null hypothesis is true.

| Example 10.15 | Cell Phones Revisited |

The paper referenced in Example 10.11 described a study of 2000 Canadians over the age of 18 that was carried out by Microsoft. Study participants were asked whether the following statement described them: "When nothing is occupying my attention, the first thing I do is reach for my phone." Of the study participants in the age group 18 to 24 years old, 77% responded "yes" to this question. In Example 10.11, it was assumed that the 77% was based on a representative sample of 800 Canadians age 18 to 24 years. A large-sample z test was used to decide if the sample data support the claim that more than 75% of all Canadians in this age group would respond "yes" when asked if the given statement describes them. In this version of the example, a randomization test will be used.

Let p represent the proportion of Canadians ages 18 to 24 years who would respond "yes" for the population represented by the sample of 800. The appropriate hypotheses for the test are $H_0: p = 0.75$ and $H_a: p > 0.75$.

Go online to statistics.cengage.com/Peck2e/Apps.html and open the Shiny app called "Randomization Test for One Proportion." Enter 800 for the Number of observations, and 616 (77% of 800) for the Number of successes. The default hypothesized value for p is 0.5, so change it to 0.75. Choose "Upper-Tailed (>)" for the form of the alternative hypothesis, and request 1000 simulated samples.

Randomization Test for One Proportion

Select number of observations:

    800

Select the number of successes

    616

Enter hypothesized value:

    0.75

Select form of alternative hypothesis:

○ Two-Tailed (Not Equal)
● Upper-Tailed (>)
○ Lower-Tailed (<)

Select number of simulated samples to generate:

○ 1　○ 10　○ 100　● 1,000　○ 10,000

| Generate Simulated Samples | Reset |

Confirm that the value of $\hat{p} = 616/800 = 0.770$ appears below "Observed Sample Proportion" in the simulation output. Notice that based on this simulation, the value of "# of Values in Tail" (shown in the accompanying figure) is 107, and that the P-Value is $107/1000 = 0.107$. This represents a randomization test approximation for the P-value for the hypothesis test.

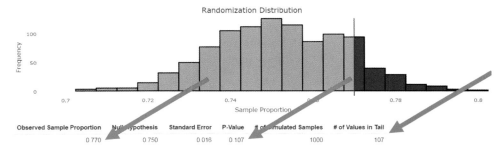

Randomization Distribution

Observed Sample Proportion	Null Hypothesis	Standard Error	P-Value	# of Simulated Samples	# of Values in Tail
0.770	0.750	0.016	0.107	1000	107

The approximate *P*-value is 0.107. Because the *P*-value is greater than 0.05, you fail to reject H_0. The sample does not provide convincing evidence that more than 75% of Canadians ages 18 to 24 would respond yes when asked if the statement "When nothing is occupying my attention the first thing I do is reach for my phone" describes them. This is consistent with the *P*-value obtained for the large-sample *z* test in Example 10.11 (which was 0.092), and so you reach the same conclusion using either method to approximate the *P*-value. This is not surprising because the sample size is large enough for the large-sample test to be appropriate. The advantage of the randomization test is that it can be used even when the sample size is not large.

Example 10.16   Vaccination Coverage

Following a two-year period when the Australian government offered free vaccinations to protect against cervical cancer to women ages 13 to 26, researchers carried out surveys of random samples of women in each of the territories in Australia **("Human Papillomavirus (HPV) Vaccination Coverage in Young Australian Women Is Higher Than Previously Estimated,"** *Vaccine* **[2014]: 592–597)**. While the sample sizes were large for most territories, for the Northern Territory, the sample size was only 12. The researchers found that 8 of the 12 women in the sample from the Northern Territory had received at least one of the three recommended doses of HPV vaccine. You can use the data in this sample to test the hypothesis that more than half of women age 13 to 26 in the Northern Territory have received at least one dose of the HPV vaccine.

Let *p* represent the population proportion of women age 13 to 26 in the Northern Territory who have received at least one dose of HPV vaccine. The hypotheses of interest are H_0: $p = 0.50$ and H_a: $p > 0.50$. The observed value of the sample proportion is $\hat{p} = 8/12 = 0.667$.

Notice that the large-sample *z* test of Section 9.3 is not an appropriate choice because the sample size condition of $np \geq 10$ is not met ($np = (12)(0.50) = 6 < 10$). However, because the sample was a random sample and the randomization test does not require a large sample, it is appropriate to proceed with a randomization test.

From the accompanying output from the Shiny app "Randomization Test for One Proportion," the approximate *P*-value for this upper-tailed randomization test is 0.197. This means that 197 of the 1000 simulated sample proportions were at least as large as 0.667. Using a significance level of 0.05, the null hypothesis would not be rejected. The sample does not provide convincing evidence that more than half of women age 13 to 26 in the Northern Territory of Australia have received at least one dose of HPV vaccine.

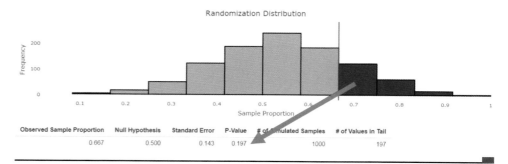

Example 10.16 illustrates that randomization tests may be used even for small random samples when the conditions for large-sample inference may not be met. For this example, had you used the large-sample *z* test, you would have obtained a *P*-value of 0.124, which is quite different than the *P*-value from the randomization test. This illustrates why you should not carry out a hypothesis test using a method for which the assumptions are not met.

An Exact Binomial Test for One Proportion

Another way to obtain a *P*-value when testing hypotheses about a population proportion is to use an exact probability approach that is based on the binomial distribution. You encountered the binomial probability distribution if you studied optional Section 6.7, Binomial and Geometric Distributions.

Example 10.17   **An Exact Binomial Test**

Suppose that you want to carry out a test of the null hypothesis H_0: $p = 0.50$. The value of p that is specified in the null hypothesis, when combined with an observed sample size, identifies a specific binomial distribution that can be used to compute an "exact" P-value. Suppose that the alternative hypothesis of interest is H_a: $p > 0.50$ and that the test will be carried out using data from a random sample of $n = 10$ independent success (S) or failure (F) observations.

Because the sample size is small, the large-sample z test is not an appropriate way to test these hypotheses. The exact binomial test does not require a large sample, so it can be used when the sample size is small and you are concerned that the sampling distribution of $\hat{p}$ may not be approximately normal.

Suppose that you observe $x = 8$ successes in the sample of size $n = 10$, which means that $\hat{p} = 8/10 = 0.8$. The binomial distribution with $n = 10$ and $p = 0.50$ (the hypothesized proportion) can be used to calculate the probability of observing a sample proportion as or more extreme than what was observed in the sample. This is the P-value for the hypothesis test.

To calculate the P-value for an exact binomial test, use the Shiny app "Exact Binomial Test for One Proportion" (statistics.cengage.com/Peck2e/Apps.html) with 10 entered for the Number of observations, and 8 for the Number of successes. The default hypothesized value is already 0.5. Specify "Upper Tail (>)" for the form of the alternative hypothesis.

Exact Binomial Test for One Proportion

Select number of observations

> 10

Select the number of successes

> 8

Enter hypothesized value:

> 0.5

Select form of alternative hypothsis:
- ◯ Two-Tailed (Not Equal)
- ◉ Upper Tail (>)
- ◯ Lower Tail (<)

The Shiny app automatically updates the probability histogram to reflect your choices, and displays the upper-tail P-value as a binomial probability. In this example, the P-value is equal to 0.055.

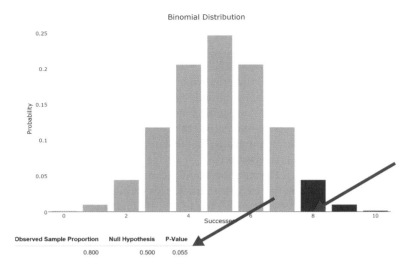

The probability that $X \geq 8$ represents the exact one-sided P-value for testing $H_0: p = 0.50$ versus $H_a: p > 0.50$. From the Shiny app, this probability is 0.055, which is the value of the P-value for the exact binomial test.

Since this P-value $= 0.055$ is greater than $\alpha = 0.05$, if you selected a significance level of 0.05 you would fail to reject H_0. Observing eight successes in $n = 10$ trials is not convincing evidence that $p > 0.50$.

The following examples illustrate the use of the exact binomial test. By revisiting the examples previously used to illustrate the randomization test, you can see how the approximate P-value from the randomization test and the P-value from the exact binomial test compare.

Example 10.18   Cell Phones Revisited Again

In the study described in Examples 10.11 and 10.15, 77% of 800 Canadians ages 18 to 24 years responded "yes" when asked if the statement "When nothing is occupying my attention the first thing I do is reach for my phone" describes them. As was done in the previous examples, this data will be used to decide whether the sample supports the claim that more than 75% of Canadians in this age group would respond "yes." In this example, an exact binomial test will be used.

For the exact binomial test, you must still assume that the Canadians sampled are representative of the population of interest, but there are no sample size conditions that must be satisfied.

Below is output from the Shiny app "Exact Binomial Test for One Proportion" that provides the binomial probability that at least 616 of the people in a random sample of size 800 would have responded yes if the null hypothesis of $p = 0.75$ is true.

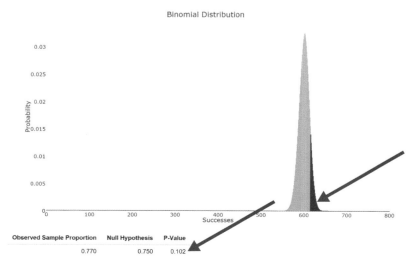

From the Shiny app, the P-value $= P(X \geq 616$ when $n = 800$ and H_0: $p = 0.75$ is true) is 0.102. This is greater than 0.05, so you fail to reject H_0. The sample does not provide convincing evidence that more than 75% of Canadians ages 18 to 24 would respond "yes" when asked if the statement, "When nothing is occupying my attention the first thing I do is reach for my phone" describes them. The P-value from the exact binomial test is similar to the approximate P-values for the large-sample z test (0.092) and the randomization test (0.107). This is not surprising because any of these three methods is appropriate for this sample size.

Example 10.19 — Vaccination Coverage Revisited

Recall that in Example 10.16, data from a random sample of 12 women from the Northern Territory of Australia was used to test the claim that more than 50% of women ages 13 to 26 had received at least one dose of the HPV vaccine. In this example, the exact binomial test will be used to test the hypotheses H_0: $p = 0.5$ and H_a: $p > 0.5$, where p represents the proportion of women age 13 to 26 in the Northern Territory who have had a least one dose of the HPV vaccine. It is reasonable to use the exact binomial test because the sample was a random sample and the exact binomial test does not have any sample size requirements.

Recall that 8 of the 12 women in the sample reported that they had received at least one dose of the vaccine. The binomial distribution with $n = 12$ and $p = 0.5$ (from the null hypothesis) can be used to determine the probability of observing 8 or more successes in a sample of size 12 when the null hypothesis is true and the actual proportion of successes is 0.5.

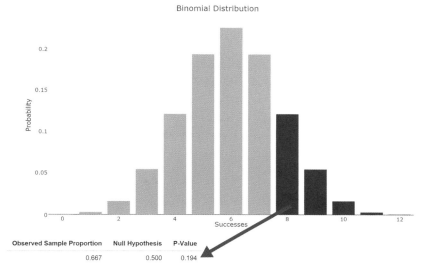

Observed Sample Proportion	Null Hypothesis	P-Value
0.667	0.500	0.194

From the accompanying Shiny app output, the probability of at least 8 successes in a sample of size 12 when $p = 0.50$ is 0.194. This is the P-value for the exact binomial test, and because the P-value is large, the null hypothesis would not be rejected.

Notice that the P-value for the exact binomial test and the approximate P-value for the randomization test (0.197) are very close, and either method would be an appropriate choice for this sample. The P-value for the large-sample z test was quite a bit smaller (0.124), and you would not want to base a conclusion on this P-value because the assumptions required for the large-sample z test are not met.

Summing It Up—Section 10.6

The following learning objectives were addressed in this section:

Mastering the Mechanics

M7: Carry out a randomization test for a population proportion.

A randomization test is a method that can be used to test hypotheses about a population proportion even if the sample size is not large enough for the large-sample z test to be appropriate. Examples 10.15 and 10.16 illustrate the use of a randomization test to carry out a hypothesis test for a population proportion.

M8: Carry out an exact binomial test for a population proportion.
The exact binomial test is a method that can be used to test hypotheses about a population proportion even if the sample size is not large enough for the large-sample z test to be appropriate. Examples 10.18 and 10.19 illustrate the use of a randomization test to carry out a hypothesis test for a population proportion.

SECTION 10.6 EXERCISES

Each Exercise Set assesses the following chapter learning objectives: M7, M8

SECTION 10.6 Exercise Set 1

10.75 We are only beginning to learn about the long-term effects of space travel on human health. A study published in 2016 (*Nature Scientific Reports 6*, Article number: 29901, www.nature.com/articles/srep29901, July 28, 2016, retrieved May 6, 2017) found that by 2014, seven of the U.S. astronauts who traveled to the moon during Apollo lunar missions of the 1960s and 1970s had died, and that three of these ($3/7 = 43\%$) had died from cardiovascular disease (CVD). The overall U.S. death rate due to CVD for adults aged 55 to 64 in 2013 was 27%. Do the data for lunar astronauts indicate that, as a group, they are at increased risk of death caused by CVD? Assume that it is reasonable to regard this sample as representative of the population of past, present, and future U.S. lunar astronauts.

a. Explain why the data in this exercise should not be analyzed using a large-sample hypothesis test for a population proportion.

b. Use the output from the Shiny app "Randomization Test for One Proportion" that appears at the bottom of the page to carry out an appropriate hypothesis test.

10.76 A study of treatment of hospitalized patients who develop pneumonia reported that 1 in 5 (20%) are readmitted to the hospital within 30 days after discharge (**"Comparison of Therapist-Directed and Physician-Directed Respiratory Care in COPD Subjects with Acute Pneumonia,"** *Respiratory Care* [2015]: 151–154).

The study reported that 15 out of $n = 162$ hospital patients who had been treated for pneumonia using a respiratory therapist protocol were readmitted to the hospital within 30 days after discharge. You would like to use this sample data to decide if the proportion readmitted is less than 0.20.

a. What hypotheses should be tested?

b. Discuss whether the conditions necessary for a large-sample hypothesis test for one proportion are satisfied.

c. The exact binomial test can be used even in cases when the sample size condition for the large-sample test is met.

Output for Exercise 10.75

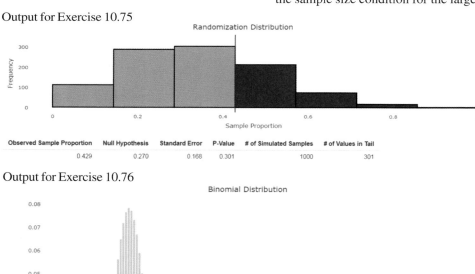

Observed Sample Proportion	Null Hypothesis	Standard Error	P-Value	# of Simulated Samples	# of Values in Tail
0.429	0.270	0.168	0.301	1000	301

Output for Exercise 10.76

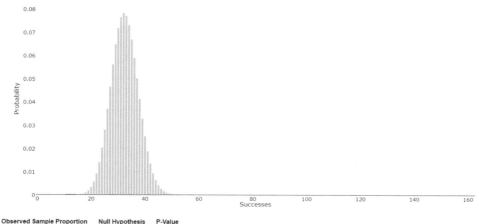

Observed Sample Proportion	Null Hypothesis	P-Value
0.093	0.200	0.000

Use the output at the bottom of the previous page from the Shiny app "Exact Binomial Test for One Proportion" to carry out an exact binomial test for one proportion, using the null hypothesis that the population proportion of subjects who will be readmitted to a hospital within 30 days after following a Respiratory Therapist protocol for treatment of pneumonia is equal to $1/5 = 0.20$.

d. Calculate the P-value for the same hypotheses using the large-sample approach for testing one population proportion. Compare this P-value to the P-value you obtained in Part (b). Would you reach the same conclusion in either case? Explain.

10.77 A sample of dogs were trained using a "Do as I do" method, in which the dog observes the trainer performing a simple task (such as climbing onto a chair or touching a chair) and is expected to perform the same task on the command "Do it!" In a separate training session, the same dogs were trained to lie down regardless of the trainer's actions.

Later, the trainer demonstrated a new simple action and said "Do it!" The dog then either repeated the new action, or repeated a previous trained action (such as lying down). The dogs were retested on the new simple action after one minute had passed, and after one hour had passed. A "success" was recorded if a dog performed the new simple action on the command "Do it!" before performing a previously trained action.

The article **"Your dog remembers more than you think" (*Science*, November 23, 2016, www.sciencemag.org/news/2016/11/your -dog-remembers-more-you-think, retrieved May 6, 2017)** reports that dogs trained using this method recalled the correct new action in 33 out of 35 trials. Suppose you want to use the data from this study to determine if more than half of all dogs trained using this method would recall the correct new action.

a. Explain why the data in this example should not be analyzed using a large-sample hypothesis test for one population proportion.

b. Perform an exact binomial test of the null hypothesis that the proportion of all dogs trained using this method who would perform the correct new action is 0.5, versus the alternative hypothesis that the proportion is greater than 0.5.

SECTION 10.6 Exercise Set 2

10.78 Data from a large study carried out in 2008 was used to estimate that 10% of all smokers who quit smoking are

smoking again after one year **("Relapse to Smoking After 1 Year of Abstinence: A Meta-analysis," www.ncbi.nlm.nih.gov/pmc /articles/PMC2577779/, June 8, 2008, retrieved May 6, 2017)**.

The outcomes of many surgical procedures are improved for patients who are not smoking. In a University of Kansas Medical Center study **("Recidivism Rates After Smoking Cessation Before Spinal Fusion," www.healio.com /orthopedics/journals/ortho/2016-3-39-2/%7B28bf5c50 -c2b8-413a-a662-f10efce6c9ef%7D/recidivism-rates-after -smoking-cessation-before-spinal-fusion, March 31, 2016, retrieved May 6, 2017)**, patients needing spinal fusion surgery were required to quit smoking before their surgery was scheduled. After one year, $n = 25$ of the patients responded to a follow-up survey, and 17 were smoking again. Assume it is reasonable to regard this sample as representative of people who quit smoking before surgery. You would like to use the data from this sample to decide if there is convincing evidence that the proportion of people who quit smoking prior to surgery who are smoking again after one year is greater than 0.10.

a. Explain why the data in this example should not be analyzed using a large-sample hypothesis test for one population proportion.

b. Use the output from the Shiny app "Randomization Test for One Proportion" at the bottpm of the page to carry out an appropriate hypothesis test.

c. Explain how the result of the hypothesis test you performed may be related to the fact that the spinal fusion patients were required to stop smoking before their surgeries.

10.79 Recall that in Exercise 10.63, a survey of 304 U.S. businesses found that 201 indicated that they monitor employees' website visits. This data was used to determine if there is convincing evidence that a majority of businesses monitor employees' website visits.

a. What hypotheses were tested?

b. Discuss whether the conditions necessary for a large-sample hypothesis test for one proportion are satisfied.

c. The exact binomial test can be used even in cases when the sample size condition for the large-sample test is met. Use the output from the Shiny app "Exact Binomial Test for One Proportion" that appears at the top of the following page to carry out an exact binomial test for one proportion, testing whether there is sufficient evidence to conclude that a majority of U.S. businesses monitor employees' website visits.

Output for Exercise 10.78

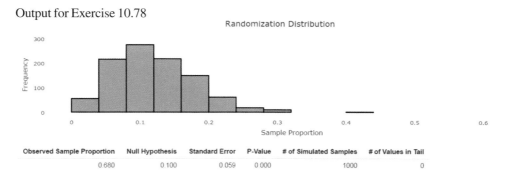

Observed Sample Proportion	Null Hypothesis	Standard Error	P-Value	# of Simulated Samples	# of Values in Tail
0.680	0.100	0.059	0.000	1000	0

Output for Exercise 10.79

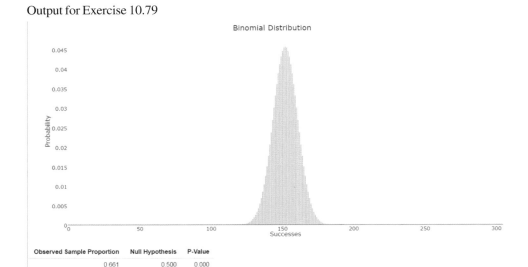

Observed Sample Proportion	Null Hypothesis	P-Value
0.661	0.500	0.000

d. Compare the P-value you obtained in Part (c) with the P-value obtained in Exercise 10.63 Part (b) in Section 10.5. Would you reach the same conclusion in either case? Explain.

10.80 At one point during the 2015 NFL season, Head Coach Bill Belichick and the New England Patriots had won 19 of their past 25 called coin flips at the beginning of NFL games (**"For Bill Belichick, Patriots' strategy is no flip of the coin," www.bostonglobe.com/sports/2015/11/04 /pnotes/vFNt235bsK8x3JLZ6FJdtK/story.html, November 4,** **2015, retrieved May 6, 2017)**. Suppose that these 25 coin toss calls can be considered as representative of all coin toss calls made by this team.

a. Perform an exact binomial test to determine if there is convincing evidence that the proportion of all coin flip calls that the Patriots win is greater than 0.5.

b. Discuss the conditions required for the exact binomial version of the hypothesis test. Write a brief explanation of why the results of the test you performed in Part (a) do not necessarily mean that Coach Belichick is able to predict the results of coin flips better than other coaches.

SECTION 10.7   Avoid These Common Mistakes

When summarizing the results of a hypothesis test, be sure to include all the relevant information. In particular, include:

1. *Hypotheses.* Whether specified in symbols or described in words, it is important that both the null and the alternative hypotheses be clearly stated. If you are using symbols to define the hypotheses, be sure to define them in the context of the problem at hand (for example, p = proportion of all students who purchase textbooks online).

2. *Test procedure.* You should be clear about what test procedure was used (for example, large-sample test for a population proportion), and why you think it was reasonable to use this procedure.

3. *Test statistic and P-value.* Be sure to include the value of the test statistic and the associated P-value. Including the P-value allows a reader who has chosen a different significance level to see whether he or she would have reached the same or a different conclusion.

4. *Conclusion in context.* Never end the report of a hypothesis test with the statement "I rejected (or did not reject) H_0." Always provide a conclusion that is in the context of the problem and that answers the question which the hypothesis test was designed to answer. Be sure to also state what significance level was used.

There are several things you should watch for when conducting a hypothesis test or when evaluating a written summary of a test:

1. A hypothesis test can never show strong support for the null hypothesis. Make sure that you don't confuse "There is no reason to believe the null hypothesis is not true" with the statement "There is convincing evidence that the null hypothesis is true." These are very different statements!

2. If you have complete information for the population, don't carry out a hypothesis test! People sometimes forget that no test is needed to answer questions about a population if you have complete information and don't need to generalize from a sample. For example, an article on prescription drug overdoses (**"Medicine Cabinet Is a Big Killer,"** *Salt Lake City Tribune,* **August 1, 2007**) reported that "In 2006, some 485 Utah deaths were attributed to poisoning—more than any previous year in the state's history. Opiate and opioid medications alone, or in combination, accounted for most of them—307." It would not be appropriate to use this information to test a hypothesis about the proportion of poisoning deaths in Utah during 2006 that involved opiate or opioid medications, because that proportion is known exactly. In this case, $p = \frac{307}{485} = 0.633$.

3. Don't confuse statistical significance with practical significance. When a null hypothesis has been rejected, be sure to step back and evaluate the result in light of its practical importance. For example, you may be convinced that the proportion who respond favorably to a proposed new medical treatment is greater than 0.4, the known proportion that respond favorably to the current treatment. But if your estimate of this proportion for the proposed new treatment is 0.405, it may not be of any practical interest.

CHAPTER ACTIVITIES

ACTIVITY 10.1 A MEANINGFUL PARAGRAPH

Write a meaningful paragraph that includes the following six terms: **hypotheses, *P*-value, reject *H*₀, Type I error, statistical significance, practical significance**. A "meaningful paragraph" is a coherent piece of writing in an appropriate context that uses all of the listed words. It is not a sequence of sentences that just define the terms. The paragraph should show that you understand the meaning of the terms and their relationship to one another. Choosing a good context will make writing a meaningful paragraph easier.

CHAPTER 10 EXPLORATIONS IN STATISTICAL THINKING

EXPLORING SAMPLING VARIABILITY IN THE CONTEXT OF HYPOTHESIS TESTING

In the exercise below, you will go online to select a random samples from a population of adults between the ages of 18 and 64.

Each person in this population was asked if he or she ever sent a text message while driving. Suppose that you would like to know if there is convincing evidence that less than half of the people in the population have sent a text message while driving. To answer this question, you will take a random sample of 30 people from the population.

Go online at statistics.cengage.com/Peck2e/Explore.html and click on the link for Chapter 10. It will take you to a web page where you can select random samples of 30 people from the population.

Click on the Select Sample button. This selects a random sample and will display the following information:

1. The ID number that identifies the person selected
2. The response to the question "Have you ever sent a text message while driving?" These responses were coded numerically—a 1 indicates a yes response and a 2 indicates a no response.

Use the sample data to answer the following questions.

a. Is this an estimation or hypothesis testing problem?

b. Are the data from sampling or from an experiment?

c. How many variables are there?

d. What type of data (categorical or numerical) do you have?

e. How many samples are there?

f. What are the appropriate null and alternative hypotheses?

g. What method might be appropriate for testing these hypotheses?

h. Are the conditions for the selected test met? Explain.

i. What is the value of the test statistic for this test?

j. What is the P-value?

k. If a significance level of 0.05 is used, what would your decision be?

l. Write a few sentences that summarize what you learned about the population proportion based on this hypothesis test.

If asked to do so by your instructor, bring your answers to Parts (a) – (l) to class. Your instructor will lead the class through the rest of this exploration.

m. Did everyone in your class get the same P-value? If not, why do you think this was the case?

n. Did everyone reach the same conclusion to reject the null hypothesis and conclude that there is evidence that the proportion who have sent a text message while driving is less than 0.50? Does this surprise you? Explain.

o. The actual proportion in the population who have sent a text while driving is 0.40. Is rejecting the null hypothesis a correct conclusion or an incorrect conclusion?

p. Did anyone in your class reach an incorrect conclusion? If so, what type of error (Type I or Type II) did they make?

ARE YOU READY TO MOVE ON?   CHAPTER 10 REVIEW EXERCISES

All chapter learning objectives are assessed in these exercises. The learning objectives assessed in each exercise are given in parentheses.

10.81 (C1, M1)
A county commissioner must vote on a resolution that would commit substantial resources to the construction of a sewer in an outlying residential area. Her fiscal decisions have been criticized in the past, so she decides to take a survey of residents in her district to find out if they favor spending money for a sewer system. She will vote to appropriate funds only if she can be reasonably sure that a majority of the people in her district favor the measure. What hypotheses should she test?

10.82 (C1, M1)
The report **"Digital Democracy Survey" (Deloitte Development LLC, 2016, www2.deloitte.com/us/en.html, retrieved November 30, 2016)** says that 69% of U.S. teens age 14 to 18 years access social media from a mobile phone. Suppose you plan to select a random sample of students at the local high school and will ask each student in the sample if he or she accesses social media from a mobile phone. You want to determine if there is evidence that the proportion of students at the high school who access social media using a mobile phone differs from the national figure of 0.69 given in the Nielsen report. What hypotheses should you test?

10.83 (C2)
The article **"How to Block Nuisance Calls" (The Guardian, November 7, 2015)** reported that in a survey of mobile phone users, 70% of those surveyed said they had received at least one nuisance call to their mobile phone in the last month. Suppose that this estimate was based on a representative sample of 600 mobile phone users. These data can be used to determine if there is evidence that more than two-thirds of all mobile phone users have received at least one nuisance call in the last month. The large-sample test for a population proportion was used to test H_0: $p = 0.667$ versus H_a: $p > 0.667$. The resulting P-value was 0.043. Using a significance level of 0.05, the null hypothesis was rejected.

a. Based on the hypothesis test, what can you conclude about the proportion of mobile phone users who received at least one nuisance call on their mobile phones within the last month?

b. Is it reasonable to say that the data provide strong support for the alternative hypothesis?

c. Is it reasonable to say that the data provide strong evidence against the null hypothesis?

10.84 (C3)
Explain why failing to reject the null hypothesis in a hypothesis test does not mean there is convincing evidence that the null hypothesis is true.

10.85 (C4, M2)
Researchers at the University of Washington and Harvard University analyzed records of breast cancer screening and diagnostic evaluations (**"Mammogram Cancer Scares**

More Frequent Than Thought," *USA TODAY*, April 16, 1998). Discussing the benefits and downsides of the screening process, the article states that although the rate of false-positives is higher than previously thought, if radiologists were less aggressive in following up on suspicious tests, the rate of false-positives would fall, but the rate of missed cancers would rise. Suppose that such a screening test is used to decide between a null hypothesis of

H_0: no cancer is present

and an alternative hypothesis of

H_a: cancer is present.

(Although these are not hypotheses about a population characteristic, this exercise illustrates the definitions of Type I and Type II errors.)

a. Would a false-positive (thinking that cancer is present when in fact it is not) be a Type I error or a Type II error?

b. Describe a Type I error in the context of this problem, and discuss the consequences of making a Type I error.

c. Describe a Type II error in the context of this problem, and discuss the consequences of making a Type II error.

d. Recall the statement in the article that if radiologists were less aggressive in following up on suspicious tests, the rate of false-positives would fall but the rate of missed cancers would rise. What aspect of the relationship between the probability of a Type I error and the probability of a Type II error is being described here?

10.86   (C4, M2)

The paper **"Living Near Nuclear Power Plants and Thyroid Cancer Risks"** (*Environmental International* **[2016]: 42–48**) investigated whether living near a nuclear power plant increases the risk of thyroid cancer. The authors of this paper concluded that there was no evidence of increased risk of thyroid cancer in areas that were near a nuclear power plant.

a. Suppose p denotes the true proportion of the population in areas near nuclear power plants who are diagnosed with thyroid cancer during a given year. The researchers who wrote this paper might have considered two rival hypotheses of the form

H_0: p is equal to the corresponding value for areas without nuclear power plants

H_a: p is greater than the corresponding value for areas without nuclear power plants

Did the researchers reject H_0 or fail to reject H_0?

b. If the researchers are incorrect in their conclusion that there is no evidence of increased risk of thyroid cancer associated with living near a nuclear power plant, are they making a Type I or a Type II error? Explain.

c. Can the result of this hypothesis test be interpreted as meaning that there is strong evidence that the risk of thyroid cancer is not higher for people living near nuclear power plants? Explain.

10.87   (C5, M5)

Suppose that you are an inspector for the Fish and Game Department and that you are given the task of determining whether to prohibit fishing along part of the Oregon coast. You will close an area to fishing if it is determined that more than 3% of fish have unacceptably high mercury levels.

a. Which of the following pairs of hypotheses would you test:

$$H_0: p = 0.03 \text{ versus } H_a: p > 0.03$$

or

$$H_0: p = 0.03 \text{ versus } H_a: p < 0.03$$

Explain the reason for your choice.

b. Would you use a significance level of 0.10 or 0.01 for your test? Explain.

10.88   (C6)

The article **"Cops Get Screened for Digital Dirt"** (*USA TODAY*, **November 12, 2010**) summarizes a report on law enforcement agency use of social media to screen applicants for employment. The report was based on a survey of 728 law enforcement agencies. One question on the survey asked if the agency routinely reviewed applicants' social media activity during background checks. For purposes of this exercise, suppose that the 728 agencies were selected at random and that you want to use the survey data to decide if there is convincing evidence that more than 25% of law enforcement agencies review applicants' social media activity as part of routine background checks.

a. Describe the shape, center, and variability of the sampling distribution of $\hat{p}$ for samples of size 728 if the null hypothesis $H_0: p = 0.25$ is true.

b. Would you be surprised to observe a sample proportion as large as $\hat{p} = 0.27$ for a sample of size 728 if the null hypothesis $H_0: p = 0.25$ were true? Explain why or why not.

c. Would you be surprised to observe a sample proportion as large as $\hat{p} = 0.31$ for a sample of size 728 if the null hypothesis $H_0: p = 0.25$ were true? Explain why or why not.

d. The actual sample proportion observed in the study was $\hat{p} = 0.33$. Based on this sample proportion, is there convincing evidence that more than 25% of law enforcement agencies review social media activity as part of background checks, or is this sample proportion consistent with what you would expect to see when the null hypothesis is true?

10.89   (C6, M6)

"Most Like It Hot" is the title of a press release issued by the Pew Research Center **(March 18, 2009, www.pewsocialtrends. org)**. The press release states that "by an overwhelming margin, Americans want to live in a sunny place." This statement is based on data from a nationally representative sample of 2260 adult Americans. Of those surveyed, 1288 indicated that they would prefer to live in a hot climate rather than a cold climate. Suppose that you want to determine if there is convincing evidence that a majority of all adult Americans prefer a hot climate over a cold climate.

a. What hypotheses should be tested in order to answer this question?

b. The P-value for this test is 0.000001. What conclusion would you reach if $\alpha = 0.01$?

For questions 10.90–10.91, answer the following four key questions (introduced in Section 7.2) and indicate whether the method that you would consider would be a large-sample hypothesis test for a population proportion.

Q Question Type	Estimation or hypothesis testing?
S Study Type	Sample data or experiment data?
T Type of Data	One variable or two? Categorical or numerical?
N Number of Samples or Treatments	How many samples or treatments?

10.90   (M3, P1)
Refer to the instructions prior to this exercise. The article **"The Average American Is in Credit Card Debt, No Matter the Economy"** (*Money*, February 9, 2016) states that in 2015, the average credit card debt for Americans between the age of 18 and 65 was $4717. You are interested in determining if there is evidence that the mean credit card debt in the current year is greater than the known average from 2015.

10.91   (M3, P1)
Refer to the instructions prior to Exercise 10.90. The paper **"I Smoke but I Am Not a Smoker"** (*Journal of American College Health* [2010]: 117–125) describes a survey of 899 college students who were asked about their smoking behavior. Of the students surveyed, 268 classified themselves as nonsmokers, but said "yes" when asked later in the survey if they smoked. These students were classified as "phantom smokers," meaning that they did not view themselves as smokers even though they do smoke at times. The authors were interested in using these data to determine if there is convincing evi-

dence that more than 25% of college students fall into the phantom smoker category.

10.92   (M4, M6)
Suppose that the sample of 899 college students described in the previous exercise can be regarded as representative of college students in the United States.

a. What hypotheses would you test to answer the question posed in the previous exercise?

b. Is the sample size large enough for a large-sample test for a population proportion to be appropriate?

c. What is the value of the test statistic and the associated P-value for this test?

d. If a significance level of 0.05 were chosen, would you reject the null hypothesis or fail to reject the null hypothesis?

10.93   (M6, P2)
Public Policy Polling conducts an annual poll on sports-related issues. In 2015, they found that in a sample of 1222 adult Americans, 794 said that they thought the designated hitter rule in professional baseball should be eliminated and that pitchers should be required to bat (**www.publicpol-icypolling.com/pdf/2015/PPP_Release_National_51216.pdf, retrieved December 1, 2016**). Suppose that this sample is representative of adult Americans. Based on the given information, is there convincing evidence that a majority of adult Americans think that the designated hitter rule should be eliminated and that pitchers should be required to bat?

10.94   (M6, P2)
Past experience is that when individuals are approached with a request to fill out and return a particular questionnaire in a provided stamped and addressed envelope, the response rate is 40%. An investigator believes that if the person distributing the questionnaire were stigmatized in some obvious way, potential respondents would feel sorry for the distributor and thus tend to respond at a rate higher than 40%. To test this theory, a distributor wore an eye patch. Of the 200 questionnaires distributed by this individual, 109 were returned. Does this provide evidence that the response rate in this situation is greater than the previous rate of 40%? State and test the appropriate hypotheses using a significance plevel of 0.05.

TECHNOLOGY NOTES

Z-Test for Proportions

JMP
Summarized data

1. Enter the data into the JMP data table with categories in the first column and counts in the second column
2. Click **Analyze** and select **Distribution**
3. Click and drag the first column name from the box under **Select Columns** to the box next to **Y, Columns**

4. Click and drag the second column name from the box under **Select Columns** to the box next to **Freq**
5. Click **OK**
6. Click the red arrow next to the column name and click **Test Probabilities**

7. Under the **Test Probabilities** section that appears in the output, click in the box across from **Yes** under the **Hypoth Prob** and type the hypothesized value for p
8. Select the appropriate option for alternative
9. Click **Done**

Note: In the two-sided case, JMP uses the square of the z test statistic, called the Chi-Square test statistic. The two methods are mathematically identical.

Note: In the one-sided cases, JMP uses the exact binomial test rather than the z test.

Raw data
1. Enter the raw data into a column

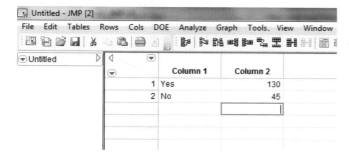

2. Click **Analyze** and select **Distribution**
3. Click and drag the first column name from the box under **Select Columns** to the box next to **Y, Columns**
4. Click **OK**
5. Click the red arrow next to the column name and click **Test Probabilities**
6. Under the **Test Probabilities** section that appears in the output, click in the box across from **Yes** under the **Hypoth Prob** and type the hypothesized value for p
7. Select the appropriate option for alternative
8. Click **Done**

Note: In the two-sided case, JMP uses the square of the z test statistic, called the Chi-Square test statistic. The two methods are mathematically identical.

Note: In the one-sided cases, JMP uses the exact binomial test rather than the z test.

Minitab
Summarized data
1. Click **Stat** then click **Basic Statistics** then click **1 Proportion...**
2. Click the radio button next to **Summarized data**
3. In the box next to **Number of Trials:** type the value for n, the sample size
4. In the box next to **Number of events:** type the value for the number of successes
5. Click **Options...**

6. Input the appropriate hypothesized value in the box next to **Test proportion:**
7. Select the appropriate alternative hypothesis from the drop-down menu next to **Alternative**
8. Check the box next to **Use test and interval based on normal distribution**
9. Click **OK**
10. Click **OK**

Raw data
1. Input the raw data into a column
2. Click **Stat** then click **Basic Statistics** then click **1 Proportion...**
3. Click in the box under **Samples in columns:**
4. Double click the column name where the raw data are stored
5. Click **Options...**
6. Select the appropriate alternative hypothesis from the drop-down menu next to **Alternative**
7. Check the box next to **Use test and interval based on normal distribution**
8. Click **OK**
9. Click **OK**

SPSS
SPSS does not have the functionality to automatically calculate a z-test for a testing a single proportion.

Excel
Excel does not have the functionality to automatically calculate a z-test for a testing a single population proportion. You may type in the formula by hand into a cell to have Excel calculate the value of the test statistic for you. Then use methods from Chapter 6 to find the P-value using the Normal Distribution.

TI-83/84
1. Press the **STAT** key
2. Highlight **TESTS**
3. Highlight **1-PropZTest...** and press **ENTER**
4. Next to **p0** type the hypothesized value for p
5. Next to **x** type the number of successes
6. Next to **n** type the sample size, n
7. Next to **prop,** highlight the appropriate alternative hypothesis
8. Highlight **Calculate** and press **ENTER**

TI-Nspire
1. Enter the Calculate Scratchpad
2. Press the **menu** key then select **6:Statistics** then select **7:Stat Tests** then select **5:1-Prop z Test...** then press **enter**
3. In the box next to **p0** type the hypothesized value for p
4. In the box next to **Successes, x** type the number of successes
5. In the box next to **n** type the sample size, n
6. In the box next to **Alternate Hyp** choose the appropriate alternative hypothesis from the drop-down menu
7. Press **OK**

11

Asking and Answering Questions About the Difference Between Two Population Proportions

Aleksandr Markin/Shutterstock.com

PREVIEW

Many statistical investigations involve comparing two populations or comparing two treatments. In Chapters 9 and 10, you saw how sample data could be used to estimate a population proportion and to test hypotheses about the value of a single population proportion. In this chapter, you will see how sample data can be used to learn about the difference between two population proportions. You will also see how data from an experiment can be used to learn about a difference in treatment proportions.

CHAPTER LEARNING OBJECTIVES

Conceptual Understanding

After completing this chapter, you should be able to

C1 Understand how a research question about the difference in two population proportions is translated into hypotheses.

C2 Understand the role that variability due to random assignment plays in drawing conclusions from experiment data.

Mastering the Mechanics

After completing this chapter, you should be able to

M1 Know the conditions for appropriate use of the large-sample confidence interval for a difference in population proportions and the large-sample test of hypotheses about a difference in population proportions.

M2 Calculate and interpret a confidence interval for a difference in population proportions.

M3 Carry out a large-sample hypothesis test for a difference in two population proportions.

M4 Distinguish between situations that involve learning from sample data and those that involve learning from experiment data.

M5 Know the conditions required for appropriate use of the large-sample z test and the large-sample z confidence interval to draw conclusions about a difference in treatment proportions.

M6 Use the large-sample z test to test hypotheses about a difference in treatment proportions.

M7 Use the large-sample z confidence interval to estimate a difference in treatment proportions.

M8 (Optional) Calculate and interpret a bootstrap confidence interval for a difference in proportions.

M9 (Optional) Carry out a randomization test for a difference in proportions.

Putting It into Practice

After completing this chapter, you should be able to

P1 Interpret a confidence interval for a difference in two population proportions in context and interpret the associated confidence level.

P2 Carry out a large-sample hypothesis test for a difference in two population proportions and interpret the conclusion in context.

P3 Carry out large-sample z test for a difference in treatment proportions and interpret the conclusion in context.

P4 Interpret a large-sample z confidence interval for a difference in treatment proportions in context and interpret the associated confidence level.

PREVIEW EXAMPLE

Cell-Phone Fundraising

After the 2010 earthquake in Haiti, many charitable organizations conducted fundraising campaigns to raise money for emergency relief. Some of these campaigns allowed a person to donate money by using their cell phone to send a text message. The amount donated was then added to the person's next cell phone bill. The report **"Early Signals on Mobile Philanthropy: Is Haiti the Tipping Point?" (Edge Research, 2010)** describes a national survey that investigated the ways in which people made donations to the Haiti relief effort. The report states that the proportion of those in a sample of Generation Y (called Gen Y and defined as those born between 1980 and 1988) who said that they had made a donation to the Haiti relief effort by cell phone was 0.17 (or 17%). This proportion was 0.14 (or 14%) for a sample of Generation X (called Gen X and defined as those born between 1968 and 1979). The proportions reported for Gen Y and Gen X are sample proportions, so you don't expect them to be exactly equal to the corresponding population proportions. But what can you learn from these sample proportions?

Suppose you plan to use the data from these two samples (Gen Y and Gen X) to draw conclusions about how these two populations might differ. Before you do this, there are a few things you would want to know. In particular, you would want to know how the samples were selected and what the sample sizes were. If these samples can be viewed as representative of the respective populations, you can then use the sample data to estimate the difference between the proportion who donated by cell phone for Gen Y and the proportion for Gen X. You might also test hypotheses about this difference. For example, you might want to decide if there is convincing evidence that the proportion who donated by cell phone is greater for the Gen Y population than for the Gen X population. This example will be revisited later in this chapter after you have seen how sample data are used to learn about the difference between two population proportions. ■

SECTION 11.1 Estimating the Difference Between Two Population Proportions

Many statistical studies are carried out in order to estimate the difference between two population proportions. For example, researchers have studied how people of different ages use mobile technology. As part of a survey conducted by Pew Research **("Americans' Views on Mobile Etiquette," August 26, 2015, www.pewinternet.org, retrieved December 12, 2016)**, people in a representative sample of 708 American adults age 18 to 29 were asked if they thought it was OK to use a cell phone while at a restaurant. The same question was asked of a representative sample of 1029 adult Americans age 30 to 49. You might expect the proportion who think it is OK to use a cell phone at a restaurant is higher for the 18 to 29 age group than for the 30 to 49 age group, but how much higher? In this section, you will see how sample data can be used to answer questions like this.

Before you get started, you need to consider some new notation. In the previous chapters, the symbol p was used to represent the proportion of "successes" in a single population. The sample proportion was denoted by $\hat{p}$. When comparing two populations, there are two samples—one from each population. You need to distinguish between the two populations and the two samples, and this is done using subscripts, as shown in the following box.

NOTATION

Proportion of successes in Population 1: p_1

Proportion of successes in Population 2: p_2

	Sample Size	Sample Proportion of Successes
Sample from Population 1	n_1	$\hat{p}_1$
Sample from Population 2	n_2	$\hat{p}_2$

In this chapter, only the case where the two samples are **independent samples** is considered. Two samples are said to be independent samples if the selection of the individuals that make up one sample does not influence the selection of the individuals in the other sample. This would be the case if a random sample is independently selected from each of the two populations.

When comparing two populations on the basis of "success" proportions, it is common to focus on the quantity $p_1 - p_2$, the difference between the two population proportions. Because $\hat{p}_1$ is an estimate of p_1 and $\hat{p}_2$ is an estimate of p_2, the obvious statistic to choose as an estimate of $p_1 - p_2$ is the difference in the sample proportions, $\hat{p}_1 - \hat{p}_2$. But the values $\hat{p}_1$ and $\hat{p}_2$ both vary from sample to sample, so the difference $\hat{p}_1 - \hat{p}_2$ will also vary. For example, a first sample from each of two populations might result in

$$\hat{p}_1 = 0.64 \qquad \hat{p}_2 = 0.61 \qquad \hat{p}_1 - \hat{p}_2 = 0.03$$

A second sample from each population might result in

$$\hat{p}_1 = 0.62 \qquad \hat{p}_2 = 0.68 \qquad \hat{p}_1 - \hat{p}_2 = -0.06$$

The sampling distribution of $\hat{p}_1 - \hat{p}_2$ describes this sample-to-sample variability.

General Properties of the Sampling Distribution of $\hat{p}_1 - \hat{p}_2$

If $\hat{p}_1 - \hat{p}_2$ is the difference in sample proportions for independently selected random samples, then the following rules hold.

Rule 1. $\mu_{\hat{p}_1-\hat{p}_2} = p_1 - p_2$

This rule says that the sampling distribution of $\hat{p}_1 - \hat{p}_2$ is centered at the actual value of the difference in population proportions. This means that the sample differences tend to cluster around the value of the actual population difference.

Rule 2. $\sigma^2_{\hat{p}_1-\hat{p}_2} = \sigma^2_{\hat{p}_1} + \sigma^2_{\hat{p}_2} = \dfrac{p_1(1-p_1)}{n_1} + \dfrac{p_2(1-p_2)}{n_2}$

and

$$\sigma_{\hat{p}_1-\hat{p}_2} = \sqrt{\dfrac{p_1(1-p_1)}{n_1} + \dfrac{p_2(1-p_2)}{n_2}}$$

This rule specifies the standard error of $\hat{p}_1 - \hat{p}_2$. The value of the standard error describes how much the $\hat{p}_1 - \hat{p}_2$ values tend to vary from the actual value of the population difference.

Rule 3. If both n_1 and n_2 are large, then the sampling distribution of $\hat{p}_1 - \hat{p}_2$ is approximately normal. The sample sizes can be considered large if $n_1 p_1 \geq 10$, $n_1(1 - p_1) \geq 10$, $n_2 p_2 \geq 10$, and $n_2(1 - p_2) \geq 10$ (or equivalently if n_1 and n_2 are large enough to expect at least 10 successes and 10 failures in each of the two samples).

In Chapter 9, the general form of a confidence interval was given as

$$\text{statistic} \pm (\text{critical value}) \left(\begin{smallmatrix} \text{standard error} \\ \text{of the statistic} \end{smallmatrix} \right)$$

Adapting this general form to the case of estimating $p_1 - p_2$ using data from large independent random samples, you use

Statistic	$\hat{p}_1 - \hat{p}_2$
Critical value	Because the sampling distribution of $\hat{p}_1 - \hat{p}_2$ is approximately normal when the sample sizes are large, a z critical value is used.
Standard error of the statistic $(\hat{p}_1 - \hat{p}_2)$	$\sqrt{\dfrac{p_1(1-p_1)}{n_1} + \dfrac{p_2(1-p_2)}{n_2}}$

Because the two population proportions are not known, the two sample proportions are used to estimate the standard error and to check sample size conditions. This results in the following large-sample confidence interval.

A Large-Sample Confidence Interval for a Difference in Population Proportions

Appropriate when the following conditions are met:
1. The samples are independent random samples from the populations of interest (or the samples are independently selected in a way that it is reasonable to regard each sample as representative of the corresponding population).
2. The sample sizes are large. This condition is met when $n_1 \hat{p}_1 \geq 10$, $n_1(1- \hat{p}_1) \geq 10$, $n_2 \hat{p}_2 \geq 10$, and $n_2(1- \hat{p}_2) \geq 10$ or (equivalently) if each sample includes at least 10 successes and 10 failures.

(continued)

When these conditions are met, a confidence interval for the difference in population proportions is

$$(\hat{p}_1 - \hat{p}_2) \pm (z \text{ critical value}) \sqrt{\frac{\hat{p}_1(1-\hat{p}_1)}{n_1} + \frac{\hat{p}_2(1-\hat{p}_2)}{n_2}}$$

The desired confidence level determines which z critical value is used. Three common confidence levels use the following critical values:

Confidence Level	z Critical Value
90%	1.645
95%	1.96
99%	2.58

Interpretation of Confidence Interval
You can be confident that the actual value of the difference in population proportions is included in the calculated interval. In a given problem, this statement should be worded in context.

Interpretation of Confidence Level
The confidence level specifies the long-run percentage of the time that this method will be successful in capturing the actual difference in population proportions.

Questions

Q Question type: estimation or hypothesis testing?

S Study type: sample data or experiment data?

T Type of data: one variable or two? Categorical or numerical?

N Number of samples or treatments: how many?

The large-sample confidence interval for a difference in population proportions is a method you should consider when the answers to the four key questions (QSTN) are *estimation, sample data, one categorical variable,* and *two samples.*

Now you're ready to look at an example, following the usual five-step process for estimation problems (EMC³).

Example 11.1 Cell Phone Etiquette

Let's return to the example introduced at the beginning of this section to answer the question, "How much greater is the proportion who think it is OK to use a cell phone in a restaurant for people age 18 to 29 than for those age 30 to 49?" The study described earlier found that 354 of 708 people in the sample of 18- to 29-year-olds and 412 of the 1029 people in the sample of 30- to 49-year-olds said that they thought it was OK to use a cell phone in a restaurant. Based on these sample data, what can you learn about the actual difference in proportions for these two populations?

Start by answering the four key questions (QSTN) for this problem.

Q Question Type	Estimation or hypothesis testing?	Estimation.
S Study Type	Sample data or experiment data?	Sample data.
T Type of Data	One variable or two? Categorical or numerical?	One categorical variable with two categories (OK to use a cell phone in a restaurant, not OK to use a cell phone in a restaurant).
N Number of Samples or Treatments	How many samples or treatments?	Two samples—one from each age group.

The answers are *estimation, sample data, one categorical variable,* and *two samples.* This combination of answers suggests that a large-sample confidence interval for a difference in population proportions should be considered.

Now you're ready to use the five-step process to learn from the data.

Steps

E   Estimate
M   Method
C   Check
C   Calculate
C   Communicate results

Step	
Estimate: Explain what population characteristic you plan to estimate.	In this example, you want to estimate the difference in the proportion of people age 18 to 29 who think it is OK to use a cell phone in a restaurant and this proportion for those age 30 to 49. This is the value of $p_1 - p_2$, **where p_1 is the proportion for the 18 to 29 age group and p_2 is the proportion for the 30 to 49 age group.**
Method: Select a potential method.	Because the answers to the four key questions are estimation, sample data, one categorical variable, and two samples, a **large-sample confidence interval for a difference in population proportions** will be considered (see Table 7.1).
	Associated with every confidence interval is a confidence level. When the method selected is a confidence interval, you will also need to specify a confidence level.
	For this example, a **confidence level of 90%** will be used.
Check: Check to make sure that the method selected is appropriate.	There are **two conditions** that need to be met in order to use the large-sample confidence interval for a difference in population proportions.
	The large-sample condition is easily verified. With "success" denoting a person who thinks it is OK to use a cell phone in a restaurant, the sample sizes are large enough because there are more than 10 successes (354 in sample 1 and 412 in sample 2) and more than 10 failures ($708 - 354 = 354$ in sample 1 and $1029 - 412 = 617$ in sample 2) in each sample.
	The requirement of independent random samples or samples that are representative of the corresponding populations is more difficult. The researchers who collected these data indicate in the report that they believe that the samples are representative of the two populations of interest.
Calculate: Use sample data to perform any necessary calculations.	$n_1 = 708 \qquad\qquad n_2 = 1029$ $\hat{p}_1 = \dfrac{354}{708} = 0.500 \qquad \hat{p}_2 = \dfrac{412}{1029} = 0.400$ For a confidence level of 90%, the appropriate z critical value is 1.645. Next, substitute the values for n_1, n_2, $\hat{p}_1$ and $\hat{p}_2$ into the confidence interval formula: $(\hat{p}_1 - \hat{p}_2) \pm (z \text{ critical value}) \sqrt{\dfrac{\hat{p}_1(1 - \hat{p}_1)}{n_1} + \dfrac{\hat{p}_2(1 - \hat{p}_2)}{n_2}}$ $(0.500 - 0.400) \pm (1.645) \sqrt{\dfrac{(0.500)(1 - 0.500)}{708} + \dfrac{(0.400)(1 - 0.400)}{1029}}$ $0.100 \pm (1.645)\sqrt{0.0006}$ $0.100 \pm (1.645)(0.024)$ 0.100 ± 0.039 $(0.061, 0.139)$
Communicate Results: Answer the research question of interest, explain what you have learned from the data, and acknowledge potential risks.	The interpretation of confidence intervals should always be worded in the context of the problem. You should always give both an interpretation of the interval itself and of the confidence level associated with the interval. **Confidence interval:** Assuming that the samples were selected in a reasonable way, you can be 90% confident that the actual difference in the proportion of people who think it is OK to use a cell phone in a restaurant for 18- to 29-year-olds and for 30- to 49-year-olds is between 0.061 and 0.139. This means that you can be confident that the proportion is higher for 18- to 29-year-olds than for 30- to 49-year-olds by somewhere between 0.061 and 0.139. **Confidence level:** The method used to construct this interval estimate will be successful in capturing the actual value of the difference in population proportions about 90% of the time.

Computing a confidence interval for a difference in population proportions can also be done using statistical software or a graphing calculator. For example, Minitab output is shown here. (Minitab has used more decimal accuracy in computing the endpoints of the confidence interval.)

CI for Two Proportions

Sample	X	N	Sample p
1	354	708	0.500000
2	412	1029	0.400389

Difference = p (1) − p (2)
Estimate for difference: 0.0996113
90% CI for difference: (0.0597794, 0.139443)

Interpreting Confidence Intervals for a Difference

Interpreting confidence intervals for a difference in proportions is a bit more complicated than interpreting a confidence interval for a single proportion. The following table shows how the interval can be interpreted in three different cases.

Case	Interpretation	Example
Both endpoints of the confidence interval for $p_1 - p_2$ are positive.	If $p_1 - p_2$ is positive, it means that you think p_1 is greater than p_2 and the interval gives an estimate of how much greater.	(0.24, 0.36) You think that p_1 is greater than p_2 by somewhere between 0.24 and 0.36.
Both endpoints of the confidence interval for $p_1 - p_2$ are negative.	If $p_1 - p_2$ is negative, it means that you think p_1 is less than p_2 and the interval gives an estimate of how much less.	(−0.14, −0.06) You think that p_1 is less than p_2 by somewhere between 0.06 and 0.14. (Or equivalently, you think that p_2 is greater than p_1 by somewhere between 0.06 and 0.14.)
0 is included in the confidence interval.	If the confidence interval includes 0, a plausible value for $p_1 - p_2$ is 0. This suggests that p_1 and p_2 could be equal.	(−0.04, 0.09) Because 0 is included in the confidence interval, it is plausible that the two population proportions could be equal.

Example 11.2 Working Parents

The report **"Raising Kids and Running a Household: How Working Parents Share the Load" (November 4, 2015, Pew Research Center, www.pewresearch.org, retrieved December 12, 2016)** described a survey of parents of children under the age of 18. Each person in a representative sample of 825 working fathers and a sample of 586 working mothers was asked if balancing the responsibilities of a job and a family was difficult. It was reported that 429 (52%) of the fathers surveyed and 352 (60%) of the mothers surveyed said that it was difficult. The two samples were independently selected and were thought to be representative of working fathers and mothers of children under 18 years old. This information can be used to estimate the difference between the proportion of working fathers who find balancing work and family difficult, p_1, and this proportion for working mothers, p_2.

Answers to the four key questions for this problem are:

Q Question Type	Estimation or hypothesis testing?	Estimation.
S Study Type	Sample data or experiment data?	Sample data.
T Type of Data	One variable or two? Categorical or numerical?	One categorical variable. The data are responses to the question about the difficulty of balancing work and family. The variable has two possible values—difficult and not difficult.
N Number of Samples or Treatments	How many samples or treatments?	Two samples—one from the population of fathers and one from the population of mothers.

So, this is *estimation, sample data, one categorical variable,* and *two samples.* This combination of answers leads you to consider a large-sample confidence interval for a difference in population proportions.

You can use the five-step process for estimation problems (EMC^3) to construct a 95% confidence interval.

E Estimate   You want to estimate $p_1 - p_2$, the difference in the proportion who find balancing work and family difficult for fathers and for mothers. Here, p_1 denotes this proportion for fathers and p_2 denotes this proportion for mothers.

M Method   Because the answers to the four key questions are estimation, sample data, one categorical variable, and two samples, a large-sample confidence interval for a difference in population proportions should be considered. A confidence level of 95% was specified for this example.

C Check   There are two conditions that need to be met in order for the large-sample confidence interval for a difference in population proportions to be appropriate. The large-sample condition is easily verified. If you define a success as someone who finds balancing work and family difficult, there are more than 10 successes in each sample and more than 10 failures in each sample.

The example specified that the samples were independently selected and that the samples could be considered as representative of the populations.

C Calculate   For a confidence level of 95%, the appropriate z critical value is 1.96.

$$\hat{p}_1 = \frac{429}{825} = 0.520 \quad \hat{p}_2 = \frac{352}{586} = 0.601$$

$$(\hat{p}_1 - \hat{p}_2) \pm (z \text{ critical value}) \sqrt{\frac{\hat{p}_1(1 - \hat{p}_1)}{n_1} + \frac{\hat{p}_2(1 - \hat{p}_2)}{n_2}}$$

$$(0.520 - 0.601) \pm (1.96) \sqrt{\frac{(0.520)(1 - 0.520)}{825} + \frac{(0.601)(1 - 0.601)}{586}}$$

$$-0.081 \pm (1.96)(0.027)$$

$$-0.081 \pm 0.053$$

$$(-0.134, -0.028)$$

Statistical software or a graphing calculator could also have been used to compute the endpoints of the confidence interval. Minitab output is shown here. The differences in the interval endpoints compared to those calculated using the formula are just due to rounding.

CI for Two Proportions

Sample	X	N	Sample p
1	429	825	0.520000
2	352	586	0.600683

Difference = p (1) − p (2)
Estimate for difference: −0.0806826
95% CI for difference: (−0.132976, −0.0283891)

C Communicate Results   **Interpret Confidence interval**
You can be 95% confident that the actual difference in the proportions of working fathers and working mothers who find balancing work and family difficult is between −0.133 and −0.028. Because both of the endpoints of the confidence interval are negative, this means that you think that the proportion of working fathers who find balancing work and family difficult is less than the proportion of working mothers who find it difficult by somewhere between 0.028 and 0.133.

Interpret Confidence level
The method used to construct this interval estimate is successful in capturing the actual difference in population proportions about 95% of the time.

Summing It Up—Section 11.1

The following learning objectives were addressed in this section:

Mastering the Mechanics

M1: Know the conditions for appropriate use of the large-sample confidence interval for a difference in population proportions and the large-sample test of hypotheses about a difference in population proportions.

There are two conditions that should be verified before a large-sample confidence interval for a difference in two population proportions is used. These are

1. The samples are independent random samples from the populations of interest, or the samples are independently selected in a way that would result in representative samples.
2. The sample sizes are large. This condition is met when $n_1\hat{p}_1$, $n_1(1 - \hat{p}_1)$, $n_2\hat{p}_2$ and $n_2(1 - \hat{p}_2)$ are all greater than or equal to 10.

M2: Calculate and interpret a confidence interval for a difference in population proportions.

The five-step process for estimation problems (EMC³) can be used to estimate a difference in two population proportions. The process of calculating and interpreting a confidence interval for a difference in two population proportions is illustrated in Examples 11.1 and 11.2.

Putting It into Practice

P1: Interpret a confidence interval for a difference in two population proportions in context and interpret the associated confidence level.

A discussion of the interpretation of confidence intervals for a difference can be found just prior to Example 11.2. If both endpoints of a confidence interval for $p_1 - p_2$ are positive, it means that you think that p_1 is greater than p_2 and the interval gives an estimate of how much greater. If both endpoints are negative, it means that you think that p_1 is less than p_2 and the interval gives an estimate of how much less. If 0 is included in the interval, it means that 0 is a plausible value for the difference in the two population proportions.

Confidence level is interpreted in the same way as it was for the large sample confidence interval for a population proportion. It specifies the long-run percentage of the time that the method will be successful in capturing the actual difference in population proportions.

SECTION 11.1 EXERCISES

Each Exercise Set assesses the following chapter learning objectives: M1, M2, P1

SECTION 11.1 Exercise Set 1

11.1 According to the **U.S. Census Bureau (www.census.gov)**, the percentage of U.S. residents living in poverty in 2015 was 12.2% for men and 14.8% for women. These percentages were estimates based on data from large representative samples of men and women. Suppose that the sample sizes were 1200 for men and 1000 for women. You would like to use the survey data to estimate the difference in the proportion living in poverty in 2015 for men and women. (Hint: See Example 11.2.)
a. Answer the four key questions (QSTN) for this problem. What method would you consider based on the answers to these questions?

b. Use the five-step process for estimation problems (EMC³) to calculate and interpret a 90% large-sample confidence interval for the difference in the proportion living in poverty in 2015 for men and women.

11.2 The U.S. Department of Transportation reported that in a large study of mobile phone use while driving conducted in 2015, 4.4% of female drivers observed and 3.5% of male drivers observed were using a handheld mobile phone while driving **("Traffic Safety Facts," September, 2016)**. Suppose that these percentages were based on observations from independent random samples of 1200 male drivers and 1200 female drivers (the actual sample sizes were much larger).

a. Are the sample sizes large enough to use the large-sample confidence interval for a difference in population proportions?

b. Assume that it is reasonable to regard these samples as representative of male drivers and female drivers in the United States. Estimate the difference in the proportion of male drivers and the proportion of female drivers who use a mobile phone while driving using a 95% confidence interval.

c. Is zero included in the confidence interval? What does this suggest about the difference in proportions for males and females?

d. Interpret the confidence interval in the context of this problem.

11.3 A Harris Poll press release dated November 1, 2016 summarized results of a survey of 2463 adults and 510 teens age 13 to 17 (**"American Teens No Longer More Likely Than Adults to Believe in God, Miracles, Heaven, Jesus, Angels, or the Devil," www.theharrispoll.com, retrieved December 12, 2016**). It was reported that 19% of the teens surveyed and 26% of the adults surveyed indicated that they believe in reincarnation. The samples were selected to be representative of American adults and teens. Use the data from this survey to estimate the difference in the proportion of teens who believe in reincarnation and the proportion of adults who believe in reincarnation. Be sure to interpret your interval in context.

SECTION 11.1 **Exercise Set 2**

11.4 The following graphical display is similar to one that appeared in *USA TODAY* (February 16, 2012).

Do you expect to get a raise or promotion this year?

Percentage who said yes:
Women **37%**
Men **52%**

Source: Adecco Workplace Insights Survey of 1,014 U.S. adults

The display summarizes data from surveys of male and female American adults. Assume that the two samples were independently selected representative samples and

that the sample size for each sample was 507. Use the given information and the five-step process for estimation problems (EMC³) to calculate and interpret a 95% large-sample confidence interval for the difference in the proportion of men who expected to get a raise or promotion in 2012 and this proportion for women.

11.5 The report **"The New Food Fights: U.S. Public Divides Over Food Science" (December 1, 2016, www.pewinternet.org, retrieved December 10, 2016)** states that younger adults are more likely to see foods with genetically modified ingredients as being bad for their health than older adults. This statement is based on a representative sample of 178 adult Americans age 18 to 29 and a representative sample of 427 adult Americans age 50 to 64. Of those in the 18 to 29 age group, 48% said they believed these foods were bad for their health, while only 38% of those in the 50 to 64 age group believed this.

a. Are the sample sizes large enough to use the large-sample confidence interval to estimate the difference in the population proportions? Explain.

b. Estimate the difference in the proportion of adult Americans age 18 to 29 who believe the foods made with genetically modified ingredients are bad for their health and the corresponding proportion for adult Americans age 50 to 64. Use a 90% confidence interval.

c. Is zero in the confidence interval? What does this suggest about the difference in the two population proportions?

11.6 A survey of high school students is described in the report **"Students on STEM" (changetheequation.org/students-stem, retrieved December 12, 2016)**. The report states that 14% of those in a sample of students in low-income households (defined as a household income less than $50,000 per year) and 24% of those in a sample of students in higher income households (defined as a household income of $50,000 or more) participated in a science club. Suppose that these samples are representative of high school students in the two income groups and that the two sample sizes were both 500. Use a 95% confidence interval to estimate the difference in the proportion participating in a science club for students in the two income groups.

ADDITIONAL EXERCISES

11.7 The article **"More Teen Drivers See Marijuana as OK; It's a Dangerous Trend" (***USA TODAY***, February 23, 2012)** describes two surveys of U.S. high school students. One survey was conducted in 2009 and the other was conducted in 2011. In 2009, 78% of the people in a representative sample of 2300 students said marijuana use is very distracting or extremely distracting to their driving. In 2011, 70% of the people in a representative sample of 2294 students answered this way. Use the five-step process for estimation problems (EMC³) to construct and interpret a 99% large-sample confidence interval for the difference in the proportion of high school students who believed marijuana was very distracting or extremely distracting in 2009 and this proportion in 2011.

11.8 The Bureau of Labor Statistics (www.bls.gov/opub /ted/2014/ted_20141112.htm, retrieved December 13, 2016) reported that 3.8% of college graduates were unemployed in October 2013 and 3.1% of college graduates were unemployed in October 2014. Suppose that the reported percentages were based on independently selected representative samples of 500 college graduates in each of these two years. Construct and interpret a 95% confidence interval for the difference in the proportions of college graduates who were unemployed in these 2 years.

11.9 The Bureau of Labor Statistics report referenced in the previous exercise also reported that 7.3% of high school graduates were unemployed in October 2013 and 5.7% of high school graduates were unemployed in October 2014. Suppose that the reported percentages were based on independently selected representative samples of 400 high school graduates in each of these 2 years.
a. Construct and interpret a 99% large-sample confidence interval for the difference in the proportions of high school graduates who were unemployed in these 2 years.
b. Is the confidence interval from Part (a) wider or narrower than the confidence interval calculated in the previous exercise? Give two reasons why it is wider or narrower.

11.10 Many people believe that they experience "information overload" in today's digital world. The report **"Information Overload" (Pew Research Center, December 7, 2016)** describes a survey in which people were asked if they feel overloaded by information. In a representative sample of 634 college graduates, 101 indicated that they suffered from information overload, while 119 people in an independent representative sample of 496 people who had never attended college said that they suffered from information overload.
a. Construct and interpret a 95% large-sample confidence interval for the proportion of college graduates who experience information overload. (Hint: This is a one-sample confidence interval.)
b. Construct and interpret a 95% large-sample confidence interval for the proportion of people who have never attended college who experience information overload.
c. Do the confidence intervals from Parts (a) and (b) overlap? What does this suggest about the two population proportions?
d. Construct and interpret a 95% large-sample confidence interval for the difference in the proportions who have experienced information overload for college graduates and for people who have never attended college.
e. Is the interval in Part (d) consistent with your answer in Part (c)? Explain.

SECTION 11.2 Testing Hypotheses About the Difference Between Two Population Proportions

In the previous section, you learned how to use data from two independent samples to construct a confidence interval estimate of a difference between two population proportions. Sample data can also be used to test hypotheses about a difference in population proportions.

Example 11.3 Detecting Plagiarism

The report **"The 2016 Inside Higher Ed Survey of Faculty Attitudes on Technology" (www.insidehighered.com/booklet/2016-survey-faculty-attitudes-technology, retrieved December 14, 2016)** describes the results of a survey of 1129 full-time college faculty and 293 part-time college faculty. Survey participants were asked if they require undergraduate students to submit papers through plagiarism-detection software; 40% of the full-time faculty and 38% of the part-time faculty said "yes."

One question of interest is whether the proportion who require students to submit papers through plagiarism-detection software is different for full-time faculty and part-time faculty. The following table summarizes what you know so far.

Population	Population Proportion	Sample Size	Sample Proportion
Full-time college faculty	p_1 = proportion of *all* full-time college faculty who require students to submit papers through plagiarism-detection software	$n_1 = 1129$	$\hat{p}_1 = 0.40$
Part-time college faculty	p_2 = proportion of *all* part-time college faculty who require students to submit papers through plagiarism-detection software	$n_2 = 293$	$\hat{p}_2 = 0.38$

Notice that the two sample proportions are not equal. But even if the two population proportions were equal, the two sample proportions wouldn't necessarily be equal because of sampling variability—differences that occur from one sample to another just by chance. The important question is whether chance is a believable explanation for the observed difference in the two sample proportions or whether this difference is large enough that it is unlikely to have occurred just by chance. A hypothesis test will help you to make this determination.

This example will be revisited a bit later in this section, after you see how the hypothesis testing procedure introduced in Chapter 10 can be adapted to answer questions about a difference in two population proportions.

The logic of hypothesis testing and a five-step process for carrying out a hypothesis test (HMC^3) were introduced in Chapter 10. The same general process is used for all hypothesis tests. What may differ from one type of test to another is

1. The form of the null hypothesis and the form of the alternative hypothesis
2. The test statistic
3. The way in which the associated P-value is determined

Consider each of these three things in the context of testing a hypothesis about a difference in population proportions:

1. Hypotheses

The hypothesis test focuses on the difference in population proportions, $p_1 - p_2$. When there is no difference between the two population proportions, $p_1 - p_2 = 0$. So

$$p_1 = p_2 \quad \text{is equivalent to} \quad p_1 - p_2 = 0$$

Similarly

$$p_1 > p_2 \quad \text{is equivalent to} \quad p_1 - p_2 > 0$$

and

$$p_1 < p_2 \quad \text{is equivalent to} \quad p_1 - p_2 < 0$$

To determine the null and alternative hypotheses, translate the question of interest into hypotheses. For example, if you want to determine whether there is evidence that p_1 is greater than p_2, you would choose $p_1 - p_2 > 0$ as the alternative hypothesis.

The form of the null hypothesis for the test is

$$H_0: p_1 - p_2 = 0$$

which states that there is no difference between the two population proportions. (While it is possible to test a null hypothesis that specifies a difference other than 0, a different test statistic would be used. Because $p_1 - p_2 = 0$ is almost always the relevant choice in applied settings, only this case is considered here.)

Null Hypothesis: $H_0: p_1 - p_2 = 0$	
If the question of interest is...	Then the alternative hypothesis is...
Is p_1 different from p_2?	$H_a: p_1 - p_2 \neq 0$
Is p_1 greater than p_2?	$H_a: p_1 - p_2 > 0$
Is p_1 less than p_2?	$H_a: p_1 - p_2 < 0$

2. Test Statistic

Recall from the previous section that when you have independent random samples, three things are known about the sampling distribution of the statistic $\hat{p}_1 - \hat{p}_2$:

1. $\mu_{\hat{p}_1 - \hat{p}_2} = p_1 - p_2$

2. $\sigma_{\hat{p}_1 - \hat{p}_2} = \sqrt{\dfrac{p_1(1 - p_1)}{n_1} + \dfrac{p_2(1 - p_2)}{n_2}}$

3. If both n_1 and n_2 are large, then the sampling distribution of $\hat{p}_1 - \hat{p}_2$ is approximately normal.

Because $\hat{p}_1 - \hat{p}_2$ has a distribution that is approximately normal, standardizing $\hat{p}_1 - \hat{p}_2$ by subtracting its mean and then dividing by its standard deviation results in a statistic that has a distribution that is approximately standard normal. This means that if both sample sizes are large and the two random samples are independently selected, then the statistic

$$z = \frac{(\hat{p}_1 - \hat{p}_2) - (p_1 - p_2)}{\sqrt{\dfrac{p_1(1 - p_1)}{n_1} + \dfrac{p_2(1 - p_2)}{n_2}}}$$

will have a distribution that is approximately the standard normal (z) distribution.

Recall that in a hypothesis test, you initially assume that the null hypothesis is true and then calculate the value of the test statistic. This is done to see if what was observed in the data is consistent with what you expect to see when the null hypothesis is true, or whether the sample data are surprising. When the null hypothesis is $H_0: p_1 - p_2 = 0$, the z statistic above simplifies to

$$z = \frac{\hat{p}_1 - \hat{p}_2}{\sqrt{\dfrac{p_1(1 - p_1)}{n_1} + \dfrac{p_2(1 - p_2)}{n_2}}}$$

You can use sample data to calculate the numerator of this statistic, but what do you do about the denominator? You can't actually calculate the denominator because you don't know the values of p_1 and p_2. The null hypothesis says that p_1 and p_2 are equal, but it doesn't specify what the common value is. This is resolved by using the sample data to estimate the common value.

When $p_1 = p_2$, both $\hat{p}_1$ and $\hat{p}_2$ are estimates of the common value of the population proportion. However, a better estimate than either $\hat{p}_1$ or $\hat{p}_2$ alone is a weighted average of the two. If the samples are not the same size, more weight is given to the value of the sample proportion that is based on the larger sample size.

If the null hypothesis $H_0: p_1 - p_2 = 0$ is true, a combined estimate of the common population proportion is

$$\hat{p}_c = \frac{n_1 \hat{p}_1 + n_2 \hat{p}_2}{n_1 + n_2} = \frac{\text{total number of successes in the two samples}}{\text{total of the two sample sizes}}$$

for combined

When the null hypothesis is true, using $\hat{p}_c$ as an estimate of the common value of the population proportion results in the following test statistic:

$$z = \frac{\hat{p}_1 - \hat{p}_2}{\sqrt{\dfrac{\hat{p}_c(1 - \hat{p}_c)}{n_1} + \dfrac{\hat{p}_c(1 - \hat{p}_c)}{n_2}}}$$

This is the test statistic you will use to test a hypothesis about a difference in population proportions.

3. Calculating a *P*-value

Once the value of the test statistic has been determined, the next step is to calculate a *P*-value, which tells you how likely it would be to observe sample data at least as extreme as what you actually observed, if the null hypothesis were true. Because this test statistic has a distribution that is approximately standard normal when the null hypothesis is true, *P*-values are calculated as areas under the standard normal curve in exactly the same way as for the one-sample test of Chapter 10.

You now have all the pieces needed to carry out the hypothesis test. The large-sample hypothesis test for a difference in two population proportions summarized in the following box is the method you should consider when the answers to the four key questions (QSTN) that lead to a recommended method are *hypothesis testing, sample data, one categorical variable,* and *two samples.*

Questions
Q Question type: estimation or hypothesis testing?
S Study type: sample data or experiment data?
T Type of data: one variable or two? Categorical or numerical?
N Number of samples or treatments: how many?

A Large-Sample Test for a Difference in Two Population Proportions

Appropriate when the following conditions are met:

1. The samples are independent random samples from the populations of interest (or the samples are independently selected in a way that results in each sample being representative of the corresponding population).
2. The sample sizes are large. This condition is met when $n_1 \hat{p}_1 \geq 10$, $n_1(1 - \hat{p}_1) \geq 10$, $n_2 \hat{p}_2 \geq 10$, and $n_2(1 - \hat{p}_2) \geq 10$ or (equivalently) if each sample includes at least 10 successes and 10 failures.

When these conditions are met, the following test statistic can be used to test the null hypothesis $H_0: p_1 - p_2 = 0$:

$$z = \frac{\hat{p}_1 - \hat{p}_2}{\sqrt{\dfrac{\hat{p}_c(1 - \hat{p}_c)}{n_1} + \dfrac{\hat{p}_c(1 - \hat{p}_c)}{n_2}}}$$

where p_c is the combined estimate of the common proportion

$$\hat{p}_c = \frac{n_1 \hat{p}_1 + n_2 \hat{p}_2}{n_1 + n_2} = \frac{\text{total number of successes in the two samples}}{\text{total of the two sample sizes}}$$

Associated *P*-value

When the alternative hypothesis is…	The *P*-value is…
$H_a: p_1 - p_2 > 0$	Area under the z curve to the right of the calculated value of the test statistic
$H_a: p_1 - p_2 < 0$	Area under the z curve to the left of the calculated value of the test statistic
$H_a: p_1 - p_2 \neq 0$	2(area to the right of z) if z is positive or 2(area to the left of z) if z is negative

---

Steps
H Hypotheses
M Method
C Check
C Calculate
C Communicate results

Now you're ready to look at an example, following the usual five-step process for hypothesis testing problems (HMC[3]).

Example 11.4   **Detecting Plagiarism Revisited**

Recall the detecting plagiarism example introduced in Example 11.3. The question posed in that example was whether the proportion of faculty members who require students to submit papers using plagiarism-detection software is different for full-time faculty and part-time faculty. From that example, we know

Population	Population Proportion	Sample Size	Sample Proportion
Full-time college faculty	p_1 = proportion of *all* full-time college faculty who require students to submit papers through plagiarism-detection software	n_1 = 1129	$\hat{p}_1$ = 0.40
Part-time college faculty	p_2 = proportion of *all* part-time college faculty who require students to submit papers through plagiarism-detection software	n_2 = 293	$\hat{p}_2$ = 0.38

As usual, begin by answering the four key questions (QSTN) in order to identify a potential method: (Q) This is a hypothesis testing problem. (S) The data are from sampling. (T) There is one categorical variable, which is the response to the question about requiring students to submit papers using plagiarism-detection software. (N) The number of samples is two. Because the answers to the four questions are *hypothesis testing, sample data, one categorical variable,* and *two samples*, you should consider a large-sample hypothesis test for a difference in population proportions.

Now you can use the five-step process for hypothesis testing problems (HMC³) to answer the question posed.

Process Step	
H Hypotheses	The claim is about the difference in proportions for the two faculty groups. You can use the following notation: p_1 = proportion of full-time college faculty who require students to submit papers through plagiarism-detection software p_2 = proportion of part-time college faculty who require students to submit papers through plagiarism-detection software The question of interest (are the population proportions different) translates into an alternative hypothesis of $p_1 - p_2 \neq 0$. The null hypothesis is that there is no difference in the population proportions. **Hypotheses:** *Null hypothesis*: H_0: $p_1 - p_2 = 0$ *Alternative hypothesis*: H_a: $p_1 - p_2 \neq 0$
M Method	Because the answers to the four key questions are hypothesis testing, sample data, one categorical variable, and two samples, consider a large-sample hypothesis test for a difference in population proportions. **Potential method:** Large-sample test for a difference in population proportions. The test statistic for this test is $$z = \frac{\hat{p}_1 - \hat{p}_2}{\sqrt{\dfrac{\hat{p}_c(1 - \hat{p}_c)}{n_1} + \dfrac{\hat{p}_c(1 - \hat{p}_c)}{n_2}}}$$ You also need to select a significance level for the test. Because no significance level was specified, you should choose a significance level based on a consideration of the consequences of a Type I error (rejecting a true null hypothesis) and a Type II error (failing to reject a false null hypothesis). In this example, a Type I error would be deciding that the proportions were not the same for the two faculty groups, when in fact the actual proportions were equal. A Type II error would be not thinking that there is a difference in the population proportions, when in fact the two proportions were not the same. In this situation, because neither type of error is much more serious than the other, you might choose a value of 0.05 for α. **Significance level:** $\alpha = 0.05$
C Check	There are two conditions that need to be met in order for the large-sample test for a difference in population proportions to be appropriate. The large samples condition is easily verified. The sample sizes are large enough because $n_1 \hat{p}_1 = 1129(0.40) = 452$ $n_1(1 - \hat{p}_1) = 1129(1 - 0.40) = 677$ $n_2 \hat{p}_2 = 293(0.38) = 111$ $n_2(1 - \hat{p}_2) = 293(1 - 0.38) = 182$ are all greater than or equal to 10. From the study description (see Example 11.3), you know that the samples were independently selected. The report states that the responses are from a survey sent to faculty members by e-mail. The survey had a large nonresponse rate, so it is possible that the samples are not representative of the populations of interest. You should keep this possibility in mind when you interpret the outcome of the hypothesis test.

(continued)

Process Step	
C Calculate	To calculate the value of the test statistic, you need to find the values of the sample proportions and the value of $\hat{p}_c$, the combined estimate of the common population proportion.

$$n_1 = 1129 \qquad \hat{p}_1 = 0.40$$

$$n_2 = 293 \qquad \hat{p}_2 = 0.38$$

$$\hat{p}_c = \frac{n_1 \hat{p}_1 + n_2 \hat{p}_2}{n_1 + n_2} = \frac{1129(0.40) + 293(0.38)}{1422} = 0.396$$

Test statistic:

$$z = \frac{\hat{p}_1 - \hat{p}_2}{\sqrt{\dfrac{\hat{p}_c(1 - \hat{p}_c)}{n_1} + \dfrac{\hat{p}_c(1 - \hat{p}_c)}{n_2}}}$$

$$= \frac{0.40 - 0.38}{\sqrt{\dfrac{(0.396)(0.604)}{1129} + \dfrac{(0.396)(0.604)}{293}}}$$

$$= \frac{0.020}{0.032} = 0.63 \text{ (rounded to two decimal places)}$$

This is a two-tailed test (the inequality in H_a is $\neq$), so the P-value is twice the area under the z curve and to the right of the calculated z value.

Associated P-value:
P-value = 2(area under z curve to the right of 0.63)

= $2 \cdot P(z > 0.63)$

= 2(0.2643)

= 0.5286

C Communicate Results

Because the P-value is greater than or equal to than the selected significance level, you fail to reject the null hypothesis.

Decision: $0.5286 \geq 0.05$, Fail to reject H_0.

The final conclusion for the test should be stated in context and answer the question posed.

Conclusion: Assuming that the samples are representative of the populations of interest, you would conclude that there is not convincing evidence that the proportion who require students to turn in papers using plagiarism-detection software differs for full-time and part-time college faculty members.

Based on this hypothesis test, you can conclude that even though the two sample proportions were different (0.40 and 0.38), this difference could have occurred just by chance and not as a result of any difference in the population proportions. So, based on the sample data, you are *not convinced* that there is a difference in the two population proportions.

It is also possible to use statistical software or a graphing calculator to carry out the calculate step in a hypothesis test. For example, Minitab output for the test of Example 11.4 is shown here.

Test and CI for Two Proportions

Sample	X	N	Sample p
1	452	1129	0.400354
2	111	293	0.378840

Difference = p (1) − p (2)
Estimate for difference: 0.0215147
95% CI for difference: (−0.0409519, 0.0839813)
Test for difference = 0 (vs ≠ 0): Z = 0.67 P-Value = 0.502

From the Minitab output, you see that $z = 0.67$ and the associated P-value is 0.502. These values are slightly different from those in Example 11.4 only because Minitab uses greater decimal accuracy in the calculations leading to the value of the test statistic.

Example 11.5 **Cell Phone Fundraising Part 2**

The Preview Example for this chapter described a study that looked at ways people donated to the 2010 Haiti earthquake relief effort. Two independently selected random samples—one of Gen Y cell phone users and one of Gen X cell phone users—resulted in the following information:

> Gen Y (those born between 1980 and 1988): 17% had made a donation via cell phone
> Gen X (those born between 1968 and 1979): 14% had made a donation via cell phone

The question posed in the Preview Example was

> Is there convincing evidence that the proportion who donated via cell phone is higher for the Gen Y population than for the Gen X population?

The report referenced in the Preview Example does not say how large the sample sizes were, but the description of the survey methodology indicates that the samples can be regarded as independent random samples. For purposes of this example, let's suppose that both sample sizes were 1200.

Now you can use the given information to answer the questions posed. Considering the four key questions (QSTN), this situation can be described as *hypothesis testing, sample data, one categorical variable (did or did not donate by cell phone), and two samples*. This combination suggests a large-sample hypothesis test for a difference in population proportions.

H Hypotheses   Using p_1 and p_2 to denote the two population proportions, define
p_1 = proportion of *all* Gen Y cell phone users who made a donation by cell phone
p_2 = proportion of *all* Gen X cell phone users who made a donation by cell phone
Because you want to determine whether there is evidence that the proportion for Gen Y (p_1) is greater than the proportion for Gen X (p_2), the alternative hypothesis will be $p_1 - p_2 > 0$.

Hypotheses:

$$H_0: p_1 - p_2 = 0$$
$$H_a: p_1 - p_2 > 0$$

M Method   Because the answers to the four key questions are hypothesis testing, sample data, one categorical variable, and two samples, a large-sample hypothesis test for a difference in population proportions will be considered. The test statistic for this test is

$$z = \frac{\hat{p}_1 - \hat{p}_2}{\sqrt{\dfrac{\hat{p}_c(1 - \hat{p}_c)}{n_1} + \dfrac{\hat{p}_c(1 - \hat{p}_c)}{n_2}}}$$

Next, choose a significance level for the test. For purposes of this example, $\alpha = 0.05$ will be used.

C Check   Checking to see if this method is appropriate, first verify that the sample sizes are large enough. Since

$$n_1\hat{p}_1 = 1200(0.17) = 204 \quad n_1(1 - \hat{p}_1) = 1200(1 - 0.17) = 996$$
$$n_2\hat{p}_2 = 1200(0.14) = 168 \quad n_2(1 - \hat{p}_2) = 1200(1 - 0.14) = 1032$$

are all greater than or equal to 10, the sample sizes are large enough. The problem description indicates that the samples are independent random samples. Because the sample sizes are large enough and an appropriate sampling method was used, it is reasonable to proceed with the hypothesis test.

C Calculate   Using Minitab to do the computations results in the following output:

Test and CI for Two Proportions

Sample	X	N	Sample p
1	204	1200	0.170000
2	168	1200	0.140000

Difference = p (1) − p (2)
Estimate for difference: 0.03
95% CI for difference: (0.00106701, 0.0589330)
Test for difference = 0 (vs > 0): Z = 2.03 P-Value = 0.021

From the Minitab output, the value of the test statistic is

$$z = 2.03$$

and the associated *P*-value is

$$P\text{-value} = 0.021$$

C Communicate Results   Because the *P*-value is smaller than the selected significance level ($0.021 < 0.05$), the null hypothesis is rejected. The sample data provide convincing evidence that the proportion donating to the Haiti relief effort by cell phone is greater for Gen Y cell phone users than for Gen X cell phone users.

Summing It Up—Section 11.2

The following learning objectives were addressed in this section:

Conceptual Understanding

C1: Understand how a research question about a difference in two population proportions is translated into hypotheses.

Sample data can be used to test hypotheses about a difference in two population proportions, p_1 and p_2. The null hypothesis is usually H_0: $p_1 - p_2 = 0$, and the research question will determine the form of the alternative hypothesis. If the question of interest is whether p_1 is greater than p_2, the alternative hypotheses will be H_a: $p_1 - p_2 > 0$. If the question of interest is whether p_1 is less than p_2, the alternative hypothesis will be H_a: $p_1 - p_2 < 0$. If the question of interest is whether p_1 and p_2 are different, the alternative hypothesis will be H_a: $p_1 - p_2 \neq 0$.

Mastering the Mechanics

M1: Know the conditions for appropriate use of the large-sample confidence interval for a difference in population proportions and the large-sample test of hypotheses about a difference in population proportions.

There are two conditions that should be verified before a large-sample hypothesis test for a difference in two population proportions is used. These are

1.   The samples are random samples from the populations of interest, or the samples are selected in a way that would result in representative samples.
2.   The sample sizes are large. This condition is met when $n_1\hat{p}_1$, $n_1(1 - \hat{p}_1)$, $n_2\hat{p}_2$ and $n_2(1 - \hat{p}_2)$ are all greater than or equal to 10.

M3: Carry out a large-sample hypothesis test for a difference in two population proportions.

The five-step process for hypothesis testing (HMC[3]) can be used to test hypotheses about a difference in two population proportions. This process of carrying out a hypothesis test for a difference in two population proportions is illustrated in Examples 11.4 and 11.5.

Putting It into Practice

P2: Carry out a large-sample hypothesis test for a difference in two population proportions and interpret the conclusion in context.

Hypothesis tests for a difference in two population proportions are illustrated in Examples 11.4 and 11.5. If the null hypothesis that the population proportions are equal is rejected, you can conclude that there is convincing evidence in support of the alternative hypothesis. If the null hypothesis is not rejected, chance differences due to sampling variability is a plausible explanation for the observed difference in sample proportions, and there is not convincing evidence that the alternative hypothesis is true.

SECTION 11.2 EXERCISES

Each Exercise Set assesses the following chapter learning objectives: C1, M1, M3, P2

SECTION 11.2 Exercise Set 1

11.11 In a survey of mobile phone owners, 53% of iPhone users and 42% of Android phone users indicated that they upgraded their phones at least every two years **("Americans Split on How Often They Upgrade Their Smartphones," July 8, 2015, www.gallup.com, retrieved December 15, 2016)**. The reported percentages were based on large samples that were thought to be representative of the population of iPhone users and the population of Android phone users. The sample sizes were 8234 for the iPhone sample and 6072 for the Android phone sample.

Suppose you want to decide if there is evidence that the proportion of iPhone owners who upgrade their phones at least every two years is greater than this proportion for Android phone users. (Hint: See Example 11.5.)
a. What hypotheses should be tested to answer the question of interest?
b. Are the two samples large enough for the large-sample test for a difference in population proportions to be appropriate? Explain.
c. Based on the following Minitab output, what is the value of the test statistic and what is the value of the associated P-value? If a significance level of 0.01 is selected for the test, will you reject or fail to reject the null hypothesis?

Test and CI for Two Proportions

Sample	X	N	Sample p
1	4364	8234	0.529998
2	2550	6072	0.419960

Difference = p (1) − p (2)
Estimate for difference: 0.110037
95% lower bound for difference: 0.0962389
Test for difference = 0 (vs > 0): Z = 13.02
P-Value = 0.000

d. Interpret the result of the hypothesis test in the context of this problem.

11.12 A hotel chain is interested in evaluating reservation processes. Guests can reserve a room by using either a telephone system or an online system that is accessed through the hotel's web site. Independent random samples of 80 guests who reserved a room by phone and 60 guests who reserved a room online were selected. Of those who reserved by phone, 57 reported that they were satisfied with the reservation process. Of those who reserved online, 50 reported that they were satisfied. Based on these data, is it reasonable to conclude that the proportion who are satisfied is greater for those who reserve a room online? Test the appropriate hypotheses using a significance level of 0.05. (Hint: See Example 11.4.)

11.13 The paper **"On the Nature of Creepiness" (*New Ideas in Psychology* [2016]: 10–15)** describes a study to investigate what people think is "creepy." Each person in a sample of women and a sample of men were asked to do the following:

Imagine a close friend of yours whose judgement you trust. Now imagine that this friend tells you that she or he just met someone for the first time and tells you that the person was creepy.

The people in the samples were then asked whether they thought the creepy person was more likely to be a male or a female. Of the 1029 women surveyed, 980 said they thought it was more likely the creepy person was male, and 298 of the 312 men surveyed said they thought it was more likely the creepy person was male.

Is there convincing evidence that the proportion of women who think the creepy person is more likely to be male is different from this proportion for men? For purposes of this exercise, you can assume that the samples are representative of the population of adult women and the population of adult men. Test the appropriate hypotheses using a significance level of 0.05.

SECTION 11.2 Exercise Set 2

11.14 The Interactive Advertising Bureau surveyed a representative sample of 1000 adult Americans and a representative sample of 1000 adults in China **("Majority of Digital Users in U.S. and China Regularly Shop and Purchase via E-Commerce," November 10, 2016, www.iab.com, retrieved December 15, 2016)**. They reported that American shoppers are much more likely to use a credit or a debit card to make an online purchase. This conclusion was based on finding that 63% of the people in the United States sample said they pay with a credit or a debit card, while only 34% of those in the China sample said that they used a credit card or a debit card to pay for online purchases. To determine if the stated conclusion is justified, you want to carry out a test of hypotheses to determine if there is convincing evidence that the proportion who pay with a credit card or a debit card is greater for adult Americans than it is for adult Chinese.
a. What hypotheses should be tested to answer the question of interest?

b. Are the two samples large enough for the large-sample test for a difference in population proportions to be appropriate? Explain.

c. Based on the following Minitab output, what is the value of the test statistic and what is the value of the associated P-value? If a significance level of 0.01 is selected for the test, will you reject or fail to reject the null hypothesis?

Test and CI for Two Proportions

Sample	X	N	Sample p
1	630	1000	0.630000
2	340	1000	0.340000

Difference = p (1) − p (2)
Estimate for difference: 0.29
95% lower bound for difference: 0.254818
Test for difference = 0 (vs > 0): Z = 12.98
P-Value = 0.000

d. Interpret the result of the hypothesis test in the context of this problem.

11.15 The article **"Most Women Oppose Having to Register for the Draft" (February 10, 2016, www.rasmussenreports.com, retrieved December 15, 2016)** describes a survey of likely voters in the United States. The article states that 36% of those in a representative sample of male likely voters and 21% of those in a representative sample of female likely voters said that they thought the United States should have a military draft. Suppose that these percentages were based on independent random samples of 500 men and 500 women.

Use a significance level of 0.01 to determine if there is convincing evidence that the proportion of male likely voters who think the United States should have a military draft is different from this proportion for female likely voters.

11.16 The article referenced in the previous exercise also reported that 53% of the Republicans surveyed indicated that they were opposed to making women register for the draft. Would you use the large-sample test for a difference in population proportions to test the hypothesis that a majority of Republicans are opposed to making women register for the draft? Explain why or why not.

ADDITIONAL EXERCISES

11.17 The report **"Young People Living on the Edge" (Greenberg Quinlan Rosner Research, 2008)** summarizes a survey of people in two independent random samples. One sample consisted of 600 young adults (age 19 to 35), and the other sample consisted of 300 parents of young adults age 19 to 35. The young adults were presented with a variety of situations (such as getting married or buying a house) and were asked if they thought that their parents were likely to provide financial support in that situation. The parents of young adults were presented with the same situations and asked if they would be likely to provide financial support in that situation. When asked about getting married, 41% of the young adults said they thought parents would provide financial support and 43% of the parents said they would provide support. Carry out a hypothesis test to determine if there is convincing evidence that the proportion of young adults who think their parents would provide financial support and the proportion of parents who say they would provide support are different.

11.18 The report referenced in the previous exercise also stated that the proportion who thought their parents would help with buying a house or renting an apartment for the sample of young adults was 0.37. For the sample of parents, the proportion who said they would help with buying a house or renting an apartment was 0.27. Based on these data, can you conclude that the proportion of parents who say they would help with buying a house or renting an apartment is significantly less than the proportion of young adults who think that their parents would help?

11.19 Gallup surveyed adult Americans about their consumer debt **("Americans' Big Debt Burden Growing, Not Evenly Distributed," February 4, 2016, www.gallup.com, retrieved December 15, 2016)**. They reported that 47% of millennials (those born between 1980 and 1996) and 61% of Gen Xers (those born between 1965 and 1971) did not pay off their credit cards each month and therefore carried a balance from month to month. Suppose that these percentages were based on representative samples of 450 millennials and 300 Gen Xers. Is there convincing evidence that the proportion of Gen Xers who do not pay off their credit cards each month is greater than this proportion for millennials? Test the appropriate hypotheses using a significance level of 0.05.

SECTION 11.3 # Inference for Two Proportions Using Data from an Experiment

In sampling situations, you often want to compare two populations to decide if there is evidence that the population proportions are different. When independent random samples are selected from each population, you know that even if the two population proportions are equal, the two sample proportions won't usually be equal. This is because of sample-to-sample variability that occurs due to the random selection process. To be convinced that there *really* is a difference between the population proportions, you need to see a difference in sample proportions that is larger than what you would expect to see just by chance due to the random selection process. This is where statistical methods help—they allow you to determine when the differences you see in the samples are unlikely to be just due to variability introduced by random selection.

Now think about what you might learn using data from an experiment. Consider an experiment to investigate whether knee replacement surgery is better than physical therapy alone as a way to eliminate pain for people with severe arthritis. In the experiment described in the article **"Study Points to Benefits of Knee Replacement Surgery Over Therapy Alone" (*The New York Times*, October 21, 2015)**, 100 adults who were considered candidates for knee replacement were followed for one year. These 100 patients were randomly assigned to one of two groups. Fifty were assigned to a group that had knee replacement surgery followed by therapy and the other half were assigned to a group that did not have surgery but did receive therapy. After one year, 86% of the patients in the surgery group and 68% of the patients in the therapy only group reported pain relief.

Of course, even if the knee replacement surgery is no better than therapy alone, you wouldn't expect the two group proportions to be exactly equal. Even if *all* 100 patients had received the same treatment, when you divide them into two groups and look at the group proportions, the two proportions wouldn't be exactly equal. In this situation, what you need to figure out is if the difference in the proportions reporting pain relief for these two groups of patients is indicating a *real* difference in the proportion experiencing pain relief for the two different treatments. You only say that there is a **significant difference** between the treatments if you can reasonably rule out the possibility that the observed difference might be due just to the way in which patients happened to be divided into the two experimental groups. To decide if a difference is significant, you need to understand differences that might result just from variability in the response and the random assignment to treatment groups, so that you can distinguish them from a difference created by a treatment effect.

In most real experimental situations, the individuals or objects receiving the treatments are not selected at random from some larger population. This means that you cannot generalize the results of the experiment to some larger population. However, if the experimental design includes random assignment of individuals to the treatments (or for random assignment of treatments to the individuals), it is possible to learn about treatment differences by testing hypotheses or calculating and interpreting confidence intervals.

Testing Hypotheses About the Difference Between Two Treatment Proportions

Sometimes the response variable in an experiment is categorical with only two possible values. For example, a ceramic tile might be classified as cracked or not cracked. A patient in a medical experiment might be classified as having improved or not improved after a particular length of time. In situations like these, you are interested in testing hypotheses about the difference in treatment "success" proportions (such as the proportion of patients who improve or the proportion of tiles that are cracked). The large-sample z test introduced in Section 11.2 can be adapted for use with experiment data by making the following modifications:

1. **Hypotheses:** The hypotheses will look the same as before, but now p_1 represents the proportion of successes for treatment 1 and p_2 represents the proportion of successes for treatment 2. The null hypothesis H_0: $p_1 - p_2 = 0$ is a statement that there is no difference in the treatment proportions (no treatment effect).

2. **Conditions:** When you previously considered the large-sample z test to test hypotheses about population proportions using sample data, there were two conditions that had to be satisfied. The first condition was that the samples had to be independent random samples from the populations of interest. The second condition was that the sample sizes had to be large. In the context of testing hypotheses about treatment proportions, these two conditions are replaced by the following two:

   1. Individuals or objects are *randomly assigned* to treatments.
   2. The number of individuals or objects in each of the treatment groups is large. This condition is met when $n_1\hat{p}_1 \geq 10$, $n_1(1 - \hat{p}_1) \geq 10$, $n_2\hat{p}_2 \geq 10$, and $n_2(1 - \hat{p}_2) \geq 10$,

(or equivalently when n_1 and n_2 are large enough to have at least 10 successes and 10 failures in each of the two treatment groups).

3. **Conclusions:** Conclusions will be worded in terms of treatment proportions. If the individuals or objects that were randomly assigned to the treatments were *also* randomly selected from some larger population, it is also reasonable to generalize conclusions about treatment effects to the larger population.

A Large-Sample *z* Test for a Difference in Two Treatment Proportions

Appropriate when the following conditions are met:

1. Individuals or objects are *randomly assigned* to treatments.
2. The number of individuals or objects in each of the treatment groups is large. This condition is met when $n_1\hat{p}_1 \geq 10$, $n_1(1 - \hat{p}_1) \geq 10$, $n_2\hat{p}_2 \geq 10$, and $n_2(1 - \hat{p}_2) \geq 10$ (or equivalently if n_1 and n_2 are large enough to have at least 10 successes and 10 failures in each of the two treatment groups).

When these conditions are met, the following test statistic can be used to test the null hypothesis $H_0: p_1 - p_2 = 0$:

$$z = \frac{\hat{p}_1 - \hat{p}_2}{\sqrt{\dfrac{\hat{p}_c(1 - \hat{p}_c)}{n_1} + \dfrac{\hat{p}_c(1 - \hat{p}_c)}{n_2}}}$$

where p_c is the combined estimate of the common proportion

$$\hat{p}_c = \frac{n_1\hat{p}_1 + n_2\hat{p}_2}{n_1 + n_2} = \frac{\text{total number of successes in the two treatment groups}}{\text{total of the two treatment group sizes}}$$

Associated *P*-value:

When the Alternative Hypothesis Is...	The *P*-Value Is...
$H_a: p_1 - p_2 > 0$	Area under the z curve to the right of the calculated value of the test statistic
$H_a: p_1 - p_2 < 0$	Area under the z curve to the left of the calculated value of the test statistic
$H_a: p_1 - p_2 \neq 0$	2(area to the right of z) if z is positive or 2(area to the left of z) if z is negative

---

Example 11.6   Duct Tape to Remove Warts?

Some people believe that you can fix anything with duct tape. Even so, many were skeptical when researchers announced that duct tape may be a more effective and less painful alternative to liquid nitrogen, which doctors routinely use to remove warts. The article **"What a Fix-It: Duct Tape Can Remove Warts" (San Luis Obispo Tribune, October 15, 2002)** described a study conducted at Madigan Army Medical Center. Patients with warts were randomly assigned to either the duct-tape treatment or the more traditional freezing treatment. Those in the duct-tape group wore duct tape over the wart for 6 days, then removed the tape, soaked the area in water, and used an emery board to scrape the area. This process was repeated for a maximum of 2 months or until the wart was gone. Data consistent with values in the article are summarized in the following table.

Treatment	*n*	Number with Wart Successfully Removed
Liquid-Nitrogen Freezing	100	60
Duct Tape	104	88

Do these data suggest that the duct tape treatment is more successful than freezing for removing warts?

H Hypotheses

You want to use data from the experiment to test the claim that duct tape is more successful than freezing for removing warts. The two treatments are duct tape and freezing, so you define the two treatment proportions as

p_1 = proportion of warts successfully removed by the duct tape treatment
p_2 = proportion of warts successfully removed by the freezing treatment

and $p_1 - p_2$ is the difference in treatment proportions. Translating the question of interest into hypotheses gives

$$H_0: p_1 - p_2 = 0$$
$$H_a: p_1 - p_2 > 0$$

The alternative hypothesis corresponds to the claim that the success proportion is greater for the duct-tape treatment.

M Method

Considering the four key questions (QSTN), this situation can be described as *hypothesis testing, experiment data, one categorical variable (with two categories—wart removed and wart not removed),* and *two treatments (duct tape and freezing).* This combination suggests a large-sample z test for a difference in treatment proportions. For purposes of this example, a significance level of 0.05 will be used.

C Check

Next, you need to verify that this method is appropriate. From the study description, you know that the participants were assigned at random to one of the two treatment groups. You also need to make sure that there are enough people in each of the two treatment groups. For these data, $\hat{p}_1 = \frac{88}{104} = 0.846$ and $\hat{p}_2 = \frac{60}{100} = 0.600$, so

$$n_1\hat{p}_1 = 104(0.846) = 87.984 \geq 10$$
$$n_1(1 - \hat{p}_1) = 104(1 - 0.846) = 16.016 \geq 10$$
$$n_2\hat{p}_2 = 100(0.600) = 60.000 \geq 10$$
$$n_2(1 - \hat{p}_2) = 100(1 - 0.600) = 40.000 \geq 10$$

Both treatment groups are large enough to proceed with the large-sample z test.

C Calculate

You need to calculate the values of the test statistic and the P-value. You can do this by hand, as shown here, or you could use a graphing calculator or a statistics software package.

Test statistic:

First you need to calculate the value of $\hat{p}_c$, the combined estimate of the common proportion.

$$\hat{p}_c = \frac{n_1\hat{p}_1 + n_2\hat{p}_2}{n_1 + n_2} = \frac{104(0.846) + 100(0.600)}{104 + 100} = 0.725$$

The value of the test statistic is then

$$z = \frac{\hat{p}_1 - \hat{p}_2}{\sqrt{\dfrac{\hat{p}_c(1 - \hat{p}_c)}{n_1} + \dfrac{\hat{p}_c(1 - \hat{p}_c)}{n_2}}} = \frac{0.846 - 0.600}{\sqrt{\dfrac{(0.725)(1 - 0.725)}{104} + \dfrac{(0.725)(1 - 0.725)}{100}}} = \frac{0.246}{0.063} = 3.91$$

P-value:

This is an upper-tailed test, so the P-value is the area under the z curve to the right of 3.91. From Appendix A Table 2, P-value ≈ 0.

C Communicate Results

Because the P-value is less that the selected significance level ($\alpha = 0.05$), the null hypothesis is rejected. The proportion of warts removed by the duct tape treatment is significantly greater than the proportion of warts removed by the freezing treatment.

Estimating the Difference in Treatment Proportions

As long as an experiment uses random assignment to create the treatment groups, the large-sample z confidence interval can be adapted for use in estimating a difference in treatment proportions. Two modifications are needed to adapt the interval introduced in Section 11.1 for use with experiment data:

1. **Conditions:** The changes to the required conditions are the same as those made for the large-sample z test.
2. **Interpretation:** The interpretation of the confidence interval estimate will now be in terms of treatment proportions rather than population proportions.

A Large-Sample Confidence Interval for a Difference in Treatment Proportions

Appropriate when the following conditions are met:

1. Individuals or objects are *randomly assigned* to treatments.
2. The number of individuals or objects in each of the treatment groups is large. This condition is met when $n_1 \hat{p}_1 \geq 10$, $n_1(1-\hat{p}_1) \geq 10$, $n_2 \hat{p}_2 \geq 10$, and $n_2(1-\hat{p}_2) \geq 10$ (or equivalently if n_1 and n_2 are large enough to have at least 10 successes and 10 failures in each of the two treatment groups).

When these conditions are met, a confidence interval for the difference in treatment proportions is

$$(\hat{p}_1 - \hat{p}_2) \pm (z \text{ critical value}) \sqrt{\frac{\hat{p}_1(1-\hat{p}_1)}{n_1} + \frac{\hat{p}_2(1-\hat{p}_2)}{n_2}}$$

The desired confidence level determines which z critical value is used. Three common confidence levels use the following critical values:

Confidence Level	z Critical Value
90%	1.645
95%	1.96
99%	2.58

Interpretation of Confidence Interval
You can be confident that the actual value of the difference in treatment proportions is included in the calculated interval. This statement should be worded in context.

Interpretation of Confidence Level:
The confidence level specifies the long-run proportion of the time that this method is expected to be successful in capturing the actual difference in treatment proportions.

Example 11.7 The Effect of Prayer

The article **"Prayer Is Little Help to Some Heart Patients, Study Shows" (*Chicago Tribune*, March 31, 2006)** described an experiment to investigate the possible effects of prayer. The following two paragraphs are from the article:

> Bypass patients who consented to take part in the experiment were divided randomly into three groups. Some patients received prayers but were not informed of that. In the second group the patients got no prayers, and also were not informed one way or the other. The third group got prayers and were told so.

> There was virtually no difference in complication rates between the patients in the first two groups. But the third group, in which patients knew they were receiving prayers, had a complication rate of 59 percent—significantly more than the rate of 52 percent in the no-prayer group.

The article also states that a total of 1800 people participated in the experiment, with 600 being assigned at random to each treatment group. The final comparison in the quote from the article was probably based on a large-sample z test for a difference in proportions, comparing the proportion with complications for the 600 patients in the no-prayer group with the proportion with complications for the 600 participants in the group that knew that someone was praying for them. You can use the given information to estimate the difference in treatment proportions for these two treatments.

E Estimate
You want to use the given information to estimate the difference between the proportion of patients with complications for the no-prayer treatment, p_1, and the proportion of patients with complications for the treatment where people knew someone was praying for them, p_2.

M Method
Because this is an estimation problem, the data are from an experiment, the one response variable (complications or no complications) is categorical, and two treatments are being compared, a method to consider is a large-sample z confidence interval for the difference in treatment proportions. For purposes of this example, a 90% confidence level will be used.

C Check
You need to check to make sure that the experiment used random assignment to create the treatment groups and that the treatment group sizes are large enough. The description of the experiment says that patients were divided randomly into treatment groups. If you define a success as someone who experienced complications, with treatment group sizes of 600 and treatment success proportions of 0.59 for the treatment group where people knew that they were receiving prayers and 0.52 for the no-prayer treatment group, there are more than 10 successes in each group and more than 10 failures in each group.

C Calculate
For a confidence level of 90%, the appropriate z critical value is 1.645.

$$\hat{p}_1 = 0.52 \quad \hat{p}_2 = 0.59$$

$$(\hat{p}_1 - \hat{p}_2) \pm (z \text{ critical value})\sqrt{\frac{\hat{p}_1(1 - \hat{p}_1)}{n_1} + \frac{\hat{p}_2(1 - \hat{p}_2)}{n_2}}$$

$$(0.52 - 0.59) \pm (1.645)\sqrt{\frac{(0.52)(1 - 0.52)}{600} + \frac{(0.59)(1 - 0.59)}{600}}$$

$$-0.070 \pm (1.645)(0.029)$$

$$-0.070 \pm 0.048$$

$$(-0.118, -0.022)$$

Statistical software or a graphing calculator could also have been used to compute the endpoints of the confidence interval. Minitab output is shown here.

CI for Two Proportions

Sample	X	N	Sample p
1	312	600	0.520000
2	354	600	0.590000

Difference = p (1) − p (2)
Estimate for difference: −0.07
90% CI for difference: (−0.117078, −0.0229225)

C Communicate Results
Based on the data from this experiment, you can be confident that the difference in the proportion of patients with complications for the no-prayer treatment and the treatment where patients knew that someone was praying for them is between −0.118 and −0.022. Because both endpoints of the interval are negative, you would conclude that the proportion with complications is greater for the treatment where patients know that someone is praying for them than for the no-prayer treatment by somewhere between 0.022 and 0.118. This is consistent with the statement in the quote from the article that says that the proportion with complications was significantly higher for the treatment where people knew someone was praying for them. The method used to construct this estimate captures the true difference in treatment proportions about 90% of the time.

Summing It Up—Section 11.3

The following learning objectives were addressed in this section:

Conceptual Understanding

C2: Understand the role that variability due to random assignment plays in drawing conclusions from experiment data.

Even when there is no difference in the actual treatment proportions, you don't expect the observed proportions for the treatments to be exactly equal. For example, if 100 people all received the same treatment and then you randomly divided them into two groups, the proportion of successes wouldn't necessarily be exactly the same for the two groups. This means that to be convinced that there is a real difference in treatment proportions, the observed difference needs to be greater than what you would expect to see just by random division into two groups when there is no real difference in treatment proportions.

Mastering the Mechanics

M4: Distinguish between situations that involve learning from sample data and those that involve learning from experiment data.

If you want to learn about characteristics of one or more populations, you collect data by selecting a sample from each population. In other situations, you want to learn about the effect of two or more treatments, and when this is a case you collect data by carrying out an experiment.

M5: Know the conditions required for appropriate use of the large-sample z test and the large-sample z confidence interval to draw conclusions about a difference in treatment proportions.

There are two conditions that must be met in order for a large-sample z test or a large-sample z confidence interval to be an appropriate way to learn about a difference in treatment proportions. They are

1. Random assignment of individuals or objects to the treatments in the experiment.
2. The treatment group sizes are large. This condition is met when $n_1 \hat{p}_1$, $n_1(1 - \hat{p}_1)$, $n_2 \hat{p}_2$ and $n_2(1 - \hat{p}_2)$ are all greater than or equal to 10.

M6: Use the large-sample z test to test hypotheses about a difference in treatment proportions.

The five-step process for hypothesis testing (HMC³) can be used to test hypotheses about the difference in two treatment proportions. The process of carrying out a hypothesis test for a difference in two treatment proportions is illustrated in Example 11.6.

M7: Use the large-sample z confidence interval to estimate a difference in treatment proportions.

The five-step process for estimation problems (EMC³) can be used to estimate the difference in two treatment proportions. The process of calculating and interpreting a confidence interval for a difference in two treatment proportions is illustrated in Example 11.7.

Putting It into Practice

P3: Carry out large-sample z test for a difference in treatment proportions and interpret the conclusion in context.

A hypothesis test for a difference in two treatment proportions is illustrated in Example 11.6. If the null hypothesis that the treatment proportions are equal is rejected, you can conclude that there is convincing evidence in support of the alternative hypothesis. If the null hypothesis is not rejected, chance differences resulting from random assignment are a plausible explanation for the observed difference in the treatment proportions, and there is not convincing evidence that the alternative hypothesis is true.

P4: Interpret a large-sample z confidence interval for a difference in treatment proportions in context and interpret the associated confidence level.

The interpretation of a confidence interval for a difference in treatment proportions is similar to the interpretation of a confidence interval for a difference in population proportions. If both endpoints of a confidence interval for $p_1 - p_2$ are positive, it means that you think that p_1 is greater than p_2 and the interval gives an estimate of how much greater. If both endpoints are negative, it means that you think that p_1 is less than p_2 and the interval gives an estimate of how much less. If zero is included in the interval, it means that zero is a plausible value for the difference in the two treatment proportions.

Confidence level is interpreted in the same way as it was for the other confidence intervals that you have seen. It specifies the long-run percentage of the time that the method will be successful in capturing the actual difference in treatment proportions.

SECTION 11.3 EXERCISES

Each Exercise set assesses the following chapter learning objectives: C2, M4, M5, M6, M7, P3, P4

SECTION 11.3 Exercise Set 1

11.20 A headline that appeared in *Woman's World* stated "Black Currant Oil Curbs Hair Loss!" (*Woman's World,* April 4, 2016). This claim was based on an experiment described in the paper **"Effect of a Nutritional Supplement on Hair Loss in Women"** (*Journal of Cosmetic Dermatology* [2015]: 76–82). In this experiment, women with stage 1 hair loss were assigned at random to one of two groups. One group was a control group who did not receive a nutritional supplement. Of the 39 women in this group, 20 showed increased hair density at the end of the study period. Those in the second group received a nutritional supplement that included fish oil, black currant oil, vitamin E, vitamin C, and lycopene. Of the 80 women in the supplement group, 70 showed increased hair density at the end of the study period.
a. Is there convincing evidence that the proportion with increased hair density is greater for the supplement treatment than for the control treatment? Test the appropriate hypotheses using a 0.01 significance level.
b. Write a few sentences commenting on the headline that appeared in *Woman's World*.
c. Based on the description of the actual experiment and the result from your hypothesis test in Part (a), suggest a more appropriate headline.

11.21 In the experiment described in the article **"Study Points to Benefits of Knee Replacement Surgery Over Therapy Alone"** (*The New York Times,* October 21, 2015), adults who were considered candidates for knee replacement were followed for one year. Suppose that 200 patients were randomly assigned to one of two groups. One hundred were assigned to a group that had knee replacement surgery followed by therapy and the other half were assigned to a group that did not have surgery but did receive therapy. After one year, 86% of the patients in the surgery group and 68% of the patients in the therapy only group reported pain relief. Is there convincing evidence that the proportion experiencing pain relief is greater for the surgery treatment than for the therapy treatment? Use a significance level of 0.05.

11.22 Use the data given in the previous exercise to construct and interpret a 95% confidence interval estimate of the difference in the proportion experiencing pain relief for the surgery treatment and this proportion for the therapy treatment.

11.23 The paper **"Passenger and Cell Phone Conversations in Simulated Driving"** (*Journal of Experimental Psychology: Applied* [2008]: 392–400) describes an experiment that investigated if talking on a cell phone while driving is more distracting than talking with a passenger. Drivers were randomly assigned to one of two groups. The 40 drivers in the cell phone group talked on a cell phone while driving in a simulator. The 40 drivers in the passenger group talked with a passenger in the car while driving in the simulator. The drivers were instructed to exit the highway when they came to a rest stop. Of the drivers talking to a passenger, 21 noticed the rest stop and exited. For the drivers talking on a cell phone, 11 noticed the rest stop and exited.
a. Use the given information to construct and interpret a 95% confidence interval for the difference in the proportions of drivers who would exit at the rest stop.
b. Does the interval from Part (a) support the conclusion that drivers using a cell phone are more likely to miss the exit than drivers talking with a passenger? Explain how you used the confidence interval to answer this question.

SECTION 11.3 Exercise Set 2

11.24 Choice blindness is the term that psychologists use to describe a situation in which a person expresses a preference and then doesn't notice when they receive something different than what they asked for. The authors of the paper **"Can Chocolate Cure Blindness? Investigating the Effect of Preference Strength and Incentives on the Incidence of Choice Blindness"** (*Journal of Behavioral and Experimental Economics* [2016]: 1–11) wondered if choice blindness would occur more often if people made their initial selection by looking at pictures of different kinds of chocolate compared with if they made their initial selection by looking as the actual different chocolate candies.

Suppose that 200 people were randomly assigned to one of two groups. The 100 people in the first group are shown a picture of eight different kinds of chocolate candy and asked which one they would like to have. After they selected, the picture is removed and they are given a chocolate candy, but not the one they actually selected. The 100 people in the second group are shown a tray with the eight different kinds of candy and asked which one they would like to receive. Then the tray is removed and they are given a chocolate candy, but not the one they selected.

If 20 of the people in the picture group and 12 of the people in the actual candy group failed to detect the switch, would you conclude that there is convincing evidence that the proportion who experience choice blindness is different for the two treatments (choice based on a picture and choice based on seeing the actual candy)? Test the relevant hypotheses using a 0.01 significance level.

11.25 The article **"Footwear, Traction, and the Risk of Athletic Injury"** (January 2016, www.lermagazine.com/article/footwear-traction-and-the-risk-of-athletic-injury, retrieved December 15, 2016) describes a study in which high school football players were given either a conventional football cleat or a swivel disc shoe. Of 2373 players who wore the conventional cleat, 372 experienced an injury during the study period. Of the 466 players who wore the swivel disc shoe, 24 experienced an injury. The question of interest is whether there is evidence that the injury proportion is smaller for the swivel disc shoe than it is for conventional cleats.
a. What are the two treatments in this experiment?
b. The article didn't state how the players in the study were assigned to the two groups. Explain why it is important to know if they were assigned to the groups at random.
c. For purposes of this example, assume that the players were randomly assigned to the two treatment groups. Carry out a hypothesis test to determine if there is evidence that the injury proportion is smaller for the swivel disc shoe than it is for conventional cleats. Use a significance level of 0.05.

11.26 Use the data given in the previous exercise to construct and interpret a 95% confidence interval estimate of the difference in the injury proportion for the traditional cleat treatment and the swivel disc shoe treatment.

11.27 Some fundraisers believe that people are more likely to make a donation if there is a relatively quick deadline given for making the donation. The paper **"Now or Never! The Effect of Deadlines on Charitable Giving: Evidence from Two Natural Field Experiments"** (*Journal of Behavioral and Experimental Economics* [2016]: 1–10) describes an experiment to investigate the influence of deadlines. In this experiment, 1.2% of those who received an e-mail request for a donation that had a three-day deadline to make a donation and 0.8% of those who received the same e-mail request but without a deadline made a donation. The people who received the e-mail request were randomly assigned to one of the two groups (e-mail with deadline and e-mail without deadline). Suppose that the given percentages are based on sample sizes of 2000 (the actual sample sizes in the experiment were much larger). Use a 90% confidence interval to estimate the difference in the proportion who donate for the two different treatments.

ADDITIONAL EXERCISES

11.28 In a study of a proposed approach for diabetes prevention, 339 people under the age of 20 who were thought to

be at high risk of developing type I diabetes were assigned at random to one of two groups. One group received twice-daily injections of a low dose of insulin. The other group (the control) did not receive any insulin but was closely monitored. Summary data (from the article **"Diabetes Theory Fails Test,"** *USA TODAY*, June 25, 2001) follow.

Group	n	Number Developing Diabetes
Insulin	169	25
Control	170	24

Do these data support the theory that the proportion who develop diabetes with the insulin treatment is significantly less than this proportion for the no-insulin treatment?

11.29 The article **"Fish Oil Staves Off Schizophrenia"** (*USA TODAY*, February 2, 2010) describes a study in which 81 patients age 13 to 25 who were considered at risk for mental illness were randomly assigned to one of two groups. Those in one group took four fish oil capsules daily. Those in the other group took a placebo. After 1 year, 5% of those in the fish oil group and 28% of those in the placebo group had become psychotic. Is it appropriate to use the large-sample z test to test hypotheses about the difference in the proportions of patients receiving the fish oil and the placebo treatments who became psychotic? Explain why or why not.

11.30 Women diagnosed with breast cancer whose tumors have not spread may be faced with a decision between two surgical treatments—mastectomy (removal of the breast) or lumpectomy (only the tumor is removed). In a long-term study of the effectiveness of these two treatments, 701 women with breast cancer were randomly assigned to one of two treatment groups. One group received mastectomies and the other group received lumpectomies and radiation. Both groups were followed for 20 years after surgery. It was reported that there was no statistically significant difference in the proportion surviving for 20 years for the two treatments (*Associated Press*, October 17, 2002). What hypotheses do you think the researchers tested in order to reach the given conclusion? Did the researchers reject or fail to reject the null hypothesis?

11.31 The article **"Spray Flu Vaccine May Work Better than Injections for Tots"** (*San Luis Obispo Tribune*, May 2, 2006) described a study that compared flu vaccine administered by injection and flu vaccine administered as a nasal spray. Each of the 8000 children under the age of 5 who participated in the study received both a nasal spray and an injection, but only one was the real vaccine and the other was salt water. At the end of the flu season, it was determined that of the 4000 children receiving the real vaccine by nasal spray, 3.9% got the flu. Of the 4000 children receiving the real vaccine by injection, 8.6% got the flu.

a. Why would the researchers give every child both a nasal spray and an injection?

b. Use a 99% confidence interval to estimate the difference in the proportion of children who get the flu after being vaccinated with an injection and the proportion of children who get the flu after being vaccinated with the nasal spray. Based on the confidence interval, would you conclude that the proportion of children who get the flu is different for the two vaccination methods? (Hint: See Example 11.7.)

11.32 Women diagnosed with breast cancer whose tumors have not spread may be faced with a decision between two surgical treatments—mastectomy (removal of the breast) or lumpectomy (only the tumor is removed). In a long-term study of the effectiveness of these two treatments, 701 women with breast cancer were randomly assigned to one of two treatment groups. One group received mastectomies, and the other group received lumpectomies and radiation. Both groups were followed for 20 years after surgery. It was reported that there was no statistically significant difference in the proportion surviving for 20 years for the two treatments (*Associated Press*, **October 17, 2002**). Suppose that this conclusion was based on a 90% confidence interval for the difference in treatment proportions. Which of the following three statements is correct? Explain why you chose this statement.

Statement 1: Both endpoints of the confidence interval were negative.
Statement 2: The confidence interval included 0.
Statement 3: Both endpoints of the confidence interval were positive.

SECTION 11.4 Simulation-Based Inference for Two Proportions (Optional)

The large-sample methods for estimating the difference between two population or treatment proportions and for testing hypotheses about the difference between two population or treatment proportions require sample sizes that are large enough to ensure that the sampling distribution of the differences in the sample proportions, $\hat{p}_1 - \hat{p}_2$, is approximately normal. When one or both sample sizes are not large enough, simulation-based methods can be used to estimate the difference in two proportions or to test hypotheses about the difference in two proportions.

Bootstrap Confidence Intervals for the Difference Between Two Population Proportions or Two Treatment Proportions

Example 11.8 Anti-Clotting Medications After Hip or Knee Surgery

In the study described in the article **"A Pilot Study Comparing Hospital Readmission Rates in Patients Receiving Rivaroxaban or Enoxaparin After Orthopedic Surgery,"** (2016, www .ptcommunity.com/system/files/pdf/ptj4106376.pdf, retrieved May 6, 2017) researchers identified patients who had received hip or knee replacement surgery and who had been given one of two types of medication to prevent blood clotting afterward. This was an observational study, conducted by reviewing the medical charts for patients who had received hip or knee replacement surgery. The outcome of interest was whether the patient was readmitted to the hospital within 30 days after leaving the hospital after surgery. The researchers were interested in estimating the difference in the proportions of patients readmitted for each of the two drugs (rivaroxaban and enoxaparin).

The study found that 8 of the 213 patients who had been given rivaroxaban to prevent blood clots were readmitted to the hospital within 30 days after their surgeries, and that 1 out of the 27 patients who had taken enoxaparin to prevent clotting was readmitted within 30 days. Based on the sample data, what can you learn about the difference in the proportion of patients who receive rivaroxaban who are readmitted and the corresponding proportion for patients who receive enoxaparin?

For this example, the answers to the four key questions (QSTN) are:

Q Question Type	Estimation or hypothesis testing	Estimation
S Study Type	Sample data or experiment data	Sample data
T Type of Data	One variable or two? Categorical or numerical?	One categorical variable with two categories (rivaroxaban or enoxaparin)
N Number of Sample or Treatments	How many sample or treatments?	Two samples (patients who received rivaroxaban and patients who received enoxaparin).

Now you can proceed with the five-step process to learn from the data.

Estimate: Explain what population characteristic you plan to estimate.

You want to estimate the difference in the proportion of patients readmitted to the hospital for those who receive rivaroxaban and the proportion readmitted for those who receive enoxaparin. If p_1 is the proportion of readmissions for patients who take rivaroxaban and p_2 is the proportion of readmissions for patients who take enoxaparin, then you will estimate the difference, $p_1 - p_2$.

Method: Select a potential method.

Only 8 patients who received rivaroxaban were readmitted, and only 1 of the patients who received enoxaparin was readmitted. These sample sizes are too small to use the large-sample z confidence interval for a difference in population proportions. Instead, you can estimate the difference in population proportions, $\hat{p}_1 - \hat{p}_2$, using a bootstrap simulation-based method. For this example, a 95% confidence level will be used.

Check: Check to make sure that the method you selected is appropriate.

The sample of patient records was not randomly selected, but the researchers believed it to be representative of the population of patients undergoing hip and knee replacements who receive rivaroxaban and the population of patients who received enoxaparin.

Calculate: Use sample data to perform any necessary calculations.

$$n_1 = 213 \qquad\qquad n_2 = 27$$
$$\hat{p}_1 = 8/213 = 0.038 \qquad \hat{p}_2 = 1/27 = 0.037$$

In order to generate bootstrap simulated differences, $\hat{p}_1 - \hat{p}_2$, first you consider a hypothetical population in which the proportion of success is 0.038, and you take a sample of size 213 from this hypothetical population. This simulated sample produces a simulated value of $\hat{p}_1$. Then you consider a hypothetical population in which the proportion of success is 0.037, and you take a random sample of size 27. This simulated sample produces a simulated value of $\hat{p}_2$. This pair of values, $\hat{p}_1$ and $\hat{p}_2$, produces a bootstrap simulated value of $\hat{p}_1 - \hat{p}_2$.

For example, a simulated value of $\hat{p}_1 = 11/213 = 0.052$ was generated from a hypothetical population with population proportion of successes $p_1 = 0.038$. A simulated value of $\hat{p}_2 = 1/27 = 0.037$ was generated from a hypothetical population with population proportion of successes $p_2 = 0.037$. The simulated difference in sample proportions is $\hat{p}_1 - \hat{p}_2 = 0.052 - 0.037 = 0.015$.

A graph produced by the Shiny app "Bootstrap Confidence Interval for Difference in Two Proportions" of 1000 simulated differences in sample proportions is shown in Figure 11.1. (This Shiny app can be found in the app collection at statistics.cengage.com/Peck2e/Apps.html)

The smallest 2.5% of the simulated differences in sample proportions were −0.097 or less, and the largest 2.5% of the simulated differences were 0.056 or greater. The 95% confidence interval based on the bootstrap resampling is (−0.097, 0.056).

FIGURE 11.1
1000 simulated differences in proportions

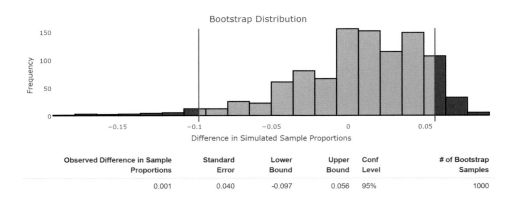

Observed Difference in Sample Proportions	Standard Error	Lower Bound	Upper Bound	Conf Level	# of Bootstrap Samples
0.001	0.040	-0.097	0.056	95%	1000

Communicate Results: Answer the research questions of interest, explain what you have learned from the data, and acknowledge potential risks.

Assuming that the samples are representative of the two populations of interest (patients that take rivaroxaban and patients who take enoxaparin), you can be 95% confident that the actual difference in the proportion of patients readmitted to the hospital after taking rivaroxaban and the proportion readmitted after taking enoxaparin is between −0.097 and 0.056. Zero is contained in this interval, indicating that it is plausible that there is no difference in the population readmission proportions for the two medications.

Randomization Tests for the Difference Between Two Population Proportions or Two Treatment Proportions

In some cases, you want to determine if two population proportions or two treatment proportions differ, or if one is smaller or larger than the other. In these situations, you would test hypotheses about the difference in the proportions.

Example 11.9 | Oxytocin Nasal Spray and Social Interaction

Oxytocin is a synthetic hormone that may improve social interaction for young children with autism. An experiment was conducted to evaluate if the use of oxytocin delivered through a nasal spray improves social interaction experiences **("The Effect of Oxytocin Nasal Spray on Social Interaction Deficits Observed in Young Children with Autism,"** *Molecular Psychology* **[2016]: 1225–1231)**.

In many studies, participants drop out for one reason or another. Sometimes participants move to a new location, sometimes they change their minds about participating, and they always have the right to just discontinue participating in the study. One concern in the oxytocin nasal spray and autism study was that young children with different diagnoses might drop out at different rates. In fact, 4 out of 23 young children with autism spectrum disorder dropped out of the study, and 4 out of 16 young children with pervasive developmental disorder dropped out. Do these data provide evidence that the proportion who drop out is different for the two different types of autism?

Here is a summary of the information in this example:

Diagnosis	Population Proportion	Sample Size	Sample Proportion
Autism spectrum disorder	p_1 = proportion of all young children with autism spectrum disorder who would drop out of the oxytocin study	$n_1 = 23$	$\hat{p}_1 = 0.174$
Pervasive developmental disorder	p_2 = proportion of all young children with pervasive developmental disorder who would drop out of the oxytocin study	$n_2 = 16$	$\hat{p}_2 = 0.250$

The answer to the question of interest, you can test the null hypothesis that the two population proportions are equal.

H Hypotheses The population proportions to be compared are p_1 = proportion of all young children with
autism spectrum disorder who would drop out of the oxytocin study, and p_2 = proportion of all young children with pervasive developmental disorder who would drop out of
the oxytocin study. The null hypothesis is $H_0\colon p_1 - p_2 = 0$. Initially, the researchers did
not know which diagnosis group would have a larger dropout rate, and so the alternative
hypothesis is two-sided, $H_a\colon p_1 - p_2 \neq 0$.

M Method A large-sample z test for a difference in population proportions is not appropriate because the
sample sizes are small. The number of young children who dropped out of the study was 4 in
each of the two diagnosis groups. Because the sample size conditions for the large-sample z
test are not met, rather than use a large-sample test, you can use a simulation-based randomization test. For this example, a significance level of $\alpha = 0.05$ will be used.

C Check The researchers assumed that the young children in the study were representative of the
population of all young children who might experience improved social interactions after
taking oxytocin nasal spray.

Because the sample sizes are not large, a randomization test will be used.

C Calculate The observed difference in the sample proportions is $\hat{p}_1 - \hat{p}_2 = 0.174 - 0.250 = -0.076$.

A simulation-based approach to a test for two proportions involves pooling the two original samples together into one hypothetical collection of $23 + 16 = 39$ young children in
which $4 + 4 = 8$ are identified as dropping out of the study. The proportion of children
who would drop out based on this hypothetical collection is $\hat{p} = 8/39 = 0.205$. This is
known as the pooled sample proportion.

A sample of $n_1 = 23$ is drawn from the combined, or pooled, collection, with replacement, and a simulated value of $\hat{p}_1$ is computed. A sample of $n_2 = 16$ is also drawn from the
pooled collection, again with replacement, and a simulated value of $\hat{p}_2$ is computed. These
are the values needed to calculate a simulated difference $\hat{p}_1 - \hat{p}_2$. This process is repeated
many times in order to construct the randomization distribution.

For example, suppose a sample of $n_1 = 23$ young children drawn from a hypothetical
population with pooled proportion $\hat{p} = 0.205$ results in $\hat{p}_1 = 6/23 = 0.261$, and a separate
sample of $n_2 = 16$ children drawn from the same population with pooled proportion $\hat{p} =
0.205$ results in $\hat{p}_2 = 4/16 = 0.250$. The simulated difference in the sample proportions
is $\hat{p}_1 - \hat{p}_2 = 0.261 - 0.250 = 0.011$.

A randomization distribution for the difference in two proportions was produced using
the Shiny app "Randomization Test for Two Proportions" and is shown in Figure 11.2. (This
Shiny app can be found in the app collection at statistics.cengage.com/Peck2e/Apps.html)

FIGURE 11.2
Randomization distribution of
difference in sample proportions

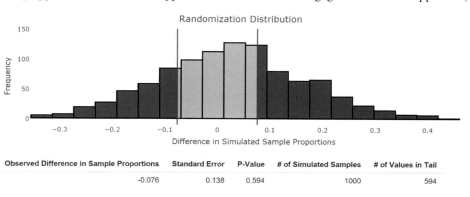

Observed Difference in Sample Proportions	Standard Error	P-Value	# of Simulated Samples	# of Values in Tail
-0.076	0.138	0.594	1000	594

Because the alternative hypothesis in this example is two-sided, the P-value is the sum
of the proportion of simulated differences in the two regions representing simulated differences that are at least as extreme as the observed difference in sample proportions. In
this example, the observed difference in sample proportions was -0.076, so the P-value
is the sum of the proportion of simulated differences that are less than -0.076 and the

proportion of simulated differences that are greater than 0.076. For the randomization distribution shown in Figure 11.2, this is 0.594. Because this P-value is greater than the specified significance level $\alpha = 0.05$, you fail to reject the null hypothesis.

C Communicate Results   Based on the sample data, there is not convincing evidence that there is a difference in population dropout proportions for young children with a diagnosis of autism spectrum disorder and those with a diagnosis of pervasive developmental disorder.

Summing It Up—Section 11.4

The following learning objectives were addressed in this section:

Mastering the Mechanics

M8 Calculate and interpret a bootstrap confidence interval for a difference in proportions.

A bootstrap confidence interval is an alternate method for calculating a confidence interval for a difference in population or treatment proportions. This method can be used even in situations where the sample size conditions necessary for the large-sample z confidence interval are not met. A bootstrap confidence interval is interpreted in the same way as the large-sample z confidence interval. Example 11.8 illustrates the calculation and interpretation of a bootstrap confidence interval for a difference in proportions.

M9 Carry out a randomization test for a difference in proportions.

A randomization test is a method that can be used to test hypotheses about a difference in population or treatment proportions even if the sample sizes are not large enough for the large-sample z test to be appropriate. Example 11.9 illustrates the use of a randomization test to carry out a hypothesis test for a difference in proportions.

SECTION 11.4 EXERCISES

Each Exercise Set assesses the following learning objectives: M8, M9

SECTION 11.4 Exercise Set 1

11.33 The article **"Rapid Evolutionary Response to a Transmissible Cancer in Tasmanian Devils" (www.nature.com /articles/ncomms12684, retrieved December 20, 2016)** describes the spread of devil facial tumor disease (DFTD), which is a fatal form of cancer that swept through the Tasmanian devil population beginning near the beginning of the 21st century. Researchers studied the genetic reaction of the Tasmanian devils by comparing the rates of occurrence of specific genetic markers of interest before and after DFTD swept across the island.

One region of Tasmania is called West Pencil Pine. Analysis of 21 tissue specimens taken from a representative sample of

Tasmanian devils living in West Pencil Pine in 2006, before DFTD swept through, revealed that 5% had a specific genetic marker. Also analyzed were 42 tissue specimens from a representative sample of devils living in the same region in 2013 and 2014, after DFTD. In this sample, 43% had the same genetic marker. A significant change in these rates would indicate a remarkably fast evolution in the genetic code of the Tasmanian devils to protect against DFTD.

a. Explain why the data from this study should not be analyzed, using a large-sample hypothesis test for the difference in two population proportions.

b. Use the output below from the Shiny app "Randomization Test for Two Proportions" to carry out a hypothesis test to

Output for Exercise 11.33

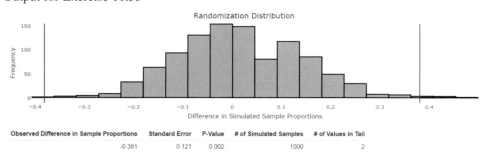

Observed Difference in Sample Proportions	Standard Error	P-Value	# of Simulated Samples	# of Values in Tail
-0.381	0.121	0.002	1000	2

Output for Exercise 11.33

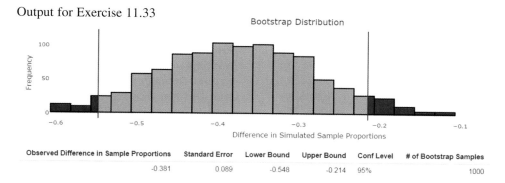

Observed Difference in Sample Proportions	Standard Error	Lower Bound	Upper Bound	Conf Level	# of Bootstrap Samples
-0.381	0.089	-0.548	-0.214	95%	1000

determine if there is convincing evidence that the proportion of Tasmanian devils with the genetic marker was greater after DFTD than before DFTD.

c. Use the output above from the Shiny app "Bootstrap Confidence Interval for Difference in Two Proportions" to identify a 95% confidence interval for the difference in the rates of occurrence of the specific genetic marker in the genes of Tasmanian devils, before and after DFTD. Interpret the confidence interval in context.

11.34 Example 11.3 describes the results of a survey of 1129 full-time college faculty and 293 part-time college faculty. Survey participants were asked if they require undergraduate students to submit papers through plagiarism-detection software; 40% of the full-time faculty and 38% of the part-time faculty said "yes." Note that the sample sizes for the two groups—full-time college faculty and

part-time college faculty—are large enough to satisfy the conditions for a large-sample test and a large-sample confidence interval for a difference in two population proportions. Even though the sample sizes are large, you can still use simulation-based methods.

a. Use the following output from the Shiny app "Randomization Test for Two Proportions" to carry out a randomization test to determine if the proportion who require students to submit papers through plagiarism-detection software is different for full-time faculty and part-time faculty.

b. Use the output from the Shiny app "Bootstrap Confidence Interval for Difference in Two Proportions" to identify a 95% confidence interval for the difference in the population proportions of full-time college faculty and part-time college faculty who require students to submit papers through plagiarism-detection software. Interpret the confidence interval in context.

Output for Exercise 11.34

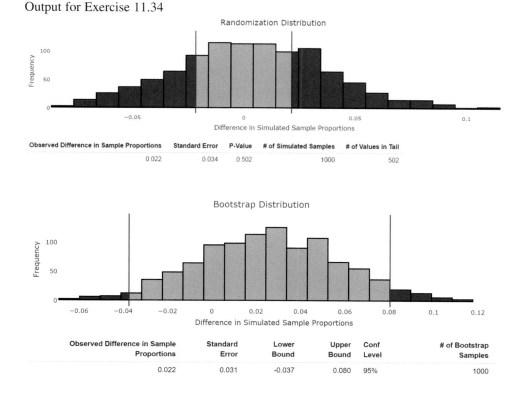

Observed Difference in Sample Proportions	Standard Error	P-Value	# of Simulated Samples	# of Values in Tail
0.022	0.034	0.502	1000	502

Observed Difference in Sample Proportions	Standard Error	Lower Bound	Upper Bound	Conf Level	# of Bootstrap Samples
0.022	0.031	-0.037	0.080	95%	1000

c. Compare your conclusion in Part (a) to the one in Example 11.3, "Detecting Plagiarism." Did you reach the same conclusion using the randomization test as when the large-sample test was used? Does this surprise you? Explain why or why not.

11.35 Researchers were interested in comparing regular-intensity exercise and high-intensity exercise for patients recovering from hospitalization due to chronic obstructive pulmonary disease (COPD). The researchers followed patients in Denmark who were enrolled in each of the two types of exercise programs **("Increased Mortality in Patients with Severe COPD Associated with High-Intensity Exercise: A Preliminary Cohort Study,"** *Journal of Chronic Obstructive Pulmonary Disease* **[2016]: 2329–2334).** Each exercise program lasted for eight weeks. The patients were followed for a total of 1.5 years. The researchers observed that 5 out of the 15 patients in the high-intensity group died within a year and a half, but that none of the 16 patients in the regular-intensity group died within a year and a half.

a. Explain why the data from this study should not be analyzed using a large-sample hypothesis test for a difference in two population proportions.

b. Carry out a hypothesis test to determine if there is convincing evidence of a difference in the population proportions who die within 1.5 years for the two exercise programs. Use the Shiny app "Randomization Test for Two Proportions" to report an approximate *P*-value and then use it to reach a decision in the hypothesis test. Remember to interpret the results of the test in context.

c. Use the Shiny app "Bootstrap Confidence Interval for Difference in Two Proportions" to obtain a 95% bootstrap confidence interval for the difference in the population proportions in patients who die within 1.5 years for the two exercise programs. Interpret the interval in the context of the research.

SECTION 11.4 Exercise Set 2

11.36 An article titled **"TCU Horned Frogs Game Preview (Part 1)" (www.uwdawgpound.com/2016/11/26/13710900 /washington-huskies-tcu-horned-frogs-game-preview-part-1, retrieved December 20, 2016)** previews a college basketball game between the University of Washington Huskies and the TCU Horned Frogs. The profile for TCU player Kenrich Williams states, "He has the statistical oddity of shooting better from behind the arc than from the free throw line but that's a marker of small sample sizes." (Note that "behind the arc" means three-point shots, which are taken behind an arc relatively far from the basket.)

At that point in the season, TCU had played five games. Kenrich Williams had made 4 of the 10 free-throws that he had attempted (40%), and 3 of the 7 three-point shots that he had attempted (42.9%). Suppose that it is reasonable to consider these shots to be representative sample of Williams's abilities to make free-throws and three-point shots for the 2016–2017 college basketball season.

a. Explain why the data in this exercise should not be analyzed using a large-sample hypothesis test for the difference in two population proportions.

b. Use the following the output from the Shiny app "Randomization Test for Two Proportions" to carry out a hypothesis test that would allow you to determine if there is evidence that supports the statement that Williams is a better shooter from behind the arc than from the free-throw line.

c. Use the output to identify a 95% confidence interval for the difference in the proportions of made free-throws and made three-point shots by Kenrich Williams for the 2016–2017 season. Interpret the confidence interval in context.

Output for Exercise 11.36

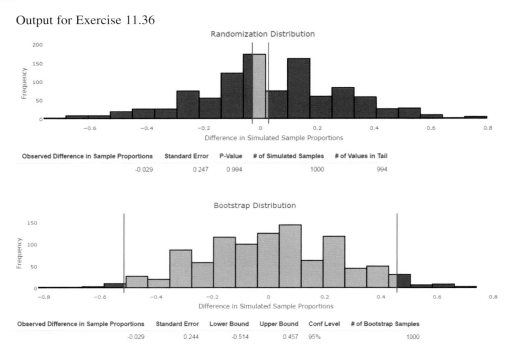

Observed Difference in Sample Proportions	Standard Error	P-Value	# of Simulated Samples	# of Values in Tail
-0.029	0.247	0.994	1000	994

Observed Difference in Sample Proportions	Standard Error	Lower Bound	Upper Bound	Conf Level	# of Bootstrap Samples
-0.029	0.244	-0.514	0.457	95%	1000

Output for Exercise 11.37

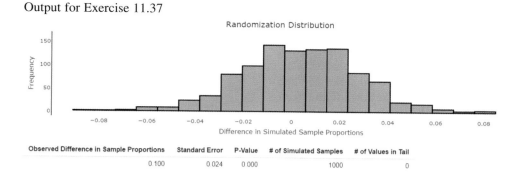

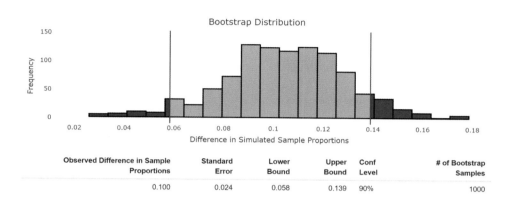

11.37 Example 11.1 describes a study in which 354 of 708 people in the sample of 18- to 29-year-olds and 412 of the 1029 people in the sample of 30- to 49-year-olds said that they thought it was OK to use a cell phone in a restaurant. Note that the sample sizes for the two groups—people ages 18 to 29 and those ages 30 to 49—are large enough to satisfy the conditions for a large-sample test and a large-sample confidence interval for the difference in two population proportions. Even though the sample sizes are large enough, you can still use simulation-based methods.

a. Use the output at the top of the page from the Shiny app "Randomization Test for Two Proportions," to carry out a randomization test to determine if there is convincing evidence that the proportion who think it is OK to use a cell phone at a restaurant is higher for the 18 to 29 age group than for the 30 to 49 age group.

b. Use the output from the Shiny app "Bootstrap Confidence Interval for Difference in Two Proportions" to identify a 90% bootstrap confidence interval for the difference in the population proportions of people ages 18 to 29 and those ages 30 to 49 who said that they think that it is acceptable to use a cell phone in a restaurant.

c. Compare your results in Part (b) to the confidence interval reported in Example 11.1 "Cell Phone Etiquette." Would your interpretation change using the bootstrap confidence interval compared with the large-sample confidence interval? Explain.

11.38 A report in **USA TODAY** described an experiment to explore the accuracy of wearable devices designed to measure heart rate (**"Wearable health monitors not always reliable, study shows," USA TODAY, October 12, 2016**).

The researchers found that when 50 volunteers wore an Apple Watch to track heart rate as they walked, jogged, and ran quickly on a treadmill for three minutes, the results were accurate compared with an EKG 92% of the time. When 50 volunteers wore a Fitbit Charge, the heart rate results were accurate 84% of the time.

a. Explain why the data from this study should not be analyzed using a large-sample hypothesis test for a difference in two population proportions.

b. Carry out a hypothesis test to determine if there is convincing evidence that the proportion of accurate results for people wearing an Apple Watch is greater than this proportion for those wearing a Fitbit Charge. Use the Shiny app "Randomization Test for Two Proportions" to report an approximate *P*-value and use it to reach a decision in the hypothesis test. Remember to interpret the results of the test in context.

c. Use the Shiny app "Bootstrap Confidence Interval for Difference in Two Proportions" to obtain a 95% bootstrap confidence interval for the difference in the population proportions of accurate results for people wearing an Apple Watch and those wearing a Fitbit Charge. Interpret the interval in the context of the research.

ADDITIONAL EXERCISES

11.39 As part of a study described in the report **"I Can't Get My Work Done!" (harmon.ie/blog/i-cant-get-my-work -done-how-collaboration-social-tools-drain-productivity, 2011, retrieved May 6, 2017)**, people in a sample of 258 cell phone users ages 20 to 39 were asked if they use their cell phones to stay connected while they are in bed, and 168 said "yes." The same question was also asked of each person in a sample of 129 cell phone users ages 40 to 49, and 61 said "yes." You might expect the proportion who stay connected while in bed to be higher for the 20 to 39 age group than for the 40 to 49 age group, but how much higher?

a. Construct and interpret a 90% large-sample confidence interval for the difference in the population proportions of cell phone users ages 20 to 39 and those ages 40 to 49 who say that they sleep with their cell phones. Interpret the confidence interval in context.

Note that the sample sizes in the two groups—cell phone users ages 20 to 39 and those ages 40 to 49—are large enough to satisfy the conditions for a large-sample test and a large-sample confidence interval for two population proportions. Even though the sample sizes are large enough, you can still use simulation-based methods.

b. Use the output below to identify a 90% bootstrap confidence interval for the difference in the population proportions of cell phone users ages 20 to 39 and those ages 40 to 49 who say that they sleep with their cell phones.

c. Compare the confidence intervals you computed in Parts (a) and (b). Would your interpretation change using the bootstrap confidence interval compared with the large-sample confidence interval? Explain.

11.40 The article **"Americans Say No to Electric Cars Despite Gas Prices"** (*USA TODAY*, May 25, 2011) describes a survey of public opinion on issues related to rising gas prices. The survey was conducted by Gallup, a national polling organization. Each person in a representative sample of low-income adult Americans (annual income less than $30,000) and each person in an independently selected representative sample of high-income adult Americans (annual income greater than $75,000) was asked whether he or she would consider buying an electric car if gas prices continue to rise. In the low-income sample, 65% said that they would not buy an electric car no matter how high gas prices were to rise. In the high-income sample, 59% responded this way. The article did not give the sample sizes, but for the purposes of this exercise, suppose the sample sizes were both 300. One question of interest is whether the proportion who would never consider buying an electric car is different for the two income groups.

Note that the sample sizes in the two groups—low-income adult Americans and high-income adult Americans—are large enough to satisfy the conditions for a large-sample test and a large-sample confidence interval for two population proportions. Even though the sample sizes are large enough, you can still use simulation-based methods.

a. Based on these data, is it reasonable to conclude that the proportions of low-income and high-income adults who would never consider buying an electric car differ? Use large-sample methods to test the appropriate hypotheses using a significance level of 0.05.

b. Use the output at the bottom of the page to carry out a randomization test of the same hypotheses tested in Part (a).

c. Compare your conclusions in Parts (a) and (b). Would you reach the same conclusion in either case? Explain.

Output for Exercise 11.39

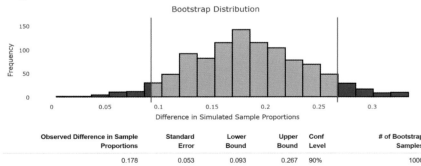

Observed Difference in Sample Proportions	Standard Error	Lower Bound	Upper Bound	Conf Level	# of Bootstrap Samples
0.178	0.053	0.093	0.267	90%	1000

Output for Exercise 11.40

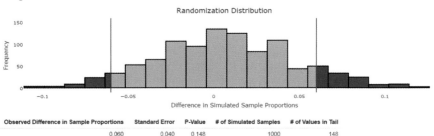

Observed Difference in Sample Proportions	Standard Error	P-Value	# of Simulated Samples	# of Values in Tail
0.060	0.040	0.148	1000	148

Avoid These Common Mistakes

The three cautions that appeared at the end of Chapter 10 apply here as well (see Chapter 10 for more detail). They were:

1. Remember that the result of a hypothesis test can never show strong support for the null hypothesis. In two-sample situations, this means that you shouldn't be *convinced* that there is *no* difference between two population proportions based on the outcome of a hypothesis test.

2. If you have complete information (a census) of both populations, there is no need to carry out a hypothesis test or to construct a confidence interval—in fact, it would be inappropriate to do so.

3. Don't confuse statistical significance and practical significance. In the two-sample setting, it is possible to be convinced that two population proportions are not equal even in situations where the actual difference between them is small enough that it is of no practical interest. After rejecting a null hypothesis of no difference (statistical significance), it is useful to look at a confidence interval estimate of the difference to get a sense of practical significance.

And here's one new caution to keep in mind when comparing two populations:

4. Correctly interpreting confidence intervals in the two-sample case is more difficult than in the one-sample case, so take particular care when providing a two-sample confidence interval interpretation. Because the two-sample confidence interval estimates a difference $(p_1 - p_2)$, the most important thing to note is whether or not the interval includes 0. If both endpoints of the interval are positive, then it is reasonable to say that, based on the interval, you believe that p_1 is greater than p_2, with the interval providing an estimate of how much greater. Similarly, if both interval endpoints are negative, you have evidence that p_1 is less than p_2, with the interval providing an estimate of the difference. If 0 is included in the interval, it is plausible that p_1 and p_2 are equal.

Drawing conclusions from experiment data requires some thought. In Chapter 1, you saw that if an experiment is carefully planned and includes random assignment to treatments, it is reasonable to conclude that observed differences in response between the experimental groups can be attributed to the treatments in the experiment. However, generalizing conclusions from an experiment that uses volunteers as subjects to a larger population is not appropriate unless a convincing argument can be made that the group of volunteers is representative of some population of interest.

Some cautions when drawing conclusions from experiment data are:

1. *Random assignment to treatments is critical.* If the design of the experiment does not include random assignment to treatments, it is not appropriate to use a hypothesis test or a confidence interval to draw conclusions about treatment differences.

2. Remember that it is not reasonable to generalize conclusions from experiment data to a larger population unless the subjects in the experiment were selected at random from the population or a convincing argument can be made that the group of volunteers is representative of the population. And even if subjects are selected at random from a population, it is still important that there be random assignment to treatments.

3. As was the case when using data from sampling to test hypotheses, remember that a hypothesis test can never show strong support for the null hypothesis. In the context of using experiment data to test hypotheses, this means you cannot say that data from an experiment provide convincing evidence that there is *no* difference between treatments.

4. Even when the data used in a hypothesis test are from an experiment, there is still a difference between statistical significance and practical significance. It is possible, especially in experiments with large numbers of subjects in each experimental group, to be convinced that two treatment proportions are not equal, even in situations where the actual difference is too small to be of any practical interest. After rejecting a null hypothesis of no difference (that is, finding statistical significance), it may be useful to look at a confidence interval estimate of the difference to get a sense of practical significance.

CHAPTER ACTIVITIES

ACTIVITY 11.1 DEFECTIVE M&MS

Materials needed: One bag of M&M's for each student (or pair of students) in the class. Half should receive bags of plain M&M's and the other half should receive peanut M&M's.

In this activity, you will use data collected by the class to determine if the proportion of defective M&M's is greater for peanut M&M's than for plain M&M's.

1. Inspect the M&M's in your bag for defects. Some possible defects are broken candies, misshapen candies, candies that do not have a visible "m" on the candy coating, and so on. Sort the M&M's into piles, with one pile for M&M's that are not defective and another pile for those that are defective. Record the type of M&M (plain or peanut) you inspected, the total number of M&M's inspected, and the total number of defective M&M's found.

2. Now combine the inspection information for the entire class to complete the following table:

Plain M&M's	Peanut M&M's
Total number inspected:	Total number inspected:
Total number defective:	Total number defective:
Proportion defective:	Proportion defective:

3. Use the class data to determine if there is convincing evidence that the population proportion of M&M's that are defective is greater for peanut M&M's than for plain M&M's.

CHAPTER 11 EXPLORATIONS IN STATISTICAL THINKING

In the exercise below, you will go online to select a random sample from a population of adults between the age of 18 and 45 and a random sample from a population of adults age 46 to 64.

Suppose that you would like to know if the proportion of people who have sent a text message while driving is greater for the younger population than for the older population.

Go online to statistics.cengage.com/Peck2e/Explore.html and click on the link for Chapter 11. This link will take you to a web page where you can select a random sample of 50 people from each population.

Click on the Select Samples button. This selects a random sample from each population and will display the following information:

1. The ID number that identifies the person selected
2. The response to the question "Have you ever sent a text message while driving?" These responses were coded numerically—a 1 indicates a yes response and a 2 indicates a no response.

Use these samples to answer the following questions.

a. Is this an estimation or a hypothesis testing problem?
b. Are the data from sampling or from an experiment?
c. How many variables are there?
d. What type of data do you have?
e. How many samples are there?
f. What are the appropriate null and alternative hypotheses?
g. What method might be appropriate for testing these hypotheses?
h. Are the conditions for the selected test met? Explain.
i. What is the value of the test statistic for this test?
j. What is the *P*-value?
k. If a significance level of 0.05 were used, what would your decision be?
l. Write a few sentences that summarize what you learned about the difference in the population proportions based on this hypothesis test.

If asked to do so by your instructor, bring your answers to Parts i through l to class. Your instructor will lead the class through a discussion of the questions that follow.

Compare the value of your test statistic and *P*-value with those obtained by another student in your class.

m. Did you both get the same test statistic value and the same *P*-value? Does this surprise you?

n. Did you both reach the same conclusion in your hypothesis test?

o. How many people in your class rejected the null hypothesis? How many did not reject the null hypothesis. Is this surprising?

ARE YOU READY TO MOVE ON? CHAPTER 11 REVIEW EXERCISES

All chapter learning objectives are assessed in these exercises. The learning objectives assessed in each exercise are given in parentheses.

11.41 (M1, M2, P1)

The Insurance Institute for Highway Safety issued a news release titled **"Teen Drivers Often Ignoring Bans on Using Cell Phones" (June 9, 2008)**. The following quote is from the news release:

> Just 1–2 months prior to the ban's Dec. 1, 2006, start, 11% of teen drivers were observed using cell phones as they left school in the afternoon. About 5 months after the ban took effect, 12% of teen drivers were observed using cell phones.

Suppose that the two samples of teen drivers (before the ban, after the ban) are representative of these populations of teen drivers. Suppose also that 200 teen drivers were observed before the ban (so $n_1 = 200$ and $\hat{p}_1 = 0.11$) and that 150 teen drivers were observed after the ban.

a. Construct and interpret a 95% large-sample confidence interval for the difference in the proportion using a cell phone while driving before the ban and the proportion after the ban.

b. Is 0 included in the confidence interval of Part (a)? What does this imply about the difference in the population proportions?

11.42 (C1, M1, M3, P2)

The news release referenced in the previous exercise also included data from independent samples of teenage drivers and parents of teenage drivers. In response to a question asking if they approved of laws banning the use of cell phones and texting while driving, 74% of the teens surveyed and 95% of the parents surveyed said they approved. The sample sizes were not given in the news release, but suppose that 600 teens and 400 parents of teens were surveyed and that these samples are representative of the two populations. Do the data provide convincing evidence that the proportion of teens who approve of banning cell phone and texting while driving is less than the proportion of parents of teens who approve? Test the relevant hypotheses using a significance level of 0.05.

11.43 (C1, M1, M3, P3)

The report titled **"Digital Democracy Survey" (2016, www.deloitte.com/us/tmttrends, retrieved December 16, 2016)** stated that 31% of the people in a representative sample of adult Americans age 33 to 49 rated a landline telephone among the three most important services that they purchase for their home. In a representative sample of adult Americans age 50 to 68, 48% rated a landline telephone as one of the top three services they purchase for their home. Suppose that the samples were independently selected and that the sample size was 600 for the 33 to 49 age group sample and 650 for the 50 to 68 age group sample. Does this data provide convincing evidence that the proportion of adult Americans age 33 to 49 who rate a landline phone in the top three is less than this proportion for adult Americans age 50 to 68? Test the relevant hypotheses using $\alpha = 0.05$.

11.44 (M3, P2)

The report **"Audience Insights: Communicating to Teens (Aged 12–17)" (2009, www.cdc.gov)** described teens' attitudes about traditional media, such as TV, movies, and newspapers. In a representative sample of American teenage girls, 41% said newspapers were boring. In a representative sample of American teenage boys, 44% said newspapers were boring. Sample sizes were not given in the report.

a. Suppose that the percentages reported were based on samples of 58 girls and 41 boys. Is there convincing evidence that the proportion of those who think that newspapers are boring is different for teenage girls and boys? Carry out a hypothesis test using $\alpha = .05$.

b. Suppose that the percentages reported were based on samples of 2000 girls and 2500 boys. Is there convincing evidence that the proportion of those who think that newspapers are boring is different for teenage girls and boys? Carry out a hypothesis test using $\alpha = 0.05$.

c. Explain why the hypothesis tests in Parts (a) and (b) resulted in different conclusions.

11.45   (C2)
In a test of hypotheses about a difference in treatment proportions, what does it mean when the null hypothesis is not rejected?

11.46   (M4, M5, M6, P3)
Many fundraisers ask for donations using e-mail and text messages. The paper **"Now or Never! The Effect of Deadlines on Charitable Giving: Evidence from Two Natural Field Experiments"** (*Journal of Behavioral and Experimental Economics* [2016]: 1–10) describes an experiment to investigate whether the proportion of people who make a donation when asked for a donation by e-mail is different from the proportion of people who make a donation when asked for a donation in a text message. In this experiment, 1.32% of those who received and opened an e-mail request for a donation and 7.77% of those who received a text message asking for a donation actually made a donation. Assume that the people who received these requests were randomly assigned to one of the two groups (e-mail or text message) and suppose that the given percentages are based on sample sizes of 2000 (the actual sample sizes in the experiment were much larger).
a. The study described is an experiment with two treatments. What are the two treatments?
b. Is there convincing evidence that the proportion who make a donation is not the same for the two different methods? Carry out a hypothesis test using a significance level of 0.05.

c. Use a 90% confidence interval to estimate the difference in the proportions who donate for the two different treatments.

11.47   (M4, M5, M7, P4)
Researchers carried out an experiment to evaluate the effectiveness of using acupuncture to treat heel pain. The experiment is described in the paper **"Effectiveness of Trigger Point Dry Needling for Plantar Heel Pain: A Randomized Controlled Trial"** (*Physical Therapy* [2014]: 1083–1094) and a follow-up response to a letter to the editor of the journal (*Physical Therapy* [2014]: 1354–1355). In this experiment, 84 patients experiencing heel pain were randomly assigned to one of two groups. One group received acupuncture and the other group received a sham treatment that consisted of using a blunt needle that did not penetrate the skin. Of the 43 patients in the sham treatment group, 17 reported pain reduction of more than 13 points on a foot pain scale (this was considered a meaningful reduction in pain). Of the 41 patients in the acupuncture group, 28 reported pain reduction of more than 13 points on the foot pain scale.
a. Use a 95% confidence interval to estimate the difference in the proportion who experience a meaningful pain reduction for the acupuncture treatment and for the sham treatment.
b. What does the interval in Part (b) suggest about the effectiveness of acupuncture in reducing heel pain?

TECHNOLOGY NOTES

Confidence Interval for p_1-p_2

JMP
Summarized data
1. Input the data into the JMP data table with categories for one variable in the first column, categories for the second variable in the second column, and counts for each combination in the third column

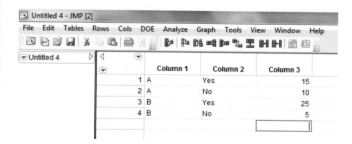

2. Click **Analyze** and select **Fit Y by X**
3. Click and drag the first column containing the response variable from the box under **Select Columns** to the box next to **X, Factor**

4. Click and drag the second column containing the group information from the box under **Select Columns** to the box next to **Y, Response**
5. Click and drag the third column containing the counts for each combination from the box under **Select Columns** to the box next to **Freq**
6. Click **OK**
7. Click the red arrow next to **Contingency Analysis of...** and select **Two Sample Test for Proportions**

Note: You can change the response of interest (i.e., "Yes" instead of "No" or "Success" instead of "Failure") by clicking the radio button at the bottom of the **Two Sample Test for Proportions** section.

Raw data
1. Input the raw data into two separate columns: one containing the response variable and one containing the group information
2. Click **Analyze** and select **Fit Y by X**

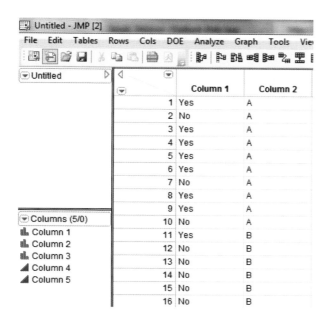

3. Click and drag the first column containing the response variable from the box under **Select Columns** to the box next to **Y, Response**

4. Click and drag the second column containing the group information from the box under **Select Columns** to the box next to **X, Factor**

5. Click **OK**

6. Click the red arrow next to **Contingency Analysis of...** and select **Two Sample Test for Proportions**

Note: You can change the response of interest (i.e., "Yes" instead of "No" or "Success" instead of "Failure") by clicking the radio button at the bottom of the **Two Sample Test for Proportions** section.

Minitab

Summarized data

1. Click **Stat** then click **Basic Statistics** then click **2 Proportions...**

2. Click the radio button next to **Summarized data**

3. In the boxes next to **First:** type the value for *n*, the total sample size in the box under the **Trials:** column and the number of successes in the box under **Events:**

4. In the boxes next to **Second:** type the value for *n*, the total sample size in the box under the **Trials:** column and the number of successes in the box under **Events:**

5. Click **Options...**

6. Input the appropriate confidence level in the box next to **Confidence Level**

7. Click **OK**

8. Click **OK**

Raw data

1. Input the raw data two separate columns

2. Click **Stat** then click **Basic Statistics** then click **2 Proportion...**

3. Select the radio button next to **Samples in different columns**

4. Click in the box next to **First:**

5. Double click the column name where the first group's raw data is stored

6. Click in the box next to **Second:**

7. Double click the column name where the second group's raw data is stored

8. Click **Options...**

9. Input the appropriate confidence level in the box next to **Confidence Level**

10. Click **OK**

11. Click **OK**

SPSS

SPSS does not have the functionality to automatically produce a confidence interval for the difference of two proportions.

Excel

Excel does not have the functionality to automatically produce a confidence interval for the difference of two proportions. However, you can type the formulas into two separate cells for the lower and upper limit to have Excel calculate these results for you.

TI-83/84

1. Press the **STAT** key

2. Highlight **TESTS**

3. Highlight **2-PropZInt...** and press **ENTER**

4. Next to **x1** type the number of successes from the first sample

5. Next to **n1** type the sample size from the first sample

6. Next to **x2** type the number of successes from the second sample

7. Next to **n2** type the sample size from the second sample

8. Next to **C-Level** type the appropriate confidence level

9. Highlight **Calculate** and press **ENTER**

TI-Nspire

1. Enter the Calculate Scratchpad

2. Press the **menu** key then select **6:Statistics** then select **6:Confidence Intervals** then **6:2-Prop z Interval...** then press **enter**

3. In the box next to **Successes, x1** type the number of successes from the first sample

4. In the box next to **n1** type the number of trials from the first sample

5. In the box next to **Successes, x2** type the number of successes from the second sample

6. In the box next to **n2** type the number of trials from the second sample

7. In the box next to **C Level** input the appropriate confidence level

8. Press **OK**

Z-Test for $p_1 - p_2$

JMP

JMP does not have the functionality to automatically provide the results of a z-test for the difference of two proportions.

Minitab

Summarized data

1. Click **Stat** then click **Basic Statistics** then click **2 Proportions...**

2. Click the radio button next to **Summarized data**

3. In the boxes next to **First:** type the value for n, the total sample size in the box under the **Trials:** column and the number of successes in the box under **Events:**
4. In the boxes next to **Second:** type the value for n, the total sample size in the box under the **Trials:** column and the number of successes in the box under **Events:**
5. Click **Options...**
6. Input the appropriate hypothesized value in the box next to **Test difference:** (this is usually 0)
7. Check the box next to **Use pooled estimate of p for test**
8. Click **OK**
9. Click **OK**

Raw data
1. Input the raw data two separate columns
2. Click **S**tat then click **B**asic **S**tatistics then click **2 P**roportion...
3. Select the radio button next to **Samples in different columns**
4. Click in the box next to **First:**
5. Double-click the column name where the first group's raw data is stored
6. Click in the box next to **Second:**
7. Double-click the column name where the second group's raw data are stored
8. Click **Options...**
9. Input the appropriate hypothesized value in the box next to **Test difference:** (this is usually 0)
10. Check the box next to **Use test and interval based on normal distribution**
11. Click **OK**
12. Click **OK**

SPSS
SPSS does not have the functionality to automatically produce a z-test for the difference of two proportions.

Excel
Excel does not have the functionality to automatically produce a z-test for the difference of two proportions. However, you can type the formulas into a cell for the test statistic in order to have Excel calculate this for you. Then use the methods from Chapter 6 to find the P-value using the Normal distribution.

TI-83/84
1. Press the **STAT** key
2. Highlight **TESTS**
3. Highlight **2-PropZTest...** and press **ENTER**
4. Next to **x1** type the number of successes from the first sample
5. Next to **n1** type the sample size from the first sample
6. Next to **x2** type the number of successes from the second sample
7. Next to **n2** type the sample size from the second sample
8. Next to **p1,** highlight the appropriate alternative hypothesis
9. Highlight **Calculate** and press **ENTER**

TI-Nspire
1. Enter the Calculate Scratchpad
2. Press the **menu** key then select **6:Statistics** then select **7:Stat Tests** then **6:2-Prop z Test...** then press **enter**
3. In the box next to **Successes, x1** type the number of successes from the first sample
4. In the box next to **n1** type the number of trials from the first sample
5. In the box next to **Successes, x2** type the number of successes from the second sample
6. In the box next to **n2** type the sample size from the second sample
7. In the box next to **Alternate Hyp** choose the appropriate alternative hypothesis from the drop-down menu
8. Press **OK**

12

Asking and Answering Questions About a Population Mean

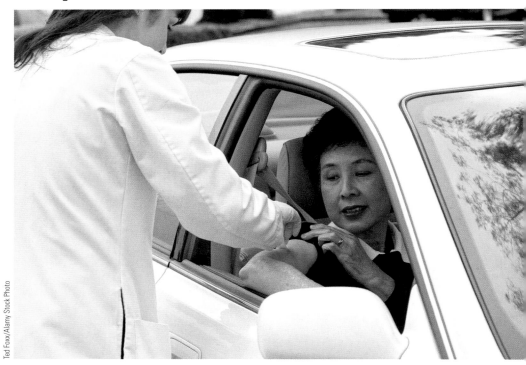

Ted Foxx/Alamy Stock Photo

PREVIEW

One of the key questions used to determine an appropriate data analysis method is whether the data are categorical or numerical. In the previous chapters, the focus has been on how categorical data can be used to learn about the value of a population proportion or a difference in proportions. In this chapter you will use numerical data from a sample to learn about the value of a population mean, such as the mean number of hours that students enrolled at a college spend studying each week or the mean weight gain of college students during their freshman year.

CHAPTER LEARNING OBJECTIVES

Conceptual Understanding

After completing this chapter, you should be able to

C1 Understand how the value of the standard deviation of the sample mean, $\bar{x}$, is related to sample size.

C2 Know what factors affect the width of a confidence interval estimate of a population mean.

Mastering the Mechanics

After completing this chapter, you should be able to

M1 Determine the mean and standard deviation of the sampling distribution of $\bar{x}$.

M2 Know when the sampling distribution of $\bar{x}$ is approximately normal.

M3 Know the conditions for appropriate use of the margin of error and confidence interval formulas when estimating a population mean.

M4 Calculate the margin of error when the sample mean, $\bar{x}$, is used to estimate a population mean μ.

M5 Use the five-step process for estimation problems (EMC^3) to calculate and interpret a confidence interval for a population mean.

M6 Calculate the sample size necessary to achieve a desired margin of error when estimating a population mean.

M7 Translate a research question or claim about a population mean into null and alternative hypotheses.

M8 Use the five-step process for hypothesis testing problems (HMC^3) to carry out a *t* test of hypotheses about a population mean.

M9 (Optional) Calculate and interpret a bootstrap confidence interval for a population mean.

M10 (Optional) Carry out a randomization test of hypotheses about a population mean.

Putting It into Practice

After completing this chapter, you should be able to

P1 Interpret a confidence interval for a population mean in context and interpret the associated confidence level.

P2 Carry out a *t* test of hypotheses about a population mean and interpret the conclusion in context.

PREVIEW EXAMPLE

Drive-Through Medicine

During a flu outbreak, many people visit emergency rooms. Before being treated, they often spend time in crowded waiting rooms where other patients may be exposed. The paper **"Drive-Through Medicine: A Novel Proposal for Rapid Evaluation of Patients During an Influenza Pandemic" (*Annals of Emergency Medicine* [2010]: 268–273)** describes a study of a drive-through model where flu patients are evaluated while they remain in their cars. The study found that not only were patients kept relatively isolated and away from each other, but also that the time to process a patient was shorter because delays related to turning over examination rooms were eliminated.

In the study, 38 people were each given a scenario for a flu case that was selected at random from the set of all flu cases actually seen in the emergency room. The scenarios provided the "patient" with a medical history and a description of symptoms that would allow the patient to respond to questions from the examining physician. These patients were processed using a drive-through procedure that was implemented in the parking structure of Stanford University Hospital. The time to process each case from admission to discharge was recorded.

Because the 38 volunteers were each representing a flu case that was selected at random from actual cases seen in the emergency room, the times were viewed as a random sample of processing times for all such cases. The researchers were interested in using the sample data to learn about the mean time to process a flu case using this new model.

This example will be revisited in Section 12.2 to see how the sample data can be used to construct a confidence interval estimate for the population mean. ∎

SECTION 12.1 The Sampling Distribution of the Sample Mean

When the purpose of a statistical study is to learn about a population mean μ, it is natural to consider the sample mean $\bar{x}$ as an estimate of μ. To understand statistical inference procedures based on $\bar{x}$, you must first study how sampling variability causes $\bar{x}$ to vary in value from one sample to another. Just as the behavior of the sample proportion $\hat{p}$ is described by its sampling distribution, the behavior of $\bar{x}$ is also described by a sampling distribution. The sample size n and characteristics of the population distribution (its shape, mean value μ, and standard deviation σ) are important in determining the sampling distribution of $\bar{x}$.

Some notation introduced in previous chapters is reviewed here.

NOTATION

n	sample size
$\bar{x}$	mean of a sample
s	standard deviation of a sample
μ	mean of the entire population
σ	standard deviation of the entire population

Suppose you are interested in learning about the time it takes students at a particular college to register for classes using a new online system. The population of interest would be all students enrolled at the college, and you would be interested in the numerical variable

$$x = \text{time to register}$$

There is variability in the values of x in the population—registration time will vary from student to student. The mean of all of the registration times in the population is denoted by μ. The population standard deviation σ is a measure of the variability in the population. A large value of σ indicates that there is a lot of student-to-student variability in the registration times. A small value of σ indicates that there is not much variability and that registration times tend to be similar.

The only way to determine the value of μ exactly is to carry out a census of the entire population. Since this isn't usually feasible, you might decide to select a sample of student registration times. These sample data can then be used to learn about the value of μ by using the sample mean, $\bar{x}$, to estimate the value of μ or to test hypotheses about the value of μ.

To learn about the sampling distribution of the statistic $\bar{x}$, begin by considering some sampling investigations. In the examples that follow, a population is specified and a sample size n is chosen. Then 500 different random samples of this size are selected. The value of $\bar{x}$ is calculated for each sample, and a histogram of these 500 $\bar{x}$ values is constructed. Because 500 is a reasonably large number of samples, the histogram of the $\bar{x}$ values should resemble the actual sampling distribution of $\bar{x}$ (which would be obtained by considering *all* possible samples). This process is repeated for several different values of n so that you can see how the choice of sample size affects the sampling distribution and so you can identify patterns that lead to important properties of the sampling distribution.

Example 12.1 Blood Platelet Volume

The paper **"Mean Platelet Volume Could Be Possible Biomarker in Early Diagnosis and Monitoring of Gastric Cancer"** (*Platelets* **[2014]: 592–594**) includes data that suggest that the distribution of

$$x = \text{platelet volume}$$

for patients who have gastric cancer is approximately normal with mean $\mu = 8.3$ and standard deviation $\sigma = 0.8$.

Figure 12.1 shows a normal curve centered at 8.3, the mean value of platelet volume. The value of the population standard deviation, 0.8, determines the extent to which the x distribution spreads out about its mean value.

FIGURE 12.1
Normal distribution of
x = platelet volume, with
μ = 8.3 and σ = 0.8

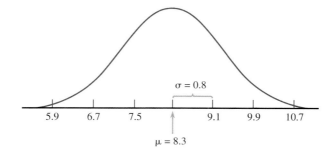

What values for the sample mean would be expected if you were to take a random sample of size 5 from this population distribution? To investigate, you can simulate sampling from this population. A statistical software package was used to select 500 random samples of size $n = 5$ from the population. The sample mean platelet volume $\bar{x}$ was calculated for each sample, and these 500 values of $\bar{x}$ were used to construct the density histogram shown in Figure 12.2. Recall from Chapter 2 that a density histogram is a histogram that uses

$$\text{density} = \frac{\text{relative frequency}}{\text{interval width}}$$ to determine the heights of the bars in the histogram.

FIGURE 12.2
Density histogram of 500 $\bar{x}$
values based on random samples
of size $n = 5$ from a normal
population with $\mu = 8.3$ and
$\sigma = 0.8$

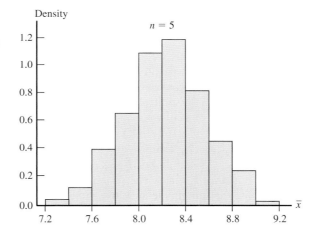

The histogram in Figure 12.2 describes the behavior of the sample mean $\bar{x}$ for samples of size $n = 5$ from the platelet volume population. Notice that there is a lot of sample-to-sample variability in the value of $\bar{x}$. For some samples $\bar{x}$ is around 7.3, and for other samples it is around 9.1. A sample of size 5 from the population of patients who have gastric cancer won't always provide precise information about the mean platelet volume in the population. What if a larger sample is selected? To investigate the effect of sample size on the behavior of $\bar{x}$, 500 samples of size 10, 500 samples of size 20, and 500 samples of size 30 were selected. Density histograms of the resulting $\bar{x}$ values, along with the density histogram for the samples of size 5, are displayed in Figure 12.3.

The first thing to notice about the histograms is that each of them is approximately normal in shape. The resemblance would be even more striking if each histogram had been based on many more than 500 $\bar{x}$ values. Second, notice that each histogram is centered at approximately 8.3, the mean of the population being sampled. Had the histograms been constructed using $\bar{x}$ values from every possible sample, they would have been centered at exactly 8.3.

The final aspect of the histograms to note is their variability relative to one another. The smaller the value of n, the more the sampling distribution spreads out about the population mean value. This is why the histograms for $n = 20$ and $n = 30$ are based on narrower class intervals than those for the two smaller sample sizes. For the larger sample

FIGURE 12.3
Density histograms for $\bar{x}$ constructed using 500 random samples of size n for the population of Example 12.1:
(a) $n = 5$; (b) $n = 10$; (c) $n = 20$; (d) $n = 30$

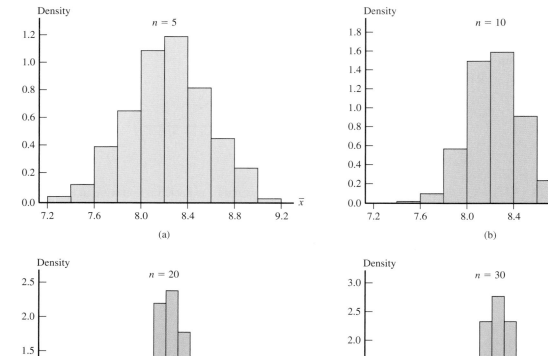

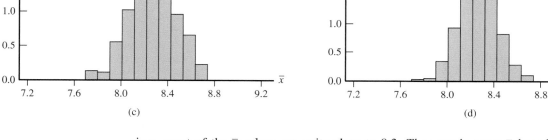

sizes, most of the $\bar{x}$ values are quite close to 8.3. *The sample mean $\bar{x}$ based on a large sample size tends to be closer to μ than $\bar{x}$ based on a small sample size.*

Example 12.2   Time to First Goal in Hockey

In this example, you will consider properties of the $\bar{x}$ distribution when the population is quite skewed (which is very unlike a normal distribution). The paper **"Is the Overtime Period in an NHL Game Long Enough?"** (*The American Statistician* **[2008]:151–154**) included data on the time (in minutes) from the start of the game to the first goal scored for the 281 regular season hockey games in the 2005–2006 season that went into overtime. Figure 12.4 displays a density histogram of the data (from a graph that appeared in the paper). The histogram has a long upper tail, indicating that the first goal is scored in the first 20 minutes of most games, but for some games, the first goal is not scored until much later in the game.

If you think of these 281 values as a population, the histogram in Figure 12.4 shows the population distribution. The skewed shape makes identification of the mean value from the histogram more difficult. The mean of the 281 values in the population was calculated to be $\mu = 13$ minutes. The median value for the population is 10 minutes, which is less than μ because the distribution is positively skewed.

For each of the sample sizes $n = 5$, 10, 20, and 30, 500 random samples of size n were selected. This was done with replacement to approximate more nearly the usual situation, in which the sample size n is only a small fraction of the population size. Density histograms of the 500 $\bar{x}$ values were constructed for each of the four sample sizes. These histograms are displayed in Figure 12.5.

FIGURE 12.4
The population distribution for Example 12.2 ($\mu = 13$)

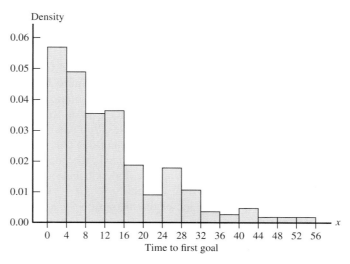

FIGURE 12.5
Four density histograms of 500 $\bar{x}$ values for Example 12.2: (a) $n = 5$; (b) $n = 10$; (c) $n = 20$; (d) $n = 30$

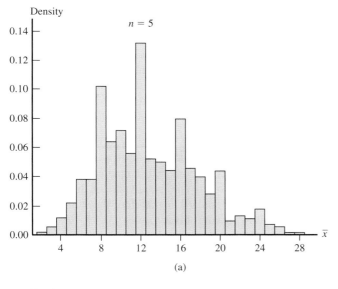

(a)

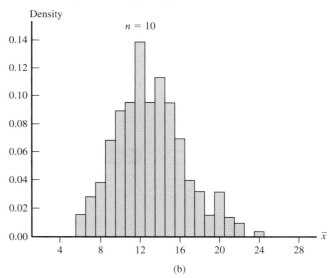

(b)

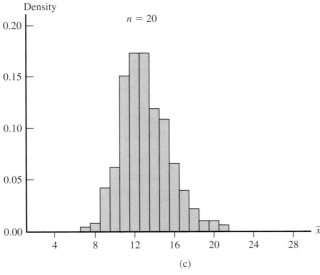

(c)

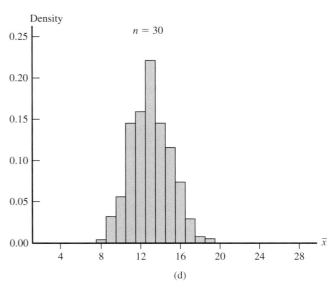

(d)

As with the samples from the normal population of Example 12.1, the means of the 500 $\bar{x}$ values for the four different sample sizes are all close to the population mean $\mu = 13$. If each histogram had been constructed using all possible samples rather than just 500 of them, they would be centered at exactly 13. Comparison of the four $\bar{x}$ histograms in Figure 12.5 also shows that as n increases, the histogram's spread about its center decreases. This was also true for the normal population of the previous example. There is less sample-to-sample variability in $\bar{x}$ for large samples than for small samples.

One aspect of the histograms in Figure 12.5 distinguishes them from those based on a random sample from a normal population (Figure 12.3). The histograms in Figure 12.5 are skewed and differ in shape more, but they become progressively more symmetric as the sample size increases. You can also see that for $n = 30$, the histogram has a shape much like a normal distribution. This is the effect of averaging. Even when n is large, one of the few large x values in the population doesn't appear in the sample very often. When one does appear, its contribution to $\bar{x}$ is outweighed by the contributions of more typical sample values.

The normal shape of the histogram for $n = 30$ is what is predicted by the *Central Limit Theorem*, which will be introduced shortly. According to this theorem, even if the population distribution does not look like a normal distribution, the sampling distribution of $\bar{x}$ is approximately normal in shape when the sample size n is reasonably large.

General Properties of the Sampling Distribution of $\bar{x}$

Examples 12.1 and 12.2 suggest that for any sample size n, the center of the $\bar{x}$ distribution (the mean value of $\bar{x}$) is equal to the value of the population mean and that the variability of the $\bar{x}$ distribution decreases as n increases. The sample histograms of Figures 12.3 and 12.5 also suggest that in some cases, the $\bar{x}$ distribution is approximately normal in shape. These observations are stated more formally in the following general properties.

> **General Properties of the Sampling Distribution of $\bar{x}$**
>
> Suppose $\bar{x}$ denotes the mean of the observations in a random sample of size n from a population with mean μ and standard deviation σ. The mean value of the sampling distribution of $\bar{x}$ is written as $\mu_{\bar{x}}$ and the standard deviation of the sampling distribution of $\bar{x}$ is written as $\sigma_{\bar{x}}$ The following rules hold:
>
> Rule 1. $\mu_{\bar{x}} = \mu$
>
> Rule 2. $\sigma_{\bar{x}} = \dfrac{\sigma}{\sqrt{n}}$. This rule is exact if the population is infinite, and is approximately correct if the population is finite and no more than 10% of the population is included in the sample.
>
> Rule 3. When the population distribution is normal, the sampling distribution of $\bar{x}$ is also normal for any sample size n.
>
> Rule 4. **(Central Limit Theorem)** When n is large, the sampling distribution of $\bar{x}$ is well approximated by a normal curve, even when the population distribution is not normal.

Rule 1, $\mu_{\bar{x}} = \mu$, states that the sampling distribution of $\bar{x}$ is always centered at the value of the population mean μ. This tells you that the $\bar{x}$ values from different random samples tend to cluster around the actual value of the population mean.

Rule 2, $\sigma_{\bar{x}} = \dfrac{\sigma}{\sqrt{n}}$, gives the relationship between $\sigma_{\bar{x}}$, the standard deviation of the sampling distribution of $\bar{x}$, and the sample size n. You can see why sample-to-sample variability in $\bar{x}$ decreases as the sample size n increases (because the sample size n is in the denominator of the expression for $\sigma_{\bar{x}}$). There is less variability in the sample means for larger samples. This means that the $\bar{x}$ values tend to cluster more tightly around the actual value of the population mean for larger samples. When $n = 4$, for example,

$$\sigma_{\bar{x}} = \frac{\sigma}{\sqrt{n}} = \frac{\sigma}{\sqrt{4}} = \frac{\sigma}{2}$$

and the sampling distribution of $\bar{x}$ has a standard deviation that is only half as large as the population standard deviation.

Rules 3 and 4 say that in some cases the shape of the sampling distribution of $\bar{x}$ is normal (when the population is normal) or approximately normal (when the population distribution is not normal but the sample size is large). Figure 12.6 illustrates these rules by showing several $\bar{x}$ sampling distributions superimposed over a graph of the population distribution.

FIGURE 12.6
Population distribution and sampling distributions of $\bar{x}$: (a) symmetric population; (b) skewed population

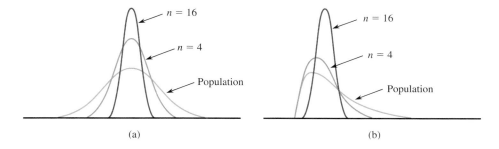

The Central Limit Theorem of Rule 4 states that when n is large, the sampling distribution of $\bar{x}$ is approximately normal for any population distribution. This result has enabled statisticians to develop large-sample methods for estimating a population mean and for testing hypotheses about a population mean that can be used even when the shape of the population distribution is unknown.

Recall that a variable is standardized by subtracting its mean value and then dividing by its standard deviation. Using Rules 1 and 2 to standardize $\bar{x}$ gives an important consequence of the last two rules.

If n is large or the population distribution is normal, the standardized variable

$$z = \frac{\bar{x} - \mu_{\bar{x}}}{\sigma_{\bar{x}}} = \frac{\bar{x} - \mu}{\dfrac{\sigma}{\sqrt{n}}}$$

has (at least approximately) a standard normal (z) distribution.

Applying the Central Limit Theorem requires a rule of thumb for deciding when n is large enough. Look back at Figure 12.5, which shows the approximate sampling distribution of $\bar{x}$ for $n = 5$, 10, 20, and 30 when the population distribution is quite skewed. The histogram for $n = 5$ is not well described by a normal curve, and this is still true of the histogram for $n = 10$. Among the four histograms, only the histogram for $n = 30$ has a shape that is reasonably well described by a normal curve. On the other hand, when the population distribution is normal, the sampling distribution of $\bar{x}$ is normal for any n.

How large n must be in order for the $\bar{x}$ distribution to be approximately normal depends on how much the population distribution differs from a normal distribution. The closer the population distribution is to being normal, the smaller the value of n necessary for the Central Limit Theorem approximation to be accurate. Many statisticians recommend the following conservative rule:

The Central Limit Theorem can safely be applied if $n \geq 30$.

| Example 12.3 | Courting Scorpion Flies |

The authors of the paper **"Should I Stay or Should I Go? Condition- and Status-Dependent Courtship Decisions in the Scorpion Fly** *Panorpa Cognate"* **(***Animal Behaviour* **[2009]: 491–497)** studied the courtship behavior of mating scorpion flies. One variable of interest was x = courtship time, which was defined as the time from the beginning of a female-male interaction until mating. Data from the paper suggest that it is reasonable to think that the population mean and standard deviation of x are $\mu = 117.1$ minutes and $\sigma = 109.1$ minutes. Notice that the population distribution of courtship times can't be normal. For a normal distribution centered at 117.1 with such a large standard deviation, it would not be uncommon to observe negative values, but courtship time can't have a negative value.

The sampling distribution of $\bar{x}$ = mean courtship time for a random sample of 20 scorpion fly mating pairs would have mean

$$\mu_{\bar{x}} = \mu = 117.1 \text{ minutes}$$

This tells you that the sampling distribution of $\bar{x}$ is centered at 117.1. The standard deviation of $\bar{x}$ is

$$\sigma_{\bar{x}} = \frac{\sigma}{\sqrt{n}} = \frac{109.1}{\sqrt{20}} = 24.40 \text{ minutes}$$

which is smaller than the population standard deviation σ. Because the population distribution is not normal and because the sample size is smaller than 30, it is not reasonable to assume that the sampling distribution of $\bar{x}$ is normal in shape.

The following examples illustrate how knowing the sampling distribution of $\bar{x}$ supports learning from sample data.

| Example 12.4 | Is the "Freshman 15" Real? |

It is a common belief that most students gain weight during their freshman year in college. The term "freshman 15" is often used to describe this weight gain. But do first-year college students really gain this much weight? The authors of the paper **"Patterns and Composition of Weight Change in College Freshmen"** **(***College Student Journal* **[2015]: 553–564)** describe a study of 103 freshmen at a college located in the Midwest region of the United States. For each student participating in the study, weight gain during the freshman year was determined. For these 103 students, the mean weight gain was 5.7 pounds, and the standard deviation was 6.8 pounds.

Although the sample wasn't actually selected at random, the researchers attempted to obtain a representative sample of students at the college. Because of this, it is reasonable to regard this sample as if it were a random sample of the freshmen at the college.

Suppose that you are interested in estimating the mean weight gain for all freshmen at the college. You don't expect the population mean to be *exactly* 5.7 pounds, but if you use 5.7 pounds as an estimate of the population mean, how accurate is this estimate likely to be? To answer this question, you can use what you know about the sampling distribution of $\bar{x}$ for random samples of size 103. You know three things that follow from the general results described earlier:

What You Know	How You Know It
The sampling distribution of $\bar{x}$ is centered at μ. This means that the $\bar{x}$ values from random samples cluster around the actual value of the population mean.	Rule 1 states that $\mu_{\bar{x}} = \mu$. This is true for random samples, and the description of the study suggests that it is reasonable to regard the sample as if it were a random sample.

(continued)

What You Know	How You Know It
Values of $\bar{x}$ will cluster fairly tightly around the actual value of the population mean. The standard deviation of $\bar{x}$, (which describes how much the $\bar{x}$ values spread out around the population mean) is $\frac{\sigma}{\sqrt{n}}$. An estimate of the standard deviation of $\bar{x}$ is $$\frac{\sigma}{\sqrt{n}} \approx \frac{s}{\sqrt{n}} = \frac{6.8}{\sqrt{103}} = 0.67 \text{ pounds}$$	Rule 2 states that $\sigma_{\bar{x}} = \frac{\sigma}{\sqrt{n}}$. In this example, $n = 103$. The value of σ is not known—this is the actual value of the population standard deviation. However, the sample standard deviation s provides an estimate of σ that can be used to estimate the standard deviation of the sampling distribution. The estimated standard deviation provides information about how tightly the $\bar{x}$ values from different random samples cluster around the value of the population mean, μ.
The sampling distribution of $\bar{x}$ is approximately normal.	Rule 4 states that the sampling distribution of $\bar{x}$ is approximately normal if n is large. Here the sample size is 103. Because the sample size is greater than 30, the normal approximation is appropriate.

Summarizing, you know that the $\bar{x}$ distribution is centered at the actual population mean, has a standard deviation of about 0.67 pounds, and is approximately normal. By using this information and what you know about normal distributions, you can now get a sense of the accuracy of the estimate $\bar{x} = 5.7$ pounds. For any variable described by a normal distribution, approximately 95% of the values are within two standard deviations of the center. Since the distribution of $\bar{x}$ is approximately normal and is centered at the actual population mean μ, you now know that about 95% of all possible random samples would produce a sample mean within about $2(0.67) = 1.34$ pounds of the actual population mean. So, a margin of error of 1.34 pounds could be reported. This tells you that the sample estimate $\bar{x} = 5.7$ pounds is likely to be within 1.34 pounds of the actual mean freshman year weight gain for students at this college. This suggests that the actual mean weight gain for students at this college is quite a bit less than the legendary freshman 15. These ideas will be formalized in Section 12.2, where you will learn about margin of error when estimating a population mean and will see how to construct confidence interval estimates for a population mean.

Example 12.5 **Fat Content of Hot Dogs**

Data from a sample can also be used to evaluate whether a claim about a population mean is believable. For example, suppose that a hot dog manufacturer claims that one of its brands of hot dogs has a mean fat content of $\mu = 18$ grams per hot dog. Consumers of this brand would probably not be unhappy if the mean is less than 18 grams but would be unhappy if it exceeds 18 grams.

In this situation, the variable of interest is

$$x = \text{fat content of a hot dog}$$

For purposes of this example, suppose that you know that σ, the standard deviation of the x distribution, is equal to 1 gram.

An independent testing organization is asked to analyze a random sample of $n = 36$ hot dogs. The fat content for each of the 36 hot dogs is measured and the sample mean is calculated to be $\bar{x} = 18.4$ grams. Does this result suggest that the manufacturer's claim that the population mean is 18 is incorrect?

The sample size, $n = 36$, is large enough to think that the sampling distribution of $\bar{x}$ will be approximately normal. The standard deviation of the $\bar{x}$ distribution is

$$\sigma_{\bar{x}} = \frac{\sigma}{\sqrt{n}} = \frac{1}{\sqrt{36}} = 0.167$$

If the manufacturer's claim is correct, you also know that

$$\mu_{\bar{x}} = \mu = 18$$

If the manufacturer's claim is correct, should you be surprised to see a sample mean of $\bar{x} = 18.4$ grams? You know that even if $\mu = 18$ grams, $\bar{x}$ will not usually be exactly 18 grams due to sampling variability. But, is it likely that you would see a sample mean at least as large as 18.4 grams when the population mean is really 18 grams?

Using the normal distribution, you can calculate the probability of observing a sample mean this large. *If the manufacturer's claim is correct*,

$$P(\bar{x} \geq 18.4) \approx P\left(z \geq \frac{18.4 - 18}{0.167}\right)$$

$$= P(z \geq 2.40)$$

$$= \text{area under the normal curve to the right of } 2.40$$

$$= 1 - 0.9918$$

$$= 0.0082$$

Values of $\bar{x}$ as large as 18.4 grams will be observed only about 0.82% of the time when a random sample of size $n = 36$ is taken from a population with mean 18 grams and standard deviation 1 gram. The value $\bar{x} = 18.4$ grams is enough greater than 18 grams that you should be skeptical of the manufacturer's claim. These ideas will be formalized in Section 12.3, where you will learn about how to test hypotheses about a population mean.

Other Cases

You now know a great deal about the sampling distribution of $\bar{x}$ in two cases: when the population distribution is normal and when the sample size is large. What happens when the population distribution is not normal and n is small? Although it is still true that $\mu_{\bar{x}} = \mu$ and $\sigma_{\bar{x}} = \frac{\sigma}{\sqrt{n}}$, unfortunately there is no general result about the shape of the sampling distribution of $\bar{x}$. When this is the case, it is possible to use a simulation-based method. Simulation-based methods are covered in Section 12.4.

Summing It Up—Section 12.1

The following learning objectives were addressed in this section:

Conceptual Understanding
C1: Understand how the value of the standard deviation of the sample mean, $\bar{x}$, is related to sample size.
The standard deviation of the sample mean describes how much the values of $\bar{x}$ tend to vary from one random sample to another. The standard deviation of $\bar{x}$ is $\sigma_{\bar{x}} = \frac{\sigma}{\sqrt{n}}$. This standard deviation decreases as n increases, so this means that the values of $\bar{x}$ don't vary as much from sample to sample as the sample size increases.

Mastering the Mechanics
M1: Determine the mean and standard deviation of the sampling distribution of $\bar{x}$.
The mean of the sampling distribution of $\bar{x}$ is $\mu_{\bar{x}} = \mu$ and the standard deviation of the sampling distribution of $\bar{x}$ is $\sigma_{\bar{x}} = \frac{\sigma}{\sqrt{n}}$. Example 12.4 illustrates the calculation of the mean and standard deviation of the sampling distribution of $\bar{x}$.

M2: Know when the sampling distribution of $\bar{x}$ is approximately normal.
The sampling distribution of $\bar{x}$ is normal if the population distribution is normal. If the population distribution is not normal, the sampling distribution of $\bar{x}$ is approximately normal if the sample size is large. This is generally the case when $n \geq 30$.

SECTION 12.1 EXERCISES

Each Exercise Set assesses the following chapter learning objectives: C1, M1, M2

SECTION 12.1 Exercise Set 1

12.1 A random sample is selected from a population with mean $\mu = 100$ and standard deviation $\sigma = 10$. Determine the mean and standard deviation of the sampling distribution of $\bar{x}$ for each of the following sample sizes:

a. $n = 9$ **d.** $n = 50$
b. $n = 15$ **e.** $n = 100$
c. $n = 36$ **f.** $n = 400$

12.2 For which of the sample sizes given in the previous exercise would it be reasonable to think that the $\bar{x}$ sampling distribution is approximately normal in shape?

12.3 The paper **"Alcohol Consumption, Sleep, and Academic Performance Among College Students"** (*Journal of Studies on Alcohol and Drugs* [2009]: 355–363) describes a study of $n = 236$ students who were randomly selected from a list of students enrolled at a liberal arts college in the northeastern region of the United States. Each student in the sample responded to a number of questions about their sleep patterns. For these 236 students, the sample mean time spent sleeping per night was reported to be 7.71 hours and the sample standard deviation of the sleeping times was 1.03 hours. Suppose that you are interested in learning about the value of μ, the population mean time spent sleeping per night for students at this college. The following table is similar to the table that appears in Example 12.4. The "what you know" information has been provided. Complete the table by filling in the "how you know it" column.

What You Know	How You Know It
The sampling distribution of $\bar{x}$ is centered at the actual (but unknown) value of the population mean.	
An estimate of the standard deviation of $\bar{x}$, which describes how the $\bar{x}$ values spread out around the population mean μ, is 0.067.	
The sampling distribution of $\bar{x}$ is approximately normal.	

12.4 Explain the difference between μ and $\mu_{\bar{x}}$

12.5 The time that people have to wait for an elevator in an office building has a uniform distribution over the interval from 0 to 1 minute. For this distribution, $\mu = 0.5$ and $\sigma = 0.289$.
a. If $\bar{x}$ is the average waiting time for a random sample of $n = 16$ waiting times, what are the values of the mean and standard deviation of the sampling distribution of $\bar{x}$?

b. Answer Part (a) for a random sample of 50 waiting times. Draw a picture of the approximate sampling distribution of $\bar{x}$ when $n = 50$.

SECTION 12.1 Exercise Set 2

12.6 A random sample is selected from a population with mean $\mu = 60$ and standard deviation $\sigma = 3$. Determine the mean and standard deviation of the sampling distribution of $\bar{x}$ for each of the following sample sizes:

a. $n = 6$ **d.** $n = 75$
b. $n = 18$ **e.** $n = 200$
c. $n = 42$ **f.** $n = 400$

12.7 For which of the sample sizes given in the previous exercise would it be reasonable to think that the $\bar{x}$ sampling distribution is approximately normal in shape?

12.8 The paper **"Alcohol Consumption, Sleep, and Academic Performance Among College Students"** (*Journal of Studies on Alcohol and Drugs* [2009]: 355–363) describes a study of $n = 236$ students that were randomly selected from a list of students enrolled at a liberal arts college in the northeastern region of the United States. Each student in the sample responded to a number of questions about their sleep patterns. For these 236 students, the sample mean additional time spent sleeping on weekend days compared to the other days of the week was reported to be 1.29 hours and the standard deviation was 1.09 hours. Suppose that you are interested in learning about the value of μ, the mean additional time spent sleeping on weekend days for students at this college. The following table is similar to the table that appears in Example 12.4. The "what you know" information has been provided. Complete the table by filling in the "how you know it" column.

What You Know	How You Know It
The sampling distribution of $\bar{x}$ is centered at the actual (but unknown) value of the population mean.	
An estimate of the standard deviation of $\bar{x}$, which describes how the $\bar{x}$ values spread out around the population mean μ, is 0.071.	
The sampling distribution of $\bar{x}$ is approximately normal.	

12.9 Explain the difference between σ and $\sigma_{\bar{x}}$.

12.10 Suppose that a random sample of size 64 is to be selected from a population with mean 40 and standard deviation 5.

a. What are the mean and standard deviation of the sampling distribution of $\bar{x}$? Describe the shape of the sampling distribution of $\bar{x}$.

b. What is the approximate probability that $\bar{x}$ will be within 0.5 of the population mean μ?

c. What is the approximate probability that $\bar{x}$ will differ from μ by more than 0.7?

ADDITIONAL EXERCISES

12.11   A random sample is selected from a population with mean $\mu = 200$ and standard deviation $\sigma = 15$. Determine the mean and standard deviation of the sampling distribution of $\bar{x}$ for each of the following sample sizes:

a. $n = 12$ **d.** $n = 40$
b. $n = 20$ **e.** $n = 90$
c. $n = 25$ **f.** $n = 300$

12.12   For which of the sample sizes given in the previous exercise would it be reasonable to think that the sampling distribution of $\bar{x}$ is approximately normal in shape?

12.13   Explain the difference between $\bar{x}$ and $\mu_{\bar{x}}$.

12.14   A sign in the elevator of a college library indicates a limit of 16 persons. In addition, there is a weight limit of 2500 pounds. Assume that the average weight of students, faculty, and staff at this college is 150 pounds, that the standard deviation is 27 pounds, and that the distribution of weights of individuals on campus is approximately normal. A random sample of 16 persons from the campus will be selected.

a. What is the mean of the sampling distribution of $\bar{x}$?

b. What is the standard deviation of the sampling distribution of $\bar{x}$?

c. What average weights for a sample of 16 people will result in the total weight exceeding the weight limit of 2500 pounds?

d. What is the probability that a random sample of 16 people will exceed the weight limit?

12.15   Suppose that the population mean value of interpupillary distance (the distance between the pupils of the left and right eyes) for adult males is 65 mm and that the population standard deviation is 5 mm.

a. If the distribution of interpupillary distance is normal and a random sample of $n = 25$ adult males is to be selected, what is the probability that the sample mean distance $\bar{x}$ for these 25 will be between 64 and 67 mm? At least 68 mm?

b. Suppose that a random sample of 100 adult males is to be selected. Without assuming that interpupillary distance is normally distributed, what is the approximate probability that the sample mean distance will be between 64 and 67 mm? At least 68 mm?

12.16   Suppose that a random sample of size 100 is to be drawn from a population with standard deviation 10.

a. What is the probability that the sample mean will be within 20 of the value of μ?

b. For this example ($n = 100$, $\sigma = 10$), complete each of the following statements by calculating the appropriate value:

  i. Approximately 95% of the time, $\bar{x}$ will be within _____ of μ.

  ii. Approximately 0.3% of the time, $\bar{x}$ will be farther than _____ from μ.

12.17   A manufacturing process is designed to produce bolts with a diameter of 0.5 inches. Once each day, a random sample of 36 bolts is selected and the bolt diameters are recorded. If the resulting sample mean is less than 0.49 inches or greater than 0.51 inches, the process is shut down for adjustment. The standard deviation of bolt diameters is 0.02 inches. What is the probability that the manufacturing line will be shut down unnecessarily? (Hint: Find the probability of observing an $\bar{x}$ in the shutdown range when the actual process mean is 0.5 inches.)

12.18   An airplane with room for 100 passengers has a total baggage limit of 6000 pounds. Suppose that the weight of baggage checked by an individual passenger, x, has a mean of 50 pounds and a standard deviation of 20 pounds. If 100 passengers will board a flight, what is the approximate probability that the total weight of their baggage will exceed the limit? (Hint: With $n = 100$, the total weight exceeds the limit when the mean weight $\bar{x}$ exceeds 6000/100.)

SECTION 12.2   A Confidence Interval for a Population Mean

In the chapter preview example, researchers wanted to use sample data to estimate the mean processing time of emergency room flu cases for a proposed new drive-through model in which flu patients are evaluated while they remain in their cars. People representing a random sample of $n = 38$ flu cases were processed using the drive-through model and the time to process each case from admission to discharge was recorded. Data read from a graph in the paper referenced in the preview were used to calculate the following summary statistics:

$$n = 38 \qquad\qquad \bar{x} = 26 \text{ minutes} \qquad\qquad s = 1.57 \text{ minutes}$$

Using the sample mean as an estimate of the population mean, you could say that you think that the mean time required to process a flu case for the drive-through model is 26 minutes.

But when you say that the population mean time is 26 minutes, you don't really believe that it is *exactly* 26 minutes. The estimate of 26 minutes is based on a sample, and the sample mean will vary from one sample to another. What you really think is that the population mean is *around* 26 minutes. To be more specific about what you mean by "around 26 minutes," you can use the sample data to calculate a confidence interval estimate of the population mean.

You have already seen confidence interval estimates in the context of estimating a population proportion, and the basic idea is the same for estimating a population mean. The general form of a confidence interval estimate is

$$\text{statistic} \pm (\text{critical value}) \left(\begin{array}{c} \text{standard error} \\ \text{of the statistic} \end{array} \right)$$

This general form can be adapted to estimating a population mean. Begin by considering the case in which σ, the population standard deviation, is known (this is not realistic, but you will see shortly how to handle the more realistic situation in which σ is unknown) and the sample size n is large enough for the Central Limit Theorem to apply ($n \geq 30$). From Section 12.1, you know that:

1. The sampling distribution of $\bar{x}$ is centered at μ, so $\bar{x}$ is an unbiased statistic for estimating μ (because $\mu_{\bar{x}} = \mu$).
2. The standard deviation of $\bar{x}$ is $\sigma_{\bar{x}} = \dfrac{\sigma}{\sqrt{n}}$.
3. As long as n is large ($n \geq 30$), the sampling distribution of $\bar{x}$ is approximately normal, even when the population distribution itself is not normal.

This suggests that a confidence interval for a population mean *when the sample size is large and σ is known* is

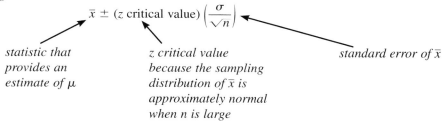

$$\bar{x} \pm (z \text{ critical value}) \left(\frac{\sigma}{\sqrt{n}} \right)$$

statistic that provides an estimate of μ   *z critical value because the sampling distribution of $\bar{x}$ is approximately normal when n is large*   *standard error of $\bar{x}$*

Example 12.6 illustrates the calculation of a confidence interval for the population mean when the population standard deviation is known.

Example 12.6   Cosmic Radiation

Cosmic radiation levels rise with increasing altitude, prompting researchers to consider how pilots and flight crews are affected by increased exposure to cosmic radiation. The Centers for Disease Control and Prevention reports that the National Council on Radiation Protection and Measurements estimates that aircrew have a mean annual radiation exposure of 3.07 millisievert (mSv) per year **(www.cdc.gov/niosh/topics/aircrew /cosmicionizingradiation.html, retrieved December 18, 2016)**. Suppose that the estimated mean exposure for aircrew is based on a random sample of $n = 100$ flight crew members.

Here, μ will represent the mean annual cosmic radiation exposure for all flight crew members. Although σ, the actual value of the population standard deviation, is not usually known, for purposes of this example, suppose that σ is known to be 0.35 mSv. Because the sample size is large and σ is known, a 95% confidence interval for μ is

z critical value for 95% confidence level

$$\bar{x} \pm (z \text{ critical value}) \frac{\sigma}{\sqrt{n}} = 3.07 \pm (1.96) \frac{0.35}{\sqrt{100}}$$
$$= 3.07 \pm 0.069$$
$$= (3.001, 3.139)$$

Based on this sample, *plausible* values of μ, the mean annual cosmic radiation exposure for all flight crew members, are those between 3.001 and 3.139 mSv. One mSv is the average annual exposure in the United States due to normal exposure to background radiation, so this means that it is estimated that the mean annual exposure of flight crew members is somewhere between about 3 and 3.14 times greater than for the general population. A confidence level of 95% is associated with the method used to produce this interval estimate.

The confidence interval formula used in Example 12.6 is straightforward, but not very useful! To calculate the confidence interval endpoints, you need to know the value of σ, the population standard deviation. It is almost never the case that you would not know the value of the population mean (which is why you would be using sample data to estimate it) but would know the value of σ for the same population. For this reason, you will probably never use the confidence interval formula used in Example 12.6. Nevertheless, it does provide a useful starting point for investigating how to estimate the population mean in more realistic situations when σ is not known.

A Confidence Interval for μ When σ Is Unknown

When the population standard deviation is unknown, there are two changes that you need to make to the confidence interval formula previously given. First, use the sample standard deviation, s, as an estimate of σ in the confidence interval formula. The second change is that you can no longer use the standard normal (z) distribution to determine the critical value used to calculate the confidence interval.

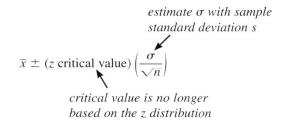

If the population distribution is at least approximately normal or if the sample size is large ($n \geq 30$), the critical value that should be used is based on a probability distribution called the t distribution. Since you have not yet studied t distributions, you need to take a short detour to learn about them.

t Distributions

Just as there are many different normal distributions, there are also many different t distributions. While normal distributions are distinguished from one another by their mean μ and standard deviation σ, t distributions are distinguished by a positive number called *degrees of freedom* (df). There is a t distribution with 1 df, another with 2 df, and so on.

Important Properties of t Distributions

1. The t distribution corresponding to any particular degrees of freedom is bell-shaped and centered at zero (just like the standard normal distribution).
2. Every t distribution has greater variability than the standard normal distribution.
3. As the number of degrees of freedom increases, the variability of the corresponding t distribution decreases.
4. As the number of degrees of freedom increases, the corresponding sequence of t distributions approaches the standard normal distribution.

The properties in the preceding box are illustrated in Figure 12.7, which shows two t curves along with the standard normal (z) curve.

FIGURE 12.7
Comparison of the standard normal (z) distribution and t distributions for 4 df and 12 df

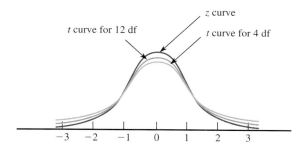

Appendix A Table 3 gives selected critical values for various *t* distributions. Critical values are given for central areas of 0.80, 0.90, 0.95, 0.98, 0.99, 0.998, and 0.999. To find a particular critical value, go down the left margin of the table to the row labeled with the desired number of degrees of freedom. Then move over in that row to the column headed by the desired central area. For example, the value in the 12-df row under the column corresponding to central area 0.95 is 2.18, so 95% of the area under the *t* curve with 12 df lies between −2.18 and 2.18. Moving over two columns, the critical value for central area 0.99 (still with 12 df) is 3.06 (see Figure 12.8). Moving down the 0.99 column to the 20-df row, you see the critical value for central area 0.99 is 2.85, so the area between −2.85 and 2.85 under the *t* curve with 20 df is 0.99.

FIGURE 12.8
t critical values illustrated

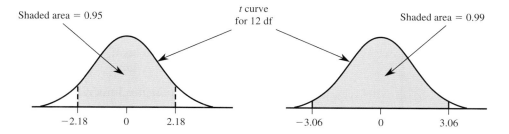

Notice that the critical values increase from left to right in each row of Appendix A Table 3. This makes sense because as you move to the right, the central area captured is larger. In each column, the critical values decrease as you move downward, reflecting decreasing variability for *t* distributions with a greater number of degrees of freedom.

The greater the number of degrees of freedom, the more closely the *t* curve resembles the *z* curve. To emphasize this, *z* critical values are included as the last row of the *t* table. Once the number of degrees of freedom exceeds 30, there is little change in the critical values as degrees of freedom increase. For this reason, Appendix A Table 3 jumps from 30 df to 40 df, then to 60 df, then to 120 df, and finally to the row of *z* critical values. If you need a critical value for a number of degrees of freedom between those tabulated, just use the critical value for the closest df. For df > 120, you can use the *z* critical values.

Many graphing calculators and statistical software packages calculate *t* critical values for any specified number of degrees of freedom, so it is not necessary to use the table or to approximate the *t* critical values if you are using technology.

A One-Sample *t* Confidence Interval for μ

Using the sample standard deviation to estimate the population standard deviation and replacing the *z* critical value with a *t* critical value results in the following confidence interval formula:

$$\bar{x} \pm (t \text{ critical value})\left(\frac{s}{\sqrt{n}}\right)$$

To find the *t* critical value, you need to know the desired confidence level and the appropriate number of degrees of freedom. When estimating a population mean μ, the number of degrees of freedom is determined by the sample size:

$$df = n - 1$$

For example, suppose that you want to use a random sample of size 25 to estimate a population mean using a 95% confidence interval. In this case, df = 25 − 1 = 24. To find the *t* critical value, you can use technology or Appendix A Table 3 (look in the 95% confidence level column and the 24 df row). The appropriate *t* critical value is 2.06.

There is one other question to answer: When is it appropriate to use this confidence interval formula? If the sample size *n* is large (*n* ≥ 30), this interval can be used regardless of the shape of the population distribution. However, when *n* is small, this interval is only appropriate if the population distribution is at least approximately normal. There are several ways that sample data can be used to assess the plausibility of normality. You can look at a normal probability plot of the sample data (looking for a plot that is reasonably straight) or you can construct a dotplot, a boxplot, or a histogram of the data (looking for approximate symmetry and no outliers).

One-Sample *t* Confidence Interval for a Population Mean μ

Appropriate when the following conditions are met:

1. The sample is a random sample from the population of interest, or the sample is selected in a way that would result in a sample that is representative of the population.
2. The sample size is large (*n* ≥ 30) *or* the population distribution is approximately normal.

When these conditions are met, a confidence interval for the population mean is

$$\bar{x} \pm (t \text{ critical value}) \left(\frac{s}{\sqrt{n}} \right)$$

The *t* critical value is based on df = *n* − 1 and the desired confidence level. It can be found using Appendix A Table 3, statistical software, or a graphing calculator.

Interpretation of Confidence Interval

You can be confident that the value of the population mean is included in the calculated interval. In a given problem, this statement should be worded in context.

Interpretation of Confidence Level

The confidence level specifies the long-run proportion of the time that this method is expected to be successful in capturing the value of population mean.

Recall the four key questions (QSTN) first introduced in Section 7.2 that guide the decision about which method to consider:

Q **Question Type**	Estimation or hypothesis testing?
S **Study Type**	Sample data or experiment data?
T **Type of Data**	One variable or two? Categorical or numerical?
N **Number of Samples or Treatments**	How many samples or treatments?

When the answers to these questions are

Q: estimation
S: sample data
T: one numerical variable
N: one sample

the method you should consider is the one-sample t confidence interval for a population mean.

Once you have selected the one-sample t confidence interval for a population mean as the method you will consider, follow the five-step process for estimation problems (EMC^3). Recall that the steps in this process are (see Section 7.3):

Step	What Is This Step?
E	**Estimate:** Explain what population characteristic you plan to estimate.
M	**Method:** Select a potential method based on the answers to the four key questions (QSTN).
C	**Check:** Check to make sure that the method selected is appropriate. Many methods for learning from data only provide reliable information under certain conditions. It is important to verify that any such conditions are met before proceeding.
C	**Calculate:** Use the sample data to perform any necessary calculations. This step is often accomplished using technology, such as a graphing calculator or statistical software.
C	**Communicate Results:** Answer the research question of interest, explain what you have learned from the data, and acknowledge potential risks.

You will see how this process is used in the examples that follow. Be sure to include all five steps when you use a confidence interval to estimate a population mean.

Example 12.7   Drive-Through Medicine Revisited

The chapter preview example described a study of a drive-through model for processing flu patients who go to the emergency room for treatment **("Drive-Through Medicine: A Novel Proposal for Rapid Evaluation of Patients During an Influenza Pandemic,"** *Annals of Emergency Medicine* **[2010]: 268–273)**. In the drive-through model, patients are evaluated while they remain in their cars. In the study, $n = 38$ people were each given a scenario for a flu case that was selected at random from the set of all flu cases actually seen in the emergency room. The scenarios provided the "patient" with a medical history and a description of symptoms that would allow the patient to respond to questions from the examining physician. These patients were processed using a drive-through procedure that was implemented in the parking structure of Stanford University Hospital. The time to process each case from admission to discharge was recorded. The following sample statistics were calculated from the data:

$$n = 38 \qquad \bar{x} = 26 \text{ minutes} \qquad s = 1.57 \text{ minutes}$$

The researchers were interested in estimating the mean processing time for flu patients using the drive-through model.

Begin by answering the four key questions:

Q Question Type	Estimation or hypothesis testing?	Estimation
S Study Type	Sample data or experiment data?	Sample data
T Type of Data	One variable or two? Categorical or numerical?	One numerical variable
N Number of Samples or Treatments	How many samples or treatments?	One sample

Because the answers are estimation, sample data, one numerical variable, and one sample, you should consider a one-sample t confidence interval for a population mean.

Now you are ready to use the five-step process to estimate the population mean. For purposes of this example, a 95% confidence level will be used.

Step	
Estimate: Explain what population characteristic you plan to estimate.	In this example, you want to estimate the value of μ, **the mean time to process a flu case using the new drive-through model.**
Method: Select a potential method based on the answers to the four key questions (QSTN).	Because the answers to the four key questions are estimation, sample data, one numerical variable, one sample, consider using a **one-sample t confidence interval for a population mean.** A confidence level of 95% was specified for this example.
Check: Check to make sure that the method selected is appropriate.	There are **two conditions** that need to be met in order to use the one-sample t confidence interval. The large sample condition is met because the sample size of 38 is greater than 30. Because the sample size is large, you do not need to worry about whether the population distribution is approximately normal. The second condition that must be met is that the sample is a random sample or one that is representative of the population. Here, the 38 flu cases were randomly selected from the population of all flu cases seen at the emergency room, so this condition is met.
Calculate: Use the sample data to perform any necessary calculations.	$n = 38$ $\bar{x} = 26$ $s = 1.57$ $df = 38 - 1 = 37$ The appropriate t critical value is found using technology or by using Appendix A Table 3 to find the t critical value that captures a central area of 0.95. (Note: When using the table, use the closest df, which is 40.) t critical value $= 2.02$ 95% confidence interval $$\bar{x} \pm (t \text{ critical value})\left(\frac{s}{\sqrt{n}}\right) = 26 \pm (2.02)\left(\frac{1.57}{\sqrt{38}}\right)$$ $$= 26 \pm 0.514$$ $$= (25.486, 26.514)$$
Communicate Results: Answer the research question of interest, explain what you have learned from the data, and acknowledge potential risks.	**Confidence interval:** You can be 95% confident that the actual mean processing time for emergency room flu cases using the new drive-through model is between 25.9 minutes and 26.51 minutes. **Confidence level:** The method used to construct this interval estimate is successful in capturing the actual value of the population mean about 95% of the time.

The researchers in this study also indicated that the average processing time for flu patients seen in the emergency room was about 90 minutes, so it appears that the drive-through procedure has promise both in terms of keeping flu patients isolated and also in reducing mean processing time.

Example 12.8 Waiting for Surgery

The authors of the paper **"Length of Stay, Wait Time to Surgery and 30-Day Mortality for Patients with Hip Fractures After Opening of a Dedicated Orthopedic Weekend Trauma Room"** (*Canadian Journal of Surgery* [2016]: 337–341) were interested in estimating the mean time that patients who broke a hip had to wait for surgery after the opening of a new hospital facility. They reported that for a representative sample of 204 people with a fractured hip, the sample mean time between arriving at the hospital and surgery to repair the hip was 28.5 hours and that the sample standard deviation of the wait times was 16.8 hours. Answers to the four key questions for this problem are:

Q Question Type	Estimation or hypothesis testing?	Estimation.
S Study Type	Sample data or experiment data?	Sample data: The data are from a sample of 204 patients.
T Type of Data	One variable or two? Categorical or numerical?	One numerical variable (wait time for surgery).
N Number of Samples or Treatments	How many samples or treatments?	One sample.

So, this is *estimation, sample data, one numerical variable,* and *one sample.* This combination of answers leads you to consider a one-sample *t* confidence interval for a population mean.

You can now use the five-step process for estimation problems (EMC³) to construct a 90% confidence interval.

E Estimate μ, the population mean wait time for surgery for patients with a fractured hip, will be estimated.

M Method   Because the answers to the four key questions are estimation, sample data, one numerical variable, one sample, a one-sample *t* confidence interval for a population mean will be considered. A confidence level of 90% was specified.

C Check   There are two conditions that need to be met in order for the one-sample *t* confidence interval to be appropriate. The sample was thought to be a representative sample of patients with a fractured hip, so the requirement of a representative sample is met.

The authors of the paper commented that there were several outliers in the data set, which suggests that the population distribution is not normal. But because the sample size is large, it is still appropriate to use the *t* confidence interval.

C Calculate   sample size: $n = 204$
sample mean wait time: $\bar{x} = 28.5$ hours
sample standard deviation: $s = 16.8$ hours
From Appendix A Table 3, the *t* critical value = 1.645 (from the *z* critical value row because df = $n - 1 = 203 > 120$, the largest number of degrees of freedom in the table). The 90% confidence interval for μ is then

$$\bar{x} \pm (t \text{ critical value})\left(\frac{s}{\sqrt{n}}\right) = 28.5 \pm (1.645)\left(\frac{16.8}{\sqrt{204}}\right)$$
$$= 28.5 \pm 1.935$$
$$= (26.565, 30.435)$$

C Communicate Results   **Interpret Confidence Interval:**
Based on this sample, you can be 90% confident that the mean wait time for surgery is between 26.57 and 30.44 hours.

Interpret Confidence Level:
The method used to construct this interval estimate is successful in capturing the actual value of the population mean about 90% of the time.

The paper referenced in Example 12.8 also gave data on surgery wait times for a representative sample of 405 patients with fractured hips who were seen at this hospital before the new facility was opened. The mean wait time for the patients in this sample was 31.5 hours and the standard deviation of wait times was 27.0 hours. Because the sample was a representative sample and the sample size was large, it is appropriate to use the one-sample t confidence interval to estimate the mean wait time for surgery before the new facility was opened.

A graphing calculator or statistical software can be used to produce a one-sample t confidence interval. Using a 90% confidence level, output from Minitab for the sample of patients who had surgery before the new facility was opened is shown here.

One-Sample T

N	Mean	StDev	SE Mean	90% CI
405	31.50	27.00	1.34	(29.29, 33.71)

The 90% confidence interval for the mean wait time before the new facility extends from 29.29 hours to 33.71 hours.

Example 12.9 **Selfish Chimps?**

Data set available

The article **"Chimps Aren't Charitable" (***Newsday*, **November 2, 2005)** summarized a research study published in the journal *Nature*. In this study, chimpanzees learned to use an apparatus that dispensed food when either of two ropes was pulled. When one of the ropes was pulled, only the chimp controlling the apparatus received food. When the other rope was pulled, food was dispensed both to the chimp controlling the apparatus and also to a chimp in the adjoining cage. The accompanying data (approximated from a graph in the paper) represent the number of times out of 36 trials that each of seven chimps chose the option that would provide food to both chimps (the "charitable" response).

$$23 \quad 22 \quad 21 \quad 24 \quad 19 \quad 20 \quad 20$$

You can use these data to estimate the mean number of times (out of 36) that chimps choose the charitable response. For purposes of this example, let's suppose it is reasonable to regard this sample of seven chimps as representative of the population of all chimpanzees.

This is an estimation problem, and you have sample data, one numerical variable (the number of times (out of 36) that the charitable response is chosen), and one sample. These are the answers to the four key questions that lead you to consider a one-sample t confidence interval for a population mean as a potential method.

The five-step process for estimation problems (EMC^3) can be used to construct a 99% confidence interval.

E Estimate The mean number of times (out of 36) that chimps choose the charitable response, μ, will be estimated.

M Method Because the answers to the four key questions are estimation, sample data, one numerical variable, and one sample, a one-sample t confidence interval for a population mean will be considered. A confidence level of 99% was specified.

C Check There are two conditions that must be met in order for the one-sample t confidence interval to be appropriate. It was stated that it is reasonable to regard the sample as representative of the population. Because the sample size is small ($n = 7$), you need to consider whether it is reasonable to think that the distribution of number of charitable responses (out of 36) for the population of all chimps is at least approximately normal. Figure 12.9 is a dotplot and Figure 12.10 is a normal probability plot of the sample data. Although it is difficult to assess with only 7 data values, the dotplot is approximately symmetric and the normal probability plot is reasonably straight, so it seems reasonable to think that the population distribution is at least approximately normal.

FIGURE 12.9
Dotplot for the sample data of Example 12.9

Number of charitable responses

FIGURE 12.10
Normal probability plot for the sample data of Example 12.9

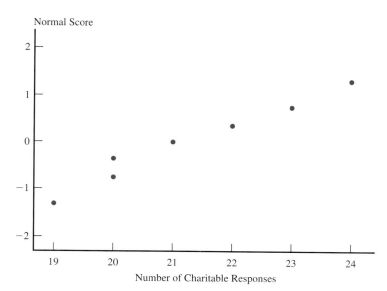

C **Calculate** Calculation of a confidence interval for the mean number of charitable responses for the population of all chimps requires $\bar{x}$ and s. From the given data,

$$\bar{x} = 21.29 \qquad\qquad s = 1.80$$

The t critical value for a 99% confidence interval based on 6 df is 3.71 (obtained from Appendix A Table 3 or using technology). The 99% confidence interval is

$$\bar{x} \pm (t \text{ critical value}) \left(\frac{s}{\sqrt{n}}\right) = 21.29 \pm (3.71)\left(\frac{1.80}{\sqrt{7}}\right)$$
$$= 21.29 \pm 2.524$$
$$= (18.766, 23.814)$$

A statistical software package or a graphing calculator could also have been used to calculate the 99% confidence interval. The following is output from SPSS.

One-Sample Statistics

	N	Mean	Std. Deviation	Std. Error Mean
CharitableResponses	7	21.2857	1.79947	.68014

	99% Confidence Interval	
	Lower	**Upper**
CharitableResponses	18.7642	23.8073

C **Communicate Results** **Interpret Confidence interval:**
Based on this sample, you can be 99% confident that the population mean number of charitable responses (out of 36 trials) is between 18.77 and 23.81.

Interpret Confidence level:
The 99% confidence level implies that if the same formula is used to calculate intervals for sample after sample randomly selected from the population of chimps, in the long run 99% of these intervals will capture μ between the lower and upper confidence limits.

Notice that based on this interval, you would conclude that, on average, chimps choose the charitable option more than half the time (more than 18 out of 36 trials). The *Newsday* headline "Chimps Aren't Charitable" was based on additional data from the study indicating that chimps' charitable behavior was not different when there was another chimp in the adjacent cage and when the adjacent cage was empty.

588 CHAPTER 12 Asking and Answering Questions About a Population Mean

Margin of Error and Choosing the Sample Size

The margin of error associated with a statistic was defined in Section 9.2 as the maximum likely estimation error expected when the statistic is used as an estimator. When $\bar{x}$ from a large random sample ($n \geq 30$) is used to estimate a population mean μ, the approximate margin of error is

$$\text{margin of error} = 1.96 \frac{\sigma}{\sqrt{n}}$$

which is usually estimated by

$$\text{estimated margin of error} = 1.96 \frac{s}{\sqrt{n}}$$

(When calculating the margin of error, it is acceptable to use either 2 or 1.96 in the calculation.) It would be unusual for the sample mean to differ from the actual value of the population mean by more than the margin of error.

Before collecting any data, an investigator may wish to determine a sample size required to achieve a particular margin of error, M. For example, with μ representing the mean fuel efficiency (in miles per gallon, mpg) for all cars of a certain model, you might want to estimate μ with a margin of error of 1 mpg. The value of n necessary to achieve this is obtained by setting the expression for the margin of error equal to 1 and then solving this equation for n. That is, you would solve

$$1 = 1.96 \frac{\sigma}{\sqrt{n}}$$

for n.

In general, if you want to estimate μ with a particular margin of error, M, you find the necessary sample size by solving the equation $M = 1.96 \frac{\sigma}{\sqrt{n}}$ for n. The result is

$$n = \left(\frac{1.96\sigma}{M}\right)^2$$

Notice that the greater the variability in the population (larger σ), the greater the required sample size. Also notice that a greater sample size is required for a smaller desired margin of error.

Use of this sample-size formula requires a value for σ, but σ is rarely known. One possible strategy for estimating σ is to carry out a preliminary study and to use the resulting sample standard deviation (or a somewhat larger value, to be conservative). Another possibility is to make an educated guess about the value of σ. For a population distribution that is not too skewed, dividing the anticipated range (the difference between the largest and the smallest values) by 4 often gives a rough estimate of the value of the standard deviation.

The sample size required to estimate a population mean μ with a specified margin of error M is

$$n = \left(\frac{1.96\sigma}{M}\right)^2$$

If the value of σ is unknown, it may be estimated based on previous information or, for a population that is not too skewed, by using $\dfrac{\text{range}}{4}$.

Example 12.10 Cost of Textbooks

A college financial aid advisor wants to estimate the mean cost of textbooks per quarter for students at the college. For the estimate to be useful, it should have a margin of error of $20 or less. How large a sample should be used to be confident of achieving this level of accuracy?

To determine the required sample size, you need a value for σ. Suppose the advisor thinks that the amount spent on books varies widely, but that most values are between $150 and $550. A reasonable estimate of σ is then

$$\frac{range}{4} = \frac{550 - 150}{4} = \frac{400}{4} = 100$$

Using this estimate of the population standard deviation, the required sample size is

$$n = \left(\frac{1.96\sigma}{M}\right)^2 = \left(\frac{(1.96)(100)}{20}\right)^2 = (9.8)^2 = 96.04$$

Rounding up, a sample size of $n = 97$ or larger is recommended.

Summing It Up—Section 12.2

The following learning objectives were addressed in this section:

Conceptual Understanding

C2: Know what factors affect the width of a confidence interval estimate of a population mean.

The width of a confidence interval estimate of a population mean is affected by the sample size, the value of the population standard deviation, and the choice of confidence level. The greater the sample size, the narrower the interval. The smaller the variability in the population (smaller population standard deviation), the narrower the interval. Decreasing the confidence level also decreases the width of the associated confidence interval.

Mastering the Mechanics

M3: Know the conditions for appropriate use of the margin of error and confidence interval formulas when estimating a population mean.

There are two conditions that need to be reasonably met in order for it to be appropriate to use the margin of error formula and the t confidence interval formula given in this section. These conditions are (1) that the sample is a random sample from the population of interest, or that the sample is selected in a way that makes it reasonable to think that the sample is representative of the population, and (2) the sample size is large. The sample size is considered to be large if $n \geq 30$.

M4: Calculate the margin of error when the sample mean, $\bar{x}$, is used to estimate a population mean μ.

The sample mean provides an estimate of the population mean. When appropriate, the margin of error can be calculated using the formula given in this section. The margin of error describes the maximum likely estimation error.

M5: Use the five-step process for estimation problems (EMC³) to calculate and interpret a confidence interval for a population mean.

The five steps for estimation problems are

> E: Estimate
>
> M: Method
>
> C: Check
>
> C: Calculate
>
> C: Communicate

This five-step process is illustrated in Examples 12.7, 12.8, and 12.9.

M6: Calculate the sample size necessary to achieve a desired margin of error when estimating a population mean.

The formula

$$n = \left(\frac{1.96\sigma}{M}\right)^2$$

can be used to calculate the sample size required to estimate a population mean with a given margin of error, M. The value of σ may be estimated using prior information or by using $\frac{range}{4}$ if you think that the population distribution is not too skewed. Example 12.10 illustrates the calculation of required sample size in a particular context.

Putting It into Practice

P1: Interpret a confidence interval for a population mean in context and interpret the associated confidence level.

A confidence interval for a population mean is interpreted as an interval of plausible values for the population mean. For examples of interpreting a confidence interval for a population mean and the associated confidence interval in context, see Examples 12.7, 12.8, and 12.9.

SECTION 12.2 EXERCISES

Each Exercise Set assesses the following chapter learning objectives: C2, M3, M4, M5, M6, P1

SECTION 12.2 Exercise Set 1

12.19 What percentage of the time will a variable that has a t distribution with the specified degrees of freedom fall in the indicated region? (Hint: See discussion on page 581.)
a. 10 df, between -1.81 and 1.81
b. 24 df, between -2.06 and 2.06
c. 24 df, outside the interval from -2.80 to 2.80
d. 10 df, to the left of -1.81

12.20 The formula used to calculate a confidence interval for the mean of a normal population is

$$\bar{x} \pm (t \text{ critical value}) \frac{s}{\sqrt{n}}$$

What is the appropriate t critical value for each of the following confidence levels and sample sizes?
a. 95% confidence, $n = 17$
b. 99% confidence, $n = 24$
c. 90% confidence, $n = 13$

12.21 The two intervals (114.4, 115.6) and (114.1, 115.9) are confidence intervals for μ = mean resonance frequency (in hertz) for all tennis rackets of a certain type. The two intervals were calculated using the same sample data.
a. What is the value of the sample mean resonance frequency?
b. The confidence level for one of these intervals is 90%, and for the other it is 99%. Which is which, and how can you tell?

12.22 The paper **"The Effects of Adolescent Volunteer Activities on the Perception of Local Society and Community Spirit Mediated by Self-Conception"** (*Advanced Science and Technology Letters* [2016]: 19–23) describes a survey of a large representative sample of middle school children in South Korea. One question in the survey asked how much time per year the children spent in volunteer activities. The sample mean was 14.76 hours and the sample standard deviation was 16.54 hours.
a. Based on the reported sample mean and sample standard deviation, explain why it is not reasonable to think that the distribution of volunteer times for the population of South Korean middle school students is approximately normal.
b. The sample size was not given in the paper, but the sample size was described as "large." Suppose that the sample size was 500. Explain why it is reasonable to use a one-sample t confidence interval to estimate the population mean even though the population distribution is not approximately normal.
c. Calculate and interpret a confidence interval for the mean number of hours spent in volunteer activities per year for South Korean middle school children. (Hint: See Example 12.7.)

12.23 Students in a representative sample of 65 first-year students selected from a large university in England participated in a study of academic procrastination (**"Study Goals and Procrastination Tendencies at Different Stages of the Undergraduate Degree,"** *Studies in Higher Education* [2016]: 2028–2043). Each student in the sample completed the Tuckman Procrastination Scale, which measures procrastination tendencies. Scores on this scale can range from 16 to 64, with scores over 40 indicating higher levels of procrastination. For the 65 first-year students in this study, the mean score on the procrastination scale was 37.02 and the standard deviation was 6.44.
a. Construct a 95% confidence interval estimate of μ, the mean procrastination scale for first-year students at this college. (Hint: See Example 12.7.)
b. Based on your interval, is 40 a plausible value for the population mean score? What does this imply about the population of first-year students?

12.24 The paper referenced in the previous exercise also reported that for a representative sample of 68 second-year students at the university, the sample mean procrastination score was 41.00 and the sample standard deviation was 6.82.
a. Construct a 95% confidence interval estimate of μ, the population mean procrastination scale for second-year students at this college.
b. How does the confidence interval for the population mean score for second-year students compare to the confidence interval for first-year students calculated in the previous exercise? What does this tell you about the difference between first-year and second-year students in terms of mean procrastination score?

12.25 The Bureau of Alcohol, Tobacco, and Firearms (ATF) has been concerned about lead levels in California wines. In a previous testing of wine specimens, lead levels ranging from 50 to 700 parts per billion were recorded. How many wine specimens should be tested if the ATF wishes to estimate the mean lead level for California wines with a margin of error of 10 parts per billion? (Hint: See Example 12.10.)

12.26 The paper **"Patterns and Composition of Weight Change in College Freshmen"** (*College Student Journal* [2015]: 553–564) reported that the freshman year weight gain for the students in a representative sample of 103 freshmen at a midwestern university was 5.7 pounds and that the standard deviation of the weight gain was 6.8 pounds. The authors also reported that 75.7% of these students gained more than 1.1 pounds, 17.4% maintained their weight within 1.1 pounds, and 6.8% lost more than 1.1 pounds.
a. Based on the given information, explain why it is not reasonable to think that the distribution of weight gains for the population of freshmen students at this university is approximately normal.
b. Explain why it is reasonable to use a one-sample *t* confidence interval to estimate the population mean even though the population distribution is not approximately normal.
c. Calculate and interpret a confidence interval for the population mean weight gain of freshmen students at this university.

SECTION 12.2 Exercise Set 2

12.27 What percentage of the time will a variable that has a *t* distribution with the specified degrees of freedom fall in the indicated region?
a. 10 df, between −2.23 and 2.23
b. 24 df, between −2.80 and 2.80
c. 24 df, to the right of 2.80

12.28 The formula used to calculate a confidence interval for the mean of a normal population is

$$\bar{x} \pm (t \text{ critical value}) \frac{s}{\sqrt{n}}$$

What is the appropriate *t* critical value for each of the following confidence levels and sample sizes?
a. 90% confidence, $n = 12$
b. 90% confidence, $n = 25$
c. 95% confidence, $n = 10$

12.29 Samples of two different models of cars were selected, and the actual speed for each car was determined when the speedometer registered 50 mph. The resulting 95% confidence intervals for mean actual speed were (51.3, 52.7) for model 1 and (49.4, 50.6) for model 2. Assuming that the two sample standard deviations were equal, which confidence interval is based on the larger sample size? Explain your reasoning.

12.30 *USA TODAY* reported that the average amount of money spent on coffee drinks each month is $78.00 (*USA Snapshot*, November 4, 2016).
a. Suppose that this estimate was based on a representative sample of 20 adult Americans. Would you recommend using the one-sample *t* confidence interval to estimate the population mean amount spent on coffee for the population of all adult Americans? Explain why or why not.

b. If the sample size had been 200, would you recommend using the one-sample *t* confidence interval to estimate the population mean amount spent on coffee for the population of all adult Americans? Explain why or why not.

12.31 Acrylic bone cement is sometimes used in hip and knee replacements to secure an artificial joint in place. The force required to break an acrylic bone cement bond was measured for six specimens, and the resulting mean and standard deviation were 306.09 Newtons and 41.97 Newtons, respectively. Assuming that it is reasonable to believe that breaking force has a distribution that is approximately normal, use a confidence interval to estimate the mean breaking force for acrylic bone cement.

12.32 A manufacturer of college textbooks is interested in estimating the strength of the bindings produced by a particular binding machine. Strength can be measured by recording the force required to pull the pages of a book from its binding. If this force is measured in pounds, how many books should be tested to estimate the average force required to break the binding with a margin of error of 0.1 pound? Assume that σ is known to be 0.8 pound.

12.33 Because of safety considerations, in May 2003, the Federal Aviation Administration (FAA) changed its guidelines for how small commuter airlines must estimate passenger weights. Under the old rule, airlines used 180 pounds as a typical passenger weight (including carry-on luggage) in warm months and 185 pounds as a typical weight in cold months. The *Alaska Journal of Commerce* (May 25, 2003) reported that Frontier Airlines conducted a study to estimate mean passenger plus carry-on weights. They found a sample mean summer weight of 183 pounds and a winter sample mean of 190 pounds. Suppose that these estimates were based on random samples of 100 passengers and that the sample standard deviations were 20 pounds for the summer weights and 23 pounds for the winter weights.
a. Construct and interpret a 95% confidence interval for the population mean summer weight (including carry-on luggage) of Frontier Airlines passengers.
b. Construct and interpret a 95% confidence interval for the population mean winter weight (including carry-on luggage) of Frontier Airlines passengers.
c. The new FAA recommendations are 190 pounds for summer and 195 pounds for winter. Comment on these recommendations in light of the confidence interval estimates from Parts (a) and (b).

ADDITIONAL EXERCISES

12.34 What percentage of the time will a variable that has a *t* distribution with the specified degrees of freedom fall in the indicated region?
a. 5 df, between −2.02 and 2.02
b. 14 df, between −2.98 and 2.98

c. 22 df, outside the interval from -1.72 to 1.72

d. 26 df, to the left of -2.48

12.35 The formula used to calculate a confidence interval for the mean of a normal population when n is small is

$$\bar{x} \pm (t \text{ critical value}) \frac{s}{\sqrt{n}}$$

What is the appropriate t critical value for each of the following confidence levels and sample sizes?

a. 95% confidence, $n = 15$

b. 99% confidence, $n = 20$

c. 90% confidence, $n = 26$

 12.36 *Consumer Reports* gave the following mileage ratings (in miles per gallon) for seven midsize hybrid 2016 model cars (**www.consumerreports.org/cro/cars/new-cars/hybrids-evs/ratings-reliability/ratings-overview.htm, retrieved December 21, 2016**). Is it reasonable to use these data and the t confidence interval of this section to construct a confidence interval for the mean mileage rating of 2016 midsize hybrid cars? Explain why or why not.

38	39	41	39	25	34	36

 12.37 Five students visiting the student health center for a free dental examination during National Dental Hygiene Month were asked how many months had passed since their last visit to a dentist. Their responses were:

6	17	11	22	29

Assuming that these five students can be considered as representative of all students participating in the free checkup program, construct and interpret a 95% confidence interval for the population mean number of months elapsed since the last visit to a dentist for the population of students participating in the program.

12.38 The paper **"The Curious Promiscuity of Queen Honey Bees (*Apis mellifera*): Evolutionary and Behavioral Mechanisms"** (*Annals of Zoology* **[2001]:255–265**) describes a study of the mating behavior of queen honeybees. The following quote is from the paper:

> Queens flew for an average of 24.2 ± 9.21 minutes on their mating flights, which is consistent with previous findings. On those flights, queens effectively mated with 4.6 ± 3.47 males (mean $\pm$ SD).

The intervals reported in the quote from the paper were based on data from the mating flights of $n = 30$ queen honeybees. One of the two intervals reported was identified as a 95% confidence interval for a population mean. Which interval is this? Justify your choice.

12.39 Use the information given in the previous exercise to construct a 95% confidence interval for the mean number of partners on a mating flight for queen honeybees. For purposes of this exercise, assume these 30 queen honeybees are representative of the population of queen honeybees.

12.40 The article **"Americans' Big Debt Burden Growing, Not Evenly Distributed"** (**www.gallup.com, retrieved December 14, 2016**) reported that for a representative sample of Americans born between 1965 and 1971 (known as Generation X), the sample mean number of credit cards owned was 4.5. Suppose that the sample standard deviation (which was not reported) was 1.0 and that the sample size was $n = 300$.

a. Construct a 95% confidence interval for the population mean number of credit cards owned by Generation X Americans.

b. The interval in Part (a) does not include 0. Does this imply that all Generation X Americans have at least one credit card? Explain.

SECTION 12.3   Testing Hypotheses About a Population Mean

In the previous section, you learned how data from a sample can be used to estimate a population mean. Sample data can also be used to test hypotheses about a population mean.

Example 12.11   Visual Aspects of Word Clouds

A word cloud is a visual representation of a text passage that uses characteristics such as font size and color to indicate the importance of a particular word. One program that will create a word cloud from text is called Wordle (found at www.wordle.net). To illustrate what a word cloud looks like, Wordle was used to create a word cloud for the text of Section 12.2, and the result is shown in Figure 12.11.

The authors of the paper **"Seeing Things in the Clouds: The Effect of Visual Features on Tag Cloud Selections"** (*Proceedings of the Nineteenth ACM Conference on Hypertext and Hypermedia*, **[2008]**) carried out a study to evaluate how people view word clouds and which characteristics of the words in a word cloud are most influential. In one part of this study, participants were shown a word cloud and asked to choose a word that they found to be visually important. The actual mean font size of all the words in the word clouds used in the study was 2.0. The researchers reasoned that if font size was not an important factor

FIGURE 12.11
A Wordle word cloud for the text of Section 12.2

in how people select visually important words, the mean font size for selected words should be equal to 2.0, the mean font size of all the words in the clouds. The researchers viewed the participants as a representative sample from the population of people who might view a word cloud and used the sample data to test the following hypotheses

$$H_0 : \mu = 2.0$$
$$H_a : \mu > 2.0$$

where μ represents the mean font size of visually important words for the population. Because they were able to reject the null hypothesis, the researchers concluded that font size was a factor in the selection of visually important words.

The researchers also studied the influence of word color and intensity (the "boldness"). Based on their analysis, they concluded that there was evidence that word intensity was also a factor in the selection of important words, but that word color was not.

The logic of hypothesis testing and a general process for carrying out a hypothesis test were developed in Chapter 10 in the context of testing hypotheses about a population proportion. The same general process (HMC³) is used for all hypothesis tests. What may differ from one type of test to another is

1. The form of the null and alternative hypotheses
2. The test statistic
3. The way in which the associated *P*-value is calculated

Consider each of these three things in the context of testing a hypothesis about a population mean:

1. Hypotheses

When testing hypotheses about a population mean, the null hypothesis will have the form

$$H_0 : \mu = \mu_0$$

where μ_0 is a particular hypothesized value. The alternative hypothesis has one of the following three forms, depending on the research question being addressed:

$$H_a : \mu > \mu_0$$
$$H_a : \mu < \mu_0$$
$$H_a : \mu \neq \mu_0$$

To determine the null and alternative hypotheses, you will need to translate the question of interest into hypotheses. For example, if you want to determine if there is evidence that the population mean is greater than 100, you would choose $H_a : \mu > 100$ as the alternative hypothesis. The null hypothesis would then be $H_0 : \mu = 100$.

2. Test Statistic

If n is large ($n \geq 30$) or if the population distribution is approximately normal, the appropriate test statistic is

$$t = \frac{\bar{x} - \mu_0}{\dfrac{s}{\sqrt{n}}}$$

If the null hypothesis is true, this test statistic has a t distribution with df $= n - 1$. This means that the P-value for a hypothesis test about a population mean will be based on a t distribution and not the standard normal (z) distribution.

3. Calculating a *P*-value

Once the value of the test statistic has been determined, the next step is to calculate a P-value. The P-value tells you how likely it would be to observe sample data as or more extreme than what was observed if the null hypothesis were true. When the null hypothesis is true, the test statistic has a distribution that is approximately a t distribution with df $= n - 1$. This means that the P-value will be calculated as an area under a t distribution curve.

To see how this is done, suppose you are carrying out a hypothesis test where the null hypothesis is $H_0 : \mu = 100$ and the alternative hypothesis is $H_a : \mu > 100$. The test statistic for this test is

$$t = \frac{\bar{x} - 100}{\dfrac{s}{\sqrt{n}}}$$

If a sample of size $n = 24$ resulted in a sample mean of $\bar{x} = 104.19$ and a sample standard deviation of $s = 7.59$, the resulting test statistic value is

$$t = \frac{104.19 - 100}{\dfrac{7.59}{\sqrt{24}}} = \frac{4.19}{1.55} = 2.70$$

Because this is an upper-tailed test, if the test statistic had been z rather than t, the P-value would be the area under the z curve to the right of 2.70. With this t statistic, the P-value is the area to the right of 2.70 under the t curve with df $= 24 - 1 = 23$. Appendix A Table 4 is a tabulation of t curve tail areas. Each column of the table is for a different number of degrees of freedom: 1, 2, 3, ... , 30, 35, 40, 60, 120, and a last column for df $= \infty$, which contains the same tail areas as the z curve. The table gives the area under each t curve to the right of values ranging from 0.0 to 4.0 in increments of 0.1. Part of this table appears in Figure 12.12.

FIGURE 12.12

Part of Appendix A Table 4: tail areas for *t* curves

df / t	1	2	...	22	23	24	...	60	120
0.0									
0.1									
⋮				⋮	⋮	⋮			
2.5			...	.010	.010	.010	...		
2.6			...	.008	.008	.008	...		
2.7			...	.007	.006	.006	...		
2.8			...	.005	.005	.005	...		
⋮				⋮	⋮	⋮			
4.0									

Area under 23-df t curve to right of 2.7

Suppose that $t = 2.7$ for an upper-tailed test based on 23 df. Then

$$P\text{-value} = \text{area to the right of } 2.7 = 0.006$$

Because the *t* curve is symmetric around 0, the *P*-value for a lower-tail test with a test statistic value of $t = -2.70$ would also be 0.006. As with *z* tests, you double a tail area to obtain the *P*-value for a two-tailed *t* test. For example, if $t = 2.6$ or if $t = -2.6$ for a two-tailed test with 23 df, then

$$P\text{-value} = 2(0.008) = 0.016$$

Once past 120 df, the tail areas don't change much, so the last column (∞) in Appendix A Table 4 provides a good approximation.

The following box summarizes how the *P*-value is obtained as a *t* curve area.

Finding *P*-Values for a *t* Test

1. Upper-tailed test:
   $H_a: \mu >$ hypothesized value

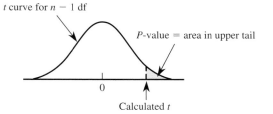

2. Lower-tailed test:
   $H_a: \mu <$ hypothesized value

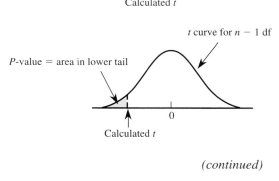

(continued)

3. **Two-tailed test:**

   $H_a: \mu \neq$ hypothesized value

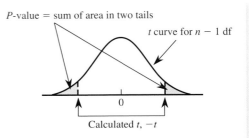

Appendix A Table 4 gives upper-tail t curve areas to the right of values 0.0, 0.1, ... , 4.0. These areas are P-values for upper-tailed tests and, by symmetry, also for lower-tailed tests. Doubling an area gives the P-value for a two-tailed test.

You now have all the pieces needed to carry out the hypothesis test. The one-sample t test for a population mean, summarized in the following box, is the method you should consider when the answers to the four key questions that lead to a recommended method are: *hypothesis testing, sample data, one numerical variable,* and *one sample.*

One-Sample *t* Test for a Population Mean

Appropriate when the following conditions are met:

1. The sample is a random sample from the population of interest or the sample is selected in a way that would result in a sample that is representative of the population.
2. The sample size is large ($n \geq 30$) *or* the population distribution is approximately normal.

When these conditions are met, the following test statistic can be used:

$$t = \frac{\bar{x} - \mu_0}{\dfrac{s}{\sqrt{n}}}$$

where μ_0 is the hypothesized value from the null hypothesis.

Form of the null hypothesis: $H_0 : \mu = \mu_0$

When the conditions above are met and the null hypothesis is true, the t test statistic has a t distribution with df $= n - 1$.

Associated *P*-value:

When the alternative hypothesis is...	The *P*-value is...
$H_a : \mu > \mu_0$	Area under the t curve to the right of the calculated value of the test statistic
$H_a : \mu < \mu_0$	Area under the t curve to the left of the calculated value of the test statistic
$H_a : \mu \neq \mu_0$	2 (area to the right of t) if t is positive or 2 (area to the left of t) if t is negative

---

Now you're ready to look at some examples, following the five-step process for hypothesis testing problems (HMC³).

Example 12.12   **Time Perception During Nicotine Withdrawal**

The authors of the paper **"Sex Differences in Time Perception During Smoking Abstinence"** (*Nicotine and Tobacco Resear*ch **[2015]: 449–454**) carried out a study to investigate how nicotine withdrawal affects time perception and decision-making. In this study, $n = 21$ male smokers were asked to abstain from smoking for 24 hours. They were shown a demo screen with a green cross that changed to a red cross after a period of time. They were then shown the green cross and asked to indicate when they thought the same amount of time had passed as in the demo. This process was repeated 15 more times with varying times and then a time discrimination score was calculated as follows for each of the 21 men.

$$time\ discrmination\ score = \frac{total\ estimated\ time}{total\ actual\ time\ of\ demos}$$

A time discrimination score greater than 1 would result for someone who tended to overestimate the actual times and a score of less than 1 would result for someone who tended to underestimate the actual times.

Suppose that the resulting data were as follows (these data are artificial but are consistent with summary quantities given in the paper):

| 1.12 | 1.03 | 1.09 | 1.03 | 1.09 | 0.97 | 0.98 | 1.20 | 1.16 | 1.03 | 1.10 |

| 1.11 | 0.98 | 1.02 | 1.20 | 0.96 | 0.78 | 1.05 | 0.90 | 1.08 | 0.95 |

These data were used to calculate the sample mean and standard deviation:

$$n = 21 \qquad \bar{x} = 1.04 \qquad s = 0.100$$

Suppose that it is reasonable to consider the people in this sample as representative of male smokers in general. This data can be used to determine if there is evidence that male smokers tend to overestimate time after having abstained from smoking for 24 hours.

Begin by answering the four key questions (QSTN) for this problem:

Q **Question Type**	Estimation or hypothesis testing?	Hypothesis testing. You want to test a claim about a population (time is overestimated after abstaining from smoking for 24 hours).
S **Study Type**	Sample data or experiment data?	Sample data. The data are from a sample of male smokers.
T **Type of Data**	One variable or two? Categorical or numerical?	One numerical variable (time discrimination score).
N **Number of Samples or Treatments**	How many samples or treatments?	One.

Because the answers are *hypothesis testing, sample data, one numerical variable,* and *one sample*, you should consider a one-sample *t* test for a population mean.

Next, use the five-step process for hypothesis testing problems (HMC³).

Process Step	
H Hypotheses	The claim is about the population of male smokers and the population characteristic of interest is a population mean: μ = mean time discrimination score for male smokers who have not smoked for 24 hours. The question of interest (do male smokers who have abstained from smoking for 24 hours tend to overestimate time?) translates into an alternative hypothesis of $\mu > 1$, because time discrimination scores greater than 1 correspond to a tendency to overestimate time. The null hypothesis is found by replacing the inequality ($>$) in the alternative hypothesis by an equal sign. **Hypotheses:** *Null hypothesis*: $H_0 : \mu = 1$ *Alternative hypothesis*: $H_a : \mu > 1$

(continued)

Process Step	
M Method	Because the answers to the four key questions are hypothesis testing, sample data, one numerical variable and one sample, consider a one-sample t test for a population mean. **Potential method:** One-sample t test for a population mean. The test statistic for this test is $$t = \frac{\bar{x} - 1}{\dfrac{s}{\sqrt{n}}}$$ The value of 1 in the test statistic is the hypothesized value from the null hypothesis. When the null hypothesis is true, this statistic will have a t distribution with df $= 21 - 1 = 20$. You also need to select a significance level for the test. In some cases, a significance level will be specified. If that is not the case, you should choose a significance level based on a consideration of the consequences of Type I and Type II errors. In this example, a Type I error is deciding that male smokers who have not smoked in 24 hours do tend to overestimate time when they really don't. A Type II error is concluding that there is no evidence that male smokers who have not smoked for 24 hours tend to overestimate time when they really do overestimate. In this situation, because neither type of error is much more serious than the other, a value for α of 0.05 is a reasonable choice. **Significance level:** $\alpha = 0.05$
C Check	There are two conditions that need to be met in order to use the one-sample t test for a population mean. The requirement of a random sample or a sample that can be regarded as representative of the population requires some thought. If the researchers conducting the study selected the sample in a way that would result in a sample that was representative of male smokers in general, then it is reasonable to proceed. The second condition is that the sample size is large or the population distribution is approximately normal. Because n is only 21 in this example, you need to verify that the normality condition is reasonable. A boxplot of the sample data is shown here. 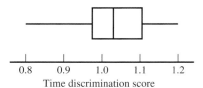 Time discrimination score Although the boxplot is not perfectly symmetric, it is not too skewed and there are no outliers. It is reasonable to think that the population distribution is at least approximately normal. Because both conditions are met, it is appropriate to use the one-sample t test.
C Calculate	$n = 21$ $\bar{x} = 1.04$ $s = 0.100$ **Test statistic:** $$t = \frac{1.04 - 1.00}{\dfrac{0.100}{\sqrt{21}}} = \frac{0.04}{0.022} = 1.82$$ This is an upper-tailed test (the inequality in H_a is $>$), so the P-value is the area under the t curve with df $= 20$ to the right of the calculated t value. From Appendix A Table 4, the area to the right of 1.82 (using 1.8, which is the closest value to 1.82 that appears in the table) is approximately 0.043. **Associated P-value:** P-value $=$ area under t curve to the right of 1.82 $\quad\quad = P(t > 1.82)$ $\quad\quad \approx 0.043$ Computer software or a graphing calculator could also be used to carry out the calculate step of the hypothesis test process. For example, in the accompanying Minitab output, you can see that the value of the test statistic is given as $t = 1.83$ and the associated P-value is given as 0.041.

(continued)

Process Step	
	One-Sample T

Test of $\mu = 1$ vs > 1

N	Mean	StDev	SE Mean	95% Lower Bound	T	P
21	1.0400	0.1000	0.0218	1.0024	1.83	0.041

The difference between what is given in the Minitab output and what was obtained by calculating the value of the test statistic and using Appendix A Table 4 is due to differences in rounding in the calculations.

C Communicate Results

Because the *P*-value is less than the selected significance level, the null hypothesis is rejected.

Decision: $0.043 < 0.05$, Reject H_0.

Conclusion: There is convincing evidence that the mean time discrimination score is greater than 1. It would be very unlikely to see a sample mean resulting in a *t* value at least this extreme just by chance when H_0 is true. This means that there is convincing evidence that the male smokers who have abstained from smoking for 24 hours do tend to overestimate time.

Example 12.13 **Goofing Off at Work**

Data set available

Many employers are concerned about employees wasting time by surfing the Internet and e-mailing friends during work hours. The article **"Who Goofs Off 2 Hours a Day? Most Workers, Survey Says" (San Luis Obispo Tribune, August 3, 2006)** summarized data from a large sample of workers. Suppose the CEO of a large company wants to determine whether the mean wasted time during an 8-hour work day for employees of her company is less than the mean of 120 minutes reported in the article. Each person in a random sample of 10 employees is asked about daily wasted time at work (in minutes). Participants would probably have to be guaranteed anonymity to obtain truthful responses. Suppose the resulting data are:

$$108 \quad 112 \quad 117 \quad 130 \quad 111 \quad 131 \quad 113 \quad 113 \quad 105 \quad 128$$

The sample mean and standard deviation calculated from these data are

$$n = 10 \qquad \bar{x} = 116.8 \qquad s = 9.45$$

Do these data provide evidence that the mean wasted time for this company is less than 120 minutes?

Considering the four key questions (QSTN), this situation can be described as *hypothesis testing, sample data, one numerical variable (wasted time),* and *one sample.* This combination suggests a one-sample *t* test for a population mean.

H Hypotheses

You want to use data from the sample to test a claim about the population of company employees. The population characteristic of interest is

μ = population mean daily wasted time for employees of this company

Translating the question of interest into hypotheses gives

$$H_0 : \mu = 120$$
$$H_a : \mu < 120$$

M Method

Because the answers to the four key questions are hypothesis testing, sample data, one numerical variable, and one sample, consider a one-sample *t* test for a population mean. The test statistic for this test is

$$t = \frac{\bar{x} - 120}{\dfrac{s}{\sqrt{n}}}$$

The value of 120 in the test statistic is the hypothesized value from the null hypothesis. When the null hypothesis is true, this statistic will have a *t* distribution with df = $10 - 1 = 9$.

You also need a significance level for the test. For this example, a significance level of $\alpha = 0.05$ will be used.

Significance level: $\alpha = 0.05$

C Check   The one-sample t test requires a random sample and either a large sample or a normal population distribution. The sample was a random sample of employees. Because the sample size is small, you must be willing to assume that the population distribution of times is at least approximately normal. The following normal probability plot appears to be reasonably straight, and although the normal probability plot and the boxplot reveal some skewness in the sample, there are no outliers.

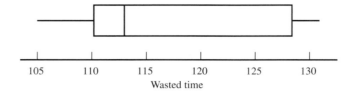

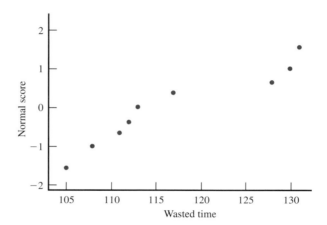

Based on these observations, it is plausible that the population distribution is approximately normal, so you can proceed with the one-sample t test.

C Calculate
$$n = 10 \qquad \bar{x} = 116.8 \qquad s = 9.45$$
$$t = \frac{116.8 - 120}{\dfrac{9.45}{\sqrt{10}}} = -1.07$$

From the df $= 9$ column of Appendix A Table 4 and rounding the test statistic value to -1.1, you get

$$P\text{-value} = \text{area to the left of } -1.1 = \text{area to the right of } 1.1 = 0.150$$

as shown:

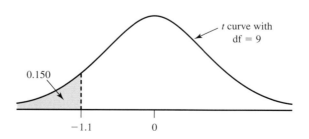

C Communicate Results   Because the P-value is greater than the selected significance level, you fail to reject the null hypothesis.

Decision:

$0.150 \geq 0.05$, Fail to reject H_0.

Interpretation:

There is not convincing evidence that the population mean wasted time per 8-hour work day for employees at this company is less than 120 minutes.

Software or a graphing calculator could also have been used to carry out the test of Example 12.13, as shown in the accompanying JMP output:

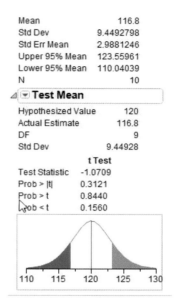

Mean	116.8
Std Dev	9.4492798
Std Err Mean	2.9881246
Upper 95% Mean	123.55961
Lower 95% Mean	110.04039
N	10

Test Mean

Hypothesized Value	120
Actual Estimate	116.8
DF	9
Std Dev	9.44928

t Test

Test Statistic	-1.0709
Prob > |t|	0.3121
Prob > t	0.8440
Prob < t	0.1560

Although you had to round the calculated t value to -1.1 to use Appendix A Table 4, JMP was able to compute the P-value corresponding to the actual value of the test statistic, which is P-value $= 0.156$ (this appears in the JMP output as Prob $< t$ just above the figure in the output; the $<$ probability is the P-value because this was a lower-tail test).

Example 12.14	**Cricket Love**

The article **"Well-Fed Crickets Bowl Maidens Over"** (*Nature Science Update*, **February 11, 1999**) reported that female field crickets are attracted to males that have high chirp rates and hypothesized that chirp rate is related to nutritional status. The chirp rates for male field crickets were reported to vary around a mean of 60 chirps per second. To investigate whether chirp rate is related to nutritional status, investigators fed male crickets a high protein diet for 8 days, and then measured chirp rate. The sample mean chirp rate for the crickets on the high-protein diet was reported to be 109 chirps per second. Is this convincing evidence that the mean chirp rate for crickets on a high-protein diet is greater than 60 (which would then imply an advantage in attracting females)? Suppose that the sample size and sample standard deviation are $n = 32$ and $s = 40$. You can test the relevant hypotheses to answer this question. A significance level of $\alpha = 0.01$ will be used.

Considering the four key questions (QSTN), this situation can be described as *hypothesis testing, sample data, one numerical variable (chirp rate),* and *one sample*. This combination suggests a one-sample t test for a population mean.

H Hypotheses   The claim to be tested is about a population mean:

$$\mu = \text{population mean chirp rate for crickets on a high protein diet}$$

The question of interest is if there is evidence that $\mu > 60$, so the hypotheses to be tested are

$$H_0 : \mu = 60$$
$$H_a : \mu > 60$$

M Method Because the answers to the four key questions are hypothesis testing, sample data, one numerical variable, and one sample, consider a one-sample t test for a population mean. The test statistic for this test is

$$t = \frac{\bar{x} - 60}{\dfrac{s}{\sqrt{n}}}$$

The value of 60 in the test statistic is the hypothesized value from the null hypothesis. When the null hypothesis is true, this statistic will have a t distribution with df $= 32 - 1 = 31$.

The significance level specified for this test is $\alpha = 0.01$.

Significance level: $\alpha = 0.01$

C Check The one-sample t test requires a random or representative sample and either a large sample or a normal population distribution. Because the sample size is large ($n = 32$), it is reasonable to proceed with the t test as long as you are willing to consider the 32 male field crickets in this study to be representative of the population of male field crickets.

C Calculate

$$n = 32$$
$$\bar{x} = 109$$
$$s = 40$$

$$t = \frac{109 - 60}{\dfrac{40}{\sqrt{32}}} = \frac{49}{7.07} = 6.93$$

This is an upper-tailed test, so the P-value is the area under the t curve with df $= 31$ and to the right of 6.93. From Appendix A Table 4 or using technology, the P-value ≈ 0.

C Communicate Results Because the P-value ≈ 0, which is less than the significance level, $\alpha = 0.01$, H_0 is rejected.

Decision:
$0 < 0.01$, Reject H_0.

Interpretation: There is convincing evidence that the population mean chirp rate for male field crickets that eat a high-protein diet is greater than 60.

Statistical Versus Practical Significance

Carrying out a hypothesis test amounts to deciding whether the value obtained for the test statistic could plausibly have resulted when H_0 is true. When the value of the test statistic leads to rejection of H_0, it is customary to say that the result is **statistically significant** at the chosen significance level α. The finding of statistical significance means that the observed deviation from what was expected when H_0 is true can't be explained by just chance variation. However, statistical significance does not necessarily mean that the true situation differs from what the null hypothesis states in any practical sense. That is, even after H_0 has been rejected, the data may suggest that there is no *practical* difference between the actual value of the population characteristic and the hypothesized value. This is illustrated in Example 12.15.

Example 12.15 "Significant" but Unimpressive Test Score Improvement

Let μ denote the true mean score on a standardized test for children in a large school district. The population mean score for all children in the United States is known to be 100. District administrators are interested in testing $H_0 : \mu = 100$ versus $H_a : \mu > 100$ using a significance level of 0.001. Data from a random sample of 2500 children in the district resulted in $\bar{x} = 101.0$ and $s = 15.00$. Minitab output from a one-sample t test is shown here:

One-Sample T
Test of mu = 100 vs > 100

N	Mean	StDev	SE Mean	95% Lower Bound	T	P
2500	101.000	15.000	0.300	100.506	3.33	0.000

From the Minitab output, the P-value ≈ 0, so H_0 is rejected. The actual mean score for this district does appear to be greater than 100. However, with $n = 2500$, you expect that $\bar{x} = 101.0$ is close to the actual value of μ. It appears that H_0 was rejected because μ is about 101 rather than 100. From a practical point of view, a 1-point difference here may not be important. A statistically significant result does not necessarily mean that there are any practical consequences.

Summing It Up—Section 12.3

The following learning objectives were addressed in this section:

Mastering the Mechanics
M7: Translate a research question or claim about a population mean into null and alternative hypotheses.
When testing hypotheses about a population mean, sample data are used to decide if there is support for a claim about the mean. The claim about the population mean or the question that you are trying to answer will determine what hypotheses are tested. Examples 12.12 and 12.13 illustrate how a research question about a population mean is translated into null and alternative hypotheses.

M8: Use the five-step process for hypothesis testing problems (HMC³) to carry out a t test of hypotheses about a population mean.
The five steps in the process for hypothesis testing are Hypotheses, Method, Check, Calculate, and Communicate (HMC³). Examples 12.12, 12.13, and 12.14 all illustrate how the steps in the process are followed to reach a conclusion in a hypothesis testing situation.

Putting It into Practice
P2: Carry out a t test of hypotheses about a population mean and interpret the conclusion in context.
Examples 12.12, 12.13, and 12.14 all illustrate hypothesis tests for a population mean, including interpreting the result of the test in context.

SECTION 12.3 EXERCISES

Each Exercise Set assesses the following chapter learning objectives: M7, M8, P2

SECTION 12.3 Exercise Set 1

12.41 Give as much information as you can about the P-value of a t test in each of the following situations. (Hint: See discussion on page 594.)
a. Upper-tailed test, df = 8, $t = 2.0$
b. Lower-tailed test, df = 10, $t = -2.4$
c. Lower-tailed test, $n = 22$, $t = -4.2$
d. Two-tailed test, df = 15, $t = -1.6$

12.42 The paper **"Playing Active Video Games Increases Energy Expenditure in Children"** (*Pediatrics* [2009]: 534–539) describes a study of the possible cardiovascular benefits of active video games. Mean heart rate for healthy boys ages 10 to 13 after walking on a treadmill at 2.6 km/hour for 6 minutes is known to be 98 beats per minute (bpm). For each of 14 boys in this age group, heart rate was measured after 15 minutes of playing Wii Bowling. The resulting sample mean and standard deviation were 101 bpm and 15 bpm, respectively. Assume that the sample of boys is representative of boys age 10 to 13 and that the distribution of heart rates after 15 minutes of Wii Bowling is approximately normal. Does the sample provide convincing evidence that the mean heart rate after 15 minutes of Wii Bowling is different from the known mean heart rate after 6 minutes walking on the treadmill?

 Data set available

Carry out a hypothesis test using $\alpha = 0.01$. (Hint: See Example 12.12.)

12.43 The paper referenced in the previous exercise also states that the known mean resting heart rate for boys in this age group is 66 bpm.
a. Is there convincing evidence that the mean heart rate after Wii Bowling for 15 minutes is higher than the known mean resting heart rate for boys of this age? Use $\alpha = 0.01$.
b. Based on the outcomes of the tests in this exercise and the previous exercise, write a paragraph comparing treadmill walking and Wii Bowling.

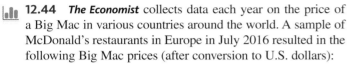 **12.44** *The Economist* collects data each year on the price of a Big Mac in various countries around the world. A sample of McDonald's restaurants in Europe in July 2016 resulted in the following Big Mac prices (after conversion to U.S. dollars):

4.44   3.15   2.42   3.96   4.35   4.51   4.17   3.69   4.62
3.80   3.36   3.85

The mean price of a Big Mac in the U.S. in July 2016 was $5.04. For purposes of this exercise, you can assume it is reasonable to regard the sample as representative of European McDonald's restaurants. Does the sample provide convincing evidence that the mean July 2016 price of a Big Mac in Europe is less than the reported U.S. price? Test the relevant hypotheses using $\alpha = 0.05$. (Hint: See Example 12.12.)

12.45 In a study of media use, each person in a large representative sample of male Canadian high school students was asked how much time they spent playing video or computer games (in minutes per day). The sample mean was 123.4 minutes and the sample standard deviation was 117.1 minutes.
a. Based on the given sample mean and standard deviation, do you think that it is reasonable to think that the distribution of time spent playing video or computer games for the population of male Canadian high school students is approximately normal? Explain why or why not.
b. Suppose you wanted to use the sample data to decide if there is evidence that the mean time spent playing video or computer games for male Canadian high school students is greater than 2 hours (120 minutes). What would you need to know to determine if the one-sample t test is an appropriate method?
c. Suppose that the sample size was 500. Carry out a hypothesis test to decide if there is evidence that the mean time spent playing video or computer games for male Canadian high school students is greater than 2 hours. Use a significance level of 0.05.
d. Now also suppose that the sample standard deviation had been 37.1 rather than 117.1. Carry out a hypothesis test to decide if there is evidence that the mean time spent playing video or computer games for male Canadian high school students is greater than 2 hours. Use a significance level of 0.05.
e. Explain why the null hypothesis was rejected in the test of Part (d) but not in the test of Part (c).

12.46 The paper titled **"Music for Pain Relief"** (*The Cochrane Database of Systematic Reviews*, **April 19, 2006**) concluded, based on a review of 51 studies of the effect of music on pain intensity, that "Listening to music reduces pain intensity levels … However, the magnitude of these positive effects is small, and the clinical relevance of music for pain relief in clinical practice is unclear." Are the authors of this paper claiming that the pain reduction attributable to listening to music is not statistically significant, not practically significant, or neither statistically nor practically significant? Explain. (Hint: See discussion on page 602.)

SECTION 12.3 **Exercise Set 2**

12.47 Give as much information as you can about the P-value of a t test in each of the following situations:
a. Two-tailed test, df $= 9$, $t = 0.73$
b. Upper-tailed test, df $= 10$, $t = 0.5$
c. Lower-tailed test, $n = 20$, $t = -2.1$
d. Two-tailed test, $n = 40$, $t = 1.7$

12.48 The report **"2016 Salary Survey Executive Summary"** (National Association of Colleges and Employers, www.naceweb .org/uploadedfiles/files/2016/publications/executive -summary/2016-nace-salary-survey-fall-executive-summary. pdf, retrieved December 24, 2016) states that the mean yearly salary offer for students graduating with mathematics and statistics degrees in 2016 was $62,985. Suppose that a random sample of 50 math and statistics graduates at a large university who received job offers resulted in a mean offer of $63,500 and a standard deviation of $3300. Do the sample data provide strong support for the claim that the mean salary offer for math and statistics graduates of this university is higher than the 2016 national average of $62,985? Test the relevant hypotheses using $\alpha = 0.05$.

12.49 The report **"Majoring in Money: How American College Students Manage Their Finances"** (SallieMae, 2016, www.news/salliemae.com, retrieved December 24, 2106) includes data from a survey of college students. Each person in a representative sample of 793 college students was asked if they had one or more credit cards and if so, whether they paid their balance in full each month. There were 500 who paid in full each month. For this sample of 500 students, the sample mean credit card balance was reported to be $825. The sample standard deviation of the credit card balances for these 500 students was not reported, but for purposes of this exercises, suppose that it was $200. Is there convincing evidence that college students who pay their credit card balance in full each month have mean balance that is lower than $906, the value reported for all college students with credit cards? Carry out a hypothesis test using a significance level of 0.01.

12.50 The authors of the paper **"Changes in Quantity, Spending, and Nutritional Characteristics of Adult, Adolescent and Child Urban Corner Store Purchases After an Environmental Intervention"** (*Preventative Medicine* [2015]: 81–85) wondered if increasing the availability of healthy food options would also increase the amount people spend at the corner store. They collected data from a representative sample of 5949 purchases at corner stores in Philadelphia after the stores increased their healthy food options. The sample mean amount spent for this sample of purchases was $2.86 and the sample standard deviation was $5.40.

a. Notice that for this sample, the sample standard deviation is greater than the sample mean. What does this tell you about the distribution of purchase amounts?

b. Before the stores increased the availability of healthy foods, the population mean total amount spent per purchase was thought to be about $2.80. Do the data from this study provide convincing evidence that the population mean amount spent per purchase is greater after the change to increased healthy food options? Carry out a hypothesis test with a significance level of 0.05.

12.51 Many consumers pay careful attention to stated nutritional contents on packaged foods when making purchases, so it is important that the information on packages be accurate. Suppose that a random sample of $n = 12$ frozen dinners of a certain type was selected and the calorie content of each one was determined. Below are the resulting observations, along with a boxplot and a normal probability plot.

255	244	239	242	265	245	259
248	225	226	251	233		

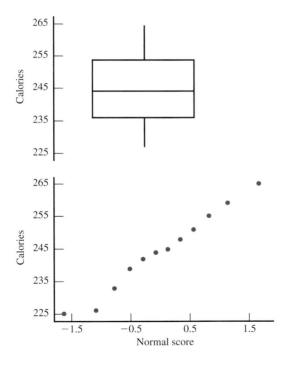

a. Is it reasonable to test hypotheses about the population mean calorie content μ using a one-sample t test? Explain why or why not.

b. The manufacturer claims that the mean calorie content for this particular type of frozen dinner is 240. Does the

boxplot suggest that actual mean content differs from the stated value? Explain your reasoning.

c. Carry out a formal test of the hypotheses suggested in Part (b).

ADDITIONAL EXERCISES

12.52 Give as much information as you can about the P-value of a t test in each of the following situations:

a. Two-tailed test, $n = 16$, $t = 1.6$
b. Upper-tailed test, $n = 14$, $t = 3.2$
c. Lower-tailed test, $n = 20$, $t = -5.1$
d. Two-tailed test, $n = 16$, $t = 6.3$

12.53 The eating habits of $n = 12$ bats were examined in the article **"Foraging Behavior of the Indian False Vampire Bat"** (*Biotropica* [1991]: 63–67). These bats consume insects and frogs. For these 12 bats, the mean time to consume a frog was $\bar{x} = 21.9$ minutes. Suppose that the standard deviation was $s = 7.7$ minutes. Is there convincing evidence that the mean supper time of a vampire bat whose meal consists of a frog is greater than 20 minutes? What assumptions must be reasonable for the one-sample t test to be appropriate?

12.54 The authors of the paper **"Mean Platelet Volume Could Be Possible Biomarker in Early Diagnosis and Monitoring of Gastric Cancer"** (*Platelets* [2014]: 592–594) wondered if mean platelet volume (MPV) might be a way to distinguish patients with gastric cancer from patients who did not have gastric cancer. MPV was recorded for 31 patients with gastric cancer. The sample mean was reported to be 8.31 femtoliters (fL) and the sample standard deviation was reported to be 0.78 fL. For healthy people, the mean MPV is 7.85 fL. Is there convincing evidence that the mean MPV for patients with gastric cancer is greater than 7.85 fL? For purposes of this exercise, you can assume that the sample of 31 patients with gastric cancer is representative of the population of all patients with gastric cancer.

12.55 People in a random sample of 236 students enrolled at a liberal arts college were asked questions about how much sleep they get each night (**"Alcohol Consumption, Sleep, and Academic Performance Among College Students,"** *Journal of Studies on Alcohol and Drugs* [2009]: 355–363). The sample mean sleep duration (average hours of daily sleep) was 7.71 hours and the sample standard deviation was 1.03 hours. The recommended number of hours of sleep for college-age students is 8.4 hours per day. Is there convincing evidence that the population mean daily sleep duration for students at this college is less than the recommended number of 8.4 hours? Test the relevant hypotheses using $\alpha = 0.01$.

12.56 *USA TODAY* (October 14, 2016) reported that Americans spend 4.1 hours per weekday checking work e-mail. This was an estimate based on a survey of 1004 white-collar workers in the United States.

a. Suppose that you would like to know if there is evidence that the mean time spent checking work e-mail for white-collar workers in the United States is more than half of the 8-hour work day. What would you need to assume about the sample in order to use the given sample data to answer this question?

b. Given that any concerns about the sample were satisfactorily addressed, carry out a test to decide if there is evidence that the mean time spent checking work e-mail for white-collar workers in the United States is more than half of the 8-hour work day. Suppose that the sample standard deviation was $s = 1.3$ hours.

12.57 An automobile manufacturer decides to carry out a fuel efficiency test to determine if it can advertise that one of its models achieves 30 mpg (miles per gallon). Six people each drive a car from Phoenix to Los Angeles. The resulting fuel efficiencies (in miles per gallon) are:

27.2   29.3   31.2   28.4   30.3   29.6

Assuming that fuel efficiency is normally distributed, do these data provide evidence against the claim that actual mean fuel efficiency for this model is (at least) 30 mpg?

12.58 A student organization uses the proceeds from a soft drink vending machine to finance its activities. The price per can was $0.75 for a long time, and the mean daily revenue during that period was $75.00. The price was recently increased to $1.00 per can. A random sample of $n = 20$ days after the price increase yielded a sample mean daily revenue and sample standard deviation of $70.00 and $4.20, respectively. Does this information suggest that the mean daily revenue has decreased from its value before the price increase? Test the appropriate hypotheses using $\alpha = 0.05$.

12.59 A hot tub manufacturer advertises that a water temperature of 100°F can be achieved in 15 minutes or less. A random sample of 25 tubs is selected, and the time necessary to achieve a 100°F temperature is determined for each tub. The sample mean time and sample standard deviation are 17.5 minutes and 2.2 minutes, respectively. Does this information cast doubt on the company's claim? Carry out a test of hypotheses using significance level 0.05.

Data set available

SECTION 12.4 Simulation-Based Inference for One Mean (Optional)

Bootstrap Confidence Intervals for One Population Mean

The sample mean $\bar{x}$ is an estimate of the population mean μ, but you don't expect that the value of $\bar{x}$ will be exactly equal to the population mean. However, for representative samples, you expect that the value of $\bar{x}$ will be near the value of μ. The notion of being near the value of a population mean can be represented by a confidence interval.

In Section 12.2 you learned about the margin of error based on the sampling distribution of $\bar{x}$, and you saw how the margin of error is used to construct confidence intervals. But the methods introduced in Section 12.2 depend on the sampling distribution of $\bar{x}$ being at least approximately normal, and this is only assured for samples known to be from normal populations, or for large sample sizes. If you are not certain that the assumptions for a large-sample confidence interval for μ are reasonable, you can use a simulation-based approach called **bootstrapping**.

Consider a hypothetical population that is similar to the population that your observed sample is actually from. By examining the distribution of sample means for samples taken from this hypothetical population, it is possible to use bootstrapping to find an confidence interval for the population mean. Bootstrapping uses random samples (called bootstrap samples) from a hypothetical population represented by the data in the observed sample, which is thought to be representative of the population. The distribution of sample means from the bootstrap samples is called a **bootstrap distribution**. The variability in this distribution indicates how far $\bar{x}$ values for samples from this hypothetical population might be from the original observed value of $\bar{x}$. Knowing how far simulated $\bar{x}$ values tend to fall from the observed value of $\bar{x}$ provides information about how far the observed value of $\bar{x}$ is likely to be from the value of μ in the actual population.

For a 95% confidence level, using the boundaries that capture the middle 95% of the bootstrap distribution is equivalent to getting a value for $\bar{x}$ minus a number, and for $\bar{x}$ plus a number (although the numbers subtracted and added won't always be the same since the bootstrap distribution may not be symmetric). This interval is a **bootstrap confidence interval estimate for μ**, and it is based on simulation rather than on knowing that the sampling distribution $\bar{x}$ is at least approximately normal.

Data set available

This method is called "bootstrapping" because it is like "pulling yourself up by the bootstraps" in the sense that you are using nothing more than the original data you collected in one sample to generate information about the sample-to-sample variability of the sample mean, $\bar{x}$.

Example 12.16   Selfish Chimps? (Revisited)

Data set available

Recall the data presented in Example 12.9 of Section 12.2, regarding the number of times out of 36 trials that each of $n = 7$ chimps chose the option to provide food to both chimps (the "charitable" response).

The original data values are displayed in the table below. Recall that the sample mean of this original sample is $\bar{x} = 21.29$.

Chimp ID	1	2	3	4	5	6	7
Trials	23	22	21	24	19	20	20

You begin by resampling from this original sample. That is, you select bootstrap samples, by selecting at random *with* replacement from the original sample. Here is one bootstrap sample from the original sample:

Resampled Chimp ID	3	2	1	4	3	5	1
Trials	21	22	23	24	21	19	23

The sample mean for this bootstrap sample is $\bar{x} = 21.86$. You can repeat this process many times, and the resulting bootstrap distribution of $\bar{x}$ values provides information about sampling variability that can be used to find confidence intervals. Here is output from the Shiny app titled "Bootstrap Confidence Interval for One Mean," used to obtain a 99% confidence interval for μ. This Shiny app can be found in the app collection at statistics .cengage.com/Peck2e/Apps.html.

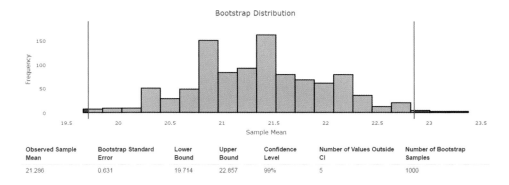

Observed Sample Mean	Bootstrap Standard Error	Lower Bound	Upper Bound	Confidence Level	Number of Values Outside CI	Number of Bootstrap Samples
21.286	0.631	19.714	22.857	99%	5	1000

The bootstrap approach produces a simulation-based bootstrap confidence interval for the population mean, μ, from 19.71 to 22.86 trials. You can be 99% confident that this interval contains the population mean. This interval represents a set of plausible values for the population mean, μ.

The EMC[3] approach applies, just as in Example 12.9.

E Estimate   The mean number of times (out of 36) that chimps choose the charitable response will be estimated.

M Method   You can use a resampling approach to determine a 99% bootstrap confidence interval for the population mean, μ.

C Check   The only condition required for the bootstrap resampling approach is that the original sample of chimps be a random sample or a sample that is representative of the population of interest. It was stated that the sample is representative, so this condition is satisfied.

C Calculate   A resampling approach was used to determine a 99% bootstrap confidence interval for the population mean number of times (out of 36) that chimps choose the charitable response. The bootstrap confidence interval was (19.71, 22.86).

C Communicate Results You can be 99% confident that the population mean number of charitable responses is between 19.71 and 22.86 trials. This is a set of plausible values for the population mean, μ. Notice that the one-sample t 99% confidence interval calcuated in Example 12.9 is (18.76, 23.81). The bootstrap confidence interval is slightly narrower than the one-sample t confidence interval.

Randomization Tests of Hypotheses About a Population Mean

A randomization approach can also be used to approximate P-values for tests of hypotheses about one population mean.

Example 12.17 Goofing Off at Work (Revisited)

Data set available

Recall the data presented in Example 12.13 of Section 12.3, regarding the amount of time a random sample of $n = 10$ employees spent wasting time at work, in minutes, on one day. The original sample data values are:

Employee ID	1	2	3	4	5	6	7	8	9	10
Minutes	108	112	117	130	111	131	113	113	105	128

The observed sample mean from the original sample is $\bar{x} = 116.8$ minutes.

Recall that the CEO wanted to determine whether the mean wasted time per day at her company is less than 120 minutes. The distance between the claimed mean of 120 minutes and the sample mean from the original sample is $120 - 116.8 = 3.2$ minutes.

The data values in the original sample can be shifted to represent a sample from a hypothetical population with mean $\mu = 120$ minutes by adding 3.2 minutes to each, as follows:

Employee ID	1	2	3	4	5	6	7	8	9	10
Minutes	108.0	112.0	117.0	130.0	111.0	131.0	113.0	113.0	105.0	128.0
Minutes + 3.2	111.2	115.2	120.2	133.2	114.2	134.2	116.2	116.2	108.2	131.2

You can then follow the HMC[3] process steps to complete the test of hypotheses.

H Hypotheses Let $\mu =$ Mean daily wasted time for employees of this company
The relevant hypotheses are

$$H_0: \mu = 120 \text{ minutes}$$
$$H_a: \mu < 120 \text{ minutes}$$

M Method A randomization approach based on resampling can be used in this example, instead of a one-sample t-test.

C Check The sample was a random sample of employees in the company, and so the only required condition is satisfied.

C Calculate Begin by resampling from the sample that has been shifted to have a mean of 120 minutes. That is, you can draw new simulated samples, selected at random with replacement from the shifted sample. Here is one bootstrap sample from the shifted sample:

Employee ID	9	2	5	9	3	2	2	2	1	4
Minutes + 3.2	108.2	115.2	114.2	108.2	120.2	115.2	115.2	115.2	111.2	133.2

The sample mean for this bootstrap sample is $\bar{x} = 115.6$ minutes.

The randomization distribution is generated by selecting many random samples from the shifted sample. This distribution can be viewed as an approximate sampling distribution for $\bar{x}$ under the assumption that the null hypothesis $\mu = 120$ is true. A P-value is the probability of obtaining a test statistic value at least as inconsistent with H_0 as what actually resulted. So, the P-value for testing the null hypothesis $H_0: \mu = 120$ minutes against the one-sided alternative $H_a: \mu < 120$ minutes is the probability of observing $\bar{x} \leq 116.8$ when H_0 is true. This probability can be approximated using the proportion of simulated values of $\bar{x}$ that fall at or below 116.8 in the randomization distribution. Below is the output from the Shiny App "Randomization Test for One Mean." This Shiny app can be found in the app collection at statistics.cengage.com/Peck2e/Apps.html.

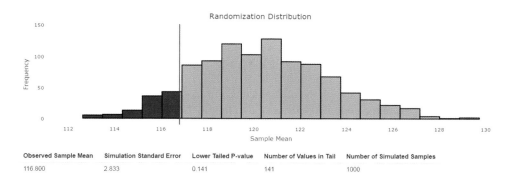

Observed Sample Mean	Simulation Standard Error	Lower Tailed P-value	Number of Values in Tail	Number of Simulated Samples
116.800	2.833	0.141	141	1000

The randomization test one-sided P-value is P-value $= 141/1000 = 0.141$. This P-value is greater than $\alpha = 0.05$, so you fail to reject H_0. The P-value obtained from the one-sample t test in Example 12.13 was 0.150, so in this case, the conclusion in the hypothesis test does not change regardless of which of these two methods is used to carry out the test.

C Communicate   Because the null hypothesis was not rejected, you would conclude that there is not convincing evidence that the mean time spent wasting time at the CEO's company is less than the value from the larger study, 120 minutes.

Summing It Up—Section 12.4

The following learning objectives were addressed in this section:

Mastering the Mechanics
M9: Calculate and interpret a bootstrap confidence interval for a population mean.
A bootstrap confidence interval is an alternate method for calculating a confidence interval for a population mean. This method can be used even in situations where the conditions necessary for the one-sample t confidence interval are not met. A bootstrap confidence interval is interpreted in the same way as the large-sample t confidence interval. Example 12.16 illustrates the calculation and interpretation of a bootstrap confidence interval for a population mean.

M10: Carry out a randomization test of hypotheses about a population mean.
A randomization-based test is a method that can be used to test hypotheses about a population mean even if the population distribution is not approximately normal and the sample size is not large enough for the one-sample t test to be appropriate. Example 12.17 illustrates the use of a randomization-based test to carry out a hypothesis test for a population mean.

SECTION 12.4 EXERCISES

Each Exercise Set assesses the following chapter learning objectives: M9, M10

SECTION 12.4 Exercise Set 1

12.60 *Consumer Reports* published the following gas mileage values (Overall MPG) for a sample of electric or plug-in hybrid car models **(www.consumerreports.org/cro/cars /new-cars/hybrids-evs/ratings-reliability/ratings-overview. htm, retrieved December 23, 2016)**.

Make and model	Overall MPG
Tesla Model S	87
BMW I3	139
Ford C-MAX	47
Nissan Leaf	106
Tesla Model X	92
Chevrolet Volt	105
Ford Focus	107
Mitsubishi i-MiEV	111

Use the output at the bottom of the page from the Shiny App "Bootstrap Confidence Interval for One Mean" to construct and interpret a 95% confidence interval for the population mean gas mileage for electric or plug-in hybrid cars. You can assume that this sample is representative of the population of electric and plug-in hybrid cars.

 12.61 The dodo was a species of flightless bird that lived on the island of Mauritius in the Indian Ocean. The first record of human interaction with the dodo occurred in 1598, and within 100 years the dodo was extinct due to hunting by humans and other newly introduced invasive species. After the extinction, the word "dodo" became synonymous with stupidity, implying that the birds lacked the intelligence to avoid or escape extinction. The closest existing relatives of the dodo are pigeons and doves.

Researchers at the American Museum of Natural History used computed tomography (CT) scans to measure the brain size ("endocranial capacity") of one of the few existing preserved dodo birds, and to measure the brain sizes in samples of eight birds that are close relatives of dodos (**"The First Endocast of the Extinct Dodo and an Anatomical Comparison Amongst Close Relatives,"** *Zoological Journal of the Linnean Society* **[2016]: 950–953**)

The brain size for the dodo was 4.17 log mm³. The following table contains the brain sizes for the sample of birds from related species (approximate values from a graph in the paper).

Brain Size
3.08
3.16
3.33
3.35
3.52
3.78
4.00

a. Use the output at the bottom of the page from the Shiny App "Randomization Test for One Mean" to help you to carry out a randomization test of the hypothesis that the population mean brain size for birds that are relatives of the dodo differs from the established dodo brain size of 4.17.

b. What does the result of your test indicate about the brain size of the dodo?

12.62 Example 12.12 provided the following 21 time discrimination scores for male smokers who had abstained from smoking for 24 hours.

1.12 1.03 1.09 1.03 1.09 0.97 0.98 1.20 1.16
1.03 1.10 1.11 0.98 1.02 1.20 0.96 0.78 1.05
0.90 1.08 0.95

a. What characteristic of the sample size indicates that the methods based on the t distribution may not be appropriate?

b. Recall that a time discrimination score of 1 indicates that there is no tendency for time to be overestimated or underestimated. Use the output at the top of the next page

Output for Exercise 12.60

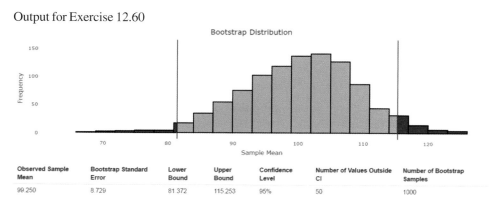

Observed Sample Mean	Bootstrap Standard Error	Lower Bound	Upper Bound	Confidence Level	Number of Values Outside CI	Number of Bootstrap Samples
99.250	8.729	81.372	115.253	95%	50	1000

Output for Exercise 12.61

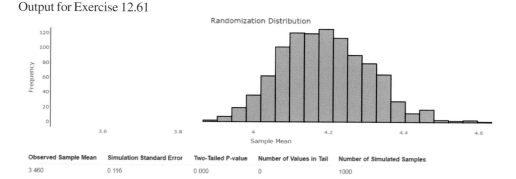

Observed Sample Mean	Simulation Standard Error	Two-Tailed P-value	Number of Values in Tail	Number of Simulated Samples
3.460	0.116	0.000	0	1000

Output for Exercise 12.62

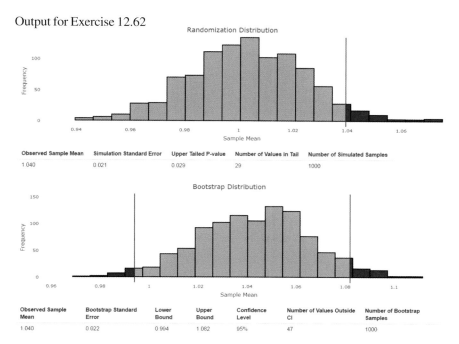

Randomization Distribution

Observed Sample Mean	Simulation Standard Error	Upper Tailed P-value	Number of Values in Tail	Number of Simulated Samples
1.040	0.021	0.029	29	1000

Bootstrap Distribution

Observed Sample Mean	Bootstrap Standard Error	Lower Bound	Upper Bound	Confidence Level	Number of Values Outside CI	Number of Bootstrap Samples
1.040	0.022	0.994	1.082	95%	47	1000

from the Shiny App "Randomization Test for One Mean" to test the hypothesis that the population mean time discrimination score for male smokers who abstain from smoking for 24 hours is significantly greater than 1.

c. Use the given output from the Shiny App "Bootstrap Confidence Interval for One Mean" to estimate the mean time discrimination score for the population of male smokers who abstain from smoking for 24 hours using a 95% confidence interval.

d. Compare the *P*-value and conclusion of the hypothesis test in Part (b) with the results of the one-sample t test in Example 12.12. Did you reach the same conclusion about whether abstaining male smokers significantly overestimate duration of time, on average? Explain.

12.63 Teams in the National Football League (NFL) are given a "bye" during one week of the season, when they can rest and not play a game. This may provide an advantage for the team in the next game they play after a bye.

In 2016, each of the 32 NFL teams was granted a bye during one of the weeks of the season. The following table contains the team name, and the number of points they won by or lost by in the game after the bye **(www.espn.com/nfl/, retrieved December 22, 2016)**. A positive value indicates that the team coming off the bye won the game, and a negative value means that they lost. You may consider these results to be a representative sample from a population of possible NFL match-ups between teams where one of the teams is coming off a bye week.

Team	Points
ARI	3
ATL	19
BAL	7
BUF	4

Team	Points
CAR	10
CHI	−26
CIN	−1
CLE	−13
DAL	6
DEN	−3
DET	7
GB	7
HOU	3
IND	7
JAX	1
KC	16
LA	−3
MIA	4
MIN	−11
NE	−7
NO	3
NYG	5
NYJ	−5
OAK	7
PHI	−1
PIT	−7
SD	8
SEA	2
SF	−18
TB	17
TEN	3
WSH	6

a. Construct a graphical display for the data. Although the sample size is at least on the borderline of being adequate for *t* distribution methods, what characteristic of the distribution indicates that the methods based on the *t* distribution may not be appropriate?

b. Use a randomization test (Shiny app: "Randomiaztion Test for One Mean") to perform a test of the hypothesis that the population mean points difference for NFL teams coming off a bye week differs from zero. Use a 0.05 significance level.

c. Use a bootstrap confidence interval (Shiny app: "Bootstrap Confidence Interval for One Mean") to estimate the population mean point difference for NFL teams coming off a bye week using a 95% confidence interval.

d. Use the results from Parts (b) and (c) to explain whether or not you believe that teams coming off a bye week have a significant advantage in points scored over their opponents.

SECTION 12.4 Exercise Set 2

 12.64 *Consumer Reports* provides ratings for televisions, including energy cost per year (in dollars) **(www.consumerreports.org/products/lcd-led-oled-tvs/ratings-overview/, retrieved December 23, 2016)**.

Energy cost data for a sample of 13 small televisions (29-inch and smaller) is displayed in the accompanying table:

Make and Model	Energy Cost (dollars)
Samsung UN28H4000	7.17
LG 28LF4520	5.17
LG 24LF4520	4.47
LG 28LH4530	5.00
Vizio D28h-C1	5.06
Vizio D28h-D1	7.00
LG 22LH4530	4.00
Element ELEFW248	4.00
Vizio D24-D1	6.00
LG 24LH4530	4.00
Insignia NS-24ER310NA17	6.00
Seiki SE23HEB2	4.00
Insignia NS-24D510NA17	5.00

Suppose that this sample is representative of the population of all small televisions. Use the output at the bottom of the page from the Shiny App "Bootstrap Confidence Interval for One Mean" to construct and interpret a 95% bootstrap confidence interval for the mean annual energy cost for the population of all small televisions.

 12.65 Researchers studied ergometer (rowing machine) time (in seconds) for international male competitors in the 2007 World Junior Rowing Championships. They found that the mean time to row 2000 meters on an ergometer for the population of international sculls competitors was 387 seconds **("Does 2000-m Rowing Ergometer Performance Time Correlate with Final Rankings at the World Junior Rowing Championship? A Case Study of 398 Elite Junior Rowers," *Journal of Sports Sciences* [2009]: 361–366)**.

A related research team studied a sample of 24 junior male sculls rowers from the United States in 2013–2014, and reported summary statistics for their 2000-meter ergometer times **("Correlates of Performance at the U.S. Rowing Youth National Championships: A Case Study of 152 Junior Rowers," *The Sport Journal*, March 3, 2014)**. Data for a representative sample of 24 rowers that are consistent with summary statistics given in the paper are shown in the accompanying table.

Rower ID	Time
1	391
2	389
3	386
4	396
5	395
6	409
7	406
8	402
9	393
10	392
11	382
12	401
13	404
14	399

(continued)

Output for Exercise 12.64

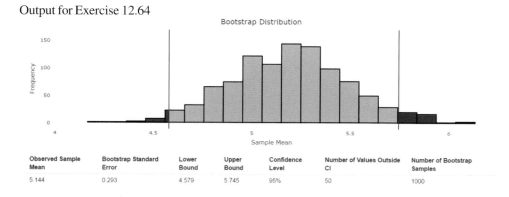

Observed Sample Mean	Bootstrap Standard Error	Lower Bound	Upper Bound	Confidence Level	Number of Values Outside CI	Number of Bootstrap Samples
5.144	0.293	4.579	5.745	95%	50	1000

Rower ID	Time
15	390
16	395
17	392
18	401
19	381
20	392
21	376
22	400
23	388
24	402

a. Use the output at the bottom of the page from the Shiny App "Randomization Test for One Mean" to carry out a randomization test of the hypothesis that the population mean 2000-meter ergometer time for U.S. junior male sculls rowers differs from the 2007 international standard of 387 seconds. Use a significance level of 0.05.

b. Based on the result of the hypothesis test, does it seem that the U.S. junior male sculls rowers have "caught up," on average, with the international championship rowers from 2007? Explain.

12.66 Exercise 12.44 asks whether a representative sample of Big Mac prices (after conversion to U.S. dollars) from countries in Europe provides evidence that the mean European price is less than the reported U.S. price of $5.04. Here are the data:

4.44 3.15 2.42 3.96 4.35 4.51 4.17 3.69 4.62 3.80 3.36 3.85

a. What characteristic of the sample indicates that the methods based on the *t* distribution may not be appropriate?

b. Use the output at the bottom of the page from the Shiny App "Randomization Test for One Mean" to test the hypothesis that the population mean price of a Big Mac in Europe is less than the reported U.S. price of $5.04. Use a significance level of 0.05.

c. Use the output at the top of the next page from the Shiny App "Bootstrap Confidence Interval for One Mean" to estimate the mean price of a Big Mac in Europe using a 95% bootstrap confidence interval.

d. Compare the P-value and conclusion of the hypothesis test in Part (b) with the results of the *t* test in Exercise 12.44. Did you reach the same conclusion about whether the mean price of a Big Mac in Europe is significantly lower than the U.S. price? Explain.

12.67 Major League Baseball (MLB) includes two groups of teams, in "leagues." There are 15 teams in each of the American League (AL) and the National League (NL). Since 1997, teams in each of the leagues play teams from the other league in "interleague" regular-season games.

One way to determine whether one league is stronger than the other is to consider the interleague winning percentages for one season for all the teams in one of the two leagues, say, the National League. For purposes of this exercise, consider the interleague games played in the 2016 season to be a represen-

Output for Exercise 12.65

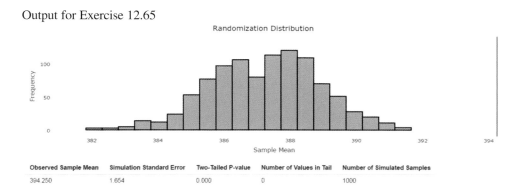

Output for Exercise 12.66(b)

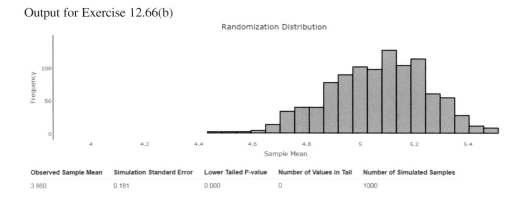

Output for Exercise 12.66(c)

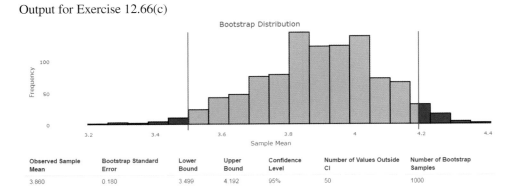

Observed Sample Mean	Bootstrap Standard Error	Lower Bound	Upper Bound	Confidence Level	Number of Values Outside CI	Number of Bootstrap Samples
3.860	0.180	3.499	4.192	95%	50	1000

tative sample of the performance of the teams in a population of potential future seasons.

Here are the 2016 interleague winning percentages for the 15 NL teams:

Team	W	L	Winning Percentage
ARI	5	15	25%
ATL	8	12	40%
CHC	15	5	75%
CIN	5	15	25%
COL	9	11	45%
LAD	10	10	50%
MIA	6	14	30%
MIL	11	9	55%
NYM	12	8	60%
PHI	11	9	55%
PIT	9	11	45%
SD	6	14	30%

Team	W	L	Winning Percentage
SF	8	12	40%
STL	8	12	40%
WSH	12	8	60%

a. What characteristic of this sample of NL team interleague winning percentages indicates that the methods based on the t distribution may not be appropriate?
b. Use a randomization test (Shiny app: "Randomization Test for One Mean") to perform a test of the hypothesis that the population mean interleague winning percentage for NL teams differs from 50%.
c. Estimate the mean interleague winning percentage for the population of NL teams using a 95% bootstrap confidence interval (Shiny app: "Bootstrap Confidence Interval for One Mean").
d. Use the results from Parts (b) and (c) to explain whether it is reasonable to say that the National League or the American League performs significantly better than the other in interleague play.

SECTION 12.5 Avoid These Common Mistakes

Now that you have seen confidence intervals and hypothesis tests for proportions and for means, you will want to think about the distinction between proportions and means when choosing an appropriate method. The best way to distinguish between situations that involve learning about a population proportion and those that involve learning about a population mean is to focus on the type of data—categorical or numerical. With categorical data, you should be thinking proportions, and with numerical data you should be thinking means.

As was the case when using sample data to learn about a population proportion, be sure to keep in mind that conditions are important. Use of the one-sample t confidence interval and hypothesis test requires that certain conditions be met. Be sure to check that these conditions are met before using these methods.

Also, remember that the result of a hypothesis test can never provide strong support for the null hypothesis. Make sure that you don't confuse "I am not convinced that the null hypothesis is false" with "I am convinced that the null hypothesis is true." These are not the same!

CHAPTER ACTIVITIES

ACTIVITY 12.1 COMPARING THE *t* AND *z* DISTRIBUTIONS

Technology Activity: Requires use of a computer or a graphing calculator.

The following instructions assume the use of Minitab. If you are using a different software package or a graphing calculator, your instructor will provide alternative instructions.

Background: Suppose a random sample will be selected from a population that is known to have a normal distribution. Then the statistic

$$z = \frac{\bar{x} - \mu}{\dfrac{\sigma}{\sqrt{n}}}$$

has a standard normal (*z*) distribution. Since it is rarely the case that σ is known, inferences for population means are usually based on the statistic

$$t = \frac{\bar{x} - \mu}{\dfrac{s}{\sqrt{n}}}$$

which has a *t* distribution rather than a *z* distribution. The informal justification for this is that the use of *s* to estimate σ introduces additional variability, resulting in a statistic whose distribution has more variability than the *z* distribution.

In this activity, you will use simulation to sample from a known normal population and then investigate how the behavior of $t = \dfrac{\bar{x} - \mu}{\dfrac{s}{\sqrt{n}}}$ compares to $z = \dfrac{\bar{x} - \mu}{\dfrac{\sigma}{\sqrt{n}}}$.

1. Generate 200 random samples of size 5 from a normal population with mean 100 and standard deviation 10.

    Using Minitab, go to the Calc Menu. Then

    Calc → Random Data → Normal
    In the "Generate" box, enter 200
    In the "Store in columns" box, enter c1-c5
    In the mean box, enter 100
    In the standard deviation box, enter 10
    Click on OK

    You should now see 200 rows of data in each of the first five columns of the Minitab worksheet.

2. Each row contains five values that have been randomly selected from a normal population with mean 100 and standard deviation 10. Viewing each row as a sample of size 5 from this population, calculate the mean and standard deviation for each of the 200 samples (the 200 rows) by using Minitab's row statistics functions, which can also be found under the Calc menu:

    Calc → Row statistics
    Choose the "Mean" button
    In the "Input Variables" box, enter c1-c5
    In the "Store result in" box, enter c7
    Click on OK

    You should now see the 200 sample means in column 7 of the Minitab worksheet. Name this column "x-bar" by typing the name in the gray box at the top of c7.

    Follow a similar process to calculate the 200 sample standard deviations, and store them in c8. Name this column "*s*."

3. Next, calculate the value of the *z* statistic for each of the 200 samples. You can calculate *z* in this example because you know that the samples were selected from a population for which $\sigma = 10$. Use the Minitab calculator function to calculate

    $$z = \frac{\bar{x} - \mu}{\dfrac{\sigma}{\sqrt{n}}} = \frac{\bar{x} - 100}{\dfrac{10}{\sqrt{5}}}$$

    as follows:

    Calc → Calculator
    In the "Store results in" box, enter c10
    In the "Expression box" type in the following:
    (c7–100)/(10/sqrt(5))
    Click on OK

    You should now see the *z* values for the 200 samples in c10. Name this column "*z*."

4. Now calculate the value of the *t* statistic for each of the 200 samples. Use the calculator function of Minitab to calculate

    $$t = \frac{\bar{x} - \mu}{\dfrac{s}{\sqrt{n}}} = \frac{\bar{x} - 100}{\dfrac{s}{\sqrt{5}}}$$

    as follows:

    Calc → Calculator
    In the "Store results in" box, enter c11
    In the "Expression box" type in the following:
    (c7–100)/(c8/sqrt(5))
    Click on OK

    You should now see the *t* values for the 200 samples in c11. Name c11 "*t*."

5. Graphs, at last! Now construct histograms of the 200 *z* values and the 200 *t* values. These two graphical displays will provide insight into how each of these two statistics behaves in repeated sampling. Use the same

scale for the two histograms so that it will be easy to compare the two distributions.

Graph → Histogram

In the "Graph variables" box, enter c10 for graph 1 and c11 for graph 2

Click the Frame dropdown menu and select multiple graphs. Then under the scale choices, select "Same X and same Y."

Now use the histograms from Step 5 to answer the following questions:

a. Write a brief description of the shape, center, and variability for the histogram of the z values. Is what you see in the histogram consistent with what you would have expected to see? Explain. (Hint: In theory, what is the distribution of the z statistic?)

b. How does the histogram of the t values compare to the z histogram? Be sure to comment on center, shape, and variability.

c. Is your answer to Part (b) consistent with what would be expected for a statistic that has a t distribution? Explain.

d. Because the z and t histograms are based on only 200 samples, they only approximate the corresponding sampling distributions. The 5th percentile for the standard normal distribution is -1.645 and the 95th percentile is 1.645. For a t distribution with df $= 5 - 1 = 4$, the 5th and 95th percentiles are -2.13 and 2.13, respectively. How do these percentiles compare to those of the distributions displayed in the histograms? (Hint: Sort the 200 z values—in Minitab, choose "Sort" from the Manip menu. Once the values are sorted, the 5th and 95th percentiles from the histogram can be found by counting in 10 [which is 5% of 200] values from either end of the sorted list. Then repeat this with the t values.)

e. Are the results of your simulation and analysis consistent with the statement that the statistic $z = \dfrac{\bar{x} - \mu}{\dfrac{\sigma}{\sqrt{n}}}$ has a standard normal (z) distribution and the statistic $t = \dfrac{\bar{x} - \mu}{\dfrac{s}{\sqrt{n}}}$ has a t distribution? Explain.

🖥 CHAPTER 12 EXPLORATIONS IN STATISTICAL THINKING

🖱 EXPLORING SAMPLING VARIABILITY IN THE CONTEXT OF CONFIDENCE INTERVALS

In this exercise, you will go online to select a random sample of 30 animated movies produced between 1980 and 2011. You will use this sample to estimate the mean length of the animated movies in this population.

Go online at statistics.cengage.com/Peck2e/Explore.html and click on the link for Chapter 12. It will take you to a web page where you can select a random sample from the animated movie population.

Click on the "sample" button. This selects a random sample of 30 movies and will display the movie name and the length (in minutes) for each movie in your sample. Each student in your class will receive a different random sample.

Use the data from your random sample to complete the following:

a. Calculate the mean length of the movies in your sample.
b. Is the mean you calculated in Part (a) the population mean or a sample mean?
c. Construct a 90% confidence interval for the mean length of the animated movies in this population.
d. Write a few sentences that provide an interpretation of the confidence interval from Part (c).
e. The actual population mean is 90.41 minutes. Did your confidence interval from Part (c) include this value?
f. Which of the following is a correct interpretation of the 90% confidence level?
   1. The probability that the actual population mean is contained in the calculated interval is 0.90.
   2. If the process of selecting a random sample of movies and then calculating a 90% confidence interval for the mean length of all animated movies made between 1980 and 2011 is repeated 100 times, exactly 90 of the 100 intervals will include the actual population mean.

3. If the process of selecting a random sample of movies and then calculating a 90% confidence interval for the mean length all animated movies made between 1980 and 2011 is repeated a very large number of times, approximately 90% of the intervals will include the actual population mean.

If asked to do so by your instructor, bring your confidence interval estimate of the mean length of animated movies made between 1980 and 2011 to class. Your instructor will lead the class through a discussion of the questions that follow.
Compare your confidence interval to a confidence interval obtained by another student in your class.
g. Are the two confidence intervals the same?
h. Do both intervals contain the actual population mean length of 90.41 minutes?
i. How many people in your class have a confidence interval that does not include the actual value of the population mean? Is this surprising given the 90% confidence level associated with the confidence intervals?

ARE YOU READY TO MOVE ON? CHAPTER 12 REVIEW EXERCISES

All chapter learning objectives are assessed in these exercises. The learning objectives assessed in each exercise are given in parentheses.

12.68 (M1, M2)
Suppose that college students with a checking account typically write relatively few checks in any given month, whereas people who are not college students typically write many more checks during a month. Suppose that 50% of a bank's accounts are held by students and that 50% are held by people who are not college students. Let x represent the number of checks written in a given month by a randomly selected bank customer.
a. Give a sketch of what the population distribution of x might look like.
b. Suppose that the mean value of x is 22.0 and that the standard deviation is 16.5. If a random sample of $n = 100$ customers is to be selected and $\bar{x}$ denotes the sample mean number of checks written during a particular month, where is the sampling distribution of $\bar{x}$ centered, and what is the standard deviation of the sampling distribution of $\bar{x}$? Sketch a rough picture of the sampling distribution.
c. What is the approximate probability that $\bar{x}$ is at most 20? At least 25?

12.69 (C1)
Let x represent the time (in minutes) that it takes a fifth-grade student to read a certain passage. Suppose that the mean and standard deviation of the x distribution are $\mu = 2$ minutes and $\sigma = 0.8$ minutes, respectively.
a. If $\bar{x}$ is the sample mean time for a random sample of $n = 9$ students, where is the sampling distribution of $\bar{x}$ centered, and what is the standard deviation of the sampling distribution of $\bar{x}$?
b. Repeat Part (a) for a sample of size of $n = 20$ and again for a sample of size $n = 100$. How do the centers and variabilities of the three $\bar{x}$ sampling distributions com-

pare to one another? Which sample size would be most likely to result in an $\bar{x}$ value close to μ, and why?

12.70 (C2)
The article **"The Association Between Television Viewing and Irregular Sleep Schedules Among Children Less Than 3 Years of Age" (Pediatrics [2005]: 851–856)** reported the accompanying 95% confidence intervals for average TV viewing time (in hours per day) for three different age groups.

Age Group	95% Confidence Interval
Less than 12 Months	(0.8, 1.0)
12 to 23 Months	(1.4, 1.8)
24 to 35 Months	(2.1, 2.5)

a. Suppose that the sample sizes for each of the three age-group samples were equal. Based on the given confidence intervals, which of the age-group samples had the greatest variability in TV viewing time? Explain your choice.
b. Now suppose that the sample standard deviations for the three age-group samples were equal, but that the three sample sizes might have been different. Which of the three age-group samples had the largest sample size? Explain your choice.
c. The interval (0.7, 1.1) is either a 90% confidence interval or a 99% confidence interval for the mean TV viewing time calculated using the sample data for children less than 12 months old. Is the confidence level for this interval 90% or 99%? Explain your choice.

12.71 (C2, P1)
Suppose that a random sample of 50 cans of a particular brand of fruit juice is selected, and the amount of juice (in ounces) in each of the cans is determined. Let μ denote the

mean amount of juice for the population of all cans of this brand. Suppose that this sample of 50 results in a 95% confidence interval for μ of (7.8, 9.4).

a. Would a 90% confidence interval have been narrower or wider than the given interval? Explain your answer.

b. Consider the following statement: There is a 95% chance that μ is between 7.8 and 9.4. Is this statement correct? Why or why not?

c. Consider the following statement: If the process of selecting a random sample of size 50 and then calculating the corresponding 95% confidence interval is repeated 100 times, exactly 95 of the resulting intervals will include μ. Is this statement correct? Why or why not?

12.72 (M4)

The authors of the paper **"Serum Zinc Levels of Cord Blood: Relation to Birth Weight and Gestational Period" (***Journal of Trace Elements in Medicine and Biology*** [2015]: 180–183)** carried out a study of zinc levels of low-birth-weight babies and normal-birth-weight babies. For a sample of 50 low-birth-weight babies, the sample mean zinc level was 17.00 and the standard error $\left(\frac{s}{\sqrt{n}}\right)$ was 0.43. For a sample of 73 normal-birth-weight babies, the sample mean zinc level was 18.16 and the standard error was 0.32. Explain why the two standard errors are not the same.

12.73 (M3, M4, M5, P1)

How much money do people spend on graduation gifts? In 2016, the **National Retail Federation (www.nrf.com)** surveyed 2511 consumers who reported that they bought one or more graduation gifts in 2016. The sample was selected to be representative of adult Americans who purchased graduation gifts in 2016. For this sample, the mean amount spent per gift was $53.73. Suppose that the sample standard deviation was $20. Construct and interpret a 98% confidence interval for the mean amount of money spent per graduation gift in 2016.

12.74 (M3, M4, M5, P1)

The authors of the paper **"Driving Performance While Using a Mobile Phone: A Simulation Study of Greek Professional Drivers" (***Transportation Research Part F*** [2016]: 164–170)** describe a study to evaluate the effect of mobile phone use by taxi drivers in Greece. Fifty taxi drivers drove in a driving simulator where they were following a lead car. The drivers were asked to carry on a conversation on a mobile phone while driving, and the following distance (the distance between the taxi and the lead car) was recorded. The sample mean following distance was 3.20 meters and the sample standard deviation was 1.11 meters.

a. Construct and interpret a 95% confidence interval for μ, the population mean following distance while talking on a mobile phone for the population of taxi drivers.

b. What assumption must be made in order to generalize this confidence interval to the population of all taxi drivers in Greece?

12.75 (M6)

Suppose that the researchers who carried out the study described in the previous exercise wanted to estimate the mean following distance with a margin of error of 0.2 meters. Using the given sample standard deviation as a preliminary estimate of the population standard deviation, calculate the required sample size.

12.76 (M7, M8, P2)

A study of fast-food intake is described in the paper **"What People Buy From Fast-Food Restaurants" (***Obesity*** [2009]:1369–1374).** Adult customers at three hamburger chains (McDonald's, Burger King, and Wendy's) in New York City were approached as they entered the restaurant at lunchtime and asked to provide their receipt when exiting. The receipts were then used to determine what was purchased and the number of calories consumed was determined. In all, 3857 people participated in the study. The sample mean number of calories consumed was 857 and the sample standard deviation was 677.

a. The sample standard deviation is quite large. What does this tell you about number of calories consumed in a hamburger-chain lunchtime fast-food purchase in New York City?

b. Given the values of the sample mean and standard deviation and the fact that the number of calories consumed can't be negative, explain why it is *not* reasonable to assume that the distribution of calories consumed is normal.

c. Based on a recommended daily intake of 2000 calories, the online Healthy Dining Finder **(www.healthydiningfinder .com)** recommends a target of 750 calories for lunch. Assuming that it is reasonable to regard the sample of 3857 fast-food purchases as representative of all hamburger-chain lunchtime purchases in New York City, carry out a hypothesis test to determine if the sample provides convincing evidence that the mean number of calories in a New York City hamburger-chain lunchtime purchase is greater than the lunch recommendation of 750 calories. Use $\alpha = 0.01$.

d. Would it be reasonable to generalize the conclusion of the test in Part (c) to the lunchtime fast-food purchases of all adult Americans? Explain why or why not.

e. Explain why it is better to use the customer receipt to determine what was ordered rather than just asking a customer leaving the restaurant what he or she purchased.

f. Do you think that asking a customer to provide his or her receipt before they ordered could have introduced a bias? Explain.

12.77 (M7, M8, P2)

Medical research has shown that repeated wrist extension beyond 20 degrees increases the risk of wrist and hand injuries. Each of 24 students at Cornell University used a proposed new computer mouse design, and while using the mouse, each student's wrist extension was recorded. Data consistent with summary values given in the paper

"Comparative Study of Two Computer Mouse Designs" (Cornell Human Factors Laboratory Technical Report RP7992) are given. Use these data to test the hypothesis that the mean wrist extension for people using this new mouse design is greater than 20 degrees. Are any assumptions required in order for it to be appropriate to generalize the results of your test to the population of all Cornell students? To the population of all university students?

27	28	24	26	27	25	25
24	24	24	25	28	22	25
24	28	27	26	31	25	28
27	27	25				

12.78 (P3)
A comprehensive study conducted by the National Institute of Child Health and Human Development tracked more than 1000 children from an early age through elementary school (*The New York Times*, **November 1, 2005**). The study concluded that children who spent more than 30 hours a week in child care before entering school tended to score higher in math and reading when they were in the third grade. The researchers cautioned that the findings should not be a cause for alarm because the differences were found to be small. Explain how the difference between the mean math score for the child care group and the overall mean for third graders could be small but the researchers could still reach the conclusion that the mean for the child care group is significantly higher than the overall mean.

TECHNOLOGY NOTES

Confidence Interval for μ Based on t-distribution

JMP
1. Input the data into a column
2. Click **Analyze** and select **Distribution**
3. Click and drag the column name from the box under **Select Columns** to the box next to **Y, Response**
4. Click **OK**
5. Click the red arrow next to the column name and select **Confidence Interval** then select the appropriate confidence level or click **Other** to specify a level

Minitab
Summarized data
1. Click **Stat** then click **Basic Statistics** then click **1-sample t…**
2. Click the radio button next to **Summarized data**
3. In the box next to **Sample size:** type the value for *n*, the sample size
4. In the box next to **Mean:** type the value for the sample mean
5. In the box next to **Standard deviation:** type the value for the sample standard deviation
6. Click **Options…**
7. Input the appropriate confidence level in the box next to **Confidence Level**
8. Click **OK**
9. Click **OK**

Raw data
1. Input the raw data into a column
2. Click **Stat** then click **Basic Statistics** then click **1-sample t…**
3. Click in the box under **Samples in columns:**
4. Double click the column name where the raw data are stored
5. Click **Options…**
6. Input the appropriate confidence level in the box next to **Confidence Level**
7. Click **OK**
8. Click **OK**

SPSS
1. Input the data into a column
2. Click **Analyze** then select **Compare Means** then select **One-Sample T Test…**
3. Highlight the column name for the variable
4. Click the arrow to move the variable to the **Test Variable(s):** box
5. Click **Options…**
6. Input the appropriate confidence level in the box next to **Confidence Interval Percentage:**
7. Click **Continue**
8. Click **OK**

Note: The confidence interval for the population means is in the **One-Sample Test** table.

Excel
Note: Excel does not have the functionality to produce a confidence interval for a single population mean automatically. However, you can use Excel to produce the estimate for the sample mean and the margin of error for the confidence interval using the steps below.
1. Input the raw data into a column
2. Click the **Data** ribbon
3. Click **Data Analysis** in the **Analysis** group

    Note: If you do not see **Data Analysis** listed on the Ribbon, see the Technology Notes for Chapter 2 for instructions on installing this add-on.

4. Select **Descriptive Statistics** from the dialog box and click **OK**
5. Click in the box next to **Input Range:** and select the data (if you used a column title, check the box next to **Labels in first row**)
6. Click the box next to **Confidence Level for Mean:**
7. In the box to the right of **Confidence Level for Mean:** type in the confidence level you are using
8. Click **OK**

Note: The margin of error can be found in the row titled Confidence Level. In order to find the lower limit of the confidence interval, subtract this from the mean shown in the output. To find the upper limit of the confidence interval, add this to the mean shown in the output.

TI-83/84
Summarized data
1. Press **STAT**
2. Highlight **TESTS**
3. Highlight **TInterval...** and press **ENTER**
4. Highlight **Stats** and press **ENTER**
5. Next to x̄ input the value for the sample mean
6. Next to **sx** input the value for the sample standard deviation
7. Next to **n** input the value for the sample size
8. Next to **C-Level** input the appropriate confidence level
9. Highlight **Calculate** and press **ENTER**

Raw data
1. Enter the data into **L1** (In order to access lists press the **STAT** key, highlight the option called **Edit...** then press **ENTER**)
2. Press **STAT**
3. Highlight **TESTS**
4. Highlight **TInterval...** and press **ENTER**
5. Highlight **Data** and press **ENTER**
6. Next to **C-Level** input the appropriate confidence level
7. Highlight **Calculate** and press **ENTER**

TI-Nspire
Summarized data
1. Enter the Calculate Scratchpad
2. Press the **menu** key and select **6:Statistics** then **6:Confidence Intervals** then **2:t Interval...** and press **enter**
3. From the drop-down menu select **Stats**
4. Press **OK**
5. Next to x̄ input the value for the sample mean
6. Next to **sx** input the value for the sample standard deviation
7. Next to **n** input the value for the sample size
8. Next to **C Level** input the appropriate confidence level
9. Press **OK**

Raw data
1. Enter the data into a data list (In order to access data lists select the spreadsheet option and press **enter**)

   Note: Be sure to title the list by selecting the top row of the column and typing a title.

2. Press the **menu** key and select **4:Statistics** then **3:Confidence Intervals** then **2:t Interval...** and press **enter**
3. From the drop-down menu select **Data**
4. Press **OK**
5. Next to **List** select the list containing your data
6. Next to **C-Level** input the appropriate confidence level
7. Press **OK**

t-Test for population mean, μ

JMP
1. Input the data into a column
2. Click **Analyze** and select **Distribution**
3. Click and drag the column name from the box under **Select Columns** to the box next to **Y, Response**
4. Click **OK**
5. Click the red arrow next to the column name and select **Test Mean**
6. In the box next to **Specify Hypothesized Mean**, type the hypothesized value of the mean, μ_0
7. Click **OK**

Note: The output provides results for all three possible alternative hypotheses.

Minitab
Summarized data
1. Click **Stat** then click **Basic Statistics** then click **1-sample t...**
2. Click the radio button next to **Summarized data**
3. In the box next to **Sample size:** type the value for *n*, the sample size
4. In the box next to **Mean:** type the value for the sample mean
5. In the box next to **Standard deviation:** type the value for the sample standard deviation
6. In the box next to **Test mean:** type the hypothesized value of the population mean
7. Click **Options...**
8. Select the appropriate alternative hypothesis from the drop-down menu next to **Alternative:**
9. Click **OK**
10. Click **OK**

Raw data
1. Input the raw data into a column
2. Click **Stat** then click **Basic Statistics** then click **1-sample t...**
3. Click in the box under **Samples in columns:**
4. Double click the column name where the raw data are stored
5. In the box next to **Test mean:** type the hypothesized value of the population mean
6. Click **Options...**
7. Select the appropriate alternative hypothesis from the drop-down menu next to **Alternative:**
8. Click **OK**
9. Click **OK**

SPSS
1. Input the data into a column
2. Click **Analyze** then select **Compare Means** then select **One-Sample T Test...**
3. Highlight the column name for the variable
4. Click the arrow to move the variable to the **Test Variable(s):** box
5. In the box next to **Test Value:** input the hypothesized test value
6. Click **OK**

Note: This procedure produces a two-sided *P*-value.

Excel

Excel does not have the functionality to automatically produce a *t* test for a single population mean. However, you may type the formula into an empty cell manually to have Excel calculate the value of the test statistic for you. You can then use the steps below to find a *P*-value for the test statistic.

1. Select an empty cell
2. Click on **Formulas**
3. Click **Insert Function**
4. Select **Statistical** from the drop-down box for category
5. Select **TDIST** and click **OK**
6. Click in the box next to **X** and select the cell containing your test statistic or type it manually
7. Click in the box next to **Deg_freedom** and type the number of degrees of freedom (*n*-1)
8. Click in the box next to **Tails** and type 1 for a one-tailed *P*-value or 2 for a two-tailed *P*-value
9. Click **OK**

Note: Choosing a one-tailed distribution in Step 8 will result in returning $P(X \geq x)$.

TI-83/84
Summarized data
1. Press **STAT**
2. Highlight **TESTS**
3. Highlight **T-Test...**
4. Highlight **Stats** and press **ENTER**
5. Next to μ_0 type the hypothesized value for the population mean
6. Next to $\bar{x}$ input the value for the sample mean
7. Next to **sx** input the value for the sample standard deviation
8. Next to **n** input the value for the sample size
9. Next to μ highlight the appropriate alternative hypothesis and press **ENTER**
10. Highlight **Calculate** and press **ENTER**

Raw data
1. Enter the data into **L1** (In order to access lists press the **STAT** key, highlight the option called **Edit...** then press **ENTER**)
2. Press **STAT**
3. Highlight **TESTS**
4. Highlight **T-Test...**
5. Highlight **Data** and press **ENTER**
6. Next to μ_0 type the hypothesized value for the population mean
7. Next to μ highlight the appropriate alternative hypothesis and press **ENTER**
8. Highlight **Calculate** and press **ENTER**

TI-Nspire
Summarized data
1. Enter the Calculate Scratchpad
2. Press the **menu** key and select **6:Statistics** then **7:Stat Tests** then **2:t test...** and press **enter**
3. From the drop-down menu select **Stats**
4. Press **OK**
5. Next to μ_0 type the hypothesized value for the population mean
6. Next to $\bar{x}$ input the value for the sample mean
7. Next to **sx** input the value for the sample standard deviation
8. Next to **n** input the value for the sample size
9. Next to **Alternate Hyp** select the appropriate alternative hypothesis from the drop-down menu
10. Press **OK**

Raw data
1. Enter the data into a data list (In order to access data lists select the spreadsheet option and press **enter**)

   Note: Be sure to title the list by selecting the top row of the column and typing a title.

2. Press the **menu** key and select **4:Statistics** then **7:Stat Tests** then **2:t test...** and press **enter**
3. From the drop-down menu select **Data**
4. Press **OK**
5. Next to μ_0 input the hypothesized value for the population mean
6. Next to **List** select the list containing your data
7. Next to **Alternate Hyp** select the appropriate alternative hypothesis from the drop-down menu
8. Press **OK**

13 Asking and Answering Questions About the Difference Between Two Means

Elena Schweitzer/Shutterstock.com

PREVIEW

In Chapter 12, you saw how sample data could be used to estimate a population mean and to test hypotheses about the value of a single population mean. In this chapter you will see how sample data can be used to learn about the difference between two population or treatment means.

CHAPTER LEARNING OBJECTIVES

Conceptual Understanding

After completing this chapter, you should be able to

C1 Understand how a research question about the difference in two population means is translated into hypotheses.

Mastering the Mechanics

After completing this chapter, you should be able to

M1 Distinguish between independent and paired samples.

M2 Know the conditions for appropriate use of the paired-samples t confidence interval and the paired-samples t test.

M3 Carry out a paired-samples t test for a difference in population means.

M4 Calculate and interpret a paired-samples t confidence interval for a difference in population means.

M5 Know the conditions for appropriate use of the two-sample t confidence interval and the two-sample t test.

M6 Carry out a two-sample t test for a difference in population means.

M7 Calculate and interpret a two-sample t confidence interval for a difference in population means.

M8 Use the two-sample t test to test hypotheses about a difference in treatment means.

M9 Use the two-sample t confidence interval to estimate a difference in treatment means.

M10 (Optional) Calculate and interpret a bootstrap confidence interval for a difference in means.

M11 (Optional) Carry out a randomization test for a difference in means.

Putting It into Practice

After completing this chapter, you should be able to

P1 Carry out a paired-samples t test for a difference in two means and interpret the conclusion in context.

P2 Interpret a paired-samples t confidence interval for a difference in two means in context and interpret the associated confidence level.

P3 Carry out a two-sample t test for a difference in two means and interpret the conclusion in context.

P4 Interpret a two-sample t confidence interval for a difference in two means in context and interpret the associated confidence level.

PREVIEW EXAMPLE

Depression and Chocolate

Is there a connection between depression and chocolate consumption? This is the question that the authors of the paper **"Mood Food: Chocolate and Depressive Symptoms in a Cross-Sectional Analysis" (*Archives of Internal Medicine* [2010]: 699–703)** set out to answer. Participants in the study were 931 adults who were not currently taking medication for depression. These participants were screened for depression with a widely used screening test. The participants were then divided into two samples based on the score on the screening test. One sample consisted of people who screened positive for depression, and the other sample consisted of people who did not. Each participant also completed a food frequency survey.

The researchers believed that the two samples were representative of the two populations of interest—adults who would screen positive for depression and adults who would not screen positive. Using methods that you will learn in this chapter, the researchers were able to conclude that the mean number of servings of chocolate per month for people who would screen positive for depression was greater than the mean number of chocolate servings per month for people who would not.

There are two other interesting things to note about this study. First, this was an observational study and not an experiment. Because of this, the researchers were not able to determine if there was a cause-and-effect relationship between depression and chocolate consumption. They also noted that even if there were a cause-and-effect relationship, this study would not have been able to identify the direction of the relationship (whether depression causes chocolate consumption or chocolate consumption causes depression). The second interesting aspect of the study is that the researchers also compared the two populations on the basis of food types other than chocolate. They did not find convincing evidence that the two populations differed with respect to mean caffeine, fat, carbohydrate, or calorie intake. The researchers believe that this makes the connection between depression and chocolate consumption even more interesting and worthy of further study. ∎

SECTION 13.1 Two Samples: Paired versus Independent Samples

Many statistical studies are carried out in order to learn about the difference between two population means. For example, many researchers have studied the ways in which college students use Facebook. As part of a study described in the paper **"Facebook Use and Academic Performance Among College Students: A Mixed-Methods Study with a Multi-Ethnic Sample"** (*Computers in Human Behavior* [2015]: 265–272), each person in a sample of 195 female Facebook users and an independent sample of 66 male Facebook users was asked to report the amount of time per day he or she spent on Facebook. The samples were chosen to be representative of female and male college students in Southern California. The authors of the paper were interested in learning whether the mean time spent by female Facebook users was greater than the mean time spent by male Facebook users. In this chapter, you will see how sample data can be used to answer questions like this that involve comparing two population means.

In the previous chapter, you used the symbol μ to represent the mean of an entire population. The sample mean was denoted by $\bar{x}$. When comparing two populations, you will have two samples—one from each population. You need to distinguish between the two populations and the two samples, and this is done using subscripts as shown in the following box.

Notation

	Population Mean	**Population Standard Deviation**
Population 1	μ_1	σ_1
Population 2	μ_2	σ_2

	Sample Size	**Sample Mean**	**Sample Standard Deviation**
Sample from Population 1	n_1	$\bar{x}_1$	s_1
Sample from Population 2	n_2	$\bar{x}_2$	s_2

Two samples are said to be independent samples if the selection of the individuals who make up one sample does not influence the selection of the individuals in the other sample. This would be the case if a random sample is selected from each of two populations. When observations from the first sample can be matched up in some meaningful way with observations in the second sample, the samples are said to be **paired**. For example, to study the effectiveness of a speed-reading course, the reading speed of subjects could be measured before they take the course and again after they complete the course. This gives rise to two related samples—one from the population of individuals who have not taken this particular course (the "before" measurements) and one from the population of individuals who have had such a course (the "after" measurements). These samples are paired. The two samples are not independently chosen, because the selection of individuals from the first (before) population completely determines which individuals make up the sample from the second (after) population.

Many studies involve comparing two population means using independent samples. However, in some situations, a study with independent samples is not the best way to obtain information about a possible difference between the populations. For example, suppose that a researcher wants to determine if there is a relationship between regular aerobic exercise and blood pressure. A random sample of people who jog regularly and a second random sample of people who do not exercise regularly are selected independently of one another. The researcher might conclude that a significant difference exists in mean blood pressure for joggers and non-joggers. But is it reasonable to conclude that the difference in mean blood pressure is attributable to jogging? It is known that blood pressure is related to body weight, and that joggers tend to be leaner than non-joggers. Based on the study described, the researcher would not be able to rule out the possibility that the observed difference in mean blood pressure is due to differences in weight.

One way to avoid this difficulty is to match subjects by weight. The researcher could match each jogger with a non-jogger who is similar in weight (although weights for different pairs might vary widely). If a difference in mean blood pressure between the two groups were still observed, the factor *weight* could then be ruled out as a possible explanation for the difference.

Studies can be designed to yield paired data in a number of different ways. Some studies involve using the same group of individuals with measurements recorded both before and after some event of interest occurs. Others might use naturally occurring pairs, such as twins or husbands and wives. Finally, as with weight in the jogging example, some studies construct pairs by matching on factors with effects that might otherwise obscure differences (or the lack of differences) between the two populations of interest.

Example 13.1 illustrates why it is important to consider whether two samples are independent or paired.

Data set available

Example 13.1   Document Word Clouds

When people are searching for information online, most people quickly scan a document and attempt to determine if it is relevant before taking the time to read the whole document. The authors of the paper **"Document Word Clouds: Visualising Web Documents as Tag Clouds to Aid Users in Relevance Decisions" (***Research and Advanced Technology for Digital Libraries*** [2009]: 94–105)** wondered if people would have an easier time determining if a document was relevant if they first saw a word cloud representation of the document. (To see an example of a word cloud, see page 593 of this text.)

The researchers chose 10 documents and created a word cloud representation for each of them. The text document version was shown to one group of people and the word cloud representation was shown to another group of people. The average time (in seconds) it took the people in the group to make a relevance decision was recorded for each version of the 10 documents and is shown in the accompanying table.

Document	Time to Relevance Decision Text Version	Time to Relevance Decision Word Cloud Version
1	3.55	2.95
2	2.94	3.04
3	3.72	2.72
4	2.63	1.97
5	2.39	2.17
6	5.36	3.64
7	1.86	1.96
8	2.49	2.04
9	2.76	2.46
10	2.77	2.63

You can view these data as consisting of two samples—a sample of times for the relevance decision for text documents and a sample of times for the relevance decision for word cloud representations.

In this case, the samples are paired rather than independent because each time in the text document sample is paired with a particular time in the word cloud sample for the same document. Notice that in 8 of the 10 data pairs, the time to relevance decision is greater for the text version of the document than for the word cloud representation of the same document. Intuitively, this suggests that the two population means may be different.

Both the text document times and the word cloud version times vary from document to document, and this variability may obscure any difference if the paired nature of the samples is ignored. To see how this might be the case, consider the two plots in Figure 13.1. The first plot (Figure 13.1(a)) ignores the pairing, and the two samples look quite similar. However, the plot in which pairs are identified (Figure 13.1(b)) does suggest that there is a difference because for 8 of the 10 pairs the text document time to relevance decision is greater than the time for the word cloud representation (these are the pairs linked by line segments that slant to the right). And in the two cases where the time for the text document is less than the time for the word cloud, the line segments are close to vertical, indicating that the two times for those documents were not very different.

FIGURE 13.1
Two plots of the paired data from Example 13.1
(a) pairing ignored
(b) pairs identified

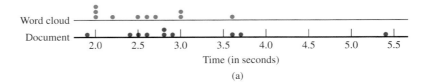

(a)

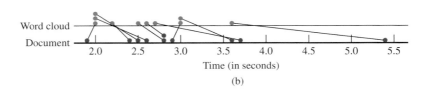

(b)

Example 13.1 suggests that when samples are paired, methods that ignore the pairing are not an appropriate way to analyze the data. This is why it is important that you determine if the samples are paired or independent before you carry out a hypothesis test or find a confidence interval estimate. Hypothesis tests and confidence intervals based on paired samples are introduced in Section 13.2. Tests and intervals based on independent samples are covered in Section 13.3.

Summing It Up—Section 13.1

The following learning objectives were addressed in this section:

Mastering the Mechanics
M1 Distinguish between independent and paired samples.
Samples are independent if the selection of the individuals in one sample does not influence the selection of the individuals in the second sample. This is the case if a random sample is selected from each of two populations. Samples are said to be paired if the observations in one sample can be paired in a meaningful way with the observations in the other sample. It is important to determine whether samples are independent or paired because the data are analyzed differently depending on how the samples were selected.

SECTION 13.1 EXERCISES

Each Exercise Set assesses the following learning objectives: M1

SECTION 13.1 Exercise Set 1

13.1 Descriptions of four studies are given. In each of the studies, the two populations of interest are the students at a particular university who live on campus and the students who live off campus. Which of these studies have samples that are independently selected?

Study 1: To determine if there is evidence that the mean amount of money spent on food each month differs for the two populations, a random sample of 45 students who live on campus and a random sample of 50 students who live off campus are selected.

Study 2: To determine if the mean number of hours spent studying differs for the two populations, a random sample students who live on campus is selected. Each student in this sample is asked how many hours he or she spends working each week. For each of these students who live on campus, a student who lives off campus and who works the same number of hours per week is identified and included in the sample of students who live off campus.

Study 3: To determine if the mean number of hours worked per week differs for the two populations, a random sample of students who live on campus and who have a brother or sister who also attends the university but who lives off campus is selected. The sibling who lives on campus is included in the on campus sample, and the sibling who lives off campus is included in the off-campus sample.

Study 4: To determine if the mean amount spent on textbooks differs for the two populations, a random sample of students who live on campus is selected. A separate random sample of the same size is selected from the population of students who live off campus.

13.2 For each of the following hypothesis testing scenarios, indicate whether or not the appropriate hypothesis test would be about a difference in two population means. If not, explain why not. (Hint: Consider the four key questions.)

Scenario 1: The international polling organization Ipsos reported data from a survey of 2000 randomly selected Canadians who carry debit cards **(Canadian Account Habits Survey, July 24, 2006)**. Participants in this survey were asked what they considered the minimum purchase amount for which it would be acceptable to use a debit card. You would like to determine if there is convincing evidence that the mean minimum purchase amount for which Canadians consider the use of a debit card to be acceptable is less than $10.

Scenario 2: Each person in a random sample of 247 male working adults and a random sample of 253 female working adults living in Calgary, Alberta, Canada, was asked how long, in minutes, his or her typical daily commute was **("Calgary Herald Traffic Study," Ipsos, September 17, 2005)**. You would like to determine if there is convincing evidence that the mean commute times differ for male workers and female workers.

Scenario 3: A hotel chain is interested in evaluating reservation processes. Guests can reserve a room using either a telephone system or an online system. Independent random samples of 80 guests who reserved a room by phone and 60 guests who reserved a room online were selected. Of those who reserved by phone, 57 reported that they were satisfied with the reservation process. Of those who reserved online, 50 reported that they were satisfied. You would like to determine if it reasonable to conclude that the proportion who are satisfied is higher for those who reserve a room online.

SECTION 13.1 Exercise Set 2

13.3 Descriptions of four studies are given. In each of the studies, the two populations of interest are female students at a particular university and male students at the university. Which of these studies have samples that are independently selected?

Study 1: To determine if there is evidence that the mean amount of time spent playing online video games differs for the two populations, a random sample of 20 female students and a random sample of 50 male students are selected.

Study 2: To determine if the mean number of hours worked per week differs for the two populations, a random sample of female students who have a brother who also attends the university is selected. The female student is included in the sample of females, and her brother is included in the sample of males.

Study 3: To determine if the mean amount of time spent using campus recreational facilities differs for the two populations, a random sample of female students is selected. A separate random sample of the same size is selected from the population of male students.

Study 4: To determine if the mean amount of money spent on housing differs for the two populations, a random sample of female students is selected. Each student in this sample is asked how far away from campus she lives. For each of these female students, a male student who lives about the same distance from campus is identified and included in the sample of male students.

13.4 For each of the following hypothesis testing scenarios, indicate whether or not the appropriate hypothesis test would be for a difference in two population means. If not, explain why not.

Scenario 1: A researcher at the Medical College of Virginia conducted a study of 60 randomly selected male soccer players and concluded that players who frequently "head" the ball in soccer have a lower mean IQ (*USA TODAY,* **August 14, 1995**). The soccer players were divided into two samples, based on whether they averaged 10 or more headers per game, and IQ was measured for each player. You would like to determine if the data support the researcher's conclusion.

Scenario 2: A credit bureau analysis of undergraduate students' credit records found that the mean number of credit cards in an undergraduate's wallet was 4.09 (**"Undergraduate Students and Credit Cards in 2004," Nellie Mae, May 2005**). It was also reported that in a random sample of 132 undergraduates, the mean number of credit cards that the

students said they carried was 2.6. You would like to determine if there is convincing evidence that the mean number of credit cards that undergraduates report carrying is less than the credit bureau's figure of 4.09.

Scenario 3: Some commercial airplanes recirculate approximately 50% of the cabin air in order to increase fuel efficiency. The authors of the paper **"Aircraft Cabin Air Recirculation and Symptoms of the Common Cold"** (*Journal of the American Medical Association* **[2002]: 483–486**) studied 1100 airline passengers who flew from San Francisco to Denver. Some passengers traveled on airplanes that recirculated air, and others traveled on planes that did not. Of the 517 passengers who flew on planes that did not recirculate air, 108 reported post-flight respiratory symptoms, while 111 of the 583 passengers on planes that did recirculate air reported such symptoms. You would like to determine if there is convincing evidence that the proportion of passengers with post-flight respiratory symptoms differs for passengers who travel on planes that do and planes that do not recirculate air.

SECTION 13.2 # Learning About a Difference in Population Means Using Paired Samples

Example 13.1 considered the time it took to make a decision about whether a document was relevant to a particular topic based on looking at the text document compared with looking at a word cloud representation of the document for a sample of 10 documents. The 10 document times can be viewed as a sample from the population of times for all documents presented in text form. The 10 word cloud times can be viewed as a sample from the population of times for all documents represented as word clouds. The time observations in the two samples are paired based on the document being considered.

When observations from one sample are paired in a meaningful way with observations in the second sample, inferences are based on the differences between the two observations in each pair. The n sample differences can then be regarded as having been selected from a large population of differences. For example, in Example 13.1, you can think of the 10 text document − word cloud time differences as having been selected from an entire population of differences (corresponding to a population of documents).

Before considering how the sample of differences is used to test hypotheses about the difference in population means, some new notation is needed:

$$\mu_d = \text{mean of the difference population}$$

and

$$\sigma_d = \text{standard deviation of the difference population}$$

The relationship between the two individual population means and the mean difference is

$$\mu_d = \mu_1 - \mu_2$$

This means that when the samples are paired, inferences about μ_d are equivalent to inferences about $\mu_1 - \mu_2$. Since inferences about μ_d can be based on the n observed sample differences, the original two-sample problem becomes a familiar one-sample problem.

The Paired t Test

To compare two population means when the samples are paired, first translate the hypotheses of interest about the value of $\mu_1 - \mu_2$ into equivalent hypotheses involving μ_d:

Hypothesis	Equivalent Hypothesis When Samples Are Paired
$H_0: \mu_1 - \mu_2 =$ hypothesized value	$H_0: \mu_d =$ hypothesized value
$H_a: \mu_1 - \mu_2 >$ hypothesized value	$H_a: \mu_d >$ hypothesized value
$H_a: \mu_1 - \mu_2 <$ hypothesized value	$H_a: \mu_d <$ hypothesized value
$H_a: \mu_1 - \mu_2 \neq$ hypothesized value	$H_a: \mu_d \neq$ hypothesized value

The general form of the null hypothesis is

$$H_0: \mu_1 - \mu_2 = \text{hypothesized value}$$

In most cases, the hypothesized value is 0, indicating that there is no difference in population means ($\mu_1 = \mu_2$). But sometimes it is something other than 0. For example, there are two different routes between San Luis Obispo, California, and Monterey, California. One is an inland route and the other is a more scenic coastal route. The scenic route is known to usually take more time, but you might be interested in deciding if the mean travel time is more than 1 hour longer. Because travel time can depend on day of the week and time of day, suppose that you collect data by having two drivers leave at the same time, with one taking the inland route and the other taking the coastal route. You repeat this process on 11 more days to get a sample of 12 travel times for each of the two routes.

With μ_1 representing the mean travel time for the scenic route and μ_2 representing the mean travel time for the inland route, the data from these two samples, which are paired by the day of the trip, can be used to test

$$H_0: \mu_1 - \mu_2 = 1 \text{ versus } H_0: \mu_1 - \mu_2 > 1$$

which is equivalent to

$$H_0: \mu_d = 1 \text{ versus } H_0: \mu_d > 1$$

Questions

Q Question type: estimation or hypothesis testing?

S Study type: sample data or experiment data?

T Type of data: one variable or two? Categorical or numerical?

N Number of samples or treatments: how many?

Sample differences (Sample 1 value – Sample 2 value) are calculated and used as the basis for testing hypotheses about μ_d. When the number of differences is large, or when it is reasonable to assume that the population of differences is approximately normal, the one-sample t test (from Chapter 12) based on the differences is the recommended test procedure. In general, if each of the two individual population distributions is normal, the population of differences also has a normal distribution. A normal probability plot or boxplot *of the differences* can be used to decide if the assumption of normality is reasonable.

The paired-samples t test, summarized in the following box, is a method you should consider when the answers to the four key questions (QSTN) are: *hypothesis testing, sample data, one numerical variable,* and *two paired samples.*

Paired-Samples t Test

Appropriate when the following conditions are met:

1. The samples are paired.
2. The n sample differences can be viewed as a random sample from a population of differences (or it is reasonable to regard the sample of differences as representative of the population of differences).
3. The *number of sample differences* is large ($n \geq 30$) *or* the population distribution *of differences* is approximately normal.

When these conditions are met, the following test statistic can be used:

$$t = \frac{\bar{x}_d - \mu_d}{\dfrac{s_d}{\sqrt{n}}}$$

where μ_d is the hypothesized value of the population mean difference from the null hypothesis, n is the number of sample differences, and $\bar{x}_d$ and s_d are the mean and standard deviation of the sample differences.

(continued)

Form of the null hypothesis: H_0: μ_d = hypothesized value

When the conditions above are met and the null hypothesis is true, this t test statistic has a t distribution with df = $n - 1$.

Associated P-value:

When the Alternative Hypothesis Is...	**The P-value Is...**
H_a: μ_d > hypothesized value	Area under the t curve to the right of the calculated value of the test statistic
H_a: μ_d < hypothesized value	Area under the t curve to the left of the calculated value of the test statistic
H_a: μ_d ≠ hypothesized value	2(area to the right of t) if t is positive or 2(area to the left of t) if t is negative

---

Example 13.2 Word Clouds Revisited

You can use the time to relevance decision data of Example 13.1 to test the claim that the mean time is greater for text documents than for word cloud representations. Because the samples are paired, the first thing to do is calculate the sample differences. These are the text – word cloud time differences for the 10 documents in the sample.

The sample data and the calculated differences are shown in the accompanying table. A positive difference indicates that the time to make a relevance decision was greater for the text document than for the word cloud representation.

Document	Time (in seconds) to Relevance Decision Text Version	Time (in seconds) to Relevance Decision Word Cloud Version	Difference (Text – Word Cloud)
1	3.55	2.95	0.60
2	2.94	3.04	−0.10
3	3.72	2.72	1.00
4	2.63	1.97	0.66
5	2.39	2.17	0.22
6	5.36	3.64	1.72
7	1.86	1.96	−0.10
8	2.49	2.04	0.45
9	2.76	2.46	0.30
10	2.77	2.63	0.14

The mean and standard deviation of these sample differences are $\bar{x}_d$ = 0.489 and s_d = 0.552. Do these data provide evidence that the mean time to make a relevance decision is greater for text documents than for word cloud representations of documents? The usual five-step process for hypothesis testing (HMC³) can be used to answer this question.

Considering the four key questions (QSTN), this situation can be described as *hypothesis testing, sample data, one numerical variable (time),* and *two paired samples.* This suggests a paired-samples t test.

H Hypotheses   You want to use data from the samples to test a claim about the difference in population means. The population characteristics of interest are:

μ_1 = mean time to make a relevance decision for text documents
μ_2 = mean mean time to make a relevance decision for word cloud representations

Because the samples are paired, you should also define μ_d:

$$\mu_d = \mu_1 - \mu_2 = \text{mean difference in time (text} - \text{word cloud)}$$

Translating the question of interest into the hypotheses gives

$$H_0: \mu_d = 0$$
$$H_a: \mu_d > 0$$

M Method   Because the answers to the four key questions are hypothesis testing, sample data, one numerical variable, and two paired samples, consider the paired-samples t test as a potential method. The test statistic for this test is

$$t = \frac{\bar{x}_d - 0}{\frac{s_d}{\sqrt{n}}}$$

When the null hypothesis is true, this statistic will have a t distribution with df $= 10 - 1 = 9$.

Next, choose a significance level for the test. For this example, $\alpha = 0.05$ will be used.

C Check   Although the 10 documents are not a random sample, the researchers chose the documents to be representative of online documents. Because the sample size is small, you must be willing to assume that the distribution of time differences for the population of online documents is at least approximately normal. The following boxplot of the 10 sample differences is not too asymmetric and there are no outliers, so it is reasonable to think that the population distribution could be at least approximately normal.

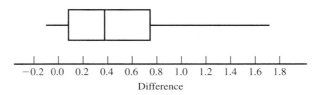

Because all conditions are met, it is appropriate to use the paired-samples t test.

C Calculate $n = 10$ $\bar{x}_d = 0.489$ $s_d = 0.552$

$$t = \frac{0.489 - 0}{\frac{0.552}{\sqrt{10}}} = \frac{0.489}{0.175} = 2.794$$

Rounding the test statistic value to 2.79 and using the df $= 9$ column of Appendix A Table 4, you get

$$P\text{-value} = \text{area to the right of } 2.79 = 0.010$$

C Communicate Results   Because the P-value (0.010) is less than α (0.05), you reject H_0. There is convincing evidence that the mean time to make a relevance decision for text documents is greater than the mean time for word cloud representations.

Statistical software could also be used for the calculate step of the test. Minitab output is shown here:

Paired T-Test
Paired T for Text − Word Cloud

	N	Mean	StDev	SE Mean
Text	10	3.047	0.975	0.308
Word Cloud	10	2.558	0.550	0.174
Difference	10	0.489	0.552	0.175

T-Test of mean difference = 0 (vs > 0): T-Value = 2.80 P-Value = 0.010

Example 13.3 Charitable Chimps

The authors of the paper **"Chimpanzees Are Indifferent to the Welfare of Unrelated Group Members"** (*Nature* [2005]: 1357–1359) concluded that "chimpanzees do not take advantage of opportunities to deliver benefits to individuals at no cost to themselves." This conclusion was based on data from a study in which chimpanzees were trained to use an apparatus that would deliver food just to themselves when one lever was pushed and would deliver food to both themselves and another chimpanzee in an adjoining cage when another lever was pushed. After training, the chimps were observed when there was no chimp in the adjoining cage and when there was another chimp in the adjoining cage.

The researchers hypothesized that if chimpanzees were motivated by the welfare of others, they would choose the option that provided food to both chimpanzees more often when there was a chimpanzee in the adjoining cage. Data on the number of times the "feed both" option was chosen out of 36 opportunities for a sample of 7 chimpanzees (approximate values from a graph in the paper) are given in the accompanying table.

	Number of Times "Feed Both" Option Was Chosen		
Chimp	Chimp in Adjoining Cage	No Chimp in Adjoining Cage	Difference
1	23	21	2
2	22	22	0
3	21	23	−2
4	23	21	2
5	19	18	1
6	19	16	3
7	19	19	0

You can use this sample data to determine if there is convincing evidence that the mean number of times the "feed both" option (the charitable response) is selected is greater when another chimpanzee is present in the adjoining cage than when there is no chimpanzee in the other cage.

Because the samples are paired, the first thing to do is to calculate the sample differences. Subtracting to obtain the (chimp – no chimp) differences results in the values shown in the difference column of the table. When a difference is positive, the trained chimp chose the charitable option more often when there was a chimp present in the other cage. A negative difference occurs when the charitable response was chosen less often when there was a chimp in the other cage.

Considering the four key questions (QSTN), this situation can be described as *hypothesis testing, sample data, one numerical variable (number of times a charitable response was chosen), and two paired samples.* This combination suggests a paired-samples *t* test.

H Hypotheses You want to use data from the samples to test a claim about the difference in population means. The population characteristics of interest are

μ_1 = mean number of charitable responses for all chimps when there is a chimp in the other cage

μ_2 = mean number of charitable responses for all chimps when there is no chimp in the other cage

Because the samples are paired,

μ_d = difference in mean number of charitable responses selections between when there is a chimp in the adjoining cage and when there is not

Translating the question of interest into hypotheses gives

$$H_0: \mu_d = 0$$
$$H_a: \mu_d > 0$$

M Method Because the answers to the four key questions are hypothesis testing, sample data, one numerical variable, and two paired samples, the paired-samples t test will be considered. The test statistic for this test is

$$t = \frac{\bar{x}_d - 0}{\dfrac{s_d}{\sqrt{n}}}$$

If the null hypothesis is true, this statistic has a t distribution with df $= 7 - 1 = 6$.

Next, choose a significance level for the test. For this example, $\alpha = 0.01$ will be used.

C Check Although the chimpanzees in this study were not randomly selected, the researchers considered them to be representative of the population of all chimpanzees. Because the sample size is small, you must be willing to assume that the distribution of differences for the population of chimpanzees is at least approximately normal. The following boxplot of the seven sample differences is reasonably symmetric, and there are no outliers, so it is reasonable to think that the population difference distribution may be approximately normal.

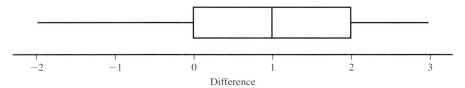

Because all conditions are met, it is appropriate to use the paired-samples t test.

C Calculate From the accompanying Minitab output, $t = 1.35$ and the associated P-value is 0.112.

Paired T-Test
Paired T for Chimp - No Chimp

	N	Mean	StDev	SE Mean
Chimp	7	20.8571	1.8645	0.7047
No Chimp	7	20.0000	2.4495	0.9258
Difference	7	0.857143	1.676163	0.633530

T-Test of mean difference = 0 (vs > 0): T-Value = 1.35 P-Value = 0.112

C Communicate Results Because the P-value (0.112) is greater than or equal to α (0.01), you fail to reject H_0. The data do not provide evidence that the mean number of times that the charitable option is chosen is greater when there is a chimpanzee in the adjoining cage. This is the basis for the statement quoted at the beginning of this example.

Confidence Interval When Samples Are Paired

When testing hypotheses about a difference in population means using paired samples, the test was based on the sample differences. The same is true when calculating a confidence interval estimate—you first calculate the differences and then use the one-sample t confidence interval introduced in Chapter 12.

The Paired-Samples t Confidence Interval for a Difference in Population Means

Appropriate when the following conditions are met:

1. The samples are paired.
2. The n sample differences can be viewed as a random sample from a population of differences (or it is reasonable to regard the sample of differences as representative of the population of differences).

(continued)

3. The *number of sample differences* is large ($n \geq 30$) *or* the population distribution *of differences* is approximately normal.

When these conditions are met, a confidence interval for the difference in population means is

$$\bar{x}_d \pm (t \text{ critical value})\left(\frac{s_d}{\sqrt{n}}\right)$$

where n is the number of sample differences, and $\bar{x}_d$ and s_d are the mean and standard deviation of the sample differences.

The t critical value is based on df $= n - 1$. The desired confidence level determines which t critical value is used. Appendix A Table 3, statistical software, or a graphing calculator can be used to obtain the t critical value.

Interpretation of Confidence Interval
You can be confident that the actual difference in population means is included in the calculated interval. This statement should be worded in context.

Interpretation of Confidence Level
The confidence level specifies the long-run proportion of the time that this method is successful in capturing the actual difference in population means.

Example 13.4 **Word Clouds One Last Time**

Example 13.2 provided data from a study to assess whether the time it takes to make a decision about the relevance of an online document topic is greater when the document is in text form than when it is represented by a word cloud. The conclusion in the hypothesis test in Example 13.2 was that there is convincing evidence that the mean time to make a relevance decision was greater for text documents than for word cloud representations. Once you have reached this conclusion, it may also be of interest to estimate how much greater the mean time is for text documents than for word clouds.

The sample data and calculated differences from Example 13.2 are shown again in the accompanying table. These data will be used to estimate the mean difference in the time it takes to make a relevance decision.

Document	Time (in seconds) to Relevance Decision Text Version	Time (in seconds) to Relevance Decision Word Cloud Version	Difference (Text – Word Cloud)
1	3.55	2.95	0.60
2	2.94	3.04	−0.10
3	3.72	2.72	1.00
4	2.63	1.97	0.66
5	2.39	2.17	0.22
6	5.36	3.64	1.72
7	1.86	1.96	−0.10
8	2.49	2.04	0.45
9	2.76	2.46	0.30
10	2.77	2.63	0.14

The answers to the four key questions are *estimation, sample data, one numerical variable (time),* and *two paired samples.* This combination leads to considering the paired-samples t confidence interval. The five-step process for estimation problems can be used to construct a 95% confidence interval for the mean difference in the time it takes to make a relevance decision.

E Estimate

You want to estimate

$$\mu_d = \mu_1 - \mu_2 = \text{mean difference in the time to make a relevance decision}$$

where

$$\mu_1 = \text{mean time to make a relevance decision for text documents}$$

and

$$\mu_2 = \text{mean time to make a relevance decision for word clouds}$$

M Method

Because the answers to the four key questions are estimation, sample data, one numerical variable, and two paired samples, a potential method is the paired-samples t confidence interval.

C Check

The conditions required for the paired-samples t confidence interval are the same as those required for the paired-samples t test. In Example 13.2, these conditions were discussed and found to be met. Therefore, it is appropriate to use the paired-samples t confidence interval to estimate the difference in means.

C Calculate

The mean and standard deviation calculated using the seven sample differences are $\bar{x}_d = 0.489$ and $s_d = 0.552$. The t critical value for df $= 9$ and a 95% confidence level is 2.26. Substituting these values into the formula for the paired-samples t confidence interval gives

$$\bar{x}_d \pm (t \text{ critical value})\left(\frac{s_d}{\sqrt{n}}\right) = 0.489 \pm (2.26)\frac{0.552}{\sqrt{10}}$$

$$= 0.489 \pm 0.395$$

$$= (0.094, 0.884)$$

A statistical software program or a graphing calculator could also have been used to compute the endpoints of the confidence interval. Minitab output is shown here.

Paired T-Test and CI
Paired T for Text – Word Cloud

	N	Mean	StDev	SE Mean
Text	10	3.047	0.975	0.308
Word Cloud	10	2.558	0.550	0.174
Difference	10	0.489	0.552	0.175

95% CI for mean difference: (0.094, 0.884)

C Communicate Results

Based on these samples, you can be 95% confident that the actual difference in mean time to make a relevance decision is somewhere between 0.094 seconds and 0.884 seconds. Because both endpoints of the interval are positive, you would estimate that the mean time to make a relevance decision is greater for text documents than for word cloud representations by somewhere between 0.094 and 0.884 seconds. The method used to construct this interval is successful in capturing the actual difference in population means about 95% of the time.

Summing It Up—Section 13.2

The following learning objectives were addressed in this section:

Conceptual Understanding
C1: Understand how a research question about the difference in two population means is translated into hypotheses.

Sample data can be used to test hypotheses about the difference in two population means, μ_1 and μ_2. The null hypothesis is usually $H_0: \mu_1 - \mu_2 = $ hypothesized value. When the samples are paired, this is equivalent to testing a null hypothesis about the mean of the differences, $H_0: \mu_d = $ hypothesized value. The research question will determine the form of the alternative hypothesis.

Mastering the Mechanics

M2: Know the conditions for appropriate use of the paired-samples *t* confidence interval and the paired-samples *t* test.

There are two conditions that need to be reasonably met in order for it to be appropriate to use the paired-samples *t* confidence interval formula and the paired-samples *t* test given in this section. These conditions are (1) that the *n* sample differences can be viewed as a random sample from a population of differences or it is reasonable to regard the sample of differences as representative of the population of differences, and (2) the sample size is large or the population difference distribution is approximately normal. The sample size is considered to be large if $n \geq 30$, where *n* is the number of sample differences.

M3: Carry out a paired-samples *t* test for a difference in population means.

The five-step process for hypothesis testing (HMC³) can be used to test hypotheses about the difference in two population means using paired samples. The process of carrying out a hypothesis test for a difference in two population means is illustrated in Examples 13.2 and 13.3.

M4: Calculate and interpret a paired-samples *t* confidence interval for a difference in population means.

The five-step process for estimation problems (EMC³) can be used to estimate the difference in two population means using paired samples. The process of calculating and interpreting a confidence interval for a difference in two population means is illustrated in Example 13.4.

Putting It into Practice

P1: Carry out a paired-samples *t* test for a difference in means and interpret the conclusion in context.

Hypothesis tests for a difference in two population means using paired samples are illustrated in Examples 13.2 and 13.3. If the null hypothesis is rejected, it is reasonable to conclude that there is convincing evidence in support of the alternative hypothesis. If the null hypothesis is not rejected, chance differences due to sampling variability is a plausible explanation for the observed difference in sample means and there is not convincing evidence that the alternative hypothesis is true.

P2: Interpret a paired-samples *t* confidence interval for a difference in means in context and interpret the associated confidence level.

A confidence interval for a difference in population means based on paired samples is interpreted as an interval of plausible values for the difference in population means. For an example of interpreting a confidence interval for a difference in population means when samples are paired and interpreting the associated confidence interval in context, see Example 13.4.

SECTION 13.2 EXERCISES

Each Exercise Set assesses the following chapter learning objectives: C1, M2, M3, M4, P1, P2

SECTION 13.2 Exercise Set 1

13.5 Many runners believe that listening to music while running enhances their performance. The authors of the paper **"Effects of Synchronous Music on Treadmill Running Among Elite Triathletes"** (*Journal of Science and Medicine in Sport* [2012]: 52–57) wondered if this is true for experienced runners. They recorded time to exhaustion for 11 triathletes while running on a treadmill at a speed determined to be near their peak running velocity. The time to exhaustion was recorded for each participant on two different days. On one day, each participant ran while listening to music that the runner selected as motivational. On a different day, each participant ran with no music playing.

For purposes of this exercise, assume that it is reasonable to regard these 11 triathletes as representative of the population of experienced triathletes. Only summary quantities were given in the paper, but the data in the table on the next page are consistent with the means and standard deviations given in the paper. Do the data provide convincing evidence that the mean time to exhaustion for experienced triathletes is greater when they run while listening to motivational music? Test the relevant hypotheses using a significance level of $\alpha = 0.05$.

TABLE FOR EXERCISE **13.5**

	Time to Exhaustion (seconds)										
Runner	1	2	3	4	5	6	7	8	9	10	11
Motivational Music	535	533	527	524	431	498	555	396	539	542	523
No Music	467	446	482	573	562	592	473	496	552	500	524

13.6 The study described in the previous exercise also measured time to exhaustion for the 11 triathletes on a day when they listened to music that the runners had classified as neutral as compared to motivational. The researchers calculated the difference between the time to exhaustion while running to motivational music and while running to neutral music. The mean difference in (motivational − neutral) was −7 seconds (the sample mean time to exhaustion was actually lower when listening to music the runner viewed as motivational than the mean when listening to music the runner viewed as neutral). Suppose that the standard deviation of the differences was $s_d = 80$. For purposes of this exercise, assume that it is reasonable to regard these 11 triathletes as representative of the population of experienced triathletes and that the population difference distribution is approximately normal. Is there convincing evidence that the mean time to exhaustion for experienced triathletes running to motivational music differs from the mean time to exhaustion when running to neutral music? Carry out a hypothesis test using $\alpha = 0.05$.

 13.7 The authors of the paper **"Statistical Methods for Assessing Agreement Between Two Methods of Clinical Measurement"** (*International Journal of Nursing Studies* [2010]: 931–936) compared two different instruments for measuring a person's ability to breathe out air. (This measurement is helpful in diagnosing various lung disorders.) The two instruments considered were a Wright peak flow meter and a mini-Wright peak flow meter. Seventeen people participated in the study, and for each person air flow was measured once using the Wright meter and once using the mini-Wright meter. The Wright meter is thought to provide a better measure of air flow, but the mini-Wright meter is easier to transport and to use. Use of the mini-Wright meter could be recommended as long as there is not convincing evidence that the mean reading for the mini-Wright meter is different from the mean reading for the Wright meter. Use the given data to determine if there is convincing evidence that the mean reading differs for the two instruments. For purposes of this exercise, you can assume that it is reasonable to consider the 17 Wright meter measurements and the 17 mini-Wright meter measurements as representative samples from

the their respective populations of measurements. (Hint: See Example 13.2.)

Subject	Mini-Wright Meter	Wright Meter
1	512	494
2	430	395
3	520	516
4	428	434
5	500	476
6	600	557
7	364	413
8	380	442
9	658	650
10	445	433
11	432	417
12	626	656
13	260	267
14	477	478
15	259	178
16	350	423
17	451	427

 13.8 To determine if chocolate milk is as effective as other carbohydrate replacement drinks, nine male cyclists performed an intense workout followed by a drink and a rest period. At the end of the rest period, each cyclist performed an endurance trial in which he exercised until exhausted, and the time to exhaustion was measured. Each cyclist completed the entire regimen on two different days. On one day, the drink provided was chocolate milk, and on the other day the drink provided was a carbohydrate replacement drink. Data consistent with summary quantities in the paper **"The Efficacy of Chocolate Milk as a Recovery Aid"** (*Medicine and Science in Sports and Exercise* [2004]: S126) are given in the table at the bottom of the page. Is there evidence that the mean time to exhaustion is greater after chocolate milk than after a carbohydrate replacement drink? Use a significance level of $\alpha = 0.05$.

TABLE FOR EXERCISE **13.8**

	Time to Exhaustion (minutes)								
Cyclist	1	2	3	4	5	6	7	8	9
Chocolate Milk	24.85	50.09	38.30	26.11	36.54	26.14	36.13	47.35	35.08
Carbohydrate Replacement	10.02	29.96	37.40	15.52	9.11	21.58	31.23	22.04	17.02

 Data set available

13.9 The paper **"Less Air Pollution Leads to Rapid Reduction of Airway Inflammation and Improved Airway Function in Asthmatic Children"** (*Pediatrics* [2009]: 1051–1058) describes a study in which children with mild asthma who live in a polluted urban environment were relocated to a less-polluted rural environment for 7 days. Various measures of respiratory function were recorded first in the urban environment and then again after 7 days in the rural environment. The accompanying graphs show the urban and rural values for three of these measures: nasal eosinophils (%), exhaled FENO concentration (ppb), and peak expiratory flow (PEF, L/min). Urban and rural values for the same child are connected by a line. The authors of the paper used paired-samples *t* tests to determine that there was a significant difference in the urban and rural means for each of these three measures. One of these tests resulted in a *P*-value less than 0.001, one resulted in a *P*-value between 0.001 and 0.01, and one resulted in a *P*-value between 0.01 and 0.05.

a. Which measure (Eosinophils, FENO, or PEF) do you think resulted in a test with the *P*-value that was less than 0.001? Explain your reasoning.

b. Which measure (Eosinophils, FENO, or PEF) do you think resulted in the test with the largest *P*-value? Explain your reasoning.

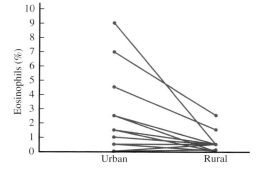

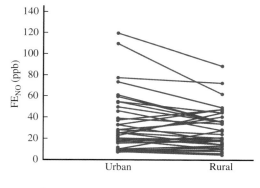

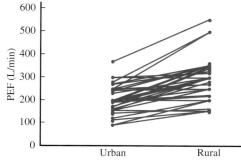

13.10 In the study described in the paper **"Exposure to Diesel Exhaust Induces Changes in EEG in Human Volunteers"** (*Particle and Fibre Toxicology* [2007]), 10 healthy men were exposed to diesel exhaust for 1 hour. A measure of brain activity (called median power frequency, or MPF in Hz) was recorded at two different locations in the brain both before and after the diesel exhaust exposure. The resulting data are given in the accompanying table. For purposes of this exercise, assume that the sample of 10 men is representative of healthy adult males.

	MPF (in Hz)			
Subject	Location 1 Before	Location 1 After	Location 2 Before	Location 2 After
1	6.4	8.0	6.9	9.4
2	8.7	12.6	9.5	11.2
3	7.4	8.4	6.7	10.2
4	8.7	9.0	9.0	9.6
5	9.8	8.4	9.7	9.2
6	8.9	11.0	9.0	11.9
7	9.3	14.4	7.9	9.1
8	7.4	11.3	8.3	9.3
9	6.6	7.1	7.2	8.0
10	8.9	11.2	7.4	9.1

Construct and interpret a 90% confidence interval estimate for the difference in mean MPF at brain location 1 before and after exposure to diesel exhaust. (Hint: See Example 13.7.)

13.11 Use the information given in the previous exercise to construct and interpret a 90% confidence interval estimate for the difference in mean MPF at brain location 2 before and after exposure to diesel exhaust.

13.12 In Exercise 13.5, a hypothesis test leads to the conclusion that there is not convincing evidence that the mean time to exhaustion for experienced triathletes is greater when they run while listening to motivational music than when they run with no music. Use the information given in that exercise to construct and interpret a 95% confidence interval for the difference in mean time to exhaustion for experienced triathletes when running to motivational music and the mean time when running with no music. Explain how this interval is consistent with the conclusion in the hypothesis test of Exercise 13.5.

SECTION 13.2 Exercise Set 2

13.13 The authors of the paper **"Ultrasound Techniques Applied to Body Fat Measurement in Male and Female Athletes"** (*Journal of Athletic Training* [2009]: 142–147) compared two different methods for measuring body fat percentage. One method uses ultrasound and the other method uses X-ray technology. The accompanying table gives body fat percentages for 16 athletes using each of these methods (a subset

Data set available

of the data given in a graph that appeared in the paper). For purposes of this exercise, you can assume that the 16 athletes who participated in this study are representative of the population of athletes. Do these data provide convincing evidence that the mean body fat percentage measurement differs for the two methods? Test the appropriate hypotheses using $\alpha = 0.05$.

Athlete	X-ray	Ultrasound
1	5.00	4.75
2	7.00	3.75
3	9.25	9.00
4	12.00	11.75
5	17.25	17.00
6	29.50	27.50
7	5.50	6.50
8	6.00	6.75
9	8.00	8.75
10	8.50	9.50
11	9.25	9.50
12	11.00	12.00
13	12.00	12.25
14	14.00	15.50
15	17.00	18.00
16	18.00	18.25

 13.14 The humorous paper **"Will Humans Swim Faster or Slower in Syrup?"** (*American Institute of Chemical Engineers Journal* [2004]: 2646–2647) investigated the fluid mechanics of swimming. Twenty swimmers each swam a specified distance in a water-filled pool and in a pool in which the water was thickened with food grade guar gum to create a syrup-like consistency. Velocity, in meters per second, was recorded. Values estimated from a graph in the paper are given. The authors of the paper concluded that swimming in guar syrup does not change mean swimming speed. Are the given data consistent with this conclusion? Carry out a hypothesis test using a 0.01 significance level.

	Velocity (m/s)	
Swimmer	Water	Guar Syrup
1	0.90	0.92
2	0.92	0.96
3	1.00	0.95
4	1.10	1.13
5	1.20	1.22
6	1.25	1.22
7	1.25	1.26
8	1.30	1.30
9	1.35	1.34
10	1.40	1.41
11	1.40	1.44

(continued)

	Velocity (m/s)	
Swimmer	Water	Guar Syrup
12	1.50	1.52
13	1.65	1.58
14	1.70	1.70
15	1.75	1.80
16	1.80	1.76
17	1.80	1.84
18	1.85	1.89
19	1.90	1.88
20	1.95	1.95

13.15 The article **"Puppy Love? It's Real, Study Says"** (*USA TODAY,* April 17, 2015) describes a study into how people communicate with their pets. The conclusion expressed in the title of the article was based on research published in *Science* (**"Oxytocin-Gaze Positive Loop and the Coevolution of Human-Dog Bonds,"** April 17, 2015). Researchers measured the oxytocin levels (in picograms per milligram, pg/mg) of 22 dog owners before and again after a 30-minute interaction with their dogs. (Oxytocin is a hormone known to play a role in parent-child bonding.) The difference in oxytocin level (before – after) was calculated for each of the 22 dog owners. Suppose that the mean and standard deviation of the differences (approximate values based on a graph in the paper) were $\bar{x}_d = 27$ pg/mg and $s_d = 30$ pg/mg.
a. Explain why the two samples (oxytocin levels before interaction and oxytocin levels after interaction) are paired.
b. Assume that it is reasonable to regard the 22 dog owners who participated in this study as representative of dog owners in general. Do the data from this study provide convincing evidence that there is an increase in mean oxytocin level of dog owners after 30 minutes of interaction with their dogs? State and test the appropriate hypotheses using a significance level of 0.05.

13.16 Two proposed computer mouse designs were compared by recording wrist extension in degrees for 24 people who each used both mouse designs (**"Comparative Study of Two Computer Mouse Designs,"** Cornell Human Factors Laboratory Technical Report RP7992). The difference in wrist extension was calculated by subtracting extension for mouse type B from the wrist extension for mouse type A for each person. The mean difference was reported to be 8.82 degrees. Assume that this sample of 24 people is representative of the population of computer users.
a. Suppose that the standard deviation of the differences was 10 degrees. Is there convincing evidence that the mean wrist extension for mouse type A is greater than for mouse type B? Use a 0.05 significance level.
b. Suppose that the standard deviation of the differences was 26 degrees. Is there convincing evidence that the mean wrist extension for mouse type A is greater than for mouse type B? Use a 0.05 significance level.

c. Briefly explain why different conclusions were reached in the hypothesis tests of Parts (a) and (b).

13.17 Breast feeding sometimes results in a temporary loss of bone mass as calcium is depleted in the mother's body to provide for milk production. The paper **"Bone Mass Is Recovered from Lactation to Postweaning in Adolescent Mothers with Low Calcium Intakes"** (*American Journal of Clinical Nutrition* [2004]: 1322–1326) gave the accompanying data on total body bone mineral content (in grams) for a representative sample of mothers both during breast feeding (B) and in the post-weaning period (P). Use a 95% confidence interval to estimate the difference in mean total body bone mineral content during post-weaning and during breast feeding.

Subject	1	2	3	4	5	6
B	1,928	2,549	2,825	1,924	1,628	2,175
P	2,126	2,885	2,895	1,942	1,750	2,184
Subject	7	8	9	10		
B	2,144	2,621	1,843	2,541		
P	2,164	2,626	2,006	2,627		

13.18 The paper **"Quantitative Assessment of Glenohumeral Translation in Baseball Players"** (*The American Journal of Sports Medicine* [2004]: 1711–1715) considered various aspects of shoulder motion for a representative sample of pitchers and a representative sample of position players. The table at the bottom of the page gives data supplied by the authors on a measure of the extent of anterior and posterior motion (in millimeters), both for the dominant arm and the nondominant arm.

Estimate the mean difference in motion between dominant and nondominant arms for pitchers using a 95% confidence interval.

13.19 Use the data given in the previous exercise to complete the following:
a. Estimate the mean difference in motion between dominant and nondominant arms for position players using a 95% confidence interval.
b. The authors asserted that pitchers have a greater difference in mean motion of their shoulders than do position players. Do you agree? Explain.

ADDITIONAL EXERCISES

13.20 The authors of the paper **"Concordance of Self-Report and Measured Height and Weight of College Students"** (*Journal of Nutrition, Education and Behavior* [2015]: 94–98) used a paired-samples *t* test to reach the conclusion that male college students

TABLE FOR EXERCISE **13.18**

Player	Position Player Dominant Arm	Position Player Nondominant Arm	Pitcher	Pitcher Dominant Arm	Pitcher Nondominant Arm
1	30.31	32.54	1	27.63	24.33
2	44.86	40.95	2	30.57	26.36
3	22.09	23.48	3	32.62	30.62
4	31.26	31.11	4	39.79	33.74
5	28.07	28.75	5	28.50	29.84
6	31.93	29.32	6	26.70	26.71
7	34.68	34.79	7	30.34	26.45
8	29.10	28.87	8	28.69	21.49
9	25.51	27.59	9	31.19	20.82
10	22.49	21.01	10	36.00	21.75
11	28.74	30.31	11	31.58	28.32
12	27.89	27.92	12	32.55	27.22
13	28.48	27.85	13	29.56	28.86
14	25.60	24.95	14	28.64	28.58
15	20.21	21.59	15	28.58	27.15
16	33.77	32.48	16	31.99	29.46
17	32.59	32.48	17	27.16	21.26

Data set available

tend to over-report both weight and height. This conclusion was based on a sample of 634 male college students selected from eight different universities. The sample mean difference between the reported weight and actual measured weight was 1.2 pounds and the standard deviation of the differences was 5.71 pounds. For purposes of this exercise, you can assume that the sample was representative of male college students.

a. Carry out a hypothesis test to determine if there is a significant difference in the mean reported weight and the mean actual weight for male college students.

b. For height, the mean difference between the reported height and actual measured height was 0.6 inches and the standard deviation of the differences was 0.8 inches. Carry out a hypothesis test to determine if there is a significant difference in the mean reported height and the mean actual height for male college students.

c. Do the conclusions reached in the hypothesis tests of Parts (a) and (b) support the given conclusion that male college students tend to over-report both height and weight? Explain.

13.21 The paper referenced in the previous exercise also compared the reported heights and weights to actual measured heights and weights for a sample of 1052 female college students selected from eight different universities. The resulting data is summarized in the accompanying table. For purposes of this exercise, you can assume that this sample is representative of female college students.

	Sample Mean Difference (Reported – Actual)	Sample Standard Deviation of Differences
Weight	−0.6	4.8
Height	−0.2	0.8

a. Carry out a hypothesis test to determine if there is a significant difference in the mean reported weight and the mean actual weight for female college students.

b. Carry out a hypothesis test to determine if there is a significant difference in the mean reported height and the mean actual height for female college students.

c. Do the conclusions reached in the hypothesis tests of Parts (a) and (b) support the conclusion that female college students tend to under-report both height and weight? Explain.

13.22 Can taking chess lessons and playing chess daily improve memory? The online article **"The USA Junior Chess Olympics Research: Developing Memory and Verbal Reasoning"** (*New Horizons for Learning*, April 2001; available at www.newhorizons.org) describes a study in which sixth-grade students who had not previously played chess participated in a program in which they took chess lessons and played chess daily for 9 months. Each student took a memory test before starting the chess program and again at the end of the 9-month period. Data (read from a graph

in the article) and calculated differences are given in the accompanying table. The author of the article wanted to know if these data support the claim that students who participated in the chess program tend to achieve higher memory scores after completion of the program. Carry out a hypothesis test to answer this question.

Student	Memory Test Score		
	Pre-test	Post-test	Difference
1	510	850	−340
2	610	790	−180
3	640	850	−210
4	675	775	−100
5	600	700	−100
6	550	775	−225
7	610	700	−90
8	625	850	−225
9	450	690	−240
10	720	775	−55
11	575	540	35
12	675	680	−5

13.23 The paper **"Driving Performance While Using a Mobile Phone: A Simulation Study of Greek Professional Drivers"** (*Transportation Research Part F* [2016]: 164–170) describes a study in which 50 Greek male taxi drivers drove in a driving simulator. In the simulator, they were asked to drive following a lead car. On one drive, they had no distractions, and the average distance between the driver's car and the lead car was recorded. In a second drive, the drivers talked on a mobile phone while driving. The authors of the paper used a paired-samples t test to determine if the mean following distance is greater when the driver has no distractions than when the driver is talking on a mobile phone. The mean of the 50 sample differences (no distraction − talking on mobile phone) was 0.47 meters and the standard deviation of the sample differences was 1.22 meters. The authors concluded that there was evidence to support the claim that the mean following distance for Greek taxi drivers is greater when there are no distractions than when the driver is talking on a mobile phone. Do you agree with this conclusion? Carry out a hypothesis test to support your answer. You may assume that this sample of 50 drivers is representative of Greek taxi drivers.

13.24 The paper referenced in the previous exercise also had the 50 taxi drivers drive in the simulator while sending and receiving text messages. The mean of the 50 sample differences (no distraction − reading text messages) was 1.3 meters and the standard deviation of the sample differences was 1.54 meters. The authors concluded that there was evidence to support the claim that the mean following distance for Greek taxi drivers is greater when there are no distractions that when the driver is texting. Do you agree with this conclusion? Carry out a hypothesis test to support your answer. You can assume that this sample of 50 drivers is representative of Greek taxi drivers.

13.25 Use the information given in the previous exercise to construct and interpret a 95% confidence interval for the difference in mean following distance for Greek taxi drivers while driving with no distractions and while driving and texting.

13.26 Babies born extremely prematurely run the risk of various neurological problems and tend to have lower IQ and verbal ability scores than babies that are not premature. The article **"Premature Babies May Recover Intelligence, Study Says"** (*San Luis Obispo Tribune*, February 12, 2003) summarized medical research that suggests that the deficits observed at an early age may decrease as children age. Children who were born prematurely were given a test of verbal ability at age 3 and again at age 8. The test is scaled so that a score of 100 would be average for normal-birth-weight children. Data for 50 children who were born prematurely were used to generate the accompanying Minitab output, where Age3 represents the verbal ability score at age 3 and Age8 represents the verbal ability score at age 8. Use the Minitab output to determine if there is convincing evidence that the mean verbal ability score for children born prematurely increases between age 3 and age 8. You can assume that it is reasonable to regard the sample of 50 children as a random sample from the population of all children born prematurely.

Paired T-Test and CI: Age8, Age3
Paired T for Age8 - Age3

	N	Mean	StDev	Se Mean
Age8	50	97.21	16.97	2.40
Age3	50	87.30	13.84	1.96
Difference	50	9.91	22.11	3.13

T-Test of mean difference = 0 (vs > 0): T-Value = 3.17
P-Value = 0.001

13.27 In a study of memory recall, 8 students from a large psychology class were selected at random and given 10 minutes to memorize a list of 20 nonsense words. Each was asked to list as many of the words as he or she could remember both 1 hour and 24 hours later. The data are given in the accompanying table. Is there convincing evidence to suggest that the mean number of words recalled after 1 hour is greater than the mean recall after 24 hours by more than 3? Use a significance level of $\alpha = 0.01$.

Student	1	2	3	4	5	6	7	8
1 hour later	14	12	18	7	11	9	16	15
24 hours later	10	4	14	6	9	6	12	12

13.28 Breast feeding sometimes results in a temporary loss of bone mass, as calcium is depleted in the mother's body to provide for milk production. The paper **"Bone Mass Is Recovered from Lactation to Postweaning in Adolescent Mothers with Low Calcium Intakes"** (*American Journal of Clinical Nutrition* [2004]: 1322–1326) gave the data at the bottom of the page on total body bone mineral content (in grams) for a sample of mothers both during breast feeding (B) and in the post-weaning period (P). Do the data suggest that actual mean bone mineral content during post-weaning is greater than that during breast feeding by more than 25 grams? State and test the appropriate hypotheses using a significance level of 0.05.

13.29 Babies born extremely prematurely run the risk of various neurological problems and tend to have lower IQ and verbal ability scores than babies who are not premature. The article **"Premature Babies May Recover Intelligence, Study Says"** (*San Luis Obispo Tribune*, February 12, 2003) summarized medical research that suggests that the deficit observed at an early age may decrease as children age. Children who were born prematurely were given a test of verbal ability at age 3 and again at age 8. The test is scaled so that a score of 100 would be average for normal-birth-weight children. Data that are consistent with summary quantities given in the paper for 50 children who were born prematurely were used to generate the accompanying Minitab output, where Age3 represents the verbal ability score at age 3 and Age8 represents the verbal ability score at age 8. Use the information in the Minitab output to construct and interpret a 95% confidence interval for the change in mean verbal ability score from age 3 to age 8. You can assume that it is reasonable to regard the sample of 50 children as a random sample from the population of all children born prematurely.

Paired T-Test and CI: Age8, Age3
Paired T for Age8 - Age3

	N	Mean	StDev	Se Mean
Age8	50	97.21	16.97	2.40
Age3	50	87.30	13.84	1.96
Difference	50	9.91	22.11	3.13

13.30 In a study of memory recall, 8 students from a large psychology class were selected at random and given 10 minutes to memorize a list of 20 nonsense words. Each was asked to list as many of the words as he or she could remember both 1 hour and 24 hours later. The data are as shown in the accompanying table. Use these data to estimate the difference in mean number of words remembered after 1 hour and after 24 hours. Use a 90% confidence interval.

Student	1	2	3	4	5	6	7	8
1 hour later	14	12	18	7	11	9	16	15
24 hours later	10	4	14	6	9	6	12	12

TABLE FOR EXERCISE **13.28**

Mother	1	2	3	4	5	6	7	8	9	10
B	1928	2549	2825	1924	1628	2175	2144	2621	1843	2541
P	2126	2885	2895	1942	1750	2184	2164	2626	2006	2627

Data set available

Learning About a Difference in Population Means Using Independent Samples

As was the case when comparing two population means using paired samples, when samples are independently selected, it is common to focus on the difference $\mu_1 - \mu_2$. Because $\bar{x}_1$ is an estimate of μ_1 and $\bar{x}_2$ is an estimate of μ_2, the obvious statistic to use as an estimate of $\mu_1 - \mu_2$ is the difference in the sample means, $\bar{x}_1 - \bar{x}_2$. But the values of $\bar{x}_1$ and $\bar{x}_2$ each vary from sample to sample, so the difference $\bar{x}_1 - \bar{x}_2$ will also vary. Because the statistic $\bar{x}_1 - \bar{x}_2$ is used to draw conclusions about the difference between two population means when the samples are independently selected, you need to know about the behavior of this statistic. The sampling distribution of $\bar{x}_1 - \bar{x}_2$ describes this behavior.

General Properties of the Sampling Distribution of $\bar{x}_1 - \bar{x}_2$

If $\bar{x}_1 - \bar{x}_2$ is the difference in sample means for independently selected random samples, then the following rules hold.

Rule 1. $\mu_{\bar{x}_1 - \bar{x}_2} = \mu_1 - \mu_2$

This rule states that the sampling distribution of $\bar{x}_1 - \bar{x}_2$ is centered at the actual value of the difference in population means. This means that differences in sample means tend to cluster around the actual difference in population means.

Rule 2. $\sigma^2_{\bar{x}_1 - \bar{x}_2} = \dfrac{\sigma_1^2}{n_1} + \dfrac{\sigma_2^2}{n_2}$

and

$$\sigma_{\bar{x}_1 - \bar{x}_2} = \sqrt{\dfrac{\sigma_1^2}{n_1} + \dfrac{\sigma_2^2}{n_2}}$$

This rule specifies the standard error of $\bar{x}_1 - \bar{x}_2$. The value of the standard error describes how much the $\bar{x}_1 - \bar{x}_2$ values tend to vary from one pair of samples to another and also from the actual difference in population means.

Rule 3.   If both n_1 and n_2 are large or if the population distributions are approximately normal, the sampling distribution of $\bar{x}_1 - \bar{x}_2$ is approximately normal. The sample sizes can be considered large if $n_1 \geq 30$ and $n_2 \geq 30$.

The logic of hypothesis testing and the general process for carrying out a hypothesis test is the same for all hypothesis tests. Recall that what may differ from one type of test to another includes

1.   The null and alternative hypotheses
2.   The test statistic
3.   The way in which the associated *P*-value is determined

Consider each of these three things in the context of testing hypotheses about a difference in population means using independent samples:

1. Hypotheses

The hypothesis test focuses on the difference in population means, $\mu_1 - \mu_2$. When there is no difference between the two population means, $\mu_1 - \mu_2 = 0$. This means that

$\mu_1 = \mu_2$ is equivalent to $\mu_1 - \mu_2 = 0$

Similarly

$\mu_1 > \mu_2$ is equivalent to $\mu_1 - \mu_2 > 0$

and

$\mu_1 < \mu_2$ is equivalent to $\mu_1 - \mu_2 < 0$

To determine the null and alternative hypotheses, you will need to translate the question of interest into hypotheses. For example, if you want to determine if there is evidence that μ_1 is greater than μ_2, you would choose $\mu_1 - \mu_2 > 0$ as the alternative hypothesis.

The general form of the null hypothesis is

$$H_0: \mu_1 - \mu_2 = \text{hypothesized value}$$

In most cases, the hypothesized value is 0, indicating no difference in the population means. But sometimes it is something other than 0. For example, suppose μ_1 and μ_2 are the mean fuel efficiencies (in miles per gallon, mpg) for cars of a particular model equipped with 4-cylinder and 6-cylinder engines, respectively. The hypotheses under consideration might be

$$H_0: \mu_1 - \mu_2 = 5 \text{ versus } H_a: \mu_1 - \mu_2 > 5$$

The null hypothesis is equivalent to the claim that mean fuel efficiency for the 4-cylinder engine is greater than the mean fuel efficiency for the 6-cylinder engine by 5 mpg. The alternative hypothesis states that the difference between the mean fuel efficiencies is greater than 5 mpg.

2. Test Statistic

Based on the properties of the sampling distribution of $\bar{x}_1 - \bar{x}_2$, when you have independent random samples you know the following three things:

1. $\mu_{\bar{x}_1 - \bar{x}_2} = \mu_1 - \mu_2$

2. $\sigma_{\bar{x}_1 - \bar{x}_2} = \sqrt{\dfrac{\sigma_1^2}{n_1} + \dfrac{\sigma_2^2}{n_2}}$

3. If both n_1 and n_2 are large or if the population distributions are approximately normal, then the sampling distribution of $\bar{x}_1 - \bar{x}_2$ is approximately normal.

This means that if the samples are independently selected, and either both samples sizes are large or the population distributions are approximately normal, the statistic

$$z = \frac{(\bar{x}_1 - \bar{x}_2) - (\mu_1 - \mu_2)}{\sqrt{\dfrac{\sigma_1^2}{n_1} + \dfrac{\sigma_2^2}{n_2}}}$$

will have a distribution that is (approximately) the standard normal (z) distribution. (This is because $\bar{x}_1 - \bar{x}_2$ has a distribution that is approximately normal and the z statistic above results from standardizing $\bar{x}_1 - \bar{x}_2$ by subtracting its mean and then dividing by its standard deviation.)

Although it is possible to base a test procedure and confidence interval on the z statistic above, the values of the population variances, σ_1^2 and σ_2^2, are almost never known. As a result, the z statistic is rarely used. When σ_1^2 and σ_2^2, are unknown, the sample variances, s_1^2 and s_2^2 are used to estimate σ_1^2 and σ_2^2 in the z statistic above, resulting in the following t statistic.

When two random samples are independently selected and when n_1 and n_2 are both large or when the population distributions are approximately normal, the t statistic

$$t = \frac{(\bar{x}_1 - \bar{x}_2) - (\mu_1 - \mu_2)}{\sqrt{\dfrac{s_1^2}{n_1} + \dfrac{s_2^2}{n_2}}}$$

has approximately a t distribution with

$$df = \frac{(V_1 + V_2)^2}{\dfrac{V_1^2}{n_1 - 1} + \dfrac{V_2^2}{n_2 - 1}} \quad \text{where } V_1 = \frac{s_1^2}{n_1} \text{ and } V_2 = \frac{s_2^2}{n_2}$$

The calculated value of df should be truncated (rounded down) to an integer value.

If one or both sample sizes are small, you can use normal probability plots, dotplots, or boxplots to evaluate whether it is reasonable to consider the population distributions to be approximately normal.

3. Calculating a *P*-value

Once the value of the test statistic and the degrees of freedom (df) have been determined, the next step is to calculate a *P*-value, which tells you how likely it would be to observe sample data as or more extreme as what was observed if the null hypothesis were true. Because this test statistic has approximately a *t* distribution, *P*-values are calculated as an area under a *t* distribution curve. If you aren't using technology, calculating the test statistic and the df can be time consuming, but once that is done, finding the *P*-value is straightforward.

You now have all the pieces needed to carry out the hypothesis test. The two-sample *t* test for a difference in populations means summarized in the following box is a method you should consider when the answers to the four key questions (QSTN) are: *hypothesis testing, sample data, one numerical variable,* and *two independently selected samples.*

Two-Sample *t* Test for a Difference in Population Means

Appropriate when the following conditions are met:

1. The samples are independently selected.
2. Each sample is a random sample from the population of interest or the samples are independently selected in a way that would result in samples that are representative of the populations.
3. Both sample sizes are large ($n_1 \geq 30$ and $n_2 \geq 30$) *or* both population distributions are approximately normal.

When these conditions are met, the following test statistic can be used:

$$t = \frac{(\bar{x}_1 - \bar{x}_2) - (\mu_1 - \mu_2)}{\sqrt{\dfrac{s_1^2}{n_1} + \dfrac{s_2^2}{n_2}}}$$

where $\mu_1 - \mu_2$ is the hypothesized value of the difference in population means from the null hypothesis (often this will be 0).

When the conditions above are met and the null hypothesis is true, the *t* test statistic has a *t* distribution with

$$df = \frac{(V_1 + V_2)^2}{\dfrac{V_1^2}{n_1 - 1} + \dfrac{V_2^2}{n_2 - 1}} \quad \text{where} \quad V_1 = \frac{s_1^2}{n_1} \text{ and } V_2 = \frac{s_2^2}{n_2}$$

The calculated value of df should be truncated (rounded down) to obtain an integer value.

Form of the null hypothesis: $H_0: \mu_1 - \mu_2 =$ hypothesized value

Associated *P*-value:

When the Alternative Hypothesis Is...	The *P*-value Is...
$H_a: \mu_1 - \mu_2 >$ hypothesized value	Area under the *t* curve to the right of the calculated value of the test statistic
$H_a: \mu_1 - \mu_2 <$ hypothesized value	Area under the *t* curve to the left of the calculated value of the test statistic
$H_a: \mu_1 - \mu_2 \neq$ hypothesized value	2(area to the right of *t*) if *t* is positive or 2(area to the left of *t*) if *t* is negative

Steps
H Hypotheses
M Method
C Check
C Calculate
C Communicate results

Now you're ready to look at some examples, following the five-step process for hypothesis testing (HMC³).

Example 13.5   Facebook and Grades

The authors of the paper **"Facebook and Academic Performance"** (*Computers in Human Behavior* **[2010]: 1237–1245**) investigated the relationship between college grade point average (GPA) and Facebook use. One question that the researchers were hoping to answer is whether the mean college GPA for students who use Facebook is less than the mean college GPA for students who do not. Two samples (141 students who were Facebook users and 68 students who were not Facebook users) were independently selected from students at a large, public midwestern university. Although the samples were not selected at random, they were selected to be representative of the two populations (students who use Facebook and students who do not use Facebook at this university). Data from these samples were used to calculate sample means and standard deviations.

Population	Population Mean	Sample Size	Sample Mean	Sample Standard Deviation
Students at the university who use Facebook	μ_1 = mean college GPA for students who use Facebook	$n_1 = 141$	$\bar{x}_1 = 3.06$	$s = 0.95$
Students at the university who do not use Facebook	μ_2 = mean college GPA for students who do not use Facebook	$n_2 = 68$	$\bar{x}_2 = 3.82$	$s = 0.41$

As always, begin by answering the four key questions (QSTN) to identify a potential method for answering the question posed: (Q) We would like to test a claim about the two populations, so this is a hypothesis testing problem. (S) The data are from samples. (T) There is one numerical variable, which is GPA. (N) There are two samples and they were independently selected. Because the answers to the four questions are *hypothesis testing, sample data, one numerical variable,* and *two independently selected samples*, you should consider a two-sample *t* test for a difference in population means.

Now you can use the five-step process for hypothesis testing problems (HMC³) to answer the question posed.

Process Step	
H Hypotheses	The question of interest is about the difference in mean GPA for the two populations. Since you are dealing with means, start by defining μ_1 and μ_2 in the context of this example.
	Population characteristics of interest:
	μ_1 = population mean GPA for students who use Facebook
	μ_2 = population mean GPA for students who do not use Facebook
	The question of interest (is the mean GPA for Facebook users lower than the mean for students who do not use Facebook) translates into an alternative hypothesis of $\mu_1 < \mu_2$. Writing this in terms of the difference in means, you get $\mu_1 - \mu_2 < 0$. The null hypothesis is that there is no difference in the population means.
	Hypotheses:
	Null hypothesis: $H_0: \mu_1 - \mu_2 = 0$
	Alternative hypothesis: $H_a: \mu_1 - \mu_2 < 0$
M Method	Because the answers to the four key questions are hypothesis testing, sample data, one numerical variable and two independently selected samples, consider a two-sample *t* test for a difference in population means.

(continued)

Process Step	
	Potential method: Two-sample t test for a difference in population means. The test statistic for this test is $$t = \frac{(\bar{x}_1 - \bar{x}_2) - (\mu_1 - \mu_2)}{\sqrt{\dfrac{s_1^2}{n_1} + \dfrac{s_2^2}{n_2}}}$$ When the null hypothesis is true, this statistic has approximately a t distribution with $$df = \frac{(V_1 + V_2)^2}{\dfrac{V_1^2}{n_1 - 1} + \dfrac{V_2^2}{n_2 - 1}} \quad \text{where} \quad V_1 = \frac{s_1^2}{n_1} \text{ and } V_2 = \frac{s_2^2}{n_2}$$ You also need to select a significance level for the test. You should choose a significance level after considering the consequences of a Type I error (rejecting a true null hypothesis) and a Type II error (failing to reject a false null hypothesis). In this example, a Type I error would be *incorrectly* concluding that the mean GPA of students who use Facebook is less than the mean GPA of students who do not. A Type II error would be not being convinced that there is a difference in the population means, when in fact the mean for students who use Facebook is less than the mean for those who do not use Facebook. In this situation, because neither type of error is much more serious than the other, a reasonable choice for α is 0.05. **Significance level:** $\alpha = 0.05$
C Check	There are three conditions that need to be met in order to use the two-sample t test for a difference in population means. The large sample condition is met because the sample sizes are both large: $$n_1 = 141 \geq 30 \text{ and } n_2 = 68 \geq 30$$ From the study description, you know that the samples were independently selected. You also know that the samples were selected to be representative of the two populations of interest.
C Calculate	$n_1 = 141 \qquad \bar{x}_1 = 3.06 \qquad s_1 = 0.95$ $n_2 = 68 \qquad \bar{x}_2 = 3.82 \qquad s_2 = 0.41$ **Test statistic:** $$t = \frac{(\bar{x}_1 - \bar{x}_2) - (\mu_1 - \mu_2)}{\sqrt{\dfrac{s_1^2}{n_1} + \dfrac{s_2^2}{n_2}}} = \frac{(3.06 - 3.82) - 0}{\sqrt{\dfrac{(0.95)^2}{141} + \dfrac{(0.41)^2}{68}}}$$ $$= \frac{-0.76}{0.094}$$ $$= -8.09$$ **Degrees of freedom:** $$V_1 = \frac{s_1^2}{n_1} = 0.0064 \qquad\qquad V_2 = \frac{s_2^2}{n_2} = 0.0025$$ $$df = \frac{(V_1 + V_2)^2}{\dfrac{V_1^2}{n_1 - 1} + \dfrac{V_2^2}{n_2 - 1}}$$ $$= \frac{(0.0064 + 0.0025)^2}{\dfrac{(0.0064)^2}{140} + \dfrac{(0.0025)^2}{67}}$$

(continued)

Process Step	
	$$= \frac{0.0000792}{0.000000386}$$ $$= 205.181$$ Truncate to df $= 205$ This is a lower-tailed test (the inequality in H_a is $<$), so the P-value is the area to the left of -8.09 under the t curve with df $= 205$. Because the t curve is symmetric and centered at 0, this area is equal to the area to the right of 8.09. Because $t = 8.09$ is greater than the largest value in the ∞ df row of Appendix A Table 4, the area to the right is approximately 0. **Associated P-value:** P-value $=$ area under t curve to the left of -8.09 $\qquad = $ area under the t curve to the right of 8.09 $\qquad = P(t > 8.09)$ $\qquad \approx 0$
C Communicate Results	Because the P-value is less than the selected significance level, you reject the null hypothesis. **Decision:** $0 < 0.05$, Reject H_0. **Conclusion:** Based on the sample data, there is convincing evidence that the mean college GPA for students at the university who use Facebook is less than the mean college GPA for students at the university who do not use Facebook.

Based on this hypothesis test, you can conclude that the sample mean GPA for students who use Facebook is enough less than the sample mean GPA for students who do not use Facebook that you don't think that this could have occurred just by chance when there is no difference in the population means.

It is also possible to use statistical software or a graphing calculator to carry out the calculate step in a hypothesis test. For example, Minitab output for the test of Example 13.5 is shown here.

Two-Sample T-Test

Sample	N	Mean	StDev	SE Mean
1	141	3.060	0.950	0.080
2	68	3.820	0.410	0.050

Difference = mu (1) − mu (2)
Estimate for difference: −0.760000
T-Test of difference = 0 (vs <): T-Value = −8.07 P-Value = 0.000 DF = 205

From the Minitab output, you see that $t = -8.07$, the associated degrees of freedom is df $= 205$, and the P-value is 0.000. The t value is slightly different from the one in Example 13.5 only because Minitab uses greater decimal accuracy in the computations leading to the value of the test statistic and the number of degrees of freedom.

Data set available

Example 13.6 **Rental Frogs?**

A frog jumping competition described in a short story written by Mark Twain inspired the real annual Calaveras County Frog Jumping Jubilee. In this competition, people enter bull frogs into a contest to see which frog jumps the farthest. Some serious competitors have frogs that they have trained, and these contestants are known as "professional frogs." Amateurs also compete with frogs that they can rent from the contest organizers, and these contestants are known as "rental frogs." The authors of the paper **"Chasing Maximal Performance: A Cautionary Tale from the Celebrated Jumping Frogs of Calaveras County"** (*Journal of Experimental Biology* [2013]: 3947–3953) wanted to compare the performance of rental frogs and professional frogs. The authors of the paper used a two-sample t test to compare the mean jump distances (in meters) for the two groups.

Data consistent with summary quantities given in the paper are shown in the accompanying table (the actual sample sizes in the study were much larger).

	Jump Distance (in meters)														
Rental Frog	0.9	0.9	1.3	0.8	1.1	1.2	0.8	0.9	1.4	1.1	1.1	1.0	0.9	1.2	1.2
Professional Frog	1.0	1.5	2.2	1.2	1.4	1.3	1.5	1.9	1.7	1.6	0.7	1.2	1.3	1.0	0.5

The authors of the paper wondered if rental frogs would not perform as well as professional frogs. Assuming that the frogs in the two samples are representative of rental frogs and professional frogs in general, the sample data can be used to determine if there is convincing evidence that the mean jumping distance for rental frogs is less than the mean distance for professional frogs.

Considering the four key questions (QSTN), this situation can be described as *hypothesis testing, sample data, one numerical variable (jump distance),* and *two independently selected samples.* This combination suggests a two-sample t test for a difference in population means.

H Hypotheses   You want to use data from the samples to test the claim that the mean jump distance for rental frogs is less than the mean jump distance for professional frogs. The population characteristics of interest are

μ_1 = population mean jump distance for rental frogs
μ_2 = population mean jump distance for professional frogs

and $\mu_1 - \mu_2$ is then the difference in population means. Translating the question of interest into hypotheses gives

$$H_0: \mu_1 - \mu_2 = 0$$
$$H_a: \mu_1 - \mu_2 < 0$$

M Method   Because the answers to the four key questions are hypothesis testing, sample data, one numerical variable, and two independently selected samples, consider a two-sample t test for a difference in population means. The test statistic for this test is

$$t = \frac{(\bar{x}_1 - \bar{x}_2) - (\mu_1 - \mu_2)}{\sqrt{\dfrac{s_1^2}{n_1} + \dfrac{s_2^2}{n_2}}}$$

Next choose a significance level for the test. For purposes of this example, $\alpha = 0.05$ will be used.

C Check   The samples were independently selected and you can assume that the samples are representative of the populations of interest. Because both of the sample sizes are small, you need to be willing to assume that the jump distance distribution is approximately normal for each of the two populations. Boxplots constructed using the sample data are shown here:

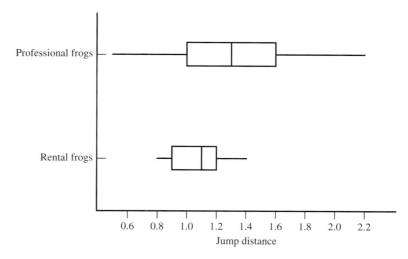

Because the boxplots are reasonably symmetric and because there are no outliers, it is reasonable to assume that the two population distributions are approximately normal. With all of the conditions met, the two-sample t test is an appropriate method.

C Calculate Using JMP to do the calculations results in the following output:

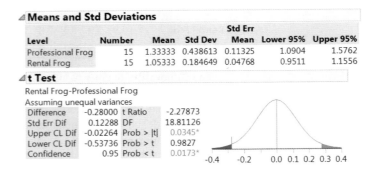

Means and Std Deviations

Level	Number	Mean	Std Dev	Std Err Mean	Lower 95%	Upper 95%
Professional Frog	15	1.33333	0.438613	0.11325	1.0904	1.5762
Rental Frog	15	1.05333	0.184649	0.04768	0.9511	1.1556

t Test

Rental Frog-Professional Frog
Assuming unequal variances

Difference	-0.28000	t Ratio	-2.27873
Std Err Dif	0.12288	DF	18.81126
Upper CL Dif	-0.02264	Prob > |t|	0.0345*
Lower CL Dif	-0.53736	Prob > t	0.9827
Confidence	0.95	Prob < t	0.0173*

From the JMP output, the value of the test statistic and df are

$t = -2.28$ (rounded from -2.27873 in the output labeled t Ratio)
df $= 18$ (rounded down from 18.81126 in the output labeled DF)

and the associated P-value is 0.0173 (from the Prob $< t$ entry in the output because this is an lower tail test).

C Communicate Results Because the P-value (0.0173) is less than the selected significance level (0.05), the null hypothesis is rejected. The sample data provide convincing evidence that the mean jump distance for rental frogs is less than the mean jump distance for professional frogs.

Suppose that the calculated value of the test statistic in Example 13.6 had been -1.28 rather than -2.28. Then the P-value would have been 0.1084 (the area to the right of 1.28 under the t curve with 18 df) and the decision would have been to not reject the null hypothesis. In this case, you would have concluded that there was not convincing evidence that the mean jump distance for rental frogs is less than the mean jump distance for professional frogs. Notice that when you fail to reject the null hypothesis of no difference between the population means, you are not saying that there is convincing evidence that the means are equal—you can only say that you are not convinced that they are different.

More on Degrees of Freedom

You have probably noticed that using the degrees of freedom formula for the two-sample t test involves quite a bit of arithmetic. An alternative approach is to calculate a conservative estimate of the P-value—one that is close to but larger than the actual P-value. If the null hypothesis is rejected using this conservative estimate, then it will also be rejected if the actual P-value is used. *A conservative estimate of the P-value for the two-sample t test can be found by using the t curve with df equal to the smaller of $(n_1 - 1)$ and $(n_2 - 1)$.*

The Pooled *t* Test

The two-sample t test is appropriate when it is reasonable to assume that the population distributions are approximately normal. If it is also known that the variances of the two populations are equal ($\sigma_1^2 = \sigma_2^2$), an alternative procedure known as the *pooled t test* can be used. This test procedure combines information from both samples to obtain a "pooled" estimate of the common variance and then uses this pooled estimate of the variance in place of s_1^2 and s_2^2 in the t test statistic. This test procedure was widely used in the past, but it has fallen into some disfavor because it is quite sensitive to departures from the assumption of equal population variances. If the population variances are equal, the pooled t test has a slightly better chance of detecting departures from the null hypothesis than does the two-sample t test of this section. However, P-values based on the pooled t test can be seriously in error if the population variances are not equal, so, in general, the two-sample t test is a better choice than the pooled t test.

Confidence Interval When Samples Are Independent

Recall that the general form of a confidence interval is

$$\text{statistic} \pm (\text{critical value})\left(\begin{array}{c}\text{standard error}\\ \text{of the statistic}\end{array}\right)$$

You can apply this general form to the case of estimating a difference in population means, $\mu_1 - \mu_2$, using data from two independent random samples. When the sample sizes are large or the population distributions are approximately normal, the appropriate confidence interval is

$$(\bar{x}_1 - \bar{x}_2) \pm (t \text{ critical value}) \sqrt{\frac{s_1^2}{n_1} + \frac{s_2^2}{n_2}}$$

To find the t critical value, you need to know the desired confidence level and the appropriate number of degrees of freedom. The df is calculated using the same formula used for the two-sample t test:

$$df = \frac{(V_1 + V_2)^2}{\dfrac{V_1^2}{n_1 - 1} + \dfrac{V_2^2}{n_2 - 1}} \quad \text{where} \quad V_1 = \frac{s_1^2}{n_1} \text{ and } V_2 = \frac{s_2^2}{n_2}$$

The Two-Sample t Confidence Interval for a Difference in Population Means

Appropriate when the following conditions are met:

1. The samples are independently selected.
2. Each sample is a random sample from the population of interest (or the samples are selected in a way that would result in samples that are representative of the populations).
3. Both sample sizes are large ($n_1 \geq 30$ and $n_2 \geq 30$) *or* the population distributions are approximately normal.

When these conditions are met, a confidence interval for a difference in population means is

$$(\bar{x}_1 - \bar{x}_2) \pm (t \text{ critical value}) \sqrt{\frac{s_1^2}{n_1} + \frac{s_2^2}{n_2}}$$

The t critical value is based on

$$df = \frac{(V_1 + V_2)^2}{\dfrac{V_1^2}{n_1 - 1} + \dfrac{V_2^2}{n_2 - 1}} \quad \text{where} \quad V_1 = \frac{s_1^2}{n_1} \text{ and } V_2 = \frac{s_2^2}{n_2}$$

The calculated value of df should be truncated (rounded down) to obtain an integer value for df. The desired confidence level determines which t critical value is used. Appendix A Table 3, statistical software, or a graphing calculator can be used to obtain the t critical value.

Interpretation of Confidence Interval
You can be confident that the actual value of the difference in population means is included in the calculated interval. This statement should be worded in context.

Interpretation of Confidence Level
The confidence level specifies the long-run proportion of the time that this method is successful in capturing the actual difference in population means.

Questions
Q Question type: estimation or hypothesis testing?
S Study type: sample data or experiment data?
T Type of data: one variable or two? Categorical or numerical?
N Number of samples or treatments: how many?

The two-sample t confidence interval for a difference in population means is a method you should consider when the answers to the four key questions are: *estimation, sample data, one numerical variable,* and *two independently selected samples.*

Now you're ready to look at an example, following the usual five-step process for estimation problems (EMC³).

Example 13.7 Teens and Texting

The report **"Teens, Social Media & Technology Overview 2015" (Pew Research Center, April 9, 2015, pewinternet.org, retrieved May 20, 2017)** states that teenage girls who own cell phones send more text messages than teenage boys who own cell phones. This conclusion was based on a survey of independent representative samples of female and male cell phone users age 13 to 17. The report stated that the mean number of text messages sent per day for the sample of girls was 79 and the mean number for the sample of boys was 56. The report also stated that a total of 929 teens were surveyed and indicated that 51% were girls. For purposes of this example, suppose that the sample sizes and sample standard deviations were as shown in the accompanying table.

Sample	Sample Size	Sample Mean	Sample Standard Deviation
Girls	$n_1 = 474$	$\bar{x}_1 = 79$	$s_1 = 10$
Boys	$n_2 = 455$	$\bar{x}_2 = 56$	$s_2 = 8$

You can use these sample data to estimate the difference in mean number of texts sent per day for girls and boys.

Start by answering the four key questions:

Q: This is an estimation problem.

S: The data are from sampling.

T: There is one numerical variable (number of text messages sent per day).

N: There are two independently selected samples (girls and boys).

This combination of answers leads to considering a two-sample t confidence interval.

You can now use the five-step process for estimation problems to construct a confidence interval estimate of the difference in mean number of text messages sent per day for girls and boys. For purposes of this example, a 90% confidence level will be used.

Steps
E Estimate
M Method
C Check
C Calculate
C Communicate results

Step	
Estimate: Explain what population characteristic you plan to estimate.	In this example, you want to estimate the difference in the mean number of text messages sent per day by girls and boys. The population characteristics of interest are μ_1 = mean number of text messages sent per day by girls μ_2 = mean number of text messages sent per day by boys $\mu_1 - \mu_2$ = difference in mean number of text messages sent per day
Method: Select a potential method based on the answers to the four key questions (QSTN).	Because the answers to the four key questions are estimation, sample data, one numerical variable, and two independently selected samples, consider constructing a two-sample t confidence interval. For this example, a confidence level of 90% was selected.
Check: Check to make sure that the method selected is appropriate.	There are three conditions that need to be met in order to use the two-sample t confidence interval. Because both sample sizes are large, you do not need to worry about whether the population distributions are approximately normal. The other two conditions that must be met have to do with how the samples were selected. Here, the study description says that the samples were representative of the two populations of interest. Also, the samples were independently selected.

(continued)

Step

Calculate: Use the sample data to perform any necessary calculations.

$n_1 = 474$ $\bar{x}_1 = 79$ $s_1 = 10$
$n_2 = 455$ $\bar{x}_2 = 56$ $s_2 = 8$

Degrees of freedom

$$V_1 = \frac{s_1^2}{n_1} = 0.211 \qquad V_2 = \frac{s_2^2}{n_2} = 0.141$$

$$\text{df} = \frac{(V_1 + V_2)^2}{\dfrac{V_1^2}{n_1 - 1} + \dfrac{V_2^2}{n_2 - 1}}$$

$$= \frac{(0.211 + 0.141)^2}{\dfrac{(0.211)^2}{474} + \dfrac{(0.141)^2}{455}}$$

$$= \frac{0.124}{0.0001}$$

$$= 1240$$

The appropriate t critical value is found using technology or by using Appendix A Table 3 to determine the t critical value that captures a central area of 0.90. (Note: When using the table, use df column ∞ because df > 120.)

t critical value $= 1.645$

90% confidence interval

$$(\bar{x}_1 - \bar{x}_2) \pm (t \text{ critical value}) \sqrt{\frac{s_1^2}{n_1} + \frac{s_2^2}{n_2}}$$

$$(79 - 56) \pm (1.645) \sqrt{\frac{10^2}{474} + \frac{8^2}{455}}$$

$$23 \pm (1.645)\sqrt{0.211 + 0.141}$$

$$23 \pm (1.645)(0.593)$$

$$23 \pm 0.976$$

$$(22.024, 23.976)$$

Communicate Results: Answer the research question of interest, explain what you have learned from the data, and acknowledge potential risks.

The interpretation of confidence intervals should always be worded in the context of the problem. You should always give interpretations of both the interval itself and the confidence level associated with the interval.

Confidence interval:

You can be 90% confident that the actual difference in mean number of text messages sent per day is between 22.024 and 23.976. Both endpoints of this interval are positive, so you estimate that the mean number of text messages sent per day by girls is greater than the mean number of text messages sent by boys by somewhere between 22.024 and 23.976.

Confidence level:

The method used to construct this interval estimate is successful in capturing the actual difference in population means about 90% of the time.

This confidence interval estimate is the basis for the statement in the report that girls send more text messages than boys.

Most statistical software packages and graphing calculators can compute the two-sample t confidence interval. Minitab was used to construct a 90% confidence interval using the data of Example 13.7, and the output is shown here:

Two-Sample T-Test and CI

Sample	N	Mean	StDev	SE Mean
1	474	79.0	10.0	0.46
2	455	56.00	8.00	0.38

Difference = μ (1) $-$ μ (2)
Estimate for difference: 23.000
90% CI for difference: (22.024, 23.976)

Data set available

Example 13.8 Freshman Year Weight Gain

The paper **"Predicting the 'Freshman 15': Environmental and Psychological Predictors of Weight Gain in First-Year University Students" (*Health Education Journal* [2010]: 321–332)** described a study conducted by researchers at Carleton University in Canada. The researchers studied a random sample of first-year students who lived on campus and a random sample of first-year students who lived off campus. Data on weight gain (in kg) during the first year, consistent with summary quantities given in the paper, are given below. A negative weight gain represents a weight loss. The researchers believed that the mean weight gain of students living on campus was higher than the mean weight gain for students living off campus and were interested in estimating the difference in means for these two groups.

On Campus	Off Campus
2.0	1.6
2.3	3.1
1.1	-2.8
-2.0	0.0
-1.9	0.2
5.6	2.9
2.6	-0.9
1.1	3.8
5.6	0.7
8.2	-0.1

The answers to the four key questions are *estimation, sample data, one numerical variable (weight gain),* and *two independently selected samples.* This combination of answers suggests using a two-sample t confidence interval. You can now use the five-step process for estimation problems to construct a 95% confidence interval for the difference in mean weight gain for students who live on campus and students who live off campus.

E Estimate You want to estimate

$$\mu_1 - \mu_2 = \text{mean difference in weight gain}$$

where

$$\mu_1 = \text{mean weight gain for first-year students living on campus}$$

and

$$\mu_2 = \text{mean weight gain for first-year students living off campus}$$

M Method Because the answers to the four key questions are estimation, sample data, one numerical variable, and two independently selected samples, consider constructing a two-sample t confidence interval. For this example, a confidence level of 95% was specified.

C Check   The samples were random samples from the two populations of interest (first-year students who live on campus and first-year students who live off campus), so the samples were independently selected. Because the sample sizes were not large, you need to be willing to assume that the population weight gain distributions are at least approximately normal. Figure 13.2 shows boxplots constructed using the data from the two samples. There are no outliers in either dataset and the boxplots are reasonably symmetric, suggesting that the assumption of approximate normality is reasonable for each of the populations.

FIGURE 13.2
Boxplots for the weight gain data of Example 13.8

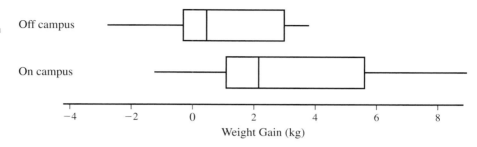

C Calculate   JMP output is shown here:

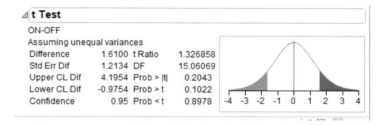

The 95% confidence interval estimate for the difference in mean weight gain is (-0.975, 4.195).

C Communicate Results   Based on these samples, you can be 95% confident that the actual difference in mean weight gain is somewhere between -0.975 kg and 4.195 kg, Because 0 is included in this interval, it is plausible that there is no difference in the means for weight gain for students who live on campus and those who live off campus. The method used to construct this interval is successful in capturing the actual difference in population means about 95% of the time.

Summing It Up—Section 13.3

The following learning objectives were addressed in this section:

Conceptual Understanding
C1: Understand how a research question about the difference in two population means is translated into hypotheses.
Sample data can be used to test hypotheses about the difference in two population means, μ_1 and μ_2. The null hypothesis is usually $H_0\!: = \mu_1 - \mu_2$. hypothesized value. The research question will determine the form of the alternative hypothesis.

Mastering the Mechanics
M5: Know the conditions for appropriate use of the two-sample t confidence interval and the two-sample t test.
There are three conditions that need to be reasonably met in order for it to be appropriate to use the two-sample t confidence interval and test. These conditions are (1) that the

samples are independently selected, (2) the two samples can be viewed as random samples from the populations of interest or it is reasonable to regard the samples as representative of the populations, and (3) the sample sizes are large or the population distributions are approximately normal. The sample sizes are considered to be large if they are both greater than or equal to 30.

M6: Carry out a two-sample *t* test for a difference in population means.
The five-step process for hypothesis testing (HMC^3) can be used to test hypotheses about the difference in two population means using independent samples. This process of carrying out a hypothesis test for a difference in two population means is illustrated in Examples 13.5 and 13.6.

M7: Calculate and interpret a two-sample *t* confidence interval for a difference in population means.
The five-step process for estimation problems (EMC^3) can be used to estimate the difference in two population means using independent samples. This process of calculating and interpreting a confidence interval for a difference in two population means is illustrated in Examples 13.7 and 13.8.

Putting It into Practice

P3: Carry out a two-sample *t* test for a difference in two means and interpret the conclusion in context.
Hypothesis tests for a difference in two population means using independent samples are illustrated in Examples 13.5 and 13.6. If the null hypothesis is rejected, you can conclude that there is convincing evidence in support of the alternative hypothesis. If the null hypothesis is not rejected, chance differences due to sampling variability is a plausible explanation for the observed difference in sample means and there is not convincing evidence that the alternative hypothesis is true.

P4: Interpret a two-sample *t* confidence interval for a difference in means in context and interpret the associated confidence level.
A confidence interval for a difference in population means based on independent samples is interpreted as an interval of plausible values for the difference in population means. For examples of interpreting a confidence interval for a difference in population means and the associated confidence interval in context, see Examples 13.7 and 13.8.

SECTION 13.3 EXERCISES

Each Exercise Set assesses the following chapter learning objectives: C1, M5, M6, M7, P3, P4

SECTION 13.3 Exercise Set 1

13.31 Example 13.5 looked at a study comparing students who use Facebook and students who do not use Facebook (**"Facebook and Academic Performance,"** *Computers in Human Behavior* **[2010]: 1237–1245**). In addition to asking the students in the samples about GPA, each student was also asked how many hours he or she spent studying each day. The two samples (141 students who were Facebook users and 68 students who were not Facebook users) were independently selected from students at a large, public midwestern university. Although the samples were not selected at random, they were selected to be representative of the two populations. For the sample of Facebook users, the mean number of hours studied per day was 1.47 hours and the standard deviation was 0.83 hours. For the sample of students who do not use Facebook, the mean was 2.76 hours and the standard deviation was 0.99 hours. Do these sample data provide convincing evidence that the mean time spent studying for Facebook users is less than the mean time spent studying for students who do not use Facebook? Use a significance level of $\alpha = 0.01$.

13.32 Many people now turn to the Internet to get information on health-related topics. The paper **"An Examination of Health, Medical and Nutritional Information on the Internet: A Comparative study of Wikipedia, WebMD and the Mayo Clinic Websites"** (*The International Journal of Communication and Health* **[2015]: 30–38**) used Flesch reading ease scores (a measure of reading difficulty based on factors such as sentence length and number of syllables in the words used) to score pages on Wikipedia and on WebMD. Higher Flesch scores correspond to more difficult reading levels. The paper reported that for a representative sample of health-related pages on Wikipedia, the mean Flesch score was 26.7 and the standard deviation of the Flesch scores was 14.1. For a representative sample of pages from WebMD, the mean score was 43.9 and the standard deviation was 19.4. Suppose that these means and standard deviations were based on samples of 40 pages from each site. Is there convincing evidence that the mean reading level for health-related pages differs for Wikipedia and WebMD? Test the relevant hypotheses using a significance level of $\alpha = 0.05$.

13.33 Use the information in the previous exercise to answer the following questions.
a. Construct a 90% confidence interval estimate of the difference in mean Flesch reading ease score for health-related pages on Wikipedia and health-related pages on WebMD.
b. What does this confidence interval imply about the readability of health-related information from these two sources? Is this consistent with the conclusion in the hypothesis test of the previous exercise?

13.34 The article **"Genetic Tweak Turns Promiscuous Animals into Loyal Mates" (***Los Angeles Times***, June 17, 2004)** summarizes a research study that appeared in the June 2004 issue of ***Nature***. In this study, 11 male meadow voles that had a single gene introduced into a specific part of the brain were compared to 20 male meadow voles that did not undergo this genetic manipulation. All of the voles were paired with a receptive female partner for 24 hours. At the end of the 24-hour period, the male was placed in a situation where he could choose either the partner from the previous 24 hours or a different female. The percentage of the time that the male spent with his previous partner during a 3-hour time period was recorded. The accompanying data are approximate values read from a graph in the *Nature* article. Do these data support the researchers' hypothesis that the mean percentage of the time spent with the previous partner is significantly greater for the population of genetically altered voles than for the population of voles that did not have the gene introduced? Test the relevant hypotheses using $\alpha = 0.05$.

Percent of Time Spent with Previous Partner							
Genetically Altered	59	62	73	80	84	85	89	92
	92	93	100					
Not Genetically Altered	2	5	13	28	34	40	48	50
	51	54	60	67	70	76	81	84
	85	92	97	99				

13.35 The National Sleep Foundation surveyed representative samples of adults in six different countries to ask questions about sleeping habits **("2013 International Bedroom Poll Summary of Findings," www.sleepfoundation.org/sites /default/files/RPT495a.pdf, retrieved May 20, 2017)**. Each person in a representative sample of 250 adults in each of these countries was asked how much sleep they get on a

typical work night. For the United States, the sample mean was 391 minutes, and for Mexico the sample mean was 426 minutes. Suppose that the sample standard deviations were 30 minutes for the U.S. sample and 40 minutes for the Mexico sample. The report concludes that on average, adults in the United States get less sleep on work nights than adults in Mexico. Is this a reasonable conclusion? Support your answer with an appropriate hypothesis test.

13.36 The report referenced in the previous exercise also gave data for representative samples of 250 adults in Canada and 250 adults in England. The sample mean amount of sleep on a work night was 423 minutes for the Canada sample and 409 minutes for the England sample. Suppose that the sample standard deviations were 35 minutes for the Canada sample and 42 minutes for the England sample.
a. Construct and interpret a 95% confidence interval estimate of the difference in the mean amount of sleep on a work night for adults in Canada and adults in England.
b. Based on the confidence interval from Part (a), would you conclude that there is evidence of a difference in the mean amount of sleep on a work night for the two countries? Explain why or why not.

SECTION 13.3 Exercise Set 2

13.37 Do female college students spend more time watching TV than male college students? This was one of the questions investigated by the authors of the paper **"An Ecological Momentary Assessment of the Physical Activity and Sedentary Behaviour Patterns of University Students"** (***Health Education Journal*** **[2010]: 116–125)**. Each student in a random sample of 46 male students at a university in England and each student in a random sample of 38 female students from the same university kept a diary of how he or she spent time over a 3-week period. For the sample of males, the mean time spent watching TV per day was 68.2 minutes, and the standard deviation was 67.5 minutes. For the sample of females, the mean time spent watching TV per day was 93.5 minutes, and the standard deviation was 89.1 minutes. Is there convincing evidence that the mean time female students at this university spend watching TV is greater than the mean time for male students? Test the appropriate hypotheses using $\alpha = 0.05$.

13.38 The paper referenced in the Preview Example of this chapter **("Mood Food: Chocolate and Depressive Symptoms in a Cross-Sectional Analysis,"** ***Archives of Internal Medicine*** **[2010]: 699–703)** describes a study that investigated the relationship between depression and chocolate consumption. Participants in the study were 931 adults who were not currently taking medication for depression. These participants were screened for depression using a widely used screening test. The participants were then divided into two samples based on their test score. One sample consisted of people who screened positive for depression, and the other sample consisted of people who did not screen positive for depression.

Each of the study participants also completed a food frequency survey. The researchers believed that the two samples were representative of the two populations of interest—adults who would screen positive for depression and adults who would not screen positive. The paper reported that the mean number of servings per month of chocolate for the sample of people that screened positive for depression was 8.39, and the sample standard deviation was 14.83. For the sample of people who did not screen positive for depression, the mean was 5.39, and the standard deviation was 8.76. The paper did not say how many individuals were in each sample, but for the purposes of this exercise, you can assume that the 931 study participants included 311 who screened positive for depression and 620 who did not screen positive. Carry out a hypothesis test to confirm the researchers' conclusion that the mean number of servings of chocolate per month for people who would screen positive for depression is greater than the mean number of chocolate servings per month for people who would not screen positive.

13.39 The paper **"Pathological Video-Game Use Among Youth Ages 8 to 18: A National Study"** (*Psychological Science* [2009]: 594–705) included the information in the accompanying table about video game playing time for representative samples of 588 males and 590 females selected from U.S. residents age 8 to 18.

Sample	Sample Size	Sample Mean (hours per week)	Sample Standard Deviation (hours per week)
Females	590	9.2	10.2
Males	588	16.4	14.1

Carry out a hypothesis test to determine if there is convincing evidence that the mean number of hours per week spent playing video games by females is less than the mean number of hours spent by males. Use a significance level of $\alpha = 0.01$.

13.40 The paper **"Facebook Use and Academic Performance Among College Students: A Mixed-Methods Study with a Multi-Ethnic Sample"** (*Computers in Human Behavior* [2015]: 265–272) describes a survey of a sample of 66 male students and a sample of 195 female students at a large university in Southern California. The authors of the paper believed that these samples were representative of male and female college students in Southern California. For the sample of males, the mean time spent per day on Facebook was 102.31 minutes. For the sample of females, the mean time was 159.61 minutes. The sample standard deviations were not given in the paper, but for purposes of this exercise, suppose that the sample standard deviations were both 100 minutes.
a. Do the data provide convincing evidence that the mean time spent of Facebook is not the same for males and for females? Test the relevant hypotheses using $\alpha = 0.05$.
b. Do you think it is reasonable to generalize the conclusion from the hypothesis test in Part (a) to the populations

of all male college students in the United States and all female college students in the United States? Explain why you think this.

13.41 Use the information in the previous exercise to answer the following questions.
a. Construct a 95% confidence interval estimate of the difference in mean time spent on Facebook for male college students and female college students in Southern California.
b. What does this confidence interval imply about the mean time spent on Facebook for these two populations of students? Is this consistent with the conclusion in the hypothesis test of the previous exercise?

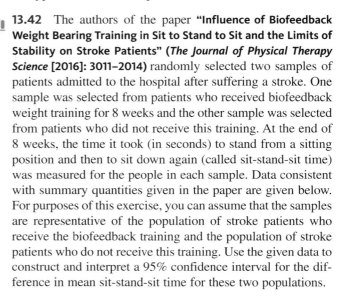 **13.42** The authors of the paper **"Influence of Biofeedback Weight Bearing Training in Sit to Stand to Sit and the Limits of Stability on Stroke Patients"** (*The Journal of Physical Therapy Science* [2016]: 3011–2014) randomly selected two samples of patients admitted to the hospital after suffering a stroke. One sample was selected from patients who received biofeedback weight training for 8 weeks and the other sample was selected from patients who did not receive this training. At the end of 8 weeks, the time it took (in seconds) to stand from a sitting position and then to sit down again (called sit-stand-sit time) was measured for the people in each sample. Data consistent with summary quantities given in the paper are given below. For purposes of this exercise, you can assume that the samples are representative of the population of stroke patients who receive the biofeedback training and the population of stroke patients who do not receive this training. Use the given data to construct and interpret a 95% confidence interval for the difference in mean sit-stand-sit time for these two populations.

Biofeedback Group
1.9 2.6 4.3 2.1 2.7 4.1 3.2 4.0 3.2 3.5 2.8
3.5 3.5 2.3 3.1

No Biofeedback Group
5.1 4.7 3.9 4.2 4.7 4.3 4.2 5.1 3.4 4.2 5.1
4.4 4.0 3.4 3.9

ADDITIONAL EXERCISES

13.43 Internet addiction has been described as excessive and uncontrolled Internet use. The authors of the paper **"Gender Difference in the Relationship Between Internet Addiction and Depression"** (*Computers in Human Behavior* [2016]: 463–470) used a score designed to measure the extent and severity of Internet addiction in a study of 836 male and 879 female sixth grade students in China. Internet Addiction was measured using Young's Internet Addiction Diagnostic Test. The lowest possible score on this test is zero, and higher scores indicate higher levels of Internet addiction. For the sample of males, the mean Internet Addiction score was 1.51 and the standard deviation was 2.03. For the sample of females, the mean was 1.07 and the standard deviation was 1.63. For purposes of this exercise, you can assume that it is reasonable to regard these two samples as representative of the population of male Chinese sixth grade students and the population of female Chinese sixth grade students, respectively.

a. The standard deviation is greater than the mean for each of these samples. Explain why it is not reasonable to think that the distribution of Internet Addiction scores would be approximately normal for either the population of male Chinese sixth grade students or the population of female Chinese sixth grade students.

b. Given your response to Part (a), would it be appropriate to use the two-sample t test to test the null hypothesis that there is no difference in the mean Internet Addiction score for male Chinese sixth grade students and female Chinese sixth grade students? Explain why or why not.

c. If appropriate, carry out a test to determine if there is convincing evidence that the mean Internet Addiction score is greater for male Chinese sixth grade students than for female Chinese sixth grade students. Use $\alpha = 0.05$.

13.44 Reduced heart rate variability (HRV) is known to be a predictor of mortality after a heart attack. One measure of HRV is the average normal-to-normal beat interval (in milliseconds) for a 24-hour time period. Twenty-two heart attack patients who were dog owners and 80 heart attack patients who did not own a dog participated in a study of the effect of pet ownership on HRV, resulting in the summary statistics shown in the accompanying table (**"Relationship Between Pet Ownership and Heart Rate Variability in Patients with Healed Myocardial Infarcts," The American Journal of Cardiology [2003]: 718–721**).

	Measure of HRV (Average Normal-to-Normal Beat Interval)	
	Mean	**Standard Deviation**
Owns Dog	873	136
Does Not Own Dog	800	134

a. The authors of this paper used a two-sample t test to test $H_0: \mu_1 - \mu_2 = 0$ versus $H_a: \mu_1 - \mu_2 \neq 0$. What conditions must be met in order for this to be an appropriate method of analysis?

b. The paper indicates that the null hypothesis in Part (a) was rejected and reported that the P-value was less than 0.05. Carry out a two-sample t test. Is your conclusion consistent with the one given in the paper?

13.45 The article **"Plugged In, but Tuned Out" (USA TODAY, January 20, 2010)** summarizes data from two surveys of kids age 8 to 18. One survey was conducted in 1999 and the other was conducted in 2009. Data on number of hours per day spent using electronic media, consistent with summary quantities in the article, are given (the actual sample sizes for the two surveys were much larger). For purposes of this exercise, you can assume that

the two samples are representative of kids age 8 to 18 in each of the 2 years when the surveys were conducted.

2009	5	9	5	8	7	6	7	9	7	9	6	9
	10	9	8									
1999	4	5	7	7	5	7	5	6	5	6	7	8
	5	6	6									

a. Because the given sample sizes are small, what assumption must be made about the distributions of electronic media use times for the two-sample t test to be appropriate? Use the given data to construct graphical displays that would be useful in determining whether this assumption is reasonable. Do you think it is reasonable to use these data to carry out a two-sample t test?

b. Do the given data provide convincing evidence that the mean number of hours per day spent using electronic media was greater in 2009 than in 1999? Test the relevant hypotheses using a significance level of $\alpha = 0.01$.

13.46 Research has shown that, for baseball players, good hip range of motion results in improved performance and decreased body stress. The article **"Functional Hip Characteristics of Baseball Pitchers and Position Players" (The American Journal of Sports Medicine, 2010: 383–388)** reported on a study of independent samples of 40 professional pitchers and 40 professional position players. For the pitchers, the sample mean hip range of motion was 75.6 degrees and the sample standard deviation was 5.9 degrees, whereas the sample mean and sample standard deviation for position players were 79.6 degrees and 7.6 degrees, respectively. Assuming that the two samples are representative of professional baseball pitchers and position players, test hypotheses appropriate for determining if there is convincing evidence that the mean range of motion for pitchers is less than the mean for position players.

13.47 What impact does fast-food consumption have on various dietary and health characteristics? The article **"Effects of Fast-Food Consumption on Energy Intake and Diet Quality among Children in a National Household Study" (Pediatrics, 2004: 112–118)** reported the accompanying summary statistics on daily calorie intake for a representative sample of teens who do not typically eat fast food and a representative sample of teens who do eat fast food.

Sample	Sample Size	Sample Mean	Sample Standard Deviation
Do not eat fast food	663	2258	1519
Eat fast food	413	2637	1138

Is there convincing evidence that the mean calorie intake for teens who typically eat fast food is greater than the mean intake for those who don't by more than 200 calories per day?

13.48 Use the information given in the previous exercise to construct and interpret a 95% confidence interval estimate of the difference in mean daily calorie intake for teens who do eat fast food on a typical day and those who do not.

13.49 Do children diagnosed with attention deficit/hyperactivity disorder (ADHD) have smaller brains than children without this condition? This question was the topic of a research study described in the paper **"Developmental Trajectories of Brain Volume Abnormalities in Children and Adolescents with Attention Deficit/Hyperactivity Disorder"** (*Journal of the American Medical Association* **[2002]: 1740–1747**). Brain scans were completed for a representative sample of 152 children with ADHD and a representative sample of 139 children without ADHD. Summary values for total cerebral volume (in cubic milliliters) are given in the following table:

	n	$\bar{x}$	s
Children with ADHD	152	1059.4	117.5
Children without ADHD	139	1104.5	111.3

Is there convincing evidence that the mean brain volume for children with ADHD is smaller than the mean for children without ADHD? Test the relevant hypotheses using a 0.05 level of significance.

13.50 Use the information given in the previous exercise to construct and interpret a 95% confidence interval estimate of the difference in mean brain volume for children with and without ADHD.

13.51 In a study of malpractice claims where a settlement had been reached, two random samples were selected: a random sample of 515 closed malpractice claims that were found not to involve medical errors and a random sample of 889 claims that were found to involve errors (*New England Journal of Medicine* **[2006]: 2024–2033**). The following statement appeared in the paper: "When claims not involving errors were compensated, payments were significantly lower on average than were payments for claims involving errors ($313,205 vs. $521,560, *P* = 0.004)."
a. What hypotheses did the researchers test to reach the stated conclusion?
b. Which of the following could have been the value of the test statistic for the hypothesis test? Explain your reasoning.
  i. $t = 5.00$ **iii.** $t = 2.33$
  ii. $t = 2.65$ **iv.** $t = 1.47$

13.52 In a study of the effect of college student employment on academic performance, the following summary statistics for GPA were reported for a sample of students who worked and for a sample of students who did not work (*University of Central Florida Undergraduate Research Journal, Spring 2005*):

	Sample Size	Mean GPA	Sandard Deviation
Students Who Are Employed	184	3.12	0.485
Students Who Are Not Employed	114	3.23	0.524

The samples were selected at random from working and nonworking students at the University of Central Florida. Does this information support the hypothesis that for students at this university, those who are not employed have a higher mean GPA than those who are employed?

13.53 Use the information given in the previous exercise to estimate the difference in mean GPA for students at the University of Central Florida who are employed and students who are not employed. Use a 90% confidence level to produce your estimate.

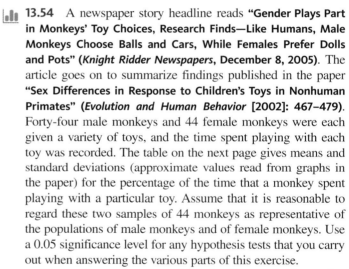 **13.54** A newspaper story headline reads **"Gender Plays Part in Monkeys' Toy Choices, Research Finds—Like Humans, Male Monkeys Choose Balls and Cars, While Females Prefer Dolls and Pots"** (*Knight Ridder Newspapers*, **December 8, 2005**). The article goes on to summarize findings published in the paper **"Sex Differences in Response to Children's Toys in Nonhuman Primates"** (*Evolution and Human Behavior* **[2002]: 467–479**). Forty-four male monkeys and 44 female monkeys were each given a variety of toys, and the time spent playing with each toy was recorded. The table on the next page gives means and standard deviations (approximate values read from graphs in the paper) for the percentage of the time that a monkey spent playing with a particular toy. Assume that it is reasonable to regard these two samples of 44 monkeys as representative of the populations of male monkeys and of female monkeys. Use a 0.05 significance level for any hypothesis tests that you carry out when answering the various parts of this exercise.
a. The police car was considered a "masculine" toy. Do these data provide convincing evidence that the mean percentage of the time spent playing with the police car is greater for male monkeys than for female monkeys?
b. The doll was considered a "feminine" toy. Do these data provide convincing evidence that the mean percentage of the time spent playing with the doll is greater for female monkeys than for male monkeys?
c. The furry dog was considered a "neutral" toy. Do these data provide convincing evidence that the mean percentage of the time spent playing with the furry dog is not the same for male and female monkeys?
d. Based on the conclusions from the hypothesis tests of Parts (a)–(c), is the quoted newspaper story headline a reasonable summary of the findings? Explain.
e. Explain why it would be inappropriate to use the two-sample *t* test to decide if there was evidence that the mean percentage of the time spent playing with the police car and the mean percentage of the time spent playing with the doll are not the same for female monkeys.

 Data set available

TABLE FOR EXERCISE **13.54**

		Percent of Time					
		Female Monkeys			Male Monkeys		
		n	Sample Mean	Sample Standard Deviation	*n*	Sample Mean	Sample Standard Deviation
Toy	**Police Car**	44	8	4	44	18	5
	Doll	44	20	4	44	9	2
	Furry Dog	44	20	5	44	25	5

13.55 A researcher at the Medical College of Virginia conducted a study of 60 randomly selected male soccer players and concluded that players who frequently "head" the ball have a lower mean IQ than those who do not (*USA TODAY*, **August 14, 1995**). The soccer players were divided into two groups, based on whether they averaged 10 or more headers per game. Mean IQs were reported in the article, but the sample sizes and standard deviations were not given. Suppose that these values were as given in the accompanying table. Do these data support the researcher's conclusion? Test the relevant hypotheses using $\alpha = 0.05$. Can you conclude that heading the ball *causes* lower IQ? Explain.

	n	Sample Mean	Sample sd
Fewer Than 10 Headers	35	112	10
10 or More Headers	25	103	8

13.56 Wayne Gretzky was one of ice hockey's most prolific scorers when he played for the Edmonton Oilers. During his last season with the Oilers, Gretzky played in 41 games and missed 17 games due to injury. The article **"The Great Gretzky"** (*Chance* **[1991]: 16–21**) looked at the number of goals scored by the Oilers in games with and without Gretzky, as shown in the accompanying table. If you view the 41 games with Gretzky as a random sample of all Oiler games in which Gretzky played and the 17 games without Gretzky as a random sample of all Oiler games in which Gretzky did not play, is there convincing evidence that the mean number of goals scored by the Oilers is higher for games when Gretzky plays? Use $\alpha = 0.01$.

	n	Sample Mean	Sample sd
Games with Gretzky	41	4.73	1.29
Games without Gretzky	17	3.88	1.18

13.57 Are girls less inclined to enroll in science courses than boys? One study **("Intentions of Young Students to Enroll in Science Courses in the Future: An Examination of Gender Differences"** (*Science Education* **[1999]: 55–76**) asked randomly selected fourth-, fifth-, and sixth-graders how many science courses they intend to take. The following data were obtained:

	n	Mean	Standard Deviation
Males	203	3.42	1.49
Females	224	2.42	1.35

Calculate a 99% confidence interval for the difference in mean number of science courses planned for males and females. Based on your interval, how would you answer the question posed at the beginning of the exercise?

13.58 Research has shown that for baseball players, good hip range of motion results in improved performance and decreased body stress. The article **"Functional Hip Characteristics of Baseball Pitchers and Position Players"** (*The American Journal of Sports Medicine*, **2010: 383–388**) reported on a study involving independent samples of 40 professional pitchers and 40 professional position players. For the sample of pitchers, the mean hip range of motion was 75.6 degrees and the standard deviation was 5.9 degrees, whereas the mean and standard deviation for the sample of position players were 79.6 degrees and 7.6 degrees, respectively. Assuming that these two samples are representative of professional baseball pitchers and position players, estimate the difference in mean hip range of motion for pitchers and position players using a 90% confidence interval.

SECTION 13.4 Inference for Two Means Using Data from an Experiment

In sampling situations, you often want to compare two populations to decide if there is evidence that their means are different. When random samples are selected from both populations, you know that even if the population means are equal, the two sample means won't usually be equal. This is because of sample-to-sample variability that occurs due to the random selection process. To be convinced that there really is a difference in population means, you need to see a difference in sample means that is greater than what you would expect to see just by chance due to the random selection process. This is where statistical methods help—they allow you to determine when the difference you see in the sample means is unlikely to be just due to variability introduced by random selection.

Now think about what you might learn using data from an experiment. An experiment might be carried out to compare two treatments or to compare a single treatment with a control. The resulting data could then be used to determine if the treatment has an effect on some response variable of interest. The treatments are "applied" to individuals (as in an experiment to compare the effect of two different diets on weight gain in fish) or to objects (as in an experiment to compare the effect of two different firing temperatures on a measure of the quality of ceramic tiles). The response that is measured in an experiment could be numerical or it could be categorical. When the response variable is numerical, you usually want to know if the treatment means differ—for example, if the mean weight gains for fish fed on two different diets are different.

In an experiment to compare two treatments, deciding if there is a significant difference between treatment means involves trying to answer the following question:

> Could the observed difference in mean response between the two experimental groups be explained just by the way the individuals or objects happened to be divided into the two groups, or is there evidence that a treatment effect is causing the difference?

You only say that there is a significant difference between the treatment means if you can reasonably rule out the explanation that says the observed difference is due just to the random assignment to treatments.

In most real experimental situations, the individuals or objects receiving the treatments are not selected at random from some larger population. In this case, you cannot generalize the results of the experiment to some larger population. However, if the experimental design included random assignment of individuals to treatments (or for random assignment of treatments to individuals), it is possible to learn about the difference in treatment means by testing hypotheses or by calculating and interpreting confidence intervals.

Example 13.9 | **Injecting Cement to Ease Pain**

Is injecting medical cement effective in reducing pain for people who have suffered fractured vertebrae? The paper **"A Randomized Trial of Vertebroplasty for Osteoporotic Spinal Fractures"** (*New England Journal of Medicine* **[2009]: 569–578**) describes a study to compare patients who underwent vertebroplasty (the injection of cement) to patients in a placebo group who underwent a fake procedure in which no cement was actually injected. Because the placebo procedure was similar to the vertebroplasty procedure except for the actual injection of cement, patients were not aware of which treatment they received. Patients participating in the study were assigned at random to one of the two experimental groups. All patients were asked to rate their pain at three different times—3 days, 14 days, and 1 month after the procedure. The summary statistics in the accompanying table were calculated from the resulting data.

	Pain Intensity			
	Vertebroplasty Group $n = 68$		Placebo Group $n = 63$	
	mean	sd	mean	sd
3 days	4.2	2.8	3.9	2.9
14 days	4.3	2.9	4.5	2.8
1 month	3.9	2.9	4.6	3.0

In this experiment, two treatments (experimental conditions) for back pain are compared. One treatment consisted of injecting cement. The second treatment consisted of a placebo, where patients underwent a procedure that appeared to be the same as the first treatment but where no cement was actually injected. Patients were randomly assigned to one of the two treatment groups. One month after the procedures, patients reported pain intensity. The mean pain intensity for the 68 patients who received the cement injection was 3.9. For the 63 patients who were in the placebo group, the mean pain intensity was 4.6. This example will be revisited in Exercise 13.65 after you have seen how the two-sample t test of Section 13.3 can be adapted for use with data from an experiment.

Testing Hypotheses About the Difference Between Two Treatment Means

It is common practice to use the two-sample t test to test hypotheses about a difference in treatment means, with the following modifications:

1. **Hypotheses:** The hypotheses will look the same as before, but now μ_1 represents the mean response for treatment 1 and μ_2 represents the mean response for treatment 2. The null hypothesis $H_0: \mu_1 - \mu_2 = 0$ is a statement that there is no difference in the treatment means (no treatment effect).
2. **Conditions:** When you previously considered the two-sample t test to test hypotheses about population means using sample data, there were three conditions that had to be satisfied. The first two conditions were that the samples had to be independently selected and they had to be random samples from the populations of interest. The third condition was that the sample sizes had to be large or that the population distributions were approximately normal. In the context of testing hypotheses about treatment means, these three conditions are replaced by the following two:

    1. Individuals or objects are *randomly assigned* to treatments.
    2. The number of individuals or objects in each of the treatment groups is large (30 or more) or the treatment response distributions (the distributions of response values that would result if the treatments were applied to a *very* large number of individuals or objects) are approximately normal.

3. **Conclusions:** Conclusions will be worded in terms of treatment means. If the individuals or objects that were randomly assigned to the treatments were *also* randomly selected from some larger population, it is also reasonable to generalize conclusions about treatment effects to the larger population.

Two-Sample t Test for a Difference in Treatment Means

Appropriate when the following conditions are met:

1. Individuals or objects are *randomly assigned* to treatments.
2. The number of individuals or objects in each of the treatment groups is large (30 or more) or the treatment response distributions (the distributions of response values that would result if the treatments were applied to a *very* large number of individuals or objects) are approximately normal.

When these conditions are met, the following test statistic can be used:

$$t = \frac{(\bar{x}_1 - \bar{x}_2) - (\mu_1 - \mu_2)}{\sqrt{\dfrac{s_1^2}{n_1} + \dfrac{s_2^2}{n_2}}}$$

where $\mu_1 - \mu_2$ is the hypothesized value of the difference in treatment means from the null hypothesis (often this will be 0).

When the conditions above are met and the null hypothesis is true, the t test statistic has a t distribution with

$$df = \frac{(V_1 + V_2)^2}{\dfrac{V_1^2}{n_1 - 1} + \dfrac{V_2^2}{n_2 - 1}} \quad \text{where } V_1 = \frac{s_1^2}{n_1} \text{ and } V_2 = \frac{s_2^2}{n_2}$$

(continued)

The calculated value of df should be truncated (rounded down) to obtain an integer value for df.

Form of the null hypothesis: $H_0: \mu_1 - \mu_2 =$ hypothesized value

Associated P-value:

When the Alternative Hypothesis Is...	The P-Value Is...
$H_a: \mu_1 - \mu_2 >$ hypothesized value	Area under the t curve to the right of the calculated value of the test statistic
$H_a: \mu_1 - \mu_2 <$ hypothesized value	Area under the t curve to the left of the calculated value of the test statistic
$H_a: \mu_1 - \mu_2 \neq$ hypothesized value	2(area to the right of t) if t is positive or 2(area to the left of t) if t is negative

The condition of normal treatment response distributions can be assessed by constructing a dotplot, boxplot, or a normal probability plot of the response values from each treatment group.

When the two-sample t test is used to test hypotheses about the difference between two treatment means, it is only an approximate test (the reported P-values are approximate). However, this is still the most common way to analyze data from experiments with two treatment groups.

Example 13.10 Fitness Trackers and Weight Loss

The article **"Activity Trackers May Undermine Weight Loss Efforts" (*The New York Times*, September 20, 2016)** describes a study published in the ***Journal of the American Medical Association*** **("The Effect of Wearable Technology Combined with a Lifestyle Intervention on Long-Term Weight Loss." [2016]: 1161–1171)**. In this study, subjects followed a low-calorie diet and exercise program for 6 months. After 6 months, the subjects were randomly assigned to one of two groups. The people in one group were provided with a website they could use to self-monitor diet and physical activity. The people in the second group were provided with a wearable fitness tracker with an accompanying web interface to monitor diet and physical activity.

The researchers were interested in learning if the mean weight loss (in kilograms) at the end of two years was different for the two treatments (self-monitoring and fitness tracker monitoring). Data from this experiment are summarized in the accompanying table.

Group	Sample Size	Mean Weight Loss	Standard Deviation
Self-Monitoring	170	5.9 kg	6.8 kg
Fitness Tracker Monitoring	181	3.5 kg	6.3 kg

Do the data from this experiment provide evidence that the mean weight loss differs for the two treatments? You can test the relevant hypotheses using a significance level of $\alpha = 0.01$.

H Hypotheses You want to use data from the experiment to determine if mean weight loss differs for people who self-monitor and people who use a fitness tracker to monitor. You can define the two treatment means as

$\mu_1 =$ mean weight loss for people who self-monitor
$\mu_2 =$ mean weight loss for people who use a fitness tracker to monitor

and $\mu_1 - \mu_2$ is the difference in treatment means. Translating the question of interest into hypotheses gives

$$H_0: \mu_1 - \mu_2 = 0$$
$$H_a: \mu_1 - \mu_2 \neq 0$$

The alternative hypothesis corresponds to the claim that the mean weight loss is not the same for the self-monitoring treatment and the fitness tracker monitoring treatment.

M Method   Considering the four key questions (QSTN), this situation can be described as *hypothesis testing, experiment data, one numerical variable (weight loss),* and *two treatments (self-monitoring and fitness tracker monitoring)*. This combination suggests a two-sample *t* test. A significance level of 0.01 was specified for this test.

C Check   Next you need to check to see if this method is appropriate. From the study description, you know that the participants were assigned at random to one of the two treatment groups. Because there are more than 30 people in each of the treatment groups, you know that both treatment groups are large enough to proceed with the two-sample *t* test. (If one or both group sizes were less than 30, you would need to consider whether it is reasonable to think that the two treatment response distributions are approximately normal.)

C Calculate   You need to calculate the value of the test statistic, the number of degrees of freedom, and the *P*-value. You can do this by hand, as shown here, or you could use a graphing calculator or a statistics software package.

   Test statistic:

$$t = \frac{(\bar{x}_1 - \bar{x}_2) - (\mu_1 - \mu_2)}{\sqrt{\dfrac{s_1^2}{n_1} + \dfrac{s_2^2}{n_2}}} = \frac{(5.9 - 3.5) - 0}{\sqrt{\dfrac{6.8^2}{170} + \dfrac{6.3^2}{181}}}$$

$$= \frac{2.4}{0.701}$$

$$= 3.42$$

   Degrees of freedom

$$V_1 = \frac{s_1^2}{n_1} = \frac{6.8^2}{170} = 0.272 \qquad\qquad V_2 = \frac{s_2^2}{n_2} = \frac{6.3^2}{181} = 0.219$$

$$df = \frac{(V_1 + V_2)^2}{\dfrac{V_1^2}{n_1 - 1} + \dfrac{V_2^2}{n_2 - 1}} = \frac{(0.272 + 0.219)^2}{\dfrac{0.272^2}{169} + \dfrac{0.219^2}{180}} = \frac{0.241}{0.0007} = 344.286$$

Truncating the calculated degrees of freedom results in df = 344.

   P-value:
   This is a two-tailed test, so the *P*-value is two times the area to the right of 3.42 under the *t* curve with df = 344. Since 3.42 is so far out in the upper tail of this *t* curve, the *P*-value ≈ 0.

C Communicate Results   Because the *P*-value is less than the selected significance level (0.01), the null hypothesis is rejected. There is convincing evidence that the mean weight loss is not the same for the self-monitoring treatment and the fitness tracker monitoring treatment. Notice that the mean weight loss for the group that used the fitness tracker was less than the mean for the group that self-monitored. This is the basis for the headline from *The New York Times* article that said that fitness trackers may undermine weight loss efforts.

   Statistics software or a graphing calculator could also have been used to complete the Calculate step. For example, Minitab output is shown on the next page. From the Minitab output, the value of the test statistic is *t* = 3.42. The reported degrees of freedom are 342 and the *P*-value is reported as 0.001. The small difference in the value of the degrees of freedom is due to differences in rounding.

Two-Sample T-Test and CI

Sample	N	Mean	StDev	SE Mean
1	170	5.90	6.80	0.52
2	181	3.50	6.30	0.47

Difference = μ (1) $-$ μ (2)
Estimate for difference: 2.400
95% CI for difference: (1.021, 3.779)
T-Test of difference = 0 (vs $\neq$): T-Value = 3.42   P-Value = 0.001   DF = 342

Estimating the Difference in Treatment Means

As long as an experiment uses random assignment to create the treatment groups, the two-sample t confidence interval can be adapted for use in estimating a difference in treatment means. For use with experiment data, two modifications are needed to adapt the interval introduced in Section 13.3.

1. **Conditions:** The changes to the required conditions are the same as those made for the two-sample t test.
2. **Interpretation:** The interpretation of the confidence interval estimate will now be in terms of treatment means rather than population means.

The Two-Sample t Confidence Interval for a Difference in Treatment Means

Appropriate when the following conditions are met:

1. Individuals or objects are *randomly assigned* to treatments.
2. The number of individuals or objects in each of the treatment groups is large (30 or more) or the treatment response distributions (the distributions of response values that would result if the treatments were applied to a *very* large number of individuals or objects) are approximately normal.

When these conditions are met, a confidence interval for the difference in treatment means is

$$(\bar{x}_1 - \bar{x}_2) \pm (t \text{ critical value}) \sqrt{\frac{s_1^2}{n_1} + \frac{s_2^2}{n_2}}$$

The t critical value is based on

$$df = \frac{(V_1 + V_2)^2}{\dfrac{V_1^2}{n_1 - 1} + \dfrac{V_2^2}{n_2 - 1}} \quad \text{where } V_1 = \frac{s_1^2}{n_1} \text{ and } V_2 = \frac{s_2^2}{n_2}$$

The calculated value of df should be truncated (rounded down) to obtain an integer value for df. The desired confidence level determines which t critical value is used. Appendix A Table 3, statistical software, or a graphing calculator can be used to obtain the t critical value.

Interpretation of Confidence Interval
You can be confident that the actual value of the difference in treatment means is included in the calculated interval. This statement should be worded in context.

Interpretation of Confidence Level
The confidence level specifies the long-run proportion of the time that this method is expected to be successful in capturing the actual difference in treatment means.

Example 13.11 **Baby Scientists**

The article **"Baby Scientists Experiment with Everything"** (*The Wall Street Journal*, April 18, 2015) argues that even very young children learn about the world by experimenting like scientists. This article references a study that appeared in *Science* ("Observing the Unexpected Enhances

Infants' Learning and Exploration," *Science* **[2015]: 91–94).** In one of the experiments described in the *Science* paper, infants were shown a video of either a car or a ball.

The infants were randomly assigned to one of two groups. Those in one group were shown a video of an object rolling down a sloped surface and then falling off the edge when it reached the end of the surface (the expected behavior). The infants in the other group were shown a video of an object rolling down the same surface but then remaining suspended in the air when it went off the end of the surface (the unexpected behavior). In some of the videos the object was a car and in others it was a ball.

After seeing the video, the infants were given both a ball and a car. The researchers calculated an "exploration preference score" by observing the time spent exploring each of the two objects and then calculating the following difference in times (measured in seconds):

$$\text{Exploration preference score} = \text{time spent on object in video} - \text{time spent on object not in video}$$

Notice that a positive score corresponds to an infant who spent more time with the object from the video and a negative score corresponds to an infant who spent more time with the object that was not in the video.

The researchers wondered if the mean exploration preference score for infants who see the video where the object behaves in an unexpected way would be different from the mean score for infants who see the video where the object behaves as expected.

The following data values are consistent with summary quantities and graphs appearing in the paper.

Exploration Preference Scores for Expected Behavior Group

$$-3 \quad 13 \quad -1 \quad -5 \quad -18 \quad 0 \quad 12 \quad 0 \quad 4 \quad 23$$
$$n_1 = 10 \qquad \bar{x}_1 = 2.5 \qquad s_1 = 11.33$$

Exploration Preference Scores for Unexpected Behavior Group

$$17 \quad 18 \quad 28 \quad -3 \quad 6 \quad 17 \quad 7 \quad 15 \quad 9 \quad 2$$
$$n_2 = 10 \qquad \bar{x}_2 = 11.6 \qquad s_2 = 9.09$$

The data from this experiment can be used to estimate the difference in mean score for the two treatments (expected behavior video and unexpected behavior video).

E Estimate You want to estimate

$$\mu_1 - \mu_2 = \text{difference in mean score}$$

where

$$\mu_1 = \text{mean score for the expected behavior treatment}$$

and

$$\mu_2 = \text{mean score for the unexpected behavior treatment}$$

M Method The answers to the four key questions are *estimation, experiment data, one numerical variable (exploration preference score),* and *two treatments*. This combination of answers leads you to consider a two-sample *t* confidence interval for a difference in treatment means. For the purposes of this example, a 90% confidence level will be used.

C Check The infants were randomly assigned to one of the two treatment groups. Because there are only 10 infants in each of the treatment groups, you need to be willing to assume that the exploration prefence score distribution for each of the two treatments is at least approximately normal. Figure 13.3 shows boxplots constructed using the data from the two treatment groups. There are no outliers in either data set and the boxplots are reasonably symmetric, suggesting that the assumption of approximate normality is reasonable.

C Calculate You need to calculate the endpoints of the confidence interval. You can do this by hand, as shown here, or you could use a graphing calculator or a statistics software package. If calculating by hand, you first need to calculate the appropriate number of degrees of freedom.

FIGURE 13.3
Boxplots for the exploration preference score data of Example 13.11

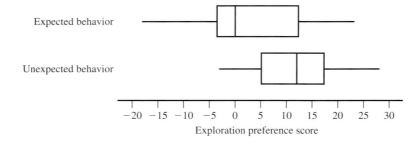

Degrees of freedom:

$$V_1 = \frac{s_1^2}{n_1} = \frac{11.33^2}{10} = 12.837 \qquad V_2 = \frac{s_2^2}{n_2} = \frac{9.09^2}{10} = 8.263$$

$$df = \frac{(V_1 + V_2)^2}{\dfrac{V_1^2}{n_1 - 1} + \dfrac{V_2^2}{n_2 - 1}} = \frac{(12.837 + 8.263)^2}{\dfrac{12.837^2}{9} + \dfrac{8.263^2}{9}} = \frac{445.210}{25.896} = 17.192$$

Truncating to an integer gives df = 17. In the 17 df row of Appendix A Table 3, the t critical value for a 90% confidence level is 1.74. The confidence interval is then

$$(\bar{x}_1 - \bar{x}_2) \pm (t \text{ critical value}) \sqrt{\frac{s_1^2}{n_1} + \frac{s_2^2}{n_2}}$$

$$= (2.5 - 11.6) \pm (1.74) \sqrt{\frac{11.33^2}{10} + \frac{9.09^2}{10}}$$

$$= -9.1 \pm 7.99$$

$$= (-17.09, -1.11)$$

C Communicate Results

Based on these samples, you can be 90% confident that the actual difference in mean exploration preference scores for the two treatments is somewhere between -17.09 and -1.11. Because 0 is not included in the interval and both endpoints of the confidence interval are negative, you would conclude that the mean exploration preference score is greater for the unexpected behavior treatment. This means that, on average, infants spend more time with the object in the video when the object in the video behaved in an unexpected way than when the object behaved as expected. Notice that the confidence interval is rather wide. This is because the two sample standard deviations are large and the sample sizes are small. The method used to construct this interval is successful in capturing the actual difference in treatment means about 90% of the time.

Many statistical computer packages can compute the two-sample t confidence interval. JMP was used to construct a 90% confidence interval using the data of this example, and the resulting output is shown here:

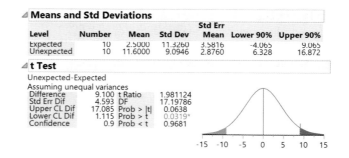

The upper and lower endpoints of the 95% confidence interval for the difference in treatment means are shown in the output labeled as "Upper CL Dif" and "Lower CL Dif." These endpoints are slightly different from those in the hand calculations because

JMP uses more decimal accuracy. Also notice that JMP has calculated the interval for the difference in means as unexpected behavior mean − expected behavior mean, so the endpoints of the interval are positive rather than negative, as in the hand calculations.

Summing It Up—Section 13.4

The following learning objectives were addressed in this section:

Mastering the Mechanics

M8:   Use the two-sample *t* test to test hypotheses about a difference in treatment means.
A hypothesis test for a difference in two treatment means is illustrated in Example 13.10. If the null hypothesis that the treatment means are equal is rejected, it is reasonable to conclude that there is convincing evidence in support of the alternative hypothesis. If the null hypothesis is not rejected, chance differences due to random assignment is a plausible explanation for the observed difference in the treatment means and there is not convincing evidence that the alternative hypothesis is true.

M9:   Use the two-sample *t* confidence interval to estimate a difference in treatment means.
The interpretation of a confidence interval for a difference in treatment means is similar to the interpretation of a confidence interval for a difference in population means. If both endpoints of a confidence interval for $\mu_1 - \mu_2$ are positive, it means that you think that μ_1 is greater than μ_2 and the interval gives an estimate of how much greater. If both endpoints are negative, it means that you think that μ_1 is less than μ_2 and the interval gives an estimate of how much less. If zero is included in the interval, it means that zero is a plausible value for the difference in the two treatment means.

Confidence level is interpreted in the same way as it was for the other confidence intervals that you have seen. It specifies the long-run percentage of the time that the method will be successful in capturing the actual difference in treatment means.

SECTION 13.4   EXERCISES

Each Exercise Set assesses the following chapter learning objectives: M8, M9

SECTION 13.4   Exercise Set 1

13.59   The paper **"The Effect of Multitasking on the Grade Performance of Business Students"** (*Research in Higher Education Journal* [2010]: 1–10) describes an experiment in which 62 undergraduate business students were randomly assigned to one of two experimental groups. Students in one group were asked to listen to a lecture but were told that they were permitted to use cell phones to send text messages during the lecture. Students in the second group listened to the same lecture but were not permitted to send text messages during the lecture. Afterwards, students in both groups took a quiz on material covered in the lecture. The researchers reported that the mean quiz score for students in the texting group was significantly lower than the mean quiz score for students in the no-texting group. In the context of this experiment, explain what it means to say that the texting group mean was *significantly lower* than the no-text group mean. (Hint: See discussion on page 662.)

13.60   The paper **"Short-Term Sleep Loss Decreases Physical Activity Under Free-Living Conditions but Does Not Increase Food Intake Under Time-Deprived Laboratory Conditions in Healthy Men"** (*American Journal of Clinical Nutrition* [2009]: 1476–1483) describes an experiment in which 30 male volunteers were assigned at random to one of two sleep conditions. Men in the 4-hour group slept 4 hours per night for two nights. Men in the 8-hour group slept 8 hours per night for two nights. On the day following these two nights, the men recorded food intake. The researchers reported that there was no significant difference in mean calorie intake for the two groups. In the context of this experiment, explain what it means to say that there is *no significant difference* in the group means. (Hint: See discussion on page 662.)

13.61   The article **"An Alternative Vote: Applying Science to the Teaching of Science"** (*The Economist*, May 12, 2011) describes an experiment conducted at the University of British Columbia. A total of 850 engineering students enrolled in a physics course participated in the experiment. These students were randomly assigned to one of two experimental groups. The two groups attended the same lectures for the first 11 weeks of the semester. In the twelfth week, one of the groups was switched to a style of teaching where students were expected to do reading assignments prior to class and then class time was used to focus on problem solving, discussion and group work. The second group continued with the traditional lecture approach. At the end of the twelfth week, the students were given a test over the course material from that week. The mean test score for students in the new teaching method group was 74 and the mean test score for students in the traditional lecture group was 41. Suppose that the two groups each consisted of 425 students and that the standard deviations of test scores for the new teaching method group and the traditional lecture method group were 20 and 24, respectively. Can you conclude that the mean test score is significantly higher for the new teaching method group than for the traditional lecture method group? Test the appropriate hypotheses using a significance level of $\alpha = 0.01$. (Hint: See Examples 13.9 and 13.10.)

13.62 Can moving their hands help children learn math? This question was investigated in the paper **"Gesturing Gives Children New Ideas About Math"** (*Psychological Science* [2009]: 267–272). Eighty-five children in the third and fourth grades who did not answer any questions correctly on a test with six problems of the form $3 + 2 + 8 = \_\_ + 8$ were participants in an experiment. The children were randomly assigned to either a no-gesture group or a gesture group. All the children were given a lesson on how to solve problems of this form using the strategy of trying to make both sides of the equation equal. Children in the gesture group were also taught to point to the first two numbers on the left side of the equation with the index and middle finger of one hand and then to point at the blank on the right side of the equation. This gesture was supposed to emphasize that grouping is involved in solving the problem. The children then practiced additional problems of this type. All children were then given a test with six problems to solve, and the number of correct answers was recorded for each child. Summary statistics are given below.

	n	$\bar{x}$	s
No Gesture	42	1.3	0.3
Gesture	43	2.2	0.4

Is there evidence that learning the gesturing approach to solving problems of this type results in a significantly higher mean number of correct responses? Test the relevant hypotheses using $\alpha = 0.05$.

13.63 Use the information in the previous exercise to estimate the difference in the mean number of correct answers for the two different methods. Use a 95% confidence level, and be sure to interpret the interval in the context of the experiment.

 13.64 The accompanying data on food intake (in Kcal) for 15 men on the day following two nights of only 4 hours of sleep each night and for 15 men on the day following two nights of 8 hours of sleep each night are consistent with summary quantities in the paper **"Short-Term Sleep Loss Decreases Physical Activity Under Free-Living Conditions But Does Not Increase Food Intake Under Time-Deprived Laboratory Conditions in Healthy Men"** (*American Journal of Clinical Nutrition* [2009]: 1476–1482). The men participating in this experiment were randomly assigned to one of the two sleep conditions.

4-Hour Sleep Group	3,585	4,470	3,068	5,338	2,221
	4,791	4,435	3,099	3,187	3,901
	3,868	3,869	4,878	3,632	4,518
8-Hour Sleep Group	4,965	3,918	1,987	4,993	5,220
	3,653	3,510	3,338	4,100	5,792
	4,547	3,319	3,336	4,304	4,057

If appropriate, estimate the difference in mean food intake for the two different sleep conditions using a 95% confidence interval. (Hint: See Example 13.11.)

13.65 Is injecting medical cement effective in reducing pain for people who have suffered fractured vertebrae? The paper **"A Randomized Trial of Vertebroplasty for Osteoporotic Spinal Fractures"** (*New England Journal of Medicine* [2009]: 569–578) describes an experiment to compare patients who underwent vertebroplasty (the injection of cement) to patients in a placebo group who underwent a fake procedure. Because the placebo procedure was similar to the vertebroplasty procedure except for the actual injection of cement, patients participating in the experiment were not aware of which treatment they received. All patients were asked to rate their pain at three different times—3 days, 14 days, and 1 month after the procedure. Summary statistics are given in the accompanying table.

	Pain Intensity			
	Vertebroplasty Group		**Placebo Group**	
	$n = 68$		$n = 63$	
	mean	sd	mean	sd
3 Days	4.2	2.8	3.9	2.9
14 Days	4.3	2.9	4.5	2.8
1 Month	3.9	2.9	4.6	3.0

a. Briefly explain why the researchers may have chosen to include a placebo group that underwent a fake procedure rather than just comparing the vertebroplasty group to a group of patients who did not receive any treatment.
b. Construct and interpret a 95% confidence interval for the difference in mean pain intensity 3 days after treatment between the vertebroplasty treatment and the fake treatment.

13.66 Use the information given in the previous exercise to complete the following:
a. Construct and interpret 95% confidence intervals for the difference in mean pain intensity between the two treatments at 14 days and at 1 month after treatment.
b. Based on the confidence intervals from Part (a) and the previous exercise, comment on the effectiveness of injecting cement as a way of reducing pain for people with fractured vertebrae.

SECTION 13.4 Exercise Set 2

13.67 The paper **"Supervised Exercise Versus Non-Supervised Exercise for Reducing Weight in Obese Adults"** (*The Journal of Sports Medicine and Physical Fitness* [2009]: 85–90) describes an experiment in which participants were randomly assigned either to a supervised exercise program or a control group. Those in the control group were told only that they should take measures to lose weight. Those in the supervised exercise group were told they should take measures to lose weight as well, but they also participated in regular supervised exercise sessions. The researchers reported that after 4 months, the mean decrease in body fat was significantly higher for the supervised exercise group than for the control group. In the context of this experiment, explain what it means to say that the exercise group mean was *significantly higher* than the control group mean.

13.68 The paper **"Effects of Caffeine on Repeated Sprint Ability, Reactive Agility Time, Sleep and Next Day Performance"** (*Journal of Sports Medicine and Physical Fitness* [2010]: 455–464) describes an experiment in which male athlete volunteers who were considered low caffeine consumers were assigned at random to one of two experimental groups. Those assigned to the caffeine group drank a beverage which contained caffeine one hour before an exercise session. Those in the no-caffeine group drank a beverage that did not contain caffeine. During the exercise session, each participant performed a test that measured reactive agility. The researchers reported that there was no significant difference in mean reactive agility for the two experimental groups. In the context of this experiment, explain what it means to say that there is *no significant difference* in the group means.

13.69 The accompanying data on food intake (in Kcal) for 15 men on the day following two nights of only 4 hours of sleep each night and for 15 men on the day following two nights of 8 hours of sleep each night is consistent with summary quantities in the paper **"Short-Term Sleep Loss Decreases Physical Activity Under Free-Living Conditions But Does Not Increase Food Intake Under Time-Deprived Laboratory Conditions in Healthy Men"** (*American Journal of Clinical Nutrition* [2009]: 1476–1482). The men participating in this experiment were randomly assigned to one of the two sleep conditions.

4-hour Sleep Group	3,585	4,470	3,068	5,338	2,221
	4,791	4,435	3,099	3,187	3,901
	3,868	3,869	4,878	3,632	4,518
8-hour Sleep Group	4,965	3,918	1,987	4,993	5,220
	3,653	3,510	3,338	4,100	5,792
	4,547	3,319	3,336	4,304	4,057

If appropriate, carry out a two-sample t test with $\alpha = 0.05$ to determine if there is a significant difference in mean food intake for the two different sleep conditions.

13.70 The paper **"If It's Hard to Read, It's Hard to Do"** (*Psychological Science* [2008]: 986–988) described an interesting study of how people perceive the effort required to do certain tasks. Each of 20 students was randomly assigned to one of two groups. One group was given instructions for an exercise routine that were printed in an easy-to-read font (Arial). The other group received the same set of instructions but printed in a font that is considered difficult to read (𝓑𝓻𝓾𝓼𝓱). After reading the instructions, subjects estimated the time (in minutes) they thought it would take to complete the exercise routine. Summary statistics follow.

	Easy Font	Difficult Font
n	10	10
$\bar{x}$	8.23	15.10
s	5.61	9.28

The authors of the paper used these data to carry out a two-sample t test and concluded at the 0.10 significance level that the mean estimated time to complete the exercise routine is significantly lower when the instructions are printed in an easy-to-read font than when printed in a font that is difficult to read. Discuss the appropriateness of using a two-sample t test in this situation.

13.71 The paper **"Does the Color of the Mug Influence the Taste of the Coffee?"** (*Flavour* [2014]: 1–7) describes an experiment in which subjects were assigned at random to one of two treatment groups. The 12 people in one group were served coffee in a white mug and were asked to rate the quality of the coffee on a scale from 0 to 100. The 12 people in the second group were served the same coffee in a clear glass mug, and they also rated the coffee. The mean quality rating for the 12 people in the white mug group was 50.35 and the standard deviation was 20.17. The mean quality rating for the 12 people in the clear glass mug group was 61.48 and the standard deviation was 16.69. For purposes of this exercise, you may assume that the distribution of quality ratings for each of the two treatments is approximately normal.
a. Use the given information to construct and interpret a 95% confidence interval for the difference in mean quality rating for this coffee when served in a white mug and when served in a glass mug.
b. Based on the interval from Part (a), are you convinced that the color of the mug makes a difference in terms of mean quality rating? Explain.

13.72 The authors of the paper **"The Empowering (Super) Heroine? The Effects of Sexualized Female Characters in Superhero Films on Women"** (*Sex Roles* [2015]: 211–220) were interested in the effect on female viewers of watching movies in which female heroines were portrayed in roles that focus on their sex appeal. They carried out an experiment in which female college students were assigned at random to one of two experimental groups. The 23 women in one group watched 13 minutes of scenes from the X-Men film series and

Data set available

then responded to a questionnaire designed to measure body esteem. Lower scores on this measure correspond to lower body satisfaction. The 29 women in the other group (the control group) did not watch any video prior to responding to the questionnaire measuring body esteem. For the women who watched the *X-Men* video, the mean body esteem score was 4.43 and the standard deviation was 1.02. For the women in the control group, the mean body esteem score was 5.08 and the standard deviation was 0.98. For purposes of this exercise, you may assume that the distribution of body esteem scores for each of the two treatments (video and control) is approximately normal.

a. Construct and interpret a 90% confidence interval for the difference in mean body esteem score for the video treatment and the no video treatment.

b. Do you think that watching the video has an effect on mean body esteem score? Explain.

ADDITIONAL EXERCISES

13.73 The article **"Dieters Should Use a Bigger Fork"** (*Food Network Magazine*, January/February 2012) described an experiment conducted by researchers at the University of Utah. The article reported that when people were randomly assigned to either eat with a small fork or to eat with a large fork, the mean amount of food consumed was significantly less for the group that ate with the large fork.

a. What are the two treatments in this experiment?

b. In the context of this experiment, explain what it means to say that the mean amount of food consumed was *significantly less* for the group that ate with the large fork.

13.74 The article **"Why We Fall for This"** (*AARP Magazine*, May/June 2011) describes an experiment investigating the effect of money on emotions. In this experiment, students at University of Minnesota were randomly assigned to one of two groups. One group counted a stack of dollar bills. The other group counted a stack of blank pieces of paper. After counting, each student placed a finger in very hot water and then reported a discomfort level. It was reported that the mean discomfort level was significantly lower for the group that had counted money. In the context of this experiment, explain what it means to say that the money group mean was *significantly lower* than the blank-paper group mean.

13.75 The article referenced in the previous exercise also described an experiment in which students at Columbia Business School were randomly assigned to one of two groups. Students in one group were shown a coffee mug and asked how much they would pay for that mug. Students in the second group were given a coffee mug identical to the one shown to the first group and asked how much someone would have to pay to buy it from them. It was reported that the mean

value assigned to the mug for the second group was significantly higher than the mean value assigned to the same mug for the first group. In the context of this experiment, explain what it means to say that the mean value was *significantly higher* for the group that was given the mug.

13.76 The paper **"The Effect of Multitasking on the Grade Performance of Business Students"** (*Research in Higher Education Journal* [2010]: 1–10) describes an experiment in which 62 undergraduate business students were randomly assigned to one of two experimental groups. Students in one group were asked to listen to a lecture but were told that they were permitted to use cell phones to send text messages during the lecture. Students in the second group listened to the same lecture but were not permitted to send text messages. Afterwards, students in both groups took a quiz on material covered in the lecture. Data from this experiment are summarized in the accompanying table.

Experimental Group	Group Size	Mean Quiz Score	Standard Deviation of Quiz Scores
Texting	31	42.81	9.91
No Texting	31	58.67	10.42

Do these data provide evidence to support the researcher's claim that the mean quiz score for the texting group is significantly lower than the mean quiz score for the no-texting group?

13.77 The paper **"Supervised Exercise Versus Non-Supervised Exercise for Reducing Weight in Obese Adults"** (*The Journal of Sports Medicine and Physical Fitness* [2009]: 85–90) describes an experiment in which participants were randomly assigned either to a supervised exercise program or a control group. Those in the control group were told that they should take measures to lose weight. Those in the supervised exercise group were told they should take measures to lose weight as well, but they also participated in regular supervised exercise sessions. Weight loss (in kilograms) at the end of four months was recorded. Data consistent with summary quantities given in the paper are shown in the accompanying table.

Experimental Group	Group Size	Weight Loss (in kg)				
Supervised Exercise	17	1.1	4.8	4.1	9.1	2.2
		10.8	9.9	2.1	5.4	10.3
		−3.7	13.4	3.5	2.9	11.2
		9.5	5.1			
Control	17	6.7	7.6	2.3	−2.1	1.9
		5.4	1.0	3.6	4.9	−0.7
		1.3	2.2	2.1	0.6	1.3
		1.3	5.4			

Data set available

Do these data provide evidence that the supervised exercise treatment results in a significantly greater mean weight loss than the treatment that just involves advising people to lose weight?

13.78 The paper **"The Effect of Multitasking on the Grade Performance of Business Students"** (*Research in Higher Education Journal* [2010]: 1–10) describes an experiment in which 62 undergraduate business students were randomly assigned to one of two experimental groups. Students in one group were asked to listen to a lecture but were told that they were permitted to use cell phones to send text messages during the lecture. Students in the second group listened to the same lecture but were not permitted to send text messages. Afterwards, students in both groups took a quiz on material covered in the lecture. Data from this experiment are summarized in the accompanying table.

Experimental Group	Group Size	Mean Quiz Score	Standard Deviation of Quiz Scores
Texting	31	42.81	9.91
No Texting	31	58.67	10.42

Use the given information to construct and interpret a 90% confidence interval for the difference in mean quiz score for the two treatments (texting allowed and texting not allowed).

 13.79 The paper **"Supervised Exercise Versus Non-Supervised Exercise for Reducing Weight in Obese Adults"** (*The Journal of Sports Medicine and Physical Fitness* [2009]: 85–90) describes an experiment in which participants were randomly assigned either to a supervised exercise program or a control group. Those in the control group were told that they should take measures to lose weight. Those in the supervised exercise group were told they should take measures to lose weight, but they also participated in regular supervised exercise sessions.

Weight loss at the end of four months was recorded. Data consistent with summary quantities given in the paper are shown in the accompanying table.

Experimental Group	Group Size	Weight Loss (in kg)				
Supervised Exercise	17	1.1	4.8	4.1	9.1	2.2
		10.8	9.9	2.1	5.4	10.3
		−3.7	13.4	3.5	2.9	11.2
		9.5	5.1			
Control	17	6.7	7.6	2.3	−2.1	1.9
		5.4	1.0	3.6	4.9	−0.7
		1.3	2.2	2.1	0.6	1.3
		1.3	5.4			

Use the given information to construct and interpret a 95% confidence interval for the difference in mean weight loss for the two treatments.

13.80 The online article **"Death Metal in the Operating Room"** (www.npr.org, December 24, 2009) describes an experiment investigating the effect of playing music during surgery. One conclusion drawn from this experiment was that doctors listening to music that contained vocal elements took more time to complete surgery than doctors listening to music without vocal elements. Suppose that μ_1 denotes the mean time to complete a specific type of surgery for doctors listening to music with vocal elements and μ_2 denotes the mean time for doctors listening to music with no vocal elements. Further suppose that the stated conclusion was based on a 95% confidence interval for $\mu_1 - \mu_2$, the difference in treatment means. Which of the following three statements is correct? Explain why you chose this statement.

Statement 1: Both endpoints of the confidence interval were negative.
Statement 2: The confidence interval included 0.
Statement 3: Both endpoints of the confidence interval were positive.

SECTION 13.5   Simulation-Based Inference for Two Means (Optional)

You have already encountered simulation-based methods for inference about one population proportion, about two population proportions, and about one population mean. In this section, you will see simulation-based methods that will allow you to make inferences about two means. These methods are especially useful when the conditions of the two-sample methods of Sections 13.2, 13.3, and 13.4 are not met (when the sample sizes are small and it is not clear that the distributions are normally distributed).

Simulation-Based Inference About Two Population Means Using Paired Samples

In Section 13.2, you saw that when the samples are paired, hypothesis tests and confidence intervals for the difference in two population means involved calculating the sample differences and then using one-sample methods with the sample of differences. This is also the case for simulation-based methods—you calculate the sample differences and then use the one-sample bootstrap confidence interval and simulation-based randomization test that were introduced in Chapter 12.

Example 13.12 **Charitable Chimps Revisited**

Example 13.3 described a study in which chimpanzees could deliver food to themselves when one lever was pushed and to both themselves and a chimp in an adjoining cage when another lever was pushed. Researchers recorded the number of times out of 16 that each of seven chimps chose the option to "feed both" when there was another chimp in the adjoining cage and when there was not a chimp in the adjoining cage. Because the samples were paired, the differences in the number of times the "feed both" option was chosen by the chimpanzees were computed.

The pairing produced the following set of differences for the seven chimps in the sample:

Chimp	1	2	3	4	5	6	7
Difference	2	0	−2	2	1	3	0

Clearly the sample size is small, and despite the fairly symmetric distribution in the boxplot presented in Example 13.3, you may still be hesitant to use the one-sample t test for the paired samples analysis. Instead, you may choose to use the simulation-based methods presented in Section 12.4.

To construct a confidence interval, begin by selecting bootstrap samples, with replacement, from the original sample of differences. Here is one bootstrap sample:

Resampled Chimp ID	5	5	5	5	1	3	6
Difference	1	1	1	1	2	−2	3

The sample mean for this bootstrap sample is $\bar{x}_d = 1.00$. You can repeat this process many times, and the resulting bootstrap distribution of $\bar{x}_d$ values provides information about sampling variability that can be used to find a confidence interval for the population mean difference, μ_d, where μ_d is the mean difference

$$\begin{pmatrix} \text{mean number of} \\ \text{charitable responses} \\ \text{when there is a chimp} \\ \text{in the adjoining cage} \end{pmatrix} - \begin{pmatrix} \text{mean number of} \\ \text{charitable responses} \\ \text{when there is not a chimp} \\ \text{in the adjoining cage} \end{pmatrix}$$

The Shiny app "Bootstrap Confidence Interval for One Mean" (which can be found in the App collection at statistics.cengage.com/Peck2e/Apps.html) produces a bootstrap distribution that can be used to obtain a bootstrap confidence interval for the mean difference, as shown here.

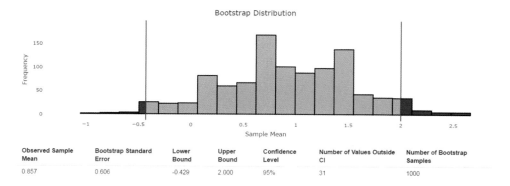

Observed Sample Mean	Bootstrap Standard Error	Lower Bound	Upper Bound	Confidence Level	Number of Values Outside CI	Number of Bootstrap Samples
0.857	0.606	-0.429	2.000	95%	31	1000

For this particular simulation, the bootstrap method produces a 95% confidence interval for the population mean difference, μ_d, of −0.43 to 2.00. You can be 95% confident that this interval contains the population mean difference.

You may also wish to perform a hypothesis test to test the claim that the population mean difference is different from zero.

To do this, the data values in the original sample can be shifted to represent a sample from a hypothetical population with mean $\mu_d = 0$ by subtracting 0.86 (the mean of the original sample of differences) from each difference, as follows:

Chimp	1	2	3	4	5	6	7
Difference	2	0	−2	2	1	3	0
Difference − 0.86	1.14	−0.86	−2.86	1.14	0.14	2.14	−0.86

Then the relevant hypotheses are:

$$H_0: \mu_d = 0$$
$$H_a: \mu_d \neq 0$$

The simulation approach begins with resampling from the original sample shifted to have a mean of 0. Here is one simulated sample from the shifted sample:

Chimp	4	4	2	1	7	3	4
Difference − 0.86	1.14	1.14	−0.86	1.14	−0.86	−2.86	1.14

The sample mean for this simulated sample is $\bar{x}_d = -0.02$.

A randomization distribution is generated by selecting many simulated samples from the shifted sample. This distribution can be viewed as the distribution of the sample mean difference if $\mu_d = 0$ is true. A P-value is the probability of obtaining a value at least as inconsistent with H_0 as what actually resulted. This means that the P-value for testing the null hypothesis $H_0: \mu_d = 0$ against the two-sided alternative $H_a: \mu_d \neq 0$ is the two times the probability of observing $\bar{x}_d \geq 0.86$ when H_0 is true. This probability can be approximated using the proportion of simulated values of $\bar{x}_d$ that fall at or above 0.86 in the distribution generated using the simulated samples.

The following randomization distribution was produced using the Shiny app "Randomization Test for One Mean." This App can be found at statistics.cengage.com/Peck2e/Apps.html.

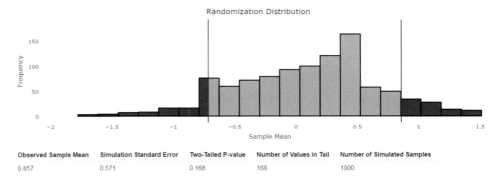

Observed Sample Mean	Simulation Standard Error	Two-Tailed P-value	Number of Values in Tail	Number of Simulated Samples
0.857	0.571	0.168	168	1000

The two-tailed P-value based on this simulation is 0.168. This P-value is greater than the 0.05 significance level, and so you fail to reject H_0. There is not convincing evidence that the mean difference is different from zero.

To use the one-sample Shiny apps with paired samples, you need to first calculate the sample differences. There are also two apps available that allow you to enter the data from the two samples without first having to calculate the differences. These apps are "Bootstrap Confidence Interval for the Difference in Two Population Means Using Paired Samples" and "Randomization Test for Difference in Two Population Means Using Paired Samples." For example, the following output was produced by the Shiny app "Randomization Test for Difference in Two Population Means Using Paired Samples" using the data from Example 13.12.

Randomization Test for Difference in Two Population Means Using Paired Samples

Choose File to Upload:

Data entry windows below must be empty.

Browse...	No file selected

Or Enter Data:

Data must be separated by a new line.

Sample 1	Sample 2
23	21
22	22
21	23
23	21
19	18
19	16
19	19

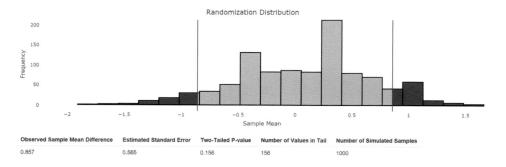

Randomization Distribution

Observed Sample Mean Difference	Estimated Standard Error	Two-Tailed P-value	Number of Values in Tail	Number of Simulated Samples
0.857	0.565	0.156	156	1000

Notice that the *P*-value based on this randomization distribution is 0.156, compared to the *P*-value of 0.168 in Example 13.12. Because simulation-based methods, such as randomization tests, are based on random sampling, the randomization distribution will vary from simulation to simulation, depending on the outcomes of the sampling process. But usually the estimated *P*-values (or the bootstrap confidence interval endpoints) are similar from one simulation to another (as was the case in this example) as long as the number of samples used in the simulation is large.

Simulation-Based Inference About the Difference in Two Treatment Means

Simulation-based randomization tests and bootstrap confidence intervals may be used to learn about the difference in two treatment means using data from a randomized experiment.

Example 13.13   **Blue Light Exposure and Blood Glucose Level**

The article **"Bright Light at Night Time Can Seriously Mess with Your Metabolism, Study Finds"** (*Science Alert,* **May 20, 2016, www.sciencealert.com/checking-your-phone -at-night-could-be-messing-with-your-metabolism, retrieved May 23, 2017)** describes research conducted to examine the effects of blue light exposure (the type of light emitted by smartphones and computer screens) on a variety of measures, including blood glucose levels and sleepiness. The study was published in the **Public Library of Science ("Morning and Evening Blue-Enriched Light Exposure Alters Metabolic Function in Normal Weight Adults," PLOS One [2016]: e0155601)**. Adult volunteers of normal weight were randomly assigned to one of two groups. The first group was exposed to blue light for three hours in the morning (30 minutes after waking), and the second group was exposed to blue light for three hours in the evening (10 hours and 30 minutes after waking).

For each subject, a baseline blood glucose level was obtained 30 minutes before the blue light exposure began. Then blood glucose level (in mg/dL) was tracked every half hour for four hours during and immediately following the blue light exposure, including when the subject ate a meal. One outcome measure of the study was the peak change from baseline in blood glucose level. A negative peak change means that the subject's blood glucose level remained below the baseline measurement during the entire time period. You can perform a hypothesis test using a significance level of $\alpha = 0.05$ to determine whether the mean peak change in blood glucose level differs for the AM and PM blue light exposure groups.

Data for the 9 subjects in the AM blue light treatment group and the 10 subjects in the PM blue light treatment group are given in the following table.

Subject ID	Group	Glucose Peak Change from Baseline (mg/dL)
1	AM	4
2	AM	−22
3	AM	7
4	AM	4
5	AM	−8
6	AM	−13
7	AM	3
8	AM	−20
9	AM	−10
10	PM	28
11	PM	12
12	PM	12
13	PM	33
14	PM	25
15	PM	22
16	PM	4
17	PM	22
18	PM	−2
19	PM	−18

The dotplots in Figure 13.4 indicate the assumption that the change distributions are approximately normal is questionable. Because both sample sizes are small, the two-sample *t* test and confidence interval might not be appropriate choices for analyzing the data from this experiment.

FIGURE 13.4
Dotplots of Peak Glucose
Change for the AM and PM
Treatments

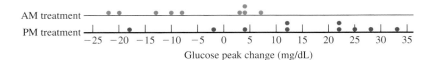

The two treatment group means can be represented as follows:

μ_1 = Mean change in glucose level for the AM treatment

μ_2 = Mean change in glucose level for the PM treatment

Translating the research question regarding whether mean change in blood glucose level differs for the AM and PM blue light exposure treatments results in the hypotheses:

$$H_0: \mu_1 - \mu_2 = 0$$
$$H_a: \mu_1 - \mu_2 \neq 0$$

Summary statistics for the data in the two samples are

	AM Treatment	**PM Treatment**
Sample size	9	10
Sample mean	−6.1	13.8
Sample standard deviation	11.02	15.61

The observed difference in means for change in glucose level, AM − PM, is $\bar{x}_1 - \bar{x}_2 = -19.9$.

A randomization test can be used to determine if there is convincing evidence that the two treatment means differ. In a randomization test using data from an experiment, you explore alternative random assignments of the subjects in a study to two (or more) groups. If the null hypothesis is true, there is no difference in the effect of the two treatments. This means that each subject would have had the same observed change whether he or she was in the AM treatment group or the PM treatment group. For example, subject 1, who was in the AM treatment group and had a change of 4, would have had a change of 4 even if that subject had been in the PM treatment group. If this is the case for every subject, the observed difference in sample means would be just due to chance in the random assignment of subjects to treatment groups.

To decide if the observed difference is consistent with what is expected due to chance alone or whether it is evidence of a real difference in treatment means, you begin by exploring what chance differences look like. Simulation is carried out by taking the original 19 observations and randomly assigning them into two groups (one of size 9 and one of size 10), and then calculating a simulated difference in means. This process is repeated many times to form a randomization distribution.

The following figure shows the simulation results from the Shiny app "Randomization Test for Difference in Two Treatment Means" for the difference in means from many different random assignments of the observed data values into two independent groups. This app can be found in the App collection at statistics.cengage.com/Peck2e/Apps.html.

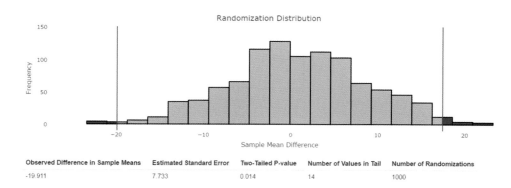

The observed difference in means was -19.9. Locating -19.9 in the distribution of simulated differences shows that it would be unusual to observe a difference this extreme if the null hypothesis of no difference in treatment means were true. Based on the randomization distribution, the probability of observing a difference at least as extreme as -19.9 is approximately 0.014. Because this two-sided P-value is less than the specified significance level of 0.05, the null hypothesis is rejected. Among many hypothetical random assignments of the changes in blood glucose levels to hypothetical AM and PM blue light exposure groups, less than 1% produce a difference in means that is at least as inconsistent with the null hypothesis as the observed difference in the means of -19.9. This is evidence that the treatment means are not equal.

You may also use the Shiny app "Bootstrap Confidence Interval for Difference in Two Treatment Means" (found in the App collection at statistics.cengage.com/Peck2e /Apps.html) to obtain a 95% confidence interval for the difference in the treatment means by examining the distribution of differences in simulated means calculated from hypothetical randomizations of the combined sample of changes in blood glucose levels into two groups.

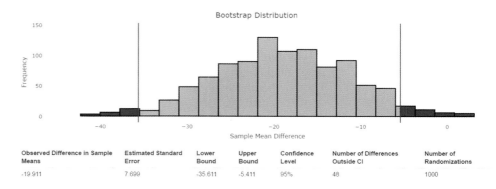

Observed Difference in Sample Means	Estimated Standard Error	Lower Bound	Upper Bound	Confidence Level	Number of Differences Outside CI	Number of Randomizations
-19.911	7.699	-35.611	-5.411	95%	48	1000

The given Shiny app output provides the bootstrap distribution for 1000 simulated differences in the means for alternative random assignments of the changes in blood glucose levels to two groups. Identifying the extreme 2.5% of differences in means on both the low and high ends of the distribution provides the endpoints for a 95% bootstrap confidence interval.

For this simulation, the 95% bootstrap confidence interval for the difference in treatment means is $(-35.6, -5.4)$. You can be 95% confident that the actual difference in the mean change in blood glucose levels, AM $-$ PM, falls between -35.6 and -5.4. Notice that this confidence interval does not include zero, and this is consistent with rejecting the null hypothesis that the difference in the mean change in glucose level for the two treatments is zero. With both endpoints negative, you would say that the mean change is greater for the PM blue light treatment than for the AM blue light treatment by somewhere between 5.4 and 35.6 mg/dL.

Simulation-Based Inference for the Difference in Two Population Means Using Independent Samples

You may wish to compare the means for two populations using independent samples selected at random from the populations, in contrast to using data from an experiment where subjects are assigned at random to treatment groups. The simulation-based methods for comparing two means using independent random samples are a little different from the methods used with data from an experiment.

Data set available

| **Example 13.14** | **Freshman Year Weight Gain Revisited** |

Example 13.8 described a study of freshman year weight gain (**"Predicting the 'Freshman 15': Environmental and Psychological Predictors of Weight Gain in First-Year University Students," *Health Education Journal* [2016]: 321–332**). In that study, weight gain (in kilograms) during the freshman year was recorded for independent random samples of

first-year students who lived on campus and first-year students who lived off campus. The data from Example 13.8 are also given here.

On Campus	Off Campus
2.0	1.6
2.3	3.1
1.1	−2.8
−2.0	0.0
−1.9	0.2
5.6	2.9
2.6	−0.9
1.1	3.8
5.6	0.7
8.2	−0.1

These samples are quite small, but the boxplots of the data from the two samples (given in Figure 13.2) were not too asymmetric and there were no outliers, so a two-sample t confidence interval was used to estimate the difference in population means. However, you might still choose to use a simulation-based method that does not depend on the assumption of normal population distributions.

The means for the populations of first-year students living on and off campus were represented by

μ_1 = mean weight gain for first-year students living on campus

μ_2 = mean weight gain for first-year students living off campus

In the case of sampling from two populations, you begin by assuming that each sample is representative of the population from which it was selected. Then, as was the case in the one-sample situation, you can get a bootstrap sample from each population by selecting a random sample, with replacement, from each of the two original samples and then calculating the difference in the means of these two bootstrap samples. This process is then repeated many times to form a bootstrap distribution for the difference in sample means.

Once you have a bootstrap distribution of simulated differences in sample means, you can use the bootstrap distribution to produce a 95% confidence interval for the difference in the population means. The endpoints of the confidence interval are the value with 2.5% of the simulated differences in the bootstrap distribution below and the value with 2.5% of the simulated differences in the bootstrap distribution above. Locating these differences in the bootstrap distribution that follows gives a 95% confidence interval for the difference in mean weight gain for students living on campus and students living off campus of (−0.64, 3.83). You can be 95% confident that the difference in population mean weight gains, On Campus – Off Campus, falls between −0.64 and 3.83 pounds. These values are not very different from the endpoints of the two-sample t confidence interval from Example 13.8, which was (−0.98, 4.20).

Here is output from the Shiny app "Bootstrap Confidence Interval for the Difference in Two Population Means Using Independent Samples." This app can be found in the App collection at statistics.cengage.com/Peck2e/Apps.html.

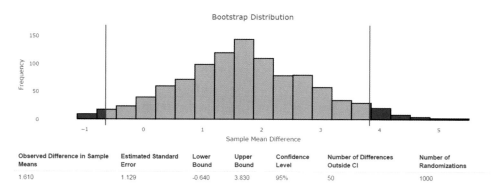

Observed Difference in Sample Means	Estimated Standard Error	Lower Bound	Upper Bound	Confidence Level	Number of Differences Outside CI	Number of Randomizations
1.610	1.129	-0.640	3.830	95%	50	1000

It is also possible to compare the population mean weight gains for students living on and off campus using a simulation-based hypothesis test, but the way the simulation is carried out is a little different. Suppose that you want to test the following hypotheses:

$$H_0: \mu_1 - \mu_2 = 0$$
$$H_a: \mu_1 - \mu_2 \neq 0$$

Summary statistics for the weight gains (in kg) calculated using the given data for students living on campus and for those living off campus follow.

	On Campus	Off Campus
Sample size	10	10
Sample mean	2.46	0.85
Sample standard deviation	3.26	2.03

The observed difference in the means for On Campus – Off Campus is $\bar{x}_1 - \bar{x}_2 = 2.46 - 0.85 = 1.61$ kg.

Recall that in the simulation-based test for a difference in two population proportions (Section 11.4), data from the two samples were combined into one group to represent the possibility that the two samples might have been taken from the same population. A similar process is used for the simulation-based test for the difference in two population means. The two samples are combined into one, and then random samples are selected with replacement from the combined samples (which represents the common population) to create a randomization distribution of the differences in sample means that is consistent with the null hypothesis of no difference in population means.

One example of a randomization distribution based on 1000 simulated differences in means using the Shiny app "Randomization Test for the Difference in Two Population Means Using Independent Samples" is shown here. This app can be found in the App collection at statistics.cengage.com/Peck2e/Apps.html.

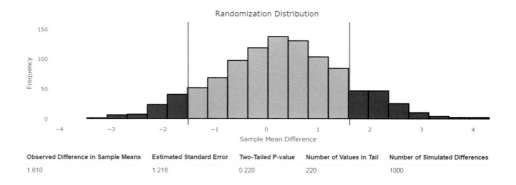

The randomization distribution in this case represents 1000 differences in two independent sample means, each computed from two random samples of size 10 taken with replacement from the combined sample of 20 weight gains from students living on and off campus.

Using the randomization distribution from this simulation, the probability of observing a difference in the sample means at least as extreme as 1.61 is 0.220. This two-tailed *P*-value is larger than the 0.05 significance level. You fail to reject the null hypothesis, and conclude that there is not convincing evidence of a difference in the population mean weight gains for students living on campus and students living off campus.

Summing It Up—Section 13.5

The following learning objectives were addressed in this section:

Mastering the Mechanics

M10: Calculate and interpret a bootstrap confidence interval for a difference in means.
A bootstrap confidence interval is an alternate method for calculating a confidence interval for a difference in population or treatment means. This method can be used even in situations where the sample size conditions necessary for the paired t or the two-sample t confidence interval are not met. A bootstrap confidence interval for a difference in means is interpreted in the same way as the paired t and the two-sample t confidence intervals. Example 13.12 illustrates the calculation and interpretation of a bootstrap confidence interval for a difference in means based on data from paired samples. Example 13.13 illustrates the calculation and interpretation of a bootstrap confidence interval for a difference in treatment means using data from an experiment, and Example 13.14 illustrates the calculation and interpretation of a bootstrap confidence interval for a difference in population means based on data from independent samples.

M11: Carry out a randomization test for a difference in means.
A randomization test is a method that can be used to test hypotheses about a difference in population or treatment means even if the sample sizes are not large enough for the paired t or the two-sample t test to be appropriate. Example 13.12 illustrates the use of a randomization test to test hypotheses about a difference in means based on data from paired samples. Example 13.13 illustrates the use of a randomization test to test hypotheses about a difference in treatment means using data from an experiment, and Example 13.14 illustrates the use of a randomization test for a difference in population means based on data from independent samples.

SECTION 13.5 EXERCISES

Each Exercise Set assesses the following learning objectives: M10, M11

SECTION 13.5 Exercise Set 1

 13.81 New "closed loop" (CL) devices have been developed to help to suppress overactive brain activity in patients with conditions such as Parkinson's disease and epilepsy (**"Conceptualization and Validation of an Open-Source Closed-Loop Deep Brain Stimulation System in Rat,"** *Scientific Reports,* **April 21, 2015, www.nature.com/articles/srep09921, retrieved May 23, 2017**). The CL device is implanted directly into a specific area of the brain, and one of the key advantages is that it can immediately apply a treatment in reaction to heightened brain activity. This may help to reduce the duration of seizures and other periods of uncontrolled movement in patients.

A study was conducted on the effectiveness of the CL device on brain activity in rats. First, each of seven rats was observed for 15 minutes with the CL device implanted but not activated (OFF). The percentage of the time that each rat was moving was recorded. Then, after each CL device was activated (CL), the percentage of the time that the rat was moving was recorded over another 15-minute period.

The OFF and CL data values given in the accompanying table are approximated from a graph in the research article. The calculated difference in movement, OFF – CL is also given.

Rat ID	OFF	CL	Difference (OFF – CL)
1	98.8	56.6	42.2
2	56.9	25.5	31.4
3	56.7	40.0	16.7
4	58.1	68.1	−10.0
5	50.8	45.1	5.7
6	53.1	35.7	17.4
7	56.8	44.5	12.3

a. Explain why two-sample t methods may not be appropriate in this context.

b. For purposes of this exercise, you can assume that these rats are representative of rats in general. Do these data support the claim that the mean difference in movement, OFF – CL, is greater than zero? Carry out a randomization test to answer this question. You can use make use of the Shiny apps in the collection at statistics.cengage.com/Peck2e/Apps.html.

13.82 Use the information given in the previous exercise to calculate a 95% bootstrap confidence interval to estimate the mean difference in movement, OFF – CL. Interpret the interval in context. You can use make use of the Shiny apps in the collection at statistics.cengage.com/Peck2e/Apps.html.

 Data set available

13.83 The Sheboygan (Wisconsin) Fire Department received a report on the potential effects of reductions in the number of firefighters it employs (**"Study of Fire Department Causes Controversy,"** *USA TODAY NETWORK-Wisconsin,* **December 22, 2016**).

In one section of the report, the average working heart rate percentage (the percentage value of the observed maximum heart rate during firefighting drills divided by an age-adjusted maximum heart rate for each firefighter) was reported for the driver of the first-arriving fire engine when only two firefighters (including the driver) were present, and for the driver of the first-arriving fire engine when more than two firefighters (up to five) were present.

The average working heart rate percentages were based on an earlier study, which included data from a sample of six drills using only two firefighters and from a sample of 18 drills using more than two firefighters. For purposes of this exercise, you can assume that these samples are representative of all drills with two firefighters and all drills with more than two firefighters.

The following data values are consistent with summary statistics given in the paper. Do these data support the claim that the mean average working heart rate percentage for the driver in a two-firefighter team is greater than the mean average working heart rate percentage for the driver in teams containing from three to five firefighters? Use a 0.05 significance level to carry out a randomization test of the given claim. You can use make use of the Shiny apps in the collection at statistics.cengage.com/Peck2e/Apps.html.

Working Heart Rate Percentage	
Two Firefighters	**More Than Two Firefighters**
97.7	73.6
87.7	63.7
89.5	72.0
97.3	83.1
81.9	54.2
79.7	73.1
	75.2
	84.5
	72.8
	70.0
	87.1
	81.1
	57.5
	68.4
	80.6
	57.8
	79.5
	75.0

13.84 Use the information given in the previous exercise to construct a 95% bootstrap confidence interval to estimate the difference in mean average working heart rates for the driver in teams of two firefighters and the driver in teams of from three to five firefighters. Interpret the interval in context. You can use make use of the Shiny apps in the collection at statistics.cengage.com/Peck2e/Apps.html.

13.85 Studies have been conducted to evaluate the effectiveness of psilocybin mushrooms on improving the quality of life for patients with cancer (**"A Dose of a Hallucinogen from a 'Magic Mushroom,' and Then Lasting Peace,"** *The New York Times,* **December 1, 2016**). In one study, patients were randomly assigned to either a low-dose psilocybin treatment or to a high-dose treatment. One outcome that was measured was a "Personal Meaning" score, collected five weeks after the psilocybin treatment. Higher scores indicated greater "Personal Meaning." The following data were estimated from a graphical display in the article. Do these data support the claim that the mean Personal Meaning score for patients with cancer taking a high dose of psilocybin is greater than the mean Personal Meaning score for patients with cancer taking a low dose? Use a randomization test to answer this question. You can use make use of the Shiny apps in the collection at statistics.cengage.com/Peck2e/Apps.html.

Personal Meaning	
Low Dose	**High Dose**
3	1
2	5
6	7
4	7
4	5
2	5
3	5
7	6
7	6
6	7
6	6
7	7
5	5
1	7
2	7
4	8
4	8
2	8
	7
	6
	4
	7
	7
	7
	6

13.86 Use the information in the previous exercise to construct a 95% bootstrap confidence interval to estimate the difference in mean Personal Meaning scores for patients with cancer in the high-dose and low-dose psilocybin groups. Interpret the interval in context. You can use make use of the Shiny apps in the collection at statistics.cengage.com/Peck2e/Apps.html.

13.87 Behavioral intervention treatments may affect perceptions of reward in human brains, and thus help people to manage weight loss. In one of the first studies relating brain activity related to rewards and weight loss, researchers recorded brain activity associated with rewards as subjects viewed images of high-calorie (HC) foods and images of low-calorie (LC) foods **("Pilot Randomized Trial Demonstrating Reversal of Obesity-Related Abnormalities in Reward System Responsivity to Food Cues with a Behavioral Intervention,"** *Nutrition and Diabetes*, **September 1, 2014, www.nature.com /nutd/journal/v4/n9/full/nutd201426a.html, retrieved May 23, 2017).** The data in the following table are consistent with graphs in the research article.

Eight adult Americans enrolled in the study had their brain activity assessed before and after completing a 24-week weight loss program. The data values represent the change in brain activity associated with high-calorie foods, the change in brain activity associated with low-calorie foods, and the difference, HC − LC, for each of the eight subjects. If the weight loss program is effective, then a greater change for the HC treatment would be anticipated, on average, compared to the change for the LC treatment. For purposes of this exercise, you can assume that these eight adults are representative of adult Americans.

Subject ID	HC	LC	Difference
1	4.3	−5.8	10.1
2	2.1	−3.1	5.2
3	4.4	−2.0	6.4
4	2.4	0.4	2.0
5	1.4	0.5	0.9
6	2.1	1.3	0.8
7	2.0	−2.8	4.8
8	2.9	−0.3	3.2

a. Explain why the two-sample t methods may not be appropriate in this context.
b. Do these data support the claim that the mean difference in change in brain activity, HC − LC, is greater than zero? Carry out a randomization test to answer this question. You can use make use of the Shiny apps in the collection at statistics.cengage.com/Peck2e/Apps.html.

13.88 Use the information in the previous exercise to construct a 95% bootstrap confidence interval to estimate the mean difference in change in brain activity, HC − LC. Interpret the interval in context. You can use make use of the Shiny apps in the collection at statistics.cengage.com /Peck2e/Apps.html.

13.89 A new set of cognitive training modules called "ONTRAC" was developed to help children with attention deficit/hyperactivity disorder (ADHD) to improve focus and to more easily dismiss distractions **("Training sensory signal-to-noise resolution in children with ADHD in a global mental health setting,"** *Translational Psychiatry*, **April 12, 2016, http://www.nature.com/tp/journal/v6/n4 /full/tp201645a.html, retrieved May 23, 2017).** Eighteen children with ADHD were randomly assigned to one of two treatment groups. One group of 11 children received the ONTRAC treatment and another group of 7 children received a control treatment.

Values for one-year improvement in ADHD Severity Score consistent with graphs and summary statistics in the research article appear in the following table:

ONTRAC	Control
11.0	4.3
8.8	3.8
12.1	8.5
10.9	4.7
8.0	1.5
8.2	7.7
11.9	8.0
11.1	
11.8	
10.8	
12.0	

a. Explain why you should be wary of using the two-sample t methods to analyze the data from this study.
b. Do these data support the claim that the mean one-year improvement in ADHD Severity Score for the ONTRAC treatment is different from the mean one-year improvement in ADHD Severity Score for the control treatment? Use a randomization test with significance level 0.05 to answer this question. You can use make use of the Shiny apps in the collection at statistics.cengage.com/Peck2e/Apps.html.

13.90 Use the information in the previous exercise to construct a 95% bootstrap confidence interval to estimate the difference in mean one-year improvement in ADHD Severity Score for the ONTRAC treatment and the control treatment. Interpret the interval in context. You can use make use of the Shiny apps in the collection at statistics.cengage.com/Peck2e /Apps.html.

 Data set available

SECTION 13.6 Avoid These Common Mistakes

The cautions that have appeared at the ends of previous chapters apply here as well. Worth repeating again are:

1. Remember that the result of a hypothesis test can never show strong support for the null hypothesis. In two-sample situations, this means that you shouldn't be *convinced* that there is *no* difference between two population means based on the outcome of a hypothesis test.

2. If you have complete information (a census) for both populations, there is no need to carry out a hypothesis test or to construct a confidence interval—in fact, it would be inappropriate to do so.

3. Don't confuse statistical significance and practical significance. In the two-sample setting, it is possible to be convinced that two population means are not equal even in situations where the actual difference between them is small enough that it is of no practical interest. After rejecting a null hypothesis of no difference (statistical significance), it is useful to look at a confidence interval estimate of the difference to get a sense of practical significance.

4. Correctly interpreting confidence intervals in the two-sample case is more difficult than in the one-sample case, so take particular care when providing a two-sample confidence interval interpretation. Because the two-sample confidence interval estimates a difference $(\mu_1 - \mu_2)$, the most important thing to note is whether or not the interval includes 0. If both endpoints of the interval are positive, then it is correct to say that, based on the interval, you think that μ_1 is greater than μ_2 and the interval provides an estimate of how much greater. Similarly, if both interval endpoints are negative, you would say that μ_1 is less than μ_2, with the interval again providing an estimate of the size of the difference. If 0 is included in the interval, it is plausible that μ_1 and μ_2 are equal.

Drawing conclusions from experiment data requires some thought. In Chapter 1, you saw that if an experiment is carefully planned and includes random assignment to treatments, it is reasonable to conclude that observed differences in response between the experimental groups can be attributed to the treatments in the experiment. However, generalizing conclusions from an experiment that uses volunteers as subjects to a larger population is not appropriate unless a convincing argument can be made that the group of volunteers is representative of some population of interest.

Some cautions when drawing conclusions from experiment data are:

1. *Random assignment to treatments is critical.* If the design of the experiment does not include random assignment to treatments, it is not appropriate to use a hypothesis test or a confidence interval to draw conclusions about treatment differences.

2. Remember that it is not reasonable to generalize conclusions from experiment data to a larger population unless the subjects in the experiment were selected at random from the population or a convincing argument can be made that the group of volunteers is representative of the population. And even if subjects are selected at random from a population, it is still important that there be random assignment to treatments.

3. As was the case when using data from sampling to test hypotheses, remember that a hypothesis test can never show strong support for the null hypothesis. In the context of using experiment data to test hypotheses, this means you cannot say that data from an experiment provide convincing evidence that there is *no* difference between treatments.

4. Even when the data used in a hypothesis test are from an experiment, there is still a difference between statistical significance and practical significance. It is possible, especially in experiments with large numbers of subjects in each experimental group, to be convinced that two treatment means are not equal, even in situations where the actual difference is too small to be of any practical interest. After rejecting a null hypothesis of no difference (statistical significance), it may be useful to look at a confidence interval estimate of the difference to get a sense of practical significance.

CHAPTER ACTIVITIES

ACTIVITY 13.1 THINKING ABOUT DATA COLLECTION

Background: In this activity you will consider two studies that allow you to investigate whether taking a keyboarding class offered as an elective course at a high school improves typing speed. You can assume that there are 2000 students at the school and that about 10% of the students at the school complete the keyboarding course.

1. Working in a group, design a study using independent samples that would allow you to compare mean typing speed for students who have not completed the keyboarding course and mean typing speed for students who have. Be sure to describe how you plan to select your samples and how you plan to measure typing speed.

2. Do you see a benefit to designing the study in a way that would result in paired samples? What would be the benefit of incorporating pairing into your study design?

ACTIVITY 13.2 A MEANINGFUL PARAGRAPH

Write a meaningful paragraph that includes the following six terms: **paired samples, significantly different, *P*-value, sample, population, alternative hypothesis.** A "meaningful paragraph" is a coherent piece of writing in an appropriate context that uses all of the listed words. The paragraph should show that you understand the meaning of the terms and their relationship to one another. A sequence of sentences that just define the terms is *not* a meaningful paragraph. When choosing a context, think carefully about the terms you need to use.

ACTIVITY 13.3 AN EXPERIMENT TO TEST FOR THE STROOP EFFECT

Background: In 1935, John Stroop published the results of his research into how people respond when presented with conflicting signals. Stroop noted that most people are able to read words quickly and that they cannot easily ignore them and focus on other attributes of a printed word, such as text color. For example, consider the following list of words:

green blue red blue yellow red

It is easy to quickly read this list of words. It is also easy to read the words even if the words are printed in color, and even if the text color is different from the color of the word. For example, people can read the words in the list

green blue red blue yellow red

as quickly as they can read the list that isn't printed in color.

However, Stroop found that if people are asked to name the text colors of the words in the list (red, yellow, blue, green, red, green), it takes them longer. Psychologists believe this is due to the reader inhibiting a natural response (reading the word) and producing a different response (naming the color of the text).

If Stroop is correct, people should be able to name colors more quickly if they do not have to inhibit the word response, as would be the case if they were shown the following:

1. Design an experiment to compare the time it takes to identify colors when they appear as text with the time it takes to identify colors when there is no need to inhibit a word response. Indicate how random assignment is incorporated into your design. What is your response variable? How will you measure it? How many subjects will you use in your experiment, and how will they be chosen?

2. When you are satisfied with your experimental design, carry out the experiment. You will need to construct a list of colored words and a corresponding list of colored bars to use in the experiment. You will also need to think about how you will implement the random assignment scheme.

3. Use the resulting data to determine if there is evidence to support the existence of the Stroop effect. Write a brief report that summarizes your findings.

ACTIVITY 13.4 QUICK REFLEXES

Background: In this activity, you will design an experiment that will allow you to investigate whether people tend to have quicker reflexes when reacting with their dominant hand than with their nondominant hand.

1. Working in a group, design an experiment that includes random assignment of participants to one of two experimental conditions. Be sure to describe what the two treatments in the experiment are, how you plan to measure quickness of reflexes, what potentially confounding variables will be directly controlled, and the role that random assignment plays in your design.

2. If assigned to do so by your instructor, carry out your experiment and analyze the resulting data. Write a brief report that describes the experimental design, includes both graphical and numerical summaries of the resulting data, and communicates any conclusions that follow from your data analysis.

💻 CHAPTER 13 EXPLORATIONS IN STATISTICAL THINKING

The Phoenix airport is considering two different expansion plans that would reduce crowding at the airport and reduce the time that planes spend on the ground with engines running prior to takeoff. As part of an environmental impact report on the proposed expansion, many different types of planes were studied. Using the expected time each plane would spend on the ground with engines running, the amount of fuel needed to taxi from the gate to the runway, takeoff, and climb to 3000 feet was calculated for each of the two plans.

Go online at statistics.cengage.com/Peck2e/Explore.html and click on the link for Chapter 13. It will take you to a web page where you can select a random sample of 30 planes from the population of all of the planes studied.

Click on the "Sample" button. This selects a random sample of 30 planes. The plane ID number and the fuel consumption for that plane for each of the two expansion plans will be displayed.

Use these data to complete the following:

(a) Construct a 95% confidence interval estimate of the difference between the mean fuel consumption for plan 1 and the mean fuel consumption for plan 2 for the population consisting of all planes studied.

(b) Write a few sentences that provide an interpretation of the confidence interval from Part (a).

(c) Which of the following statements is most appropriate given the confidence interval in Part (a)?

1. You can be 95% confident that the mean fuel consumption for planes in this population is greater for plan 1 than for plan 2.
2. You can be 95% confident that the mean fuel consumption for planes in this population is greater for plan 2 than for plan 1.
3. The interval provides convincing evidence that there is no difference in the mean fuel consumption for the two plans.
4. Based on the interval, it is possible that there is no difference in the mean fuel consumption for the two plans.

(d) Which of the following is a correct interpretation of the 95% confidence level? (Select all correct interpretations.)

1. The probability that the true mean difference in fuel consumption is contained in the calculated interval is 0.95.
2. If the process of selecting a random sample of planes and then calculating a 95% confidence interval for the mean difference in fuel consumption is repeated 100 times, 95 of the 100 intervals will include the population mean difference in fuel consumption.
3. If the process of selecting a random sample of planes and then calculating a 95% confidence interval for the mean difference in fuel consumption is repeated a very large number of times, approximately 95% of the calculated intervals will include the population mean difference in fuel consumption.

ARE YOU READY TO MOVE ON?   CHAPTER 13 REVIEW EXERCISES

All chapter learning objectives are assessed in these exercises. The learning objectives assessed in each exercise are given in parentheses.

13.91 (C1)
An individual can take either a scenic route to work or a nonscenic route. She decides that use of the nonscenic route can be justified only if it reduces the mean travel time by more than 10 minutes.

a. If μ_1 refers to the mean travel time for nonscenic route and μ_2 to the mean travel time for scenic route, what hypotheses should be tested?

b. If μ_1 refers to the mean travel time for scenic route and μ_2 to the mean travel time for nonscenic route, what hypotheses should be tested?

13.92 (M1)
Descriptions of three studies are given. In each of the studies, the two populations of interest are students majoring in science at a particular university and students majoring in liberal

arts at this university. For each of these studies, indicate whether the samples are independently selected or paired.

Study 1: To determine if there is evidence that the mean number of hours spent studying per week differs for the two populations, a random sample of 100 science majors and a random sample of 75 liberal arts majors are selected.

Study 2: To determine if the mean amount of money spent on textbooks differs for the two populations, a random sample of science majors is selected. Each student in this sample is asked how many units he or she is enrolled in for the current semester. For each of these science majors, a liberal arts major who is taking the same number of units is identified and included in the sample of liberal arts majors.

Study 3: To determine if the mean amount of time spent using the campus library differs for the two populations, a random sample of science majors is selected. A separate random sample of the same size is selected from the population of liberal arts majors.

13.93 (C1)

For each of the following hypothesis testing scenarios, indicate whether or not the appropriate hypothesis test would be for a difference in population means. If not, explain why not.

Scenario 1: The authors of the paper **"Adolescents and MP3 Players: Too Many Risks, Too Few Precautions"** (*Pediatrics* [2009]: e953–e958) studied independent random samples of 764 Dutch boys and 748 Dutch girls age 12 to 19. Of the boys, 397 reported that they almost always listen to music at a high volume setting. Of the girls, 331 reported listening to music at a high volume setting. You would like to determine if there is convincing evidence that the proportion of Dutch boys who listen to music at high volume is greater than this proportion for Dutch girls.

Scenario 2: The report **"Highest Paying Jobs for 2009–10 Bachelor's Degree Graduates"** (National Association of Colleges and Employers, February 2010) states that the mean yearly salary offer for students graduating with accounting degrees in 2010 is $48,722. A random sample of 50 accounting graduates at a large university resulted in a mean offer of $49,850 and a standard deviation of $3300. You would like to determine if there is strong support for the claim that the mean salary offer for accounting graduates of this university is higher than the 2010 national average of $48,722.

Scenario 3: Each person in a random sample of 228 male teenagers and a random sample of 306 female teenagers was asked how many hours he or she spent online in a typical week **(Ipsos, January 25, 2006)**. The sample mean and standard deviation were 15.1

hours and 11.4 hours for males and 14.1 and 11.8 for females. You would like to determine if there is convincing evidence that the mean number of hours spent online in a typical week is greater for male teenagers than for female teenagers.

13.94 (C1, M5, M6, P3)

Do male college students spend more time studying than female college students? This was one of the questions investigated by the authors of the paper **"An Ecological Momentary Assessment of the Physical Activity and Sedentary Behaviour Patterns of University Students"** (*Heath Education Journal* [2010]: 116–125). Each student in a random sample of 46 male students at a university in England and each student in a random sample of 38 female students from the same university kept a diary of how he or she spent time over a 3-week period. For the sample of males, the mean time spent studying per day was 280.0 minutes, and the standard deviation was 160.4 minutes. For the sample of females, the mean time spent studying per day was 184.8 minutes, and the standard deviation was 166.4 minutes. Is there convincing evidence that the mean time male students at this university spend studying is greater than the mean time for female students? Test the appropriate hypotheses using $\alpha = 0.05$.

13.95 (C1, M5, M6, P3)

The paper **"Sodium content of Lunchtime Fast Food Purchases at Major U.S. Chains"** (*Archives of Internal Medicine* [2010]: 732–734) reported that for a random sample of 850 meal purchases made at Burger King, the mean sodium content was 1685 mg, and the standard deviation was 828 mg. For a random sample of 2107 meal purchases made at McDonald's, the mean sodium content was 1477 mg, and the standard deviation was 812 mg. Based on these data, is it reasonable to conclude that there is a difference in mean sodium content for meal purchases at Burger King and meal purchases at McDonald's? Use $\alpha = 0.05$.

13.96 (C1, M5, M6, P3)

The paper referenced in the previous exercise also gave information on calorie content. For the sample of Burger King meal purchases, the mean number of calories was 1008, and the standard deviation was 483. For the sample of McDonald's meal purchases, the mean number of calories was 908, and the standard deviation was 624. Based on these samples, is there convincing evidence that the mean number of calories in McDonald's meal purchases is less than the mean number of calories in Burger King meal purchases? Use $\alpha = 0.01$.

13.97 (M2, M3, P1)

The article **"A Shovel with a Perforated Blade Reduces Energy Expenditure Required for Digging Wet Clay"** (*Human Factors*, 2010: 492–502) described a study in which each of 13 workers performed a task using a conventional shovel and using a shovel with a blade that was perforated with small holes. The authors of the cited article provided the following data on energy expenditure (in Kcal/kg(subject)/lb(clay)):

Worker:	1	2	3	4	5	6	7
Conventional	0.0011	0.0014	0.0018	0.0022	0.0010	0.0016	0.0028
Perforated	0.0011	0.0010	0.0019	0.0013	0.0011	0.0017	0.0024

Worker:	8	9	10	11	12	13
Conventional	0.0020	0.0015	0.0014	0.0023	0.0017	0.0020
Perforated	0.0020	0.0013	0.0013	0.0017	0.0015	0.0013

Do these data provide convincing evidence that the mean energy expenditure using the conventional shovel exceeds that using the perforated shovel? Test the relevant hypotheses using a significance level of $\alpha = 0.05$.

13.98 (M2, M3, P1)
Head movement evaluations are important because disabled individuals may be able to operate communications aids using head motion. The paper **"Constancy of Head Turning Recorded in Healthy Young Humans" (*Journal of Biomedical Engineering* [2008]: 428–436)** reported the accompanying data on neck rotation (in degrees) both in the clockwise direction (CL) and in the counterclockwise direction (CO) for 14 subjects. For purposes of this exercise, you may assume that the 14 subjects are representative of the population of adult Americans. Based on these data, is it reasonable to conclude that mean neck rotation is greater in the clockwise direction than in the counterclockwise direction? Carry out a hypothesis test using a significance level of $\alpha = 0.01$.

Subject:	1	2	3	4	5	6	7
CL:	57.9	35.7	54.5	56.8	51.1	70.8	77.3
CO:	44.2	52.1	60.2	52.7	47.2	65.6	71.4

Subject:	8	9	10	11	12	13	14
CL:	51.6	54.7	63.6	59.2	59.2	55.8	38.5
CO:	48.8	53.1	66.3	59.8	47.5	64.5	34.5

13.99 (M2, M3, M4, P1, P2)
The paper **"The Truth About Lying in Online Dating Profiles" (*Proceedings, Computer-Human Interactions* [2007]: 1–4)** describes an investigation in which 40 men and 40 women with online dating profiles agreed to participate in a study. Each participant's height (in inches) was measured and the actual height was compared to the height given in that person's online profile. The differences between the online profile height and the actual height (profile – actual) were used to calculate the values in the accompanying table.

Men	Women
$\bar{x}_d = 0.57$	$\bar{x}_d = 0.03$
$s_d = 0.81$	$s_d = 0.75$
$n = 40$	$n = 40$

For purposes of this exercise, assume that the two samples are representative of male online daters and female online daters.
a. Use the paired-samples t test to determine if there is convincing evidence that, on average, male online daters overstate their height in online dating profiles. Use $\alpha = 0.05$.
b. Construct and interpret a 95% confidence interval for the difference between the mean online dating profile height and mean actual height for female online daters.
c. Use the two-sample t test of Section 13.3 to test $H_0: \mu_m - \mu_f = 0$ versus $H_a: \mu_m - \mu_f > 0$, where μ_m is the mean height difference (profile – actual) for male online daters and μ_f is the mean height difference (profile – actual) for female online daters.
d. Explain why a paired-samples t test was used in Part (a) but a two-sample t test was used in Part (c).

13.100 (M5, M6, P2)
Here's one to sink your teeth into: The authors of the article **"Analysis of Food Crushing Sounds During Mastication: Total Sound Level Studies" (*Journal of Texture Studies* [1990]: 165–178)** studied the nature of sounds generated during eating. Peak loudness was measured (in decibels at 20 cm away) for both open-mouth and closed-mouth chewing of potato chips and of tortilla chips. A sample of size 10 was used for each of the four possible combinations (such as closed-mouth potato chip, and so on). We are not making this up! Summary values taken from plots given in the article appear in the accompanying table. For purposes of this exercise, suppose that it is reasonable to regard the peak loudness distributions as approximately normal.

	n	$\bar{x}$	s
Potato Chip			
Open mouth	10	63	13
Closed mouth	10	54	16
Tortilla Chip			
Open mouth	10	60	15
Closed mouth	10	53	16

a. Construct a 95% confidence interval for the difference in mean peak loudness between open-mouth and closed-mouth chewing of potato chips. Be sure to interpret the resulting interval.
b. For closed-mouth chewing (the recommended method!), construct a 95% confidence interval for the difference in mean peak loudness between potato chips and tortilla chips.

13.101 (M2, M4, P2)

Samples of both surface soil and subsoil were taken from eight randomly selected agricultural locations in a particular county. The soil samples were analyzed to determine surface pH and subsoil pH, with the results shown in the accompanying table.

Location	1	2	3	4	5	6	7	8
Surface pH	6.55	5.98	5.59	6.17	5.92	6.18	6.43	5.68
Subsoil pH	6.78	6.14	5.80	5.91	6.10	6.01	6.18	5.88

a. Calculate a 90% confidence interval for the mean difference between surface and subsoil pH for agricultural land in this county.

b. What conditions must be met for the interval in Part (a) to be valid?

13.102 (M8)

The paper **"Effects of Caffeine on Repeated Sprint Ability, Reactive Agility Time, Sleep and Next Day Performance"** (*Journal of Sports Medicine and Physical Fitness* [2010]: 455–464) describes an experiment in which male athlete volunteers who were considered low caffeine consumers were assigned at random to one of two experimental groups. Those assigned to the caffeine group drank a beverage which contained caffeine 1 hour before an exercise session. Those in the no-caffeine group drank a beverage that did not contain caffeine 1 hour before an exercise session. That night, participants wore a device that measures sleep activity. The researchers reported that there was no significant difference in mean sleep duration for the two experimental groups. In the context of this experiment, explain what it means to say that there is *no significant difference* in the group means. In particular, explain if this means that the mean sleep durations for the two groups are equal.

13.103 (M8)

In the paper **"Happiness for Sale: Do Experiential Purchases Make Consumers Happier than Material Purchases?"** (*Journal of Consumer Research* [2009]: 188–197), the authors distinguish between spending money on experiences (such as travel) and spending money on material possessions (such as a car). In an experiment to determine if the type of purchase affects how happy people are after the purchase has been made, 185 college students were randomly assigned to one of two groups. The students in the "experiential" group were asked to recall a time when they spent about $300 on an experience. They rated this purchase on three different happiness scales that were then combined into an overall measure of happiness. The students assigned to the "material" group recalled a time that they spent about $300 on an object and rated this purchase in the same manner. The mean happiness score was 5.75 for the experiential group and 5.27 for the material group. Standard deviations and sample sizes were not given in the paper, but for purposes of this exercise, suppose that they were as follows:

Experiential	Material
$n_1 = 92$	$n_2 = 93$
$s_1 = 1.2$	$s_2 = 1.5$

Using the following Minitab output, carry out a hypothesis test to determine if these data support the authors' conclusion that, on average, "experiential purchases induced more reported happiness." Use $\alpha = 0.05$

Two-Sample T-Test and CI

Sample	N	Mean	StDev	SE Mean
1	92	5.75	1.20	0.13
2	93	5.27	1.50	0.16

Difference = mu (1) − mu (2)

Estimate for difference: 0.480000

95% lower bound for difference: 0.149917

T-Test of difference = 0 (vs >): T-Value = 2.40 P-Value = 0.009 DF = 175

13.104 (M9)

The article **"An Alternative Vote: Applying Science to the Teaching of Science"** (*The Economist*, May 12, 2011) describes an experiment conducted at the University of British Columbia. A total of 850 engineering students enrolled in a physics course participated in the experiment. Students were randomly assigned to one of two experimental groups. Both groups attended the same lectures for the first 11 weeks of the semester. In the twelfth week, one of the groups was switched to a style of teaching where students were expected to do reading assignments prior to class, and then class time was used to focus on problem solving, discussion, and group work. The second group continued with the traditional lecture approach. At the end of the twelfth week, students were given a test over the course material from that week. The mean test score for students in the new teaching method group was 74, and the mean test score for students in the traditional lecture group was 41. Suppose that the two groups each consisted of 425 students. Also suppose that the standard deviations of test scores for the new teaching method group and the traditional lecture method group were 20 and 24, respectively. Estimate the difference in mean test score for the two teaching methods using a 95% confidence interval. Be sure to give an interpretation of the interval.

TECHNOLOGY NOTES

Confidence Interval for $\mu_1 - \mu_2$

1. Input the raw data for both groups into the first column
2. Input the group information into the second column

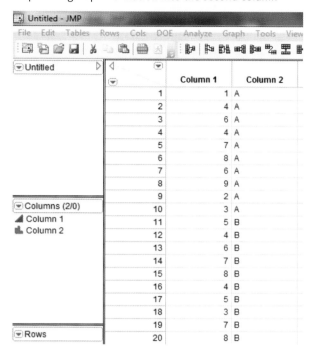

3. Click **Analyze** and select **Fit Y by X**
4. Click and drag the first column's name from the box under **Select Columns** to the box next to **Y, Response**
5. Click and drag the second column's name from the box under **Select Columns** to the box next to **X, Factor**
6. Click **OK**
7. Click the red arrow next to **Oneway Analysis of...** and select *t* **Test**

Note: The 95% confidence interval is automatically displayed. To change the confidence level, click the red arrow next to **Oneway Analysis of...** and select **Set α Level** then click the appropriate α-level or select **Other** and type the appropriate level.

Minitab

Summarized data

1. Click **S̲tat** then click **Basic Statistics** then click **2̲-sample t...**
2. Click the radio button next to **Summarized data**
3. In the boxes next to **First:** For the first sample, type the value for *n*, the sample size in the box under **Sample size:** and type the value for the sample mean in the box under **Mean:** and finally type the value for the sample standard deviation in the box under **Standard deviation:**
4. In the boxes next to **Second:** For the second sample, type the value for *n*, the sample size in the box under **Sample size:** and type the value for the sample mean in the box under **Mean:** and finally type the value for the sample standard deviation in the box under **Standard deviation:**
5. Click **Options...**

6. Input the appropriate confidence level in the box next to **Confidence Level**
7. Click **OK**
8. Click **OK**

Raw data

1. Input the raw data for each group into a separate column
2. Click **S̲tat** then click **Basic Statistics** then click **2̲-sample t...**
3. Click the radio button next to **Samples in different columns:**
4. Click in the box next to **First:**
5. Double click the column name where the first group's data are stored
6. Click in the box next to **Second:**
7. Double-click the column name where the second group's data are stored
8. Click **Options...**
9. Input the appropriate confidence level in the box next to **Confidence Level**
10. Click **OK**
11. Click **OK**

SPSS

1. Input the raw data for BOTH groups into the first column
2. Input the data for groups into the second column (input A for the first group and B for the second group)
3. Click **A̲nalyze** then click **Compare Means** then click **Independent̲-Samples T Test**
4. Click the name of the column containing the raw data and click the arrow to move this variable to the **Test Variable(s):** box
5. Click the name of the column containing the group information and click the arrow to move this variable to the **Grouping Variable:** box
6. Click the **D̲efine Groups...** button
7. In the box next to **Group 1:** type A
8. In the box next to **Group 2:** type B
9. Click **Continue**
10. Click **Options...**
11. Input the confidence level in the box next to **Confidence Interval Percentage:**
12. Click **Continue**
13. Click **OK**

Note: This procedure produces confidence intervals under the assumption of equal variances AND also when equal variances are not assumed.

Excel

Excel does not have the functionality to produce a confidence interval automatically for the difference of two population means. However, you can manually type the formulas for the lower and upper limits into two separate cells and have Excel calculate the results for you. You may also use Excel to find the *t* critical value based on the confidence level using the following steps:

1. Click in an empty cell
2. Click **Formulas**

Data set available

3. Click **Insert Function**
4. Select **Statistical** from the drop-down box for the category
5. Select **TINV** and click **OK**
6. In the box next to **Probability** type in the value representing one minus your selected confidence level
7. In the box next to **Deg_freedom** type in the degrees of freedom ($n - 1$)
8. Click **OK**

TI-83/84
Summarized data

1. Press **STAT**
2. Highlight **TESTS**
3. Highlight **2-SampTInt…** and press **ENTER**
4. Highlight **Stats** and press **ENTER**
5. Next to $\bar{x}1$ input the value for the sample mean from the first sample
6. Next to **sx1** input the value for the sample standard deviation from the first sample
7. Next to **n1** input the value for the sample size from the first sample
8. Next to $\bar{x}2$ input the value for the sample mean from the second sample
9. Next to **sx2** input the value for the sample standard deviation from the second sample
10. Next to **n2** input the value for the sample size from the second sample
11. Next to **C-Level** input the appropriate confidence level
12. Highlight **Calculate** and press **ENTER**

Raw data

1. Enter the data into **L1** and **L2** (In order to access lists press the **STAT** key, then press **ENTER**)
2. Press **STAT**
3. Highlight **TESTS**
4. Highlight **2-SampTInt…** and press **ENTER**
5. Highlight **Data** and press **ENTER**
6. Next to **C-Level** input the appropriate confidence level
7. Highlight **Calculate** and press **ENTER**

TI-Nspire
Summarized data

1. Enter the Calculate Scratchpad
2. Press the **menu** key and select **6:Statistics** then **6:Confidence Intervals** then **4:2-Sample t Interval…** and press **enter**
3. From the drop-down menu select **Stats**
4. Press **OK**
5. Next to $\bar{x}1$ input the value for the sample mean from the first sample
6. Next to **sx1** input the value for the sample standard deviation from the first sample
7. Next to **n1** input the value for the sample size from the first sample
8. Next to $\bar{x}2$ input the value for the sample mean from the second sample
9. Next to **sx2** input the value for the sample standard deviation from the second sample
10. Next to **n2** input the value for the sample size from the second sample

11. Next to **C Level** input the appropriate confidence level
12. Press **OK**

Raw data

1. Enter the data into two separate data lists (In order to access data lists select the spreadsheet option and press **enter**)

Note: Be sure to title the lists by selecting the top row of the column and typing a title.

2. Press the **menu** key and select **4:Statistics** then **3:Confidence Intervals** then **2:t Interval…** and press **enter**
3. From the drop-down menu select **Data**
4. Press **OK**
5. Next to **List 1** select the list containing the first data sample from the drop-down menu
6. Next to **List 2** select the list containing the second data sample from the drop-down menu
7. Next to **C-Level** input the appropriate confidence level
8. Press **OK**

Two-sample t-Test for $\mu_1 - \mu_2$

JMP

1. Input the raw data for both groups into the first column
2. Input the group information into the second column
3. Click **Analyze** and select **Fit Y by X**
4. Click and drag the first column's name from the box under **Select Columns** to the box next to **Y, Response**
5. Click and drag the second column's name from the box under **Select Columns** to the box next to **X, Factor**
6. Click **OK**
7. Click the red arrow next to **Oneway Analysis of…** and select **t Test**

Minitab
Summarized data

1. Click **Stat** then click **Basic Statistics** then click **2-sample t…**
2. Click the radio button next to **Summarized data**
3. In the boxes next to **First:** For the first sample, type the value for n, the sample size in the box under **Sample size:** and type the value for the sample mean in the box under **Mean:** and finally type the value for the sample standard deviation in the box under **Standard deviation:**
4. In the boxes next to **Second:** For the second sample, type the value for n, the sample size in the box under **Sample size:** and type the value for the sample mean in the box under **Mean:** and finally type the value for the sample standard deviation in the box under **Standard deviation:**
5. Click **Options…**
6. Input the appropriate hypothesized value for the difference of the population means in the box next to **Test difference:**
7. Select the appropriate alternative from the drop-down menu next to **Alternative:**
8. Click **OK**
9. Click **OK**

Note: You may also run this test with the assumption of equal variances by clicking the checkbox next to **Assume equal variances** after Step 8 in the above sequence.

Raw data

1. Input the raw data into two separate columns
2. Click **Stat** then click **Basic Statistics** then click **2-sample t...**
3. Click the radio button next to **Samples in different columns:**
4. Click in the box next to **First:**
5. Double-click the column name where the first group's data is stored
6. Click in the box next to **Second:**
7. Double-click the column name where the second group's data is stored
8. Click **Options...**
9. Input the appropriate hypothesized value for the difference of the population means in the box next to **Test difference:**
10. Select the appropriate alternative from the drop-down menu next to **Alternative:**
11. Click **OK**
12. Click **OK**

Note: You may also run this test with the assumption of equal variances by clicking the checkbox next to **Assume equal variances** after Step 11 in the above sequence.

SPSS

1. Input the raw data for BOTH groups into the first column
2. Input the data for groups into the second column (input A for the first group and B for the second group)
3. Click **Analyze** then click **Compare Means** then click **Independent-Samples T Test**
4. Click the name of the column containing the raw data and click the arrow to move this variable to the **Test Variable(s):** box
5. Click the name of the column containing the group information and click the arrow to move this variable to the **Grouping Variable:** box
6. Click the **Define Groups...** button
7. In the box next to **Group 1:** type A
8. In the box next to **Group 2:** type B
9. Click **Continue**
10. Click **OK**

Note: This procedure produces two-sample t-tests under the assumption of equal variances AND also when equal variances are not assumed. It also outputs a two-tailed *P*-value.

Excel

1. Input the raw data for each group into two separate columns
2. Click on the **Data** ribbon
3. Click **Data Analysis** in the **Analysis** group

Note: If you do not see **Data Analysis** listed on the Ribbon, see the Technology Notes for Chapter 2 for instructions on installing this add-on.

If you are performing a test where you are assuming equal variances, continue with Steps 4–9 below. If you are performing a test where you are not assuming equation variances, skip to Step 10.

4. Select **t-Test: Two Samples Assuming Equal Variances** from the dialog box and click **OK**
5. Click in the box next to **Variable 1 Range:** and select the first column of data

6. Click in the box next to **Variable 2 Range** and select the second column of data (if you have used and selected column titles for BOTH variables, select the check box next to **Labels**)
7. Click in the box next to **Hypothesized Mean Difference** and type your hypothesized value (in general, this will be 0)
8. Click in the box next to **Alpha:** and type in the significance level
9. Click **OK**
10. Select **t-Test: Two Samples Assuming Unequal Variances** from the dialog box and click **OK**
11. Click in the box next to **Variable 1 Range:** and select the first column of data
12. Click in the box next to **Variable 2 Range** and select the second column of data (if you have used and selected column titles for BOTH variables, select the check box next to **Labels**)
13. Click in the box next to **Hypothesized Mean Difference** and type your hypothesized value (in general, this will be 0)
14. Click in the box next to **Alpha:** and type in the significance level
15. Click **OK**

Note: This procedure outputs *P*-values for both a one-sided and two-sided test.

TI-83/84
Summarized data

1. Press **STAT**
2. Highlight **TESTS**
3. Highlight **2-SampT-Test...**
4. Highlight **Stats** and press **ENTER**
5. Next to $\bar{x}1$ input the value for the sample mean from the first sample
6. Next to **sx1** input the value for the sample standard deviation from the first sample
7. Next to **n1** input the value for the sample size from the first sample
8. Next to $\bar{x}2$ input the value for the sample mean from the second sample
9. Next to **sx2** input the value for the sample standard deviation from the second sample
10. Next to **n2** input the value for the sample size from the second sample
11. Next to $\mu 1$ highlight the appropriate alternative hypothesis and press **ENTER**
12. Highlight **Calculate** and press **ENTER**

Raw data

1. Enter the data into **L1** and **L2** (In order to access lists press the **STAT** key, highlight the option called **Edit...** then press **ENTER**)
2. Press **STAT**
3. Highlight **TESTS**
4. Highlight **2-SampT-Test...**
5. Highlight **Data** and press **ENTER**
6. Next to $\mu 1$ highlight the appropriate alternative hypothesis and press **ENTER**
7. Highlight **Calculate** and press **ENTER**

TI-Nspire
Summarized data

1. Enter the Calculate Scratchpad

2. Press the **menu** key and select **6:Statistics** then **7:Stat Tests** then **4:2-Sample t test...** and press **enter**
3. From the drop-down menu select **Stats**
4. Press **OK**
5. Next to $\bar{x}1$ input the value for the sample mean for the first sample
6. Next to **sx1** input the value for the sample standard deviation for the first sample
7. Next to **n1** input the value for the sample size for the first sample
8. Next to $\bar{x}2$ input the value for the sample mean for the second sample
9. Next to **sx2** input the value for the sample standard deviation for the second sample
10. Next to **n2** input the value for the sample size for the second sample
11. Next to **Alternate Hyp** select the appropriate alternative hypothesis from the drop-down menu
12. Press **OK**

Raw data

1. Enter the data into a data list (In order to access data lists select the spreadsheet option and press **enter**)

   Note: Be sure to title the list by selecting the top row of the column and typing a title.
2. Press the **menu** key and select **4:Statistics** then **4:Stat Tests** then **4:2-Sample t test...** and press **enter**
3. From the drop-down menu select **Data**
4. Press **OK**
5. Next to **List 1** select the list containing your data from the first sample
6. Next to **List 2** select the list containing your data from the second sample
7. Next to **Alternate Hyp** select the appropriate alternative hypothesis from the drop-down menu
8. Press **OK**

Confidence Interval for Paired Data

JMP
1. Enter the data for one group in the first column
2. Enter the paired data from the second group in the second column

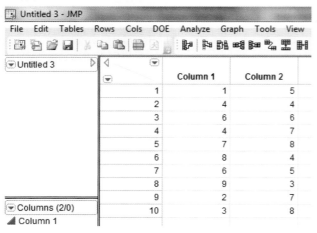

3. Click **Analyze** and select **Matched Pairs**
4. Click and drag the first column name from the box under **Select Columns** to the box next to **Y, Paired Response**
5. Click and drag the second column name from the box under **Select Columns** to the box next to **Y, Paired Response**
6. Click **OK**

Note: The 95% confidence interval is automatically displayed. To change the confidence level, click the red arrow next to **Oneway Analysis of...** and select **Set α Level** then click the appropriate α-level or select **Other** and type the appropriate level.

Minitab
Summarized data
1. Click **Stat** then click **Basic Statistics** then click **Paired t...**
2. Click the radio button next to **Summarized data**
3. In the box next to **Sample size:** type the sample size, n
4. In the box next to **Mean:** type the sample mean for the DIFFERENCE of each pair of data values
5. In the box next to **Standard deviation:** type the sample standard deviation for the DIFFERENCE of each pair of data values
6. Click **Options...**
7. Input the appropriate confidence level in the box next to **Confidence Level**
8. Click **OK**
9. Click **OK**

Raw data
1. Input the raw data into two separate columns
2. Click **Stat** then click **Basic Statistics** then click **Paired t...**
3. Click in the box next to **First sample:**
4. Double click the column name where the first group's data are stored
5. Click in the box next to **Second sample:**
6. Double click the column name where the second group's data are stored
7. Click **Options...**
8. Input the appropriate confidence level in the box next to **Confidence Level**
9. Click **OK**
10. Click **OK**

SPSS
1. Input the data for each group into two separate columns
2. Click **Analyze** then click **Compare Means** then click **Paired-Samples T Test**
3. Select the first column and click the arrow to move it into the Pair 1 row, Variable 1 column
4. Select the second column and click the arrow to move it to the Pair 1 row, Variable 2 column
5. Click **Options...**
6. Input the appropriate confidence level in the box next to **Confidence Interval Percentage:**
7. Click **Continue**
8. Click **OK**

Excel

Excel does not have the functionality to produce a confidence interval automatically for the difference of two population means. However, you can manually type the formulas for the lower and upper limits into two separate cells and have Excel calculate the results for you. You may also use Excel to find the t critical value based on the confidence level using the following steps:

1. Click in an empty cell
2. Click **Formulas**
3. Click **Insert Function**
4. Select **Statistical** from the drop-down box for the category
5. Select **TINV** and click **OK**
6. In the box next to **Probability** type in the value representing one minus your selected confidence level
7. In the box next to **Deg_freedom** type in the degrees of freedom ($n - 1$)
8. Click **OK**

TI-83/84

The TI-83/84 does not provide the option for a paired t confidence interval. However, one can be found by entering the difference data into a list and following the procedures in Chapter 12.

TI-Nspire

The TI-Nspire does not provide the option for a paired t confidence interval. However, one can be found by entering the difference data into a list and following the procedures in Chapter 12.

Paired *t*-Test for Difference of Population Means

JMP

1. Enter the data for one sample in the first column
2. Enter the paired data from the second sample in the second column
3. Click **Analyze** and select **Matched Pairs**
4. Click and drag the first column name from the box under **Select Columns** to the box next to **Y, Paired Response**
5. Click and drag the second column name from the box under **Select Columns** to the box next to **Y, Paired Response**
6. Click **OK**

Minitab

Summarized data

1. Click **Stat** then click **Basic Statistics** then click **Paired t...**
2. Click the radio button next to **Summarized data**
3. In the box next to **Sample size:** type the sample size, n
4. In the box next to **Mean:** type the sample mean for the DIFFERENCE of each pair of data values
5. In the box next to **Standard deviation:** type the sample standard deviation for the DIFFERENCE of each pair of data values
6. Click **Options...**
7. Input the appropriate hypothesized value for the difference of the paired population means in the box next to **Test mean:**
8. Select the appropriate alternative from the drop-down menu next to **Alternative:**
9. Click **OK**
10. Click **OK**

Raw data

1. Input the raw data into two separate columns
2. Click **Stat** then click **Basic Statistics** then click **Paired t...**
3. Click in the box next to **First:**
4. Double click the column name where the first group's data are stored
5. Click in the box next to **Second:**
6. Double click the column name where the second group's data are stored
7. Click **Options...**
8. Input the appropriate hypothesized value for the difference of the paired population means in the box next to **Test mean:**
9. Select the appropriate alternative from the drop-down menu next to **Alternative:**
10. Click **OK**
11. Click **OK**

SPSS

1. Input the data for each group into two separate columns
2. Click **Analyze** then click **Compare Means** then click **Paired-Samples T Test**
3. Select the first column and click the arrow to move it into the Pair 1 row, Variable 1 column
4. Select the second column and click the arrow to move it to the Pair 1 row, Variable 2 column
5. Click **OK**

Note: This procedure produces a two-sided *P*-value.

Excel

1. Input the raw data for each group into two separate columns
2. Click on the **Data** ribbon
3. Click **Data Analysis** in the **Analysis** group

Note: If you do not see **Data Analysis** listed on the Ribbon, see the Technology Notes for Chapter 2 for instructions on installing this add-on.

4. Select **t-Test: Paired Two Sample for Means** from the dialog box and click **OK**
5. Click in the box next to **Variable 1 Range:** and select the first column of data
6. Click in the box next to **Variable 2 Range** and select the second column of data (if you have used and selected column titles for BOTH variables, select the check box next to **Labels**)
7. Click in the box next to **Hypothesized Mean Difference** and type your hypothesized value (in general, this will be 0)
8. Click in the box next to **Alpha:** and type in the significance level
9. Click **OK**

Note: This procedure outputs *P*-values for both a one-sided and two-sided test.

TI-83/84

The TI-83/84 does not provide the option for a paired t test. However, one can be found by entering the difference data into a list and following the procedures in Chapter 12.

TI-Nspire

The TI-Nspire does not provide the option for a paired t test. However, one can be found by entering the difference data into a list and following the procedures in Chapter 12.

14 Learning from Categorical Data

yichaoMA/Shutterstock.com

PREVIEW

This chapter introduces three additional methods for learning from categorical data. Sometimes a categorical data set consists of observations on a single variable of interest (univariate data). When the categorical variable has only two possible categories, the methods introduced in Chapters 9, 10, and 11 can be used to learn about the proportion of "successes." For example, suppose calls made to the 9-1-1 emergency number are classified according to whether they are for true emergencies or not. You can estimate the proportion of calls that are for true emergencies or you can use data from two different cities to determine if there is evidence of a difference in the proportions of true emergency calls. But the methods of Chapters 9, 10, and 11 are only appropriate when the categorical variable of interest has two possible categories. In this chapter, you will see how to analyze data on a categorical variable with more than two possible categories.

You will also see how to compare two or more populations on the basis of a categorical variable. This is illustrated in the following example.

CHAPTER LEARNING OBJECTIVES

Conceptual Understanding
After completing this chapter, you should be able to

C1 Understand how a chi-square goodness-of-fit test can be used to answer a question of interest about a categorical variable.

C2 Understand that the way in which data summarized in a two-way table were collected determines which chi-square test (independence or homogeneity) is appropriate.

Mastering the Mechanics
After completing this chapter, you should be able to

M1 Determine which chi-square test (goodness-of-fit, independence, or homogeneity) is appropriate in a given situation.

M2 Determine appropriate null and alternative hypotheses for chi-square tests.

M3 Know the conditions necessary for the chi-square goodness-of-fit test to be appropriate.

M4 Know the conditions necessary for the chi-square tests of independence or homogeneity to be appropriate.

M5 Calculate the value of the test statistic and find the associated *P*-value for chi-square tests.

Putting It into Practice
After completing this chapter, you should be able to

P1 Carry out a chi-square goodness-of-fit test and interpret the result in context.

P2 Carry out a chi-square test of homogeneity and interpret the result in context.

P3 Carry out a chi-square test of independence and interpret the result in context.

PREVIEW EXAMPLE

Heart Attacks in High-Rise Buildings

Does the chance of surviving a heart attack depend on whether you live in a house or a high-rise apartment building? If you live in a high-rise apartment building, does the chance of survival depend on how high up in the building the apartment is? The authors of the paper **"Out-of-Hospital Cardiac Arrest in High-Rise Buildings: Delays to Patient Care and Effect on Survival (*Canadian Medical Association Journal* [2016]: 413–419)** investigated these questions using survival data from representative samples of 5531 heart attacks that occurred in a house or townhouse, 667 heart attacks that occurred in an apartment building on the first or second floor, and 1696 heart attacks that occurred in an apartment building on the third or higher floor. The resulting data are summarized in the following table:

	Survived	Did Not Survive	Total
House or Townhouse	217	5,314	**5,531**
Apartment First or Second Floor	35	632	**667**
Apartment Third or Higher Floor	46	1,650	**1,696**
Total	**298**	**7,596**	**7,894**

Notice that three different populations are being compared (heart attacks that occur in a house or townhouse, heart attacks that occur in an apartment building on the first or second floor, and heart attacks that occur in an apartment building on the third or higher floor).

Had the researchers been interested in comparing only houses and apartments, they might have used the two-sample test of Chapter 11 to determine if the proportion surviving is significantly different for heart attacks that occur in houses and for heart attacks that occur in apartments. However, because there are more than two populations to be compared, the researchers analyzed the data using one of the methods you will learn in this chapter. This example will be revisited in Section 14.2 to see what can be learned from these data. ■

A categorical data set might also be bivariate, consisting of observations on two categorical variables, such as political affiliation and support for a particular ballot initiative. A method for determining if there is an association between two categorical variables is also introduced in this chapter.

SECTION 14.1 Chi-Square Tests for Univariate Categorical Data

Univariate categorical data sets arise in a variety of settings. If each student in a sample of 100 is classified according to whether he or she is enrolled full-time or part-time, data on a categorical variable with two categories result. Each person in a sample of 100 registered voters in a particular city might be asked which of five city council members he or she favors for mayor. This would result in observations on a categorical variable with five categories. Univariate categorical data are usually summarized in a one-way frequency table, as illustrated in the following example.

Example 14.1 Tasty Dog Food?

The article **"Can People Distinguish Pâté from Dog Food?" (American Association of Wine Economists, April 2009, www.wine-economics.org)** describes a study that investigated whether people can tell the difference between dog food, pâté (a spread made of finely chopped liver, meat, or fish), and processed meats (such as Spam and liverwurst). Researchers used a food processor to make spreads that had the same texture and consistency as pâté from Newman's Own brand dog food and from the processed meats. Each participant in the study tasted five spreads (duck liver pâté, Spam, dog food, pork liver pâté, and liverwurst). After tasting all five spreads, each participant was asked to choose the one that they thought was the dog food. The first few observations were

> liverwurst   pork liver pâté   liverwurst   dog food

After the number of observations was counted for each category, the data for 50 participants were summarized in the following table. This table is an example of a **one-way frequency table**.

	Spread Chosen as Dog Food				
	Duck Liver Pâté	Spam	Dog Food	Pork Liver Pâté	Liverwurst
Frequency	3	11	8	6	22

(Note: The frequencies in the table are consistent with summary values given in the paper. However, the sample size in the study was not actually 50.)

In general, for a categorical variable with k possible values (k different categories), sample data are summarized in a one-way frequency table consisting of k cells, which may be displayed either horizontally or vertically. In this section, you will consider testing hypotheses about the proportions of the population that fall into the possible categories.

NOTATION

k = number of categories of a categorical variable
p_1 = population proportion for Category 1
p_2 = population proportion for Category 2
⋮
p_k = population proportion for Category k
    (Note: $p_1 + p_2 + \cdots + p_k = 1$)

The hypotheses to be tested have the form

H_0: p_1 = hypothesized proportion for Category 1
    p_2 = hypothesized proportion for Category 2
⋮
    p_k = hypothesized proportion for Category k

H_a: H_0 is not true. At least one of the population category proportions differs from the corresponding hypothesized value.

For the dog food study of Example 14.1, you could define

p_1 = proportion of all people who would choose duck liver pâté as the dog food
p_2 = proportion of all people who would choose Spam as the dog food
p_3 = proportion of all people who would choose dog food as the dog food
p_4 = proportion of all people who would choose pork liver pâté as the dog food
p_5 = proportion of all people who would choose liverwurst as the dog food

One possible null hypothesis of interest might be

$$H_0: p_1 = 0.20, \ p_2 = 0.20, \ p_3 = 0.20, \ p_4 = 0.20, \ p_5 = 0.20$$

This hypothesis specifies that any of the five spreads is equally likely to be the one identified as dog food. If this is true, you would expect $\frac{1}{5}$ or 20% of the participants in the sample to have selected each of the five spreads. In this case, the expected count would be $50(0.20) = 10$ for each category.

In general, the expected counts are:

Expected cell count for Category 1 = np_1
Expected cell count for Category 2 = np_2
$\vdots$
Expected cell count for Category k = np_k

The expected counts for a particular null hypothesis are calculated using the hypothesized proportions from the null hypothesis.

A null hypothesis of the form just described can be tested by first selecting a random sample of size n and then classifying each response into one of the k possible categories. To decide whether the sample data are compatible with the null hypothesis, you compare each observed count (the observed category frequency) to an expected count.

Example 14.2   *Jeopardy!* Nerds

The article **"Memo to Alex Trebeck: Your Viewers Are Nerds" (October 21, 2016, www.today .yougov.com/news/2016/10/21/jeopardy-fans-are-nerds/, retrieved May 26, 2017)** reports that viewers of the popular game show *Jeopardy!* tend to label themselves as nerds and that *Jeopardy!* viewers are more educated than the average American. These conclusions were based on a survey of a representative sample of people who said that they had watched the show in the past year. Data on the highest level of education completed for the sample of *Jeopardy!* viewers that is consistent with percentages given in the article are given in the accompanying table. The article did not give the sample size, so the numbers in the table are based on assuming a sample size of 1000.

Highest Level of Education Completed	Observed Frequency
Less than High School	10
High School	270
Some College	180
2-Year Degree	70
4-Year College Degree	300
Post-Graduate Study	170
Total	**1,000**

The article also indicated that for the general population in the United States, the percentages of the population falling into each of the education categories are as shown in the following table.

Highest Level of Education Completed	Percentage of U.S. Population
Less than High School	7
High School	36
Some College	22
2-Year Degree	9
4-Year College Degree	17
Post-Graduate Study	9
Total	**100**

The given information can be used to investigate the claim that *Jeopardy!* viewers tend to be more educated than the general population. You can represent the education category proportions as

p_1 = proportion of *Jeopardy!* viewers with a highest level of education completed of less than high school

p_2 = proportion of *Jeopardy!* viewers with a highest level of education completed of high school

p_3 = proportion of *Jeopardy!* viewers with a highest level of education completed of some college

p_4 = proportion of *Jeopardy!* viewers with a highest level of education completed of 2-year degree

p_5 = proportion of *Jeopardy!* viewers with a highest level of education completed of 4-year degree

p_6 = proportion of *Jeopardy!* viewers with a highest level of education completed of post-graduate study

If *Jeopardy!* viewers are like the population in general in terms of education level, you would expect that the *Jeopardy!* viewer proportions would match those of the general population. This means that they would be

$$p_1 = 0.07$$
$$p_2 = 0.36$$
$$p_3 = 0.22$$
$$p_4 = 0.09$$
$$p_5 = 0.17$$
$$p_6 = 0.09$$

The hypotheses of interest are then

$H_0: p_1 = 0.07, p_2 = 0.36, p_3 = 0.22, p_4 = 0.09, p_5 = 0.17, p_6 = 0.09$
$H_a: H_0$ is not true.

There were a total of 1000 *Jeopardy!* viewers in the sample. If the null hypothesis is true, the expected counts for the first two categories are

$$\left(\begin{array}{c}\text{expected count for} \\ \text{less than high school}\end{array}\right) = n\left(\begin{array}{c}\text{hypothesized proportion} \\ \text{for less than high school}\end{array}\right) = 1000(0.07) = 70$$

$$\left(\begin{array}{c}\text{expected count for} \\ \text{high school}\end{array}\right) = n\left(\begin{array}{c}\text{hypothesized proportion} \\ \text{high school}\end{array}\right) = 1000(0.36) = 360$$

Expected counts for the other four categories are calculated in a similar way. The observed and expected counts are given in the following table.

Highest Level of Education Completed	Observed Count	Expected Count
Less than High School	10	70
High School	270	360
Some College	180	220
2-Year Degree	70	90
4-Year College Degree	300	170
Post-Graduate Study	170	90
Total	**1,000**	**1,000**

Because the observed counts are based on a sample of *Jeopardy!* viewers, it would be surprising to see exactly 7% of the sample falling into the first education category, exactly 36% falling into the second category, and so on, even when H_0 is true. If the differences between the observed and expected cell counts can reasonably be attributed to sampling variability, the data are considered compatible with H_0. On the other hand, if the differences between the observed and the expected cell counts are too large to be explained by chance differences from one sample to another, H_0 should be rejected in favor of H_a. To make a decision, you need to assess how different the observed and expected counts are.

The goodness-of-fit statistic, denoted by X^2, is a quantitative measure of the extent to which the observed counts differ from those expected when H_0 is true. (The Greek letter χ (chi) is often used in place of X. The symbol X^2 is referred to as the chi-square statistic. Using X^2 rather than χ^2 follows the convention of denoting sample statistics by Roman letters.)

The **goodness-of-fit statistic, X^2**, results from first calculating the quantity

$$\frac{(\text{observed count} - \text{expected count})^2}{\text{expected count}}$$

for each category, where for a sample of size n,

$$\frac{\text{expected}}{\text{count}} = n \left(\begin{array}{c} \text{hypothesized value of corresponding} \\ \text{population proportion} \end{array} \right)$$

The X^2 statistic is the sum of these quantities for all k categories:

$$X^2 = \sum_{\text{all categories}} \frac{(\text{observed count} - \text{expected count})^2}{\text{expected count}}$$

The value of the X^2 statistic reflects the size of the differences between observed and expected counts. When the differences are big, the value of X^2 tends to be large, which suggests H_0 should be rejected. A small value of X^2 (it can never be negative) occurs when the observed cell counts are quite similar to those expected when H_0 is true, and so would be considered consistent with H_0.

As with previous test procedures, a conclusion is reached by comparing a P-value to the significance level for the test. The P-value is the probability of observing a value of X^2 at least as large as the observed value if H_0 were true. Calculating this P-value requires information about the sampling distribution of X^2 when H_0 is true.

When the null hypothesis is true and the sample size is large, the behavior of X^2 is described approximately by a **chi-square distribution**. A chi-square distribution curve has no area associated with negative values and is not symmetric, with a longer tail on the right. There are many different chi-square distributions, and each one has a different number of degrees of freedom (df). Curves corresponding to several chi-square distributions are shown in Figure 14.1.

FIGURE 14.1
Chi-square curves

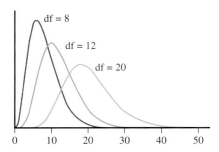

For a test procedure based on X^2, the associated *P*-value is the area under the appropriate chi-square curve and to the right of the calculated X^2 value. Appendix A Table 5 gives upper-tail areas for chi-square distributions with 2 to 20 df.

To find the area to the right of a particular X^2 value, locate the appropriate df column in Appendix A Table 5. Determine which listed value is closest to the X^2 value of interest, and read the right-tail area corresponding to this value from the left-hand column of the table. For example, for a chi-square distribution with df = 4, the area to the right of $X^2 = 8.18$ is 0.085, as shown in Figure 14.2.

FIGURE 14.2
A chi-square upper-tail area

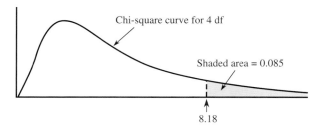

For this same chi-square distribution (df = 4), the area to the right of 9.70 is approximately 0.045 (the area to the right of 9.74, the closest entry in the table for df = 4).

It is also possible to use computer software or a graphing calculator to determine areas under a chi-square curve.

Chi-Square Goodness-of-Fit Test

When a null hypothesis that specifies *k* category proportions is true, the X^2 goodness-of-fit statistic has approximately a chi-square distribution with df = $k - 1$, as long as none of the expected cell counts is too small. When expected counts are small, and especially when an expected count is less than 1, the value of

$$\frac{(\text{observed count} - \text{expected count})^2}{\text{expected count}}$$

can be inflated because it involves dividing by a small number. *The use of the chi-square distribution is appropriate when the sample size is large enough for every expected count to be at least 5.* If any of the expected counts are less than 5, categories can be combined in a sensible way to create acceptable expected cell counts. If you do this, remember to calculate the number of degrees of freedom based on the reduced number of categories.

Chi-Square Goodness-of-Fit Test

Appropriate when the following conditions are met:

1. Observed counts are based on a random sample or a sample that is representative of the population.
2. The sample size is large. The sample size is large enough for the chi-square goodness-of-fit test to be appropriate if every expected count is at least 5.

When these conditions are met, the following test statistic can be used:

$$X^2 = \sum_{\text{all categories}} \frac{(\text{observed count} - \text{expected count})^2}{\text{expected count}}$$

When the null hypothesis is true, the X^2 statistic has a chi-square distribution with df $= k - 1$, where k is the number of category proportions specified in the null hypothesis.

Hypotheses
H_0: p_1 = hypothesized proportion for Category 1

p_2 = hypothesized proportion for Category 2

$\vdots$

p_k = hypothesized proportion for Category k

H_a: H_0 is not true. At least one of the population category proportions differs from the corresponding hypothesized value.

Associated *P*-value
The *P*-value is the area to the right of X^2 under the chi-square curve with df $= k - 1$. Upper-tail areas for chi-square distributions can be found in Appendix A Table 5.

Questions

Q Question type: estimation or hypothesis testing?

S Study type: sample data or experiment data?

T Type of data: one variable or two? Categorical or numerical?

N Number of samples or treatments: how many?

The chi-square goodness-of-fit test is a method you should consider when the answers to the four key questions are *hypothesis testing, sample data, one categorical variable (with more than two categories),* and *one sample.*

The following examples illustrate the use of the chi-square goodness-of-fit test.

Example 14.3   Tasty Dog Food Continued

You can use the dog food taste data of Example 14.1 to test the hypothesis that the five different spreads (duck liver pâté, Spam, dog food, pork liver pâté, and liverwurst) are chosen equally often when people who have tasted all five spreads are asked to identify the one they think is the dog food. The data from Example 14.1 are also shown here:

	Spread Chosen as Dog Food				
	Duck Liver Pâté	Spam	Dog Food	Pork Liver Pâté	Liverwurst
Frequency	3	11	8	6	22

Start by considering the four key questions (QSTN). You would like to use the sample data to test a claim about the corresponding population, so this is a hypothesis testing problem. The data are from a sample. There is one categorical variable with five categories corresponding to the five types of spread that could be identified as dog food. There is one sample. This combination of answers leads you to consider the chi-square goodness-of-fit test.

Steps
H Hypotheses
M Method
C Check
C Calculate
C Communicate results

Now you can use the five-step process for hypothesis testing problems (HMC³).

Process Step	
H Hypotheses	The question of interest is about the population category proportions, so you need to define these proportions in the context of this example.
	Population characteristics of interest
	p_1 = proportion of all people who would choose duck liver pâté as the dog food
	p_2 = proportion of all people who would choose Spam as the dog food
	p_3 = proportion of all people who would choose dog food as the dog food
	p_4 = proportion of all people who would choose pork liver pâté as the dog food
	p_5 = proportion of all people who would choose liverwurst as the dog food
	The question of interest (Are the five spreads identified equally often as the one thought to be dog food?) results in a null hypothesis that specifies that all five category proportions are 0.20.
	Hypotheses
	Null hypothesis:
	H_0: $p_1 = 0.20$, $p_2 = 0.20$, $p_3 = 0.20$, $p_4 = 0.20$, $p_5 = 0.20$
	Alternative hypothesis:
	H_a: At least one of the population proportions is not 0.20
M Method	Because the answers to the four key questions are hypothesis testing, sample data, one variable, categorical with more than two categories, and one sample, a chi-square goodness-of-fit test is considered.
	The test statistic for this test is
	$$X^2 = \sum_{\text{all categories}} \frac{(\text{observed count} - \text{expected count})^2}{\text{expected count}}$$
	When the null hypothesis is true, this statistic has approximately a chi-square distribution with
	$$\text{df} = k - 1 = 5 - 1 = 4$$
	A significance level of $\alpha = 0.05$ will be used for this test.
C Check	In order to use the chi-square goodness-of-fit test, you must be willing to assume that the participants in this study can be regarded as a random or representative sample. If this assumption is not reasonable, you should be cautious in generalizing results from this analysis to any larger population.
	To check the large sample condition, you need to calculate the expected count for each of the five categories. For the first category, the hypothesized proportion (from the null hypothesis) is 0.20. Because the sample size was 50, the expected count for the duck liver category is $50(0.20) = 10$. Because the other hypothesized proportions are also 0.20, all of the expected counts are equal to 10. All expected counts are at least 5, so the sample size is large enough for the chi-square goodness-of-fit test to be appropriate.
C Calculate	**Test statistic**
	$$X^2 = \sum_{\text{all categories}} \frac{(\text{observed count} - \text{expected count})^2}{\text{expected count}}$$
	$$= \frac{(3 - 10)^2}{10} + \frac{(11 - 10)^2}{10} + \frac{(8 - 10)^2}{10} + \frac{(6 - 10)^2}{10} + \frac{(22 - 10)^2}{10}$$
	$$= 4.9 + 0.1 + 0.4 + 1.6 + 14.4$$
	$$= 21.4$$

(continued)

Process Step		
	Degrees of freedom: $k - 1 = 5 - 1 = 4$	
	The P-value is the area under the chi-square curve with df = 4 and to the right of 21.4. Because 21.4 is greater than the largest value in the 4 df column of Appendix A Table 5, the area to the right is less than 0.001.	
	Associated P-value	
	P-value = area under chi-square curve to the right of 21.4 < 0.001	
C Communicate results	Because the P-value is less than the selected significance level, the null hypothesis is rejected.	
	$$P\text{-value} < 0.001 < 0.05, \text{ Reject } H_0$$	
	Interpretation	
	Based on these sample data, there is convincing evidence that the proportion identifying a spread as dog food is not the same for all five spreads. Here, it is interesting to note that the largest differences between the observed counts and the counts that would have been expected if the null hypothesis of equal proportions were true are in the duck liver pate and the liverwurst categories. This indicates that fewer than expected chose duck liver and many more than expected chose liverwurst as the one they thought was dog food. So, although you reject the hypothesis that the proportion choosing each of the five spreads is the same, it is not because people were actually able to identify which one was really dog food.	

It is also possible to use statistical software to carry out the calculate step in a goodness-of-fit test. For example, Minitab output for the test of this example is shown here.

Chi-Square Goodness-of-Fit Test

Category	Observed	Test Proportion	Expected	Contribution to Chi-Sq
1	3	0.2	10	4.9
2	11	0.2	10	0.1
3	8	0.2	10	0.4
4	6	0.2	10	1.6
5	22	0.2	10	14.4

N	DF	Chi-Sq	P-Value
50	4	21.4	0.000

From the Minitab output, you see that $X^2 = 21.4$ and that the P-value is approximately 0. This is consistent with the values from the hand calculations shown above. Minitab also constructs a graphical display comparing the observed and expected counts for each category. This display is shown in Figure 14.3. From this graph, it is easy to see the two spread categories where there were large differences between observed and expected values.

FIGURE 14.3
Minitab graph of observed and expected counts for Example 14.3

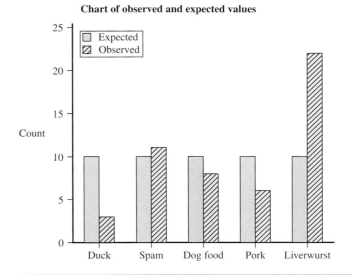

Chart of observed and expected values

Example 14.4 *Jeopardy!* Viewers Revisited

You can use the *Jeopardy!* data in Example 14.2 to test the hypothesis that *Jeopardy!* viewers are like the population in general in terms of education level. A 0.05 significance level will be used. The observed and expected counts calculated in Example 14.2 were

Highest Level of Education Completed	Observed Count	Expected Count
Less than High School	10	70
High School	270	360
Some College	180	220
2-Year Degree	70	90
4-Year College Degree	300	170
Post-Graduate Study	170	90
Total	1,000	1,000

H Hypotheses   If *Jeopardy!* viewers are like the population in terms of general education, then you would expect the proportions falling into the education categories to match those for the general population. This leads to the hypotheses given in Example 14.2:

$$H_0: p_1 = 0.07, p_2 = 0.36, p_3 = 0.22, p_4 = 0.09, p_5 = 0.17, p_6 = 0.09$$

$H_a: H_0$ is not true. At least one of the education category proportions is different from the value specified in H_0.

M Method   Consider the four key questions. You want to use the sample data to test a claim about the education level proportions, so this is a hypothesis-testing problem. The data are from a sample of *Jeopardy!* viewers. There is one categorical variable, which is education level. This variable has six categories. Finally, there is one sample. Because the answers to the four questions are hypothesis testing, sample data, one categorical variable with more than two categories, and one sample, you should consider a chi-square goodness-of-fit test.

The test statistic for the chi-square goodness-of-fit test is

$$X^2 = \sum_{\text{all categories}} \frac{(\text{observed count} - \text{expected count})^2}{\text{expected count}}$$

When the null hypothesis is true, this statistic has a chi-square distribution with df $= 6 - 1 = 5$.

A significance level of 0.05 was specified for this test.

C Check   Next you need to check to see if this method is appropriate. The study description in Example 14.2 states that the sample of viewers was representative of the population of all *Jeopardy!* viewers. The sample size is large enough for the chi-square test to be appropriate because all six expected counts are much greater than 5.

C Calculate $$X^2 = \sum_{\text{all categories}} \frac{(\text{observed count} - \text{expected count})^2}{\text{expected count}}$$

$$= \frac{(10 - 70)^2}{70} + \frac{(270 - 360)^2}{360} + \frac{(180 - 220)^2}{220} + \frac{(70 - 90)^2}{90} + \frac{(300 - 170)^2}{170}$$

$$+ \frac{(170 - 90)^2}{90}$$

$$= 51.429 + 22.500 + 7.273 + 4.444 + 99.412 + 71.111$$

$$= 256.169$$

P-value: the *P*-value is the area to the right of 256.169 under a chi-square curve with df $= 5$. The calculated value of X^2 is greater than 20.51 (the largest entry in the df $= 5$ column of Appendix A Table 5), so *P*-value < 0.001.

C Communicate Results   Because the *P*-value is less than the significance level of $\alpha = 0.05$, H_0 is rejected. There is sufficient evidence to conclude that *Jeopardy!* viewers are not like the population in general with respect to education level. This is consistent with the conclusion in the article that said that *Jeopardy!* viewers are "more educated than the average American."

Statistics software could also have been used to carry out the Calculate step. Minitab output for the data and hypothesized proportions of this example is shown here.

Chi-Square Goodness-of-Fit Test for Observed Counts in Variable: Number of Viewers
Using category names in Education Level

Category	Observed	Test Proportion	Expected	Contribution to Chi-Sq
Less than high school	10	0.07	70	51.4286
High school	270	0.36	360	22.5000
Some college	180	0.22	220	7.2727
2-year degree	70	0.09	90	4.4444
4-year degree	300	0.17	170	99.4118
Post-graduate study	170	0.09	90	71.1111

N	DF	Chi-Sq	P-Value
1000	5	256.169	0.000

Example 14.5   Hybrid Car Purchases

Green Car reported the top three states for sales of hybrid cars between 2010 and 2013 as California, Texas, and Florida **("Where Do Hybrids, Diesels Sell Best? State-by-State Data Shows Answers," June 4, 2014, www.greencarreports.com, retrieved May 26, 2017).** Suppose that each car in a random sample of hybrid car sales in 2010 from these three states is classified by the state where the sale took place, resulting in the accompanying table.

State	Observed Frequency
California	697
Texas	153
Florida	150
Total	**1,000**

(The given observed counts are based on a hypothetical sample of 1000 hybrid car sales, but they are consistent with hybrid sales figures given in the article.)

A X^2 goodness-of-fit test and a significance level of $\alpha = 0.01$ will be used to test the hypothesis that hybrid sales for these three states are proportional to their 2010 populations. Population sizes in 2010 from the Census Bureau web site are given in the following table. The population proportion for each state was calculated by dividing each state population by the total population of the three states.

State	2010 Population	Population Proportion
California	37,253,956	0.459
Texas	25,145,561	0.310
Florida	18,801,310	0.231
Total	**81,200,827**	

If these same population proportions hold for hybrid car sales, the expected counts are

Expected count for California $= 1000(0.459) = 459$
Expected count for Texas $= 1000(0.310) = 310$
Expected count for Florida $= 1000(0.231) = 231$

These expected counts have been entered in Table 14.1.

TABLE 14.1 Observed and Expected counts for Example 14.5

State	Observed Counts	Expected Counts
California	697	459
Texas	153	310
Florida	150	231

H Hypotheses You are interested in learning about the proportion of hybrid car sales for each of the given states. You can define these proportions as

p_1 = proportion of hybrid car sales for California
p_2 = proportion of hybrid car sales for Texas
p_3 = proportion of hybrid car sales for Florida

If sales of hybrid cars in these three states are proportional to population size, then

$$p_1 = 0.459 \qquad p_2 = 0.310 \qquad p_3 = 0.231$$

The hypotheses of interest are

$$H_0: p_1 = 0.459 \qquad p_2 = 0.310 \qquad p_3 = 0.231$$

H_a: H_0 is not true. At least one of the state hybrid sales proportions is different from the value specified in H_0.

M Method For this example, answers to the four key questions are hypothesis testing, sample data, one categorical variable with three categories, and one sample. These answers lead you to consider a chi-square goodness-of-fit test. The test statistic for the chi-square goodness-of-fit test is

$$X^2 = \sum_{\text{all categories}} \frac{(\text{observed count} - \text{expected count})^2}{\text{expected count}}$$

When the null hypothesis is true, this statistic has a chi-square distribution with df = $3 - 1 = 2$.

A significance level of $\alpha = 0.01$ was specified for this example.

C Check Next you need to check to see if this method is appropriate. The study description indicates that the sample of car sales was a random sample. Because all three expected counts are at least 5, the sample size is large enough for the chi-square test to be appropriate.

C Calculate From Minitab:

Chi-Square Goodness-of-Fit Test for Observed Counts in Variable: Hybrid Sales

Using category names in State

Category	Observed	Test Proportion	Expected	Contribution to Chi-Sq
California	697	0.459	459	123.407
Texas	153	0.310	310	79.513
Florida	150	0.231	231	28.403

N	DF	Chi-Sq	P-Value
1000	2	231.323	0.000

From the Minitab output, the values of the test statistic and df are

$$X^2 = 231.323 \qquad df = 2$$

and the associated *P*-value is 0.000.

C Communicate Results Because the P-value is less than the significance level of $\alpha = 0.01$, H_0 is rejected. There is convincing evidence that hybrid sales are not proportional to population size.

Looking back at the Minitab output, notice that there is a column labeled "Contribution to Chi-Sq." This column shows the individual values of $\dfrac{\text{(observed count} - \text{expected count})^2}{\text{expected count}}$, which are summed to produce the value of the chi-square statistic. Notice that the two states with the largest contribution to the chi-square statistic are California and Texas. For Texas, observed hybrid sales were much smaller than expected (observed = 153, expected = 310), whereas for California, observed sales were higher than expected (observed = 697, expected = 459).

Summing It Up—Section 14.1

The following learning objectives were addressed in this section:

Conceptual Understanding

C1: Understand how a chi-square goodness-of-fit test can be used to answer a question of interest about a categorical variable.

A chi-square goodness-of-fit test can be used to analyze data on a categorical variable with more than two categories. The question of interest specifies or implies category proportions that determine the form of the null hypothesis. By comparing the observed category counts to the counts that would be expected if the null hypothesis were true, you can determine if there is convincing evidence that the null hypothesis should be rejected.

Mastering the Mechanics

M1: Determine which chi-square test (goodness-of-fit, independence, or homogeneity) is appropriate in a given situation.

A chi-square goodness-of-fit test is used to test hypotheses about the distribution of a single categorical variable in one population. In this situation, the sample data are usually summarized in a one-way frequency table.

M2: Determine appropriate null and alternative hypotheses for chi-square tests.

For a chi-square goodness-of-fit test, the null hypothesis specifies particular values for the proportions of the population falling into each of the categories of a categorical variable. Example 14.2 illustrates how a question of interest is translated into a null and alternative hypothesis for a chi-square goodness-of-fit test.

M3: Know the conditions necessary for the chi-square goodness-of-fit test to be appropriate.

There are two conditions that are necessary in order for a chi-square goodness-of-fit test to be appropriate: (1) The observed counts are based on a random sample from the population of interest or the sample is representative of the population, and (2) the sample size is large. The sample size is considered to be large if all of the expected counts are greater than or equal to 5.

M5: Calculate the value of the test statistic and find the associated P-value for chi-square tests.

The chi-square test statistic is

$$X^2 = \sum_{all\ categories} \frac{(\text{observed count} - \text{expected count})^2}{\text{expected count}}$$

When the conditions for the chi-square goodness-of-fit test are met and the null hypothesis is true, this statistic has a chi-square distribution with df $= k - 1$, where k is the number of category proportions specified in the null hypothesis. The P-value for a chi-square goodness-of-fit test is the area under the appropriate chi-square curve and to the right of the calculated value of the test statistic. The P-value can be determined using Appendix A Table 5 or by using statistical software or a graphing calculator. Example 14.3 illustrates that calculation of the value of the chi-square statistic and associated P-value.

Putting It into Practice

P1: Carry out a chi-square goodness-of-fit test and interpret the result in context.

For examples of chi-square goodness-of-fit tests and the interpretation of the results in context, see Examples 14.3, 14.4, and 14.5.

SECTION 14.1 EXERCISES

Each Exercise Set assesses the following chapter learning objectives: C1, M1, M2, M3, M5, P1

SECTION 14.1 Exercise Set 1

14.1 A particular cell phone case is available in a choice of four different colors. A store sells all four colors. To test the hypothesis that sales are equally divided among the four colors, a random sample of 100 purchases is identified.

a. If the resulting X^2 value were 6.4, what conclusion would you reach when using a test with significance level 0.05?

b. What conclusion would be appropriate at significance level 0.01 if $X^2 = 15.3$?

c. If there were six different colors rather than just four, what would you conclude if $X^2 = 13.7$ and a test with $\alpha = 0.05$ was used?

14.2 What is the approximate *P*-value for the following values of X^2 and df?

a. $X^2 = 6.62$, df $= 3$
b. $X^2 = 16.97$, df $= 10$
c. $X^2 = 30.19$, df $= 17$

14.3 The authors of the paper **"Is It Really About Me? Message Content in Social Awareness Streams"** (*Computer Supported Cooperative Work 2010*) studied a random sample of 350 Twitter users. For each Twitter user in the sample, the tweets sent during a particular time period were analyzed and the Twitter user was classified into one of the following categories based on the type of messages they usually sent:

Category	Description
IS	Information sharing
OC	Opinions and complaints
RT	Random thoughts
ME	Me now (what I am doing now)
O	Other

The accompanying table gives the observed counts for the five categories (approximate values from a graph in the paper).

Twitter Category	IS	OC	RT	ME	O
Observed Count	51	61	64	101	73

Carry out a hypothesis test to determine if there is convincing evidence that the proportions of Twitter users falling into each of the five categories are not all the same. Use a significance level of $\alpha = 0.05$. (Hint: See Example 14.3.)

14.4 The article **"In Bronx, Hitting Home Runs Is a Breeze"** (*USA TODAY*, June 2, 2009) included a classification of 87 home runs hit at the new Yankee Stadium

according to the direction that the ball was hit, resulting in the accompanying data.

Direction	Left Field	Left Center	Center	Right Center	Right Field
Number of Home Runs	18	10	7	18	34

a. Assume that this sample of 87 home runs is representative of home runs hit at Yankee Stadium. Carry out a hypothesis test to determine if there is convincing evidence that the proportions of home runs hit are not the same for all five directions.

b. Write a few sentences describing how the observed counts for the five directions differ from what would have been expected if the proportion of home runs were the same for all five directions.

14.5 In 2014, the University of Houston carried out a study for the Texas Lottery Commission **("Demographic Survey of Texas Lottery Players," www.uh.edu/class/hobby/_docs /Texas%20Lottery%20Study%202014.pdf, retrieved May 27, 2017).** The accompanying table gives the age distribution for a representative sample of 375 Texas Lottery players.

Age Group	Frequency
18 to 24	7
25 to 34	32
35 to 44	34
45 to 54	60
55 to 64	106
65 and older	136
Total	**375**

Using data from the U.S. Census Bureau (www.census.gov) for 2014, the age distribution of adults in Texas was: 14% between age 18 and 24, 20% between age 25 and 34, 19% between age 35 and 44, 18% between age 45 and 54, 14% between age 55 and 64, and 15% age 65 or older. Is it reasonable to conclude that one or more of the age groups buys a disproportionate share of Texas Lottery tickets? Use a chi-square goodness-of-fit test with $\alpha = 0.05$. (Hint: See Example 14.5.)

SECTION 14.1 Exercise Set 2

14.6 What is the approximate *P*-value for the following values of X^2 and df?

a. $X^2 = 14.44$, df $= 6$
b. $X^2 = 16.91$, df $= 9$
c. $X^2 = 32.32$, df $= 20$

14.7 Packages of mixed nuts made by a certain company contain four types of nuts. The percentages of nuts of Types 1, 2, 3, and 4 are advertised to be 40%, 30%, 20%, and 10%, respectively. A random sample of nuts is selected, and each one is categorized by type.

a. If the sample size is $n = 200$ and the resulting test statistic value is $X^2 = 19.0$, what conclusion would be appropriate for a significance level of 0.001?

b. If the random sample had consisted of only 40 nuts, would you use the chi-square goodness-of-fit test? Explain your reasoning.

14.8 The authors of the paper **"Talking Smack: Verbal Aggression in Professional Wrestling"** (*Communication Studies* **[2008]: 242–258)** analyzed the content of 36 hours of televised professional wrestling. Each instance of verbal aggression was classified according to the type of aggression, resulting in the frequency table at the bottom of the page. Assume that this sample of 804 verbal aggression incidents is representative of all verbal aggression incidents in televised professional wrestling. Carry out a hypothesis test to determine if there is convincing evidence that the proportions of verbal aggression incidents are not the same for all five types of aggression. Use a significance level of 0.01.

14.9 The **"Global Automotive 2016 Color Popularity Report"** (Axalta Coating Systems, www.axaltacs.com, retrieved May 27, 2017) included data on the colors for a sample of new cars sold in North America. The report stated that 25% of the cars in the sample were white, 21% were black, 16% were grey, 11% were silver, 10% were red, and 17% were some other color. Suppose that these percentages were based on a random sample of 1200 new cars sold in North America. Is there convincing evidence that the proportions of new cars sold are not the same for all of the six color categories?

14.10 A popular urban legend is that more babies than usual are born during certain phases of the lunar cycle, especially near the full moon. The paper **"The Effect of the Gravitation of the Moon on Frequency of Births"** (*Environmental Health Insights* **[2010]: 65–69)** classified a random sample of 1007 births at a large hospital in Japan according to lunar phase. In each lunar cycle (27.32 days), the moon moves 360 degrees relative to the earth. To determine lunar phase, the researchers divided the 360 degrees in one lunar cycle into 12 phases of 30 degrees. The sample data are summarized in the accompanying frequency table.

Lunar Phase (degrees)	Number of Births
0–30	90
31–60	81
61–90	76

(continued)

Lunar Phase (degrees)	Number of Births
91–120	87
121–150	90
151–180	76
181–210	94
211–240	79
241–270	76
271–300	80
301–330	93
331–360	85

The researchers concluded that the frequency of births is not related to lunar cycle. Carry out a chi-square goodness-of-fit test to determine if the data are consistent with the researchers' claim. Use a significance level of 0.05 for your test.

ADDITIONAL EXERCISES

14.11 What is the approximate P-value for the following values of X^2 and df?

a. $X^2 = 34.52$, df $= 13$
b. $X^2 = 39.25$, df $= 16$
c. $X^2 = 26.00$, df $= 19$

14.12 For which of the X^2 and df pairs in the previous exercise would the null hypothesis be rejected if a significance level of $\alpha = 0.01$ were used?

14.13 Think about how you would answer the following question.

> Next Wednesday's meeting has been moved forward two days. What day is the meeting now that it has been rescheduled?

This question is ambiguous, as "moved forward" can be interpreted in two different ways. Would you have answered Monday or Friday? The authors of the paper **"Even Abstract Motion Influences the Understanding of Time"** (*Metaphor and Symbol* **[2011]: 260–271)** wondered if the answers Monday and Friday would be provided an equal proportion of the time. Each student in a random sample of students at Stanford University was asked this question, and the responses are summarized in the following table.

Response	Frequency
Monday	11
Friday	33

The authors of the paper used a chi-square goodness-of-fit test to test the null hypothesis H_0: $p_1 = 0.50$, $p_2 = 0.50$, where p_1 is the proportion who would respond Monday, and p_2 is the proportion who would respond Friday. They reported $X^2 = 11.00$ and P-value < 0.001. What conclusion can be drawn from this test?

TABLE FOR EXERCISE 14.8

Type of Aggression	Swearing	Competence Attack	Character Attack	Physical Appearance Attack	Other
Observed Count	219	166	127	75	217

Data set available

14.14 A certain genetic characteristic of a particular plant can appear in one of three forms (phenotypes). A researcher has developed a theory, according to which the hypothesized proportions are $p_1 = 0.25$, $p_2 = 0.50$, and $p_3 = 0.25$. A random sample of 200 plants yields $X^2 = 4.63$.

a. Carry out a test of the null hypothesis that the theory is correct, using level of significance $\alpha = 0.05$.

b. Suppose that a random sample of 300 plants had resulted in the same value of X^2. How would your analysis and conclusion differ from those in Part (a)?

14.15 The article **"Linkage Studies of the Tomato"** (*Transactions of the Royal Canadian Institute* [1931]: 1–19) reported the accompanying data on phenotypes resulting from crossing tall cut-leaf tomatoes with dwarf potato-leaf tomatoes. There are four possible phenotypes: (1) tall cut-leaf, (2) tall potato-leaf, (3) dwarf cut-leaf, and (4) dwarf potato-leaf.

	Phenotype			
	1	**2**	**3**	**4**
Frequency	926	288	293	104

Mendel's laws of inheritance imply that $p_1 = \dfrac{9}{16}$, $p_2 = \dfrac{3}{16}$, $p_3 = \dfrac{3}{16}$, and $p_4 = \dfrac{1}{16}$. Are the data from this experiment consistent with Mendel's laws? Use a 0.01 significance level.

14.16 Suppose that each observation in a random sample of 100 fatal bicycle accidents in 2015 was classified according to the day of the week on which the accident occurred. Data consistent with information on the Insurance Institute for Highway Safety website **(www.iihs.org/iihs/topics/t/general-statistics/fatalityfacts/overview-of-fatality-facts)** are given in the following table.

Day of Week	Frequency
Sunday	16
Monday	12
Tuesday	12
Wednesday	13
Thursday	14
Friday	15
Saturday	18

Based on these data, is it reasonable to conclude that the proportion of fatal bicycle accidents in 2015 was not the same for all days of the week? Use $\alpha = 0.05$.

14.17 Birds use color to select and avoid certain types of food. The authors of the article **"Colour Avoidance in Northern Bobwhites: Effects of Age, Sex, and Previous Experience"** (*Animal Behaviour* [1995]: 519–526) studied the pecking behavior of 1-day-old bobwhites. In an area painted white, they inserted four pins with different colored heads. The color of the pin chosen on the bird's first peck was noted for each of 33 bobwhites, resulting in the accompanying table.

Color	First Peck Frequency
Blue	16
Green	8
Yellow	6
Red	3

Do these data provide evidence of a color preference? Test using $\alpha = 0.01$.

14.18 The authors of the paper **"External Factors and the Incidence of Severe Trauma: Time, Date, Season and Moon"** (*Injury* [2014]: S93–S99) classified admissions to hospitals in Germany according to season. They wondered if severe trauma injuries were more common in some seasons than others. For purposes of this exercise, assume that there were 1200 trauma cases in the sample and that the sample is representative of severe trauma injuries in Germany. The data in the accompanying table are consistent with summary quantities given in the paper. Do these data support the theory that the proportion of severe trauma cases is not the same for the four seasons? Test the relevant hypotheses using a significance level of 0.05.

Season				
Winter	**Spring**	**Summer**	**Fall**	**Total**
228	332	352	288	1,200

SECTION 14.2 Tests for Homogeneity and Independence in a Two-Way Table

Bivariate categorical data (resulting from observations made on two different categorical variables) can also be summarized in a table. For example, suppose that residents of a particular city can watch national news on networks ABC, CBS, NBC, or FOX. A researcher wishes to know whether there is any relationship between political philosophy (liberal, moderate, or conservative) and preferred news network for people who regularly watch the national news. The two variables of interest are *political philosophy* and *preferred network*. If a random sample of 300 regular watchers is selected and these two variables are recorded for each person in the sample, the data set is bivariate and might initially be displayed as follows:

Observation	Political Philosophy	Preferred Network
1	Liberal	CBS
2	Conservative	ABC
3	Conservative	FOX
⋮	⋮	⋮
299	Moderate	NBC
300	Liberal	NBC

Bivariate categorical data are usually summarized in a **two-way frequency table**. This is a rectangular table that consists of a row for each possible category of one of the variables and a column for each possible category of the other variable. There is a cell in the table for each possible combination of category values. The number of times each particular combination occurs in the data set is entered in the corresponding cell of the table. These numbers are called **observed cell counts**. For example, one possible table for the political philosophy and preferred network example is given in Table 14.2. This table has three rows and four columns (because political philosophy has three possible categories and preferred network has four possible categories). Two-way frequency tables are often described by the number of rows and columns in the table (specified in that order: rows first, then columns). Table 14.2 is called a 3 × 4 table. The smallest two-way frequency table is a 2 × 2 table, which has only two rows and two columns, resulting in four cells.

TABLE 14.2 A 3 × 4 Frequency Table

	ABC	CBS	NBC	FOX	Row Marginal Total
Liberal	20	20	25	15	**80**
Moderate	45	35	50	20	**150**
Conservative	15	40	10	5	**70**
Column Marginal Total	**80**	**95**	**85**	**40**	**300**

Marginal totals are obtained by adding the observed cell counts in each row and also in each column of the table. The row and column marginal totals, along with the total of all observed cell counts in the table—the **grand total**, have been included in Table 14.2. The marginal totals provide information on the distribution of observed values for each variable separately. In this example, the row marginal totals reveal that the sample consisted of 80 liberals, 150 moderates, and 70 conservatives. Similarly, column marginal totals indicate how often each of the preferred network categories occurred: 80 preferred ABC, 95 preferred CBS, and so on. The grand total, 300, is the total number of observations in the bivariate data set.

Table 14.2 was constructed using the values of two different categorical variables for all individuals in a single sample. This table could be used to investigate any association between political philosophy and preferred network. In this type of bivariate categorical data set, only the grand total (the sample size) is known before the data are collected.

Two-way tables are also used when data are collected to compare two or more populations or treatments on the basis of a single categorical variable. In this situation, independent samples are selected from each population or treatment. For example, data could be collected at a university to compare students, faculty, and staff on the basis of primary mode of transportation to campus (car, bicycle, motorcycle, bus, or by foot). One random sample of 200 students, another random sample of 100 faculty members, and a third random sample of 150 staff members might be chosen. The selected individuals could be interviewed to obtain the necessary transportation information. Data from such a study could be summarized in a 3 × 5 two-way frequency table with row categories of student, faculty, and staff and column categories corresponding to the five possible modes of transportation. The observed cell counts could then be used to learn about differences and similarities among the three groups. In this situation, one set of marginal totals (the sample sizes for the different groups) is known before the data are collected. In the 3×5 situation just discussed, the row totals would be known (200, 100, and 150).

Comparing Two or More Populations or Treatments: A Test of Homogeneity

When the purpose of a study is to compare two or more populations or treatments on the basis of a categorical variable, the question of interest is whether the category proportions are the same for all the populations or treatments. The test procedure uses a chi-square statistic to compare the observed counts to those that would be expected if there were no differences among proportions.

Example 14.6 **Heart Attacks in High-Rise Buildings Revisited**

The paper referenced in the Preview Example (**"Out-of-Hospital Cardiac Arrest in High-Rise Buildings: Delays to Patient Care and Effect on Survival,"** *Canadian Medical Association Journal* **[2016]: 413–419)** compared people who lived in a house or townhouse, people who lived on the first or second floor of an apartment building, and people who lived on the third or a higher floor in an apartment building on the basis of heart attack survival rates. Table 14.3, a 3×4 two-way frequency table, is the result of classifying each heart attack victim in independently selected representative samples of 5531 heart attacks that occurred in a house or townhouse, 667 heart attacks that occurred in an apartment building on the first or second floor, and 1696 heart attacks that occurred in an apartment building on the third or higher floor into one of two categories (survived and did not survive).

TABLE 14.3 Observed Counts for Example 14.6

	Survived	Did Not Survive	Row Marginal Total
House or Townhouse	217	5,314	**5,531**
Apartment First or Second Floor	35	632	**667**
Apartment Third or Higher Floor	46	1,650	**1,696**
Column Marginal Total	**298**	**7,596**	**7,894**

Notice that there were a total of 7894 heart attack victims in the three samples combined. Of these, 298 survived. The proportion of the total who survived is then

$$\frac{298}{7894} = 0.03775$$

If there were no difference in survival for the three different groups, you would expect about 3.775% of the heart attack victims who live in a house or townhouse to survive, about 3.775% of the heart attack victims who live in an apartment on the first or second floor to survive, and about 3.775% of the heart attack victims who live in an apartment on the third or a higher floor to survive. This means that if there is no difference in survival, the expected number surviving for each of the three cells in the "Survived" column are

Expected count for live in a house or townhouse and survive = 0.03775(5531) = 208.795

Expected count for live in an apartment on first or second floor and survive = 0.03775(667) = 25.179

Expected count for live in an apartment on third or higher floor and survive = 0.03775(1696) = 64.024

Notice that the expected cell counts do not need to be whole numbers.

The expected cell counts for the remaining cells can be calculated in a similar manner. The proportion of the total who did not survive is

$$\frac{7596}{7894} = 0.96225$$

This proportion can be used to calculate expected counts for the cells in the "Did Not Survive" column:

Expected count for live in a house or townhouse and did not survive = 0.96225(5531) = 5322.205

Expected count for live in an apartment on first or second floor and did not survive = 0.96225(667) = 641.821

Expected count for live in an apartment on third or higher floor and did not survive = 0.96225(1696) = 1631.976

It is common practice to display the observed cell counts and the corresponding expected cell counts in the same table, with the expected cell counts enclosed in parentheses. Table 14.4 gives the observed cell counts and the expected cell counts. Notice that each marginal total for the expected cell counts is equal to the corresponding marginal total for the observed cell counts. (Sometimes there will be small differences in the marginal totals due to rounding when calculating the expected cell counts.)

TABLE 14.4 Observed and Expected Counts for Example 14.6

	Survived	Did Not Survive	Row Marginal Total
House or Townhouse	217 (208.795)	5,314 (5,322.205)	**5,531**
Apartment First or Second Floor	35 (25.179)	632 (641.821)	**667**
Apartment Third or Higher Floor	46 (64.024)	1,650 (1,631.976)	**1,696**
Column Marginal Total	**298**	**7,596**	**7,894**

A quick comparison of the observed and expected cell counts in Table 14.4 reveals some large differences, suggesting that the survival proportions may not be the same for the three groups considered. This will be explored further in Example 14.7.

In Example 14.6, the expected count for a cell corresponding to a particular group–response combination was calculated in two steps. First, the response *marginal proportion* was calculated (for example, 298/7894 for the "survived" response). Then this proportion was multiplied by a marginal group total (for example, 5531(298/7894) for the house or townhouse group). Algebraically, this is equivalent to first multiplying the row and column marginal totals and then dividing by the grand total:

$$\frac{(5531)(298)}{7894}$$

To compare two or more populations or treatments on the basis of a categorical variable, calculate an **expected cell count** for each cell by selecting the corresponding row and column marginal totals and then using the formula

$$\text{expected cell count} = \frac{(\text{row marginal total})(\text{column marginal total})}{\text{grand total}}$$

These expected cell counts represent what is expected if there are no differences in the category proportions for the groups under study.

The X^2 statistic, introduced in Section 14.1, is used to compare the observed cell counts to the expected cell counts. When there are large differences between the observed and expected counts, this leads to a large value of X^2 and suggests that the hypothesis of no differences between the populations or treatments should be rejected. A formal test procedure is described in the accompanying box.

Chi-Square Test for Homogeneity

Appropriate when the following conditions are met:

1. Observed counts are from independently selected random samples, *or* subjects in an experiment are randomly assigned to treatment groups.
2. The sample sizes are large. The sample size is large enough for the chi-square test for homogeneity to be appropriate if every expected count is at least 5. If some expected counts are less than 5, rows or columns of the table may be combined to achieve a table with satisfactory expected counts.

When these conditions are met, the following test statistic can be used:

$$X^2 = \sum_{\text{all cells}} \frac{(\text{observed count} - \text{expected count})^2}{\text{expected count}}$$

The expected cell counts are estimated from the sample data using the formula

$$\text{expected cell count} = \frac{(\text{row marginal total})(\text{column marginal total})}{\text{grand total}}$$

When the conditions above are met and the null hypothesis is true, the X^2 statistic has a chi-square distribution with df = (number of rows $-$ 1)(number of columns $-$ 1).

Hypotheses

H_0: The population (or treatment) category proportions are the same for all of the populations (or treatments).

H_a: The population (or treatment) category proportions are not all the same for all of the populations (or treatments).

Associated *P*-value

The *P*-value associated with the calculated value of the test statistic is the area to the right of X^2 under the chi-square curve with

$$\text{df} = (\text{number of rows} - 1)(\text{number of columns} - 1)$$

Upper-tail areas for chi-square distributions can be found in Appendix A Table 5.

Example 14.7 **Heart Attacks in High-Rise Buildings One Last Time**

The following table of observed and expected cell counts appeared in Example 14.6.

	Survived	Did Not Survive	Row Marginal Total
House or Townhouse	217 (208.795)	5,314 (5,322.205)	**5,531**
Apartment First or Second Floor	35 (25.179)	632 (641.821)	**667**
Apartment Third or Higher Floor	46 (64.024)	1,650 (1,631.976)	**1,696**
Column Marginal Total	**298**	**7,596**	**7,894**

You are interested in learning if there is a difference in the survival category proportions for the three populations (heart attack victims who live in a house or townhouse, heart attack victims who live on the first or second floor of an apartment building, and heart attack victims who live on the third or a higher floor in an apartment building).

H Hypotheses The hypotheses of interest are

H_0: Proportions in each survival category are the same for all three groups.
H_a: The survival category proportions are not all the same for all three groups.

M Method This is a hypothesis testing problem. Representative samples from three different populations were independently selected. The response is categorical. In this situation, you should consider a chi-square test of homogeneity.

The test statistic for the chi-square test of homogeneity is

$$X^2 = \sum_{\text{all cells}} \frac{(\text{observed count} - \text{expected count})^2}{\text{expected count}}$$

A significance level of $\alpha = 0.05$ will be used in this example.

C Check Next you need to check to see if this method is appropriate.

Because the samples were independent representative samples and the expected cell counts are all at least 5, the chi-square test of homogeneity is appropriate.

C Calculate The two-way table has three rows and two columns, so the appropriate df is

$$\text{df} = (\text{number of rows} - 1)(\text{number of columns} - 1) = (3 - 1)(2 - 1) = 2$$

$$X^2 = \sum_{\text{all cells}} \frac{(\text{observed count} - \text{expected count})^2}{\text{expected count}}$$

$$= \frac{(217 - 208.795)^2}{208.795} + \cdots + \frac{(1650 - 1631.976)^2}{1631.976} = 9.59$$

P-value: The *P*-value is the area to the right of 9.59 under a chi-square curve with df = 2. The calculated value of X^2 is between 9.21 and 10.59 in the df = 2 column of Appendix A Table 5, so $0.005 < P\text{-value} < 0.010$.

C Communicate Results Because the *P*-value is less than $\alpha = 0.05$, H_0 is rejected. There is strong evidence that the proportions in the survival categories are not the same for the three groups compared. Notice that there are more people who survived in the house or townhouse and first or second floor apartment categories and fewer people who survived in the third and higher floor apartment category than would have been expected if the survival proportions were the same for all three groups. This led the researchers who collected these data to conclude that there is a smaller chance of survival for people who suffer a heart attack in an apartment that is on the third or higher floor.

Most statistical computer packages can calculate expected cell counts, the value of the X^2 statistic, and the associated *P*-value. This is illustrated in the following example.

Example 14.8 **Keeping the Weight Off**

The article **"Daily Weigh-Ins Can Help You Keep Off Lost Pounds, Experts Say"** (**Associated Press, October 17, 2005**) describes an experiment in which 291 people who had lost at least 10% of their body weight in a medical weight-loss program were assigned at random to one of three groups for follow-up. One group met monthly in person, one group "met" online monthly in a chat room, and one group received a monthly newsletter by mail. After 18 months, participants in each group were classified according to whether or not they had regained more than 5 pounds, resulting in the counts given in Table 14.5.

TABLE 14.5 Observed Counts for Example 14.8

	Amount of Weight Gained		
	Regained 5 Lb or Less	Regained More Than 5 Lb	Row Marginal Total
In-Person	52 (41.0)	45 (56.0)	97
Online	44 (41.0)	53 (56.0)	97
Newsletter	27 (41.0)	70 (56.0)	97

Does there appear to be a difference in the weight gain category proportions for the three follow-up methods?

H Hypotheses The hypotheses of interest are

H_0: Proportions for the two weight gain categories are the same for the three follow-up methods.

H_a: The weight gain category proportions are not the same for all three follow-up methods.

M Method This is a hypothesis testing problem. The data are from an experiment. There is one categorical variable and there are three treatments. In this situation, you should consider a chi-square test of homogeneity.

The test statistic for the chi-square test of homogeneity is

$$X^2 = \sum_{\text{all cells}} \frac{(\text{observed count} - \text{expected count})^2}{\text{expected count}}$$

This table has three rows and two columns, so the appropriate df is

$$\text{df} = (\text{number of rows} - 1)(\text{number of columns} - 1) = (3 - 1)(2 - 1) = 2$$

A significance level of $\alpha = 0.01$ will be used for this test.

C Check Next, you need to determine if this method is appropriate. Table 14.5 includes the expected cell counts, all of which are greater than 5, so the large sample condition is satisfied. The subjects in this experiment were assigned at random to the treatment groups.

C Calculate JMP output follows. For each cell, this output includes the observed cell count, the expected cell count, and the value of $\frac{(\text{observed count} - \text{expected count})^2}{\text{expected count}}$ for that cell (its contribution to the X^2 statistic). From the output, $X^2 = 13.773$ (in the last row of the output), df $= 2$ (from the row in the output that is just below the table) and the associated P-value $= 0.001$ (from the last row of the output).

Contingency Analysis of Weight Gain By Follow-up Method

Mosaic Plot

Freq: Count

Contingency Table

Follow-up Method: Count / Expected / Cell Chi^2	Weight Gain: 5 Lb. or less	Weight Gain: More than 5 Lb.	
In-Person	52 41 2.9512	45 56 2.1607	97
Newsletter	27 41 4.7805	70 56 3.5000	97
Online	44 41 0.2195	53 56 0.1607	97
	123	168	291

Tests

N	DF	-LogLike	RSquare (U)
291	2	7.0481148	0.0356

Test	ChiSquare	Prob>ChiSq
Likelihood Ratio	14.096	0.0009*
Pearson	13.773	0.0010*

C Communicate Results Because the P-value is less than $\alpha = 0.01$, H_0 is rejected. The data indicate that the proportions who have regained more than 5 pounds are not the same for the three follow-up methods. Comparing the observed and expected cell counts, you can see that the observed number in the newsletter group who had regained more than 5 pounds was higher than would have been expected, and the observed number in the in-person group who had regained more than 5 pounds was lower than would have been expected if there were no differences in the three follow-up methods.

Testing for Independence of Two Categorical Variables

The X^2 test statistic and test procedure can also be used to investigate association between two categorical variables in a single population. When there is an association, knowing the value of one variable provides information about the value of the other variable. When there is *no* association between two categorical variables, they are said to be independent.

With bivariate categorical data, the question of interest is whether there is an association, and the null and and alternative hypotheses are

> H_0: The two variables are independent (or equivalently, there is no association between the two variables).
>
> H_a: The two variables are not independent (or equivalently, there is an association between the two variables).

To see how expected counts are obtained in this situation, recall from Chapter 5 that if two outcomes A and B are independent, then

$$P(A \text{ and } B) = P(A)P(B)$$

This means that the proportion of time that the two outcomes occur together in the long run is the product of the two individual long-run relative frequencies. Similarly, two categorical variables are independent in a population if, for each particular category of the first variable and each particular category of the second variable,

$$\begin{pmatrix} \text{proportion of individuals} \\ \text{in a particular category} \\ \text{combination} \end{pmatrix} = \begin{pmatrix} \text{proportion in} \\ \text{specified category of} \\ \text{first variable} \end{pmatrix} \begin{pmatrix} \text{proportion in} \\ \text{specified category} \\ \text{of second variable} \end{pmatrix}$$

Multiplying the right-hand side of this expression by the sample size gives the expected number of individuals in the sample who are in the specified category combination if the variables are independent. However, these expected counts cannot be calculated, because the individual population proportions are not known. The solution is to estimate each population proportion using the corresponding sample proportion:

$$\begin{pmatrix} \text{estimated expected number} \\ \text{in a partciular category} \\ \text{combination} \end{pmatrix} = (\text{sample size}) \begin{pmatrix} \dfrac{\begin{pmatrix} \text{observed number} \\ \text{in category of} \\ \text{first variable} \end{pmatrix}}{\text{sample size}} \end{pmatrix} \begin{pmatrix} \dfrac{\begin{pmatrix} \text{observed number} \\ \text{in category of} \\ \text{second variable} \end{pmatrix}}{\text{sample size}} \end{pmatrix}$$

$$= \dfrac{\begin{pmatrix} \text{observed number} \\ \text{in category of} \\ \text{first variable} \end{pmatrix} \begin{pmatrix} \text{observed number} \\ \text{in category of} \\ \text{second variable} \end{pmatrix}}{\text{sample size}}$$

Suppose that the observed counts are displayed in a rectangular table in which the rows correspond to the categories of the first variable and the columns to the categories of the second variable. The numerator in the preceding expression for expected counts is just the product of the row and column marginal totals. This is exactly how expected counts were calculated in the test for homogeneity of several populations, even though the reasoning that leads to the formula is different.

Chi-Square Test for Independence

Appropriate when the following conditions are met:

1. Observed counts are from a random sample or a sample that is selected in a way that would result in it being representative.
2. The sample size is large. The sample size is large enough for the chi-square test for independence to be appropriate if every expected count is at least 5. If some expected counts are less than 5, rows or columns of the table can be combined to achieve a table with satisfactory expected counts.

When these conditions are met, the following test statistic can be used:

$$X^2 = \sum_{\text{all cells}} \frac{(\text{observed count} - \text{expected count})^2}{\text{expected count}}$$

The expected cell counts are estimated from the sample data using the formula

$$\text{expected cell count} = \frac{(\text{row marginal total})(\text{column marginal total})}{\text{grand total}}$$

When these conditions are met and the null hypothesis is true, the X^2 statistic has a chi-square distribution with

$$df = (\text{number of rows} - 1)(\text{number of columns} - 1).$$

Hypotheses

H_0: The two variables are independent (or equivalently, there is no association between the two variables).

H_a: The two variables are not independent (or equivalently, there is an association between the two variables).

Associated *P*-value

The *P*-value associated with the calculated value of the test statistic is the area to the right of X^2 under the chi-square curve with

$$df = (\text{number of rows} - 1)(\text{number of columns} - 1)$$

Upper-tail areas for chi-square distributions can be found in Appendix A Table 5.

Example 14.9 A Pained Expression

The paper **"Facial Expression of Pain in Elderly Adults with Dementia"** (*Journal of Undergraduate Research* **[2006]**) examined the relationship between a nurse's assessment of a patient's facial expression and the patient's self-reported level of pain. Data for 89 patients are summarized in Table 14.6. Because patients with dementia do not always give a verbal indication that they are in pain, the authors of the paper were interested in determining if there is an association between a facial expression that reflects pain and self-reported pain.

TABLE 14.6 Observed Counts for Example 14.9

Facial Expression	Self-Report	
	No Pain	Pain
No Pain	17	40
Pain	3	29

H Hypotheses   The question of interest is whether there is an association between facial expression and self-reported pain. The hypotheses of interest are then

H_0: Facial expression and self-reported pain are independent.

H_a: Facial expression and self-reported pain are not independent.

M Method You should consider a chi-square test of independence because the answers to the four key questions are hypothesis testing, sample data, two categorical variables, and one sample.

The test statistic for the chi-square test of independence is

$$X^2 = \sum_{\text{all cells}} \frac{(\text{observed count} - \text{expected count})^2}{\text{expected count}}$$

The data table has two rows and two columns, so the appropriate df is

$$\text{df} = (\text{number of rows} - 1)(\text{number of columns} - 1) = (2-1)(2-1) = 1$$

A significance level of $\alpha = 0.05$ will be used for this test.

C Check Next you need to determine if this method is appropriate. To do this, the expected cell counts must be calculated.

	Cell	
Row	Column	Expected Cell Count
1	1	$\frac{(57)(20)}{89} = 12.81$
1	2	$\frac{(57)(69)}{89} = 44.19$
2	1	$\frac{(32)(20)}{89} = 7.19$
2	2	$\frac{(32)(69)}{89} = 24.81$

All of the expected counts are greater than 5. Although the participants in the study were not randomly selected, they were thought to be representative of the population of nursing home patients with dementia. The observed and expected counts (shown in parentheses) are given together in Table 14.7.

TABLE 14.7 Observed and Expected Counts for Example 14.9

	Self-Report	
Facial Expression	No Pain	Pain
No Pain	17 (12.81)	40 (44.19)
Pain	3 (7.19)	29 (24.81)

C Calculate

$$X^2 = \sum_{\text{all cells}} \frac{(\text{observed count} - \text{expected count})^2}{\text{expected count}}$$

$$= \frac{(17 - 12.81)^2}{12.81} + \cdots + \frac{(29 - 24.81)^2}{24.81} = 4.92$$

P-value: The *P*-value is the area to the right of 4.92 under a chi-square curve with df = 1. The entry closest to 4.92 in the 1-df column of Appendix A Table 5 is 5.02, so the approximate *P*-value for this test is *P*-value ≈ 0.025.

C Communicate Results Because the *P*-value is less than $\alpha = 0.05$, H_0 is rejected. There is convincing evidence of an association between a nurse's assessment of facial expression and self-reported pain.

Example 14.10 **Exercise and Sleep Quality**

The National Sleep Foundation asked each person in a representative sample of 1000 adult Americans about activity level and sleep quality (**"2013 Sleep in America Poll," February 20, 2013, www.sleepfoundation.org/sites/default/files/RPT336%20Summary%20of%20 Findings%2002%2020%202013.pdf, retrieved May 27, 2017**). Survey participants were classified into one of four activity levels (none, light, moderate, and vigorous). Each participant was also classified into one of two sleep categories. Data consistent with summary quantities given in the paper are given in Table 14.8. Expected cell counts (calculated under the assumption of no association between activity level and sleep quality) are also shown in Table 14.8.

TABLE 14.8 Observed and Expected Counts for Example 14.10

	Poor Sleep Quality	Good Sleep Quality
No activity	40 (21.96)	50 (68.04)
Light	116 (117.12)	364 (362.88)
Moderate	57 (61.00)	193 (189.00)
Vigorous	31 (43.92)	149 (136.08)

The Sleep Foundation was interested in using these sample data to determine whether there was an association between quality of sleep and activity level.

H Hypotheses   The hypotheses of interest are

H_0: Quality of sleep and activity level are independent.
H_a: Quality of sleep and activity level are not independent.

M Method   The answers to the four key questions are hypothesis testing, sample data, two categorical variables, and one sample. A chi-square test of independence will be considered. The test statistic for the chi-square test of independence is

$$X^2 = \sum_{\text{all cells}} \frac{(\text{observed count} - \text{expected count})^2}{\text{expected count}}$$

This table has four rows and two columns, so the appropriate df is

$$\text{df} = (\text{number of rows} - 1)(\text{number of columns} - 1) = (4 - 1)(2 - 1) = 3$$

A significance level of $\alpha = 0.01$ will be used for this test.

C Check   Next you need to check to see if this method is appropriate. Table 14.8 includes the calculated expected cell counts, all of which are greater than 5, so the large sample condition is satisfied. The sample was a representative sample of adult Americans, so it is appropriate to use the chi-square test of independence.

C Calculate   Minitab output follows. For each cell, the Minitab output includes the observed cell count, the expected cell count, and the value of $\frac{(\text{observed count} - \text{expected count})^2}{\text{expected count}}$ for that cell (its contribution to the X^2 statistic). From the output, $X^2 = 24.991$, df $= 3$ and the associated P-value $= 0.000$.

Chi-Square Test for Association: Activity Level, Sleep Quality

Rows: Activity Level	Columns: Sleep Quality		
	Poor Sleep Quality	Good Sleep Quality	All
None	40 21.96	50 68.04	90
Light	116 117.12	364 362.88	480
Moderate	57 61.00	193 189.00	250
Vigorous	31 43.92	149 136.08	180
All	244	756	1000
Cell Contents:	Count		
	Expected count		

Pearson Chi-Square = 24.991, DF = 3, P-Value = 0.000

C Communicate Results   Because the P-value is less than $\alpha = 0.01$, H_0 is rejected at the 0.01 significance level. There is convincing evidence that an association exists between quality of sleep and activity level.

Summing It Up—Section 14.2

The following learning objectives were addressed in this section:

Conceptual Understanding

C2: Understand that the way in which data summarized in a two-way table were collected determines which chi-square test (independence or homogeneity) is appropriate.

Two-way frequency tables are used to summarize categorical data in two situations. Sometimes independent samples are selected from two or more groups and data on one categorical variable are obtained for each of these samples. This results in a two-way table where the rows represent the different populations or treatments and the columns represent the different categories of the categorical variable. In this situation, the question of interest is whether the category proportions are the same for the different groups and the appropriate test is a chi-square test for homogeneity.

Another way two-way tables are used is to summarize data from a single sample where data on two categorical variables are obtained for the individuals in the sample. In this case, the rows of the table represent the categories of one of the variables and the columns represent the categories of the second variable. The question of interest is whether the two categorical variables are independent (or equivalently, there is no association between the two variables) and the appropriate test is the chi-square test for independence.

Mastering the Mechanics

M1: Determine which chi-square test (goodness-of-fit, independence, or homogeneity) is appropriate in a given situation.

A chi-square test for homogeneity is appropriate when you want to determine if the distributions of responses on a categorical variable are the same for more than one population or treatment. This test uses data on a single categorical variable that have been collected for independent samples from the populations or treatment of interest.

A chi-square test for independence is appropriate when you want to determine if there is an association between two categorical variables. This test uses data on two categorical variables that have been collected from a single sample from the population of interest.

M2: Determine appropriate null and alternative hypotheses for chi-square tests.

For a chi-square test for homogeneity, the null hypothesis specifies that the category proportions are the same for all of the groups that are being compared. The alternative hypothesis is that the category proportions are not all the same for all of the groups.

For a chi-square test for independence, the null hypothesis specifies that there is no association between two categorical variables. This is equivalent to saying that the two categorical variables are independent. The alternative hypothesis is that there is an association (the variables are not independent).

M4: Know the conditions necessary for the chi-square tests of independence or homogeneity to be appropriate.

There are two conditions that are necessary in order for a chi-square test for homogeneity to be appropriate: (1) The observed counts are based on independently selected random (or representative) samples from the populations or subjects were randomly assigned to treatment groups in the case of an experiment, and (2) the sample sizes are large. The sample sizes are considered to be large if all of the expected counts are greater than or equal to 5.

There are also two conditions that are necessary in order for a chi-square test for independence to be appropriate: (1) The observed counts are based on a random (or representative) sample from the population of interest, and (2) the sample size is large. The sample size is considered to be large if all of the expected counts are greater than or equal to 5.

M5: Calculate the value of the test statistic and find the associated *P*-value for chi-square tests.

The chi-square test statistic is

$$X^2 = \sum_{all\ categories} \frac{(\text{observed count} - \text{expected count})^2}{\text{expected count}}$$

When the conditions for the chi-square test of homogeneity or the chi-square test for independence are met and the null hypothesis is true, this statistic has a chi-square distribution with degrees of freedom based on the number of rows and the number of columns in the two-way table used to summarize the data. The degrees of freedom are calculated as df = (number of rows − 1)(number of columns − 1). The *P*-value for a chi-square test of homogeneity or the chi-square test for independence is the area under the appropriate chi-square curve and to the right of the calculated value of the test statistic. The *P*-value can be determined using Appendix A Table 5 or by using statistical software or a graphing calculator.

Putting It into Practice

P2: Carry out a chi-square test of homogeneity and interpret the result in context.

For examples of a chi-square test for homogeneity and the interpretation of the results in context, see Examples 14.7 and 14.8.

P3: Carry out a chi-square test of independence and interpret the result in context.

For examples of a chi-square test for independence and the interpretation of the results in context, see Examples 14.9 and 14.10.

SECTION 14.2 EXERCISES

Each Exercise Set assesses the following chapter learning objectives: C2, M1, M2, M4, M5, P2, P3.

SECTION 14.2 Exercise Set 1

14.19 Some colleges now allow students to pay their tuition using a credit card. The report **"Credit Card Tuition Payment Survey 2014" (www.creditcards.com/credit-card -news/tuition-charge-fee-survey.php, retrieved May 27, 2017)** includes data from a survey of 100 public four-year colleges, 100 private four-year colleges, and 100 community colleges. The accompanying table gives information on credit card acceptance for each of these samples of colleges. For purposes of this exercise, suppose that these three samples are representative of the populations of public four-year colleges, private four-year colleges, and community colleges in the United States. Is there convincing evidence that the proportions in each of the two credit card categories are not the same for all three types of colleges? Test the relevant hypotheses using a 0.05 significance level. (Hint: See Example 14.7.)

	Accepts Credit Cards for Tuition Payment	Does Not Accept Credit Cards for Tuition Payment
Public Four-Year Colleges	92	8
Private Four-Year Colleges	68	32
Community Colleges	100	0

14.20 The Knight Foundation asked each person in a representative sample of high school students and in a representative sample of high school teachers which of the rights guaranteed by the First Amendment they thought was the most important **("Future of the First Amendment 2014 Survey of High School Students and Teachers," www.knightfoundation.org/media /uploads/publication_pdfs/Future_of_the_First_Amendment _cx2.pdf, retrieved May 27, 2017).** Suppose that the sample size for each sample was 1000. Data consistent with summary values given in the paper are summarized in the accompanying table.

Table for exercise 14.20

	Most Important First Amendment Right				
	Freedom of Speech	Freedom of the Press	Freedom of Religion	Freedom to Peacefully Assemble	Freedom to Petition the Government
Students	650	30	250	20	50
Teachers	400	60	420	50	70

a. Carry out a hypothesis test to determine if there is convincing evidence that the proportions falling into the five First Amendment rights categories are not the same for teachers and students. Use a significance level of $\alpha = 0.01$.

b. Based on your test in Part (a) and a comparison of observed and expected cell counts, write a brief description of how teachers and students differ with respect to what they view as the most important of the First Amendment rights.

14.21 The authors of the paper **"The Relationship of Field of Study to Current Smoking Status Among College Students"** (*College Student Journal* [2009]: 744–754) carried out a study to investigate if smoking rates were different for college students in different majors. Each student in a large random sample of students at the University of Minnesota was classified according to field of study and whether or not they had smoked in the past 30 days. The data are given in the accompanying table.

Field of Study	Smoked in the Last 30 Days	Did Not Smoke in Last 30 Days
1. Undeclared	176	489
2. Art, design, performing arts	149	336
3. Humanities	197	454
4. Communication, languages	233	389
5. Education	56	170
6. Health sciences	227	717
7. Mathematics, engineering, sciences	245	924
8. Social science, human services	306	593
9. Individualized course of study	134	260

a. Is there evidence that field of study and smoking status are not independent? Use the accompanying Minitab output to test the relevant hypotheses using $\alpha = 0.01$. (Hint: See Example 14.10.)

Chi-Square Test: Smoked, Did Not Smoke
Expected counts are printed below observed counts
Chi-Square contributions are printed below expected counts

	Smoked	Did Not Smoke	Total
1	176	489	665
	189.23	475.77	
	0.925	0.368	
2	149	336	485
	138.01	346.99	
	0.875	0.348	
3	197	454	651
	185.25	465.75	
	0.746	0.297	
4	233	389	622
	177.00	445.00	
	17.721	7.048	
5	56	170	226
	64.31	161.69	
	1.074	0.427	

(continued)

	Smoked	Did Not Smoke	Total
6	227	717	944
	268.62	675.38	
	6.449	2.565	
7	245	924	1169
	332.65	836.35	
	23.094	9.185	
8	306	593	899
	255.82	643.18	
	9.844	3.915	
9	134	260	394
	112.12	281.88	
	4.272	1.699	
Total	1723	4332	6055

Chi-Sq = 90.853, DF = 8, P-Value = 0.000

b. Write a few sentences describing how smoking status is related to field of study. (Hint: Focus on cells that have large values of

$$\frac{(\text{observed cell count} - \text{expected cell count})^2}{\text{expeted cell count}}$$

14.22 The paper **"Credit Card Misuse, Money Attitudes, and Compulsive Buying Behavior: Comparison of Internal and External Locus of Control Consumers"** (*College Student Journal* [2009]: 268–275) describes a survey of college students at two midwestern public universities. Based on the survey responses, students were classified into two "locus of control" groups (internal and external) based on whether they believe that they control what happens to them. Those in the internal locus of control group believe that they are usually in control of what happens to them, whereas those in the external locus of control group believe that factors outside their control usually determine what happens to them. Each student was also classified according to a measure of compulsive buying. The resulting data are summarized in the accompanying table. Can the researchers conclude that there is an association between locus of control and compulsive buying behavior? Carry out a test using $\alpha = 0.01$. Assume it is reasonable to regard the sample as representative of college students at midwestern public universities. (Hint: See Example 14.9.)

		Locus of Control	
		Internal	External
Compulsive Buyer?	Yes	3	14
	No	52	57

SECTION 14.2 Exercise Set 2

14.23 The data in the accompanying table are from the paper **"Gender Differences in Food Selections of Students at a Historically Black College and University"** (*College Student Journal* [2009]: 800–806). Suppose that each person in a random sample of 48 male students and in a random sample of 91 female students at a particular college was classified according to gender and whether they usually or rarely eat three meals a day.

	Usually Eat 3 Meals a Day	Rarely Eat 3 Meals a Day
Male	26	22
Female	37	54

a. Is there evidence that the proportions who would fall into each of the two response categories are not the same for males and females? Use the X^2 statistic to test the relevant hypotheses with a significance level of $\alpha = 0.05$.

b. Are your calculations and conclusions from Part (a) consistent with the accompanying Minitab output? Explain.

Expected counts are printed below observed counts
Chi-Square contributions are printed below expected counts

	Usually	Rarely	Total
Male	26	22	48
	21.76	26.24	
	0.828	0.686	
Female	37	54	91
	41.24	49.76	
	0.437	0.362	
Total	63	76	139

Chi-Sq = 2.314, DF = 1, P-Value = 0.128

c. Because the response variable in this exercise has only two categories (usually and rarely), you could have also answered the question posed by carrying out a large-sample z test of $H_0: p_1 - p_2 = 0$ versus $H_a: p_1 - p_2 \neq 0$, where p_1 is the proportion who usually eat three meals a day for males and p_2 is the proportion who usually eat three meals a day for females. Minitab output from the large-sample z test is shown. Using a significance level of $\alpha = 0.05$, does the large-sample z test lead to the same conclusion as in Part (a)?

Test for Two Proportions

Sample	X	N	Sample p
Male	26	48	0.541667
Female	37	91	0.406593

Difference = p (1) − p (2)
Test for difference = 0 (vs not = 0): Z = 1.53 P-Value = 0.128

d. How do the *P*-values from the tests in Parts (a) and (c) compare? Does this surprise you? Explain.

14.24 The Knight Foundation investigated whether high school students agreed with the statement that people should be allowed to burn or deface the American flag as a political statement. This question was asked in a survey of a representative sample of high school students in 2004 and also in a survey of a representative sample of high school students in 2014 (**"Future of the First Amendment**

2014 Survey of High School Students and Teachers," www .knightfoundation.org/media/uploads/publication_pdfs /Future_of_the_First_Amendment_cx2.pdf, retrieved May 27, 2017). Suppose that the sample size was 1000 in each of the two years. Data consistent with summary values given in the paper are summarized in the accompanying table. Is there convincing evidence that the proportions falling into each of the response categories were not the same for high school students in 2004 and 2014?

People should be allowed to burn or deface the American flag as a political statement					
	Strongly Agree	Mildly Agree	Mildly Disagree	Strongly Disagree	Don't know
2004	80	80	110	630	100
2014	70	70	110	660	90

14.25 The paper **"Contemporary College Students and Body Piercing"** (*Journal of Adolescent Health* [2004]: 58–61) described a survey of 490 undergraduate students at a state university in the southwestern region of the United States. Each student in the sample was classified according to class standing (freshman, sophomore, junior, or senior) and body art category (body piercings only, tattoos only, both tattoos and body piercings, no body art). Use the data in the accompanying table to determine if there is evidence of an association between class standing and body art category. Assume that it is reasonable to regard the sample as representative of the students at this university. Use $\alpha = 0.01$.

	Body Piercings Only	Tattoos Only	Both Body Piercing and Tattoos	No Body Art
Freshman	61	7	14	86
Sophomore	43	11	10	64
Junior	20	9	7	43
Senior	21	17	23	54

14.26 Each person in a large sample of German adolescents was asked to indicate which of 50 popular movies he or she had seen in the past year. Based on the response, the amount of time (in minutes) of alcohol use contained in the movies the person had watched was estimated. Each person was then classified into one of four groups based on the amount of movie alcohol exposure (groups 1, 2, 3, and 4, with 1 being the lowest exposure and 4 being the highest exposure). Each person was also classified according to school performance. The resulting data are given in the table on the next page (from **"Longitudinal Study of Exposure to Entertainment Media and Alcohol Use Among German Adolescents," *Pediatrics* [2009]: 989–995**). For purposes of this exercise, assume this sample is a random sample of German adolescents. Is there evidence of an association between school performance and movie exposure to alcohol? Carry out a hypothesis test using $\alpha = 0.05$.

		Alcohol Exposure Group			
		1	2	3	4
School Performance	Excellent	110	93	49	65
	Good	328	325	316	295
	Average/Poor	239	259	312	317

ADDITIONAL EXERCISES

14.27 Give an example of a situation where it would be appropriate to use a chi-square test of homogeneity. Describe the populations that would be sampled and the variable that would be recorded.

14.28 Give an example of a situation where it would be appropriate to use a chi-square test of independence. Describe the population that would be sampled and the two variables that would be recorded.

14.29 Explain the difference between situations that would lead to a chi-square goodness-of-fit test and those that would lead to a chi-square test of homogeneity.

14.30 Explain the difference between situations that would lead to a chi-square test for homogeneity and those that would lead to a chi-square test for independence.

14.31 Are babies born to mothers who use assistive reproduction technology (ART) more likely to be born prematurely than babies conceived naturally? The data in the accompanying table are from the paper **"Child Growth from Birth to 18 Months After Assisted Reproduction Technology"** (*International Journal of Nursing* [2010]: 1159–1166). The data are from a random sample of 19,614 births. Each birth was classified according to whether the mother used ART and whether the baby was premature. Use these data to decide if there is convincing evidence of an association between the use of ART and whether or not a baby is premature. Use $\alpha = 0.01$.

	Conceived Using ART	Conceived Naturally
Premature	154	2,405
Not Premature	212	18,843

14.32 The report **"Smartphone Ownership and Internet Usage Continues to Climb in Emerging Economies" (Pew Research Center, February 22, 2016, www.pewglobal .org/2016/02/22/smartphone-ownership-and-internet -usage-continues-to-climb-in-emerging-economies/, retrieved May 27, 2017)** provided the following information on smartphone ownership for representative samples of adults in several different countries.

Country	Percent Who Own a Smartphone	Percent Who Do Not Own a Smartphone
United States	72%	28%
Spain	71%	29%
Italy	60%	40%
India	17%	83%

a. Suppose the sample sizes were 1000 for the United States and for India and 500 for Spain and Italy. Complete the following two-way table by entering the observed counts.

Country	Own a Smartphone	Do Not Own a Smartphone
United States		
Spain		
Italy		
India		

b. Carry out a hypothesis test to determine if there is convincing evidence that the smartphone ownership proportions are not the same for all four countries. Use a significance level of $\alpha = 0.05$.

14.33 Each person in a representative sample of 445 college students age 18 to 24 was classified according to age and to the response to the following question: "How often have you used a credit card to buy items knowing you wouldn't have money to pay the bill when it arrived?" Possible responses were never, rarely, sometimes, or frequently **("Majoring in Money: How American College Students Manage Their Finances," June 28, 2016, salliemae.newshq.businesswire.com /sites/salliemae.newshq.businesswire.com/files/doc_library /file/SallieMae_MajoringinMoney_2016.pdf, retrieved May 27, 2017)**. The responses are summarized in the accompanying table. Do these data provide evidence that there is an association between age group and the response to the question? Test the relevant hypotheses using $\alpha = 0.01$.

	Age 18 to 20	Age 21 to 22	Age 23 to 24
Never	72	62	29
Rarely	36	34	32
Sometimes	30	42	40
Frequently	12	24	32

14.34 Does viewing angle affect a person's ability to tell the difference between a female nose and a male nose? This important (?) research question was examined in the article **"You Can Tell by the Nose: Judging Sex from an Isolated Facial Feature"** (*Perception* [1995]: 969–973). Eight Caucasian males and eight Caucasian females posed for nose photos. The article states that none of the volunteers wore nose studs or had prominent nasal hair. Each person placed a black Lycra tube over his or her head in such a way that only the nose protruded through a hole in the material. Photos were then taken from three different angles: front view, three-quarter view, and profile. These photos were shown to a sample of undergraduate students. Each student in the sample was shown one of the nose photos and asked whether it was a photo of a male or a female. The response was classified as either correct or incorrect. The accompanying table was constructed using summary values reported in the article. Is there evidence that the proportion of correct sex identifications differs for the three different nose views?

	View		
Sex ID	Front	Profile	Three-Quarter
Correct	23	26	29
Incorrect	17	14	11

14.35 The report **"Majoring in Money: How American College Students Manage Their Finances" (June 28, 2016, salliemae.newshq.businesswire.com/sites/salliemae.newshq.businesswire.com/files/doc_library/file/SallieMae_MajoringinMoney_2016.pdf, retrieved May 27, 2017)** included data from a study in which 792 people in a representative sample of college students age 18 to 24 were asked how they perceive their money management skills. Possible responses were excellent, good, average, not very good, and poor. Each student in the sample was also classified by sex, resulting in the data in the accompanying table. Is there convincing evidence that there is an association between sex and how students perceive their money management skills? Test the relevant hypotheses using a significance level of $\alpha = 0.05$.

	Perception of Money Management Skills				
	Excellent	Good	Average	Not Very Good	Poor
Male	103	139	89	17	8
Female	83	192	135	22	4

14.36 The report referenced in the previous exercise also provided data on perception of money management skills by age group. Use the data from 788 people in the accompanying table to determine if there is evidence of an association between age and perception of money management skills. Because some of the expected values are less than 5, construct a new table that combines the not very good and the poor categories. Use a significance level of $\alpha = 0.05$.

	Perception of Money Management Skills				
	Excellent	Good	Average	Not Very Good	Poor
Age 18 to 20	73	146	104	17	8
Age 21 to 22	54	103	85	10	3
Age 23 to 24	58	84	36	7	0

14.37 The authors of the paper **"Risk of Malnutrition Is an Independent Predictor of Mortality, Length of Hospital Stay, and Hospitalization Costs in Stroke Patients" (*Journal of Stroke and Cerebrovascular Diseases* [2016]: 799–806)** describe a sample of patients admitted to a hospital after suffering a stroke. Each of 537 patients was classified according to a measure of risk of malnutrition (with possible categories low, medium, and high) and whether or not the patient was alive at 6 months following the stroke. The authors concluded that there was an association between survival and the risk of malnutrition. Do you agree? Support your answer with evidence based on that the given data. For purposes of this exercise, you may assume that the sample of 537 patients is representative of stroke patients.

	Survived	Did Not Survive
Low Risk of Malnutrition	322	20
Medium Risk of Malnutrition	29	10
High Risk of Malnutrition	91	65

14.38 Jail inmates can be classified into one of the following four categories according to the type of crime committed: violent crime, crime against property, drug crime, and public-order offenses. Suppose that random samples of 500 male inmates and 500 female inmates are selected, and each inmate is classified according to type of crime. The data in the accompanying table are based on summary values given in the article **"Profile of Jail Inmates" (*USA TODAY*, April 25, 1991)**. You would like to know whether male and female inmates differ with respect to crime type proportions.

	Gender	
Type of Crime	Male	Female
Violent	117	66
Property	150	160
Drug	109	168
Public–Order	124	106

a. Is this a test of homogeneity or a test of independence?
b. Test the relevant hypotheses using a significance level of $\alpha = 0.05$.

SECTION 14.3 Avoid These Common Mistakes

Keep the following in mind when analyzing categorical data using one of the chi-square tests presented in this chapter:

1. Don't confuse tests for homogeneity with tests for independence. The hypotheses and conclusions are different for the two types of test. Tests for homogeneity are used when the individuals in each of two or more independent samples are classified according to a single categorical variable. Tests for independence are used when individuals in a *single* sample are classified according to two categorical variables.

2. As was the case for the hypothesis tests in earlier chapters, remember that you can never say you have strong support for the null hypothesis. For example, if you do not reject the null hypothesis in a chi-square test for independence, you cannot conclude that there is convincing evidence that the variables are independent. You can only say that you were not convinced that there is an association between the variables.

3. Be sure that the conditions for the chi-square test are met. *P*-values based on the chi-square distribution are only approximate, and if the large sample condition is not met, the actual *P*-value may be quite different from the approximate one based on the chi-square distribution. This can lead to incorrect conclusions. Also, for the chi-square test of homogeneity, the assumption of *independent* samples is particularly important.

4. Don't jump to conclusions about causation. Just as a strong correlation between two numerical variables does not mean that there is a cause-and-effect relationship between them, an association between two categorical variables does not imply a causal relationship.

CHAPTER ACTIVITIES

ACTIVITY 14.1 PICK A NUMBER, ANY NUMBER ...

Background: There is evidence to suggest that people are not very good random number generators. In this activity, you will investigate this phenomenon by collecting and analyzing a set of human-generated "random" digits.

For this activity, work in a group with four or five other students.

1. Each member of your group should ask 25 different people to pick a digit from 0 to 9 at random and record the responses.

2. Combine the responses you collected with those of the other group members to form a single sample. Summarize the resulting data in a one-way frequency table.

3. If people are good at picking digits at random, what would you expect for the proportion of the responses in the sample that are 0? That are 1?

4. State a null hypothesis and an alternative hypothesis that could be tested to determine whether there is evidence that the 10 digits from 0 to 9 are not selected an equal proportion of the time when people are asked to pick a digit at random.

5. Carry out the appropriate hypothesis test, and write a few sentences indicating whether or not the data support the theory that people are not good random number generators.

ARE YOU READY TO MOVE ON? CHAPTER 14 REVIEW EXERCISES

All chapter learning objectives are assessed in these exercises. The learning objectives assessed in each exercise are given in parentheses.

14.39 (M1, M2, M3, M5)
The authors of the paper **"Racial Stereotypes in Children's Television Commercials"** (*Journal of Advertising Research* **[2008]: 80–93)** counted the number of times that characters of different ethnicities appeared in commercials aired on Philadelphia television stations, resulting in the data in the accompanying table.

	African American	Asian	Caucasian	Hispanic
Frequency	57	11	330	6

Based on the 2000 Census, the proportion of the U.S. population falling into each of these four ethnic groups are 0.177 for African American, 0.032 for Asian, 0.734 for Caucasian, and 0.057 for Hispanic. Do these data provide sufficient evidence to conclude that the proportions appearing in commercials are not the same as the census proportions? Test the relevant hypotheses using a significance level of 0.01.

14.40 (C1, P1)
The report **"Fatality Facts 2004: Bicyclists 2015"** (Insurance Institute, 2015, www.iihs.org/iihs/topics/t/pedestrians-and

-bicyclists/fatalityfacts/bicycles, retrieved May 27, 2017) included the following table classifying 814 fatal bicycle accidents that occurred in 2015 according to the time of day the accident occurred.

Time of Day	Number of Accidents
Midnight to 3 A.M.	46
3 A.M. to 6 A.M.	50
6 A.M. to 9 A.M.	87
9 A.M. to Noon	70
Noon to 3 P.M.	76
3 P.M. to 6 P.M.	158
6 P.M. to 9 P.M.	192
9 P.M. to Midnight	135

For purposes of this exercise, assume that these 814 bicycle accidents are a random sample of fatal bicycle accidents. Do these data support the hypothesis that fatal bicycle accidents are not equally likely to occur in each of the 3-hour time periods used to construct the table? Test the relevant hypotheses using a significance level of $\alpha = 0.05$.

14.41 (C1, P1)
Suppose a safety officer proposes that bicycle fatalities are twice as likely to occur between noon and midnight as during midnight to noon and suggests the following hypothesis: $H_0: p_1 = \frac{1}{3}, p_2 = \frac{2}{3}$, where p_1 is the proportion of accidents occurring between midnight and noon and p_2 is the proportion occurring between noon and midnight. Do the data given in the previous exercise provide evidence against this hypothesis, or are the data compatible with it? Justify your answer with an appropriate test.

14.42 (C1, P1)
The report referenced in the previous two exercises also classified 817 fatal bicycle accidents according to the month in which the accidents occurred, resulting in the accompanying table.

Month	Number of Accidents
January	47
February	47
March	40
April	61
May	66
June	92
July	94
August	78
September	89
October	85
November	66
December	52

a. To determine if some months are riskier than others, use the given data to test the null hypothesis H_0: $p_1 = \frac{1}{12}, p_2 = \frac{1}{12}, \ldots, p_{12} = \frac{1}{12}$, where p_1 is the proportion of fatal bicycle accidents that occur in January, p_2 is the proportion for February, and so on. Use a significance level of $\alpha = 0.01$.

b. The null hypothesis in Part (a) specifies that fatal accidents are equally likely to occur in any of the 12 months. But not all months have the same number of days. What null and alternative hypotheses would you test if you wanted to take differing month lengths into account?

c. Test the hypotheses proposed in Part (b) using a 0.05 significance level.

14.43 (M2, M4, M5)
In a study of high-achieving high school graduates, the authors of the report **"High-Achieving Seniors and the College Decision" (Lipman Hearne, October 2009)** surveyed 828 high school graduates who were considered "academic superstars" and 433 graduates who were considered "solid performers." One question on the survey asked the distance from their home to the college they attended. Assuming these two samples are random samples of academic superstars and solid performers nationwide, use the accompanying data to determine if it is reasonable to conclude that the distribution of responses over the distance from home categories is not the same for academic superstars and solid performers. Use $\alpha = 0.05$.

	Distance of College from Home (in miles)				
Student Group	Less than 40	40 to 99	100 to 199	200 to 399	400 or More
Academic Superstars	157	157	141	149	224
Solid Performers	104	95	82	65	87

14.44 (C2, P2)
The accompanying data on degree of spirituality for a random sample of natural scientists and a random sample of social scientists working at research universities appeared in the paper **"Conflict Between Religion and Science Among Academic Scientists" (Journal for the Scientific Study of Religion [2009]: 276–292).** Is there evidence that the spirituality category proportions are not the same for natural and social scientists? Test the relevant hypotheses using a significance level of $\alpha = 0.01$.

	Degree of Spirituality			
	Very	Moderate	Slightly	Not at All
Natural Scientists	56	162	198	211
Social Scientists	56	223	243	239

14.45　(M2, M4, M5)

The authors of the paper **"Movie Character Smoking and Adolescent Smoking: Who Matters More, Good Guys or Bad Guys?" (Pediatrics [2009]: 135–141)** studied characters who were depicted smoking in movies released between 2000 and 2005. The smoking characters were classified according to sex and whether the character type was positive, negative, or neutral. The resulting data are given in the accompanying table. Assume that this sample is representative of smoking movie characters. Do these data provide evidence of an association between sex and character type for movie characters who smoke? Use $\alpha = 0.05$.

Sex	Character Type		
	Positive	Negative	Neutral
Male	255	106	130
Female	85	12	49

14.46　(M2, M4, M5)

The report **"Education Pays 2016" (The College Board, trends.collegeboard.org/sites/default/files/education-pays -2016-full-report.pdf, retrieved May 27, 2017)** provided information on education level and earnings for a sample of adult Americans who are employed full-time. Data consistent with summary percentages given in the report are summarized in the accompanying table. Suppose this data resulted from a representative sample of 1001 working adults whose highest level of education was either a high school diploma, an Associate degree, or a Bachelor's degree. Each person in the sample was classified according to education level (high school diploma, Associate degree, or Bachelor's degree) and yearly income (with possible categories of less than \$20,000, \$20,000 to \$39,999, \$40,000 to \$59,999, and \$60,000 or more). Is there evidence of an association between income category and education level? Test the appropriate hypotheses using a 0.05 significance level.

	Income Level			
	Less than \$20,000	\$20,000 to \$39,999	\$40,000 to \$59,999	\$60,000 or more
High School Diploma	8	68	106	243
Associate Degree	11	56	56	63
Bachelor's Degree	47	160	101	82

14.47　(C2, P3)

The report **"Consumer Revolving Credit and Debt Over the Life Cycle and Business Cycle"** describes a study conducted by the Federal Reserve Bank of Boston **(www.bostonfed. org, October 2015, retrieved May 27, 2017)**. Data consistent with summary values given in the report are summarized

in the accompanying table. Suppose that this data resulted from a random sample of 800 adult Americans age 20 to 39 years old who have at least one credit card. Each person in the sample was classified according to age (with possible categories of 20 to 24 years, 25 to 29 years, 30 to 34 years, and 35 to 39 years). The people in the sample were also classified according to whether or not they pay the full balance on their credit cards each month or sometimes or always carry over a balance from month to month.

	Pay Full Balance Each Month	Carry Balance from Month to Month
Age 20 to 24 years	75	95
Age 25 to 29 years	74	126
Age 30 to 34 years	75	145
Age 35 to 39 years	67	143

a. To investigate whether or not people pay their balance in full each month is related to age, which chi-square test (homogeneity or independence) would be the appropriate test? Explain your choice.

b. Carry out an appropriate test to determine if these data provide convincing evidence that whether or not people pay their balance in full each month is related to age.

c. To what population would it be reasonable to generalize the conclusion from the test in Part (b)?

14.48　(C2, P2)

The following passage is from the paper **"Gender Differences in Food Selections of Students at a Historically Black College and University" (College Student Journal [2009]: 800–806)**:

> Also significant was the proportion of males and their water consumption (8 oz. servings) compared to females ($X^2 = 8.166$, $P = .086$). Males came closest to meeting recommended daily water intake (64 oz. or more) than females (29.8% vs. 20.9%).

This statement was based on carrying out a chi-square test of homogeneity using data in a two-way table where rows corresponded to sex (male, female) and columns corresponded to number of servings of water consumed per day, with categories none, one, two to three, four to five, and six or more.

a. What hypotheses did the researchers test? What is the number of degrees of freedom associated with the reported value of the X^2 statistic?

b. The researchers based their statement on a test with a significance level of 0.10. Would they have reached the same conclusion if a significance level of 0.05 had been used? Explain.

14.49　(C2, P2)

The paper referenced in the previous exercise also included the accompanying data on how often students said they had consumed fried potatoes (French fries or potato chips) in the past week.

		Number of Times Consumed Fried Potatoes in the Past Week					
		0	1 to 3	4 to 6	7 to 13	14 to 20	21 or more
Sex	Male	2	10	15	12	6	3
	Female	15	15	10	20	19	12

Use the accompanying Minitab output to carry out a chi-square test of homogeneity. Do you agree with the authors' conclusion that there was a significant difference in consumption of fried potatoes for males and females? Explain.

Expected counts are printed below observed counts

Chi-Square contributions are printed below expected counts

	0	1-3	4-6	7-13	14-20	21 or more	Total
M	2	10	15	12	6	3	48
	5.87	8.63	8.63	11.05	8.63	5.18	
	2.552	0.216	4.696	0.082	0.803	0.917	
F	15	15	10	20	19	12	91
	11.13	16.37	16.37	20.95	16.37	9.82	
	1.346	0.114	2.477	0.043	0.424	0.484	
Total	17	25	25	32	25	15	139

Chi-Sq = 14.153, DF = 5, P-Value = 0.015

14.50 (C2, P3)

The press release titled **"Nap Time"** (July 2009, pewresearch.org, retrieved May 27, 2017) described results from a nationally representative survey of 1488 adult Americans. The survey asked several demographic questions (such as sex, age, and income) and also included a question asking respondents if they had taken a nap in the past 24 hours. The press release stated that 38% of the men surveyed and 31% of the women surveyed reported that they had napped in the past 24 hours. For purposes of this exercise, suppose that men and women were equally represented in the sample.

a. Use the given information to fill in observed cell counts for the following table:

	Napped	Did Not Nap	Row Total
Men			744
Women			744

b. Use the data in the table from Part (a) to carry out a hypothesis test to determine if there is an association between sex and napping.

c. The press release states that more men than women nap. Although this is true for the people in the sample, based on the result of your test in Part (b), is it reasonable to conclude that this holds for adult Americans in general? Explain.

TECHNOLOGY NOTES

X² Goodness-of-Fit Test

JMP

1. Enter each category name into the first column
2. Enter the count for each category into the second column

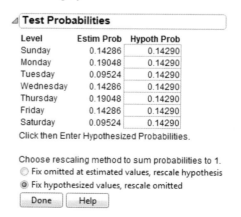

3. Click **Analyze** then select **Distribution**
4. Click and drag the name for the first column from the box under **Select Columns** to the box next to **Y, Columns**
5. Click and drag the name for the second column from the box under **Select Columns** to the box next to **Freq**
6. Click **OK**

7. Click the red arrow next to the column name and select **Test Probabilities**
8. Under **Test Probabilities** input the hypothesized probabilities for each category

Test Probabilities		
Level	Estim Prob	Hypoth Prob
Sunday	0.14286	0.14290
Monday	0.19048	0.14290
Tuesday	0.09524	0.14290
Wednesday	0.14286	0.14290
Thursday	0.19048	0.14290
Friday	0.14286	0.14290
Saturday	0.09524	0.14290

Click then Enter Hypothesized Probabilities.

Choose rescaling method to sum probabilities to 1.
○ Fix omitted at estimated values, rescale hypothesis
◉ Fix hypothesized values, rescale omitted

Done   Help

9. Click **Done**

Note: The test statistic and P-value for the chi-squared test will appear in the row called **Pearson**.

Minitab

MINITAB Student Version 14 does not have the functionality to produce a x^2 goodness-of-fit test.

SPSS

1. Input the observed data into one column
2. Click **Analyze** then click **Nonparametric Tests** then click **One Sample...**
3. Click the **Settings** tab
4. Select the radio button next to **Customize tests**
5. Check the box next to **Compare observed probabilities to hypothesized (Chi-Square test)**
6. Click the **Options...** button
7. Select the appropriate option (to test with equal probabilities for each category or to input expected probabilities manually)
8. Once you have selected the appropriate option and input expected probabilities if necessary, click **OK**
9. Click **Run**

Excel

Excel does not have the functionality to produce the χ^2 goodness-of-fit test automatically. However, you can use Excel to find the *P*-value for this test once you have found the value of the test statistic by using the following steps.

1. Click on an empty cell
2. Click **Formulas**
3. Click **Insert Function**
4. Select **Statistical** from the drop-down box for category
5. Select **CHIDIST** and press **OK**
6. In the box next to **X**, type the value of the test statistic
7. In the box next to **Deg_freedom** type the value for the degrees of freedom
8. Click **OK**

Note: This outputs the value for $P(X \geq x)$.

TI-83/84

1. Enter the observed cell counts into **L1** and the expected cell counts into **L2** (in order to access lists press the **STAT** key, highlight the option called **Edit...** then press **ENTER**)
2. Press **STAT**
3. Highlight **TESTS**
4. Highlight **X2GOF-Test** and press **ENTER**
5. Next to Observed enter **L1**
6. Next to Expected enter **L2**
7. Next to df enter the appropriate df (this will be the number of categories - 1)
8. Highlight **Calculate** and press **ENTER**

TI-Nspire

1. Enter the observed data into a data list (In order to access data lists select the spreadsheet option and press **enter**)

   Note: Be sure to title the lists by selecting the top row of the column and typing a title.

2. Enter the expected data into a data list
3. Press **menu** and select **4:Statistics** then **4:Stat Tests** then **7: χ^2 GOF...** and press **enter**

4. For **Observed List** select the title of the list that contains the observed data from the drop-down menu
5. For **Expected List** select the title of the list that contains the expected data from the drop-down menu
6. For **Deg of Freedom** input the degrees of freedom for this test
7. Press **OK**

X^2 Tests for Independence and Homogeneity

JMP

1. Input the data table into the JMP data table
2. Click **Analyze** then select **Fit Y by X**
3. Click and drag the first column name from the box under **Select Columns** to the box next to **Y, Response**

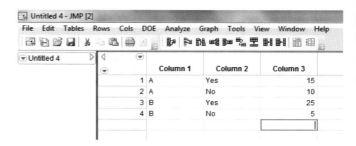

4. Click and drag the second column name from the box under **Select Columns** to the box next to **X, Factor**
5. Click and drag the third column name from the box under **Select Columns** to the box next to **Freq**
6. Click **OK**

Minitab

1. Input the data table into MINITAB

	C1-T	C2	C3	C4	C5
		Male	Female		
1	Small	145	165		
2	Medium	120	145		
3	Large	300	75		
4	XL	45	6		
5					
6					

2. Click **Stat** then **Tables** then **Chi-Square Test (Table in Worksheet)**
3. Select all columns containing data (do NOT select the column containing the row labels)
4. Click **OK**

Note: This output returns expected cell counts as well as the chi-square test statistic and *P*-value.

SPSS

1. Enter the row variable data into one column
2. Enter the column variable data into a second column
3. Click **Analyze** then click **Descriptive Statistics** then click **Crosstabs...**

4. Select the name of the row variable and press the arrow to move the variable to the box under **Row(s):**
5. Select the name of the column variable and press the arrow to move the variable to the box under **Column(s):**
6. Click **Statistics...**
7. Check the box next to Chi-square
8. Click **Continue**
9. Click **Cells...**

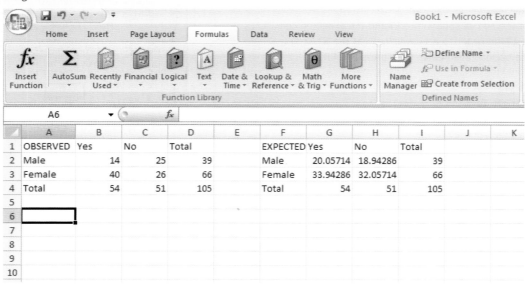

10. Check the box next to Expected
11. Click **Continue**
12. Click **OK**

Note: The *P*-value for this test can be found in the **Chi-Square Tests** table in the Pearson Chi-Square row.

Excel
1. Input the observed contingency table
2. Input the expected table
3. Click on an empty cell
4. Click **Formulas**
5. Click **Insert Function**
6. Select **Statistical** from the drop-down box for category
7. Select **CHITEST** and press **OK**
8. Click in the box next to **Actual_range** and select the data values from the actual table (do NOT select column or row labels or totals)
9. Click in the box next to **Expected_range** and select the data values from the expected table (do NOT select column or row labels or totals)
10. Click **OK**

TI-83/84
1. Input the observed contingency table into matrix **A** (To access and edit matrices, press **2nd** then **x⁻¹**, then highlight **EDIT** and press **ENTER**. Then highlight **[A]** and press **ENTER**. Type the value for the number of rows and press **ENTER**; type the value for the number of columns and press **ENTER**. Type the data values into the matrix.)

 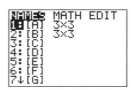

2. Press **STAT**
3. Highlight **TESTS**
4. Highlight **χ²-Test...** and press **ENTER**
5. Highlight **Calculate** and press **ENTER**

Image for Head "Excel"

	A	B	C	D	E	F	G	H	I	J	K
1	OBSERVED	Yes	No	Total		EXPECTED	Yes	No	Total		
2	Male	14	25	39		Male	20.05714	18.94286	39		
3	Female	40	26	66		Female	33.94286	32.05714	66		
4	Total	54	51	105		Total	54	51	105		
5											
6											
7											
8											
9											
10											

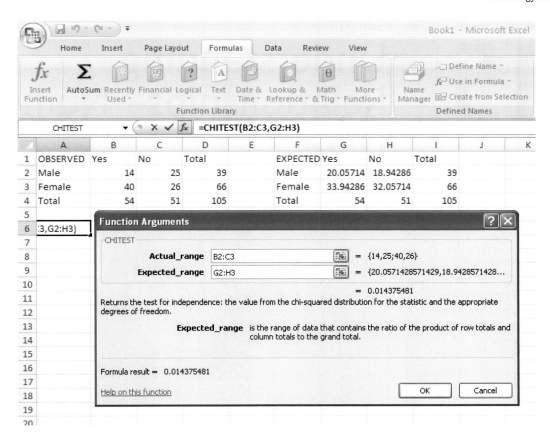

TI-Nspire

1. Enter the Calculator Scratchpad
2. Press the **menu** key then select **7:Matrix & Vector** then select **1:Create** then select **1:Matrix...** and press **enter**
3. Next to **Number of rows** enter the number of rows in the contingency table (do not include title rows or total rows)
4. Next to **Number of columns** enter the number of columns in the contingency table (do not include title columns or total columns)
5. Press **OK**
6. Input the values into the matrix (pressing tab after entering each value and press enter when you are finished)
7. Press **ctrl** then press **var**
8. Type in **amat** and press **enter**

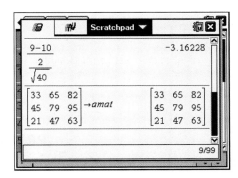

9. Press the **menu** key then select **6:Statistics** then **7:Stat tests** then **8: χ² 2-way Test...**
10. For **Observed Matrix**, select amat from the drop-down list
11. Press **OK**

15 Understanding Relationships—Numerical Data

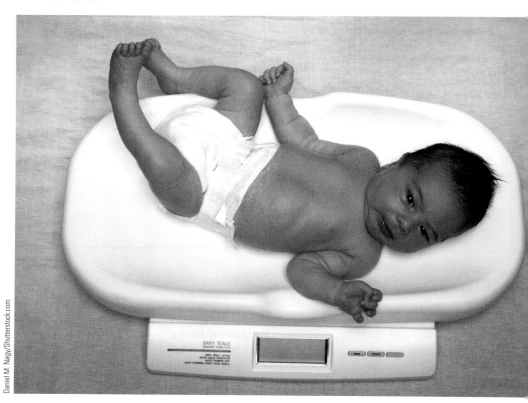

Daniel M. Nagy/Shutterstock.com

PREVIEW

In Chapter 4, you learned how to describe relationships between two numerical variables. When the relationship was judged to be linear you found the equation of the least squares regression line and assessed the quality of the fit using the scatterplot, the residual plot, and the values of the coefficient of determination (r^2) and the standard deviation about the least squares regresssion line (s_e). In this chapter you will learn how to make inferences about the slope of the population regression line.

CHAPTER LEARNING OBJECTIVES

Conceptual Understanding

After completing this chapter, you should be able to

C1 Understand how probabilistic and deterministic models differ.

C2 Understand that the simple linear regression model provides a basis for making inferences about linear relationships.

Mastering the Mechanics

After completing this chapter, you should be able to

M1 Interpret the parameters of the simple linear regression model in context.

M2 Use scatterplots, residual plots, and normal probability plots or boxplots to assess the credibility of the assumptions of the simple linear regression model.

M3 Know the conditions for appropriate use of methods for making inferences about the slope of a population regression line, β.

M4 Use the five-step process for estimation problems (EMC³) and computer output to construct and interpret a confidence interval estimate for the slope of a population regression line.

M5 Use the five-step process for hypothesis testing (HMC³) to test hypotheses about the slope of a population regression line.

M6 Use graphs to identify outliers and potentially influential points.

Putting It into Practice

After completing this chapter, you should be able to

P1 Interpret a confidence interval for the slope of a population regression line in context.

P2 Carry out the model utility test and interpret the result in context.

PREVIEW EXAMPLE

Premature Babies

Babies born prematurely (before the 37th week of pregnancy) often have low birth weights. Is a low birth weight related to factors that affect brain function? The authors of the paper **"Intrauterine Growth Restriction Affects the Preterm Infant's Hippocampus"(*Pediatric Research* [2008]: 438-43)** hoped to use data from a study of premature babies to answer this question. They measured x = birth weight (in grams) and y = hippocampus volume (in mL) for 26 premature babies. The hippocampus is a part of the brain that is important in the development of both short- and long-term memory. The sample correlation coefficient for their data is $r = 0.4722$ and the equation of the least squares regression line is $\hat{y} = 1.67 + 0.0026x$. The pattern in the scatterplot (Figure 15.1) suggests there may be a positive linear relationship. However, the value of the correlation coefficient is not very large, and the value of the slope is close to zero. Could the pattern observed in the scatterplot—and the nonzero slope—be plausibly explained by chance? That is, is it plausible that there is no relationship between birth weight and hippocampus volume in the population of all premature babies? Or does the sample provide convincing evidence of a linear relationship between these two variables? If there is evidence of a meaningful relationship between these two variables, the regression line could be used to predict the hippocampus volume. If the predicted volume was sufficiently small, early cognitive therapy could be recommended. On the other hand, if there is no meaningful relationship between these variables, low birth weight should not automatically trigger potentially expensive therapy. ∎

FIGURE 15.1
Scatterplot of birth weight
versus hippocampus volume

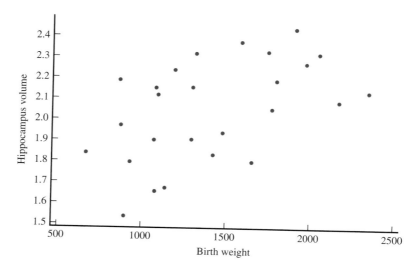

In this chapter, you will learn methods that will help you determine if there is a real and useful linear relationship between two variables or if the pattern in the data could be simply due to chance differences that occur when a sample is selected from a population.

SECTION 15.1 The Simple Linear Regression Model

A *deterministic relationship* between two variables x and y is one in which the value of y is completely determined by the value of the independent variable x. A deterministic relationship can be described, or "modeled," using mathematical notation, such as $y = f(x)$ where $f(x)$ is a particular function of x. This relationship is deterministic in the sense that the value of the independent variable is all that is needed to determine the value of the dependent variable. For example, you might convert x = temperature in degrees centigrade to y = temperature in degrees Fahrenheit using $y = f(x)$, where $f(x) = \frac{9}{5}x + 32$. Once the centigrade temperature is known, the Fahrenheit temperature is completely determined. Or you might determine y = amount of money in a savings account after x years, using the compound interest formula, $y = P\left(1 + \frac{r}{n}\right)^{nx}$, where P is the principal (the amount of money deposited), r is the interest rate, and n is the number of times each year the interest is compounded. The number of years you leave the principal in the bank determines the amount in the account.

In many situations the variables of interest are not deterministically related. For example, the value of y = first-year college grade point average is not determined solely by x = high school grade point average, and y = crop yield is determined partly by factors other than x = amount of fertilizer used. The relationship between two variables, x and y, that are not deterministically related can be described by extending the deterministic model to specify a probabilistic model. The general form of a **probabilistic model** allows y to be larger or smaller than $f(x)$ by a random amount e. The model equation for a probabilistic model has the form

$$y = \text{deterministic function of } x + \text{random deviation}$$
$$= f(x) + e$$

In a scatterplot of y versus x, some of the data points will fall above the graph of $f(x)$ and some will fall below. Thinking geometrically, if $e > 0$, the corresponding point in the scatterplot will lie above the graph of the function $y = f(x)$. If $e < 0$, the corresponding point will fall below the graph of $f(x)$.

For example, consider the probabilistic model

$$y = \underbrace{50 - 10x + x^2}_{f(x)} + e$$

The graph of the function $y = 50 - 10x + x^2$ is shown as the orange curve in Figure 15.2. The observed point $(4, 30)$ is also shown in the figure. Because $f(4) = 50 - 10(4) + 4^2 =$

$50 - 40 + 16 = 26$ for this point, you can write $y = f(x) + e$, where $e = 4$. The point $(4, 30)$ falls 4 units above the graph of the function, $y = 50 - 10x + x^2$.

FIGURE 15.2
A deviation from the deterministic part of a probabilistic model

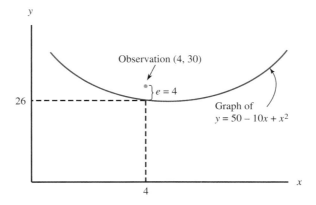

Simple Linear Regression Model

The simple linear regression model is a special case of the general probabilistic model in which the deterministic function, $f(x)$, is linear (so its graph is a straight line).

> **DEFINITION**
>
> The **simple linear regression model** assumes that there is a line with vertical or y intercept α and slope β, called the **population regression line.** When a value of the independent variable x is fixed and an observation on the dependent variable y is made,
>
> $$y = \alpha + \beta x + e$$
>
> Without the random deviation e, all observed (x, y) points would fall exactly on the population regression line. The inclusion of e in the model equation recognizes that points will deviate from the line by a random amount.

Figure 15.3 shows two observations in relation to the population regression line.

FIGURE 15.3
Two observations and deviations from the population regression line

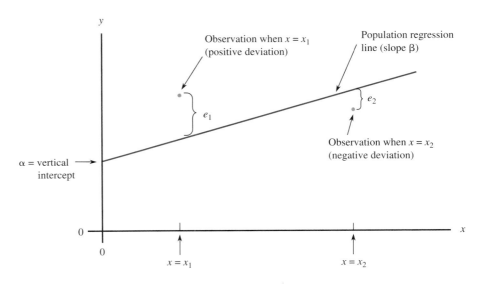

Before you actually observe a value of y for any particular value of x, you are uncertain about the value of e. It could be negative, positive, or even 0. Also, e might be quite large in magnitude (resulting in a point far from the population regression line) or quite small (resulting in a point very close to the line). The simple linear regression model makes some assumptions about the distribution of e at any particular x value in the population.

BASIC ASSUMPTIONS OF THE SIMPLE LINEAR REGRESSION MODEL

1. The distribution of e at any particular x value has mean value 0. That is, $\mu_e = 0$.
2. The standard deviation of e (which describes the spread of its distribution) is the same for any particular value of x. This standard deviation is denoted by σ_e.
3. The distribution of e at any particular x value is normal.
4. The random errors $e_1, e_2, ..., e_n$ associated with different observations are independent of one another.

The simple linear regression model assumptions about the variability in the values of e in the population imply that there is also variability in the y values observed at any particular value of x. Consider y when x has some fixed value x^*, so that

$$y = \alpha + \beta x^* + e.$$

Because α and β are fixed (they are unknown population values), $\alpha + \beta x^*$ is also a fixed number. The sum of a fixed number and a normally distributed variable (e) is also a normally distributed variable (the bell-shaped curve is simply shifted), so y itself has a normal distribution. Furthermore, $\mu_e = 0$ implies that the mean value of y is $\alpha + \beta x^*$, the height of the population regression line for the value $x = x^*$. Finally, because there is no variability in the fixed number $\alpha + \beta x^*$, the standard deviation of y is the same as the standard deviation of e. These properties are summarized in the following box.

At any fixed value x^*, y has a normal distribution, with

$$\begin{array}{c}\text{mean } y \text{ value}\\ \text{for } x^*\end{array} = \left(\begin{array}{c}\text{height of the population}\\ \text{regression line above } x^*\end{array}\right) = \alpha + \beta x^*$$

and

$$\text{standard deviation of } y \text{ for a fixed value } x^* = \sigma_e$$

The slope, β, of the population regression line is the *mean* or *expected* change in y associated with a 1-unit increase in x. The y intercept, α, is the height of the population line when $x = 0$.

The value of σ_e determines how much the (x, y) observations deviate vertically from the population line; when σ_e is small, most observations will be close to the line, but when σ_e is large, the observations will tend to deviate more from the line.

The key features of the model are illustrated in Figures 15.4 and 15.5. Notice that the three normal curves in Figure 15.4 have identical spreads. This is a consequence of σ_e being the same at any value of x, which implies that the variability in the y values at a particular value of x is constant—the variability does not depend on the value of x.

FIGURE 15.4
Illustration of the simple linear regression model

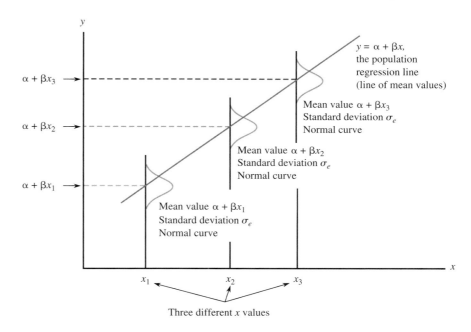

FIGURE 15.5
The simple linear regression model: (a) small σ_e ; (b) large σ_e

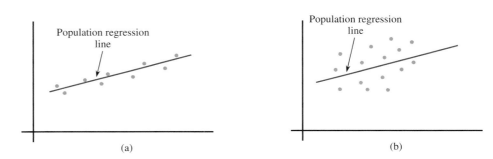

Example 15.1 **Stand on Your Head to Lose Weight?**

The authors of the article **"On Weight Loss by Wrestlers Who Have Been Standing on Their Heads"** (paper presented at the Sixth International Conference on Statistics, Combinatorics, and Related Areas, Forum for Interdisciplinary Mathematics, 1999, with the data also appearing in *A Quick Course in Statistical Process Control*, Mick Norton, 2005) state that "amateur wrestlers who are overweight near the end of the weight certification period, but just barely so, have been known to stand on their heads for a minute or two, get on their feet, step back on the scale, and establish that they are in the desired weight class. Using a headstand as the method of last resort has become a fairly common practice in amateur wrestling."

Does this really work? Data were collected in an experiment where weight loss was recorded for each wrestler after exercising for 15 minutes and then doing a headstand for 1 minute 45 sec. Based on these data, the authors of the article concluded that there was in fact a demonstrable weight loss that was greater than that for a control group that exercised for 15 minutes but did not do the headstand. (The authors give a plausible explanation for why this might be the case based on the way blood and other body fluids collect in the head during the headstand and the effect of weighing while these fluids are draining immediately after standing.) The authors also concluded that a simple linear regression model was a reasonable way to describe the relationship between the variables

$$y = \text{weight loss (in pounds)}$$

and

$$x = \text{body weight prior to exercise and headstand (in pounds)}$$

Suppose that the actual model equation has $\alpha = 0$, $\beta = 0.001$, and $\sigma_e = 0.09$ (these values are consistent with the findings in the article). The population regression line is shown in Figure 15.6.

FIGURE 15.6

The population regression line for Example 15.1

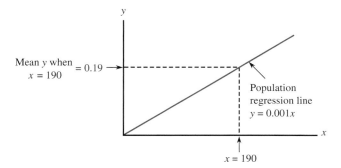

If the distribution of the random errors at any fixed weight (x value) is normal, then the variable y = weight loss is normally distributed with

$$\mu_y = 0 + 0.001x$$
$$\sigma_y = \sigma_e = 0.09$$

For example, when $x = 190$ (corresponding to 190-pound wrestlers), weight loss has mean value

$$\mu_y = 0 + 0.001(190) = 0.19 \text{ pounds}$$

Because the standard deviation of y is $\sigma_y = 0.09$, the interval $0.19 \pm 2(0.09) = (0.01, 0.37)$ includes y values that are within 2 standard deviations of the mean value for y when $x = 190$. Roughly 95% of the weight loss observations made for 190-lb wrestlers will be in this range. The slope $\beta = 0.001$ can be interpreted as the mean change in weight associated with each additional pound of body weight.

More insight into model properties can be gained by thinking of the population of all (x, y) pairs as consisting of many smaller subpopulations. Each subpopulation contains pairs for which x has a fixed value. Suppose, for example, that in a large population of college students the variables

$$x = \text{grade point average in major courses}$$

and

$$y = \text{starting salary after graduation}$$

are related according to the simple linear regression model. Then you can think about the subpopulation of all pairs with $x = 3.20$ (corresponding to all students with a grade point average of 3.20 in major courses), the subpopulation of all pairs having $x = 2.75$, and so on. The model assumes that for each of these subpopulations, y is normally distributed with the same standard deviation, and that the *mean y value* (rather than y itself) is linearly related to x.

In practice, the judgment of whether the simple linear regression model is appropriate—that is, the judgments about the credibility of the assumptions underlying the linear regression model—must be based on knowledge of how the data were collected, as well as an inspection of various plots of the data and the residuals. The sample observations should be independent of one another, which will be the case if the data are from a random sample. In addition, the scatterplot should show a linear rather than a curved pattern, and the vertical spread of points should be very similar throughout the range of x values. Figure 15.7 shows plots with three different patterns; only the first pattern is consistent with the simple linear regression model assumptions.

FIGURE 15.7
Some commonly encountered patterns in scatter plots: (a) Consistent with the simple linear regression model; (b) Suggests a nonlinear probabilistic model; (c) Suggests that variability in y changes with x

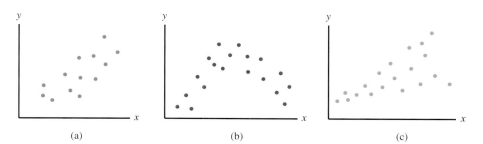

(a) (b) (c)

Estimating the Population Regression Line

In Section 15.3, you will see how to check whether the basic assumptions of the simple linear regression model are reasonable. When this is the case, the values of α and β (y intercept and slope of the population regression line) can be estimated from sample data.

The estimates of α and β are denoted by a and b, respectively. These estimates are the values of the intercept and slope of the least squares regression line. Recall that that the least squares regression line is the line for which the sum of squared vertical deviations of points in the scatterplot from the line is smaller than for any other line.

The estimates of the slope and the y intercept of the population regression line are the slope and y intercept, respectively, of the least squares regression line. That is,

$$b = \text{estimate of } \beta = \frac{\Sigma(x - \bar{x})(y - \bar{y})}{\Sigma(x - \bar{x})^2}$$

$$a = \text{estimate of } \alpha = \bar{y} - b\bar{x}$$

The values of a and b are usually obtained using statistical software or a graphing calculator. If the slope and intercept are calculated by hand, you can use the following computational formula:

$$b = \frac{\Sigma xy - \dfrac{(\Sigma x)(\Sigma y)}{n}}{\Sigma x^2 - \dfrac{(\Sigma x)^2}{n}}$$

The estimated regression line is the familiar least squares regression line

$$\hat{y} = a + bx$$

Let x^* denote a specified value of the independent variable x. Then $a + bx^*$ has two different interpretations:

1. It is a point estimate of the mean y value when $x = x^*$.
2. It is a point prediction of an individual y value to be observed when $x = x^*$.

Data set available

Example 15.2 **Mother's Age and Baby's Birth Weight**

Medical researchers have noted that adolescent females are much more likely to deliver low-birth-weight babies than are adult females. (Low birth weight in humans is generally defined as a weight below 2500 grams.) Because low-birth-weight babies have higher mortality rates, a number of studies have examined the relationship between birth weight and mother's age for babies born to young mothers.

One such study is described in the article **"Body Size and Intelligence in 6-Year-Olds: Are Offspring of Teenage Mothers at Risk?"** (*Maternal and Child Health Journal* [2009]: 847–856). The following data on

$$x = \text{maternal age (in years)}$$

and

$$y = \text{birth weight of baby (in grams)}$$

are consistent with summary values given in the article and also with data published by the National Center for Health Statistics.

Observation

	1	2	3	4	5	6	7	8	9	10
x	15	17	18	15	16	19	17	16	18	19
y	2,289	3,393	3,271	2,648	2,897	3,327	2,970	2,535	3,138	3,573

A scatterplot of the data is given in Figure 15.8. The scatterplot shows a linear pattern, and the spread in the y values appears to be similar across the range of x values. This supports the appropriateness of the simple linear regression model.

FIGURE 15.8
Scatterplot of birth weight versus maternal age for Example 15.2

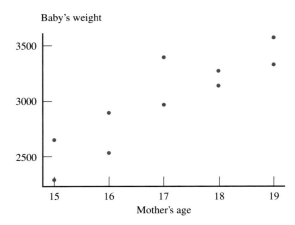

For these data, the equation of the estimated regression line was found using statistical software, resulting in

$$\hat{y} = a + bx = -1163.45 + 245.15x$$

An estimate of the mean birth weight of babies born to 18-year-old mothers results from substituting $x = 18$ into the estimated equation:

estimated mean y for 18-year-old mothers $= a + bx$
$$= -1163.45 + 245.15(18)$$
$$= 3249.25 \text{ grams}$$

Similarly, you would predict the birth weight of a baby to be born to a particular 18-year-old mother to be

$\hat{y} =$ predicted y value when $x = 18$
$$= a + b(18)$$
$$= 3249.25 \text{ grams}$$

The estimate of the mean weight and the prediction of an individual baby weight are identical, because the same x value was used in each calculation. However, their interpretations differ. One is the prediction of the weight of a single baby whose mother is 18, whereas the other is an estimate of the mean weight of *all* babies born to 18-year-old mothers.

In Example 15.2, the x values in the sample ranged from 15 to 19. The estimated regression equation should not be used to make an estimate or prediction for any x value much outside this range. Without sample data for such values, or some clear theoretical reason for expecting the relationship to be linear outside the observed range of x values, you have no reason to believe that the estimated linear relationship continues outside the range from 15 to 19. Making predictions outside this range can be misleading, and statisticians refer to this as the **danger of extrapolation**.

Estimating σ_e^2 and σ_e

The value of σ_e describes the extent to which observed points (x, y) tend to fall close to or far away from the population regression line. A point estimate of σ_e is based on

$$SSResid = \Sigma(y - \hat{y})^2$$

where $\hat{y}_1 = a + bx_1, \cdots, \hat{y}_n = a + bx_n$ are the fitted or predicted y values and the residuals are $y_1 - \hat{y}_1, \cdots y_n - \hat{y}_n$. SSResid is a measure of the extent to which the sample data spread out around the estimated regression line.

DEFINITION

The statistic for estimating the variance σ_e^2 is

$$s_e^2 = \frac{SSResid}{n - 2}$$

where

$$SSResid = \Sigma(y - \hat{y})^2 = \Sigma y^2 - a\Sigma y - b\Sigma xy$$

The subscript in σ_e^2 and s_e^2 is a reminder that you are estimating the variance of the "errors" or residuals.

The estimate of σ_e is the **estimated standard deviation**

$$s_e = \sqrt{s_e^2}$$

The number of degrees of freedom associated with estimating σ_e^2 or σ_e in simple linear regresssion is $n - 2$.

The estimates and number of degrees of freedom here have analogs in previous work involving a single sample $x_1, x_2, \ldots, x_n$. The sample variance s^2 had a numerator of $\Sigma(x - \bar{x})^2$, a sum of squared deviations (residuals), and denominator $n - 1$, the number of degrees of freedom associated with s^2 and s. The use of $\bar{x}$ as an estimate of μ in the formula for s^2 reduces the number of degrees of freedom by 1, from n to $n - 1$. In simple linear regression, estimation of two quantities, α and β, results in a loss of 2 degrees of freedom, leaving $n - 2$ as the number of degrees of freedom associated with SSResid, s_e^2 and s_e.

Once the estimated regression equation has been found, the usefulness of this model is evaluated using a residual plot and the values of s_e and the coefficient of determination, r^2. Recall from Chapter 4 that the values of s_e and r^2 are interpreted as described in the following box.

The coefficient of determination, r^2, is the proportion of variability in y that can be explained by the approximate linear relationship between x and y.

The value of s_e, the estimated standard deviation about the population regression line, is interpreted as the typical amount by which an observation deviates from the population regression line.

Example 15.3 **Predicting Elk Weight**

Wildlife biologists monitor the ecological health of animals. For large animals whose habitat is relatively inaccessible, this can present some practical problems. The Rocky Mountain elk is the fourth largest deer species. Males range up to 7.5 feet in length and over 500 pounds in weight. The equipment, manpower, and time needed to weigh these creatures make direct measurement of weight difficult and expensive. The authors of the paper **"Estimating Elk Weight From Chest Girth"** *(Wildlife Society Bulletin* **[1996]: 58–61)** found they could reliably estimate elk weights by a much more practical method: measuring the chest girth and then using linear regression to estimate the weight. They measured the chest girth (in cm) and weight (in kg) of 19 Rocky Mountain elk in Custer State Park, South Dakota. The resulting data (from a scatterplot in the paper) is given in the accompanying table.

Girth (cm)	Weight (kg)
96	87
105	196
108	163
109	196
110	183
114	171
121	230
124	225
131	211
135	231
137	225
138	266
140	241
142	264
157	284
157	292
159	300
155	337
162	339

The scatterplot (Figure 15.9) provides evidence of a strong positive linear relationship between

$$x = \text{chest girth}$$

and

$$y = \text{weight}$$

FIGURE 15.9
Scatterplot of weight versus chest girth for Example 15.3

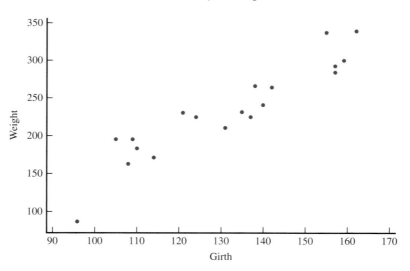

Partial Minitab regression output is shown here.

Regression Analysis: Weight versus Girth
The regression equation is
Weight = − 136 + 2.81 Girth

Predictor	Coef	SE Coef	T	P
Constant	−135.51	35.75	−3.79	0.001
Girth	2.8063	0.2686	10.45	0.000

S = 23.6626   R-Sq = 86.5%   R-Sq(adj) = 85.7%

From the output,

$$\hat{y} = -136 + 2.81x$$
$$r^2 = 0.865$$
$$s_e = 23.6626$$

Approximately 86.5% of the observed variation in elk weight y can be attributed to the linear relationship between weight and chest girth. The magnitude of a typical deviation from the population regression line is about 23.7 kg, which is relatively small in comparison to the y values themselves.

Another important assumption of the simple linear regression model is that the random deviations at any particular x value are normally distributed. In Section 15.3, you will see how the residuals can be used to determine whether this assumption is plausible.

Summing It Up—Section 15.1

The following learning objectives were addressed in this section:

Conceptual Understanding
C1: Understand how probabilistic and deterministic models differ.
A deterministic model of the relationship between two variables x and y is one in which the value of y is completely determined by the value of x. A probabilistic model extends a deterministic model by incorporating a random deviation term. The general form of a probabilistic model is $y = f(x) + e$, where $f(x)$ is a function of x and e is a random deviation.

Mastering the Mechanics
M1: Interpret the parameters of the simple linear regression model in context.
The slope β of the population regression line is the mean, or expected, change in y associated with a 1-unit change in x. The y intercept α is the height of the population line when $x = 0$.

SECTION 15.1 EXERCISES

Each exercise set assesses the following chapter learning objectives: C1, M1

SECTION 15.1 Exercise Set 1

15.1 Identify the following relationships as deterministic or probabilistic:
a. The relationship between the length of the sides of a square and its perimeter.
b. The relationship between the height and weight of an adult.
c. The relationship between SAT score and college freshman GPA.
d. The relationship between tree height in centimeters and tree height in inches.

15.2 Let x be the size of a house (in square feet) and y be the amount of natural gas used (therms) during a specified period. Suppose that for a particular community, x and y are related according to the simple linear regression model with

$$\beta = \text{slope of population regression line} = .017$$
$$\alpha = y \text{ intercept of population regression line} = -5.0$$

Houses in this community range in size from 1000 to 3000 square feet.
a. What is the equation of the population regression line?
b. Graph the population regression line by first finding the point on the line corresponding to $x = 1000$ and then the point corresponding to $x = 2000$, and drawing a line through these points.
c. What is the mean value of gas usage for houses with 2100 sq. ft. of space?
d. What is the average change in usage associated with a 1 sq. ft. increase in size?
e. What is the average change in usage associated with a 100 sq. ft. increase in size?
f. Would you use the model to predict mean usage for a 500 sq. ft. house? Why or why not?

15.3 Suppose that a simple linear regression model is appropriate for describing the relationship between $y =$ house price (in dollars) and $x =$ house size (in square feet) for houses in a large city. The population regression line is $y = 23{,}000 + 47x$ and $\sigma_e = 5000$.
a. What is the average change in price associated with one extra square foot of space? With an additional 100 sq. ft. of space?
b. Approximately what proportion of 1800 sq. ft. homes would be priced over $110,000? Under $100,000?

SECTION 15.1 Exercise Set 2

15.4 Identify the following relationships as deterministic or probabilistic:
a. The relationship between height at birth and height at one year of age.
b. The relationship between a positive number and its square root.
c. The relationship between temperature in degrees Fahrenheit and degrees centigrade.
d. The relationship between adult shoe size and shirt size.

15.5 The flow rate in a device used for air quality measurement depends on the pressure drop x (inches of water) across the device's filter. Suppose that for x values between 5 and 20, these two variables are related according to the simple linear regression model with population regression line $y = -0.12 + 0.095x$.
a. What is the mean flow rate for a pressure drop of 10 inches? A drop of 15 inches?
b. What is the average change in flow rate associated with a 1 inch increase in pressure drop? Explain.

15.6 The paper **"Predicting Yolk Height, Yolk Width, Albumen Length, Eggshell Weight, Egg Shape Index, Eggshell Thickness, Egg Surface Area of Japanese Quails Using Various Egg Traits as Regressors"** (*International Journal of Poultry Science* [2008]: 85–88) suggests that the simple linear regression model is reasonable for describing the relationship between y = eggshell thickness (in micrometers) and x = egg length (mm) for quail eggs. Suppose that the population regression line is $y = 0.135 + 0.003x$ and that $\sigma_e = 0.005$. Then, for a fixed x value, y has a normal distribution with mean $0.135 + 0.003x$ and standard deviation 0.005.
a. What is the mean eggshell thickness for quail eggs that are 15 mm in length? For quail eggs that are 17 mm in length?
b. What is the probability that a quail egg with a length of 15 mm will have a shell thickness that is greater than 0.18 μm?
c. Approximately what proportion of quail eggs of length 14 mm have a shell thickness of greater than 0.175? Less than 0.178?

ADDITIONAL EXERCISES

15.7 Tom and Ray are managers of electronics stores with slightly different pricing strategies for USB drives. In Tom's store, customers pay the same amount, c, for each USB drive. In Ray's store, it is a little more exciting. The customer pays an up-front cost of $1.00. Ray charges the same price per USB drive, c, but at the register the customer flips a coin. If the coin lands heads up, the customer gets his or her $1.00 back, plus another dollar off the total cost of the USB drives purchased.
a. Which of these pricing strategies can be expressed as a deterministic model?
b. Using mathematical notation, specify a model using Tom's pricing strategy that relates y = total cost to x = number of USB drives purchased.

c. Using mathematical notation, specify a model using Ray's pricing strategy that relates y = total cost to x = number of USB drives purchased.
d. Describe the distribution of e for the probabilistic model described above. What is the mean of the distribution of e? What is the standard deviation of e?

15.8 Identify the following relationships as deterministic or probabilistic:
a. The relationship between the speed limit and a driver's speed.
b. The relationship between the price in dollars and the price in Euros of an object.
c. The relationship between the number of pages and the number of words in a text book.
d. The relationship between the possible numbers of pennies and the nickels in a pile if no other coins are in the pile and the amount of money in the pile is $3.00.

15.9 Hormone replacement therapy (HRT) is thought to increase the risk of breast cancer. The accompanying data on x = percent of women using HRT and y = breast cancer incidence (cases per 100,000 women) for a region in Germany for 5 years appeared in the paper **"Decline in Breast Cancer Incidence after Decrease in Utilisation of Hormone Replacement Therapy"** (*Epidemiology* [2008]: 427–430). The authors of the paper used a simple linear regression model to describe the relationship between HRT use and breast cancer incidence.

HRT Use	Breast Cancer Incidence
46.30	103.3
40.60	105.0
39.50	100.0
36.60	93.8
30.00	83.5

a. What is the equation of the estimated regression line?
b. What is the estimated average change in breast cancer incidence associated with a 1 percentage point increase in HRT use?
c. What would you predict the breast cancer incidence to be in a year when HRT use was 40%?
d. Should you use this regression model to predict breast cancer incidence for a year when HRT use was 20%? Explain.
e. Calculate and interpret the value of r^2.
f. Calculate and interpret the value of s_e.

15.10 Consider the accompanying data on x = advertising share and y = market share for a particular brand of soft drink during 10 randomly selected years.

x 0.103 0.072 0.071 0.077 0.086 0.047 0.060 0.050 0.070 0.052
y 0.135 0.125 0.120 0.086 0.079 0.076 0.065 0.059 0.051 0.039

a. Construct a scatterplot for these data. Do you think the simple linear regression model would be appropriate for describing the relationship between x and y?

 Data set available

b. Calculate the equation of the estimated regression line and use it to obtain the predicted market share when the advertising share is 0.090.

c. Compute r^2. How would you interpret this value?

d. Calculate a point estimate of σ_e. How many degrees of freedom is associated with this estimate?

15.11 The paper **"Depression, Body Mass Index, and Chronic Obstructive Pulmonary Disease—A Holistic Approach"** (*International Journal of COPD* [2016]:239–249) gives data on

x = change in Body Mass Index (BMI, in kilograms/meter^2)

and

y = change in a measure of depression

for patients suffering from depression who participated in a pulmonary rehabilitation program. The table below contains a subset of the data given in the paper and are approximate values read from a scatterplot in the paper.

BMI Change (kg/m²)	Depression Score Change
0.5	−1
−0.5	9
0	4
0.1	4
0.7	6
0.8	7
1	13
1.5	14
1.1	17
1	18
0.3	12
0.3	14

The accompanying computer output is from Minitab.

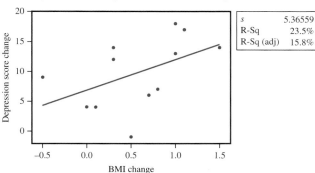

Fitted Line Plot
Depression score change = 6.873 + 5.078 BMI change

s	5.36559
R-Sq	23.5%
R-Sq (adj)	15.8%

```
       S      R-sq
5.36559    23.48%

Coefficients
Term         Coef   SE Coef   T-Value   P-Value   VIF
Constant     6.87   2.26      3.04      0.012
BMI change   5.08   2.90      1.75      0.110     1.00

Regression Equation
Depression score change = 6.87 + 5.08 BMI change
```

Data set available

a. What percentage of observed variation in depression score change can be explained by the simple linear regression model?

b. Give a point estimate of σ_e and interpret this estimate.

c. Give an estimate of the average change in depression score change associated with a 1 kg/m^2 increase in BMI change.

d. Calculate a point estimate of the mean depression score change for a patient whose BMI change was 1.2 kg/m^2.

15.12 The production of pups and their survival are the most significant factors contributing to gray wolf population growth. The causes of early pup mortality are unknown and difficult to observe. The pups are concealed within their dens for 3 weeks after birth, and after they emerge it is difficult to confirm their parentage. Researchers recently used portable ultrasound equipment to investigate some factors related to reproduction (**"Diagnosing Pregnancy, in Utero Litter Size, and Fetal Growth with Ultrasound in Wild, Free-Ranging Wolves,"** *Journal of Mammology* [2006]: 85–92). A scatterplot of y = length of an embryonic sac diameter (in cm) and x = gestational age (in days) is shown below. Computer output from a regression analysis is also given.

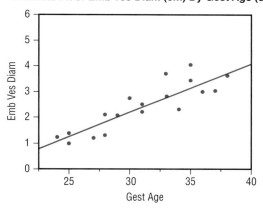

Bivariate Fit of Emb Ves Diam (cm) By Gest Age (days)

—— Linear Fit

Linear Fit

Emb Ves Diam (cm) = −3.497279 + 0.1903121*Gest Age (days)

Summary of Fit

RSquare	0.792803
RSquare Adj	0.780615
Root Mean Square Error	0.450587
Mean of Response	2.482526
Observations (or Sum Wgts)	19

Lack of Fit

Analysis of Variance

Parameter Estimates

Term	Estimate	Std Error	t Ratio	Prob>|t|
Intercept	−3.497279	0.748605	−4.67	0.0002*
Gest Age (days)	0.1903121	0.023597	8.07	<.0001*

a. What is the equation of the estimated regression line?

b. What is the estimated embryonic sac diameter for a gestational age of 30 days?

c. What is the average change in sac diameter associated with a 1-day increase in gestational age?

d. What is the average change in sac diameter associated with a 5-day increase in gestational age?

e. Would you use this model to predict the mean embryonic sac diameter for all gestation ages from conception to birth? Why or why not?

SECTION 15.2 # Inferences Concerning the Slope of the Population Regression Line

The slope coefficient β in the simple linear regression model represents the average or expected change in the response variable y that is associated with a 1-unit increase in the value of the independent variable x. For example, consider $x =$ the size of a house (in square feet) and $y =$ selling price of the house. If the simple linear regression model is appropriate for the population of houses in a particular city, β would be the average increase in selling price associated with a 1-square-foot increase in size. As another example, if $x =$ amount of time per week a computer system is used and $y =$ the resulting annual maintenance cost, then β would be the expected change in the maintenance cost associated with using the computer system one additional hour per week.

The value of β is nearly always unknown, but it can be estimated from sample data. The slope of the least squares regression line, b, provides an estimate. In some situations, the value of the statistic b may vary greatly from sample to sample, and the value of b computed from a single sample may be quite different from the value of the population slope, β. In other situations, almost all possible samples result in a value of b that is quite close to β. The sampling distribution of b provides information about the behavior of this statistic.

Properties of the Sampling Distribution of b

When **the four basic assumptions of the simple linear regression model** are satisfied

1. The mean value of the sampling distribution of b is β. That is, $\mu_b = \beta$. This means that the sampling distribution of b is centered at the value of β and that b is an unbiased statistic for estimating β.

2. The standard deviation of the sampling distribution of the statistic b is

$$\sigma_b = \frac{\sigma_e}{\sqrt{\Sigma(x - \bar{x})^2}}$$

3. The statistic b has a normal distribution (a consequence of the model assumption that the random deviation e is normally distributed).

The fact that b is unbiased tells you that the sampling distribution is centered at the right place, but it gives no information about variability. If σ_b is large, the sampling distribution of b will be quite spread out around β and an estimate far from the value of β could result. For $\sigma_b = \dfrac{\sigma_e}{\sqrt{\Sigma(x - \bar{x})^2}}$ to be small, the numerator σ_e should be small (little variability about the population line) and/or the denominator $\sqrt{\Sigma(x - \bar{x})^2}$ should be large. Because $\Sigma(x - \bar{x})^2$ is a measure of how much the observed x values spread out, β tends to be more precisely estimated when the x values in the sample are spread out rather than when they are close together.

The normality of the sampling distribution of b implies that the standardized variable

$$z = \frac{b - \beta}{\sigma_b}$$

has a standard normal distribution. However, inferential methods cannot be based on this statistic, because the value of σ_b is not known (because the unknown σ_e appears in the numerator of σ_b). One way to proceed is to estimate σ_e with s_e to obtain an estimate of σ_b.

The **estimated standard deviation of the statistic b** is

$$s_b = \frac{s_e}{\sqrt{\Sigma(x - \bar{x})^2}}$$

When the four basic assumptions of the simple linear regression model are satisfied, the probability distribution of the standardized variable

$$t = \frac{b - \beta}{s_b}$$

is the t distribution with df $= n - 2$.

In the same way that $t = \dfrac{\bar{x} - \mu}{\dfrac{s}{\sqrt{n}}}$ was used in Chapter 12 to develop a confidence interval for μ, the t variable in the preceding box can be used to obtain a confidence interval for β.

Confidence Interval for β

When the four basic assumptions of the simple linear regression model are satisfied, a confidence interval for β, the slope of the population regression line, has the form

$$b \pm (t \text{ critical value})s_b$$

where the t critical value is based on df $= n - 2$. Appendix A Table 3 gives critical values corresponding to the most frequently used confidence levels.

The interval estimate of β is centered at b and extends out from the center by an amount that depends on the sampling variability of b. When s_b is small, the interval is narrow, implying that you have relatively precise knowledge of the value of β. Calculation of a confidence interval for the slope of a population regression line is illustrated in Example 15.4.

In Section 7.2, you learned four key questions that guide the decision about what statistical inference method to consider in any particular situation. In Section 7.3, a five-step process for estimation problems was introduced.

The four key questions of Section 7.2 were

Q Question Type	Estimation or hypothesis testing?
S Study Type	Sample data or experiment data?
T Type of Data	One variable or two? Categorical or numerical?
N Number of Samples or Treatments	How many samples or treatments?

When the answers to these questions are

Q: estimation
S: sample data
T: two numerical variables
N: one sample

the method you will want to consider in a regression setting is the confidence interval for the slope of a population regression line.

Once you have selected the confidence interval for the slope of a population regression line as the method you want to consider, because this is an estimation problem you would follow the five-step process for estimation problems (EMC³).

 Example 15.4   **The Bison of Yellowstone Park**

Data set available

The dedicated work of conservationists for over 100 years has brought the bison in Yellowstone National Park from near extinction to a herd of over 3000 animals. This recovery is a mixed blessing. Many bison have been exposed to the bacteria that cause brucellosis, a disease that infects domestic cattle, and there are many domestic cattle herds near Yellowstone. Because of concerns that free-ranging bison can infect nearby cattle, it is important to monitor and manage the size of the bison population and, if possible, keep bison from transmitting this bacteria to ranch cattle. The article **"Reproduction and Survival of Yellowstone Bison" (*The Journal of Wildlife Management* [2007]: 2365–2372)** described a large multiyear study of the factors that influence bison movement and herd size. The researchers studied a number of environmental factors to better understand the relationship between bison reproduction and the environment. One factor thought to influence reproduction is stress due to accumulated snow, which makes foraging more difficult for the pregnant bison. Data on

$$y = \text{spring calf ratio (SCR)}$$

and

$$x = \text{previous fall snow-water equivalent (SWE)}$$

for 17 years are shown in the accompanying table. Spring calf ratio is the ratio of calves to adults, a measure of reproductive success. Snow water equivalent is the depth of water that would result if the snow pack were melted. The SWE measurements in the table are in thousands of centimeters (so 1.93 represents 1930 cm). The researchers were interested in estimating the mean change in spring calf ratio associated with each additional 1000 cm in snow-water equivalent.

SCR	SWE	SCR	SWE
0.19	1.933	0.22	3.317
0.14	4.906	0.22	3.332
0.21	3.072	0.18	3.511
0.23	2.543	0.21	3.907
0.26	3.509	0.25	2.533
0.19	3.908	0.19	4.611
0.29	2.214	0.22	6.237
0.23	2.816	0.17	7.279
0.16	4.128		

For this example, the answers to the four key questions are *estimation, sample data, two numerical variables,* and *one sample.*

Q **Question Type**	Estimation or hypothesis testing?	Estimation
S **Study Type**	Sample data or experiment data?	Sample data
T **Type of Data**	One variable or two? Categorical or numerical?	Two numerical values
N **Number of Samples or Treatments**	How many samples or treatments?	One sample (regression)

This combination of answers suggests considering a confidence interval for the slope of a population regression line. You can now use the five-step process (EMC³) to estimate the slope of the population regression line.

Step	
Estimate	In this example, the value of β, the mean increase in spring calf ratio for each additional 1000 cm of snow-water equivalent, will be estimated.
Method	Because the answers to the four key questions are estimation, sample data, two numerical values, and one sample, a confidence interval for β, the slope of the population regression line, will be considered.
	For this example, a 95% confidence level will be used.
Check	The four basic assumptions of the simple linear regression model need to be met in order to use this confidence interval.
	The investigators collected data from 17 successive years. To proceed, you would need to assume that these years are representative of yearly circumstances at Yellowstone, and that each year's reproduction and snowfall is independent of previous years. You should keep this in mind when you get to the step that involves interpretation.
	A scatterplot of the data is shown here. The pattern in the plot looks linear and the spread does not seem to be different for different values of x.

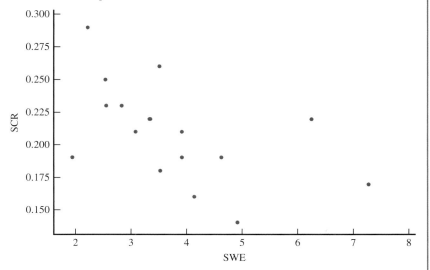

In Section 15.3, you will see how the assumption that the distribution of the random errors is normal can be evaluated. For now, you can assume that this assumption is reasonable and proceed with the rest of this example.

Calculate	JMP regression output is shown here:

◿ **Linear Fit**

SCR = 0.2606709 - 0.0136667*SWE

◿ **Summary of Fit**

RSquare	0.257696
RSquare Adj	0.208209
Root Mean Square Error	0.033512
Mean of Response	0.209412
Observations (or Sum Wgts)	17

◿ **Parameter Estimates**

Term	Estimate	Std Error	t Ratio	Prob>|t|
Intercept	0.260609	0.023888	10.91	<.0001*
SWE	-0.013667	0.005989	-2.28	0.0375*

$df = n - 2 = 17 - 2 = 15$

The t critical value for a 95% confidence level and df = 15 is 2.13.

$b \pm (t \text{ critical value})s_b$

$= -0.0137 \pm (2.13)(0.00599)$

$= (-0.0265, -0.0009)$

(continued)

Step	
Communicate Results	**Confidence interval:**
	You can be 95% confident that the true average change in spring calf ratio associated with an increase of 1000 cm in the snow-water equivalent is between -0.027 and -0.001. This means that you think the average spring calf ratio decreases by somewhere between 0.001 and 0.027 with each additional 1000 cm of snow-water equivalent.
	Confidence level:
	The method used to construct this interval estimate is successful in capturing the actual value of the slope of the population regression about 95% of the time.

Hypothesis Tests Concerning β

Hypotheses about β can be tested using a t test similar to the t tests introduced in Chapter 12. The null hypothesis states that β has a specified hypothesized value. The t statistic results from standardizing b, the estimate of β, under the assumption that H_0 is true. When H_0 is true, the sampling distribution of this statistic is the t distribution with df $= n - 2$.

Hypothesis Test for the Slope of the Population Regression Line, β

Appropriate when the four basic assumptions of the simple linear regression model are reasonable:

1. The distribution of e at any particular x value has mean value 0 (that is $\mu_e = 0$).
2. The standard deviation of e is σ_e, which does not depend on x.
3. The distribution of e at any particular x value is normal.
4. The random deviations $e_1, e_2, e_3, \ldots e_n$ associated with different observations are independent of one another.

When these conditions are met, the following test statistic can be used:

$$t = \frac{b - \beta_0}{s_b}$$

where β_0 is the hypothesized value from the null hypothesis.

Form of the null hypothesis: $H_0 : \beta = \beta_0$

When the assumptions of the simple linear regression model are reasonable and the null hypothesis is true, the t test statistic has a t distribution with df $= n - 2$.

Associated P-value:

When the alternative hypothesis is...	The P-value is...
$H_a : \beta > \beta_0$	Area to the right of the computed t under the appropriate t curve
$H_a : \beta < \beta_0$	Area to the left of the computed t under the appropriate t curve
$H_a : \beta \neq \beta_0$	2(area to the right of t) if t is positive
	or
	2(area to the left of the t) if t is negative

This test is a method you should consider when the answers to the four key questions are *hypothesis testing, sample data, two numerical variables,* and *one sample.* You would carry out this test using the five-step process for hypothesis testing problems (HMC³).

Inference for a population slope generally focuses on two questions:

(1) Is the population slope different from zero?
(2) What are plausible values for the population slope?

The question of plausible values can be addressed by calculating a confidence interval for the population slope. The question of whether a population slope is equal to zero can be answered by using the hypothesis testing procedure with a null hypothesis H_0: $\beta = 0$. This test of H_0: $\beta = 0$ versus H_a: $\beta \neq 0$ is called the *model utility test for simple linear regression*. The default computer output for inference for a regression slope is for the model utility test.

When the null hypothesis of the model utility test is true, the population regression line is a horizontal line, and the value of y in the simple linear regression model does not depend on x. That is,

$$y = \alpha + \beta x + e$$
$$= \alpha + 0x + e$$
$$= \alpha + e$$

If β is in fact equal to 0, knowledge of x will be of no use — it will have no "utility" for predicting y. On the other hand, if β is different from 0, there is a useful linear relationship between x and y, and knowledge of x may be useful for predicting y. This is illustrated by the scatterplots in Figure 15.10.

FIGURE 15.10
(a) $\beta = 0$; (b) $\beta \neq 0$

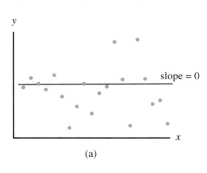
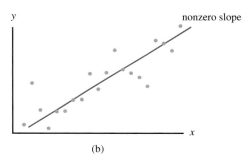

(a)   (b)

The Model Utility Test for Simple Linear Regression

The **model utility test for simple linear regression** is the test of

$$H_0: \beta = 0$$

versus

$$H_a: \beta \neq 0$$

The null hypothesis specifies that there is *no* useful linear relationship between x and y, whereas the alternative hypothesis specifies that there *is* a useful linear relationship between x and y. If H_0 is rejected, you can conclude that the simple linear regression model is useful for predicting y.

The test statistic is the t ratio

$$t = \frac{b - 0}{s_b} = \frac{b}{s_b}.$$

It is recommended that the model utility test be carried out before using an estimated regression line to make predictions or to estimate a mean y value.

 Example 15.5   The British (Musical) Invasion

Data set available   Have you ever recognized a song from your past when scanning from station to station on your car radio? After hearing just a very short segment of the song, maybe you could remember the title of the song, the artist, or even when the song was released. The article **"Plink: 'Thin slices' of Music" (_Krumhansl, C. Music Perception_ [2010]:337–354)** describes a study of this phenomenon. The investigator compiled a list of hit songs from _Rolling Stone_, _Billboard_, and Blender lists of songs, and added some more recent songs familiar to college students. Twenty-three college students were then exposed to 56 clips of songs. Most of these students had had musical training, and they listened to popular music for an average of 21.7 hours per week. After hearing three short clips from a song (only 400 ms in duration), the students were asked what year they thought each of the songs was released. The accompanying table shows the actual release year and the average of the release years given by the students. The actual release years ranged from 1965 (The Beatles, "Help") to 2008 (Katy Perry, "I Kissed a Girl").

Actual and Judged Release Years							
Actual Release	**Judged Release**	**Actual Release**	**Judged Release**	**Actual Release**	**Judged Release**	**Actual Release**	**Judged Release**
1998	1997.2	1976	1983.3	1976	1988.0	1970	1985.4
1967	1973.7	2008	1995.0	2006	1996.7	1975	1985.9
1998	1996.3	1971	1979.8	1974	1985.4	1991	1993.3
1999	1993.3	1965	1976.8	2007	1999.8	2008	1995.4
1983	1985.4	1967	1975.0	1976	1987.2	1965	1977.6
1982	1988.0	1971	1978.0	1974	1977.6	1987	1990.7
1965	1970.2	1967	1978.0	1970	1982.8	1975	1986.3
1991	1992.8	1984	1983.3	1971	1976.3	1968	1986.7
1983	1984.1	1984	1989.8	1999	1988.5	1987	1988.0
1976	1979.3	1968	1976.7	1997	1994.1	2008	1990.2
1971	1975.4	1965	1978.5	2006	1995.4	1982	1991.1
1981	1984.6	1965	1977.2	1981	1989.3	1979	1983.7
1967	1973.7	1979	1986.7	2008	1993.7	2000	1989.8
2007	1997.2	1997	1996.3	1965	1981.1	2000	1991.1

Is there a relationship between the average judged release year and the actual release year for these songs? A scatterplot of the data (Figure 15.11) suggests that there is a linear relationship between these two variables, but this can be confirmed this using the model utility test.

With x = Actual release year and y = Judged release year (the average of the student responses for judged release year), the equation of the estimated regression line is $\hat{y} = 1095 + 0.449x$. The five-step process for hypothesis testing can be used to carry out the model utility test.

FIGURE 15.11
Scatterplot of Judged release year versus Actual release year

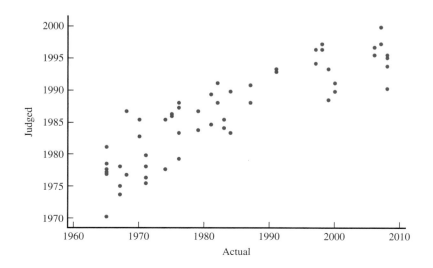

Process Step	
H Hypotheses	In the model utility test, the null hypothesis is that there is no useful relationship between the actual year and the judged release year: H_0: $\beta = 0$. The alternative hypothesis specifies that there is a useful relationship: $\beta \neq 0$. **Hypotheses:** *Null hypothesis*: H_0: $\beta = 0$ *Alternative hypothesis*: H_a: $\beta \neq 0$
M Method	Because the answers to the four key questions are hypothesis testing, sample data, two numerical variables in a regression setting, and one sample, a hypothesis test for the slope of a population regression line will be considered. The test statistic for this test is $$ t = \frac{b - 0}{s_b} = \frac{b}{s_b} $$ The value of 0 in the test statistic is the hypothesized value from the null hypothesis. For this example, a significance level of 0.05 will be used. **Significance level:** $\alpha = 0.05$
C Check	In Section 15.3, you will see how to assess whether the four assumptions of the simple linear regression model are reasonable. For this example, you can assume that these assumptions are reasonable and proceed with the model utility test.
C Calculate	JMP output is shown here: 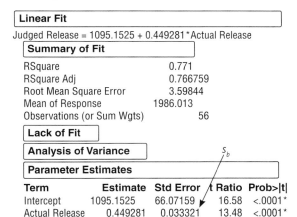 **Test statistic:** $$ t = \frac{b - 0}{s_b} = \frac{0.449 - 0}{0.0333} = 13.48 $$ **Associated *P*-value:** $P - \text{value} = $ twice area under t curve to the right of 13.48 $\qquad = 2P(t > 13.48)$ $\qquad \approx 0$
C Communicate results	Because the *P*-value is less than the selected significance level, the null hypothesis is rejected. **Decision:** Reject H_0. **Conclusion:** The sample data provide convincing evidence that there is a useful linear relationship between the actual release year and the judged release year.

Because the model utility test confirms that there is a useful linear relationship between judged release year and actual release year, it would be reasonable to use the estimated regression line to predict the judged release year for a given song based on its actual release year. Of course, before you do this, you would also want to evaluate the accuracy of predictions by looking at the value of s_e.

If H_0: $\beta = 0$ cannot be rejected using the model utility test at a reasonably small significance level, the search for a useful model could continue. One possibility is to relate y to x using a nonlinear model—an appropriate strategy if the scatterplot shows curvature. Another possibility is to consider models that include more than one predictor variable.

Summing It Up—Section 15.2

The following learning objectives were addressed in this section:

Conceptual Understanding

C2: Understand that the simple linear regression model provides a basis for making inferences about linear relationships.

The simple linear regression model assumes that there is a population regression line that describes how the mean value of y changes as x changes. The simple linear regression model has the form $y = \alpha + \beta x + e$, where e represents a random deviation from the line. The value of β specifies the change in the mean value of y associated with a 1-unit increase in the value of x. If $\beta = 0$, the mean value of y does not change with x.

Mastering the Mechanics

M3: Know the conditions for appropriate use of methods for making inferences about the slope of a population regression line, β.

There are four assumptions about the distribution of the random errors in the simple linear regression model that must be reasonable in order for the confidence interval formula and hypothesis test procedure of this section to be an appropriate way to answer questions about the value of the slope of the population regression line. They are

1. The distribution of e at any particular x value has mean value 0.
2. The standard deviation of e, s_e, is the same for any particular value of x.
3. The distribution of e at any given x value is normal.
4. The random errors associated with different observations are independent of one another.

M4: Use the five-step process for estimations problems (EMC³) and computer output to construct and interpret a confidence interval estimate for the slope of a population regression line.

Assuming the basic assumptions of the simple linear regression model are reasonable, a confidence interval for the slope of a population regression line can be calculated as $b \pm (t$ critical value$)s_b$, where s_b is the estimated standard deviation of b. Example 15.4 illustrates that calculation and interpretation of a confidence interval for the slope of a population regression line.

M5: Use the five-step process for hypothesis testing (HMC³) to test hypotheses about the slope of a population regression line.

Assuming the basic assumptions of the simple linear regression model are reasonable, hypotheses about the slope of a population regression line can be tested using the test statistic $t = \dfrac{b - \beta_0}{s_b}$, where β_0 is the hypothesized value from the null hypothesis and s_b is the estimated standard deviation of b. Example 15.5 illustrates the process of testing a hypothesis about the slope of a population regression line.

Putting It into Practice

P1: Interpret a confidence interval for the slope of a population regression line in context.

For an example of interpreting a confidence interval for the slope of a population regression line in context, see Example 15.4.

P2: Carry out the model utility test and interpret the result in context.
The model utility test is a test of the null hypothesis $H_0: \beta = 0$ versus the alternative hypothesis $H_a: \beta \neq 0$. If the null hypothesis is rejected, you can conclude that there is a useful linear relationship. Example 15.5 provides an example of the model utility test.

SECTION 15.2 EXERCISES

Each exercise set assesses the following chapter learning objectives: C2, M3, M4 , M5, P1, P2

SECTION 15.2 Exercise Set 1

15.13 The standard deviation of the errors, σ_e, is an important part of the linear regression model.
a. What is the relationship between the value of σ_e and the value of the test statistic in a test of a hypotheses about β?
b. What is the relationship between the value of σ_e and the width of a confidence interval for β?

15.14 A journalist is reporting about some research on appropriate amounts of sleep for people 9 to 19 years of age. In that research, a linear regression model is used to describe the relationship between alertness and number of hours of sleep the night before. The researchers reported a 95% confidence interval, but newspapers usually report an estimate and a margin of error. Explain how the journalist could determine the margin of error from the reported confidence interval.

15.15. A nursing student has completed his final project, and is preparing for a meeting with his project advisor. The subject of his project was the relationship between systolic blood pressure (SBP) and body mass index (BMI). The last time he met with his advisor he had completed his measurements, but only entered half his data into his statistical software. For the data he had entered, the necessary conditions for inference for β were met. In a short paragraph, explain, using appropriate statistical terminology, which of the conditions below must be rechecked.

1. The standard deviation of e is the same for all values of x.
2. The distribution of e at any particular x value is normal.

 15.16 Consider the accompanying data on $x = $ research and development expenditure (thousands of dollars) and $y = $ growth rate (% per year) for eight different industries.

| x | 2024 | 5038 | 905 | 3572 | 1157 | 327 | 378 | 191 |
| y | 1.90 | 3.96 | 2.44 | 0.88 | 0.37 | −0.90 | 0.49 | 1.01 |

a. Would a simple linear regression model provide useful information for predicting growth rate from research and development expenditure? Use a 0.05 level of significance.
b. Use a 90% confidence interval to estimate the average change in growth rate associated with a $1000 increase in expenditure. Interpret the resulting interval

15.17 The paper **"The Effects of Split Keyboard Geometry on Upper Body Postures"** (Ergonomics [2009]: 104–111) describes a study to determine the effects of several keyboard characteristics on typing speed. One of the variables considered was the front-to-back surface angle of the keyboard. Minitab output resulting from fitting the simple linear regression model with $x = $ surface angle (degrees) and $y = $ typing speed (words per minute) is given below.

Regression Analysis: Typing Speed versus Surface Angle

The regression equation is
Typing Speed = 60.0 + 0.0036 Surface Angle

Predictor	Coef	SE Coef	T	P
Constant	60.0286	0.2466	243.45	0.000
Surface Angle	0.00357	0.03823	0.09	0.931

$S = 0.511766$ R-Sq = 0.3%   R-Sq(adj) = 0.0%

Analysis of Variance

Source	DF	SS	MS	F	P
Regression	1	0.0023	0.0023	0.01	0.931
Residual Error	3	0.7857	0.2619		
Total	4	0.7880			

a. Suppose that the basic assumptions of the simple linear regression model are met. Carry out a hypothesis test to decide if there is a useful linear relationship between x and y.
b. Are the values of s_e and r^2 consistent with the conclusion from Part (a)? Explain.

15.18 Do taller adults make more money? The authors of the paper **"Stature and Status: Height, Ability, and Labor Market Outcomes"** (*Journal of Political Economics* [2008]: 499–532) investigated the association between height and earnings. They used the simple linear regression model to describe the relationship between $x = $ height (in inches) and $y = $ log(weekly gross earnings in dollars) in a very large sample of men. The logarithm of weekly gross earnings was used because this transformation resulted in a relationship that was approximately linear. The paper reported that the slope of the estimated regression line was $b = 0.023$ and the standard deviation of b was $s_b = 0.004$. Carry out a hypothesis test to decide if there is convincing evidence of a useful linear relationship between height and the logarithm of weekly earnings. You can assume that the basic assumptions of the simple linear regression model are met.

 15.19 Acrylamide is a chemical that is sometimes found in cooked starchy foods and which is thought to increase the risk of certain kinds of cancer. The paper **"A Statistical Regression Model for the Estimation of Acrylamide Concentrations in French Fries for Excess Lifetime Cancer Risk Assessment"** (*Food and Chemical Toxicology* [2012]: 3867–3876) describes a study to investigate the effect of frying time (in seconds) and acrylamide concentration (in micrograms per kilogram) in

 Data set available

french fries. The data in the accompanying table are approximate values read from a graph that appeared in the paper.

Frying Time	Acrylamide Concentration
150	155
240	120
240	190
270	185
300	140
300	270

a. For these data, the estimated regression line for predicting y = acrylamide concentration based on x = frying time is $\hat{y} = 87 + 0.359x$. What is an estimate of the average change in acrylamide concentration associated with a 1-second increase in frying time?
b. What would you predict for acrylamide concentration for a frying time of 250 seconds?
c. Use the given Minitab output to decide if there is convincing evidence of a useful linear relationship between acrylamide concentration and frying time. You may assume that the necessary conditions have been met.

S	R-sq	R-sq(adj)	R-sq(pred)
54.7108	14.38%	0.00%	0.00%

Coefficients

Term	Coef	SE Coef	T-Value	P-Value	VIF
Constant	87	112	0.78	0.480	
x	0.359	0.438	0.82	0.459	1.00

Regression Equation
$y = 87 + 0.359\, x$

SECTION 15.2 **Exercise Set 2**

15.20 Consider a test of hypotheses about, β the population slope in a linear regression model.
a. If you reject the null hypothesis, $\beta = 0$, what does this mean in terms of a linear relationship between x and y?
b. If you fail to reject the null hypothesis, $\beta = 0$, what does this mean in terms of a linear relationship between x and y?

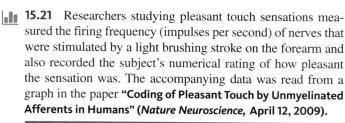 **15.21** Researchers studying pleasant touch sensations measured the firing frequency (impulses per second) of nerves that were stimulated by a light brushing stroke on the forearm and also recorded the subject's numerical rating of how pleasant the sensation was. The accompanying data was read from a graph in the paper **"Coding of Pleasant Touch by Unmyelinated Afferents in Humans"** (*Nature Neuroscience,* April 12, 2009).

Firing Frequency	Pleasantness Rating	Firing Frequency	Pleasantness Rating
23	0.2	28	2.0
24	1.0	34	2.3
22	1.2	33	2.2
25	1.2	36	2.4
27	1.0	34	2.8

 Data set available

a. Estimate the mean change in pleasantness rating associated with an increase of 1 impulse per second in firing frequency using a 95% confidence interval. Interpret the resulting interval.
b. Carry out a hypothesis test to decide if there is convincing evidence of a useful linear relationship between firing frequency and pleasantness rating.

15.22 The largest commercial fishing enterprise in the southeastern United States is the harvest of shrimp. In a study described in the paper **"Long-term Trawl Monitoring of White Shrimp, Litopenaeus setiferus (Linnaeus), Stocks within the ACE Basin National Estuariene Research Reserve, South Carolina"** (*Journal of Coastal Research* [2008]:193–199), researchers monitored variables thought to be related to the abundance of white shrimp. One variable the researchers thought might be related to abundance is the amount of oxygen in the water. The relationship between mean catch per tow of white shrimp and oxygen concentration was described by fitting a regression line using data from ten randomly selected offshore sites. (The "catch" per tow is the number of shrimp caught in a single outing.) Computer output is shown below.

The regression equation is
Mean catch per tow = −5859 + 97.2 O2 Saturation

Predictor	Coef	SE Coef	T	P
Constant	−5859	2394	−2.45	0.040
O2 Saturation	97.22	34.63	2.81	0.023

S = 481.632   R-Sq = 49.6%   R-Sq(adj) = 43.3%

a. Is there convincing evidence of a useful linear relationship between the shrimp catch per tow and oxygen concentration density? Explain.
b. Would you describe the relationship as strong? Why or why not?
c. Construct a 95% confidence interval for β and interpret it in context.
d. What margin of error is associated with the confidence interval in Part (c)?

15.23 The authors of the paper **"Decreased Brain Volume in Adults with Childhood Lead Exposure"** (*Public Library of Science Medicine* [May 27, 2008]: e112) studied the relationship between childhood environmental lead exposure and a measure of brain volume change in a particular region of the brain. Data were given for x = mean childhood blood lead level (μg/dL) and y = brain volume change (BVC, in percent). A subset of data read from a graph that appeared in the paper was used to produce the accompanying Minitab output.

Regression Analysis: BVC versus Mean Blood Lead Level
The regression equation is
BVC = −0.00179 − 0.00210 Mean Blood Lead Level

Predictor	Coef	SE Coef	T	P
Constant	−0.001790	0.008303	−0.22	0.830
Mean Blood Lead Level	−0.0021007	0.0005743	−3.66	0.000

Carry out a hypothesis test to decide if there is convincing evidence of a useful linear relationship between x and y. You can assume that the basic assumptions of the simple linear regression model are met.

ADDITIONAL EXERCISES

15.24 a. Explain the difference between the line $y = \alpha + \beta x$ and the line $\hat{y} = a + bx$.
b. Explain the difference between β and b.
c. Let x^* denote a particular value of the independent variable. Explain the difference between $\alpha + \beta x^*$ and $a + bx^*$.

15.25 What is the distinction between σ_e and s_e?

15.26 The accompanying data are a subset of data from the report **"Great Jobs, Great Lives" (Gallup-Purdue Index 2015 Report, www.gallup.com/reports/197144/gallup-purdue -index-report-2015.aspx, retrieved May 27, 2017)**. The values are approximate values read from a scatterplot. Students at a number of universities were asked if they agreed that their education was worth the cost. One variable in the table is the *U.S. News and World Report* ranking of the university in 2015. The other variable in the table is the percentage of students at the university who responded "strongly agree."

University Ranking	Percentage of Alumni Who Strongly Agree
28	53
29	58
30	62
37	55
45	54
47	62
52	55
54	62
57	70
60	58
65	66
66	55
72	65
75	57
82	67
88	59
98	75

a. Fit a linear regression model that would allow you to predict the the percentage of alumni who strongly agree that their education was worth the cost, using 2015 university ranking as a predictor.
b. Do the sample data support the hypothesis that there is a useful linear relationship between the percentage of

alumni who strongly agree that their education was worth the cost and 2015 university ranking? Test the appropriate hypotheses using $\alpha = .05$.

15.27 The paper referenced in Exercise 15.11 (**"Depression, Body Mass Index, and Chronic Obstructive Pulmonary Disease—A Holistic Approach,"** *International Journal of COPD* **[2016]:239–249**) gave data on

x = change in Body Mass Index (BMI in kilograms/meter^2)

and

y = change in a measure of depression

for patients suffering from depression who participated in a pulmonary rehabilitation program. JMP output for these data is shown below.

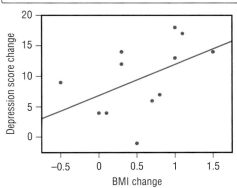

Bivariate Fit of Depression Score Change by BMI Change

— Linear Fit

Linear Fit

Depression score change = 6.8725681 + 5.077821*BMI Change

Summary of Fit

RSquare	0.234828
RSquare Adj	0.158311
Root Mean Square Error	5.365593
Mean of Response	9.75
Observations (or Sum Wgts)	12

Analysis of Variance

Source	DF	Sum of Squares	Mean Square	F Ratio
Model	1	88.35409	88.3541	3.0690
Error	10	287.89591	28.7896	**Prob > F**
C.Total	11	376.25000		0.1104

Parameter Estimates

Term	Estimate	Std Error	t Ratio	Prob>|t|
Intercept	6.8725681	2.257651	3.04	0.0124*
BMI Change	5.077821	2.898557	1.75	0.1104

a. What does the scatterplot suggest about the relationship between depression score change and BMI change?
b. What is the equation of the estimated regression line?
c. Is there is a useful linear relationship between the two variables? Carry out an appropriate test using a significance level of $\alpha = 0.05$.

15.28 In anthropological studies, an important characteristic of fossils is cranial capacity. Frequently skulls are at least partially decomposed, so it is necessary to use other characteristics to obtain information about capacity (in cm³). One measure that has been used is the length of the lambda-opisthion chord (in mm). The article **"Vertesszollos and the Presapiens Theory" (*American Journal of Physical Anthropology* [1971])** reported the accompanying data for $n = 7$ Homo erectus fossils.

x (chord length)	78	75	78	81	84	86	87
y (capacity)	850	775	750	975	915	1015	1030

Suppose that from previous evidence, anthropologists had believed that for each 1-mm increase in chord length, cranial capacity would be expected to increase by 20 cm³. Do these new experimental data provide convincing evidence against this prior belief?

 Data set available

15.29 Suppose you are given the computer output shown. You are interested in testing the null hypothesis $\beta = 1.0$ versus an alternative hypothesis of $\beta > 1.0$. Describe how you would use the given computer output to test these hypotheses.

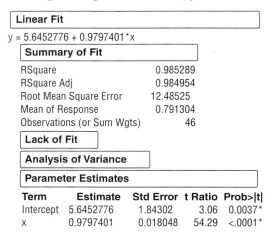

Linear Fit

$y = 5.6452776 + 0.9797401 * x$

Summary of Fit	
RSquare	0.985289
RSquare Adj	0.984954
Root Mean Square Error	12.48525
Mean of Response	0.791304
Observations (or Sum Wgts)	46

Lack of Fit

Analysis of Variance

Parameter Estimates

Term	Estimate	Std Error	t Ratio	Prob>|t|
Intercept	5.6452776	1.84302	3.06	0.0037*
x	0.9797401	0.018048	54.29	<.0001*

SECTION 15.3 Checking Model Adequacy

Section 15.2 introduced methods for estimating and testing hypotheses about β, the slope in the simple linear regression model

$$y = \alpha + \beta x + e$$

In this model, e represents the random deviation of a y value from the population regression line $\alpha + \beta x$. The methods presented in Section 15.2 require that some assumptions about the random deviations in the simple linear regression model be met in order for inferences to be valid. These assumptions include:

1. At any particular x value, the distribution of e is normal.
2. At any particular x value, the standard deviation of e is σ_e, which is constant over all values of x (that is, σ_e does not depend on x).

Inferences based on the simple linear regression model are still appropriate if model assumptions are slightly violated (for example, mild skew in the distribution of e). However, interpreting a confidence interval or the result of a hypothesis test when assumptions are seriously violated can result in misleading conclusions. For this reason, it is important to be able to detect any serious violations.

Residual Analysis

If the deviations $e_1, e_2, \ldots, e_n$ from the population line were available, they could be examined for any inconsistencies with model assumptions. For example, a normal probability plot of these deviations would suggest whether or not the normality assumption was plausible. However, because these deviations are

$$e_1 = y_1 - (\alpha + \beta x_1)$$
$$\vdots$$
$$e_n = y_n - (\alpha + \beta x_n)$$

they can be calculated only if α and β are known. In practice, this will not be the case. Instead, diagnostic checks must be based on the residuals

$$y_1 - \hat{y}_1 = y_1 - (a + b x_1)$$
$$\vdots$$
$$y_n - \hat{y}_n = y_n - (a + b x_n)$$

which are the deviations from the *estimated* regression line. When all model assumptions are met, the mean value of the residuals at any particular x value is 0. Any observation that gives a large positive or negative residual should be examined carefully for any unusual circumstances, such as a recording error or nonstandard experimental condition. Identifying residuals with unusually large magnitudes is made easier by inspecting **standardized residuals**.

Recall that a quantity is standardized by subtracting its mean value (0 in this case) and dividing by its actual or estimated standard deviation:

$$\text{standardized residual} = \frac{\text{residual}}{\text{estimated standard deviation of residual}}$$

The value of a standardized residual tells you the distance (in standard deviations) of the corresponding residual from its expected value, 0.

Because residuals at different x values have different standard deviations (depending on the value of x for that observation),[1] computing the standardized residuals can be tedious. Fortunately, many computer regression programs provide standardized residuals.

Example 15.6   Revisiting the Elk

Data set available

Example 15.3 introduced data on

$$x = \text{chest girth (in cm)}$$

and

$$y = \text{weight (in kg)}$$

for a sample of 19 Rocky Mountain elk. (See Example 15.3 for a more detailed description of the study.)

Inspection of the scatterplot in Figure 15.12 suggests the data are consistent with the assumptions of the simple linear regression model.

FIGURE 15.12
Scatterplot for the elk data

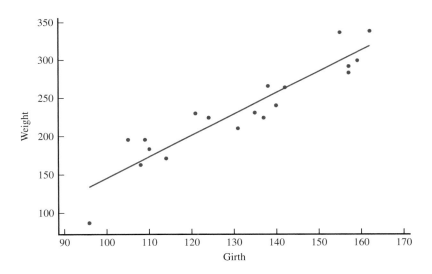

The data, residuals, and the standardized residuals (computed using Minitab) are given in Table 15.1. For the residual with the largest magnitude, 38.1397, the standardized residual is 1.813. That is, the residual is approximately 1.8 standard deviations above its expected value of 0. This value is not particularly unusual in a sample of this size. Also notice that for the negative residual with the largest magnitude, −38.2661, the standardized residual is −1.923, still not unusual in a sample of this size. On the standardized scale, no residual here is surprisingly large.

---

[1] The estimated standard deviation of the ith residual, $y_i - \hat{y}_i$, is $s_e \sqrt{1 + \frac{1}{n} + \frac{(x_i - \bar{x})^2}{\Sigma(x - \bar{x})^2}}$

TABLE 15.1 Data, residuals, and standardized residuals for the elk data

Observation	Girth (cm) x	Weight (kg) y	Residual	Standardized Residual	$\hat{y}$
1	96	98	−38.2661	−1.923	136.266
2	105	196	34.9314	1.680	161.069
3	108	163	−6.3361	−0.301	169.336
4	109	196	23.9080	1.133	172.092
5	110	183	8.1522	0.385	174.848
6	114	171	−14.8711	−0.695	185.871
7	121	230	24.8380	1.145	205.162
8	124	225	11.5705	0.531	213.429
9	131	211	−21.7203	−0.993	232.720
10	135	231	−12.7436	−0.583	243.744
11	137	225	−24.2553	−1.111	249.255
12	138	266	13.9889	0.641	252.011
13	140	241	−16.5228	−0.759	257.523
14	142	264	0.9655	0.044	263.034
15	157	284	−20.3720	−0.975	304.372
16	157	292	−12.3720	−0.592	304.372
17	159	300	−9.8837	−0.477	309.884
18	155	337	38.1397	1.813	298.860
19	162	339	20.8488	1.020	318.151

Next, consider the assumption of the normality of e's. Figure 15.13 shows box plots of the residuals and standardized residuals. The box plots are approximately symmetric and there are no outliers, so the assumption of normally distributed errors seems reasonable.

FIGURE 15.13
Boxplots of residuals and standardized residuals for the elk data

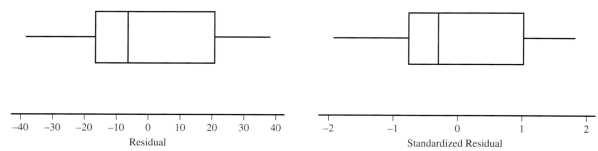

Notice that the boxplots of the residuals and standardized residuals are nearly identical. While it is preferable to work with the standardized residuals, if you do not have access to a computer package or calculator that will produce standardized residuals, a plot of the unstandardized residuals should suffice.

A normal probability plot of the standardized residuals (or the residuals) is another way to assess whether it is reasonable to assume that $e_1, e_2, ..., e_n$ all come from the same normal distribution. An advantage of the normal probability plot, shown in Figure 15.14, is that the value of each residual can be seen, which provides more information about the distribution. The pattern in the normal probability plot of the standardized residuals and pattern in the normal probability plot of the the residuals for the elk data are reasonably straight, confirming that the assumption of normality of the error distribution is reasonable. Also notice that the pattern in both normal probability plots is similar, so you don't need to construct both—either plot could be used.

FIGURE 15.14
Normal probability plots of residuals and standardized residuals for the elk data

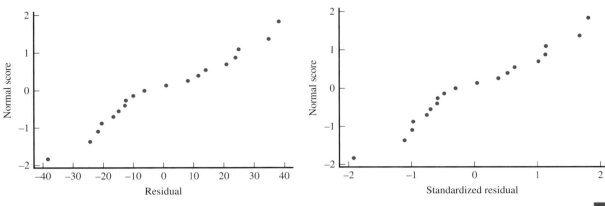

Plotting the Residuals

A plot of the (x, residual) pairs is called a **residual plot**, and a plot of the (x, standardized residual) pairs is a **standardized residual plot**. Residual and standardized residual plots typically exhibit the same general shapes. If you are using a computer package or graphing calculator that calculates standardized residuals, the standardized residual plot is recommended. If not, it is acceptable to use the unstandardized residual plot instead.

A standardized residual plot or a residual plot is often helpful in identifying unusual or highly influential observations and in checking for violations of model assumptions. A desirable plot is one that exhibits no particular pattern (such as curvature or a much greater spread in one part of the plot than in another) and that has no point that is far removed from all the others. A point in the residual plot falling far above or far below the horizontal line at height 0 corresponds to a large residual, which can indicate unusual behavior, such as a recording error, a nonstandard experimental condition, or an atypical experimental subject. A point with an x value that differs greatly from others in the data set could have exerted excessive influence in determining the estimated regression line.

A standardized residual plot, such as the one pictured in Figure 15.15(a) is desirable, because no point lies much outside the horizontal band between -2 and 2 (so there is no unusually large residual corresponding to an outlying observation). There is no point far to the left or right of the others (which could indicate an observation that might greatly influence the estimated line), and there is no pattern to indicate that the model should somehow be modified.

When the plot has the appearance of Figure 15.15(b), the fitted model should be changed to incorporate curvature (a nonlinear model). The increasing spread from left to right in Figure 15.15(c) suggests that the variance of y is not the same at each x value but rather increases with x. A straight-line model may still be appropriate, but the best-fit line should be obtained by using *weighted least* squares rather than ordinary least squares. This involves giving more weight to observations in the region exhibiting low variability and less weight to observations in the region exhibiting high variability. A specialized regression analysis textbook or a statistician should be consulted for more information on using weighted least squares.

The standardized residual plots of Figures 15.15(d) and 15.15(e) show an outlier (a point with a large standardized residual) and a potentially influential observation, respectively. Consider deleting the observation corresponding to such a point from the data set and refitting a line. Substantial changes in estimates and various other quantities are a signal that a more careful analysis should be carried out before proceeding.

FIGURE 15.15
Examples of residual plots:
(a) satisfactory plot;
(b) plot suggesting that a curvilinear regression model is needed;
(c) plot indicating nonconstant variance;
(d) plot showing a large residual;
(e) plot showing a potentially influential observation

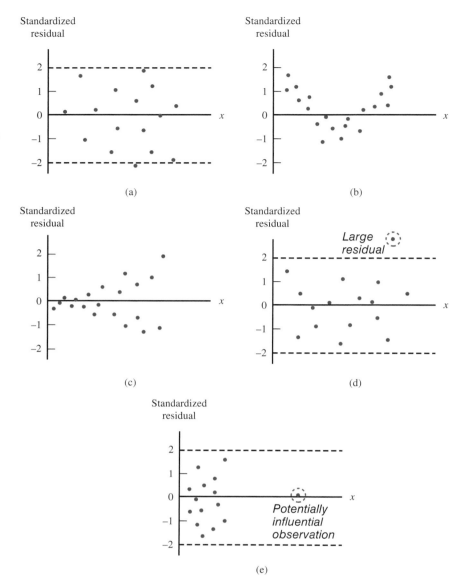

Example 15.7   The Business of Baseball

Data set available

The article **"The Business of Baseball" (www.forbes.com/mlb-valuations/list/#tab:overall, retrieved May 27, 2017)** ranked the 30 Major League Baseball teams based on their 2016 value (in millions of dollars). Also included in the article are data on annual operating income (in millions of dollars). A positive value for operating income indicates a profit for the year, and a negative operating income represents a loss for the year.

Team	2016 Value (millions of dollars)	Operating Income (millions of dollars)
New York Yankees	3,400	13.0
Los Angeles Dodgers	2,500	−73.2
Boston Red Sox	2,300	43.2
San Francisco Giants	2,250	72.6
Chicago Cubs	2,200	50.8
New York Mets	1,650	46.8
St. Louis Cardinals	1,600	59.8

(continued)

Team	2016 Value (millions of dollars)	Operating Income (millions of dollars)
Los Angeles Angels of Anaheim	1,340	41.7
Washington Nationals	1,300	22.5
Philadelphia Phillies	1,235	−8.9
Texas Rangers	1,225	−4.7
Seattle Mariners	1,200	16.8
Atlanta Braves	1,175	27.8
Detroit Tigers	1,150	11.0
Houston Astros	1,100	66.6
Chicago White Sox	1,050	20.2
Baltimore Orioles	1,000	8.8
Pittsburgh Pirates	975	35.3
Arizona Diamondbacks	925	17.4
Minnesota Twins	910	18.5
Cincinnati Reds	905	9.0
Toronto Blue Jays	900	1.2
San Diego Padres	890	32.9
Milwaukee Brewers	875	27.0
Kansas City Royals	865	39.0
Colorado Rockies	860	5.5
Cleveland Indians	800	18.0
Oakland Athletics	725	32.7
Miami Marlins	675	15.8
Tampa Bay Rays	650	8.2

To investigate whether there is a relationship between y = 2016 value and x = annual operating income, a simple linear regression model was fit. Figure 15.16 shows a scatterplot of the data and the least squares regression line. Notice that there are two teams that stand out in the plot. One has an unusually low operating income (the L.A. Dodgers with an operating income of −73.2 million dollars). The other team that stands out is a team with an unusually high 2016 value (the New York Yankees, with a 2016 value of 3,400 million dollars (3.4 billion dollars)).

FIGURE 15.16

Scatterplot of 2016 value versus annual operating income for 30 Major League Baseball teams

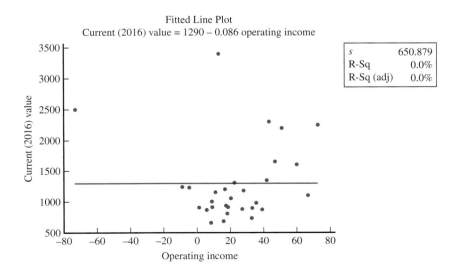

For this model, r^2 is approximately 0 and the model utility test does not reject the hypothesis that $\beta = 0$. This would lead you to think that there is not a useful linear relationship between 2016 value and operating income. But looking back at the scatterplot, you can see that if you were to ignore the two unusual teams, there does appear to be a positive linear relationship between 2016 value and operating income for the data set that consists of the remaining 28 teams. To investigate the potential influence of these two teams on the model, you can delete these two teams from the data set and then fit a regression model to the remaining data.

Figure 15.17 shows a scatterplot and the least squares regression line for the 28 major league baseball teams that remain after the Dodgers and the Yankees are excluded from the data set. Notice that the slope of the line has changed dramatically (from 0.086 to 12.60) and that the r^2 value is now 0.334. The model utility test confirms that there is a useful linear relationship between 2016 value and operating income for these 28 teams.

FIGURE 15.17

Scatterplot of 2016 value versus annual operating income for 28 Major League Baseball teams (L.A. Dodgers and N.Y. Yankees excluded)

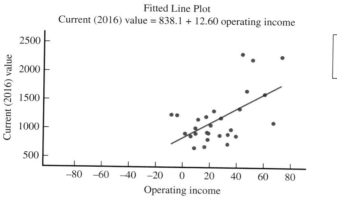

Data set available

Example 15.8 Competitive Cross-Country Skiing

The paper **"Time Trials Predict the Competitive Performance Capacity of Junior Cross-Country Skiers"** (*International Journal of Sports Physiology and Performance* **[2014]: 12–18**) described a study to investigate whether scores in trials performed by junior cross-country skiers during the preseason could be used to predict performance during the competitive season. Data on

$$x = \text{preseason score}$$

and

$$y = \text{competition score}$$

consistent with a scatterplot that appeared in the paper are shown in Table 15.2.

TABLE 15.2 Data, Residuals, and Standardized Residuals for Example 15.8

Skier	Preseason Score	Competition Score	Residual	Standardized Residual
1	93	100	−9.107	−0.160
2	150	105	−60.221	−0.958
3	140	200	44.624	0.719
4	210	200	−24.288	−0.368
5	215	310	80.790	1.221
6	260	290	16.490	0.248
7	300	220	−92.888	−1.417
8	330	410	67.578	1.056
9	380	300	−91.644	−1.537
10	400	480	68.666	1.201

Is it reasonable to use the given data to construct a confidence interval or test hypotheses about the average change in competition score associated with a 1-point increase in the preseason score? It depends on whether the assumptions that the distribution of the deviations from the population regression line at any fixed x value is approximately normal and that the variance of the distribution does not depend on x are reasonable. Constructing a plot of the standardized residuals will provide insight into whether these assumptions are in fact reasonable.

Figure 15.18 shows a box plot of the standardized residuals and a standardized residual plot. The boxplot is reasonably symmetric with no outliers, and the standardized residual plot does not show evidence of any patterns or increasing spread.

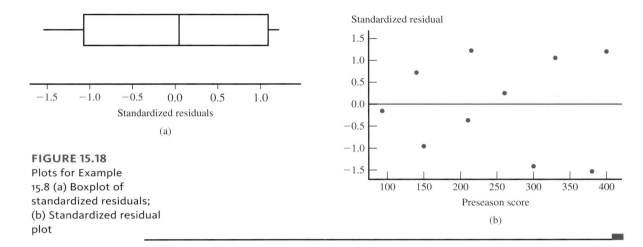

FIGURE 15.18
Plots for Example 15.8 (a) Boxplot of standardized residuals; (b) Standardized residual plot

Example 15.9 **A New Pediatric Tracheal Tube**

The article **"Appropriate Placement of Intubation Depth Marks in a New Cuffed, Paediatric Tracheal Tube" (*British Journal of Anaesthesia* [2004]: 80–87)** describes a study of the use of tracheal tubes in newborns and infants. Newborns and infants have small trachea, and there is little margin for error when inserting tracheal tubes. Using X-rays of a large number of children aged 2 months to 14 years, the researchers examined the relationships between appropriate trachea tube insertion depth and other variables such as height, weight, and age. A scatterplot and a standardized residual plot constructed using data on the insertion depth and height of the children (both measured in cm) are shown in Figure 15.19.

FIGURE 15.19
(a) Scatterplot for insertion depth vs. height data of Example 15.9; (b) standardized residual plot

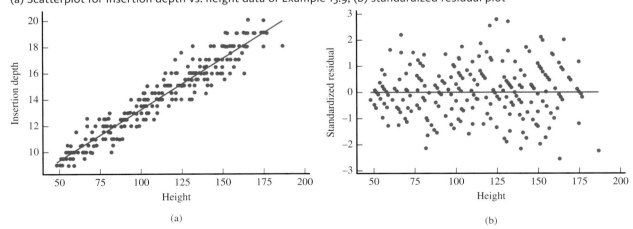

Residual plots like the ones pictured in Figure 15.19(b) are desirable. No point lies much outside the horizontal band between -2 and 2 (so there are no unusually large residuals corresponding to outliers). There is no point far to the left or right of the others (no observation that might be influential), and there is no pattern of curvature or differences in the variability of the residuals for different height values to indicate that the model assumptions are not reasonable.

But consider what happens when the relationship between insertion depth and weight is examined. A scatterplot of insertion depth and weight (kg) is shown in Figure 15.20(a), and a standardized residual plot in Figure 15.20(b). While some curvature is evident in the original scatterplot, it is even more clearly visible in the standardized residual plot. A careful inspection of these plots suggests that along with curvature, the residuals may be more variable at larger weights. When plots have this curved appearance and increasing variability in the residuals, the linear regression model is not appropriate.

FIGURE 15.20

(a) Scatterplot for insertion depth vs. weight data of Example 15.9; (b) standardized residual plot

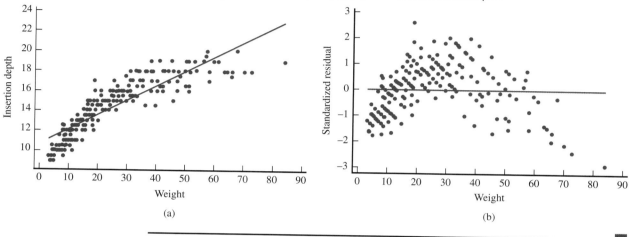

Example 15.10   Looking for Love in All the Right... Trees?

Treefrogs' search for mating partners was the examined in the article, **"The Cause of Correlations Between Nightly Numbers of Male and Female Barking Treefrogs (*Hyla gratiosa*) Attending Choruses" (*Behavioral Ecology* [2002]: 274–281)**. A lek, in the world of animal behavior, is a cluster of males gathered in a relatively small area to exhibit courtship displays. The "female preference" hypothesis asserts that females will prefer larger leks over smaller leks, presumably because there are more males to choose from. The scatterplot and residual plot in Figure 15.21 show the relationship between the number of females and the number of males in observed leks of barking treefrogs. You can see that the unequal variance, which is noticeable in the scatterplot, is even more evident in the residual plot. This indicates that the assumptions of the linear regression model are not reasonable in this situation.

FIGURE 15.21

(a) Scatterplot for treefrog data of Example 15.10; (b) residual plot

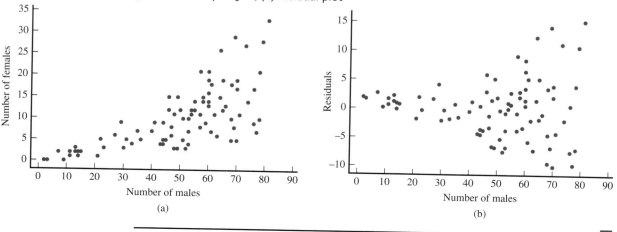

Summing It Up—Section 15.3

The following learning objectives were addressed in this section:

Mastering the Mechanics

M2: Use scatterplots, residual plots, and normal probability plots or boxplots to assess the credibility of the assumptions of the simple linear regression model.

To assess whether the assumptions of the simple linear regression model are reasonable, you can look at graphs based on the residuals or standardized residuals. To decide if it is reasonable to think that the distribution of the random deviations is normal, you can look at a boxplot or normal probability plot of the residuals. If you use a boxplot, look for a boxplot that is approximately symmetric and that has no outliers. If you use a normal probability plot, look for a linear pattern in the plot to indicate that the normality assumption is reasonable. In order to decide if it is reasonable to think that the simple linear regression model is appropriate and that the standard deviation of the random deviations is the same for different values of x, look at a plot of the residuals or the standardized residuals versus x. You want to see a plot that does not show changing variability and that has no apparent patterns. Example 15.8 illustrates the use of plots to assess the appropriateness of the simple linear regression model in a particular context.

M6: Use graphs to identify outliers and potentially influential observations.

An outlier is an observation with a large residual (a point that is far away from the estimated regression line in the y direction). A potentially influential observation is one that is far removed from the rest of the data in the x direction. Example 15.7 considers how to assess the effect of outliers and potentially influential observations on the equation of the regression line.

SECTION 15.3 EXERCISES

Each exercise set assesses the following chapter learning objectives: M2, M6

SECTION 15.3 Exercise Set 1

15.30 The following graphs are based on data from an experiment to assess the effects of logging on a squirrel population in British Columbia (**"Effects of Logging Pattern and Intensity on Squirrel Demography,"** *The Journal of Wildlife Management* **[2007]: 2655–2663).** Plots of land, each nine hectares in area, were subjected to different percentages of logging, and the squirrel population density for each plot was measured after 3 years. The scatterplot, residual plot, and a boxplot of the residuals are shown here.

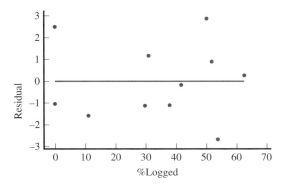

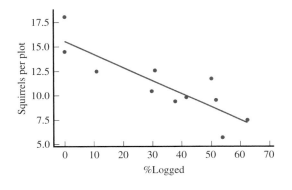

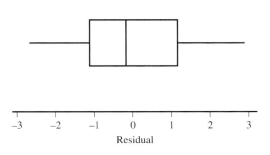

Does it appear that the assumptions of the simple linear regression model are plausible? Explain your reasoning in a few sentences.

15.31 The clutch size (number of eggs laid) for turtles is known to be influenced by body size, latitude, and average environmental temperature. Researchers gathered data on Gopher tortoises in Okeeheelee County Park in Florida to further understand the factors that affect reproduction in these animals **("Geographic Variation in Body and Clutch Size of Gopher Tortoises," *Copeia* [2007]: 355–363)**. The scatterplot, residual plot, and a normal probability plot of the residuals for the least squares regression line with x = body length and y = clutch size are shown here.

Does it appear that the assumptions of the simple linear regression model are plausible? Explain your reasoning in a few sentences.

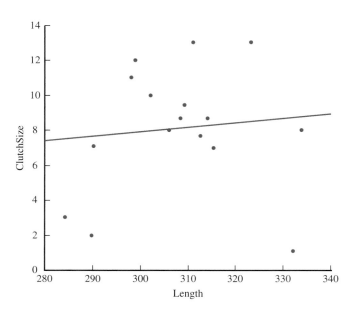

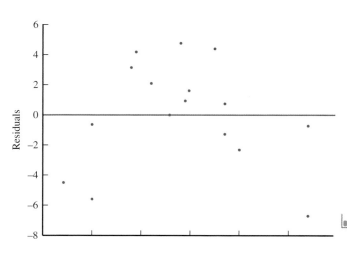

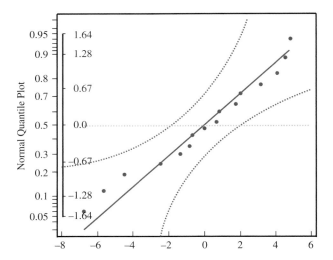

15.32 Carbon aerosols have been identified as a contributing factor in a number of air quality problems. In a chemical analysis of diesel engine exhaust, x = mass (μg/cm^2) and y = elemental carbon (μg/cm^2) were recorded **("Comparison of Solvent Extraction and Thermal Optical Carbon Analysis Methods: Application to Diesel Vehicle Exhaust Aerosol" *Environmental Science Technology* [1984]: 231–234)**. The estimated regression line for this data set is $\hat{y} = 31 + .737x$. The accompanying table gives the observed x and y values and the corresponding standardized residuals.

x	164.2	156.9	109.8	111.4	87.0
y	181	156	115	132	96
Std resid	2.52	0.82	0.27	1.64	0.08
x	161.8	230.9	106.5	97.6	79.7
y	170	193	110	94	77
Std resid	1.72	−0.73	0.05	−0.77	−1.11
x	118.7	248.8	102.4	64.2	89.4
y	106	204	98	76	89
Std resid	−1.07	−0.95	−0.73	−0.20	−0.68
x	108.1	89.4	76.4	131.7	100.8
y	102	91	97	128	88
Std resid	−0.75	−0.51	0.85	0.00	−1.49
x	78.9	387.8	135.0	82.9	117.9
y	86	310	141	90	130
Std resid	−0.27	−0.89	0.91	−0.18	1.05

a. Construct a standardized residual plot. Are there any unusually large residuals? Do you think that there are any influential observations?
b. Is there any pattern in the standardized residual plot that would indicate that the simple linear regression model is not appropriate?
c. Based on your plot in Part (a), do you think that it is reasonable to assume that the variance of y is the same at each x value? Explain.

15.33 The article **"Vital Dimensions in Volume Perception: Can the Eye Fool the Stomach?"** (*Journal of Marketing Research* [1999]: 313–326) gave the accompanying data on the dimensions (in cm) of the containers for 27 representative

Data set available

food products (Gerber baby food, Cheez Whiz, Skippy Peanut Butter, and Ahmed's tandoori paste, to name a few).

Product	Maximum Width (cm)	Minimum Width (cm)
1	2.50	1.80
2	2.90	2.70
3	2.15	2.00
4	2.90	2.60
5	3.20	3.15
6	2.00	1.80
7	1.60	1.50
8	4.80	3.80
9	5.90	5.00
10	5.80	4.75
11	2.90	2.80
12	2.45	2.10
13	2.60	2.20
14	2.60	2.60
15	2.70	2.60
16	3.10	2.90
17	5.10	5.10
18	10.20	10.20
19	3.50	3.50
20	2.70	1.20
21	3.00	1.70
22	2.70	1.75
23	2.50	1.70
24	2.40	1.20
25	4.40	1.20
26	7.50	7.50
27	4.25	4.25

a. Fit the simple linear regression model that would allow prediction of the maximum width of a food container based on its minimum width.

b. Calculate the standardized residuals (or just the residuals if you don't have access to a computer program that gives standardized residuals) and make a residual plot to determine whether there are any outliers.

c. The data point with the largest residual is for a 1-liter Coke bottle. Delete this data point and refit the regression. Did deletion of this point result in a large change in the equation of the estimated regression line?

d. For the regression line of Part (c), interpret the estimated slope and, if appropriate, the estimated intercept.

e. For the data set with the Coke bottle deleted, do you think that the assumptions of the simple linear regression model are reasonable? Give statistical evidence for your answer.

15.34 Models of climate change predict that global temperatures and precipitation will increase in the next 100 years, with the largest changes occurring during winter in northern latitudes. Researchers gathered data on the potential effects of climate change for flowering plants in Norway (**"Climatic Variability, Plant Phenology, and Northern Ungulates,"** *Ecology* **[1999]: 1322–1339**). The table below gives data for one flower species. Range of flowering dates and elevation for different sites in Norway were used to construct the given scatterplot. A potentially influential point is indicated on the scatterplot.

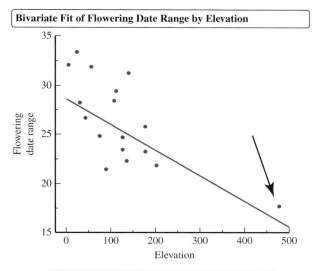

Bivariate Fit of Flowering Date Range by Elevation

Flowering Range versus Elevation: *Tussilago Farfara*	
Elevation (Meters Above Sea Level)	Flowering Date Range
23.3	33.4
5.6	32.0
55.6	31.9
140.0	31.3
31.1	28.1
112.2	29.3
106.7	28.4
42.2	26.6
75.6	24.9
176.7	25.7
126.7	24.7
126.7	23.5
176.7	23.2
201.1	21.8
133.3	22.3
90.0	21.4
41.1	19.7
125.6	17.6
477.8	17.6

a. Fit a linear regression model using all 19 observations. What are the values of a, b, r^2, s_e?

 Data set available

b. Fit a linear regression model with the indicated point omitted. What are the values of a, b, r^2, s_e?

c. In a few sentences, describe any differences you found in Parts (a) and (b).

d. The researchers could use the estimated regression equation based on all 19 observations to make predictions for elevations ranging from 0 to 500 meters; or they could use the estimated regression equation based on the 18 observations (omitting the observation identified by the arrow) to make predictions for elevations ranging from 0 to 200 meters. Which strategy would you recommend, and why?

SECTION 15.3 **Exercise Set 2**

15.35 In the study described in Exercise 15.31, the effect of latitude on mean clutch size was investigated. Data from various locations in Florida, Georgia, Alabama, and Mississippi on y = mean clutch size and x = latitude were measured. The scatterplot, standardized residual plot, and several graphs of the standardized residuals are shown below.

Does it appear that the assumptions of the simple linear regression model are plausible? Explain your reasoning in a few sentences.

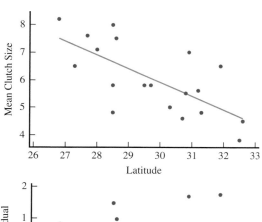

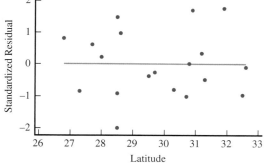

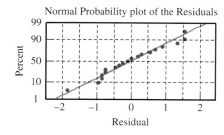

Normal Probability plot of the Residuals

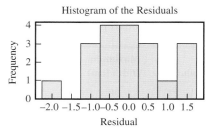

Histogram of the Residuals

15.36 Exercise 15.21 gave data on x = nerve firing frequency and y = pleasantness rating when nerves were stimulated by a light brushing stoke on the forearm. The x values and the corresponding residuals from a simple linear regression are as follows:

Firing Frequency, x	Standardized Residual
23	−1.83
24	0.04
22	1.45
25	0.20
27	−1.07
28	1.19
34	−0.24
33	−0.13
36	−0.81
34	1.17

a. Construct a standardized residual plot. Does the plot exhibit any unusual features?

b. A normal probability plot of the standardized residuals follows. Based on this plot, do you think it is reasonable to assume that the error distribution is approximately normal? Explain.

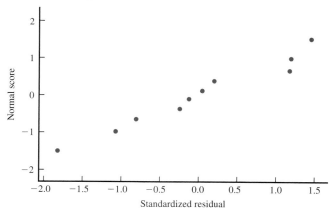

15.37 The accompanying scatterplot, based on 34 sediment samples with x = sediment depth (cm) and y = oil and grease content (mg/kg), appeared in the article **"Mined Land Reclamation Using Polluted Urban Navigable Waterway Sediments"** (*Journal of Environmental Quality* **[1984]: 415–422)**. Discuss the effect that the observation (20, 33,000) will have on the estimated regression line. If this point were omitted, what do you think will happen to the slope of the estimated regression line compared to the slope when this point is included?

 Data set available

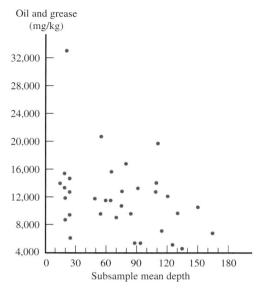

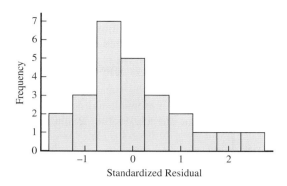

15.38 Investigators in northern Alaska periodically monitored radio collared wolves in 25 wolf packs over 4 years, keeping track of the packs' home ranges (**"Population Dynamics and Harvest Characteristics of Wolves in the Central Brooks Range, Alaska,"** *Wildlife Monographs*, **[2008]: 1–25**). The home range of a pack is the area typically covered by its members in a specified amount of time. The investigators noticed that wolf packs with larger home ranges tended to be located more often by monitoring equipment. The investigators decided to explore the relationship between home range and the number of locations per pack. A scatterplot and standardized residual plot of the data are shown below, as well as plots of the standardized residuals.

Does it appear that the assumptions of the simple linear regression model are plausible? Explain your reasoning in a few sentences.

ADDITIONAL EXERCISES

15.39 Carbon acrosols have been identified as a contributing factor in a number of air quality problems. In a chemical analysis of diesel engine exhaust, x = mass ($\mu g/cm^2$) and y = elemental carbon ($\mu g/cm^2$) were recorded (**"Comparison of Solvent Extraction and Thermal Optical Carbon Analysis Methods: Application to Diesel Vehicle Exhaust Aerosol"** *Environmental Science Technology* **[1984]: 231–234**). The estimated regression line for this data set is $\hat{y} = 31 + 0.737x$.

A scatterplot of the data and a standardized residual plot are shown below.

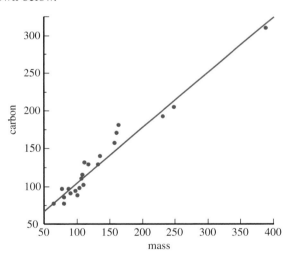

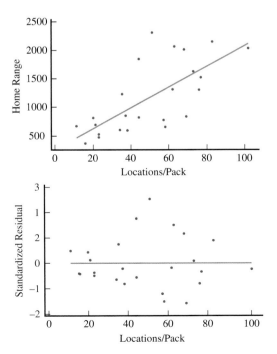

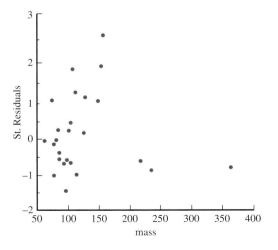

a. Are there any unusually large residuals? Do you think that there are any influential observations?

b. Is there any pattern in the standardized residual plot that would indicate that the simple linear regression model is not appropriate?

c. Based on the scatterplot and the standardized residual plot, do you think that it is reasonable to assume that the variance of y is the same at each x value? Explain.

15.40 Models of climate change predict that global temperatures and precipitation will increase in the next 100 years, with the largest changes occurring during winter in northern latitudes. Researchers recently gathered data on the potential effects of climate change for flowering plants in Norway **("Climatic Variability, Plant Phenology, and Northern Ungulates," *Ecology* [1999]: 1322–1339)**. The table below gives data for one flower species. A scatterplot of the "range of flowering dates" versus latitude for different sites in Norway is also shown. Two points that are potentially influential are indicated on the scatterplot.

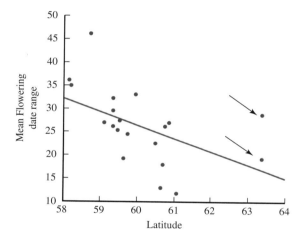

Flowering Range Versus Latitude: *Anemone Hepatica*	
Latitude (N)	Flowering Date Range
58.7	46.1
58.2	35.9
58.2	34.7
59.4	32.3
60.0	33.0
59.4	29.7
59.1	26.9
59.3	26.2
59.5	25.6
59.5	27.6
59.7	19.1

(continued)

Flowering Range Versus Latitude: *Anemone Hepatica*	
Latitude (N)	Flowering Date Range
59.8	24.4
60.8	26.2
60.9	26.8
63.4	28.7
63.4	19.2
60.5	22.5
60.7	17.9
60.7	12.9
61.1	11.8

a. Fit a linear regression model using all 20 observations. What are the values of a, b, r^2, and s_e?

b. Fit a linear regression model with the two observations identified by arrows omitted. What are the values of a, b, r^2, and s_e?

c. In a few sentences, describe any differences you found in Parts (a) and (b).

d. The researchers could use the estimated regression equation based on all 20 observations to make predictions for latitudes ranging from 58 to 64, or they could use the estimated regression equation based on the 18 observations (omitting the two observations identified by arrows) to make predictions for latitudes ranging from 58 to 61. Which strategy would you recommend, and why?

15.41 The sand scorpion is a predator that always hunts from a motionless resting position outside its own burrow. When prey appears on the horizon, within say 20 cm, the scorpion assumes an alert posture; it determines the angular position of the prey, makes a quick rotation, and runs after it. In a recent study of the scorpion's accuracy, the angular position (0 degrees = right in front) of the prey, and the turning angle of the scorpion was recorded for 23 attacks. A simple regression model relating the response angle (r) of the predator to the target angle (t) position of the prey, $\hat{r} = a + bt$, was fit. The resulting residual plot is shown. Describe the locations of any outliers you see in the residual plot.

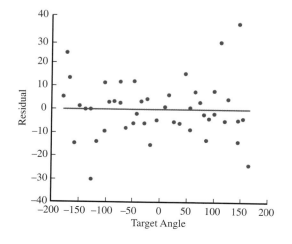

15.42 The production of pups and their survival are the most significant factors contributing to gray wolf population growth. The causes of early pup mortality are unknown, and difficult to observe. The pups are concealed within their dens for 3 weeks after birth, and after they emerge it is difficult to confirm their parentage. Researchers recently used portable ultrasound equipment to investigate some factors related to reproduction (**"Diagnosing Pregnancy, in Utero Litter Size, and Fetal Growth with Ultrasound in Wild, Free-Ranging Wolves,"** *Journal of Mammology* **[2006]: 85–92)**.

A scatterplot and linear regression of the length of a wolf fetus (in centimeters, measured from crown to rump) and gestational age (in days) is shown below. Identify the point that has the largest residual by giving its approximate coordinates.

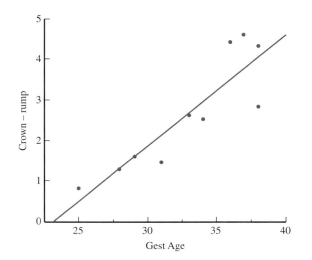

15.43 The authors of the article **"Age, Spacing and Growth Rate of Tamarix as an Indication of Lake Boundary Fluctuations at Sebkhet Kelbia, Tunisia"** (*Journal of Arid Environments* **[1982]: 43–51)** used a simple linear regression model to describe the relationship between y = vigor (average width in centimeters of the last two annual rings) and x = stem density (stems/m^2). The estimated model was based on the following data. Also given are the standardized residuals.

x	4	5	6	9	14
y	0.75	1.20	0.55	0.60	0.65
Std resid	-0.28	1.92	-0.90	-0.28	0.54

x	15	15	19	21	22
y	0.55	0.00	0.35	0.45	0.40
Std resid	0.24	-2.05	-0.12	0.60	0.52

a. What assumptions are required for the simple linear regression model to be appropriate?

b. Construct a normal probability plot of the standardized residuals. Does the assumption that the random deviation distribution is normal appear to be reasonable? Explain.

c. Construct a standardized residual plot. Are there any unusually large residuals?

d. Is there anything about the standardized residual plot that would cause you to question the use of the simple linear regression model to describe the relationship between x and y?

ARE YOU READY TO MOVE ON?   CHAPTER 15 REVIEW EXERCISES

All chapter learning objectives are assessed in these exercises. The learning objectives assessed in each exercise are given in parentheses.

15.44 (C1)
Explain what distinguishes a deterministic model from a probabilistic model.

15.45 (C2)
In the context of the simple linear regression model, explain the difference between α and a. Between β and b. Between σ_e and s_e.

15.46 (M1)
The SAT and ACT exams are often used to predict a student's first-term college grade point average (GPA). Different formulas are used for different colleges and majors. Suppose that a student is applying to State U with an intended major in civil engineering. Also suppose that for this college and this major, the following model is used to predict first term GPA.

$$GPA = a + b(ACT)$$
$$a = 0.5$$
$$b = 0.1$$

a. In this context, what would be the appropriate interpretation of the value of a?

b. In this context, what would be the appropriate interpretation of the value of b?

15.47 (M2)
Theropods were carnivorous dinosaurs, characterized by short forelimbs, living in the Jurassic and Cretaceous periods. (*Tyrannosaurus rex* is classified as a Theropod.) What scientists know about therapods is based on studying incomplete skeletal remains. In a study described in the paper **"My Theropod is Bigger than Yours…or not: Estimating Body**

 Data set available

Size from Skull Length in Theropods" (*Journal of Vertebrate Paleontology* [2007]: 108–115), researchers used data from skeletons to develop a model describing the relationship between body length and skull length. JMP was used to produce the following graphical displays and computer output. When you evaluate the fit of an estimated regression line, all of the information below is considered as a whole. However, the summary statistics in the computer output and the different plots each convey some specific information.

a. Using only the scatterplot, do you think a linear model does a good job of describing the relationship? Explain why or why not.

b. Using only the residual plot, what can you determine about whether the basic assumptions of the linear regression model are met?

c. Using only the normal probability plot and boxplot of the residuals, what can you determine about whether the basic assumptions of the linear regression model are met?

d. Using only the values of r^2 and s_e, what can you say about the quality of the fit of the linear model for these data?

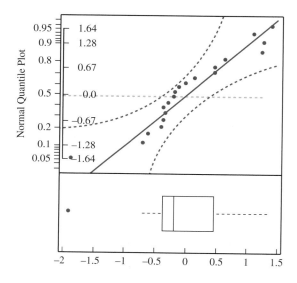

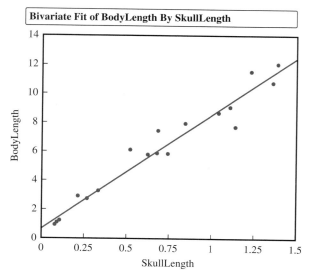

Bivariate Fit of BodyLength By SkullLength

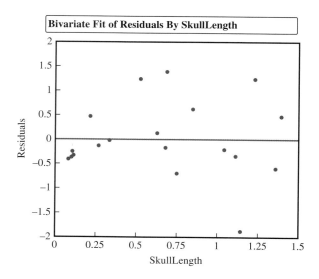

Bivariate Fit of Residuals By SkullLength

Linear Fit

BodyLength = 0.7061088 + 7.791973*SkullLength

Summary of Fit

RSquare	0.953929
RSqureAdj	0.951218
Root Mean Square Error	0.801042
Mean of Response	5.859474
Observations(or Sum Wgts)	19

Analysis Of Variance

Parameter Estimates

Term	Estimate	Std Error	t Ratio	Prob>|t|
Intercept	0.7061088	0.330485	2.14	0.0475*
SkullLength	7.791973	0.415318	18.76	<.0001*

15.48 (M3)

There are 4 basic assumptions necessary for making inferences about β, the slope of the population regression line.

a. What are the four assumptions?

b. Which assumptions can be checked using sample data?

c. What statistics or graphs would be used to check each of the assumptions you listed in Part (b)?

15.49 (M3, M4, P1, P2)

Ruffed grouse are a species of birds that nest on the ground. Because of this, chick survival at night in the first few weeks of life depends on avoiding predators. Biologists have theorized that protection from predators might be supplied by the mother hen's choice of brooding sites. One variable that biologists thought might be related to survival is the density of vegetation in the vicinity of the nest. Dense vegetation would possible reduce the ability of predators to detect the nests. The paper **"Nocturnal Roost Habitat Selection by Ruffed Grouse Broods"** (*Journal of Field Ornithology* [2005]:168–174) describes a study in which researchers monitored the survival of the brood (number of chicks surviving /number of eggs hatched) in 23 nests in different vegetation densities (thousands of stems / hectare).

Computer output (from JMP) is shown below.

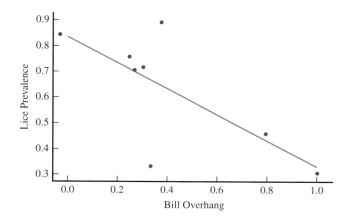

Linear Fit

BroodSurvival = 0.9468008 − 0.0261902*StemDensity

Summary of Fit

RSquare	0.193788
RSqureAdj	0.155397
Root Mean Square Error	0.287538
Mean of Response	0.436043
Observations(or Sum Wgts)	23

Parameter Estimates

Term	Estimate	Std Error	t Ratio	Prob>|t|
Intercept	0.9468008	0.235108	4.03	0.0006*
StemDensity	−0.02619	0.011657	−2.25	0.0355*

---

a. Is there convincing evidence of a useful linear relationship between brood survival and stem density? Explain.

b. Would you describe the relationship as strong? Why or why not?

c. Construct a 95% confidence interval for β and interpret it in context.

d. What margin of error is associated with the confidence interval in part (c)?

15.50 (M6)

Researchers in Hawaii have recently documented a large increase in the prevalence of a bird parasite known as chewing lice (**"Explosive Increase in Ectoparasites in Hawaiian Forest Birds,"** *The Journal of Parasitology* **[2008]: 1009–1021**). Current data suggest that the prevalence of chewing lice may be less for bird species with a high degree of bill overhang. A species is said to have bill overhang when the upper bill extends downward in front of the end of the lower bill. The following scatterplot shown shows the relationship between the prevalence of chewing lice and bill overhang for 8 bird species in the Hawaiian Islands. A residual plot is also shown. Use these plots to identify any outliers or potentially influential observations. For each point you identify, assess its influence on the estimated slope of the regression line.

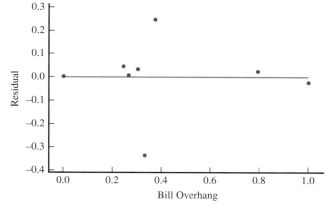

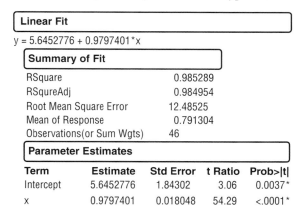

15.51 (M5)

Suppose you are given the computer output shown. You want to test the hypothesis, $\beta = 1.0$. Describe how you would use the computer output to test this hypothesis.

Linear Fit

y = 5.6452776 + 0.9797401*x

Summary of Fit

RSquare	0.985289
RSqureAdj	0.984954
Root Mean Square Error	12.48525
Mean of Response	0.791304
Observations(or Sum Wgts)	46

Parameter Estimates

Term	Estimate	Std Error	t Ratio	Prob>|t|
Intercept	5.6452776	1.84302	3.06	0.0037*
x	0.9797401	0.018048	54.29	<.0001*

TECHNOLOGY NOTES

Regression Test

TI-83/84

1. Enter the data for the independent variable into **L1** (In order to access lists press the **STAT** key, highlight the option called **Edit...** then press **ENTER**)
2. Enter the data for the dependent variable into **L2**
3. Press **STAT**

4. Highlight **TESTS**
5. Highlight **LinRegTTest...** and press **ENTER**
6. Next to β & ρ select the appropriate alternative hypothesis
7. Highlight **Calculate**

TI-Nspire

1. Enter the data into two separate data lists (In order to access data lists select the spreadsheet option and press **enter**) **Note:** Be sure to title the lists by selecting the top row of the column and typing a title.
2. Press the **menu** key and select **4:Stat Tests** then **4:Stats Tests** then **A:Linear Reg t Test...** and press **enter**
3. In the box next to **X List** choose the list title where you stored your independent data from the drop-down menu
4. In the box next to **Y List** choose the list title where you stored your dependent data from the drop-down menu
5. In the box next to **Alternate Hyp** choose the appropriate alternative hypothesis from the drop-down menu
6. Press **OK**

JMP

1. Input the data for the dependent variable into the first column
2. Input the data for the independent variable into the second column
3. Click **Analyze** and select **Fit Y by X**
4. Select the dependent variable (Y) from the box under **Select Columns** and click on **Y, Response**
5. Select the independent variable (X) from the box under **Select Columns** and click on **X, Factor**
6. Click the red arrow next to **Bivariate Fit of...** and select **Fit Line**

MINITAB

1. Input the data for the dependent variable into the first column
2. Input the data for the independent variable into the second column
3. Select **Stat** then **Regression** then **Regression...**
4. Highlight the name of the column containing the dependent variable and click **Select**
5. Highlight the name of the column containing the independent variable and click **Select**
6. Click **OK**

Note: You may need to scroll up in the Session window to view the t-test results for the regression analysis.

SPSS

1. Input the data for the dependent variable into one column
2. Input the data for the independent variable into a second column
3. Click **Analyze** then click **Regression** then click **Linear...**
4. Select the name of the dependent variable and click the arrow to move the variable to the box under **Dependent:**
5. Select the name of the independent variable and click the arrow to move the variable to the box under **Independent(s):**
6. Click **OK**

Note: The p-value for the regression test can be found in the **Coefficients** table in the row with the independent variable name.

Excel

1. Input the data for the dependent variable into the first column
2. Input the data for the independent variable into the second column
3. Select **Analyze** then choose **Regression** then choose **Linear...**
4. Highlight the name of the column containing the dependent variable
5. Click the arrow button next to the **Dependent** box to move the variable to this box
6. Highlight the name of the column containing the independent variable
7. Click the arrow button next to the **Independent** box to move the variable to this box
8. Click **OK**

Note: The test statistic and p-value for the regression test for the slope can be found in the third table of output. These values are listed in the row titled with the independent variable name and the columns entitled *t Stat and P-value.*

16

Asking and Answering Questions About More Than Two Means

James Woodson/Digital Vision/Getty Images

PREVIEW

In Chapter 13, you learned methods for testing $H_0: \mu_1 - \mu_2 = 0$ (or equivalently, $\mu_1 = \mu_2$), where μ_1 and μ_2 are the means of two different populations or the mean responses when two different treatments are applied. However, many investigations involve comparing more than two population or treatment means, as illustrated in the following example.

Conceptual Understanding

After completing this chapter, you should be able to

C1 Understand how a research question about differences between three or more population or treatment means is translated into hypotheses.

Mastering the Mechanics

After completing this chapter, you should be able to

M1 Translate a research question or claim about differences between three or more population or treatment means into null and alternative hypotheses.

M2 Know the conditions for appropriate use of the ANOVA F test

M3 Carry out an ANOVA F test.

M4 Use a multiple comparison procedure to identify differences in population or treatment means.

Putting It into Practice

After completing this chapter, you should be able to

P1 Carry out an ANOVA F test and interpret the conclusion in context.

P2 Interpret the results of a multiple comparison procedure in context.

PREVIEW EXAMPLE

Risky Soccer

In a study to see if the high incidence of head injuries among soccer players might affect memory recall, researchers collected data from three samples of college students (**"No Evidence of Impaired Neurocognitive Performance in Collegiate Soccer Players," *The American Journal of Sports Medicine* [2002]: 157–162**). One sample consisted of soccer athletes, one sample consisted of athletes whose sport was not soccer, and one sample was a comparison group consisting of students who did not participate in sports. The following information on scores from the Hopkins Verbal Learning Test (which measures memory recall) was given in the paper.

Group	Soccer Athletes	Nonsoccer Athletes	Comparison Group
Sample Size	86	95	53
Sample Mean Score	29.90	30.94	29.32
Sample Standard Deviation	3.73	5.14	3.78

Notice that the three sample means are different. But even if the population means were equal, you would not expect the three sample means to be exactly equal. Are the differences in sample means consistent with what is expected simply due to chance differences from one sample to another when the population means are equal, or are the differences large enough that you should conclude that the three population means are not all equal? This is the type of problem considered in this chapter ■

SECTION 16.1   The Analysis of Variance—Single-Factor ANOVA and the *F* Test

When more than two populations or treatments are being compared, the characteristic that distinguishes the populations or treatments from one another is called the **factor** under investigation. For example, an experiment might be carried out to compare three different methods for teaching reading (three different treatments), in which case the factor of interest would be *teaching method*, a qualitative factor. If the growth of the fish raised in waters having different salinity levels—0%, 10%, 20%, and 30%—is of interest, the factor *salinity level* is quantitative.

A **single-factor analysis of variance (ANOVA)** problem involves a comparison of k population or treatment means $\mu_1, \mu_2, \ldots, \mu_k$. The objective is to test

$$H_0: \mu_1 = \mu_2 = \cdots = \mu_k$$

against

$$H_a: \text{At least two of } \mu\text{'s are different}$$

When comparing populations, the analysis is based on independently selected random samples, one from each population. When comparing treatment means, the data are from an experiment, and the analysis assumes random assignment of the experimental units (subjects or objects) to treatments. If, in addition, the experimental units are chosen at random from a population of interest, it is also possible to generalize the results of the analysis to this population.

Whether the null hypothesis in a single-factor ANOVA should be rejected depends on how much the samples from the different populations or treatments differ from one another. Figure 16.1 displays observations that might result when random samples are selected from each of three populations. Each dotplot displays five observations from the first population, four observations from the second population, and six observations from the third population. For both displays, the three sample means are located by arrows. The means of the two samples from Population 1 are equal, as are the means for the two samples from Population 2 and for the two samples from Population 3.

FIGURE 16.1

Two possible data sets when three populations are compared: green circle = observation from Population 1; orange circle = observation from Population 2; blue circle = observation from Population 3

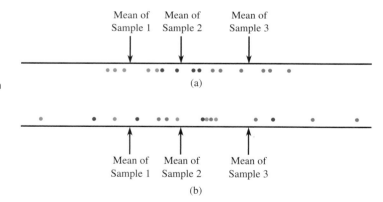

After looking at the data in Figure 16.1(a), you would probably think that the claim $\mu_1 = \mu_2 = \mu_3$ appears to be false. Not only are the three sample means different, but also the three samples are clearly separated. The differences between the three sample means are quite large relative to the variability within each sample.

The situation pictured in Figure 16.1(b) is much less clear-cut. The sample means are as different as they were in the first data set, but now there is considerable overlap among the three samples. The separation between sample means might be due to the substantial variability in the populations (and therefore the samples) rather than to differences between μ_1, μ_2, and μ_3. The phrase *analysis of variance* comes from the idea of analyzing variability in the data to see how much can be attributed to differences in the μ's and how much is due to variability in the individual populations. In Figure 16.1(a), the *within-sample variability* is small relative to the *between-sample variability*, whereas in Figure 16.1(b), a great deal more of the total variability is due to variation within each sample. If differences between the sample means can be explained entirely by within-sample variability, there is no compelling reason to reject H_0: $\mu_1 = \mu_2 = \mu_3$.

Notation and Assumptions

Notation in single-factor ANOVA is a natural extension of the notation used in earlier chapters for comparing two population or treatment means.

ANOVA Notation

k = number of populations or treatments being compared

Population or treatment	1	2	...	k
Population or treatment mean	μ_1	μ_2	...	μ_k
Population or treatment variance	σ_1^2	σ_2^2	...	σ_k^2
Sample size	n_1	n_2	...	n_k
Sample mean	$\bar{x}_1$	$\bar{x}_2$	...	$\bar{x}_k$
Sample variance	s_1^2	s_2^2	...	s_k^2

$N = n_1 + n_2 + \cdots + n_k$ (the total number of observations in the data set)

T = grand total = sum of all N observations in the data set = $n_1\bar{x}_1 + n_2\bar{x}_2 + \cdots + n_k\bar{x}_k$

$\bar{\bar{x}}$ = grand mean = $\dfrac{T}{N}$

A decision between

$$H_0: \mu_1 = \mu_2 = \cdots = \mu_k$$

and

$$H_a: \text{At least two of } \mu\text{'s are different}$$

is based on examining the $\bar{x}$ values to see whether observed differences are small enough to be explained by sampling variability alone or whether an alternative explanation for the differences is more plausible.

Example 16.1　An Indicator of Heart Attack Risk

The article **"Could Mean Platelet Volume Be a Predictive Marker for Acute Myocardial Infarction?"** (*Medical Science Monitor* **[2005]: 387–392)** described a study in which four groups of patients seeking treatment for chest pain were compared with respect to the mean platelet volume (MPV, measured in fL). The four groups considered were based on the clinical diagnosis: (1) noncardiac chest pain, (2) stable angina, (3) unstable angina, and (4) heart attack. The purpose of the study was to determine if the mean MPV differed for the four groups, and in particular if the mean MPV was different for the heart attack group, because then MPV could be used as an indicator of heart attack risk.

To carry out this study, patients seen for chest pain were divided into groups according to diagnosis. The researchers then selected a random sample of 35 from each of the resulting $k = 4$ groups. The researchers believed that this sampling process would result in samples that were representative of the four populations of interest and that could be regarded as if they were random samples from these four populations. Table 16.1 presents summary values given in the paper.

TABLE 16.1 Summary Values for MPV Data of Example 16.1

Group Number	Group Description	Sample Size	Sample Mean	Sample Standard Deviation
1	Noncardiac chest pain	35	10.89	0.69
2	Stable angina	35	11.25	0.74
3	Unstable angina	35	11.37	0.91
4	Heart attack	35	11.75	1.07

With μ_i denoting the true mean MPV for group i ($i = 1, 2, 3, 4$), consider the null hypothesis $H_0: \mu_1 = \mu_2 = \mu_3 = \mu_4$. Figure 16.2 shows a comparative boxplot of the data from the four samples (based on data consistent with summary values given in the paper). The mean MPV for the heart attack sample is larger than for the other three samples, and the boxplot for the heart attack sample appears to be shifted a bit higher than the boxplots for the other three samples. However, because the four boxplots show substantial overlap, it is not obvious whether H_0 is plausible or should be rejected. In situations like this, a formal test procedure is helpful.

FIGURE 16.2
Boxplots for Example 16.1

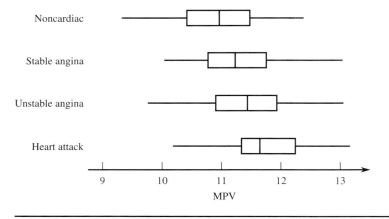

As with the inferential methods of previous chapters, the validity of the ANOVA test for $H_0: \mu_1 = \mu_2 = \cdots = \mu_k$ requires that some conditions be met.

Conditions for ANOVA

1. Each of the k population or treatment response distributions is normal.
2. $\sigma_1 = \sigma_2 = \cdots = \sigma_k$ (The k normal distributions have equal standard deviations.)
3. The observations in the sample from any particular one of the k populations or treatments are independent of one another.
4. When comparing population means, the k random samples are selected independently of one another. When comparing treatment means, experimental units are assigned at random to treatments.

In practice, the test based on these assumptions works well as long as the conditions are not too badly violated. If the sample sizes are reasonably large, normal probability plots or boxplots of the data in each sample are helpful in checking the condition of normality. Often, however, sample sizes are so small that a separate normal probability plot or boxplot for each sample is of little value in checking normality. In this case, a single combined plot can be constructed by first subtracting $\bar{x}_1$ from each observation in the first sample, $\bar{x}_2$ from each value in the second sample, and so on, and then constructing a normal probability or boxplot of all *N* deviations from their respective means. Figure 16.3 shows such a normal probability plot for the data of Example 16.1.

FIGURE 16.3
A normal probability plot using the combined data of Example 16.1

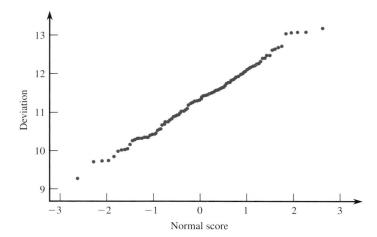

There is a formal procedure for testing the equality of population standard deviations. Unfortunately, it is quite sensitive to even a small violation of the normality condition. However, the equal population or treatment standard deviation condition can be considered reasonably met if the largest of the sample standard deviations is at most twice the smallest one. For example, the largest standard deviation in Example 16.1 is $s_4 = 1.07$, which is less than 2 times the smallest standard deviation ($s_1 = 0.69$).

The analysis of variance test procedure is based on the following measures of variation in the data.

DEFINITION

A measure of differences among the sample means is the **treatment sum of squares,** denoted by **SSTr** and given by

$$\text{SSTr} = n_1(\bar{x}_1 - \bar{\bar{x}})^2 + n_2(\bar{x}_2 - \bar{\bar{x}})^2 + \cdots + n_k(\bar{x}_k - \bar{\bar{x}})^2$$

A measure of variation within the *k* samples, called **error sum of squares** and denoted by **SSE**, is

$$\text{SSE} = (n_1 - 1)s_1^2 + (n_2 - 1)s_2^2 + \cdots + (n_k - 1)s_k^2$$

Each sum of squares has an associated degrees of freedom (df):

   treatment df $= k - 1$ error df $= N - k$

A **mean square** is a sum of squares divided by its df.

A **mean square for treatments** and a **mean square for error** are calculated as follows:

$$\text{mean square for treatments} = \text{MSTr} = \frac{\text{SSTr}}{k - 1}$$

$$\text{mean square for error} = \text{MSE} = \frac{SSE}{N - k}$$

The number of error degrees of freedom comes from adding the number of degrees of freedom associated with each of the sample variances:

$$(n_1 - 1) + (n_2 - 1) + \cdots (n_k - 1) = n_1 + n_2 + \cdots n_k - 1 - 1 - \cdots 1$$

$$= N - k$$

Example 16.2   Heart Attack Calculations

For the mean platelet volume (MPV) data of Example 16.1, the grand mean $\bar{\bar{x}}$ was calculated to be 11.315. Notice that because the sample sizes are all equal, the grand mean is just the average of the four sample means (this will not usually be the case when the sample sizes are unequal). With $\bar{x}_1 = 10.89$, $\bar{x}_2 = 11.25$, $\bar{x}_3 = 11.37$, $\bar{x}_4 = 11.75$, and $n_1 = n_2 = n_3 = n_4 = 35$,

$$\text{SSTr} = n_1(\bar{x}_1 - \bar{\bar{x}})^2 + n_2(\bar{x}_2 - \bar{\bar{x}})^2 + \cdots + n_k(\bar{x}_k - \bar{\bar{x}})^2$$

$$= 35(10.89 - 11.315)^2 + 35(11.25 - 11.315)^2 + 35(11.37 - 11.315)^2$$
$$+ 35(11.75 - 11.315)^2$$

$$= 6.322 + 0.148 + 0.106 + 6.623$$

$$= 13.199$$

Because $s_1 = 0.69$, $s_2 = 0.74$, $s_3 = 0.91$, $s_4 = 1.07$,

$$\text{SSE} = (n_1 - 1)s_1^2 + (n_2 - 1)s_2^2 + \cdots + (n_k - 1)s_k^2$$

$$= (35 - 1)(0.69)^2 + (35 - 1)(0.74)^2 + (35 - 1)(0.91)^2 + (35 - 1)(1.07)^2$$

$$= 101.888$$

The numbers of degrees of freedom are

$$\text{treatment df} = k - 1 = 3$$

$$\text{error df} = N - k = 35 + 35 + 35 + 35 - 4 = 136$$

from which

$$\text{MSTr} = \frac{\text{SSTr}}{k - 1} = \frac{13.199}{3} = 4.400$$

$$\text{MSE} = \frac{\text{SSE}}{N - k} = \frac{101.888}{136} = 0.749$$

Both MSTr and MSE are statistics whose values can be calculated once sample data are available. Both statistics MSTr and MSE have sampling distributions, and these sampling distributions have mean values. The following box describes the relationship between the mean values of MSTr and MSE.

When H_0 is true ($\mu_1 = \mu_2 = \cdots = \mu_k$),

$$\mu_{\text{MSTr}} = \mu_{\text{MSE}}$$

However, when H_0 is false,

$$\mu_{\text{MSTr}} > \mu_{\text{MSE}}$$

and the greater the differences among the μ's, the larger μ_{MSTr} will be relative to μ_{MSE}.

According to this result, when H_0 is true, you would expect the values of the two mean squares to be close. However, you would expect MSTr to be substantially greater than MSE when some μ's differ greatly from others. This means that a calculated MSTr that is much larger than MSE is inconsistent with the null hypothesis. In Example 16.2, MSTr = 4.400 and MSE = 0.749, so MSTr is about six times as large as MSE. Can this be attributed solely to sampling variability, or is the ratio MSTr/MSE large enough to suggest that the null hypothesis should be rejected? Before a formal test procedure can be described, you need to learn about a new family of probability distributions called F distributions.

An F distribution always arises in connection with a ratio. A particular F distribution is obtained by specifying both numerator degrees of freedom (df_1) and denominator degrees of freedom (df_2). Figure 16.4 shows an F curve for a particular choice of df_1 and

df_2. The ANOVA test of this section is an upper-tailed test, so a P-value is the area under an appropriate F curve to the right of the calculated value of the test statistic.

FIGURE 16.4
An F curve and P-value for an upper-tailed test

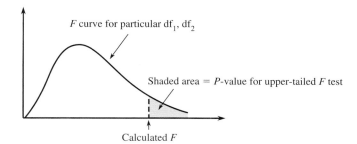

Constructing tables of these upper-tail areas is cumbersome, because there are two degrees of freedom rather than just one (as in the case of t distributions). For selected (df_1, df_2) pairs, the F table (Appendix Table 7) gives only the four numbers that capture tail areas 0.10, 0.05, 0.01, and 0.001, respectively. Here are the four numbers for $df_1 = 4$, $df_2 = 10$ along with the statements that can be made about the P-value:

Tail area	0.10	0.05	0.01	0.001
Value	2.61	3.48	5.99	11.28
	↑	↑	↑	↑	↑
	a	b	c	d	e

a. $F < 2.61 \rightarrow$ tail area = P-value > 0.10
b. $2.61 < F < 3.48 \rightarrow 0.05 < P$-value < 0.10
c. $3.48 < F < 5.99 \rightarrow 0.01 < P$-value < 0.05
d. $5.99 < F < 11.28 \rightarrow 0.001 < P$-value < 0.01
e. $F > 11.28 \rightarrow P$-value < 0.001

For example, if $F = 7.12$, then $0.001 < P$-value < 0.01. If a test with $\alpha = 0.05$ is used, H_0 should be rejected, because P-value $< \alpha$. The most frequently used statistics software packages can provide exact P-values for F tests.

Single Factor ANOVA F Test for Equality of Three or More Means

Appropriate when the following conditions are met:

1. Each of the k population or treatment response distributions is normal.
2. $\sigma_1 = \sigma_2 = \cdots = \sigma_k$ (The k normal distributions have equal standard deviations.)
3. The observations in the sample from any particular one of the k populations or treatments are independent of one another.
4. When comparing population means, the k random samples are selected independently of one another. When comparing treatment means, experimental units are assigned at random to treatments.

When these conditions are met, the following test statistic can be used:

$$F = \frac{\text{MSTr}}{\text{MSE}}$$

When the conditions above are met and the null hypothesis is true, the F statistic has an approximate F distribution with

$$df_1 = k - 1 \quad \text{and} \quad df_2 = N - k$$

Form of the null hypothesis: H_0: $\mu_1 = \mu_2 = \cdots = \mu_k$

Form of the alternative hypothesis: H_a: At least two of the μ's are different

The P-value is: Area under the F curve to the right of the calculated value of the test statistic

| **Example 16.3** | **Heart Attacks Revisited** |

Recall that the two mean squares for the MPV data given in Example 16.1 were calculated in Example 16.2 to be

$$\text{MSTr} = 4.400 \qquad \text{MSE} = 0.749$$

You can now use the five-step process for hypothesis testing problems (HMC³) to test the hypotheses of interest.

Process Step	
H Hypotheses	The question of interest is whether there are differences in mean MPV for the four different diagnosis groups. **Population characteristics of interest:** μ_1 = mean MPV for patients with noncardiac chest pain μ_2 = mean MPV for patients with stable angina μ_3 = mean MPV for patients with unstable angina μ_4 = mean MPV for patients with a heart attack **Hypotheses:** *Null hypothesis*: $H_0: \mu_1 = \mu_2 = \mu_3 = \mu_4$ *Alternative hypothesis*: H_a: At least two of the μ's are diffrent
M Method	Because the answers to the four key questions are hypothesis testing, sample data, one numerical variable and four independently selected samples, consider an ANOVA *F* test. **Potential method:** ANOVA *F* test. The test statistic for this test is $$F = \frac{\text{MSTr}}{\text{MSE}}$$ When the null hypothesis is true, this statistic has approximately an *F* distribution with $$df_1 = k - 1 \quad \text{and} \quad df_2 = N - k$$ Once you have decided to proceed with the test, you need to select a significance level for the test. In this example, a significance level of 0.05 will be used. **Significance level:** $\alpha = 0.05$
C Check	The samples were independently selected random samples. The largest sample standard deviation (from Table 16.1, $s_4 = 1.07$) is not more than twice as large as the smallest sample standard deviation ($s_1 = 0.69$), so the equal population standard deviations condition is reasonably met. A normal probability plot (see Figure 16.3) indicates that the normality condition is also reasonably met.
C Calculate	MSTr = 4.400 MSE = 0.749 (from Example 16.2) **Test statistic:** $$F = \frac{\text{MSTr}}{\text{MSE}} = \frac{4.400}{0.749} = 5.87$$ **Degrees of freedom** $$df_1 = k - 1 = 4 - 1 = 3$$ $$df_2 = N - k = 140 - 4 = 136$$

(continued)

Process Step	
	Associated *P*-value:
	P-value = area under *F* curve to the right of 5.87
	Using $df_1 = 3$ and $df_2 = 120$ (the closest value to 136 that appears in the table), Appendix Table 7 shows that the area to the right of 5.78 is 0.001. Since $5.87 > 5.78$ it follows that the *P*-value is less than 0.001.
C Communicate results	Because the *P*-value is less than the selected significance level, you reject the null hypothesis.
	Decision: Reject H_0.
	The final conclusion for the test should be stated in context and answer the question posed.
	Conclusion: There is convincing evidence that at least two of the treatment means for MPV are different.

Techniques for determining which means differ are introduced in Section 16.2.

Example 16.4   Hormones and Body Fat

The article **"Growth Hormone and Sex Steroid Administration in Healthy Aged Women and Men"** (*Journal of the American Medical Association* [2002]: 2282–2292) described an experiment to investigate the effect of four treatments on various body characteristics. In this double-blind experiment, each of 57 female subjects age 65 or older was assigned at random to one of the following four treatments: (1) placebo "growth hormone" and placebo "steroid" (denoted by P + P); (2) placebo "growth hormone" and the steroid estradiol (denoted by P + S); (3) growth hormone and placebo "steroid" (denoted by G + P); and (4) growth hormone and the steroid estradiol (denoted by G + S).

The following table gives data on change in body fat mass over the 26-week period following the treatments that are consistent with summary quantities given in the article.

Treatment	Change in Body Fat Mass (kg)			
	P + P	**P + S**	**G + P**	**G + S**
	0.1	−0.1	−1.6	−3.1
	0.6	0.2	−0.4	−3.2
	2.2	0.0	0.4	−2.0
	0.7	−0.4	−2.0	−2.0
	−2.0	−0.9	−3.4	−3.3
	0.7	−1.1	−2.8	−0.5
	0.0	1.2	−2.2	−4.5
	−2.6	0.1	−1.8	−0.7
	−1.4	0.7	−3.3	−1.8
	1.5	−2.0	−2.1	−2.3
	2.8	−0.9	−3.6	−1.3
	0.3	−3.0	−0.4	−1.0
	−1.0	1.0	−3.1	−5.6
	−1.0	1.2		−2.9
				−1.6
				−0.2
n	14	14	13	16
$\bar{x}$	0.064	−0.286	−2.023	−2.250
s	1.545	1.218	1.264	1.468
s^2	2.387	1.484	1.598	2.155

For this example, $N = 57$, grand total $T = -65.4$, and $\bar{\bar{x}} = \dfrac{-65.4}{57} = -1.15$.

Process Step	
H Hypotheses	The question of interest is whether there are differences in mean change in body fat mass for the four treatments. **Population characteristics of interest:** μ_1 = mean change in body fat mass for the P + P treatment μ_2 = mean change in body fat mass for the P + S treatment μ_3 = mean change in body fat mass for the G + P treatment μ_4 = mean change in body fat mass for the G + S treatment **Hypotheses:** *Null hypothesis*: $H_0: \mu_1 = \mu_2 = \mu_3 = \mu_4$ *Alternative hypothesis*: H_a: At least two of the μ's are different
M Method	Because the answers to the four key questions are hypothesis testing, experiment data, one numerical variable, and four treatments, consider an ANOVA F test. **Potential method:** ANOVA F test. The test statistic for this test is $$F = \frac{\text{MSTr}}{\text{MSE}}$$ When the null hypothesis is true, this statistic has approximately an F distribution with $$df_1 = k - 1 \quad \text{and} \quad df_2 = N - k$$ Once you have decided to proceed with the test, you need to select a significance level for the test. For this example, a significance level of 0.01 will be used. **Significance level:** $\alpha = 0.01$
C Check	The subjects in the experiment were randomly assigned to treatments. The largest sample standard deviation ($s_1 = 1.545$) is not more than twice as large as the smallest sample standard deviation ($s_2 = 1.218$), so the equal population standard deviations condition is reasonably met. Boxplots of the data from each of the four samples are shown in Figure 16.5. The boxplots are roughly symmetric, and there are no outliers, so the normality condition is also reasonably met.
C Calculate	$\text{SSTr} = n_1(\bar{x}_1 - \bar{\bar{x}})^2 + n_2(\bar{x}_2 - \bar{\bar{x}})^2 + \cdots + n_k(\bar{x}_k - \bar{\bar{x}})^2$ $\quad\quad = 14(0.064 - (-1.15))^2 + 14(-0.286 - (-1.15))^2$ $\quad\quad\quad + 13(-2.023 - (-1.15))^2 + 16(-2.250 - (-1.15))^2$ $\quad\quad = 60.35$ treatment df $= k - 1 = 3$ $\text{SSE} = (n_1 - 1)s_1^2 + (n_2 - 1)s_2^2 + \cdots + (n_k - 1)s_k^2$ $\quad\quad = 13(2.387) + 13(1.484) + 12(1.598) + 15(2.155)$ $\quad\quad = 101.82$ **Test statistic:** $$F = \frac{\text{MSTr}}{\text{MSE}} = \frac{\text{SSTr/treatment df}}{\text{SSE/error df}} = \frac{60.35/3}{101.82/53} = \frac{20.12}{1.92} = 10.48$$ **Degrees of freedom** $$df_1 = k - 1 = 4 - 1 = 3$$ $$df_2 = N - k = 57 - 4 = 53$$ **Associated P-value:** P-value = area under F curve to the right of 10.48 Using $df_1 = 3$ and $df_2 = 60$ (the closest value to 53 that appears in the table), Appendix Table 7 shows that the area to the right of 6.17 is 0.001. Since $10.48 > 6.17$ it follows that the P-value is less than 0.001.

(continued)

Process Step	
C Communicate results	Because the *P*-value is less than the selected significance level, reject the null hypothesis.
	Decision: Reject H_0.
	Conclusion: There is convincing evidence that the mean changes in body fat mass are different for at least two patient groups.

FIGURE 16.5
Boxplots for the data of
Example 16.4

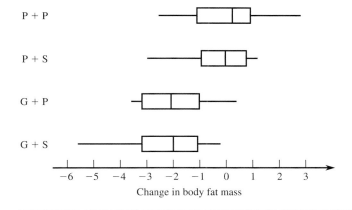

Change in body fat mass

Summarizing an ANOVA

ANOVA calculations are often summarized in a tabular format called an ANOVA table. To understand such a table, one more sum of squares must be defined.

Total sum of squares, denoted by **SSTo**, is given by

$$SSTo = \sum_{\text{all } N \text{ obs.}} (x - \bar{\bar{x}})^2$$

with associated df $= N - 1$

The relationship between the three sums of squares SSTo, SSTr, and SSE is

$$SSTo = SSTr + SSE$$

which is called the *fundamental identity for single-factor ANOVA*

The quantity SSTo, the sum of squared deviations about the grand mean, is a measure of total variability in the data set consisting of all *k* samples. The quantity SSE results from measuring variability separately within each sample and then combining. Such within-sample variability is present regardless of whether or not H_0 is true. The magnitude of SSTr, on the other hand, depends on whether the null hypothesis is true or false. The more the μ's differ, the larger SSTr will tend to be. SSTr represents variation that can (at least to some extent) be explained by differences among the means. An informal paraphrase of the fundamental identity for single-factor ANOVA is

total variation = explained variation + unexplained variation

Once any two of the sums of squares have been calculated, the remaining one is easily obtained from the fundamental identity. Often SSTo and SSTr are calculated first (using computational formulas given in the appendix to this chapter), and then SSE is obtained by subtraction: SSE = SSTo − SSTr. All the degrees of freedom, sums of squares, and mean squares are entered in an ANOVA table, as displayed in Table 16.2.

The *P*-value usually appears to the right of *F* when the analysis is done by a statistical software package.

TABLE 16.2 General Format for a Single-Factor ANOVA Table

Source of Variation	df	Sum of Squares	Mean Square	F
Treatments	$k - 1$	SSTr	$\text{MSTr} = \dfrac{\text{SSTr}}{k - 1}$	$F = \dfrac{\text{MSTr}}{\text{MSE}}$
Error	$N - k$	SSE	$\text{MSE} = \dfrac{\text{SSE}}{N - k}$	
Total	$N - 1$	SSTo		

An ANOVA table from Minitab for the change in body fat mass data of Example 16.4 is shown in Table 16.3. The reported *P*-value is 0.000, consistent with the previous conclusion that *P*-value < 0.001.

TABLE 16.3 An ANOVA Table from Minitab for the Data of Example 16.4

```
Analysis of Variance
Source      DF        Adj SS      Adj MS      F-Value     P-Value
Group        3         60.37      20.123      10.48       0.000
Error       53        101.81       1.921
Total       56        162.18
```

Summing It Up—Section 16.1

The following learning objectives were addressed in this section:

Conceptual Understanding

C1: Understand how a research question about differences between three or more population or treatment means is translated into hypotheses.

When data on a numerical variable are used to compare three or more means, the null hypothesis states that the population or treatment means are all equal. The alternative hypothesis states that the population or treatment means are not all equal. If the null hypothesis is rejected, it means that you have evidence that at least two of the population or treatment means differ.

Mastering the Mechanics

M1: Translate a research question or claim about differences between three or more population or treatment means into null and alternative hypotheses.

Example 16.1 illustrates how a research question is translated into hypotheses that can be tested using the ANOVA *F* test.

M2: Know the conditions for appropriate use of the ANOVA *F* test.

The conditions that must be satisfied in order for the ANOVA *F* test to be appropriate are:

1. Each of the *k* population or treatment response distributions is normal.
2. $\sigma_1 = \sigma_2 = \cdots = \sigma_k$ (The *k* normal distributions have equal standard deviations.)
3. The observations in the sample from any particular one of the *k* populations or treatments are independent of one another.
4. When comparing population means, the *k* random samples are selected independently of one another. When comparing treatment means, experimental units are assigned at random to treatments.

M3: Carry out an ANOVA *F* test.

Examples 16.3 and 16.4 illustrate the steps in the ANOVA *F* test.

Putting It into Practice

P1: Carry out an ANOVA *F* test and interpret the conclusion in context.

When the null hypothesis is not rejected, you conclude that there is not convincing evidence that the population or treatment means differ. In this situation, the observed differences in sample means can be explained by sampling variability, and the population or treatment means might be equal. If the null hypothesis is rejected, there is convincing evidence that at least two of the means differ. Examples 16.3 and 16.4 illustrate interpreting conclusions of ANOVA *F* tests in context.

SECTION 16.1 EXERCISES

Each Exercise Set assesses the following chapter learning objectives: C1, M1, M2, M3, P1

SECTION 16.1 Exercise Set 1

16.1 Give as much information as you can about the *P*-value for an upper-tailed *F* test in each of the following situations.
a. $df_1 = 4$, $df_2 = 15$, $F = 5.37$
b. $df_1 = 4$, $df_2 = 15$, $F = 1.90$
c. $df_1 = 4$, $df_2 = 15$, $F = 4.89$
d. $df_1 = 3$, $df_2 = 20$, $F = 14.48$
e. $df_1 = 3$, $df_2 = 20$, $F = 2.69$
f. $df_1 = 4$, $df_2 = 50$, $F = 3.24$

16.2 Employees of a certain state university system can choose from among four different health plans. Each plan differs somewhat from the others in terms of hospitalization coverage. Four random samples of recently hospitalized individuals were selected, each sample consisting of people covered by a different health plan. The length of the hospital stay (number of days) was determined for each individual selected.
a. What hypotheses would you test to decide whether the mean lengths of stay are not the same for all four health plans?
b. If each sample consisted of eight individuals and the value of the ANOVA *F* statistic was $F = 4.37$, what conclusion would be appropriate for a test with $\alpha = 0.01$?
c. Answer the question posed in Part (b) if the *F* value given there resulted from sample sizes $n_1 = 9$, $n_2 = 8$, $n_3 = 7$, and $n_4 = 8$.

16.3 The authors of the paper **"Age and Violent Content Labels Make Video Games Forbidden Fruits for Youth"** (*Pediatrics* **[2009]: 870–876**) carried out an experiment to determine if restrictive labels on video games actually increased the attractiveness of the game for young game players. Participants read a description of a new video game and were asked how much they wanted to play the game. The description also included an age rating. Some participants read the description with an age restrictive label of 7+, indicating that the game was not appropriate for children under the age of 7. Others read the same description, but with an age restrictive label of 12+, 16+, or 18+. The following data for 12- to 13-year-old boys are consistent with summary statistics given in the paper. (The sample sizes in the actual experiment were larger.) For purposes of this exercise, you can assume that the boys were assigned at random to one of the four age label treatments

(7+, 12+, 16+, and 18+). Data shown are the boys' ratings of how much they wanted to play the game on a scale of 1 to 10. Do the data provide convincing evidence that the mean rating associated with the game description by 12- to 13-year-old boys is not the same for all four restrictive rating labels? Test the appropriate hypotheses using a significance level of 0.05.

7+ label	12+ label	16+ label	18+ label
6	8	7	10
6	7	9	9
6	8	8	6
5	5	6	8
4	7	7	7
8	9	4	6
6	5	8	8
1	8	9	9
2	4	6	10
4	7	7	8

16.4 The authors of the paper **"Reading Subtitles and Taking Enotes While Learning Scientific Materials in a Multimedia Environment"** (*Educational Technology and Society* **[2016]: 47–58**) were interested in determining if including subtitles and providing opportunities to take electronic notes while listening to online materials would enhance learning for students whose first language was not English. Students were randomly assigned to one of four groups. In the first group, subtitles were included in the online materials that the students were asked to study, but the ability to take electronic notes was not provided. For the second group, no subtitles were provided but students were able to take electronic notes. For the third group, both subtitles and the ability to take electronic notes were available. Students in the fourth group did not have access to either subtitles or the ability to take electronic notes. After studying the online materials, all students took a 14-question test on the material studied.

Minitab output based on data consistent with summary quantities in the paper is shown on the next page. Is there evidence to conclude that the mean test score differs for at least two of the treatments? Use the given computer output to test the appropriate hypotheses with a significance level of 0.05.

📊 Data set available

Analysis of Variance

```
Source    DF    Adj SS    Adj MS    F-Value  P-Value
Factor     3    37.66    12.558     2.39     0.077
Error     63   332.36     5.276
Total     66   370.13
```

SECTION 16.1 Exercise Set 2

16.5 Give as much information as you can about the P-value of the single-factor ANOVA F test in each of the following situations.
a. $k = 5$, $n_1 = n_2 = n_3 = n_4 = n_5 = 4$, $F = 5.37$
b. $k = 5$, $n_1 = n_2 = n_3 = 5$, $n_4 = n_5 = 4$, $F = 2.83$
c. $k = 3$, $n_1 = 4$, $n_2 = 5$, $n_3 = 6$, $F = 5.02$
d. $k = 3$, $n_1 = n_2 = 4$, $n_3 = 6$, $F = 15.90$
e. $k = 4$, $n_1 = n_2 = 15$, $n_3 = 12$, $n_4 = 10$, $F = 1.75$

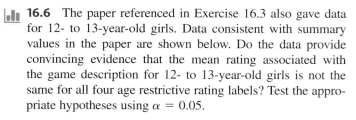 **16.6** The paper referenced in Exercise 16.3 also gave data for 12- to 13-year-old girls. Data consistent with summary values in the paper are shown below. Do the data provide convincing evidence that the mean rating associated with the game description for 12- to 13-year-old girls is not the same for all four age restrictive rating labels? Test the appropriate hypotheses using $\alpha = 0.05$.

7+ label	12+ label	16+ label	18+ label
4	4	6	8
7	5	4	6
6	4	8	6
5	6	6	5
3	3	10	7
6	5	8	4
4	3	6	10
5	8	6	6
10	5	8	8
5	9	5	7

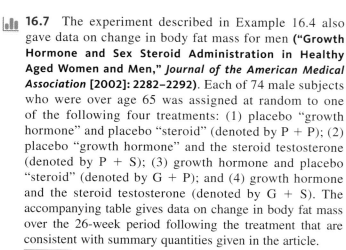 **16.7** The experiment described in Example 16.4 also gave data on change in body fat mass for men (**"Growth Hormone and Sex Steroid Administration in Healthy Aged Women and Men,"** *Journal of the American Medical Association* **[2002]: 2282–2292**). Each of 74 male subjects who were over age 65 was assigned at random to one of the following four treatments: (1) placebo "growth hormone" and placebo "steroid" (denoted by P + P); (2) placebo "growth hormone" and the steroid testosterone (denoted by P + S); (3) growth hormone and placebo "steroid" (denoted by G + P); and (4) growth hormone and the steroid testosterone (denoted by G + S). The accompanying table gives data on change in body fat mass over the 26-week period following the treatment that are consistent with summary quantities given in the article.

Treatment	Change in Body Fat Mass (kg)			
	P + P	P + S	G + P	G + S
	0.3	−3.7	−3.8	−5.0
	0.4	−1.0	−3.2	−5.0
	−1.7	0.2	−4.9	−3.0
	−0.5	−2.3	−5.2	−2.6
	−2.1	1.5	−2.2	−6.2
	1.3	−1.4	−3.5	−7.0
	0.8	1.2	−4.4	−4.5
	1.5	−2.5	−0.8	−4.2
	−1.2	−3.3	−1.8	−5.2
	−0.2	0.2	−4.0	−6.2
	1.7	0.6	−1.9	−4.0
	1.2	−0.7	−3.0	−3.9
	0.6	−0.1	−1.8	−3.3
	0.4	−3.1	−2.9	−5.7
	−1.3	0.3	−2.9	−4.5
	−0.2	−0.5	−2.9	−4.3
	0.7	−0.8	−3.7	−4.0
		−0.7		−4.2
		−0.9		−4.7
		−2.0		
		−0.6		
n	17	21	17	19
$\bar{x}$	0.100	−0.933	−3.112	−4.605
s	1.139	1.443	1.178	1.122
s^2	1.297	2.082	1.388	1.259

Also, $N = 74$, grand total $= -158.3$, and the mean of all 74 observations is $\bar{\bar{x}} = \dfrac{-158.3}{74} = -2.139$ Carry out an F test to see whether mean change in body fat mass differs for the four treatments.

16.8 In an experiment to investigate the performance of four different brands of spark plugs intended for use on a 125-cc motorcycle, five plugs of each brand were tested, and the number of miles (at a constant speed) until failure was observed. A partially completed ANOVA table is given. Fill in the missing entries, and test the relevant hypotheses using a 0.05 level of significance.

Source of Variation	df	Sum of Squares	Mean Square	F
Treatments				
Error		235,419.04		
Total		310,500.76		

ADDITIONAL EXERCISES FOR SECTION 16.1

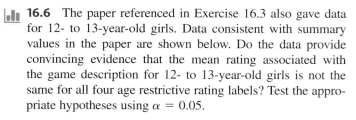 **16.9** Do people feel hungrier after sampling a healthy food? The authors of the paper **"When Healthy Food Makes You Hungry"** (*Journal of Consumer Research* [2010]: S34–S44)

carried out a study to answer this question. They randomly assigned volunteers into one of three groups. The people in the first group were asked to taste a snack that was billed as a new health bar containing high levels of protein, vitamins, and fiber. The people in the second group were asked to taste the same snack but were told it was a tasty chocolate bar with a raspberry center. After tasting the snack, participants were asked to rate their hunger level on a scale from 1 (not at all hungry) to 7 (very hungry). The people in the third group were asked to rate their hunger but were not given a snack.

The data in the accompanying table are consistent with summary quantities given in the paper (although the sample sizes in the actual study were larger).

a. Do these data provide evidence that the mean hunger rating differs for at least two of the treatments ("healthy" snack, "tasty" snack, no snack)? Test the relevant hypotheses using a significance level of 0.05.
b. Is it reasonable to conclude that the mean hunger rating is greater for people who do not get a snack? Explain.

Treatment Group	Hunger Rating									Sample Mean	Sample Standard Deviation
Healthy	4	7	7	4	5	3	4	7	6	5.2	1.56
Tasty	4	1	3	2	6	2	5	3	4	3.3	1.58
No Snack	3	4	5	6	5	4	2	4	4	4.1	1.17
								Overall mean = 4.2			

16.10 The accompanying summary statistics for a measure of social marginality for samples of youths, young adults, adults, and seniors appeared in the paper **"Perceived Causes of Loneliness in Adulthood"** (*Journal of Social Behavior and Personality* [2000]: 67–84). The social marginality score measured actual and perceived social rejection, with higher scores indicating greater social rejection. For purposes of this exercise, assume that it is reasonable to regard the four samples as representative of the U.S. population in the corresponding age groups and that the distributions of social marginality scores for these four groups are approximately normal with the same standard deviation. Is there evidence that the mean social marginality score differs for at least two of the four age groups? Test the relevant hypotheses using $\alpha = 0.05$.

Age Group	Youths	Young Adults	Adults	Seniors
Sample Size	106	255	314	36
$\bar{x}$	2.00	3.40	3.07	2.84
s	1.56	1.68	1.66	1.89

16.11 The chapter Preview Example described a study comparing three groups of college students (soccer athletes, nonsoccer athletes, and a comparison group consisting of students who did not participate in intercollegiate sports). The following is information on scores from the Hopkins Verbal Learning Test (which measures immediate memory recall).

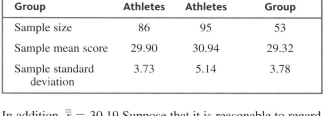

Group	Soccer Athletes	Nonsoccer Athletes	Comparison Group
Sample size	86	95	53
Sample mean score	29.90	30.94	29.32
Sample standard deviation	3.73	5.14	3.78

In addition, $\bar{\bar{x}} = 30.19$ Suppose that it is reasonable to regard these three samples as random samples from the three student populations of interest. Is there sufficient evidence to conclude that the mean Hopkins score is not the same for the three student populations? Use $\alpha = 0.05$.

16.12 It is common for baseball pitchers to use stretching to prepare for a game. But does this make a difference? The authors of the paper **"The Acute Effects of Upper Extremity Stretching on Throwing Velocity in Baseball Throwers"** (*Journal of Sports Medicine* [2013]: 1–7) carried out an experiment to compare two different types of stretching and a control treatment consisting of no stretching. Participants were adult males with varying levels of baseball throwing experience and who were not professional or collegiate baseball players. Participants in the two stretching treatments went through a warm-up that included 8 minutes of stretching. Each participant (all three groups) then threw 10 pitches, and the average speed (km/hour) was calculated.

a. Explain why it is important that the participants be assigned at random to the three different treatment groups (Stretching Method 1, Stretching Method 2, and No Stretching).
b. The following computer output and summary values are based on simulated data that are consistent with information and conclusions given in the paper. Use the given output to determine if there is evidence to support the claim that the mean average speed is not the same for all three treatments. Use a significance level of 0.05 for your test.

```
Analysis of Variance

Source   DF   Adj SS   Adj MS   F-Value  P-Value
Factor    2     1097    548.7     2.44     0.094
Error    78    17570    225.3
Total    80    18667

Means

Factor              N   Mean  StDev       95% CI
Stretching         27  86.26 15.95  (80.51, 92.01)
  Method 1
Stretching         27  90.30 17.06  (84.55, 96.05)
  Method 2
No Stretching      27  95.26 11.42  (89.51, 101.01)
```

c. Previous research on the effect of stretching on performance in other sports, such as running, has concluded that stretching can improve performance. Why do you think that the authors of this paper were surprised by the results of this study?

16.13 Parents are frequently concerned when their child seems slow to begin walking (although when the child finally walks, the resulting havoc sometimes has the parents wishing they could turn back the clock!). The article **"Walking in the Newborn" (***Science***, 176 [1972]: 314–315)** reported on an experiment in which the effects of several different treatments on the age at which a child first walks were compared. Children in the first group were given special walking exercises for 12 minutes per day beginning at age 1 week and lasting 7 weeks. The second group of children received daily exercises but not the walking exercises administered to the first group. The third and fourth groups were control groups. They received no special treatment and differed only in that the third group's progress was checked weekly, whereas the fourth group's progress was checked just once at the end of the study. Observations on age (in months) when the children first walked are shown in the accompanying table. Also given is the ANOVA table, obtained from the SPSS computer package.

a. Verify the entries in the ANOVA table.

b. State and test the relevant hypotheses using a significance level of 0.05.

	Age			n	Total
Treatment 1	9.00	9.50	9.75	6	60.75
	10.00	13.00	9.50		
Treatment 2	11.00	10.00	10.00	6	68.25
	11.75	10.50	15.00		
Treatment 3	11.50	12.00	9.00	6	70.25
	11.50	13.25	13.00		
Treatment 4	13.25	11.50	12.00	5	61.75
	13.50	11.50			

Analysis of Variance					
Source	df	Sum of sq.	Mean Sq.	F Ratio	F Prob
Between Groups	3	14.779	4.926	2.142	.129
With in Group	19	43.690	2.299		
Total	22	58.467			

SECTION 16.2 Multiple Comparisons

When H_0: $\mu_1 = \mu_2 = \cdots = \mu_k$ is rejected by the F test, you believe that there are differences among the k population or treatment means. A natural question to ask at this point is, which means differ? For example, with $k = 4$, it might be the case that $\mu_1 = \mu_2 = \mu_4$, with μ_3 different from the other three means. Another possibility is that $\mu_1 = \mu_4$ and $\mu_2 = \mu_3$. Still another possibility is that all four means are different from one another. A **multiple comparisons procedure** is a method for identifying differences among the μ's once the hypothesis that all of the means are equal has been rejected. The **Tukey-Kramer** (T-K) multiple comparisons procedure is one of these methods.

The T-K procedure is based on calculating a confidence interval for the difference in each possible pair of μ's. For example, for $k = 3$, there are three differences to consider:

$$\mu_1 - \mu_2 \qquad \mu_1 - \mu_3 \qquad \mu_2 - \mu_3$$

(The difference $\mu_2 - \mu_1$ is not considered, because the interval for $\mu_1 - \mu_2$ provides the same information. Similarly, intervals for $\mu_3 - \mu_1$ and $\mu_3 - \mu_2$ are not necessary.) Once all confidence intervals have been calculated, each is examined to determine whether the interval includes 0. If a particular interval does not include 0, the two means are declared "significantly different" from one another. If an interval includes 0, there is no evidence of a significant difference between the means involved.

Suppose, for example, that $k = 3$ and that the three confidence intervals are

Difference	T-K Confidence Interval
$\mu_1 - \mu_2$	$(-0.9, 3.5)$
$\mu_1 - \mu_3$	$(2.6, 7.0)$
$\mu_2 - \mu_3$	$(1.2, 5.7)$

Because the interval for $\mu_1 - \mu_2$ includes 0, you would say that μ_1 and μ_2 do not differ significantly. The other two intervals do not include 0, so you would conclude that $\mu_1 \neq \mu_3$ and $\mu_2 \neq \mu_3$.

The T-K intervals are based on critical values for a probability distribution called the *Studentized range distribution*. These critical values appear in Appendix A Table 8. To find a critical value, enter the table at the column corresponding to the number of populations or treatments being compared, move down to the row corresponding to the number of error degrees of freedom $(N - k)$, and select either the value for a 95% confidence level or the one for a 99% level.

The Tukey–Kramer Multiple Comparison Procedure

When there are k populations or treatments being compared, $\dfrac{k(k-1)}{2}$ confidence intervals must be calculated. Denoting the relevant Studentized range critical value (from Appendix A Table 8) by q, the intervals are as follows:

$$\text{For } \mu_i - \mu_j:\qquad (\bar{x}_i - \bar{x}_j) \pm q\sqrt{\frac{\text{MSE}}{2}\left(\frac{1}{n_i} + \frac{1}{n_j}\right)}$$

Two means are judged to differ significantly if the corresponding interval does not include zero.

If the sample sizes are all the same, you can use n to denote the common value of $n_1, n_2, \ldots, n_k$. In this case, the $\pm$ term for each interval is the same quantity

$$q\sqrt{\frac{\text{MSE}}{n}}$$

Example 16.5   Hormones and Body Fat Revisited

Example 16.4 introduced the accompanying data on change in body fat mass resulting from a double-blind experiment designed to compare the following four treatments: (1) placebo "growth hormone" and placebo "steroid" (denoted by P + P); (2) placebo "growth hormone" and the steroid estradiol (denoted by P + S); (3) growth hormone and placebo "steroid" (denoted by G + P); and (4) growth hormone and the steroid estradiol (denoted by G + S). From Example 16.4, MSTr = 20.12, MSE = 1.92, and $F = 10.48$ with an associated P-value < 0.001. It was concluded that the mean change in body fat mass is not the same for all four treatments.

	Change in Body Fat Mass (kg)			
Treatment	P + P	P + S	G + P	G + S
	0.1	−0.1	−1.6	−3.1
	0.6	0.2	−0.4	−3.2
	2.2	0.0	0.4	−2.0
	0.7	−0.4	−2.0	−2.0
	−2.0	−0.9	−3.4	−3.3
	0.7	−1.1	−2.8	−0.5
	0.0	1.2	−2.2	−4.5
	−2.6	0.1	−1.8	−0.7
	−1.4	0.7	−3.3	−1.8
	1.5	−2.0	−2.1	−2.3
	2.8	−0.9	−3.6	−1.3
	0.3	−3.0	−0.4	−1.0
	−1.0	1.0	−3.1	−5.6
	−1.0	1.2		−2.9
				−1.6
				−0.2
n	14	14	13	16
$\bar{x}$	0.064	−0.286	−2.023	−2.250
s	1.545	1.218	1.264	1.468
s^2	2.387	1.484	1.598	2.155

Appendix A Table 8 gives the 95% Studentized range critical value $q = 3.74$ (using $k = 4$ and error df $= 60$, the closest tabled value to df $= N - k = 53$). The first two T-K intervals are

$$\mu_1 - \mu_2: \quad (0.064 - (-0.286)) \pm 3.74 \sqrt{\left(\frac{1.92}{2}\right)\left(\frac{1}{14} + \frac{1}{14}\right)}$$

$$= 0.35 \pm 1.39$$

$$= (-1.04, 1.74) \quad \leftarrow Includes\ 0$$

$$\mu_1 - \mu_3: \quad (0.064 - (-2.023)) \pm 3.74 \sqrt{\left(\frac{1.92}{2}\right)\left(\frac{1}{14} + \frac{1}{13}\right)}$$

$$= 2.09 \pm 1.41$$

$$= (0.68, 3.50) \quad \leftarrow Does\ not\ include\ 0$$

The remaining intervals are

$\mu_1 - \mu_4$	$(0.97, 3.66)$	$\leftarrow$ *Does not include 0*
$\mu_2 - \mu_3$	$(0.32, 3.15)$	$\leftarrow$ *Does not include 0*
$\mu_2 - \mu_4$	$(0.62, 3.31)$	$\leftarrow$ *Does not include 0*
$\mu_3 - \mu_4$	$(-1.14, 1.60)$	$\leftarrow$ *Includes 0*

You would conclude that μ_1 is not significantly different from μ_2 and that μ_3 is not significantly different from μ_4. You would also conclude that μ_1 and μ_2 are significantly different from both μ_3 and μ_4. Note that Treatments 1 and 2 were treatments that administered a placebo in place of the growth hormone and Treatments 3 and 4 were treatments that included the growth hormone. This analysis was the basis of the researchers' conclusion that growth hormone, with or without steroids, decreased body fat mass.

Minitab can be used to calculate T-K intervals if raw data are available. Typical output (based on Example 16.5) is shown in Figure 16.6. From the output, you can see that the confidence interval for $\mu_1 (P + P) - \mu_2 (P + S)$ is $(-1.039, 1.739)$, that for $\mu_2 (P + S) - \mu_4 (G + S)$ is $(0.619, 3.309)$, and so on.

FIGURE 16.6
The T-K intervals for Example 16.5 (from Minitab)

Tukey 95% Simultaneous Confidence Intervals
All Pairwise Comparisons

Individual confidence level = 98.95%

G + S subtracted from:

	Lower	Center	Upper	
G + P	-1.145	0.227	1.599	(------*------)
P + S	0.619	1.964	3.309	(------*------)
P + P	0.969	2.314	3.659	(------*-----)

```
        --------+---------+---------+---------+-
             -2.0       0.0       2.0       4.0
```

G + P subtracted from:

	Lower	Center	Upper	
P + S	0.322	1.737	3.153	(------*------)
P + P	0.672	2.087	3.503	(------*-------)

```
        --------+---------+---------+---------+-
             -2.0       0.0       2.0       4.0
```

P + S subtracted from:

	Lower	Center	Upper	
P + P	-1.039	0.350	1.739	(------*------)

```
        --------+---------+---------+---------+-
             -2.0       0.0       2.0       4.0
```

Why calculate the T-K intervals rather than use the two-sample t confidence interval for a difference between μ's from Chapter 13? The answer is that the T-K intervals control the **simultaneous confidence level** at approximately 95% (or 99%, if chosen as the overall confidence level). That is, if the procedure is used repeatedly on many different data sets, in the long run only about 5% (or 1%) of the time would at least one of the intervals not include the actual value that the interval is estimating. Consider using separate 95% t intervals, each one having a 5% error rate. In those instances, the chance that at least one interval would make an incorrect statement about a difference in μ's increases dramatically with the number of intervals calculated. The Minitab output in Figure 16.6 shows that to achieve a simultaneous confidence level of about 95% (experimentwise or "family" error rate of 5%) when $k = 4$ and error df $= 76$, the individual interval confidence levels must be 98.95% (individual error rate 1.05%).

An effective display for summarizing the results of any multiple comparisons procedure involves listing the $\bar{x}$'s and underscoring pairs judged to be not significantly different. The process for constructing such a display is described in the following box.

Summarizing the Results of the Tukey–Kramer Procedure

1. List the sample means in increasing order, identifying the corresponding population or treatment just above the value of each $\bar{x}$.
2. Use the T-K intervals to determine the group of means that do not differ significantly from the first mean in the list. Draw a horizontal line extending from the smallest mean to the last mean in the group identified. For example, if there are five means, arranged in order,

Population	3	2	1	4	5
Sample mean	$\bar{x}_3$	$\bar{x}_2$	$\bar{x}_1$	$\bar{x}_4$	$\bar{x}_5$

and μ_3 is judged to be not significantly different from μ_2 or μ_1, but is judged to be significantly different from μ_4 and μ_5, draw the following line:

Population	3	2	1	4	5
Sample mean	$\underline{\bar{x}_3}$	$\underline{\bar{x}_2}$	$\underline{\bar{x}_1}$	$\bar{x}_4$	$\bar{x}_5$

3. Use the T–K intervals to determine the group of means that are not significantly different from the second smallest mean. (You need consider only means that appear to the right of the mean under consideration.) If there is already a line connecting the second smallest mean with all means in the new group identified, no new line need be drawn. If this entire group of means is not underscored with a single line, draw a line extending from the second smallest to the last mean in the new group. Continuing with our example, if μ_2 is not significantly different from μ_1 but is significantly different from μ_4 and μ_5, no new line need be drawn. However, if μ_2 is not significantly different from either μ_1 or μ_4 but is judged to be different from μ_5, a second line is drawn as shown:

Population	3	2	1	4	5
Sample mean	$\underline{\bar{x}_3}$	$\underline{\bar{x}_2}$	$\underline{\bar{x}_1}$	$\underline{\bar{x}_4}$	$\bar{x}_5$

4. Continue considering the means in the order listed, adding new lines as needed.

To illustrate this summary procedure, suppose that four samples with $\bar{x}_1 = 19$, $\bar{x}_2 = 27$, $\bar{x}_3 = 24$, and $\bar{x}_4 = 10$ are used to test $H_0\colon \mu_1 = \mu_2 = \mu_3 = \mu_4$ and that this hypothesis is rejected. Suppose the T-K confidence intervals indicate that μ_2 is significantly different from both μ_1 and μ_4, and that there are no other significant differences. The resulting summary display would then be

Population	4	1	3	2
Sample mean	$\underline{10}$	$\underline{19}$	$\underline{24}$	27

Example 16.6 **Sleep Time**

Data set available

Suppose that a biologist wanted to study the effects of ethanol on sleep time. Twenty rats, matched for age and other characteristics, were used in the study. Each rat was given an oral injection having a particular concentration of ethanol per body weight. The rapid eye movement (REM) sleep time for each rat was then recorded for a 24-hour period, with the results shown in the following table:

Treatment	Observations					$\bar{x}$
1. 0 (control)	88.6	73.2	91.4	68.0	75.2	79.28
2. 1 g/kg	63.0	53.9	69.2	50.1	71.5	61.54
3. 2 g/kg	44.9	59.5	40.2	56.3	38.7	47.92
4. 4 g/kg	31.0	39.6	45.3	25.2	22.7	32.76

Table 16.4 (an ANOVA table from SAS) leads to the conclusion that actual mean REM sleep time is not the same for all four treatments (the *P*-value for the *F* test is 0.0001).

TABLE 16.4 SAS ANOVA Table for Example 16.6
Analysis of Variance Procedure
Dependent Variable: TIME

Source	DF	Sum of Squares	Mean Square	F Value	Pr > F
Model	3	5882.35750	1960.78583	21.09	0.0001
Error	16	1487.40000	92.96250		
Total	**19**	**7369.75750**			

The T-K intervals are

Difference	Interval	Includes 0?
$\mu_1 - \mu_2$	17.74 ± 17.446	no
$\mu_1 - \mu_3$	31.36 ± 17.446	no
$\mu_1 - \mu_4$	46.24 ± 17.446	no
$\mu_2 - \mu_3$	13.08 ± 17.446	yes
$\mu_2 - \mu_4$	28.78 ± 17.446	no
$\mu_3 - \mu_4$	15.16 ± 17.446	yes

The only T-K intervals that include zero are those for $\mu_2 - \mu_3$ and $\mu_3 - \mu_4$. The corresponding underscoring pattern is

$\bar{x}_4$	$\bar{x}_3$	$\bar{x}_2$	$\bar{x}_1$
32.76	47.92	61.54	79.28

Figure 16.7 displays the SAS output that agrees with the underscoring shown above; letters are used to indicate groupings in place of the underscoring.

FIGURE 16.7
SAS output for Example 16.6

Alpha = 0.05 df = 16 MSE = 92.9625
Critical Value of Studentized Range = 4.046
Minimum Significant Difference = 17.446
Means with the same letter are not significantly different.

Tukey Grouping		Mean	N	Treatment
	A	79.280	5	0 (control)
	B	61.540	5	1 g/kg
C	B	47.920	5	2 g/kg
C		32.760	5	4 g/kg

 Data set available

Example 16.7 **Mug Color and the Taste of Coffee**

The paper **"Does the Colour of the Mug Influence the Taste of the Coffee?"** (*Flavour* **[2014]: 1–7)** describes an experiment to investigate whether the color of a coffee mug has an effect on how people perceive the flavor intensity of coffee. In this experiment, 18 volunteers were randomly assigned to one of three groups. All were served the same coffee, but those in the first group received their coffee in a white mug, those in the second group received their coffee in a blue mug, and those in the third group received their coffee in a clear glass mug. After tasting the coffee, each person rated the flavor intensity of the coffee. The mean intensity rating for the three treatment groups (approximate values from a graph in the paper) were 50 for the white mug group, 42 for the blue mug group, and 23 for the clear glass mug group.

The researchers used an ANOVA F test to test the null hypothesis of no difference in the mean intensity rating for the three treatments. They reported an F statistic value of 4.78 and a P-value of 0.025. Based on this P-value, they rejected the null hypothesis and concluded that the means for intensity rating were different for at least two of the groups. A multiple comparison procedure was used to identify differences in treatment means, and the following conclusions were reached:

- Coffee was rated as significantly more intense when served in a white mug than when serve in a clear glass mug.
- None of the other differences were found to be significant.

These results are summarized in the following underscoring pattern. The treatment group means are arranged in order from smallest to largest.

Clear	Blue	White
23	42	50

Based on this analysis, you would conclude that there is a difference between the mean intensity rating for coffee served in a white mug and coffee served in a clear glass mug. The effect of a blue mug is unclear. There is not convincing evidence of a difference between the mean intensity rating of coffee served in a blue mug and coffee served in a white mug, and there is also not convincing evidence of a difference for coffee served in a blue mug and coffee served in a clear glass mug.

Summing It Up—Section 16.2

The following learning objectives were addressed in this section:

Mastering the Mechanics

M4: Use a multiple comparison procedure to identify differences in population or treatment means.

A multiple comparison procedure is a method for determining which means differ. When the null hypothesis of equal population or treatment means has been rejected, a multiple comparison procedure can help to identify which means are different from one another. The Tukey-Kramer (T-K) method described in this section is one such procedure. This method involves calculating confidence intervals for all possible differences in population means. Two means are then judged as significantly different if the corresponding interval for their difference does not include 0. Examples 16.5 and 16.6 illustrate the use of the Tukey-Kramer multiple comparison method.

Putting It into Practice

P2: Interpret the results of a multiple comparison procedure in context.

Examples 16.6 and 16.7 illustrate how the results from a multiple comparison procedure are interpreted in context.

SECTION 16.2 EXERCISES

Each Exercise Set assesses the following chapter learning objectives: M4, P2

SECTION 16.2 Exercise Set 1

16.14 Leaf surface area is an important variable in plant gas-exchange rates. Dry matter per unit surface area (mg/cm³) was measured for trees raised under three different growing conditions. Let μ_1, μ_2, and μ_3 represent the mean dry matter per unit surface area for the growing conditions 1, 2, and 3, respectively. Suppose that the given 95% T-K confidence intervals are:

Difference	$\mu_1 - \mu_2$	$\mu_1 - \mu_3$	$\mu_2 - \mu_3$
Interval	$(-3.11, -1.11)$	$(-4.06, -2.06)$	$(-1.95, 0.05)$

Which of the following four statements do you think describes the relationship between μ_1, μ_2, and μ_3? Explain your choice.
a. $\mu_1 = \mu_2$, and μ_3 differs from μ_1 and μ_2.
b. $\mu_1 = \mu_3$, and μ_2 differs from μ_1 and μ_3.
c. $\mu_2 = \mu_3$, and μ_1 differs from μ_2 and μ_3.
d. All three μ's are different from one another.

16.15 The accompanying underscoring pattern appears in the article **"Women's and Men's Eating Behavior Following Exposure to Ideal-Body Images and Text" (*Communications Research* [2006]: 507–529)**. Women either viewed slides depicting images of thin female models with no text (treatment 1); viewed the same slides accompanied by diet and exercise-related text (treatment 2); or viewed the same slides accompanied by text that was unrelated to diet and exercise (treatment 3). A fourth group of women did not view any slides (treatment 4). Participants were assigned at random to the four treatments. Participants were then asked to complete a questionnaire in a room where pretzels were set out on the tables. An observer recorded how many pretzels participants ate while completing the questionnaire. Write a few sentences interpreting this underscoring pattern.

Treatment:	2	1	4	3
Mean number of pretzels consumed:	0.97	1.03	2.20	2.65

16.16 The following data resulted from a flammability study in which specimens of five different fabrics were tested to determine burn times (in seconds).

Fabric					
1	17.8	16.2	15.9	15.5	
2	13.2	10.4	11.3		
3	11.8	11.0	9.2	10.0	
4	16.5	15.3	14.1	15.0	13.9
5	13.9	10.8	12.8	11.7	

$$\text{MSTr} = 23.67$$
$$\text{MSE} = 1.39$$
$$F = 17.08$$
$$P\text{-value} = 0.000$$

The accompanying output gives the T-K intervals as calculated by Minitab. Identify significant differences and give the underscoring pattern.

Data set available

Individual error rate = 0.00750
Critical value = 4.37
Intervals for (column level mean) − (row level mean)

	1	2	3	4
	1.938			
2	7.495			
	3.278	−1.645		
3	8.422	3.912		
	−1.050	−5.983	−6.900	
4	3.830	−0.670	−2.020	
	1.478	−3.445	−4.372	0.220
5	6.622	2.112	0.772	5.100

SECTION 16.2 Exercise Set 2

16.17 The paper **"Trends in Blood Lead Levels and Blood Lead Testing among U.S. Children Aged 1 to 5 Years" (*Pediatrics* [2009]: e376–e385)** gave data on blood lead levels (in mg/dL) for samples of children living in homes that had been classified either at low, medium, or high risk of lead exposure, based on when the home was constructed. After using a multiple comparison procedure, the authors reported the following:

1. The difference in mean blood lead level between low-risk housing and medium-risk housing was significant.
2. The difference in mean blood lead level between low-risk housing and high-risk housing was significant.
3. The difference in mean blood lead level between medium-risk housing and high-risk housing was significant.

Which of the following sets of T-K intervals (Set 1, 2, or 3) is consistent with the authors' conclusions? Explain your choice.

μ_L = mean blood lead level for children living in low-risk housing
μ_M = mean blood lead level for children living in medium-risk housing
μ_H = mean blood lead level for children living in high-risk housing

Difference	Set 1	Set 2	Set 3
$\mu_L - \mu_M$	$(-0.6, 0.1)$	$(-0.6, -0.1)$	$(-0.6, -0.1)$
$\mu_L - \mu_H$	$(-1.5, -0.6)$	$(-1.5, -0.6)$	$(-1.5, -0.6)$
$\mu_M - \mu_H$	$(-0.9, -0.3)$	$(-0.9, 0.3)$	$(-0.9, -0.3)$

16.18 The paper referenced in the Exercise 16.15 also gave the following underscoring pattern for men.

Treatment:	4	3	1	2
Mean number of pretzels consumed:	2.70	3.8	5.96	6.61

a. Write a few sentences interpreting this underscoring pattern.
b. Using your answers from Part (a) and from the Exercise 16.15, write a few sentences describing the differences between how men and women respond to the treatments.

16.19 Do lizards play a role in spreading plant seeds? Some research carried out in South Africa would suggest so (**"Dispersal of Namaqua Fig [*Ficus cordata cordata*] Seeds by the Augrabies Flat Lizard [*Platysaurus broadleyi*]," *Journal of Herpetology* [1999]: 328–330**). The researchers collected 400 seeds of a particular type of fig, 100 of which were from each treatment: lizard dung, bird dung, rock hyrax dung, and uneaten figs. They planted these seeds in batches of 5, and for each group of 5 they recorded how many of the seeds germinated. This resulted in 20 observations for each treatment. The treatment means and standard deviations are given in the accompanying table.

Treatment	*n*	$\bar{x}$	*s*
Uneaten figs	20	2.40	0.30
Lizard dung	20	2.35	0.33
Bird dung	20	1.70	0.34
Hyrax dung	20	1.45	0.28

a. Construct the appropriate ANOVA table, and test the hypothesis that there is no difference between mean number of seeds germinating for the four treatments.
b. Is there evidence that seeds eaten and then excreted by lizards germinate at a higher rate than those eaten and then excreted by birds? Give statistical evidence to support your answer.

ADDITIONAL EXERCISES

 16.20 Suppose that samples of six different brands of diet or imitation margarine were analyzed to determine the level of physiologically active polyunsaturated fatty acids (PAPUFA, in percent), resulting in the accompanying data.

Imperial	14.1	13.6	14.4	14.3	
Parkay	12.8	12.5	13.4	13.0	12.3
Blue Bonnet	13.5	13.4	14.1	14.3	
Chiffon	13.2	12.7	12.6	13.9	
Mazola	16.8	17.2	16.4	17.3	18.0
Fleischmann's	18.1	17.2	18.7	18.4	

a. Carry out a test to determine if there is evidence of differences among the true mean PAPUFA percentages for the different brands. Use $\alpha = 0.05$.
b. Use the T-K procedure to calculate 95% simultaneous confidence intervals for all differences between pairs of means and give the corresponding underscoring pattern.

16.21 In an experiment to investigate the effect of the portrayal of female characters in superhero movies, researchers randomly assigned female college students to one of three groups (**"The Empowering (Super) Heroine? The Effects of Sexualized Female Characters in Superhero Films on Women,"**

Sex Roles [2015]: 211–220). One group was a control group, one group watched 13 minutes of video scenes from the movie *Spider-Man* (where a sexy female character was portrayed as a victim), and one group watched 13 minutes of video scenes from the movie *X-Men* (where a sexy female character was portrayed as a heroine). The women in the control group did not watch a video. The women in all three groups then completed a questionnaire and their answers were used to calculate a measure of gender stereotyping, with lower values indicating attitudes more accepting of equality of women and men.

The researchers used a one-way ANOVA to analyze the data. The following Minitab ANOVA output and summary statistics are based on data consistent with information and conclusions from the paper.

```
Analysis of Variance

Source   DF   Adj SS  Adj MS  F-Value  P-Value
Factor    2    3.264  1.6321   3.34     0.041
Error    79   38.632  0.4890
Total    81   41.896

Means

Factor        N    Mean   StDev      95% CI
Control      29   2.668   0.673  (2.409, 2.926)
Spider Man   30   3.120   0.800  (2.866, 3.374)
X-Men        23   3.020   0.580  (2.730, 3.310)
```

a. Use the given output to test the null hypothesis of no difference in mean gender stereotyping score for the three different treatment groups. Use a significance level of 0.05.
b. Minitab reported the following T-K intervals:

Control – X-Men:	$(-0.819, 0.114)$
Control – Spider Man:	$(-0.887, -0.017)$
X-Men – Spider Man :	$(-0.563, 0.363)$

Use this information to construct the corresponding underscore pattern.
c. Write a few sentences describing what you learned from the results of the Tukey-Kramer procedure.

 16.22 Consider the accompanying data on plant growth after the application of five different types of growth hormone.

	1	13	17	7	14
	2	21	13	20	17
Hormone	**3**	18	14	17	21
	4	7	11	18	10
	5	6	11	15	8

a. Carry out the ANOVA *F* test using a significance level of $\alpha = 0.05$.
b. What happens when the T-K procedure is applied? (*Note:* This "contradiction" can occur when H_0 is "barely" rejected. It happens because the test and the multiple comparison method are based on different distributions. Consult your friendly neighborhood statistician for more information.)

 Data set available

ANOVA Computations (Optional)

Single-Factor ANOVA

Let T_1 denote the sum of the observations in the sample from the first population or treatment, and let $T_2, \ldots, T_k$ denote the other sample totals. Also let T represent the sum of all N observations—the **grand total**—and

$$CF = \text{correction factor} = \frac{T^2}{N}$$

Then

$$SSTo = \sum_{\text{all } N \text{ obs.}} x^2 - CF$$

$$SSTr = \frac{T_1^2}{n_1} + \frac{T_2^2}{n_2} + \cdots + \frac{T_k^2}{n_k} - CF$$

$$SSE = SSTo - SSTr$$

Example 15A.1

Treatment 1	4.2	3.7	5.0	4.8	$T_1 = 17.7$	$n_1 = 4$
Treatment 2	5.7	6.2	6.4		$T_2 = 18.3$	$n_2 = 3$
Treatment 3	4.6	3.2	3.5	3.9	$T_3 = 15.2$	$n_3 = 4$
					$T = 51.2$	$N = 11$

$$CF = \text{correction factor} = \frac{T^2}{N} = \frac{(51.2)^2}{11} = 238.31$$

$$SSTr = \frac{T_1^2}{n_1} + \frac{T_2^2}{n_2} + \cdots + \frac{T_k^2}{n_k} - CF$$

$$= \frac{(17.7)^2}{4} + \frac{(18.3)^2}{3} + \frac{(15.2)^2}{4} - 238.31$$

$$= 9.40$$

$$SSTo = \sum_{\text{all } N \text{ obs.}} x^2 - CF = (4.2)^2 + (3.7)^2 + \cdots + (3.9)^2 - 238.31 = 11.81$$

$$SSE = SSTo - SSTr = 118.1 - 9.40 = 2.41$$

ARE YOU READY TO MOVE ON? **CHAPTER 16 REVIEW EXERCISES**

All chapter learning objectives are assessed in these exercises. The learning objectives assessed in each exercise are given in parentheses for each exercise.

16.23 (C1, M1, M2, M3)
The paper **"Women's and Men's Eating Behavior Following Exposure to Ideal-Body Images and Text"** (*Communication Research* [2006]: 507–529) describes an experiment in which 74 men were assigned at random to one of four treatments:
1. Viewed slides of fit, muscular men
2. Viewed slides of fit, muscular men accompanied by diet and fitness-related text
3. Viewed slides of fit, muscular men accompanied by text not related to diet and fitness
4. Did not view any slides

The participants then went to a room to complete a questionnaire. In this room, bowls of pretzels were set out on the tables. A research assistant noted how many pretzels were consumed by each participant while completing the questionnaire. Data consistent with summary quantities given in the paper are given in the accompanying table. Do these data provide convincing evidence that the mean number of pretzels consumed is not the same for all four treatments? Test the relevant hypotheses using a significance level of 0.05.

Data set available

Treatment 1	Treatment 2	Treatment 3	Treatment 4
8	6	1	5
7	8	5	2
4	0	2	5
13	4	0	7
2	9	3	5
1	8	0	2
5	6	3	0
8	2	4	0
11	7	4	3
5	8	5	4
1	8	5	2
0	5	7	4
6	14	8	1
4	9	4	1
10	0	0	
7	6	6	
0	3	3	
12	12		
	5		
	6		
	10		
	8		
	6		
	2		
	10		

16.24 (P1)

Can use of an online plagiarism-detection system reduce plagiarism in student research papers? The paper **"Plagiarism and Technology: A Tool for Coping with Plagiarism"** (*Journal of Education for Business* [2005]: 149–152) describes a study in which randomly selected research papers submitted by students during five semesters were analyzed for plagiarism. For each paper, the percentage of plagiarized words in the paper was determined by an online analysis. In each of the five semesters, students were told during the first two class meetings that they would have to submit an electronic version of their research papers and that the papers would be reviewed for plagiarism. Suppose that the number of papers sampled in each of the five semesters and the means and standard deviations for percentage of plagiarized words are as given in the accompanying table. For purposes of this exercise, assume that the conditions necessary for the ANOVA F test are reasonable. Do these data provide evidence to support the claim that mean percentage of plagiarized words is not the same for all five semesters? Test the appropriate hypotheses using $\alpha = 0.05$.

Semester	n	Mean	Standard deviation
1	39	6.31	3.75
2	42	3.31	3.06
3	32	1.79	3.25
4	32	1.83	3.13
5	34	1.50	2.37

Data set available

16.25 (M4, P2)

The paper referenced in Exercise 16.3 described an experiment to determine if restrictive age labeling on video games increased the attractiveness of the game for boys ages 12 to 13. In that exercise, the null hypothesis was H_0: $\mu_1 = \mu_2 = \mu_3 = \mu_4$, where μ_1 is the population mean attractiveness rating for the game with the 7+ age label, and μ_2, μ_3, and μ_4 are the population mean attractiveness scores for the 12+, 16+, and 18+ age labels, respectively. The sample data are given in the accompanying table.

7+ label	12+ label	16+ label	18+ label
6	8	7	10
6	7	9	9
6	8	8	6
5	5	6	8
4	7	7	7
8	9	4	6
6	5	8	8
1	8	9	9
2	4	6	10
4	7	7	8

a. Calculate the 95% T-K intervals and then use the underscoring procedure described in this section to identify significant differences among the age labels.

b. Based on your answer to Part (a), write a few sentences commenting on the theory that the more restrictive the age label on a video game, the more attractive the game is to 12- to 13-year-old boys.

16.26 (M4)

The authors of the paper **"Beyond the Shooter Game: Examining Presence and Hostile Outcomes among Male Game Players"** (*Communication Research* [2006]: 448–466) studied how video game content might influence attitudes and behavior. Male students at a large midwestern university were assigned at random to play one of three action-oriented video games. Two of the games involved some violence—one was a shooting game and one was a fighting game. The third game was a nonviolent race car driving game. After playing a game for 20 minutes, participants answered a set of questions. The responses were used to determine values of three measures of aggression: (1) a measure of aggressive behavior; (2) a measure of aggressive thoughts; and (3) a measure of aggressive feelings. The authors hypothesized that the means for the three measures of aggression would be greatest for the fighting game and lowest for the driving game.

a. For the measure of aggressive behavior, the paper reports that the mean score for the fighting game was significantly higher than the mean scores for the shooting and driving game, but that the mean scores for the shooting and driving games were not significantly different. The three sample means were:

	Driving	Shooting	Fighting
Sample mean	3.42	4.00	5.30

Use the underscoring procedure of this section to construct a display that shows any significant differences in mean aggressive behavior score among the three games.

b. For the measure of aggressive thoughts, the three sample means were:

	Driving	Shooting	Fighting
Sample mean	2.81	3.44	4.01

The paper states that the mean score for the fighting game only significantly differed from the mean score for the driving game, and that the mean score for the shooting game did not significantly differ from either the fighting or driving games. Use the underscoring procedure of this section to construct a display that shows any significant differences in mean aggressive thoughts score among the three games.

TECHNOLOGY NOTES

ANOVA

JMP

1. Input the raw data into the first column
2. Input the group information into the second column

Untitled - JMP		
File Edit Tables Rows Cols DOE Analyze Graph Tools View		
Untitled	Column 1	Column 2
1	1	A
2	4	B
3	6	C
4	4	A
5	7	B
6	8	C
7	6	A
8	9	B
9	2	C
Columns (5/0) — 10	3	A

3. Click **Analyze** then select **Fit Y by X**
4. Click and drag the first column name from the box under **Select Columns** to the box next to **Y, Response**
5. Click and drag the second column name from the box under **Select Columns** to the box next to **X, Factor**
6. Click **OK**
7. Click the red arrow next to **Oneway Analysis of...** and select **Means/ANOVA**

MINITAB

Data stored in separate columns

1. Input each group's data in a separate column

	C1	C2	C3	C4
	Group 1	Group 2	Group 3	
1	1	4	1	
2	2	5	5	
3	3	6	6	
4	4	1	7	
5	5	7	5	
6	6	8	6	
7	7	9	4	
8	8	5	5	
9	9	6	6	
10	10	4	8	

2. Click **Stat** then **ANOVA** then **One-Way (Unstacked)...**
3. Click in the box under **Responses (in separate columns):**
4. Double-click the column name containing each group's data
5. Click **OK**

Data stored in one column

1. Input the data into one column
2. Input the group information into a second column

Worksheet 3

↓	C1	C2	C3
	Data	Group	
7	7	1	
8	8	1	
9	9	1	
10	10	1	
11	4	2	
12	5	2	
13	6	2	
14	1	2	
15	7	2	
16	8	2	
17	9	2	

3. Click **Stat** then **ANOVA** then **One-Way...**
4. Click in the box next to **Response:** and double-click the column name containing the raw data values
5. Click in the box next to **Factor:** and double-click the column name containing the group information
6. Click **OK**

SPSS

1. Input the raw data for all groups into one column
2. Input the group information into a second column (use group numbers)

*Untitled1 [DataSet0] - PASW Statistics Student Version Data Editor

File Edit View Data Transform Analyze Graphs Utilities

1 : VAR00002   1.00

	VAR00001	VAR00002	var	var	
1	1.00	1.00			
2	2.00	1.00			
3	3.00	1.00			
4	4.00	1.00			
5	5.00	1.00			
6	6.00	1.00			
7	7.00	1.00			
8	8.00	1.00			
9	9.00	1.00			
10	10.00	1.00			
11	2.00	2.00			
12	4.00	2.00			
13	3.00	2.00			
14	5.00	2.00			

3. Click **Analyze** then click **Compare Means** then click **One-Way ANOVA...**
4. Click the name of the column containing the raw data and click the arrow to move it to the box under **Dependent List:**
5. Click the name of the column containing the group data and click the arrow to move it to the box under **Factor:**
6. Click **OK**

Excel

1. Input the raw data for each group into a separate column
2. Click the **Data** ribbon
3. Click **Data Analysis** in the **Analysis** group

Note: If you do not see **Data Analysis** listed on the Ribbon, see the Technology Notes for Chapter 2 for instructions on installing this add-on.

4. Select **Anova: Single Factor** and click **OK**
5. Click on the box next to **Input Range** and select ALL columns of data (if you typed and selected column titles, click the box next to **Labels in First Row**)
6. Click in the box next to **Alpha** and type the significance level
7. Click **OK**

Note: The test statistic and p-value can be found in the first row of the table under *F* and *P-value*, respectively.

TI-83/84

1. Enter the data for each group into a separate list starting with **L1** (In order to access lists press the **STAT** key, highlight the option called **Edit...** then press **ENTER**)
2. Press **STAT**
3. Highlight **TESTS**
4. Highlight **ANOVA** and press **ENTER**
5. Press **2ⁿᵈ** then **1**
6. Press **,**
7. Press **2ⁿᵈ** then **2**
8. Press **,**
9. Continue to input lists where data is stored separated by commas until you input the final list
10. When you are finished entering all lists, press **)**
11. Press **ENTER**

TI-Nspire

Summarized Data

1. Enter the summary information for the first group in a list in the following order: the value for *n* followed by a comma then the value of $\bar{x}$ followed by a comma then the value of *s* (In order to access data lists select the spreadsheet option and press **enter**)

Note: Be sure to title the lists by selecting the top row of the column and typing a title.

2. Enter the summary information for the first group in a list in the following order: the value for *n* followed by a comma then the value of $\bar{x}$ followed by a comma then the value of *s*
3. Continue to enter summary information for each group in this manner

4. When you are finished entering data for each group, press **menu** then **4:Statistics** then **4:Stat Tests** then **C:ANOVA...** then press **enter**
5. For **Data Input Method** choose **Stats** from the drop-down menu
6. For **Number of Groups** enter the number of groups, k
7. In the box next to **Group 1 Stats** select the list containing group one's summary statistics
8. In the box next to **Group 2 Stats** select the list containing group one's summary statistics
9. Continue entering summary statistics in this manner for all groups
10. Press **OK**

Raw data
1. Enter each group's data into separate data lists (In order to access data lists, select the spreadsheet option and press **enter**)

Note: Be sure to title the lists by selecting the top row of the column and typing a title.
2. Press the **menu** key and select **4:Statistics** then **4:Stat Tests** then **C:ANOVA...** and press **enter**
3. For **Data Input Method** choose **Data** from the drop-down menu
4. For **Number of Groups** input the number of groups, k
5. Press **OK**
6. For **List 1** select the list title that contains group one's data from the drop-down menu
7. For **List 2** select the list title that contains group two's data from the drop-down menu
8. Continue to select the appropriate lists for all groups
9. When you are finished inputting lists press **OK**

Appendix A

APPENDIX A   Statistical Tables

TABLE 1 Random Numbers

Row																				
1	4	5	1	8	5	0	3	3	7	1	2	8	4	5	1	1	0	9	5	7
2	4	2	5	5	8	0	4	5	7	0	7	0	3	6	6	1	3	1	3	1
3	8	9	9	3	4	3	5	0	6	3	9	1	1	8	2	6	9	2	0	9
4	8	9	0	7	2	9	9	0	4	7	6	7	4	7	1	3	4	3	5	3
5	5	7	3	1	0	3	7	4	7	8	5	2	0	1	3	7	7	6	3	6
6	0	9	3	8	7	6	7	9	9	5	6	2	5	6	5	8	4	2	6	4
7	4	1	0	1	0	2	2	0	4	7	5	1	1	9	4	7	9	7	5	1
8	6	4	7	3	6	3	4	5	1	2	3	1	1	8	0	0	4	8	2	0
9	8	0	2	8	7	9	3	8	4	0	4	2	0	8	9	1	2	3	3	2
10	9	4	6	0	6	9	7	8	8	2	5	2	9	6	0	1	4	6	0	5
11	6	6	9	5	7	4	4	6	3	2	0	6	0	8	9	1	3	6	1	8
12	0	7	1	7	7	7	2	9	7	8	7	5	8	8	6	9	8	4	1	0
13	6	1	3	0	9	7	3	3	6	6	0	4	1	8	3	2	6	7	6	8
14	2	2	3	6	2	1	3	0	2	2	6	6	9	7	0	2	1	2	5	8
15	0	7	1	7	4	2	0	0	0	1	3	1	2	0	4	7	8	4	1	0
16	6	6	5	1	6	1	8	1	5	5	2	6	2	0	1	1	5	2	3	6
17	9	9	6	2	5	3	5	9	8	3	7	5	0	1	3	9	3	8	0	8
18	9	9	9	6	1	2	9	3	4	6	5	6	4	6	5	8	2	7	4	0
19	2	5	6	3	1	9	8	1	1	0	3	5	6	7	9	1	4	5	2	0
20	5	1	1	9	8	1	2	1	1	6	9	8	1	8	1	9	9	1	2	0
21	1	9	8	0	7	4	6	8	4	0	3	0	8	1	1	0	6	2	3	2
22	9	7	0	9	6	3	8	9	9	7	0	6	5	4	3	6	5	0	3	2
23	1	7	6	4	8	2	0	3	9	6	3	6	2	1	0	7	7	3	1	7
24	6	2	5	8	2	0	7	8	6	4	6	6	8	9	2	0	6	9	0	4
25	1	5	7	1	1	1	9	5	1	4	5	2	8	3	4	3	0	7	3	5
26	1	4	6	6	5	6	0	1	9	4	0	5	2	7	6	4	3	6	8	8
27	1	8	5	0	2	1	6	8	0	7	7	2	6	2	6	7	5	4	8	7
28	7	8	7	4	6	5	4	3	7	9	3	9	2	7	9	5	4	2	3	1
29	1	6	3	2	8	3	7	3	0	7	2	4	8	0	9	9	9	4	7	0
30	2	8	9	0	8	1	6	8	1	7	3	1	3	0	9	7	2	5	7	9
31	0	7	8	8	6	5	7	5	5	4	0	0	3	4	1	2	7	3	7	9
32	8	4	0	1	4	5	1	9	1	1	2	1	5	3	2	8	5	5	7	5
33	7	3	5	9	7	0	4	9	1	2	1	3	2	5	1	9	3	3	8	3
34	4	7	2	6	7	6	9	9	2	7	8	7	5	5	5	2	4	4	3	4
35	9	3	3	7	0	7	0	5	7	5	6	9	5	4	3	1	4	6	6	8
36	0	2	4	9	7	8	1	6	3	8	7	8	0	5	6	7	2	7	5	0
37	7	1	0	1	8	4	7	1	2	9	3	8	0	0	8	7	9	2	8	6
38	9	7	9	4	4	5	3	1	9	3	4	5	0	6	3	5	9	6	9	8
39	0	4	2	5	0	0	9	9	6	4	0	6	9	0	3	8	3	5	7	2
40	0	7	1	2	3	6	1	7	9	3	9	5	4	6	8	4	8	8	0	6
41	3	5	6	6	2	4	4	5	6	3	7	8	7	6	5	2	0	4	3	2
42	6	6	8	5	5	2	9	7	9	3	3	1	6	9	5	9	7	1	1	2
43	9	5	0	4	3	1	1	7	3	9	2	7	7	4	7	0	3	1	2	8
44	5	1	7	8	9	4	7	2	9	2	8	9	9	8	0	6	3	7	2	1
45	1	6	3	9	4	1	3	2	1	1	8	5	6	3	4	1	9	3	1	7
46	4	4	8	6	4	0	3	8	3	8	3	5	9	5	9	4	8	3	9	4
47	7	7	6	6	4	5	4	4	8	4	4	0	3	9	8	5	2	0	2	3
48	2	5	6	6	3	7	0	6	5	6	9	0	1	9	5	2	6	9	1	2

TABLE 1 Random Numbers *(Continued)*

Row																			
49	9	4	0	4	7	5	3	2	8	7	2	7	4	9	3	9	6	5	5	6
50	7	3	1	5	6	6	5	0	3	5	3	7	2	8	6	2	4	1	8	7
51	7	5	8	2	8	8	8	7	6	4	1	1	0	2	3	1	9	3	6	0
52	3	3	6	0	9	1	1	0	3	2	7	8	2	0	5	3	4	8	9	8
53	0	2	9	6	9	8	9	3	8	1	5	3	9	9	7	0	7	7	1	6
54	8	5	9	6	2	9	6	8	2	1	2	4	7	0	6	8	3	4	6	1
55	5	4	7	6	1	0	0	1	0	4	6	1	4	1	5	0	9	6	5	5
56	5	0	3	6	4	1	9	8	4	4	1	2	0	2	5	1	8	1	2	1
57	0	2	6	3	7	5	1	1	6	6	0	5	8	1	2	3	3	6	1	3
58	3	8	1	6	3	8	1	4	5	2	9	4	2	5	7	3	2	3	1	8
59	9	1	5	6	0	6	5	6	6	3	6	2	3	0	0	0	1	8	5	9
60	5	3	5	6	3	9	5	4	7	3	6	6	7	5	0	1	5	6	7	3
61	9	6	6	4	5	7	7	6	1	5	4	4	8	0	6	5	7	6	3	0
62	6	3	0	6	7	9	5	5	4	6	2	2	8	4	4	0	0	9	9	8
63	8	5	8	3	5	2	0	6	6	0	0	6	0	6	3	0	1	7	0	5
64	3	8	2	4	9	0	9	2	6	2	9	5	1	9	1	9	0	8	3	3
65	1	4	4	1	1	7	4	6	3	6	5	6	5	5	7	7	0	3	5	8
66	5	9	9	5	3	7	2	5	1	7	1	1	0	7	1	0	9	2	8	8
67	8	7	1	7	5	2	5	6	8	7	9	9	1	3	9	6	4	9	3	0
68	6	7	2	3	1	4	9	2	1	7	0	8	6	7	8	9	9	4	7	4
69	2	3	2	8	7	0	9	7	1	1	1	2	8	2	9	1	0	6	7	7
70	2	9	5	7	8	4	7	9	0	3	6	9	2	0	6	0	6	2	6	8
71	4	8	9	8	3	2	7	6	9	1	9	8	6	9	5	2	4	9	9	9
72	1	5	6	5	7	7	5	4	3	4	3	8	1	8	9	9	4	4	1	1
73	1	8	1	1	7	2	8	5	5	8	9	9	9	6	2	0	1	6	6	7
74	5	7	7	0	9	5	5	6	8	6	8	2	2	6	0	5	5	1	8	7
75	1	8	6	0	5	4	8	3	4	5	3	5	8	7	7	7	8	5	7	0
76	2	6	6	7	9	4	2	2	8	7	4	3	4	9	6	1	9	4	3	9
77	3	6	6	4	5	7	8	3	0	2	8	4	6	7	2	1	4	5	2	3
78	0	7	8	0	1	2	1	1	3	4	2	1	6	9	3	3	5	4	0	4
79	8	3	6	0	5	7	7	9	1	5	8	8	4	9	5	7	2	2	7	6
80	5	3	6	9	0	6	3	8	7	5	9	5	9	7	4	2	5	6	2	9
81	0	9	3	7	7	2	8	6	4	3	2	9	4	8	2	9	9	6	9	9
82	9	4	7	4	0	0	0	3	5	4	6	6	2	6	2	3	6	1	1	4
83	5	5	4	1	7	8	6	4	2	3	2	9	8	4	6	3	8	3	0	5
84	5	3	0	0	5	4	8	0	7	4	7	6	2	1	1	2	1	2	6	9
85	3	3	0	9	3	2	9	4	0	5	5	4	8	7	5	7	5	3	8	8
86	3	0	5	7	1	9	5	8	0	0	4	5	3	0	3	0	2	7	6	7
87	5	0	8	6	0	8	1	6	2	0	8	6	5	4	0	7	2	9	1	0
88	3	6	4	7	8	2	3	5	7	9	8	5	2	7	6	9	0	2	4	9
89	9	0	4	4	9	1	6	8	5	2	8	9	0	7	5	7	2	5	1	8
90	9	5	2	6	9	3	9	6	5	1	8	8	7	8	2	0	4	4	7	9
91	9	4	5	7	0	3	4	6	4	2	5	4	8	6	1	1	9	1	8	8
92	8	1	1	8	0	5	4	2	8	5	3	3	3	0	1	1	4	4	8	3
93	6	9	4	7	8	3	3	9	1	2	5	0	1	2	3	0	1	1	2	5
94	0	0	6	8	8	7	2	4	4	7	6	6	0	3	4	7	5	6	8	2
95	5	3	3	9	3	8	4	9	1	9	1	7	8	4	5	2	2	5	4	4
96	2	5	6	2	7	6	0	3	8	1	4	4	2	6	8	3	6	3	2	8
97	7	4	3	7	9	6	8	6	2	8	3	8	4	2	2	0	7	0	5	3
98	1	9	0	8	8	0	1	2	2	2	7	5	6	5	5	7	8	7	2	6
99	2	4	8	0	2	5	2	7	0	5	9	6	6	1	5	8	7	9	7	5
100	4	1	7	8	6	7	1	1	5	8	9	4	8	9	8	3	0	9	0	7

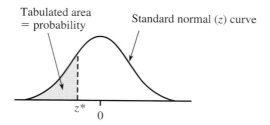

Tabulated area = probability

Standard normal (z) curve

z^* 0

TABLE 2 Standard Normal Probabilities (Cumulative z Curve Areas)

Z*	.00	.01	.02	.03	.04	.05	.06	.07	.08	.09
−3.8	.0001	.0001	.0001	.0001	.0001	.0001	.0001	.0001	.0001	.0000
−3.7	.0001	.0001	.0001	.0001	.0001	.0001	.0001	.0001	.0001	.0001
−3.6	.0002	.0002	.0001	.0001	.0001	.0001	.0001	.0001	.0001	.0001
−3.5	.0002	.0002	.0002	.0002	.0002	.0002	.0002	.0002	.0002	.0002
−3.4	.0003	.0003	.0003	.0003	.0003	.0003	.0003	.0003	.0003	.0002
−3.3	.0005	.0005	.0005	.0004	.0004	.0004	.0004	.0004	.0004	.0003
−3.2	.0007	.0007	.0006	.0006	.0006	.0006	.0006	.0005	.0005	.0005
−3.1	.0010	.0009	.0009	.0009	.0008	.0008	.0008	.0008	.0007	.0007
−3.0	.0013	.0013	.0013	.0012	.0012	.0011	.0011	.0011	.0010	.0010
−2.9	.0019	.0018	.0018	.0017	.0016	.0016	.0015	.0015	.0014	.0014
−2.8	.0026	.0025	.0024	.0023	.0023	.0022	.0021	.0021	.0020	.0019
−2.7	.0035	.0034	.0033	.0032	.0031	.0030	.0029	.0028	.0027	.0026
−2.6	.0047	.0045	.0044	.0043	.0041	.0040	.0039	.0038	.0037	.0036
−2.5	.0062	.0060	.0059	.0057	.0055	.0054	.0052	.0051	.0049	.0048
−2.4	.0082	.0080	.0078	.0075	.0073	.0071	.0069	.0068	.0066	.0064
−2.3	.0107	.0104	.0102	.0099	.0096	.0094	.0091	.0089	.0087	.0084
−2.2	.0139	.0136	.0132	.0129	.0125	.0122	.0119	.0116	.0113	.0110
−2.1	.0179	.0174	.0170	.0166	.0162	.0158	.0154	.0150	.0146	.0143
−2.0	.0228	.0222	.0217	.0212	.0207	.0202	.0197	.0192	.0188	.0183
−1.9	.0287	.0281	.0274	.0268	.0262	.0256	.0250	.0244	.0239	.0233
−1.8	.0359	.0351	.0344	.0336	.0329	.0322	.0314	.0307	.0301	.0294
−1.7	.0446	.0436	.0427	.0418	.0409	.0401	.0392	.0384	.0375	.0367
−1.6	.0548	.0537	.0526	.0516	.0505	.0495	.0485	.0475	.0465	.0455
−1.5	.0668	.0655	.0643	.0630	.0618	.0606	.0594	.0582	.0571	.0559
−1.4	.0808	.0793	.0778	.0764	.0749	.0735	.0721	.0708	.0694	.0681
−1.3	.0968	.0951	.0934	.0918	.0901	.0885	.0869	.0853	.0838	.0823
−1.2	.1151	.1131	.1112	.1093	.1075	.1056	.1038	.1020	.1003	.0985
−1.1	.1357	.1335	.1314	.1292	.1271	.1251	.1230	.1210	.1190	.1170
−1.0	.1587	.1562	.1539	.1515	.1492	.1469	.1446	.1423	.1401	.1379
−0.9	.1841	.1814	.1788	.1762	.1736	.1711	.1685	.1660	.1635	.1611
−0.8	.2119	.2090	.2061	.2033	.2005	.1977	.1949	.1922	.1894	.1867
−0.7	.2420	.2389	.2358	.2327	.2296	.2266	.2236	.2206	.2177	.2148
−0.6	.2743	.2709	.2676	.2643	.2611	.2578	.2546	.2514	.2483	.2451
−0.5	.3085	.3050	.3015	.2981	.2946	.2912	.2877	.2843	.2810	.2776
−0.4	.3446	.3409	.3372	.3336	.3300	.3264	.3228	.3192	.3156	.3121
−0.3	.3821	.3783	.3745	.3707	.3669	.3632	.3594	.3557	.3520	.3483
−0.2	.4207	.4168	.4129	.4090	.4052	.4013	.3974	.3936	.3897	.3859
−0.1	.4602	.4562	.4522	.4483	.4443	.4404	.4364	.4325	.4286	.4247
−0.0	.5000	.4960	.4920	.4880	.4840	.4801	.4761	.4721	.4681	.4641

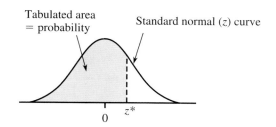

Tabulated area = probability

Standard normal (z) curve

TABLE 2 Standard Normal Probabilities (Cumulative z Curve Areas) *(Continued)*

z*	.00	.01	.02	.03	.04	.05	.06	.07	.08	.09
0.0	.5000	.5040	.5080	.5120	.5160	.5199	.5239	.5279	.5319	.5359
0.1	.5398	.5438	.5478	.5517	.5557	.5596	.5636	.5675	.5714	.5753
0.2	.5793	.5832	.5871	.5910	.5948	.5987	.6026	.6064	.6103	.6141
0.3	.6179	.6217	.6255	.6293	.6331	.6368	.6406	.6443	.6480	.6517
0.4	.6554	.6591	.6628	.6664	.6700	.6736	.6772	.6808	.6844	.6879
0.5	.6915	.6950	.6985	.7019	.7054	.7088	.7123	.7157	.7190	.7224
0.6	.7257	.7291	.7324	.7357	.7389	.7422	.7454	.7486	.7517	.7549
0.7	.7580	.7611	.7642	.7673	.7704	.7734	.7764	.7794	.7823	.7852
0.8	.7881	.7910	.7939	.7967	.7995	.8023	.8051	.8078	.8106	.8133
0.9	.8159	.8186	.8212	.8238	.8264	.8289	.8315	.8340	.8365	.8389
1.0	.8413	.8438	.8461	.8485	.8508	.8531	.8554	.8577	.8599	.8621
1.1	.8643	.8665	.8686	.8708	.8729	.8749	.8770	.8790	.8810	.8830
1.2	.8849	.8869	.8888	.8907	.8925	.8944	.8962	.8980	.8997	.9015
1.3	.9032	.9049	.9066	.9082	.9099	.9115	.9131	.9147	.9162	.9177
1.4	.9192	.9207	.9222	.9236	.9251	.9265	.9279	.9292	.9306	.9319
1.5	.9332	.9345	.9357	.9370	.9382	.9394	.9406	.9418	.9429	.9441
1.6	.9452	.9463	.9474	.9484	.9495	.9505	.9515	.9525	.9535	.9545
1.7	.9554	.9564	.9573	.9582	.9591	.9599	.9608	.9616	.9625	.9633
1.8	.9641	.9649	.9656	.9664	.9671	.9678	.9686	.9693	.9699	.9706
1.9	.9713	.9719	.9726	.9732	.9738	.9744	.9750	.9756	.9761	.9767
2.0	.9772	.9778	.9783	.9788	.9793	.9798	.9803	.9808	.9812	.9817
2.1	.9821	.9826	.9830	.9834	.9838	.9842	.9846	.9850	.9854	.9857
2.2	.9861	.9864	.9868	.9871	.9875	.9878	.9881	.9884	.9887	.9890
2.3	.9893	.9896	.9898	.9901	.9904	.9906	.9909	.9911	.9913	.9916
2.4	.9918	.9920	.9922	.9925	.9927	.9929	.9931	.9932	.9934	.9936
2.5	.9938	.9940	.9941	.9943	.9945	.9946	.9948	.9949	.9951	.9952
2.6	.9953	.9955	.9956	.9957	.9959	.9960	.9961	.9962	.9963	.9964
2.7	.9965	.9966	.9967	.9968	.9969	.9970	.9971	.9972	.9973	.9974
2.8	.9974	.9975	.9976	.9977	.9977	.9978	.9979	.9979	.9980	.9981
2.9	.9981	.9982	.9982	.9983	.9984	.9984	.9985	.9985	.9986	.9986
3.0	.9987	.9987	.9987	.9988	.9988	.9989	.9989	.9989	.9990	.9990
3.1	.9990	.9991	.9991	.9991	.9992	.9992	.9992	.9992	.9993	.9993
3.2	.9993	.9993	.9994	.9994	.9994	.9994	.9994	.9995	.9995	.9995
3.3	.9995	.9995	.9995	.9996	.9996	.9996	.9996	.9996	.9996	.9997
3.4	.9997	.9997	.9997	.9997	.9997	.9997	.9997	.9997	.9997	.9998
3.5	.9998	.9998	.9998	.9998	.9998	.9998	.9998	.9998	.9998	.9998
3.6	.9998	.9998	.9999	.9999	.9999	.9999	.9999	.9999	.9999	.9999
3.7	.9999	.9999	.9999	.9999	.9999	.9999	.9999	.9999	.9999	.9999
3.8	.9999	.9999	.9999	.9999	.9999	.9999	.9999	.9999	.9999	1.0000

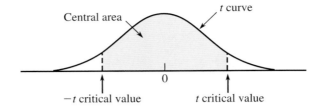

TABLE 3 *t* Critical Values

Central area captured:		.80	.90	.95	.98	.99	.998	.999
Confidence level:		80%	90%	95%	98%	99%	99.8%	99.9%
	1	3.08	6.31	12.71	31.82	63.66	318.31	636.62
	2	1.89	2.92	4.30	6.97	9.93	23.33	31.60
	3	1.64	2.35	3.18	4.54	5.84	10.21	12.92
	4	1.53	2.13	2.78	3.75	4.60	7.17	8.61
	5	1.48	2.02	2.57	3.37	4.03	5.89	6.86
	6	1.44	1.94	2.45	3.14	3.71	5.21	5.96
	7	1.42	1.90	2.37	3.00	3.50	4.79	5.41
	8	1.40	1.86	2.31	2.90	3.36	4.50	5.04
	9	1.38	1.83	2.26	2.82	3.25	4.30	4.78
	10	1.37	1.81	2.23	2.76	3.17	4.14	4.59
	11	1.36	1.80	2.20	2.72	3.11	4.03	4.44
	12	1.36	1.78	2.18	2.68	3.06	3.93	4.32
	13	1.35	1.77	2.16	2.65	3.01	3.85	4.22
	14	1.35	1.76	2.15	2.62	2.98	3.79	4.14
	15	1.34	1.75	2.13	2.60	2.95	3.73	4.07
	16	1.34	1.75	2.12	2.58	2.92	3.69	4.02
Degrees of	17	1.33	1.74	2.11	2.57	2.90	3.65	3.97
freedom	18	1.33	1.73	2.10	2.55	2.88	3.61	3.92
	19	1.33	1.73	2.09	2.54	2.86	3.58	3.88
	20	1.33	1.73	2.09	2.53	2.85	3.55	3.85
	21	1.32	1.72	2.08	2.52	2.83	3.53	3.82
	22	1.32	1.72	2.07	2.51	2.82	3.51	3.79
	23	1.32	1.71	2.07	2.50	2.81	3.49	3.77
	24	1.32	1.71	2.06	2.49	2.80	3.47	3.75
	25	1.32	1.71	2.06	2.49	2.79	3.45	3.73
	26	1.32	1.71	2.06	2.48	2.78	3.44	3.71
	27	1.31	1.70	2.05	2.47	2.77	3.42	3.69
	28	1.31	1.70	2.05	2.47	2.76	3.41	3.67
	29	1.31	1.70	2.05	2.46	2.76	3.40	3.66
	30	1.31	1.70	2.04	2.46	2.75	3.39	3.65
	40	1.30	1.68	2.02	2.42	2.70	3.31	3.55
	60	1.30	1.67	2.00	2.39	2.66	3.23	3.46
	120	1.29	1.66	1.98	2.36	2.62	3.16	3.37
z critical values	∞	1.28	1.645	1.96	2.33	2.58	3.09	3.29

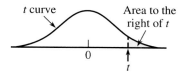

t curve / Area to the right of t

0   t

TABLE 4 Tail Areas for *t* Curves

t \ df	1	2	3	4	5	6	7	8	9	10	11	12
0.0	.500	.500	.500	.500	.500	.500	.500	.500	.500	.500	.500	.500
0.1	.468	.465	.463	.463	.462	.462	.462	.461	.461	.461	.461	.461
0.2	.437	.430	.427	.426	.425	.424	.424	.423	.423	.423	.423	.422
0.3	.407	.396	.392	.390	.388	.387	.386	.386	.386	.385	.385	.385
0.4	.379	.364	.358	.355	.353	.352	.351	.350	.349	.349	.348	.348
0.5	.352	.333	.326	.322	.319	.317	.316	.315	.315	.314	.313	.313
0.6	.328	.305	.295	.290	.287	.285	.284	.283	.282	.281	.280	.280
0.7	.306	.278	.267	.261	.258	.255	.253	.252	.251	.250	.249	.249
0.8	.285	.254	.241	.234	.230	.227	.225	.223	.222	.221	.220	.220
0.9	.267	.232	.217	.210	.205	.201	.199	.197	.196	.195	.194	.193
1.0	.250	.211	.196	.187	.182	.178	.175	.173	.172	.170	.169	.169
1.1	.235	.193	.176	.167	.162	.157	.154	.152	.150	.149	.147	.146
1.2	.221	.177	.158	.148	.142	.138	.135	.132	.130	.129	.128	.127
1.3	.209	.162	.142	.132	.125	.121	.117	.115	.113	.111	.110	.109
1.4	.197	.148	.128	.117	.110	.106	.102	.100	.098	.096	.095	.093
1.5	.187	.136	.115	.104	.097	.092	.089	.086	.084	.082	.081	.080
1.6	.178	.125	.104	.092	.085	.080	.077	.074	.072	.070	.069	.068
1.7	.169	.116	.094	.082	.075	.070	.066	.064	.062	.060	.059	.057
1.8	.161	.107	.085	.073	.066	.061	.057	.055	.053	.051	.050	.049
1.9	.154	.099	.077	.065	.058	.053	.050	.047	.045	.043	.042	.041
2.0	.148	.092	.070	.058	.051	.046	.043	.040	.038	.037	.035	.034
2.1	.141	.085	.063	.052	.045	.040	.037	.034	.033	.031	.030	.029
2.2	.136	.079	.058	.046	.040	.035	.032	.029	.028	.026	.025	.024
2.3	.131	.074	.052	.041	.035	.031	.027	.025	.023	.022	.021	.020
2.4	.126	.069	.048	.037	.031	.027	.024	.022	.020	.019	.018	.017
2.5	.121	.065	.044	.033	.027	.023	.020	.018	.017	.016	.015	.014
2.6	.117	.061	.040	.030	.024	.020	.018	.016	.014	.013	.012	.012
2.7	.113	.057	.037	.027	.021	.018	.015	.014	.012	.011	.010	.010
2.8	.109	.054	.034	.024	.019	.016	.013	.012	.010	.009	.009	.008
2.9	.106	.051	.031	.022	.017	.014	.011	.010	.009	.008	.007	.007
3.0	.102	.048	.029	.020	.015	.012	.010	.009	.007	.007	.006	.006
3.1	.099	.045	.027	.018	.013	.011	.009	.007	.006	.006	.005	.005
3.2	.096	.043	.025	.016	.012	.009	.008	.006	.005	.005	.004	.004
3.3	.094	.040	.023	.015	.011	.008	.007	.005	.005	.004	.004	.003
3.4	.091	.038	.021	.014	.010	.007	.006	.005	.004	.003	.003	.003
3.5	.089	.036	.020	.012	.009	.006	.005	.004	.003	.003	.002	.002
3.6	.086	.035	.018	.011	.008	.006	.004	.004	.003	.002	.002	.002
3.7	.084	.033	.017	.010	.007	.005	.004	.003	.002	.002	.002	.002
3.8	.082	.031	.016	.010	.006	.004	.003	.003	.002	.002	.001	.001
3.9	.080	.030	.015	.009	.006	.004	.003	.002	.002	.001	.001	.001
4.0	.078	.029	.014	.008	.005	.004	.003	.002	.002	.001	.001	.001

(continued)

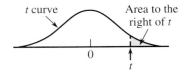

TABLE 4 Tail Areas for *t* Curves *(Continued)*

t \ df	13	14	15	16	17	18	19	20	21	22	23	24
0.0	.500	.500	.500	.500	.500	.500	.500	.500	.500	.500	.500	.500
0.1	.461	.461	.461	.461	.461	.461	.461	.461	.461	.461	.461	.461
0.2	.422	.422	.422	.422	.422	.422	.422	.422	.422	.422	.422	.422
0.3	.384	.384	.384	.384	.384	.384	.384	.384	.384	.383	.383	.383
0.4	.348	.347	.347	.347	.347	.347	.347	.347	.347	.347	.346	.346
0.5	.313	.312	.312	.312	.312	.312	.311	.311	.311	.311	.311	.311
0.6	.279	.279	.279	.278	.278	.278	.278	.278	.278	.277	.277	.277
0.7	.248	.247	.247	.247	.247	.246	.246	.246	.246	.246	.245	.245
0.8	.219	.218	.218	.218	.217	.217	.217	.217	.216	.216	.216	.216
0.9	.192	.191	.191	.191	.190	.190	.190	.189	.189	.189	.189	.189
1.0	.168	.167	.167	.166	.166	.165	.165	.165	.164	.164	.164	.164
1.1	.146	.144	.144	.144	.143	.143	.143	.142	.142	.142	.141	.141
1.2	.126	.124	.124	.124	.123	.123	.122	.122	.122	.121	.121	.121
1.3	.108	.107	.107	.106	.105	.105	.105	.104	.104	.104	.103	.103
1.4	.092	.091	.091	.090	.090	.089	.089	.089	.088	.088	.087	.087
1.5	.079	.077	.077	.077	.076	.075	.075	.075	.074	.074	.074	.073
1.6	.067	.065	.065	.065	.064	.064	.063	.063	.062	.062	.062	.061
1.7	.056	.055	.055	.054	.054	.053	.053	.052	.052	.052	.051	.051
1.8	.048	.046	.046	.045	.045	.044	.044	.043	.043	.043	.042	.042
1.9	.040	.038	.038	.038	.037	.037	.036	.036	.036	.035	.035	.035
2.0	.033	.032	.032	.031	.031	.030	.030	.030	.029	.029	.029	.028
2.1	.028	.027	.027	.026	.025	.025	.025	.024	.024	.024	.023	.023
2.2	.023	.022	.022	.021	.021	.021	.020	.020	.020	.019	.019	.019
2.3	.019	.018	.018	.018	.017	.017	.016	.016	.016	.016	.015	.015
2.4	.016	.015	.015	.014	.014	.014	.013	.013	.013	.013	.012	.012
2.5	.013	.012	.012	.012	.011	.011	.011	.011	.010	.010	.010	.010
2.6	.011	.010	.010	.010	.009	.009	.009	.009	.008	.008	.008	.008
2.7	.009	.008	.008	.008	.008	.007	.007	.007	.007	.007	.006	.006
2.8	.008	.007	.007	.006	.006	.006	.006	.006	.005	.005	.005	.005
2.9	.006	.005	.005	.005	.005	.005	.005	.004	.004	.004	.004	.004
3.0	.005	.004	.004	.004	.004	.004	.004	.004	.003	.003	.003	.003
3.1	.004	.004	.004	.003	.003	.003	.003	.003	.003	.003	.003	.002
3.2	.003	.003	.003	.003	.003	.002	.002	.002	.002	.002	.002	.002
3.3	.003	.002	.002	.002	.002	.002	.002	.002	.002	.002	.002	.001
3.4	.002	.002	.002	.002	.002	.002	.002	.001	.001	.001	.001	.001
3.5	.002	.002	.002	.001	.001	.001	.001	.001	.001	.001	.001	.001
3.6	.002	.001	.001	.001	.001	.001	.001	.001	.001	.001	.001	.001
3.7	.001	.001	.001	.001	.001	.001	.001	.001	.001	.001	.001	.001
3.8	.001	.001	.001	.001	.001	.001	.001	.001	.001	.000	.000	.000
3.9	.001	.001	.001	.001	.001	.001	.000	.000	.000	.000	.000	.000
4.0	.001	.001	.001	.001	.000	.000	.000	.000	.000	.000	.000	.000

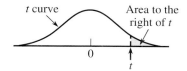

TABLE 4 Tail Areas for *t* Curves *(Continued)*

df t	25	26	27	28	29	30	35	40	60	120	∞(= z)
0.0	.500	.500	.500	.500	.500	.500	.500	.500	.500	.500	.500
0.1	.461	.461	.461	.461	.461	.461	.460	.460	.460	.460	.460
0.2	.422	.422	.421	.421	.421	.421	.421	.421	.421	.421	.421
0.3	.383	.383	.383	.383	.383	.383	.383	.383	.383	.382	.382
0.4	.346	.346	.346	.346	.346	.346	.346	.346	.345	.345	.345
0.5	.311	.311	.311	.310	.310	.310	.310	.310	.309	.309	.309
0.6	.277	.277	.277	.277	.277	.277	.276	.276	.275	.275	.274
0.7	.245	.245	.245	.245	.245	.245	.244	.244	.243	.243	.242
0.8	.216	.215	.215	.215	.215	.215	.215	.214	.213	.213	.212
0.9	.188	.188	.188	.188	.188	.188	.187	.187	.186	.185	.184
1.0	.163	.163	.163	.163	.163	.163	.162	.162	.161	.160	.159
1.1	.141	.141	.141	.140	.140	.140	.139	.139	.138	.137	.136
1.2	.121	.120	.120	.120	.120	.120	.119	.119	.117	.116	.115
1.3	.103	.103	.102	.102	.102	.102	.101	.101	.099	.098	.097
1.4	.087	.087	.086	.086	.086	.086	.085	.085	.083	.082	.081
1.5	.073	.073	.073	.072	.072	.072	.071	.071	.069	.068	.067
1.6	.061	.061	.061	.060	.060	.060	.059	.059	.057	.056	.055
1.7	.051	.051	.050	.050	.050	.050	.049	.048	.047	.046	.045
1.8	.042	.042	.042	.041	.041	.041	.040	.040	.038	.037	.036
1.9	.035	.034	.034	.034	.034	.034	.033	.032	.031	.030	.029
2.0	.028	.028	.028	.028	.027	.027	.027	.026	.025	.024	.023
2.1	.023	.023	.023	.022	.022	.022	.022	.021	.020	.019	.018
2.2	.019	.018	.018	.018	.018	.018	.017	.017	.016	.015	.014
2.3	.015	.015	.015	.015	.014	.014	.014	.013	.012	.012	.011
2.4	.012	.012	.012	.012	.012	.011	.011	.011	.010	.009	.008
2.5	.010	.010	.009	.009	.009	.009	.009	.008	.008	.007	.006
2.6	.008	.008	.007	.007	.007	.007	.007	.007	.006	.005	.005
2.7	.006	.006	.006	.006	.006	.006	.005	.005	.004	.004	.003
2.8	.005	.005	.005	.005	.005	.004	.004	.004	.003	.003	.003
2.9	.004	.004	.004	.004	.004	.003	.003	.003	.003	.002	.002
3.0	.003	.003	.003	.003	.003	.003	.002	.002	.002	.002	.001
3.1	.002	.002	.002	.002	.002	.002	.002	.002	.001	.001	.001
3.2	.002	.002	.002	.002	.002	.002	.001	.001	.001	.001	.001
3.3	.001	.001	.001	.001	.001	.001	.001	.001	.001	.001	.000
3.4	.001	.001	.001	.001	.001	.001	.001	.001	.001	.000	.000
3.5	.001	.001	.001	.001	.001	.001	.001	.001	.000	.000	.000
3.6	.001	.001	.001	.001	.001	.001	.000	.000	.000	.000	.000
3.7	.001	.001	.000	.000	.000	.000	.000	.000	.000	.000	.000
3.8	.000	.000	.000	.000	.000	.000	.000	.000	.000	.000	.000
3.9	.000	.000	.000	.000	.000	.000	.000	.000	.000	.000	.000
4.0	.000	.000	.000	.000	.000	.000	.000	.000	.000	.000	.000

TABLE 5 Upper-Tail Areas for Chi-Square Distributions

Right-tail area	df = 1	df = 2	df = 3	df = 4	df = 5
> 0.100	< 2.70	< 4.60	< 6.25	< 7.77	< 9.23
0.100	2.70	4.60	6.25	7.77	9.23
0.095	2.78	4.70	6.36	7.90	9.37
0.090	2.87	4.81	6.49	8.04	9.52
0.085	2.96	4.93	6.62	8.18	9.67
0.080	3.06	5.05	6.75	8.33	9.83
0.075	3.17	5.18	6.90	8.49	10.00
0.070	3.28	5.31	7.06	8.66	10.19
0.065	3.40	5.46	7.22	8.84	10.38
0.060	3.53	5.62	7.40	9.04	10.59
0.055	3.68	5.80	7.60	9.25	10.82
0.050	3.84	5.99	7.81	9.48	11.07
0.045	4.01	6.20	8.04	9.74	11.34
0.040	4.21	6.43	8.31	10.02	11.64
0.035	4.44	6.70	8.60	10.34	11.98
0.030	4.70	7.01	8.94	10.71	12.37
0.025	5.02	7.37	9.34	11.14	12.83
0.020	5.41	7.82	9.83	11.66	13.38
0.015	5.91	8.39	10.46	12.33	14.09
0.010	6.63	9.21	11.34	13.27	15.08
0.005	7.87	10.59	12.83	14.86	16.74
0.001	10.82	13.81	16.26	18.46	20.51
< 0.001	>10.82	>13.81	>16.26	>18.46	>20.51

Right-tail area	df = 6	df = 7	df = 8	df = 9	df = 10
>0.100	<10.64	<12.01	<13.36	<14.68	<15.98
0.100	10.64	12.01	13.36	14.68	15.98
0.095	10.79	12.17	13.52	14.85	16.16
0.090	10.94	12.33	13.69	15.03	16.35
0.085	11.11	12.50	13.87	15.22	16.54
0.080	11.28	12.69	14.06	15.42	16.75
0.075	11.46	12.88	14.26	15.63	16.97
0.070	11.65	13.08	14.48	15.85	17.20
0.065	11.86	13.30	14.71	16.09	17.44
0.060	12.08	13.53	14.95	16.34	17.71
0.055	12.33	13.79	15.22	16.62	17.99
0.050	12.59	14.06	15.50	16.91	18.30
0.045	12.87	14.36	15.82	17.24	18.64
0.040	13.19	14.70	16.17	17.60	19.02
0.035	13.55	15.07	16.56	18.01	19.44
0.030	13.96	15.50	17.01	18.47	19.92
0.025	14.44	16.01	17.53	19.02	20.48
0.020	15.03	16.62	18.16	19.67	21.16
0.015	15.77	17.39	18.97	20.51	22.02
0.010	16.81	18.47	20.09	21.66	23.20
0.005	18.54	20.27	21.95	23.58	25.18
0.001	22.45	24.32	26.12	27.87	29.58
<0.001	>22.45	>24.32	>26.12	>27.87	>29.58

TABLE 5 Upper-Tail Areas for Chi-Square Distributions *(Continued)*

Right-tail area	df = 11	df = 12	df = 13	df = 14	df = 15
>0.100	<17.27	<18.54	<19.81	<21.06	<22.30
0.100	17.27	18.54	19.81	21.06	22.30
0.095	17.45	18.74	20.00	21.26	22.51
0.090	17.65	18.93	20.21	21.47	22.73
0.085	17.85	19.14	20.42	21.69	22.95
0.080	18.06	19.36	20.65	21.93	23.19
0.075	18.29	19.60	20.89	22.17	23.45
0.070	18.53	19.84	21.15	22.44	23.72
0.065	18.78	20.11	21.42	22.71	24.00
0.060	19.06	20.39	21.71	23.01	24.31
0.055	19.35	20.69	22.02	23.33	24.63
0.050	19.67	21.02	22.36	23.68	24.99
0.045	20.02	21.38	22.73	24.06	25.38
0.040	20.41	21.78	23.14	24.48	25.81
0.035	20.84	22.23	23.60	24.95	26.29
0.030	21.34	22.74	24.12	25.49	26.84
0.025	21.92	23.33	24.73	26.11	27.48
0.020	22.61	24.05	25.47	26.87	28.25
0.015	23.50	24.96	26.40	27.82	29.23
0.010	24.72	26.21	27.68	29.14	30.57
0.005	26.75	28.29	29.81	31.31	32.80
0.001	31.26	32.90	34.52	36.12	37.69
<0.001	>31.26	>32.90	>34.52	>36.12	>37.69

Right-tail area	df = 16	df = 17	df = 18	df = 19	df = 20
>0.100	<23.54	<24.77	<25.98	<27.20	<28.41
0.100	23.54	24.76	25.98	27.20	28.41
0.095	23.75	24.98	26.21	27.43	28.64
0.090	23.97	25.21	26.44	27.66	28.88
0.085	24.21	25.45	26.68	27.91	29.14
0.080	24.45	25.70	26.94	28.18	29.40
0.075	24.71	25.97	27.21	28.45	29.69
0.070	24.99	26.25	27.50	28.75	29.99
0.065	25.28	26.55	27.81	29.06	30.30
0.060	25.59	26.87	28.13	29.39	30.64
0.055	25.93	27.21	28.48	29.75	31.01
0.050	26.29	27.58	28.86	30.14	31.41
0.045	26.69	27.99	29.28	30.56	31.84
0.040	27.13	28.44	29.74	31.03	32.32
0.035	27.62	28.94	30.25	31.56	32.85
0.030	28.19	29.52	30.84	32.15	33.46
0.025	28.84	30.19	31.52	32.85	34.16
0.020	29.63	30.99	32.34	33.68	35.01
0.015	30.62	32.01	33.38	34.74	36.09
0.010	32.00	33.40	34.80	36.19	37.56
0.005	34.26	35.71	37.15	38.58	39.99
0.001	39.25	40.78	42.31	43.81	45.31
<0.001	>39.25	>40.78	>42.31	>43.81	>45.31

TABLE 6 Binomial Probabilities

n = 5

							p						
x	0.05	0.1	0.2	0.25	0.3	0.4	0.5	0.6	0.7	0.75	0.8	0.9	0.95
0	.774	.590	.328	.237	.168	.078	.031	.010	.002	.001	.000	.000	.000
1	.204	.328	.410	.396	.360	.259	.156	.077	.028	.015	.006	.000	.000
2	.021	.073	.205	.264	.309	.346	.313	.230	.132	.088	.051	.008	.001
3	.001	.008	.051	.088	.132	.230	.313	.346	.309	.264	.205	.073	.021
4	.000	.000	.006	.015	.028	.077	.156	.259	.360	.396	.410	.328	.204
5	.000	.000	.000	.001	.002	.010	.031	.078	.168	.237	.328	.590	.774

n = 10

							p						
x	0.05	0.1	0.2	0.25	0.3	0.4	0.5	0.6	0.7	0.75	0.8	0.9	0.95
0	.599	.349	.107	.056	.028	.006	.001	.000	.000	.000	.000	.000	.000
1	.315	.387	.268	.188	.121	.040	.010	.002	.000	.000	.000	.000	.000
2	.075	.194	.302	.282	.233	.121	.044	.011	.001	.000	.000	.000	.000
3	.010	.057	.201	.250	.267	.215	.117	.042	.009	.003	.001	.000	.000
4	.001	.011	.088	.146	.200	.251	.205	.111	.037	.016	.006	.000	.000
5	.000	.001	.026	.058	.103	.201	.246	.201	.103	.058	.026	.001	.000
6	.000	.000	.006	.016	.037	.111	.205	.251	.200	.146	.088	.011	.001
7	.000	.000	.001	.003	.009	.042	.117	.215	.267	.250	.201	.057	.010
8	.000	.000	.000	.000	.001	.011	.044	.121	.233	.282	.302	.194	.075
9	.000	.000	.000	.000	.000	.002	.010	.040	.121	.188	.268	.387	.315
10	.000	.000	.000	.000	.000	.000	.001	.006	.028	.056	.107	.349	.599

TABLE 6 Binomial Probabilities *(Continued)*

n = 15

x	0.05	0.1	0.2	0.25	0.3	0.4	0.5	0.6	0.7	0.75	0.8	0.9	0.95
0	.463	.206	.035	.013	.005	.000	.000	.000	.000	.000	.000	.000	.000
1	.366	.343	.132	.067	.031	.005	.000	.000	.000	.000	.000	.000	.000
2	.135	.267	.231	.156	.092	.022	.003	.000	.000	.000	.000	.000	.000
3	.031	.129	.250	.225	.170	.063	.014	.002	.000	.000	.000	.000	.000
4	.005	.043	.188	.225	.219	.127	.042	.007	.001	.000	.000	.000	.000
5	.001	.010	.103	.165	.206	.186	.092	.024	.003	.001	.000	.000	.000
6	.000	.002	.043	.092	.147	.207	.153	.061	.012	.003	.001	.000	.000
7	.000	.000	.014	.039	.081	.177	.196	.118	.035	.013	.003	.000	.000
8	.000	.000	.003	.013	.035	.118	.196	.177	.081	.039	.014	.000	.000
9	.000	.000	.001	.003	.012	.061	.153	.207	.147	.092	.043	.002	.000
10	.000	.000	.000	.001	.003	.024	.092	.186	.206	.165	.103	.010	.001
11	.000	.000	.000	.000	.001	.007	.042	.127	.219	.225	.188	.043	.005
12	.000	.000	.000	.000	.000	.002	.014	.063	.170	.225	.250	.129	.031
13	.000	.000	.000	.000	.000	.000	.003	.022	.092	.156	.231	.267	.135
14	.000	.000	.000	.000	.000	.000	.000	.005	.031	.067	.132	.343	.366
15	.000	.000	.000	.000	.000	.000	.000	.000	.005	.013	.035	.206	.463

n = 20

x	0.05	0.1	0.2	0.25	0.3	0.4	0.5	0.6	0.7	0.75	0.8	0.9	0.95
0	.358	.122	.012	.003	.001	.000	.000	.000	.000	.000	.000	.000	.000
1	.377	.270	.058	.021	.007	.000	.000	.000	.000	.000	.000	.000	.000
2	.189	.285	.137	.067	.028	.003	.000	.000	.000	.000	.000	.000	.000
3	.060	.190	.205	.134	.072	.012	.001	.000	.000	.000	.000	.000	.000
4	.013	.090	.218	.190	.130	.035	.005	.000	.000	.000	.000	.000	.000
5	.002	.032	.175	.202	.179	.075	.015	.001	.000	.000	.000	.000	.000
6	.000	.009	.109	.169	.192	.124	.037	.005	.000	.000	.000	.000	.000
7	.000	.002	.055	.112	.164	.166	.074	.015	.001	.000	.000	.000	.000
8	.000	.000	.022	.061	.114	.180	.120	.035	.004	.001	.000	.000	.000
9	.000	.000	.007	.027	.065	.160	.160	.071	.012	.003	.000	.000	.000
10	.000	.000	.002	.010	.031	.117	.176	.117	.031	.010	.002	.000	.000
11	.000	.000	.000	.003	.012	.071	.160	.160	.065	.027	.007	.000	.000
12	.000	.000	.000	.001	.004	.035	.120	.180	.114	.061	.022	.000	.000
13	.000	.000	.000	.000	.001	.015	.074	.166	.164	.112	.055	.002	.000
14	.000	.000	.000	.000	.000	.005	.037	.124	.192	.169	.109	.009	.000
15	.000	.000	.000	.000	.000	.001	.015	.075	.179	.202	.175	.032	.002
16	.000	.000	.000	.000	.000	.000	.005	.035	.130	.190	.218	.090	.013
17	.000	.000	.000	.000	.000	.000	.001	.012	.072	.134	.205	.190	.060
18	.000	.000	.000	.000	.000	.000	.000	.003	.028	.067	.137	.285	.189
19	.000	.000	.000	.000	.000	.000	.000	.000	.007	.021	.058	.270	.377
20	.000	.000	.000	.000	.000	.000	.000	.000	.001	.003	.012	.122	.358

(continued)

TABLE 6 **Binomial Probabilities** *(Continued)*

n = 25													
							p						
x	0.05	0.1	0.2	0.25	0.3	0.4	0.5	0.6	0.7	0.75	0.8	0.9	0.95
0	.277	.072	.004	.001	.000	.000	.000	.000	.000	.000	.000	.000	.000
1	.365	.199	.024	.006	.001	.000	.000	.000	.000	.000	.000	.000	.000
2	.231	.266	.071	.025	.007	.000	.000	.000	.000	.000	.000	.000	.000
3	.093	.226	.136	.064	.024	.002	.000	.000	.000	.000	.000	.000	.000
4	.027	.138	.187	.118	.057	.007	.000	.000	.000	.000	.000	.000	.000
5	.006	.065	.196	.165	.103	.020	.002	.000	.000	.000	.000	.000	.000
6	.001	.024	.163	.183	.147	.044	.005	.000	.000	.000	.000	.000	.000
7	.000	.007	.111	.165	.171	.080	.014	.001	.000	.000	.000	.000	.000
8	.000	.002	.062	.124	.165	.120	.032	.003	.000	.000	.000	.000	.000
9	.000	.000	.029	.078	.134	.151	.061	.009	.000	.000	.000	.000	.000
10	.000	.000	.012	.042	.092	.161	.097	.021	.001	.000	.000	.000	.000
11	.000	.000	.004	.019	.054	.147	.133	.043	.004	.001	.000	.000	.000
12	.000	.000	.001	.007	.027	.114	.155	.076	.011	.002	.000	.000	.000
13	.000	.000	.000	.002	.011	.076	.155	.114	.027	.007	.001	.000	.000
14	.000	.000	.000	.001	.004	.043	.133	.147	.054	.019	.004	.000	.000
15	.000	.000	.000	.000	.001	.021	.097	.161	.092	.042	.012	.000	.000
16	.000	.000	.000	.000	.000	.009	.061	.151	.134	.078	.029	.000	.000
17	.000	.000	.000	.000	.000	.003	.032	.120	.165	.124	.062	.002	.000
18	.000	.000	.000	.000	.000	.001	.014	.080	.171	.165	.111	.007	.000
19	.000	.000	.000	.000	.000	.000	.005	.044	.147	.183	.163	.024	.001
20	.000	.000	.000	.000	.000	.000	.002	.020	.103	.165	.196	.065	.006
21	.000	.000	.000	.000	.000	.000	.000	.007	.057	.118	.187	.138	.027
22	.000	.000	.000	.000	.000	.000	.000	.002	.024	.064	.136	.226	.093
23	.000	.000	.000	.000	.000	.000	.000	.000	.007	.025	.071	.266	.231
24	.000	.000	.000	.000	.000	.000	.000	.000	.002	.006	.024	.199	.365
25	.000	.000	.000	.000	.000	.000	.000	.000	.000	.001	.004	.072	.277

TABLE 7 Values That Capture Specified Upper-Tail *F* Curve Areas

df$_2$	Area	1	2	3	4	5	6	7	8	9	10
1	.10	39.86	49.50	53.59	55.83	57.24	58.20	58.91	59.44	59.86	60.19
	.05	161.40	199.50	215.70	224.60	230.20	234.00	236.80	238.90	240.50	241.90
	.01	4052.00	5000.00	5403.00	5625.00	5764.00	5859.00	5928.00	5981.00	6022.00	6056.00
2	.10	8.53	9.00	9.16	9.24	9.29	9.33	9.35	9.37	9.38	9.39
	.05	18.51	19.00	19.16	19.25	19.30	19.33	19.35	19.37	19.38	19.40
	.01	98.50	99.00	99.17	99.25	99.30	99.33	99.36	99.37	99.39	99.40
	.001	998.50	999.00	999.20	999.20	999.30	999.30	999.40	999.40	999.40	999.40
3	.10	5.54	5.46	5.39	5.34	5.31	5.28	5.27	5.25	5.24	5.23
	.05	10.13	9.55	9.28	9.12	9.01	8.94	8.89	8.85	8.81	8.79
	.01	34.12	30.82	29.46	28.71	28.24	27.91	27.67	27.49	27.35	27.23
	.001	167.00	148.50	141.10	137.10	134.60	132.80	131.60	130.60	129.90	129.20
4	.10	4.54	4.32	4.19	4.11	4.05	4.01	3.98	3.95	3.94	3.92
	.05	7.71	6.94	6.59	6.39	6.26	6.16	6.09	6.04	6.00	5.96
	.01	21.20	18.00	16.69	15.98	15.52	15.21	14.98	14.80	14.66	14.55
	.001	74.14	61.25	56.18	53.44	51.71	50.53	49.66	49.00	48.47	48.05
5	.10	4.06	3.78	3.62	3.52	3.45	3.40	3.37	3.34	3.32	3.30
	.05	6.61	5.79	5.41	5.19	5.05	4.95	4.88	4.82	4.77	4.74
	.01	16.26	13.27	12.06	11.39	10.97	10.67	10.46	10.29	10.16	10.05
	.001	47.18	37.12	33.20	31.09	29.75	28.83	28.16	27.65	27.24	26.92
6	.10	3.78	3.46	3.29	3.18	3.11	3.05	3.01	2.98	2.96	2.94
	.05	5.99	5.14	4.76	4.53	4.39	4.28	4.21	4.15	4.10	4.06
	.01	13.75	10.92	9.78	9.15	8.75	8.47	8.26	8.10	7.98	7.87
	.001	35.51	27.00	23.70	21.92	20.80	20.03	19.46	19.03	18.69	18.41
7	.10	3.59	3.26	3.07	2.96	2.88	2.83	2.78	2.75	2.72	2.70
	.05	5.59	4.74	4.35	4.12	3.97	3.87	3.79	3.73	3.68	3.64
	.01	12.25	9.55	8.45	7.85	7.46	7.19	6.99	6.84	6.72	6.62
	.001	29.25	21.69	18.77	17.20	16.21	15.52	15.02	14.63	14.33	14.08
8	.10	3.46	3.11	2.92	2.81	2.73	2.67	2.62	2.59	2.56	2.54
	.05	5.32	4.46	4.07	3.84	3.69	3.58	3.50	3.44	3.39	3.35
	.01	11.26	8.65	7.59	7.01	6.63	6.37	6.18	6.03	5.91	5.81
	.001	25.41	18.49	15.83	14.39	13.48	12.86	12.40	12.05	11.77	11.54
9	.10	3.36	3.01	2.81	2.69	2.61	2.55	2.51	2.47	2.44	2.42
	.05	5.12	4.26	3.86	3.63	3.48	3.37	3.29	3.23	3.18	3.14
	.01	10.56	8.02	6.99	6.42	6.06	5.80	5.61	5.47	5.35	5.26
	.001	22.86	16.39	13.90	12.56	11.71	11.13	10.70	10.37	10.11	9.89

(continued)

TABLE 7 Values That Capture Specified Upper-Tail F Curve Areas (*continued*)

df₂	Area	1	2	3	4	5	6	7	8	9	10
10	.10	3.29	2.92	2.73	2.61	2.52	2.46	2.41	2.38	2.35	2.32
	.05	4.96	4.10	3.71	3.48	3.33	3.22	3.14	3.07	3.02	2.98
	.01	10.04	7.56	6.55	5.99	5.64	5.39	5.20	5.06	4.94	4.85
	.001	21.04	14.91	12.55	11.28	10.48	9.93	9.52	9.20	8.96	8.75
11	.10	3.23	2.86	2.66	2.54	2.45	2.39	2.34	2.30	2.27	2.25
	.05	4.84	3.98	3.59	3.36	3.20	3.09	3.01	2.95	2.90	2.85
	.01	9.65	7.21	6.22	5.67	5.32	5.07	4.89	4.74	4.63	4.54
	.001	19.69	13.81	11.56	10.35	9.58	9.05	8.66	8.35	8.12	7.92
12	.10	3.18	2.81	2.61	2.48	2.39	2.33	2.28	2.24	2.21	2.19
	.05	4.75	3.89	3.49	3.26	3.11	3.00	2.91	2.85	2.80	2.75
	.01	9.33	6.93	5.95	5.41	5.06	4.82	4.64	4.50	4.39	4.30
	.001	18.64	12.97	10.80	9.63	8.89	8.38	8.00	7.71	7.48	7.29
13	.10	3.14	2.76	2.56	2.43	2.35	2.28	2.23	2.20	2.16	2.14
	.05	4.67	3.81	3.41	3.18	3.03	2.92	2.83	2.77	2.71	2.67
	.01	9.07	6.70	5.74	5.21	4.86	4.62	4.44	4.30	4.19	4.10
	.001	17.82	12.31	10.21	9.07	8.35	7.86	7.49	7.21	6.98	6.80
14	.10	3.10	2.73	2.52	2.39	2.31	2.24	2.19	2.15	2.12	2.10
	.05	4.60	3.74	3.34	3.11	2.96	2.85	2.76	2.70	2.65	2.60
	.01	8.86	6.51	5.56	5.04	4.69	4.46	4.28	4.14	4.03	3.94
	.001	17.14	11.78	9.73	8.62	7.92	7.44	7.08	6.80	6.58	6.40
15	.10	3.07	2.70	2.49	2.36	2.27	2.21	2.16	2.12	2.09	2.06
	.05	4.54	3.68	3.29	3.06	2.90	2.79	2.71	2.64	2.59	2.54
	.01	8.68	6.36	5.42	4.89	4.56	4.32	4.14	4.00	3.89	3.80
	.001	16.59	11.34	9.34	8.25	7.57	7.09	6.74	6.47	6.26	6.08
16	.10	3.05	2.67	2.46	2.33	2.24	2.18	2.13	2.09	2.06	2.03
	.05	4.49	3.63	3.24	3.01	2.85	2.74	2.66	2.59	2.54	2.49
	.01	8.53	6.23	5.29	4.77	4.44	4.20	4.03	3.89	3.78	3.69
	.001	16.12	10.97	9.01	7.94	7.27	6.80	6.46	6.19	5.98	5.81
17	.10	3.03	2.64	2.44	2.31	2.22	2.15	2.10	2.06	2.03	2.00
	.05	4.45	3.59	3.20	2.96	2.81	2.70	2.61	2.55	2.49	2.45
	.01	8.40	6.11	5.18	4.67	4.34	4.10	3.93	3.79	3.68	3.59
	.001	15.72	10.66	8.73	7.68	7.02	6.56	6.22	5.96	5.75	5.58
18	.10	3.01	2.62	2.42	2.29	2.20	2.13	2.08	2.04	2.00	1.98
	.05	4.41	3.55	3.16	2.93	2.77	2.66	2.58	2.51	2.46	2.41
	.01	8.29	6.01	5.09	4.58	4.25	4.01	3.84	3.71	3.60	3.51
	.001	15.38	10.39	8.49	7.46	6.81	6.35	6.02	5.76	5.56	5.39
19	.10	2.99	2.61	2.40	2.27	2.18	2.11	2.06	2.02	1.98	1.96
	.05	4.38	3.52	3.13	2.90	2.74	2.63	2.54	2.48	2.42	2.38
	.01	8.18	5.93	5.01	4.50	4.17	3.94	3.77	3.63	3.52	3.43
	.001	15.08	10.16	8.28	7.27	6.62	6.18	5.85	5.59	5.39	5.22

TABLE 7 Values That Capture Specified Upper-Tail *F* Curve Areas (*continued*)

df$_2$	Area	1	2	3	4	5	6	7	8	9	10
						df$_1$					
20	.10	2.97	2.59	2.38	2.25	2.16	2.09	2.04	2.00	1.96	1.94
	.05	4.35	3.49	3.10	2.87	2.71	2.60	2.51	2.45	2.39	2.35
	.01	8.10	5.85	4.94	4.43	4.10	3.87	3.70	3.56	3.46	3.37
	.001	14.82	9.95	8.10	7.10	6.46	6.02	5.69	5.44	5.24	5.08
21	.10	2.96	2.57	2.36	2.23	2.14	2.08	2.02	1.98	1.95	1.92
	.05	4.32	3.47	3.07	2.84	2.68	2.57	2.49	2.42	2.37	2.32
	.01	8.02	5.78	4.87	4.37	4.04	3.81	3.64	3.51	3.40	3.31
	.001	14.59	9.77	7.94	6.95	6.32	5.88	5.56	5.31	5.11	4.95
22	.10	2.95	2.56	2.35	2.22	2.13	2.06	2.01	1.97	1.93	1.90
	.05	4.30	3.44	3.05	2.82	2.66	2.55	2.46	2.40	2.34	2.30
	.01	7.95	5.72	4.82	4.31	3.99	3.76	3.59	3.45	3.35	3.26
	.001	14.38	9.61	7.80	6.81	6.19	5.76	5.44	5.19	4.99	4.83
23	.10	2.94	2.55	2.34	2.21	2.11	2.05	1.99	1.95	1.92	1.89
	.05	4.28	3.42	3.03	2.80	2.64	2.53	2.44	2.37	2.32	2.27
	.01	7.88	5.66	4.76	4.26	3.94	3.71	3.54	3.41	3.30	3.21
	.001	14.20	9.47	7.67	6.70	6.08	5.65	5.33	5.09	4.89	4.73
24	.10	2.93	2.54	2.33	2.19	2.10	2.04	1.98	1.94	1.91	1.88
	.05	4.26	3.40	3.01	2.78	2.62	2.51	2.42	2.36	2.30	2.25
	.01	7.82	5.61	4.72	4.22	3.90	3.67	3.50	3.36	3.26	3.17
	.001	14.03	9.34	7.55	6.59	5.98	5.55	5.23	4.99	4.80	4.64
25	.10	2.92	2.53	2.32	2.18	2.09	2.02	1.97	1.93	1.89	1.87
	.05	4.24	3.39	2.99	2.76	2.60	2.49	2.40	2.34	2.28	2.24
	.01	7.77	5.57	4.68	4.18	3.85	3.63	3.46	3.32	3.22	3.13
	.001	13.88	9.22	7.45	6.49	5.89	5.46	5.15	4.91	4.71	4.56
26	.10	2.91	2.52	2.31	2.17	2.08	2.01	1.96	1.92	1.88	1.86
	.05	4.23	3.37	2.98	2.74	2.59	2.47	2.39	2.32	2.27	2.22
	.01	7.72	5.53	4.64	4.14	3.82	3.59	3.42	3.29	3.18	3.09
	.001	13.74	9.12	7.36	6.41	5.80	5.38	5.07	4.83	4.64	4.48
27	.10	2.90	2.51	2.30	2.17	2.07	2.00	1.95	1.91	1.87	1.85
	.05	4.21	3.35	2.96	2.73	2.57	2.46	2.37	2.31	2.25	2.20
	.01	7.68	5.49	4.60	4.11	3.78	3.56	3.39	3.26	3.15	3.06
	.001	13.61	9.02	7.27	6.33	5.73	5.31	5.00	4.76	4.57	4.41
28	.10	2.89	2.50	2.29	2.16	2.06	2.00	1.94	1.90	1.87	1.84
	.05	4.20	3.34	2.95	2.71	2.56	2.45	2.36	2.29	2.24	2.19
	.01	7.64	5.45	4.57	4.07	3.75	3.53	3.36	3.23	3.12	3.03
	.001	13.50	8.93	7.19	6.25	5.66	5.24	4.93	4.69	4.50	4.35
29	.10	2.89	2.50	2.28	2.15	2.06	1.99	1.93	1.89	1.86	1.83
	.05	4.18	3.33	2.93	2.70	2.55	2.43	2.35	2.28	2.22	2.18
	.01	7.60	5.42	4.54	4.04	3.73	3.50	3.33	3.20	3.09	3.00
	.001	13.39	8.85	7.12	6.19	5.59	5.18	4.87	4.64	4.45	4.29

(*continued*)

TABLE 7 Values That Capture Specified Upper-Tail *F* Curve Areas (*continued*)

df$_2$	Area	1	2	3	4	5	6	7	8	9	10
10	.10	3.29	2.92	2.73	2.61	2.52	2.46	2.41	2.38	2.35	2.32
	.05	4.96	4.10	3.71	3.48	3.33	3.22	3.14	3.07	3.02	2.98
	.01	10.04	7.56	6.55	5.99	5.64	5.39	5.20	5.06	4.94	4.85
	.001	21.04	14.91	12.55	11.28	10.48	9.93	9.52	9.20	8.96	8.75
11	.10	3.23	2.86	2.66	2.54	2.45	2.39	2.34	2.30	2.27	2.25
	.05	4.84	3.98	3.59	3.36	3.20	3.09	3.01	2.95	2.90	2.85
	.01	9.65	7.21	6.22	5.67	5.32	5.07	4.89	4.74	4.63	4.54
	.001	19.69	13.81	11.56	10.35	9.58	9.05	8.66	8.35	8.12	7.92
12	.10	3.18	2.81	2.61	2.48	2.39	2.33	2.28	2.24	2.21	2.19
	.05	4.75	3.89	3.49	3.26	3.11	3.00	2.91	2.85	2.80	2.75
	.01	9.33	6.93	5.95	5.41	5.06	4.82	4.64	4.50	4.39	4.30
	.001	18.64	12.97	10.80	9.63	8.89	8.38	8.00	7.71	7.48	7.29
13	.10	3.14	2.76	2.56	2.43	2.35	2.28	2.23	2.20	2.16	2.14
	.05	4.67	3.81	3.41	3.18	3.03	2.92	2.83	2.77	2.71	2.67
	.01	9.07	6.70	5.74	5.21	4.86	4.62	4.44	4.30	4.19	4.10
	.001	17.82	12.31	10.21	9.07	8.35	7.86	7.49	7.21	6.98	6.80
14	.10	3.10	2.73	2.52	2.39	2.31	2.24	2.19	2.15	2.12	2.10
	.05	4.60	3.74	3.34	3.11	2.96	2.85	2.76	2.70	2.65	2.60
	.01	8.86	6.51	5.56	5.04	4.69	4.46	4.28	4.14	4.03	3.94
	.001	17.14	11.78	9.73	8.62	7.92	7.44	7.08	6.80	6.58	6.40
15	.10	3.07	2.70	2.49	2.36	2.27	2.21	2.16	2.12	2.09	2.06
	.05	4.54	3.68	3.29	3.06	2.90	2.79	2.71	2.64	2.59	2.54
	.01	8.68	6.36	5.42	4.89	4.56	4.32	4.14	4.00	3.89	3.80
	.001	16.59	11.34	9.34	8.25	7.57	7.09	6.74	6.47	6.26	6.08
16	.10	3.05	2.67	2.46	2.33	2.24	2.18	2.13	2.09	2.06	2.03
	.05	4.49	3.63	3.24	3.01	2.85	2.74	2.66	2.59	2.54	2.49
	.01	8.53	6.23	5.29	4.77	4.44	4.20	4.03	3.89	3.78	3.69
	.001	16.12	10.97	9.01	7.94	7.27	6.80	6.46	6.19	5.98	5.81
17	.10	3.03	2.64	2.44	2.31	2.22	2.15	2.10	2.06	2.03	2.00
	.05	4.45	3.59	3.20	2.96	2.81	2.70	2.61	2.55	2.49	2.45
	.01	8.40	6.11	5.18	4.67	4.34	4.10	3.93	3.79	3.68	3.59
	.001	15.72	10.66	8.73	7.68	7.02	6.56	6.22	5.96	5.75	5.58
18	.10	3.01	2.62	2.42	2.29	2.20	2.13	2.08	2.04	2.00	1.98
	.05	4.41	3.55	3.16	2.93	2.77	2.66	2.58	2.51	2.46	2.41
	.01	8.29	6.01	5.09	4.58	4.25	4.01	3.84	3.71	3.60	3.51
	.001	15.38	10.39	8.49	7.46	6.81	6.35	6.02	5.76	5.56	5.39
19	.10	2.99	2.61	2.40	2.27	2.18	2.11	2.06	2.02	1.98	1.96
	.05	4.38	3.52	3.13	2.90	2.74	2.63	2.54	2.48	2.42	2.38
	.01	8.18	5.93	5.01	4.50	4.17	3.94	3.77	3.63	3.52	3.43
	.001	15.08	10.16	8.28	7.27	6.62	6.18	5.85	5.59	5.39	5.22

TABLE 7 Values That Capture Specified Upper-Tail *F* Curve Areas (*continued*)

df₂	Area	1	2	3	4	5	6	7	8	9	10
							df₁				
20	.10	2.97	2.59	2.38	2.25	2.16	2.09	2.04	2.00	1.96	1.94
	.05	4.35	3.49	3.10	2.87	2.71	2.60	2.51	2.45	2.39	2.35
	.01	8.10	5.85	4.94	4.43	4.10	3.87	3.70	3.56	3.46	3.37
	.001	14.82	9.95	8.10	7.10	6.46	6.02	5.69	5.44	5.24	5.08
21	.10	2.96	2.57	2.36	2.23	2.14	2.08	2.02	1.98	1.95	1.92
	.05	4.32	3.47	3.07	2.84	2.68	2.57	2.49	2.42	2.37	2.32
	.01	8.02	5.78	4.87	4.37	4.04	3.81	3.64	3.51	3.40	3.31
	.001	14.59	9.77	7.94	6.95	6.32	5.88	5.56	5.31	5.11	4.95
22	.10	2.95	2.56	2.35	2.22	2.13	2.06	2.01	1.97	1.93	1.90
	.05	4.30	3.44	3.05	2.82	2.66	2.55	2.46	2.40	2.34	2.30
	.01	7.95	5.72	4.82	4.31	3.99	3.76	3.59	3.45	3.35	3.26
	.001	14.38	9.61	7.80	6.81	6.19	5.76	5.44	5.19	4.99	4.83
23	.10	2.94	2.55	2.34	2.21	2.11	2.05	1.99	1.95	1.92	1.89
	.05	4.28	3.42	3.03	2.80	2.64	2.53	2.44	2.37	2.32	2.27
	.01	7.88	5.66	4.76	4.26	3.94	3.71	3.54	3.41	3.30	3.21
	.001	14.20	9.47	7.67	6.70	6.08	5.65	5.33	5.09	4.89	4.73
24	.10	2.93	2.54	2.33	2.19	2.10	2.04	1.98	1.94	1.91	1.88
	.05	4.26	3.40	3.01	2.78	2.62	2.51	2.42	2.36	2.30	2.25
	.01	7.82	5.61	4.72	4.22	3.90	3.67	3.50	3.36	3.26	3.17
	.001	14.03	9.34	7.55	6.59	5.98	5.55	5.23	4.99	4.80	4.64
25	.10	2.92	2.53	2.32	2.18	2.09	2.02	1.97	1.93	1.89	1.87
	.05	4.24	3.39	2.99	2.76	2.60	2.49	2.40	2.34	2.28	2.24
	.01	7.77	5.57	4.68	4.18	3.85	3.63	3.46	3.32	3.22	3.13
	.001	13.88	9.22	7.45	6.49	5.89	5.46	5.15	4.91	4.71	4.56
26	.10	2.91	2.52	2.31	2.17	2.08	2.01	1.96	1.92	1.88	1.86
	.05	4.23	3.37	2.98	2.74	2.59	2.47	2.39	2.32	2.27	2.22
	.01	7.72	5.53	4.64	4.14	3.82	3.59	3.42	3.29	3.18	3.09
	.001	13.74	9.12	7.36	6.41	5.80	5.38	5.07	4.83	4.64	4.48
27	.10	2.90	2.51	2.30	2.17	2.07	2.00	1.95	1.91	1.87	1.85
	.05	4.21	3.35	2.96	2.73	2.57	2.46	2.37	2.31	2.25	2.20
	.01	7.68	5.49	4.60	4.11	3.78	3.56	3.39	3.26	3.15	3.06
	.001	13.61	9.02	7.27	6.33	5.73	5.31	5.00	4.76	4.57	4.41
28	.10	2.89	2.50	2.29	2.16	2.06	2.00	1.94	1.90	1.87	1.84
	.05	4.20	3.34	2.95	2.71	2.56	2.45	2.36	2.29	2.24	2.19
	.01	7.64	5.45	4.57	4.07	3.75	3.53	3.36	3.23	3.12	3.03
	.001	13.50	8.93	7.19	6.25	5.66	5.24	4.93	4.69	4.50	4.35
29	.10	2.89	2.50	2.28	2.15	2.06	1.99	1.93	1.89	1.86	1.83
	.05	4.18	3.33	2.93	2.70	2.55	2.43	2.35	2.28	2.22	2.18
	.01	7.60	5.42	4.54	4.04	3.73	3.50	3.33	3.20	3.09	3.00
	.001	13.39	8.85	7.12	6.19	5.59	5.18	4.87	4.64	4.45	4.29

(*continued*)

TABLE 7 Values That Capture Specified Upper-Tail *F* Curve Areas (*continued*)

df$_2$	Area	1	2	3	4	5	6	7	8	9	10
30	.10	2.88	2.49	2.28	2.14	2.05	1.98	1.93	1.88	1.85	1.82
	.05	4.17	3.32	2.92	2.69	2.53	2.42	2.33	2.27	2.21	2.16
	.01	7.56	5.39	4.51	4.02	3.70	3.47	3.30	3.17	3.07	2.98
	.001	13.29	8.77	7.05	6.12	5.53	5.12	4.82	4.58	4.39	4.24
40	.10	2.84	2.44	2.23	2.09	2.00	1.93	1.87	1.83	1.79	1.76
	.05	4.08	3.23	2.84	2.61	2.45	2.34	2.25	2.18	2.12	2.08
	.01	7.31	5.18	4.31	3.83	3.51	3.29	3.12	2.99	2.89	2.80
	.001	12.61	8.25	6.59	5.70	5.13	4.73	4.44	4.21	4.02	3.87
60	.10	2.79	2.39	2.18	2.04	1.95	1.87	1.82	1.77	1.74	1.71
	.05	4.00	3.15	2.76	2.53	2.37	2.25	2.17	2.10	2.04	1.99
	.01	7.08	4.98	4.13	3.65	3.34	3.12	2.95	2.82	2.72	2.63
	.001	11.97	7.77	6.17	5.31	4.76	4.37	4.09	3.86	3.69	3.54
90	.10	2.76	2.36	2.15	2.01	1.91	1.84	1.78	1.74	1.70	1.67
	.05	3.95	3.10	2.71	2.47	2.32	2.20	2.11	2.04	1.99	1.94
	.01	6.93	4.85	4.01	3.53	3.23	3.01	2.84	2.72	2.61	2.52
	.001	11.57	7.47	5.91	5.06	4.53	4.15	3.87	3.65	3.48	3.34
120	.10	2.75	2.35	2.13	1.99	1.90	1.82	1.77	1.72	1.68	1.65
	.05	3.92	3.07	2.68	2.45	2.29	2.18	2.09	2.02	1.96	1.91
	.01	6.85	4.79	3.95	3.48	3.17	2.96	2.79	2.66	2.56	2.47
	.001	11.38	7.32	5.78	4.95	4.42	4.04	3.77	3.55	3.38	3.24
240	.10	2.73	2.32	2.10	1.97	1.87	1.80	1.74	1.70	1.65	1.63
	.05	3.88	3.03	2.64	2.41	2.25	2.14	2.04	1.98	1.92	1.87
	.01	6.74	4.69	3.86	3.40	3.09	2.88	2.71	2.59	2.48	2.40
	.001	11.10	7.11	5.60	4.78	4.25	3.89	3.62	3.41	3.24	3.09
∞	.10	2.71	2.30	2.08	1.94	1.85	1.77	1.72	1.67	1.63	1.60
	.05	3.84	3.00	2.60	2.37	2.21	2.10	2.01	1.94	1.88	1.83
	.01	6.63	4.61	3.78	3.32	3.02	2.80	2.64	2.51	2.41	2.32
	.001	10.83	6.91	5.42	4.62	4.10	3.74	3.47	3.27	3.10	2.96

TABLE 8 Critical Values of q for the Studentized Range Distribution

Error df	Confidence level	Number of populations, treatments, or levels being compared							
		3	4	5	6	7	8	9	10
5	95%	4.60	5.22	5.67	6.03	6.33	6.58	6.80	6.99
	99%	6.98	7.80	8.42	8.91	9.32	9.67	9.97	10.24
6	95%	4.34	4.90	5.30	5.63	5.90	6.12	6.32	6.49
	99%	6.33	7.03	7.56	7.97	8.32	8.61	8.87	9.10
7	95%	4.16	4.68	5.06	5.36	5.61	5.82	6.00	6.16
	99%	5.92	6.54	7.01	7.37	7.68	7.94	8.17	8.37
8	95%	4.04	4.53	4.89	5.17	5.40	5.60	5.77	5.92
	99%	5.64	6.20	6.62	6.96	7.24	7.47	7.68	7.86
9	95%	3.95	4.41	4.76	5.02	5.24	5.43	5.59	5.74
	99%	5.43	5.96	6.35	6.66	6.91	7.13	7.33	7.49
10	95%	3.88	4.33	4.65	4.91	5.12	5.30	5.46	5.60
	99%	5.27	5.77	6.14	6.43	6.67	6.87	7.05	7.21
11	95%	3.82	4.26	4.57	4.82	5.03	5.20	5.35	5.49
	99%	5.15	5.62	5.97	6.25	6.48	6.67	6.84	6.99
12	95%	3.77	4.20	4.51	4.75	4.95	5.12	5.27	5.39
	99%	5.05	5.50	5.84	6.10	6.32	6.51	6.67	6.81
13	95%	3.73	4.15	4.45	4.69	4.88	5.05	5.19	5.32
	99%	4.96	5.40	5.73	5.98	6.19	6.37	6.53	6.67
14	95%	3.70	4.11	4.41	4.64	4.83	4.99	5.13	5.25
	99%	4.89	5.32	5.63	5.88	6.08	6.26	6.41	6.54
15	95%	3.67	4.08	4.37	4.59	4.78	4.94	5.08	5.20
	99%	4.84	5.25	5.56	5.80	5.99	6.16	6.31	6.44
16	95%	3.65	4.05	4.33	4.56	4.74	4.90	5.03	5.15
	99%	4.79	5.19	5.49	5.72	5.92	6.08	6.22	6.35
17	95%	3.63	4.02	4.30	4.52	4.70	4.86	4.99	5.11
	99%	4.74	5.14	5.43	5.66	5.85	6.01	6.15	6.27
18	95%	3.61	4.00	4.28	4.49	4.67	4.82	4.96	5.07
	99%	4.70	5.09	5.38	5.60	5.79	5.94	6.08	6.20
19	95%	3.59	3.98	4.25	4.47	4.65	4.79	4.92	5.04
	99%	4.67	5.05	5.33	5.55	5.73	5.89	6.02	6.14
20	95%	3.58	3.96	4.23	4.45	4.62	4.77	4.90	5.01
	99%	4.64	5.02	5.29	5.51	5.69	5.84	5.97	6.09
24	95%	3.53	3.90	4.17	4.37	4.54	4.68	4.81	4.92
	99%	4.55	4.91	5.17	5.37	5.54	5.69	5.81	5.92
30	95%	3.49	3.85	4.10	4.30	4.46	4.60	4.72	4.82
	99%	4.45	4.80	5.05	5.24	5.40	5.54	5.65	5.76
40	95%	3.44	3.79	4.04	4.23	4.39	4.52	4.63	4.73
	99%	4.37	4.70	4.93	5.11	5.26	5.39	5.50	5.60
60	95%	3.40	3.74	3.98	4.16	4.31	4.44	4.55	4.65
	99%	4.28	4.59	4.82	4.99	5.13	5.25	5.36	5.45
120	95%	3.36	3.68	3.92	4.10	4.24	4.36	4.47	4.56
	99%	4.20	4.50	4.71	4.87	5.01	5.12	5.21	5.30
∞	95%	3.31	3.63	3.86	4.03	4.17	4.29	4.39	4.47
	99%	4.12	4.40	4.60	4.76	4.88	4.99	5.08	5.16

Answers

Collecting Data in Reasonable Ways

SECTION 1.2
Exercise Set 1

1.1 observational study; the person conducting the study merely recorded whether or not the boomers sleep with their phones within arm's length and whether or not people used their phones to take photos.

1.2 observational study; researchers recorded responses to questions on a survey.

1.3 experiment; the researchers assigned different toddlers to experimental conditions (adult played with/talked to the robot or the adult ignored the robot).

1.4 observational study; researchers surveyed U.S. adults and drew a conclusion from the survey results

1.5 experiment; because the researchers assigned study participants to one of three experimental groups (meditation, distraction task, or relaxation technique).

Additional Exercises

1.11 observational study; there was no assignment of subjects to experimental groups.

1.13 experiment; the study participants were assigned to one of the two experimental groups (how much would you pay for the mug or how much would you sell the mug for).

SECTION 1.3
Exercise Set 1

1.15 **(a)** census
(b) population characteristic

1.16 The sample is the 100 San Fernando Valley residents, and the population of interest is all San Fernando Valley residents.

1.17 Not appropriate because it only surveyed 100 residents from those who attended a community forum in Van Nuys on a Monday. The residents who attended the forum are likely those who feel strongly about the issue. A more appropriate headline might be "Over two-thirds of those attending a community forum OK with 1-cent transit tax."

1.18 There are several reasonable approaches. One possibility is to use a list of all the students at the school and to write all of the names on otherwise identical slips of paper. Thoroughly mix the slips of paper, and select 150 slips. Include the individuals whose names are on the slips of paper in the sample.

1.19 **(a)** all U.S. women
(b) only women from Maryland, Minnesota, Oregon, and Pennsylvania were included in the sample.
(c) Given that only women from four states were included in the sample, the sample is not likely to be representative of the population of interest.
(d) Selection bias is present because the selection method excluded women from all states other than Maryland, Minnesota, Oregon, and Pennsylvania.

Additional Exercises

1.25 Answers will vary.
(a) One example of a leading question is "Knowing that there are health problems associated with consuming too much sugar, and that soft drinks contain large amounts of added sugar, should there be a tax on soft drinks to encourage people to consume less?"

(b) For example, a double-barreled question is "How satisfied are you with the food and service at this restaurant?" It would be better to split the bad question into two questions, such as "How satisfied are you with the food at this restaurant" and "How satisfied are you with the service at this restaurant."

1.27 The population is the 5000 bricks in the lot. The sample is the 100 bricks chosen for inspection.

1.29 Bias introduced through the two different sampling methods may have contributed to the different results. The online sample could suffer from voluntary response bias in that perhaps only those who feel very strongly would take the time to go to the web site and register their vote. In addition, younger people might be more technologically savvy, and therefore the web site might over-represent the views of younger people (particularly students) who support the parade. The telephone survey responses might over-represent the view of permanent residents (as students might only use cell phones and not have a local phone number).

Exercise Set 1

1.30 Random assignment allows the researcher to create groups that are equivalent, so that the subjects in each experimental group are as much alike as possible. This ensures that the experiment does not favor one experimental condition (playing Unreal Tournament 2004 or Tetris) over another.

1.31 (a) Allowing participants to choose which group they want to be in could introduce systematic differences between the two experimental conditions (compression socks or regular socks), resulting in potential confounding. Those who would choose compression socks might, in some way, be different from those who would choose regular socks.
(b) It would be good to have the runners be blind to the type of socks they were given to eliminate the possible psychological advantage the runners might have if they knew they were wearing compression socks.

1.32 (a) The attending nurse was responsible for administering medication after judging the degree of pain and nausea, so the researchers did not want the nurse's personal beliefs about the different surgical procedures to influence measurements.
(b) Because the children who had the surgery could easily determine whether the surgical procedure was laparoscopic repair or open repair based on the type of incision.

1.33 There are several possible approaches. One possibility is to write the subjects names on otherwise identical slips of paper. Mix the slips of paper thoroughly and draw out slips one at a time. The names on the first 15 slips are assigned to the experimental condition of listening to a Mozart piano sonata for 24 minutes. The names on the next 15 slips are assigned to the experimental condition of listening to popular music for the same length of time. The remaining 15 names are assigned to the relaxation with no music experimental condition.

1.34 (1) Does a dietary supplement consisting of Omega 3, Omega 6, and antioxidants reduce hair loss in women with stage 1 hair loss? (2) The experimental conditions are the supplement and control. (3) The response variable is the change in hair density. (4) The experimental units are the 120 women who volunteered for the study. (5) Yes, the design incorporates random assignment of women to either the supplement group or the control group. (6) Yes, there was a control group. (7) Yes, there was blinding. The expert who determined the change in hair density did not know which of the women had taken the supplement. There is no indication that the women were blinded as to which treatment they received, although this could be incorporated.

Additional Exercises

1.41 (a) Some surgical procedures are more complex and require a greater degree of concentration; music with a vocal component might be more distracting when the surgical procedure is more complex.

(b) The temperature of the room might affect the comfort of the surgeon; if the surgeon is too hot or too cold, she or he might be uncomfortable, and therefore more easily distracted by the vocal component.
(c) If the music is too loud, the surgeon might be distracted and unable to focus, regardless of the presence or absence of the vocal component. If the music is too soft, the surgeon might try to concentrate on listening to the vocal component and therefore pay more attention to the music rather than to the surgical procedure.

1.43 This experiment could not have been double-blind because the surgeon would know whether or not there was a vocal component to the music.

1.45 (a) Probably not, because the judges might not believe that Denny's food is as good as that of other restaurants.
(b) Experiments are often blinded in this way to eliminate preconceptions about particular experimental treatments.

Exercise Set 1

1.46 (a) This was most likely an observational study.

(b) It is not reasonable to conclude that pushing a shopping cart causes people to be less likely to purchase junk food because the results of observational studies cannot be used to draw cause-and-effect conclusions.

1.47 (a) It is not reasonable to conclude that owning a dog causes higher heart rate variability. This was an observational study, so cause-and-effect conclusions cannot be drawn.

(b) It is not reasonable to generalize the results of this survey to all adult Americans because the study participants were not randomly selected from the population of all adult Americans.

1.48 The researcher would have had to assign the nine cyclists at random to one of the three experimental conditions (chocolate milk, Gatorade, or Endurox).

1.49
Study 1: This is an observational study; random selection was used; this was not an experiment so there were no experimental groups; no, because this was not an experiment, cause-and-effect cannot be concluded; it is reasonable to generalize to the population of students at this particular large college.

Study 2: This study was an experiment; random selection was not used; there was no random assignment to experimental conditions (the grouping was based on gender); the conclusion is not appropriate because of confounding of gender and treatment (women ate pecans, and men did not eat pecans); it is not reasonable to generalize to a larger population.

Study 3: This is an observational study; no random selection; no random assignment to experimental groups;

the conclusion is not appropriate because this was an observational study, and therefore cause-and-effect conclusions cannot be drawn; cannot generalize to any larger population.

Study 4: This is an experiment; no random selection; there was random assignment to experimental groups; yes, because this was an experiment with random assignment of subjects to experimental groups, we can draw cause-and-effect conclusions; cannot generalize to a larger population.

Study 5: This is an experiment; there was random selection from students enrolled at a large college; random assignment of subjects to experimental groups was used; because this was a simple comparative experiment with random assignment of subjects to experimental groups, we can draw cause-and-effect conclusions; there was random selection of students, so we can generalize conclusions from this study to the population of all students enrolled at the large college.

Additional Exercises

1.55 It might be that people who live in the South have a less healthy diet and exercise less than those in other parts of the country. As a result, the higher percentage of Southerners with high blood pressure might have nothing to do with living in the South.

1.57 There was no random selection from some population.

1.59 Yes, because this was an experiment and there was random assignment of subjects to experimental groups.

ARE YOU READY TO MOVE ON?
Chapter 1 Review Exercises

1.61 (a) experiment, there is random assignment of subjects to experimental conditions.
(b) observational study, there was no assignment of subjects to experimental conditions
(c) observational study, there was no assignment of subjects to experimental conditions.
(d) experiment, there was random assignment of study participants to experimental conditions.

1.63 (a) population characteristic
(b) statistic
(c) population characteristic
(d) statistic
(e) statistic

1.65 The council president could assign a unique identifying number to each of the names on the petition, numbered from 1 to 500. On identical slips of paper, write the numbers 1 to 500, with each number on a single slip of paper. Thoroughly mix the slips of paper and select 30 numbers. The 30 numbers correspond to the unique numbers assigned to names on the petition. These 30 are the names that would be in the sample.

1.67 Without random assignment of the study participants to experimental conditions, confounding could impact the conclusions of the study. For example, people who would choose an attractive avatar might be more outgoing and willing to engage than someone who would choose an unattractive avatar.

1.69 (a) The alternate assignment to the experimental groups (large serving bowls, small serving bowls) would probably produce groups that are similar.
(b) Blinding ensures that individuals do not let personal beliefs influence their measurements. The research assistant who weighed the plates and estimated the calorie content of the food might (intentionally or not) have let personal beliefs influence the estimate of the calorie content of the food on the plate.

1.71 (a) (1) Does using hand gestures help children learn math? (2) Using hand gestures and not using hand gestures. (3) Number correct on the six-problem test. (4) The 128 children in the study; they were selected because they were the children who answered all six questions on the pretest incorrectly. (5) Yes, the children were assigned randomly to one of the two experimental groups. (6) Yes, the control group is the experimental condition of not using any hand gestures. (7) There was no blinding. It would not be possible to include blinding of subjects in this experiment (the children would know whether or not they were using hand gestures), and there is no need to blind the person recording the response because the test was graded with each answer correct or incorrect, so there is no subjectivity in recording the responses.
(b) The conclusions are reasonable because the subjects were assigned to the treatment groups at random.

1.73 (a) No, the 60 games selected were the 20 most popular (by sales) for each of three different gaming systems. The study excluded the games that were not in the top 20 most popular (by sales).
(b) It is not reasonable to generalize to all video games because of the exclusion of those games not in the top 20 (by sales).

1.75

Study 1: observational study; no random selection; no random assignment to experimental groups; not reasonable to conclude that taking calcium supplements is the cause of the increased heart attack risk; not reasonable to generalize conclusions from this study to a larger population.

Study 2: observational study; there was random selection from the population of people living in Minneapolis who receive Social Security; no random assignment of subjects to experimental groups; not reasonable to conclude that taking calcium supplements is the cause of the increased heart attack risk; it is reasonable to generalize the results of this study to the population of people living in Minneapolis who receive Social Security.

Study 3: experiment; there was random selection from the population of people living in Minneapolis who receive Social Security; no random assignment of subjects to experimental groups; not reasonable to conclude that taking calcium supplements is the cause of the increased risk of heart attack because the participants in this study who did not have a previous history of heart problems were given the calcium supplement, and those with a history of heart problems were not given the supplement. It is not possible to determine the role of the calcium supplement because only those study participants who did not have a history of heart problems were given the supplement; it is possible to generalize the results from this study to the population of all people living in Minneapolis who receive Social Security. However, it is unclear (due to the confounding described in Question 4) what the conclusion would be.

Study 4: experiment; no random selection from some larger population; there was random assignment of study participants to experimental groups; it is reasonable to conclude that taking calcium supplements is the cause of the increased risk of heart attack; it is not reasonable to generalize conclusions from this study to some larger population.

CHAPTER 2
Graphical Methods for Describing Data Distributions

SECTION 2.1
Exercise Set 1

2.1 (a) numerical, discrete (b) categorical (c) numerical, continuous (d) numerical, continuous (e) categorical

2.2 (a) discrete (b) continuous (c) discrete (d) discrete

2.3
Data Set 1: one variable; categorical; summarize the data distribution; bar chart.

Data Set 2: one variable; numerical; compare groups; comparative dotplot or comparative stem-and-leaf display.

Data Set 3: two variables; numerical; investigate relationship; scatterplot.

Data Set 4: one variable; categorical, compare groups, comparative bar chart.

Data Set 5: one variable; numerical; summarize the data distribution; dotplot, stem-and-leaf display or histogram.

Additional Exercises

2.7 (a) numerical (b) numerical (c) categorical (d) numerical (e) categorical

2.9 (a) categorical (b) numerical (c) numerical (d) categorical

2.11 (a) numerical (b) numerical (c) numerical (d) categorical (e) categorical (f) numerical (g) categorical

2.13 one variable; numerical; compare groups, comparative dotplot or comparative stem-and-leaf display.

2.15 one variable; numerical, summarize the data distribution; dotplot, stem-and-leaf plot or histogram.

SECTION 2.2
Exercise Set 1

2.16 (a)

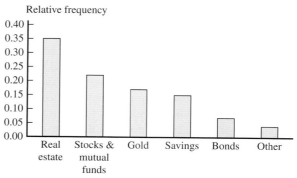

(b) Over half of the responses (57%) were from people who indicated that the best long-term investments were real estate (35%) and stocks & mutual funds (22%). The remaining 43% of respondents indicated that gold (17%), savings (15%), bonds (7%), and other (4%) were the best long-term investments.

2.17

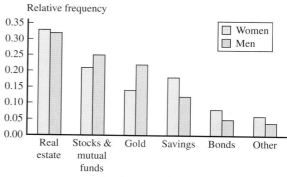

(b) In general, women and men have similar rankings for the best long-term investments. However, one notable difference is that women rank savings higher than gold, and men rank gold higher than savings.

Additional Exercises

2.21

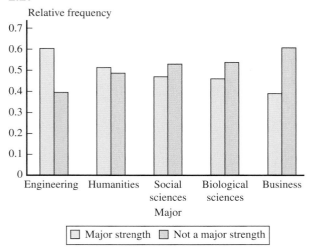

Relative frequency

The majors in which a majority of students indicated that critical thinking is "a major strength" are engineering and humanities. A majority of students in other majors (social sciences, biological sciences, and business) indicated that critical thinking is "not a major strength." The greatest disparities between "a major strength" and "not a major strength" occurred with engineering and business majors, and humanities majors had the smallest difference between the two responses.

2.23   The relative frequency distribution is:

Rating	Relative Frequency
G	1/25 = 0.04
PG	6/25 = 0.24
PG-13	13/25 = 0.52
R	5/25 = 0.20

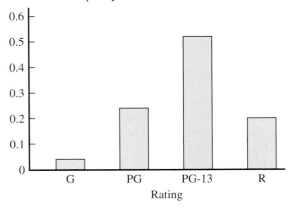

Relative frequency

PG-13 is the rating with the highest relative frequency (0.52), followed by PG (0.24), R (0.20), and G (0.04). Seventy-two percent (72%) of the top 25 movies of 2015 are PG-13 or R, and the remaining 28% are rated G or PG.

SECTION 2.3

Exercise Set 1

2.24 **(a)** The dotplot below shows the concussion rate (concussions per 10,000 athletes).

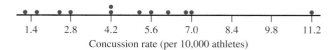

Concussion rate (per 10,000 athletes)

(b) The dotplot below shows the concussion rate (concussions per 10,000 athletes), with different symbols for boys and girls.

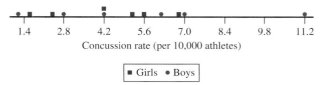

Concussion rate (per 10,000 athletes)

■ Girls • Boys

The sport with an unusually high concussion rate (compared to all the other sports) is football. Without considering football, the distribution of concussion rates for girls' sports is essentially the same as the distribution for boys' sports. However, if you consider football, the concussion rates for girls' sports tend to be lower than the rates for boys' sports.

2.25 **(a)**

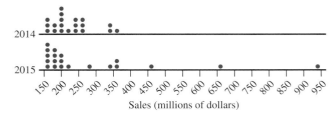

Sales (millions of dollars)

(b) The shape of both distributions is positively skewed. The distribution of 2015 ticket sales is centered at about $280 million, which is higher than the center of the distribution of 2014 ticket sales, which is centered around $230 million. The lowest ticket sales for both 2014 and 2015 are approximately $150 million. Ticket sales for 2014 have a maximum value of approximately $350 million, which is much lower than the highest ticket sales for 2015 of $937 million. The difference between the lowest and highest values for 2014 is approximately $200 million, which is less than the difference for 2015 of a little under $800 million.

2.26

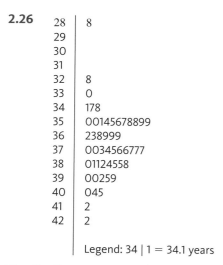

```
28 | 8
29 |
30 |
31 |
32 | 8
33 | 0
34 | 178
35 | 00145678899
36 | 238999
37 | 0034566777
38 | 01124558
39 | 00259
40 | 045
41 | 2
42 | 2
```

Legend: 34 | 1 = 34.1 years

The distribution of median ages is centered at approximately 37 years old, with values ranging from 28.8 to 42.2 years. The distribution is approximately symmetric, with one unusual value of 28.8 years.

2.27 (a)

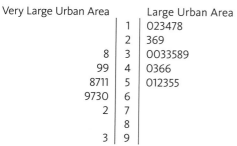

```
Very Large Urban Area        Large Urban Area
                      1 | 023478
                      2 | 369
                 8 |  3 | 0033589
                99 |  4 | 0366
              8711 |  5 | 012355
              9730 |  6 |
                 2 |  7 |
                   |  8 |
                 3 |  9 |
```

Legend: 4|6 = 46 extra hours per year

(b) The statement "The larger the urban areas, the greater the extra travel time during peak period travel" is generally consistent with the data. Although there is overlap between the times for the very large and large urban areas, the extra travel times for the very large urban areas are generally greater than those for the large urban areas.

2.28 (a)

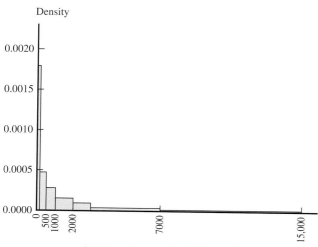

Credit card balance (Credit Bureau data)

(b)

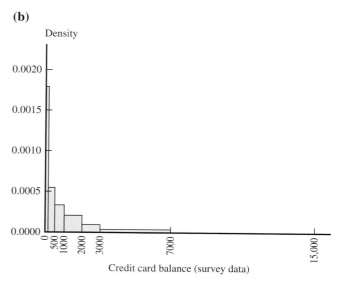

Credit card balance (survey data)

(c) The histograms are similar in shape. A notable difference is that the Credit Bureau data show that 7% of students have credit card balances of at least $7000, but no survey respondent indicated a balance of at least $7000.

(d) Yes, because students with credit card balances of $7000 or more might be too embarrassed to admit that they have such a high balance.

2.29 (a) If the exam is quite easy, the scores would be clustered at the high end of the scale, with a few low scores for the students who did not study. The histogram would be negatively skewed.

(b) If the exam is difficult, the scores would be clustered around a much lower value, with only a few high scores. The histogram would be positively skewed.

(c) In this case, the histogram would be bimodal, with a cluster of high scores and a cluster of low scores.

Additional Exercises

2.37

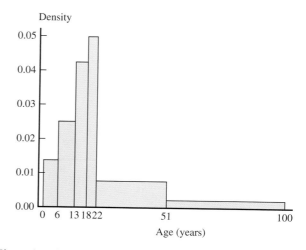

Age (years)

Given that the eye is naturally drawn to large areas, the incorrect histogram exaggerates the proportion of clients in the wider age groups.

2.39 The distribution of wind speed is positively skewed and bimodal. There are peaks in the 35–40 m/s and 60–65 m/s intervals.

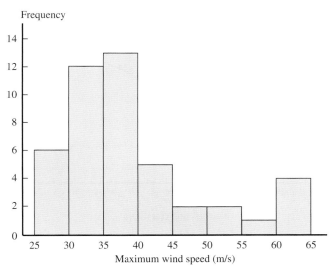

2.41

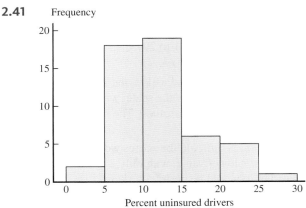

The distribution of the percent of uninsured drivers is positively skewed, centered in the 10–15% interval. The percent uninsured drivers varies between a low of 3.9% and a high of 25.9%. Approximately 75% of states have 15% or less uninsured drivers.

SECTION 2.4

Exercise Set 1

2.42 (a)

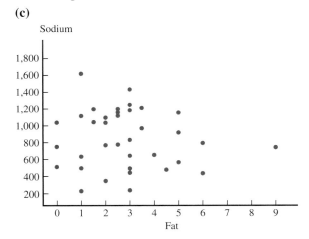

The scatterplot shows the expected positive relationship between grams of fat and calories. The relationship is weak.

(b)

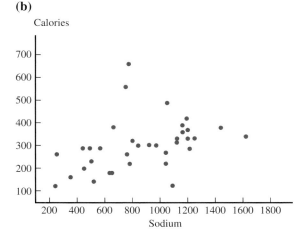

As was observed in the calories versus fat scatterplot, there is also a weak, positive relationship between calories and sodium. The relationship between calories and sodium appears to be a little stronger than the calories versus fat relationship.

(c)

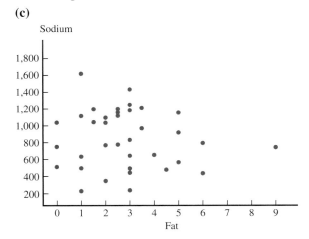

There is no apparent relationship between sodium and fat.

(d)

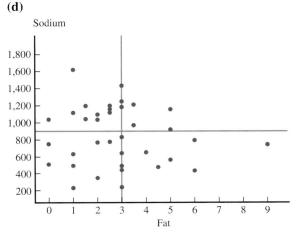

The lower-left region corresponds to healthier fast-food choices. This region corresponds to food items with fewer than 3 grams of fat and fewer than 900 milligrams of sodium.

2.43 **(a)**

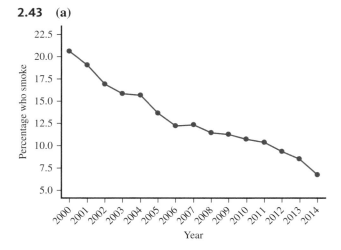

There has been a downward trend in the percent of grade 12 students who smoke daily, from a high of 20.6% in 2000 down to 6.7% in 2014.

(b)

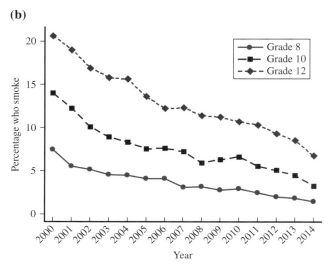

(c) There has been a downward trend in the percent of students who smoke regardless of grade level. For all the years in the data set, grade 8 students had a lower percentage of smokers than the other grades. The percentage of grade 10 students who smoke was between the percentages of grade 8 and grade 12, and the percentage of grade 12 students who smoke was the highest for all years in the data set.

Additional Exercises

2.47 **(a)**

Percent of households with a computer

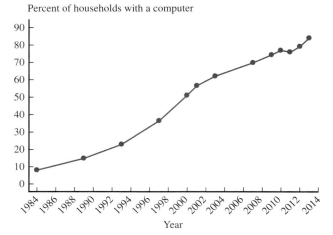

(b) The percentage of households with computers has increased over time, from a low of approximately 8% in 1984 to nearly 84% in 2013. The rate of increase of the percentage of households with a computer increased over time from 1984 to 2001, and the rate of increase remained roughly constant from 2001 to 2013.

2.49

Average cost

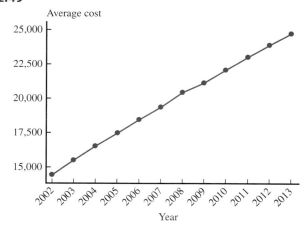

There is a strong, positive trend in the average cost per year for tuition, fees, and room and board for four-year public institutions in the United States. The average cost has steadily increased from a low of $14,439 per year in 2002 to a high of $24,706 per year in 2013.

SECTION 2.5

Exercise Set 1

2.51 **(a)** categorical

(b) A bar chart was used because the response is a categorical variable, and dotplots are used for numerical responses.

(c) This is not a correct representation of the response data because the percent values add up to over 100% (they add to 107%).

2.52 **(a)** Overall score is numerical. Grade is categorical. **(b)** The figure is equivalent to a segmented bar graph because the bar is divided into segments, with different shaded regions representing the different grades ("Top of the Class," "Passing," "Barely Passing," and "Failing"), and the height of each segment is equal to the frequency for that category (for example, there are five school districts in the Top of the Class category, three in the Passing category, and so on), making the area of each shaded region proportional to the relative frequencies for each grade.

(c)

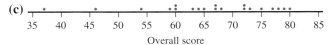
Overall score

One alternate assignment of grades is to require that "Top of the Class" schools earn grades of 72 or higher, "Passing" schools earn between 66 and 71, "Barely Passing" schools earn between 61 and 65, and "Failing" schools earn 60 or below. This alternative is suggested because there appear to be clusters of dots on the dotplot that correspond to the suggested ranges.

2.53 Answers may vary. Students and teachers differ in their opinions regarding use of Facebook and students being allowed to report on controversial issues in their student newspaper. For both questions, 61% of students agree with the statements given. However, the percentages of teachers who agree are much lower (39% and 29%, respectively, for the two statements). Similarly, a larger percentage of teachers disagree with the two statements (57% and 67%, respectively).

Additional Exercises

2.57 **(a)** The areas in the display are not proportional to the values they represent. The "no" category seems to represent more than 68%.

(b)

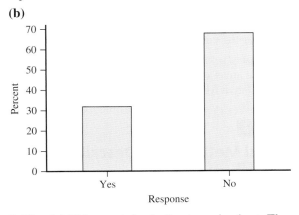
Response

2.59 **(a)** This graph is similar to a pie chart. The portions of the house add up to 100%.
(b) Chart III is the segmented bar chart that is a graph of the data used to create the given graph. The height of the "Yes" portion of the bar in Chart III has a height that corresponds to 23%. The "No" portion of the bar in Chart III has a height that corresponds to 48%, which gives a total

height of 71% for the combined "Yes" and "No" responses. Finally, the "Not sure" portion of the bar makes up the remaining 29%, making the total to 100%.

ARE YOU READY TO MOVE ON?
Chapter 2 Review Exercises

2.61

Data Set 1: one variable; numerical; summarize the data distribution; a dotplot, stem-and-leaf display or histogram would be an appropriate graphical display.

Data Set 2: two variables; numerical; investigate the relationship; a scatterplot would be an appropriate graphical display.

Data Set 3: two variables; categorical; compare groups; comparative bar chart.

Data Set 4: one variable; categorical; summarize the data distribution; a bar chart would be an appropriate graphical display.

Data Set 5: one variable; numerical; compare groups; a comparative dotplot or comparative stem-and-leaf display would be an appropriate graphical display.

2.63 **(a)**

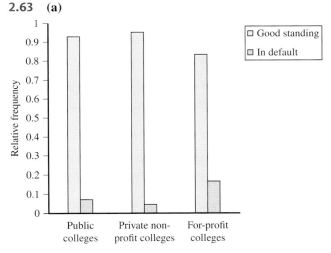

(b) The "In Default" bar in the "For-Profit Colleges" category is taller than either of the other "In Default" bars.

2.65 **(a)**

0	9
1	4667788899
2	0011333333334556667789
3	0000000011223333457
4	4
5	1

Legend: 1|4 = 14 cents per gallon

(b) The center is approximately 26 cents per gallon, and most states have a tax that is near the center value, with tax values ranging from 9 cents per gallon to 51 cents per gallon. The distribution is approximately symmetric.

(c) The only value that might be considered unusual is the 51 cents per gallon tax in Pennsylvania. Although not an obvious outlier, it is approximately 7 cents per gallon higher than the next lower gasoline tax. There are no other states that make such a big jump to the next higher tax.

2.67 (a)

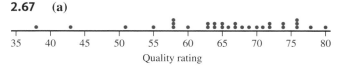

(b) A typical value for quality rating is approximately 66.

(c) Quality rating varies between a low score of 38 defects per 100 vehicles and a high score of 80 defects per 100 vehicles. Noting the minimum and maximum values on the dotplot supports this answer.

(d) Two brands (Fiat and Jeep) seem to stand out as having much lower quality ratings (defects per 100 vehicles) than the others, with values of 38 and 43, respectively.

(e)

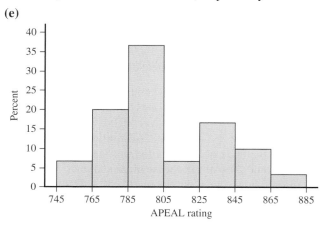

The histogram is centered in the 785–805 range, with values that range between approximately 745 and 885. The distribution is bimodal.

(f)

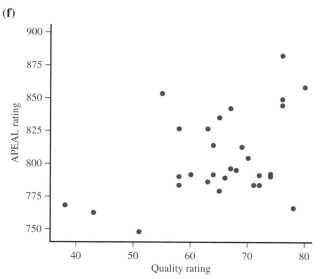

There is a weak positive association between customer satisfaction (as measured by the APEAL rating) and quality rating. Brands with higher quality ratings (higher number of defects per 100 vehicles) tend to have higher APEAL ratings.

2.69 (a)

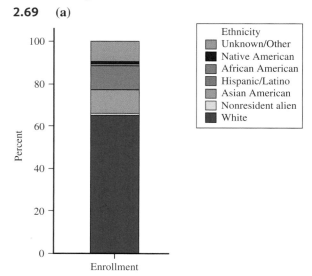

(b) The segmented bar graph in part (a) is more informative because it is easier to get a sense of the percentages of each ethnicity enrolled. Specifically, in the original graphical display, with the "Nonwhite" category further subdivided, it is difficult to compare the "Nonwhite" breakout categories with the other categories represented in the pie chart.

(c) The pie chart combined with the segmented bar graph could have been chosen because some of the pie slices might be very thin and hard to see, and too many pieces could be difficult to visually process.

(d) I would recommend the combination pie chart with segmented bar chart because of the number of categories. In the single pie chart, several categories are too small to distinguish from each other. The same is true of the segmented pie chart. The combination pie chart with segmented bar chart gives a good sense of the overall percentages for the categories.

2.71 The display is misleading because the area principle is violated. The areas of the cocaine mounds are not proportional to the relative frequencies being represented.

CHAPTER 3
Numerical Methods for Describing Data Distributions

SECTION 3.1
Exercise Set 1

3.1 The distribution is approximately symmetric with no outliers, so the mean and standard deviation should be used to describe the center and spread, respectively.

3.2 The distribution is skewed with an outlier, so the median and interquartile range should be used.

3.3 The distribution is approximately symmetric, so the mean and standard deviation should be used to describe center and spread, respectively.

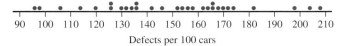

Defects per 100 cars

3.4 The average may not be the best measure of a typical value for this data set because the distribution is clearly skewed.

Additional Exercises

3.9 The distribution is skewed, so median and interquartile range should be used.

3.11 The distribution is roughly symmetric with no obvious outliers, so the mean and standard deviation should be used.

SECTION 3.2

Exercise Set 1

3.12 $\bar{x} = 51.33$ ounces; this is a typical value for the amount of alcohol poured. $s = 15.22$ ounces; this represents how much, on average, the values in the data set spread out, or deviate, from the mean.

3.13 **(a)** $\bar{x} = 59.23$ ounces; this is a typical value for the amount of alcohol poured. $s = 16.71$ ounces; this represents how much, on average, the values in the data set spread out, or deviate, from the mean.

(b) Individuals pouring alcohol into short, wide glasses pour, on average, more alcohol than when pouring into tall, slender glasses.

3.14 **(a)** $\bar{x} = 5.69$ millions of dollars, $s = 3.68$ millions of dollars
(b) $\bar{x} = 4.182$ millions of dollars, $s = 1.168$ millions of dollars. When New York and Los Angeles were excluded from the data set, the mean and standard deviation both decreased. This suggests that using the mean and standard deviation as measures of center and variability for data sets with outliers present can be risky, because outliers can have a significant impact on those measures.

3.15 Answers will vary. One possible answer: The mean is $444, but it is likely that some parents spend only a small amount. There is probably a lot of variability in amount spent, so we would expect a large value for the standard deviation.

Additional Exercises

3.21 **(a)** $\bar{x} = 299$
(c) $s = \sqrt{s^2} = \sqrt{1630} = 40.37$

3.23 The deviations are exactly the same as the corresponding deviations for the original data set. Since the deviations are the same, the new variance and standard deviation are also the same as the old variance and standard deviation. Subtracting the same number from

or adding the same number to every value in a data set does not change the value of the variance or standard deviation.

SECTION 3.3

Exercise Set 1

3.25 **(a)** $median = \dfrac{411,960 + 412,669}{2} = 412,314.5$. This value, 412,314.5, is the value that divides the ordered data set into two halves. This tells you that half of the values in the data set had average weekly circulations of less than 412,314.5, and the other half had average weekly circulations of more than 412,314.5.

(b) The median is preferable to the mean because the distribution is positively skewed and contains outliers.

(c) It is not reasonable to generalize from this sample to the population of daily newspapers in the United States because these newspapers were not randomly selected. Rather, they are the top 20 newspapers in weekday circulation.

3.26 Lower quartile = 440.5 milligrams/slice; upper quartile = 667 milligrams/slice. The lower quartile of 440.5 milligrams/slice is the value such that 25% of the cheese pizzas have sodium contents lower than this value, and 75% are higher. The upper quartile of 667 milligrams/slice is the value such that 75% of the cheese pizzas have sodium contents lower than this value, and 25% are higher. The interquartile range is $iqr = 667 - 440.5 = 226.5$ milligrams/slice. The interquartile range of 226.5 milligrams/slice is the range of the middle 50% of the cheese pizza sodium contents. It tells you how spread out the middle 50% of the data values are.

3.27 Median is 318 minutes. The lower quartile is 175.5 minutes, and the upper quartile is 497 minutes. The interquartile range is $iqr = 497 - 175.5 = 321.5$ minutes. Half of the values of number of minutes used in cell phone calls in one month are less than or equal to 318 minutes, and half of the data values of number of minutes used in cell phone calls are greater than or equal to 318 minutes. The middle 50% of the data values have a range of 321.5 minutes.

3.28 The median tipping percentage is 15.15%. The lower quartile is 12.3%, and the upper quartile is 18.95%. The interquartile range is $iqr = 18.95 - 12.3 = 6.65\%$. The median tipping percentage of 15.15% indicates that half of the tips were below 15.15%, and the remaining half were above 15.15%. The interquartile range indicates that the middle 50% of tips had a range of 6.65%.

Additional Exercises

3.33 **(a)** The large difference between the mean cost and the median cost along with the fact that the mean is greater than the median tells us that there are some large outliers in the distribution of wedding costs in 2012.

(b) I agree that the average cost is misleading because 50% of the weddings in 2012 cost less than $18,086, which is much lower than the mean wedding cost.

(c) I agree with this statement. The mean is strongly influenced by outliers, and large outliers will pull the mean toward larger values. Given that the mean is so much larger than the median (which divides the distribution of wedding costs in half), less than 50% of the wedding costs must be greater than or equal to the mean.

3.35 **(a)** mean will be greater than the median.

(b) $\bar{x}$ = 370.69 seconds, *median* = 369.5 seconds.

(c) The largest time could be increased by any amount and not affect the sample median because the position of the middle value will not change if the largest value is increased. The largest time could be decreased to 370 seconds without changing the value of the median.

SECTION 3.4

Exercise Set 1

3.36 Minimum = 2 tons; lower quartile = 12,423 tons; median = 38,533 tons; upper quartile = 80,541 tons; maximum = 365,507 tons.

3.37

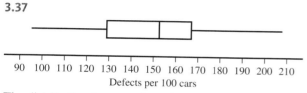

The distribution is approximately symmetric with no outliers. The minimum value is 95, the lower quartile is approximately 129, the median is approximately 153, the upper quartile is approximately 167, and the maximum value is 208 (all units are defects per 100 cars).

3.38 **(a)** Alaska (9 cents per gallon) is not an outlier, but Pennsylvania (51.4 cents per gallon) is an outlier. Values greater than upper quartile + 1.5(*iqr*) or less than lower quartile − 1.5(*iqr*) are considered outliers. For this data set, values are outliers if they are greater than 31.4 + 1.5(31.4 − 20.9) = 47.15 cents per gallon or less than 20.9 − 1.5(31.4 − 20.9) = 5.15 cents per gallon. The largest value of 51.4 cents per gallon is greater than 47.15, and the smallest value is not less than 5.15.

(b)

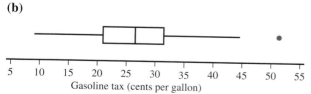

The boxplot is approximately symmetric, but there is one outlier on the high side.

3.39 **(a)** lower quartile = 15.8 inches, upper quartile = 21.15 inches. *iqr* = 21.15 − 15.8 = 5.35 inches.

(b) Any values greater than upper quartile + 1.5(*iqr*) or less than lower quartile − 1.5(*iqr*) are considered outliers. For this data set, values are outliers if they are greater than 21.15 + 1.5(5.35) = 29.175 inches or less than

15.8 − 1.5(5.35) = 7.775 inches. The values 30.2 inches and 31.4 inches are outliers.

(c)

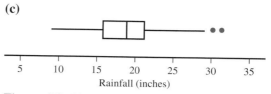

The modified boxplot shows two outliers at the high end of the scale. The distribution of inches of rainfall is slightly positively skewed.

3.40

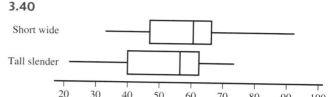

Both distributions (short, wide and tall, slender) are skewed, although the direction of skew is different for the two distributions. The amount of alcohol poured into short, wide glasses tends to be more than the amount poured into tall, slender glasses.

Additional Exercises

3.47 **(a)**

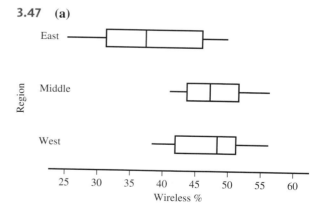

(b) The Eastern region has the smallest median (37.2%), and the Middle and Western regions have medians that are much closer to each other (47.1% and 48.3%, respectively). The distribution of wireless percent for the West is negatively skewed, and the other two region distributions are positively skewed. The Eastern region has the most variability in wireless percent, and the Middle region has the least variability.

3.49 **(a)** Lower quartile: 233.5 million dollars; Upper quartile: 458.5 million dollars; Interquartile range = 458.5 – 233.5 = 225.0 million dollars. Outliers are observations that are smaller than 233.5 − 1.5(225.0) = −104 million dollars and larger than 458.5 + 1.5(225.0) = 796.0. There are no values below zero or above 796.0 million dollars, so there are no outliers.

(b)

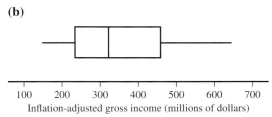

100   200   300   400   500   600   700
Inflation-adjusted gross income (millions of dollars)

(c) The boxplot shows that the distribution of inflation-adjusted gross incomes is positively skewed. You would expect the mean to be greater than the median.

SECTION 3.5

Exercise Set 1

3.50 First national aptitude test: $z = 1.5$. Second national aptitude test: $z = 1.875$. The student performed better on the second national aptitude test relative to the other test takers because the z-score for the second test is higher than for the first test.

3.51 **(a)** 40 minutes is 1 standard deviation above the mean; 30 minutes is 1 standard deviation below the mean. The values that are 2 standard deviations away from the mean are 25 and 45 minutes.
(b) Approximately 95% of times are between 25 and 45 minutes; approximately 0.3% of times are less than 20 minutes or greater than 50 minutes; approximately 0.15% of times are less than 20 minutes.

3.52 The 10th percentile of $0 indicates that 10% of students have $0 or less of student debt. The 25th percentile (which is the lower quartile) indicates that 25% of students have $0 or less of student debt. The 50th percentile (the median) indicates that 50% of students have $11,000 or less of student debt. The 75th percentile (the upper quartile) indicates that 75% of students have $24,600 or less of student debt. The 90th percentile indicates that 90% of students have $39,300 or less of student debt.

3.53 **(a)** Because $(132)(0.54) = 71.28$, we are looking for the data value that is in approximately the 71st position in the ordered list of observations. Therefore, the 54th percentile is approximately 100 meters.
(b) Because $(132)(0.80) = 105.6$, we are looking for the data value that is in approximately the 106th position in the ordered list of observations. Therefore, the 80th percentile is approximately 2000 meters.
(c) Because $(132)(0.92) = 121.44$, we are looking for the data value that is in approximately the 121st position in the ordered list of observations. Therefore, the 92nd percentile is approximately 5000 meters.

Additional Exercises

3.59 **(a)** 1100 gallons; **(b)** 1400 gallons; **(c)** 1700 gallons

3.61 **(a)** 120
(b) 20
(c) -0.5
(d) 97.5

(e) Since a score of 40 is 3 standard deviations below the mean, that corresponds to a percentile of 0.15%. Therefore, there were relatively few scores below 40.

ARE YOU READY TO MOVE ON?
Chapter 3 Review Exercises

3.63 The mean APEAL rating is $\bar{x} = 804.27$, which is a typical or representative value for the APEAL ratings in the sample. The standard deviation is $s = 32.10$ and represents how much, on average, the values in the data set spread out, or deviate, from the mean APEAL rating.

3.65 **(a)** $\bar{x} = 27.31\%$, $s = 23.83\%$
(b) After removing the 105% tip, the new mean and standard deviation are $\bar{x}_{new} = 23.23\%$ and $s_{new} = 15.70\%$. These values are much smaller than the mean and standard deviation computed with 105 included. This suggests that the mean and standard deviation can change dramatically when outliers are present (or removed) from the data set and, therefore, are probably not the best measures of center and spread to use in this situation.

3.67 **(a)** median = 140 seconds; half the values are less than 140 seconds and half the values greater than 140 seconds. $iqr = 100$ seconds; the middle 50% of the data values have a range of 100 seconds.
(b) There is an outlier in the data set.

3.69 **(a)** Median = 8 g/serving; lower quartile = 7 g/serving; upper quartile = 12 g/serving; interquartile range = $12 - 7 = 5$ g/serving
(b) Median = 10 g/serving; lower quartile = 6 g/serving; upper quartile = 13 g/serving; interquartile range = $13 - 6 = 7$ g/serving
(c) There are no outliers in the sugar content data.
(d) The minimum value and lower quartile are the same because the smallest five values in the data set are all equal to 7.

(e)

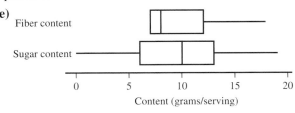

Fiber content

Sugar content

0   5   10   15   20
Content (grams/serving)

The sugar content in grams/serving is much more variable than the fiber content in grams/serving. The boxplot of fiber content shows that the minimum and lower quartiles are equal to each other, which is not observed in the sugar content. The distribution of sugar content values is approximately symmetric, which is different from the skewed fiber content distribution.

3.71 **(a)** The 25th percentile indicates that 25% of full-time female workers age 25 or older with an Associate degree earn $26,900 or less. The 50th percentile indicates that 50% of full-time female workers age 25 or older with an Associate degree earn $39,300 or less. The 75th percentile indicates that 75% of full-time female workers age 25 or older with an Associate degree earn $53,400 or less.

(b) The 25th, 50th, and 75th percentile values for men are all greater than the corresponding percentiles for female workers, indicating that full-time employed men age 25 or older with an Associate degree, in general, earn more than full-time employed women age 25 or older with an Associate degree.

CHAPTER 4
Describing Bivariate Numerical Data

SECTION 4.1
Exercise Set 1

4.1   Scatterplot 1:   (i) Yes (ii) Yes (iii) Negative
Scatterplot 2:   (i) Yes (ii) No (iii) –
Scatterplot 3:   (i) Yes (ii) Yes (iii) Positive
Scatterplot 4:   (i) Yes (ii) Yes (iii) Positive

4.2   **(a)** Negative correlation, because as interest rates rise, the number of loan applications might decrease.
(b) Close to zero, because there is no reason to believe that height and IQ should be related.
(c) Positive correlation, because taller people tend to have larger feet.
(d) Positive correlation, because as the minimum daily temperature increases, the cooling cost would also increase.

4.3   **(a)** The value of the correlation coefficient is negative, which suggests that students who use a cell phone for more hours per day tend to have lower GPAs.
(b) The relationship between texting and GPA has the same direction as the direction for cell phone use and GPA, so the correlations must have the same sign. The relationship between texting and GPA is not as strong as the relationship between cell phone use and GPA, so the correlation coefficient must be closer to zero. Therefore, $r = -0.10$ is the only option that satisfies both of the criteria.

(c) Since it is reasonable to believe that texts sent would be approximately equal to texts received, there would be a positive association between these items. In addition, given that the two texting items are nearly perfectly correlated, the correlation coefficient must be close to $+1$ or -1. Therefore, the correlation coefficient would be close to $+1$.

4.4   **(a)**

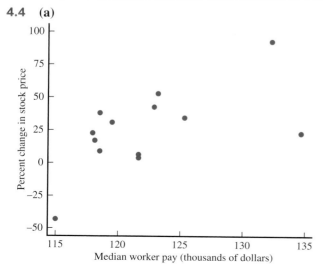

(b) $r = 0.578$; There is a moderately strong, positive relationship.

(c) The conclusion is justified based on these data because there is a positive association between the variables. In general, as median worker pay increases, so does the percent change in stock price.

(d) It is not reasonable to generalize conclusions based on these data to all U.S. companies because these companies were not randomly selected. These are the top 13 highest paying companies in the United States.

4.5   **(a)**

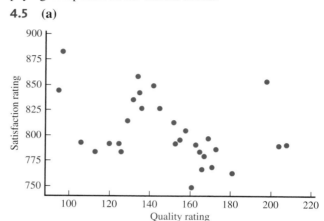

There is a weak, negative linear relationship between satisfaction rating and quality rating.

(b) The correlation of $r = -0.4$ indicates that there is a weak, negative linear relationship between satisfaction rating and quality rating.

4.6   No, the statement is not correct. A correlation of 0 indicates that there is not a linear relationship between two variables. There could be a strong nonlinear relationship (for example, a quadratic relationship) between the two variables.

4.7   No, it is not reasonable to conclude that increasing alcohol consumption will increase income. Correlation measures the strength of association, but association does not imply causation.

4.8   The correlation between college GPA and academic self-worth ($r = 0.48$) indicates that there is a weak or moderate positive linear relationship between those variables. This tells us that athletes with higher a GPA tend to feel better about themselves academically than those with lower grades. The correlation between college GPA and high school GPA ($r = 0.46$) indicates that there is a weak or moderate positive relationship between those variables as well. This tells us that those athletes with higher high school GPA tend to also have a higher college GPA. Finally, the correlation between college GPA and a measure of tendency to procrastinate ($r = -0.36$) indicates that there is a weak negative linear relationship between those variables. Athletes with a lower college GPA tend to procrastinate more than athletes with a higher college GPA.

Additional Exercises

4.15 **(a)** $r = 0.394$

(b) $r = -0.664$

(c) There is a stronger relationship between Happiness Index and response to Statement 2.

(d) There is a weak, positive association between the Happiness Index and the response to Statement 1. In contrast, there is a moderate, negative association between the Happiness Index and the response to Statement 2.

4.17 $r = 0.975$; this value is consistent with the previous answer, because the correlation coefficient is large and positive (close to 1), which indicates a strong positive association between the amount spent on science and the amount spent on pets.

4.19 The sample correlation coefficient would be closest to -0.9. Cars traveling at a faster rate of speed will travel the length of the highway segment more quickly than those who are traveling more slowly, and the correlation would be strong.

SECTION 4.2

Exercise Set 1

4.21 It makes sense to use the least squares regression line to summarize the relationship between x and y for Scatterplot 1, but not for Scatterplot 2. Scatterplot 1 shows a linear relationship between x and y, but Scatterplot 2 shows a curved relationship between x and y.

4.22 It would be larger because the least squares regression line is *the* line with the minimum value for the sum of the squared vertical deviations from the line. All other lines would have larger values for the sum of the squared vertical deviations.

4.23 **(a)** $\hat{y} = -5.0 + 0.017x$

(b) 30.7 therms.

(c) 0.017 therms.

(d) No, because the regression line was determined based on house sizes between 1000 and 3000 square feet. There is no guarantee that the linear relationship will continue outside this range of house sizes.

4.24 **(a)** The response variable (y) is birth weight, and the predictor variable (x) is mother's age.
(b) It is reasonable to use a line to summarize the relationship because the scatterplot shows a clear linear relationship between birth weight and mother's age.
(c) $\hat{y} = -1163.4 + 245.15x$

(d) The slope of 245.15 is the amount, on average, by which the birth weight increases when the mother's age increases by one year.

(e) It is not appropriate to interpret the intercept of the least squares regression line. The intercept is the birth weight for a mother who is zero years old, which is impossible. In addition, the intercept is negative, indicating a negative birth weight, which is also impossible.

(f) 3249.3 g
(g) 2513.85 g
(h) No, because 23 years is well outside the range of ages used in determining the least squares regression line.

4.25 **(a)** The response variable is the acrylamide concentration, and the predictor variable is the frying time.

(b)

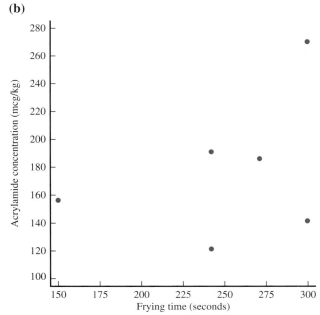

There is a weak, positive association ($r = 0.379$) between frying time and acrylamide concentration.

(c) $\hat{y} = 86.9 + 0.359x$, where y is the acrylamide concentration (in micrograms per kg), and x is the frying time in seconds.

(d) The slope is positive, and it is consistent with the observed positive association in the scatterplot, as well as with the positive correlation coefficient.

(e) Yes, the scatterplot and least squares regression line equation do support the researcher's conclusion that higher frying times tend to be paired with higher acrylamide concentrations.

(f) $\hat{y} = 86.9 + 0.359(225) = 167.675$ mcg per kg

(g) No, I would not use the least squares regression equation to predict the acrylamide concentration for a frying time of 500 seconds. The data set that was used to create the least squares regression line was based on frying times that vary between 150 and 300 seconds; the value 500 seconds is far outside that range of values. There is no guarantee that the observed trend will continue as far as 500 seconds.

4.26 **(a)** $\hat{y} = 11.48 + 0.970x$
(b) 496.48.
(c) 302.48

Additional Exercises

4.33 $\hat{y} = 13.450 - 0.195x$

4.35 Age is a better predictor of number of cell phone calls. The linear relationship between age and number of cell phone calls is stronger than the relationship between age and number of text messages sent.

4.37 The slope is the change in predicted price for each additional mile from the Bay, so the slope would be −4000.

Exercise Set 1

4.39 (a) $r^2 = 0.303$; approximately 30.3% of the variability in the percentage of alumni who strongly agree can be explained by the linear relationship between the percentage of alumni who strongly agree and ranking.
(b) $s_e = 5.347$ (obtained using software); s_e tells us that the percentage of alumni who strongly agree will deviate from the least squares regression line by 5.347, on average.
(c) The relationship between the percentage of alumni who strongly agree and ranking is moderate and positive. It is positive because the slope of the least squares regression line is positive, and it is moderate because the correlation coefficient (which has the same sign as the slope) is $r = \sqrt{r^2} = \sqrt{0.303} = 0.550$

4.40 A small value of s_e indicates that residuals tend to be small. Because residuals represent the difference between an observed y value and a predicted y value, the value of s_e tells us how much accuracy we can expect when using the least squares regression line to make predictions.

4.41 It is important to consider both r^2 and s_e when evaluating the usefulness of the least squares regression line because a large r^2 (which indicates the proportion of variability in y that can be explained by the linear relationship between x and y) tells us that knowing the value of x is helpful in predicting y, and a small s_e indicates that residuals tend to be small.

4.42 (a) Yes, the scatterplot looks reasonably linear.

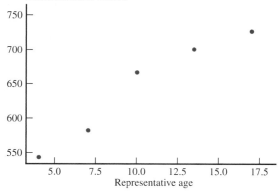

(b) $\hat{y} = 492.80 + 14.763x$
(c) The residuals are −7.55, −12.14, 26.87, 9.00, and −16.17. There is a curved pattern in the residual plot. The curvature indicates that the relationship between median distance walked and representative age is not linear.

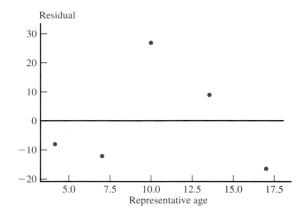

4.43 (a) The pattern for girls differs from boys in that the girls' scatterplot shows more apparent nonlinearity in the pattern.
(b) $\hat{y} = 480 + 12.525x$
(c) The residuals are −37.70, 10.63, 50.56, 8.52, −32.01. The curvature of the residual plot indicates that a curve is more appropriate than a line for describing the relationship between median distance walked and representative age.

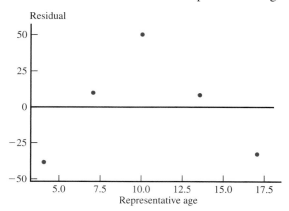

4.44 (a)

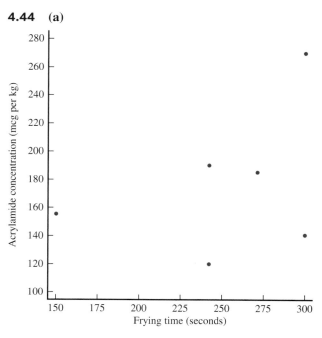

(b) $\hat{y} = 86.9 + 0.359x$, where y is the acrylamide concentration (in mcg/kg), and x is the frying time in seconds. The predicted acrylamide concentration for a frying time of 270 seconds is $\hat{y} = 86.9 + 0.359(270) = 183.83$ mcg/kg. The residual associated with the observation (270, 185) is $y - \hat{y} = 185 - 183.83 = 1.17$.
(c) The observation (150,155) is potentially influential because that point has an x value that is far away from the rest of the data set.
(d) $\hat{y} = -44 + 0.83(270) = 180.1$; this prediction is smaller than that made in part (b).

4.45 **(a)** There appears to be a linear relationship.
(b) $\hat{y} = 18.483 + 0.0029x$
(c) The observation (3928, 46.8) is not influential, because the x-value for that observation is not far from the rest of the data. In addition, removal of the potentially influential point produces a least squares regression line with a y-intercept and slope similar to the original line.
(d) Those points are not considered influential even though they are far from the rest of the data because they follow the trend of the remaining data points. Removal of those points would produce a least squares regression line similar to the line found using the full data set.
(e) $s_e = 9.16217$; a typical deviation from the least squares regression line is 9.16217 percentage points.
(f) $r^2 = 0.832$; approximately 83.2% of the variability in percentage transported can be explained by the linear relationship between percentage transported and number of salmon.

4.55 **(a)** $\hat{y} = 6.873 + 5.078x$, where $\hat{y}$ is the predicted depression score change and x is the BMI change.
(b) Using software, we find that $r^2 = 0.235$ and $s_e = 5.366$. Therefore, only 23.5% of the variation in depression score change can be explained by the linear relationship between these variables. In addition, the value of s_e is large relative to the y values. The values of r^2 and s_e indicate that the least squares regression line does not do a good job describing the relationship between depression score change and BMI change.
(c) The point $(0.5, -1)$ is the only obvious outlier. No, the observation with the largest residual does not correspond to the patient with the largest change in BMI.

ARE YOU READY TO MOVE ON?

Chapter 4 Review Exercises

4.57 Scatterplot 1: (i) Yes (ii) Yes (iii) Negative
Scatterplot 2: (i) Yes (ii) No (iii) –
Scatterplot 3: (i) Yes (ii) Yes (iii) Positive
Scatterplot 4: (i) Yes (ii) Yes (iii) Positive

4.59 **(a)** $r = -0.10$; there is a weak, negative linear relationship. Because the relationship is negative, larger arch heights tend to be paired with smaller, average hopping heights.

(b) The correlations coefficients support the conclusion since they are all fairly close to 0.

4.61 **(a)** $r = 0.944$; there is a strong positive linear relationship between sugar consumption and depression rate.
(b) No, because you can't conclude that a cause-and-effect relationship exists just based on a strong correlation.
(c) These countries were not a random sample of all countries, and it is unlikely that they are representative.

4.63 **(a)** negative, because it is likely that the work is more stressful and less enjoyable for nurses with high patient-to-nurse ratios.
(b) negative, because it is likely that the patients at hospitals where the patient-to-nurse ratio is high will not get as much individual attention and will be less satisfied with their care.
(c) negative, because it is likely that the quality of patient care suffers when patient-to-nurse ratio is high.

4.65 There is no evidence that the form of the relationship between x and y remains the same outside the range of the data.

4.67 **(a)** size, because the relationship between price and size ($r = 0.700$) is stronger than the relationship between price and land-to-building ratio ($r = -0.332$).
(b) $\hat{y} = 1.33 + 0.00525x$

4.69 **(a)** $\hat{y} = 2014.73 + 66.376x$, where $\hat{y}$ is the predicted number of heart transplants and x is the year using 1 to represent 2006, 2 to represent 2007, and so on. The number of transplants has increased over time, with a predicted increase of about 66 transplants per year.
(b) The residual plot shows a curved pattern, suggesting that the relationship between year and number of heart transplants is not linear.

4.71 **(a)** Removing Jupiter Networks would have the greatest impact on the equation of the least squares regression line.

(b) An outlier in a bivariate data set is a point that has a large residual. Outliers fall far away from the least squares regression line. An influential observation has an x value that is far away from the rest of the data.

CHAPTER 5
Probability

SECTION 5.1

Exercise Set 1

5.1 **(a)** In the long run, about 0.1% of eggs are double-yolk.
(b) We would expect to find 5 double-yolk eggs because 0.1% of 5000 is 5.

5.2 In the long run, 86% of the time this particular flight that flies between Phoenix and Atlanta will arrive on time.

5.3 **(a)** 0.83 **(b)** 0.83 **(c)** 0.17 **(d)** 0.48 **(e)** 0.52

Additional Exercises

5.7 Approximately 167; P(observing six when a fair die is rolled) $= \frac{1}{6}$

5.9 (a) 0.17 (b) 0.85 (c) 0.15 (d) 0.75

SECTION 5.2

Exercise Set 1

5.11 (a) 0.225 (b) 0.775 (c) 0.06 (d) 0.715 (e) 0.234
(f) The airline should not be particularly worried. The probability of selecting one person who was delayed overnight is approximately $\frac{75}{8000} = 0.0094$, and the probability of selecting additional people who were delayed overnight decreases from 0.0094.

5.12 (a) $S = \{AB, AC, AD, AE, BC, BD, BE, CD, CE, DE\}$
(b) Yes. (c) 0.3
(d) No, the probability does not change. (e) The probability increases to 0.6.

5.13 0.11

Additional Exercises

5.17 (a) 0.611
(b) 0.389

5.19 (a) 0.034
(b) 0.000062
(c) 0.035

5.21 (a) $S = \{AB, AC, AD, AE, AF, BC, BD, BE, BF, CD, CE, CF, DE, DF, EF\}$ (b) Yes. (c) 0.0667 (d) 0.4
(e) 0.533

SECTION 5.3

Exercise Set 1

5.22 (a) Not mutually exclusive, because there are seniors who are computer science majors.
(b) Not mutually exclusive, because there are female students who are computer science majors.
(c) Mutually exclusive, because a student cannot have a college residence that is more than 10 miles from campus and live in a college dormitory that is on campus.
(d) Mutually exclusive, because college football teams are male teams and females would not be on the team.

5.23 The events are dependent because the probability that a male favors affirmative action for women is not equal to the probability that a female favors affirmative action for women.

5.24 (a) 0.70 (b) 0.189 (c) 0.5 (d) 0.811 (e) 0.799
(f) 0.101

5.25 (a)

	E (stop at first light)	Not E (does not stop at first light)	Total
F (stop at second light)	150	150	300
Not F (does not stop at second light)	250	450	700
Total	400	600	1,000

(b) (i) 0.55 (ii) 0.45 (iii) 0.40 (iv) 0.25

5.26 (a)

	Cable TV	Not Cable TV	Total
Internet Service	250	170	420
Not Internet Service	550	30	580
Total	800	200	1,000

(b) (i) 0.25 (ii) 0.72

5.27 (a) (i) 0 (ii) 0.64
(b) No, events A and B are not mutually exclusive. For mutually exclusive events, $P(A \cup B) = P(A) + P(B)$. If A and B were mutually exclusive, then $P(A \cup B) = P(A) + P(B) = 0.26 + 0.34 = 0.60$, which is different from the value of $P(A \cup B)$ that is given.

5.28 0.000625

5.29 (a)

	Cash (M)	Credit Card (not M)	Total
Extended Warranty (E)	75	85	160
No Extended Warranty (not E)	395	445	840
Total	470	530	1,000

(b) 0.555. In the long run, approximately 55.5% of customers will either pay with cash or purchase an extended warranty, or both.

Additional Exercises

5.39 If the events F and I were mutually exclusive, we could add the probabilities to obtain $P(F \cup I)$. Since $P(F) + P(I) = 0.71 + 0.52 = 1.23$ is greater than 1 (and probabilities cannot be greater than 1), we know that F and I cannot be mutually exclusive.

5.41 (a) No, the events O and N are not independent. If they were independent, $P(O \cap N) = P(O) \cdot P(N) = (0.7)(0.07) = 0.049$, which is not equal to the value of $P(O \cap N)$ given in the exercise.

(b)

	N	Not N	Total
O	40	660	**700**
Not O	30	270	**300**
Total	**70**	**930**	**1,000**

(c) 0.730. In the long run, 73% of the time airline ticket purchasers will buy their ticket online or not show up for a flight, or both.

SECTION 5.4

Exercise Set 1

5.43 **(a)** The conditional probability is the 80%, because you are told that 80% *of those receiving such a citation* attended traffic school. The key phrase is "of those receiving such a citation," which is what indicates that the 80% is a conditional probability.
(b) $P(F) = 0.20$ and $P(E|F) = 0.80$

5.44 **(a)** **(i)** 0.62 **(ii)** 0.36 **(iii)** 0.40 **(iv)** 0.29 **(v)** 0.31
(b) **(i)** The probability that a randomly selected Honda Civic buyer is male. **(ii)** The probability that a randomly selected Honda Civic buyer purchased a hybrid. **(iii)** The probability that, of the males, a randomly selected buyer purchased a hybrid. **(iv)** The probability that, of the females, a randomly selected buyer purchased a hybrid. **(v)** The probability that, of the hybrid purchasers, a randomly selected buyer is female.
(c) These probabilities are not equal.

5.45 **(a)** $P(L) = 0.493$: The probability that the goalkeeper jumps to the left.

$P(C) = 0.063$: The probability that the goalkeeper stays in the center.

$P(R) = 0.444$: The probability that the goalkeeper jumps to the right.

$P(B|L) = 0.142$: The probability that, given that the goalkeeper jumps to the left, the goal was blocked.

$P(B|C) = 0.333$: The probability that, given that the goalkeeper stays in the center, the goal was blocked.

$P(B|R) = 0.126$: The probability that, given that the goalkeeper jumps to the right, the goal was blocked.
(b)

	Blocked (B)	Not Blocked (not B)	Total
Goalkeeper Jumped Left (*L*)	70	423	**493**
Goalkeeper Stayed Center (*C*)	21	42	**63**
Goalkeeper Jumped Right (*R*)	56	388	**444**
Total	**147**	**853**	**1,000**

(c) 0.147.
(d) It seems as if the best strategy is for the goalkeeper to stay in the center when defending a kick.

5.46 **(a)**

	Completely Satisfied	Not Completely Satisfied	Total
Public School	836	2,150	**2,986**
Private School	377	231	**608**
Total	**1,213**	**2,381**	**3,594**

(b) 0.3375
(c) 0.1692
(d) 0.3799
(e) 0.2326
(f) For E and F to be independent, $P(E \cap F) = P(E) \cdot P(F)$. The necessary probabilities are $P(E) = \frac{1213}{3594} = 0.3375$, $P(F) = \frac{608}{3594} = 0.1692$, and $P(E \cap F) = \frac{377}{3594} = 0.1049$. Finally, $P(E) \cdot P(F) = (0.3375)(0.1692) = 0.05711$, which is not equal to 0.1049. Therefore, the events E and F are not independent.

Additional Exercises

5.51 $P(A|B)$ would be larger than $P(B|A)$ because it is more likely that a professional basketball player is over 6 feet tall than for an individual who is over 6 feet tall to be a professional basketball player.

5.53 **(a)** 0.490 **(b)** 0.3125 **(c)** 0.240 **(d)** 0.100

5.55 **(a)** 0.46
(b) It more likely that he or she is in the first priority group.
(c) It is not particularly likely that a student in the third priority group would get more than nine units during the first attempt to register because, of those in the third priority group, only 36% of students get more than nine units.

SECTION 5.5

Exercise Set 1

5.57 **(a)** 0.55
(b) 0.45
(c) 0.40
(d) 0.25

5.58 **(a)** 0.25
(b) 0.72

5.59 **(a)** 0.90
(b) 0.77

5.60: **(a)** 0.60
(b) 0.56
(c) 0.33

Additional Exercises

5.67
(a) 0.49
(b) 0.3125
(c) 0.24
(d) 0.10

5.69 The reason that $P(T)$ is not the average of the two given conditional probabilities is because there are different numbers (or different proportions) of people in the two given age groups (19 to 36 and 37 or older).

SECTION 5.6

Exercise Set 1

5.70 There are 18,639 participants in the study, and of those, 12,392 reported their weight within 3 pounds of their actual weight, so P(weight reported within 3 pounds) $\approx$ $\frac{12,392}{18,639} = 0.665$. Therefore, approximately 66.5% of women reported their weight within 3 pounds of their actual weight. Define an "older woman" as age 65+ years, and a "younger woman" as age <45 years. The proportion of older women who under-report their weight by at least 4 pounds is P(older women under report weight) $\approx \frac{263 + 444}{3292} = 0.215$. The proportion of younger women who under-report their weight by at least 4 pounds is P(younger women under report weight) $\approx \frac{297 + 373}{2471} = 0.271$. The proportion of older women who over-report their weight by at least 4 pounds is P(older women over report weight) $\approx \frac{300}{3292} = 0.091$. The proportion of younger women who over-report their weight by at least 4 pounds is P(younger women over report weight) $\approx \frac{207}{2471} = 0.084$. These probabilities are consistent with the conclusion given in the paper.

Exercise Set 2

5.71 The following probabilities can be used to justify the report's conclusion that females are more likely to wear seatbelts than males in both urban and rural areas.

P(urban wear seat belt | male) $\approx \frac{871}{871 + 129} = 0.871$

P(urban wear seat belt | female) $\approx \frac{928}{928 + 72} = 0.928$

P(rural wear seat belt | male) $\approx \frac{770}{770 + 230} = 0.770$

P(rural wear seat belt | female) $\approx \frac{837}{837 + 163} = 0.837$

The differences in percentages of females and males who wear seatbelts are below, which support the report's conclusion that the difference in the proportion of females and the proportion of males who wear seatbelts is greater for rural areas.

Urban: $0.928 - 0.871 = 0.057$

Rural: $0.837 - 0.770 = 0.067$

SECTION 5.7

Exercise Set 1

5.72 (a) 0.143
(b) 0.639

(c) 0.130
(d) 0.484

5.73 (a) Answers will vary depending on outcome of simulation.
(b) This is not a fair way of distributing licenses because those companies/individuals who are requesting multiple licenses are given approximately the same chance to get two or three licenses as an individual has to get a single license. Perhaps companies who request multiple licenses should be required to submit one application per license.

Additional Exercises

5.77 (a) Results from the simulation will vary. The exact answer is 0.6504.
(b) Jacob's decrease in the probability of on-time completion resulted in the bigger change in the probability that the project is completed on time.

5.79 (a) Answers will vary. One possibility is to assign digits 0 through 7 to represent a win for Seed 1, and digits 8 and 9 a win for Seed 4.
(b) Answers will vary. One possibility is to assign digits 0 through 5 to represent a win for Seed 2, and digits 6 through 9 a win for Seed 3.

(c) Answers will vary. One set of possible digit assignments is provided here.

If Seed 1 won game 1 and Seed 2 won game 2, then assign digits 0 through 5 to represent a win for Seed 1, and digits 6 through 9 a win for Seed 2.

If Seed 1 won game 1 and Seed 3 won game 2, then assign digits 0 through 6 to represent a win for Seed 1, and digits 7 through 9 a win for Seed 3.

If Seed 4 won game 1 and Seed 2 won game 2, then assign digits 0 through 2 to represent a win for Seed 4, and digits 3 through 9 a win for Seed 2.

If Seed 4 won game 1 and Seed 3 won game 2, then assign digits 0 through 3 to represent a win for Seed 4, and digits 4 through 9 a win for Seed 3.

(d) Answers will vary. One possible answer is given here. The first three digits of Row 15 in the Random Numbers Table (Appendix A Table 1) are 0 7 1. Choose one digit at a time, with the first digit representing the winner of game 1, the second digit representing the winner of game 2, and the third digit representing the winner of game 3. The first digit selected, 0, indicates that the winner of game 1 was Seed 1. The second digit selected, 7, indicates that the winner of game 2 was Seed 3. Because Seed 1 and Seed 3 won, the third digit, 1, indicates that Seed 1 won the tournament.
(e) Answers will vary.
(f) Answers will vary.
(g) The estimated probabilities from parts (e) and (f) differ because they are based on different runs of the simulations, and the estimated probability in part (f) is based on more repetitions. The estimate in part (f) is most likely a better estimate because it is based on more repetitions of the simulation than the estimate in part (e).

ARE YOU READY TO MOVE ON?

Chapter 5 Review Exercises

5.81 In the long run, 18% of all calls for assistance will be to help someone who is locked out of his or her car.

5.83 **(a)** The 10 possible outcomes are: *BC, BM, BP, BS, CM, CP, CS, MP, MS,* and *PS.*
(b) The probability of each outcome is 1/10 = 0.1.
(c) There are 4 of the 10 possible outcomes that have the statistics department representative on the committee. Therefore, the probability that one of the committee members is the statistics department representative is 4/10 = 0.4.
(d) There are 3 of the 10 possible outcomes that have both committee members from laboratory science departments. Therefore, the probability that both committee members come from laboratory science departments is 3/10 = 0.3.

5.85 **(a)** Answers will vary. For example, let event *A* be the car purchaser is a male, and event *B* be the car purchaser is a female.
(b) Answers will vary. For example, let event *C* be the car purchaser is female, and event *D* be the car purchaser is under age 65.

5.87
(a) 0.184
(b) 0.374
(c) 0.035
(d) 0.469

5.89 **(a)**

	Internet	No Internet	Total
Phone	230	90	**320**
No Phone	190	490	**680**
Total	**420**	**580**	**1,000**

(b) **(i)** $P(\text{both Internet and phone service}) = \frac{230}{1000} = 0.230$
(ii) $P(\text{exactly one of the two services}) = \frac{90 + 190}{1000} = 0.280$

5.91 $P(\text{called for jury duty in both of next two years}) = (0.15)(0.15) = 0.0225$
$P(\text{called for jury duty in all of next three years}) = (0.15)(0.15)(0.15) = 0.003375$

5.93 **(a)** 0.875
(b) 0.846
(c) 0.096
(d) 0.384
(e) No, the probabilities in parts (c) and (d) are not equal. Part (c) is the proportion of female drivers who do not use a seat belt, and part (d) is the proportion of drivers who do not use a seat belt who are female. There is no reason to believe that these proportions should be equal.

5.95 **(a)** **(i)** $P(TD|D) = 0.99$
(ii) $P(TD|C) = 0.01$
(iii) $P(C) = 0.99$
(iv) $P(D) = 0.01$

(b)

	TD	TC	Total
D	10	0	10
C	10	980	990
Total	20	980	1,000

(c) $P(TD) = \frac{20}{1000} = 0.02$
(d) $P(C|TD) = \frac{10}{20} = 0.5$; yes, this value is consistent with the argument given in the quote, namely, half of the dirty tests are false.

5.97 The following conditional probabilities can be used to assess whether one can be equally confident in a positive test result and a negative test result. $P(\text{not pregnant} \mid \text{positve test result}) = \frac{0}{202 + 0} = 0.000$, and $P(\text{pregnant} \mid \text{negative test result}) = \frac{9}{416 + 9} = 0.021$. These conditional probabilities are not equal, so one cannot be equally confident in a positive test result and a negative test result.

5.99 **(a)** 0.5625
(b) 0.340
(c) 0.105

CHAPTER 6

Random Variables and Probability Distributions

NOTE: Your numerical answers may sometimes differ slightly from those in the answer section, depending on whether a table, a graphing calculator, or statistical software is used to compute probabilities.

SECTION 6.1

Exercise Set 1

6.1 **(a)** discrete **(b)** continuous **(c)** discrete **(d)** discrete **(e)** continuous

6.2 The possible values for *x* are *x* = 1, 2, 3, … (the positive integers).

Answers will vary. One possible answer is

Outcomes	x
S	1
LS	2
RLS	3
RRS	3
LRLRS	5

6.3 **(a)** 3, 4, 5, 6, 7
(b) −3, −2, −1, 1, 2, 3
(c) 0, 1, 2
(d) 0, 1

Additional Exercises

6.7 (a) discrete (b) continuous (c) continuous (d) discrete

6.9 (a) 2, 3, 4, 5, 6, 7, 8, 9, 10, 11, 12
(b) $-5, -4, -3, -2, -1, 0, 1, 2, 3, 4, 5$
(c) 1, 2, 3, 4, 5, 6

SECTION 6.2

Exercise Set 1

6.10 (a) $p(4) = 0.01$
(b) It is the probability of randomly selecting a carton of one dozen eggs and finding exactly 1 broken egg.

(c) $P(y \leq 2) = 0.95$; the probability that a randomly selected carton of eggs contains 0, 1, or 2 broken eggs is 0.95.

(d) $P(y < 2) = 0.85$; it is smaller because it does not include the possibility of 2 broken eggs.
(e) Exactly 10 unbroken eggs is equivalent to exactly 2 broken eggs; $p(2) = 0.10$.
(f) At least 10 unbroken eggs is equivalent to 10, 11, or 12 unbroken eggs, or 2, 1, or 0 broken eggs; $P(y \leq 2) = 0.95$.

6.11 (a) For 1000 graduates, you would expect to see approximately 450 graduates who contributed nothing, 300 graduates to contribute $10, 200 graduates to contribute $25, and 50 graduates to contribute $50.
(b) $0 (c) 0.25 (d) 0.55

6.12 (a) (1, 2), (1, 3), (1, 4), (2, 3), (2, 4), (3, 4)
(b) Because the bottles are randomly selected, the outcomes are equally likely and each has probability 1/6.

(c)

x	0	1	2
p(x)	$\frac{1}{6}$	$\frac{4}{6}$	$\frac{1}{6}$

6.13 (a)

x	0	1	2	3	4
p(x)	0.4096	0.4096	0.1536	0.0256	0.0016

(b) The most likely outcomes are 0 and 1.
(c) $P(x \geq 2) = 0.1808$

Additional Exercises

6.19 (a) $k = \dfrac{1}{15}$ (b) $P(y \leq 3) = 0.4$

(c) $P(2 \leq y \leq 4) = 0.6$

SECTION 6.3

Exercise Set 1

6.21 (a)

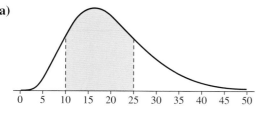

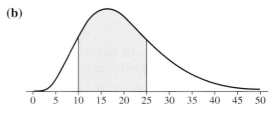

(c)

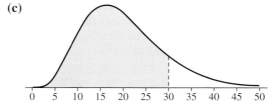

(d)

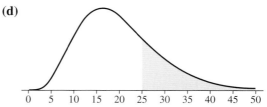

6.22 (a) $P(4 \leq x \leq 7)$
(b) The probability that a randomly selected individual waits between 4 and 7 minutes for service at a bank is 0.26.
6.23 (a) $h = 2$ (b) $P(x > 0.5) = 0.25$
(c) $P(x \leq 0.25) = 0.4375$
6.24 (a) $\frac{1}{2}(0.40)(5) = 1$
(b) $P(x < 0.20) = 0.5$; $P(x < 0.1) = 0.125$;
$P(x > 0.3) = 0.125$
(c) $P(0.10 < x < 0.20) = 0.375$

Additional Exercises

6.29 The probability $P(x < 1)$ is the smallest. $P(x > 3)$ and $P(2 < x < 3)$ are equal, and larger than the other two probabilities.
6.31 $P(2 < x < 3) = P(2 \leq x \leq 3) < P(x < 2) < P(x > 7)$. The smallest two probabilities are both equal to 1/10, the third probability is equal to 2/10, and the fourth probability is equal to 3/10.

SECTION 6.4

Exercise Set 1

6.32 (a) $\mu_y = 0.56$; this is the mean value of the number of broken eggs in the population of egg cartons.
(b) $P(y < 0.56) = P(y = 0) = 0.65$. This is not particularly surprising because, in the long run, 65% of egg cartons contain no broken eggs.
(c) This computation of the mean is incorrect because it assumes that the numbers of broken eggs (0, 1, 2, 3, or 4) are all equally likely.

6.33 (a) $\mu_x = 16.38$, $\sigma_x = 1.9984$
(b) The mean, $\mu_x = 16.38$ cubic feet represents the long-run average storage space of freezers sold by this particular

appliance dealer. The standard deviation, $\sigma_x = 1.9984$ cubic feet, represents a typical amount by which the storage space in freezers purchased deviates from the mean.

6.34 Answers may vary. Two possible probability distributions are shown below.

Probability Distribution 1 ($\mu_x = 3$ and $\sigma_x = 1.265$):

x	1	2	3	4	5
p(x)	0.15	0.20	0.30	0.20	0.15

Probability Distribution 2 ($\mu_x = 3$ and $\sigma_x = 1.643$):

x	1	2	3	4	5
p(x)	0.30	0.15	0.10	0.15	0.30

Additional Exercises

6.39 **(a)** $\mu_x = 2.3$ represents the long-run average number of lots ordered per customer.
(b) $\sigma_x^2 = 0.81$, $\sigma_x = 0.9$ lots. A typical deviation from the mean is about 0.9 lots.

SECTION 6.5

Exercise Set 1

6.41 **(a)** 0.9599
(b) 0.2483
(c) 0.1151
(d) 0.9976
(e) 0.6887
(f) 0.6826
(g) approximately 1

6.42 **(a)** 0.9909
(b) 0.9909
(c) 0.1093
(d) 0.1267
(e) 0.0706
(f) 0.0228
(g) 0.9996
(h) approximately 1

6.43 **(a)** 0.5
(b) 0.9772
(c) 0.9772
(d) 0.8185
(e) 0.9938
(f) approximately 1

6.44 **(a)** At most 60 wpm: 0.5; Less than 60 wpm: 0.5
(b) 0.8185
(c) 0.0013; it would be surprising, because the probability of finding such a typist is very small.
(d) The probability that a randomly selected typist has a typing speed that exceeds 75 wpm is 0.1587. The probability that both typists have typing speeds that exceed 75 wpm is $(0.1587)(0.1587) = 0.0252$.
(e) typing speeds of 47.376 wpm or less.

6.45 0.3173

6.46 The proportion of corks produced by this machine that are defective is approximately 0. The second machine produces fewer defective corks.

Additional Exercises

6.53 It is not reasonable to think that time playing video or computer games is approximately normal because time cannot be negative, and 0 minutes is 1.054 standard deviations below the mean $\left(z = \frac{0 - 123.4}{117.1} = -1.054\right)$, which indicates that approximately 14.6% of playing times would be negative if the playing times actually followed a normal distribution.

6.55 Since these values are times, they must all be positive. In the normal distribution with mean 9.9 and standard deviation 6.2, approximately 5.5% of processing times would be less than or equal to zero.

6.57 **(a)** 0.1359
(b) 0.228
(c) 0.5955

6.59 **(a)** 0.9332
(b) 72.82 minutes

6.61 $P(x < 4.9) = 0.0228$; $P(x \geq 5.2) \approx 0$

6.63 To get an A, your score must be greater than 85.252, so you received an A.

6.65 The bulbs should be replaced after 657.9 hours.

SECTION 6.6

Exercise Set 1

6.67 **(a)** The plot does not look linear. This supports the author's statement.

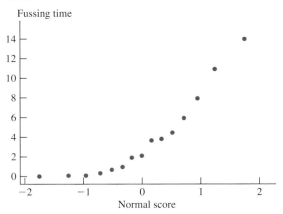

(b) $r = 0.921$; *critical r* (from Table 6.2) is 0.911; it is reasonable to think that the population distribution is normal.

6.68 **(a)**

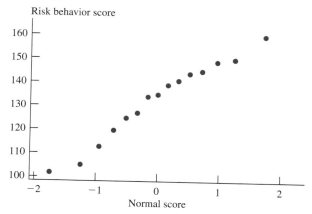

Risk behavior score

(b)

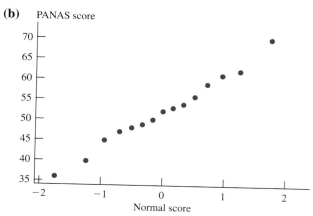

PANAS score

(c) Because both normal probability plots are approximately linear, it seems reasonable that both risk behavior scores and PANAS scores are approximately normally distributed.

Additional Exercises

6.71 **(a)** Yes, the normal probability plot appears linear.
(b) $r = 0.994$; *critical r* for $n = 6$ lies between 0.832 (*critical r* for $n = 5$) and 0.880 (*critical r* for $n = 10$). The computed correlation coefficient is larger than the *critical r* for $n = 6$, so it is reasonable to think that the fuel efficiency distribution is approximately normal.

6.73

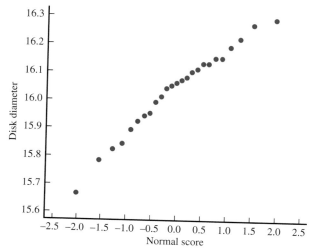

Yes, the normal probability plot appears to be linear, so it is reasonable to think that disk diameter is approximately normally distributed.

SECTION 6.7

Exercise Set 1

6.74 **(a)** 0, 1, 2, 3, 4, and 5.
(b)

x	0	1	2	3	4	5
p(x)	0.2373	0.3955	0.2637	0.0879	0.0146	0.0010

6.75 **(a)** $P(x = 2) = 0.0486$; the probability that exactly two of the four randomly selected households have cable TV is 0.0486.
(b) $P(x = 4) = 0.6561$
(c) $P(x \leq 3) = 0.3439$

6.76 **(a)** $P(x = 2) = 0.2637$
(b) $P(x \leq 1) = 0.6328$
(c) $P(2 \leq x) = 0.3672$
(d) $P(x \neq 2) = 0.7363$

6.77 **(a)** 0.7359
(b) 0.3918
(c) 0.0691

6.78 **(a)** There is not a fixed number of trials, which is required for the binomial distribution. This setting is geometric.
(b) **(i)** $p(4) = 0.0623$
　　(ii) $P(x \leq 4) = 0.2836$
　　(iii) $P(x > 4) = 0.7164$
　　(iv) $P(x \geq 4) = 0.7787$

(c) The differences between the four probabilities are shown in bold font.

　　(i) The probability that it takes **exactly four songs** until the first song by the particular artist is played is 0.0623.
　　(ii) The probability that it takes **at most four songs** until the first song by the particular artist is played is 0.2836.
　　(iii) The probability that it takes **more than four songs** until the first song by the particular artist is played is 0.7164.
　　(iv) The probability that it takes **at least four songs** until the first song by the particular artist is played is 0.7787.

6.79 **(a)** The probability distribution of x is geometric, with success probability $p = 0.44$. The distribution is geometric because we are waiting until we find someone that washes sheets at least once a week.
(b) 0.138
(c) 0.824
(d) 0.176

Additional Exercises

6.87 The binomial probability distribution with $n = 5$ and $p = 0.5$:

x	0	1	2	3	4	5
p(x)	0.03125	0.15625	0.31250	0.31250	0.15625	0.03125

6.89 This scenario is sampling without replacement. Since more than 5% of the population is being sampled (the percentage of the population being sampled is 2,000/10,000 = 20%), the binomial distribution will not give a good approximation of the probability distribution of the number of invalid signatures.

6.91 **(a)** $P(x \leq 7) + P(x \geq 18) = 0.0216 + 0.0216 = 0.0432$.
(b) 0.8542, 0.8542
(c) 0.1548; 0.1548; the probabilities are large compared to the probabilities in Part (b) because p (the probability of a head) in Part (c) is closer to 0.5 than in Part (b).
(d) Changing the rule for fair to $7 \leq x \leq 18$ increases the likelihood that the coin will be judged fair and decreases the probability that the coin is judged to not be fair. Although the new rule is more likely to judge a fair coin fair, it is also more likely to judge a biased coin as fair.

SECTION 6.8

Exercise Set 1

6.93 **(a)** 0.026
(b) 0.7580
(c) 0.7366
(d) 0.9109

6.94 **(a)** 0.8849
(b) 0.8739
(c) 0.0110
(d) 0.5434

6.95 **(a)** 0.3248
(b) 0.2763
(c) 0.0012

Additional Exercises

6.99 **(a)** Both $np = (60)(0.7) = 42$ and $n(1 - p) = 60(1 - 0.7) = 18$ are at least 10.
(b) **(i)** 0.1121
    (ii) 0.4440
    (iii) 0.5560
(c) The probability in (i) represents the probability of exactly 42 correct; the probability in (ii) is the probability of less than 42 correct, the probability in (iii) represents the probability of 42 or fewer correct.
(d) The normal approximation is not appropriate here because $n(1 - p) = 60(1 - 0.96) = 2.4$, which is less than 10.
(e) The small probability in Part (d) indicates that it is extremely unlikely for someone who is not faking the test to correctly answer 42 or fewer questions, compared with the probability that a person who is faking the test correctly answers 42 or fewer questions (0.556).

ARE YOU READY TO MOVE ON?

Chapter 6 Review Exercises

6.101 The depth, x, can take on any value between 0 and 100 ($0 \leq x \leq 100$), inclusive. x is continuous.

6.103 **(a)** The plane can accommodate 100 passengers, so if 100 or fewer passengers show up for the flight, everyone can be accommodated; $P(x \leq 100) = 0.82$.
(b) $P(x > 100) = 0.18$.
(c) The first person on the standby list: $P(x \leq 99) = 0.65$. The third person on the standby list: $P(x \leq 97) = 0.27$.

6.105 **(a)** $P(x < 10) = 0.5$; $P(x > 15) = 0.25$
(b) $P(7 < x < 12) = 0.25$
(c) $c = 18$ minutes

6.107 **(a)** Supplier 1 **(b)** Supplier 2
(c) I would recommend Supplier 1 because the bulbs will last, on average, longer than those from Supplier 2. Additionally, the bulbs from Supplier 1 have less variability in their lifetimes, so there is more consistency in the bulb lifetimes when compared with Supplier 2.
(d) Approximately 1000 hours.
(e) Approximately 100 hours.

6.109 **(a)** 0.284
(b) 0.091
(c) 0.435
(d) 29.93 mm

6.111 **(a)** 0.159
(b) 51.4 minutes
(c) 41.6 minutes

6.113 No, it is not reasonable to think that the distribution of 2015 Honda Accord prices in this area is approximately normal because of the curvature in the pattern of the points apparent in the normal probability plot.

CHAPTER 7
An Overview of Statistical Inference— Learning from Data

SECTION 7.1

Exercise Set 1

7.1 The inferences made involve estimation.

7.2 **(a)** American teenagers between the ages of 12 and 17.
(b) The percentage of teens who own a cell phone, the percentage of teens who use a cell phone to send and receive text messages, and the percentage of teens ages 16–17 who have used a cell phone to text while driving.
(c) No, the actual percentage of teens owning a cell phone is probably not exactly 75%. The value of a sample statistic won't necessarily be equal to the population value.
(d) I would expect the estimate of the percentage of teens who own a cell phone to be more accurate. The sample contained teens aged 12–17; however, only those teens aged 16 and 17 were asked about texting while driving. The number of teens aged 16 and 17 is a subset of the overall sample, and so the estimate is based on a smaller sample.

7.3 The inference made is one that involves hypothesis testing.

7.4 **(a)** People driving along stretches of highway that have digital billboards.

(b) The time required to respond to road signs is greater when digital billboards are present.
(c) Answers will vary. One possible answer is: In addition to the information provided, I would like to know if the subjects were randomly selected or not, and whether or not the subjects were randomly assigned to any experimental groups. I would also like to know if there was a control group in which the subjects did not see digital billboards or a group in which subjects did not see any change in the display on the digital billboard.
(d) lower

7.5 **(a)** Answers will vary. One possible answer is "What proportion of students approve of a recent decision made by the university to increase athletic fees in order to upgrade facilities?"
(b) Answers will vary. One possible answer is "Do more than half of students approve of the recent decision made by the university to increase athletic fees in order to up-grade facilities?"

Additional Exercises

7.11 The inference made is one that involves hypothesis testing.

SECTION 7.2
Exercise Set 1
7.13 Estimate a population mean. The data are numerical, rather than categorical.

7.14 The type of data determines what graphical, numerical, and inferential methods are appropriate.

7.15 The other two questions are (1) **Q: Question Type** (estimation problem or hypothesis testing problem), and (2) **S: Study Type** (sample data or experiment data).

7.16 Q: Estimation
S: Sample data
T: One variable, categorical data
N: One sample

7.17 Q: Hypothesis testing
S: Sample data
T: One variable, numerical data
N: Two samples

7.18 Q: Estimation
S: Sample data
T: Two variables, both variables are numerical
N: One sample

Additional Exercises

7.25 Q: Estimation
S: Sample data
T: One variable, categorical
N: One sample

7.27 Q: Hypothesis testing
S: Experiment
T: One variable, categorical
N: Two treatments

7.29 Q: Hypothesis testing
S: Experiment data
T: One variable, numerical
N: Two treatments

ARE YOU READY TO MOVE ON?
Chapter 7 Review Exercises

7.31 Estimation, because the researchers estimated that "8 in 10" consider appearance when shopping for fresh produce. The researchers did not indicate that they had a claim that they wanted to test.

7.33 **(a)** Answers will vary. One possible answer is: "What proportion of people who purchased season tickets for home games of the New York Yankees purchased alcoholic beverages during the game?"
(b) Answers will vary. One possible answer is: "Do fewer than 50% of people who purchased season tickets for home games of the New York Yankees drive to the game?"

7.35 **(a)** The proportion of people who take a garlic supplement who get a cold is lower than the proportion of those who do not take a garlic supplement and who get a cold.
(b) Yes, it is possible that the conclusion is incorrect. The observed difference in treatment effects may be due to chance variability in the response variable and the random assignment to treatments, and not due to the treatment.
(c) greater

7.37 The type of data collected determines not only the type of numerical and graphical methods that can be applied but also the particular inferential method or methods that can be used. Numerical data require different inferential methods than categorical data, and univariate data require different inferential methods than bivariate data.

7.39 Q: Hypothesis testing
S: Experiment data
T: One variable, categorical
N: Two treatments

7.41 **(a)** Estimate, Method, Check, Calculate, Communicate results
(b) The difference is that the "estimate" step is replaced with "hypotheses," where you determine the hypotheses you wish to test, rather than defining the population characteristic or treatment effect you wish to estimate.

CHAPTER 8
Sampling Variability and Sampling Distributions

SECTION 8.1
Exercise Set 1
8.1 No, because the value of $\hat{p}$ will vary from sample to sample.

8.2 **(a)** The histogram on the left; it has values that tend to deviate more from the center, and it is more spread out than the histogram on the right.

(b) $n = 75$, because the histogram on the right seems to have less sample-to-sample variability, is centered at 0.55, and has tall bars close to 0.55.

8.3 **(a)** a population proportion.
(b) $p = 0.22$.

8.4 **(a)** a sample proportion.
(b) $\hat{p} = 0.38$.

8.5 p is the proportion of successes in the entire population, and $\hat{p}$ is the proportion of successes in the sample.

Additional Exercises

8.11 Sample statistics are computed from a sample. Since one sample from the population is likely to differ from other possible samples taken from the same population, the sample statistics computed from different samples are likely to be different.

8.13 **(a)** a sample proportion.
(b) $\hat{p} = 0.45$.

SECTION 8.2

Exercise Set 1

8.15 **(a)** $\mu_{\hat{p}} = 0.65$; $\sigma_{\hat{p}} = 0.151$
(b) $\mu_{\hat{p}} = 0.65$; $\sigma_{\hat{p}} = 0.107$
(c) $\mu_{\hat{p}} = 0.65$; $\sigma_{\hat{p}} = 0.087$
(d) $\mu_{\hat{p}} = 0.65$; $\sigma_{\hat{p}} = 0.067$
(e) $\mu_{\hat{p}} = 0.65$; $\sigma_{\hat{p}} = 0.048$
(f) $\mu_{\hat{p}} = 0.65$; $\sigma_{\hat{p}} = 0.034$

8.16 For $p = 0.65$: 30, 50, 100, and 200. For $p = 0.2$: 50, 100, and 200.

8.17 **(a)** mean: $\mu_{\hat{p}} = 0.03$; standard deviation: $\sigma_{\hat{p}} = \sqrt{\frac{0.03(1 - 0.03)}{100}} = 0.017$
(b) The sampling distribution is not approximately normal because $np = (100)(0.03) = 3$ is less than the required 10.
(c) The change in sample size does not change the mean of the sampling distribution. However, the standard deviation will decrease to 0.009. The mean does not change when the sample size is increased because the sampling distribution is always centered at the population value (in this case, $\mu_{\hat{p}} = 0.03$) regardless of the sample size. The standard deviation of the sampling distribution will decrease as the sample size increases because the sample size (n) is in the denominator of the formula for standard deviation. As sample size increases, standard deviation of the sampling distribution of $\hat{p}$ decreases.
(d) The sampling distribution of $\hat{p}$ is approximately normal because $np = (400)(0.03) = 12$ and $n(1 - p) = (400)(1 - 0.03) = 388$. Both of these values are at least 10.

8.18 **(a)** mean: $\mu_{\hat{p}} = 0.37$; standard deviation: $\sigma_{\hat{p}} = \sqrt{\frac{0.37(1 - 0.37)}{100}} = 0.048$
(b) The sampling distribution is approximately normal because $np = (100)(0.37) = 37$ and $n(1 - p) = (100)(1 - 0.37) = 63$ are both at least 10.
(c) The change in sample size does not change the mean of the sampling distribution. However, the standard deviation

will decrease to 0.024. The mean does not change when the sample size is increased because the sampling distribution is always centered at the population value (in this case, $\mu_{\hat{p}} = 0.37$) regardless of the sample size. The standard deviation of the sampling distribution will decrease as the sample size increases because the sample size (n) is in the denominator of the formula for standard deviation.
(d) Yes, the sampling distribution of $\hat{p}$ is approximately normal because $np = (400)(0.37) = 148$ and $n(1 - p) = (400)(1 - 0.37) = 252$ are both at least 10.

Additional Exercises

8.23 $n = 100$ and $p = 0.5$.

8.25 For $p = 0.2$: 50 and 100. For $p = 0.8$: 50 and 100. For $p = 0.6$: 25, 50, and 100.

8.27 For samples of size $n = 40$: $p = 0.45$ and $p = 0.70$. For samples of size $n = 75$: $p = 0.20$, $p = 0.45$, and $p = 0.70$.

SECTION 8.3

Exercise Set 1

8.29 **(a)** $s_{\hat{p}} = \sqrt{\frac{0.48(1 - 0.48)}{500}} = 0.0223$
(b) More sample-to-sample variability in the sample proportions because $\sigma_{\hat{p}}$ is now greater than $\sigma_{\hat{p}}$ when $n = 500$.
(c) The sample size is smaller because, in order for $\sigma_{\hat{p}}$ to be greater, the denominator of $\sigma_{\hat{p}} = \sqrt{\frac{p(1 - p)}{n}}$ must be smaller than when $n = 500$.

8.30

What You Know	How You Know It
The sampling distribution of $\hat{p}$ is centered at the actual (but unknown) value of the population proportion.	Rule 1 states that $\mu_{\hat{p}} = p$. This is true for random samples, and the description of the study says that the sample was selected at random.
An estimate of the standard deviation of $\hat{p}$, which describes how much the $\hat{p}$ values spread out around the population proportion p is 0.015.	Rule 2 states that $\sigma_{\hat{p}} = \sqrt{\frac{p(1-p)}{n}}$. In this exercise, $n = 1000$. The value of p is not known. However, $\hat{p}$ provides an estimate of p that can be used to estimate the standard deviation of the sampling distribution. Specifically, $\hat{p} = 0.35$, so $\sigma_{\hat{p}} = \sqrt{\frac{0.35(1 - 0.35)}{1000}} = 0.015$. This standard deviation provides information about how tightly the $\hat{p}$ values from different random samples will cluster around p.
The sampling distribution of $\hat{p}$ is approximately normal.	Rule 3 states that the sampling distribution of $\hat{p}$ is approximately normal if n is large and p is not too close to 0 or 1. Here the sample size is 1000. The sample includes 350 successes and 650 failures, which are both much greater than 10. So, we conclude that the sampling distribution of $\hat{p}$ is approximately normal.

8.31 **(a)** $\hat{p} = \frac{27}{46} = 0.587$
(b) No, it is not reasonable to think that this estimate is within 0.05 of the actual value of the population proportion. From Rule 1, we know that the sampling distribution is centered at p. From Rule 2, the standard deviation of the sampling distribution of $\hat{p}$ is $\sigma_{\hat{p}} = \sqrt{\frac{0.587(1 - 0.587)}{46}} = 0.073$. By Rule 3, the sampling distribution of $\hat{p}$ is approximately normally distributed because the sample includes 27 successes and 19 failures, which are both greater than 10. For any variable described by a normal distribution, about 95% of the values are within two standard deviations of the center. Since the sampling distribution of $\hat{p}$ is approximately normal and is centered at the actual population proportion p, we now know that about 95% of all possible random samples of size $n = 46$ will produce a sample proportion that is within $2(0.073) = 0.146$ of the actual value of the population proportion. This tells us that the sample estimate $\hat{p} = 0.587$ is likely to be within 0.146 of the actual proportion of all young adults with pierced tongues that have receding gums. This margin of error of 0.146 is over twice the value of 0.05, so it is not reasonable to think that this estimate is within 0.05 of the actual value of the population proportion.

8.32 From Rule 1, we know that the sampling distribution of $\hat{p}$ is centered at the population value $p = 0.46$.
Rule 2 tells us that $\sigma_{\hat{p}} = \sqrt{\frac{0.46(1 - 0.46)}{1500}} = 0.013$.
Finally, Rule 3 allows us to determine that the shape of the sampling distribution of $\hat{p}$ is approximately normal because $np = 1500(0.46) = 690 \geq 10$ and $n(1 - p) = 1500(1 - 0.46) = 810 \geq 10$. We can use the results of Rules 1, 2, and 3 to find the probability of observing a sample proportion of 0.62 or larger just by chance (due to sampling variability). This probability is $P(\hat{p} > 0.62) = P\left(z > \frac{0.62 - 0.46}{0.013}\right) = P(z > 12.31) \approx 0$. Therefore, it seems likely that the proportion of California four-year degree graduates who attended a two-year college in the previous 10 years is different from the national figure.

Additional Exercises

8.37 No, it is not likely that this estimate is within 0.05 of the actual value of the population proportion. From Rule 1, we know that the sampling distribution is centered at p. From Rule 2, the standard deviation of the sampling distribution of $\hat{p}$ is $\sigma_{\hat{p}} = 0.049$. By Rule 3, the sampling distribution of $\hat{p}$ is approximately normal because $np = 100(0.38) = 38 \geq 10$ and $n(1 - p) = 100(1 - 0.38) = 62 \geq 10$. Since the sampling distribution of $\hat{p}$ is approximately normal and is centered at the actual population proportion p, we now know that about 95% of all possible random samples of size $n = 100$ will produce a sample proportion that is within $2(0.049) = 0.098$ of the actual value of the population proportion. This margin of error is nearly twice the value of 0.05, so it is unlikely that this estimate is within 0.05 of the population proportion.

8.39 By Rule 1, we know that the sampling distribution of $\hat{p}$ is centered at the unknown value of p, the true proportion of social network users who believe that it is not OK to "friend" your boss. Rule 2 says that the standard deviation of the sampling distribution of $\hat{p}$ is $\sigma_{\hat{p}} = \sqrt{\frac{p(1 - p)}{n}}$. The value of p is not known. However, $\hat{p}$ provides an estimate of p that can be used to estimate the standard deviation of the sampling distribution. In this case, $\hat{p} = 0.56$, so the estimate of the standard deviation of the sampling distribution is $\sigma_{\hat{p}} = 0.014$. Finally, Rule 3 says that the sampling distribution of $\hat{p}$ is approximately normal if n is large and p is not too close to 0 or 1. Here the sample size is 1200. The sample includes 672 successes (56% of 1,200) and 528 failures (44% of 1,200), which are both much greater than 10. Therefore, we can conclude that the sampling distribution of $\hat{p}$ is approximately normal. The probability of observing a sample proportion at least as large as what we actually observed ($\hat{p} = 0.56$) if the true value were 0.5 is $P(\hat{p} \geq 0.56) = P\left(z \geq \frac{0.56 - 0.5}{0.014}\right) = P(z \geq 4.29) \approx 0.000$. It is highly unlikely that a sample proportion as large as 0.56 would be observed if the true value were 0.5 (or less). Therefore, it seems plausible that the proportion of social network users who believe that it is not OK to "friend" your boss is greater than 0.5.

ARE YOU READY TO MOVE ON?
Chapter 8 Review Exercises

8.41 No, because the value of $\hat{p}$ varies from sample to sample.

8.43 Different samples will likely yield different values of $\hat{p}$, which is the concept of sampling variability (or sample-to-sample variability). However, there is only one true value for the population proportion.

8.45 **(a)** Population proportion.
(b) $p = 0.21$.

8.47 For both $p = 0.70$ and $p = 0.30$: $n = 50$, $n = 100$, and $n = 200$.

8.49 **(a)** Mean: $\mu_{\hat{p}} = p = 0.48$; Standard Deviation: $\sigma_{\hat{p}} = \sqrt{\frac{p(1 - p)}{n}} = \sqrt{\frac{0.48(1 - 0.48)}{200}} = 0.035$
(b) $np = 200(0.48) = 96 \geq 10$ and $n(1 - p) = 200(1 - 0.48) = 104 \geq 10$; Since both np and $n(1 - p)$ are both at least 10, we know that the sampling distribution of $\hat{p}$ is approximately normal.
(c) The change in sample size does not affect the mean of the sampling distribution of $\hat{p}$ because $\mu_{\hat{p}} = p$, regardless of sample size. However, the change in sample size does affect the standard deviation of the sampling distribution of $\hat{p}$ because $\sigma_{\hat{p}}$ depends on sample size. The new standard deviation is $\sigma_{\hat{p}} = \sqrt{\frac{0.48(1 - 0.48)}{50}} = 0.071$.
(d) $np = 50(0.48) = 24 \geq 10$ and $n(1 - p) = 50(1 - 0.48) = 26 \geq 10$; Since both np and $n(1 - p)$

are both at least 10, we know that the sampling distribution of $\hat{p}$ is approximately normal.

8.51

What You Know	How You Know It
The sampling distribution of $\hat{p}$ is centered at the actual (but unknown) value of the population proportion.	Rule 1 states that $\mu_{\hat{p}} = p$. This is true for random samples, and the description of the study says that the sample was selected at random.
An estimate of the standard deviation of $\hat{p}$, which describes how much the $\hat{p}$ values spread out around the population proportion p is 0.016.	Rule 2 states that $\sigma_{\hat{p}} = \sqrt{\frac{p(1-p)}{n}}$. In this exercise, $n = 1000$. The value of p is not known. However, $\hat{p}$ provides an estimate of p that can be used to estimate the standard deviation of the sampling distribution. Specifically, $\hat{p} = 0.46$, so $\sigma_{\hat{p}} = \sqrt{\frac{0.46(1-0.46)}{1000}} = 0.016$. This standard deviation provides information about how tightly the $\hat{p}$ values from different random samples will cluster around p.
The sampling distribution of $\hat{p}$ is approximately normal.	Rule 3 states that the sampling distribution of $\hat{p}$ is approximately normal if n is large and p is not too close to 0 or 1. Here the sample size is 1000. The sample includes 460 successes and 540 failures, which are both much greater than 10. So, we conclude that the sampling distribution of $\hat{p}$ is approximately normal.

8.53 By Rule 1, we know that the sampling distribution of $\hat{p}$ is centered at the unknown value of p, the true proportion of adult Americans who own smartphones. Rule 2 says that the standard deviation of the sampling distribution of $\hat{p}$ is $\sigma_{\hat{p}} = \sqrt{\frac{p(1-p)}{n}}$. The value of p is not known. However, $\hat{p}$ provides an estimate of p that can be used to estimate the standard deviation of the sampling distribution. In this case, $\hat{p} = \frac{1297}{1907} = 0.68$, so the estimate of the standard deviation of the sampling distribution is $\sigma_{\hat{p}} = \sqrt{\frac{0.68(1-0.68)}{1907}} = 0.011$. Finally, Rule 3 says that the sampling distribution of $\hat{p}$ is approximately normal if n is large and p is not too close to 0 or 1. Here the sample size is 1907. The sample includes 1297 successes and 610 failures, which are both greater than 10. Therefore, we can conclude that the sampling distribution of $\hat{p}$ is approximately normal. The probability of observing a sample proportion at least as large as what we observed ($\hat{p} = 0.68$) if the true value was 0.6 is equal to $P(\hat{p} \geq 0.68) = P\left(z \geq \frac{0.68 - 0.6}{0.011}\right) = P(z \geq 7.27) \approx 0.000$. It is highly unlikely that a sample proportion as large as 0.68 would be observed if the true value was 0.6. Therefore, it seems plausible that the proportion of adult Americans who own smartphones is greater than 0.6.

CHAPTER 9

Estimating a Population Proportion

NOTE: Answers may vary slightly if you are using statistical software, a graphing calculator or depending on how the values of sample statistics are rounded when performing hand calculations. Don't worry if you don't match these numerical answers exactly (but your answers should be similar).

SECTION 9.1

Exercise Set 1

9.1 An unbiased statistic with a smaller standard error is preferred because it is likely to result in an estimate that is closer to the actual value of the population characteristic than an unbiased statistic that has a larger standard error.

9.2 Statistics II and III

9.3 Statistic I, because it has a smaller bias than Statistics II and III.

9.4 $n = 200$.

9.5 **(a)** The formula for the standard error of $\hat{p}$ is $\sigma_{\hat{p}} = \sqrt{\frac{p(1-p)}{n}}$. The quantity $p(1-p)$ reaches a maximum value when $p = 0.5$.
(b) The standard error of $\hat{p}$ is the same when $p = 0.2$ as when $p = 0.8$ because when $p = 0.2$, $(1-p) = (1-0.2) = 0.8$. Similarly, when $p = 0.8$, $(1-p) = (1-0.8) = 0.2$. So, the quantity $p(1-p)$ is the same in both cases.

9.6 $n = 400$ and $p = 0.8$

Additional Exercises

9.13 A biased statistic might be chosen over an unbiased statistic if the bias is not too large, and the standard error of the biased statistic is much smaller than the standard error of the unbiased statistic. In this case, the observed value of the biased statistic might be closer to the actual value than the value of an unbiased statistic.

9.15 $n = 200$.

SECTION 9.2

Exercise Set 1

9.17 Statement 1: Incorrect, because the value 0.0157 is the standard error of $\hat{p}$, and therefore approximately 32% of all possible values of $\hat{p}$ would differ from the value of the actual population proportion by more than 0.0157 (using properties of the normal distribution).

Statement 2: Correct

Statement 3: Incorrect, because the phrase "will never differ from the value of the actual population proportion" is wrong. The value 0.0307 is the margin of error and indicates that in about 95% of all possible random samples, the estimation error will be less than the margin of error. In about 5% of the random samples, the estimation error will be greater than the margin of error.

9.18 **(a)** 0.049

(b) $n = 100$

(c) $1/\sqrt{2} \approx 0.707$

9.19 **(a)** $\hat{p} = 0.299$; the sample proportion ($\hat{p}$) is the statistic that was used.

(b) $\sigma_{\hat{p}} = 0.026$

(c) margin of error = 0.052; this estimate of the proportion of all businesses that have fired workers for misuse of the Internet is unlikely to differ from the actual population proportion by more than 0.052.

9.20 **(a)** yes

(b) no

(c) no

(d) no

9.21 **(a)** $\hat{p} = \frac{125}{734} = 0.170$

(b) The sample was selected in such a way that makes it representative of the population of U.S. college students. Additionally, there are 125 successes and 609 failures in the sample, which are both at least 10. These conditions together verify that the margin of error formula is appropriate.

(c) margin of error = 0.027

(d) It is unlikely that the estimated proportion of U.S. college students who check their cell phones for something other than the time at least twice during the night ($\hat{p} = 0.170$) will differ from the true proportion by more than 0.027.

9.22 **(a)** $\hat{p} = 0.350$

(b) The sample is representative of the population of adult Americans. Additionally, there are 584 successes and 1084 failures in the sample, which are both greater than 10. These conditions together verify that the margin of error formula is appropriate.

(c) margin of error = 0.023

(d) It is unlikely that the estimated proportion of adult Americans who carry credit card debt from month to month ($\hat{p} = 0.350$) will differ from the true proportion by more than 0.023.

Additional Exercises

9.29 margin of error = 0.023; It is unlikely that the estimated proportion of Americans who prefer cheese on their burgers ($\hat{p} = 0.84$) will differ from the true proportion by more than 0.023.

9.31 **(a)** $\hat{p} = 0.38$

(b) margin of error = 0.030; It is unlikely that the estimated proportion of adults who have traveled by air at least once in the previous year who have yelled at a complete stranger while traveling ($\hat{p} = 0.38$) will differ from the true proportion by more than 0.030.

9.33 In this case, margin of error = 0.032, which rounds to 3%. The margin of error tells you that it is unlikely

that the estimate will differ from the actual population proportion by more than 0.03.

Exercise Set 1

9.34 **(a)** The confidence intervals are centered at $\hat{p}$. In this case, the intervals are not centered in the same place because two different samples were taken, each yielding a different value of $\hat{p}$.

(b) Interval 2 conveys more precise information about the value of the population proportion because Interval 2 is narrower than Interval 1.

(c) A smaller sample size produces a larger margin of error. In this case, Interval 1 (being wider than Interval 2) was based on the smaller sample size.

(d) Interval 1 would have the higher confidence level, because the z critical value for higher confidence is larger, resulting in a wider confidence interval.

9.35 **(a)** 95%

(b) $n = 100$

9.36 The method used to construct this interval estimate is successful in capturing the actual value of the population proportion about 95% of the time.

9.37 **(a)** yes

(b) no

(c) no

(d) no

9.38 **(a)** 1.645

(b) 2.58

(c) 1.28

9.39 Question type (Q): Estimation; Study type (S): Sample data; Type of data (T): One categorical variable; Number of samples or treatments (N): One sample.

9.40 Estimate (E): The proportion of hiring managers and human resources professionals who use social networking sites to research job applicants, p, will be estimated.

Method (M): Because the answers to the four key questions are estimation, sample data, one categorical variable, and one sample (see Exercise 9.39), consider a 95% confidence interval for the proportion of hiring managers and human resource professionals who use social networking sites to research job applicants.

Check (C): The sample is representative of hiring managers and human resource professionals. In addition, the sample includes 1200 successes and 1467 failures, which are both greater than 10. The two required conditions are satisfied.

Calculations (C): (0.4311,0.4689)

Communicate Results (C):

Interpret confidence interval: You can be 95% confident that the actual proportion of hiring managers and human resources professionals is somewhere between 0.4311 and 0.4689.

Interpret confidence level: The method used to construct this interval estimate is successful in capturing the actual value of the population proportion about 95% of the time.

9.41 **(a)** (0.302, 0.318). You can be 90% confident that the true proportion of students applying to college who want to attend a college within 250 miles of home is somewhere between 0.302 and 0.318.
(b) (0.492, 0.528). You can be 90% confident that the true proportion of parents of students applying to college who want their child to attend a college within 250 miles from home is somewhere between 0.492 and 0.528.
(c) The two confidence intervals from (a) and (b) do not have the same width because the two sample sizes are different, and also because the margins of error are based on two different values for $\hat{p}$.

9.42 (0.689, 0.751). You can be 95% confident that the true proportion of U.S. college students who believe that a student or faculty member on campus who uses language considered racist, sexist, homophobic, or offensive should be subject to disciplinary action is somewhere between 0.689 and 0.751.

9.43 **(a)** (0.155, 0.269). You can be 95% confident that the actual proportion of all adult Americans who planned to purchase a Valentine's Day gift for their pet is somewhere between 0.155 and 0.269.
(b) The 95% confidence interval computed for the actual sample size would have been narrower than the confidence interval computed in Part (a) because the standard error, and hence the margin of error, would have been smaller.

Additional Exercises

9.55 **(a)** (0.494, 0.546). You can be 90% confident that the actual proportion of adult Americans who would say that lying is never justified is somewhere between 0.494 and 0.546.
(b) (0.6252, 0.6748). You can be 90% confident that the actual proportion of adult Americans who would say that it is often or sometimes OK to lie to avoid hurting someone's feelings is somewhere between 0.6252 and 0.6748.
(c) The confidence interval in Part (a) indicates that it is plausible that at least 50% of adult Americans would say that lying is never justified, and the confidence interval in part (b) indicates that it is also plausible that well over 50% of adult Americans would say that it is often or sometimes OK to lie to avoid hurting someone's feelings. These are contradictory responses.

9.57 Intervals constructed when the sample proportions ($\hat{p}$) are closer to 0.5 are wider than those farther from 0.5 because the product $\hat{p}(1 - \hat{p})$ is largest when $\hat{p} = 0.5$. In Exercise 9.56, $\hat{p} = \frac{851}{1001} = 0.850$ and in this exercise, $\hat{p} = \frac{811}{1001} = 0.810$. The sample proportion in this exercise is closer to 0.5 than that in Exercise 9.56. Therefore, the interval in this exercise will be wider than the interval in Exercise 9.56.

9.59 (0.477, 0.563). You can be 90% confident that the true proportion of all homeowners in western states who have considered installing solar panels is between 0.477 and 0.563.

SECTION 9.4

Exercise Set 1

9.61 Assuming a 95% confidence level, and using a conservative estimate of $p = 0.5$, $n = 2401$.

9.62 Assuming a 95% confidence level, and using the preliminary estimate of $p = 0.63$, $n = 358.191$. The sample should contain 359 individuals. Using the conservative estimate of $p = 0.5$, $n = 384.16$. The sample should contain 385 individuals. As expected, the sample size computed using the conservative estimate is larger than the sample size computed using the preliminary estimate. The sample size 385 should be used for this study because it should result in a margin of error of no greater than 0.05 because the margin of error is largest with $p = 0.5$. The smaller sample size (359) is only likely to result in a margin of error no greater than 0.05 if $p > 0.63$ or if $p < 1 - 0.63 = 0.37$. If $\hat{p}$ lies between 0.37 and 0.63, which could happen with a different sample, then the margin of error would be greater than 0.05.

9.63 Assuming a 95% confidence level, and using a conservative estimate of $p = 0.5$, $n = 97$.

Additional Exercises

9.67 Assuming a 95% confidence level, and using the preliminary estimate of $p = 0.32$, $n = 335$. Using the conservative estimate of $p = 0.5$, $n = 385$.

SECTION 9.5

Exercise Set 1

9.69 **(a)** A 95% bootstrap confidence interval for the population proportion of those who would reply "No" to the question is (0.774, 0.829). Based on this sample, you can be 95% confident that the actual proportion who would reply "No" to the question "Should you be friends with your boss on Facebook?" is somewhere between 0.774 and 0.829.
(b) In this case, one could argue that this situation would tend to overestimate the true proportion because individuals who respond to anonymous and voluntary surveys tend to have strong feelings about the subject, and therefore might artificially inflate the proportion who would reply "No."

9.70 **(a)** No, it is not appropriate to use the large-sample confidence interval for a population proportion to estimate the proportion of the transportation services customers who have tried Uber or Lyft at least once. We don't have at least 10 successes in the sample, which is one of the necessary conditions.
(b) Yes, it is appropriate to use a bootstrap confidence interval for a population proportion to estimate the proportion of the transportation services customers who have tried Uber or Lyft because a random sample of regular customers who are 55 or older was taken.
(c) A 95% bootstrap confidence interval for the population proportion of customers 55 or older who have used Uber or Lyft at least once is (0.000, 0.333). Based on this sample,

you can be 95% confident that the actual proportion of customers 55 or older who have used Uber or Lyft at least once is somewhere between 0.000 and 0.333.
(d) Yes, the value obtained in the study (21%) is contained within the bootstrap confidence interval. Confidence intervals provide a range of plausible values for the true proportion. The interval contains 0.21, so we know that 0.21 is a plausible value for the true proportion of customers 55 or older who have used Uber or Lyft at least once.

9.71 Different simulations will produce different results, so answers will vary. For one simulation, a 95% bootstrap confidence interval for the population proportion of U.S. businesses who have fired workers for e-mail misuse was (0.230, 0.332). Based on this sample, you can be 95% confident that the actual proportion of U.S. businesses who have fired workers for e-mail misuse is somewhere between 0.230 and 0.332.

9.72 **(a)** It would not be appropriate to use a large-sample confidence interval for one proportion to estimate Kevin Love's success rate for three-point shots during the 2016 season because there are only 5 successes in the sample, which is fewer than the required 10 successes to use the large-sample interval.
(b) Different simulations will produce different results, so answers will vary. For one simulation, a 90% bootstrap confidence interval for the population proportion of Kevin Love's three-point shot success rate during the 2016 NBA season was (0.105, 0.421). Based on this sample, you can be 90% confident that the actual proportion of Kevin Love's three-point shot success rate during the 2016 NBA season is somewhere between 0.105 and 0.421.

ARE YOU READY TO MOVE ON?

Chapter 9 Review Exercises

9.77 Statistic II is preferred because, although both statistics are unbiased, the standard error of Statistic II is smaller, which is likely to result in an estimate that is closer to the actual value of the population characteristic than Statistic I.

9.79 The sample statistic $\hat{p}$ for a random sample of size $n = 800$ would tend to be closer to the actual value of 0.6 than would a random sample of size $n = 400$ because larger sample sizes correspond to smaller standard errors. Statistics with smaller standard errors tend to produce estimates that are closer to the actual value of the population characteristic than statistics with larger standard errors.

9.81 The situation with the smallest standard error will generally yield estimates closer to the actual value of p. The standard error for situation I is $\sqrt{\frac{0.5(1 - 0.5)}{1000}} = 0.016$, the standard error for situation II is $\sqrt{\frac{0.6(1 - 0.6)}{200}} = 0.035$, and the standard error for situation III is $\sqrt{\frac{0.7(1 - 0.7)}{100}} = 0.046$. Situation I has the smallest standard error, so estimates will tend to be closest to the actual value of p.

9.83 **(a)** $\sigma_{\hat{p}} = 0.0217$
(b) The standard error of $\hat{p}$ would be smaller for samples of size $n = 400$.
(c) No, cutting the sample size in half from 400 to 200 will increase the standard error by a factor of $\sqrt{2} \approx 1.4142$.

9.85 **(a)** Confidence intervals are centered at $\hat{p}$. In this case, the intervals are not centered in the same place because two different samples were taken, each yielding a different value of $\hat{p}$.
(b) Interval 2 conveys more precise information about the value of the population proportion because Interval 2 is narrower than Interval 1.
(c) Interval 1 is the confidence interval that was based on the smaller sample size because smaller samples convey less precise information about the value of the population proportion, and result in a wider confidence interval.
(d) Interval 1 has the higher confidence level because it is wider than Interval 2. The z critical value for higher confidence levels is larger than z critical values for lower confidence levels, which would result in a wider confidence interval.

9.87 The meaning of the 90% confidence level refers to the fact that the method used to construct this interval estimate is successful in capturing the actual value of the population proportion 90% of the time.

9.89 (0.697, 0.743) You can be 90% confident that the actual proportion of adult American Internet users who use Facebook is between 0.697 and 0.743.

9.91 Assuming a 95% confidence level, and using a conservative estimate of $p = 0.5$, $n = 384.16$. The sample should include 385 packages of ground beef.

CHAPTER 10

Asking and Answering Questions About a Population Proportion

NOTE: Answers may vary slightly if you are using statistical software, a graphing calculator or depending on how the values of sample statistics are rounded when performing hand calculations. Don't worry if you don't match these numerical answers exactly (but your answers should be similar).

SECTION 10.1

Exercise Set 1

10.1 $\hat{p}$ is a sample statistic. Hypotheses are about population characteristics.

10.2 $H_0: p = \frac{2}{3}$ versus $H_a: p > \frac{2}{3}$, where p is the proportion of employers who perform background checks.

10.3 $H_0: p = 0.7$ versus $H_a: p \neq 0.7$, where p is the proportion of college students who are Facebook users and log into their Facebook profile at least six times a day.

10.4 **(a)** There is convincing evidence that the proportion of American adults who favor drafting women is less than 0.5.

(**b**) Yes

(**c**) Yes

10.5 (**a**) The conclusion is consistent with testing H_o: HGH in addition to IVF does not increase the chance of getting pregnant versus H_a: HGH in addition to IVF does increase the chance of getting pregnant.

(**b**) A statistical hypothesis test is capable of demonstrating strong support for the alternative hypothesis (by rejecting the null hypothesis). The statement of "no strong evidence" is referring to no strong evidence in support of the alternative hypothesis. Therefore, the null hypothesis was not rejected.

10.6 The sample data provide convincing evidence against the null hypothesis. If the null hypothesis were true, the sample data would be very unlikely.

Additional Exercises

10.13 (**a**) legitimate

(**b**) not legitimate

(**c**) not legitimate

(**d**) legitimate

(**e**) not legitimate

10.15 H_0: $p = 0.6$ versus H_a: $p > 0.6$. In order to make the change, the university requires evidence that *more than* 60% of the faculty are in favor of the change.

SECTION 10.2

Exercise Set 1

10.17 Not rejecting the null hypothesis when it is not true.

10.18 A small significance level, because α is the probability of a Type I error.

10.19 (**a**) Before filing charges of false advertising against the company, the consumer advocacy group would require convincing evidence that more than 10% of the flares are defective.

(**b**) A Type I error is thinking that more than 10% of the flares are defective when in fact 10% (or fewer) of the flares are defective. This would result in the expensive and time-consuming process of filing charges of false advertising against the company when the company advertising is not false. A Type II error is not thinking that more than 10% of the flares are defective when in fact more than 10% of the flares are defective. This would result in the consumer advocacy group not filing charges when the company advertising was false.

10.20 (**a**) A Type I error would be thinking that less than 90% of the TV sets need no repair when in fact (at least) 90% need no repair. The consumer agency might take action against the manufacturer when the manufacturer is not at fault. A Type II error would be *not* thinking that less than 90% of the TV sets need no repair when in fact less than 90% need no repair. The consumer agency would not take action against the manufacturer when the manufacturer is making untrue claims about the reliability of the TV sets.

(**b**) Taking action against the manufacturer when the manufacturer is not at fault could involve large and unnecessary legal costs to the consumer agency. $\alpha = 0.01$ should be recommended.

10.21 (**a**) This is a Type II error because the statement describes the result of failing to reject the null hypothesis when the null hypothesis is actually false (concluding that the woman does not have breast cancer when, in actuality, she does). This probability is approximately P(Type II error) $\approx \frac{1}{13} = 0.077$.

(**b**) The other error that is possible is a Type I error, in which the null hypothesis is rejected when it should not be. In this scenario, a Type I error is concluding that a woman has cancer when she really does not have cancer. This probability is approximately P(Type I error) $\approx \frac{90}{637} = 0.141$.

Additional Exercises

10.27 A Type I error is rejecting a true null hypothesis and a Type II error is not rejecting a false null hypothesis.

10.29 Answers will vary.

10.31 It is not necessary to carry out a hypothesis test to determine if the proportion of registered voters in California who voted in the 2016 presidential election is less than the national proportion of 0.600 because the 57.8% stated in the article is the proportion of all registered voters who voted in the 2016 election, which is a population characteristic. This tells us that the population proportion is less than 0.600, so there is no need to carry out a hypothesis test.

SECTION 10.3

Exercise Set 1

10.32 (**a**) Approximately normal with mean 0.341 and standard deviation 0.015.

(**b**) You would not be surprised to observe a sample proportion of $\hat{p} = 0.33$ for a sample of size 1000 if the null hypothesis H_0: $p = 0.341$ is true. $\hat{p} = 0.33$ is less than one standard deviation below what you would expect it to be if the null hypothesis were true.

(**c**) You would be surprised to observe a sample proportion of $\hat{p} = 0.31$ for a sample of size 1000 if the null hypothesis H_0: $p = 0.341$ is true. This value is more than 2 standard deviations below the mean.

(**d**) $z = \dfrac{\hat{p} - p}{\sqrt{\dfrac{p(1-p)}{n}}} = \dfrac{0.307 - 0.341}{\sqrt{\dfrac{0.341(1-0.341)}{1000}}} = -2.27$ This z statistic is 2.27 standard deviations below what you would expect it to be if the null hypothesis were true. $P(z \leq -2.27$ when H_0 is true) $= 0.012$. This probability indicates that it is unlikely we would observe a $\hat{p}$ at least as extreme as what we observed if H_0 is true. Because the computed probability is so small, there is convincing evidence that the goal is not being met.

10.33 (**a**) Approximately normal with mean 0.5 and standard deviation 0.017.

(**b**) I would not be particularly surprised to observe a sample proportion of $\hat{p} = 0.52$ for a sample of size 844 if the null

hypothesis $H_0: p = 0.5$ is true. $z = \dfrac{\hat{p} - p}{\sqrt{\frac{p(1-p)}{n}}} = \dfrac{0.52 - 0.5}{\sqrt{\frac{0.5(1 - 0.5)}{844}}} =$
1.16. This z statistic is not very large in magnitude, and indicates that $\hat{p} = 0.52$ is just over one standard deviation above what you would expect it to be if the null hypothesis were true. $P(z \geq 1.16$ when H_0 is true) = 0.123. This probability indicates that it is not unlikley you would observe a $\hat{p}$ at least as extreme as what we observed if H_0 is true.
(c) You would be surprised to observe a sample proportion of $\hat{p} = 0.54$ for a sample of size 844 if the null hypothesis $H_0: p = 0.5$ is true. $z = \dfrac{\hat{p} - p}{\sqrt{\frac{p(1-p)}{n}}} = \dfrac{0.54 - 0.5}{\sqrt{\frac{0.5(1 - 0.5)}{844}}} = 2.32$. This z statistic indicates that $\hat{p} = 0.54$ is 2.32 standard deviations above what you would expect it to be if the null hypothesis were true. $P(z \geq 2.32$ when H_0 is true) = 0.010. This probability indicates that it is unlikely that we would observe a $\hat{p}$ at least as extreme as what we observed if H_0 is true.
(d) $z = \dfrac{\hat{p} - p}{\sqrt{\frac{p(1-p)}{n}}} = \dfrac{0.59 - 0.5}{\sqrt{\frac{0.5(1 - 0.5)}{844}}} = 5.23$. This z statistic indicates that $\hat{p} = 0.59$ is 5.23 standard deviations above what you would expect it to be if the null hypothesis were true. $P(z \geq 5.23$ when H_0 is true) ≈ 0. Because this computed probability is essentially zero, there is convincing evidence that the null hypothesis $H_0: p = 0.5$ is not true.
(e) No, it is not reasonable to generalize this conclusion to adults living in the United States because the sample was of adults living in Belgium. We can therefore only generalize to the population of adults living in Belgium.

Additional Exercises

10.37 **(a)** Approximately normal with mean 0.20 and standard deviation = 0.013.
(b) The sample proportion $\hat{p} \geq 0.221$ would convince me that more than 20% of adults have sent a love letter via e-mail.

SECTION 10.4
Exercise Set 1

10.38 **(a)** A P-value of 0.0003 means that it is very unlikely (probability = 0.0003), assuming that H_0 is true, that you would get a sample result at least as inconsistent with H_0 as the one obtained in the study. H_0 would be rejected.
(b) A P-value of 0.350 means that it is not particularly unlikely (probability = 0.350), assuming that H_0 is true, that you would get a sample result at least as inconsistent with H_0 as the one obtained in the study. There is no reason to reject H_0.

10.39 **(a)** $H_0: p = 0.5$ versus $H_a: p > 0.5$, where p represents the proportion of adult Americans who prefer name-brand frozen vegetables over store brand frozen vegetables.
(b) Because the P-value of 0.173 is greater than α of 0.05, we fail to reject H_0. There is insufficient evidence to conclude that the proportion of adult Americans who prefer name-brand frozen vegetables over store brand frozen vegetables is greater than 0.5.

10.40 This step involves using the answers to the four key questions (QSTN) to identify an appropriate method.

Additional Exercises

10.45 A P-value of 0.002 means that it is very unlikely (probability = 0.002), assuming that H_0 is true, that you would get a sample result at least as inconsistent with H_0 as the one obtained in the study. This is strong evidence against the null hypothesis.

SECTION 10.5
Exercise Set 1

10.47 Estimation, sample data, one numerical variable, one sample. A hypothesis test for a population proportion would not be appropriate.

10.48 Hypothesis testing, sample data, one categorical variable, one sample. A hypothesis test for a population proportion would be appropriate.

10.49 Estimation, sample data, one categorical variable, one sample. A hypothesis test for a population proportion would not be appropriate.

10.50 **(a)** Large-sample z test is not appropriate.
(b) Large-sample z test is appropriate.
(c) Large-sample z test is appropriate.
(d) Large-sample z test is not appropriate.

10.51 **(a)** 0.2912
(b) 0.1788
(c) 0.0233
(d) 0.0125
(e) 0.9192

10.52 **(a)** $H_0: p = 0.4$, $H_a: p < 0.4$, $z = -2.969$, P-value = 0.001, reject H_0. There is convincing evidence that the proportion of all adult Americans who would answer the question correctly is less than 0.4.
(b) $H_0: p = \frac{1}{3}$, $H_a: p > \frac{1}{3}$, $z = 2.996$, P-value = 0.001, reject H_0. There is convincing evidence that more than one-third of adult Americans would select a wrong answer.

10.53 $H_0: p = 0.25$, $H_a: p > 0.25$, $z = 0.45$, P-value = 0.328, fail to reject H_0. There is no convincing evidence that more than 25% of Americans age 16 to 17 have sent a text message while driving.

10.54 $H_0: p = 0.5$ versus $H_a: p > 0.5$; $z = 2.73$; P-value $= P(z \geq 2.73) = 0.003$; reject the null hypothesis. There is convincing evidence that a majority of U.S. consumers prefer to stream TV shows rather than to watch them live.

10.55 **(a)** $H_0: p = 0.5$ versus $H_a: p < 0.5$; $z = -2.24$; P-value $= P(z \leq -2.24) = 0.013$; reject the null hypothesis. There is convincing evidence that less than half of Americans ages 14 to 18 years usually use social media while watching TV.
(b) $H_0: p = 0.5$ versus $H_a: p < 0.5$; $z = -1.00$. P-value $= P(z \leq -1.00) = 0.159$; fail to reject the null hypothesis. There is not convincing evidence that less than half of Americans age 14 to 18 years usually use social media while watching TV.

(c) Both results *suggest* that fewer than half of Americans age 14 to 18 years usually use social media while watching TV. However, getting 225 out of 500 people responding this way (as opposed to 45 out of 100) provides much *stronger* evidence of this fact.

10.56 H_0: $p = 0.5$, H_a: $p < 0.5$, $z = -2.536$, P-value $= 0.006$, reject H_0. There is convincing evidence that the proportion of all adult Americans who want car web access is less than 0.5. The marketing manager is not correct in his claim.

Additional Exercises

10.67 **(a)** $z = 2.530$, P-value $= 0.0057$, reject H_0
(b) No. The survey only included women ages 22 to 35.

10.69 H_0: $p = 0.25$ versus H_a: $p > 0.25$; $z = 1.03$; P-value $= P(z \geq 1.03) = 0.15$; fail to reject the null hypothesis. There is not convincing evidence that more than one-quarter of all adult Americans ages 26 to 32 years own a fitness band.

10.71 $z = 1.069$, P-value $= 0.143$, fail to reject H_0

10.73 H_0: $p = \frac{1}{3}$ versus H_a: $p > \frac{1}{3}$; $z = 3.08$; P-value $= P(z \geq 3.08) = 0.001$; reject the null hypothesis. There is convincing evidence that more than one-third of adult Americans who have checked their credit scores within the past 12 months did so as part of regular financial planning.

SECTION 10.6

Exercise Set 1

10.75 **(a)** The large-sample hypothesis test for a population proportion cannot be used in this example because the large sample size conditions (np_0 and $n(1 - p_0)$ both at least 10) are not satisfied.
(b) The randomization test was used to test the hypotheses H_0: $p = 0.27$ versus H_a: $p > 0.27$, where p is the proportion of lunar astronauts who are at increased risk of CVD. A P-value of 0.301 was obtained. Because the P-value of 0.301 is greater than any reasonable significance level, we fail to reject the null hypothesis. There is not sufficient evidence to conclude that, as a group, lunar astronauts are at increased risk of death caused by CVD.

10.76 **(a)** H_0: $p = 0.20$ versus H_a: $p < 0.20$, where p is the proportion of hospital patients who had been treated for pneumonia using a respiratory therapist protocol who were readmitted to the hospital within 30 days after discharge.
(b) It is not stated that the sample was randomly selected or that the sample is representative of the population of all hospital patients who had been treated for pneumonia using a respiratory therapist protocol. As such, you must use caution and assume the sample was selected in a reasonable way. The large sample size condition has been satisfied because there are at least 10 successes and failures in the sample.

(c) The output for the exact binomial test indicates that the P-value is 0.000. Since the P-value of 0.000 is less than any reasonable significance level, we reject the null hypothesis. We have sufficient evidence to conclude that the proportion of subjects who will be readmitted to a hospital within 30 days after following a respiratory therapist protocol for treatment of pneumonia is less than 0.20.
(d) In this case, $\hat{p} = \frac{15}{162} = 0.093$, and $z = \frac{0.093 - 0.20}{\sqrt{\frac{0.20(1 - 0.20)}{162}}} = -3.42$. This is a lower-tailed test (the inequality in H_a is $<$), so the P-value is the area under the z curve and to the left of -3.42. Therefore, the P-value is $P(z \leq -3.42) = 0.0003$. This P-value is larger than the one obtained in part (c), but the conclusion is the same.

10.77 **(a)** These data should not be analyzed using a large-sample hypothesis test for one proportion because the number of successes and failures are not both at least 10. In this case, the number of successes is 33, but the number of failures is 2, which is less than 10.
(b) The exact binomial test gives a P-value of approximately 0. Therefore, you reject the null hypothesis. You have sufficient evidence to conclude that the proportion of all dogs trained using this method who would perform the correct new action is greater than 0.5.

ARE YOU READY TO MOVE ON?
Chapter 10 Review Exercises

10.81 H_0: $p = 0.5$ versus H_a: $p > 0.5$, where p is the proportion of the people in the district who favor the measure.

10.83 **(a)** Because the null hypothesis was rejected, there is convincing evidence that the proportion of mobile phone users who received at least one nuisance call on their mobile phones within the past month is greater than 0.667.
(b) Yes, it is reasonable to say that the data provide strong support for the alternative hypothesis.
(c) Yes, it is reasonable to say that the data provide strong evidence against the null hypothesis.

10.85 **(a)** A false positive would be a Type I error.
(b) A Type I error is rejecting the null hypothesis when, in fact, the null hypothesis is true. In this case, rejecting the null hypothesis of "no cancer is present" in favor of the alternative "cancer is present" means that a person would be diagnosed with cancer when the person is actually cancer free. In this case, a person would undergo cancer treatment unnecessarily.
(c) A Type II error is failing to reject the null hypothesis when, in fact, the null hypothesis is false. In this case, the null hypothesis of "no cancer is present" is not rejected, so a person with cancer is declared cancer-free, and does not undergo necessary cancer treatment.
(d) If false positives fall, then the probability of a Type I error (a "false alarm"), α, decreases, and the probability of

a Type II error (missed cancers) would increase. In order to decrease the probability of a Type II error by adjusting the significance level, the probability of a Type I error must increase. The probability of a false positive must increase in order for the probability of a missed cancer ("false negative") to decrease.

10.87 (a) Test the hypotheses $H_0: p = 0.03$ versus $H_a: p > 0.03$. This pair of hypotheses was chosen because the researcher is interested in determining of there is sufficient evidence to conclude that more than 3% of fish have unacceptably high mercury levels. A hypothesis test can only show strong support for the alternative hypothesis, never the null hypothesis.
(b) To choose the significance level, consider the consequences of Type I and Type II errors. A Type I error would result in rejecting the null hypothesis when the null hypothesis is actually true. In this case, we would close an area to fishing when, in fact, the area has acceptable mercury levels. A Type II error would result in not rejecting the null hypothesis when the null hypothesis is actually false. In this case, we would not close an area to fishing when, in fact, the area should be closed due to high mercury levels. A Type II error is more serious than a Type I error, so choose a significance level of $\alpha = 0.10$.

10.89 (a) The hypotheses are $H_0: p = 0.5$ versus $H_a: p > 0.5$, where p is the proportion of adult Americans who prefer a hot climate over a cold climate.
(b) Because the P-value is less than the significance level α, you reject the null hypothesis. There is convincing evidence that a majority of adult Americans prefer a hot climate over a cold climate.

10.91 Question Type (Q): Hypothesis testing
Study Type (S): Sample data
Type of Data (T): One categorical variable
Number of Samples or Treatments (N): One sample
A hypothesis test for a population proportion is appropriate.

10.93 $H_0: p = 0.5$ versus $H_a: p > 0.5$; $z = 10.49$; P-value $= P(z \geq 10.49) \approx 0$; reject the null hypothesis. There is convincing evidence that a majority of adult Americans think that the designated hitter rule should be eliminated and that pitchers should have to bat.

CHAPTER 11
Asking and Answering Questions About the Difference Between Two Population Proportions

NOTE: Answers may vary slightly if you are using statistical software, a graphing calculator or depending on how the values of sample statistics are rounded when performing hand calculations. Don't worry if you don't match these numerical answers exactly (but your answers should be similar).

SECTION 11.1
Exercise Set 1

11.1 (a) Estimation, sample data, one categorical variable, two samples. A large-sample confidence interval for a difference in proportions should be considered.
(b) $(-0.050, -0.002)$. You can be 90% confident that the actual difference between the proportion of male U.S. residents living in poverty and this proportion for females is between -0.050 and -0.002. Because both endpoints of this interval are negative, you would estimate that the proportion of men living in poverty is smaller than the proportion of women living in poverty by somewhere between 0.002 and 0.050.

11.2 (a) Yes. $\hat{p}_1 = 0.035$ $\hat{p}_2 = 0.044$. $n_1 \hat{p}_1 = 1200(0.035) = 42$, $n_1(1 - \hat{p}_1) = 1200(1 - 0.035) = 1,158$, $n_2 \hat{p}_2 = 1200(0.044) = 52.8$ and $n_2(1 - \hat{p}_2) = 1200(1 - 0.044) = 1147.2$ are all at least 10
(b) $(-0.025, 0.007)$
(c) Zero is included in the confidence interval. This means that it is plausible that the two population proportions could be equal.
(d) You can be 95% confident that the actual difference in proportions of males who use a mobile phone while driving and females who use a mobile phone while driving is somewhere between -0.025 and 0.007. Since zero is contained in the confidence interval, it is plausible that the proportion of males who use a mobile phone while driving and the proportion of females who use a mobile phone while driving could be equal.

11.3 $(0.032, 0.108)$. You can be 95% confident that the actual difference in the proportion of adults and teens age 13–17 who believe in reincarnation is somewhere between 0.032 and 0.108. Because both endpoints of the confidence interval are positive, you believe that the proportion of adults who believe in reincarnation is greater than the proportion of teens age 13–17 who believe in reincarnation by somewhere between 0.032 and 0.108.

Additional Exercises

11.7 $(0.047, 0.113)$. You can be 99% confident that the proportion of high school students who believed marijuana use is very distracting in 2009 was greater than this proportion in 2011 by somewhere between 0.047 and 0.114.

11.9 (a) $(-0.029, 0.061)$. You can be 99% confident that the actual difference in the proportion of high school graduates who were unemployed in October 2013 and high school graduates who were unemployed in October 2014 is somewhere between -0.029 and 0.061. Because the endpoints of the confidence interval have opposite signs, zero is included in the interval, and there may be no difference in the proportion of high school graduates that were unemployed in October 2013 and the proportion that were unemployed in October 2014.
(b) Wider, because the confidence level in Part (a) is greater than the confidence level in the previous exercise, and the sample sizes are smaller.

SECTION 11.2

Exercise Set 1

11.11 **(a)** $H_0: p_1 - p_2 = 0$ versus $H_a: p_1 - p_2 > 0$, where p_1 is the actual proportion of iPhone owners who upgrade their phones at least every two years, and p_2 is the actual proportion of Android owners who upgrade their phones at least every two years.
(b) Yes, because there are more than 10 successes and 10 failures in each sample: $n_1\hat{p}_1 = 8234(0.53) = 4364.02$, $n_1(1 - \hat{p}_1) = 8234(1 - 0.53) = 3869.98$, $n_2 p_2 = 6072(0.42) = 2550.24$, and $n_2(1 - \hat{p}_2) = 6072(1 - 0.42) = 3521.76$, which are all at least 10.
(c) The test statistic is $z = 13.02$ and the P-value $= 0.000$. Because the P-value of 0.000 is less than α of 0.01, we reject H_0.
(d) There is convincing evidence that the proportion of iPhone owners who upgrade their phones at least every two years is greater than this proportion for Android phone users.

11.12 $H_0: p_T - p_O = 0$, $H_a: p_T - p_O < 0$, $z = -1.667$, P-value $= 0.048$, reject H_0. There is convincing evidence that the proportion who are satisfied is higher for those who reserve a room online.

11.13 $H_0: p_1 - p_2 = 0$, $H_a: p_1 - p_2 \neq 0$. $z = -0.22$, P-value $= 0.8258$, fail to reject H_0. There is not convincing evidence that the proportion of women who think the creepy person is more likely to be male is different from this proportion for men.

Additional Exercises

11.17 $H_0: p_1 - p_2 = 0$, $H_a: p_1 - p_2 \neq 0$, $z = -0.574$, P-value $= 0.566$. Fail to reject H_0. There is not convincing evidence of a difference between the proportion of young adults who think that their parents would provide financial support for marriage and the proportion of parents who say they would provide financial support for marriage.

11.19 $H_0: p_1 - p_2 = 0$, $H_a: p_1 - p_2 > 0$. $z = 3.76$, P-value ≈ 0, reject H_0. There is convincing evidence that the proportion of Gen-Xers who do not pay off their credit cards each month is greater than this proportion for millennials.

SECTION 11.3

Exercise Set 1

11.20 **(a)** $H_0: p_1 - p_2 = 0$, $H_a: p_1 - p_2 > 0$. $z = 4.32$, P-value ≈ 0, reject H_0. There is convincing evidence that the proportion with increased hair density is greater for the supplement treatment than for the control treatment.
(b) There are a couple of problems with the headline. First, there was more than just black currant oil in the supplement treatment. Second, the headline is too definitive. The result of a statistical test does not prove a result. Rather, it provides evidence that supports a particular conclusion.
(c) A more appropriate headline might be "Black Currant Oil, Combined with Other Supplements, May Contribute to Less Hair Loss."

11.21 $H_0: p_1 - p_2 = 0$, $H_a: p_1 - p_2 > 0$. $z = 3.02$, P-value $= 0.001$, reject H_0. There is convincing evidence that the proportion experiencing pain relief is greater for the surgery treatment than for the therapy treatment.

11.22 $(0.066, 0.294)$. You can be 95% confident that the difference in proportions experiencing pain relief for the surgery treatment and the proportion for the therapy treatment is between 0.0661 and 0.294. Both endpoints of the confidence interval are positive, so you can estimate that the proportion of those experiencing pain relief for the surgery treatment is greater than the proportion of those experiencing pain relief for the therapy treatment by somewhere between 6.61 and 29.39 percentage points.

11.23 **(a)** $(0.042, 0.458)$. You can be 95% confident that the difference in proportions of drivers who would exit at the rest stop while talking to a passenger and who would exit while talking on a cell phone is between 0.042 and 0.458. Both endpoints of the confidence interval are positive, so you can estimate that the proportion of drivers talking with a passenger who would exit at the rest stop is greater than the proportion of drivers talking on a cell phone who would exit at the rest stop by somewhere between 0.042 and 0.458.
(b) Yes, the interval in Part (a) does support the conclusion that drivers using a cell phone are more likely to miss the exit than drivers talking with a passenger.

Additional Exercises

11.29 No. It is not appropriate to use the two-sample z test because the groups are not large enough. You are not told the sizes of the groups, but you know that each is, at most, 81. The sample proportion for the fish oil group is 0.05, and $81(0.05) = 4.05$, which is less than 10. So, the conditions for the two-sample z test are not satisfied.

11.31 **(a)** If people believe that one of the treatments is more effective than the other (for example, that the injection is more effective than the spray), it might affect how they interpret and report symptoms.
(b) $(0.033, 0.061)$. You can be 99% confident that the proportion of children who get sick with the flu after being vaccinated with a nasal spray minus the proportion of children who get sick with the flu after being vaccinated with an injection is between 0.033 and 0.061. Because both endpoints of the confidence interval are positive, you believe that the proportion of children who get sick with the flu after being vaccinated with the nasal spray is greater than the proportion of children who get sick with the flu after being vaccinated with an injection by somewhere between 0.033 and 0.061.

SECTION 11.4

Exercise Set 1

11.33 **(a)** In order to use the large-sample hypothesis test for the difference in two population proportions, the number of successes and failures in each sample must be at

least 10. In this case, the number of successes in one of the groups is $n_1\hat{p}_1 = (21)(0.05) = 1.05$, which is less than 10.
(b) $H_0: p_1 - p_2 = 0$, $H_a: p_1 - p_2 < 0$. For the given simulation, the P-value is 0.002. Because this P-value of 0.002 is less than any reasonable significance level, you reject the null hypothesis. There is convincing evidence that the proportion of Tasmanian devils with the genetic marker was greater after DFTD than before DFTD.
(c) The given simulation produced a confidence interval of $(-0.548, -0.214)$. You can be 95% confident that the true difference in the rates of occurrence of the specific genetic marker in the genes of Tasmanian devils, before and after DFTD, lies somewhere between -0.548 and -0.214. Because both endpoints of the confidence interval are negative, you believe that the rate of occurrence of the genetic marker in the genes of Tasmanian devils after DFTD is greater than the rate before DFTD by somewhere between 0.214 and 0.548.

11.34   **(a)** $H_0: p_1 - p_2 = 0$, $H_a: p_1 - p_2 \neq 0$. The given randomization test resulted in a P-value of 0.502. Because this P-value of 0.502 is greater than any reasonable significance level, you fail to reject the null hypothesis. There is not convincing evidence that the proportion who require students to submit papers through plagiarism-detection software is different for full-time faculty and part-time faculty.
(b) For the given simulation, a confidence interval of $(-0.037, 0.080)$ was obtained. You can be 95% confident that the actual difference in the population proportions of full-time college faculty and part-time college faculty who require students to submit papers through plagiarism-detection software is somewhere between -0.037 and 0.080. Because the endpoints of the confidence interval have opposite signs, zero is included in the interval, and there may be no difference in in the population proportions of full-time college faculty and part-time college faculty who require students to submit papers through plagiarism-detection software.
(c) The same conclusion was reached using the randomization test and when the large-sample test was used.

11.35   **(a)** Data from this study should not be analyzed using a large-sample hypothesis test for a difference in two population proportions because the number of successes in each group (5 in the high-intensity group and 0 in the regular-intensity group) are both less than 10.
(b) $H_0: p_1 - p_2 = 0$ versus $H_a: p_1 - p_2 \neq 0$. Different simulations will produce different results, so answers will vary. For one randomization test, the P-value was 0.017. Because this P-value was 0.017 is less than a significance level of 0.05, the null hypothesis is rejected. There is convincing evidence of a difference in the population proportions of patients in the two exercise groups (high-intensity and regular-intensity) who die within 1.5 years.
(c) Different simulations will produce different results, so answers will vary. One simulation resulted in a confidence interval of $(0.123, 0.133)$. You can be 95% confident

that the actual difference in the population proportions of patients who die within 1.5 years for the two exercise groups is somewhere between 0.123 and 0.133. Because both endpoints of the confidence interval are positive, you believe that the proportion of patients who die within 1.5 years for the high-intensity group is greater than the proportion of patients who die within 1.5 years for the regular-intensity group by somewhere between 0.123 and 0.133.

Additional Exercises

11.39   **(a)** $(0.091, 0.265)$. You can be 90% confident that the difference in the population proportions of cell phone users age 20 to 39 and those age 40 to 49 who say that they sleep with their cell phones is between 0.091 and 0.265.
(b) The given simulation resulted in a bootstrap confidence interval of $(0.093, 0.267)$. You can be 90% confident that the difference in the population proportions of cell phone users age 20 to 39 and those age 40 to 49 who say that they sleep with their cell phones is between 0.093 and 0.267.
(c) The confidence intervals in Parts (a) and (b) are very similar.

ARE YOU READY TO MOVE ON?
Chapter 11 Review Exercises

11.41   **(a)** $(-0.078, 0.058)$. You can be 95% confident that the actual difference in proportion of teens using a cell phone while driving before the ban and the proportion after the ban is somewhere between -0.078 and 0.058. Because the endpoints of the confidence interval have opposite signs, zero is included in the interval, and there may be no difference in the proportion of teens using a cell phone while driving before the ban and the proportion after the ban.
(b) Yes, zero is included in the confidence interval. This implies that there may be no difference in the proportion of teens using a cell phone while driving before the ban and the proportion after the ban.

11.43 $H_0: p_1 - p_2 = 0$, $H_a: p_1 - p_2 < 0$. $z = -6.134$, P-value ≈ 0, reject H_0. There is convincing evidence that the proportion of adult Americans age 33 to 49 who rate a landline phone in the top three is less than this proportion for adult Americans age 50 to 68.

11.45   When the null hypothesis is not rejected, you do not have convincing evidence that the alternative hypothesis is true.

11.47   **(a)** $(0.084, 0.492)$. You can be 95% confident that the actual difference in the proportion who experience a meaningful pain reduction for the acupuncture treatment and for the sham treatment is somewhere between 0.084 and 0.492.
(b) The confidence interval suggests that acupuncture is more effective in reducing heel pain than the sham treatment.

CHAPTER 12

Asking and Answering Questions About a Population Mean

NOTE: Answers may vary slightly if you are using statistical software or a graphing calculator, or depending on how the values of sample statistics are rounded when performing hand calculations. Don't worry if you don't match these numerical answers exactly (but your answers should be similar).

SECTION 12.1

Exercise Set 1

12.1 (a) 100, 3.333
(b) 100, 2.582
(c) 100, 1.667
(d) 100, 1.414
(e) 100, 1.000
(f) 100, 0.500

12.2 The sampling distribution of $\bar{x}$ will be approximately normal for the sample sizes in Parts (c)–(f), since those sample sizes are all greater than or equal to 30.

12.3 (1) The mean of the sampling distribution of $\bar{x}$ is equal to the population mean μ. (2) $\sigma_{\bar{x}} \leq \dfrac{s}{\sqrt{n}} = \dfrac{1.03}{\sqrt{236}} =$ 0.067. (3) The sampling distribution of $\bar{x}$ is approximately normal because the sample size ($n = 236$) is greater than 30.

12.4 The quantity μ is the population mean, while $\mu_{\bar{x}}$ is the mean of the $\bar{x}$ distribution. It is the mean value of $\bar{x}$ for all possible random samples of size n.

12.5 (a) $\mu_{\bar{x}} = \mu = 0.5$, $\sigma_{\bar{x}} = \dfrac{\sigma}{\sqrt{n}} = \dfrac{0.289}{\sqrt{16}} = 0.07225$.
(b) When $n = 50$, $\mu_{\bar{x}} = \mu = 0.5$ and $\sigma_{\bar{x}} = \dfrac{\sigma}{\sqrt{n}} = \dfrac{0.289}{\sqrt{50}} =$ 0.041. Since $n \geq 30$ the distribution of $\bar{x}$ is approximately normal.

Additional Exercises

12.11 (a) 200, 4.330
(b) 200, 3.354
(c) 200, 3.000
(d) 200, 2.371
(e) 200, 1.581
(f) 200, 0.866

12.13 $\bar{x}$ is the mean of a single sample, while $\mu_{\bar{x}}$ is the mean of the $\bar{x}$ distribution. It is the mean value of $\bar{x}$ for all possible random samples of size n.

12.15 (a) 0.8185, 0.0013
(b) 0.9772, 0.0000

12.17 $P(0.49 < \bar{x} < 0.51) = 0.9974$; the probability that the manufacturing line will be shut down unnecessarily is $1 - 0.9974 = 0.0026$.

Exercise Set 1

12.19 (a) 90%
(b) 95%
(c) 1%
(d) 5%

12.20 (a) 2.12
(b) 2.81
(c) 1.78

12.21 (a) 115.0
(b) The 99% confidence interval is wider than the 90% confidence interval. The 90% interval is (114.4, 115.6) and the 99% interval is (114.1, 115.9).

12.22 (a) If the distribution of volunteer times is approximately normal, for the sample standard deviation of $s = $ 16.54 hours and the sample mean of $\bar{x} = 14.76$ hours, approximately 18.6% of volunteer times would be negative, which is impossible. Therefore, it is not reasonable to think that the distribution of volunteer times is approximately normal.
(b) The two conditions required to use a one-sample t distribution are that the sample was either randomly selected or is representative of the population, and the population distribution is either normally distributed or that the sample size is at least 30. We are told that the sample is representative of the population, and that the sample size is 500. Both conditions are satisfied.
(c) (13.307, 16.213). You can be 95% confident that the mean number of hours spent in volunteer activities per year for South Korean middle school children is between 13.307 and 16.213 hours.

12.23 (a) (35.424, 38.616). You can be 95% confident that the mean procrastination scale for first-year students at this college is between 35.424 and 38.616.
(b) Forty (40) is not a plausible value for the mean population procrastination scale score because that value is not contained within the confidence interval.

12.24 (a) (39.349, 42.651). You can be 95% confident that the mean procrastination scale for second-year students at this college is between 39.349 and 42.651.
(b) The confidence interval for second-year students does not overlap, and has values that are entirely above, the confidence interval for first-year students. Therefore, it seems plausible that second-year students tend to procrastinate more than first-year students at this university.

12.25 Using (sample range)/4 = 162.5 as an estimate of the population standard deviation, a sample size of 1015 is needed.

12.26 (a) If the distribution of weight gains is approximately normal, then, for the given sample standard deviation of $s = 6.8$ pounds and the sample mean of $\bar{x} = 5.7$ pounds, you should see approximately 16% of the weight gain observations be less than -1.1 pounds (one standard deviation below the mean). It is reported that 6.8% lost more than 1.1 pounds, which is too small a percentage if the normal distribution model was correct. Therefore, it is not reasonable to think that the distribution of weight gains is approximately normal.

(b) The two conditions required to use a one-sample t distribution are that the sample was either randomly selected or is representative of the population, and the population distribution is either normally distributed or that the sample size is at least 30. We are told that the sample is representative, and that the sample size is 103. Therefore, both conditions are satisfied in the exercise.

(c) (4.371, 7.029). You can be 95% confident that the mean weight gain of freshmen students at this university is between 4.371 and 7.029 pounds.

Additional Exercises

12.35   (a) 2.15
(b) 2.86
(c) 1.71

12.37   (5.776, 28.224). You can be 95% confident that the mean number of months elapsed since the last visit for the population of students participating in the program is between 5.776 and 28.224.

12.39   (3.301, 5.899). You can be 95% confident that the mean number of partners on a mating flight is between 3.301 and 5.899.

SECTION 12.3

Exercise Set 1

12.41   (a) 0.040
(b) 0.019
(c) 0.000
(d) 0.130

12.42   (a) H_0: $\mu = 98$, H_a: $\mu \neq 98$, $t = 0.748$, P-value $= 0.468$, fail to reject H_0. There is not convincing evidence that the mean heart rate after 15 minutes of Wii Bowling is different from the mean after 6 minutes of walking on a treadmill.

12.43   (a) H_0: $\mu = 66$, H_a: $\mu > 66$, $t = 8.731$, P-value ≈ 0, reject H_0. There is convincing evidence that the mean heart rate after 15 minutes of Wii Bowling is higher than the mean resting heart rate.

12.44 H_0: $\mu = 5.07$ H_a: $\mu < 5.07$, $t = -6.36$, P-value ≈ 0, reject H_0. There is convincing evidence that the mean price of a Big Mac in Europe is less than \$5.07.

12.45   (a) It is not reasonable to think the distribution of time spent playing video or computer games for the population of male Canadian high school students is approximately normal. If the distribution were approximately normal with the given mean and standard deviation, roughly 15% of the observed times would be negative, which is not possible.

(b) We would need to know that the sample was randomly selected or is representative of the population of male

Canadian high school students. Additionally, we would need to know that the distribution of the population of times is normal or that the sample is large (at least size 30).

(c) H_0: $\mu = 120$, H_a: $\mu > 120$; $t = -0.649$, P-value $= 0.258$; fail to reject H_0. There is not convincing evidence that the average time spent playing video or computer games for male Canadian high school students is greater than 2 hours (120 minutes).

(d) H_0: $\mu = 120$, H_a: $\mu > 120$ minutes; $t = 2.049$, P-value $= 0.020$; reject H_0. There is convincing evidence that the average time spent playing video or computer games for male Canadian high school students is greater than 2 hours (120 minutes).

(e) The sample standard deviation of 117.1 minutes in Part (c) indicates that there is likely to be more variability in the number of minutes of video or computer game playing than when the sample standard deviation was 37.1 minutes, as in Part (d). If the mean of the population of average time spent playing video or computer games for male Canadian high school students was indeed 120 minutes, it is less likely that a sample mean of 123.4 minutes would be obtained when the standard deviation was 37.1 minutes than when the standard deviation was 117.1 minutes. Therefore, the larger standard deviation in Part (c) makes it more difficult to detect a difference, resulting in a failure to reject the null hypothesis.

12.46   By saying that listening to music reduces pain levels, the authors are telling you that the study resulted in convincing evidence that pain levels are reduced when music is being listened to. (In other words, the results of the study were statistically significant.) By saying, however, that the magnitude of the positive effects was small, the authors are telling you that the effect was not practically significant.

Additional Exercises

12.53 H_0: $\mu = 20$, H_a: $\mu > 20$, $t = 0.85$, P-value $= 0.205$, fail to reject H_0. There is not convincing evidence that the mean suppertime is greater than 2 minutes.

12.55 H_0: $\mu = 8.4$, H_a: $\mu < 8.4$; $t = -10.29$, P-value ≈ 0, reject H_0. There is convincing evidence that the population mean sleep duration for students at this college is less than the recommended number of 8.4 hours.

12.57 H_0: $\mu = 30$, H_a: $\mu < 30$, $t = -1.160$, P-value $= 0.149$, fail to reject H_0. There is not convincing evidence that the mean fuel efficiency under these circumstances is less than 30 miles per gallon.

12.59 H_0: $\mu = 15$, H_a: $\mu > 15$, $t = 5.682$, P-value $= 0.000$, reject H_0. There is convincing evidence that the mean time to 100°F is greater than 15 minutes.

SECTION 12.4

Exercise Set 1

12.60   The given simulation produced a confidence interval of (81.372, 115.253). You can be 95% confident that the population mean gas mileage for electric or plug-in hybrid cars lies somewhere between 81.372 and 115.253 mpg.

12.61   **(a)** $H_0: \mu = 4.17$, $H_a: \mu \neq 4.17$. The given randomization test resulted in a P-value of 0.000. Because this P-value of 0.000 is less than any reasonable significance level, you reject the null hypothesis. There is convincing evidence that the population mean brain size for birds that are relatives of the dodo differs from the established dodo brain size of 4.17.
(b) The results of this test provide evidence that the population mean brain size for birds that are relatives of the dodo differs from the established dodo brain size of 4.17.

12.62   **(a)** The sample size of $n = 21$ is smaller than 30, so the methods based on the t distribution may not be appropriate.
(b) $H_0: \mu = 1$ versus $H_a: \mu > 1$. The given randomization test resulted in a P-value of 0.029. Because this P-value of 0.029 is less than a significance level of 0.05, you reject the null hypothesis. There is convincing evidence that the population mean time discrimination score for male smokers who abstain from smoking for 24 hours is significantly greater than 1.
(c) The given simulation produced a confidence interval of (0.994, 1.082). You can be 95% confident that the population mean time discrimination score for the population of male smokers who abstain from smoking for 24 hours lies somewhere between 0.994 and 1.082.
(d) Yes, the same conclusion was reached.

12.63   **(a)**

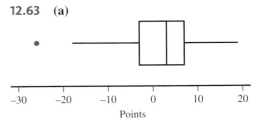

The boxplot indicates an outlier in the distribution, so the methods based on the t distribution might not be appropriate.
(b) $H_0: \mu = 0$ versus $H_a: \mu \neq 0$. Different simulations will produce different results, so answers will vary. One randomization test resulted in a P-value of 0.355. Because this P-value of 0.355 is greater than a significance level of 0.05, you fail to reject the null hypothesis. There is not convincing evidence that the population mean points difference for NFL teams coming off a bye week differs from zero.
(c) Different simulations will produce different results, so answers will vary. One simulation produced a confidence interval of (-1.906, 4.688). You can be 95% confident that the population mean points difference for NFL teams coming off a bye week lies somewhere between -1.906 and 4.688.
(d) Based on the results from Parts (b) and (c), you would not believe that teams coming off a bye week have a significant advantage in points scored over their opponents. In the hypothesis test in Part (b), the null hypothesis was not rejected, concluding that you do not have convincing evidence that the population mean points difference for NFL teams coming off a bye week differs from zero. The

confidence interval in Part (c) includes zero, so zero is a plausible value for the true mean points difference. Both Parts (b) and (c) lead to the same conclusion.

ARE YOU READY TO MOVE ON?
Chapter 12 Review Exercises

12.69   **(a)** $\mu_{\bar{x}} = 2$, $s_{\bar{x}} = 0.267$
(b) In each case $\mu_{\bar{x}} = 2$. When $n = 20$, $\sigma_{\bar{x}} = 0.179$, and when $n = 100$, $\sigma_{\bar{x}} = 0.08$. All three centers are the same, and the larger the sample size, the smaller the standard deviation of $\bar{x}$. Since the distribution of $\bar{x}$ when $n = 100$ is the one with the smallest standard deviation of the three, this sample size is most likely to result in a value of $\bar{x}$ close to μ.

12.71   **(a)** Narrower.
(b) The statement is not correct. The population mean, μ, is a constant, and it is not appropriate to talk about the probability that it falls within a certain interval.
(c) The statement is not correct. You can say that *in the long run* about 95 out of every 100 samples will result in confidence intervals that will contain μ, but we cannot say that in 100 such samples, *exactly* 95 will result in confidence intervals that contain μ.

12.73   (52.80, 54.66). You can be 98% confident that the mean amount of money spent per graduation gift in 2016 was between \$52.80 and \$54.66.

12.75   119

12.77 $H_0: \mu = 20$, $H_a: \mu > 20$, $t = 14.836$, P-value = 0.000, reject H_0. There is convincing evidence that the mean wrist extension for all people using the new mouse design is greater than 20 degrees. To generalize the result to the population of Cornell students, you need to assume that the 24 students used in the study are representative of all students at the university. To generalize the result to the population of all university students, you need to assume that the 24 students used in the study are representative of all university students.

CHAPTER 13
Asking and Answering Questions About the Difference Between Two Means

NOTE: Answers may vary slightly if you are using statistical software or a graphing calculator, or depending on how the values of sample statistics are rounded when performing hand calculations. Don't worry if you don't match these numerical answers exactly (but your answers should be similar).

SECTION 13.1
Exercise Set 1

13.1   Studies 1 and 4 have samples that are independently selected.

13.2 Scenario 1: The appropriate test is not about a difference in population means. There is only one sample. Scenario 2: The appropriate test is about a difference in population means. Scenario 3: The appropriate test is not about a difference in population means. There are two samples, but the variable is categorical rather than numerical.

SECTION 13.2

Exercise Set 1

13.5 $H_0: \mu_d = 0$, $H_a: \mu_d > 0$, $t = -0.247$, P-value = 0.595, fail to reject H_0. There is not convincing evidence that the mean time to exhaustion for experienced triathletes is greater when they run while listening to motivational music.

13.6 $H_0: \mu_d = 0$, $H_a: \mu_d \neq 0$, $t = -0.290$, P-value = 0.778, fail to reject H_0. There is not convincing evidence that the mean time to exhaustion for experienced triathletes is different when they run while listening to motivational music than when they run listening to neutral music.

13.7 $H_0: \mu_d = 0$, $H_a: \mu_d \neq 0$, $t = 0.23$, P-value = 0.825, fail to reject H_0. There is not convincing evidence that the mean reading differs for the Mini-Wright meter and the Wright meter.

13.8 $H_0: \mu_d = 0$, $H_a: \mu_d > 0$, $t = 4.458$, P-value = 0.001, reject H_0. There is convincing evidence that the mean time to exhaustion is greater after chocolate milk than after carbohydrate replacement drink.

13.9 (a) Eosinophils. Virtually every child showed a reduction in eosinophils percentage, with many of the reductions by large amounts.
(b) FE_{NO}. While the majority seems to have shown a reduction, there are several who showed increases, and some of those by non-negligible amounts.

13.10 $(-3.069, -0.791)$. You can be 90% confident that the mean difference in MPF at brain location 1 is between -3.069 and -0.791.

13.11 $(-2.228, -0.852)$. You can be 90% confident that the mean difference in MPF at brain location 1 is between -2.228 and -0.852.

13.12 $(-58.198, 46.598)$; you can be 95% confident that the difference in mean time to exhaustion for experienced triathletes when running to motivational music and the mean time when running with no music is somewhere between -58.198 and 46.598 seconds. This interval is consistent with the conclusion in the hypothesis test in 13.5 because in 13.5, the null hypothesis of no difference in the means was not rejected, and the confidence interval included zero, which also leads us to conclude no difference.

Additional Exercises

13.21 (a) $H_0: \mu_d = 0$, $H_a: \mu_d \neq 0$, $t = -4.054$, P-value ≈ 0, reject H_0. There is evidence of a significance difference in

the mean reported weight and the mean actual weight for female college students.
(b) $H_0: \mu_d = 0$, $H_a: \mu_d \neq 0$, $t = -8.109$, P-value ≈ 0, reject H_0. There is evidence of a significance difference in the mean reported height and the mean actual height for female college students.
(c) Both of the hypothesis tests in Parts (a) and (b) resulted in rejecting the null hypothesis. Additionally, the t test statistics are both negative, which tells us that the sample mean differences are less than zero for reported value minus actual value. This means that there is evidence that female college students tend to under-report both height and weight.

13.23 $H_0: \mu_d = 0$, $H_a: \mu_d > 0$, $t = 2.724$, P-value = 0.004, reject H_0. There is convincing evidence to support the claim that the mean following distance for Greek taxi drivers is greater when there are no distractions than when the driver is talking on a mobile phone. This conclusion is consistent with the claim made by the authors.

13.25 (0.862, 1.738); you can be 95% confident that the mean following distance for Greek taxi drivers while driving with no distractions and while driving and texting is between 0.862 and 1.738 meters. Both endpoints of this interval are positive, so you would think that the mean following distance for Greek taxi drivers while driving with no distractions is greater than the mean while driving and texting by somewhere between 0.862 and 1.738 meters.

13.27 $H_0: \mu_d = 3$, $H_a: \mu_d > 3$, $t = 0.856$, P-value = 0.210, fail to reject H_0. There is not convincing evidence that the mean number of words recalled after 1 hour is greater than the mean number of words recalled after 24 hours by more than 3.

13.29 (3.626, 16.194); you can be 95% confident that the mean change in verbal ability score of children born prematurely from age 3 to age 8 is between 3.626 and 16.194. Both endpoints of this interval are positive, so you would think that the mean verbal ability score at age 8 is greater than the mean verbal ability score at age 3 by somewhere between 3.626 and 16.194.

SECTION 13.3

Exercise Set 1

13.31 $H_0: \mu_F - \mu_{NF} = 0$, $H_a: \mu_F - \mu_{NF} < 0$, $t = -9.29$, P-value < 0.000, reject H_0. There is convincing evidence that the mean time spent studying is less for Facebook users than for those who do not use Facebook.

13.32 $H_0: \mu_1 - \mu_2 = 0$, $H_a: \mu_1 - \mu_2 \neq 0$, $t = -4.536$, P-value ≈ 0, reject H_0. There is convincing evidence that the mean reading level for health-related pages differs for Wikipedia and WebMD.

13.33 (a) $(-23.520, -10.881)$, you can be 90% confident that the difference in mean Flesch reading ease score for health-related pages on Wikipedia and health-related pages on WebMD is between -23.520 and -10.881. Both endpoints of this interval are negative, so you would think that

the mean Flesch reading ease score for WebMD is greater than the mean Flesch reading ease score for Wikipedia by somewhere between 10.8805 and 23.5195.
(b) This interval indicates that, on average, WebMD pages have more difficult reading levels. This is consistent with the conclusion reached in the previous exercise.

13.34 $H_0: \mu_{GA} - \mu_{NGA} = 0, H_a: \mu_{GA} - \mu_{NGA} > 0, t = 3.34$, P-value $= 0.001$, reject H_0. There is convincing evidence that the mean percentage of the time spent with the previous partner is greater for genetically altered voles than for voles that were not genetically altered.

13.35 $H_0: \mu_1 - \mu_2 = 0, H_a: \mu_1 - \mu_2 > 0, t = 11.068$, P-value ≈ 0, reject H_0. There is convincing evidence that, on average, adults in the United States get less sleep on work nights than adults in Mexico.

13.36 **(a)** (7.206, 20.794), you can be 95% confident that the difference in the mean amount of sleep on a work night for adults in Canada and adults in England is between 7.206 and 20.794 minutes. Both endpoints of this interval are positive, so you would think that the mean amount of sleep on a work night for adults in Canada is greater than the mean amount of sleep on a work night for adults in England by somewhere between 7.206 and 20.794 minutes.
(b) Yes, because zero is not contained in the confidence interval.

Additional Exercises

13.43 **(a)** If the distributions were both approximately normal, then 22.8% of the male Internet Addiction scores would be negative, and 25.6% of the female Internet Addiction scores would be negative. Assuming that the scores must be positive, these percentages are too large to make it reasonable to believe that the distributions are approximately normal.
(b) Yes, because the sample sizes are both much larger than 30 and the samples were said to be representative.
(c) $H_0: \mu_1 - \mu_2 = 0, H_a: \mu_1 - \mu_2 < 0, t = -4.934$, P-value ≈ 0, reject H_0. There is convincing evidence that the mean Internet Addiction score is greater for male Chinese sixth grade students than for female Chinese sixth grade students.

13.45 **(a)** Since boxplots are roughly symmetric and since there are no outliers in either sample, the assumption of normality is plausible, and it is reasonable to carry out a two-sample t test.
(b) $H_0: \mu_{2009} - \mu_{1999} = 0, H_a: \mu_{2009} - \mu_{1999} > 0, t = 3.332$, P-value $= 0.001$, reject H_0. There is convincing evidence that the mean time spent using electronic media was greater in 2009 than in 1999.

13.47 $H_0: \mu_1 - \mu_2 = 200, H_a: \mu_1 - \mu_2 > 200, t = 2.201$, P-value $= 0.014$, reject H_0. There is convincing evidence to conclude that the mean calorie intake for teens who typically eat fast food is greater than the mean intake for those who don't eat fast food by more than 200 calories per day.

13.49 $H_0: \mu_W - \mu_{WIO} = 0, H_a: \mu_W - \mu_{WIO} < 0, t = -3.36$, P-value ≈ 0.000, reject H_0. There is convincing evidence

that the mean brain volume is smaller for children with ADHD than for children without ADHD.

13.51 **(a)** μ_1 = mean payment for claims not involving errors; μ_2 = mean payment for claims involving errors; $H_0: \mu_1 - \mu_2 = 0; H_a: \mu_1 - \mu_2 < 0$.
(b) Answer: (ii) 2.65. Since the samples are large, a t distribution with a large number of degrees of freedom is used, which can be approximated with the standard normal distribution. $P(z > 2.65) = 0.004$, which is the P-value given. None of the other possible values of t gives the correct P-value.

13.53 (0.010, 0.210), you can be 90% confident that the difference in mean GPA for students at the University of Central Florida who are employed and students who are not employed is between 0.010 and 0.210.

13.55 $H_0: \mu_1 - \mu_2 = 0, H_a: \mu_1 - \mu_2 < 0, t = -3.867$, P-value ≈ 0, reject H_0. There is convincing evidence to conclude that male soccer players who frequently "head" the ball have a lower mean IQ than those who do not. Yes, these data support the researcher's conclusion.

13.57 (0.643, 1.357), you can be 99% confident that the difference in mean number of science courses planned for males and females is between 0.643 and 1.357. This interval indicates that, on average, girls are somewhat less inclined to enroll in science courses. However, not all girls will enroll in fewer courses than boys.

SECTION 13.4
Exercise Set 1

13.59 The mean quiz score for those in the texting group is enough lower than the mean for the no-texting group that a difference this large is not likely to have occurred by chance (due just to the random assignment) when there is no real difference between the treatment means.

13.60 The difference in mean calorie intake for the 4-hour and 8-hour sleep groups could have occurred by chance just due to the random assignment.

13.61 $H_0: \mu_N - \mu_T = 0, H_a: \mu_N - \mu_T > 0, t = 21.78$, P-value ≈ 0.000, reject H_0. There is convincing evidence that the mean test score is higher for the new teaching method.

13.62 $H_0: \mu_G - \mu_{NG} = 0, H_a: \mu_G - \mu_{NG} > 0, t = 11.753$, P-value ≈ 0.000, reject H_0. There is convincing evidence that learning the gesturing approach to solving problems of this type results in a higher mean number of correct responses.

13.63 (0.748, 1.053), you can be 95% confident that the difference in the mean numbers of correct answers for the two different methods (gesture and no gesture) is between 0.748 and 1.053. Both endpoints of this interval are positive, so you would think that the mean number of correct answers for the gesture method is greater than the mean number of correct answers for the no gesture method by somewhere between 0.748 and 1.053.

13.64 (−814.629, 524.096). You can be 95% confident that the difference in mean food intake for the two sleep conditions is between −814.629 and 524.096. Because 0 is included in the confidence interval, it is possible that there is no difference in the treatment means.

13.65 **(a)** If the vertebroplasty group had been compared to a group of patients who did not receive any treatment, and if, for example, the people in the vertebroplasty group experienced a greater pain reduction on average than the people in the "no treatment" group, then it would be impossible to tell whether the observed pain reduction in the vertebroplasty group was caused by the treatment or merely by the subjects' knowledge that some treatment was being applied. By using a placebo group, it is ensured that the subjects in both groups have the knowledge of some "treatment," so that any differences between the pain reduction in the two groups can be attributed to the nature of the vertebroplasty treatment.
(b) (−0.687, 1.287). You can be 95% confident that the difference in mean pain intensity three days after treatment for the vertebroplasty treatment and the fake treatment is between −0.687 and 1.287.

13.66 **(a)** At 14 days: (−1.186, 0.786). You can be 95% confident that the difference in mean pain intensity 14 days after treatment for the vertebroplasty treatment and the fake treatment is between −1.186 and 0.786. At one month: (−1.722, 0.322). You can be 95% confident that the difference in mean pain intensity one month after treatment for the vertebroplasty treatment and the fake treatment is between −1.722 and 0.322.
(b) The fact that all of the intervals contain zero tells you that you do not have convincing evidence of a difference in the mean pain intensity for the vertebroplasty treatment and the fake treatment at any of the three times.

Additional Exercises

13.73 **(a)** The two treatments are small fork and large fork.
(b) The mean amount of food for those in the small fork group is enough larger than the mean for the small fork group that a difference this large is not likely to have occurred by chance (due just to the random assignment) when there is no real difference between the treatment means.

13.75 The mean value assigned to the mug for those in the group given the mug is enough higher than the mean for the other group that a difference this large is not likely to have occurred by chance (due just to the random assignment) when there is no real difference between the treatment means.

13.77 $H_0: \mu_S - \mu_C = 0$, $H_a: \mu_S - \mu_C > 0$, $t = 2.633$, P-value = 0.007, reject H_0. There is convincing evidence that the supervised exercise treatment results in a significantly greater mean weight loss than the treatment that just involves advising people to lose weight.

13.79 (0.731, 5.959), you can be 95% confident that difference in mean weight loss for the two treatments

is between 0.731 and 5.959 Kg. Both endpoints of the confidence interval are positive, so you would think that the mean weight loss for those who would have supervised exercise is greater than the mean weight loss for those who are merely told to take measures to lose weight by somewhere between 0.731 and 5.959 Kg.

SECTION 13.5
Exercise Set 1

13.81 **(a)** The sample size is not greater than 30 and a dotplot of the difference data suggests that it may not be reasonable to assume that the difference distribution is normal.
(b) $H_0: \mu_d = 0$ versus $H_a: \mu_d > 0$. Different simulations will produce different results, so answers will vary. For one randomization test, the P-value was 0.003. Because this P-value was 0.003 is less than any reasonable significance level, the null hypothesis is rejected. There is convincing evidence that the mean difference in movement is greater than 0.

13.82 Different simulations will produce different results, so answers will vary. One simulation resulted in a confidence interval of (5.120, 29.659). You can be 95% confident that the actual mean difference in movement is somewhere between 5.120 and 29.659.

13.83 $H_0: \mu_1 - \mu_2 = 0$ versus $H_a: \mu_1 - \mu_2 > 0$. Different simulations will produce different results, so answers will vary. For one randomization test, the P-value was 0.001. Because this P-value of 0.001 is less than the significance level of 0.05, the null hypothesis is rejected. There is convincing evidence that the difference in the population mean heart rate percentage for drivers in a two-firefighter team is greater than the mean heart rate for drivers in a team with more than two firefighters.

13.84 Different simulations will produce different results, so answers will vary. One simulation resulted in a confidence interval of (9.678, 23.450). You can be 95% confident that the actual difference in population mean heart rate percentage for drivers in a two-firefighter team is greater than the mean heart rate for drivers in a team with more than two firefighters by somewhere between 9.678 and 23.450.

13.85 $H_0: \mu_1 - \mu_2 = 0$ versus $H_a: \mu_1 - \mu_2 < 0$. Different simulations will produce different results, so answers will vary. For one randomization test, the P-value was 0.001. Because this P-value of 0.001 is less than a significance level of 0.05 or 0.01, the null hypothesis is rejected at either of these significance levels. There is convincing evidence that the mean Personal Meaning Score for patients taking a high dose is greater than the mean score for patients taking a low dose.

13.86 Different simulations will produce different results, so answers will vary. One simulation resulted in a confidence interval of (−3.222, −0.738). You can be 95% confident that the actual difference (Low − High) in mean

Personal Meaning Score is greater for patients taking a high dose than for patients taking a low dose by somewhere between 0.738 and 3.222.

ARE YOU READY TO MOVE ON?

Chapter 13 Review Exercises

13.91 (a) $H_0: \mu_1 - \mu_2 = 10$ versus $H_a: \mu_1 - \mu_2 > 10$
(b) $H_0: \mu_1 - \mu_2 = -10$ versus $H_a: \mu_1 - \mu_2 < -10$

13.93 Scenario 1: no. Scenario 2: no. Scenario 3: yes.

13.95 $H_0: \mu_B - \mu_M = 0, H_a: \mu_B - \mu_M \neq 0, t = 6.22$, P-value ≈ 0.000, reject H_0. There is convincing evidence that the mean sodium content is not the same for meal purchases at Burger King and McDonalds.

13.97 $H_0: \mu_d = 0, H_a: \mu_d > 0, t = 2.68, P$-value $= 0.010$, reject H_0. There is convincing evidence that the mean energy expenditure is greater for the conventional shovel than for the perforated shovel.

13.99 (a) $H_0: \mu_d = 0, H_a: \mu_d > 0, t = 4.451, P$-value ≈ 0, reject H_0. There is convincing evidence that, on average, male online daters overstate their height.
(b) $(-0.210, 0.270)$. You can be 95% confident that for female online daters, the mean difference in reported height and actual height is between -0.210 and 0.270 inches.
(c) $H_0: \mu_M - \mu_F = 0, H_a: \mu_M - \mu_F > 0, t = 3.094$, P-value $= 0.001$, reject H_0. There is convincing evidence that the mean height difference is greater for males than for females.
(d) In Part (a), the male profile heights and the male actual heights are paired (according to which individual has the actual height and the height stated in the profile), and with paired samples, you use the paired t test. In Part (c), you were dealing with two independent samples (the sample of males and the sample of females), and therefore, the two-sample t test was appropriate.

13.101 (a) $(-0.186, 0.111)$. You can be 90% confident that the mean difference in pH for surface and subsoil is between -0.186 and 0.111. Because 0 is included in the interval, it is possible that there is no difference in mean pH.
(b) You must assume that the distribution of differences across all locations is normal.

13.103 $H_0: \mu_1 - \mu_2 = 0, H_a: \mu_1 - \mu_2 > 0, t = 2.40$, P-value $= 0.009$, reject H_0. There is convincing evidence that the mean measure of happiness for experiential purchasers is greater than the mean measure of happiness for material purchasers. These data do support the authors' conclusion.

CHAPTER 14
Learning from Categorical Data

NOTE: Answers may vary slightly if you are using statistical software or a graphing calculator, or depending on how the values of sample statistics are rounded when performing hand calculations. Don't worry if you don't match these numerical answers exactly (but your answers should be similar).

SECTION 14.1

Exercise Set 1

14.1 (a) fail to reject H_0
(b) reject H_0
(c) reject H_0

14.2 (a) 0.085
(b) 0.075
(c) 0.025

14.3 $H_0: p_1 = p_2 = p_3 = p_4 = p_5 = 0.2, X^2 = 20.686$, P-value ≈ 0, reject H_0. There is convincing evidence that the Twitter category proportions are not all equal.

14.4 (a) $H_0: p_1 = p_2 = p_3 = p_4 = p_5 = 0.2, X^2 = 25.241$, P-value ≈ 0, reject H_0. There is convincing evidence that the proportions of home runs hit are not the same for all five directions.
(b) For home runs going to left center and center field, the observed counts are significantly lower than the numbers that would have been expected if the proportion of home runs hit was the same for all five directions; while for right field, the observed count is much high than the number that would have been expected.

14.5 $H_0: p_1 = 0.14, p_2 = 0.20, p_3 = 0.19, p_4 = 0.18$, $p_5 = 0.14, p_6 = 0.14, X^2 = 251.981, P$-value ≈ 0, reject H_0. There is convincing evidence that one or more of the age groups buys a disproportionate share of lottery tickets.

Additional Exercises

14.11 (a) 0.001
(b) 0.001
(c) P-value > 0.100

14.13 Reject H_0. There is convincing evidence that the response proportions are not each 0.5.

14.15 $p_1 = \frac{9}{16}, p_2 = \frac{3}{16}, p_3 = \frac{3}{16}, p_4 = \frac{1}{16}, X^2 = 1.469$, P-value $= 0.690$, fail to reject H_0. The data are consistent with Mendel's laws.

14.17 $H_0: p_1 = p_2 = p_3 = p_4 = 0.25, X^2 = 11.242$, P-value $= 0.0105$, fail to reject H_0. There is not convincing evidence that there is a color preference.

SECTION 14.2

Exercise Set 1

14.19 H_0: The proportions falling into the two credit card response categories are the same for all three types of colleges, H_a: The proportions falling into the two credit card response categories are not all the same for all three types of colleges, $X^2 = 54.477, P$-value ≈ 0, reject H_0. There is convincing evidence that the proportions falling into the two credit card response categories are not the same for all three types of colleges.

14.20 (a) H_0: The proportions falling into the First Amendment rights categories are the same for teachers and students, H_a: The proportions falling into the First Amendment rights categories are not all the same for teachers and

students, $X^2 = 128.849$, P-value ≈ 0, reject H_0. There is convincing evidence that the proportions falling into the First Amendment rights categories are not all the same for teachers and students.
(b) The largest contributions to the chi-square statistics come from the columns corresponding to freedom of speech and freedom of religion. More students and fewer teachers than expected chose freedom of speech. Fewer students and more teachers than expected chose freedom of religion.

14.21   **(a)** H_0: Field of study and smoking status are independent, H_a: Field of study and smoking status are not independent, $X^2 = 90.853$, P-value ≈ 0, reject H_0. There is convincing evidence that field of study and smoking status are not independent.
(b) The particularly high contributions to the chi-square statistic (in order of importance) come from the field of communication, languages, in which there was a disproportionately high number of smokers; from the field of mathematics, engineering, and sciences, in which there was a disproportionately low number of smokers; and from the field of social science and human services, in which there was a disproportionately high number of smokers.

14.22 H_0: Locus of control and compulsive buyer behavior are independent, H_a: Locus of control and compulsive buyer behavior are not independent, $X^2 = 5.402$, P-value $= 0.020$, fail to reject H_0. There is not convincing evidence that there is an association between locus of control and compulsive buyer behavior.

Additional Exercises

14.27   Answers will vary.

14.29   In a chi-square goodness-of-fit test, one population is compared to fixed category proportions. In a chi-square test of homogeneity, two or more populations are compared.

14.31 H_0: There is no association between use of ART and whether baby is premature, H_a: There is an association between use of ART and whether baby is premature, $X^2 = 326.111$, P-value ≈ 0, reject H_0. There is convincing evidence that there is an association between use of ART and whether the baby is premature.

14.33 H_0: There is no association between age group and the response to the question, H_a: There is an association between age group and the response to the question, $X^2 = 28.646$, P-value ≈ 0, reject H_0. There is convincing evidence that there is an association between age group and the response to the question.

14.35 H_0: There is no association between age and perception of money management skills, H_a: There is an association between age and perception of money management skills, $X^2 = 12.146$, P-value $= 0.007$, reject H_0. There is convincing evidence that there is an association between age and perception of money management skills.

14.37 H_0: There is no association between survival and the risk of malnutrition, H_a: There is an association between survival and the risk of malnutrition, $X^2 = 96.219$, P-value ≈ 0, reject H_0. There is convincing evidence that there is an association between survival and the risk of malnutrition. This is consistent with the authors' conclusion.

ARE YOU READY TO MOVE ON?
Chapter 14 Review Exercises

14.39 H_0: $p_1 = 0.177$, $p_2 = 0.032$, $p_3 = 0.734$, $p_4 = 0.057$, H_a: At least one of the population proportions is not as stated in H_0, $X^2 = 19.599$, P-value ≈ 0, reject H_0. There is convincing evidence that the proportions of characters of different ethnicities appearing in commercials are not all the same as the census proportions.

14.41 H_0: $p_1 = \frac{1}{3}$, $p_2 = \frac{2}{3}$, H_a: At least one of the population proportions is not as stated in H_0, $X^2 = 5.052$, P-value $= 0.025$, reject H_0. There is convincing evidence that bicycle fatalities are not twice as likely to occur between noon and midnight as during midnight to noon, so the data provide evidence against the hypothesis.

14.43 H_0: Proportions are the same for academic superstars and solid performers, H_a: Proportions are not the same for academic superstars and solid performers, $X^2 = 12.438$, P-value $= 0.014$, reject H_0. There is convincing evidence that the distribution of responses over the distance from home categories is not the same for academic superstars and solid performers.

14.45 H_0: There is no association between sex and character type, H_a: There is an association between sex and character type, $X^2 = 13.702$, P-value $= 0.001$, reject H_0. There is convincing evidence that there is an association between sex and character type.

14.47   **(a)** test of independence, there was a single sample, and two categorical variables were recorded for each person in the sample.
(b) H_0: There is no association between whether or not people pay their balance in full each month and age, H_a: There is an association between whether or not people pay their balance in full each month and age, $X^2 = 6.746$, P-value $= 0.080$, fail to reject H_0. There is not convincing evidence that there is an association between whether or not people pay their balance in full each month and age.
(c) Because it was a random sample, it is reasonable to generalize the conclusion to adult Americans age 20 to 39 years old who have at least one credit card.

14.49 H_0: Proportions are the same for males and females, H_a: Proportions are not the same for males and females, $X^2 = 14.153$, P-value $= 0.015$, reject H_0. There is convincing evidence that the distribution of the consumption of fried potatoes is not the same for males and females. This agrees with the authors' conclusion.

CHAPTER 15

Understanding Relationships—Numerical Data

NOTE: Answers may vary slightly if you are using statistical software or a graphing calculator, or depending on how the values of sample statistics are rounded when performing hand calculations. Don't worry if you don't match these numerical answers exactly (but your answers should be similar).

SECTION 15.1

Exercise Set 1

15.1 **(a)** deterministic **(b)** probabilistic **(c)** probabilistic **(d)** deterministic

15.2 **(a)** $y = -5.0 + 0.017x$

(b)

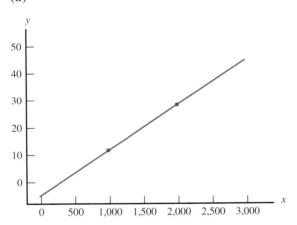

(c) 30.7 therms
(d) 0.017 therms
(e) 1.7 therms
(f) No. The given relationship only applies to houses whose sizes are between 1,000 and 3,000 square feet. The size of this house, 500 square feet, lies outside this range.

15.3 **(a)** $47, $4,700 **(b)** 0.3156, 0.0643

15.7 **(a)** Tom's pricing strategy
(b) $y = cx$
(c) $y = cx + e$
(d) e will be -2 for about half of the purchases and $+1$ for about half of the purchases.

15.9 **(a)** $\hat{y} = 45.572 + 1.335x$
(b) 1.335 cases per 100,000 women.
(c) 98.990 cases per 100,000 women.
(d) No, 20% is outside the range of the data.
(e) $r^2 = 0.830$ Eighty-three percent of the variability in breast cancer incidence can be explained by the approximate linear relationship between breast cancer incidence and percent of women using HRT.

(f) $s_e = 4.154$ A typical deviation of a breast cancer incidence value in this data set from the value predicted by the estimated regression line is about 4.154.

15.11 **(a)** 0.335
(b) $s_e = 5.366$; This is a typical deviation of a depression score change value in the sample from the value predicted by the least-squares line.
(c) 5.08
(d) 12.966

SECTION 15.2

Exercise Set 1

15.13 **(a)** The larger σ_e is, the larger σ_b will be. Because σ_b is in the denominator of the test statistic, when σ_e is large, the value of the test statistic will tend to be small.
(b) The larger σ_e is, the larger σ_b will be. This means that when σ_e is large, the confidence interval for β will tend to be wide.

15.14 The width of the confidence interval is twice the margin of error. The journalist could calculate the width of the confidence interval and then divide by 2.

15.15 Both conditions should be re-checked to make sure that they are still reasonable.

15.16 **(a)** $H_0: \beta = 0, H_a: \beta \neq 0, t = 2.284$, P-value $= 0.062$, fail to reject H_0. There is not convincing evidence that the simple linear regression model would provide useful information for predicting growth rate from research and development expenditure.
(b) (0.000086, 0.001064) You can be 90% confident that the mean change in growth rate associated with a $1000 increase in research and development expenditure is between 0.000086 and 0.001064.

15.17 **(a)** $H_0: \beta = 0, H_a: \beta \neq 0, t = 0.09$, P-value $= 0.931$, fail to reject H_0. There is not convincing evidence of a useful linear relationship between typing speed and surface angle.
(b) $s_e = 0.512$ This value is relatively small when compared to typing speeds of around 60, and tells you that the y values typically differ from the values predicted by the virtually horizontal estimated regression line by about 0.5. However, $r^2 = 0.003$, which means that virtually none of the variation in typing speed can be attributed to the approximate linear relationship between typing speed and surface angle. This is consistent with the conclusion that there is not convincing evidence of a useful linear relationship between the two variables.

15.18 $H_0: \beta = 0, H_a: \beta \neq 0, t = 5.75$, P-value ≈ 0, reject H_0. There is convincing evidence of a useful linear relationship between height and the logarithm of weekly earnings.

15.19 **(a)** 0.359
(b) 176.750
(c) $H_0: \beta = 0, H_a: \beta \neq 0, t = 0.820$, P-value $= 0.459$, fail to reject H_0. There is not convincing evidence of a useful linear relationship between acrylamide concentration and frying temperature.

Additional Exercises

15.25 The simple linear regression model assumes that for any fixed value of x, the distribution of y is normal with mean $\alpha + \beta x$ and standard deviation σ_e. Thus σ_e is the shared standard deviation of these y distributions. The quantity s_e is an estimate of σ_e obtained from a particular set of (x, y) observations.

15.27 (a) The scatterplot suggests that the simple linear regression model might be appropriate.
(b) $\hat{y} = 6.873 + 5.078x$
(c) $H_0: \beta = 0$, $H_a: \beta \neq 0$, $t = 1.75$, P-value $= 0.1104$, fail to reject H_0. There is not convincing evidence of a useful linear relationship.

15.29 The test statistic for this test is $t = \dfrac{b - 1}{s_b}$. From the computer output, $b = 0.9797401$ and $s_b = 0.018048$, so $t = -1.123$. Because this is an upper-tail test, the null hypothesis would not be rejected.

SECTION 15.3

Exercise Set 1

15.30 Yes. The pattern in the scatterplot looks linear and the spread around the line does not appear to be changing with x. There is no pattern in the residual plot, and the boxplot of the residuals is approximately symmetric, and there are no outliers.

15.31 Yes, but the linear relationship is not very strong.

15.32 (a) There is one unusually large standardized residual, 2.52, for the point (164.2, 181). The point (387.8, 310) may be an influential point.

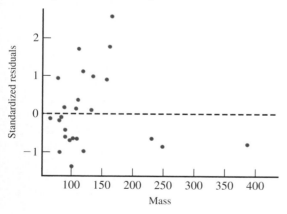

(b) No, the plot seems consistent with the simple linear regression model.
(c) If you include the point with the unusually large standardized residual you might suspect that the variances of the y distributions decrease as the x values increase. However, from the relatively small number of points included there is not strong evidence that the assumption of constant variance does not apply.

15.33 (a) $\hat{y} = 0.939 + 0.873x$
(b) The standardized residual plot shows that there is one point that is a clear outlier (the point whose standardized residual is 3.721). This is the point for product 25.

(c) $\hat{y} = 0.703 + 0.918x$, removal of the point resulted in a reasonably substantial change in the equation of the estimated regression line.
(d) For every 1-cm increase in minimum width, the mean maximum width increases by about 0.918 cm. It does not make sense to interpret the intercept because it is clearly impossible to have a container whose minimum width is zero.

(e) The pattern in the standardized residual plot suggests that the variances of the y distributions decrease as x increases, and therefore that the assumption of constant variance is not valid.

15.34 (a) $\hat{y} = 28.516 - 0.0257x$
$a = 28.516$
$b = -0.0257$
$r^2 = 0.303$
$s_e = 4.418$
(b) $\hat{y} = 29.371 - 0.0351x$
$a = 29.371$
$b = -0.0351$
$r^2 = 0.195$
$s_e = 4.225$
(c) Deleting the potentially influential data point did not have a big impact on the slope and intercept of the estimated regression line.

Additional Exercises

15.39 (a) There is one point that has a standardized residual of about 2.6. The data point with an x value of about 400 is far away from the rest of the data points in the x direction and might be influential.
(b) No
(c) Yes. The spread of points around the line does not appear to be different for different values of x.

15.41 Three points that sand out as outliers are those that have residuals of about -30, 30, and 38.

ARE YOU READY TO MOVE ON?
Chapter 15 Review Exercises

15.45 α and β are the intercept and slope of the population regression line. a and b are the least squares estimates of α and β.

15.47 (a) Yes. The pattern in the scatterplot is approximately linear.
(b) There are no obvious patterns in the residual plot that would indicate that the basic assumptions are not met.
(c) Because the normal probability plot of the residuals is roughly linear and the boxplot of the residuals is approximately symmetric with no outliers, the assumption that the distribution of e is approximately normal is reasonable.
(d) $r^2 = 0.953929$ and $s_e = 0.801042$

15.49 (a) $H_0: \beta = 0$, $H_a: \beta \neq 0$, $t = -2.25$, P-value $= 0.0355$. For $\alpha = 0.05$, reject H_0, there is convincing evidence of a linear relationship.

(b) $r^2 = 0.193788$, the relationship is not strong.
(c) $(-0.051, 0.001)$
(d) 0.025

15.51 The test statistic for this test is $t = \dfrac{b - 1}{s_b}$. From the computer output, $b = 0.9797401$ and $s_b = 0.018048$, so $t = -1.123$. Because this is an upper-tail test, the null hypothesis would not be rejected.

CHAPTER 16
Asking and Answering Questions About More Than Two Means

NOTE: Answers may vary slightly if you are using statistical software or a graphing calculator, or depending on how the values of sample statistics are rounded when performing hand calculations. Don't worry if you don't match these numerical answers exactly (but your answers should be similar).

SECTION 16.1

Exercise Set 1

16.1 **(a)** $0.001 < P\text{-value} < 0.01$ **(b)** $P\text{-value} > 0.10$
(c) $P\text{-value} = 0.01$ **(d)** $P\text{-value} < 0.001$ **(e)** $0.05 < P\text{-value} < 0.10$ **(f)** $0.01 < P\text{-value} < 0.05$ (using $\text{df}_1 = 4$ and $\text{df}_2 = 60$)

16.2 **(a)** $H_0: \mu_1 = \mu_2 = \mu_3 = \mu_4$, H_a: At least two of the four μ_i's are different.
(b) $P\text{-value} = 0.012$, fail to reject H_0
(c) $P\text{-value} = 0.012$, fail to reject H_0

16.3 $F = 6.687$, $P\text{-value} = 0.001$, reject H_0

16.4 $H_0: \mu_1 = \mu_2 = \mu_3 = \mu_4$, H_a: at least two of the four μ's are different $F = 2.39$, $P\text{-value} = 0.077$, fail to reject H_0.

Additional Exercises

16.9 **(a)** $H_0: \mu_1 = \mu_2 = \mu_3$, H_a: at least two of the three μ's are different $F = 3.86$, $P\text{-value} = 0.035$, reject H_0.
(b) No, the sample mean hunger rating is highest for the healthy snack group.

16.11 $F = 2.62$, $0.05 < P\text{-value} < 0.10$, fail to reject H_0

16.13 **(a)** See solutions manual for detailed computations.
(b) $F = 2.142$, $P\text{-value} > 0.10$, fail to reject H_0

SECTION 16.2

Exercise Set 1

16.14 Since the interval for $\mu_2 - \mu_3$ is the only one that contains zero, we have evidence of a difference between μ_1 and μ_2, and between μ_1 and μ_3, but not between μ_2 and μ_3. Thus, statement **c** is the correct choice.

16.15 In increasing order of the resulting mean numbers of pretzels eaten, the treatments were: slides with related text, slides with no text, no slides, and slides with unrelated text. There were no significant differences between the results for slides with related text and slides with no text, or for no slides and slides with unrelated text. However, there was a significant difference between the mean numbers of pretzels eaten for no slides and slides with no text (and also between the results for no slides and slides with related text). Likewise, there was a significant difference between the mean numbers of pretzels eaten for slides with unrelated text and slides with no text (and also between the results for slides with unrelated text and slides with related text).

16.16

	Fabric 3	Fabric 2	Fabric 5	Fabric 4	Fabric 1
Sample mean	10.5	11.633	12.3	14.96	16.35

Additional Exercises

16.21 **(a)** $F = 3.34$, $P\text{-value} = 0.041$, reject H_0.
(b)

Control	X-Men	Spider Man
2.668	3.020	3.120

(c) The mean measure of gender stereotyping is significantly different for the Control group and the Spider Man group. The mean for X-Men was not significantly different from either of the other two groups. Because lower values indicate attitudes more accepting of equality, you can conclude that those who did not watch a video were more accepting of equality than those who watched Spider Man.

ARE YOU READY TO MOVE ON?
Chapter 16 Review Exercises

16.23 $F = 5.273$, $P\text{-value} = 0.002$, reject H_0

16.25 **(a)**

Difference	Interval	Includes 0?
$\mu_1 - \mu_2$	$(-4.027, 0.027)$	Yes
$\mu_1 - \mu_3$	$(-4.327, -0.273)$	No
$\mu_1 - \mu_4$	$(-5.327, -1.273)$	No
$\mu_2 - \mu_3$	$(-2.327, 1.727)$	Yes
$\mu_2 - \mu_4$	$(-3.327, 0.727)$	Yes
$\mu_3 - \mu_4$	$(-3.027, 1.027)$	Yes

	7+ label	12+ label	16+ label	18+ label
Sample mean	4.8	6.8	7.1	8.1

(b) The more restrictive the age label on the video game, the higher the sample mean rating given by the boys used in the experiment. However, according to the T-K intervals, the only significant differences were between the means for the 7 + label and the 16 + label and between the means for the 7 + label and the 18 + label.

Index

Note: Hyphenated page numbers in *italics* refer to online-only material.

Note: Hyphenated page numbers in *italics* refer to online-only material.

Note: Hyphenated page numbers in *italics* refer to online-only material.